Accessing subject matter on the CD-ROM, Mineralolgy Tutorials (version 2.1)

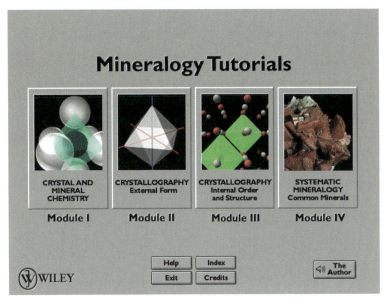

the first menu screen ▶
followed by the first subject
screen ▼ in Module I

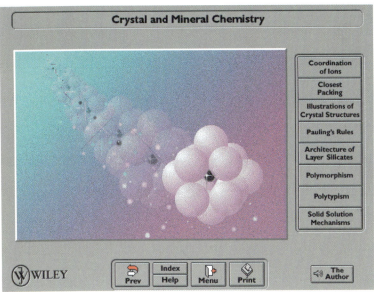

▼ EXAMPLES OF INDEX SCREENS ▼

the main index screen,
showing major subject
categories in the four
modules ▶

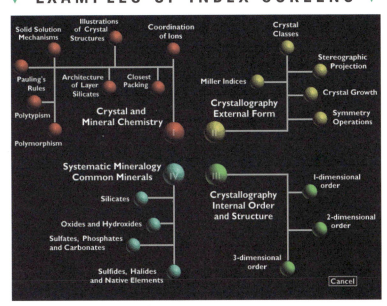

The colored spheres are "hot" and can be clicked with the mouse.

The index screen for module II shows a more detailed screen when the "Crystal Classes" button is clicked. ▶

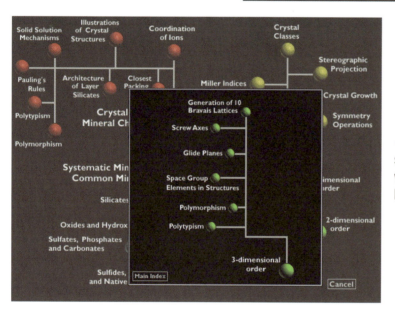

◀ The index screen for module III shows a more detailed screen when the "3-dimensional order" button is clicked.

The index screen for module IV shows the subdivision of the silicates when the "Silicates" button is clicked. ▶

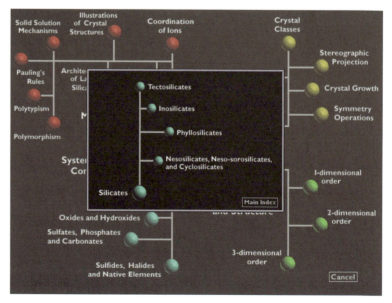

The colored spheres are "hot" and can be clicked with the mouse.

MINERAL SCIENCE

COVER

Close-up photograph of a polished slab of malachite from Zimbabwe (Harvard Mineralogical Museum).

SUPPLEMENTARY MATERIALS

For laboratory and homework assignments: Klein, C., 1994, *Minerals and Rocks: Exercises in Crystallography, Mineralogy, and Hand Specimen Petrology*, revised edition, John Wiley & Sons, Inc., New York, 405 pp. A *Solutions Manual* with completely worked answers to most of the assignments in *Minerals and Rocks: Exercises* is available to instructors who have adopted *Minerals and Rocks: Exercises* from John Wiley & Sons.

For classroom instruction: A 35-mm slide set that incorporates approximately 150 of the most important illustrations from the text will be available to instructors upon request from the publisher.

The *Manual of Mineralogy* was written by James D. Dana in 1848 and revised by him in 1857, in 1878 as the *Manual of Mineralogy and Lithology*, and in 1887 as the *Manual of Mineralogy and Petrography*. An edition number was given to some, but not all, reprintings of each revision. For example, the 1887 revision was reprinted in 1891 as the 10th Edition, but in 1893, 1895, and 1900 as the 12th Edition.* Each subsequent revision, as *Dana's Manual of Mineralogy*, has been given an edition number as follows: *13*, 1912, and *14*, 1929 by William E. Ford. *15*, 1941; *16*, 1952; *17*, 1959; and *18*, 1971 by C. S. Hurlbut, Jr. *19*, 1977 by C. S. Hurlbut, Jr. and C. Klein. *20*, 1985 by C. Klein and C. S. Hurlbut, Jr. *21*, 1993 by C. Klein and C. S. Hurlbut, Jr. *22*, 2002 by C. Klein.

The information regarding revisions by James D. Dana was supplied by Clifford J. Awald, Buffalo Museum of Science.

THE 22ND EDITION OF THE MANUAL OF

MINERAL SCIENCE

(after JAMES D. DANA)

CORNELIS KLEIN
The University of New Mexico

with continued contribution of
CORNELIUS S. HURLBUT, Jr.
Harvard University

JOHN WILEY & SONS, INC.

Acquisitions Editor   **Clifford Mills**
Marketing Manager   **Clay Stone**
Sr. Production Editor   **Kelly Tavares**
Cover   **Marc Klein**
Illustration Coordinator   **Sandra Rigby**

This book was set in Optima by UG / GGS Information Services, Inc.
and printed and bound by Courier.

ISBN: 0-471-25177-1

Printed in the United States of America

10 9 8 7 6 5 4 3

PREFACE

This 22nd edition was prepared during a time of change in many Geology (or Earth Science, or Earth and Planetary Science, or Geoscience) departments, their undergraduate courses, and students. To reflect the extent of the revision, this text for the introductory mineralogy course even has a new title, *Manual of Mineral Science*, succeeding many earlier editions with the name *Manual of Mineralogy*. Some of the changes in the teaching of mineralogy were emphasized for me in June, 1996, when I was among the participants in a National Science Foundation-sponsored workshop on "Teaching Mineralogy" at Smith College. In review sessions about "goals in the college course in mineralogy," the majority of participants concurred that the subject of crystal chemistry and the crystal chemical treatment of the rock-forming minerals were a top priority. The identification of rock-forming minerals by hand specimen study was considered a very important component of the accompanying laboratory. However, no clear consensus emerged on how much time should be spent on crystallographic concepts, and to what extent such concepts should be pursued.

I also learned that students are more comfortable if the course begins with chemical and crystallochemical subject matter, instead of crystallography. The main reason is that almost all undergraduate mineralogy courses have a college-level chemistry course as a prerequisite. Several reviewers of the 21st edition of the *Manual of Mineralogy* pointed this out as well. Last year, for the first time in a long teaching career in mineralogy, I myself began the course with subject matter in chemistry and crystal chemistry and followed with crystallographic concepts. It worked well.

As I began the revision of this text, several reviewers made it clear that the new edition should be made more topical (instead of exhaustive or encyclopedic), more accessible to the average mineralogy student, and more "user-friendly." I've made every

attempt to make the text more readable, understandable, and "user-friendly" through reordering of subject matter, adding new chapter introductions and "interest boxes," streamlining and simplifying aspects of crystallography, and refining illustrations. At the same time, every effort has been made not to compromise the level of topic discussions. Mineralogical concepts are not always easy, and indeed, learning in general is hard work.

A NEW, FLEXIBLE ORGANIZATION

The first seven chapters have now been reorganized to reflect my own successful experience in covering chemistry and crystal chemistry early. Although the order of subject coverage in the present text is quite different from that of earlier editions, every effort was made to keep the first seven chapters, that deal with conceptual material, as independent of each other as possible. Instructors can feel confident that they may cover these chapters in any order they prefer.

In this much-revised text, the first, Introductory chapter is now followed by a second chapter on the "Physical Properties of Minerals in Hand Specimen." Chapter 3, "Elements of Crystal Chemistry," can be viewed either as an introduction to or an overview of that subject, depending on the students' familiarity with college-level chemistry. Chapter 4, "Mineral Reactions, Stability and Behavior," covers a broad array of mineral reactions and behavior and briefly discusses some very basic aspects of thermodynamics before introducing the concept of mineral stability and the diagrams that illustrate such stabilities. Crystallographic subjects follow in Chapters 5 and 6. The treatment in these two chapters is very different from that in prior editions of this text. Chapter 5, "Overview of Crystallographic Concepts," covers the most important crystallographic concepts without losing the student in too much detail. Chapter 6, "Selected Point

Groups and Space Groups," discusses first the concept of stereographic projection and then proceeds with the detailed description of 19 of the most common point groups and examples of some space group representations. Some instructors might consider the material in this chapter more advanced and too detailed for an undergraduate mineralogy course. However, I still find instruction in the use of the stereonet in crystallography a very helpful pedagogical exercise for thinking in three dimensions. Introducing this projection in mineralogy is useful also because it recurs, from a somewhat different perspective, in a subsequent course in structural geology.

Chapter 7, "Analytical Methods in Mineral Science," gives an introduction to the most commonly used analytical techniques, with most extensive coverage on optical methods and X-ray powder diffraction. Chapters 8 through 12 have been re-arranged in comparison with similar chapters on rock-forming minerals in earlier editions. In keeping with the emphasis on crystal chemistry, Chapters 8, 9 and 10 begin with an overview section on the crystal chemistry of specific mineral groups. That is followed, in each of these chapters, by systematic descriptions of the most important (common) rock-forming minerals. Chapter 11 discusses the crystal chemistry of the main rock-forming silicates and Chapter 12 provides the systematic mineralogic descriptions of members of these same silicate groups. Chapter 13 is a brief introduction to gem minerals, and Chapter 14 provides three mineral identification tables. A chapter entitled "Mineral Assemblages" (Chapter 14 in the prior, 21st edition of this text) was eliminated because very few instructors can find time for this subject in a one-semester mineralogy course.

A MORE ACCESSIBLE APPROACH

In addition to a very different order, this 22nd edition now includes new "introductory statements" at the beginning of each chapter. These set the stage for what follows. Eight new color plates with photographs of 72 of the most common minerals are new. Also new are thirteen one-page "interest boxes" that highlight mineralogic (and/or crystallographic) concepts, and their connection with subjects of more general interest:

- Boxes 5.1 and 5.2 highlight symmetry and patterns, respectively.
- Box 8.1 discusses diamond synthesis.
- Boxes 8.2 and 8.3 address some geologic and environmental aspects of sulfides and their common occurrence in vein systems.

- Boxes 9.1, 9.2 and 10.1 discuss some economically important minerals and their applications in manufactured products.
- Boxes 12.1, 12.2, 12.4, and 12.5 highlight geologic occurrences of some major silicate groups as well as the unique industrial applications of some of them.
- Box 12.3 addresses the issue of whether natural silica dust in the environment is indeed a health hazard.

A MINERALOGY EDUCATION

Although this edition has streamlined and simplified some concepts, to my mind it continues to provide what should be taught in a basic, one-semester mineralogy course. I addressed the topic coverage by asking "what should geology (or environmental science) B.A. or B.S. recipients be prepared to deal with?", be it in subsequent graduate school, or as professionals in the work force. My answer is that students who have taken a single undergraduate mineralogy course, should be reasonably comfortable in consulting much of what appears in *Reviews in Mineralogy* (published by the Mineralogical Society of America) and in such reference volumes as those that are part of *Rock-forming Minerals* (the various volumes written by W. A. Deer, R. A. Howie, and J. Zussman), in *Dana's New Mineralogy*, by R. V. Gaines, H. C. Skinner, E. E. Foord, B. Mason and A. Rosenzweig, or in the four-volume set entitled *Handbook of Mineralogy*, by J.W. Anthony, R.A. Bideaux, K.W. Bladh, and M.C. Nichols. To be reasonably comfortable in consulting any of these volumes, new graduates must have had considerable grounding in crystal chemistry, crystal structure, mineral behavior, and crystallography, and they must have some fundamental understanding of mineral stability and/or mineral assemblage diagrams. As part of this background, I think, new graduates should have some familiarity with point and space groups and their notation, statements about unit cell dimensions and content, atomic coordination, and so on, all of which is interwoven in every mineral description in the reference volumes mentioned above.

AN EXPANDED AND MORE USABLE CD-ROM

Included with this edition of the *Manual of Mineral Science* is a revised version (2.1) of the CD-ROM entitled *Mineralogy Tutorials*. Designed to be used by students and/or instructors, the CD-ROM is subdivided into four modules:

I, Crystal and Mineral Chemistry;
II, Crystallography: external form;
III, Crystallography, internal order and structure; and
IV, Systematic Mineralogy.

The first three modules provide many animations that deal with three-dimensional concepts (in crystal chemistry and crystallography) and which are difficult to explain or visualize from a two-dimensional, static book illustration. I would suggest that if students are required to review concepts on the CD-ROM in these three modules, at the same time as the instructor covers such material in class, the learning process can be speeded up. Furthermore, the CD-ROM serves a broader spectrum of learning styles than the text on its own.

Module IV contains brief text pages for 104 of the most common minerals. On these text pages are entries in green (hot) text that provide instant, clickable linkages to crystal structure illustrations, compositional and assemblage diagrams, stability and phase diagrams, solid solution mechanisms, and so on. This fourth module, therefore, is an excellent alternative way to review one's understanding of the crystallography, crystal chemistry, and paragenesis of rock-forming minerals by various instant pathways, through the click of a mouse on hot text. By such quickly branching pathways the student can call up images and concepts that have been developed in a much more linear sequence in the text.

This version of the CD-ROM also provides audio explanations by the author on about 50 screens so as to aid the user's understanding of presentations and/or animations that follow.

ACCESSING OF THE CD-ROM (VERSION 2.1) AND THE INDEX SCREENS

You may access subject material on the CD-ROM in two ways. A click on one of the four "buttons" for modules I, II, III, and IV, found on the first menu screen, leads directly into subject matter, as shown by the first screen for module I that follows the very first menu screen. These two screens are printed in color in the frontpiece of this text.

You may also access subject matter on the CD-ROM more systematically via the index screens that link each subject "button" directly, through the click of a mouse, to the discussion of that subject. In several cases, subjects in the graphical index can be pursued in more detail with a graphical display of subcategories that is superimposed on the original index menu. Examples of four index screens are given in the frontpiece of the text.

Throughout the printed text of the *Manual of Mineral Science*, icons appear along **Subject Headings**, where concepts discussed in the text can also be located on the CD-ROM. In many places, written references specify locations within one of the four modules of the CD-ROM.

LABORATORY MANUAL

An accompanying laboratory manual is available from John Wiley & Sons for use in the mineralogy laboratory: C. Klein, *Minerals and Rocks: Exercises in Crystallography, Mineralogy, and Hand Specimen Petrology*, 1994 revised edition, 405 pp., ISBN 0-471-00042-6

SLIDE SET

The Instructor's slide set has been updated for the 22nd edition of the text. Including 20 new slides, this collection of 150 figures is a great resource for class lectures. ISBN: 0-471-07580-9.

SPECIAL RECOGNITION

I wish to express my gratitude to my co-author (from the 19th through the 21st editions of this text), Cornelius S. Hurlbut, Jr., for inviting me to join him as a junior author for the 19th edition of the *Manual of Mineralogy*, published in 1977. Ever since then, we have jointly contributed to the three more recent editions of this text. Our cooperative efforts have been wonderful in every way; highly professional and respectful of each other. Having been able to be a partner in the development of several mineralogy texts over the last 25 years has added much to my professional development as an instructor and I much cherish my continued friendship with Connie Hurlbut. As a tribute to Connie, I will now re-introduce a poem, written by him, that was part of the front matter of the 19th edition of the *Manual of Mineralogy*, 1977.

A Mineral

A mineral is a wondrous thing,
At least it is to me,
For in its ordered structure
Lies a world of mystery.
The secrets that it has withheld
For countless ages past
And clung to most tenaciously
Are being learned at last.
Each year using new techniques
Or with a new device,
We make our knowledge more complete,
Our data more precise.
But let us not in trying to solve
A mineral mystery
Forget that minerals are a part
Of natural history.
Nor in our quest for more detail
When probing an unknown,
Forget that every mineral
Has a beauty of its own.
With progress in technology
Each year sees new machines
That try to copy nature
By sophisticated means.
But for all these modern methods
We can not yet compete
With the world of ordered beauty
That lies beneath our feet.

C.S.H.

ACKNOWLEDGMENTS

I wish to express my sincere appreciation to several reviewers who provided me with in-depth assessments of the 21st edition of the *Manual of Mineralogy* (revised edition with CD-ROM). These are:

Richard W. Berry, San Diego State University, San Diego, California,

James G. Brophy, Indiana University, Bloomington,

Robert L. Hooper, University of Wisconsin, Eau Claire,

Theodore C. Labotka, University of Tennessee, Knoxville,

Jonathan B. Martin, University of Florida, Gainesville,

Jill D. Pasteris, Washington University, St. Louis, Missouri, and

Erich U. Petersen, The University of Utah, Salt Lake City.

Their valuable commentary has resulted in this extensively revised text. Jim Brophy and Jill Pasteris subsequently reviewed the first six chapters of the present text as well. Their additional comments were helpful and supportive.

CD-ROM version 2.0 was reviewed by John de Martino and Jennifer Whitney, both of whom are graduate students at California State University, Northridge. Their views on the incorporation of the CD-ROM in the text and on the use of images for lectures by overhead projection were enthusiastically supportive.

I also wish to thank a number of my departmental colleagues at the University of New Mexico for reviewing short sections of new text and/or for providing input as well as illustrations. These are:

Roger Y. Anderson,
Adrian Brearley,
John Husler,
Les D. McFadden,
Frans Rietmeijer,
Charles Shearer,
Michael Spilde, and
Huifang Xu.

Others, who reviewed aspects of the text and/or provided input are:

Charles W. Burnham, Durango, CO.,
John M. Hayes, Woods Hole Oceanographic Institution, Woods Hole, MA.,
J. A. Mandarino, Toronto, Ontario, Canada,
Malcolm Ross, Washington, D.C.,
L. Stixrude, University of Michigan, Ann Arbor, and
J. B. Thompson, Jr., Harvard University, Cambridge, MA.

I am grateful to Carl A. Francis and William C. Metropolis, of the Harvard Mineralogical Museum, for their assistance in photographing specimens from the collection for the eight new color plates in this text. They also provided for the loan of the malachite specimen used for the cover.

I wish to thank Senior Geology editor at John Wiley & Sons, Clifford W. Mills, for his continued support and understanding; Sandra Rigby, Illustration Coordinator; Kelly Tavares, who was a most efficient Senior Production Editor; and Madelyn Lesure, Design Director, who provided much input into the design and layout of the text. I am grateful to my son, Marc A. Klein, of New York City, who did much of the original lay-out of the design for the cover.

The word processing of the revisions was most efficiently and enthusiastically accomplished by Mabel T. Chavez of Santo Domingo Pueblo, New Mexico.

Cornelis Klein,
Albuquerque, New Mexico

CONTENTS

CHAPTER 1

INTRODUCTION

"Minerals are the basic stuff of the Earth, and their study will always remain at the core of the Earth Sciences." Frank C. Hawthorne, 1993

Mineral science (traditionally referred to as mineralogy) is the study of naturally occurring solid substances that constitute the solid portions of the universe. Minerals are the products of complex earth and planetary processes over a wide range of temperatures and pressures, and as such, they offer a key to the understanding of the origin and evolution of the Earth and planets. Minerals are also the raw materials on which much of society's technological development and economic productivity are based. Minerals and organic matter are the main constituents of soils. Gem minerals and the gemstones cut therefrom are the exceptionally beautiful and generally highly durable representatives of the mineral world.

Mineral science, the study and understanding of naturally occurring, generally inorganic, solid substances, called minerals, plays a central role in earth science curricula. Earth science is a broad field of study that aims at understanding the origin, evolution, and behavior of the Earth; it is also concerned with comprehending the place of the Earth in the solar system and the universe. The Earth and surrounding planets are made of solid materials called *rocks*, which generally consist of an aggregate of one or more minerals. The study of these *rock-forming minerals*—be it their chemistry, their atomic structure, their physical properties, the chemical reactions that created them, or a combination of these scientific assessments—has provided insight into many aspects of earth science. For example, *geophysics* is the study of the physics of the Earth, emphasizing its physical nature and dynamics. In conjunction with geophysicists, mineralogists have explored mineral behavior under experimentally produced extremely high pressure and temperature conditions, and have thus provided new insights into the mineralogical makeup of the Earth's mantle and core.

THE STUDY OF MINERALS AS PART OF EARTH SCIENCE

The relationship of mineral science to several branches of the earth sciences is shown schematically in Fig. 1.1. Additional subdisciplines of earth science could have been added to the diagram, such as soil science, geochronology, and sedimentology, but the main reason for the schematic is to argue that a very large part of what is done in the earth sciences involves minerals in one way or another. Beginning with the subject of petrology (Fig. 1.1) and going clockwise from there, one may ask, "What is the mineralogical basis for each of these branches of earth science?"

Petrology. The scientific study of rocks, their overall composition, their mineralogy, texture, structure and conditions of origin. *Experimental petrology* involves synthesizing rocks and minerals in the laboratory to evaluate the physical and chemical conditions under which they form.

Geochemistry. Much of geochemistry deals with the relative abundance, distribution, and migration of chemical elements (and their isotopes) in

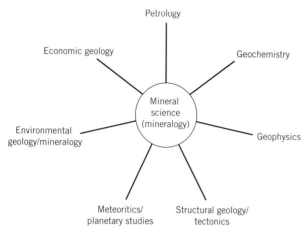

FIG. 1.1. The central role of mineral science in the earth sciences.

Earth and planetary materials that are represented by minerals, rocks, and soils.

Geophysics. This branch of earth science deals with such geologic phenomena as the temperature distribution of the Earth's interior and the source, configuration, and variations of the geomagnetic field. In all such investigations the physical properties of minerals and rocks are basic parameters to understanding the physical phenomena that are measured.

Structural Geology and Tectonics. A scientific discipline that evaluates rock deformation on both small and large scales. The scale of its investigation ranges from submicroscopic lattice defects in crystals to faulting and folding associated with mountain building and the large-scale upward and downward movements of the crust.

Meteoritics and Planetary Studies. The field of meteoritics is dedicated to the study of the chemistry and mineralogy of meteorite samples that have been collected on Earth. Planetary studies involve the study of rocks and soil (e.g., lunar regolith) returned from scientific missions to other planets. Such studies aid in an assessment of the geologic history of the planets from which the materials originated: asteroids, Mars, and the moon.

Environmental Geology and Mineralogy. Environmental geology is a scientific field that applies geologic research to the problems of land use and civil engineering. This includes reclaiming mined lands (with, e.g., sulfide minerals or radioactive minerals in waste dumps) and identifying geologically stable sites for

housing of nuclear waste and the construction of nuclear power plants. Environmental mineralogy is concerned with the interaction of minerals (their surfaces, their fracture patterns, their particle size) with biological systems. An example of this is the role of natural and anthropogenic mineral dust and the occurrence of pulmonary diseases.

Economic Geology. This field of study is concerned with the distribution of mineral deposits, the economic considerations involved in their recovery, and the assessment of available reserves. Economic geology covers the extraction of all materials from the Earth, including metal-rich ores, fossil fuels, and industrial materials such as salt, gypsum, building stone, and sand and gravel.

MINERAL SCIENCE

Mineral science, also referred to as mineralogy, is the title as well as the subject of this text. It concerns itself mainly with the study of naturally occurring inorganic solids called *minerals*; these may be of terrestrial or extraterrestrial origin. The subject is most closely related to the discipline of inorganic chemistry, but in mineral science the focus is specifically on naturally occurring solid substances.

The subject of mineral science encompasses five subdisciplines (see Fig. 1.2). Here we will briefly introduce these categories, beginning with descriptive mineralogy and followed by the next subject in a clockwise order of Fig. 1.2.

Descriptive Mineralogy. This involves the measurement and recording of physical properties (parameters) that help identify and describe a specific mineral. Some gross features of a mineral such as crystal form, hardness, color, and specific gravity ("heft") can be evaluated in hand specimen (i.e., using the five senses and some basic testing tools). Other, more objective criteria such as optical properties (e.g., refractive indices) and X-ray diffraction data (e.g., dimensions of the smallest building block of the atomic structure, the size of the unit cell) require specialized techniques and equipment such as a petrographic microscope and an X-ray diffraction system, respectively. Other measurable physical properties of minerals include their magnetic behavior, electrical properties, radioactivity, and their mechanical response to applied compressive or tensile loading.

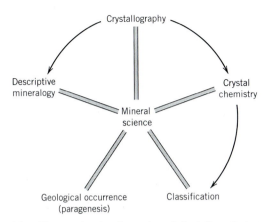

FIG. 1.2. Diagram showing the major subdisciplines that constitute mineral science. Arrows imply major contribution of one subfield to another.

Crystallography. This branch of science is very broad. Prior to the discovery of X-radiation by Wilhelm Conrad Roentgen in 1895 and the subsequent X-ray diffraction experiment by Max von Laue (observing the regular pattern of X-ray diffraction effects when a crystalline substance is properly positioned in an X-ray beam), crystallography concerned itself mainly with the geometric form, external symmetry, and optical properties of crystals. Since 1912 its main emphasis has been, and still is, the investigation of the internal structure of crystalline materials, whether organic or inorganic in origin. Most commonly an X-ray beam is used as the energy source for X-ray diffraction experiments. However, electron beams are also used for producing electron diffraction patterns. The ultimate aim of modern crystallographic techniques is the determination of the *crystal structure.* It provides information on the location of all the atoms, bond positions and bond types, internal symmetry (space group symmetry), and the chemical content of the unit cell. All such information is fundamental to concepts of *crystal chemistry,* which concerns itself with the interrelation of variability in chemistry and structure.

Crystal Chemistry. The field of crystal chemistry relates the chemical composition, the internal structure, and the physical properties of crystalline materials. A specific mineral is defined on the basis of its crystal structure, chemical composition, and related physical properties. In many mineral groups the overall pattern of the structure is relatively constant, whereas the chemical composition of such a group may be

highly variable. This is commonly called a *specific structure type,* showing extensive chemical substitution (i.e., solid solution). The assessment of structure type, of the atomic bonding arrangement, of the variability in overall chemistry, and of related changes in physical properties of a crystalline subtance are the domain of crystal chemistry.

Classification. There are approximately 3800 mineral species, and each has a distinctive name. To make sense of the divergent chemistries and structures represented by these minerals, it is customary to classify them according to a rational crystallochemical scheme. This means that minerals are first classified by their anionic group. Such classifications would include elements, sulfides, oxides, carbonates, silicates, and so on. Second, in groups with many species and complex structures, as in the silicate group, further subclassifications are made, mainly on the basis of the structural (atomic) arrangements of the silicate tetrahedra.

Geologic Occurrence. A common synonymous term is paragenesis. Geologic occurrence refers to the characteristic association or occurrence of a mineral or a mineral assemblage in a well-defined geologic setting. For example, a common occurrence (or paragenesis) for the relatively common sulfide *sphalerite,* ZnS, would be reported as "in ore deposits of hydrothermal origin." Garnet, a chemically complex silicate, is especially characteristic of metamorphic rocks. Its occurrence would be reported as "mainly in Al-rich rock types that are the product of regional metamorphism."

DEFINITION OF MINERAL

Although it is difficult to formulate a succinct and universally acceptable definition for the word *mineral,* the following applies well to the subject coverage in this text:

> A mineral is a naturally occurring solid with a highly ordered atomic arrangement and a definite (but not fixed) chemical composition. It is usually formed by inorganic processes.

A step-by-step analysis of this somewhat restrictive definition will aid in its understanding. At the end of the discussion of the implications of this definition there will be a brief overview of how a broader definition of *mineral* is now used in many research aspects of mineral and materials science.

The qualification *naturally occurring* distinguishes between substances formed by natural processes and those made in the laboratory. Industrial and research laboratories routinely produce synthetic equivalents of many naturally occurring materials, including valuable gemstones such as emeralds, rubies, and diamonds. Since the beginning of the twentieth century, mineralogic studies have relied heavily on the results from synthetic systems in which the products are given the names of their naturally occurring counterparts. Such practice is generally accepted, although it is at variance with the strict interpretation of *naturally occurring.* In this textbook *mineral* means a naturally occurring substance, and its name may be qualified by *synthetic* if purposely produced by laboratory techniques. One might now ask how to refer to $CaCO_3$ (calcite) that sometimes forms in concentric layers in city water mains. The material is precipitated from water by natural processes but in a manmade system. Most mineralogists would refer to it by its mineral name, calcite, because humanity's part in its formation was inadvertent.

The qualification *solid* excludes gases and liquids. Thus, H_2O as ice in a glacier is a mineral, but water is not. Likewise liquid mercury, found in some mercury deposits, must be excluded by a strict interpretation of the definition. However, in a classification of natural materials, such substances that otherwise are like minerals in chemistry and occurrence are called *mineraloids* and fall in the domain of the mineralogist.

A *highly ordered atomic arrangement* indicates an internal structural framework of atoms (or ions) arranged in a regular geometric pattern. Because this is the criterion of a crystalline solid, minerals are *crystalline.* Solids that lack an ordered atomic arrangement are called *amorphous* (e.g., glass). Several natural solids are amorphous. Examples are volcanic glass (which is not classified as a mineral because of its highly variable composition and lack of ordered atomic structure), limonite (a hydrous iron oxide), and allophane (a hydrous aluminum silicate); also several metamict minerals qualify—for example, microlite, gadolinite, and allanite (in metamict minerals the original crystallinity has been destroyed, to various degrees, by radiation from radioactive elements present in the original structure). They, with the liquids water and mercury, which also lack internal order, are classified as *mineraloids.*

The statement that a mineral has a *definite chemical composition* implies that it can be expressed by a specific chemical formula. For example, the chemical composition of quartz is expressed as SiO_2. Because quartz contains no chemical elements other than silicon and oxygen, its formula is definite. Quartz is, therefore, often referred to as a pure substance. Most minerals, however, do not have such well-defined compositions. Dolomite, $CaMg(CO_3)_2$, is not always a pure Ca-Mg-carbonate. It may contain considerable amounts of Fe and Mn in place of Mg. Because these amounts vary, the composition of dolomite is said to range between certain limits and is, therefore, *not fixed.* The chemical formula of a pure dolomite is given as $CaMg(CO_3)_2$; such a formula is commonly referred to as an *end-member composition.* The formula of a variety of dolomite with variable Fe and Mn in its structure would be expressed as $Ca(Mg,Fe,Mn)(CO_3)_2$ without specific subscripts for Mg, Fe, or Mn. The pure end-member formula and the one with greater chemical variability have a set of atomic ratios in common. In the end-member formula Ca : Mg : CO_3 is 1 : 1 : 2, and in the iron, and manganese-containing variety of dolomite, Ca : (Mg + Fe + Mn) : CO_3 is also 1 : 1 : 2. In other words, the overall atomic ratios in these formulas remain the same (*fixed*) even though one of the two dolomite types indicates a range in chemical composition (*its chemistry is not fixed*).

According to the traditional definition, a mineral is formed *by inorganic processes.* This is prefaced with *usually* and thus includes in the realm of mineralogy those organically produced compounds that answer all the other requirements of a mineral. The outstanding example is the calcium carbonate of mollusk shells. The oyster's shell and the pearl that may be within it are composed in large part of aragonite, identical to the inorganically formed mineral.

Although several forms of $CaCO_3$ (calcite, aragonite, vaterite) and monohydrocalcite, $CaCO_3 \cdot H_2O$, are the most common biogenic minerals (meaning "mineral formed by organisms"), many other biogenic species have been recognized. Opal (an amorphous form of SiO_2), magnetite (Fe_3O_4), fluorite (CaF_2), several phosphates, some sulfates, Mn-oxides, and pyrite (FeS_2), as well as elemental sulfur, are all examples of minerals that can be precipitated by organisms (see Lowenstam 1981). The human body also produces essential minerals. Apatite, $Ca_5(PO_4)_3(OH)$, is the principal constituent of bones and teeth. The body can also produce mineral matter (calculi) in the urinary system. Such calculi consist predominantly of calcium phosphates (such as hydroxylapatite, carbonate-apatite, and whitlockite), calcium oxalates that are very uncommon in the mineral world, and magnesium phosphates (see Gibson 1974).

However, petroleum and coal, frequently referred to as *mineral fuels,* are excluded. Although naturally formed, they have neither a definite chemical composition nor an ordered atomic arrangement. An exception is graphite, formed when coal beds have been subjected to high temperatures that drive off the volatile hydrocarbons and crystallize the remaining carbon.

Having evaluated in detail the implications of the definition of mineral that is most applicable to the subject and purposes of this text, it is instructive to discuss the broader definition accepted in much of ongoing mineral research. There is a great diversity in the scientific research venues that directly relate to mineral science. Taken in part from R. J. Hemly (1999), this includes: (1) the synthesis of new "minerals" at high pressures and temperatures to simulate materials that are expected to reside in the Earth's mantle and core (*Ultrahigh-Pressure Mineralogy* 1998; see reference list at the end of this chapter; see also p. 111); (2) the investigations of the transition from the crystalline to the amorphous (noncrystalline, disordered) state by application of extremely high pressures or by electron beam irradiation; (3) research on microorganisms that cause mineral precipitation and dissolution and control the distribution of elements in diverse environments at and below the surface of the Earth (*Geomicrobiology* 1997); (4) the study of mineral surfaces and their involvement in controlling reactions that occur near the Earth's surface; and (5) the synthetic production of zeolitic structures with spaceous channels that can be used in industrial applications as molecular sieves, ion exchangers, and catalysts. These wide-ranging research endeavors are part of mineral and materials science. With time, further results from such investigations will provide the mineralogist and earth scientist with a fuller understanding of the complex and heterogeneous makeup of the Earth and other planets.

HISTORY OF MINERALOGY

Although it is impossible in a few paragraphs to trace systematically the development of mineralogy, some of the highlights of its development can be singled out. The emergence of mineralogy as a science is relatively recent, but the practice of mineralogical arts is as old as human civilization. Natural

FIG. 1.3. Prospecting with a forked stick (A) and trenching (B) in the fifteenth century. (From Agricola, *De Re Metallica*, translated into English. Dover Publications, New York, 1950.)

FIG. 1.4. Portrait of Niels Stensen (Latinized to Nicolaus Steno). Steno was born in Copenhagen, Denmark, in 1638 and died in 1686. (From Scherz, G., *Steno, Geological Papers.* Odense University Press, 1969.)

weighing malachite and precious metals, smelting mineral ores, and making delicate gems of lapis lazuli and emerald. As the Stone Age gave way to the Bronze Age, other minerals were sought from which metals could be extracted.

We are indebted to the Greek philosopher Theophrastus (372–287 B.C.) for the first written work on minerals and to Pliny, who 400 years later recorded the mineralogical thought of his time. During the following 1300 years, the few works that were published on minerals contained much lore and fable with little factual information. If one were to select a single event signaling the emergence of mineralogy as a science, it would be the publication in 1556 of *De Re Metallica* by the German physician Georgius Agricola. This work gives a detailed account of the mining practices of the time and includes the first factual account of minerals. The book was translated into English from the Latin in 1912[1] by the former president of the United States, Herbert Hoover, and his wife, Lou Henry Hoover. An illustration from this book is reproduced in Fig. 1.3. In 1669 an important contribution was made to crystallography by Nicolas Steno through his study of quartz crystals. (A portrait of Nicolas Steno is reproduced in Fig. 1.4.) He noted that despite their differences in origin, size, or habit, the angles between corresponding faces were constant (see Fig. 1.5). More than a century passed before the next major contributions were made. In 1780 Carangeot invented a device (contact goniometer) for the measurement of interfacial crystal angles (see Fig. 1.7a). In 1783 Romé de l'Isle made angular measurements on crystals confirming Steno's work and formu-

pigments made of red hematite and black manganese oxide were used in cave paintings by early humans, and flint tools were prized possessions during the Stone Age. Tomb paintings in the Nile Valley executed nearly 5000 years ago show busy artificers

---

[1]Published in 1950 by Dover Publications, Inc., New York.

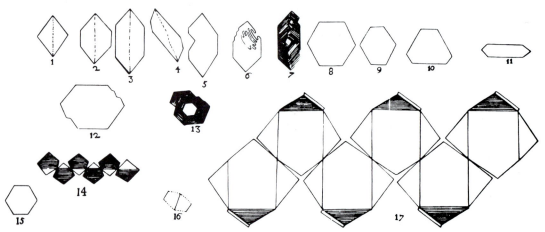

FIG. 1.5. Steno's drawings of various quartz and hematite crystals, illustrating the constancy of angles among crystals of different habits. (From Schafkranovski, J. J., 1971, *Die Kristallographischen Entdeckungen N. Stenens*, in *Steno as Geologist.* Odense University Press.)

lated the law of the constancy of interfacial angles. The following year, 1784, René J. Haüy showed that crystals were built by stacking together tiny identical building blocks, which he called integral molecules (see Fig. 1.6). The concept of integral molecules survives almost in its original sense in the unit cells of modern crystallography. Later (1801) Haüy, through his study of hundreds of crystals, developed the theory of rational indices for crystal faces.

In the early nineteenth century rapid advances were made in the field of mineralogy. In 1809 Wollaston invented the reflecting goniometer, which permitted highly accurate and precise measurements of the positions of crystal faces. Where the contact goniometer had provided the necessary data for studies on crystal symmetry, the reflecting goniometer (see Figs. 1.7c and d) would provide extensive, highly accurate measurements on naturally occurring and artificial crystals. These data made crystallography an exact science. Between 1779 and 1848 Berzelius, a Swedish chemist, and his students studied the chemistry of minerals and developed the principles of our present chemical classification of minerals.

In 1815 the French naturalist Cordier, whose legacy to mineral science is honored in the name of the mineral *cordierite*, turned his microscope on crushed mineral fragments in water. He thereby initiated the *immersion method* that others, later in the century, developed into an important technique for the study of the optical properties of mineral fragments. The usefulness of the microscope in the study of minerals was greatly enhanced by the invention in 1828 by the Scotsman, William Nicol, of a polarizing device that permitted the systematic study of the behavior of light in crystalline substances. The polarizing microscope became, and still is, a powerful determinative tool in mineralogical studies. An early model is illustrated in Fig. 1.8. In the latter part of the nineteenth century Fedorov, Schoenflies, and Barlow, working independently, almost simultaneously developed theories for the internal symmetry and order within crystals which became the foundations for later work in X-ray crystallography.

The most far-reaching discovery of the twentieth century must be attributed to Max von Laue of the University of Munich. In 1912 in an experiment performed by Friedrich and Knipping at the suggestion of von Laue, it was demonstrated that crystals could diffract X-rays. Thus was proved for the first time the regular and ordered arrangement of atoms in crystalline material. Almost immediately X-ray diffraction became a powerful method for the study of minerals

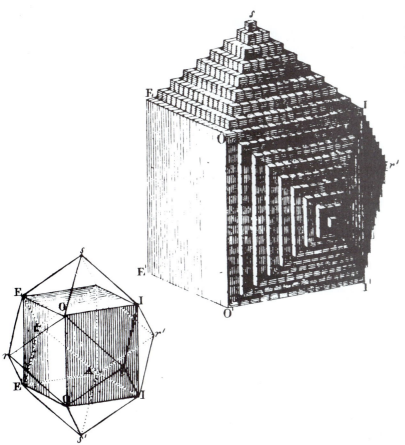

FIG. 1.6. Illustration of the concept developed by R. J. Haüy (1743–1826) of tiny identical building blocks underlying the external form of crystals. In this figure the development of a dodecahedron of garnet is shown. (From Marr, G. M., 1970, *Geschichte der Kristallkunde*. Reprinted by Sandig, Walluf, Germany.)

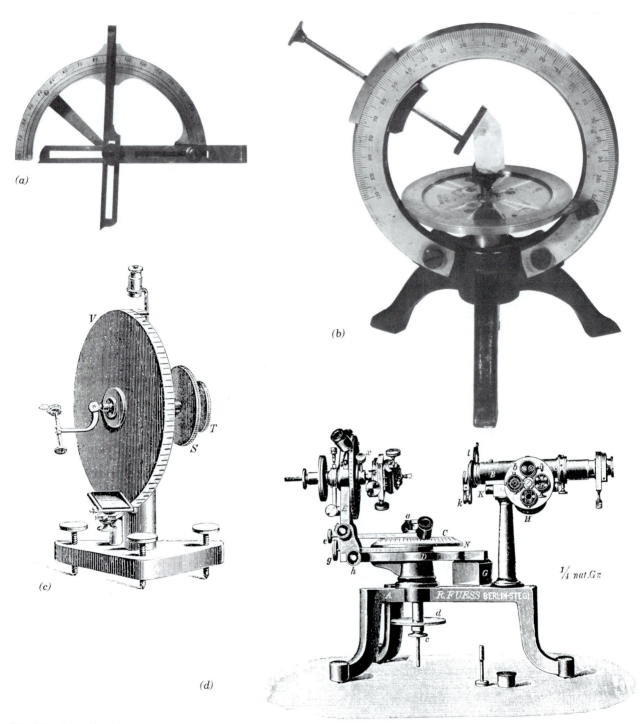

FIG. 1.7. Examples of instruments used for the measurement of angles between crystal faces.
(a) Brass contact goniometer of the Carangeot type. This was used at Harvard University in 1797 (see Frondel, 1983). *(b)* A two-circle contact goniometer based on the 1896 design of Victor Goldschmidt.
(c) The earliest reflecting one-circle goniometer as invented by W. H. Wollaston in 1809. (From Tschermak, G. and Becke, F., 1921, *Lehrbuch der Mineralogie*. Hölder-Pichler-Tempsky, Vienna.)
(d) A two-circle reflecting goniometer as developed in the latter part of the nineteenth century. (From Groth, P., 1895, *Physikalische Krystallographie*, Leipzig.) Compare with Fig. 6.1.

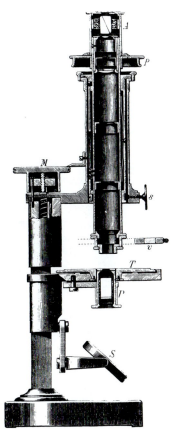

FIG. 1.8. Polarizing microscope as available in the mid-nineteenth century. (From Tschermak, G. and Becke, F., 1921, *Lehrbuch der Mineralogie*. Hölder-Pichler-Tempsky, Vienna.) Compare with Fig. 7.12.

and all other crystalline substances, and in 1914 the earliest crystal structure determinations were published by W. H. Bragg and W. L. Bragg in England. (Their photographs are given in Fig. 1.9.) Modern X-ray diffraction equipment with online, dedicated computers has made possible the relatively rapid determination of highly complex crystal structures. The advent of the electron microprobe in the early 1960s, for the study of the chemistry of minerals on a microscale, has provided yet another powerful tool that is now routinely used for the study of the chemistry of minerals, synthetic compounds, and glasses. Electron microprobes (see Figs. 1.10 and 1.11) can provide accurate, many-element analyses of solid

FIG. 1.9. Portraits of *(a)* Sir William Henry Bragg (1862–1942) and *(b)* his son Sir William Lawrence Bragg (1890–1971). Father and son received the Nobel Prize for Physics in 1915. Both men are eminently known for their researches in the field of crystal structure by X-ray methods. (*a* from Godfrey Argent, London, photograph by Walter Stoneman; *b* from Times Newspapers, Ltd., London.)

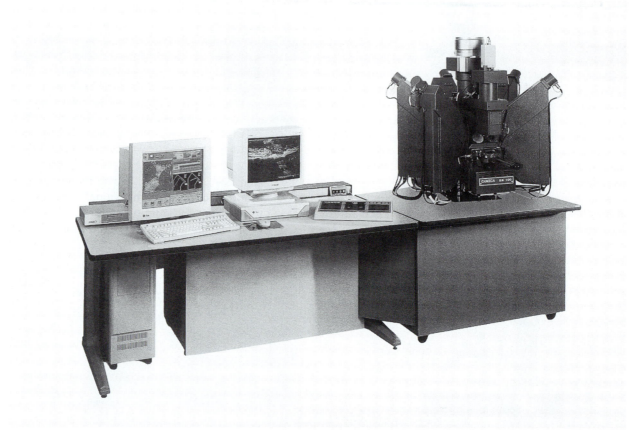

FIG. 1.10. A computer-automated electron microprobe manufactured by Cameca, Courbevoie, France. The electron beam column and X-ray spectrometers are on the right with computer control and read-out systems on the left. This is a Cameca SX 100 Electron Probe Microanalyzer. (Courtesy of Cameca Instruments, Inc., Trumbull, Ct).

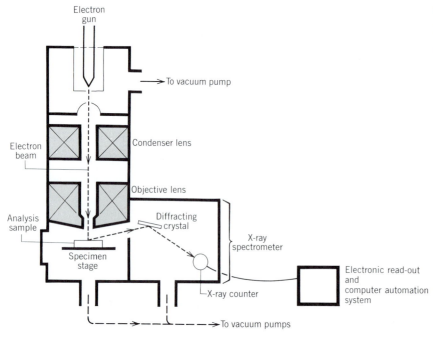

FIG. 1.11. Schematic cross section through the electron optical column and X-ray spectrometer of an electron micro-probe.

materials in a grain size as small as about one micrometer (0.001mm). The majority of mineral analyses now produced are made by electron microprobe not only because of the fine focus of the electron beam of the instrument, but because the analyses can be made *in situ*—on specific mineral grains in polished sections and polished thin sections of rocks. This has eliminated the laborious process of mineral separation and concentration, which is a requirement for several other mineral analysis techniques (see Chapter 7 for further discussion of analytical techniques).

Since early 1970 another electron beam instrument, which can magnify the internal architecture

of minerals many millions of times, has produced elegant and powerful visual images of atomic structures. This instrument, the transmission electron microscope, is illustrated in Fig. 1.12, and a schematic representation of the magnification process of an object into a final and highly enlarged image is shown in Fig. 1.13. The most visually instructive application of this technique is known as high-resolution transmission electron microscopy (HRTEM), which allows the study of crystalline materials at resolutions approaching the scale of atomic distances (see P. R. Buseck 1983). The technique can produce projected two-dimensional images of three-dimensional crystalline structures. These images show that many minerals have infinitely extending, periodic (meaning: perfectly repeating) internal structural arrangements; an example of such a "perfect" structure is illustrated by the HRTEM image in Fig. 1.14 for the chemically complex mineral tourmaline. The HRTEM images have also shown that minerals may contain defects that are deviations from idealized ("perfect") structures (see Figs. 4.53 and 4.54).

FIG. 1.12.   Transmission electron microscope, Philips Tecnai TEM manufactured by the FEI Company. The tall vertical feature is the electron column with control mechanisms on either side and a viewing screen at table-top level; the viewing screen is used for transmission electron microscope (TEM) imaging and for display of electron diffraction patterns. In the middle of the electron column and slightly to the right is the sample assembly holder. An energy dispersive X-ray detector (EDX) with liquid nitrogen dewar for cooling is affixed to the right of the column. Computer control of the whole system and a monitor for data display are to the far right. (Courtesy FEI Company, Hillsboro, Or.)

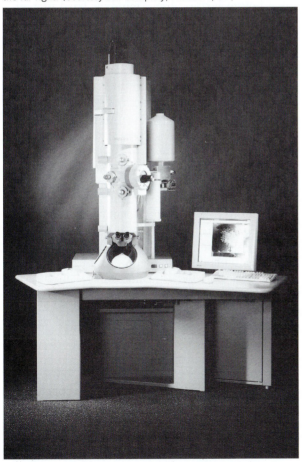

FIG. 1.13.   Schematic cross section through the column of a transmission electron microscope, showing the electron beam path for structural imaging. The four lenses are electromagnetic lenses.

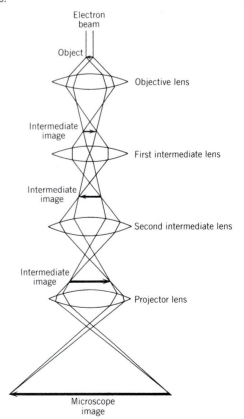

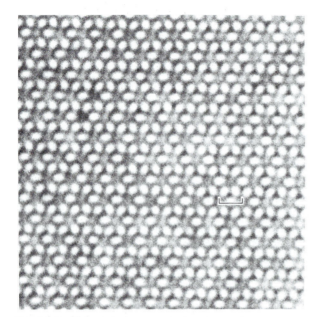

FIG. 1.14. High-resolution transmission electron microscope (HRTEM) image of the structure of tourmaline. (From Iijima, S., Cowley, J. M., and Donnay, G., 1973, High-resolution electron microscopy of tourmaline crystals. *Tschermaks Mineralogische Petrographische Mitteilungen*, v. 20, pp. 216–224.) The white areas in the photograph correspond to regions of low electron density in the structure of tourmaline. The six-fold pattern is the image of the six-fold Si_6O_{18} ring in tourmaline (compare with Fig. 11.14). The bar scale represents 15 angstroms.

The field of mineral science now encompasses a very broad area of study that includes X-ray, electron, and neutron diffraction by minerals, mineral synthesis, crystal physics, the evaluation of the thermodynamic stability of minerals, petrography (the study of rocks and minerals in thin section), petrology (the study of rocks), experimental petrology, and aspects of metallurgy and ceramics. Because it is difficult to predict which contributions (made in relatively recent times) to the science of mineralogy will prove to be most enduring and important, Table 1.1 provides a tabulation of the Roebling medalists and some of their major research contributions. This list illustrates the diversity of internationally recognized professional contributions, and it is a reasonable assumption that it includes those mineralogists whom future historians will regard as the mineralogical giants of our time.

The Roebling medal was established in 1937 by the Mineralogical Society of America in memory of Colonel Washington A. Roebling (1837–1926), who had made a generous financial gift to the society in 1926. Colonel Roebling, the designer of such well-known suspension bridges as those over the Niagara River at Niagara Falls, over the Allegheny River at Pittsburgh, over the Ohio River in Cincinnati, and the Brooklyn Bridge over the East River in New York City, had a deep, life-long interest in the study of minerals. The Roebling medal signifies the highest recognition of achievement American mineralogy can bestow on outstanding investigators in the United States or abroad. The presentation and acceptance speeches for each Roebling award can be found in *American Mineralogist*.

In 1927, Roebling's son John donated his father's mineral collection of about 16,000 specimens to the National Museum of Natural History (Smithsonian Institution) in Washington, D.C. This collection, known as the Washington A. Roebling mineral collection, was undoubtedly one of the largest and finest private collections of its time. This acquisition, together with another collection of about 9100 specimens (the Canfield collection), made the Smithsonian mineral collection one of the best in the world (A. Roe 1990, Washington A. Roebling, his life and his mineral collection, *Mineralogical Record*, v. 21, pp. 13–30).

ECONOMIC IMPORTANCE OF MINERALS

Since before historic time, minerals have played a major role in humanity's way of life and standard of living. With each successive century they have become increasingly important, and today we depend on them in countless ways—from the construction of skyscrapers to the manufacture of computers. Modern civilization depends on and necessitates the prodigious use of minerals. A few minerals such as talc, asbestos, and sulfur are used essentially as they come from the ground, but most are first processed to obtain a usable material. Some of the more familiar of these products are bricks, glass, cement, plaster, and a score of metals ranging from iron to gold. Metallic ores and industrial minerals are mined on every continent wherever specific minerals are sufficiently concentrated to be economically extracted.

According to the U.S. Bureau of Mines (as reported in *Geotimes*, 1989, v. 34, p. 19), "each year, every American requires 40,000 pounds of new minerals. At that level of consumption, the average newborn infant will need a lifetime supply of 795 pounds of lead (mainly for car batteries, solder, and electronic components); 757 pounds of zinc (as an alloy of copper to make brass, as protective coatings on steel, and as chemical compounds in rubber and paints); 1500 pounds of copper (mainly used in electrical motors, generators, communication equipment, and wiring); 3593 pounds of aluminum (for all

TABLE 1.1   Recipients of the Roebling Medal Awarded by the Mineralogical Society of America

Award Year	Recipient and Institutional Affiliation	Examples of Outstanding Contributions to Mineralogy
1937	Charles Palache, Harvard University	goniometric studies in crystallography and mineral paragenesis; minerals of Franklin Furnace, N.J.
1938	Waldemar T. Schaller, U.S. Geological Survey	chemical mineralogy; crystallography and paragenesis of pegmatite minerals
1940	Leonard James Spencer, British Museum (Natural History)	mineral systematics; long-time editor of *Mineralogical Magazine* and *Mineralogical Abstracts*
1941	Esper S. Larsen, Jr., Harvard University	mineralogy and petrology; the determination of the optical properties of nonopaque minerals
1945	Edward H. Kraus, University of Michigan	occurrence and origin of minerals; crystallographic forms; development of apparatus for mineral testing
1946	Clarence S. Ross, U.S. Geological Survey	petrography and petrology, particularly clay minerals; microscopic techniques; geochemistry of ore deposits
1947	Paul Niggli, Technische Hochschule, Zurich	crystallography and structure of minerals; igneous and metamorphic rocks; minerals of the Swiss Alps
1948	William Lawrence Bragg, Cavendish Laboratory, University of Cambridge	crystal structure determinations by X-ray diffraction techniques; recipient (jointly with his father, William Henry Bragg) of the Nobel Prize for Physics in 1915
1949	Herbert E. Merwin, Geophysical Laboratory, Carnegie Institution	crystal optics; characterization of synthetic, transparent, and opaque phases
1950	Norman L. Bowen, Geophysical Laboratory, Carnegie Institution	application of experimental physico-chemical data and principles to petrological problems
1952	Frederick E. Wright, Geophysical Laboratory, Carnegie Institution	optical properties of minerals; design of improved petrographic microscope and test plates
1953	William F. Foshag, U.S. National Museum	characterization of new minerals; minerals of the United States and Mexico; gemology
1954	Cecil Edgar Tilley, University of Cambridge	application of physico-chemical principles to the study of mineral assemblages in metamorphic and igneous rocks
1955	Alexander N. Winchell, University of Wisconsin	evaluation of major rock-forming silicate groups; textbooks on optical mineralogy
1956	Arthur F. Buddington, Princeton University	quantitative petrologic studies of many rock types and of ore deposits
1957	Walter F. Hunt, University of Michigan	mineralogy; editor for 35 years of *The American Mineralogist*
1958	Martin J. Buerger, Massachusetts Institute of Technology	structural crystallography; development of single crystal X-ray techniques; textbooks in crystallography
1959	Felix L. Machatschki, University of Vienna	atomic arrangement of major silicate groups; atomic substitution; relation of crystal structure to paragenesis
1960	Thomas F. W. Barth, Oslo University	petrology and X-ray crystallography; petrogenetic relations of rocks in regions of Norway (e.g., the Oslo region) and North America
1961	Paul Ramdohr, University of Heidelberg	reflected light microscopy; mineralogy and genesis of ore deposits; major text on ore mineralogy
1962	John W. Gruner, University of Minnesota	X-ray crystallography of clay minerals; mineralogy and petrology of iron-formations; uranium mineralization
1963	John Frank Schairer, Geophysical Laboratory, Carnegie Institution	experimental studies of silicate phase equilibria
1964	Clifford Frondel, Harvard University	structural mineralogy; characterization of many new minerals; mineralogy of uranium and thorium; co-author of *Dana's System*
1965	Adolph Pabst, University of California Berkeley	X-ray crystallographic and structural mineralogy; evaluation of the metamict state and of sheet silicates
1966	Max H. Hey, British Museum (Natural History)	mineral chemistry of zeolites, the chlorite group, and meteorites; author of *Index of Mineral Species*; long-time editor of *Mineralogical Magazine*
1967	Linus Pauling, University of California, San Diego	crystal and molecular structures; quantum chemistry; theory of chemical bonding; author of *The Nature of the Chemical Bond*; recipient of the Nobel Prize in Chemistry in 1954 and the Nobel Peace Prize in 1962
1968	Tei-ichi Ito, University of Tokyo	structural crystallography; polymorphism; X-ray powder diffraction
1969	Fritz Laves, Technische Hochschule, Zurich	structure and paragenesis of feldspars; crystal chemistry of metallic compounds
1970	George W. Brindley, Pennsylvania State	structural crystallography of layer silicates; clay mineralogy; crystallography;
1971	J. D. H. Donnay, Johns Hopkins University	crystal optics; relationship of morphology and structure; twinning; author of *Crystal Data*
1972	Elburt F. Osborn, Bureau of Mines, U.S. Department of Interior	experimental petrology in rock-forming systems; crystallization and differentiation trends in magmas
1973	George Tunell, University of California, Santa Barbara	experimental investigations of oxidized ore minerals; physical-chemical evaluations of ore-forming processes

(continued)

TABLE 1.1 (continued)

Award Year	Recipient and Institutional Affiliation	Examples of Outstanding Contributions to Mineralogy
1974	Ralph E. Grim, University of Illinois, Champaign-Urbana	clay mineralogy; author of *Clay Mineralogy* and *Applied Clay Mineralogy*
1975	O. Frank Tuttle, Stanford University	experimental petrology; development of hydrothermal research equipment; experimental studies on the origin of granite
1975	Michael Fleischer, U.S. Geological Survey	trace element geochemistry; evaluation of new mineral species; long-time chairman of international commission on new minerals; long-time editor of section on mineralogy in *Chemical Abstracts*
1976	Carl W. Correns, University of Göttingen	physical-chemical evaluations of sedimentology; chemical and physical oceanography; mineralogy of sediments
1977	Raimond Castaing, University of Paris	inventor of the electron microprobe and pioneering work on theory of quantitative analysis; development of ion beam microprobe
1978	James B. Thompson, Jr., Harvard University	theoretical evaluation of petrologic systems; thermodynamics of mineral systems; metamorphic reactions in pelitic schists; crystal chemistry of amphiboles
1979	William H. Taylor, Cavendish Laboratory, University of Cambridge	structural crystallography; structural features of feldspar, zeolite, and aluminosilicate minerals
1980	D. S. Korzhinskii, Academy of Sciences, Moscow	application of chemical thermodynamics to petrology; author of *Physicochemical Basis for the Analysis of the Paragenesis of Minerals* and of *Theory of Metasomatic Zoning*
1981	Robert M. Garrels, University of South Florida	theoretical studies of ore formation; chemical thermodynamics of mineral systems; phase diagrams for minerals at low temperature; chemical evolution of the ocean and atmosphere; co-author of *Solutions, Minerals and Equilibria*
1982	Joseph V. Smith, University of Chicago	structural crystallography of rock-forming minerals; zeolites and feldspars; lunar mineralogy and petrology; author of *Feldspars* (2 volumes)
1983	Hans P. Eugster, Johns Hopkins University	solid-fluid equilibria in hydrothermal systems; chemical sedimentation and water chemistry such as in active salt lakes
1984	Paul B. Barton, Jr., U.S. Geological Survey	petrology of ores; the chemistry and physics of the ore-forming process
1985	Francis J. Turner, University of California, Berkeley	metamorphic petrology
1986	Edwin Roedder, U.S. Geological Survey	fluid inclusions in minerals
1987	Gerald V. Gibbs, Virginia Polytechnic Institute and State University	mathematical foundations of crystallography; application of molecular orbital theory to chemical bonding
1988	Julian R. Goldsmith, University of Chicago	order–disorder in feldspars; phase equilibria of carbonates; oxygen-isotope fractionations among minerals
1989	Helen D. Megaw, University of Cambridge	X-ray crystal structure of feldspars; origin of ferroelectricity in oxides
1990	Sturges W. Bailey, University of Wisconsin	structural and crystal chemical studies of layer silicate minerals
1991	E-an Zen, U.S. Geological Survey	application of thermodynamics to petrology; the temperature-pressure regime of the Appalachian mountain belt; igneous thermobarometry
1992	Hatten S. Yoder, Jr., Geophysical Laboratory, Carnegie Institution	experimental petrology and its application to mineral paragenesis; study of the role of water in metamorphism and the petrogenesis of igneous rocks; author of *Generation of Basaltic Magma*
1993	Brian Mason, National Museum of Natural History (Smithsonian Institution)	characterization of a broad range of terrestrial and meteoritic mineralogy; author of *Principles of Geochemistry*, and of *Meteorites*; co-author of *Mineralogy*
1994	William A. Bassett, Cornell University	experimental studies in high-pressure mineral physics using the diamond cell that he developed
1995	William S. Fyfe, University of Western Ontario	experimental phase equilibria, crystal chemistry, aqueous geochemistry, mineral spectroscopy, high temperature processes, and environmental geochemistry
1996	Donald H. Lindsley, State University of New York at Stony Brook	experimental and theoretical work on phase relations in Fe-Ti oxides, pyroxenes, and olivines; thermometry, barometry, and oxybarometry of rock types with these ferromagnesian minerals
1997	Ian S. E. Carmichael, University of California, Berkeley	studies of volcanic regions in the Pacific rim; tectonic and volcanic history of Western Mexico; thermodynamic properties of silicate liquids
1998	C. Wayne Burnham, Pennsylvania State University	experimental and thermodynamic evaluations of the behavior of volatiles in igneous petrogenesis
1999	Ikuo Kushiro, University of Tokyo	experimental studies on the genesis of basaltic and andesitic magmas and on the properties of magmas at high pressures
2000	Robert C. Reynolds, Jr., Dartmouth College	X-ray diffraction studies of mixed-layered clay minerals and computer simulations of diffraction patterns of disordered minerals
2001	Peter J. Wyllie, California Institute of Technology	experimental petrology on various aspects of igneous processes, producing a long series of phase diagrams that have elucidated many magmatic processes

sorts of things such as beverage cans, folding lawn chairs, and aircraft); 32,700 pounds of pig iron (for kitchen utensils, automobiles, ships, and large buildings); 28,213 pounds of salt (for cooking, highway de-icing, and detergents); and 1,238,101 pounds of stone, sand, gravel, and cement (for building roads, homes, schools, offices, and factories)." A nationwide commitment to recycling will lower many of the above estimates.

The location of mineable metal and industrial mineral deposits, and the study of the origin, size, and ore grade of these deposits is the domain of economic geologists. But a knowledge of the chemistry, occurrence, and physical properties of minerals is basic to pursuits in economic geology.

NAMING OF MINERALS

Minerals are most commonly *classified* on the basis of the presence of a major chemical component (an anion or anionic complex) into oxides, sulfides, silicates, carbonates, phosphates, and so forth. This is especially convenient because most minerals contain only one major anion. However, *naming* minerals is not based on such a logical chemical scheme.

The careful description and identification of minerals commonly require highly specialized techniques such as chemical analysis and measurement of physical properties, among which are the specific gravity, optical properties, and X-ray parameters that relate to the atomic structure of minerals. However, the names of minerals are not arrived at in an analogous scientific manner. Minerals may be given names on the basis of some physical property or chemical aspect, or they may be named after a locality, a public figure, a mineralogist, or almost any other subject considered appropriate. Some examples of mineral names and their derivations are as follows:

Albite ($NaAlSi_3O_8$) from the Latin *albus* (white), an allusion to its color.

Rhodonite ($MnSiO_3$) from the Greek *rhodon* (a rose), an allusion to its characteristically pink color.

Chromite ($FeCr_2O_4$) because of the presence of a large amount of chromium in the mineral.

Magnetite (Fe_3O_4) because of its magnetic properties.

Franklinite ($ZnFe_2O_4$) after a locality, Franklin, New Jersey, where it occurs as the dominant zinc mineral.

Sillimanite (Al_2SiO_5) after Professor Benjamin Silliman of Yale University (1779–1864).

An international committee, the Commission on New Minerals and New Mineral Names of the Inter-

national Mineralogical Association, now reviews all new mineral descriptions and judges the appropriateness of new mineral names as well as the scientific characterization of newly discovered mineral species. The *Glossary of Minerals Species*, published in 1991 by Michael Fleischer (see end of this chapter for complete reference), lists the internationally recommended names for nearly 3500 mineral entries. This text will use the names given in that listing. An alphabetical listing of about 3800 mineral species is given in *Encyclopedia of Mineral Names*, 1997 (see end of this chapter for complete reference).

REFERENCES AND LITERATURE OF MINERALOGY

The first comprehensive book on mineralogy in English, *A System of Mineralogy*, was written by James D. Dana in 1837. Since then, through subsequent revisions, it has remained a standard reference work. The last complete edition (the sixth) was published in 1892, with supplements in 1899, 1909, and 1915. Parts of a seventh edition, known as *Dana's System of Mineralogy*, appeared as three separate volumes in 1944, 1951, and 1962. The first two volumes cover the nonsilicate minerals, and volume three deals with silica (quartz and its polymorphs). A more recent reference is the five-volume work, *Rock-Forming Minerals*, by W. A. Deer, R. A. Howie, and J. Zussman; additional volumes are now being published sequentially (see complete reference at the end of this chapter). The treatment of the physical properties of all minerals in *Dana's System* is exhaustive. The coverage in *Rock-Forming Minerals*, however, is more topical and expansive in the areas of chemistry, structure, and experimental studies, but is essentially confined to the rock-forming minerals. Another handy reference on the chemistry and nomenclature of minerals is *Mineralogische Tabellen* (in German) by H. Strunz. A useful tabulation of mineral data can be found in the four-volume set titled *Handbook of Mineralogy* by J. Anthony et al. (see end of this chapter for complete reference). An in-depth treatment of topical subjects in mineralogy is provided by *Reviews in Mineralogy*, vols. 1 to 40, published by the Mineralogical Society of America. Example titles are *Orthosilicates* (vol. 5), *Pyroxenes* (vol. 7), *Amphiboles and Other Hydrous Pyriboles—Mineralogy* (vol. 8), *Micas* (vol. 13), and *Hydrous Phyllosilicates (Exclusive of Micas)* (vol. 19).

The diverse literature of mineralogy is found in papers published in scientific journals all over the world. The most widely circulated mineralogical journals in the English-speaking world are *American Mineralogist*, published by the Mineralogical Society of America, *The Canadian Mineralogist*, published by the Mineralogical Association of Canada, and the *Mineralogical Magazine*, published by the Mineralogical Society of Great Britain. In 1989 a new mineralogical journal, *European Journal of Mineralogy*, was created, following the merged publications of the mineralogical societies of Germany, France, and Italy. *The Mineralogical Record*, first published in 1970, devotes itself to mineralogical subjects that are often of more general appeal to hobbyists or mineral collectors than those found in the other above-mentioned journals.

REFERENCES AND SELECTED READING

Standard Mineralogical Reference Works

Anthony, J. W., R. A. Bideaux, K. W. Bladh, and M. C. Nichols. *Handbook of mineralogy*. Vol. 1, *Elements, sulfides, sulfosalts*, 1990. Vol. 2, *Silica, silicates*, 1995. Vol. 3, *Halides, hydroxides, oxides*, 1997. Vol. 4, *Arsenates, phosphates and vanadates*, 2000. Tucson, Ariz.: Mineral Data Publishing.

Blackburn, W. H., and W. H. Dennen. 1997. *Encyclopedia of mineral names*. The Canadian Mineralogist, Special Publication 1.

Dana, J. D. 1944–1962. *A System of mineralogy*. 7th ed. 3 vols. New York: Wiley. Rewritten by C. Palache, H. Berman, and C. Frondel.

Deer, W. A., R. A. Howie, and J. Zussman. 1962. *Rock-forming minerals*. 5 vols. New York: Wiley. Five completely revised volumes (nos. 1A, 1B, 2A, 2B, and 5B) were published in 1982, 1986, 1978, 1997, and 1996, respectively, and are available from the Geological Society, London, Great Britain.

Fleischer, M. 1991. *Glossary of mineral species*. Tucson, Ariz.: The Mineralogical Record.

Gaines, R. B., H. C. W. Skinner, E. E. Foord, B. Mason, and A. Rosenzweig. 1997. *Dana's new mineralogy*. New York: Wiley.

Nickel, E. H., and M. C. Nichols. 1991. *Mineral reference manual*. New York: Van Nostrand Reinhold.

Reviews in mineralogy. 40 vols. Washington, D.C.: *Mineralogical Society of America*, 1974–2000.

Strunz, H. 1970. *Mineralogische Tabellen*. 5th ed. Leipzig: Akademische Verlagsgesellschaft.

Historical Accounts

Agricola, G. 1950. *De Re Metallica*. Translated from the first Latin edition of 1556 by Herbert C. Hoover and Lou H. Hoover. New York: Dover Publications.

Ford, W. E. 1918. The growth of mineralogy from 1818 to 1918. *American Journal of Science* 46: 240–54.

Frondel, C. 1983. An overview of crystallography in North America. Pp. 1–24 in *Crystallography in North America*. Edited by D. McLachlan and J. P. Glusker. New York: American Crystallographic Association.

Greene, F. C., and J. G. Burke. 1978. The science of minerals in the age of Jefferson. *American Philosophical Society* 68, pt. 4: 113 pp.

Hazen, R. M. 1984. Mineralogy: A historical review. *Journal of Geological Education* 32: 288–98.

Knopf, A. 1941. Petrology. *Geology 1888–1938*. Geological Society of America, 50th anniversary vol.: 333–64.

Kraus, E. H. 1941. Mineralogy. *Geology 1888–1938*. Geological Society of America, 50th anniversary vol.: 307–32.

Pirsson, L. V. 1918. The rise of petrology as a science. *American Journal of Science* 46: 222–40.

Economic Geology/Natural Resources

Cameron, E. G. 1986. *At the crossroads: The mineral problems of the United States*. New York: Wiley.

Carr, D. C., ed. 1994. *Industrial minerals and rocks*. 6th ed. Littleton, Colo.: Society for Mining, Metallurgy, and Exploration, Inc.

Evans, A. M. 1987. *Introduction to ore geology*, 2nd ed. London: Blackwell Scientific Publications.

Guilbert, J. M., and C. F. Park Jr. 1986. *The geology of ore deposits*. New York: W. H. Freeman.

Holland, H. D., and U. Petersen. 1995. *Living dangerously: The Earth, its resources, and the environment*. Princeton, N.J.: Princeton University Press.

Kessler, S. E. 1994. *Mineral resources, economics, and the environment*. New York: MacMillan.

U.S. Bureau of Mines. Various years. *Minerals yearbook*. Washington, D.C.: U.S. Government Printing Office.

Additional References

Banfield, J. F., and K. H. Nealson, eds. 1997. Geomicrobiology: Interactions between rocks and minerals. *Reviews in Mineralogy* 35, Mineralogical Society of America.

Buseck, P. R. 1983. Electron microscopy of minerals. *American Scientist* 71, no. 2: 175–85.

Gibson, R. I. 1974. Descriptive Human Pathological Mineralogy. *American Mineralogist* 59: 1177–82.

Hawthorne, F. C. 1993. Minerals, mineralogy, and mineralogists: Past, present, and future. *The Canadian Mineralogist* 31: 253–96.

Hemley, R. J. 1999. Mineralogy at the Crossroads. *Science* 285: 1026–27.

Hemley, R. J., ed. 1998. Ultrahigh-pressure mineralogy: Physics and chemistry of the Earth's deep interior. *Reviews in Mineralogy* 37, Mineralogical Society of America.

Lowenstam, H. A. 1981. Minerals formed by organisms. *Science* 211: 1126–31.

CHAPTER 2

PHYSICAL PROPERTIES OF MINERALS IN HAND SPECIMEN

The physical properties of a mineral are an expression of the interrelation of a mineral's crystal structure and its chemical composition. Physical properties are generally more easily determined than the chemical composition or the structure of a mineral. A combination of visual observations followed by some simple physical tests may be enough to uniquely characterize a particular mineral. In this chapter the main emphasis is on properties that are easily and quickly determined in hand specimen. A subsequent chapter will deal with more objective physical parameters such as those obtained from optical and X-ray diffraction studies.

In most mineralogy courses the student is required to enroll in one or more related laboratory sessions. One common aspect of such laboratory instruction is the study of minerals in hand specimen. This involves learning to use the eyes and other unaided senses as well as some very basic testing tools to arrive at a reasonably correct (or even unique) identification. The order of subject discussion in this chapter is broadly that which should be adopted by the student in the laboratory or in the field. Those properties that depend solely on visual inspection will be discussed first. These include the following:

Crystal form and *habit*
Intergrowths, twins, and *striations*
State of aggregation
Luster, color, and *streak*
Other properties depending on light
Cleavage

This is followed by the evaluation of properties that require some relatively simple testing equipment:

Hardness
Specific gravity (or *"heft"*)
Magnetism
Radioactivity
Solubility in HCl

The chapter concludes with a brief introduction to an electrical property, *piezoelectricity*, which has found enormous application in the control of radio frequencies, such as in quartz watches.

CRYSTAL FORM AND CRYSTAL HABIT

One of the most aesthetically pleasing aspects of mineral specimens in museum exhibits, in mineral shows, or in private collections is their occurrence in well-formed crystals or crystal groupings. *Crystals are bounded by smooth plane surfaces and assume regular geometric forms* (see Fig. 2.1). The regular geometric shape of a crystal, its *crystal form*, is not only pleasing to the eye but is also a diagnostic physical property. Mineral specimens that exhibit good crystal outline have formed under favorable geological conditions and are commonly the result of chemical deposition from a solution (or melt) in an open space, such as a vug or cavity in a rock formation.

The external form of crystals, or their *morphology*, is the outward expression of their internal ordered atomic arrangement. Figure 2.2 shows a packing model of the structure of halite, NaCl, with Na^+ as small and Cl^- as larger spheres. The outline of the model shows the presence of two crystal forms: a cube and an octahedron. The cube is the form with the six square faces perpendicular to the three axes (a_1, a_2, and a_3); the octahedron is the form with eight

FIG. 2.1. Scanning electron microscope (SEM) photograph of three beautifully formed crystals of the mineral analcime, $NaAlSi_2O_6 \cdot H_2O$, from Ischia, Italy. Each crystal displays a single form, a trapezohedron that reflects a high symmetry content in the isometric system. The trapezohedron is composed of 24 trapezium-shaped faces (from Gottardi, G., and Galli, E., 1985, *Natural Zeolites*, Springer-Verlag, New York; with permission).

triangular faces, each of which cuts off one of the eight corners of the cube. Upon careful inspection of the atomic packing shown in Fig. 2.2 it becomes obvious that the cube faces are underlain by planes that contain half Na^+ and half Cl^- ions. On the other hand, the octahedral faces at the corners are underlain by planes containing only Cl^-. As such, different crystal planes represent differences in the underlying atomic environment. Figure 2.2 also shows the geometric appearance of two independent geometric forms, the cube and the octahedron; combinations of these forms are shown as well. In this early chapter in the text, without a basis in crystallography, suffice it to say that the study of crystal forms leads the investigator to the overall assessment of the symmetry content of the crystal at hand and subsequently to a classification of its symmetry as part of a crystal class (or point group).[1] The term *form* has a specific meaning in crystallography. In its most familiar sense, *form* is used to indicate general outward appearance. In crystallography, however, external shape is denoted by the word *habit*, whereas *form* is used in a special and restricted sense. *A form consists of a group of crystal faces, all of which have the same relation to the elements of symmetry* (inherent in the crystal). *Crystal habit*, therefore, includes its general shape and irregularities of growth.

---

[1]An introductory overview of elements of crystallography is given in Chapter 5. A systematic treatment of morphology is the subject of Chapter 6.

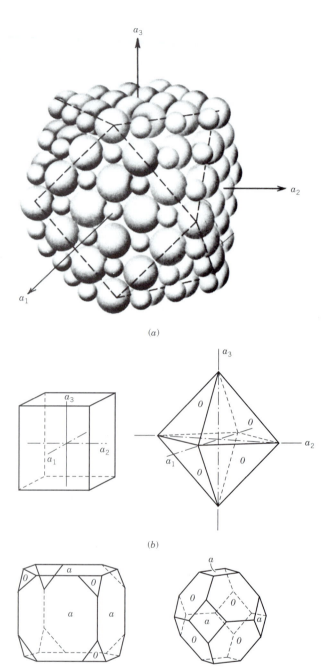

FIG. 2.2. *(a)* Packing model of halite with cubo-octahedral outline. a_1, a_2, and a_3 are reference axes that apply to the isometric system. *(b)* The geometric expression, in the external form of crystals, of the cube and the octahedron. *(c)* Two geometric combinations of a cube and an octahedron. The difference in these two crystals is the predominance of the cube in the left drawing and the predominance of the octahedron in the right drawing. The letters *a* and *o* are standard labels for cube and octahedral faces, respectively. See text for discussion.

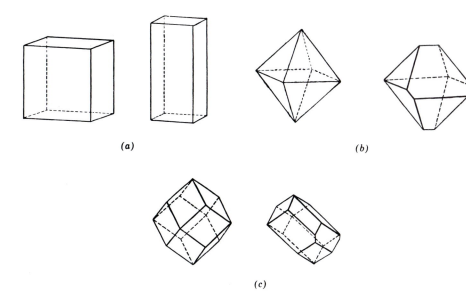

FIG. 2.3. Examples of perfect as well as distorted forms in the isometric system *(a)* Perfect cube and malformed cube. *(b)* Octahedron and asymmetric octahedron. *(c)* Dodecahedron and malformed dodecahedron.

Although "perfect" crystals are shown in idealized crystal drawings in Chapter 6 and elsewhere in this text, most real crystals tend to be somewhat malformed. As such, their true symmetry is not immediately apparent; furthermore, the size of the equivalent faces may be unequal, and therefore the shape of the crystal as a whole appears distorted. Figure 2.3 shows some examples of crystal habit in the isometric system, and Fig. 2.4 illustrates how various divergent shapes and crystallographic forms may arise in the isometric system. The small cubic blocks that are used to build up the various shapes (in Fig. 2.4) are a schematic representation of the unit cells of the underlying regular atomic arrangement. A *unit cell* is defined as *the smallest unit of structure that if stacked indefinitely in three dimensions would form the whole structure.*

Three commonly used terms express the quality of the development of the external faces on a crystal:

Euhedral. From the Greek roots *eu*, meaning good, and *hedron*, meaning plane; describing a mineral that is completely bounded by crystal

FIG. 2.4. Different external shapes are the result of the systematic stacking of cubic unit cells. *(a)* Perfect cube. *(b)* Distorted cubes. *(c)* Perfect octahedron, and *(d)* Perfect dodecahedron.

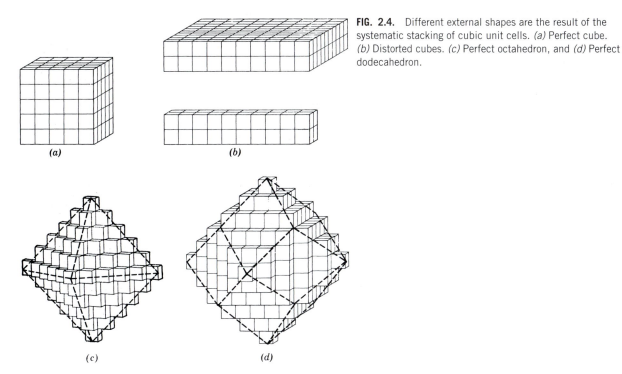

faces and whose growth during crystallization was not restrained or interfered with by adjacent crystals or mineral grains.

Subhedral. From the Latin root *sub*, meaning less than; describing a crystal or mineral grain that is partly bounded by crystal faces and partly by surfaces formed against preexisting grains.

Anhedral. From the Greek root *an*, meaning without; for minerals that lack crystal faces and that may show rounded or irregular surfaces produced by the crowding of adjacent minerals during crystallization.

All the crystal drawings in this text are therefore of *euhedral* crystals, whereas the minerals that are intergrown, for example, in a granite, will tend to be *subhedral* and *anhedral*.

If mineral specimens display well-developed crystal forms, the form names are used to describe their outward appearance,[2] as the following examples illustrate:

prismatic—for a crystal with one dimension markedly longer than the other two

rhombohedral—with the external form of a rhombohedron

cubic—with the external form of a cube

octahedral—with the external form of an octahedron

pinacoidal—with the pronounced development of one or more two-sided forms, the pinacoid

INTERGROWTHS, TWINS, AND STRIATIONS

Most minerals occur as random aggregates of grains in the rocks of the Earth. These grains are generally anhedral (lacking external faces), but, being crystalline, they possess an internal order evidenced by cleavage, optical properties, and X-ray diffraction.

There are, however, some relatively common intergrowth patterns of well-formed crystals (as well as anhedral grains) that are *not* random in nature. Such as *parallel* growths of the same crystalline substance, and crystallographically controlled (nonrandom) intergrowths of two or more crystals of the same substance that are related by a symmetry element (that is not normally present in either individual crystal). Such crystallographically controlled intergrowths are called *twins* or *twinned crystals*.

An aggregate of similar crystals with their crystallographic axes and faces parallel is called a *parallel growth*. Such aggregates, although they may at first appear to represent several crystals, are a single crystal because the internal (atomic) structure remains unchanged in orientation throughout the specimen. Figure 2.5 illustrates some types of parallel growth as found in quartz and barite.

A twin is a symmetrical intergrowth of two (or more) crystals of the same substance. Such crystallographic intergrowths are called *twinned crystals*. The two or more individuals of the twinned aggregate are related by a symmetry element[3] that is absent in the original (untwinned) crystal (see pp. 208 to 213). Twinning is common in crystals, and the size of the twinned units can range from an almost atomic scale (with twin lamellae or twin domains on the order of tens to hundreds of angstroms in size) to such a large scale that the individuals are easily seen by the naked eye. Generally the twins that are easiest to recognize in hand specimen are *contact twins* or *penetration twins*. Although the twinned relationship in such symmetric intergrowths is easily recognized, the twin law that underlies the twinned relationship

---

[2]In addition to the form name, a *Miller index* is commonly given to describe the orientation of the form with respect to crystallographic axes. See pp. 197 and 200.

[3]The orientation of the twin element (most commonly a mirror plane or a rotation axis) is generally given by a Miller index (see p. 209).

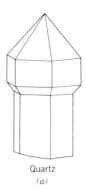

Quartz
(a)

Quartz
(b)

Barite
(c)

FIG. 2.5. Examples of parallel growth. *(a)* Overgrowth of a larger crystal of quartz on a smaller one, forming the shape of a scepter. *(b)* The termination of a large quartz crystal in a collection of smaller crystals, all in parallel orientation. *(c)* Parallel intergrowth of barite crystals.

Contact Twin   Penetration Twins

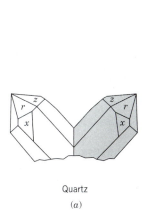

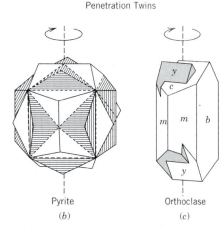

Quartz
(a)

Pyrite
(b)

Orthoclase
(c)

FIG. 2.6. *(a)* Two quartz crystals twinned across a mirror plane. This is referred to as a contact twin because there is a definite surface separating the two individuals. *(b)* and *(c)* are examples of penetration twins in which the individuals are joined along an irregular surface and the twin axis is parallel to the vertical axis. *(b)* Pyrite crystals in which the two individuals are related by 90° rotation about the vertical axis. *(c)* Orthoclase crystals that are related by 180° rotation about the vertical axis.

may not be so obvious. Examples of some common contact and penetration twins are shown in Fig. 2.6.

A twin relationship that is more subtle in its appearance is *polysynthetic twinning*. In a polysynthetic twin the successive composition planes of the twin are parallel to each other. When a large number of individuals in a polysynthetic twin are closely spaced, crystal faces or cleavage surfaces cutting across the composition planes show striations owing to the reversed positions of adjacent individuals. A highly diagnostic polysynthetic twin is albite twinning in the plagioclase feldspar series. The individual twin lamellae that can be seen by the naked eye are commonly quite thin, ranging from 0.1 to several millimeters in thickness. This twin is evidenced by paral-

lel lines or striations seen on cleavage directions. The striations resulting from this polysynthetic twinning are shown in Figures 2.7 *a* and *b*. Polysynthetic twinning in a magnetite crystal is shown in Figure 2.7*c*.

Striations, however, as seen on crystal faces, are by no means always the result of polysynthetic twinning. Figures 2.7 *d* and *e* show striations that result from the intergrowths of two forms. Pyrite cubes (Fig. 2.7*d*) typically show striations that are the result of successive combinations of other faces or of another form (called a pyritohedron) in narrow lines with the cube. The magnetite crystal in Figure 2.7*e* shows striations on the faces of a dodecahedron caused by the stepwise growth of octahedral faces (with triangular outline).

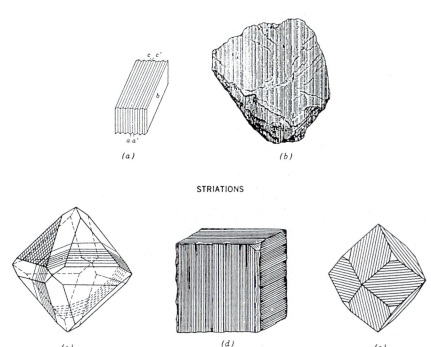

(a)

(b)

STRIATIONS

(c)

(d)

(e)

FIG. 2.7. Polysynthetic twinning and striations. *(a)* Albite polysynthetically twinned parallel to the vertical plane marked *b*; this plane is identified by the Miller index (010). *(b)* The appearance of albite twinning as striations or as parallel groovings across a cleavage surface or crystal face. *(c)* Octahedral crystal of magnetite with twinning lamellae appearing as striations on an octahedral face. *(d)* Striations on a cube of pyrite. *(e)* Striations on the faces of a dodecahedron of magnetite caused by the presence of octahedral faces *(o)* (From C. Klein, 1994, *Minerals and Rocks: Exercises in Crystallography, Mineralogy and Hand Specimen Petrology*, rev. ed. New York: Wiley, p. 308).

STATE OF AGGREGATION

Most mineral specimens, instead of exhibiting well-formed crystals, will tend to be *aggregates* of many smaller grains, ranging in outline from euhedral through subhedral to anhedral. These crystalline aggregates are traditionally defined by descriptive terms, such as the following (see also Fig. 2.8):

Massive. Applied to a mineral specimen totally lacking crystal faces.

Cleavable. Applied to a specimen exhibiting one or several well-developed cleavage directions.

Granular. Made up of mineral grains that are of approximately equal size. The term is mainly applied to minerals whose grains range in size

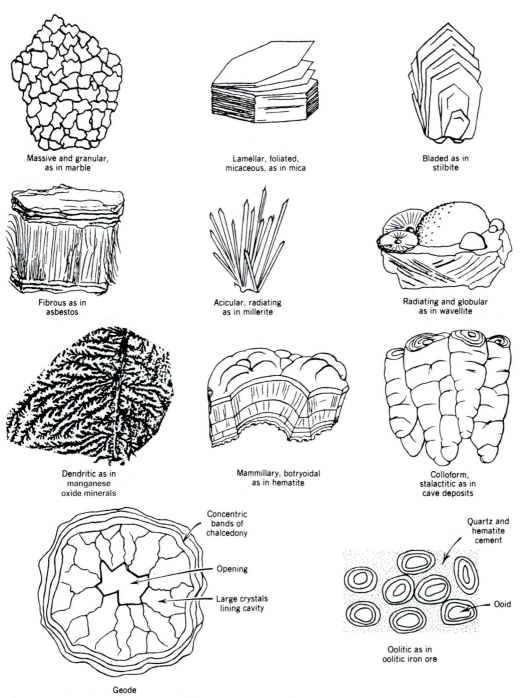

Massive and granular, as in marble

Lamellar, foliated, micaceous, as in mica

Bladed as in stilbite

Fibrous as in asbestos

Acicular, radiating as in millerite

Radiating and globular as in wavellite

Dendritic as in manganese oxide minerals

Mammillary, botryoidal as in hematite

Colloform, stalactitic as in cave deposits

Concentric bands of chalcedony

Opening

Large crystals lining cavity

Geode

Quartz and hematite cement

Ooid

Oolitic as in oolitic iron ore

FIG. 2.8.   Some common mineral habits and occurrences (From C. Klein, 1994, *Minerals and Rocks: Exercises in Crystallography, Mineralogy and Hand Specimen Petrology*, rev. ed. New York: Wiley, p. 305).

from about 2 to 10 mm. If the individual grains are larger, the aggregate is described as *coarse-granular*, if smaller, it is *fine-granular*.

Compact. Applied to a specimen so fine-grained that the state of aggregation is not obvious to the eye.

Lamellar. Made up of layers like the leaves in a book.

Foliated. Made up of thin leaves or plates that can be separated from each other, as in graphite or mica.

Micaceous. Applied to a mineral whose separation into thin plates occurs with great ease, as in mica.

Bladed. With individual crystals (or grains) that are flattened blades or flattened elongate crystals.

Fibrous. Having a tendency to crystallize in needlelike grains or fibers, as in some amphiboles and in asbestos. In asbestos the fibers are *separable*; that is, they are easy to pull apart.

Acicular. From the Latin root *acicula*, meaning needle; describing a mineral with a needlelike habit.

Radiating (or radiated). Describing a mineral in which acicular crystals radiate from a central point.

Dendritic. From the Greek root *dendron*, meaning tree; applied to a mineral exhibiting a branching pattern.

Banded. Describing a mineral aggregate in which a single species may show thin and roughly parallel banding (as in banded malachite), or in which two or more minerals form a finely banded intergrowth (as in chert and hematite bands in banded iron-formation).

Concentric. With bands or layers arranged in parallel positions about one or more centers (as in malachite).

Mammillary. From the Latin word *mamma*, meaning breast; with an external form made up of rather large, rounded prominences. Commonly shown by massive hematite or goethite.

Botryoidal. From the Greek root *botrys*, meaning bunch or cluster of grapes; having the form of a bunch of grapes. The rounded prominences are generally smaller than those described as mammillary. Botryoidal forms are common in smithsonite, chalcedony, and prehnite.

Globular. Having a surface made of little spheres or globules.

FIG. 2.9. Reniform hematite, Cumbria, England. (Harvard Mineralogical Museum)

Reniform. From the Latin *renis*, meaning kidney; with a rounded, kidney-shaped outer surface as in some massive hematite specimens (see Fig. 2.9).

Colloform. From the Greek root *collo*, meaning cementing or welding; because there is often no clear distinction between the four previous descriptive terms (mammillary, botryoidal, globular, and reniform), the term *colloform* includes them all.

Stalactitic. From the Greek *stalaktos*, meaning dripping; made up of small stalactites, which are conical or cylindrical in form as is common on the ceilings of caves.

Concretionary. Clustering about a center, as in calcium carbonate concretions in clay. Some concretions are roughly spherical, whereas others assume a great variety of shapes.

Geode. A rock cavity lined with mineral matter but not wholly filled. Geodes may be banded as in agate, through successive depositions of material, and the central part is commonly filled with crystals projecting into an open space.

Oolitic. From the Greek *oön*, meaning egg; made up of oolites, which are small, round, or ovate (meaning egg-shaped) accretionary bodies, resembling the roe of fish. This texture is common in some iron-rich specimens, made of hematite, known as oolitic iron ore.

Pisolitic. From the Greek *pisos*, meaning pea; therefore, pea-sized; having a texture similar to that of an oolitic aggregate but somewhat coarser in grain size. Bauxite, the major source of aluminum ore, is commonly pisolitic.

LUSTER, COLOR, AND STREAK

These three properties—luster, color, and streak—are easily observed, and they may serve as distinguishing criteria. We will consider each property separately.

Luster

The term *luster* refers to the general appearance of a mineral surface in reflected light. The two distinct types of luster are *metallic* and *nonmetallic*, but there is no sharp division between them. Although the difference between these types of luster is not easy to describe, the eye discerns it easily and, after some experience, seldom makes a mistake. *Metallic* is the luster of a metallic surface such as chrome, steel, copper, and gold. These materials are quite opaque to light; no light passes through even at very thin edges. Galena, pyrite, and chalcopyrite are common minerals with metallic luster. *Nonmetallic* luster is generally shown by light-colored minerals that transmit light, if not through thick portions at least through their edges. Minerals with an intermediate luster are said to be *submetallic*. The following terms are used to describe further the luster of *nonmetallic minerals.*

> *Vitreous.* With the luster of a piece of broken glass. This is commonly seen in quartz and many nonmetallic minerals.
>
> *Resinous or waxy.* With the luster of a piece of resin. This is common in samples of sphalerite.
>
> *Pearly.* With the luster of mother-of-pearl. An iridescent pearlike luster. This is characteristic of mineral surfaces that are parallel to well-developed cleavage planes. The cleavage surface of talc and the basal plane cleavage of apophyllite show pearly luster.
>
> *Greasy.* Appears as though covered with a thin layer of oil. This luster results from light scattered by a microscopically rough surface. Some milky quartz and nepheline specimens may show this.
>
> *Silky, or silklike.* Describing the luster of a skein of silk or a piece of satin. This is characteristic of some minerals in fibrous aggregates. Examples are fibrous gypsum, known as satin spar, and chrysotile asbestos.
>
> *Adamantine.* From the Greek word *adamas*, meaning diamond; with the luster of the diamond. This is the brilliant luster shown by some minerals that also have a high refractive index and as such refract light strongly, as does dia-

mond. Examples are the carbonate of lead, *cerussite*, and the sulfate of lead, *anglesite*.

Color

The color of a mineral is easily observed and instantly evaluated. For some minerals it is characteristic and serves as a distinguishing property. Indeed, most people will recognize a large number of gem minerals and gemstones shown in Plates IX to XII (Chapter 13) on the basis of their color alone. Yet in many minerals color is one of the most changeable and unreliable diagnostic properties. The paragraphs that follow will address only the practical aspects of using color in the recognition of minerals; the origin of color is discussed on pp. 157 to 164.

Because color varies not only from one mineral to another, but also within the same mineral (or mineral group), the mineralogy student must learn in which minerals it is a constant property and can therefore be relied upon as a distinguishing criterion. Most minerals with a metallic luster vary little in color, but most nonmetallic minerals vary widely in color. Although the color of a freshly broken surface of metallic minerals is often highly diagnostic, these same minerals may become tarnished with time. Such a tarnish dulls some minerals such as galena, which has a bright, bluish lead-gray color on a fresh surface, but may become dull upon long exposure to air. Bornite, which on a freshly broken surface has a brownish-bronze color, may be so highly tarnished on an older surface that it shows variegated purples and blues; hence, it is called *peacock ore.* In other words, in the identification of minerals with a metallic luster, it is important to have a freshly broken surface to which a more tarnished surface can be compared.

In contrast to metallic minerals, most mineral with nonmetallic luster, vary widely in color. However, a few have such a *constant color* that their coloration can be used as a truly diagnostic property. Examples are malachite, which is green; azurite, which is blue; rhodonite, which is rose-red to pink; and turquoise, which gives its name to the turquoise color, a greenish blue to blue-green. Most nonmetallic minerals have a relatively narrow *range in colors*, although some show an unusually large range. Members of the plagioclase feldspar series range from almost pure white in albite, through light gray to darker gray toward the anorthite end-member. Most common garnets show various shades of red to red-brown to brown. Members of the clinopyroxene group range from almost white in pure diopside, to light green in diopside with some iron in substitution for magnesium in the structure, through dark green in hedenbergite, to almost black in many augites. Members of the orthopyroxene series (enstatite to ferrosilite) range from light beige to darker brown. On the other hand, tourmaline may show many colors (red, blue, green, brown, and black) as well as distinct color zonation (from colorless through pink to green) within a single crystal. Similarly, gem minerals such as corundum, beryl, quartz, and numerous others occur in many colors; the gemstones cut from them are given varietal names. In short, in most nonmetallic minerals color is a helpful property, but not commonly a truly diagnostic (and therefore unique) property.

Streak

It is commonly useful, especially in metallic minerals, to test the color of the fine powder, or the color of the streak. To determine the color of the streak of a mineral, you will use a piece of unglazed white porcelain called a *streak plate*. Minerals with a hardness higher than that of the streak plate will not powder by rubbing them against it. The hardness of a streak plate is about 7 (see discussion of hardness, later in the chapter). This test will show that black hematite with a silvery luster has a red streak. Most minerals with a nonmetallic luster will have a whitish streak, even through the minerals themselves are colored.

◤ OTHER PROPERTIES DEPENDING ON LIGHT

Minerals are commonly described in terms of the amount of light they can transmit. Such properties are grouped under the term *diaphaneity*, meaning the light-transmitting qualities of a mineral, from the Greek word *diaphanes*, meaning transparent. Examples follow.

Transparent. Describing a mineral that is capable of transmitting light, and through which an object may be seen. Quartz and calcite are commonly transparent. Most gem materials are highly transparent and are commonly priced on the basis of the quality of their transparency.

Translucent. Said of a mineral that is capable of transmitting light diffusely, but is not transparent. Although a translucent mineral allows light to be transmitted, it will not show a sharp outline of an object seen through it. Some varieties of gypsum are commonly translucent.

Opaque. Describing a mineral that is impervious to visible light, even on the outer edges of the mineral. Most metallic minerals are opaque.

Play of Colors

Interference of light either at the surface or in the interior of a mineral may produce a series of colors as the angle of the incident light changes. The striking flashes of varied color against a white or black background, as seen in precious opal, is called *play of colors* (see Plate X, no. 7, Chapter 13). This phenomenon was originally thought to be the result of thin film interference, but electron microscopic study of opal has revealed that the underlying reason for the color play is the presence (in precious opal) of a regular three-dimensional array of equal-size spheres. These spheres consist of amorphous silica, SiO_2, with small amounts of water; they are cemented together by amorphous silica with a slightly different water content (see Fig. 2.10). In precious opal the uniformly packed spheres occur in patches (domains) ranging from less than a millimeter to more than a centimeter across. These regularly arrayed domains act as diffraction gratings for white light, and resolve white light into its spectral colors in accordance with the modified Bragg equation (see also Chapter 7):

$$n\lambda = \mu 2d \sin \theta$$

where n is a small number (1, 2, or 3) and is known as the order of reflection, λ is the wavelength of a specific color, μ is the refractive index of opal (which must be considered because the process of diffraction takes place within the SiO_2 of the opal), d is the spacing between the spheres in the precious opal (in the sample in Fig. 2.10 this is about 3000 Å), and θ is the angle of incidence and reflection. The wavelength of the diffracted spectral color is determined by the value of d (interplanar spacing) and varies with the angle θ (see Fig. 2.11). Common opal lacks this regular internal stacking of spheres, and the scattered white light produces a milky *opalescence*.

An internal iridescence is caused by light diffracted and reflected from closely spaced fractures, cleavage planes, twin lamellae, exsolution lamellae, or minute foreign inclusions in parallel orientation. Some specimens of labradorite (member of the plagioclase series) show colors ranging from blue to green or yellow and red with changing thickness of lamellar pairs. This *iridescence*, also called *schiller* and *labradorescence*, is the result of light scattered by extremely fine, less than $\frac{1}{10}$ micron or thinner in width, exsolution

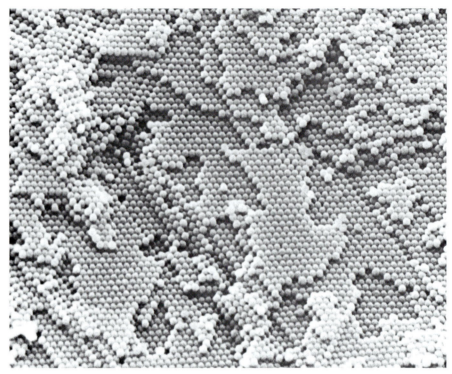

FIG. 2.10. Scanning electron micrograph of an opal with chalky appearance, showing hexagonal closest packing of silica spheres (diameter of spheres is approximately 3000 Å). Because of the weak bonding between the spheres they are completely intact; in typical precious opal samples many of the spheres are cleaved. (Courtesy of Darragh, P. J., Gaskin, A. J. and Sanders, J. V., 1976, Opals. *Scientific American*, v. 234, no. 4, pp. 84–95.)

lamellae in the compositional range of An_{47} to An_{58} (see Fig. 2.12). The delicate bluish, grayish sheen of some albitic feldspars (peristerite), ranging in composition from 2 to 16% Ca/(Ca + Na), is caused by scatter from exsolution lamellae of An_0 and An_{25} composition.

A surface iridescence similar to that produced by soap bubbles or thin films of oil on water is caused by interference of light as it is reflected from thin surface films produced by oxidation or alteration. It is most commonly seen on metallic minerals, particularly hematite, bornite, limonite, and sphalerite.

Chatoyancy and Asterism

In reflected light some minerals have a silky appearance, which results from closely packed parallel

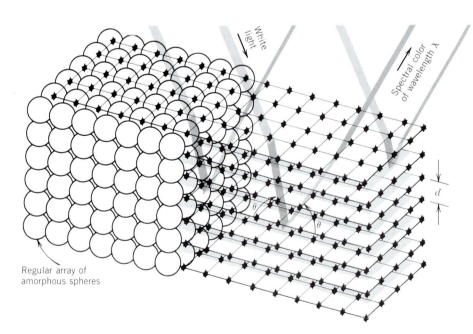

Regular array of amorphous spheres

FIG. 2.11. The spectral colors of precious opal are the result of diffraction by regularly spaced lattice planes. These planes result from amorphous spheres in a regular close-packed array. The spacing of the lattice planes is shown as *d*. The λ of the diffracted spectral line is a function of *d* and the angle θ. (Redrawn after Darragh, P. J., Gaskin, A. J. and Sanders, J. V., 1976, Opals. *Scientific American*, v. 234, pp. 84–95. Copyright Scientific American Inc., George V. Kelvin.)

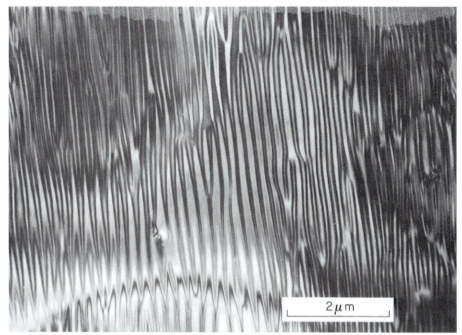

FIG. 2.12. Microstructure in labradorite showing very fine essentially parallel lamellae. These lamellae act as a diffraction grating for white light, producing spectral colors known as *labradorescence*. This photograph was taken with a transmission electron microscope. (From Champness, P. E. and Lorimer, G. W., 1976, Exsolution in silicates, chapter 4.1 in *Electron Microscopy in Mineralogy*, H. R. Wenk, ed. Springer-Verlag, New York.)

fibers or from a parallel arrangement of inclusions or cavities. When a cabochon gemstone is cut from such a mineral, it shows a band of light at right angles to the length of the fibers or direction of the inclusions. This property, known as *chatoyancy*, is

FIG. 2.13. Asterism in a sphere of rose quartz. Sphere diameter is 5.5 cm. The 6-rayed star is caused by microscopic needlelike inclusions of rutile (TiO_2) that are oriented in three directions (at 120° to each other) by the quartz structure. These inclusions reflect a spotlight source yielding the six-rayed star. (Harvard Mineralogical Museum.)

shown particularly well by "satin spar" gypsum, *cat's eye*, a gem variety of chrysoberyl, and *tiger's eye*, fibrous crocidolite replaced by quartz.

In some crystals, particularly those of the hexagonal system, inclusions may be arranged in three crystallographic directions at 120° to each other. A cabochon stone cut from such a crystal shows what might be called a triple chatoyancy, that is, one beam of light at right angles to each direction of inclusions producing a six-pointed star. The phenomenon, seen in star rubies and sapphires, is termed *asterism* (see Fig. 2.13) and is the result of scattering of light from inclusions of rutile arranged in three crystallographic directions. Some phlogopite mica containing rutile needles oriented in a pseudohexagonal pattern shows a striking asterism in transmitted light.

Luminescence

Any emission of light by a mineral that is not the direct result of incandescence is *luminescence*. This phenomenon may be brought about in several ways and is usually observed in minerals containing impurity ions called *activators*. Most luminescence is faint and can be seen only in the dark.

Fluorescence and Phosphorescence

Minerals that luminesce during exposure to ultraviolet light, X-rays, or cathode rays are *fluorescent*. If the luminescence continues after the exciting rays are cut off, the mineral is said to be *phosphorescent*. There is no sharp distinction between fluorescence and phosphorescence, because some minerals that appear only to fluoresce can be shown by refined methods to continue to glow for a small fraction of a second after the removal of the exciting radiation.

The cause of fluorescence is similar to the cause of color (see pp. 157 to 162), and ions of the transition metals are effective activators. Electrons, excited by the invisible short radiation, are raised to higher energy levels. When they fall back to their initial (ground) state, they emit visible light of the same wavelength. However, these excited electrons may fall back to an energy level intermediate between their excited state and the ground state (Fig. 2.14). They then emit a photon of light of lower energy (longer wavelength) than that which provided the original excitation. If the original excitation is produced by ultraviolet (uv) light, the fluorescence is commonly in the visible range.

In phosphorescent minerals there is a time lag between the excitation of electrons to a higher energy level and their return to the ground state. Minerals vary in their ability to absorb uv light at a given wavelength. Thus some fluoresce only in shortwave uv, whereas others may fluoresce only in longwave uv, and still others will fluoresce under either wavelength of uv. The color of the emitted light varies considerably with the wavelengths or source of uv light.

Fluorescence is an unpredictable property, for some specimens of a mineral show it, whereas other apparently similar specimens, even from the same locality, do not. Thus, only some fluorite, the mineral from which the property receives its name, will fluoresce. Its usual blue fluorescence may result from the presence of organic material or rare earth ions. Other minerals that frequently but by no means invariably fluoresce are scheelite, willemite, calcite, eucryptite, scapolite, diamond, hyalite, and autunite. The pale blue fluorescence of most scheelite is ascribed to molybdenum substituting for tungsten. And the brilliant fluorescence of willemite and calcite from Franklin, New Jersey, is attributed to the presence of manganese.

With the development of synthetic phosphors, fluorescence has become a commonly observed phenomenon in fluorescent lamps, paints, cloth, and tapes. The fluorescent property of minerals also has practical applications in prospecting and ore dressing. With a portable uv light one can at night detect scheelite in an outcrop, and underground the miner can quickly estimate the amount of scheelite on a freshly blasted surface. At Franklin, New Jersey, uv light has long been used to determine the amount of willemite that goes into the tailings. Eucryptite is an ore of lithium in the great pegmatite at Bikita, Zimbabwe. In white light it is indistinguishable from quartz, but under uv light it fluoresces a salmon pink and can be easily separated.

CLEAVAGE, PARTING, AND FRACTURE

The properties of *cleavage*, *parting*, and *fracture* are the response of a crystalline material to an external force. When such a force is applied, a mineral is subjected to stress. If the internal structure of the crystalline substance is deformed, due to the stress, it is said to have undergone *strain*. As such, stress relates to the force applied, and strain to the resultant deformation. The strength of a crystalline material is a function of its bonding mechanism(s) and the presence (or absence) of structural defects. The type of bonding (see Chapter 3) is of major importance in a mineral's reaction to an applied force. If a mineral contains structural defects along specific planes or directions, it will tend to deform along such features more easily than a mineral with a truly perfect (or better-ordered) structure. If the strain in the mineral is so great as to exceed its overall strength, it will break. Many minerals have planar directions in their structure that are systematically weaker than other directions. This is the result of planes in the crystal structure that are joined by fewer bonds per unit volume than are

FIG. 2.14. Schematic energy level diagram for the absorption (Abs) of ultraviolet radiation, and resulting fluorescence in the visible light region.

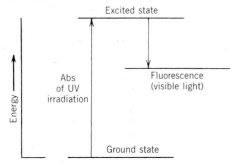

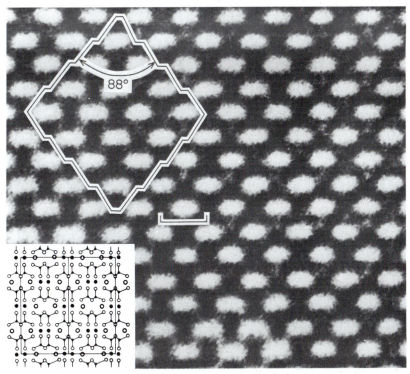

FIG. 2.15. High-resolution transmission electron microscope (HRTEM) image of orthopyroxene showing possible cleavage surfaces a 88° to each other. The length of the bar is 8.8 Å, which is the length of the b axis in orthopyroxene. The white regions in the image correspond to areas between M_2 sites in the pyroxene structure. These light regions have relatively fewer atoms than the rest of the structure and hence have a low electron density. The insert shows the pyroxene structure at the same scale as the structure image. (From Buseck, P. R. and Iijima, S., 1974, High resolution electron microscopy of silicates. *American Mineralogist*, v. 59, pp. 1–21.)

other planes in the structure, or are joined by weaker bonds.

Cleavage

Cleavage is the tendency of minerals to break parallel to atomic planes. Cleavage is a reflection of internal structure because within a structure the strength of chemical bonding is commonly different in different directions. This is especially well shown by layer structures (such as the various mica types and also graphite) in which the bonding within layers is very strong but between layers is much weaker. In these structures a perfect cleavage exists parallel to the layering.

Cleavage may be very well developed (perfect) in some crystals, as shown by the basal cleavage of micas, or it may be fairly obscure, as in beryl and apatite. In some minerals it is completely absent, as, for example, in quartz. Graphite, for example, has a well-developed platelike cleavage parallel to the basal plane. Within these cleavage plates there is a strong covalent bond among the carbon atoms, but across the plates there are weak van der Waals bonds giving rise to the cleavage (see Fig. 3.26). A weak bond is usually accompanied by a large interplanar spacing because the attractive force cannot

hold the planes closely together. Diamond has only one bond type, the covalent bond, and its excellent cleavage parallel to the faces of the octahedron takes place along atomic planes having the largest interplanar spacing. The relationship between internal atomic structure and resultant cleavage directions is well shown by the structure image of a pyroxene in Fig. 2.15. The planes of lowest bond density (i.e., the planes of lowest relative electron density) coincide with the two directions of the prism, the intersection of which is parallel to the c axis. The internal angle between these two cleavage directions is approximately 88°.

In describing a cleavage, its quality and crystallographic direction should be given. The quality is expressed as perfect, good, fair, and so forth. The direction is expressed by the name or Miller index (see p. 198) of the form that the cleavage parallels, such as cubic {001} (Fig. 2.16), octahedral {111}, rhombohedral {10$\bar{1}$1}, prismatic {110}, or pinacoidal {001}. Cleavage is always consistent with the symmetry; thus, if one octahedral cleavage direction is developed, it implies that there must be three other symmetry-related directions. If one dodecahedral cleavage direction is present, it similarly implies five other symmetry-related directions. Not all minerals show cleavage, and only a comparatively few show

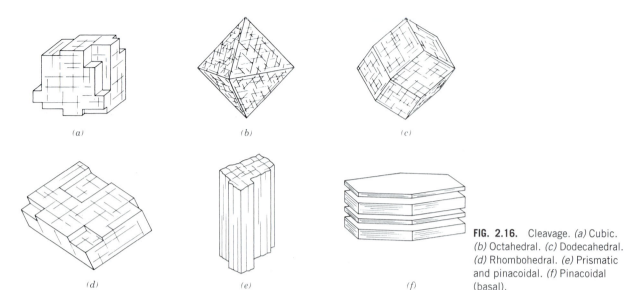

FIG. 2.16. Cleavage. *(a)* Cubic. *(b)* Octahedral. *(c)* Dodecahedral. *(d)* Rhombohedral. *(e)* Prismatic and pinacoidal. *(f)* Pinacoidal (basal).

it in an eminent degree, but in these it serves as an outstanding diagnostic criterion.

Parting

When minerals break along planes of structural weakness, they have parting. The weakness may result from pressure or twinning or exsolution; and, because it is parallel to rational crystallographic planes, it resembles cleavage. However, parting, unlike cleavage, is not shown by all specimens but only by those that are twinned or have been subjected to the proper pressure. Even in these specimens there are a limited number of planes in a given direction along which the mineral will break. For example, twinned crystals part along composition planes, but between these planes they fracture irregularly. Familiar examples of parting are found in the octahedral parting of magnetite, the basal parting of pyroxene, and the

rhombohedral parting of corundum (see Figs. 2.17*a* and *b*).

Fracture

In some crystal structures the strength of the bonds is approximately the same in all directions. Breaking of such crystals generally will not follow a particular crystallographic direction. The way minerals break when they do not yield along cleavage or parting surfaces is their *fracture*. Fracture patterns can be distinctive and highly diagnostic in mineral identification. Different kinds of fracture are designated as follows:

1. *Conchoidal*. The smooth, curved fracture resembling the interior surface of a shell (see Fig. 2.18). This is most commonly observed in such substances as glass and quartz.
2. *Fibrous* and *splintery*.

FIG. 2.18. Conchoidal fracture, obsidian.

FIG. 2.17. *(a)* Basal parting, pyroxene. *(b)* Rhombohedral parting, corundum.

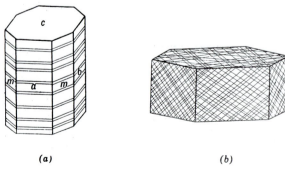

(a)   (b)

3. *Hackly.* Jagged fractures have sharp edges.
4. *Uneven* or *irregular.* Fractures produce rough and irregular surfaces.

HARDNESS

The resistance that a smooth surface of a mineral offers to scratching is its *hardness* (designated by **H**). The degree of hardness is determined by observing the comparative ease or difficulty with which one mineral is scratched by another, or by a file or knife. The hardness of a mineral might then be said to be its "scratchability." The evaluation of hardness is the assessment of the reaction of a crystal structure to stress without *rupture* (cleavage, parting, and fracture are various forms of rupture). In metallic-bonded crystals that can flow plastically, scratching results in a groove. However, brittle materials with ionic or covalent bonds react to a hardness test by microfracturing (rupture on a very fine scale). The effect of ionic size and charge in ionically bonded structures is discussed in Chapter 3 (see Fig. 3.19). This illustrates how different chemical compounds with the same internal structure increase in hardness with decreasing ionic size and increasing ionic charge. In relating the hardness of a crystal structure to its bonding, it must be noted that the structure's overall strength is a composite of all of its bond types, whereas the hardness of that same structure is an expression of its weakest bonding. For example, in silicates, all of which are based on various arrangements of SiO_4 tetrahedra, the hardness ranges from 1, as in talc, to 7, as in quartz, and to 8, as in topaz. Such a variation suggests that hardness is not a function of Si-O bonding, but rather of the other bond types present in the structure. In talc the basal silicate layers are held together by weak van der Waals and/or hydrogen bonds; in quartz there is a uniform bond strength within a relatively dense network of SiO_4 tetrahedra; and in topaz there are somewhat weaker Al-(F, OH) bonds.

A series of 10 common minerals was chosen by the Austrian mineralogist F. Mohs in 1824 as a scale, by comparison with which the relative hardness of any mineral can be told. The following minerals, arranged in order of increasing hardness, comprise what is known as the *Mohs scale of hardness:*

1. Talc
2. Gypsum
3. Calcite
4. Fluorite
5. Apatite
6. Orthoclase
7. Quartz
8. Topaz
9. Corundum
10. Diamond

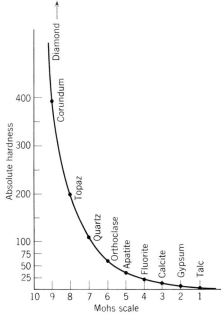

FIG. 2.19. Comparison of Mohs relative hardness scale and absolute measurements of hardness.

These minerals are arranged in an order of increasing relative hardness. Their hardness can also be measured by more quantitative techniques than a scratch test, and this leads to an absolute hardness scale as shown in Fig. 2.19. The relative position of the minerals in the Mohs scale is preserved, but corundum, for example, is two times as hard as topaz and four times harder than quartz.

Talc, number 1 in the Mohs scale, has a structure made up of plates so weakly bound to one another that the pressure of the fingers is sufficient to slide one plate over the other. At the other end of the scale is diamond, with its constituent carbon atoms so firmly bound to each other that no other mineral can force them apart to cause a scratch.

In order to determine the relative hardness of any mineral in terms of this scale, it is necessary to find which of these minerals it can and which it cannot scratch. In making the determination, the following should be observed:

1. Sometimes when one mineral is softer than another, portions of the first will leave a mark on the second that may be mistaken for a scratch. Such a mark can be rubbed off, whereas a true scratch will be permanent.
2. The surfaces of some minerals are frequently altered to material that is much softer than the original mineral. A *fresh surface* of the specimen to be tested must therefore be used.

TABLE 2.1   Mohs Hardness Scale and Additional Observations (From C. Klein, 1994, *Minerals and Rocks: Exercises in Crystallography, Mineralogy, and Hand Specimen Petrology*, rev. ed. New York: Wiley, p. 315).

Mineral	Mohs Hardness	Other Materials	Observations on the Minerals
Talc	1		Very easily scratched by the fingernail; has a greasy feel
Gypsum	2	~2.2 fingernail	Can be scratched by the fingernail
Calcite	3	~3.2 copper penny	Very easily scratched with a knife and just scratched by a copper coin
Fluorite	4		Easily scratched with a knife but not as easily as calcite
Apatite	5	~5.1 pocketknife ~5.5 glass plate	Scratched with a knife with difficulty
Orthoclase	6	~6.5 steel file	Cannot be scratched with a knife, but scratches glass with difficulty
Quartz	7	~7.0 streak plate	Scratches glass easily
Topaz	8		Scratches glass very easily
Corundum	9		Cuts glass
Diamond	10		Used as a glass cutter

3. The physical nature of a mineral may prevent a correct determination of its hardness. For instance, if a mineral is pulverulent, granular, or splintery, it may be broken down and apparently scratched by a mineral much softer than itself. It is always advisable when making the hardness test to confirm it by reversing the order of procedure; that is, do not only try to scratch mineral *A* by mineral *B*, but also try to scratch *B* by *A*.

Several materials serve in addition to the above scale (see Table 2.1): the hardness of the fingernail is a little over 2, a copper coin about 3, the steel of a pocketknife a little over 5, window glass $5\frac{1}{2}$, and the steel of a file $6\frac{1}{2}$. With a little practice, the hardness of minerals under 5 can be quickly estimated by the ease with which they can be scratched with a pocketknife.

Hardness is a vectorial property. Thus, crystals may show varying degrees of hardness, depending on the directions in which they are scratched. The directional hardness differences in most common minerals are so slight that, if they can be detected at all, it is only through the use of delicate instruments. Two exceptions are kyanite and calcite. In kyanite, $\mathbf{H} = 5$ parallel to the length, but $\mathbf{H} = 7$ across the length of the crystal. The hardness of calcite is 3 on all surfaces except the basal pinacoid {0001}; on this form it can be scratched by the fingernail and has a hardness of 2.

Because there is a general link between hardness and chemical composition, the following generalizations can be made.
1. Most hydrous minerals are relatively soft ($\mathbf{H} < 5$).
2. Halides, carbonates, sulfates, and phosphates are also relatively soft ($\mathbf{H} < 5\frac{1}{2}$).
3. Most sulfides are relatively soft ($\mathbf{H} < 5$) with pyrite being an exception ($\mathbf{H} < 6$ to $6\frac{1}{2}$).

4. Most anhydrous oxides and silicates are hard ($\mathbf{H} > 5\frac{1}{2}$).

Because hardness is a highly diagnostic property in mineral identification, most determinative tables use relative hardness as a primary sorting parameter, as is done in the Determinative Table 14.1 in Chapter 14.

◆ TENACITY

The resistance that a mineral offers to breaking, crushing, bending, or tearing—in short, its cohesiveness—is known as *tenacity*. The following terms are used to describe tenacity in minerals:
1. *Brittle.* A mineral that breaks and powders easily. This is characteristic of crystals with dominant ionic bonding.
2. *Malleable.* A mineral that can be hammered out into thin sheets.
3. *Sectile.* A mineral that can be cut into thin shavings with a knife.
4. *Ductile.* A mineral that can be drawn into a wire.

The characteristics described in 2, 3, and 4 (malleability, sectility, ductility) are diagnostic of materials held together by metallic bonding. The metallic bond conveys to crystalline substances the unique property of yielding to applied stress by plastic deformation. As shown in Fig. 3.24, metals are regarded as cations surrounded by a dense cloud of mobile electrons. These cations can, under stress, move past each other without setting up repulsive electrostatic forces. This atomic property is responsible for the physical behavior of metals under stress.
5. *Flexible.* A mineral that bends but does not resume its original shape when the pressure is released. Cleavage sheets of chlorite and talc are

flexible, but they do not snap back to their original position after having been bent. In other words, their deformation is permanent. The bonding between the OH-rich layers in these silicates is by a combination of van der Waals and hydrogen bonds, whereas the bonding within the tetrahedral (Si-Al-O) sheets is a mixture of covalent and ionic bonding. The flexibility of the sheets is the result of slippage along OH-layers in the structure (see Fig. 11.35).

6. *Elastic.* A mineral that, after being bent, will resume its original position upon the release of the pressure. Sheets of mica can be bent, and they will snap back into their original position after the bending has stopped. In contrast to the structures of talc and chlorite, the mica structure contains K^+-rich layers that exert a much stronger force on the sheets of Si–Al tetrahedra than do the hydrogen or van der Waals bonding mechanisms. The ionic bonding between K^+ ions and the Si–Al tetrahedral sheets is responsible for the elasticity of mica (see Figs. 11.33 and 11.34).

SPECIFIC GRAVITY

Specific gravity (**G**) *or relative density*[4] *is a number that expresses the ratio between the weight of a substance and the weight of an equal volume of water at 4°C* (this temperature is coincident with the maximum density of water). Thus, a mineral with a specific gravity of 2 weighs twice as much as the same volume of water. The specific gravity of a mineral is frequently an important aid in its identification, particularly in working with fine crystals or gemstones, when other tests would injure the specimens. A listing of minerals arranged according to increasing specific gravity is given in Table 14.2 of Chapter 14.

The specific gravity of a crystalline substance depends on (1) the kind of atoms of which it is composed, and (2) the manner in which the atoms are packed together. In isostructural compounds (substances with identical structures) in which the packing is constant, those with elements of higher atomic weight will usually have higher specific gravities. This is well illustrated by the orthorhombic carbonates listed in Table 2.2, in which the chief differences are the atomic weight of the various cations.

---

[4]Density and specific gravity are sometimes used interchangeably. However, density requires the citation of units, for example, grams per cubic centimeter or pounds per cubic foot.

TABLE 2.2   Specific Gravity Increase with Increasing Atomic Weight of Cation In Orthorhombic Carbonates

Mineral	Composition	Atomic Weight of Cation	Specific Gravity
Aragonite	$CaCO_3$	40.08	2.94
Strontianite	$SrCO_3$	87.62	3.78
Witherite	$BaCO_3$	137.34	4.31
Cerussite	$PbCO_3$	207.19	6.58

In a solid solution series (see page 90), there is a continuous change in specific gravity (or density) with change in chemical composition. For example, the mineral olivine, $(Mg,Fe)_2SiO_4$, is a solid solution series between forsterite, Mg_2SiO_4 (**G** 3.3), and fayalite, Fe_2SiO_4 (**G** 4.4). Thus, from determination of specific gravity one can obtain a close approximation of the chemical composition of an Mg-Fe olivine (see Fig. 12.4b). A similar relationship in an amphibole series is shown in Fig. 2.20.

The influence of the packing of atoms on specific gravity is well illustrated in polymorphous compounds (substances of identical chemical composition but with different structures; see Table 4.2). In these compounds the composition remains constant, but the packing of the atoms varies. The most dramatic example is given by diamond and graphite, both elemental carbon. Diamond—with specific gravity 3.5—has a closely packed structure, giving a high density of atoms per unit volume; whereas in graphite—with specific gravity 2.23—the layers of carbon atoms are loosely packed.

Average Specific Gravity

Most people from everyday experience have acquired a sense of relative weight even in regard to minerals. For example, ulexite (**G** 1.96) seems light,

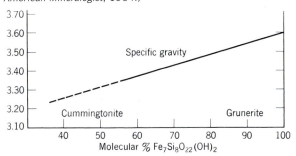

FIG. 2.20. Variation of specific gravity with composition in the monoclinic cummingtonite-grunerite series ranging in composition from $Fe_2Mg_5Si_8O_{22}(OH)_2$ to $Fe_7Si_8O_{22}(OH)_2$. (After Klein, C., *American Mineralogist*, 1964.)

whereas barite (**G** 4.5) seems heavy for nonmetallic minerals. This means that one has developed an idea of an average specific gravity or a feeling of what a nonmetallic mineral of a given size should weigh. This average specific gravity can be considered to be between 2.65 and 2.75. The reason for this is that the specific gravities of quartz (**G** 2.65), feldspar (**G** 2.60–2.75), and calcite (**G** 2.72), the most common and abundant nonmetallic minerals, fall mostly within this range. The same sense may be developed in regard to metallic minerals: graphite (**G** 2.23) seems light, whereas silver (**G** 10.5) seems heavy. The average specific gravity for metallic minerals is about 5.0, that of pyrite. Thus, with a little practice, one can, by merely lifting specimens, distinguish minerals that have comparatively small differences in specific gravity.

Determination of Specific Gravity

In order to determine specific gravity accurately, the mineral must be homogeneous and pure, requirements frequently difficult to fulfill. It must also be compact with no cracks or cavities within which bubbles or films of air could be imprisoned. For normal mineralogic work, the specimen should have a volume of about one cubic centimeter. If these conditions cannot be met, a specific gravity determination by any rapid and simple method means little.

The necessary steps in making an ordinary specific gravity determination are, briefly, as follows: the mineral is first weighed in air. Let this weight be represented by W_a. It is then immersed in water and weighed again. Under these conditions it weighs less, because in water it is buoyed up by a force equivalent to the weight of the water displaced. Let the weight in water be represented by W_w. Then $W_a - W_w$ equals the apparent loss of weight in water, or the weight of an equal volume of water. The expression $W_a/(W_a - W_w)$ will therefore yield a number which is the specific gravity.

Jolly Balance

Because specific gravity is merely a ratio, it is not necessary to determine the absolute weight of the specimen but merely values proportional to the weights in air and in water. This can be done by means of a *Jolly balance* (Fig. 2.21),[5] with which the data for making the calculations are obtained by the stretching of a spiral spring. In using the balance, a fragment is first placed on the upper scale pan and

[5]Manufactured by Eberbach and Son, Ann Arbor, Michigan.

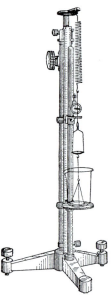

FIG. 2.21. Jolly balance.

the elongation of the spring noted. This is proportional to the weight in air (W_a). The fragment is then transferred to the lower pan and immersed in water. The elongation of the spring is now proportional to the weight of the fragment in water (W_w).

When the mineral is weighed immersed in water, it is buoyed up and weighs less than it does in air; this weight loss is equal to the weight of water it displaces. Hence, if one finds first the weight of a mineral fragment on a pan of the balance in air, and subsequently its weight while immersed in water (it being suspended on a pan by a thin wire thread), one subtracts the two weights. The difference is the weight of the equal volume of water. For example, the weight of a small quartz fragment is 4.265 grams in air; in water it is 1.609 grams. The loss of weight, or weight of an equal volume of water exactly to it, is therefore 2.656 grams; hence the specific gravity is

$$\frac{4.265}{4.265 - 2.656} = \frac{4.265}{1.609} = 2.65$$

which is indeed that of quartz.

In other words, the specific gravity of a mineral (**G**) can be expressed as follows,

$$\mathbf{G} = \frac{W_a}{W_a - W_w}$$

in which W_a is the weight in air and W_w is the weight in water.

Heavy Liquids

Several liquids with relatively high densities are sometimes used in the determination of the specific gravity of minerals. The two liquids most easily used are bromoform (**G** 2.89) and methylene iodide (**G** 3.33).[6] These liquids are miscible with acetone (**G** 0.79), and thus, by mixing, a solution of any intermediate specific gravity may be obtained. A mineral grain is introduced into the heavy liquid and the liquid is diluted with acetone until the mineral neither rises nor sinks. The specific gravity of the liquid and the mineral are then the same, and that of the liquid may be quickly determined by means of a Westphal balance.

Heavy liquids are frequently used in the separation of grains from mixtures composed of several constituents. For example, a separation of the constituent mineral grains of a sand composed of quartz (**G** 2.65), tourmaline (**G** 3.20), and garnet (**G** 4.25) could be quickly made. In bromoform the quartz would float and the tourmaline and garnet would sink; they can be separated from quartz using a separatory funnel. After removing and washing these "heavy minerals" in acetone, they could be separated from each other in methylene iodide; the tourmaline would float and the garnet would sink.

MAGNETISM

Many minerals experience no attraction for a magnetic field (these are referred to as *diamagnetic*), whereas other minerals may be drawn to a magnetic field (these are known as *paramagnetic*). The origin of magnetism in minerals is discussed on pp. 164 to 167. Because minerals show a wide range of magnetic susceptibility they can be separated from each other by strong electromagnets. Magnetic separation by a Franz Isodynamic Separator is a common procedure in industrial and research laboratories. Such a technique separates minerals that are paramagnetic from those that are diamagnetic. On a commercial scale, electromagnetic separation is used to separate ore minerals from gangue (waste).

In a mineralogy laboratory the student will encounter two common highly magnetic (paramagnetic) minerals, magnetite and pyrrhotite. That is, both are easily attracted to a small hand magnet. Both are opaque and may occur as minor constituents in a wide range of mineral associations and rock types. Even if present in small quantities, or with small-sized grains, removing one or several grains from a specimen with a needle or a pock-

etknife will allow you to test the magnetism of individual grains. Magnetite is very strongly attracted to a magnet, pyrrhotite less so.

RADIOACTIVITY

Minerals containing uranium and thorium will continually undergo *decay reactions* in which radioactive isotopes of U and Th form various daughter elements and also release energy in the form of alpha and beta particles and gamma radiation. The radiation produced can be measured in the laboratory or in the field using a Geiger counter or a scintillation counter. A radiation counter, therefore, is helpful in the identification of U- and Th-containing minerals. Examples are uraninite, pitchblende, thorianite, and autunite. The origin of radioactivity is discussed on pp. 167 to 168.

SOLUBILITY IN HCl

The positive identification of carbonate minerals is much aided by the fact that the carbon–oxygen bond of the (CO_3) group in carbonates becomes unstable and breaks down in the presence of hydrogen ions available in acids. This is expressed by the reaction

$$2H^+ + CO_3 \rightarrow H_2O + CO_2 \nearrow ,$$

which is the basis for the "fizz" test with dilute hydrochloric acid. Calcite, aragonite, witherite, and strontianite as well as Cu-carbonates show bubbling or effervescence (fizz) when a drop of dilute HCl is placed on the mineral. The fizz is the result of the release of CO_2. Other carbonates such as dolomite, rhodochrosite, magnesite, and siderite will show effervescence only in hot HCl.

PIEZOELECTRICITY

The conduction of electricity in crystals is related to the type of chemical bonding. Minerals with pure metallic bonding, such as the native metals, are excellent electrical *conductors*, whereas those in which the bonding is partially metallic, as in some sulfide minerals, are *semiconductors*. Ionically or covalently bonded minerals are usually *nonconductors*. In some nonconducting minerals it is possible to induce electrical charges using directed pressure (piezoelectricity) or temperature (pyroelectricity).

Piezoelectricity is extensively used in industrial applications for the control of radio frequencies in

---

[6]Although bromoform and methylene iodide are miscible, *do not mix them*. The mixture will become black.

electronic circuits. Piezoelectricity occurs only in those crystalline substances that lack a center of symmetry (see Table 5.3). Of the 32 crystal classes (point groups; see Chapter 6), 21 have no center of symmetry, and of these all but one, the gyroidal class (with point group symmetry 432), has at least one *polar* crystallographic axis with different crystal forms at opposite ends of that specific axis. (A *polar* axis is a crystallographic concept. A crystal is said to have a polar axis if one direction of an axis is not related by symmetry to the opposite direction along that same axis. Such a crystal is referred to as *noncentrosymmetric*, meaning that it lacks a center of symmetry). If pressure is exerted at the ends of a polar axis, a flow of electrons toward one end produces a negative electrical charge, whereas a positive charge is induced at the opposite end. This is *piezoelectricity*, and any mineral crystallizing in one of the 21 classes with polar axes should show it. However, in some minerals the charge developed is too weak to be detected.

The property of piezoelectricity was first detected in quartz in 1881 by Pierre and Jacques Curie, but nearly 40 years passed before it was used in a practical way. Toward the end of World War I it was found that sound waves produced by a submarine could be detected by the piezoelectric current generated when they impinged on a submerged quartz plate. The device was developed too late to have great value during the war, but it pointed the way to other applications. In 1921 the piezoelectric property of quartz was first used to control radio frequencies, and since then millions of quartz plates have been used for this purpose. When subjected to an alternating current, a properly cut slice of quartz is mechanically deformed and vibrates by being flexed first one way and then the other; the thinner the slice, the greater the frequency of vibration. By placing a quartz plate in the electric field generated by a radio circuit, the frequency of transmission or reception is controlled when the frequency of the quartz coincides with the oscillations of the circuit. The tiny quartz plate used in digital and analog quartz watches serves the same function as quartz oscillators used to control radio frequencies. That is, it mechanically vibrates at a constant frequency that is a function of the plate's thickness and the crystallographic orientation of the slicing within the original quartz crystal. This quartz frequency controls accurately the radio frequency of the electronic circuit in the watch. This circuit counts the crystal frequency and provides the digital time display of the watch. Figure 2.22 is an illustration of the basic schematic of a liquid crystal watch for the display of hours and minutes. The quartz crystal controls an oscillator circuit, which, in turn, generates pulses of one second in time. These "second" pulses are counted to produce "minute" and "hour" pulses. Each of these pulses is decoded to provide proper outputs for the digital watch display. Powered by a 1.5 V silver oxide battery, a quartz plate vibrates approximately 100,000 times per second. An inexpensive quartz

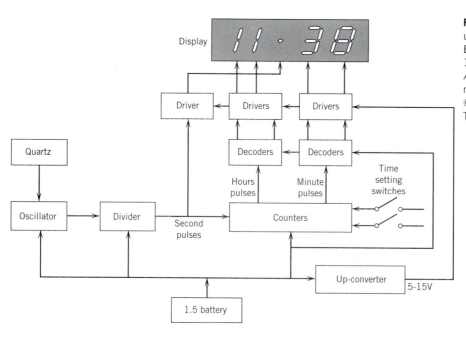

FIG. 2.22. Block diagram of a liquid crystal watch. (Redrawn after Burfoot, J. C. and Taylor, G. W., 1979, *Polar Dielectrics and Their Applications*. University of California Press, Berkeley, Calif. Copyright © 1979, Jack Burfoot and George Taylor.)

watch today is more accurate than the best-made mechanical watch, and precision-manufactured quartz clocks are accurate to within one second per ten years.

The piezoelectric property of tourmaline has been known almost as long as that of quartz, but compared with quartz, tourmaline is a less effective radio oscillator and is rare in occurrence. Nevertheless, small amounts of it are used today in piezoelectric pressure gauges. In tourmaline, which is hexagonal in symmetry, the vertical c axis is a polar axis. Plates cut normal to this direction will generate an electrical current when subjected to a transient pressure. The current generated is proportional to the area of the plate and to the pressure. Tourmaline gauges were developed to record the blast pressure of the first atomic bomb in 1945 and since then have been used by the United States with each atomic explosion. Lesser pressures also can be recorded by them, however, such as those generated by firing a rifle or by surf beating on a sea wall.

REFERENCES AND SELECTED READING

Hochleitner, R. 1994. *Minerals, classifying and collecting them.* Barron's Educational Series Inc. New York: Hauppauge.

Hurlbut, C. S., Jr., and W. E. Sharp. 1998. *Dana's minerals and how to study them.* 4th ed. New York: Wiley.

Keffer, F. 1967. The magnetic properties of materials. *Scientific American* 217: 222–38.

Klein, C. 1994. *Minerals and rocks: Exercises in crystallography, mineralogy, and hand specimen petrology.* Rev. ed. New York: Wiley.

Loeffler, B. M., and R. G. Burns. 1976. Shedding light on the color of gems and minerals. *American Scientist* 64: 636–47.

Simon and Schuster's guide to rocks and minerals. 1978. New York: Simon and Schuster.

Zussman, J. 1977. *Physical methods in determinative mineralogy.* New York: Academic Press.

CHAPTER 3
ELEMENTS OF CRYSTAL CHEMISTRY

The definition of a mineral includes the following phrase: "a solid with a highly ordered atomic arrangement and a definite (but not fixed) chemical composition." This implies that the chemical elements that make up a mineral are arranged in a three-dimensional structure with inherent symmetry. Furthermore, the chemical bonds that hold the structure together are also influenced by this same structural symmetry. The assessment of structure (and its symmetry), of the atomic bonding arrangement, of the variability in its chemistry, and of the related physical properties of a crystalline substance are the domain of crystal chemistry. A discussion of some basic aspects of atoms and ions is part of the field of inorganic chemistry. However, in view of its importance to crystal chemistry this chapter will review some basic concepts that have already been covered in a general chemistry course. Aspects of the external symmetry of crystals, as an expression of their internal atomic arrangement, are covered in Chapters 5 and 6.

Crystal chemistry is a scientific discipline that is closely tied to inorganic chemistry. The term *inorganic chemistry* originally meant nonliving chemistry because it arose from the arts and recipes dealing with minerals and ores. It evolved from the finding of naturally occurring substances (minerals and rocks) that had useful properties, such as chert and flint that were worked into tools during the middle Pleistocene period (about 500,000 years ago or less). As the Stone Age gave way to the Bronze Age, other minerals were sought from which metals could be extracted. Paintings in tombs constructed 5000 years ago in ancient Egypt depict craftsmen smelting ores for metals and weighing gem materials. This led to the recognition of nine chemical elements already in the ancient world: Au, Ag, Cu, Fe, Pb, Sn, Hg, S, and C. Modern inorganic chemistry grew out of the alchemy of the Middle Ages. By the eighteenth century highly original observations about the chemistry and physical behavior of substances were interpreted by the French Chemist Antoine Lavoisier in essentially modern terms. In a publication in 1789 ("Traité elementaire de Chimie") he listed 31 chemical elements, in addition to the 9

identified earlier. The discovery of chemical elements has continued through the nineteenth and twentieth centuries.

The domain of inorganic chemistry embraces all the elements that are now known (see periodic table as an endpaper at the back of this text); this includes 109 elements, 6 of which are inert gases, and 20 of which are not found naturally. The subject of the crystal chemistry of common, naturally occurring substances (minerals) is, for the most part, concerned with a much smaller number of chemical elements, namely those that are abundant constituents of the rock-forming minerals. These are the minerals that make up the common rock types found in the Earth's crust, such as granite, basalt, and gabbro. To set the overall "chemical stage," this chapter begins with a chemical overview of the Earth's crust before reviewing some basic concepts that are part of inorganic chemistry. Specific crystal chemical concepts—such as bonding forces in crystals, coordination of ions, aspects of crystal structure (and some common structure types), and compositional variation in minerals—follow. The chapter ends with the recalculation of mineral analyses and the graphical representation of their composition.

CHEMICAL COMPOSITION OF THE EARTH'S CRUST

Geophysical investigations indicate a division of the Earth into a crust, mantle, and core (see Fig. 3.1a). The crust is approximately 36 km thick under the continents and 10 to 13 km thick under the oceans (Fig. 3.1b). The boundary between the crust and the underlying upper mantle is referred to as the Mohorovičić discontinuity, commonly known as the *Moho*. The upper part of the crust, which consists of the materials immediately underfoot, is composed of a relatively large percentage of sedimentary rocks and unconsolidated materials. However, this sedimentary cover forms but a thin veneer on an underlying basement of igneous and metamorphic rocks. Clarke and Washington (1924) estimated that the upper 10 miles (16 km) of the crust consist of 95% igneous rocks (or their metamorphic equivalents), 4% shale, 0.75% sandstone, and 0.25% limestone. The average composition of igneous rocks, therefore, would closely approximate the average crustal composition. To this effect, Clarke and Washington compiled 5159 "superior" analyses of igneous rocks, the average of which represents the average composition of the continental crust (see Table 3.1). It turns out that this average composition is intermediate between that of granite and basalt (or its coarser grained equivalents, diabase and gabbro), which are the two most common igneous rock types. If the sampling by Clarke and Washington had included an appropriate number of basalts from the deep ocean basins, their average would have been more representative of the average crust than of just the average continental crust. Consideration of the extensive basalts on the ocean floor would lower the average Si, K, and Na values but would increase the proportion of Fe, Mg, and Ca.

It should be noted that eight elements make up approximately 99 weight percent of the Earth's crust; of these, oxygen is by far the most abundant. This predominance is even more apparent when the figures are recalculated from the original weight

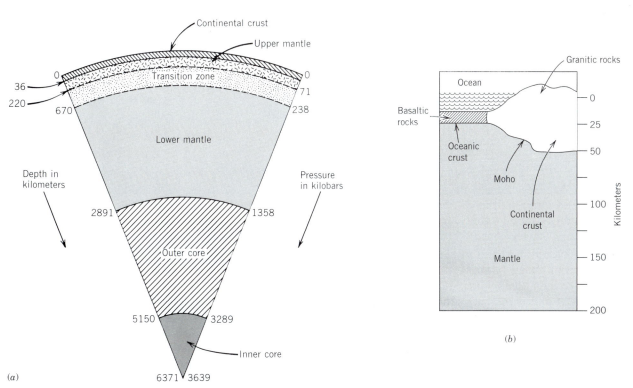

FIG. 3.1 *(a)* Major subdivisions of the Earth's interior. The pressure is expressed in kilobars, where 1 kilobar = 1000 bars, and 1 bar = 0.987 atmosphere. (From Liu, L. and Bassett, W. A., 1986, *Elements, Oxides, Silicates: High-Pressure Phases with Implications for the Earth's Interior.* Oxford University Press, New York.). *(b)* Schematic representation of the Earth's crust. Crustal rocks are less dense than the rocks of the mantle below. The crust is thicker beneath the continents than under the oceans.

TABLE 3.1   Average Amounts of the Elements in Crustal Rocks, in Weight Percent for the Common Elements (as Indicated By %) and in Parts Per Million for the Less Abundant Elements*

Atomic Number	Element	Crustal Average	Granite (G-1)	Diabase (W-1)	Atomic Number	Element	Crustal Average	Granite (G-1)	Diabase (W-1)
1	H	0.14%	0.04%	0.06%	45	Rh	0.005		<0.001
3	Li	20	22	15	46	Pd	0.01	0.002	0.025
4	Be	2.8	3	0.8	47	Ag	0.07	0.05	0.08
5	B	10	1.7	15	48	Cd	0.2	0.03	0.15
6	C	200	200	100	49	In	0.1	0.02	0.07
7	N	20	59	52	50	Sn	2	3.5	3.2
8	O	46.60%	48.50%	44.90%	51	Sb	0.2	0.31	1.01
9	F	625	700	250	52	Te	0.01	<1	<1
11	Na	2.83%	2.46%	1.60%	53	I	0.5	<0.03	<0.03
12	Mg	2.09%	0.24%	3.99%	55	Cs	3	1.5	0.9
13	Al	8.13%	7.43%	7.94%	56	Ba	0.04%	0.12%	0.02%
14	Si	27.72%	33.96%	24.61%	57	La	30	101	9.8
15	P	0.10%	0.04%	0.06%	58	Ce	60	170	23
16	S	260	58	123	59	Pr	8.2	19	3.4
17	Cl	130	70	200	60	Nd	28	55	15
19	K	2.59%	4.51%	0.53%	62	Sm	6.0	8.3	3.6
20	Ca	3.63%	0.99%	7.83%	63	Eu	1.2	1.3	1.1
21	Sc	22	2.9	35	64	Gd	5.4	5	4
22	Ti	0.44%	0.15%	0.64%	65	Tb	0.9	0.54	0.65
23	V	135	17	264	66	Dy	3.0	2.4	4
24	Cr	100	20	114	67	Ho	1.2	0.35	0.69
25	Mn	0.09%	0.02%	0.13%	68	Er	2.8	1.2	2.4
26	Fe	5.00%	1.37%	7.76%	69	Tm	0.5	0.15	0.30
27	Co	25	2.4	47	70	Yb	3.4	1.1	2.1
28	Ni	75	1	76	71	Lu	0.5	0.19	0.35
29	Cu	55	13	110	72	Hf	3	5.2	2.7
30	Zn	70	45	86	73	Ta	2	1.5	0.50
31	Ga	15	20	16	74	W	1.5	0.4	0.5
32	Ge	1.5	1.1	1.4	75	Re	0.001	<0.002	<0.002
33	As	1.8	0.5	1.9	76	Os	0.005	0.00007	0.0003
34	Se	0.05	0.007	0.3	77	Ir	0.001	0.00001	0.003
35	Br	2.5	0.4	0.4	78	Pt	0.01	0.0019	0.0012
37	Rb	90	220	21	79	Au	0.004	0.004	0.004
38	Sr	375	250	190	80	Hg	0.08	0.1	0.2
39	Y	33	13	25	81	Tl	0.5	1.2	0.11
40	Zr	165	210	105	82	Pb	13	48	7.8
41	Nb	20	24	9.5	83	Bi	0.2	0.07	0.05
42	Mo	1.5	6.5	0.57	90	Th	7.2	50	2.4
44	Ru	0.01			92	U	1.8	3.4	0.58

*From Mason, B. and Moore, C. B., 1982, *Principles of Geochemistry.* Copyright © 1982 by John Wiley & Sons, Inc., New York.

percent column (Fig. 3.2) to atom percent and volume percent. As such, the Earth consists almost entirely of oxygen compounds, especially silicates but also oxides and carbonates. Thus, the minerals referred to as the "rock-forming" minerals are, with few exceptions, members of these groups. In terms of number of atoms, oxygen exceeds 60%. If the volumes of the most common ions are considered, oxygen is found to constitute about 94% of the total volume of the crust. In other words, the Earth's crust, on an atomic scale, consists essentially of a close packing of oxygen anions with interstitial metal ions, chiefly Si.

It is noteworthy that many elements important to our economy have very low values for their average abundances in the crust (see Table 3.1). For example, Cu (atomic number, $Z = 29$) = 55 parts per million (ppm), Pb ($Z = 82$) = 13 ppm, and Hg ($Z = 80$) = 0.08 ppm. On the other hand, the less commonly used element Zr ($Z = 40$) is more abundant (165 ppm) than Cu. Similarly, Ga ($Z = 31$) is more abundant (15 ppm) than Hg. Clearly, in order to produce metals needed for our economy, one must locate areas of high concentrations to make the mining profitable. Copper is not extracted from rocks of average composition, but from ore deposits in which

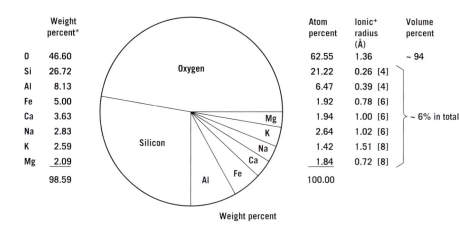

	Weight percent*			Atom percent	Ionic+ radius (Å)	Volume percent
O	46.60			62.55	1.36	~ 94
Si	26.72			21.22	0.26 [4]	
Al	8.13			6.47	0.39 [4]	
Fe	5.00			1.92	0.78 [6]	
Ca	3.63			1.94	1.00 [6]	~ 6% in total
Na	2.83			2.64	1.02 [6]	
K	2.59			1.42	1.51 [8]	
Mg	2.09			1.84	0.72 [8]	
	98.59			100.00		

FIG. 3.2 The eight most common elements in the Earth's crust. From B. Mason, and C. B. Moore, 1982, *Principles of Geochemistry*, 4th ed. (New York: Wiley). †Ionic radii taken from Table 3.11. Numbers in square brackets refer to coordination number.

copper has been concentrated and is present in specific copper-bearing minerals (ore minerals).

Some elements, for example rubidium (Rb) ($Z = 37$), are dispersed throughout common minerals and are never concentrated. Rb does not form specific Rb compounds, but is housed in K-rich minerals, and therefore is an example of what is called a *dispersed element*. Other elements are strongly concentrated in specific minerals. For example, Zr is concentrated in zircon ($ZrSiO_4$), and Ti in rutile (TiO_2) and ilmenite ($FeTiO_3$).

Although we cannot directly determine the average composition of the Earth as a whole, assumptions and calculations can be made on the basis of meteorite samples. These are generally concluded to represent materials analogous to materials within the Earth. The composition of iron meteorites (mainly FeNi alloy; see Fig. 3.3) is believed to be very similar to the composition of the Earth's core. Stony (silicate) meteorites with little metal are probably similar to materials in the mantle. On the basis of the known average compositions of these types of meteorites and the known volumes of the core, mantle, and crust, one can arrive at an average estimate for the composition of the Earth as a whole. Mason and Moore (1982) made such a calculation, with the following results, in weight percent: Fe, 34.63%; O, 29.53%; Si, 15.20%; Mg, 12.70%; Ni, 2.39%; S, 1.93%; Ca, 1.13%; Al 1.09%; and seven other elements (Na, K, Cr, Co, Mn, P, and Ti) each in amounts from 0.1% to 1%. Although all of these elements in the calculated average Earth abundance (except for Ni and S) are also major elements in the average crustal abundance listing (Table 3.1 and Fig. 3.2), they appear in a different order because of the consideration of materials from the Fe-Ni-S–rich core and the silicate-rich mantle.

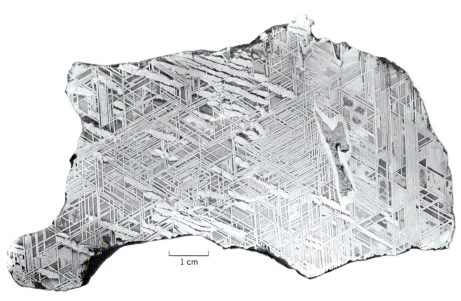

FIG. 3.3 A polished and chemically etched slab of the Edmonton (Kentucky) iron meteorite. The octahedral pattern seen throughout this meteorite consists of a crystallographically controlled intergrowth of two types of Fe-Ni alloy: *kamacite*, which contains about 5.5 weight percent Ni, and *taenite*, which tends to have a variable Ni content ranging from about 27 to 65 weight percent. This texture is known as Widmanstätten pattern. The irregular inclusion at the right center consists of troilite, FeS. (Smithsonian Astrophysical Observatory, courtesy of J. A. Wood; Harvard Mineralogical Museum.)

1 cm

THE ATOM

An atom is the smallest subdivision of matter that retains the characteristics of the element. It consists of a
very small, massive nucleus composed of *protons* and
neutrons surrounded by a much larger region thinly
populated by *electrons* (see Table 3.2). Although
atoms are so small that their images can barely be
resolved even with the highest magnification of
the transmission electron microscope, their sizes can
be deduced from measurements of interatomic distances. The atomic radii are expressed in nanometer
or angstrom units (1 nanometer = 10 angstroms). For
example, the smallest atom, hydrogen, has a radius of
only 0.46 Å, whereas the largest, cesium, has a radius
of 2.72 Å (see Table 3.10). Each proton carries a unit
positive charge; the neutron, as the name implies, is
electrically neutral; and each electron carries a unit
negative charge (see Table 3.2). Because the atom as
a whole is electrically neutral, there must be as many
electrons as protons. The weight of the atom is concentrated in the nucleus, because the mass of an
electron is only 1/1837 that of the proton. Although
the electrons and nuclei are both extremely small, the
diameter of the atom is approximately 10,000 times
the diameter of its nucleus, such that almost all the
volume of an atom is occupied by its electrons that
fill the space around the nucleus.

The simplest atom is that of hydrogen, which
consists of one proton and one electron (see Fig.
3.4). Atoms of the other natural elements have from
two protons (helium) to 92 protons (uranium).

The fundamental difference between atoms of
the different elements lies in the electrical charge of
the nucleus. This positive charge is the same as the
number of protons, and this number, equal to the
number of electrons in an uncharged atom, is called
the *atomic number* (= Z). The sum of the number of
protons and the number of neutrons determines the
characteristic mass, or mass number of an element.
Atoms of the same element but with differing numbers of neutrons are called *isotopes*. For example,
oxygen (Z = 8) has three isotopes, the most common

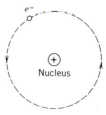

FIG. 3.4 Schematic representation of the hydrogen atom as
based on the model of Niels Bohr. A single electron (e⁻) moves
in an orbit around the nucleus, as in a planetary system.

of which has a nucleus with eight protons and eight
neutrons (Z = 16); this is known as ^{16}O. Rarer and
heavier isotopes of oxygen carry eight protons and
nine or ten neutrons; these are ^{17}O and ^{18}O, respectively. Similarly, hydrogen, H, can exist in several
isotopic forms. The element H (Z = 1) consists of
one proton and one electron; the H isotope with one
neutron in the nucleus is ^2H, known as deuterium
(D), and the isotope with two neutrons in the nucleus
is ^3H, tritium (T). A schematic illustration of some
isotopes is given in Fig. 3.5.

Chemical Elements and the Periodic Table

The chemical elements are listed in alphabetical order
in Table 3.3. In this table their chemical symbols,
atomic number (Z), and atomic weights are also

FIG. 3.5 Schematic illustration of the content of protons and
neutrons in the nucleus of several elements. The isotopes in the
first column have equal numbers of protons and neutrons; those
in the second column have an extra neutron; the isotope in the
third column has two extra neutrons.

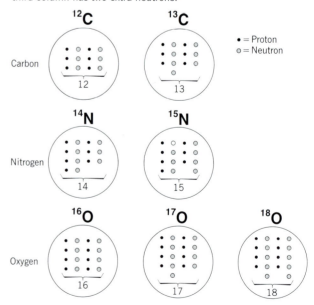

TABLE 3.2   Some Atomic Particles*

Particle	Symbol	Atomic Mass Units	Relative Charge
Electron	*e*	0.0005486	−1
Proton	*p*	1.007276	+1
Neutron	*n*	1.008665	0

*Consideration of other particles discovered in studies of high-energy
physics is unnecessary in this context.

TABLE 3.3   Alphabetical Listing of the Elements, Their Symbols, Atomic Number, and Atomic Weights

Name	Symbol	Atomic Number, Z	Atomic Weight*	Name	Symbol	Atomic Number, Z	Atomic Weight*
Actinium	Ac	89	227.0278	Mendelevium	Md	101	(258)
Aluminum	Al	13	26.98154	Mercury	Hg	80	200.59
Americium	Am	95	(243)	Molybdenum	Mo	42	95.94
Antimony	Sb	51	121.75	Neodymium	Nd	60	144.24
Argon	Ar	18	39.948	Neon	Ne	10	20.179
Arsenic	As	33	74.9216	Neptunium	Np	93	237.0482
Astatine	At	85	~210	Nickel	Ni	28	58.70
Barium	Ba	56	137.33	Niobium	Nb	41	92.9064
Berkelium	Bk	97	(247)	Nitrogen	N	7	14.0067
Beryllium	Be	4	9.01218	Nobelium	No	102	(259)
Bismuth	Bi	83	208.9804	Osmium	Os	76	190.2
Bohrium	Bh	107	(264.12)	Oxygen	O	8	15.9994
Boron	B	5	10.81	Palladium	Pd	46	106.4
Bromine	Br	35	79.904	Phosphorus	P	15	30.97376
Cadmium	Cd	48	112.41	Platinum	Pt	78	195.09
Calcium	Ca	20	40.08	Plutonium	Pu	94	(244)
Californium	Cf	98	(251)	Polonium	Po	84	(209)
Carbon	C	6	12.011	Potassium	K	19	39.0983
Cerium	Ce	58	140.12	Praeseodymium	Pr	59	140.9077
Cesium	Cs	55	132.9054	Promethium	Pm	61	(145)
Chlorine	Cl	17	35.453	Protoactinium	Pa	91	231.0359
Chromium	Cr	24	51.996	Radium	Ra	88	226.0254
Cobalt	Co	27	58.9332	Radon	Rn	86	(222)
Copper	Cu	29	63.546	Rhenium	Re	75	186.207
Curium	Cm	96	(247)	Rhodium	Rh	45	102.9055
Dubnium	Db	105	(262.1144)	Rubidium	Rb	37	85.4678
Dysprosium	Dy	66	162.50	Ruthenium	Ru	44	101.07
Einsteinium	Es	99	(252)	Rutherfordium	Rf	104	(261.1089)
Erbium	Er	68	167.26	Samarium	Sm	62	150.4
Europium	Eu	63	151.96	Scandium	Sc	21	44.9559
Fermium	Fm	100	(257)	Seaborgium	Sg	106	(263.1186)
Fluorine	F	9	18.998403	Selenium	Se	34	78.96
Francium	Fr	87	(223)	Silicon	Si	14	28.0855
Gadolinium	Gd	64	157.25	Silver	Ag	47	107.868
Gallium	Ga	31	69.72	Sodium	Na	11	22.98977
Germanium	Ge	32	72.59	Strontium	Sr	38	87.62
Gold	Au	79	196.9665	Sulfur	S	16	32.064
Hafnium	Hf	72	178.49	Tantalum	Ta	73	180.9479
Hassium	Hs	108	(265.1306)	Technetium	Tc	43	98.906
Helium	He	2	4.00260	Tellurium	Te	52	127.60
Holmium	Ho	67	164.9304	Terbium	Tb	65	158.9254
Hydrogen	H	1	1.0079	Thallium	Tl	81	204.37
Indium	In	49	114.82	Thorium	Th	90	232.0381
Iodine	I	53	126.9045	Thulium	Tm	69	168.9342
Iridium	Ir	77	192.22	Tin	Sn	50	118.69
Iron	Fe	26	55.847	Titanium	Ti	22	47.90
Krypton	Kr	36	83.80	Tungsten	W	74	183.85
Lanthanum	La	57	138.9055	Uranium	U	92	238.029
Lawrencium	Lr	103	(260)	Vanadium	V	23	50.9415
Lead	Pb	82	207.2	Xenon	Xe	54	131.30
Lithium	Li	3	6.941	Ytterbium	Yb	70	173.04
Lutetium	Lu	71	174.967	Yttrium	Y	39	88.9059
Magnesium	Mg	12	24.305	Zinc	Zn	30	65.38
Manganese	Mn	25	54.9380	Zirconium	Zr	40	91.22
Meitnerium	Mt	109	(268)				

*A value given in parantheses denotes the mass of the most stable known isotope.

given. Although such a table is useful as a reference for the information it contains, it provides no insight into the well-known regularity of the atomic build-up. Instead, the *periodic table* (see endpaper at the back of this text) is the widely used chemical listing of all the elements. In this table the elements are arranged in order of increasing atomic number (*Z*) because this order exhibits best the periodic repetition of chemical and physical properties of the elements. In other words, the charge on the nucleus (which is reflected in the atomic number) and the number of electrons in the neutral atom are what determines the order in which the elements occur. The table is organized into *vertical columns*, also called *groups*, with Roman numerals (I, II, . . . VIII); these numbers are equal to the numbers of electrons contained in their outermost shell (this is discussed in more detail on page 50); the *horizontal rows*, also known as *periods*, are designated by Arabic numerals (1, 2, . . . , 7).

In the periodic table (see endpaper at the back of this text), most of the elements are *metals*, but those that are shaded, in the right-hand part, are *non-metals*. Metals have distinct properties such as high electrical conductivity, metallic luster, generally high melting points, ductility (can be drawn into wires), and malleability (can be hammered into sheets). Nonmetals, in contrast, show poor electrical conductivity, do not have the characteristic luster of metals, and, as solids, are brittle. The far right-hand column (VIII) contains the inert, noble gases.

The two long rows below the main part of the table are known as the *lanthanide series*, or *rare earths* (*Z* 58 through 71), and the *actinide series* (*Z* 90 through 103). These elements belong in the body of the table but are placed below simply to conserve space. Most of the elements listed in the table occur naturally; however, elements with atomic numbers 43, 61, and 93 through 109 are known only from synthesis.

The Bohr Model of the Atom

The first widely accepted picture of the atom was that developed in 1913 by the Danish physicist Niels

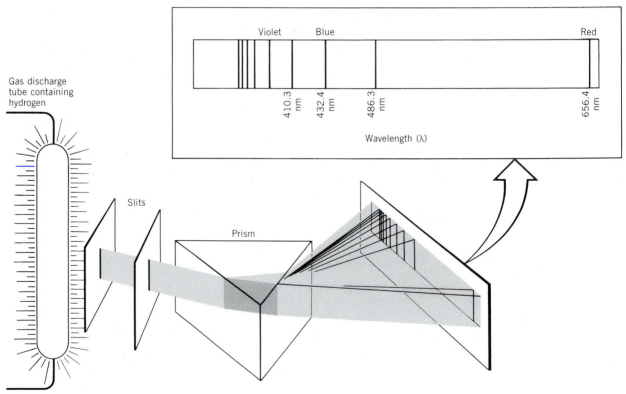

FIG. 3.6   The line spectrum of hydrogen obtained by refraction through a glass prism. The wavelengths of the lines in the visible spectrum are given in nanometers (nm). The lines to the left of these are in the ultraviolet region of the spectrum. The distinct line separation is due to the fact that the light energy released has only distinct quantities of energy, and none in between. (From J. E. Brady, and G. E. Humiston, 1982, *General Chemistry: Principles and Structure*, 3rd ed. New York: Wiley; copyright © John Wiley & Sons; reprinted by permission.)

Bohr. This model is based on the fact that when an electrical charge is passed through a tube containing hydrogen, light is emitted, and the light's spectrum is found to consist of several sharp lines with specific wavelength (λ) values (see Fig. 3.6). Four of these lines, in the visible part of the spectrum, can be seen with the naked eye, and the others, in the ultraviolet region, can be recorded on photographic film. Because of the specific and characteristic wavelength (λ) values of the light lines emitted by all chemical elements heated to a high temperature, Niels Bohr concluded that electrons of the elements occur in specific energy levels at various distances away from the nucleus. He postulated that when an electron absorbs energy it jumps to a higher energy level, and when it loses energy it drops to a lower energy level (see Fig. 3.8). This led Niels Bohr to conclude that electrons occur in discrete, or *quantized*, energy levels. The energy of the emitted radiation is proportional to the wavelength of the radiation, in accordance with the Einstein equation

$$E = \frac{hc}{\lambda}$$

where E is the energy, c is the velocity of light, λ is the wavelength of the radiation emitted, and h is a proportionality constant known as Planck's constant. This is a universal constant of nature that relates the energy of a quantum of radiation to the frequency of the oscillator that emitted it. Its numerical value is 6.62517×10^{-27} erg sec.

In the Bohr model of the atom the electrons are visualized as circling the nucleus in "orbits" (as in a planetary system), or energy levels, at fixed distances from the nucleus, depending on their energies. Figure 3.4 is a schematic illustration of the hydrogen atom as based on the Bohr theory. At the center is the *nucleus*, which, except in the hydrogen atom, is made up of *protons* and *neutrons* (the *hydrogen* nucleus is made up of a single proton).

In chemical elements with more than one electron, the electrons are distributed in shells designated by n (e.g., $n = 1$, $n = 2$, $n = 3$, etc.) and known as the *principal quantum numbers*. The corresponding electron shells (or orbits) are designated K, L, M, N, etc. (see Fig. 3.7). For this model, Bohr derived an equation for the energy of the electron, as follows:

$$E = -A \frac{1}{n^2}$$

where A is a constant involving knowledge of the mass and charge of the electron and Planck's constant, and n is the principal quantum number, which identifies the electron orbit. From this equation it be-

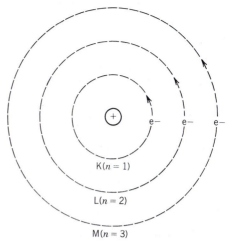

FIG. 3.7 The Bohr model of the atom. Electrons travel along specific orbits of fixed energy levels (known as K, L, M, N, . . . shells, with principal quantum number $n = 1, 2, 3, 4, . . . , \infty$).

comes clear that the energy of an electron in a particular orbit depends on the value of n. The greatest absolute value of energy is represented by $n = 1$, because this yields the largest value for the fraction $1/n^2$. Because of the negative sign in the equation this is the lowest (most negative) energy, E (see Fig. 3.8). In Fig. 3.8 the energy (E) scale is expressed in negative units, in accordance with the above equation. Although this may appear unusual, the important fact to note is that the electron orbits represent

FIG. 3.8 Schematic representation of the energy levels of electrons (for $n = 1, 2, 3, 4, . . . , \infty$) in the Bohr atomic model. The two vertical "up" arrows show two possibilities for the increase in energy level of an electron as a result of the absorption of energy; the three "down" arrows show three possibilities for how electron energy is lost as a result of emission of energy.

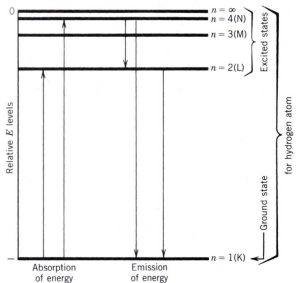

energy levels with various distinct energy differences between them.

The Schrödinger Model of the Atom

Although the concept of electrons circling the nucleus in well-defined orbits (as in the Bohr model) gained wide acceptance, it did not satisfactorily explain a number of important observations. It was not applicable to the line spectra of complex atoms (with atomic numbers higher than those of hydrogen). Additional spectral lines were interpreted as the result of the axial spin (clockwise or counterclockwise) of the electrons, which was not accounted for in the Bohr model. A single quantum number, n, furthermore, did not account for possible elliptical electron orbits. It also gave no basic understanding for the quantization of energy between various orbital levels, and it failed to explain why an orbiting electron did not radiate energy.

In 1923 the French physicist Louis-Victor de Broglie demonstrated that electrons, instead of behaving solely as particles whose positions can be determined in space (as in the Bohr model), had properties identical to those of waves. The wavelength (λ) of a particle with mass m and velocity v is expressed as

$$\lambda = \frac{h}{mv}$$

where h is Planck's constant.

With electrons having wavelike properties, it becomes impossible to visualize them as being in a specific place at a particular time. This notion, expressed as the *uncertainty principle*, was introduced by the German scientist Werner Heisenberg. It implies that the motion of an electron around the nucleus of an atom cannot be satisfactorily described in terms of orbits, be they circular or elliptical.

In 1926 all of the above developments were incorporated into a new atomic model by Erwin Schrödinger, and were expressed as a wave equation. In this equation the electrons are described by wave functions, and the theoretical model is founded on the quantum properties of energy, that is, the theory of *quantum mechanics*. The Schrödinger equation relates the probability of finding an electron at a given time, in a specific place, to the mass and potential energy of the particle at that time and place.

One form of the Schrödinger equation is as follows:

$$\frac{\partial^2\psi}{\partial x^2} + \frac{\partial^2\psi}{\partial y^2} + \frac{\partial^2\psi}{\partial z^2} + \frac{8\pi^2 m}{h^2}(E - V)\,\psi = 0$$

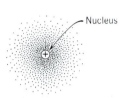

FIG. 3.9 Schematic illustration of the electron probability density distribution (ψ^2) of an electron around a nucleus (an electron cloud). The dots do not show the location of the electron but only the probability of finding the electron in that location. The greater the density of dots, the greater the probability of finding the electron in that region.

where ψ is a wave function in terms of three coordinate axial directions (x, y, z), m is the mass of the electron, E is the total energy of the electron, V is the potential energy of the electron at some specified point, and h is Planck's constant.

The Schrödinger equation is also stated as follows:

$$\nabla^2\psi + \frac{8\pi^2 m}{h^2}(E - V)\,\psi = 0$$

where the ∇^2 symbol is a differential operator with

$$\nabla^2\psi = \frac{\partial^2\psi}{\partial x^2} + \frac{\partial^2\psi}{\partial y^2} + \frac{\partial^2\psi}{\partial z^2}$$

The unique contribution of the Schrödinger equation is its representation of physical observations in a way that had not been possible in earlier developed models. The function $\psi(x, y, z)$, which occurs in the wave equation, describes the behavior of an electron. The square of this function, $\psi^2(x, y, z)$, defines a region in space (x, y, z) where the electron may be found with a certain probability. This is shown schematically in Fig. 3.9. This illustration might be considered the final print of millions of superimposed photographs of the position of an electron in three-dimensional space, defined by x, y, and z axial directions. With the electron in rapid motion, the final composite photograph would produce an array of dots (like a cloud) with dense and more openly spaced regions. The dense regions would be those of a high probability of finding an electron, and those of low density would represent a lower probability of locating an electron. The wave theory, therefore, portrays the motion of electrons only in terms of the probability of finding a certain electron within a small volume unit. It does not describe the movement of electrons in well-defined, simple orbits.

The electron behavior summarized by the Schrödinger wave equation can be compared to the vibrations of a string stretched between two fixed points. Figure 3.10 shows some of the ways in which a string can vibrate. In Fig. 3.10a the string vibrates in the simplest way, that is, between two fixed end points. In Figure 3.10b the string distorts like a standing sinusoidal

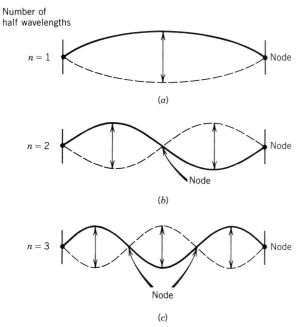

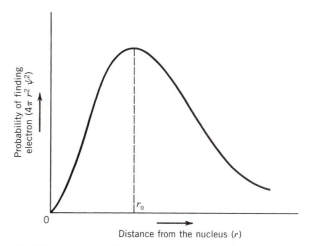

FIG. 3.11 The radial distribution function ($4\pi r^2 \psi^2$) for the s-orbital plotted against increasing distance (r) from the nucleus. The vertical axis is a measure of the probability of finding the electron at a specific distance from the nucleus. The maximum in this function coincides with r_0, the radius of the smallest orbit in the Bohr atomic model.

FIG. 3.10 Relationship between principal quantum number n and the number of half wavelengths in a standing wave. In solutions to the wave equation, nodes represent regions with no electron density (see Fig. 3.12).

wave, with a central point where there is no displacement. The points of no displacement are known as *nodes*. Figure 3.10c shows a string vibrating between two fixed end nodes, and with two nodes at one-third and two-thirds of the length of the string. Each of these different vibration patterns can be defined by a fundamental audio frequency, ν audio (where ν audio = v/λ; v = velocity of sound and λ = wavelength of the string) and a principal quantum number, n. In the one-dimensional example of a string, the quantum number n gives the number of half-wavelengths in the vibrations, and $(n + 1)$ is the number of nodes (including the nodes at the ends of the string). In the case of a three-dimensional electron wave, nodes may form along each of the three axial directions.

In order to completely specify the position of an electron in three-dimensional space, *three quantum numbers* are needed:

1. The *principal quantum number, n*
2. The *azimuthal quantum number* (or *orbital shape quantum number*)
3. The *magnetic quantum number, m*.

These three quantum numbers follow from the solution of the Schrödinger wave equation and represent specified parameters in the mathematical formulation of ψ.

In three dimensions, the *principal quantum number, n*, is a function of the distance r of the electron from the nucleus. The probability of finding the elec-

tron at a distance r from the nucleus is given by $4\pi r^2\psi^2$. For the s-orbital[1] in the hydrogen atom, this is plotted in Fig. 3.11. This curve shows that the electron occupies all of the small specified volume around the nucleus, but that it is most frequently found at distance r_0, the radius predicted in the Bohr model of the atom.

The principal quantum number, n, reflects the effective volume (or mean radius) of an electron orbital and can have any positive integral value from 1 to infinity, 1, 2, 3, . . . , ∞. It also reflects the energy levels, or shells, in an atom. The larger the value of n, the greater is the average energy of the levels belonging to the shell. As in the Bohr theory, $n = 1$ defines the K shell, $n = 2$ defines the L shell, $n = 3$ defines the M shell, and so on. n determines the position of the horizontal rows of the periodic table (see endpaper at the back of this text).

The *azimuthal quantum number* (or *orbital shape quantum number*), l, determines the general shape of the region in which an electron moves (i.e., the shape of the orbital), and to some degree its energy. For a given shell, l may have values of 0, 1, 2, 3, . . . , to a maximum of $(n - 1)$ for that shell. This means that for the K shell, with $n = 1$, the only value of l that is possible is $l = 0$. When $n = 2$, two values of l are possible, 0 and 1, resulting in two subshells for the L shell. The values of l that are possible for each value of n are given in Table 3.4. It follows

---

[1]The wave functions that describe the motions of an electron are known as "orbitals," to distinguish them from the "orbits" in the Bohr atomic model.

TABLE 3.4   Summary of the Three Quantum Numbers

Principal Quantum Number, n (Shell)	Azimuthal Quantum Number, l (Subshell)	Subshell Designation	Magnetic Quantum Number, m (Orbital)	Number of Orbitals in Subshell	Maximum Number of Electrons	
1 (K)	0	1s	0	1	2	2
2 (L)	0	2s	0	1	2	}8
	1	2p	−1, 0, +1	3	6	
3 (M)	0	3s	0	1	2	
	1	3p	−1, 0, +1	3	6	}18
	2	3d	−2, −1, 0, +1, +2	5	10	
4 (N)	0	4s	0	1	2	
	1	4p	−1, 0, +1	3	6	}32
	2	4d	−2, −1, 0, +1, +2	5	10	
	3	4f	−3, −2, −1, 0, +1, +2, +3	7	14	

from this table that the number of subshells in any given shell is equal to its value of n.

The various states of l (0, 1, 2, 3, ..., $n-1$) have been given letter designations as follows:

value of l	0	1	2	3	4	5	6...
subshell designation	s	p	d	f	g	h	i...

The letters s, p, d, and f are abbreviations of the spectroscopic terms *sharp*, *principal*, *diffuse*, and *fundamental*, respectively. The following discussion will be limited to the s, p, d, and f subshells, because they are the ones populated by electrons in atoms in their lowest energy state, or ground state. The s orbital is spherical in shape, the p orbital is quasi-dumbbell-shaped and axially directed, and the d orbital has various shapes (see Fig. 3.12).

To designate a subshell within a given shell, the value of n (for the shell) is followed by the letter designation of the subshell. That is, the 2s subshell is a subshell of the second shell ($n = 2$), with $l = 0$. The 3d subshell is a subshell of the third shell ($n = 3$), with $l = 2$.

The *magnetic quantum number*, m, restricts the orientation and shape of each type of orbital. It has integer values that range from $-l$ to $+l$. When $l = 0$, only one value of m is permitted, $m = 0$. This means that the s subshell has only one orbital (the s orbital).

Table 3.4 gives a summary of the interrelationship of the n, l, and m quantum numbers and the number of subshell orbitals.

In addition to the three quantum numbers, n, l, and m, which follow from the solution of the wave equation, there is a fourth quantum number, the *spin quantum number*, s, which defines the direction of spin of the electron in space. Because an electron can spin only in one of two directions, it has only two values, namely $+\frac{1}{2}$ and $-\frac{1}{2}$. A spinning electron

behaves as a small magnet and will produce a magnetic field while moving around its orbit, both from its orbital motion and from its spin. An orbiting electron produces a magnetic field, just like the magnetic field produced by an electric current moving through a coiled wire. In addition, the electron's axial spin can produce a magnetic field. In Fig. 3.13, which illustrates an electron orbital and the spinning motion of an electron, the overall magnetic field is represented by H. Because the spinning electron behaves as a small magnet, there will be an interaction between H (the magnetic field strength) and the field produced by the axial spin of the electron. The axial spin of the electron will either reinforce or oppose the field strength (H), depending on whether the spin is clockwise or anticlockwise. Note that in this picture of the electron's spin, the electron is depicted as a charged particle, not a wave function.

It is useful, as in the discussion of magnetism (see pp. 164 to 167), to keep account of the electron spin directions in an atom. This is commonly done by representing an electron with its associated spin (in one direction) by an arrow pointing up, ↑, and an electron with a reverse spin (in the opposite direction) by an arrow pointing down, ↓. To indicate the distribution of electrons among orbitals, the arrows are placed over bars that symbolize orbitals, such as for

H $\frac{\uparrow}{1s}$ He $\frac{\uparrow\downarrow}{1s}$

Li $\frac{\uparrow\downarrow}{1s}\frac{\uparrow}{2s}$ Be $\frac{\uparrow\downarrow}{1s}\frac{\uparrow\downarrow}{2s}$

and

Ne $\frac{\uparrow\downarrow}{1s}\frac{\uparrow\downarrow}{2s}\underbrace{\frac{\uparrow\downarrow}{}\frac{\uparrow\downarrow}{}\frac{\uparrow\downarrow}{}}_{2p}$

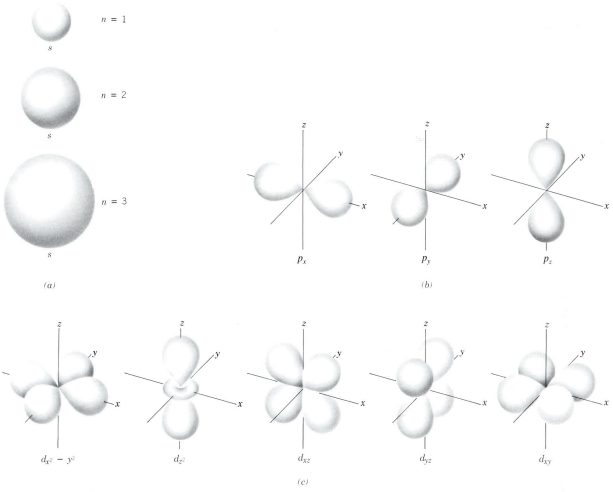

FIG. 3.12 Surfaces showing the angular dependence of the function ψ^2 for s, p, and d-orbitals of the hydrogen atom. These angular wave functions can be regarded as probability distributions of electrons.

Two electrons paired in the same orbital have spin directions opposing each other (see discussion of the Pauli exclusion principle, below). The magnetic moments of such paired electrons nullify each other (see Table 4.7), whereas an effective magnetic moment is

FIG. 3.13 The orbital and spin motions of an electron. The magnetic field H is the result of the electron moving along the orbital. The spin of the electron can reinforce or oppose the magnetic field H, depending on the rotation direction of the spin (clockwise or anticlockwise).

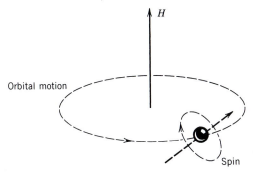

the result of unpaired electrons in outer orbitals. The spin of unpaired electrons is a major factor in the magnetic properties of atoms. This will be further discussed under "Origin of Magnetic Properties" in Chapter 4.

In conclusion, each electron in an atom can be assigned values for four quantum numbers, n, l, m, and s, which determine the orbital in which the electron occurs and the direction in which the electron spins.

There is, however, a restriction on the values these quantum numbers may have. This is known as the *Pauli exclusion principle*, which states that no two electrons in any one atom may have all four quantum numbers the same. This has the effect of limiting the number of electrons in any given orbital to two, and it also requires that the spins of the two electrons be in opposite directions (with s values of $+\frac{1}{2}$ and $-\frac{1}{2}$). Because of the restriction of only two electrons per orbital, the maximum number of electrons that can occur in the various (s, p, d, and f)

TABLE 3.5   Quantum Notation and Electron Distribution

Shell and Main Energy Levels (n)	Energy Sublevels	Number of Orbitals		Maximum Number of Electrons	
K ($n = 1$)	$1s$ ($l = 0$)	1	1	2	2
L ($n = 2$)	$2s$ ($l = 0$)	1	} 4	2	} 8
	$2p$ ($l = 1$)	3		6	
M ($n = 3$)	$3s$ ($l = 0$)	1		2	
	$3p$ ($l = 1$)	3	} 9	6	} 18
	$3d$ ($l = 2$)	5		10	
N ($n = 4$)	$4s$ ($l = 0$)	1		2	
	$4p$ ($l = 1$)	3	} 16	6	} 32
	$4d$ ($l = 2$)	5		10	
	$4f$ ($l = 3$)	7		14	
O ($n = 5$)	$5s$ ($l = 0$)	1		2	
	$5p$ ($l = 1$)	3	} 16	6	} 50*
	$5d$ ($l = 2$)	5		10	
	$5f$ ($l = 3$)	7		14	
P ($n = 6$)	$6s$ ($l = 0$)	1		2	
	$6p$ ($l = 1$)	3		6	} 72*
	$6d$ ($l = 2$)	5		10	
Q ($n = 7$)	$7s$ ($l = 0$)	1		2	98*

*This number is not reached in naturally occurring atoms.

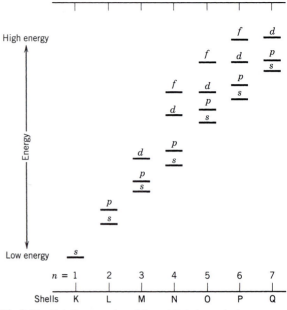

FIG. 3.14 Relative energies of the orbitals in neutral, many-electron isolated atoms.

subshells is also limited. For example, the s subshell has only one orbital and therefore can house only a maximum of two electrons; the p subshell has three orbitals and therefore can house a maximum of six electrons. This is illustrated in Table 3.5.

The relative energies of shells, subshells, and orbitals in atoms with more than one electron are illustrated in Fig. 3.14. This shows that the energy of a shell increases with increasing value of the principal quantum number, n. It also shows that, as the value of n increases, there is an overlap in the energy levels of subshells for $n = 3$ and higher. That is, the $4s$ subshell has a lower relative energy level than the $3d$ subshell. Such overlap becomes even more common in the higher shells. This sequence of energy levels is critical in the determination of the arrangement of electrons in the atom. A complete table of electron configurations of the elements is given in Table 3.6.

Electron Configuration and the Periodic Table

In the earlier, introductory discussion to the periodic table, we noted that the table is arranged by increasing atomic number (Z). Now that we have discussed the orbital structure of the atom, we can address this periodicity in greater detail. In general, electrons tend to occupy higher energy orbitals only after lower energy levels have been filled (see Table 3.6).

The periodic table (the endpaper at the back of this text) is organized such that the vertical columns (identified by I, II, III, IV, . . . , VIII) list atoms whose outer shells contain the number of electrons equal to the Roman numeral at the top of the column. For example, all elements in column I ($Z = 1$, 3, 11, 19, 37, 55, and 87) have only one electron in the s orbital of the outer shell (see also Table 3.6). All elements in column II have two s orbital electrons in the outer shell. Elements in column III contain three electrons (two s and one p). Those in column IV have four electrons (two s and two p), and so forth, until column VIII, which contains all those elements (except He) with eight outer shell electrons (two s and six p). Helium has two electrons in the s orbital, completely filling the K shell.

The horizontal *rows* in the periodic table are numbered 1 through 7; these are equivalent to the K, L, M, N, . . . shells. In a left-to-right sequence in any given row, the outer shell of an atom is progressively filled, beginning with the s orbital and ending with the p orbitals. Atoms with $Z = 21$ to 30 (in row 4), with $Z = 39$ to 48 (in row 5), and with $Z = 57$ to 80 (in row 6) are called the *transition elements* because the orbital electrons, in excess of those present in calcium ($Z = 20$), fill in inner shells. For example, in row 4, elements with $Z = 21$ to 30 fill the $3d$ orbital of the more interior M shell; in row 6, elements with $Z = 57$ to 80 first fill $4f$ and subsequently $5d$ orbitals on the interior side of the P shell.

TABLE 3.6   Electron Configurations of the Atoms

Element	K 1s	L 2s	L 2p	M 3s	M 3p	M 3d	N 4s	N 4p	N 4d	N 4f	O 5s	O 5p	O 5d	O 5f	O 5g	P 6s	P 6p	P 6d	Q 7s
1. H	1																		
2. He	2																		
3. Li	2	1																	
4. Be	2	2																	
5. B	2	2	1																
6. C	2	2	2																
7. N	2	2	3																
8. O	2	2	4																
9. F	2	2	5																
10. Ne	2	2	6																
11. Na	2	2	6	1															
12. Mg	2	2	6	2															
13. Al	2	2	6	2	1														
14. Si	2	2	6	2	2														
15. P	2	2	6	2	3														
16. S	2	2	6	2	4														
17. Cl	2	2	6	2	5														
18. Ar	2	2	6	2	6														
19. K	2	2	6	2	6		1												
20. Ca	2	2	6	2	6		2												
21. Sc	2	2	6	2	6	1	2												
22. Ti	2	2	6	2	6	2	2												
23. V	2	2	6	2	6	3	2												
24. Cr	2	2	6	2	6	5	1												
25. Mn	2	2	6	2	6	5	2												
26. Fe	2	2	6	2	6	6	2												
27. Co	2	2	6	2	6	7	2												
28. Ni	2	2	6	2	6	8	2												
29. Cu	2	2	6	2	6	10	1												
30. Zn	2	2	6	2	6	10	2												
31. Ga	2	2	6	2	6	10	2	1											
32. Ge	2	2	6	2	6	10	2	2											
33. As	2	2	6	2	6	10	2	3											
34. Se	2	2	6	2	6	10	2	4											
35. Br	2	2	6	2	6	10	2	5											
36. Kr	2	2	6	2	6	10	2	6											
37. Rb	2	2	6	2	6	10	2	6			1								
38. Sr	2	2	6	2	6	10	2	6			2								
39. Y	2	2	6	2	6	10	2	6	1		2								
40. Zr	2	2	6	2	6	10	2	6	2		2								
41. Nb	2	2	6	2	6	10	2	6	4		1								
42. Mo	2	2	6	2	6	10	2	6	5		1								
43. Tc	2	2	6	2	6	10	2	6	5		2								
44. Ru	2	2	6	2	6	10	2	6	7		1								
45. Rh	2	2	6	2	6	10	2	6	8		1								
46. Pd	2	2	6	2	6	10	2	6	10										
47. Ag	2	2	6	2	6	10	2	6	10		1								
48. Cd	2	2	6	2	6	10	2	6	10		2								
49. In	2	2	6	2	6	10	2	6	10		2	1							
50. Sn	2	2	6	2	6	10	2	6	10		2	2							
51. Sb	2	2	6	2	6	10	2	6	10		2	3							
52. Te	2	2	6	2	6	10	2	6	10		2	4							
53. I	2	2	6	2	6	10	2	6	10		2	5							

(continued)

TABLE 3.6   (continued)

Shell / Element	K 1s	L 2s	2p	M 3s	3p	3d	N 4s	4p	4d	4f	O 5s	5p	5d	5f	5g	P 6s	6p	6d	Q 7s
54. Xe	2	2	6	2	6	10	2	6	10		2	6							
55. Cs	2	2	6	2	6	10	2	6	10		2	6				1			
56. Ba	2	2	6	2	6	10	2	6	10		2	6				2			
57. La	2	2	6	2	6	10	2	6	10		2	6	1			2			
*58. Ce	2	2	6	2	6	10	2	6	10	1	2	6	1			2			
*59. Pr	2	2	6	2	6	10	2	6	10	2	2	6	1			2			
*60. Nd	2	2	6	2	6	10	2	6	10	3	2	6	1			2			
*61. Pm	2	2	6	2	6	10	2	6	10	4	2	6	1			2			
*62. Sm	2	2	6	2	6	10	2	6	10	5	2	6	1			2			
*63. Eu	2	2	6	2	6	10	2	6	10	6	2	6	1			2			
*64. Gd	2	2	6	2	6	10	2	6	10	7	2	6	1			2			
*65. Tb	2	2	6	2	6	10	2	6	10	8	2	6	1			2			
*66. Dy	2	2	6	2	6	10	2	6	10	9	2	6	1			2			
*67. Ho	2	2	6	2	6	10	2	6	10	10	2	6	1			2			
*68. Er	2	2	6	2	6	10	2	6	10	11	2	6	1			2			
*69. Tm	2	2	6	2	6	10	2	6	10	12	2	6	1			2			
*70. Yb	2	2	6	2	6	10	2	6	10	13	2	6	1			2			
*71. Lu	2	2	6	2	6	10	2	6	10	14	2	6	1			2			
72. Hf	2	2	6	2	6	10	2	6	10	14	2	6	2			2			
73. Ta	2	2	6	2	6	10	2	6	10	14	2	6	3			2			
74. W	2	2	6	2	6	10	2	6	10	14	2	6	4			2			
75. Re	2	2	6	2	6	10	2	6	10	14	2	6	5			2			
76. Os	2	2	6	2	6	10	2	6	10	14	2	6	6			2			
77. Ir	2	2	6	2	6	10	2	6	10	14	2	6	7			2			
78. Pt	2	2	6	2	6	10	2	6	10	14	2	6	9			1			
79. Au	2	2	6	2	6	10	2	6	10	14	2	6	10			1			
80. Hg	2	2	6	2	6	10	2	6	10	14	2	6	10			2			
81. Tl	2	2	6	2	6	10	2	6	10	14	2	6	10			2	1		
82. Pb	2	2	6	2	6	10	2	6	10	14	2	6	10			2	2		
83. Bi	2	2	6	2	6	10	2	6	10	14	2	6	10			2	3		
84. Po	2	2	6	2	6	10	2	6	10	14	2	6	10			2	4		
85. At	2	2	6	2	6	10	2	6	10	14	2	6	10			2	5		
86. Rn	2	2	6	2	6	10	2	6	10	14	2	6	10			2	6		
87. Fr	2	2	6	2	6	10	2	6	10	14	2	6	10			2	6		1
88. Ra	2	2	6	2	6	10	2	6	10	14	2	6	10			2	6		2
89. Ac	2	2	6	2	6	10	2	6	10	14	2	6	10			2	6	1	2
*90. Th	2	2	6	2	6	10	2	6	10	14	2	6	10			2	6	2	2
*91. Pa	2	2	6	2	6	10	2	6	10	14	2	6	10	2		2	6	1	2
*92. U	2	2	6	2	6	10	2	6	10	14	2	6	10	3		2	6	1	2
*93. Np	2	2	6	2	6	10	2	6	10	14	2	6	10	4		2	6	1	2
*94. Pu	2	2	6	2	6	10	2	6	10	14	2	6	10	6		2	6		2
*95. Am	2	2	6	2	6	10	2	6	10	14	2	6	10	7		2	6		2
*96. Cm	2	2	6	2	6	10	2	6	10	14	2	6	10	7		2	6	1	2
*97. Bk	2	2	6	2	6	10	2	6	10	14	2	6	10	8		2	6	1	2
*98. Cf	2	2	6	2	6	10	2	6	10	14	2	6	10	10		2	6		2
*99. Es	2	2	6	2	6	10	2	6	10	14	2	6	10	11		2	6		2
*100. Fm	2	2	6	2	6	10	2	6	10	14	2	6	10	12		2	6		2
*101. Mv	2	2	6	2	6	10	2	6	10	14	2	6	10	13		2	6		2
*102. No	2	2	6	2	6	10	2	6	10	14	2	6	10	14		2	6		2

*Lanthanide and actinide elements; some configurations uncertain.

The symbolism for summarizing an atom's electron configuration is as follows: the symbol of each orbital is followed by an exponent indicating the number of electrons present in the orbital. The symbol for atomic silicon ($Z = 14$) is $1s^2 2s^2 2p^6 3s^2 3p^2$. This signifies two electrons in the $1s$ and $2s$ orbitals, six in the $2p$ orbital, and two in the $3s$ and $3p$ orbitals. A listing of the symbolic notation of the electronic structure for elements $Z = 1$ to 37 is given in Table 3.7; this should be compared with Table 3.6.

The preceding discussion has shown that the periodic table of elements is not just a tabulation, but rather the ordered result of basic chemical properties that depend on the nature of the outer electrons, the *valence electrons*. These are the electrons available for chemical bonding. As a result of the similarity of the chemical character of certain elements (because of their similarity in outer electron configurations), such elements often have similar chemical behavior and can be found in similar crystallographic sites within mineral structures.

THE ION

Until this point, we have concerned ourselves with atoms in which the number of protons and electrons is equal. However, elements in the periodic table can be divided into two groups: those that have a tendency to give up electrons, and those that are capable of acquiring electrons. Those that are electron donors are the *metals* (in the left-hand part of the periodic table, see endpaper at the back of this text) and those that are electron acceptors are the *nonmetals* (in the right-hand part of the periodic table). When one or more electrons are lost from the electron configuration of an atom, a *cation* is formed, and when electrons are added, an *anion* results. This can be expressed as:

$$X_{atom} - e^- \rightarrow X^+_{cation}$$

and

$$X_{atom} + e^- \rightarrow X^-_{anion}$$

In either process, energy is involved. The energy required to remove the most weakly held electron from a neutral atom (to infinity) is known as the *first ionization potential*. This value (listed for a representative number of elements in Table 3.7) expresses how strongly the nucleus of a neutral atom attracts an electron in a partially filled orbital. Figure 3.15a illustrates that the first ionization potential values increase with increasing atomic number, within each period. This increase coincides with the progressive filling of electron orbitals and expresses the reluctance of atoms to lose electrons from orbitals that are nearly completely filled. It further shows that inert gases (He, Ne, Ar, Kr) have maximal values and alkali metals (Li, Na, K, Rb) have minimal values. This means that the electron configurations of the inert

TABLE 3.7 First Ionization Potentials, Electronegativity Values, and Electronic Structure of the Elements Through Atomic Number 37

Z	Element	First Ionization Potential, in Electron Volts (e.v.)*,†	Electro-negativity‡	Electronic Structure
1	H	13.598	2.1	$1s^1$
2	He	24.587	0	$1s^2$
3	Li	5.392	1.0	$1s^2 2s^1$
4	Be	9.322	1.5	$1s^2 2s^2$
5	B	8.298	2.0	$1s^2 2s^2 2p^1$
6	C	11.260	2.5	$1s^2 2s^2 2p^2$
7	N	14.534	3.1	$1s^2 2s^2 2p^3$
8	O	13.618	3.5	$1s^2 2s^2 2p^4$
9	F	17.422	4.1	$1s^2 2s^2 2p^5$
10	Ne	21.564	0	$1s^2 2s^2 2p^6$
11	Na	5.139	1.0	$[Ne]3s^1$
12	Mg	7.646	1.3	$[Ne]3s^2$
13	Al	5.986	1.5	$[Ne]3s^2 3p^1$
14	Si	8.151	1.8	$[Ne]3s^2 3p^2$
15	P	10.486	2.1	$[Ne]3s^2 3p^3$
16	S	10.360	2.4	$[Ne]3s^2 3p^4$
17	Cl	12.967	2.9	$[Ne]3s^2 3p^5$
18	Ar	15.759	0	$[Ne]3s^2 3p^6$
19	K	4.341	0.9	$[Ar]4s^1$
20	Ca	6.113	1.1	$[Ar]4s^2$
21	Sc	6.54	1.2	$[Ar]3d^1 4s^2$
22	Ti	6.82	1.3	$[Ar]3d^2 4s^2$
23	V	6.74	1.5	$[Ar]3d^3 4s^2$
24	Cr	6.766	1.6	$[Ar]3d^5 4s^1$
25	Mn	7.435	1.6	$[Ar]3d^5 4s^2$
26	Fe	7.870	1.7	$[Ar]3d^6 4s^2$
27	Co	7.86	1.7	$[Ar]3d^7 4s^2$
28	Ni	7.635	1.8	$[Ar]3d^8 4s^2$
29	Cu	7.726	1.8	$[Ar]3d^{10} 4s^1$
30	Zn	9.394	1.7	$[Ar]3d^{10} 4s^2$
31	Ga	5.999	1.8	$[Ar]3d^{10} 4s^2 4p^1$
32	Ge	7.899	2.0	$[Ar]3d^{10} 4s^2 4p^2$
33	As	9.81	2.2	$[Ar]3d^{10} 4s^2 4p^3$
34	Se	9.752	2.5	$[Ar]3d^{10} 4s^2 4p^4$
35	Br	11.814	2.8	$[Ar]3d^{10} 4s^2 4p^5$
36	Kr	13.999		$[Ar]3d^{10} 4x^2 4p^6$
37	Rb	4.177	0.9	$[K]5s^1$

*e.v. = electron volt = 23 kilocalories/mole.
†From Lide, D. R., ed., 1991, *CRC Handbook of Chemistry and Physics*, 72nd ed. CRC Press, Boca Raton, Fla.
‡From J. E. Brady, J. W. Russell, and J. R. Holum, 2000, *Chemistry, matter and its changes*, 3rd ed. (New York: Wiley).

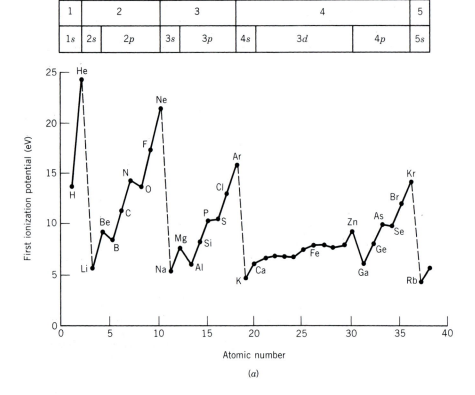

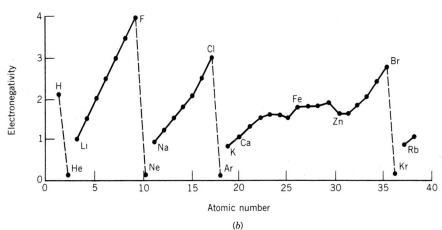

FIG. 3.15 *(a)* Variation of the first ionization potential as a function of increasing atomic number (*Z*) for the first 37 elements. *(b)* Variation of the electronegativity for the same elements as shown in *(a)*.

gases are more stable than those of the alkali metals. As such, a relatively small amount of energy is needed to remove an electron from an alkali metal, thus producing a stable monovalent (1+) ion. Indeed, the stability of alkali and halide ions and the lack of chemical reactivity of the noble (inert) gases is attributable to the *extraordinary stability of configuration of 2, 10, 18, 36, 54, and 86 electrons about an atomic nucleus* (see Table 3.6). As such, the alkali metals (in column I of the periodic table; see endpaper at the back of this text) carry one more elec-

tron than the noble-gas atom, and this electron is easily lost. The halogens in column VII of the periodic table contain one less electron than a noble-gas atom, and they easily gain an electron. Similarly, the elements of column II of the periodic table can by losing two electrons produce ions (Be^{2+}, Mg^{2+}, Ca^{2+}, Sr^{2+}, Ba^{2+}) with a noble-gas structure.

Figure 3.15a also shows that it takes more energy to remove an electron from beryllium than it does to remove an electron from lithium or boron. An unpaired electron, whether it is in a 2s orbital (as in lithium) or in a $2p$ orbital (as in boron; see Table 3.7), is evidently less well bonded to the atom than a paired electron, as in the filled $2s$ orbital of beryllium. A similar effect is seen in period 3 in the first ionization potential of magnesium versus that of sodium and aluminum. The ionization potentials of elements in period 4 with partially filled d orbitals (see Table 3.6) vary little with atomic number. The general trend, within each period, is one of low ionization potentials (elements acting as electron donors, and as such metallic in their character) to high ionization potentials (elements acting as electron acceptors, and as such nonmetallic in character).

Ionization potentials that express the energy needed to remove additional electrons (i.e., more than one electron) are very much larger than those of the first ionization reported in Table 3.7. Such higher values reflect the much greater energy needed to remove an electron from an atom that has already acquired a positive charge, as well as the higher energy needed to remove an additional electron from remaining electrons that fill the orbital in which they occur. Because of these energy barriers, elements involved in chemical reactions tend to lose only their *valence* electrons, which are those that reside outside filled orbitals.

Elements in column I in the periodic table, which include Li, Na, K, Rb, easily lose the one and only outer electron, as seen in Fig. 3.15. This results in the formation of monovalent cations such as Li^+, Na^+, K^+, and Rb^+. For elements in column II (Be, Mg, Ca, Sr, Ba), there are similarly low ionization potential values (see Fig. 3.15 and Table 3.7), suggesting that relatively little energy is needed to make these atoms into divalent cations such as Be^{2+}, Mg^{2+}, Ca^{2+}, and so on. Similar considerations hold for the formation of trivalent (e.g., Al^{3+}) and tetravalent (e.g., Si^{4+}) cations.

Several elements are found in more than one *valence* or *oxidation state*. For example, iron (Fe) can occur in a divalent state (*ferrous* iron, Fe^{2+}) or in an even more oxidized trivalent (*ferric* iron, Fe^{3+}) state. The electronic configuration of atomic Fe is $1s^2 2s^2 2p^6 3s^2 3p^6 3d^6 4s^2$ (Table 3.6). In Fe^{2+}, the atom loses the two $4s$ electrons, but it can lose one $3d$ electron as well, making Fe^{3+}, which then has a half-filled subshell. Because in a crystalline substance d orbitals have energy differences that are similar to wavelengths in visible light, transition elements, such as iron, tend to play a major role in the coloration of minerals (see pp. 157 to 164). Table 3.8 lists the valence states of some common ions.

TABLE 3.8 The Valence States of Ions (and of Ionic Groups) that Occur Abundantly in Rock-Forming Minerals (see also Table 3.12).

Cations	
Na^+	Si^{4+}
K^+	Ti^{4+}
	C^{4+}
Mg^{2+}	
Fe^{2+} (also Fe^{3+})	P^{5+}
Ca^{2+}	
Mn^{2+} (also Mn^{3+})	S^{6+}
Al^{3+}	

Anions and Anionic Groups		
O^{2-}	$(SiO_4)^{4-}$	$(PO_4)^{3-}$
$(OH)^{1-}$	$(CO_3)^{2-}$	$(SO_4)^{2-}$

Although ionization potentials are useful predictors of some chemical properties, Linus Pauling developed an additional concept, known as *electronegativity*. This is a measure of the ability of an atom *in a crystal structure or molecule* to attract electrons to its outer shell. It is represented as a dimensionless number (see Table 3.7 and Fig. 3.22) that is calculated from the known bond strengths between atoms in molecules. Elements with low electronegativity are electron donors and those with high values are electron acceptors. Ranges of electronegativity values are illustrated in Fig. 3.15*b*. Here it can be seen that, within a specified period, electronegativity values rise as a function of increasing atomic number. It is also clear from Fig. 3.15*b* that the electronegativity values of elements within column I (H, Li, Na, K, Rb, and so on; see Table 3.7) or within column II (Be, Mg, Ca, Sr, and so on) decrease with increasing atomic number. The same holds for their values of ionization potential. This leads to the conclusion that the bond strength (or binding energy) between the nucleus and the first valence electron of an element (in a specified group) decreases as the volume of the atom in the group increases. This implies that the large atoms hold their outer valence electrons more loosely than do smaller atoms. The electronegativity values of the noble gases are set at zero because these atoms do not attract electrons. The concept of electronegativity is especially useful in assessing the type of bond formed between different atoms. It is well known that elements with very different electronegativity values will tend to form bonds that are *ionic* in character, whereas those with similar electronegativities form *covalent* bonds. As such, the *difference* in electronegativity values is a very useful parameter in the estimation of the chemical bonding mechanism.

BONDING FORCES IN CRYSTALS

The forces that bind together the atoms (or ions, or ionic groups) of crystalline solids are electrical in nature. Their type and intensity are largely responsible for the physical and chemical properties of minerals. Hardness, cleavage, fusibility, electrical and thermal conductivity, compressibility, and the coefficient of thermal expansion are directly related to binding forces. In general, the stronger the average bond, the harder is the crystal, the higher its melting point, and the smaller its coefficient of thermal expansion. The great hardness of a diamond is attributed to the very strong electrical forces linking its constituent carbon atoms. The minerals periclase, MgO, and halite, NaCl, are isostructural (i.e., their structures have the same arrangement), yet periclase melts at 2820°C, whereas halite melts at 801°C. The greater amount of heat energy required to separate the atoms in periclase indicates that it has a stronger electrical bond than halite.

These electrical forces are chemical bonds and can be described as belonging to one of five principal bond types: ionic, covalent, metallic, van der Waals, and hydrogen bonding. It should be understood that this classification is one of expediency and that transitions, or hybrids, may exist between all types. The electrical interaction of the ions or atoms constituting the structural units determines the properties of the resulting crystal. It is the similarity in properties among crystals having similar types of electrical interaction that justifies the use of the classification of bonding mechanisms.

The bonding forces linking the atoms of silicon and oxygen in quartz display in almost equal amount the characteristics of the ionic and the covalent bond. The strong bonding between Si^{4+} and O^{2-} ions in SiO_2 causes the electron densities of the two ions to be localized between the nuclei of Si and O. Such distortions of the electron clouds from generally spherical to more ellipsoidal shapes have been determined directly from X-ray diffraction intensity measurements. Figure 3.16 shows a considerable build-up of charge density in the Si–O bond of coesite (SiO_2), with the strongest charge concentration in the vicinity of the more electronegative oxygen atoms. Another example of a crystal displaying two bond types is galena, PbS. This mineral exhibits characteristics related to the metallic bond (Pb–Pb),

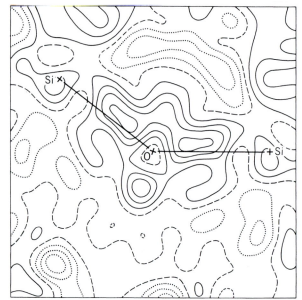

FIG. 3.16 A map of the electron density distribution of an Si–O–Si bond in coesite, one of the high-pressure polymorphs of SiO_2. This map displays the difference between the total electron density and the electron density prior to formation of the bond. The solid contours represent positive electron density, the dotted contours negative density, and the long dashed contour is the zero density contour. The contour interval is 0.1 electrons per Å^3. The distance between O and Si is 1.61 Å. (From K. L. Geisinger and G. V. Gibbs, 1983, An X-ray diffraction study of the electron distribution in coesite. *Geological Society of America, Abstracts with Programs*, 15:580.)

as expressed by good electrical conductivity and bright metallic luster, and of the ionic bond (Pb–S), as shown by excellent cleavage and brittleness. Furthermore, many minerals (such as mica) contain two or more bond types of different character and strength.

Ionic Bond

An assessment of the chemical activity of elements in terms of the occupancy of their outer orbitals with valence electrons leads to the conclusion that all atoms have a strong tendency to achieve an inert gas configuration with a completely filled valence shell. The noble gases—helium, neon, argon, krypton, and xenon—have completely filled valence shells and are almost completely inert (see Table 3.6). An ionic bond is achieved when one or more electrons in the valence shell of an atom are transferred to the valence shell of another, so that they both achieve an inert gas configuration. Sodium, for example, has a single valence electron in its outer orbital that it loses readily, leaving the atom with an unbalanced positive charge and the noble gas configuration of neon. Atomic chlorine, on the other hand, needs to acquire one electron to achieve the noble gas structure of argon (see Tables 3.6 and 3.7).

These electron losses and gains can be stated as

$$Na(1s^2 2s^2 2p^6 3s^1) \rightarrow Na^+(1s^2 2s^2 2p^6) + e^-$$

$$Cl(1s^2 2s^2 2p^6 3s^2 3p^5) + e^- \rightarrow Cl^-(1s^2 2s^2 2p^6 3s^2 3p^6)$$

The electron lost by the sodium is picked up by the chlorine. Once formed, the Na^+ and Cl^- attract each other because of their opposite charges. *The attraction between oppositely charged ions constitutes the ionic (or electrostatic) bond* (see Fig. 3.17). The formation of this bond is the result of *the exchange of electron(s) of the metal atom (forming a cation) to the nonmetal atom (forming an anion).*

In a crystal of sodium chloride characteristic properties may be recognized: cubic crystal habit, cleavage, specific gravity, index of refraction, and so forth. These properties in no way resemble those of the shining metal (Na) or the greenish, acrid gas (Cl$_2$), which are the elemental constituents of the substance. Touching the crystal to the tongue yields the taste of the solution. In other words, the properties conveyed into the crystal by its constituent elements are the properties of the *ions*, not the elements.

Physically, ionically bonded crystals are generally of moderate hardness and specific gravity, have fairly high melting points, and are poor conductors of electricity and heat. The lack of electrical conductivity in the ionic bonding of crystals is due to the stability of the ions, which neither gain nor lose electrons easily. Because the electrostatic charge constituting the ionic bond is evenly spread over the ion, a cation will tend to surround itself with as many anions as can be fitted around it. This means that the ionic bond is nondirectional, and the symmetry of the resultant crystals is generally high (see Fig. 3.17 and Table 3.9 on page 62).

FIG. 3.17   An idealized ionic structure in two dimensions.

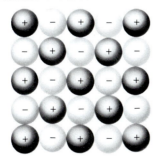

The strength of the ionic bond, u, depends on two factors: (1) the center-to-center spacing between the ions (r), and (2) the product of their charges (q):

$$u = (A \, q_1 \, q_2)/r$$

where A is a numerical quantity, the *Madelung constant.* This expression is similar to that for the Coulombic force (F) operating over a distance d, where F is defined by the equation on page 65 (see also Fig. 3.29b).

The effect of increased interionic distance on the strength of the ionic bond is readily seen when we compare the halides of sodium. Figure 3.18a shows the melting points and interionic distances for these compounds. It is apparent that the strength of the bond, as measured by the melting temperature, is inversely proportional to the bond length. The melting temperatures of the fluorides of the alkali metals illustrate that it does not matter whether the size of the anion or cation is varied: the shorter the bond length, the stronger is the bond (see Fig. 3.18c). LiF is an exception to this generalization, which is explained by anion–anion (F–F) repulsion in a structure having a very small cation.

The charge of the coordinated ions has an even more powerful effect on the strength of the bond. A comparison of the absolute values of the melting temperatures for the alkali-earth oxides (Fig. 3.18b), which are divalent compounds ($q = 2$), with the absolute values for the monovalent alkali fluorides ($q = 1$), in which the interionic spacings are closely comparable (Fig. 3.18c), reveals the magnitude of the effect. Although the interionic distance is almost the same for corresponding oxides and fluorides, the bonds uniting the more highly charged ions are much stronger. Figure 3.19 illustrates the effect of interionic spacing and charge on hardness. All substances cited as examples in these two figures have the same structure and may be regarded as ionically bonded.

Covalent Bond

We have seen that ions of chlorine may enter into ionically bonded crystals as stable units of structure because, by taking an electron from a metal such as Na, they achieve a filled outer orbital. A single atom of chlorine with an incomplete valence orbital is in a highly reactive condition. It seizes upon and combines with almost anything in its neighborhood. Generally, its nearest neighbor is another chlorine atom, and the two unite in such a way that two electrons, one from each chlorine atom, do double duty

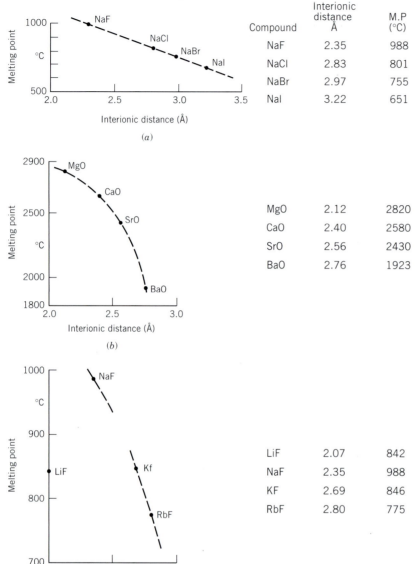

Compound	Interionic distance Å	M.P (°C)
NaF	2.35	988
NaCl	2.83	801
NaBr	2.97	755
NaI	3.22	651
MgO	2.12	2820
CaO	2.40	2580
SrO	2.56	2430
BaO	2.76	1923
LiF	2.07	842
NaF	2.35	988
KF	2.69	846
RbF	2.80	775

FIG. 3.18 Melting point versus interionic distance in ionic-bonded compounds. (Data from *Handbook of chemistry and physics*, 72nd ed. Boca Raton, Fla.: CRC Press, 1991.)

in the outer orbitals of both atoms, and both thus achieve the stable inert gas configuration. As a result of this sharing of an electron, the two atoms of chlorine are strongly bound together.

The bonding mechanism between two chlorine atoms can be symbolically stated as follows:

$$\cdot\ddot{\underset{..}{C}}l\colon + \cdot\ddot{\underset{..}{C}}l\colon \rightarrow \colon\ddot{\underset{..}{C}}l\colon\ddot{\underset{..}{C}}l\colon$$

Here the small dots represent the outer-shell (valence) electrons in the *s* and *p* orbitals of the M shell in chlorine (this notation is known as Lewis symbols, after the American chemist Gilbert N. Lewis, 1875–1946). The valency shell is completed by the sharing of electrons between the two Cl on the right

side of the equation. A schematic representation of this type of covalent bonding is shown in Fig. 3.20.

The number of covalent bonds that an atom may form can commonly be predicted by counting the number of electrons required to achieve a stable electron configuration (such as that of a noble gas). Carbon, for example, has four electrons in its valence shell; through sharing four additional electrons, it achieves the noble gas configuration of neon.

This *electron-sharing* or *covalent* bond is the strongest of the chemical bonds. Minerals so bonded are characterized by general insolubility, great stability, and very high melting points. They do not yield

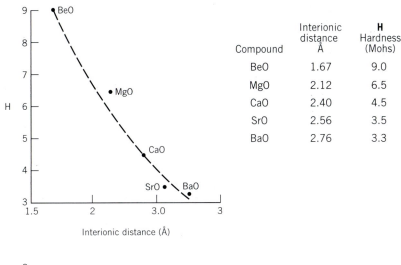

Compound	Interionic distance Å	**H** Hardness (Mohs)
BeO	1.67	9.0
MgO	2.12	6.5
CaO	2.40	4.5
SrO	2.56	3.5
BaO	2.76	3.3

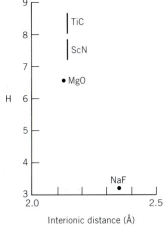

$Na^+ F^-$	2.35	3.2
$Mg^{2+} O^{2-}$	2.12	6.5
$Sc^{3+} N^{3-}$	2.23	7-8
$Ti^{4+} C^{4-}$	2.23	8-9

FIG. 3.19 Hardness versus interionic distance and charge in ionic-bonded compounds. (Data from R. C. Evans, 1952, *Crystal chemistry*. London: Cambridge University Press.)

ions in the dissolved state and are nonconductors of electricity both in the solid state and in solution. Because the electrical forces constituting the bond are sharply localized in the vicinity of the shared electron, the bond is highly directional, and the symmetry of the resulting crystals is likely to be lower than where ionic bonding occurs (see Table 3.9). In chlorine, the bonding energy of the atom is entirely consumed in linking to one neighbor, and stable Cl_2 molecules result that show little tendency to link one to another. Certain other elements—in general, those near the middle of the periodic table, such as car-

bon, silicon, aluminum, and sulfur—have two, three, and four vacancies in their outer orbitals. They therefore can form up to four covalent bonds with neighboring atoms. In some cases, these are multiple shared pairs of electrons between two atomic centers. This may result in very stable groups, which in turn link together to form larger aggregates or groups.

Carbon is an outstanding example of such an atom. The four valence electrons in each carbon are sufficient to fill the bonding orbitals by electron sharing with four other carbon atoms, forming a very stable, firmly bonded configuration having the shape of a tetrahedron with a central carbon atom bonded to four others at the apices (see Fig. 3.21). Every carbon atom is linked in this way to four others, forming a continuous network. The energy of the bonds is strongly localized in the vicinity of the shared electrons, producing a very rigid structure—that of diamond, the hardest natural substance. Because all the valence electrons in diamond are used for bonding, none can move freely to conduct electricity.

FIG. 3.20 Schematic representation of the electron distribution between two covalently bonded atoms.

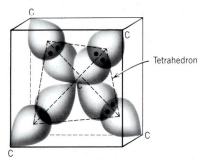

FIG. 3.21 Schematic representation of the overlap of orbitals in a C atom (at the center of the cube) with similar orbitals from four carbon atoms at the corners. This represents the covalent bonding in diamond.

In covalently bonded structures, the interatomic distance is generally equal to the arithmetic mean of the interatomic distances in crystals of the elemental substances. Thus, in diamond, the C–C spacing is 1.54 Å; in metallic silicon Si–Si distance is 2.34 Å. We may therefore suppose that if these atoms unite to form a compound, SiC, the silicon–carbon distance will be near 1.94 Å, the arithmetic mean of the elemental spacings. X-ray measurement determines this spacing in the familiar synthetic abrasive, silicon carbide, as 1.93 Å.

Estimation of the Character of the Bonding Mechanism

It is now generally recognized that there is some electron sharing in most ionically bonded crystals, whereas atoms in covalently bonded substances often have some electrostatic charges. Assessment of the relative proportions of ionic to covalent character is based in part on the polarizing power and polarizability of the ions involved. Compounds of a highly polarizing cation with an easily polarized anion, such as AgI, may show strongly covalent character. In contrast, AgF, because of the lower polarizability of the smaller fluorine ion, is a dominantly ionically bonded compound.

Bonds between elements of the first and seventh columns of the periodic table and between the second and the sixth columns are dominantly ionic. Examples are the alkali halides and the alkali-earth oxides. Bonds between like atoms or atoms close together in the periodic table will be covalent.

Linus Pauling, in 1939, provided a method by which the percentage of ionic character of a chemical bond can be estimated. The basis for this method is his scale of electronegativities of the elements (see Table 3.7 for a partial listing of these values; see also Fig. 3.22). *Electronegativity* is a measure of the ability of an atom to attract electrons to itself, and is expressed in a set of dimensionless numbers. Elements with low electronegativity are electron donors, and those with high values act as electron acceptors. The *differences* in the electronegativity values of the elements are an expression of the ionic character of the bond formed by their atoms. This difference is expressed as $X_A - X_B$, where X_A is the electronegativity of element A bonded to element B with its own electronegativity value X_B. Linus Pauling used this differ-

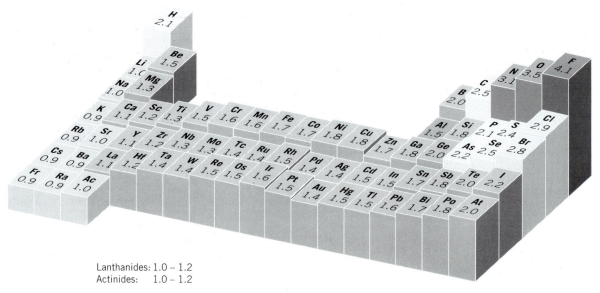

Lanthanides: 1.0 – 1.2
Actinides: 1.0 – 1.2

The noble gases are assigned electronegativities of zero and are omitted from this illustration.

FIG. 3.22 The electronegativities of the elements. (From J. E. Brady, J. W. Russell, and J. R. Holum, 2000, *Chemistry, matter and its changes*, 3rd ed., New York: Wiley.)

ence in electronegativity values in an equation to estimate the ionic character of a single bond:

amount of ionic character $= 1 - e^{-1/4(X_A - X_B)^2}$

This function is shown as a curve in Fig. 3.23. The usefulness of this function (and curve) can be illustrated by plotting, as an example, the differences in electronegativity values for the elements of the four compounds listed in the lower half of Fig. 3.19. These are NaF, MgO, ScN, and TiC. The electronegativity values of the elements are given in Fig. 3.22; the values for $(X_A - X_B)$ for these compounds are as follows:

NaF	3.1	ScN	1.9
MgO	2.3	TiC	1.2

The plot of these values on Fig. 3.23 shows that NaF is essentially ionic in its bond type (~92%) and that TiC is about 30% ionic in its bond type; the two other compounds are intermediate between these values. Similarly, one can evaluate the ionic character of bond types between silicon–oxygen, aluminum–oxygen, and boron–oxygen. The values for $(X_A - X_B)$ for these turn out to be 1.7 (for Si–O), 2.0 (for Al–O), and 1.5 (for B–O) (see Fig. 3.22). The 1.7 value for the Si–O bond is plotted in Fig. 3.23 and shows the bond type to be 50% ionic in character. The Al–O bond is 63% ionic, and the B–O bond only 44% ionic.

This shows that electronegativity, along with ionic size and valence, is helpful in predicting the chemical behavior of elements. It follows from Fig. 3.23 that compounds made of elements with very different values of electronegativity are more ionic than compounds made up of elements close to each other in electronegativity.

Metallic Bond

Metallic sodium is soft, lustrous, opaque, and sectile, and conducts heat and electrical current well. X-ray diffraction analysis reveals that it has the regular repetitive pattern of a true crystalline solid. The properties of the metal differ so from those of its salts that we are led to suspect a different mechanism of bonding. Because sodium, like all true metals, conducts electricity, electrons are free to move readily through the structure. So prodigal with their electrons are sodium and its close relatives, cesium, rubidium, and potassium, that the impact of the radiant energy of light knocks a considerable number entirely free of the structure. This photoelectric effect, on which such instruments as exposure meters depend, shows that the electrons are very weakly tied into the metal structure.

We may thus postulate that the structural units of true metals are actually the atomic nuclei plus nonvalence electron orbitals bound together by the aggregate electrical charge of a cloud of valence electrons that surrounds the nuclei. Many of the electrons owe no affinity to any particular nucleus and are free to drift through the structure or even out of it entirely, without disrupting the bonding mechanism. This is schematically illustrated in Fig. 3.24, which shows positively charged ions (from which the valence electrons have been removed) in a dense cloud of valence electrons. The attractive force between the nuclei with their filled electron orbitals (but lacking valence electrons) and the cloud of negative electrons holds such structures together. This type of bond is fittingly called the *metallic bond*. To it metals owe their high plasticity, tenacity, ductility, and conductivity, as well as their generally low hardness. Among minerals, only the native metals display

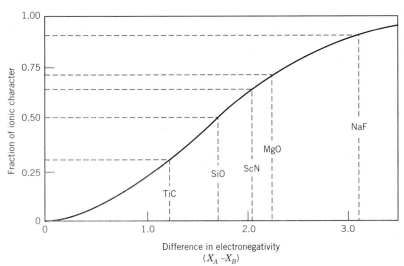

FIG. 3.23 Curve relating the amount of ionic character of a bond A–B to the difference in electronegavity $(X_A–X_B)$ of the atoms. Several compounds listed in Fig. 3.19 are plotted as well (see text). (After L. Pauling, 1960, *The nature of chemical bond.* Ithaca, N.Y.: Cornell University Press.)

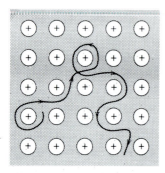

FIG. 3.24 A schematic cross section through the structure of a metal. Each circle with a positive charge represents a nucleus with filled, nonvalence electron orbitals of the metal atoms. The mobile electrons are represented by the cloud around the atoms (light gray shading). A possible electron path between the nuclei is shown.

pure metallic bonding. Table 3.9 gives a brief listing of some of the properties related to the metallic bond type in crystalline materials.

van der Waals Bond

Molecules such as N_2, O_2, and Cl_2 form molecular solids despite the fact that all the valence orbitals are occupied either by nonbonding electrons or by electrons used in covalent bonding to form *dimers* (a dimer is a molecule created from two identical simpler units). For example, if we take energy away from a Cl_2 gas by cooling it, the molecules ultimately will collapse into the close-packed, chaotic liquid state. If still more heat energy is removed, the amplitude of vibrations of the Cl_2 molecules is further reduced, and ultimately the minute, stray electrical fields existing about the essentially electrically neutral atoms will serve to lock the sluggishly moving molecules into the orderly structure of the solid state. This phenomenon of the solidification of chlorine takes place at very low temperatures, and warming above $-102°C$ will permit the molecules to break the very weak bonds and return to the disordered state of a liquid.

Neutral molecules such as Cl_2 may develop a small concentration of positive charge at one end, with a corresponding dearth of positive charge (which results in a small negative charge) at another end. This is the result of electrons in the occupied orbitals of the interacting atoms synchronizing their motions to avoid each other as much as possible. Figure 3.25 is a schematic illustration of electrons in

TABLE 3.9   Examples of Properties Conferred by the Principal Types of Chemical Bond

Property	Bond Type			
	Ionic (Electrostatic)	Covalent (Electron-shared)	Metallic	van der Waals (Residual)
Bond strength	Strong	Very strong	Variable strength, generally moderate	Weak
Mechanical	Hardness moderate to high, depending on interionic distance and charge; brittle	Hardness great Brittle	Hardness low to moderate; gliding common; high plasticity; sectile, ductile, malleable	Crystals soft and somewhat plastic
Electrical	Poor conductors in the solid state; melts and solutions conduct by ion transport	Insulators in solid state and melt	Good conductors; conduction by electron transport	Insulators in both solid and liquid state
Thermal (melting point = m.p.; coefficient of thermal expansion = coef.)	m.p. moderate to high depending on interionic distance and charge; low coef.	m.p. high; low coef.; atoms and molecules in melt	Variable m.p. and coef.; atoms in melt	Low m.p.; high coef.; liquid crystal molecules in melt
Solubility	Soluble in polar solvents to yield solutions containing ions	Very low solubilities	Insoluble, except in acids or alkalis by chemical reaction	Soluble in organic solvents to yield solutions
Structure	Nondirected; gives structures of high coordination and symmetry	Highly directional; gives structures of lower coordination and symmetry	Nondirected; gives structures of very high coordination and symmetry	Nondirected; symmetry low because of shape of molecules
Examples	Halite, NaCl; Fluorite, CaF_2; most minerals	Diamond, C; sphalerite, ZnS; molecules of O_2; organic molecules; graphite (strong bond)	Copper, Cu; silver, Ag; gold, Au; electrum, (Au, Ag); most metals	Sulfur (weak bond); organic compounds; graphite (weak bond)

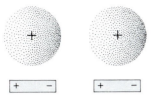

FIG. 3.25 Polarization of an atom, because of an increase in the concentration of electrons on one side of the atom, causes a dipole effect. The weak dipole attraction is that of the van der Waals bond.

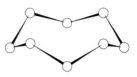

FIG. 3.27 S_8 rings occur in the crystal structure of sulfur. These rings are linked to each other by van der Waals bonds. Compare with Fig. 8.3.

orbitals having synchronized their motions such that an instantaneous and weak dipole attraction is produced between the two atoms. This weak dipole can induce a similar effect in neighboring atoms, which will cause the whole molecular structure to be bonded together by this weak dipole effect. Such bonding is especially effective over large distances in molecular structures. In the formation of crystals these molecules are aligned with negative poles against positive poles of neighboring molecules. Such is the mechanism that bonds the Cl_2 molecules in solid Cl_2. This weak bond, which ties neutral molecules and essentially uncharged structure units into a cohesive structure by virtue of small residual charges on their surfaces, is called the *van der Waals* (or *residual*) bond and is the weakest of the chemical bonds. Common in organic compounds and solidified gases, it is not often encountered in minerals, but when it is, it generally defines a zone of ready

cleavage and low hardness. An example is the mineral graphite, which consists of covalently bonded sheets of carbon atoms linked only by van der Waals bonds (see Fig. 3.26 and Table 3.9).

The most common form of crystalline sulfur is made up of discrete S_8 molecules with a cyclic structure (see Figs. 3.27 and 8.3). Within the ring there is pure covalent bonding, but adjacent rings are held together by van der Waals forces, which account for the low hardness ($H = 1\frac{1}{2}$ to $2\frac{1}{2}$) and melting point (at 112.8°C) of sulfur.

Hydrogen Bond

Polar molecules can form crystalline structures by the attraction between the oppositely charged ends of molecules (see Fig. 3.28a). The *hydrogen bond* is an electrostatic bond between a positively charged hydrogen ion and a negatively charged ion such as O^{2-} and N^{3-}. Because hydrogen has only one electron, when it transfers this electron to another more electronegative ion in ionic bonding, the remaining proton of the hydrogen nucleus becomes unshielded. This positive ion has the ability to form weak hydrogen bonds with other negative ions or with the negative ends of polar molecules, such as H_2O. The closeness of approach allows the formation of a dipole–dipole bond which is relatively weak when compared with an ionic or covalent bonding mechanism. However, it is considerably stronger than the van der Waals bond.

Ice is an especially good example of bonding in an intermolecular structure. The shape of an H_2O molecule is polar (see Fig. 3.28c), and because of it the two hydrogen atoms in the H_2O molecule provide the bonding to two other neighboring H_2O molecules. Two additional neighboring H_2O molecules in turn provide H atoms to provide two more hydrogen bonds. Therefore each oxygen atom is bound to four neighboring oxygen atoms, in a tetrahedral arrangement, by intervening hydrogen bonds. As in the case of the van der Waals bonds, the hydrogen bonds are weak, but there are many of them per unit volume of structure. Figure 3.28d is an illustration of the tetrahedrally bonded structure of one of the poly-

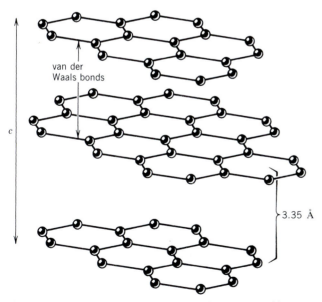

FIG. 3.26 Perspective sketch of the graphite structure with covalent bonding between carbon atoms within layers and residual (van der Waals) bonding between layers. Note large separation (3.35 Å) between layers.

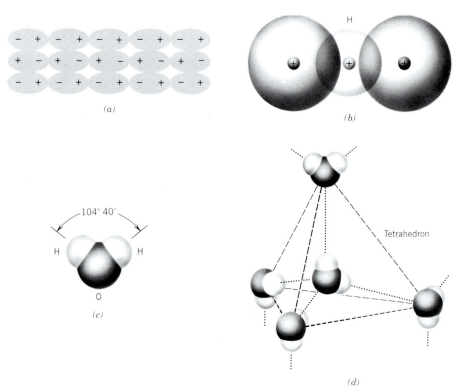

FIG. 3.28 *(a)* Schematic representation of the packing of polar molecules in a crystalline solid. Charges of opposite sign are arranged as closest neighbors. *(b)* Model of a hydrogen bond. *(c)* A water molecule and the bond angle between H–O–H. *(d)* Hydrogen bonding as shown by one of the polymorphs of ice. The coordination is tetrahedral and similar to that in diamond.

morphs of ice. It is a well-known fact that ice is less dense than water at the melting temperature. At the onset of melting this relatively open network structure collapses and in the resulting liquid the H_2O molecules are more densely packed than in the solid.

Hydrogen bonding is common in hydroxides in which the $(OH)^-$ group does not behave strictly as a spherical anionic group but is more realistically represented by an asymmetrical coordination, which produces a dipole effect. Hydrogen bonding is also present in many of the layer silicates such as micas and clay minerals, which contain hydroxyl groups.

Crystals with More than One Bond Type

Among naturally occurring substances, with their tremendous diversity and complexity, the presence of only one type of bonding is rare, and two or more bond types coexist in most minerals. Where this is so, the crystal shares in the properties of the different bond types represented, and often strongly directional properties result. Thus, in the mineral graphite, the cohesion of the thin sheets that make up the mineral structure is the result of the strong covalent bonding in the plane of sheets, whereas the excellent cleavage between the sheets reflects the weak van der Waals bonds joining the sheets together (see Fig.

3.26). The layer silicates, which consist of sheets of strongly bonded silica tetrahedra with a relatively weak ionic and/or hydrogen bond joining the sheets together through the cations, similarly reflect, in their remarkable basal cleavage between the sheets, the difference in strength of the two bond types (see Fig. 11.33). As we shall see, the prismatic habit and cleavage of the pyroxenes and amphiboles, and the chunky, blocky habit and cleavage of the feldspars, may similarly be traced to the influence of relatively weak bonds joining together more strongly bonded structure units having a chain, band, or block shape (see also Chapter 2).

◆ ATOMIC AND IONIC RADII

The sizes of atoms or ions are difficult to define but even more difficult to measure experimentally. The *radius* of an atom is defined by the radius of the maximum radial charge density of the outermost shells of the atom, but the *effective radius* of an atom (or ion) is also dependent on the type and number of neighboring atoms and/or ions, and on the charge of the ion. In a crystal of a pure metal, where identical atoms are bonded to each other, the radius of the individual atom is assumed to be one-half the bond length. Such measurements provide sizes for atomic

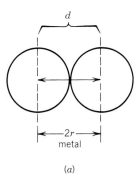

(a)

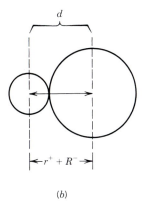

(b)

FIG. 3.29 *(a)* Metallic radii are half the distance between the centers of two adjoining metal atoms. *(b)* Ionic radii are defined as the distance between centers of the cation and anion, with the radius of e.g., the anion well known (rO^{2-} in 3-coordination = 1.36 Å; rF^- in 3-coordination = 1.30 Å). Therefore, the radius of Mg^{2+} is obtained by subtracting 1.36 Å from the internuclear distance between Mg^{2+} and O^{2-} in MgO.

radii (see Fig. 3.29*a* and Table 3.10). In ionic crystals, where two oppositely charged ions are bonded together, however, the distance between a positive and a negative ion is the sum of two different radii (see Fig. 3.29*b*). This distance is determined by electrostatic forces. There is, between any pair of oppositely charged ions, an attractive electrostatic force

that is directly proportional to the product of their charges and inversely proportional to the square of the distance between their centers. This is known as Coulomb's law, formulated in 1787 by the French physicist Charles Coulomb. It is stated as follows:

$$F = k\frac{(q^+)(q^-)}{d^2}$$

where *F* is the force of attraction between two oppositely charged ions, q^+ and q^- are the charges on the ions, *d* is the distance between them, and *k* is a proportionality constant. A force operating over a certain distance can also be expressed as energy (*E*). In Fig. 3.30 the curve representing the attractive force is the lower one, with (−) energy values. When ions approach each other under the influence of these forces, repulsive forces are also set up. These repulsive forces arise from the interaction of the negatively charged electron clouds and from the opposition of the positively charged nuclei; they increase rapidly with diminishing internuclear distance. In Fig. 3.30 this is represented by the upper curve with (+) energy values. The distance at which these repulsive forces balance the attractive forces is the characteristic interionic spacing (bond length) for a pair of ions. This is shown in Fig. 3.30 by the minimum value in the curve that is the resultant of the attractive and repulsive forces.

In the simplest case, when both cations and anions are fairly large and of low charge and both have numerous symmetrically disposed neighbors of opposite sign, ions may be regarded as spheres in contact. Sodium chloride, in which both cation and anion are monovalent, fairly large, and surrounded by six neighbors of opposite polarity, is a good example. In such crystals the interionic distance may be regarded as the sum of the radii of the two ions in contact.

If one of the ionic radii in an internuclear distance is well known from prior experimental mea-

TABLE 3.10   Metallic Radii (In Å)*

Li	Be												
1.57	1.12												
Na	Mg	Al											
1.91	1.60	1.43											
K	Ca	Sc	Ti	V	Cr	Mn	Fe	Co	Ni	Cu	Zn	Ga	Ge
2.35	1.97	1.64	1.47	1.35	1.29	1.37	1.26	1.25	1.25	1.28	1.37	1.53	1.39
Rb	Sr	Y	Zr	Nb	Mo	Tc	Ru	Rh	Pd	Ag	Cd	In	Sn	Sb
2.50	2.15	1.82	1.60	1.47	1.40	1.35	1.34	1.34	1.37	1.44	1.52	1.67	1.58	1.61
Cs	Ba	La	Hf	Ta	W	Re	Os	Ir	Pt	Au	Hg	Ti	Pb	Bi
2.72	2.24	1.88	1.59	1.47	1.41	1.37	1.35	1.36	1.39	1.44	1.55	1.71	1.75	1.82

*The values refer to 12-coordination. From Wells, A. F., 1991, *Structural Inorganic Chemistry*, 5th ed. Clarendon Press, Oxford, England, 1382 pp.

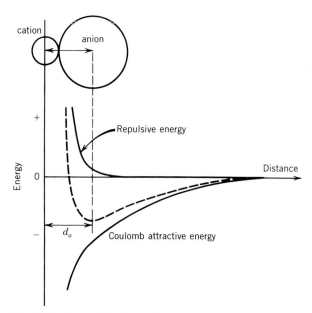

FIG. 3.30 Electrostatic interaction between a cation and anion. The attractive force acts over longer distances than the repulsion. The attractive and repulsive forces add to produce a resultant (dashed curve) in which the minimum value corresponds to the equilibrium distance (d_0) between the centers of the cation and anion.

surements, the radius of the other ion can be obtained (see Fig. 3.29b). For example, once Linus Pauling (in 1927) determined the radius of 6-coordinated O^{2-} to be 1.40 Å, the radii of many cations that are bonded ionically to oxygen were obtained by subtracting the value of 1.40 Å from the measured bond length between cation–oxygen pairs. Such measurements provide ionic radii.[2] However, the ionic radius of an ion may not be constant from one crystal structure to another. This is due to possible changes in bond type and coordination number (= the number of closest neighbors around a specific atom or ion). Because of such influences, the values of ionic radii given in tabulations generally represent averaged values.

Shannon and Prewitt (1969; revised by Shannon, 1976) evaluated in great detail the variation of ionic size as a function of coordination number. For example, they found considerable variation in the ionic size of O^{2-} as a function of coordination number,

ranging from 1.35 Å in 2-fold coordination to 1.42 Å in 8-fold coordination (see Table 3.11). The ionic radius values reported by Shannon and Prewitt (1969) and Shannon (1976) (many of which are listed in Table 3.11; for a complete listing see Shannon's Table 1) are referred to as *effective* ionic radii. The term effective is used because they were determined empirically from highly accurate data for a very large number of oxide structures. According to Shannon and Prewitt (1969) and Shannon (1976), these radii may well represent the best possible fit because they reproduce interatomic distances in a wide variety of crystalline solids. It is clear from Table 3.11 that many of the radii (be they for anions or cations) vary as a function of *coordination number* (**C.N.** = the number of atoms that surround a particular atom or ion in a structure; coordination number is discussed in detail in a subsequent section). For examples of anion radius changes see O^{2-} and F^- in Table 3.11. Radius changes in cations as a function of coordination number are clear for many of those listed in Table 3.11. For example, the K^+ radius = 1.38 Å for 6-coordination, 1.51 Å for 8-coordination, and 1.59 Å for 10-coordination. The increase in the cation radius reflects the expansion of the cation into the space (or void) provided by the surrounding anions. Figure 3.31 graphically illustrates the expansion of the radius of some selected cations as a function of coordination number.

It is instructive to compare the metallic radii, given in Table 3.10, with the ionic radii for the same elements, given in Table 3.11. In all cases the ionic radius of cations is considerably smaller than the metallic radius, for the same element. This is the result of the loss of one or more outer electrons, and the resultant reduction in the overall size of the electron cloud. Anions, on the other hand, because they gain electrons, are larger than the corresponding neutral atom. In a specific crystal structure, the measured radius of a given element may be somewhere between the element's atomic and ionic radii because the bond type in that specific structure is a mixture of several bond types, such as ionic, covalent, and/or metallic.

A regular change in ionic size is reflected by the arrangement of the elements in the periodic table. For elements of the same column, the ionic radii increase as the atomic number increases. For example, in Table 3.11, column II, the smallest ion is Be^{2+}, with a radius of 0.16 to 0.45 Å. The last ion listed in column II is Ba^{2+}, with a radius ranging from 1.35 to 1.61 Å. While ionic radii generally increase with increasing atomic number, the trivalent ions of the lanthanide elements decrease in radius with increasing

[2]In practice, the value of one anion, usually oxygen, is assumed. Cation radii are then calculated from interatomic distances in oxides. These cation radii are subsequently used to calculate other anion radii (from interatomic measurements of, for example, chlorides or sulfides). These are then used to calculate other cation radii, which in turn are used to calculate oxygen radii. The process is repeated iteratively so as to arrive at a self-consistent set of ionic radii.

TABLE 3.11   Effective Ionic Radii (In Å) for Ions Commonly Found In Minerals*

Main group elements (← Column / Row ↓):

Row	I	II	III	IV	V	VI	VII
2	Li⁺ 0.59 [4], 0.74 [6], 0.92 [8]	Be²⁺ 0.16 [3], 0.27 [4], 0.45 [6]	B³⁺ 0.11 [4], 0.27 [6]	C⁴⁺ −0.08 [3], 0.15 [6], 0.16 [6]	N⁵⁺ −0.10 [3], 0.13 [6]	O²⁻ 1.36 [3], 1.38 [4], 1.40 [6], 1.42 [8]	F⁻ 1.31 [4], 1.33 [6]
3	Na⁺ 0.99 [4], 1.02 [6], 1.18 [8], 1.24 [9], 1.39 [12]	Mg²⁺ 0.57 [4], 0.72 [6], 0.89 [8]	Al³⁺ 0.39 [4], 0.48 [5], 0.54 [6]	Si⁴⁺ 0.26 [4], 0.40 [6]	P⁵⁺ 0.17 [4], 0.29 [5], 0.38 [6]	S²⁻ 1.84 [4]; S⁶⁺ 0.12 [4], 0.29 [6]	Cl⁻ 1.81 [6]
4	K⁺ 1.38 [6], 1.51 [8], 1.55 [9], 1.59 [10], 1.64 [12]	Ca²⁺ 1.00 [6], 1.12 [8], 1.18 [9], 1.23 [10], 1.34 [12]	Ga³⁺ 0.47 [4], 0.55 [5], 0.62 [6]	Ge⁴⁺ 0.39 [4], 0.53 [6]	As³⁺ 0.58 [6]; As⁵⁺ 0.34 [4], 0.46 [6]	Se²⁻ 1.98 [6]	Br⁻ 1.96 [6]
5	Rb⁺ 1.52 [6], 1.61 [8], 1.66 [10], 1.72 [12]	Sr²⁺ 1.18 [6], 1.26 [8], 1.36 [10], 1.44 [12]	In³⁺ 0.62 [4], 0.80 [6], 0.92 [8]	Sn⁴⁺ 0.69 [6], 0.81 [8]	Sb³⁺ 0.76 [6]; Sb⁵⁺ 0.60 [6]	Te²⁻ 2.21 [6]	I⁻ 2.20 [6]
6	Cs⁺ 1.67 [6], 1.74 [8], 1.81 [10], 1.85 [11], 1.88 [12]	Ba²⁺ 1.35 [6], 1.42 [8], 1.47 [9], 1.52 [10], 1.61 [12]		Pb²⁺ 1.19 [6], 1.29 [8], 1.35 [9], 1.40 [10]	Bi³⁺ 0.96 [5], 1.03 [6], 1.17 [8]		

Transition elements:

Row	Sc / Y / La	Ti / Zr / Hf	V / Nb / Ta	Cr / Mo / W	Mn / Re	Fe / Rh	Co	Ni / Pd / Pt	Cu / Ag	Zn / Cd / Hg
4	Sc³⁺ 0.75 [6], 0.87 [8]	Ti⁴⁺ 0.42 [4], 0.61 [6], 0.74 [8]	V⁵⁺ 0.36 [4], 0.46 [5], 0.54 [6]	Cr³⁺ 0.62 [6]; Cr⁴⁺ 0.41 [4], 0.55 [6]; Cr⁶⁺ 0.26 [4]	Mn²⁺ 0.83 [6], 0.96 [8]; Mn³⁺ 0.65 [6]; Mn⁴⁺ 0.53 [6]	Fe²⁺ 0.63 [4], 0.78 [6], 0.92 [8]; Fe³⁺ 0.65 [6], 0.78 [8]	Co²⁺ 0.74 [6], 0.90 [8]	Ni²⁺ 0.55 [4], 0.69 [6]	Cu⁺ 0.46 [2], 0.77 [6]; Cu²⁺ 0.57 [4], 0.65 [5], 0.73 [6]	Zn²⁺ 0.60 [4], 0.74 [6], 0.90 [8]
5	Y³⁺ 0.90 [6], 1.02 [8]	Zr⁴⁺ 0.72 [6], 0.78 [7], 0.84 [8], 0.89 [9]	Nb⁵⁺ 0.64 [6], 0.69 [7], 0.74 [8]	Mo⁴⁺ 0.65 [6]; Mo⁶⁺ 0.41 [4], 0.59 [6]	Re⁴⁺ 0.63 [6]; Re⁷⁺ 0.38 [4], 0.53 [6]	Rh⁴⁺ 0.60 [6]		Pd²⁺ 0.64 [4], 0.86 [6]	Ag⁺ 1.15 [6], 1.28 [8]	Cd²⁺ 0.58 [4], 0.74 [6], 0.90 [8]
6	La³⁺ 1.03 [6], 1.16 [8], 1.22 [9], 1.27 [10]	Hf⁴⁺ 0.71 [6], 0.76 [7], 0.83 [8]	Ta⁵⁺ 0.64 [6], 0.69 [7], 0.74 [8]	W⁶⁺ 0.42 [4], 0.51 [5], 0.60 [6]				Pt²⁺ 0.80 [6]		Hg²⁺ 0.94 [4], 1.01 [6], 1.14 [8]

Th⁴⁺	U⁴⁺ / U⁶⁺
0.94 [6], 1.05 [8], 1.09 [9], 1.13 [10]	U⁴⁺ 0.89 [6], 1.00 [8]; U⁶⁺ 0.52 [4], 0.73 [6]

*Numbers in square brackets are the coordination numbers of the ions. Radii in upright digits are from Shannon (1976). Radii in italics are from Pauling (1960), revised and supplemented by Ahrens (1952). For complete references see reference list at end of chapter.

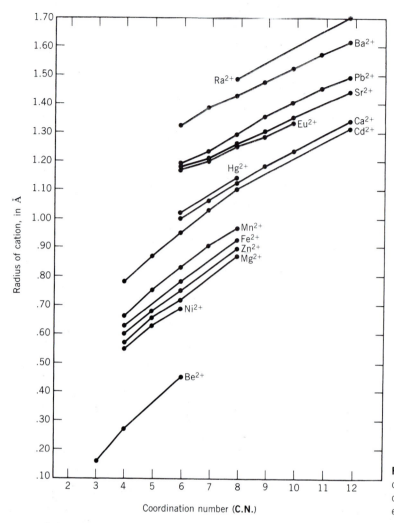

FIG. 3.31 The change in ionic radius as a function of coordination number (**C.N.**) for some selected cations (From Shannon, 1976; for complete reference see list at end of chapter.)

atomic number, from La^{3+} ($Z = 57$), with a radius of 1.16 Å (8-coordination), to Lu^{3+} ($Z = 71$), with a radius of 0.98 Å (8-coordination). This feature, known as the *lanthanide contraction*, is the result of building up inner electron orbitals before adding to a new outer orbital (see Table 3.6). As a result of the increasing nuclear charge, and the fairly weak "shielding" of this positive charge by inner-shell 4f electrons, an increased attraction is exerted on the outer-shell 5s, 5p, and 5d electrons, which draws them in more tightly, causing a decrease in ionic radius.

For positive ions with the same electronic structure, the radii decrease with increasing charge. For example, the ionic radii of the metallic elements in the third horizontal row, all of which have two electrons in the first shell and eight in the second shell, decrease (for 6-coordination) from Na$^+$, with a radius of approximately 1.02 Å, to P^{5+}, with a radius of

0.38 Å. The size of these ions with identical electron configurations decreases because the increased nuclear charge exerts a greater pull on the electrons, thus decreasing the effective radius of the ion.

For an element that can exist in several valence states (ions of the same element with different charges), the higher the charge on the positive ion, the smaller is its radius. For example, Mn^{2+} = 0.83 Å, Mn^{3+} = 0.65 Å, and Mn^{4+} = 0.53 Å. This decrease in size is due to the greater pull exerted by the nucleus on a reduced electron cloud.

In addition to variations in size as a function of coordination number and bond type, there can also be considerable change in the *shape* of some atoms and ions. Atoms and ions are not rigid bodies but respond to external electrical forces by dilation and deformation. A larger number of neighboring ions tends to distend the central ion as a function of increasing coordination number; a smaller number al-

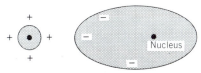

FIG. 3.32 Polarization effect of a small, highly charged cation on a large anion.

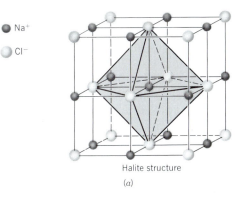

Halite structure

(a)

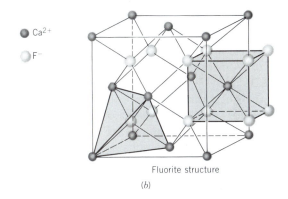

Fluorite structure

(b)

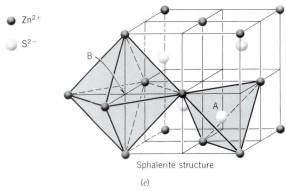

Sphalerite structure

(c)

lows it to collapse a little. Some distortion of shape may accompany the distention of ions. These effects are collectively called *polarization* and are of considerable importance in crystal structures. If the apparent shape and size are strongly affected by its structural environment, the ion is said to have a high polarizability; if, on the other hand, it behaves essentially as a rigid sphere in all environments, it is said to have a low polarizability. Generally, large monovalent anions with a noble gas electronic structure are most easily polarized. The greater the polarization between two neighboring ions, the more the electron density is localized between the two nuclei and the more covalent is the nature of the bond between them. Figure 3.32 is a schematic illustration of polarization of a large monovalent anion by a highly charged smaller cation.

COORDINATION OF IONS

When oppositely charged ions unite to form a crystal structure in which the binding forces are dominantly electrostatic (i.e., ionic), each ion tends to gather to itself, or to *coordinate*, as many ions of opposite sign as size permits. When the atoms are linked by simple electrostatic (ionic) bonds, they may be regarded as approximately spherical in shape, and the resulting geometry is simple. The coordinated ions always cluster about the central coordinating ion in such a way that their centers lie at the apices of a polyhedron. Thus, in a stable crystal structure, each cation lies at the approximate center of a *coordination polyhedron* of anions. The number of anions in the polyhedron is the *coordination number* (**C.N.**) of the cation with respect to the given anion, and is determined by their relative sizes. Thus, in NaCl each Na^+ has six closest Cl^- neighbors and is said to be in 6-coordination with Cl (**C.N.** 6). In fluorite, CaF_2, each calcium ion is at the center of a coordination polyhedron consisting of eight fluorine ions and hence is in 8-coordination with respect to fluorine (**C.N.** 8). See Fig. 3.33 for illustrations of both structures.

FIG. 3.33 Visualization of coordination polyhedra in various mineral structures represented by "ball and stick" structure drawings. *(a)* The halite, NaCl, structure where the ions are positioned in a cubic arrangement (with overall isometric symmetry). Both Na^+ and Cl^- are in 6-coordination (**C.N.** 6) with each other. A coordination polyhedron about Na^+ is shown; it is an octahedron. *(b)* The fluorite, CaF_2, structure with overall isometric symmetry. Each Ca^{2+} is coordinated to eight neighboring F^- (**C.N.** 8); this arrangement is cubic in shape. F^-, however, is coordinated only to four Ca^{2+} cations (**C.N.** 4); this is tetrahedral in shape. *(c)* The sphalerite, ZnS, structure with overall isometric symmetry. Each S^{2-} (in position marked *A*) is surrounded by four Zn^{2+} ions, in tetrahedral coordination (**C.N.** 4). The position marked *B* is empty and has octahedral surroundings.

Anions may also be regarded as occupying the centers of coordination polyhedra formed by cations. In NaCl each chloride ion has six sodium neighbors and hence is in 6-coordination with respect to sodium. Because both sodium and chlorine are in 6-coordination, there must be equal numbers of both, in agreement with the formula, NaCl. On the other hand, examination of the fluorite structure reveals that each fluorine ion has four closest calcium neighbors and hence is in 4-coordination with respect to calcium (**C.N.** 4). Although these four calcium ions do not touch each other, they form a definite coordination polyhedron about the central fluorine ion in such a way that the calcium ions lie at the apices of a regular tetrahedron (see Fig. 3.33b). Because each calcium ion has eight fluorine neighbors (**C.N.** 8), whereas each fluorine ion has only four calcium neighbors, it is obvious that there are twice as many fluorine as calcium ions in the structure. This accords with the formula CaF_2, and with the valences for calcium and fluorine.

Consideration of the relative sizes of the Ca^{2+} and F^{1-} ions (they are reasonably similar in size and both fairly large; $Ca^{2+[8]} = 1.12$ Å and $F^{-[6]} = 1.33$ Å) might lead one to conclude that the structure of fluorite, CaF_2, could consist of equal numbers of both ions in 8-coordination. The fact that in fluorite only half the possible calcium sites are filled calls attention to an important restriction of crystal structures; namely, *the total numbers of ions of all kinds in any stable ionic crystal structure must be such that the crystal as a whole is electrically neutral.* That is, the total number of positive charges must equal the total number of negative charges; hence, in fluorite there can be only half as many divalent positive calcium ions as there are monovalent fluorine ions.

Figure 3.33c illustrates the sphalerite, ZnS, structure in which Zn^{2+} is in tetrahedral coordination with S^{2-}.

Radius Ratio

Although each ion in a crystal affects every other ion to some extent, the strongest forces exist between ions that are nearest neighbors. These are said to constitute the *first coordination shell.* The geometric arrangement of this shell, and hence the coordination number, is a function of the relative sizes of the coordinating ions. However, we have already seen that the effective size of an ion is not constant but depends to varying degrees on the total number of closest neighboring ions, its coordination number (**C.N.**), its polarizability, and the type of bonding involved. *Because of these influences, it is unrealistic*

to regard ions and atoms as rigid spheres with well-established constant radii. Nonetheless, the average sizes of ions are useful as a first approximation in the prediction of approximate interatomic distances in unknown structures. Furthermore, the ratio of the radius of a cation to that of an adjoining anion can be used reasonably successfully in predicting the number of closest neighbors.

The relative size of ions is generally expressed as a radius ratio: radius ratio = $R_A : R_X$, where R_A is the radius of the cation and R_X the radius of the anion, in angstrom units. The radius ratio of sodium and chlorine in halite, NaCl, is therefore:

$$R_{Na^+} = 1.02 \text{ Å (\textbf{C.N.} 6)} \qquad R_{Cl^-} = 1.81 \text{ Å (\textbf{C.N.} 6)}$$
$$R_{Na^+} : R_{Cl^-} = 1.02/1.81 = 0.56$$

The radius ratio of calcium and fluorine in fluorite, CaF_2, is

$$R_{Ca^{2+}} = 1.12 \text{ Å (\textbf{C.N.} 8)} \qquad R_{F^-} = 1.31 \text{ Å (\textbf{C.N.} 4)}$$
$$R_{Ca^{2+}} : R_{F^-} = 1.12/1.31 = 0.86$$

When two or more cations are present in a structure, coordinated with the same anion, separate radius ratios must be computed for each.

To illustrate the usefulness of the radius ratio in predicting coordination in crystal structures, we will need to regard ions as perfect spheres of constant radius. Using a cation of which the radius size is gradually increased with respect to that of surrounding anions (of constant size), it is possible to calculate geometric limits for certain coordination types. This is illustrated in Figure 3.34 (this same illustration is presented as an animation in module I of the CD-ROM, and viewing that will clarify the concept). Figure 3.34a illustrates *linear* or 2-coordination, which is very rare in ionic bonded crystals. Examples are the uranyl group $(UO_2)^{2+}$, the nitrite group $(NO_2)^{2-}$, and copper with respect to oxygen in cuprite, Cu_2O. *Triangular* or 3-coordination is stable between the limits of 0.155 and 0.225 (see Fig. 3.34b) and is common in nature in CO_3, NO_3, and BO_3 groups. *Tetrahedral* or 4-coordination (see Fig. 3.34c) occurs when the R_A (cation): R_X (anion) ratio becomes 0.225 (as a result of the changeover from triangular to tetrahedral coordination). At this geometric ratio (0.225) four coordinated anions touch each other and the central cation. With an expanding cation (in tetrahedral coordination) the anions will continue to touch the central cation but not each other. Tetrahedral coordination therefore is possible over the range of radius ratios from 0.225 to 0.414. Tetrahedral coordination is typified by the SiO_4 group in silicates, by the *A*-type ion in spinel, and by the sphalerite (ZnS) and diamond structures. *Octahe-*

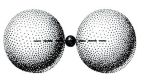

(a) Linear

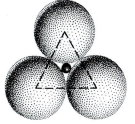

(b) Triangular

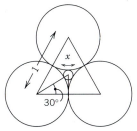

$$\cos 30° = \frac{½}{½ + ½ x}$$

$$½ + ½ x = \frac{½}{\cos 30°} = \frac{½}{0.8660} = 0.5774$$

$$½ x = 0.5774 - 0.50 = 0.0774$$

$$x = 0.155$$

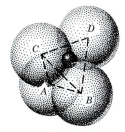

(c) Tetrahedral

C
G
D
F
A E B

G is location of
center of small ion,
in center of tetrahedron

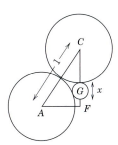

In base triangle

$$\cos 30° = \frac{AE}{AF}$$

$$\therefore AF = \frac{AE}{\cos 30°} = \frac{½}{\cos 30°}$$

$$= ½ \cdot \frac{2}{\sqrt{3}} = \frac{1}{\sqrt{3}}$$

In vertical triangle CAF

$$CF = \sqrt{AC^2 - AF^2} =$$

$$\sqrt{(1)^2 - \left(\frac{1}{\sqrt{3}}\right)^2} = \sqrt{1 - \frac{1}{3}}$$

$$\sqrt{\frac{2}{3}} = .81649$$

Also $CG = ¾\ CF$, because center of
tetrahedron G is $¼$ up from the base.

Furthermore $CG = ½ + ½ x$

$$\therefore ½ + ½ x = ¾ \cdot .81649 = .6124$$

$$\therefore ½ x = .612 - .5 - .1124$$

$$x = 0.225$$

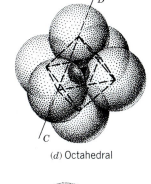

(d) Octahedral

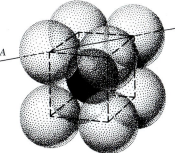

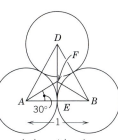

(e) Cubic

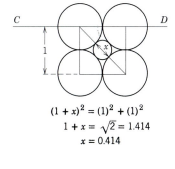

$$(1 + x)^2 = (1)^2 + (1)^2$$

$$1 + x = \sqrt{2} = 1.414$$

$$x = 0.414$$

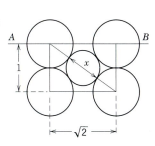

$$(1 + x)^2 = (1)^2 + (\sqrt{2})^2$$

$$1 + x = \sqrt{1 + 2} = 1.732$$

$$x = 0.732$$

FIG. 3.34 Illustration of the geometric derivation of limiting ratios for R_A (cation): R_X (anion) as a function of coordination (number). Note that the central cation gradually increases in size in going from (a) to (e), whereas the surrounding anions remain constant in size. *(a) Linear* or 2-coordination of anions around a cation. The radius of the cation can be minute and can range upward to a maximal size predicted by the $R_A : R_X < 0.155$, as derived for triangular coordination. *(b) Triangular* or 3-coordination of anions about a cation. The limiting value for the changeover from linear to triangular coordination is shown to be at $R_A : R_X = 0.155$. *(c) Tetrahedral* or 4-coordination of anions about a cation. The limiting value for the changeover from triangular to tetrahedral coordination is shown to be at $R_A : R_X = 0.225$. *(d) Octahedral* or 6-coordination of anions about a cation. The limiting value for the changeover from tetrahedral to octahedral coordination is shown to be at $R_A : R_X = 0.414$. *(e) Cubic* or 8-coordination of anions about a cation. The limiting value for the changeover from octahedral to cubic coordination is shown to be at $R_A : R_X = 0.732$.

dral or 6-coordination (see Fig. 3.34*d*) occurs when the R_A (cation): R_X (anion) ratio reaches 0.414, which is the lower limit of radius ratio for stable 6-coordination. The centers of the coordinated ions lie at the apices of a regular octahedron; at the radius ratio value of 0.414, the six coordinated anions touch each other and the central cation. The upper limit for the radius ratio for 6-coordination turns out to be 0.732 (see Fig. 3.34*e*). Hence, one expects 6 to be the common coordination number when the radius ratio lies between 0.414 and 0.732. Na and Cl in halite, Ca and oxygen in calcite, the *B*-type cations in spinel, and many cations in silicates are examples of 6-coordination. When the cation size is increased further (past the radius ratio value 0.732) *cubic* or 8-coordination results. In this case the cation is approaching the size of the coordinating anions but it is still somewhat smaller (than the anions). The coordinating ions lie at the eight corners of a cube (see Fig. 3.34*e*). At the radius ratio value of 0.732 all eight anions touch each other as well as the central cation. Allowing the radius of the cation to increase further such that its radius becomes equal to that of the anions, gives rise to a radius ratio value of 1.0. Hence, cubic coordination occurs most commonly for radius ratios between 0.732 and 1.0. An example of 8-coordination (as a distorted cube) is found in the *M2* crystallographic site of monoclinic pyroxenes; this site commonly houses relatively large cations such as Ca^{2+} and Na^+.

Figure 3.35 illustrates the change from 6- to 8-coordination in the alkali chlorides with increasing ionic radius of the cation. It is interesting to note that rubidium may be in either 6- or 8-coordination, and thus rubidium chloride is polymorphous (i.e., it is able to occur in two different structural arrangements). Li^+, Na^+, and K^+ have radius ratios with Cl^- between 0.414 and 0.732, and therefore go into

6-coordination with Cl^-. Cs^+ has a radius ratio with Cl^- between 0.732 and 1.00, and thus goes into 8-coordination with Cl^-. Rb^+ has a radius ratio with Cl^- close to 0.732, and therefore can have either 6- or 8-coordination with Cl^-; this is the origin of its polymorphism.

Figure 3.36 illustrates the various regular coordination polyhedra as a function of radius ratio and coordination number.

When coordinating and coordinated ions (or atoms) are the same size, the radius ratio is 1.0. The structures that result are known as *closest packed* and are discussed on p. 73.

Examples of 5-, 7-, 9-, and 10-coordination are also known. Such coordination numbers are possible only in complex structures in which the anions are not closely packed.

The coordination polyhedra in experimentally determined structures are almost always distorted. The smaller and more strongly polarizing the coordinating cation, or the larger and more polarizable the anion, the greater is the distortion and the wider the departure from the theoretical radius ratio limits. Also, *if the bonding mechanism is not dominantly ionic, radius ratio considerations may not be safely used to determine the coordination number.*

Obviously, every ion in a crystal structure has some effect on every other ion—it is attractive if the charges are opposite, repulsive if the same. Hence, ions tend to group themselves in crystal structures in such a way that cations are as far apart as possible yet consistent with the coordination of the anions that will result in electrical neutrality. Thus, when cations share anions between them, they do so in such a way as to place themselves as far apart as possible. Hence, the coordination polyhedra formed around each are linked commonly through corners

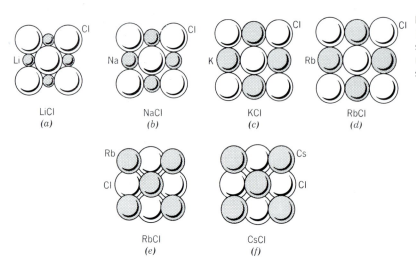

FIG. 3.35 Change in coordination from 6 to 8 in alkali chlorides. *(a)*, *(b)*, *(c)*, and *(d)* have sodium chloride type structures with 6-coordination; *(e)* and *(f)* have cesium chloride type structures with 8-coordination.

LiCl
(a)

NaCl
(b)

KCl
(c)

RbCl
(d)

RbCl
(e)

CsCl
(f)

Minimum Radius Ratio $R_A : R_X$	Coordination Number C. N.	Packing Geometry		
< 0.155	2	Linear		
0.155	3	Corners of an equilateral triangle (triangular coordination)		
0.225	4	Corners of a tetrahedron (tetrahedral coordination)		
0.414	6	Corners of an octahedron (octahedral coordination)		
0.732	8	Corners of a cube (cubic coordination)		
1.0	12	Corners of a cuboctahedron (close packing)		

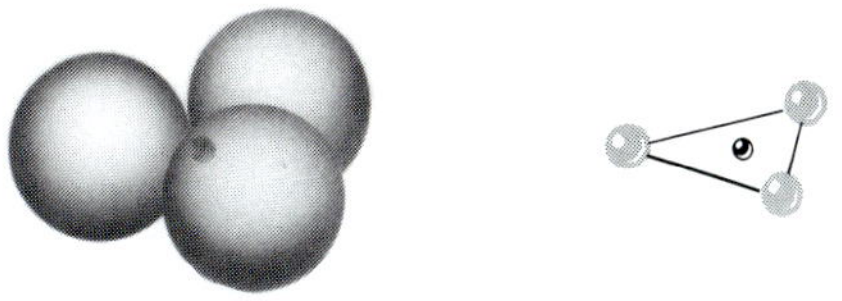

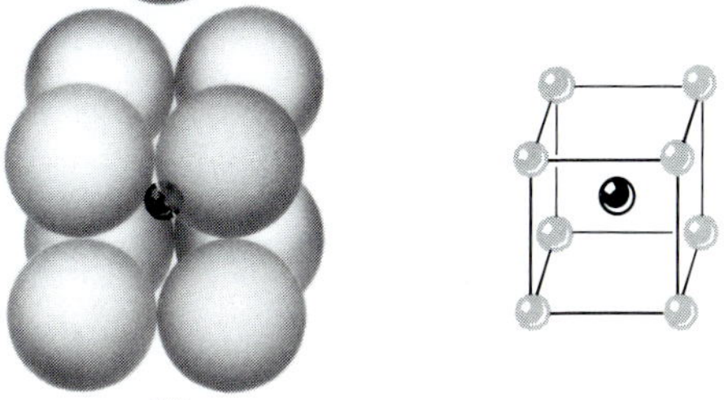

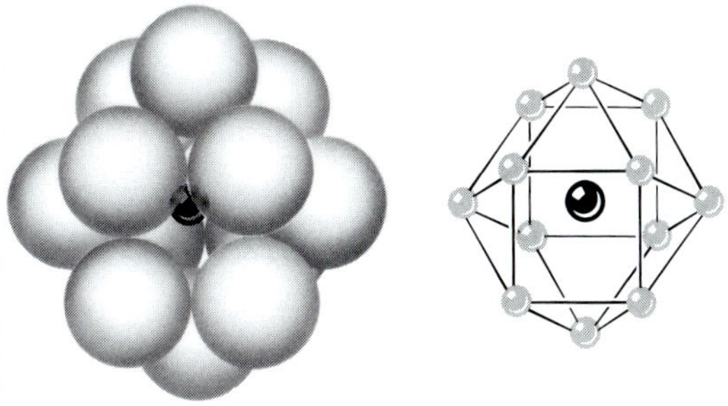

FIG. 3.36 Atomic packing schemes.

and edges but generally not through faces (Fig. 3.42). Cations tend to share as small a number of anions as possible, and sharing of as many as three or four anions is relatively rare.

At this stage, it may be instructive to list the most common ionic charges, the most common coordination polyhedra, and ionic sizes of the 11 most common elements in the Earth's crust. This information is given in Table 3.12. Knowledge of this listing will be most helpful in the subsequent discussion of silicate

and oxide structures, as well as in understanding aspects of the compositional variation in minerals.

When coordinating and coordinated ions (or atoms) are the same size, the radius ratio is 1. Trial with a tray of identical spheres, such as Ping-Pong balls, reveals that spherical units may be arranged in three dimensions in either of two ways, called *hexagonal closest packing* (HCP) and *cubic closest packing* (CCP). In either arrangement, each sphere is in contact with 12 closest neighbors (**C.N.** 12). With reference to

TABLE 3.12   Some Common Ions (Exclusive of Hydrogen) that Occur in Rock-Forming Minerals, Arranged in Decreasing Ionic Size

Ion	Coordination Number with Oxygen		Ionic Radius Å
O^{2-}			1.36 [3]
K^+	8–12		1.51 [8]–1.64 [12]
Na^+	8–6 ⎫	cubic to	1.18 [8]–1.02 [6]
Ca^{2+}	8–6 ⎬	octahedral	1.12 [8]–1.00 [6]
Mn^{2+}	6 ⎫		0.83 [6]
Fe^{2+}	6 ⎪		0.78 [6]
Mg^{2+}	6 ⎬	octahedral	0.72 [6]
Fe^{3+}	6 ⎪		0.65 [6]
Ti^{4+}	6 ⎪		0.61 [6]
Al^{3+}	6 ⎭		0.54 [6]
Al^{3+}	4 ⎫		0.39 [4]
Si^{4+}	4 ⎪	tetrahedral	0.26 [4]
P^{5+}	4 ⎬		0.17 [4]
S^{6+}	4 ⎭		0.12 [4]
C^{4+}	3	triangular	−0.08 [3]

*The first column lists the most common ionic (valency) states of the elements. The second column lists their most common coordination with respect to oxygen, and the third column lists ionic sizes for specific coordinations (the number in square brackets is **C.N.**). A complete listing of elemental abundances is given in Table 3.1.

Fig. 3.37a, it is clear that a sphere can be surrounded by six spheres of equal size, all of which lie in the same plane and touch each other. This is referred to as a *hexagonal closest packed layer* because of the hexagonal arrangement of the spheres (marked *A* in Fig. 3.37a). Between the spheres one can distinguish two types of voids, *B* and *C*, on the basis of the orientation of their triangular shapes. In Fig. 3.37b a part of a second hexagonal closest packed layer is superimposed on top of the *B* voids. This sequence of stacking can be represented by the combination of letters *AB*. If a third layer is now put on top of the second layer but with the spheres resting in the dimples of the second layer above the *A* spheres in the first layer, we obtain a sequence *ABA*, which can be extended upward by another layer of spheres on top of the *B* voids, giving rise to an indefinite stacking sequence, *ABABAB* ..., which is known as *hexagonal closest packing* (HCP). If, however, we choose to stack the third layer on top of the second layer of the *AB* sequence in the dimples that directly overlie the *C* voids in the first layer, we form a three-layer sequence *ABC*, as in Fig. 3.37c. We can continue this type of stacking sequence indefinitely, resulting in an *ABCABCABC* ... sequence, or *cubic closest packing* (CCP). Figure 3.38 illustrates the hexagonal lattice that underlies the HCP arrangement and the cubic lattice that underlies the CCP arrangement. 12-coordination is rare in minerals, with the exception of the native metals; they, and many alloys, have structures based on HCP or CCP. Some

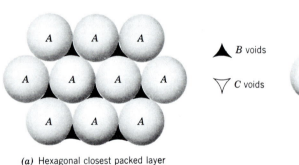

(a) Hexagonal closest packed layer

▲ *B* voids

▽ *C* voids

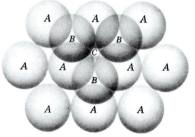

(b) Hexagonal closest packing sequence: *AB AB*.....

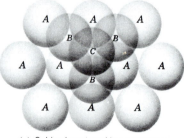

(c) Cubic closest packing sequence: *ABC ABC*.....

The plane of this drawing turns out to be perpendicular to the 3-fold axis at the corner of a cube

FIG. 3.37 Closest packing of like spheres.

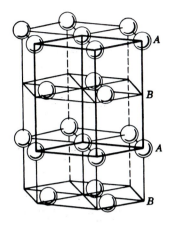
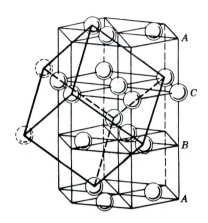
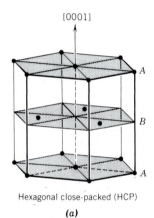
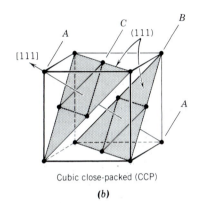

[0001]

A

B

A

Hexagonal close-packed (HCP)

[111]

A

C
(111)

B

A

Cubic close-packed (CCP)

(a)

(b)

FIG. 3.38 (a) Two representations of a lattice based on hexagonal closest packing (HCP). (b) Two representations of a lattice based on cubic closest packing (CCP). Upon inspection of the location of nodes in this lattice, it is found to be all-face-centered (F; or FCP).

examples of hexagonal closest packing are given in Fig. 3.39. Animations that build both closest packing types (HCP and CCP) are given in module I of the CD-ROM.

Pauling's Rules

Every stable crystal structure bears witness to the operation of some broad generalizations that determine the structure of solid matter. These principles were enunciated in 1929 by Linus Pauling in the form of the following five rules (animations of these are presented in module I of the CD-ROM):

Rule 1. A coordination polyhedron of anions is formed about each cation, the cation–anion distance being determined by the radius sum and the coordination number (i.e., number of nearest neighbors) of the cation by the radius ratio. This was discussed in the previous section.

Rule 2. The electrostatic valency principle. In a stable crystal structure the total strength of the valency bonds that reach an anion from all the neighboring cations is equal to the charge of the anion.

This needs some further examination. The strength of an electrostatic bond (e.v.) may be defined as an ion's valence charge (z) divided by its coordination number (n): e.v. $= z/n$, expressed in absolute values. The resulting number, called the *electrostatic valency* (e.v.), is a measure of the strength of any of the bonds reaching the coordinating ion from its nearest neighbors. For instance, in NaCl, the Cl^- is surrounded by six Na^+ neighbors in octahedral coordination, and each of the bonds reaching Na^+ has a strength (e.v.) of $\frac{1}{6}$. This means that six bonds between a central Na^+ and six octahedrally coordinated closest neighbors of Cl^- completely and exactly neutralize the charge on the central Na^+ (see Fig. 3.40a). The Cl^- also has six neighbors (Na^+), so that the e.v. for each of the bonds reaching Cl^- is $\frac{1}{6}$. As such, the charge on the Cl^- is neutralized by six bonds of $\frac{1}{6}$ reaching this central ion from six (octahedrally coordinated) Na^+ ions. Figures 3.40b and c give examples of electrostatic valencies in additional coordination polyhedra. The electrostatic valency rule is very helpful in the evaluation of the polyhedral nature of crystal structures. In a stable structure the sum of the electrostatic valencies from

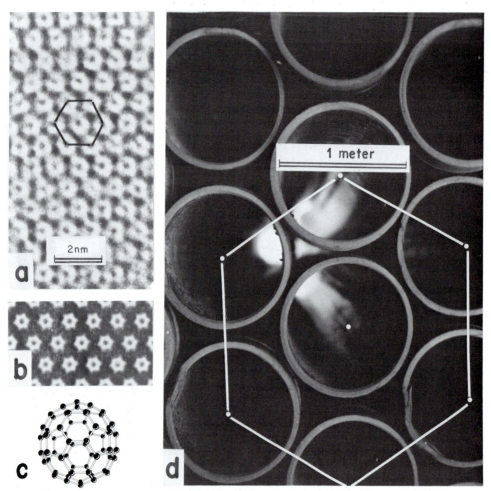

FIG. 3.39 Examples of hexagonal closest packing (HCP). *(a)* The hexagonal closest packing of C_{60} molecules ("buckminsterfullerene") as revealed by high-resolution transmission electron microscopy (HRTEM). These complex carbon molecules are of synthetic origin and have not been found in nature. Scale is in nanometers (nm): 1 nm = 10 Å. (For further discussion, see R. F. Curt and R. E. Smalley, 1991, Fullerenes. *Scientific American*, 265: 54–63.) *(b)* A computer calculated image that replicates the observed structure image in *(a)*. (*a* and *b* from S. Wang and P. R. Buseck, 1991, Packing of C_{60} molecules and related fullerenes in crystals: A direct view. *Chemical Physics Letters*, 182: 1–4). *(c)* Illustration of the structure of C_{60} "buckminsterfullerene," named after the American engineer and philosopher R. Buckminster Fuller because of his invention of the geodesic dome which underlies the structure of C_{60}. *(d)* Hexagonal closest packing on a very large scale, as seen in everyday life when steel, concrete, or plastic pipes are stacked.

cations in coordination polyhedra exactly balances the charge on the anion that is shared among these coordination polyhedra. For example, in the structure of grossular garnet ($Ca_3Al_2Si_3O_{12}$), Ca^{2+} is in cubic coordination (**C.N.** = 8), Al^{3+} is in octahedral coordination (**C.N.** = 6), and Si^{4+} is in tetrahedral coordination (**C.N.** = 4), all of them with oxygen. Their bonds, therefore, have electrostatic valencies of $\frac{2}{8} = \frac{1}{4}$, $\frac{3}{6} = \frac{1}{2}$, and $\frac{4}{4} = 1$, respectively. In order to satisfy (neutralize) the 2^- charge on a shared oxygen, that oxygen ion must belong to *two* cubic Ca^{2+} polyhedra, *one* Al^{3+} octahedron, and *one* Si^{4+} tetrahedron ($2 \times \frac{1}{4} + \frac{1}{2} + 1 = 2$). This is the fundamental **linkage of coordination polyhedra** in grossular garnet.

Crystals in which all bonds are of equal strength are called *isodesmic*. This generalization is so simple that it seems trivial, but in some cases unexpected results emerge from calculation of electrostatic valencies. For example, minerals of the spinel group have formulas of the type AB_2O_4, where A is a divalent cation such as Mg^{2+} or Fe^{2+} and B is a trivalent cation such as Al^{3+} or Fe^{3+}. Such compounds have been called aluminates and ferrates, by analogy with such compounds as borates and oxalates. This nomenclature suggests that ionic clusters or radicals are present in the structure. X-ray data reveal that the A ions are in 4-coordination, whereas the B ions are in 6-coordination with oxygen. Hence, for the A

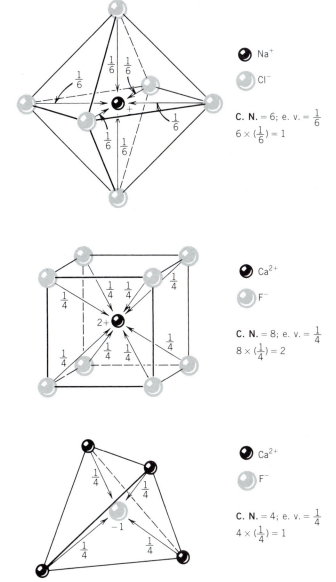

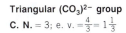

such groups, the anions are more strongly bonded to the central coordinating cation than they can possibly be bonded to any other ion. For example, in the carbonate group, C^{4+} is in 3-coordination with O^{2-}, and hence we may compute the e.v. $= \frac{4}{3} = 1\frac{1}{3}$. This is greater than one-half the charge on the oxygen ion, and hence a functional group, or radical, exists. This is the carbonate triangle, the basic structural unit of carbonate minerals (see Fig. 3.41a). Another example is the sulfate group. O^{2-} is in 4-coordination with S^{6+}; hence the e.v. $= \frac{6}{4} = 1\frac{1}{2}$. Because this is greater than one-half the charge on the oxygen ion, the sulfate radical forms a tightly knit group, and oxygen is more strongly bonded to sulfur than it can be bound

FIG. 3.41 Examples of polyhedral anionic groupings about a central cation, in which the electrostatic valency that the central cation contributes to each coordinating anion is greater than half the anion's charge. Therefore, the anion is more tightly bonded to the central cation than it can be to any other cation in the structure. These discrete bonding units (or complex ions) occur in what are known as anisodesmic structures.

Triangular $(CO_3)^{2-}$ group
C. N. = 3; e. v. $= \frac{4}{3} = 1\frac{1}{3}$

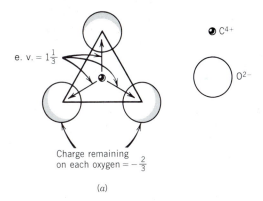

Charge remaining on each oxygen $= -\frac{2}{3}$

(a)

Tetrahedral $(SO_4)^{2-}$ group
C. N. = 4; e. v. $= \frac{6}{4} = 1\frac{1}{2}$

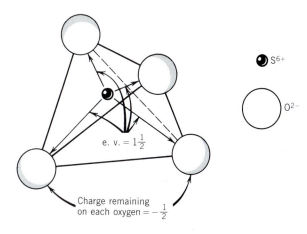

Charge remaining on each oxygen $= -\frac{1}{2}$

(b)

FIG. 3.40 Illustration of the neutralization of a central ion by bonds from the nearest neighbors. Each of these bonds has an electrostatic valency (e.v.). The total of all bonds with specific e.v.'s neutralizes the central ion. *(a)* Octahedral coordination in the halite structure (see Fig. 3.33a). *(b)* Cubic coordination of F^{-1} around Ca^{2+} in the fluorite structure (see Fig. 3.33b). *(c)* Tetrahedral coordination of Ca^{2+} around F^- in the fluorite structure (see Fig. 3.33b).

ions, e.v. $= \frac{2}{4} = \frac{1}{2}$, and for B ions, e.v. $= \frac{3}{6} = \frac{1}{2}$. All bonds are the same strength, and such crystals are isodesmic multiple oxides.

When small, highly charged cations coordinate larger and less strongly charged anions, compact, firmly bonded groups result, as in the carbonates and nitrates. If the strength of the bonds within such groupings is computed, the numerical value of the electrostatic valency is always greater than one-half the total charge on the anion. This means that in

to any other ion in the structure. This is the tetrahedral unit (see Fig. 3.41*b*) that is the fundamental basis of the structure of all sulfates. Compounds such as sulfates and carbonates are said to be *anisodesmic.*

Of course, it must be understood that if the tightly bonded groups are regarded as single structural units, then in a compound such as $CaCO_3$, calcite, all Ca to $(CO_3)^{2-}$ group bonds are of equal strength and resemble those of an isodesmic crystal. Similarly, in simple sulfates like barite, all Ba to $(SO_4)^{2-}$ group bonds are equal in strength. The crystals are, however, called anisodesmic because of the presence of the strong C–O and S–O bonds in addition to the weaker Ca to $(CO_3)^{2-}$ group and Ba to $(SO_4)^{2-}$ group bonds.

Logic requires yet another case: that in which the strength of bonds joining the central coordinating

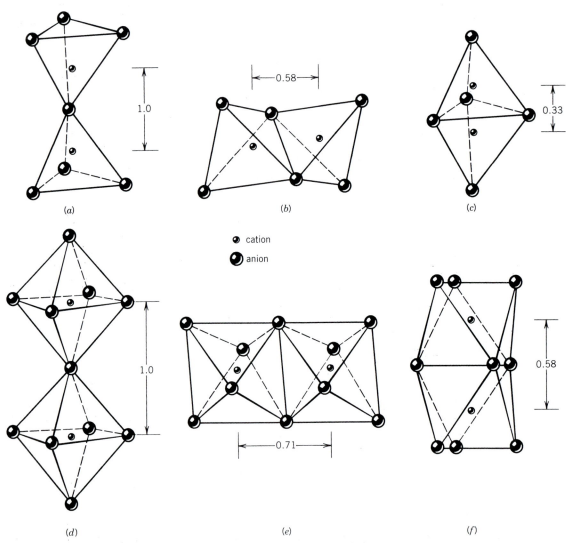

FIG. 3.42 *(a)* Tetrahedrons sharing corners, as is commonly found in crystal structures. Here the cation–cation distance is given as 1.0. *(b)* Tetrahedrons sharing edges; very uncommon. The cation–cation distance is reduced from 1.0 to 0.58. In this case cation–cation repulsion would occur, causing severe polyhedral distortion. *(c)* Tetrahedrons sharing faces; this is never found when both tetrahedra are occupied by a cation. The cation–cation distance is reduced to 0.33 from 1.0 (in *a*) and 0.58 (in *b*). *(d)* Octahedrons sharing corners, as is common in crystal structures. The cation–cation distance is given as 1.0. *(e)* Octahedrons sharing edges is also common. The cation–cation distance is reduced to 0.71 (but the cations are still considerably farther apart than in the case of tetrahedrons sharing edges; see *b*). *(f)* Octahedrons sharing faces; this is not uncommon in crystal structures. The cation–cation distance is reduced to 0.58 from 1.0 (in *d*) and 0.71 (in *f*). Face-sharing of octahedrons is possible because the cation–cation distance is larger for octahedrons than tetrahedrons. Furthermore, cations in octahedral coordination tend to have lower charges (e.g., Mg^{2+}, Fe^{2+}) than cations in tetrahedral coordination (e.g., Si^{4+}, Al^{3+}); such, the repulsive force between cations inside octahedrons is generally less than between cations inside tetrahedrons.

cation to its coordinated anions equals exactly half the charge of the anion. In this case each anion may be bonded to some other unit in the structure just as strongly as it is to the coordinating cation. The other unit may be an identical cation, and the anion shared between two cations may enter into the coordination polyhedra of both. Let us consider the case of silicon–oxygen groupings, in which the oxygens are in tetrahedral coordination about the central Si^{4+}. The e.v. of the bonds between oxygen and Si^{4+} is $\frac{4}{4} = 1$. This is half the bonding strength of the oxygen ion. Consequently, an SiO_4 tetrahedron may link to some other ion just as strongly as to the central Si^{4+} ion. If this ion is another Si^{4+}, two tetrahedra may combine, linked through a common oxygen to form a single $(Si_2O_7)^{6-}$ group. In similar fashion, SiO_4 tetrahedra may join, or *polymerize*, to form chains, sheets, or networks by sharing oxygen ions. Such crystals are called *mesodesmic*. The most important example is the silicates.

All ionic crystals may be classified on the basis of the relative strengths of their bonds into isodesmic and anisodesmic crystals.

Rule 3. The existence of edges, and particularly of faces, common to two anion polyhedra in a coordinated structure decreases its stability (see Fig. 3.42). This effect is large for cations with high valency and small coordination number and is especially large when the radius ratio approaches the lower limit of stability of the polyhedron.

Rule 4. In a crystal containing different cations, those of high valency and small coordination number tend *not* to share polyhedral elements with each other, and when they do, the shared edges contract (to put more negative charge between the cations), and the cations are displaced from their polyhedral centers, away from the shared edge or face (to minimize cation–cation repulsion).

Rules 1 through 4 are all designed to maximize cation–anion attractions and to minimize anion-anion and cation–cation repulsion.

Rule 5. The principle of parsimony. The number of essentially different kinds of constituents in a crystal tends to be small, because, characteristically, there are only a few types of contrasting cation and anion sites. Thus, in structures with complex compositions, a number of different ions may occupy the same structural position (site). These ions must be considered as a single "constituent."

This can be illustrated with respect to the polyhedral representation of the amphibole structure (see Fig. 3.43), where the number of different crystallographic

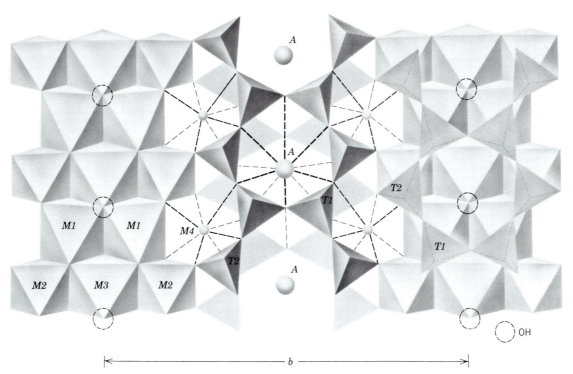

FIG. 3.43 Polyhedral representation of the structure of monoclinic amphibole projected down the *a* axis. Polyhedral sites are marked as follows: T = tetrahedral; $M1$, $M2$, and $M3$ = regular octahedral; $M4$ = distorted cubic; A is irregular coordination to 10 to 12 neighboring oxygens and (OH); (OH) notes the specific location of hydroxyl in the formula. (After J. J. Papike, et al., 1969, *Mineralogical Society of America, Special Paper*, 2: 120.)

sites is limited but the range of chemical constituents in the amphibole composition is large. Figure 3.43 shows the presence of tetrahedral *T* sites (labeled *T1* and *T2*), octahedral *M* sites (labeled *M1*, *M2*, and *M3*, because of slightly different octahedral size and shape), an *M4* site (with irregular 8-coordination), a very large and irregularly coordinated *A* site, and the location of (OH)⁻ groups. The tetrahedral sites will accommodate Si^{4+} and Al^{3+}. The *M1*, *M2*, and *M3* sites house Mg^{2+}, Fe^{2+}, Mn^{2+}, Al^{3+}, Fe^{3+}, and Ti^{4+}. The *M4* sites can be the locations for Mg^{2+}, Mn^{2+}, Fe^{2+}, Ca^{2+}, and Na^{2+}. The *A* site may be occupied by Na^+ or K^+, and the (OH)⁻ location also houses Cl^- and F^-. As such, as many as 13 different ions are distributed among five distinctly different crystallographic sites (tetrahedral, regular octahedral, distorted cubic, the very large *A* site, and the (OH) location). An interactive approach to the cation distribution in an amphibole structure is provided in module I of the CD-ROM.

CRYSTAL STRUCTURE

A knowledge of the structure of crystalline materials is basic to the understanding of properties such as cleavage, hardness, density, melting point, refractive index, X-ray diffraction pattern, and solid solution. Furthermore, our knowledge of atomic and ionic sizes and of the nature of the chemical bonds in

crystals derives from the precise determinations of crystal structures.

A structural crystallographer determines the structure of a crystalline material through the interpretation of the interrelationship of various properties such as external symmetry, X-ray diffraction effects, electron diffraction patterns, density, and chemical composition. The resulting crystal structure provides us with a knowledge of the geometric arrangement of all of the atoms (or ions) in the unit cell, and the bonding and coordination between them. The symmetry of the internal, atomic arrangement of a crystal is expressed by its *space group* (see Chapter 5), whereas the morphology of the crystal is reflected in the *point group symmetry* (see Chapter 5).

Let us illustrate briefly the chemical and physical interrelationships in the derivation of the simple structure of sodium chloride, NaCl. The external morphology of halite is consistent with isometric symmetry (point group: $4/m\,\overline{3}\,2/m$). X-ray diffraction data indicate that the unit cell has an edge dimension (*a*) of 5.64 Å. A chemical analysis yields 39.4 weight percent Na and 60.6 weight percent Cl. Division of these weight percentage figures by the appropriate atomic weights yields an atomic ratio for Na : Cl or 1 : 1, which is expressed as NaCl. The density of halite is 2.165 g/cm³. If the density (*D*) and the volume (*V*) of the unit cell (*a³*) are known, then the number of formula units per unit cell (*Z*) can be calculated from $D = (Z \times M)/(N \times V)$,

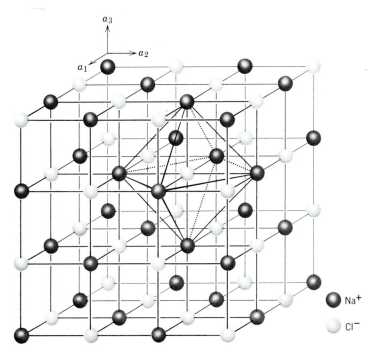

FIG. 3.44 Sodium chloride (NaCl) structure based on a cubic lattice type. Note that each ion is surrounded by six neighboring ions of opposite charge outlining octahedral polyhedra (see also Fig. 3.48).

Na^+

Cl^-

where M is molecular weight and N is Avogadro's number (6.02338×10^{23}). The number of formula units per unit cell for halite is 4, which means that the unit cell contains 4 NaCl units, or four Na^+ and four Cl^- ions. It is now possible to assess the structural arrangement of the ions in NaCl. The radius ratio of Na^+ to Cl^- would predict a coordination of 6 about each of the ions. A reasonable and indeed correct interpretation of the crystal structure is given in Fig. 3.44.

Illustration of Crystal Structures

The illustration of a three-dimensional crystal structure on a two-dimensional page is commonly the result of projection, using *fractional coordinates*, x, y, and z. For example, as in Fig. 3.45a, the location of an atom within the outlined cell with edges a, b, and c can be described in terms of distances from the origin whose components are parallel with the edges. These components are x', y', and z', where $x = x'/a$, $y = y'/b$, and $z = z'/c$; x, y, and z range from 0 to 1. A listing of the fractional coordinates of all atoms (and/or ions) in a unit cell (this is the smallest unit of the structure that, when repeated regularly and indefinitely in three dimensions, yields the entire structure) defines the crystal structure. From that listing a two-dimensional projection can be constructed. Usually such projections are made onto one of the cell faces or onto planes perpendicular to a specific direction in the crystal (e.g., the crystallographic axes). The fractional heights of atoms

above the plane of projection are noted adjacent to them, although values of 0 to 1 corresponding to positions on the bottom and top faces of the unit cell are often omitted. The coordinates 0, 0, 0 specify an atom at the origin of the unit cell.

As an example, a structure projection onto a horizontal plane of high tridymite, SiO_2, is shown in Fig. 3.45b. Two unit cells are shown. The x and y coordinates of the oxygen and silicon ions in the structure are shown by their locations in the a_1–a_2 plane. The z coordinates of the ions are shown by fractions indicating their location above the page of the projection. Instead of using fractional coordinates, structure projections may be represented by atomic positions accompanied by numbers ranging from 0 to 100. An atom in the lower face of the unit cell is indicated by 0, one on the top face by 100, and intermediate heights are given accordingly. An atom marked 75, for example, is located three-quarters of the way up the unit cell. Two examples of such illustrations are given in Fig. 3.46.

For many purposes it is useful to represent a crystal structure in terms of coordination polyhedra instead of the locations of the atoms or ions. Figure 3.47 illustrates the structure of high tridymite in terms of the linkage of SiO_4 tetrahedra (compare with Fig. 3.45b). To aid in the visualization of complex crystal structures, crystal structure models can be built or obtained commercially. Such models reproduce the internal atomic arrangement on an enormously magnified scale (e.g., 1 Å might be represented by 1 cm in a model). Several types of

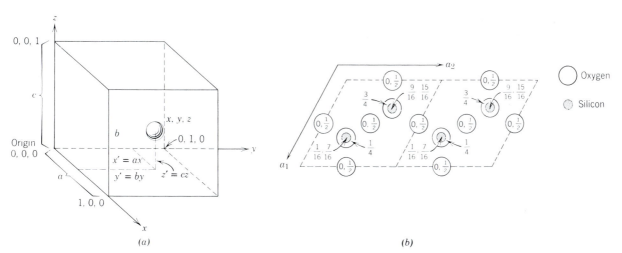

FIG. 3.45 *(a)* Fractional cell coordinates (x, y, z) of an atom in a unit cell as outlined. *(b)* Atomic arrangement in the high tridymite polymorph of SiO_2, projected on (0001). Oxygens occur in the base of the cell (0), one quarter ($\frac{1}{4}$), halfway ($\frac{1}{2}$), and three-quarters of the way up ($\frac{3}{4}$). In this projection some oxygen positions superimpose, as shown by the z coordinates 0, $\frac{1}{2}$. Silicon positions, above and below oxygen locations, occur at $\frac{1}{16}$, $\frac{7}{16}$, $\frac{9}{16}$, and $\frac{15}{16}$ of the z axis.

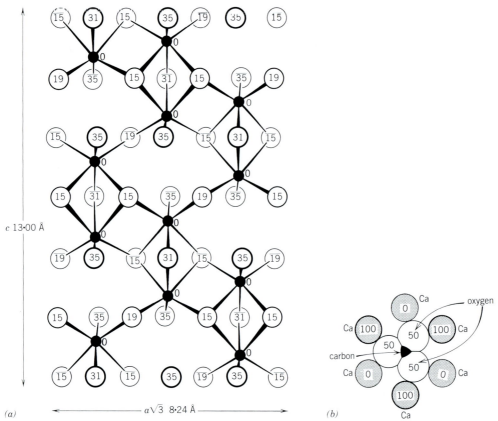

(a)   ←——————— $a\sqrt{3}$ 8·24 Å ———————→

(b)

FIG. 3.46 Examples of projections of atomic structures. *(a)* The structure of corundum, Al_2O_3, projected on $(2\overline{1}\overline{1}0)$. The heights of oxygens relate to the plane of aluminum ions at zero. Compare this illustration with the polyhedral representation in Fig. 9.2*b*. *(b)* Projection of some of the ion positions in calcite, $CaCO_3$, onto (0001). The upper layer of Ca^{2+} ions is shown by the number 100; the Ca^{2+} ions in the lower layer by 0. Note the triangular outline of the CO_3 group.

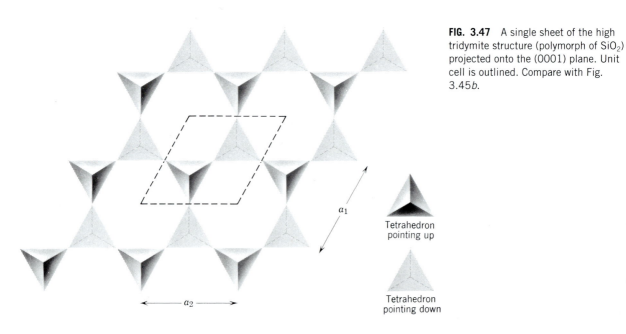

FIG. 3.47 A single sheet of the high tridymite structure (polymorph of SiO_2) projected onto the (0001) plane. Unit cell is outlined. Compare with Fig. 3.45*b*.

Tetrahedron pointing up

Tetrahedron pointing down

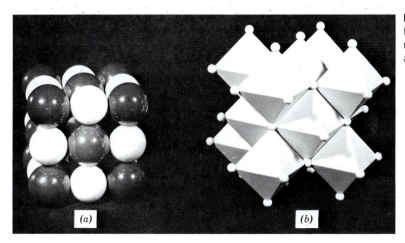

FIG. 3.48 *(a)* Close-packed model of NaCl structure. Na^+ ions light; Cl^- ions dark. *(b)* Polyhedral representation of octahedral coordination of Cl^- about Na^+ (as well as Na^+ about Cl^-) in NaCl.

models are illustrated in Figs. 3.48 and 3.49. A *close-packed model* is most realistic in terms of the representation of the filling of space on an atomic scale (see Fig. 3.48*a*). It is difficult to visualize the interior of such a model, however. A *polyhedral model* (Figs. 3.48*b* and 3.49*a*) represents the coordination polyhedra instead of specific atomic locations. Note the different appearances of two models in Fig. 3.48, both of which represent the NaCl structure. A polyhedral model of a complex structure, such as in Fig. 3.49*a*, is most useful in visualizing the overall symmetry and the regular repeat units of the structure. This often-idealized regularity of the polyhedral arrangement also allows relatively easy identification of the space group elements in the structure. An *open "ball and stick" model* (Fig. 3.49*b*) allows a view of the coordination inside the model and portrays the irregularly coordinated environment of ions in a complex structure.

Isostructuralism

Although it might seem that uraninite, UO_2, and fluorite, CaF_2, would have little in common, their X-ray powder diffraction patterns show analogous lines, although different in spacing and intensity. Structure analysis reveals that there are four U^{4+} ions in uraninite around each oxygen, whereas eight O^{2-} are grouped about each uranium. In fluorite, four Ca^{2+} are grouped about each fluorine, and eight F^{1-} are packed about each calcium. Uraninite and fluorite have structures that are analogous in every respect, although their unit cell dimensions are different and other properties are, of course, totally dissimilar. These two substances are said to be *isostructural* or *isotypous*, and belong to the same structure type. Occasionally the term *isomorphism* is used instead

of *isostructuralism*. Crystals in which the centers of the constituent atoms occupy geometrically similar positions, regardless of the size of the atoms or the absolute dimensions of the structure, are said to belong to the same structure type. For example, all isometric crystals in which there are equal numbers of cations and anions in 6-coordination belong to the NaCl (halite) structure type. A few of the large number of minerals of diverse composition belonging to this structure type are KCl, sylvite; MgO, periclase; NiO, bunsenite; PbS, galena; MnS, alabandite; AgCl, chlorargyrite; and TiN, osbornite. The relative sizes of cations and surrounding anions are of major importance in the type of packing of atoms, and the subsequent resulting structure. This was covered extensively in an earlier part of this chapter, under "Coordination of Ions."

The two minerals stishovite, SiO_2, and rutile, TiO_2, are also isostructural. In both structures the cation (Si^{4+} or Ti^{4+}) is surrounded by six oxygen neighbors, in octahedral packing about the cation. In all other known forms of SiO_2 (including quartz), as well as all other silicates found in the Earth's crust, silicon is surrounded by four neighboring oxygens, in tetrahedral packing. Stishovite is a very high pressure polymorphic form of quartz.

Of great importance in mineralogy is the concept of the *isostructural group*: a group of minerals related to each other by analogous structure, generally having a common anion and frequently displaying extensive ionic substitution. Many groups of minerals are isostructural, of which the barite group of sulfates, the calcite group of carbonates, and the aragonite group of carbonates are perhaps the best examples. The extremely close relationship that exists among the members of many groups is illustrated by the aragonite group listed in Table 3.13.

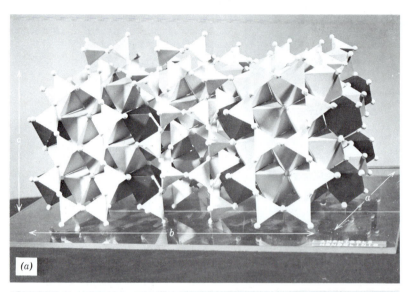

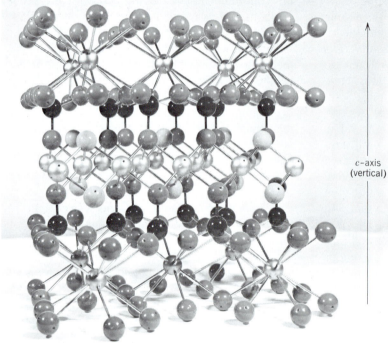

FIG. 3.49 (a) Polyhedral structure model representing a monoclinic amphibole. Tetrahedral linking of SiO_4 (Si_4O_{11} chains parallel to c axis) and octahedral coordination of cations between chains. Dark octahedra represent *M4* positions in structure. (b) Expanded "ball and stick" model of the structure of biotite $K(Mg,Fe)_3$-$(AlSi_3O_{10})(OH)_2$. Top and bottom layers represent K^+ ions in 12 coordination with oxygen. Central layer represents Mg^{2+} and Fe^{2+} in 6-coordination with O^{2-} and $(OH)^-$.

c–axis (vertical)

TABLE 3.13 **Aragonite Group of Orthorhombic Isostructural Carbonates**

Mineral	Chemical Composition	Cation	Size (Å)	Unit Cell Dimensions (Å)			Volume (Å³)	Specific Gravity (**G**)	Cleavage Angle 110 ∧ 1$\bar{1}$0
				a	b	c			
Aragonite	$CaCO_3$	Ca^{2+}	1.18	4.96	7.97	5.74	226.91	2.94	63°48′
Strontianite	$SrCO_3$	Sr^{2+}	1.34	5.11	8.41	6.03	259.14	3.78	62°41′
Cerussite	$PbCO_3$	Pb^{2+}	1.35	5.19	8.44	6.15	269.39	6.58	62°46′
Witherite	$BaCO_3$	Ba^{2+}	1.47	5.31	8.90	6.43	303.87	4.31	62°12′

*Because the metal ions are coordinated to nine oxygens, the ionic radii (see Table 3.11) are given for nine-fold coordination.

EXAMPLES OF SOME COMMON STRUCTURE TYPES

In the visualization of inorganic crystal structures it is often useful to view the larger anions as a closest packed array with the smaller cations housed in the interstitial positions (interstices). It was shown in Fig. 3.37 that close-packed layers of spheres can be stacked into a hexagonal closest packed sequence (HCP: $ABAB$ . . .) or a cubic closest packed sequence (CCP: $ABCABC$ . . .). In Fig. 3.50a the interstices that arise in such close-packed sequences are shown as tetrahedral and octahedral in coordination. These tetrahedral and octahedral interstitial sites may or may not be fully occupied by cations, as we will see in various of the following structures. In structures in which the cations are larger than can be housed in octahedral coordination ($R_A : R_X$ values between 0.73 and 1.0; see Fig. 3.36) they commonly occur in *simple cubic packing* (SCP), with the cations at the center of a cube and the anions at the eight corners (see Fig. 3.50b).

The important structure types that will be reviewed here are those of NaCl (halite), CsCl (cesium chloride), ZnS (sphalerite), CaF_2 (fluorite), TiO_2 (rutile, $CaTiO_3$ (perovskite), $MgAl_2O_4$ (spinel), and the silicates.

NaCl Structure

This structure type is adopted by a large number of AX compounds in the appropriate range of radius ratio. In this structure type the anions are in CCP, with the cations filling all the octahedral sites. All tetrahedral interstices are empty. Values, of $R_A : R_X$ in the range of 0.73 to 0.41 favor octahedral coordination (see Fig. 3.36). The cations and anions in this structure occur in edge-sharing octahedra (see Fig.

3.51), with each of the 12 edges of an octahedron shared with a neighboring octahedron. Some examples are listed.

Halides:	LiF, LiCl, LiBr, LiI
	NaF, NaCl, NaBr, NaI
	KF, KCl, KBr, KI
	RbF, RbCl, RbBr, RbI
Oxides:	MgO, CaO, SrO, BaO, NiO
Sulfides:	MgS, CaS, MnS, PbS

The structure of pyrite (FeS_2) can be considered as a derivative of the NaCl structure (see Fig. 8.8) in which Fe^{2+} is in the Na position and covalently bonded S_2 (in pairs) in the Cl position.

CsCl Structure

The cesium chloride structure instead of the NaCl structure is adopted by AX compounds when the radius ratio $R_A : R_X$ is greater than 0.73 (see Fig. 3.36). Therefore, the anions (X) are in simple cubic packing (SCP) and the cations fill the large interstices between them. The overall structure (see Fig. 3.52) is made up of centered cubes that share faces with six other neighboring cubes. This feature makes this structure an unattractive choice for highly charged cations (see Pauling's Rule 3).

Examples of AX compounds that exhibit this structure are:

$$CsCl, CsBr, CsI, (NH_4)Cl, (NH_4)Br$$

Sphalerite (ZnS) Structure

The radius ratio, $R_{Zn} : R_S = 0.32$ (radius of $Zn^{2+} = 0.60$ Å in 4-coordination; $S^{2-} = 1.84$ Å in 4-coordination; see Table 3.11) predicts that Zn^{2+} is in tetrahedral

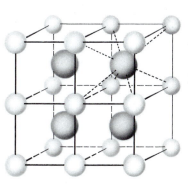

B voids are tetrahedral
C voids are octahedral

(a) (b)

FIG. 3.50 *(a)* Hexagonal closest packing sequence (HCP) with tetrahedral interstices (or voids, marked B) and octahedral interstices (marked C). *(b)* Simple cubic packing (SCP) results when $R_A : R_X$ lies between 0.73 and 1.0. The relatively large cation is surrounded by eight nearest neighbors in cubic coordination.

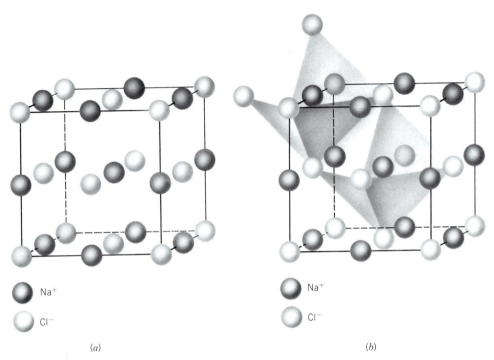

Na⁺

Cl⁻

Na⁺

Cl⁻

(a)   (b)

FIG. 3.51   *(a)* The structure of NaCl, halite. The Na⁺ and Cl⁻ ions are arranged in a face-centered cubic lattice. *(b)* The same structure showing the edge-sharing octahedrons about the Na⁺. Similar edge-sharing octahedrons could be drawn about Cl⁻.

coordination with the neighboring S^{2-} (see Fig. 3.53*a*). The sphalerite structure may be considered as a derivative of the diamond, C, structure (see Fig. 3.53*b*) in which half the carbon atoms in the diamond structure are replaced by Zn and the other half by S. In comparison with the sphalerite structure, diamond may be viewed as having C atoms in the cation positions of the *AX* compound, as well as C atoms in the anion positions. Derivatives of the sphalerite structure, such as chalcopyrite, $CuFeS_2$, and tetrahedrite, $Cu_{12}Sb_4S_{13}$, are illustrated in Fig. 8.5.

FIG. 3.52   The structure of CsCl. The ions are distributed in a primitive cubic lattice. Each cation is surrounded by eight neighbors, and so is each anion.

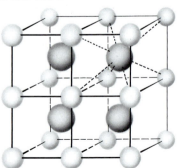

SiC (silicon carbide; carborundum) is isostructural with sphalerite.

CaF₂ Structure

For AX_2 compounds in which the radius ratio ($R_A : R_X$) exceeds 0.73 (see Fig. 3.36), the fluorite structure may be adopted. In this structure the Ca^{2+} ions are arranged at the corners and face centers of a cubic unit cell and F^- are at the centers of the eight equal cubelets into which the cell may be mentally divided (see Fig. 3.54). Each Ca^{2+} is surrounded by eight F^- in cubic coordination, and each F^- is surrounded by four Ca^{2+} at the corners of a tetrahedron. The fluorite structure may be derived from the CsCl structure by replacing Cl^- with F^- and *every other* Cs^+ with Ca^{2+}. This leaves alternate cubic interstices vacant and results in the octahedral cleavage of fluorite and isostructural minerals. The cubic coordination of F^- about a central Ca^{2+} is the result of the radius ratio of $R_{Ca} : R_F \sim .75$. These cubic coordination polyhedra share edges only; similarly, the tetrahedral coordination polyhedra share only edges. This arrangement allows for the maximum separation of the Ca^{2+} cations from each other (see Fig. 3.54). Because the *A* cation has double the charge of the *X* anion, the

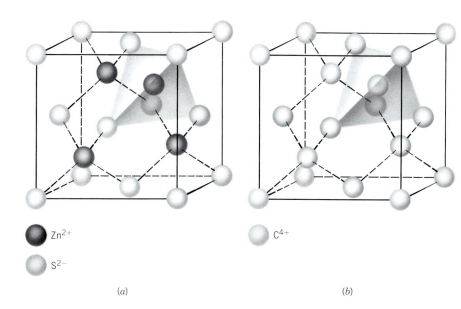

Zn²⁺

S²⁻

(a)

C⁴⁺

(b)

FIG. 3.53 *(a)* The structure of sphalerite, ZnS. Both Zn^{2+} and S^{2-} are in a face-centered cubic array. Zn^{2+} is in tetrahedral coordination with four S^{2-} neighbors. *(b)* The structure of diamond, C. The carbons are arranged in a face-centered cubic lattice. The coordination of C by four nearest carbon neighbors is tetrahedral.

number of anions in the structure must be double that of the cations in order to achieve electrostatic neutrality. As such, the resulting general formula is AX_2. This structure type is adopted by large number of AX_2 halides and oxides:

Halides: CaF_2, SrF_2, BaF_2, CuF_2, CdF_2, $SrCl_2$, $BaCl_2$

Oxides: ZrO_2, HfO_2, CeO_2, UO_2

Rutile (TiO₂) Structure

The rutile structure is based on HCP packing, with Ti filling half the octahedral interstitial positions. AX_2 compounds that exhibit $R_A : R_X$ ratios between about 0.73 and 0.41 (see Fig. 3.36) may adopt the rutile structure in which the A cation is octahedrally coordinated (**C.N.** = 6) to the X anion (see Fig. 3.55). The

FIG. 3.54 The structure of fluorite, CaF_2. Ca^{2+} ions are arranged in a face-centered cubic lattice. The F^- ions are in simple cubic packing (SCP) with Ca^{2+} occupying the voids at the centers of *alternating* cubic interstices.

Ca²⁺

F⁻

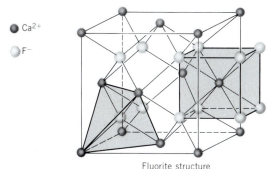

Fluorite structure

oxygen anions are coordinated by three cations in a triangular array (**C.N.** = 3). The structure consists of octahedrons that are linked along horizontal edges. This linking pattern forms strips of octahedrons parallel to the c axis, and these bands are cross-linked to each other by corner-sharing of neighboring octahedrons. This structure results in the prismatic cleavage of rutile, parallel to c. A large number of inorganic compounds assume the rutile structure. Examples are:

MgF_2, NiF_2, CoF_2, FeF_2, MnF_2, ZnF_2

TiO_2, MnO_2, SnO_2, WO_2, PbO_2, SiO_2 (stishovite, a high-pressure polymorph of quartz; see Fig. 4.32a)

Perovskite (ABO₃) Structure

This structure type is based on CCP of oxygen, with one-quarter of the oxygens replaced by a large A cation. This large cation position is in 12-coordination with the surrounding oxygens. The B cations occur in octahedrons that share only apices (see Fig. 3.56). The valence of the A and B ions is not specified; however, the total valence of both ions ($A + B$) must be equal to 6 (to balance the O_3^{2-} in the formula ABO_3). The perovskite structure type is adopted by many compounds. These examples are arranged in columns for the charge of the cations in the A and B sites:

$A^{1+}B^{5+}$	$A^{2+}B^{4+}$	$A^{3+}B^{3+}$
$NaNbO_3$	$CaTiO_3$	$LaCrO_3$
$KNbO_3$	$SrTiO_3$	$YAlO_3$
$KTaO_3$	$BaZrO_3$	$LaAlO_3$
	$BaTiO_3$	

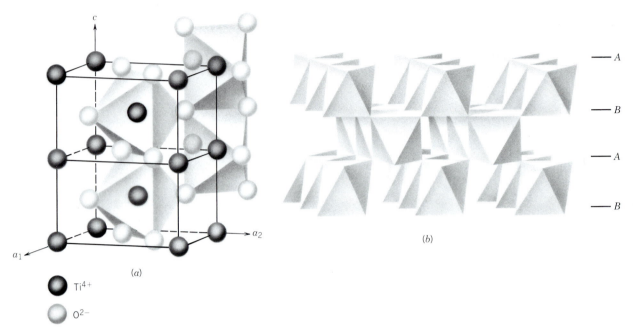

Ti^{4+}

O^{2-}

FIG. 3.55   Two views of the structure of rutile, TiO_2. *(a)* Standard orientation of two unit cells of rutile stacked in the *c* direction. The octahedrons share two horizontal edges with adjacent octahedrons, forming bands parallel to the vertical *c* axis. *(b)* The array of bands of edge-sharing octahedrons (parallel to *c*) running into the page. Chains of edge-sharing octahedrons running parallel to *c* are clearly seen, cross-linked by corner-sharing octahedrons. The HCP stackings are shown by *ABAB*. . . . (From G. A. Waychunas, 1991, Crystal chemistry of oxides and hydroxides, in *Oxide Minerals. Reviews in Mineralogy*, 25: 11–68.)

FIG. 3.56   The structure of perovskite, $CaTiO_3$ in a perspective projection looking down the *c* axis. Modified after J. R. Symth and D. L. Bish, 1988, *Crystal structures and cation sties of the rock-forming minerals*. Boston: Allen and Unwin. Layers of anion-sharing octahedra (containing Ti^{4+}) are oriented perpendicular to the *c* axis. Ca^{2+} is in 12-coordination.

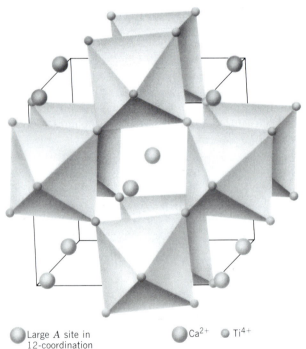

Large *A* site in 12-coordination

Ca^{2+} Ti^{4+}

The perovskite structure type is of especial interest because it is considered to be a common structure type under very high-pressure conditions, as in the central and deep mantle (see page 111).

Spinel (*AB*O₄) Structure

The spinel structure type consists of a CCP array of oxygens in which one-eighth of the tetrahedral interstices (*A*) and half of the octahedral interstices (*B*) are occupied by cations. All spinels contain two differing cations, or at least two different valences of the same cation in the ratio of 2 : 1. Spinels are classified as *normal* or *inverse* spinels, depending on where the more abundant of the cations is housed. If it occurs in the octahedral site, it is classified as *normal*. If it is equally split between the octahedral and tetrahedral sites, it is *inverse*.

The CCP layers of oxygen are stacked parallel to octahedral planes, resulting in alternating layers of octahedral sites and tetrahedral sites (see Fig. 3.57). Occupied octahedra are joined along edges to form rows and planes in the structure, and tetrahedra provide cross-links between layers of octahedra (Fig. 3.57*a*). A plan view of an oxygen layer parallel to the {111} plane and its coordination with cations is given in Fig. 3.57*b*.

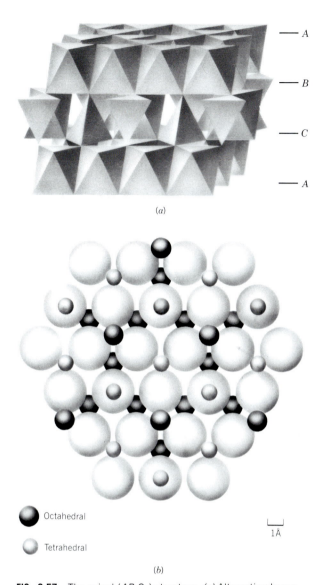

— A

— B

— C

— A

(a)

● Octahedral

○ Tetrahedral

|___|
1Å

(b)

FIG. 3.57 The spinel (AB_2O_4) structure. *(a)* Alternating layers parallel to {111} of octahedral and octahedral-tetrahedral polyhedra, as based upon approximate cubic closest packing. *(b)* A close-packed layer of oxygen in the spinel structure, projected onto the {111} plane. The large spheres are oxygen, and the cation layers on either side of the oxygen layer are shown as well. (*a* and *b* redrawn after G. A. Waychunas, 1991, Crystal chemistry of oxides and hydroxides, in *Oxide Minerals. Reviews in Mineralogy*, 25.)

In the general formula of spinel (ABO_4), the smaller tetrahedral *A* site is commonly occupied by Mg^{2+}, Fe^{2+}, Mn^{2+}, Zn^{2+} and the larger octahedral *B* site by Al^{3+}, Cr^{3+}, or Fe^{3+}. The coordination polyhedra about the various cations in spinel are not what might be predicted on the basis of the ionic sizes of the cations. Because Mg^{2+} is larger than Al^{3+}, one would expect Mg to occur in the octahedral *B* site and Al in the tetrahedral *A* site. In the normal spinel structure (e.g., $MgAl_2O_4$), however, the general con-

cepts of radius ratio do not apply; indeed, the larger cation is in the smaller polyhedron, and vice versa. Only when crystal field stabilization energies are considered instead of geometric aspects of the ions does it become clear why the larger cation may occupy tetrahedral sites.

The spinel structure has a coordination scheme similar to that of the silicates of the olivine series, Mg_2SiO_4 to Fe_2SiO_4. This compositional series can be represented as $X_2^{2+}Y^{4+}O_4$. Although this is not the same as $A^{2+}B_2^{3+}O_4$, as in spinel, in both cases the overall cation charge is identical. If one compares the structure of an Mg–Fe olivine with that of a possible Mg–Fe spinel, one finds that the spinel structure is about 12% denser than the olivine structure of the same composition. This leads to the conclusion that the spinel form of Mg_2SiO_4 must be abundant in the mantle as a result of very high confining pressures (see Fig. 4.9). Some examples of spinel composition follow:

Spinel	$MgAl_2^{3+}O_4$
Hercynite	$FeAl_2^{3+}O_4$
Gahnite	$ZnAl_2^{3+}O_4$
Chromite	$Fe^{2+}Cr_2^{3+}O_4$
Magnesiochromite	$Mg^{2+}Cr_2^{3+}O_4$

For further discussion of spinel, see pages 387–392.

Silicate Structures

Silicates, compounds consisting of abundant oxygen and silicon, are the major mineral components of the Earth's crust. The oxygen in these structures is in close packing with the cations in various coordination polyhedra (see Table 3.12) between the oxygens. In *all* crustal silicates, silicon is in tetrahedral coordination with oxygen. Only in high-pressure phases, such as stishovite (a polymorph of SiO_2), is Si found to occur in 6-coordination with oxygen. As such, the tetrahedral $(SiO_4)^{4-}$ is a fundamental "packing unit" of silicates. The various ways in which this $(SiO_4)^{4-}$ group can link itself to other $(SiO_4)^{4-}$ groups by sharing one, two, three, or all four corner oxygens of the tetrahedron are fundamental to any classification of silicates. An overview of these tetrahedral linking (or polymerization) schemes is given in Fig. 3.58. The basic underlying reason for the ability of the $(SiO_4)^{4-}$ tetrahedron to link itself in so many ways to other $(SiO_4)^{4-}$ tetrahedra is the fact that the electrostatic valency (e.v.) of the bonds between Si^{4+} and oxygen is $\frac{4}{4} = 1$. This is exactly half the bonding strength of the oxygen ion. Consequently, the SiO_4 tetrahedron may link itself through a "bridging" oxygen to another SiO_4 tetrahedron.

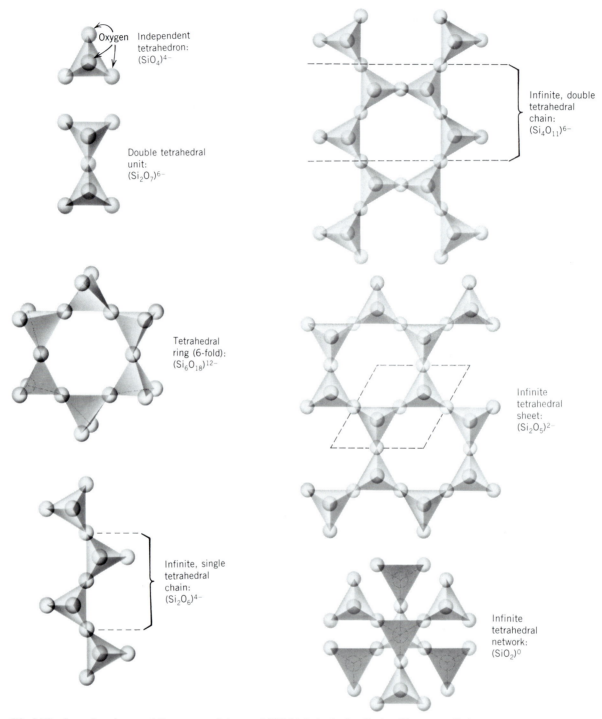

FIG. 3.58 Examples of some of the common linkages of (SiO_4) tetrahedra in silicates. The oxygen that links two tetrahedra is known as the "bridging" oxgyen.

Individual silicate structures are discussed in detail in Chapter 11.

In addition to the classification of silicates used in this text, there are other classifications proposed by Liebau (1985) and Zoltai (1960) (see references at the end of this chapter).

COMPOSITIONAL VARIATION IN MINERALS

As already noted, it is the exceptional mineral that is a *pure substance*, and most minerals display extensive variation in chemical composition. Compositional variation is the result of substitution, in a given structure, of an ion or ionic group for another ion or ionic group. This type of process, referred to as *ionic substitution*, or *solid solution*, occurs among minerals that are isostructural. A concise definition of solid solution is as follows: *A solid solution is a mineral structure in which specific atomic site(s) are occupied in variable proportions by two or more different chemical elements (or groups).* The main factors that determine the amount of solid solution taking place in a crystal structure are:

1. *The comparative sizes of the ions, atoms, or ionic groups that are substituting for each other.* Generally a wide range of substitution is possible if the *size difference* between the ions (or atoms) is *less than about 15%.* If the radii of the two elements differ by 15 to 30%, substitution is limited or rare, and if the radii differ by more than 30%, little substitution is likely.

2. *The charges of the ions involved in the substitution.* If the charges are the same, as in Mg^{2+} and Fe^{2+}, the structure in which the ionic replacement occurs will remain electrically neutral. If the charges are not the same, as in the case of Al^{3+} substituting for Si^{4+}, additional ionic substitutions elsewhere in the structure must take place in order to maintain overall electrostatic neutrality.

3. *The temperature at which the substitution takes place.* There is, in general, a greater tolerance toward atomic substitution at higher temperatures when thermal vibrations (of the overall structure) are greater and the sizes of available atomic sites are larger. Therefore, in a given structure, one expects a greater variability in its composition at higher temperature than at lower temperature.

The types of solid solution can be discussed in terms of *substitutional*, *interstitial*, and *omission* solid solution mechanisms. There is extensive coverage of these concepts under the heading "Solid Solution Mechanisms" in module I of the CD-ROM.

Substitutional Solid Solution

The simplest types of ionic substitutions are *simple cationic* or *anionic* substitutions. In a compound of the type A^+X^-, A^+ may be partly or wholly replaced by B^+. In this instance there is no valence change. Such a substitution is illustrated by the substitution of Rb^+ in the K^+ position of KCl or biotite. A simple anionic substitution can be represented in an A^+X^- compound in which part or all of X^- can be replaced by Y^-. An example is the incorporation of Br^- in the structure of KCl in place of Cl^-. An example of a

complete binary solid solution series (meaning substitution of one element by another over the total possible compositional range, as defined by two end member compositions) is provided by olivine $(Mg,Fe)_2SiO_4$. Mg^{2+} can be replaced in part or completely by Fe^{2+}; the *end members* of the olivine series between which there is complete solid solution are Mg_2SiO_4 (forsterite) and Fe_2SiO_4 (fayalite). Another example of a complete solid solution series is given by $(Mn,Fe)CO_3$, which occurs as a series from $MnCO_3$ (rhodochrosite) to $FeCO_3$ (siderite). An example of a complete anionic series between two compounds is given by KCl and KBr. The size of the two anions is within 10% of each other, allowing for complete substitution of Cl^- and Br^-, and vice versa (see Fig. 3.59). Later in the chapter, Table 3.15 presents an example of extensive, but not complete, simple cation solid solution in the mineral sphalerite, in which Fe^{2+} is the main replacement for Zn^{2+} in the structure.

If in a general composition of $A^{2+}X^{2-}$ a cation B^{3+} substitutes for some of A^{2+}, electrical neutrality can be maintained if an identical amount of A^{2+} is at the same time replaced by cation C^+. This can be represented by

$$2A^{2+} \rightleftarrows 1B^{3+} + 1C^+$$

with identical total electrical charges on both sides of the equation. This type of substitution is known as *coupled substitution.* The substitution of Fe^{2+} and Ti^{4+} for $2Al^{3+}$ in the corundum (Al_2O_3) structure of the gem sapphire is an example of such coupled substitution. The plagioclase feldspar series can be represented in terms of two end members, $NaAlSi_3O_8$ (albite) and $CaAl_2Si_2O_8$ (anorthite). The complete solid solution between these two end member compositions illustrates coupled substitution:

$$Na^{1+}Si^{4+} \rightleftarrows Ca^{2+}Al^{3+}$$

This means that for each Ca^{2+} that replaces Na^{1+} in the feldspar structure, one Si^{4+} is replaced by Al^{3+} in the Si–O framework. The equation shows that the total electrical charges on both sides of the equation are identical; as such the structure remains neutral. An example of only limited coupled solid solution is provided by two pyroxenes, diopside $(CaMgSi_2O_6)$ and jadeite $(NaAlSi_2O_6)$. The coupled replacement can be represented as: $Ca^{2+}Mg^{2+} \rightleftarrows Na^+Al^{3+}$. Although the pyroxene structure is neutral with either type of cation pair, the atomic sites and coordination polyhedra do not allow for a complete solid solution of this type (see Fig. 3.59).

Atomic sites that in some structure are *unfilled (vacant)* may become partly or wholly occupied due to a substitutional scheme in which a vacancy is

Simple cationic:

$Mg^{2+} \rightleftarrows Fe^{2+}$ (complete)	Mg^{2+} [6] = 0.72Å
$Fe^{2+} \rightleftarrows Mn^{2+}$ (complete)	Fe^{2+} [6] = 0.78Å
$Na^{+} \rightleftarrows K^{+}$ (partial)	Mn^{2+} [6] = 0.83Å
$Mg^{2+} \rightleftarrows Mn^{2+}$ (partial)	Na^{+} [8] = 1.18Å
	K^{+} [8] = 1.51Å

Simple anionic:

$Br^{-} \rightleftarrows Cl^{-}$ (complete)	Br^{-} [6] = 1.96Å
	Cl^{-} [6] = 1.81Å
$I^{-} \rightleftarrows Cl^{-}$ (partial)	I^{-} [6] = 2.20Å

Coupled cationic: *Extent of solid solution:*

$Na^{+}Si^{4+}$	$\rightleftarrows Ca^{2+}Al^{3+}$	complete at high
2 cations:	2 cations:	temperature in
total charge 5^{+}	total charge 5^{+}	plagioclase
$Ca^{2+}Mg^{2+}$	$\rightleftarrows Na^{+}Al^{3+}$	limited, as in omphacite,
2 cations:	2 cations:	a member of the
total charge 4^{+}	total charge 4^{+}	pyroxene group
$Mg^{2+}Al_2^{3+}$	$\rightleftarrows Fe^{2+}_2 Ti^{4+}$	extensive, as in the
3 cations:	3 cations:	spinel group
total charge 8^{+}	total charge 8^{+}	
$\square^{\circ} + Si^{4+}$	$\rightleftarrows Na^{+}Al^{3+}$	extensive, as in
2 sites:	2 cations in 2	arfvedsonite, a sodium
total charge 4^{+}	sites:	amphibole
$\square^{\circ}$ = vacancy	total charge 4^{+}	

FIG. 3.59 *Examples of various types of substitutional solid solution. Ionic radii are taken from Table 3.11. Square brackets after the ions enclose the coordination number.*

involved. For example, a partial coupled substitution in the amphibole tremolite, $\square Ca_2Mg_5Si_8O_{22}(OH)_2$, where $\square$ is the normally vacant *A* site in the structure, may result from

$$\square\ Si^{4+} \leftrightarrows Na^{1+}Al^{3+}$$

In this coupled substitutional scheme Al^{3+} replaces Si^{4+} in the tetrahedral position, and the additional Na^{1+} is housed in a site that is normally vacant (shown as $\square$) (see Fig. 3.59).

Interstitial Solid Solution

Between atoms, or ions, or ionic groups of a crystal structure interstices exist, which normally are considered voids. When ions or atoms are located in these structural voids, we speak of *interstitial substitution* or *interstitial solid solution*. In some crystal structures these voids may be channel-like cavities, as in beryl, $Be_3Al_2Si_6O_{18}$. In this ring silicate, large ions as well as molecules can occupy the tubular cavities of the superimposed rings (see Fig. 3.60). Considerable amounts of K^{+}, Rb^{+}, Cs^{+}, and H_2O as well as CO_2 are reported in beryl analyses, housed interstitially in the hexagonal channels. H_2O and CO_2 molecules are weakly bonded to the internal oxygens of the six-fold Si_6O_{18} rings. The large monovalent alkalis K^{+}, Rb^{+}, and Cs^{+} are also located in-

side these rings, but they are much more strongly bonded, because of the following coupled substitutional mechanisms:

$$Si^{4+} \rightleftarrows Be^{2+} + 2R^{+}$$

and

$$Si^{4+} \rightleftarrows Al^{3+} + R^{+}$$

where *R* represents K^{+}, Rb^{+}, or Cs^{+}. In the first substitution two monovalent alkali cations are housed inside the interstice of the hexagonal Si_6O_{18} chain; in the second, only one cation is positioned there.

Another common example of interstitial solid solution is found in the zeolite group of silicates (see Chapter 11). The zeolites constitute a group of tectosilicates in which (SiO_4) and (AlO_4) tetrahedra are linked together in a very open framework. Within this skeletal framework there may be large holes and continuous channels, with openings ranging from 2 to 9 Å in diameter. These channels provide easy access to the interior of the crystals, and they also accommodate water molecules. Because the water molecules are weakly tied, by hydrogen bonding, to the surrounding Al–Si framework, mild heating will remove them from these interstitial positions without collapse of the framework. Upon removal of the heat and in the presence of liquid water or steam, zeolites rehydrate by re-incorporation of H_2O molecules in

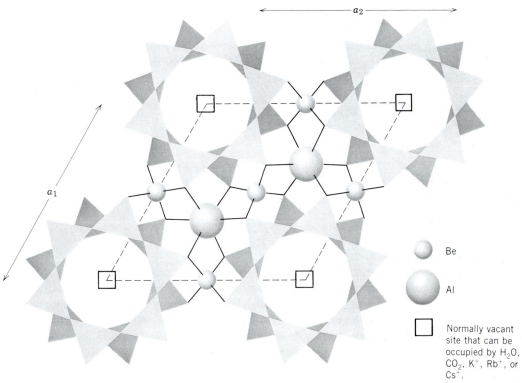

FIG. 3.60 Plan of the hexagonal structure of beryl, $Be_3Al_2Si_6O_{18}$, projected onto the basal plane (0001). The Si_6O_{18} rings at two different heights are shown. The hexagonal channels are the locus for large alkali ions and neutral molecules such as H_2O and CO_2. A unit cell is outlined by dashed lines.

Be

Al

Normally vacant site that can be occupied by H_2O, CO_2, K^+, Rb^+, or Cs^+.

the interstices from which the molecules were originally removed.

Omission Solid Solution

Omission solid solution occurs when a more highly charged cation replaces two or more other cations for charge balance. The substitution can take place only in one atomic site, leaving other sites vacant or omitted. For instance, when Pb^{2+} substitutes for K^+ in the blue-green variety of microcline feldspar ($KAlSi_3O_8$), known as amazonite, it replaces *two* K^+ ions, but occupies only one site:

$$K^+ + K^+ \leftrightarrows Pb^{2+} + \square$$

creating a lattice vacancy, $\square$, which can become a color center.

The best-known mineralogical example of this type of solid solution is provided by pyrrhotite, $Fe_{(1-x)}S$. In pyrrhotite, sulfur occurs in hexagonal closest packing and iron in 6-coordination with sulfur. If each octahedral site were occupied by Fe^{2+}, the formula of pyrrhotite would be FeS. In pyrrhotites, however, there is a variation in the percentage of *vacancies* in the octahedral sites, causing

the composition to range from Fe_6S_7 through $Fe_{11}S_{12}$, close to FeS. The formula is generally expressed as $Fe_{(1-x)}S$, with x ranging from 0 to 0.2. Minerals, such as pyrrhotite, in which a particular structural site is incompletely filled are known as *defect structures* (see Fig. 3.61 for a structural image of pyrrhotite). When Fe^{2+} is absent from some of the octahedral sites in pyrrhotite, with the sulfur net completely intact, the structure is not electrically neutral. It is very likely that some of the Fe is in the Fe^{3+} state to compensate for the deficiency in Fe^{2+}. If this is so, a neutral pyrrhotite formula can be written as $(Fe^{2+}_{1-3x}Fe^{3+}_{2x})\square_xS$, where $\square$ represents vacancies in the cation position. If x in this example were 0.1, there would be 0.1 vacancies, 0.7 Fe^{2+}, and 0.2 Fe^{3+} in the structure of the pyrrhotite. In other words, Fe^{3+} is accommodated in the pyrrhotite structure by the following substitution: $Fe^{2+} + Fe^{2+} + Fe^{2+} \leftrightarrows Fe^{3+} + Fe^{3+} + \square$, which creates a vacancy $\square$.

Another mineral example of omission solid solution (with defects, or cation vacancies) is maghemite, γFe_2O_3, an oxidation product of magnetite, Fe_3O_4 (or $Fe^{3+}_2Fe^{2+}O_4$). Maghemite is an iron-deficient spinel-type structure in which the tetrahedral and octahedral cation sites show considerable vacancies. The

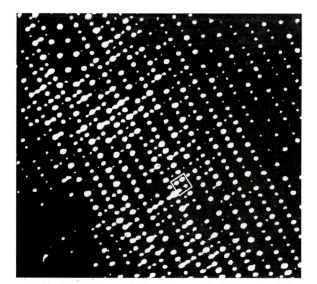

FIG. 3.61 High-resolution transmission electron microscope (HRTEM) structure image of pyrrhotite in which the white spots correspond to columns of iron atoms; these columns are aligned perpendicular to the plane of the photograph. The less intense spots represent columns from which some of the iron atoms are missing (omission solid solution). The columns are alternately more or less fully occupied by iron atoms. The white square is 3 Å to a side. (From L. Pierce and P. R. Buseck, 1974, Electron imaging of pyrrhotite superstructures. *Science* 186: 1209–1212; copyright 1974 by the AAAS. See also P. R. Buseck, 1983, Electron microscopy of minerals. *American Scientist* 7: 175–185.)

maghemite composition may be stated as follows: $Fe^{3+}Fe^{3+}_{1.67}\ \square_{0.33}\ O_4$, where $\square$ denotes a cation vacancy.

RECALCULATION OF CHEMICAL ANALYSES

Elements such as gold, arsenic, and sulfur occur in the native state and their formulas are the chemical symbols of the elements. Most minerals, however, are compounds composed of two or many more elements; their formulas, recalculated from the results of quantitative chemical analyses, indicate the atomic proportions of the elements present. Thus, in galena, PbS, there is one atom of sulfur for each atom of lead, and in chalcopyrite, $CuFeS_2$, there is one atom of sulfur for each atom of copper and iron. Such specific atomic proportions underlie the definition of a mineral, which states that "a mineral . . . has a definite (but generally not fixed) chemical composition." Very few minerals, however, have a truly constant composition; examples of specific and essentially constant compositions are provided by quartz, SiO_2, and kyanite, Al_2SiO_5. Such minerals are often referred to as *pure substances*. The great majority of

minerals show large variations in composition within specific atomic sites of their structures. For example, ZnS, sphalerite, shows a large range of Fe content. The Fe substitutes for Zn in the sphalerite structure, and in general terms an Fe-rich (ferroan) sphalerite composition can be expressed as (Zn, Fe)S. Here the total cation–anion ratio is still fixed, 1 : 1, however the Zn and Fe contents are variable.

As already noted, a quantitative chemical analysis provides the basic information for the atomic formula of a mineral. A weight percent analysis in terms of metals, or oxides, gives us a listing of what elements are present and in what concentrations, but it gives us no direct information about how the elements (or ions) occur in the structure of the mineral. A quantitative weight percent analysis should add up close to 100%. Slight variations above or below 100% occur because of cumulative small errors that are part of the analytical procedure. As an example, let us consider the analysis of chalcopyrite in Table 3.14. The percentages in column 1, reported by the chemist, represent the percentages by weight of the various elements. Because the elements have different atomic weights, these percentages do not represent the ratios of the different atoms. To arrive at their relative proportions, the weight percentage in each case is divided by the atomic weight of the element. This gives a series of numbers (column 3), the *atomic proportions*, from which the *atomic ratios* can be quickly derived. In the analysis of chalcopyrite, these ratios are Cu : Fe : S = 1 : 1 : 2; thus $CuFeS_2$ is the chemical formula.

The reverse procedure, that of calculating the percentage composition from the formula, is shown

TABLE 3.14   Chalcopyrite ($CuFeS_2$) Recalculations

	1 Weight Percent	2 Atomic Weights	3 Atomic Proportions	4 Atomic Ratios	
Cu	34.30	63.55	0.53973	1	
Fe	30.59	55.85	0.54772	1	approx.
S	34.82	32.06	1.08575	2	
Total	99.71				
	Atomic Weights				

Cu	63.55	$Cu(in\%) = \dfrac{63.55}{183.52} \times 100\% =$	34.63
Fe	55.85	$Fe(in\%) = \dfrac{55.85}{183.52} \times 100\% =$	30.43
S($2\times$)	$(32.06) \times 2$	$S(in\%) = \dfrac{64.12}{183.52} \times 100\% =$	34.94
Total	183.52		Total 100.00

in the lower half of Table 3.14. The formula of chalcopyrite is $CuFeS_2$, which makes for a gram-formula weight of 183.52. The calculated Cu, Fe, and S weight percentage values are very similar to the measured percentages reported in column 1 of Table 3.14. The calculated values are slightly different from the determined values, probably because of a slight experimental error in the analysis technique.

Let us now examine the recalculation of some sphalerite analyses, which show a considerable range in Fe, and of troilite, FeS (Table 3.15). The upper half of the table shows the weight percentages for the various elements and the lower part of the table provides atomic proportions obtained by dividing each weight percentage figure by the appropriate atomic weight. One of the lower lines in Table 3.15 shows that the ratio of (Zn + Fe + Mn + Cd) : S is constant in the sphalerite analyses, namely 1 : 1, and that it is 1 : 1 in the troilite as well. However, the Zn in the sphalerite structure is partially substituted for by variable amounts of Fe, Mn, and Cd. The maximum amount of Fe reported in Table 3.15 is in column 4, which shows 18.25 weight percent Fe. This recalculates to 0.327 atomic proportions out of a total of 1.060 atomic proportions for (Zn + Fe + Mn + Cd). These analyses of naturally occurring sphalerites lead to the conclusion that the representation of sphalerite compositions as ZnS is a great oversimplification. Only analysis 1 (Table 3.15) is pure ZnS and analysis 5 is pure FeS. Analyses 2, 3, and 4 contain variable amounts of Fe, Mn, and Cd. Analysis 4, which contains the largest amounts of elements other than Zn and S, can serve to illustrate some of the useful atomic proportion recalcu-

lations. Its total cation content (in atomic percentage) is 1.060. Fe is $0.327/1.060 \times 100\% = 30.8\%$; similarly, Mn, Cd, and Zn values are 4.5, 0.2, and 64.4%, respectively. These cation proportion values can be used as subscripts to represent the specific composition of the sample of sphalerite in column 4; this leads to the formulation $(Zn_{0.644}Fe_{0.308}Mn_{0.045}Cd_{0.002})S$. The other specific formulas in Table 5.3 were obtained in a similar manner. It is obvious from the analyses that Zn and Fe are the main variables. Often it is necessary to obtain the cation distribution of such *major* elements only. For analysis 4 in Table 3.15 we then must eliminate from consideration the quantities of Cd and Mn. The total of (Fe + Zn) in the atomic proportions column = 1.010, rather than 1.060 for (Fe + Mn + Cd + Zn). If now we wish to know the Fe versus Zn percentage in this analysis, we factor $0.327/1.010 \times 100\% = 32.4\%$. This value, and those similarly obtained for the other analyses, are listed in the last line of Table 3.15. These Fe : Zn ratios show that a selection of sphalerites may show *a compositional range from pure* ZnS to $(Zn_{0.68}Fe_{0.32})S$. The mineral troilite is found only in meteorites and shows no Zn in its chemical composition; it can therefore be considered as a compound of constant composition, FeS. The large variation of Fe in sphalerites is due to the substitution of Fe for Zn in the same structural site in the sphalerite structure; this is also known as *solid solution* of Fe in ZnS. The largest amount of Fe solid solution in the analyses of Table 3.15 is shown by analysis 4, with composition $(Zn_{0.676}Fe_{0.324})S$.

The majority of minerals such as silicates, oxides, carbonates, phosphates, sulfates, and so on, are

TABLE 3.15 Recalculation of Several Sphalerite Analyses and of Troilite

	Weight Percentage*				
	1	2	3	4	5
Fe	0	0.15	7.99	18.25	63.53
Mn	0	0	0	2.66	0
Cd	0	0	1.23	0.28	0
Zn	67.10	66.98	57.38	44.67	0
S	32.90	32.78	32.99	33.57	36.47
Total	100.00	99.91	99.59	99.43	100.00

	Recalculated in Terms of Atomic Proportions				
Fe	0	0.003 ⎫	0.143 ⎫	0.327 ⎫	1.137
Mn	0	0 ⎪	0 ⎪	0.048 ⎪	0
Cd	0	0 ⎬ 1.027	0.011 ⎬ 1.032	0.002 ⎬ 1.060	0
Zn	1.026	1.024 ⎭	0.878 ⎭	0.683 ⎭	0
S	1.026	1.022	1.029	1.047	1.137
(Zn + Fe + Mn + Cd):S	1 : 1	1 : 1	1 : 1	1 : 1	1 : 1
Formulas	ZnS	$(Zn_{0.997}Fe_{0.003})S$	$(Zn_{0.851}Fe_{0.138}Cd_{0.011})S$	$(Zn_{0.644}Fe_{0.308}Mn_{0.045}Cd_{0.002})S$	FeS
Fe : Zn	0 : 100	0.3 : 99.7	14.0 : 86.0	32.4 : 67.6	100 : 0

*Analysis 1–4 from *Dana's System*, v. 1, 1944; analysis 5 for pure troilite FeS.

compounds containing large amounts of oxygen. By convention the analyses of such minerals are reported as percentages of oxides, rather than as percentages of metals. Calculations similar to those outlined above are performed with the oxide components to arrive at molecular proportions of the oxides, rather than atomic proportions of the elements. Table 3.16 gives an example of the recalculation of an analysis of gypsum. The analytically determined oxide components in column 1 are divided by the molecular weights of the corresponding oxides (column 2) to arrive at molecular proportions (column 3). From the molecular ratios in column 4 we see that $CaO : SO_3 : H_2O = 1 : 1 : 2$, and the composition can be written as $CaO \cdot SO_3 \cdot 2H_2O$, or as $CaSO_4 \cdot 2H_2O$. The latter formula is preferred because it avoids creating the erroneous impression that the mineral is composed of discrete oxide molecules.

In the lower part of Table 3.16 the chemical analysis of a member of the olivine series is given. The desired result of a recalculation of the analysis again is a statement of the molecular proportions of the oxide components, or of the cation components. The steps in going from columns 1 to 3 are the same as in the gypsum analysis. Column 4 lists the values for the atomic proportions of the various atoms. The atomic proportions of the metal atoms are the same as the corresponding molecular proportions in column 3, because, for example, one "molecule" of SiO_2 contributes 1 Si and one "molecule" of FeO contributes 1 Fe. However, the number of oxygens contributed by each molecular proportion is not the same as the numbers listed in column 3. For each of

the divalent metals there is one oxygen atom, but for each silicon there are two oxygens. In other words, one molecular proportion of FeO contributes 1 oxygen, but one molecular proportion of SiO_2 contributes two oxygens. This is reflected in the numbers of column 5.

The total number of oxygens contributed by the atomic proportions in column 5 is 2.3535. From crystal structure data we know that olivine, with the general formula of $(Mg, Fe)_2SiO_4$, has four oxygens per formula unit. In order to arrive at the cation proportions in terms of four oxygens, we multiply each of the cation numbers in column 4 with the 4/2.3535 ratio. This leads to the numbers in columns 6 and 7. The final formulation for this olivine is $(Mg_{1.12}Fe_{.86}Mn_{.01})SiO_4$, obtained by assuming 2.0 for total cations, instead of 2.022 as calculated. Such a formula is often expressed in terms of *end member compositions*. In olivine these are Mg_2SiO_4, forsterite, and Fe_2SiO_4, fayalite. The amounts of forsterite (Fo) and fayalite (Fa) in the general formula are directly proportional to the atomic proportions of Mg and Fe, or molecular proportions of MgO and FeO in the analysis. For example, in Table 3.16, this leads to Fo_{57} and Fa_{43}. If the olivine composition ranges mainly between Mg and Fe compositions, it suffices to state the final recalculation as Fo_{57}.

The olivine structure is relatively simple, with Mg and Fe substituting for each other in the same structural position (site). Many silicates, however, have various structural sites among which elements can be distributed. An example would be a member of the diopside-hedenbergite series $(CaMgSi_2O_6$ —

TABLE 3.16   Recalculation of Gypsum and Olivine Analyses

Gypsum	1 Weight Percent	2 Molecular Weights	3 Molecular Proportions	4 Molecular Ratios			
CaO	32.44	56.08	0.57846	1 ⎫			
SO₃	46.61	80.06	0.58219	1 ⎬ approx.			
H₂O	20.74	18.0	1.15222	2 ⎭			
Total	99.79						

Olivine	1 Weight Percent	2 Molecular Weights	3 Molecular Proportions	4 Atomic Proportions Cations	5 Oxygens	6 On Basis of 4 Oxygens	7 Atomic Ratios
SiO₂	34.96	60.09	0.58179	Si 0.5818	1.1636	0.989	1
FeO	36.77	71.85	0.51176	Fe²⁺ 0.5118	0.5118	0.870 ⎫	
MnO	0.52	70.94	0.00733	Mn 0.0073	0.0073	0.012 ⎬ 2.022	2
MgO	27.04	40.31	0.67080	Mg 0.6708	0.6708	1.140 ⎭	
Total	99.29				2.3535		

Olivine in terms of end member compositions: Mg = 1.140, Fe = 0.870; total; 2.010. Percentage Mg = 56.7; percentage Fe = 43.3; percentage Mg_2SiO_4 = % forsterite = % Fo = 56.7%; percentage Fe_2SiO_4 = % fayalite = % Fa = 43.3%.

TABLE 3.17   Recalculation of a Pyroxene Analysis

	1	2	3	4	5		6			7
	Weight Percent	Molecular Proportions	Cation Proportions	Number of Oxygens	Cations on Basis of 6 Oxygens		Cation Assignment			End Member Recalculation
SiO_2	50.38	0.8384	0.8384	1.6768	1.875	Si	1.875	} 2.0		MgO as $MgSiO_3$
Al_2O_3	3.01	0.0295	0.0590	0.0885	0.132	→Al	0.125	}		(enstatite = En)
										FeO as $FeSiO_3$
					↘	Al	0.007	}		(ferrosilite = Fs)
TiO_2	0.45	0.0056	0.0056	0.0112	0.012	Ti	0.012	}		CaO as $CaSiO_3$
Fe_2O_3	1.95	0.0122	0.0244	0.0366	0.055	Fe^{3+}	0.055	} 1.033		(wollastonite = Wo)
FeO	4.53	0.0630	0.0630	0.0630	0.141	Fe^{2+}	0.141	} ≈ 1		Using molecular
MnO	0.09	0.0013	0.0013	0.0013	0.003	Mn	0.003	}		proportions:
MgO	14.69	0.3643	0.3643	0.3643	0.815	Mg	0.815	}		
										MgO = 0.3644
CaO	24.32	0.4321	0.4321	0.4321	0.966	Ca	0.966	} 1.006		FeO = 0.0630
Na_2O	0.46	0.0074	0.0148	0.0074	0.033	Na	0.033	} ≈ 1		CaO = 0.4321
K_2O	0.15	0.0016	0.0032	0.0016	0.007	K	0.007	}		Total = 0.8595
Total	99.94		Total O = 2.6828							
										%En = 42.39

$$\text{Oxygen factor } \frac{6}{2.6828} = 2.236469$$

%Fs = 7.33

%Wo = 50.27

$CaFeSi_2O_6$) in the pyroxenes. Table 3.17 gives a chemical analysis, as well as the various recalculation steps for pyroxene similar to that outlined for olivine. In column 1 of Table 3.17 percentages are reported for FeO as well as Fe_2O_3. In analyses made by instrumental techniques Fe^{3+} and Fe^{2+} cannot be distinguished, and total Fe is then reported as FeO only, or Fe_2O_3 only. Column 2 provides molecular proportions, obtained by dividing the weight percent figures by the appropriate molecular weights. Column 3 lists the cation proportions for each oxide "molecule." Note, for example, that Al_2O_3 contains two Al, therefore 0.0295 is multiplied by 2. Column 4 lists the number of oxygens contributed by each oxide "molecule." Note that SiO_2 contains two oxygens per "molecule"; the molecular proportion figure (0.8384) is therefore multiplied by 2. The total number for all oxygens contributed by all the oxide "molecules" in column 4 is 2.6828. From crystal structure data we know that a pyroxene of this type has a formula such as $Ca(Mg,Fe)(Si,Al)_2O_6$. The formula is therefore recalculated on the basis of six oxygens, and the numbers in column 3 are multiplied by 2.23647 to obtain the figures in column 5. These cations are subsequently assigned to specific atomic sites in the pyroxene structure. If necessary, sufficient Al is added to Si so that the sum Si + Al = 2.0 in column 6; the remainder of the Al is added to the sum of the intermediate size cations. Intermediate size cations (Al, Ti, Fe^{3+}, Fe^{2+}, Mn, Mg) are assigned to the M1 cation position (see Fig. 11.16) in the structure; these cation

numbers together amount to 1.033, in agreement with the general formula $Ca_1(Mg, Fe)_1(Si, Al)_2O_6$. The remaining larger cations (Ca, Na, K) are assigned to the M2 position in the pyroxene structure, and their total is 1.006, again approximately 1.

Pyroxene analyses are often recalculated in terms of end member compositions. The analysis in Table 3.17 contains major amounts of CaO, MgO, and FeO and can therefore be recalculated in terms of $CaSiO_3$ (wollastonite), $MgSiO_3$ (enstatite), and $FeSiO_3$ (ferrosilite) end members. The calculations are given in column 7, showing that the weight percent analysis can be recast as $Wo_{50.3}En_{42.4}Fs_{7.3}$.

Complex hydrous silicates such as amphiboles are recalculated by the same sequence of steps as illustrated for olivine and pyroxene, but the H_2O content is evaluated as (OH) groups in the amphibole structure. Table 3.18 gives an example of a Na-containing actinolite analysis. A general actinolite formula is $Ca_2(Mg,Fe)_5Si_8O_{22}(OH)_2$. Column 1 lists weight percentages for $H_2O(+)$ and $H_2O(-)$. The $H_2O(+)$ is considered part of the actinolite structure but the $H_2O(-)$ is not and is therefore neglected in subsequent calculations (see footnote to Table 3.18). Column 2 lists the molecular proportions, column 3 the cation proportions, and column 4 gives the total contribution of (O,OH) for each of the "molecules" in column 2. The sum in column 4 divided into 24 gives the ratio by which the whole analysis must be multiplied in order to put it on a basis of 24(O,OH). Column 3 is now multiplied by the factor 8.4803,

TABLE 3.18   Recalculation of an Amphibole Analysis

	1 Weight Percent	2 Molecular Proportions	3 Number of Cations	4 Total Number of (O, OH)	5 Cations on Basis of 24 (O, OH)	6 Cations on Basis of 23 Oxygens
SiO_2	56.16	0.9346	0.9346	1.8692	7.926 ⎱ 7.958	7.931
Al_2O_3	0.20	0.0019	0.0038	0.0057	0.032 ⎰ ≈8	0.032
Fe_2O_3	1.81	0.0113	0.0226	0.0339	0.192 ⎫	0.192
FeO	6.32	0.0880	0.0880	0.0880	0.746 ⎬ 5.111	0.747
MgO	19.84	0.4921	0.4921	0.4921	4.173 ⎭ ≈5	4.176
MnO	2.30	0.0324	0.0324	0.0324	0.275 ⎫	0.275
CaO	9.34	0.1665	0.1665	0.1665	1.412 ⎬ 2.068	1.413
Na_2O	1.30	0.0210	0.0420	0.0210	0.356 ⎭ ≈2	0.356
K_2O	0.14	0.0015	0.0030	0.0015	0.025	0.025
$H_2O(+)$	2.16	0.1198	0.2396	0.1198	2.032 ≈2	
$H_2O(-)$*	0.48	—		—		
Total	100.05			2.8301		

$$\frac{24}{2.8301} = 8.4803; \quad 2.8301 - 0.1198 = 2.7103; \quad \frac{23}{2.7103} = 8.4861$$

*This H_2O is weakly held by the powdered sample and is removed by drying at low temperature. It is assumed to be absorbed from the atmosphere, not structurally bound.

giving the results in column 5 on the basis of 24(O,OH). In general it will be found that Si is less than 8.0 and Al is added to bring (Si,Al) to 8.0. The remaining Al, if any, is added to the cation group immediately below consisting of Fe^{3+}, Fe^{2+}, and Mg. The total for this group of intermediate size cations is 5.111, close to 5.0, and the total for the remaining larger size cations is 2.068, close to 2.0. The intermediate size cations enter the M1, M2, and M3 sites in the amphibole structure and the larger cations enter M4 (see Fig. 11.22). Mn could be distributed among M1, M2, and M3 as well as M4 because its ionic size lies between that of Ca^{2+} and Mg^{2+}. The (OH) content turns out to be about 2. The recalculated analysis now has the general form (Ca, Na, Mn)$_2$(Fe, Mg)$_5$(Si, Al)$_8$O$_{22}$(OH)$_2$. Instrumental analysis results of hydrous minerals such as amphiboles do not provide information on the oxidation state of iron (Fe^{2+} vs. Fe^{3+}), for example, or on the presence of structural water. Such anhydrous analytical results are frequently recalculated on an anhydrous basis. For the analysis in Table 3.18 such a basis would be 23 oxygens, replacing two (OH)$^-$ by one O^{2-}. Although this is not correct in terms of the structure of amphiboles, it does allow for comparison among analytical results in which H_2O was not determined. In column 6 the actinolite is recalculated on an anhydrous basis and the cation results are similar to, but not identical with, the results in column 5.

Additional examples of recalculations of mineral analyses followed by recalculation exercises are given in Klein's *Mineral and Rocks: Exercises in Crys-*

tallography, Mineralogy, and Hand Specimen Petrology (see full reference at the end of the chapter).

GRAPHIC REPRESENTATION OF MINERAL COMPOSITION

Because most minerals show solid solution features ranging from partial to complete, it is frequently useful to show such variations in composition in graphic form. (This subject is covered extensively under the heading "Solid Solution Mechanisms" in module I of the CD-ROM.) A composition graph can be constructed after we have chosen the chemical components pertaining to the mineral compositions to be graphed. The mineral kyanite, Al_2SiO_5, is known to be essentially constant in composition. A chemical analysis of a pure kyanite yields the following data, in terms of weight percent: Al_2O_3, 62.91, and SiO_2, 37.08. Selecting Al_2O_3 and SiO_2 as the two end components of a bar graph, we can position the kyanite composition on it in terms of weight percent of the oxides (see Fig. 3.62a). In order to plot a mineral composition in terms of the weight percent of its components we need to obtain its chemical analysis or we can recalculate the weight percentages of the components from the formula as shown in Table 3.14. However, formulas that provide direct information about atomic proportions of elements or molecular proportions of components can be used directly for graphing of compositions. The formula of kyanite can be written as $1Al_2O_3 + 1SiO_2 = 1Al_2SiO_5$. In

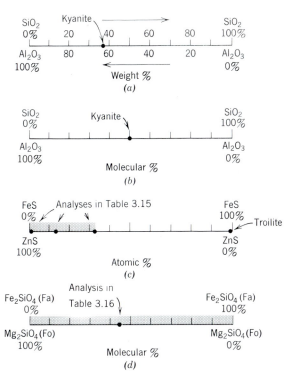

FIG. 3.62 Graphic representations of chemical compositions in terms of bar diagrams. *(a)* Kyanite (Al_2SiO_5) composition in terms of weight percentage of the oxides. *(b)* Kyanite in terms of molecular percentage of the oxides. *(c)* Sphalerite analyses from Table 3.15 in terms of atomic percentages of Zn and Fe. *(d)* Complete solid solution series of olivine and composition of olivine from Table 3.16 in terms of molecular percentages of forsterite and fayalite.

other words, the composition of kyanite can be expressed in terms of the molecular proportions of the oxide components Al_2O_3 and SiO_2. This is shown in Fig. 3.62*b*.

In Table 3.15 we recalculated several sphalerite analyses. These compositions can be shown graphically (see Fig. 3.62*c*) in terms of two end members, ZnS and FeS. Intermediate compositions can be obtained directly from the Zn : Fe ratios in the last line of Table 3.15. After these numbers are plotted, and with the addition of further literature analyses as well, the solid solution extent of FeS in sphalerite can be generalized on the basis of a population of analysis points. Although only very few points are shown on the bar graph in Fig. 3.62*c*, it can be stated that sphalerite shows a partial solid solution series from ZnS to at least $Zn_{0.676}Fe_{0.324}S$. This statement is shown graphically by the shading. Troilite, on the other hand, shows no Zn content and must appear as a single point on the bar graph, indicating a total lack of Zn substitution. The graph shows that between troilite and the most Fe-rich sphalerite plotted there is a region in which homogeneous composi-

tions of Fe-rich sphalerites are not found. In Table 3.16 we recalculated an olivine analysis in terms of molecular percentages of MgO and FeO; such percentages can also be expressed as Mg_2SiO_4(Fo) and Fe_2SiO_4(Fa), the two end member compositions of the olivine solid solution series. Figure 3.62*d* shows a bar graph depicting the complete solid solution series between Fo and Fa as well as the specific composition of the olivine of Table 3.16. Bar diagrams of this type form the horizontal axis in *variation diagrams* that relate changes in physical properties to variation in composition (see Figs. 2.20 and 7.48). Composition bars also form the horizontal axis in temperature-composition diagrams (see Fig. 4.44 and Chapter 4).

As we have already seen in Tables 3.16 to 3.18, mineral analyses can be very complex and may show substitution of several elements in a specific atomic site of the mineral structure. In order to portray the variation of at least three components instead of just two, as in Fig. 3.62, a *triangular diagram* is frequently used. Triangular coordinate graph paper is available commercially[3] with ruling for each 1% division. Let us first illustrate the use of a triangular composition diagram by plotting some fairly simple end member compositions. In Figure 3.63*a* we list the mineral names, formulas, and compositions in terms of weight percentages of the oxides for two end members of the orthopyroxene series, enstatite and ferrosilite (also known as orthoferrosilite to reflect the orthorhombic nature of the mineral) and the two end members of the olivine series, forsterite and fayalite. On the basis of the data given in Fig. 3.63*a* we have two choices for a triangular plot, namely, on the basis of weight percentages, or on the basis of the relative cation proportions in the formulas. We will do the first plot on the basis of the weight percentage values.

In Fig. 3.63*b* we have chosen the corners of the triangle as 100 weight percent SiO_2, 100 weight percent MgO, and 100 weight percent FeO. This allows us to plot all four compositions because they are made up only of these three oxide components. Enstatite and forsterite contain 40.16 and 57.30 weight percent MgO, respectively. The MgO scale extends from 0 to 100% on the left side of the triangle, and all we need to do is locate the MgO values (we could have chosen the SiO_2 values as well, and we would have plotted the identical compositional points). The right-hand side of the triangle extends between 100% SiO_2 and 100% FeO, and the direction of increasing FeO (from 0% to 100%) is along

[3]From Keuffel and Esser Co.

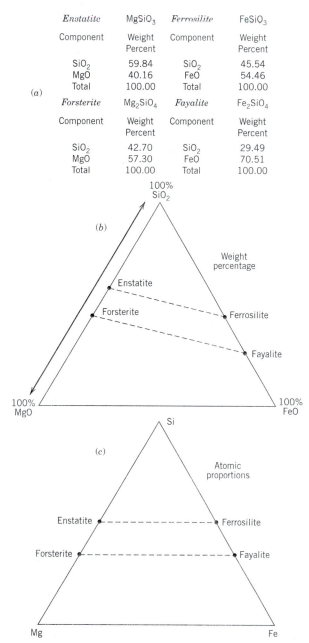

Enstatite	MgSiO₃	*Ferrosilite*	FeSiO₃
Component	Weight Percent	Component	Weight Percent
SiO₂	59.84	SiO₂	45.54
MgO	40.16	FeO	54.46
Total	100.00	Total	100.00
Forsterite	Mg₂SiO₄	*Fayalite*	Fe₂SiO₄
Component	Weight Percent	Component	Weight Percent
SiO₂	42.70	SiO₂	29.49
MgO	57.30	FeO	70.51
Total	100.00	Total	100.00

FIG. 3.63 Plotting of some simple silicate compositions on triangular coordinate paper.

the right side from top to bottom. Here we have located the compositions of the minerals in (a) on the basis of the given FeO weight percentage values. We have joined the end members within each series by a dashed line to show where intermediate compositions would plot. The two lines are not quite parallel because of the slight differences in MgO content (57.30 − 40.16 = 17.14 weight percent) and the FeO content (70.51 − 54.46 = 16.05 weight percent) of the Mg and Fe end member compositions. The final plot, in weight percent, is easy enough to

achieve *if* indeed we have the weight percentage values at hand. In Fig. 3.63c we have chosen the elements Si, Mg, and Fe for the corners. In order to plot the four minerals in terms of cation proportions we might conclude that each of the mineral analyses would have to be recalculated from weight percentage, via molecular proportions, to cation proportions, and so on, as shown in Tables 3.17 and 3.18 for two different silicates. This is the correct procedure for complex mineral compositions; for simple end member compositions for which we know the exact formulas, however, that is unnecessary. Let us inspect the formulas.

Enstatite contains one Mg and one Si atom (or ion) per formula. The same ratio holds for ferrosilite, with one Fe and one Si. In other words, enstatite contains one Mg out of a total of two cations (Mg + Si = 2); the situation is analogous in ferrosilite with one Si to one Fe. On such a scale our compositions will plot at one-half for Mg, Fe, or Si along the two slanting sides of the triangle. If the components had been expressed as 100% Si, 100% Mg, and 100% Fe, the points would have been at the 50% : 50% locations. The plotted positions are shown and have been joined by a dashed line that is parallel to the bottom edge of the triangle.

Now let us plot the two compositions in the olivine series. Forsterite contains two Mg and one Si, which is a total of three cations; fayalite contains two Fe and one Si, again a total of three cations. This means that the plotted compositions must lie at two-thirds of Mg (or Fe) from the 1 Mg (or 1 Fe) corners. In terms of percentages, the Mg and Fe values would locate at 66.6%. These two points are shown and have also been connected by a dashed line, which again is parallel to the base, and as such also parallel to the line for intermediate compositions in the orthopyroxene series. A weight percentage diagram, as in Fig. 3.63b, is commonly used in the plotting of rock and mineral compositions in igneous rocks, because the investigator is mainly interested in the possible changes in composition of the melt from which an igneous rock formed. However, mineralogists and petrologists who deal with sedimentary and metamorphic rocks much prefer the atomic representation. Such a diagram is especially easy to construct if end member formulas are known or are easily located. The regularity of the diagram (with parallel lines) is also easy to remember.

Now that we have plotted simple compositions let us see what is involved if we wish to plot the composition of a pyroxene as listed in Table 3.17. This analysis has been recalculated in terms of three components, CaSiO₃ (wollastonite), MgSiO₃ (enstatite), and $FeSiO_3$ (ferrosilite). The ratios of these three components are the same as they would be for CaO, MgO, and FeO (these two sets of alternate components are completely interchangeable). In Fig. 3.64 we will use the three mineral components, wollastonite (Wo), enstatite (En), and ferrosilite (Fs), which mark the corners of the triangular graph paper. Any pyroxene composition that involves only two of the three components can be plotted along an edge of the triangle, identical to the procedure in the bar graphs of Fig. 3.62. For example, diopside ($CaMgSi_2O_6$) can be plotted along the left edge of the triangle: $1CaSiO_3 + 1MgSiO_3 = 1CaMgSi_2O_6$. Similarly, hedenbergite ($CaFeSi_2O_6$) can be plotted along the right edge, according to $1CaSiO_3 + 1FeSiO_3 = 1CaFeSi_2O_6$. Compositions along the bottom edge can be expressed in general as $(Mg, Fe)SiO_3$. A specific compositional location can be obtained from a formulation $(Fe_{0.80}Mg_{0.20})SiO_3$, which can be restated as 80 molecular percent ferrosilite (Fs) and 20 molecular percent enstatite (En). This composition is shown in Fig. 3.64.

A more general pyroxene composition that shows the presence of all three components may be expressed as $Wo_{45}En_{20}Fs_{35}$. This is shorthand for a pyroxene containing 45 molecular percent CaO(Wo), 20 molecular percent MgO(En), and 35 molecular percent FeO(Fs). In order to locate this composition on the triangular graph paper, note that the join En-Wo represents 0% Fs; similarly the En-Fs join represents 0% Wo, and so forth. The distance between the corner and the opposite side of the triangle is graduated in "%" lines from 100% at the corner to 0% along the join. For example, the $CaSiO_3$ corner represents 100% $CaSiO_3$ and horizontal lines between this corner and the base of the triangle mark the change from 100% to 0% $CaSiO_3$. If one wishes to plot $Wo_{45}En_{20}Fs_{35}$, locate first the Wo_{45} line (marked on Fig. 3.64), then locate the En_{20} line (also marked). Where the two lines intersect, the composition is located on the Fs_{35} line. The position of the composition, $Wo_{50}En_{43}Fs_7$, from Table 3.17 is also shown in Fig. 3.64.

Like bar graphs, triangular diagrams are frequently used to depict the extent of solid solution in minerals. Figure 3.65 shows the extent of solid solution among minerals in the chemical system SiO_2-MgO-FeO. In this representation we can show the olivine series $(Mg,Fe)_2SiO_4$ and the orthorpyroxene series $(Mg,Fe)SiO_3$, both of which show an essentially complete solid solution between the end member compositions. SiO_2 (quartz and its various polymorphs) shows no substitution in terms of FeO

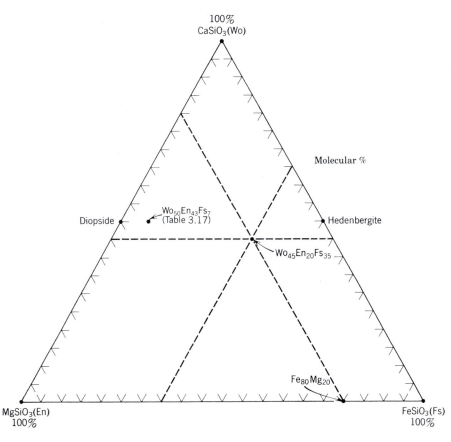

FIG. 3.64 Plotting of chemical compositions on triangular coordinate paper. The specific compositions represented are for the pyroxene family of silicates. See text for explanation of plotting procedure.

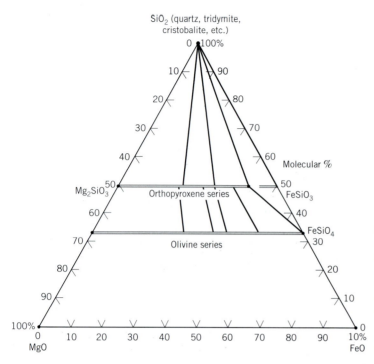

FIG. 3.65 Triangular representation in terms of molecular percent of minerals in the chemical system MgO–FeO–SiO_2. The olivine and orthopyroxene series show complete solid solution series between Mg and Fe end members, as shown by the continuous bars. Lines connecting SiO_2 to various orthopyroxene compositions, and between orthopyroxene and olivine compositions, are *tielines*; these connect coexisting minerals in rocks. The triangle outlines a cristobalite-orthopyroxene (Fs_{83})-fayalite coexistence.

or MgO and is drawn as a point, indicating its constant composition. Forsterite (Mg_2SiO_4) can be written as $2MgO + 1SiO_2 = 1Mg_2SiO_4$; of the three oxide "molecules," one is SiO_2 and the other two are MgO. On the left side of the triangle, Fo is therefore located at $\frac{1}{3}$ away from the MgO corner. Enstatite ($MgSiO_3$) can be written as $1MgO + 1SiO_2 =$

$1MgSiO_3$, which is located halfway between the MgO and SiO_2 corners. The plotting process here is identical to that in Fig. 3.63*c*.

In addition to depicting graphically the extent of solid solution, triangular diagrams are useful in showing the minerals that make up (coexist in) a specific rock type. In such diagrams, known as *assemblage*

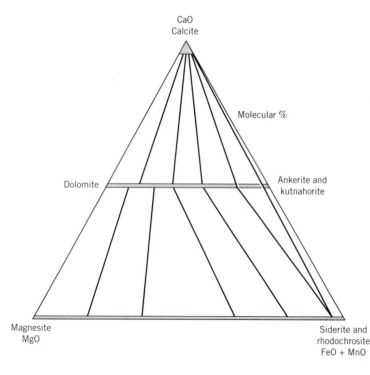

FIG. 3.66 Compositions of carbonates in the system CaO-MgO-FeO-MnO-CO_2. The CO_2 component is not considered in this diagram. ($FeO + MnO$) is a combined component. A complete solid solution series exists between dolomite and ankerite. A complete series also exists between magnesite and siderite. Tielines connect possible coexisting carbonate members. The three-mineral triangle portrays the coexistence of calcite-ankerite-siderite.

diagrams, the minerals that coexist with each other (i.e., touch each other along the borders of their grains) are connected by *tielines*. Such tielines denote the fact that two, three, or four minerals are found next to each other in a rock. In Fig. 3.65 some tielines have been drawn for possible olivine-orthopyroxene pairs, as well as for an assemblage, in a different rock, of cristobalite, Fe-rich orthopyroxene, and fayalite. This mineral assemblage could be found in a high-temperature, Fe-rich basalt. Such a three-mineral coexistence is denoted by an *assemblage triangle*, of which all of the three sides are tielines.

Triangular diagrams would tend to limit us to representation of only three components, which may be simple elements, compound oxides, or more complex components as expressed by mineral formulas. In order to represent more than three components on one triangle, some components are frequently combined and some may not be considered in the graphic representation. For example, in an attempt to represent carbonate compositions in the system $CaO-MgO-FeO-MnO-CO_2$, we must somehow reduce the possible number of compositional variables. Because all carbonates contain CO_2 there will be little gained in using the CO_2 component to show the small CO_2 variations that do exist. We will therefore ignore CO_2 in the graphic representation. We may further decide to combine the FeO and MnO because Fe^{2+} and Mn^{2+} substitute freely for each other in the carbonate structure and because Mn^{2+} typically is far less abundant in most environments than Fe^{2+}. This leaves three components, CaO, MgO, and (FeO + MnO). Figure 3.66 shows that a complete solid solution series may exist between magnesite ($MgCO_3$) and siderite + rhodochrosite ($FeCO_3$ + $MnCO_3$); the completeness of this series is well documented (see also Fig. 10.16). It also shows an almost complete series between dolomite, $CaMg(CO_3)_2$, and ankerite, $CaFe(CO_3)_2$ + kutnahorite, $CaMn(CO_3)_2$. Calcite ($CaCO_3$) shows a very limited ionic substitution by Mg, Fe, and Mn. Tielines show coexistences of magnesite-dolomite, calcite-dolomite, and calcite-ankerite-siderite in the triangle. The three-mineral assemblage is not uncommon in banded Precambrian iron-formations. Additional discussion of the procedures in plotting mineral compositions on triangular diagrams, and exercises on the graphic representa-

tion of mineral compositions, are given in Klein (1994).

REFERENCES AND SELECTED READING

Ahrens, L. H. 1952. The use of ionization potentials. *Geochimica et Cosmochimica Acta* 2: 155–69.

Bloss, F. D. 1994. *Crystallography and crystal chemistry: An introduction.* Reprint of original text of 1971. Washington, D.C.: Mineralogical Society of America.

Brady, J. E., J. W. Russell, and J. R. Holum. 2000. *Chemistry, matter, and its changes.* 3rd ed. New York: Wiley.

Bragg, W. L., and G. F. Claringbull. 1965. *Crystal structure of minerals.* Ithaca, N.Y.: Cornell University Press.

Clarke, F. W., and H. S. Washington. 1924. *The composition of Earth's crust.* U.S. Geological Survey Professional Paper 127.

Cotton, F. A., G. Wilkinson, and P. L. Gaus. 1995. *Basic inorganic chemistry.* 3rd ed. New York: Wiley.

Douglas, B., D. McDaniel, and J. Alexander. 1994. *Concepts and models of inorganic chemistry.* 3rd ed. New York: Wiley.

Evans, R. C. 1966. *An introduction to crystal chemistry.* 2nd ed. Cambridge, England: Cambridge University Press.

Griffen, D. T. 1992. *Silicate crystal chemistry.* New York: Oxford University Press.

Huheey, J. E., E. A. Keiter, and R. L. Keiter. 1993. *Inorganic chemistry: Principles of structure and reactivity.* 4th ed. New York: Harper Collins.

Klein, C. 1994. *Minerals and rocks: Exercises in crystallography, mineralogy, and hand specimen petrology.* Rev. ed. New York: Wiley.

Kotz, J. C., and P. Triechel, Jr. 1999. *Chemistry and chemical reactivity.* 4th ed. Orlando, Fla.: Saunders College Publishing.

Liebau, F. 1985. *Structural chemistry of silicates, structure, bonding and classification.* New York: Springer-Verlag.

Mason, B., and C. B. Moore. 1982. *Principles of geochemistry.* 4th ed. New York: Wiley.

Pauling, L. 1960. *The nature of the chemical bond.* 3rd ed. Ithaca, N.Y.: Cornell University Press.

Pimentel, G. C., and R. D. Spratley. 1969. *Chemical bonding clarified through quantum mechanics.* San Francisco: Holden Day.

Shannon, R. D., and C. T. Prewitt. 1969. Effective ionic radii in oxides and flourides. *Acta Crystallographica* 25: 925–46.

Shannon, R. D. 1976. Revised effective ionic radii and systematic studies of interatomic distances in halides and chalcogenides. Acta Crystallographica A32: 751–767.

Silberberg, M. S. 2000. *Chemistry, the molecular nature of matter and change.* 2nd ed. New York: McGraw Hill.

Smyth, J. R., and D. L. Bish. 1988. *Crystal structures and cation sites of the rock-forming minerals.* Boston: Allen & Unwin.

Wells, A. G. 1991. *Structural inorganic chemistry.* 5th ed. Oxford, England: Clarendon Press.

Zoltai, T. 1960. Classification of silicates and other minerals, with tetrahedral structures. *American Mineralogist* 45: 960–73.

CHAPTER 4

MINERAL REACTIONS, STABILITY, AND BEHAVIOR

An important objective in the study of minerals is the deduction of the processes they have undergone as a result of their geologic past. Such processes may have involved changes in temperature or in pressure, or both, as well as changes in their chemical environment. Minerals that have formed from the solidification of molten material (a melt) are referred to as igneous in origin; those that are the result of reactions among minerals in preexisting rock are known as metamorphic; and those that originate from the deposition of material from water or air are referred to as sedimentary. Igneous minerals commonly crystallize from a magma as a result of lowering of temperature in the magma chamber. These minerals react with the melt, as well as with each other, as the temperature continues to fall. Internal structural changes may occur within these same igneous minerals as well as a function of changing (generally falling) temperature. Metamorphic minerals are the result of chemical reactions among solids and of transformations within minerals in the solid state in response to changes in temperature or pressure, or both, in the metamorphic environment. Sedimentary minerals are the result of reactions of preexisting minerals with the atmosphere or hydrosphere and reflect a low and rather narrow range of temperature for their origin.

The objective of this chapter is to introduce *crystallization* and *mineral reactions* (chemical as well as structural in nature) and to assess some of these reactions in terms of quantitative evaluations of temperature and pressure as presented in mineral *stability* (or *phase*) *diagrams*. This, by necessity, leads to a brief glimpse of the important and extensive subject of thermodynamics. That is followed by a discussion of *polymorphism*, which is the process by which a mineral of constant composition rearranges its internal structure in response to changes in temperature, or pressure, or both, in the surrounding geologic environment. This leads to the subject of *polytypism*, which represents a much more restrictive type of reorganization of a structure than in polymorphism. It is common in layered structures.

A variety of other features indicate specific processes that minerals have undergone. For instance, in the discussion of solid solution in minerals, in Chapter 3, it was pointed out that an increase in temperature commonly leads to an increase in the extent of solid solution. Conversely, a decrease in temperature applied to the crystal structure of a mineral may reduce the extent of solid solution that a specific structure can tolerate. This leads to the process of *exsolution*. Although almost all minerals show highly ordered internal atomic structures, some, due to internal "self irradiation" or external irradiation by radioactive elements, become partly or wholly amorphous, also known as *metamict*. The process responsible for such structural changes is discussed under the *metamict state*. Several processes produce atomic defects in crystalline substances, which are now directly resolvable by the use of electron microscope structure images. Such *defect structures* and their origin are evaluated in this chapter as well.

The chapter concludes with discussion of the origin of *twinning*, of *color*, of *magnetic properties*, and of *radioactivity*.

CRYSTALLIZATION

Crystals are formed from solutions, melts, and vapors. The atoms in these disordered states have a random distribution, but with changing temperature, pressure, and concentration they may join in an ordered arrangement characteristic of the crystalline state.

As an example of crystallization from solution consider sodium chloride dissolved in water (see Fig. 4.1). If the water is allowed to evaporate, the solution contains more and more Na^+ and Cl^- per unit volume. Ultimately, the point is reached when the remaining water can no longer retain all the salt in solution, and solid salt begins to precipitate. If the evaporation of the water is very slow, the ions of sodium and chlorine will group themselves together to form one or a few crystals with characteristic

FIG. 4.1. Schematic representation of a *nucleus* of NaCl in an evaporating saline lake. The nucleus, as drawn, consists of only 125 regularly packed ions (Na^+ and Cl^-). Additional ions will accrete, in an ordered array, on the outside of the nucleus cube, thus allowing it to grow to a larger crystal. The scale used for ionic size is the angstrom unit ($1\text{Å} = 10^{-8}$ cm). A cubic crystal of NaCl, 1 cm along each edge, would contain approximately 10^{23} ions or atoms.

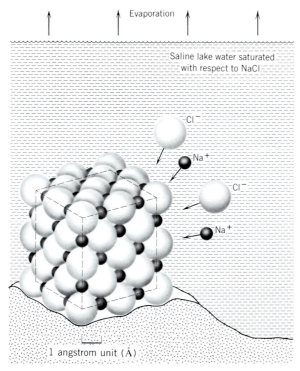

shapes and often with a common orientation. If evaporation is rapid, many centers of crystallization will be set up, usually resulting in many small, randomly oriented crystals.

Crystals also form from solution by lowering the temperature or pressure. Hot water will dissolve sightly more salt, for instance, than cold water; and, if a hot solution is allowed to cool, a point may be reached where the solution becomes sufficiently concentrated that salt will crystallize. Again, the higher the pressure, the more salt water can hold in solution. Thus, lowering the pressure of a saturated solution will result in supersaturation and crystals will form. *Therefore, in general, crystals may form from a solution by evaporation of the solvent, by lowering the temperature, or by decreasing the pressure.*

A crystal is formed from a melt in much the same way as from a solution. The most familiar example of crystallization from fusion is the formation of ice crystals when water freezes. Although it is not ordinarily considered in this way, water is fused ice. When the temperature is lowered sufficiently, the H_2O molecules, which were free to move in any direction in the liquid state, become fixed and arrange themselves in a definite order to build up a solid, crystalline mass. The formation of igneous rocks from molten magmas, though more complicated, is similar to the freezing of water. In a magma many of the ions of the elements are in an uncombined state, although there is considerable cross-linking of ions and ionic groups. The crystal growth in a cooling magma is the result of two competing tendencies: (1) thermal vibrations that tend to destroy the nuclei of potential minerals, and (2) attractive forces that tend to aggregate atoms (and/or ions) into crystal structures. As the temperature falls, the effect of the first tendency diminishes, which allows the effect of the attractive forces to dominate.

Although crystallization from a vapor is less common than from a solution or a melt, the underlying principles are much the same. As the vapor is cooled, the dissociated atoms or molecules are brought closer together, eventually locking themselves into a crystalline solid. Familiar examples of this mode of crystallization are the formation of snowflakes from air laden with water vapor, and the formation of sulfur crystals at the base of fumaroles, or the neck of volcanoes.

Crystal Growth

Anyone who has collected minerals or has viewed mineralogical exhibits in museums, in rock shops, or in jeweler's display cases knows the aesthetic beauty and attraction of well-formed crystals. Most of these crystals are the result of chemical deposition from a solution (or a melt) in an open space, such as a vug or cavity in a rock formation.

The question that arises is, how do such well-formed crystals grow from small to larger ones? Or one might rephrase the same question in chemical terms, such as, how are the chemical building blocks (atoms, ions, or ionic clusters) incorporated into a well-ordered crystalline pattern?

Here we will briefly discuss some of the most basic aspects of crystal growth. The first stage in the growth of a crystal is that of *nucleation*, which implies that growth can commence only after a *nucleus* (or *seed*) has been formed. In most cases, the *nuclei* are the initial products of precipitation (in a water-rich environment) or of crystallization (as in a melt). The *nucleus* is the result of the coming together of various ions (in the solution or melt) to form the initial regular structural pattern of a crystalline solid. For example, in an evaporating salt lake, the conditions might be appropriate for the random precipitation of nuclei of NaCl. This means that Na^+ and Cl^- ions in the lake water are combining with each other in a regular cubic array of alternating ions of Na^+ and Cl^- as required by the structure of NaCl (halite, or rock salt; see Fig. 4.1). The formation of a single crystal of halite is generally preceded by the random formation of large numbers of potential nuclei. Most nuclei may not lead to further crystal growth, because in a saturated solution (saturated with respect to Na^+ and Cl^- ions) there is also a tendency for nuclei to go back into solution (to be redissolved). This is because these very small beginnings of an ordered structure have a very large surface area with respect to volume. A high surface area implies that there are many atoms (on the outer surface of the crystal) with unsatisfied chemical bonds (see Fig. 4.2). Such a crystal (or mineral grain) with a high surface area is more soluble than a crystal (or mineral grain) with a high volume, in which most of the atoms are internal with completely satisfied chemical bonds.

If a nucleus is to survive, it must grow rapidly enough to reduce its surface energy (calculated as the ratio of surface area to volume) and thus its solubility. If a nucleus reaches a *critical size* through rapid deposition of further layers of ions, it will have a high chance of surviving as a larger crystal. An idealized picture of crystal growth is one of enlargement of the nucleus by the ordered accretion of additional ions to its outer surfaces (see Fig. 4.1). Such a picture, we now know, is too simple. The outer solid surface of a nucleus (or crystal) in contact with a saturated solution represents a surface of unsatisfied chemical bonds (see Fig. 4.2). The energy of such a surface is lowered when an atom attaches itself to it, and the amount of energy released by such an attachment depends on where such an attachment occurs. For example, in Fig. 4.2*b* the "step" location on the surface of the crystal is where it can lose energy through the attachment by additional ions. This is because in ionically bonded crystals, as in halite (NaCl), the energy of attachment is greatest at corners, intermediate at edges, and least in the middle of faces. The greater attraction of atoms to the corners of ionic crystals commonly leads to rapid growth in these directions. This may result in the for-

FIG. 4.2. *(a)* A section through a corner of an NaCl crystal showing well-bonded, closely packed ions in the internal part of the crystal and unsatisfied chemical bonds at the outer surfaces of the crystal (∼ represents unsatisfied and = represents satisfied chemical bonds). *(b)* A crystal surface showing a submicroscopic step. Attachment of ions at the location of such a step lowers the energy of the crystal surface. This energy is the cumulative result of the unsatisfied bonds. *(c)* Submicroscopic clumps of atoms, shown as blocks, attached to the outer three surfaces of a crystal. Such blocks create steps for the attachment of new layers of ions on the outer surfaces of the crystal.

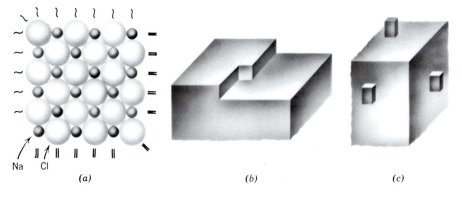

Na   Cl
(a)   (b)   (c)

mation of *dendrites* with branches radiating from a central core. In some types of crystals, which differ from NaCl because of their nonionic bonding, it is thought that atoms accrete on the outer surface as clumps of atoms. Such clumps, shown as blocks in Fig. 4.2c, provide the outer surface with steps along which a new outer layer of crystal can be built up.

In this brief introduction to the subject of crystal growth it is implied that the addition of ions (or atoms, or clumps of atoms) to the outside of a crystal occurs in a regular and continuous pattern. It is now known that this is commonly not the case. Indeed, natural and synthetic crystals commonly contain *imperfections*. These will be discussed more fully on p. 151.

Intergrowths of Crystals

The previous discussion has concentrated on the growth of individual crystals in an open (unconstrained) space, which may lead to the formation of well-formed (euhedral), relatively large single crystals. However, most minerals occur as *random aggregates* of anhedral grains (lacking external crystal faces) in the rocks of the Earth. There are, however, some relatively common intergrowth patterns of well-formed crystals (as well as anhedral grains) that are *not* random in nature. Such are *parallel growths* (see Fig. 2.5) and crystallographically controlled intergrowths of the same substance known as *twins* or *twinned crystals* (see Fig. 2.6). A third type of nonrandom (crystallographically oriented) overgrowth of one crystalline substance on another of different composition is known as *epitaxis*. Although the two intergrown crystals will tend to have different structures (and unit cell sizes) because they are compositionally distinct, there may be planes in their internal structures where there is a good fit (or the least amount of misfit) between the two individuals. As shown in Fig. 4.3a staurolite (with monoclinic symmetry) occurs in parallel growth with kyanite (with triclinic symmetry) with the (010)[1] plane of staurolite parallel with the (100) plane of kyanite. This has come about because the atomic spacing in the (010) plane in staurolite is similar to that in the (100) plane in kyanite. Figure 4.3b illustrates the epitaxial overgrowth of a plagioclase feldspar (oligoclase) on a crystal of microcline. The common direction between the two different structures is identified by the forward sloping arrows that are parallel to a crystal-

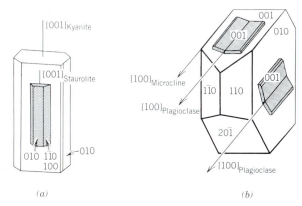

FIG. 4.3. Examples of epitaxis. *(a)* Parallel growth of staurolite ($Fe_2Al_9O_6(SiO_4)_4(O,OH)_2$) and kyanite ($Al_2SiO_5$). *(b)* Oligoclase ($NaAlSi_3O_8$ with about 13% substitution by $CaAl_2Si_2O_8$) overgrowths on microcline ($KAlSi_3O_8$). (This figure from R. Kern and R. Gindt, 1958, *Bulletin Societé Française Min. Cryst.*, 81: 264.)

lographic direction identified as [100]. The planes of attachment are (001) in plagioclase and (001) and (010) in microcline. These three planes have a fairly good (but not perfect) fit between them in terms of the internal structure and atomic spacing for the two minerals involved.

MINERAL REACTIONS

This section will present a brief overview of some of the mineral reactions that take place in igneous melts as a function of decreasing temperature; in metamorphic settings; in sedimentary rocks; and of reactions that are suggested for the ultrahigh-pressure conditions in the Earth's mantle.

Reactions in an Igneous Regime

Igneous rocks make up approximately 95 percent of the upper 10 miles (16 km) of the Earth's crust, but their great abundance is hidden on the Earth's surface by a relatively thin but widespread layer of sedimentary and metamorphic rocks. Igneous rocks have crystallized from a silicate melt, known as a *magma*. The high temperature (about 900° to 1600°C) necessary to generate magmas, as concluded from melting and cooling experiments in the laboratory and from the similarly high temperatures measured in the boiling magmas of active volcanoes, suggests that their source lies deep in the Earth's crust. A magma is principally made up of O, Si, Al, Fe, Ca, Mg, Na, and K, but it also contains considerable amounts of H_2O and CO_2, as well as lesser gaseous components such as H_2S, HCl, CH_4, and CO. As a magma cools there

[1]These triple digit symbols inside parentheses are Miller indices and are discussed on p. 198.

is generally a definite order of crystallization of the various mineral constituents. In a magma consisting mainly of O, Si, Mg, and Fe, for example, the mineral with the highest melting point, Mg-rich olivine, Mg_2SiO_4, would crystallize first, followed by more Fe-rich olivine approaching fayalite, Fe_2SiO_4. The forsterite–fayalite series is a total solid solution series. Many other rock-forming minerals are total solid solution series as well. In such compositional series, crystallization takes place over a range of temperatures and the solids that separate from the melt may have a considerable range in composition. In general, the crystallization of a magma is the result of two types of reactions: *continuous reactions*, in solid solution series, whereby the composition of early-formed crystals changes continuously due to interaction with the surrounding melt, and *discontinuous reactions*, in which early-formed crystals react with the melt to give rise to new minerals with different crystal structures and chemical compositions. Such continuous and discontinuous reactions, as well as separation of magma and crystals, lead to what is known as *magmatic differentiation*.

This concept was first developed by N. L. Bowen, an American petrologist, on the basis of his studies of the textures and mineralogical makeup of many igneous rock types and his correlative experimental studies. Mg-Fe-rich igneous silicates constitute a series of mineral groups that are related to

each other by discontinuous reactions. For example, early-formed olivine may be rimmed by pyroxene; aphiboles may form rims at the expense of the earlier formed pyroxene; and biotite may form as a reaction product of the earlier crystallized amphiboles. On the other hand, members of the plagioclase feldspar series represent a continuous reaction series in which Ca-rich plagioclase crystallizes early from the melt, enriching the residual melt in alkalies (Na and K).

Figure 4.4 illustrates schematically what is known as *Bowen's reaction series*. As a magma of basaltic composition cools, olivine and calcic plagioclase (with a high anorthite component) may crystallize first. If these minerals remain in contact with the melt, they will tend to form pyroxene and more sodic plagioclase (with an albite component). The resulting rock will be a gabbro or a basalt. If, however, the bulk of the early olivine and the calcic plagioclase is removed from the melt, as by crystal settling, the bulk composition of the remaining melt will tend to become enriched in Si, Al, Fe^{2+}, alkalies, and H_2O and CO_2. Such a melt could produce a mineral assemblage consisting mainly of amphiboles, some mica, alkali feldspar, and SiO_2. Figure 4.4 shows that amphiboles, micas, alkali feldspars, and quartz are relatively lower-temperature, late-stage crystallization products. It is of interest to note that the discontinuous reaction series reflect increasing complexity in the linkage of (SiO_4) tetrahedra in the silicates, be-

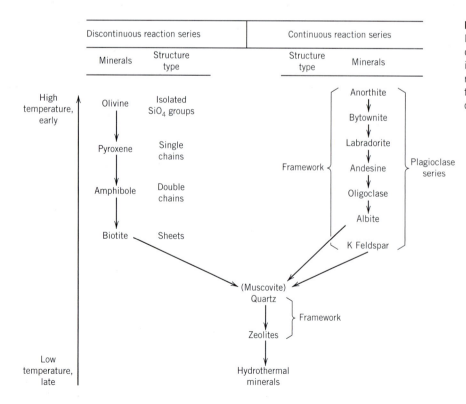

FIG. 4.4. Bowen's reaction series, relating changes in mineral groups (in discontinuous reactions) and changes in mineral composition (in continuous reactions) to a general decrease in temperature during the crystallization of a cooling magma.

ginning (at high temperature) with isolated tetrahedra as in olivine and ending (at low temperature) with framework structures such as quartz and zeolites. Intermediate architectural schemes are represented by pyroxenes with single chain structures, amphiboles with double chain structures, and micas with sheet structures (see also Chapter 11).

Many of the mineral groups depicted in Fig. 4.4 are anhydrous: olivine, pyroxene, feldspar, and quartz. Hydrous mineral groups are amphibole, mica, and zeolites. Therefore, if during the cooling of a melt pyroxenes crystallize first but subsequently give way, in part, to amphiboles, the crystallizing amphiboles have acquired (OH) groups from the remaining melt. Upon further crystallization and chemical interaction with the remaining melt, the amphiboles may give way, in part, to hydrous layer silicates such as mica or talc.

Reactions Under Metamorphic Conditions

Metamorphic rocks are derived from preexisting rocks (igneous, sedimentary, or metamorphic) by mineralogical, textural, and structural changes. Such changes may be the result of marked variations in temperature, pressure, and shearing stress at considerable depth in the Earth's crust. Weathering effects, at atmospheric conditions, are not considered part of metamorphism, nor are chemical reactions involving partial melting, because these are part of igneous processes. Metamorphic changes such as recrystallization and chemical reactions among mineral constituents take place essentially in the solid state, although the solids may exchange chemical species with small amounts of a liquid phase consisting mainly of H_2O (as water, steam, or supercritical fluid, depending on the temperature and pressure at which the reactions take place). The general conditions of formation of metamorphic rocks lie between those of sedimentary rocks that form at essentially atmospheric conditions of T and P and those of igneous rocks that are the result of crystallization from a melt at high T. Metamorphic rocks may be the result also of very large changes in pressure in conjunction with increasing metamorphic temperature; mineral assemblages in the Earth's mantle have formed in response to very high confining pressures (see page 111). Excluding gain or loss of H_2O, many metamorphic reactions are generally considered to be essentially *isochemical*; this implies that during recrystallization and the process of chemical reactions the bulk chemistry of rocks has remained essentially constant. If this is not the case, and if additional elements have been introduced into the

rock—by circulating fluids, for example—it is said to have undergone *metasomatism*.

In general, metamorphic rocks can be divided into two groups: (1) those formed by contact metamorphism and (2) those formed by regional metamorphism. *Contact metamorphic rocks* occur as concentric zones (aureoles) around hot igneous intrusive bodies. Such metamorphic rocks lack schistosity, and the relatively large temperature gradient from the intrusive contact to the unaffected country rock gives rise to zones that may differ greatly in mineral assemblages. Sandstone are converted to *quartzites*, and shales are changed to *hornfels*, a fine-grained dense rock. *Regional metamorphic rocks* are the result of increases in T or P, or both, on a regional scale (areas a few hundred to thousand of miles in extent) in response to mountain building or deep burial of rocks. Contact and regionally metamorphosed rocks generally reflect an increase in temperature (*progressive* or *prograde* metamorphism) in their assemblages. However, when assemblages of high-temperature origin (e.g., igneous or high-temperature metamorphic rocks) fail to survive conditions of lower-temperature metamorphism, the process is referred to as *retrograde* (or *retrogressive*) metamorphism.

Although metamorphic reactions take place without addition of chemical species to the rock, some chemical components may be lost, especially during the higher-temperature ranges of metamorphism. For example, a shale—largely composed of hydrous silicates such as clays—may convert into a slate and subsequently into a schist at increasing metamorphic temperatures, during which processes the mineral assemblage becomes less and less hydrous due to progressive *dehydration*. In other words, with an increase in metamorphic temperatures water is generally lost from the rock. Similarly, carbonate-rich sedimentary rocks while undergoing increasing temperatures of metamorphism will tend to lose CO_2, which is known as *decarbonation*. Such processes can be illustrated as follows:

$$Al_2Si_2O_5(OH)_4 \rightarrow Al_2SiO_5 + SiO_2 + H_2O,$$

kaolinite   andalusite quartz

where andalusite is one of three structural forms (polymorphs, see p. 134) of Al_2SiO_5, and

$$CaCO_3 + SiO_2 \rightarrow CaSiO_3 + CO_2$$

calcite   chert   wollastonite

Although such losses of H_2O and CO_2 to a thin layer of fluid phase surrounding the mineral grains are common during the conditions of progressive metamorphism, additions of chemical species from the fluid phase to the minerals are considered part of metasomatic processes.

A transmission electron microscope (TEM) photograph of an arrested prograde metamorphic reaction in which muscovite gives way to several reaction products that include potassium feldspar, biotite, spinel, and corundum is shown in Figure 4.5. This arrested reaction shows the original reactant, muscovite (with some Mg and Fe replacing octahedral Al in the structure), and two of the products, potassium feldspar and biotite. The complete reaction is as follows:

$$\text{muscovite (containing some Mg and Fe)} \rightarrow$$
$$\rightarrow \text{K-feldspar} + \text{biotite} + \text{spinel} + \text{corundum} + H_2O$$

An example of an arrested retrograde metamorphic reaction of an original high-temperature pyroxene (of enstatite composition, $MgSiO_3$, but with monoclinic symmetry, referred to as clinoenstatite) altering to a complex mixture of hydrous silicate products (including *biopyriboles*, see p. 460), is shown in a

FIG. 4.5. Transmission electron micrograph of an original muscovite (Mu) that has partially reacted to biotite platelets (Bi) and potassium feldspar (kf) as a result of metamorphism between 750° and 900°C. The other reaction products mentioned in the text (spinel and corundum) are not present in this image (From A. J. Brearley, 1986, An electron optical study of muscovite breakdown in pelitic xenoliths during pyrometamorphism, *Mineralogical Magazine* 50: 385–97).

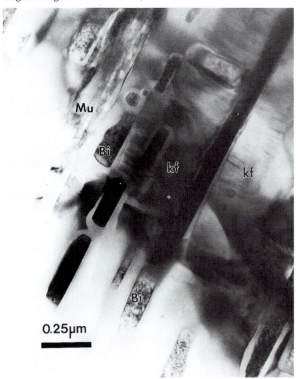

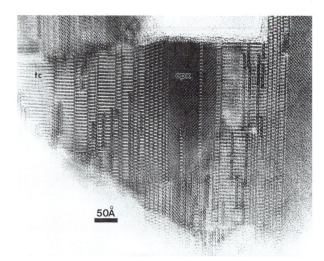

FIG. 4.6. High resolution transmission electron microscope (HRTEM) image of an arrested retrograde reaction in which the original, high-temperature pyroxene has transformed into a mixture of amphiboles, "biopyriboles" as well as talc. The original pyroxene is identified by cpx (clinopyroxene; it is a monoclinic form with enstatite composition). The amphiboles are the thinnest vertical strips; the "biopyriboles" are the various "zipper-like" strips, with variable widths; and talc (tc) is present in the far left corner. This mixture of silicates occurs in the Allende meteorite. (Courtesy A. J. Brearley, Department of Earth and Planetary Sciences, University of New Mexico.)

high-resolution transmission electron microscope (HRTEM) image in Figure 4.6. The original clinoenstatite (marked as cpx, which is an abbreviation for the generic term *clinopyroxene*) is riddled with lamellae that represent silicate structures with chains of variable width (these are wider than the structural chain units that are found in the structure of the original pyroxene). The homogeneous wide band at the left is the ultimate, stable alteration product talc. A qualitative equation for this reaction is as follows:

$$x MgSiO_3 + n H_2O \rightarrow y Mg_7Si_8O_{22}(OH)_2$$
$$\text{clinoenstatite} \qquad\qquad \text{anthopyllite}$$
$$+ \text{"biopyriboles"} + z Mg_3Si_4O_{10}(OH)_2$$
$$\text{talc}$$

Reactions in a Weathering Environment

The minerals of sedimentary rocks can be divided into two major groups: minerals that are resistant to the mechanical and chemical breakdown of the weathering cycle and minerals that are newly formed from the products of chemical weathering. The relative stabilities of minerals to weathering are shown in Fig. 4.7. This listing means that quartz is one of the chemically and mechanically most resistant minerals, whereas olivine is easily altered chemically. The

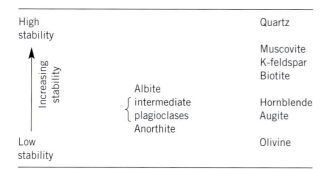

FIG. 4.7. Relative stabilities of some rock-forming silicates in the weathering cycle.

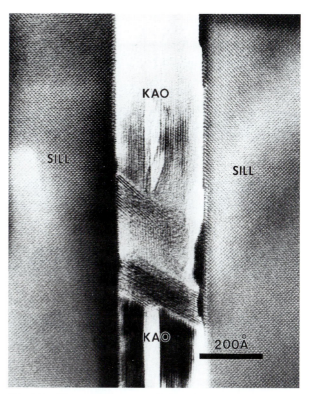

FIG. 4.8. High resolution transmission electron microscope (HRTEM) image of the arrested reaction of sillimanite (sill) altering to kaolinite (kao) during low-temperature, hydrous conditions. (Courtesy of A. J. Brearley, Department of Earth and Planetary Sciences, University of New Mexico.)

positions of minerals listed between quartz and olivine represent intermediate stabilities ("survival rates") during the weathering process. It is of interest to note that this sequence is very similar to Bowen's reaction series in reverse (see Fig. 4.4). This similarity indicates that the minerals formed at the lowest temperature in the crystallization of a melt are also the most stable at ordinary temperatures and pressures (atmospheric conditions).

Important chemical breakdown reactions involving high-temperature igneous (or metamorphic) silicates have reaction products that fall into three categories: (1) layer silicates such as kaolinite and montmorillonite, (2) silica in solution as H_4SiO_4, and (3) Na^+, K^+, Ca^{2+}, and Mg^{2+} ions in solution. Examples of two such reactions are as follows:

$$3KAlSi_3O_8 + 12H_2O + 2H^+ \rightarrow KAl_3Si_3O_{10}(OH)_2$$
orthoclase   muscovite
$$+ 6H_4SiO_4 + 2K^+,$$

in which the muscovite continues its alteration to kaolinite by the following reaction:

$$2KAl_3Si_3O_{10}(OH)_2 + 3H_2O + 2H^+$$
muscovite
$$\rightarrow 3Al_2Si_2O_5(OH)_4 + 2K^+$$
kaolinite

Fig. 4.8 is a high resolution transmission electron microscope (HRTEM) image of an arrested reaction in which sillimanite is giving way to kaolinite according to the following low-temperature hydration reaction:

$$Al_2SiO_5 + SiO_2 + H_2O \rightarrow Al_2Si_2O_5(OH)_4$$
sillimanite quartz   kaolinite

Detrital sedimentary rocks consist mainly of the most resistant rock-forming minerals, quartz, K-feldspar, mica and lesser plagioclase, as well as small amounts of garnet, zircon, and spinel (mag-

netite). The detrital rock types may be regarded as accidental, mechanical mixtures of genetically unrelated resistant minerals.

Chemical sedimentary rocks that result from the inorganic or organic precipitation of minerals can be interpreted in large part in terms of chemical and physico-chemical principles that apply at low temperatures (25°C) and atmospheric pressure. Such chemical sedimentary assemblages, therefore, are not random or accidental but reflect the concentrations of ions in solution, as well as conditions such as temperature, pressure, and salinity of the sedimentary basin.

Ultrahigh-Pressure Reactions

Having concerned ourselves with mineral reactions that occur in the Earth's crust, we may now ask the question: What type of structural arrangements are found in minerals that occur below the Earth's crust, in various parts of the mantle?

Figure 3.1a is a cross-section of the Earth, giving the depth locations of the upper and lower mantle

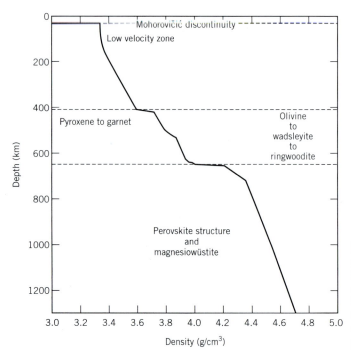

FIG. 4.9. Density profile in the upper mantle as a function of depth. Changes in the structural packing of minerals and in mineral assemblages as a function of the changing density of the mantle. Below 660 km, it is likely that all silicon is in 6-coordination with oxygen. (From L. Stixrude, Unified description of the phase equilibria and physical properties of mantle assemblages, *Journal of Geophysical Research*, in preparation; courtesy L. Stixrude.)

and the transition zone between them. On the left-hand side, this cross-section gives depth in kilometers, and on the right the pressures associated with these depths. Clearly, the overall density of materials in the deeper parts of the Earth must be considerably greater than in the Earth's crust. As such, the atomic packing schemes of common rock-forming minerals, deeper down, must be different from those observed in minerals in crustal rocks.

The average mean density of rocks in the crust is only about 2.8 g/cm^3, whereas the mean density of the Earth as a whole is 5.52 g/cm^3. This means that rocks below the crust (i.e., below the Mohorovičić discontinuity) must increase their density as a function of changing composition or atomic packing of the elements that constitute the minerals. The Earth's core is generally accepted to consist mainly of Fe and Ni, with minor amounts of S, C, and Si. Such a composition satisfies the geophysical parameters of density and magnetism of the core. The mantle, between the Mohorovičić discontinuity (at 36 km below continents) and the core (at 2900 km; see Fig. 3.1*a*), is thought to consist of mixtures of silicates, oxides, minor sulfides, and lesser metal. A density profile for the upper part of the mantle as a function of distance below the Earth's surface is given in Fig. 4.9.

The makeup of the upper part of the mantle (between 36 and 220 km) has been sampled directly in places where rocks have been brought to the Earth's surface by violent volcanic events. Such rocks occur in what are known as *kimberlite pipes*. These kimberlite pipes consist of a mixture of rock types and contain in them several minerals that are the result of reactions at high to ultrahigh pressures. Among them

FIG. 4.10. Photomicrograph of an association of coesite (c), kyanite (k), omphacite (o), and garnet (g). Coesite is a high-pressure polymorph of SiO$_2$; kyanite is the high-pressure polymorph of Al$_2$SiO$_5$; and omphacite is a high-pressure form of clinopyroxene. This association is part of a nodule collected in the Roberts Victor kimberlite pipe in South Africa. It is concluded to have originated in the upper parts of the Earth's mantle. From J. R. Smyth, and C. J. Hatton, 1977, A coesite-sanidine grospydite from the Roberts Victor kimberlite, *Earth and Planetary Science Letters*, 34: 284–90.

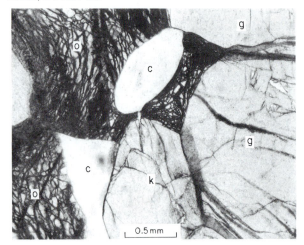

are diamond (the high-pressure structural form of graphite; both are carbon in composition); coesite, SiO_2, a high-pressure form of quartz; kyanite, a high-pressure form of Al_2SiO_5; and omphacite, a complex pyroxene composition indicative of a high-pressure origin. An illustration of a high-pressure mineral assemblage from a kimberlite is given in Fig. 4.10. Most kimberlite pipes have their root zones in the upper mantle in a depth range of about 150 to 200 km down. In recent years diamond and coesite-bearing rocks have also been discovered in the rock masses of some deep mountain roots that have become exposed at the Earth's surface due to exhumation. These mountain roots were originally subducted to a depth of about 100 km. Even though these direct samplings of tiny parts of the upper mantle give us some understanding of the mineral reactions in that very upper part of our planet, geologists have not found samples that represent deeper parts of the mantle.

The most direct way to answer questions about the makeup of rocks at much greater depths in the mantle is to simulate high pressure and temperature conditions in the laboratory. One of the most versatile

instruments that allows for heating and squeezing of minerals in the laboratory is known as the *diamond anvil cell*, in which a tiny specimen is compressed between the tips of two gem-quality diamonds (see Fig. 4.11). Because the surface area of each diamond tip is so small, very high pressures can be achieved with relatively modest forces pushing the diamonds together. Diamond makes an ideal anvil for exerting ultrahigh pressures on mineral samples because (1) it is the strongest material known and (2) it is transparent, allowing the sample to be observed right through the diamond anvils (with a petrographic microscope). Furthermore, high-power laser beams can be focused through the diamonds, thereby heating the sample to several thousand degrees while it is under very high pressure. Such experiments have allowed researchers to duplicate in the laboratory the conditions that are expected to exist throughout the Earth's mantle, and to observe and define the mineral reactions that occur under such extreme conditions.

High-pressure experiments in the diamond anvil cells—as well as with other high-pressure instrumentation—show that the minerals we are familiar with

FIG. 4.11. The diamond anvil cell used for high pressure studies, manufactured by High Pressure Diamond Optics, Inc., Tucson, Arizona. *(a)* The small size of this instrument allows it to be mounted on the stage of a petrographic microscope for direct observation of phase transitions. The large knurled screw applies the pressure. The two opposing diamonds in the anvil are located in the center of the circular metal housing. *(b)* Enlarged cross-section through the center of the diamond anvil. A = diamond supports. B = diamond anvils. C = gasket that holds the study sample in place. D = sample cavity. Courtesy of High Pressure Diamond Optics, Inc.

(a)

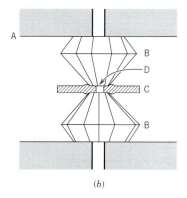

(b)

in the Earth's crust transform to new mineral structure types when subjected to the conditions of the deeper mantle. As shown in Fig. 4.9 olivine, (Mg, Fe)$_2$SiO$_4$ will transform to a denser structure (wadsleyite) at a depth of about 410 km. Further down, at about 520 km, wadsleyite transforms to an even denser structure known as ringwoodite. These reactions are:

$$Mg_2^{[6]}Si^{[4]}O_4 \quad\rightarrow Mg_2^{[6]}Si^{[4]}O_4$$

olivine structure   wadsleyite
with D = 3.22 gm/cm^3 with D = 3.47 gm/cm^3

$$\rightarrow Mg_2^{[6]}Si^{[4]}O_4$$

ringwoodite
with D = 3.55 gm/cm^3

where the density values are given for the Mg-end-member compositions. The superscripts in square bracket represent coordination numbers: [6] = **C.N.6** and [4] = **C.N.4**.

Similarly, the pyroxene structure adopts a denser, garnet type structure at about 410 km depth. At depths below 660 km it is likely that *all silicon is in 6-coordination* instead of 4-coordination, as in silicates in crustal rocks. The most abundant types of densely packed silicate structures to be found below the 660 km zone are the perovskite and rutile structure types. SiO$_2$ would be expected to occur as stishovite, with Si in 6-coordination with oxygen, isostructural with rutile.

Common mineral compositions rich in Mg and Ca, such as olivines and pyroxenes, are expected to occur in a perovskite structure type. Small amounts of the oxide mineral magnesiowüstite, (Mg, Fe)O, is expected to form as well as a high-pressure product from ringwoodite (see Fig. 4.9). Examples of reactions in the upper part of the mantle are as follows (from Liu and Bassett 1986; see also *Ultrahigh-Pressure Mineralogy* 1998; for complete references see listing at end of this chapter):

$$Mg^{[6]}Si^{[4]}O_3 \rightarrow \tfrac{1}{2}Mg^{[6]}_2Si^{[4]}O_4 + \tfrac{1}{2}Si^{[6]}O_2 \rightarrow$$
enstatite   wadsleyite   stishovite
structure   structure   (rutile structure)

$$\rightarrow \tfrac{1}{2}Mg^{[6]}_2Si^{[4]}O_4 + \tfrac{1}{2}Si^{[6]}O_2 \rightarrow$$
ringwoodite   stishovite
(spinel structure)   (rutile structure)

$$\rightarrow Mg^{[6]}Si^{[6]}O_3 \rightarrow Mg^{[8]}Si^{[6]}O_3$$
ilmenite*   perovskite
structure   structure

For a discussion of the ilmenite, FeTiO$_3$, structure, see page 372.

Examples of the greatly increased densities, reflecting much denser packing of elements on an atomic scale, of a few common silicate compositions are given in Table 4.1.

TABLE 4.1   Compositions and Measured Densities of Three Common Crustal Mineral Types as Compared with Their Calculated Densities in High-Pressure Polymorphs with SiO$_6$ Octahedra*

Mineral	Density	C.N. (Si)
SiO$_2$ (quartz)	2.65	4
SiO$_2$ (stishovite)	4.29*	6 (rutile structure type)
CaSiO$_3$ (wollastonite)	2.91	4
CaSiO$_3$	4.25*	6 (perovskite structure type)
MgSiO$_3$ (enstatite)	3.21	4
MgSiO$_3$	4.10*	6 (perovskite structure type)

*High-pressure polymorphic data from Hazen, R. M. and Finger, L. W., 1991, Predicted High-Pressure Mineral Structures with Octahedral Silicon, *Geophysical Laboratory Annual Report*, Carnegie Institution of Washington, no. 2250, pp. 101–107.

It is now clear that many silicate minerals transform to the perovskite structure at or below about 660 km depth. In this structure (see Fig. 4.12) six oxygens surround each silicon cation, and eight such octahedra are grouped around a magnesium cation. Such a structure is in sharp contrast to that of silicate minerals in the Earth's crust, where four oxygens surround each silicon in tetrahedral coordination. Because of the high predicted abundance of perovskite-structured material, with an end-member composition of Mg$^{[8]}$Si$^{[6]}$O$_3$, and a more general range of composition of (Mg, Fe)$^{[8]}$Si$^{[6]}$O$_3$ to (Mg, Fe, Ca)$^{[8]\ to\ [12]}$(Si, Al)$^{[6]}$O$_3$ in the

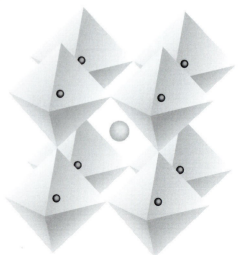

FIG. 4.12. The perovskite structure type. The Si^{4+} (small black spheres) are located at the center of octahedra, where each octahedral corner represents one oxygen (**C.N.** = 6). The large cation in the center of the structure may represent Mg^{2+}, Fe^{2+} and/or Ca^{2+} and is coordinated to eight SiO$_6$ octahedra packed around it. The coordination of this site is variable from **C.N.** = 8 to **C.N.** = 12.

lower parts of the mantle, it is therefore concluded that perovskite is the most abundant structural type within the Earth, although the mineral perovskite ($CaTiO_3$) is very uncommon in the crust.

MINERAL STABILITY

In the prior discussion of mineral reactions in this chapter, the concept of mineral stability has not yet been mentioned. To progress further in our understanding of mineral behavior, some basic aspects of mineral stability and *mineral stability diagrams* must be discussed.

Phase Diagrams

The behavior of solids, liquids, and gases under variable external conditions, such as those of tempera-

ture and pressure, is commonly expressed in what is known as a *phase diagram* (or a *stability diagram*). A *phase* is a homogeneous substance with a well-defined set of physical and chemical properties. As such, the term *phase* can be used interchangeably with the term *mineral* only if the mineral is homogeneous (i.e., exhibits no compositional variation). For example, low quartz (SiO_2) is a low-temperature phase in the chemical system $Si–O_2$ (or SiO_2); kyanite (Al_2SiO_5) is a high-pressure phase in the chemical system $Al_2O_3–SiO_2$ (or Al_2SiO_5). When a mineral exhibits solid solution, as in the complete solid solution series between forsterite (Mg_2SiO_4) and fayalite (Fe_2SiO_4), we speak of a *phase region*. A phase may be solid, liquid, or gaseous, as in the case of H_2O with three distinct phases—ice, water, and steam.

Figure 4.13 is a phase diagram for the chemical system H_2O; specifically it is a *pressure-temperature stability diagram*, commonly known as a *P–T diagram*.

FIG. 4.13. *(a)* P-T diagram for H_2O. Six different structural types (polymorphs) of ice are indicated by I, II, III, V, VI, and VII. (After P. W. Bridgeman, 1937, *Jour. Chemical Physics* 5, p. 965, and *Phase Diagrams for Ceramists*, © American Ceramic Society, Columbus, Ohio, 1964.) For clarity the water/water vapor curve has been offset slightly toward higher pressure. The shaded region in figure *(a)* is enlarged, but not to scale, in *(b)*.

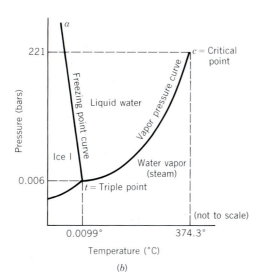

It indicates the state (i.e., solid, liquid, or gas) at a specific set of P-T conditions. The diagram in Figure 4.13 shows that at low temperatures and high pressures gas condenses to liquid, and that at even lower temperatures both gas and liquid give way to various different structures (polymorphs) of ice. Along the various curves in this diagram two phases can *coexist stably* (*in equilibrium*, see below). For example, ice and water can coexist along the various *T–P* curves (freezing point curves) on the left side of the diagram. In Fig. 4.13, at the point where all three curves meet, three phases, ice, water, and steam, can coexist; this is known as the *triple point t*. Along the curve *t–c*, water and steam can coexist but with increasing *P* and *T* (going in the upper right direction along the curve) the water phase becomes less dense (expands due to increasing temperature) and the steam phase more compressed (due to increasing *P*). At point *c* (*critical point*) the two phases become identical, hence indistinguishable. In *P–T* space, to the upper right of *c* we no longer speak of water or steam but we speak of *supercritical* aqueous fluid.

Stability, Activation Energy, and Equilibrium

Phase, or mineral stability, diagrams are very useful in providing a visual image of what mineral or group of minerals is stable with respect to some other mineral or mineral groups at a specific set of external (e.g., specified *P* and *T*) conditions. For example, in Fig. 4.13, different polymorphs of ice are stable in different parts of the diagram (i.e., at different conditions of *P* and *T*). Each numbered ice field outlines the *P–T* space in which a specific polymorph of ice is stable. Ice VI has a much larger *P–T* region over which it is stable than, for example, ice III. Furthermore, one can say that ice VII is stable at the highest pressures and a range of low temperatures, whereas ice I is stable only over a range of relatively low *P* and *T* conditions. This leads to the conclusion that ice VII is unstable in the *P–T* space of ice I; the reverse statement is also true.

The concept of *stability* in a chemical system is related to the energy of the system (see discussion of Gibbs free energy, later in this chapter), but it can be intuitively appreciated in terms of the situation of some mechanical blocks with respect to their resting surface. In Fig. 4.14a three of the same blocks are shown in different positions. In Fig. 4.14a the block is barely stable, even though the bottom edge on which it stands was beveled off. This position is clearly the least stable of the three block positions, because as a result of a slight displacement the block in (a) will change to either position (b) or (c). This is also referred to as the block in (a) being *metastable* with respect to the other two blocks (in *b* and *c*). Note also that the center of gravity (g) of the block in (a) is in the highest position above the resting surface. The block in (b) is in a more stable condition because a reasonably large displacement is needed to change it to the position of the block in (c). If only a small disturbance is applied to the block in (b), it will remain in the same orientation. Note also that the center of gravity (g) of block (b) is lower than that in (a) but considerably higher than that in (c). In (c) the orientation of the block is said to be *stable* because even after a reasonably large disturbance (or displacement) it will return to the orientation in which it is shown. The center of gravity (g) is now at the lowest position of the three orientations. The physical displacements (or disturbances) applied to the various blocks can be expressed in energy terms, as the *activation energy*. The amount of energy necessary to change the position of block (b) to that of block (c) is the activation energy needed to cause the change in position; in a chemical system the activation energy is the energy needed to cause a chemical reaction to occur. In applying the concept of stability to phase diagrams, we are generally concerned with differences in stability (i.e., differences in energy val-

FIG. 4.14. Illustration of various degrees of stability for the same block in different orientations. *(a)* is less stable than *(b)* or *(c)*, or *metastable* with respect to *(b)* and *(c)*. The most stable position is *(c)*.

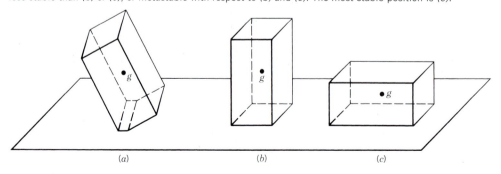

ues) and not the absolute values. For example, with reference to Fig. 4.13, we would say that liquid water is unstable with respect to ice I, in the *P–T* of ice I stability.

The concept of *stability* is also related to time. If, for example, water and ice coexist in constant amounts indefinitely, that is, no water is forming at the expense of ice, or vice versa (as along curve *a–t* in Fig. 4.13*b*), we can say that under these specific conditions water and ice are *in equilibrium*. In rocks, where the constituent minerals have coexisted since their formation, perhaps several million years ago, one cannot always conclude unambiguously whether the mineral constituents are in equilibrium with each other or not. If no *reaction rims* are observed between minerals that touch each other in the rock, we may assume that the minerals were in equilibrium at the time of formation; this assumption, however, must generally be supported by further detailed chemical and textural information. If, however, megascopically or microscopically visible rims exist between minerals, we may tend to conclude that some of the minerals were *not* in equilibrium with each other. For example, garnet may be separated by a chlorite rim from a coexisting biotite. Here we would conclude that the garnet and biotite were *not* in equilibrium with each other, as they are separated by a reaction product, chlorite. In experimental studies, the experimentalist will conclude that the phases under investigation are in equilibrium when no further change takes place between them during a certain time interval. This may range from a few hours, to several months, or even years, depending on the speed or sluggishness of the reactions studied, and on the patience of the investigator.

Components

In Fig. 4.13 we have illustrated stability fields, in terms of bordering equilibrium curves, for phases in a system that can be described chemically by one compound component, H_2O. Phases in a system are described by independent chemical species known as *components*; a minimum number of chemical variables is generally chosen. This can be rephrased as *components are the chemical entities necessary to define the compositions of all the phases in a system.* For example, in the case of H_2O, we chose the command component H_2O, instead of defining the chemical system in terms of two components, H_2 and O_2. In the system Al_2SiO_5 (andalusite-sillimanite-kyanite), Al_2SiO_5 is generally chosen as the compound component although three elements, Al, Si, and O, or two oxide components, Al_2O_3 and SiO_2,

could have been selected to define the system chemically. If one is interested in the stability relations of wollastonite, $CaSiO_3$, one might choose to represent one's findings in terms of the compound component, $CaSiO_3$. If, however, one wishes to know the stability fields of pyroxenes in the system CaO-MgO-FeO-SiO_2, one generally chooses the three compound components $CaSiO_3$-$MgSiO_3$-$FeSiO_3$ to define the system chemically (see Fig. 4.23).

Introductory Thermodynamics

The phase diagram for H_2O, as given in Fig. 4.13, can be delineated by a large number of pressure-temperature experiments on H_2O and very careful characterization of the resultant reaction products. However, such a diagram can also be calculated on the basis of known parameters that relate to the various configurations of atoms, ions, and molecules as a function of physical and chemical conditions. This scientific method, which allows us to make quantitative assessments of mineral and other phase equilibria, is known as *thermodynamics*. Here we will give only a very cursory introduction to this subject, enough to aid in the understanding of some aspects of the phase diagrams that follow.

A fundamental and universal observation is that all organizations of matter drive toward a minimal energy state (or arrangement). Minerals and rocks tend toward the lowest energy state that is, the most stable state for their constituents.

The *first law of thermodynamics* states that "the internal energy (*E*) of an isolated system is constant," in which a system is defined as any part of the universe being considered (e.g., it may be a hand specimen of a specific rock type, a specific mineral assemblage, or a specific chemical mix in a platinum crucible ready for experimental study). In a *closed system* (one in which there is no addition or subtraction of material; the mass remains constant, but there can be a loss or gain of energy) the change in internal energy (the differential of *E*, *dE*) will be the difference between heat (a form of energy, defined as *Q*) added to the system (expressed as *dQ*) and work (another form of energy, defined as *W*) done by the system (expressed as *dW*). The first law can be stated as:

$$dE = dQ - dW$$

Because work (*W*) = force × distance, and because force = pressure (*P*) × surface area, *W* = *P* × surface area × distance, or *W* = pressure (*P*) × volume (*V*).

At constant pressure, and in the absence of electrical, etc., work, this results in *dW* = *PdV*. When this is substituted in the above equation it yields the most familiar form of the first law of thermodynamics as

$$dE = dQ - PdV$$

The effect of this equation on a mineral can be qualitatively assessed. When energy, as heat, is added to a mineral, the increase in internal energy (dE) of the mineral is proportional to, but less than, the heat added because part of the added energy is transformed into the work of thermal expansion of the mineral.

The *second law of thermodynamics* relates a change in thermal energy of a system (at constant pressure, P, and constant temperature, T) to a change in the degree of order (or disorder) in that system. *Entropy* (S) is a quantity that represents the degree of disorder in a system. As discussed on p. 138 (see "Order-Disorder Polymorphism"), the state of greatest order in a crystalline material is at the lowest temperature, with increasing disorder of atoms in the structure the result of increasing temperatures. Relating this to entropy, a rigorously ordered structure has a lower entropy than one that is disordered. The concept of entropy can be illustrated with respect to Fig. 4.13. When heat is added to ice, some of it will convert to water, which, because of its much less regularly ordered structure than ice, will have a higher entropy. As a consequence of the second law,

$$\frac{dQ}{T} = dS$$

where dQ is the absorption of a quantity of heat. The concept of entropy can be further illustrated with respect to the system H_2O, as shown in Fig. 4.13. Because entropy is related to the amount of disorder (or chaos) in a system, there will be increases in entropy in the following reactions in the system H_2O: (1) ice → vapor, (2) ice → liquid, and (3) liquid → vapor.

The *third law of thermodynamics* states that "at absolute zero (0 Kelvin, which is equivalent to $-273°C$), a crystalline structure approaches perfect order, and the entropy of such a perfect crystal is zero."

This leads to the introduction of another important thermodynamic function, namely, *Gibbs free energy*, G, as:

$$G = E + PV - TS$$

where E = internal energy, P = pressure, V = volume, T = temperature, and S = entropy. The term *free energy* expresses the energy in excess of the internal energy; this is the excess energy needed to drive a chemical reaction. When the Gibbs free energy equation is differentiated and combined with

the equations for the first and second laws, the following relation results:

$$dG = VdP - SdT$$

For a system in equilibrium, at constant P and T, it can be shown that the Gibbs free energy (G) is a minimum. Furthermore, at equilibrium, the Gibbs free energies of the reactant (r) and product (p) are equal, that is $G_r = G_p$, and $dG = 0$.

The above equation of the Gibbs free energy can be differentiated with respect to T, at constant P, or with respect to P, at constant T. This results in two important relations that express the change in free energy (dG) with respect to pressure and temperature. These are:

$$\left(\frac{\partial G}{\partial P}\right)_T = V \qquad \text{where } V = \text{volume}$$

and

$$\left(\frac{\partial G}{\partial T}\right)_P = -S \qquad \text{where } S = \text{entropy}$$

The first of these expressions tells us that dense phases (i.e., those with small volumes) are favored at high pressures, and the second shows that high entropy states (with greater atomic disorder) are favored at high temperatures. Basic aspects of the Gibbs free energy equation can be shown graphically. The equations given above state that dG, a change in the Gibbs free energy, is a function of only P (stated as dP) and of T (stated as dT). Because the three variables G, P, and T, are interrelated, the function G can be represented graphically in terms of two variables, P and T. This is shown in Fig. 4.15a for two minerals (or phases) marked A and B. Each phase has its distinct G surface. Where the two G surfaces intersect, the two minerals (or phases) are in equilibrium because the condition of $G_A = G_B$ is satisfied. Figures 4.15b and c illustrate two cross sections through the three-dimensional picture in a. Figure 4.15b shows a cross-over of the entropy of B with respect to A (as a function of G and T at constant P). Figure 4.15c shows a cross-over in volume of B with respect to that of A (as a function of G and P at constant T). Figure 4.15b is a cross section of the diagram in a at constant pressure; this is also known as an *isobaric* section. Figure 4.15c is a cross section at constant temperature, also known as an *isothermal* section. Figure 4.15d is the standard $P–T$ diagram obtained from Fig. 4.15a by projection of the equilibrium curve (line of intersection) along which phases A and B coexist onto the basal plane of $P–T$. This shows that for any phases (minerals) to be at equilibrium, they must be at the same P and T.

Figure 4.13, for the system H_2O, illustrates the $P–T$ regions in which the specifically labeled state

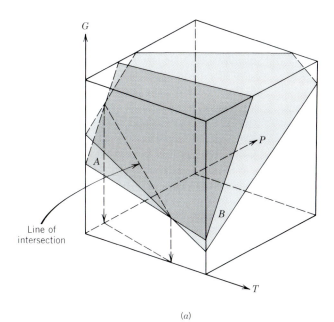

(a)

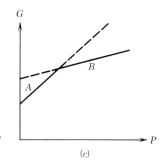 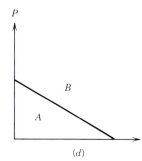

(b)   (c)   (d)

FIG. 4.15. *(a)* Three-dimensional representation of *G–T–P* space, with free energy surfaces for two minerals, *A* and *B*. Where the two surfaces intersect (along the line of intersection) the free energies of the two minerals are equal. Two sections through this space are shown: *(b)* a *G–T* section and *(c)* a *G–P* section. *(d)* is a projection of the line of intersection from above onto the *P–T* plane.

(e.g., ice, or liquid, or vapor) has a lower free energy than other possible states. Along the curves (phase boundaries), the adjoining states have equal free energies and are in equilibrium.

The last relationship that we will mention is the *Clapeyron equation*, which allows us to determine the *P–T* trajectory of equilibrium states (on a *P–T* diagram) as a function of entropy and volume changes. When ice and water are in equilibrium along a specific curve, as in Fig. 4.13, the changes in the Gibbs free energy along the equilibrium curve must be equal. This allows us to equate two Gibbs free energy expressions (i.e., one for the reactant, ice, and one for the product, water). As such the one equation must equal the other, as follows: For the reactant (*r*):

$$dG_r = V_r dP - S_r dT$$

and for the product (*p*):

$$dG_p = V_p dP - S_p dT$$

At equilibrium:

$$V_r dP - S_r dT = V_p dP - S_p dT$$

Rearrangement results in:

$$(V_p - V_r)dP = (S_p - S_r)dT$$

which results in:

$$\frac{dP}{dT} = \frac{\Delta S}{\Delta V}$$

the *Clapeyron equation*. In this equation ΔS is the total entropy of the products minus the total entropy of the reactants. Similarly, ΔV is the total volume of the products minus the total volume of the reactants. The *dP/dT* function, as expressed on a *P–T* diagram, is a function of both changes in entropy and volume of the system.

Examples of Mineral Stability (Phase) Diagrams

In the following discussion and illustrations of some representative mineral stability diagrams, we will concentrate on those that represent conditions pertaining to the *solid state*. This decision is made because this is a mineralogy text in which almost all of our concerns center on solids. A few diagrams, used in igneous petrology, may involve a high-temperature melt, and some, used in low-temperature processes (such as in chemical sedimentation in the ocean), may involve a low-temperature fluid phase.

One-Component Diagrams

Figure 4.16 shows four stability diagrams for various common minerals. The diagrams are defined by pressure and temperature axes (P–T diagrams) identical to those used in Fig. 4.13. As such, all four diagrams depict the stability fields of various different structural arrangements (*polymorphs*, see p. 134), in a specific chemical system, as a function of P and T. Figure 4.16a shows the stability fields of diamond, graphite, carbon III, and melt for a system that is composed only of carbon (C). Diamond and graphite are the two common polymorphs of carbon. Diamond has a very large stability field in the high-pressure region of the diagram. The diamond field also extends to high

FIG. 4.16. Examples of mineral stability diagrams (P–T diagrams) for one-component systems. Each of these has been determined by experiments. When curves, or lines, are solid their locations have been determined with certainty; when they are dashed, their location is less well known. (a) The system C based on experimental data from various sources. (b) The system Al$_2$SiO$_5$ as based on research results from Holdaway, 1971; complete reference is given at the end of this chapter. (c) The system SiO$_2$ from various sources. (d) The system CaCO$_3$. A kilobar = 1000 bars; 1 bar = 0.987 atmosphere.

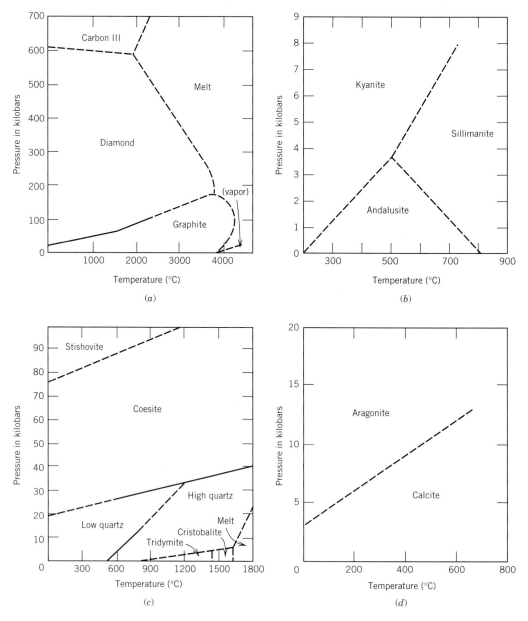

temperature at certain pressures (up to about 4000°C at about 150 kilobars). Graphite is stable over a wide temperature range but only at relatively low pressures. Diamond has a much more densely packed structure than graphite, with a molecular volume that is about 36 percent smaller than that of graphite. As such, diamond would be expected to be the high-pressure phase, as shown in the stability diagram. The polymorphic reaction in transforming diamond to graphite, and vice versa, is reconstructive. In such reconstructions, chemical bonds must be broken, and the structure must be reassembled. This requires large amounts of energy (activation energy), and because of this, diamond is very persistent under ordinary pressures and temperatures (atmospheric conditions), even though thermodynamically the diagram shows that graphite is the stable phase under atmospheric conditions. The conversion rate of diamond to graphite, at atmospheric conditions, must be infinitely slow and cannot be detected.

Figure 4.16*b* is a stability diagram for the three different structural arrangements (polymorphic forms) of Al_2SiO_5: andalusite, sillimanite, and kyanite. This diagram is regarded as a one-component system because Al_2SiO_5 is the compound but single component that represents the composition of all three polymorphs. Because kyanite occupies the high-pressure region of the P–T diagram, one would conclude kyanite to be the densest polymorph of Al_2SiO_5. This is so, as shown by the densities and structural studies of the three Al_2SiO_5 polymorphs. The three structures may be represented as follows:

$[Al]^6[Al]^6SiO_5$ kyanite
$[Al]^5[Al]^6SiO_5$ andalusite
$[Al]^4[Al]^6SiO_5$ sillimanite

where the square brackets with superscripts denote the coordination polyhedron about Al ($[Al]^6$ = octahedral; $[Al]^5$ = irregular 5-fold; $[Al]^4$ = tetrahedral). It is clear from this notation that kyanite represents the closest atomic packing of the three polymorphs.

In the case of the experimental study of the stability fields of the three polymorphs of Al_2SiO_5, very high purity minerals may be used for the experiments although most commonly the starting materials are high purity chemicals. For example, gem grade and inclusion-free kyanite may be found at increasing T to react to form sillimanite (see Fig. 4.16*b*). The position of the curve separating the kyanite and sillimanite fields is based on the first evidence of sillimanite forming at the expense of kyanite with increasing T and on evidence of the reverse reaction, namely sillimanite giving way to kyanite at decreasing T or increasing P. The determination of the beginning of such reactions is generally based on a combination

of X-ray powder diffraction and optical microscopic techniques. Because of uncertainties in various aspects of the experimental techniques in locating specific reaction boundaries, the reaction curves, defined in P and T values, may represent relatively broad reaction zones, although they are commonly shown as narrow lines or curves.

Kyanite, or sillimanite, or andalusite, without any textural indication of reaction to another Al_2SiO_5 polymorph, can be found in metamorphic rocks of all geological ages, including early Precambrian. This observation signifies that the activation energy necessary to transform a high-temperature or high-pressure polymorph into the lower P–T polymorph stable at atmospheric conditions has not been provided. However, in many other metamorphic rocks that contain Al_2SiO_5 polymorphs, the following rimming reactions have been observed:

a center of kyanite rimmed by sillimanite
a center of sillimanite rimmed by kyanite
a center of sillimanite rimmed by andalusite
a center of andalusite rimmed by kyanite

All of the above textural occurrences delineate various P–T paths across the different equilibrium boundaries. An example of such P–T paths interpreted as the result of various geologic processes, is shown in Fig. 4.17.

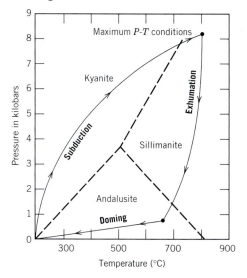

FIG. 4.17. The Al_2SiO_5 stability diagram superimposed on which are arrows for various geologic processes. Crustal material was originally subducted to conditions of maximum P–T of about 8 kilobars and 800°C. Subsequent to this, these metamorphic rocks were exhumed and finally domed, returning the subducted material right back to the Earth's surface. In addition to the evaluation of the textural reactions among the Al_2SiO_5 minerals, a geologist (or petrologist) must assess the P–T conditions for other minerals in these same rocks, as well as the tectonic parameters of the region.

Figure 4.16c outlines the stability fields of phases in the system SiO_2. Quartz, as low quartz, is the SiO_2 phase in plutonic, metamorphic, and sedimentary rocks, reflecting its general temperature of origin below 1000°C. Tridymite and cristobalite are found in volcanic assemblages of many geological ages. This means that these minerals exist metastably for long geological time periods. In other volcanic occurrences, the original cristobalite or tridymite may have been converted to low quartz, but still preserving the crystal form of cristobalite or tridymite (these occurrences, therefore, represent pseudomorphs of low quartz after higher-temperature SiO_2 polymorphs). Such pseudomorphic occurrences suggest that the activation energy necessary for the reconstructive transformations (from cristobalite or tridymite to low quartz) was possibly provided by some reheating due to later metamorphism of the original volcanics. The high-pressure polymorphs of SiO_2, coesite and stishovite, have been found in meteorite impact craters. Coesite also occurs as inclusions in diamonds, and inside pyrope garnets in very pyrope-rich high-grade metamorphic rocks from Parigi, Northern Italy (Chopin 1984; for complete reference see reference list at end of this chapter). Coesite has also been found in a xenolith nodule (a xenolith is a foreign inclusion) in a kimberlite pipe (see Fig. 4.10). This kimberlite pipe originated at a depth of about 170 to 200 km in the mantle, at pressures of approximately 60 to 70 kilobars (a pressure region in Figs. 4.16a and c where both diamond and coesite are stable). Coesite-diamond coexistences have also been reported from very high-pressure metamorphism of crustal rocks in eastern China (Shutong et al. 1992; for complete reference see reference list at end of this chapter). The coesite–diamond–jadeite (another high-pressure mineral) assemblage is considered the result of burial of the crust to great depths, metamorphism, and subsequent exhumation back toward the Earth's surface (see Fig. 4.17).

Figure 4.16d shows the stability diagram for $CaCO_3$, with two polymorphic forms, calcite and aragonite. By far the most common carbonate, be it sedimentary, metamorphic, or igneous in origin, is calcite. The phase diagram in Fig. 4.16d suggests that $CaCO_3$ that is formed at normal temperatures and pressures (at essentially atmospheric conditions) should be calcite. However, this is not what is observed. Primary precipitation of $CaCO_3$ from seawater is commonly in the form of aragonite, giving rise to aragonite muds. Many organisms, furthermore, build their shells of aragonite, and the main constituent of pearls is aragonite. These observations imply that aragonite is formed and precipitated as a

metastable phase that, over time, will convert to calcite. This polymorphic transformation can be very slow, as shown by the preservation of aragonite in Pennsylvanian reefs, which are about 300 million years old. However, the aragonite to calcite polymorphic reaction is more easily accomplished than the transformations mentioned above for diamond-graphite or coesite-quartz. If energy is applied in the form of increased temperature, as a result of burial, aragonite will transform to calcite more quickly.

Two-Component Diagrams

Two-component stability diagrams are most commonly constructed with a horizontal composition bar and a vertical temperature axis. In these diagrams pressure is a constant and does not change. Such *temperature–composition* (*T–X*) *diagrams* may exhibit very different features as a result of: (1) complete solid solution between two end members, (2) partial solid solution between two end members as expressed by the presence of a miscibility gap, and (3) no solid solution between various mineral species that can be represented along the two-component composition bar.

Let us begin with some examples of two-component *T–X* diagrams in which there is complete solid solution over the whole temperature range of the diagram. This is shown by the olivine series, between Mg_2SiO_4 (forsterite) and Fe_2SiO_4 (fayalite), and by the high-temperature region in the plagioclase feldspar series ($NaAlSi_3O_8$ to $CaAl_2Si_2O_8$). Both diagrams, depicted in Figs. 4.18a and b, are the result of experimental studies at atmospheric pressure. That is, the pressure conditions of the experiments are fixed at 1 atmosphere (= 1.01325 bars). These two diagrams are also known as *liquidus diagrams* because they involve a liquid phase, or melt. Such liquidus diagrams are instructive in evaluating the melting relations of igneous rock compositions, as well as for the study of a crystallization sequence from a melt. The upper curve in Fig. 4.18a is known as the *liquidus*, a line or surface along which compositions of melt are in equilibrium with a crystalline phase. The lower curve is known as the *solidus*, a line or surface along which compositions of a crystalline phase are in equilibrium with the melt. Everything above the liquidus is liquid; everything below the solidus is solid. With the aid of the schematic diagram in Fig. 4.18c the crystallization behavior of solids from a melt in such diagrams can be illustrated. This diagram shows that the pure end member A melts at T_A and pure end member B melts at T_B, but intermediate compositions (AB), which consist of a single phase (part of the solid solution series), melt through a range of

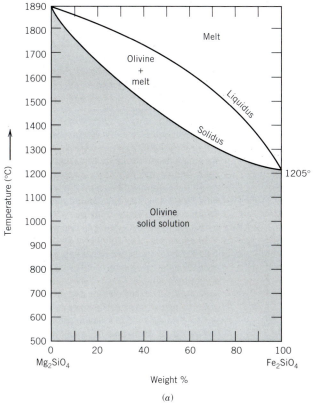

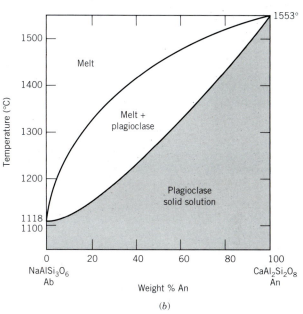

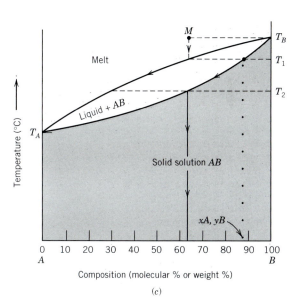

FIG. 4.18. Examples of two-component systems with complete or extensive solid solution between end members. *(a)* Equilibrium temperature–composition (*T–X*) diagram for the two-component system Mg₂SiO₄–Fe₂SiO₄ (olivine). This is based on experiments above 1100°C by Bowen and Schairer (1932). This version of the diagram has been extended to a temperature of 500°C to illustrate the extensive field of homogeneous solid solution. *(b)* Equilibrium temperature–composition diagram for the high-temperature region of the two-component system NaAlSi₃O₈ (albite; Ab)–CaAl₂Si₂O₈ (anorthite; An) (After Bowen 1913). *(c)* Schematic temperature–composition diagram for a two-component chemical system that shows a complete solid solution series between two end member components.

temperatures intermediate between T_A and T_B. A melt of composition *M* at temperature T_B will be entirely melt. When it cools to T_1 it will start to crystallize out a member of the solid solution series *AB* with specific composition *xA, yB*. These crystals are enriched in the *B* component with respect to the melt composition *M*, and their growth will deplete the melt in component *B*. The melt composition will,

as a result of this depletion, move along the liquidus curve toward *A* as indicated by the upper arrow. As a result of the continual lowering of the temperature, the solid phase of original composition *xA, yB* will react with the melt in the direction of the lower arrow along the solidus. As such, both the melt and crystalline products will increase in content of *A* with decreasing *T* and the ratio of solid to melt will increase. Finally, at T_2 the crystallized products have a composition which is that of the original melt M, and the amount of melt in equilibrium with the crystals will reach zero. Now, because only a solid phase remains (i.e., we are in the subsolidus region), with continued lowering of the temperature the com-

position of the crystalline product will remain constant at the bulk composition M of the original melt. The above observations apply to both diagrams, Figs. 4.18a and b.

Although the original equilibrium melting and crystallization experiments necessary to construct

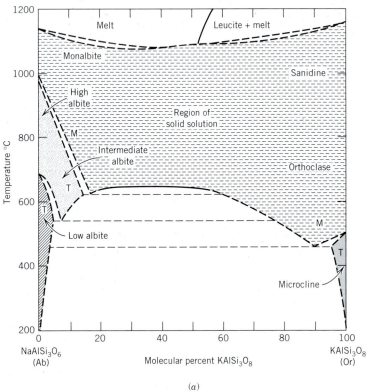

FIG. 4.19. (a) Schematic phase diagram for the system $NaAlSi_3O_8$ (Ab)-$KAlSi_3O_8$ (K-spar) showing a large miscibility gap at temperatures below approximately 650°C. M and T mean monoclinic and triclinic, respectively. (Modified with permission after J. V. Smith and W. L. Brown, 1988, *Feldspar minerals*, v. 1, *Crystal structures, physical, chemical, and microtextural properties*. New York: Springer-Verlag, Fig. 1.2, 828 pp.) (b) Approximate and schematic phase diagram for the same chemical system, except for addition of H_2O to the melt, at 5 kilobars of P_{H_2O}. Now the solidus and miscibility gap intersect at point e, the eutectic point; see text for discussion. S.S. = solid solution. (Modified after S. A. Morse, 1970, Alkali feldspars with water at 5 kb pressure, *Journal of Petrology* 11: 221–51.)

(a)

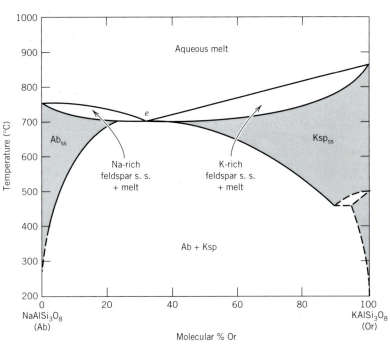

Fig. 4.18*b* indicated a complete solid solution series for plagioclase at high temperature, subsequent detailed single-crystal X-ray diffraction and transmission electron microscope studies have shown this to be untrue at lower temperatures. This is further discussed as part of "Exsolution Processes" on p. 143.

The two-component system for the alkali feldspars series ($NaAlSi_3O_8$-$KAlSi_3O_8$) is an example of only very limited solid solution at low temperature with an almost complete range of solid solution at high temperature. This is shown in a *T–X* diagram in Fig. 4.19*a*. The lack of solid solution at low temperatures is mainly the result of the large difference in ionic sizes of Na^+ (1.18 Å in 8-coordination) and K^+ (1.51 Å in 8-coordination; see Table 3.11). Lamellar exsolution textures (see page 144) are common in this system (e.g., perthite). The mineral names listed inside the diagram are the names for polymorphs of $NaAlSi_3O_8$ (monalbite = monoclinic albite; high, intermediate, and low albite) and of $KAlSi_3O_8$ (sanidine, orthoclase, and microcline).

It is of considerable geological interest to see what happens to the configuration of the diagram in Fig. 4.19*a* when H_2O (at high water pressure = P_{H_2O}) is added to the chemical system as in Fig. 4.19*b*. A high aqueous-fluid pressure considerably lowers the melting temperatures, while increased pressure applied to the system raises the maximum temperature of the miscibility gap such that now the solidus and the miscibility gap intersect. Point *e*, the *eutectic point* (see further discussion below), is the lowest temperature point on the liquidus at which a unique melt of fixed composition is in equilibrium with two feldspar compositions (i.e., a Na-rich feldspar solid solution and a K-rich feldspar solid solution). The crystallization of a melt composition at *e*, therefore, allows for the direct crystallization from the liquid of coexisting albite and K-spar, without the need for an exsolution process (in the solid state) as is required in the anhydrous system of Fig. 4.19*a*. Coarse-grained coexistences of albite and K-spar without exsolution textures are common in granites and pegmatites; such occurrences reflect crystallization of both minerals from a melt. Perthitic intergrowths, however, represent exsolution in the solid state (see p. 144).

A very different temperature–composition diagram results when there is *no* solid solution between various minerals in a two-component system. Figure 4.20*a* is a two-component phase diagram with a *eutectic* relationship. Phases *A* and *B* are pure substances and as such there is no solid solution between them. For composition *A* the melting temperature is T_A. Similarly, for composition *B* the melting temperature is T_B. The addition of some of

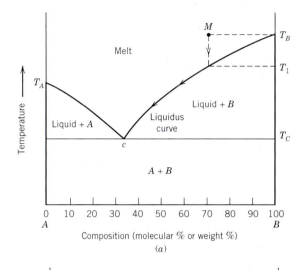

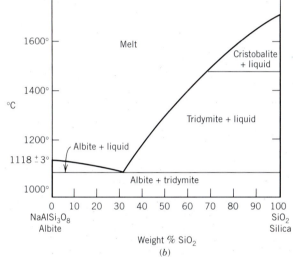

FIG. 4.20. *(a)* Schematic temperature–composition section showing eutectic crystallization of components *A* and *B*; both components are pure substances (no solid solution between them). *(b)* The system $NaAlSi_3O_8$(albite)–SiO_2. (After J. F. Schairer and N. L. Bowen, 1956, *American Journal of Science* 254: 161.)

composition *B* to a melt of *A* lowers the temperature of the liquid that can coexist with *A*, along the curve between T_A and *e* (*eutectic point*, which is the minimum temperature point of the liquid field). Similarly the melting temperature of the melt that can coexist with *B* is lowered by the addition of some of *A*, as shown by the curve between T_B and *e*. The lowest temperature at which crystals and melt are in equilibrium is T_C, the temperature of the eutectic. Let us now look at the crystallization sequence of a melt of composition *M*. Upon lowering of the temperature of the melt at *M* (original temperature = T_B) the melt will start to crystallize some of the pure substance *B* at T_1. The crystallization of *B* from the melt will continue along the *liquidus curve* (T_B to e), continually increasing the content of substance A in the melt. At the eutectic point (e) phase A will join B as a crystallization product. At this point there is no further change in composition of the melt because both A and B crystallize from the melt in the same proportions as are present in the melt. With continued crystallization the melt will disappear and the final crystalline products will be B and A in the proportions represented by the original bulk composition M. Figure 4.20b is an illustration of a eutectic melting relationship in the high-temperature part of the system NaAlSi$_3$O$_8$-SiO$_2$.

Three- or More-Component Diagrams

Because most rock types (igneous, metamorphic, and sedimentary) consist on the order of six to ten, or more, chemical components, one-component and two-component stability diagrams generally have only limited applicability to more complex naturally occurring mineral assemblages. However, easy graphic representation of multicomponent chemical systems is limited mainly to three-component (triangular) diagrams. For the methods of plotting mineral formulas on triangular diagrams, see pages 99–103. One may also consider four-component systems with chemical components at each of the four corners of a tetrahedron, but such tetrahedral representations become graphically complex.

Triangular phase diagrams are commonly used in igneous petrology to represent the experimentally studied melting relations of igneous rock compositions, as well as the crystallization sequence from a melt. An example of such a diagram is given in Fig. 4.21 in terms of three compound components SiO$_2$, KAlSiO$_4$, and Mg$_2$SiO$_4$. The contours in this diagram represent melting temperatures and are known as *isotherms*; the surface defined by these isotherms is the *liquidus surface*. The arrows along the boundaries of the various phase fields indicate crystallization paths, with decreasing temperature.

In the study of the mineralogy of assemblages and rock types, triangular diagrams that depict mineral stabilities below the liquidus and solidus surfaces—that is, *subsolidus* triangular phase diagrams—are most instructive. A schematic, composite liquidus-subsolidus diagram is shown in Fig. 4.22a for the triangular composition space of feldspar. The bottom triangle is based on three major feldspar end members: KAlSi$_3$O$_8$–NaAlSi$_3$O$_8$–CaAl$_2$Si$_2$O$_8$. This compositional triangle alone is commonly used to depict the extent of solid solution in the feldspars. Figure 4.22b shows the experimentally determined extent of solid solution between end member compositions as a function of the temperature of the experiments at $P_{H_2O} = 1$ kilobar. This diagram illustrates that at the highest temperature (900°C) the feldspars show the most extensive solid solution field. At the lowest temperature studied, 650°C, there is the least extent of solid solution. Indeed, along the join KAlSi$_3$O$_8$–NaAlSi$_3$O$_8$ there is a considerable break in the central part of the solid solution series at 650°C. This reflects the intersection of the miscibility gap in the system at about that temper-

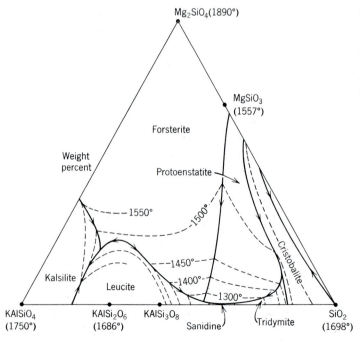

FIG. 4.21. Melting relations among minerals in the system SiO$_2$–KAlSiO$_4$–Mg$_2$SiO$_4$. Temperatures next to compositions refer to the melting temperatures of those compositions in °C. Protoenstatite instead of enstatite is produced in experiments of this type. The compositions of protoenstatite and enstatite are identical, but their structures differ in detail. (After W. C. Luth, 1967, *Journal of Petrology*, 373, reprinted by permission of Oxford University Press.)

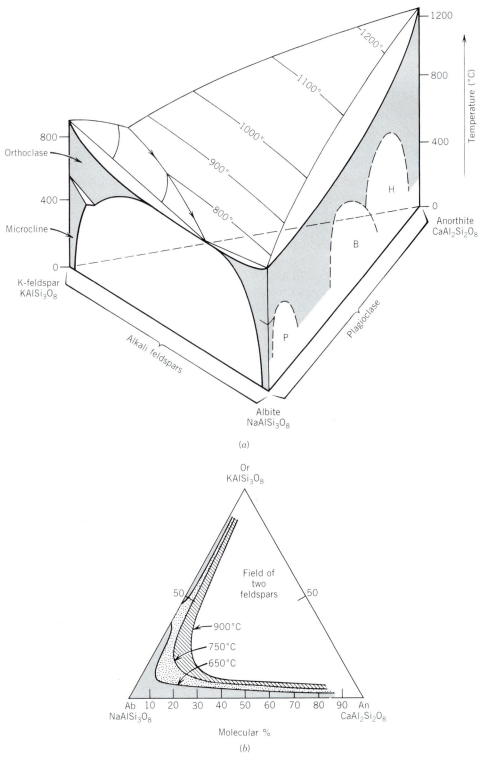

FIG. 4.22. *(a)* Highly schematic temperature–composition diagram for the three-component system KAlSi$_3$O$_8$ (K-feldspar)-NaAlSi$_3$O$_8$ (albite)-CaAl$_2$Si$_2$O$_8$ (anorthite), at water pressure of about 5 kilobars. Details of the interior are complex and have been omitted. The upper, contoured surface of the diagram is the liquidus surface. The three intergrowth regions, as a result of miscibility gaps, in the lower temperature range of the plagioclase series are: P = peristerite, B = Bøggild intergrowth, H = Huttenlocher intergrowth. Solid solution is shown by shading. Compare the left vertical side of this diagram with more detailed information given in Fig. 4.19. (Adapted from P. H. Ribbe, 1987, Feldspars, in *McGraw-Hill Encyclopedia of Science and Technology*, 6th ed., v. 7, p. 45; reproduced with permission of McGraw-Hill.) *(b)* Experimentally determined extent of solid solution plotted on the triangular base of the diagram in *(a)*, at P_{H2O} = 1 kilobar. (After P. H. Ribbe, 1975, The chemistry, structure, and nomenclature of feldspar. *Reviews in Mineralogy, Feldspar Mineralogy*, 2: R1–R72, fig. R1.)

ature (see Fig. 4.19a), which is the T–X section for $KAlSi_3O_8$–$NaAlSi_3O_8$, without H_2O present in the system (see also the discussion on exsolution, later in this chapter). The plagioclase join in Fig. 4.22b suggests that solid solution at 650°C is continuous between $NaAlSi_3O_8$–$CaAl_2Si_2O_8$. It is known (see Fig. 4.46) that three miscibility gaps exist in this series, below about 800°C. These gaps are not reflected in the experimental results of Fig. 4.22b, because the chemical analytical techniques (used to determine compositions of experimental products) cannot resolve the extremely fine-scale nature of the exsolution lamellae across the three gaps. In order to ascertain that the plagioclase grains obtained experimentally are homogeneous (or inhomogeneous, i.e., consisting of lamellar intergrowths), single-crystal X-ray or transmission electron microscope techniques must be used, in addition to the commonly used technique of electron probe microanalysis (see Chapter 7).

Another very commonly used triangular diagram is one that represents the main compositional variation among members of the pyroxene group. This is shown in Fig. 4.23 as a function of three components, $CaSiO_3$ (wollastonite, a pyroxenoid), $MgSiO_3$ (enstatite), and $FeSiO_3$ (ferrosilite). The compositional extent between end members of various pyroxene series, in common igneous and metamorphic rocks, is shown by shading. This diagram is the result of the compilation and graphic plotting of thousands of pyroxene analyses from the published literature.

Such a compilation of analyses from natural occurrences does not distinguish compositional regions that specifically represent high-temperature pyroxenes (as in basalt flows) from lower-temperature pyroxenes (as in gabbros). Such a diagram is useful only as a graphic representation of the average extent of solid solution of common pyroxenes. The open space between the augite region and that of pigeonite/orthopyroxene is a miscibility gap (see p. 148).

Triangular composition diagrams similar to Fig. 4.23 are commonly used in the graphic representation of other mineral groups, such as members of the olivine (see Fig. 12.1) and amphibole groups (Fig. 11.21). Three-component triangular diagrams may also be used to illustrate the common coexistence of possible mineral pairs, or of groups of three minerals. Such diagrams are known as *assemblage diagrams*.

In the construction of such a diagram, one of the first steps is the listing of the minerals that compose a specific rock under study. A simple igneous granite may consist of orthoclase, albite, quartz, and biotite. The texture of this specific granite may indicate that all four minerals formed as crystallization products at about the same elevated temperature. A petrologist, therefore, might conclude that orthoclase-albite-quartz-biotite is the *mineral assemblage* (also referred to as the *mineral paragenesis*) of this granite. Although the term *assemblage* (or *paragenesis*) is often loosely used to include all the minerals that

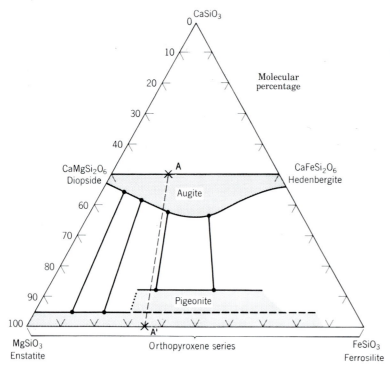

FIG. 4.23. *(a)* The extent of pyroxene solid solution in the system $CaSiO_3$–$MgSiO_3$–$FeSiO_3$. Representative tielines across the miscibility gap between augite and the more Mg-Fe-rich pyroxenes are shown. AA' locates the section across this diagram shown in Fig. 4.47.

compose a rock, it should be restricted to those minerals in a rock that appear to have been *in equilibrium* (*the equilibrium mineral assemblage*). The above granite may be found to contain vugs in outcrop that are lined by several clay minerals, bauxite as well as some limonite. Clearly the minerals in the vugs represent relatively low-temperature alteration products of the original much higher-temperature granite. The high-temperature granitic assemblage consists of orthoclase-albite-quartz-biotite, and a separate, later formed, low-temperature assemblage consists of clays-bauxite-limonite. In short, an assemblage consists of minerals that formed under the same, or very similar, conditions of pressure and temperature. In practice all minerals that coexist (physically touch each other) and show no alteration or rimming relations are commonly considered to constitute the assemblage. In a distinctly banded rock the mineral assemblages in the various bands are often very different from each other because the bands may represent major differences in bulk chemistries.

The evaluation of whether all minerals in a rock are in *equilibrium* with each other is generally not straightforward. If the texture does not show reaction rims and alterations, it is possible that the minerals are in equilibrium; nevertheless, further detailed chemical tests are often needed to define the equilibrium assemblage without ambiguity. Similarly, the

evaluation of equilibrium in an experimental study is also not straightforward. However, in spite of the problems inherent in the assessment of an equilibrium assemblage, triangular mineral assemblage diagrams are commonly used to depict the observed pairs, and groupings of three possible minerals in a specific chemical system. Figure 4.24 shows commonly observed coexistences (as deduced from natural assemblage occurrences as well as synthetic studies) of minerals in the system SiO_2 (quartz, tridymite, cristobalite)–$NaAlSiO_4$(nepheline)–$KAlSiO_4$ (kalsilite). This chemical system also includes the alkali fieldspar series and leucite ($KAlSi_2O_6$). The diagram shows that there is complete solid solution in the alkali feldspar series and extensive solid solution in nepheline, kalsilite, and leucite. The diagram was determined experimentally and is drawn for high-temperature conditions of about 1000°C. The SiO_2 phase, as would be expected at this temperature, is tridymite. The assemblages depicted, therefore, are analogous to those found in high-temperature volcanic occurrences. Tielines connect mineral compositions that represent equilibrium coexistences. Triangles consist of three sets of tielines (outlining the triangle) depicting three-phase equilibrium coexistences. The diagram illustrates that feldspathoids do not coexist with tridymite because of the intervening alkali feldspar series. This specific triangular diagram is of importance in igneous

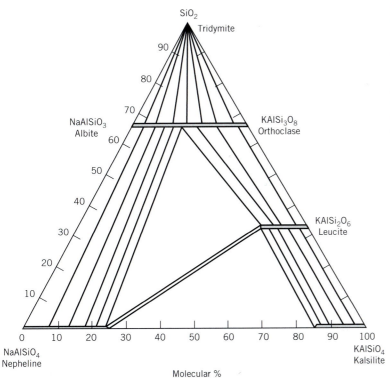

FIG. 4.24. The system SiO_2–$NaAlSiO_4$ (nepheline)–$KAlSiO_4$ (kalsilite) at approximately 1000°C and atmospheric pressure, as determined experimentally. The extent of solid solution for several phase regions is shown, as are tielines between coexisting pairs of minerals. Triangles represent three-phase coexistences. (Adapted from *Petrologic Phase Equilibria*, 21st ed., Fig. 4.39, by W. G. Ernst. Copyright © 1976 by W. H. Freeman and Company.)

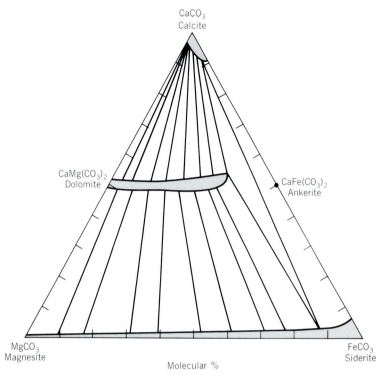

FIG. 4.25. *(a)* The system $CaCO_3$–$MgCO_3$–$FeCO_3$ showing extent of solid solution for phase regions and coexistences for natural assemblages metamorphosed to about 400°C (the biotite zone of the greenschist facies). A three-phase assemblage is shown by a triangle. In this study the extent of solid solution is based on experimental data as well as on analytical results of natural assemblages. The composition of coexisting carbonates was based on electron microprobe analyses of carbonates over a wide compositional range in low- to medium-grade metamorphic rocks. The open regions across which tielines are drawn are miscibility gaps. Extent of solid solution is shaded. (Adapted from Anovitz, L. M. and Essene, E. J., 1987, Phase equilibria in the system $CaCO_3$–$MgCO_3$–$FeCO_3$, *Journal of Petrology*, v. 28, pp. 389–415.)

petrology because it outlines mineral compositions in SiO_2-rich and SiO_2-poor rocks. All assemblages above the albite-orthoclase series contain an SiO_2 phase, whereas all assemblages below the alkali feldspar series are free of SiO_2 minerals.

Another illustration of an assemblage diagram for a carbonate system with very common carbonate minerals is shown in Fig. 4.25. This diagram is based upon naturally occurring compositional ranges, and coexistences of metamorphic carbonates in a temperature range of about 400°C (the biotite zone of the greenschist facies). It shows a very small solid solution region for calcite, $CaCO_3$, an extensive (but not complete) series between dolomite, $CaMg(CO_3)_2$, and ankerite, $CaFe(CO_3)_2$, and a complete series between magnesite, $MgCO_3$, and siderite, $FeCO_3$. Common tielines between coexisting pairs are shown, as well as one three-phase coexistence.

Triangular assemblages diagrams are most commonly determined from experimental results, or from carefully studied and chemically analyzed natural assemblages. In order to delineate the extent of solid solution of any of the minerals in such diagrams, one of several chemical analytical techniques must be used (see Chapter 7). However, for mineral assemblages that represent a relatively simple chemical system, and where the minerals involved are relatively easily identifiable, mineral assemblage diagrams can be constructed on the basis of hand

specimen study of a large suite of assemblages that relate to the system in question. In hand specimen study no knowledge can be gained about solid solution; therefore, the mineral formula for each mineral is plotted as a point on the triangular diagram, without reference to any possible solid solution extent. Coexistences (and as such tielines) are based upon the visual observation of two or three minerals touching each other in the hand specimen. An example of such an assemblage diagram in the system Cu–Fe–S is shown in Fig. 4.26. Here all minerals are represented as points on the diagram, reflecting the fact that any possible solid solution within any of the minerals was ignored.

Until now we have discussed and illustrated triangular diagrams with only three components (one at each corner). Petrologists may also use triangular diagrams for assemblage representations in which two components are grouped at a specific corner, so as to represent more complex chemical systems that are more alike to those found in rocks. Figure 4.27 illustrates two assemblage diagrams in which two oxide components have been grouped together in the right-hand corner. This allows for a more complete representation of the mineral compositions in this instance. In Fig. 4.27 the basalt assemblage may have crystallized at a constant and relatively low *P* (e.g., 1 to 2 kilobars) and a temperature of somewhere between 1000 and 1200°C. The chemically

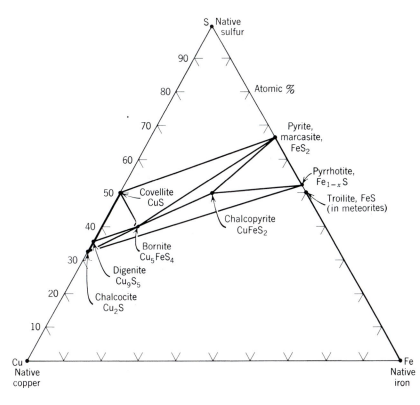

FIG. 4.26. Some of the most common sulfides represented in the Cu–Fe–S system. Many of these sulfides (e.g., bornite and chalcopyrite) show some solid solution of especially Cu and Fe; this is not shown in the diagram. Tielines connect commonly occurring pairs of minerals. Triangles indicate coexistences of three sulfides. The Fe-FeS coexistence is common in iron meteorites.

equivalent eclogite assemblage formed at a much higher pressure, possibly over a range of 12 to 30 kilobars, and a temperature range of 400 to 800°C. In other words, the two assemblage diagrams represent very different P and T conditions for identical bulk chemical compositions. In assemblage diagrams obtained from experimental studies the temperatures and pressures can be closely controlled and isobaric and isothermal phase diagrams can be obtained for

much narrower ranges in P and T than is generally possible for natural rock or mineral systems.

Diagrams for Mineral Reactions Involving H_2O or CO_2

Pressure–temperature diagrams are used to outline the stability field of a mineral that, along some $P–T$ curve, gives way to another mineral by a chemical reaction. The only $P–T$ diagrams that we have discussed until now are those that have involved

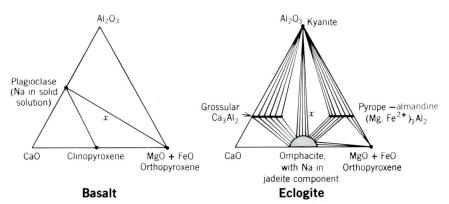

FIG. 4.27. Coexisting minerals in the system Al_2O_3–CaO–(MgO + FeO)–SiO_2–Na_2O. In *basalt* a common assemblage is plagioclase-clinopyroxene-orthopyroxene, as shown by the triangle of connecting tielines (the *x* represents a possible bulk composition of basalt with this mineral assemblage). In *eclogite*, of the same bulk composition, marked by *x*, the assemblage kyanite-omphacite-pyrope (as a component in almandine) occurs. Composition bars and omphacite field in the eclogite diagram outline approximate extent of solid solution. Eclogitic rocks may form in the mantle of the Earth from basaltic compositions.

structural (polymorphic) changes (see Figs. 4.13 and 4.16). However, experiments can also be done that delineate the stability fields of, for example, a hydrous mineral and its anhydrous (higher temperature) reaction products. Figure 4.28a shows the stability region of anthophyllite, $Mg_7Si_8O_{22}(OH)_2$, in the low-temperature part of the diagram and the stability of the higher-temperature reaction products, enstatite ($MgSiO_3$), and quartz (SiO_2), and fluid. This chemical system can be defined as $MgO–SiO_2–H_2O$. The curve delineates the $P–T$ region over which the breakdown of Mg–amphibole occurs. According to

$$Mg_7Si_8O_{22}(OH)_2 \rightleftharpoons 7MgSiO_3 + SiO_2 + H_2O \nearrow$$

FIG. 4.28. *(a)* Schematic $P–T$ diagram for the system $MgO–SiO_2–H_2O$. The stability fields of anthophyllite, $Mg_7Si_8O_{22}(OH)_2$, and reaction products, enstatite, $MgSiO_3$ + quartz + H_2O are shown. (After H. J. Greenwood, 1963, *Journal of Petrology* 4: 325.) *(b)* $P–T$ diagram for the system $CaO–SiO_2–CO_2$, showing the stability fields of calcite in the presence of SiO_2, and without SiO_2 in the assemblage. (Modified from *Petrologic Phase Equilibria*, 21st ed. Fig. 6.14a, by W. G. Ernst. Copyright © 1976 by W. H. Freeman and Company. Reprinted by permission.)

The H_2O enters the fluid phase produced in the reaction.

A pressure–temperature diagram involving mineral stabilities as a function of CO_2 is shown in Fig. 4.28b. In the system $CaO–CO_2$, calcite is stable to high temperatures, even at relatively low CO_2 pressures (in Fig. 4.28b calcite is still stable at about 1000°C and P_{CO_2} pressure of less than 10 bars). At temperatures above the curve, calcite decomposes as follows:

$$CaCO_3 \rightleftharpoons CaO + CO_2 \nearrow$$

However, the addition of SiO_2 (quartz) to the system, resulting in the system $CaO-SiO_2–CO_2$, produces a large stability field for wollastonite, $CaSiO_3$, on account of the reaction:

$$CaCO_3 + SiO_2 \rightleftharpoons CaSiO_3 + CO_2 \nearrow$$

Figure 4.28b illustrates that $CaCO_3$ decomposes at much lower temperatures in the presence of SiO_2 than in the simple system $CaO–CO_2$. The diagrams shown in Fig. 4.28 are commonly used in metamorphic petrology to evaluate mineral stabilities as a function of increasing temperature.

In the evaluation of the conditions that metamorphic rocks may have undergone, composite $P–T$ diagrams with reaction curves for several minerals are commonly used. Figure 4.29 is such a diagram. The reaction curves for Al_2SiO_5 and for muscovite + quartz = K-feldspar + sillimanite + H_2O are especially relevant to metamorphic rocks with high Al_2O_3 contents relative to other components such as CaO,

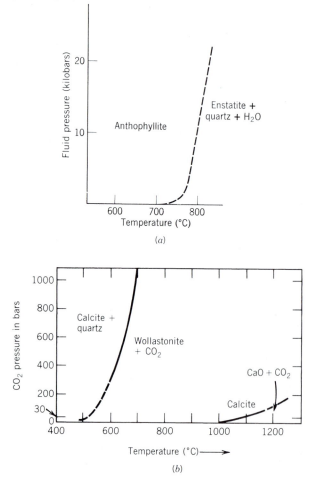

(a)

(b)

FIG. 4.29. Reaction curves for some common metamorphic minerals. For significance of shaded area, see text.

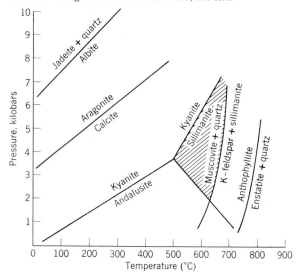

MgO, and FeO. Shales with abundant clay minerals have relatively high Al_2O_3 contents and during metamorphism mineral reactions take place in the shale, as well as recrystallization. The metamorphic equivalents of shales, known collectively as pelitic schists, may contain abundant sillimanite, muscovite, and quartz. This type of schist, on the basis of the reaction curves, and stability fields of minerals in Fig. 4.29 may have equilibrated under P–T conditions as outlined by the shaded area in Fig. 4.29. In metamorphic rocks that may have undergone very high pressures, such as eclogites, the originally present albite may, during metamorphism, have reacted to form jadeite + quartz, and any original calcite may have transformed to aragonite. The presence of jadeite + quartz, for example, would indicate a minimum pressure during metamorphism of about 10 kilobars at 300°C (see upper left curve in Fig. 4.29).

Eh–pH Diagrams

Until now we have discussed various types of phase diagrams that are applicable to minerals that have formed in high-temperature and/or high-pressure environments, as may be the case for igneous and metamorphic rocks. At low temperatures (essentially room temperature; 25°C) and low pressure (atmospheric pressure), such as prevail under atmospheric conditions, it is often useful to express the stability fields of phases (or minerals) in terms of Eh (oxidation potential) and pH (the negative logarithm of the hydrogen ion concentration; it allows for the definition of acidic conditions with pH = 0–7, for basic conditions with pH = 7–14, and for neutral solutions at pH = 7). The construction of such diagrams, also known as Pourbaix diagrams, on the basis of thermodynamic parameters is outlined in Garrels and Christ (1965; see reference list). Figure 4.30 illustrates the stability fields of some iron oxides and iron sulfides as calculated for atmospheric conditions (25°C, 1 atmosphere pressure). It shows at a glance that hematite is stable in an oxidizing region (high Eh). In hematite, $Fe_2^{3+}O_3$, all iron is present in the most oxidized, trivalent state. Magnetite, on the other hand, is stable under more reducing conditions, lower Eh. Magnetite, Fe_3O_4, which can be rewritten as $FeO·Fe_2O_3$, consists of $\frac{1}{3}Fe^{2+}$ and $\frac{2}{3}Fe^{3+}$. The two sulfides, pyrite and pyrrhotite, occur under reducing conditions and pH values between 4 and 9. In both of these, the oxidation state of iron is Fe^{2+}. Along the lines that separate the various mineral fields, two of the minerals can coexist in equilibrium. For example, hematite and magnetite can coexist under conditions of variable pH and Eh (line *ab*, Fig. 4.30);

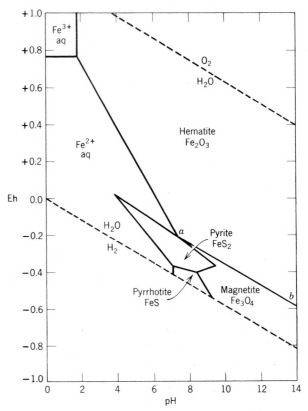

FIG. 4.30. Stability fields of iron oxides and iron sulfides in water at 25°C and 1 atmosphere total pressure, with an activity of total dissolved sulfur = 10^{-6}. The pyrrhotite field is calculated on the basis of the formula FeS in this diagram; more correctly it would be $Fe_{1-x}S$. (Adapted from R. M. Garrels and C. L. Christ, 1965, *Solutions, minerals, and equilibria.* San Francisco: Freeman, Cooper and Co., Fig. 7.20.)

indeed, hematite and magnetite commonly occur together in Precambrian sedimentary iron-formations. Diagrams of this type are especially useful in evaluating some of the physical chemical parameters that prevail during conditions of atmospheric weathering and of chemical sedimentation and diagenesis of water-laid sediments (at essentially atmospheric pressure conditions and temperatures ranging from 25°C to about 100°C). The size of the sulfide fields (pyrite and pyrrhotite) depends on the amount of sulfur in solution. If sulfur were lower and the CO_2 content high there would be a stability field of siderite ($FeCO_3$) instead. On the basis of the frequent coexistence of hematite and magnetite in Proterozoic banded iron-formations one may be able to evaluate to some extent parameters such as Eh and pH of the original Proterozoic sedimentary basin.

Another example of an Eh–pH diagram is given in Fig. 4.31. This is for the stability fields of three manganese oxides, one manganese hydroxide, and

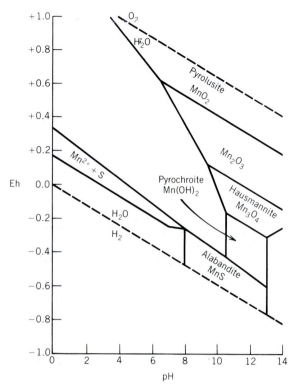

FIG. 4.31. Stability fields of manganese oxides, hydroxide, and sulfide in water at 25°C and 1 atmosphere total pressure, with activity of total dissolved sulfur = 10^{-1}. (Adapted from D. D. Runnels, 1962, in H. H. Schmitt, ed., *Equilibrium diagrams for minerals at low temperatures and pressure*, Cambridge, Mass.: Geology Club of Harvard, 199 pp.)

one manganese sulfide. As in Fig. 4.30, the most oxidized oxide is stable at the highest Eh values. Pyrolusite, MnO_2, is the most oxidized naturally occurring oxide, with all Mn in the Mn^{4+} state. Manganite, Mn_2O_3, with all Mn in the Mn^{3+} state, is stable over a slightly less highly oxidizing Eh range. Hausmannite, Mn_3O_4, which can be rewritten as $MnO \cdot Mn_2O_3$, contains $\frac{1}{3}Mn^{2+}$ and $\frac{2}{3}Mn^{3+}$. As a consequence, it is stable at Eh values below manganite. In $Mn(OH)_2$ and MnS, both of which show stability fields at the bottom of the Eh scale, Mn is present at Mn^{2+}. This type of diagram is useful not only in the prediction of what Mn-rich minerals may precipitate from an Mn-rich system at atmospheric conditions, but also for the interpretation of the conditions of formation of naturally occurring Mn compounds in manganese formations.

POLYMORPHIC REACTIONS

Polymorphism occurs whenever a specific chemical compound exists in more than one structural or atomic arrangement. Therefore, *the ability of a specific chemical substance to crystallize with more than one type of structure* (as a function of changes in temperature, pressure, or both) is known as *polymorphism* (from the Greek meaning "many forms"). The various structures of such a chemical element or compound are known as *polymorphic forms*, or *polymorphs*. Examples of some polymorphous minerals are given in Table 4.2; Fig. 4.32 shows the stability regions of various polymorphs, in terms of

TABLE 4.2   Examples of Polymorphous Minerals

Composition	Mineral Name	Crystal System and Space Group	Hardness	Specific Gravity
C	Diamond	Isometric—$Fd3m$	10	3.52
	Graphite	Hexagonal–$P6_3/mmc$	1	2.23
FeS_2	Pyrite	Isometric—$Pa3$	6	5.02
	Marcasite	Orthorhombic—$Pnnm$	6	4.89
$CaCO_3$	Calcite	Rhombohedral—$R\bar{3}c$	3	2.71
	Aragonite	Orthorhombic—$Pnam$	$3\frac{1}{2}$	2.94
SiO_2	Low quartz	Hexagonal—$P3_121$	7	2.65
	High quartz	Hexagonal—$P6_222$		2.53
	High tridymite	Hexagonal—$P6_3/mmc$	7	2.20
	Low tridymite	Orthorhombic—$C222_1$		2.26
	High cristobalite	Isometric—$Fd3m$	$6\frac{1}{2}$	2.20
	Low cristobalite	Tetragonal—$P4_12_12$		2.32
	Coesite	Monoclinic—$C2/c$	$7\frac{1}{2}$	3.01
	Stishovite	Tetragonal—$P4_2/mnm$		4.30

*Crystal systems and space group notation are discussed in Chapters 5 and 6, respectively.

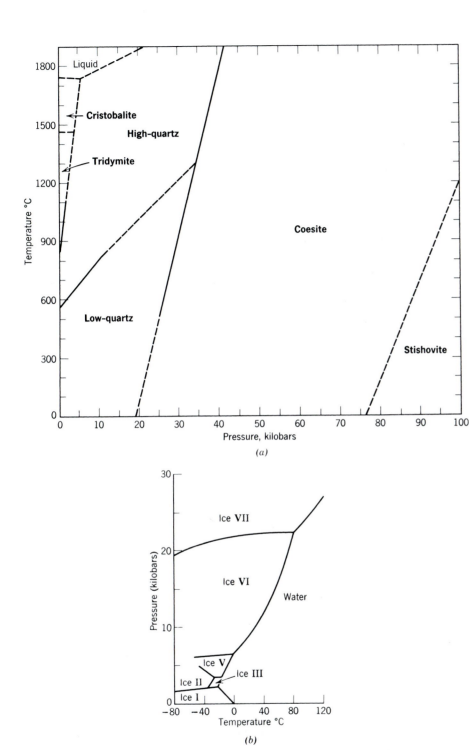

FIG. 4.32. *(a)* Stability relations of the SiO_2 polymorphs. Pressure is expressed in kilobars = 1000 bars, where 1 bar = 0.987 atmosphere. *(b)* P–T diagram for H_2O. Six polymorphic types are indicated by I, II, III, V, VI, and VII. (After P. W. Bridgman, 1935, *Jour. Chem. Phys.* 5, p. 965, and *Phase Diagrams for Ceramists*, copyright 1964 by the American Ceramic Society.)

temperature (*T*) and pressure (*P*), for the chemical systems SiO_2 and H_2O (Fig. 4.16 shows additional diagrams for other polymorphic systems).

Three major types of mechanisms are recognized by which one polymorphic form of a substance can change to another. These are: *reconstructive, displacive,* and *order-disorder polymorphism.* Each of

these is illustrated with animations in module I of the CD-ROM.

The reason why a constant chemical composition may have different structural arrangements is that some structural configurations represent greater (or lesser) internal (structural) energies (*E*) than others. The relative internal energy of a specific poly-

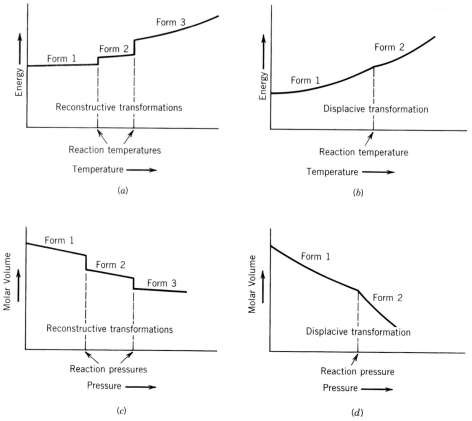

FIG. 4.33. Variation in internal energy (*E*) as a function of temperature for *(a)* three reconstructive polymorphs and *(b)* two displacive polymorphic forms. (Adapted from M. J. Buerger, 1961, Polymorphism and phase transformations, *Fortschr. Miner.* 39, pp. 9–14.) Variation in molar volume as a function of pressure for *(c)* three reconstructive polymorphs and *(d)* two displacive polymorphic forms.

morph may be a function of temperature, pressure, or both. A higher internal energy, as a function of increasing temperature, is caused by higher frequencies of vibrations of the atoms. Figure 4.33*a* shows the abrupt differences in the relative internal energy levels as a function of temperature for three polymorphs that are related by reconstructive transformations; these three structures are completely different from each other. Figure 4.33*b* shows the continuous increase in internal energy level in one polymorph (form 1) as a function of increasing temperature, until at a specific temperature the transformation takes place. This is what occurs in a displacive transformation, as a function of increasing temperature. Pressure can also be a major driving force in polymorphic transformations. Increasing pressure favors the development of structural arrangements that result in an increase in the density of atomic packing (as reflected in increased density (*D*), or specific gravity (**G**)), and a decrease in molar volume (molar volume is the volume occupied by one mole of sub-

stance). Figure 4.33*c* shows abrupt discontinuities in the molar volume values, whereas Fig. 4.33*d* shows no discontinuities; here, however, the slopes of the molar volumes are discontinuous.

Reconstructive Polymorphism

In a *reconstructive* polymorphic reaction the internal rearrangement in going from one form to another is extensive. It involves the breaking of atomic bonds and a reassembly of the structural units in a different arrangement (see Fig. 4.34). This type of transformation requires a large amount of energy, is not readily reversed, and is sluggish. An example of such a polymorphic reaction is the change from tridymite or cristobalite to low quartz. All three are polymorphs of SiO_2. Cristobalite and tridymite are formed at high temperatures and relatively low pressures (see Fig. 4.32*a*), such as in SiO_2-rich lava flows, from the devitrification of volcanic glasses, such as obsidian. Table 4.2 lists the specific gravity values as 2.20 for

(a)

(b)

FIG. 4.34. Schematic representation of reconstructive polymorphism. The transformation of a hypothetical structure *(a)* made of octahedral coordination polyhedra to *(b)* requires the breaking of bonds and complete rearrangement of the octahedral units. (Adapted from M. J. Buerger, 1961, Polymorphism and phase transformations, *Fortschr. Miner.* 39, pp. 9–14.)

both forms; these are the lowest values for any of the SiO_2 polymorphs listed. A high activation energy is needed to change the cristobalite (or tridymite) SiO_2 network into the arrangement of the low quartz structure. Cristobalite and tridymite are metastable in terms of atmospheric conditions; however, both minerals are abundantly present in very old terrestrial volcanic flows as well as in Precambrian lunar lavas. Such persistence of metastable minerals testifies to the fact that high energy is required to activate a reconstructive polymorphic transformation. Coesite and stishovite are forms of SiO_2 that are stable in the high to very high pressure part of the stability diagram in Fig. 4.32a. Coesite and stishovite both occur in meteorite craters, as a result of meteorite impacts on Earth. Coesite is also found in kimberlites, which are high pressure host rocks for diamond and which originate in the upper part of the Earth's mantle. In Table 4.2 note the high specific gravity for coesite and the very high value for stishovite, as compared to those of the other SiO_2 polymorphs. Stishovite is unusually densely packed, on an atomic level, for a silicate structure. In its structure each Si is surrounded by six neighboring oxygens (in octahedral coordination), whereas all other SiO_2 polymorphs contain Si with four oxygen neighbors in tetrahedral coordination. The importance of variable pressure regimes in the formation of some polymorphs is also seen in Fig. 4.32b, where the stability fields of the various polymorphs of ice are strongly dependent on the pressure applied to the system H_2O. The differences between the polymorphs for C, for FeS_2, and for $CaCO_3$ (Table 4.2) are such that a major reworking and rearrangement of the structure is needed to

go from one structure type to the other; each of these examples represents reconstructive polymorphism.

Displacive Polymorphism

In a *displacive* polymorphic reaction the internal adjustment in going from one form to another is very small and requires little energy. The structure is generally left completely intact, and no bonds between ions are broken; only a slight displacement of atoms (or ions) and readjustment of bond angles ("kinking") between ions is needed. This type of transformation occurs instantaneously and is easily reversible. Figure 4.35a is a schematic representation of a possible tetrahedral structure (with relatively high symmetry) "kinking" to produce two structural arrangements with lesser symmetry. These may be related by a symmetry element such as a twin plane (a mirror plane) as shown in Fig. 4.35b. A similar displacive transformation occurs when the high quartz form of SiO_2 is cooled to below 537°C (at atmospheric pressure; see also Fig. 4.32a) and rearranges its structure to that of low quartz. The difference between the two forms of quartz is expressed by their space groups (high quartz, $P6_222$; low quartz, $P3_221$; for discussion of space groups, see Chapters 5 and 6) and can be portrayed in a basal projection of the SiO_2 framework for both forms (see Fig. 4.36). The structural arrangement in the low-temperature form is slightly less symmetric and somewhat more dense than in the high-temperature form. The transition from high to low quartz can be viewed as the result of "kinking" of atomic bonds in the original high quartz structure. Because the high-temperature form of

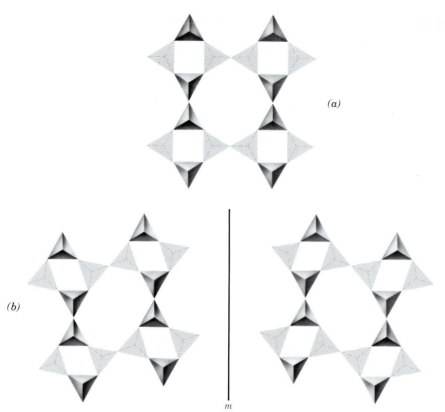

FIG. 4.35. Schematic representation of displacive polymorphism. *(a)* A hypothetical structure consisting of an infinite network of tetrahedra. It represents a fairly open atomic arrangement with relatively high symmetry. This would be the high-temperature polymorphic form. *(b)* The same infinite network of tetrahedra as shown in *(a)* but in a less symmetrical and somewhat collapsed ("kinked") arrangement. This represents the lower temperature polymorph. The two oppositely "kinked" structures are mirror images of each other, with *m* representing the mirror.

quartz (high quartz) shows a higher symmetry than the low-temperature form, twinning ("transformation twinning") may result in the transition from high to low quartz. Such twins, known as *Dauphiné twins* (see Fig. 4.56), represent a megascopic expression of the presence of obversed and reversed units of structure in a low quartz crystal.

Order–Disorder Polymorphism

Yet another type of polymorphism is referred to as an *order–disorder transformation*. This is commonly observed in metallic alloys but also occurs in minerals. To appreciate this transformation, it should be noted that perfect order occurs only at absolute zero (0 Kelvin, which is equivalent to −273.15°C). An increase in temperature disturbs the perfect order of a structure, until at some high temperature a totally disordered (totally random) state is obtained (see Fig. 4.37). As such, there is no definite transition point between perfect order and complete disorder. In a state of perfect order, the atoms in a structure are arranged in specific crystallographic site locations. At high temperatures, close to but below the melting point of a substance, atoms (or ions) tend to become completely disordered and are ready to break away from the structure. Slow cooling of a mineral structure will permit the original randomized ions (at high temperatures) to select specific sites in the structure and become more ordered. Figure 4.38*a* is a qualitative illustration of various degrees of disorder.

A more quantitative statement of order-disorder is shown in Fig. 4.38*b*. This shows an alloy of composition *AB* with 50% of *A* and 50% of *B* existing in various states of disorder, of which a totally disordered and a perfectly ordered state are two extreme conditions. In a perfectly ordered state atoms of *A* are arranged in perfect and regular repeat with respect to *B* atoms (see Fig. 4.38*b*-I). Atom *A* is always

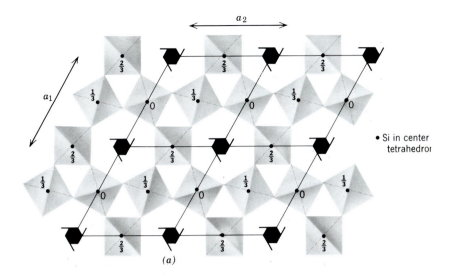

a_2

a_1

$\frac{2}{3}$ $\frac{2}{3}$ $\frac{2}{3}$

$\frac{1}{3}$ 0 $\frac{1}{3}$ 0 $\frac{1}{3}$ 0

$\frac{2}{3}$ $\frac{2}{3}$ $\frac{2}{3}$

$\frac{1}{3}$ 0 $\frac{1}{3}$ 0 $\frac{1}{3}$ 0 $\frac{1}{3}$

$\frac{2}{3}$ $\frac{2}{3}$ $\frac{2}{3}$ $\frac{2}{3}$

(a)

• Si in center
  tetrahedror

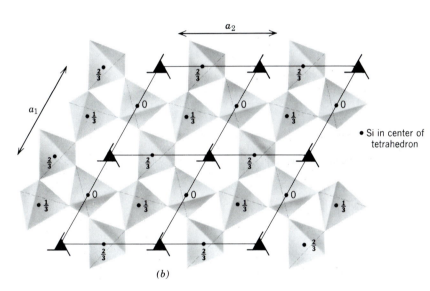

a_2

a_1

$\frac{2}{3}$ $\frac{2}{3}$ $\frac{2}{3}$ $\frac{2}{3}$

0   0   0

$\frac{1}{3}$ $\frac{1}{3}$ $\frac{1}{3}$

$\frac{2}{3}$ $\frac{2}{3}$ $\frac{2}{3}$

$\frac{1}{3}$ 0 $\frac{1}{3}$ 0 $\frac{1}{3}$ 0 $\frac{1}{3}$

$\frac{2}{3}$ $\frac{2}{3}$ $\frac{2}{3}$

(b)

• Si in center of
  tetrahedron

FIG. 4.36. *(a)* Projection of the tetrahedral SiO_2 framework of high quartz onto (0001). Four unit cells and sixfold screw axes (denoted as 6_2) are shown. Fractional heights represent locations of centers (Si) of tetrahedra. *(b)* Projection of the tetrahedral SiO_2 framework of low quartz onto (0001). Four unit cells and some threefold screw axes (denoted as 3_2) are shown. Fractional heights represent locations of centers (Si) of tetrahedra. These two figures illustrate a displacive transformation.

FIG. 4.37. The relationship of structural order and temperature. The higher the temperature the more disordered the distribution of atoms in specific structural sites. This would apply, for example, to the distribution of Al^{3+} versus Si^{4+} among tetrahedrally coordinated sites in the feldspar structure (see Fig. 4.39).

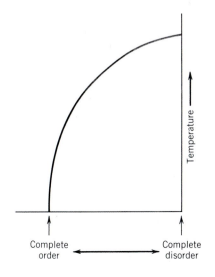

Temperature

Complete order ⟷ Complete disorder

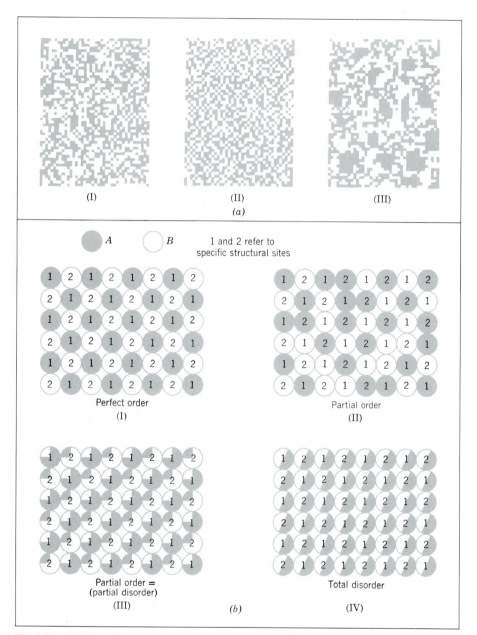

FIG. 4.38. (a) Distribution patterns of black-and-white squares. I: statistically random; II: somewhat less random, toward a chessboardlike pattern; III: even less random with segregations of larger black-and-white domains. (From F. Laves, reproduced in C. W. Correns, 1967, *Introduction to mineralogy.* New York: Springer-Verlag, p. 92. Original reference to Laves could not be located.) (b) Schematic illustration of order-disorder polymorphism in alloy *AB* (see text for explanation).

in structural site 1 and atom *B* in structural site 2. Two examples of somewhat disordered states for the atomic arrangement of such an alloy are shown in Figs. 4.38*b*-II and III. Here the distributions are not perfectly ordered and yet also are not random. In Fig. 4.38*b*-II the ratio *A* : *B* is still 1 : 1, but the atoms *A* and *B* are in a partially disordered array. In Fig. 4.38*b*-III, out of every four no. 1 sites three are occu-

pied by *A* (on the average) and one is occupied by *B* (on the average). The opposite holds for site no. 2. In other words, in Fig. 4.38*b*-III the probability ratio for site 1 being occupied by atoms of *A* rather than *B* is 3 : 1. A state of total disorder for the alloy *AB* is shown schematically in Fig. 4.38*b*-IV. Total disorder, on an atomic scale, implies equal probability of finding *A* or *B* in a specific site in the structure. In other

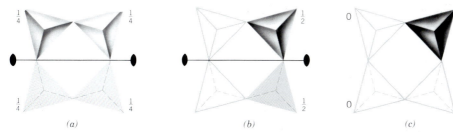

FIG. 4.39. Schematic representation of order-disorder of Al^{3+} and Si^{4+} in four linked tetrahedra that may be part of a tetrahedral framework structure (the composition of the ring is $AlSi_3O_8$.) *(a)* The probability of finding Al^{3+} is equal for all four tetrahedra, that is, $\frac{1}{4}$ Al is statistically distributed over the four tetrahedral sites (*total disorder*). This is compatible with a twofold rotation axis as shown. *(b)* The Al-Si distribution is such that each of two tetrahedra (on average) contains $\frac{1}{2}$ Al, whereas the other two tetrahedra contain Si only. This *partial order* is still compatible with twofold rotation symmetry. *(c)* All of the Al (one Al^{3+} per ring) is concentrated in one tetrahedron. This is a state of *complete order*. The ring now has lost the twofold rotation symmetry.

words, the probability of a given atomic site being occupied by one type of atom instead of another equals 1. This means that for a graphic representation, as in Fig. 4.38*b*-IV, each atomic position is indicated to represent *A* as well as *B* occupancy (in a statistical sense).

An example of order-disorder polymorphism in a mineral is shown by potassium feldspar ($KAlSi_3O_8$), in which Al occupies a structural position identical with and replacing Si in the mineral. The high-temperature form, sanidine, shows a disordered distribution of Al in the aluminosilicate framework. The low-temperature K-feldspar, microcline, however, shows an ordered distribution of Al in the aluminosilicate framework. States of intermediate order (which are equivalent to intermediate disorder) are present between that of high-temperature sanidine and low-temperature microcline. Figure 4.39 is a schematic illustration of Al-Si order-disorder in a tetrahedral ring of a structure; this is similar but not identical to the order-disorder in $KAlSi_3O_8$.

Polytypism

A special kind of polymorphism, known as *polytypism*, occurs when two polymorphs differ only in the stacking of identical, two-dimensional sheets or layers. As a consequence, the unit cell dimensions parallel to the sheets will be identical in the two polytypes. However, the atomic spacings between the sheets (or layers) will be related to each other as multiples or submultiples. Polytypism is a well-known feature of SiC, ZnS, the micas, and other layer silicates. The only difference between sphalerite (ZnS) and its polymorph wurtzite (ZnS) is that the S atoms in sphalerite are in cubic closest pack-

ing, whereas in wurtzite they are in hexagonal closest packed arrays. However, wurtzite shows extensive polytypism as reflected in *c* dimensions of the unit cell (the "thickness" of the basic ZnS layer is 3.12 Å). Examples of wurtzite polytypes and their *c* dimensions are 4*H*, 12.46 Å; 6*H*, 18.73 Å; 8*H*, 24.96 Å; and 10*H*, 31.20 Å (*H* refers to hexagonal cell). The Å dimensions show multiples of the basic unit along the *c* axis.

Micas and other layer silicates consist of infinitely extending tetrahedral silicate sheets (of composition $Si_2O_5(OH)$) that are stacked in various ways along the *c* axis. The inherent symmetry of such sheets is monoclinic. Because of the hexagonal symmetry about the (OH) group in the plan view of the $Si_2O_5(OH)$ tetrahedral sheets, there are six alternate directions (see Fig. 4.40*a*) in which $Si_2O_5(OH)$ sheets can be stacked. These six directions can be represented by three vectors (1, 2, and 3, and opposite negative directions $\bar{1}$, $\bar{2}$, and $\bar{3}$) at 120° to each other. If the stacking of $Si_2O_5(OH)$ sheets is always in the same direction (see Fig. 4.40*b*), the resulting structure will have monoclinic symmetry; this is referred to as the 1*M* (*M* = monoclinic) polytype with stacking sequence [1, 1, 1, . . .]. If the stacking sequence of $Si_2O_5(OH)$ sheets consists of alternating 1 and $\bar{1}$ directions (in opposing directions along the same vector), a structure results that is best described as orthorhombic. This stacking sequence can be expressed as [1, $\bar{1}$, 1, $\bar{1}$, . . .] and the polytype is known as 2*O* (*O* = orthorhombic). If the stacking sequence consists of two vectors at 60° to each other, that is [1, $\bar{2}$, 1, $\bar{2}$, . . .], another monoclinic polytype results; this is known as 2*M*₁ (see Fig. 4.40*c*). When all three vectors come into play, such as in [1, 2, 3, . . .], a trigonal polytype, 3*T* (*T* = trigonal) results. Module I

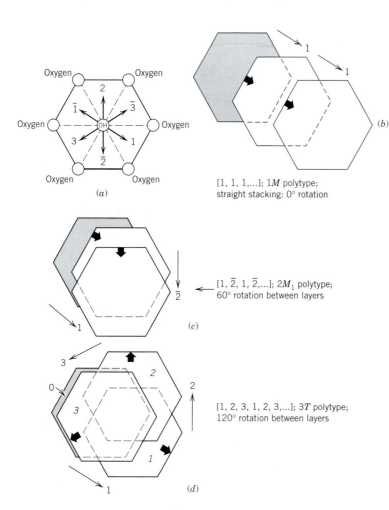

FIG. 4.40. Schematic illustration of some possible stacking polymorphs (*polytypes*) in mica. *(a)* Three vectorial directions for possible location of the (OH) group in an $Si_2O_5(OH)$ sheet that is stacked above or below the hexagonal ring shown. *(b)* Stacking of $Si_2O_5(OH)$ sheets in the same direction. *(c)* Stacking in two directions at 60° to each other. *(d)* Stacking in three directions at 120° to each other. Here, sheet 3 should lie directly above sheet 0 but is offset slightly for illustrative purposes.

of the CD-ROM has extensive animations that illustrate these various polytype stackings.

Table 4.3 lists the stacking sequence, its rotation angle, the number of layer repeats and the resulting repeat distance (along a vertical c axis), as well as the polytype symbol for six different stacking arrangements. A polytype is represented by a *Ramsdell structure symbol* consisting of a number to indicate the number of layers in the repeating unit along the vertical (c) axis and letters to designate the resulting symmetry (M = monoclinic, O = orthorhombic, T = trigonal, H = hexagonal). Subscripts 1 and 2 are

TABLE 4.3   Aspects of Stacking in Sheet Silicates; the Vertical Dimension of the Layer in Such Structures in Approximately 10 Å.

Stacking sequence	[1, 1]	[1, $\bar{2}$]	[1, 2]	[1, $\bar{1}$]	[1, 2, 3]	[1, $\bar{3}$, 2, $\bar{1}$, 3, $\bar{2}$]
Layer rotation	0°	60°	120°	180°	120°	60°
Number of layers in the final repeat	1	2	2	2	3	6
Original layer thickness	10 Å	10 Å	10 Å	10 Å	10 Å	10 Å
Final repeat distance	10 Å	20 Å	20 Å	20 Å	30 Å	60 Å
Ramsdell symbol for polytype	1M	2M_1	2M_2	2O	3T	6H

(Based on Figure 6 in Hollocher, K. 1997. Building crystal structure ball models using pre-drilled templates: sheet structures, tridymite, and cristobalite. In *Teaching Mineralogy*, Mineralogical Society of America: 255–283.)

used to differentiate structures that have the same unit repeat and symmetry.

EXSOLUTION PROCESSES

In the discussion of solid solution in minerals in Chapter 3 it was pointed out that if two ions (or atoms) have very different sizes, the amount of substitution for each other would be expected to be limited. However, it was also pointed out that increased temperature would aid the substitution of ions of divergent size. That is, at elevated temperatures, the structure of a mineral expands overall and the amplitudes of vibration of the atoms become larger. As temperature increases, previously distinct structural sites become more similar on average, and finally the sites may become indistinguishable. The interchange of cations between sites leads to chemical disorder, where the average chemical content of each site may become the same. In other words, where there might have been only limited solid solution between, for example, Na^+ (radius = 1.18 Å for

C.N. = 8) and K^+ (radius = 1.51 Å for C.N. = 8; from Table 3.11) at low temperature in a silicate (e.g., K-Na feldspar series), it is well known that at high temperature, such as about 1000°C, this same silicate series shows complete solid solution (or miscibility) between the $NaAlSi_3O_8$ and $KAlSi_3O_8$ end members.

Here we will address what happens when an originally homogeneous high-temperature mineral, containing cations of considerably different size, is allowed to cool to low temperature. The term *exsolution* refers to *the process whereby an initially homogeneous solid solution separates into two (or possibly more) distinct crystalline minerals without the addition or removal of material to or from the system. This means that no change in the bulk composition takes place* (see Fig. 4.41). Exsolution or "unmixing" is analogous to a well-known phenomenon that occurs when a mixture of oil and vinegar (vinaigrette) is made. The oil and vinegar are energetically stirred or shaken so as to produce a homogeneous but cloudy liquid suspension of very fine particles of both (oil and vinegar). When this mixture

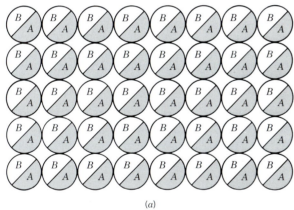

(a)

Low temperature

(b)

Decreasing temperature

FIG. 4.41. *(a)* Schematic two-dimensional representation of elements *A* and *B* in a completely disordered array (at high temperature). *(b)* Separation of elements *A* and *B* to *A*-rich and *B*-rich regions = exsolution (at lower temperature).

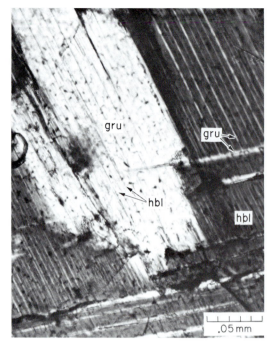

FIG. 4.42. Photomicrograph of relatively coarse exsolution lamellae of hornblende (hbl) in grunerite (gru) and grunerite (gru) lamellae in hornblende (hbl). The lamellae are oriented parallel to (001) and (100), respectively.

FIG. 4.43. *(a)* Bright-field transmission electron microscope (TEM) image of cummingtonite exsolution lamellae in a glaucophane host. The narrow lamellae of ferromagnesian amphibole occur in two symmetrically related orientations, nearly parallel to $(2\,8\,\bar{1}$ and $(2\,\bar{8}\,\bar{1})$ of the clinoamphibole structure. These un-

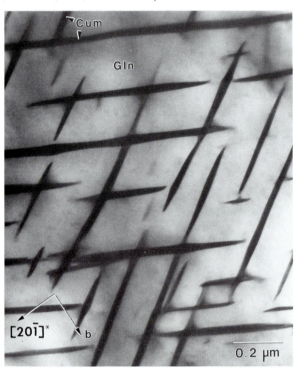

(a)

is allowed to stand for a short time, the original components will separate out ("unmix") and produce two clear liquids, oil and vinegar, as the end members.

Exsolution generally, although not necessarily, occurs on cooling. Some relatively coarse exsolution features can be observed under a high-power microscope when the scale of exsolution is about 1 μm (1 μm = 10^{-3} mm). An example of microscope exsolution in amphiboles is shown in a photomicrograph in Fig. 4.42. Exsolution lamellae that separate from the originally homogeneous host mineral are generally crystallographically oriented, as seen in Fig. 4.42. In amphiboles the orientation of exsolved lamellae is commonly parallel to {001} and {100}. In alkali (Na-K) feldspars, exsolution lamellae can often be seen in hand specimens to be approximately parallel to {100}. Such coarse-grained intergrowths in alkali feldspars consist of Na-rich feldspar lamellae exsolved from a K-rich host; they are known as *perthite* (see Fig. 12.73). If exsolution is on such a scale that it can be resolved only microscopically, it is referred to as *microperthite*; and if X-ray diffraction techniques are needed to resolve extremely fine (submicro-

usual orientations represent planes of best dimensional fit between the glaucophane and cummingtonite structures. This amphibole appeared perfectly homogeneous in the optical microscope and in the electron microprobe. (From E. A. Smelik and D. R. Veblen, 1991, Exsolution of cummingtonite from glaucophane: A new direction for exsolution lamellae in clinoamphiboles, *American Mineralogist* 76: 971–84.) *(b)* High-resolution transmission electron microscope (HRTEM) image showing a sodic pyroxene lamella (Napx) parallel to (100) in an augite matrix. (From M. T. Otten and P. R. Buseck, 1987, TEM study of the transformation of augite to sodic pyroxene in eclogitized ferrogabbro, *Contributions to Mineralogy and Petrology* 96: 529–38).

(b)

scopic) lamellae, the intergrowth is known as *cryptoperthite*. These types of crystallographically controlled intergrowths resulting from exsolution are common in many mineral systems: alkali feldspars, pyroxenes, amphiboles, and iron oxides, to name a few examples. Examples of extremely fine exsolution lamellae in amphiboles and pyroxenes are given in Fig. 4.43.

The origin of exsolution textures is best illustrated with reference to a schematic temperature-composition (*T-X*) phase diagram in Fig. 4.44. The horizontal axis represents the compositional variation, in terms of molecular percentages, between two possible silicates that are isostructural. The hypothetical silicates are monoclinic and are represented as ASi_xO_y and BSi_xO_y (for pyroxene or amphibole groups, for example). The ionic sizes of A and B differ by about 25%. If we were to collect from the literature the chemical analyses for all low-temperature geological occurrences of these two minerals, we might conclude that there is very limited solid solution between the two end members, as shown graphically by the low-temperature (T_1) composition bar. These same two hypothetical silicates might also be

present in high-temperature basalt flows that cooled very rapidly. At such high temperature (T_2) we may find a complete solid solution series (shown by shading in Fig. 4.44). What happens to the extent of solid solution between T_2 and T_1 is outlined by the *miscibility gap* in the diagram . This gap represents a temperature–composition field in which solid solution between the end members decreases gradually from higher to lower *T*. For specific compositions in this region two minerals coexist, rather than one homogeneous mineral, as above the gap.

Let us take a specific composition (X_1) in the region above the miscibility gap at T_2. At this temperature, the structure of the silicate is in a high-energy state and will allow relatively easy accommodation of A and B (although these differ by as much as 25% in ionic radius) in the same atomic sites. In other words, the A and B ions randomly occupy the various cationic sites in the structure (disorder; see Fig. 4.41*a*). On very slow cooling of this mineral a temperature (T_4) is reached at which the structure is unable to maintain a complete solid solution between A and B ions. Stresses in the structure resulting from the difference in size of the cations cause

FIG. 4.44.   *(a)* Schematic *T–X* (*X* = composition and is expressed as molecular or atomic %) diagram for hypothetical silicates ASi_xO_y and BSi_xO_y. At temperature T_2 a complete solid solution exists between the two end members. At lower temperatures a miscibility gap occurs. The arrows along the edge of the gap indicate the change in compositions of the coexisting silicates with decreasing *T*. The process of exsolution is discussed in the text. *(b)* Cross-sections of mineral (with composition X_1 in FIG. 4.44*a*) illustrating the sequence of appearance of possible exsolution lamellae. (1) homogeneous, at T_2. (2) exsolution lamellae parallel to (001) at T_4. (3) additional set of exsolution lamellae parallel to (100) at T_3, and coarsened lamellae parallel to (001).

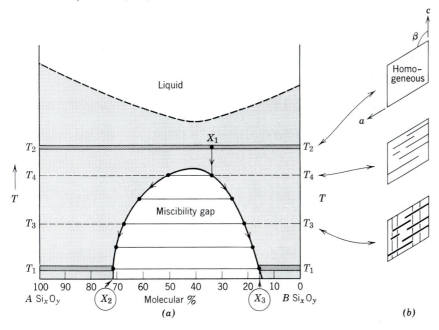

(a)   (b)

diffusion of A^+ ions to produce silicate regions with a preponderance of A^+; simultaneous B^+ ion diffusion results in silicate regions rich in B^+. This is the beginning of the process known as *unmixing*, which means that an originally homogeneous mineral segregates into two chemically different minerals (see Fig. 4.41*b*). The miscibility gap continually widens as the temperature is lowered from T_4. Under equilibrium conditions, the compositions of lamellae of the exsolution intergrowth that form as a function of decreasing temperature are given by points along the edge of the miscibility gap; such compositional points are joined by tielines. At the lowest temperature, T_1, we may find a rather coarse-grained, crystallographically well-oriented intergrowth of (A,B) Si$_x$O$_y$ lamellae with the specific composition (X_2) in a host of (B,A)Si$_x$O$_y$ with specific composition (X_3). Although the segregation of ions within the silicate structure, which began at T_4, is complex on an atomic level, it is due mainly to the inability of a single structure to house ions of disparate sizes in a random distribution when decreasing temperature lowers the energy of the crystal. Illustrations on exsolution can be found in module I of the CD-ROM as part of the title "Solid Solution Mechanisms." Figure 4.45 is an illustration of the free energy (G)–composition curve for the exsolution process. It shows the high free energy content of the high-temperature disordered compositional region, and the much lower free energy states of the two exsolved minerals, whose compositions are marked as x_2 and x_3 on the compositional bar. The free energy, also known as the Gibbs free energy (G), is defined as $G = E + PV - TS$, where E = internal energy, PV is a pressure-volume term (which for solids at atmospheric pressure is negligible compared to other thermodynamic quantities; see also p. 117), and TS is a temperature-entropy term.

Although exsolution intergrowths are common in many mineral groups, the resulting textures are rarely visible in hand specimens. Perthite in the alkali feldspar series is an example of exsolution on a large scale, visible in hand specimen. The schematic sequence of cross sections in Fig. 4.44*b* illustrates what may be seen under the microscope as a result of the processes depicted in Fig. 4.44*a*. An originally homogeneous crystal of composition X_1 develops sets of differently oriented exsolution lamellae between temperatures T_4 and T_3.

These segregation processes all take place within the silicate structure in the solid state. The T–X diagram depicting this sequence of events (Fig. 4.44) is,

therefore, in general referred to as a *subsolidus phase diagram*.

The process of unmixing, on an atomic scale, is one of atomic migration (diffusion). An increase in temperature (and an increase in kinetic energy) will increase the mobility of an atom (or ion) and its chances of breaking away from its neighbors and moving to a new position. This means that diffusion rate is strongly temperature dependent. As the temperature is lowered, as in the unmixing process depicted in Fig. 4.44, the mobility of the atoms may become so limited that the original homogeneous mineral cannot achieve the pure end member compositions shown in Fig. 4.41*b*. A composition may result that is a compromise between lowering the thermodynamic free energy of the system, and the decrease in the mobility of the atoms. The phase or assemblage of phases with the lowest free energy will be the most stable. In the exsolution process, the exsolved phases (the two minerals) have a much lower free energy than the original homogeneous, high-temperature and disordered composition (see Fig. 4.45).

Exsolution lamellae may range from visible in hand specimens (as in microcline perthite) to so small that they can be resolved only by very high-resolution transmission electron microscopy (see Fig. 4.43). The scale of the exsolution lamellae is a good indicator of the cooling rate of the rock in which this intergrowth texture occurs. In slowly cooled rocks, more time is available for diffusion (and unmixing), and as a result the exsolution texture is coarser. In rocks that are cooled extremely quickly ("quenched"), the high-

FIG. 4.45. Free energy (G)-composition curve illustrating the free energy changes during exsolution. The dashed line, which is tangential to the two minima in the free energy curve, gives the composition of the coexisting minerals at a specific temperature.

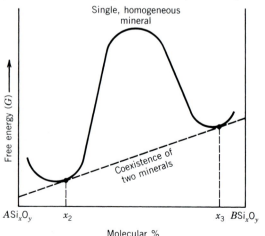

temperature, homogeneous, disordered mineral may be preserved, although in a metastable state. In a slightly slower but still rapid cooling process, very fine exsolution textures may develop.

As mentioned earlier, exsolution phenomena and their resulting exsolution textures are common in many mineral systems. Examples of two such systems with miscibility gaps are given in Figs. 4.46 and 4.47. Figure 4.46a shows a highly schematic diagram with three broadly outlined miscibility gaps in the plagioclase feldspar series. The locations of these gaps were originally established by single crystal

FIG. 4.46. *(a)* Approximate location of three miscibility gaps in the low-temperature region of the plagioclase feldspar series. *(b)* Schematic phase diagram for the plagioclase feldspar series showing a range of almost complete subsolidus solid solution at high temperatures, and the various miscibility regions at lower temperatures. (Simplified with permission after J. V. Smith and W. L. Brown, 1988, *Feldspar minerals*, v. 1, *Crystal structures, physical, chemical, and microtextural properties*. New York: Springer-Verlag, Fig. 1.4, 828 pp.)

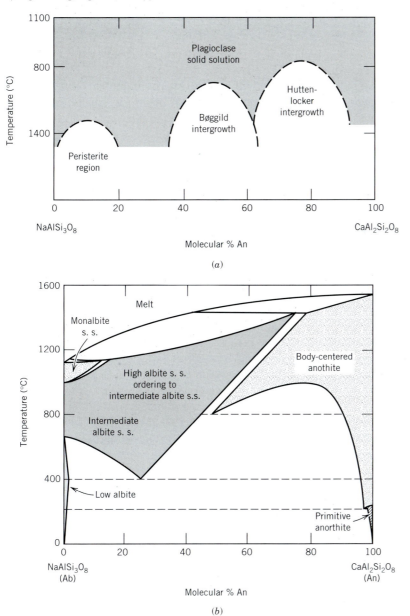

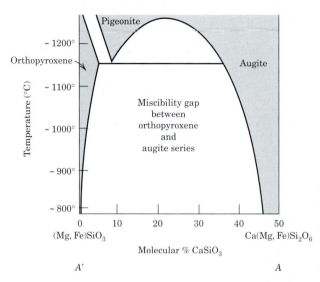

FIG. 4.47. A temperature–composition section across the pyroxene compositional regions in the Wo (wollastonite)–En (enstatite)–Fs (ferrosilite) diagram; see Fig. 4.23. This section (labeled *A'A*) is at about $En_{65}Fs_{35}$ (see location as marked *A'A* in Fig. 4.23). A broad miscibility gap exists between orthopyroxene (opx) and augite (clinopyroxene = cpx) compositions. Pigeonite is stable only at high-temperature conditions.

X-ray diffraction studies. The three regions have been named, in increasing An content, the *peristerite gap*, the *Bøggild intergrowth region*, and the *Huttenlocker intergrowth region*. These regions represent exsolution phenomena on an extremely fine scale. Exsolution lamellae are responsible for the *iridescence* (see page 25) of moonstone, as a result of the presence of closely spaced lamellae of An_0 and An_{25} composition (peristerite gap). Similarly, the *schiller* or *labradorescence* of labradorite is caused by very fine exsolution lamellae of An_{47} to An_{58} composition (see Fig. 2.12) across the gap described as Bøggild intergrowth. A temperature–composition diagram that incorporates the data in Fig. 4.46*a* for the plagioclase series is given in Fig. 4.46*b*. This figure represents the most up-to-date interpretation of data showing three miscibility gaps at temperatures below about 800°C and an almost complete solid solution series above about 800°C, up to the solidus of the diagram. Figure 4.47 is a pyroxene subsolidus section (note that the solidus and/or liquidus surfaces are not even shown) across pyroxene compositional space as shown in Fig. 4.23. This temperature composition section shows the slight expansion of solid solution at elevated temperatures in orthopyroxene, and the much enlarged solid solution field of augite at high temperatures. It also demonstrates that pigeonite is specifically a high-temperature phase region that must invert at about 1150°C to an orthopyroxene matrix with augite exsolution lamellae (due to presence of the miscibility gap and the fact that the structure of pigeonite is different from that of orthopyroxene). Pigeonite occurs only in quickly quenched high-temperature basalts. If the basalt has

cooled slowly the original pigeonite is now represented by orthopyroxene with augite exsolution lamellae.

METAMICT MINERALS

Minerals described as metamict originally formed as crystalline solids, but their crystal structure has been destroyed, to various degrees, by radiation from radioactive elements present in the original structure.

All metamict minerals are radioactive, and it is believed that their structural breakdown results mainly from the bombardment of alpha particles emitted from radioactive uranium or thorium contained in these minerals. The decay of a radioactive element can result in the complete destruction of the periodic structure around it. Usually this is caused by daughter elements of different size and charge from the parent element. However, the process is not well understood, because some minerals containing less than 0.5% of these elements may be metamict, whereas other minerals containing high percentages of them may be crystalline.

The various stages of metamictization (the various degrees to which the original crystalline structure has been destroyed) are determined by a combination of X-ray diffraction and high resolution transmission electron microscopy (HRTEM) techniques. A well-ordered structure gives well-defined X-ray diffraction patterns, shows optical microscopic interference effects, and has ordered electron beam images (electron diffraction and structural projections). Different stages of destruction of the original

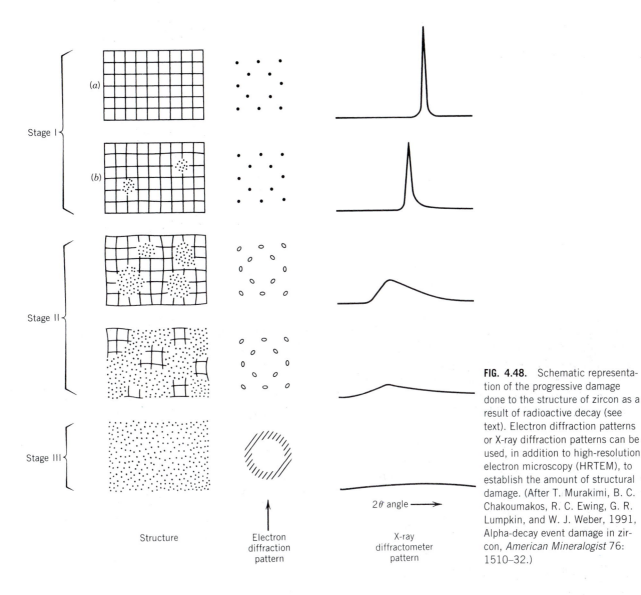

Stage I

(a)

(b)

Stage II

Stage III

Structure

Electron
diffraction
pattern

2θ angle ———►

X-ray
diffractometer
pattern

FIG. 4.48. Schematic representation of the progressive damage done to the structure of zircon as a result of radioactive decay (see text). Electron diffraction patterns or X-ray diffraction patterns can be used, in addition to high-resolution electron microscopy (HRTEM), to establish the amount of structural damage. (After T. Murakimi, B. C. Chakoumakos, R. C. Ewing, G. R. Lumpkin, and W. J. Weber, 1991, Alpha-decay event damage in zircon, *American Mineralogist* 76: 1510–32.)

structure are reflected in changes in the diffraction and structural patterns. Figure 4.48 shows several such stages of destruction of the original crystal structure of zircon ($ZrSiO_4$) owing to the presence of radioactive uranium and thorium. Stage 1(a) shows a well-ordered structure with well-established periodicities throughout; this is the unaltered structure. Stage 1(b) shows some destruction of the structure and the appearance of some aperiodic domains (amorphous regions). Stage II shows large increases in the volume percentage of aperiodic (amorphous) domains as a result of further destruction of the structure. In the final stage, III, all periodicities of the original structure are lacking and the material is totally amorphous. The changes in the electron diffraction and X-ray diffraction patterns are the evidence on

which the structural pictures are based. Figure 4.49 is an HRTEM illustration of metamict domains in an otherwise crystalline matrix of olivine (($Mg,Fe)_2 SiO_4$). In this instance the damage was artificially induced by high-voltage ion beam irradiation of the originally completely crystalline olivine.

The original nonmetamict mineral may well have shown good cleavage, in addition to the well-defined electron and X-ray diffraction patterns in Fig. 4.48. The metamict product will not show cleavage but instead will show conchoidal fracture. Many metamict minerals are bounded by crystal faces and are thus amorphous pseudomorphs after an earlier crystalline mineral. When a metamict mineral is heated, its crystal structure may be reconstituted and its density increased.

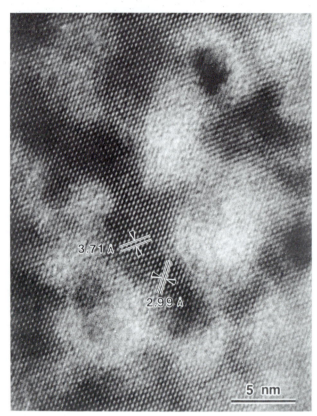

FIG. 4.49. High-resolution transmission electron microscopic (HRTEM) structural image of radiation-induced metamict domains inside a well-crystallized host of olivine ($(Mg, Fe)_2 SiO_4$). The damage was artificially produced by subjecting the original olivine crystal to a high-kilovoltage ion beam. Scale bar is in nanometers (nm): 1 nm = 10 angstroms. (From L. M. Wang, M. L. Miller, and R. C. Ewing, 1991, High-resolution TEM observation of displacement cascades in krypton-irradiated silicate minerals, in *Proceedings of the 49th Annual Meeting of the Electron Microscopy Society of America*, EMSA, pp. 910–11).

MINERALOIDS (NONCRYSTALLINE MINERALS)

The definition of a mineral states that "a mineral . . . has a highly ordered atomic arrangement." There are, however, a number of noncrystalline, natural solids classed as amorphous. Figure 4.50*a* is a schematic illustration, on an atomic scale, of the amorphous state.

Amorphous minerals include gel minerals and glasses. Gel minerals (mineraloids) are usually formed under conditions of low temperature and low pressure and commonly originate during weathering processes. They characteristically occur in mammillary, botryoidal, and stalactitic masses. The ability of amorphous materials to absorb all kinds of ionic species accounts for their often wide variations in chemical composition. Commonly amorphous minerals are limonite, $FeO \cdot OH \cdot nH_2O$, and allophane, a hydrous aluminum silicate.

The structure of a silica-rich glass, such as volcanic glass, is said to have *short-range order* but lacks *long-range order*. The Si^{4+} and Al^{3+} ions in such a glass occur in tetrahedral coordination, as they do in crystalline compounds. In glasses, how-

FIG. 4.50. Schematic illustration of the atomic arrangement in an amorphous alloy *(a)* and a crystalline alloy *(b)*.

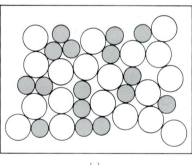

(a)

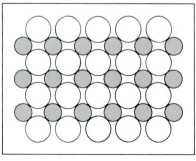

(b)

ever, such tetrahedral coordination polyhedra do not repeat their pattern over more than a few angstrom units. In other words, *long-range order* is absent, although *short-range order* (as shown by the presence of coordination tetrahedra) is present.

A well-known example of a partly amorphous material is opal. Its chemical composition can be represented as $SiO_2 \cdot nH_2O$ with an average range of H_2O content from 4 to 9 weight percent. Opal was originally considered to be completely without internal structure; however, careful electron beam studies show that it contains an ordered arrangement of very small SiO_2 spheres (see Fig. 2.10).

PSEUDOMORPHISM

The existence of a mineral with the outward crystal form of another mineral species is known as *pseudomorphism*. If a crystal of a mineral is altered so that the internal structure or chemical composition is changed but the external form is preserved, it is called a *pseudomorph* (from the Greek meaning "false form"). The chemical composition and structure of a pseudomorph belong to one mineral species, whereas the crystal form corresponds to another. For example, pyrite, FeS_2, may change to limonite, $FeO \cdot OH \cdot nH_2O$, but it will preserve all the external features of the pyrite. Such a crystal is described as a pseudomorph of *limonite after pyrite*. Pseudomorphs usually are further defined according to the manner in which they were formed:

1. *Substitution*. In this type of pseudomorph there is a gradual removal of the original material and a corresponding and simultaneous replacement of it by another with no chemical reaction between the two. A common example of this is the substitution of silica for wood fiber to form petrified wood. Another example is quartz, SiO_2, after fluorite, CaF_2.

2. *Encrustation*. In the formation of this type of pseudomorph a crust of one mineral is deposited over crystals of another. A common example is quartz encrusting cubes of fluorite. The fluorite may later be carried away entirely by solution, but its former presence is indicated by the casts left in the quartz.

3. *Alteration*. In this type of pseudomorph there has been only a partial addition of new material, or a partial removal of the original materials. The change of anhydrite, $CaSO_4$, to gypsum, $CaSO_4 \cdot 2H_2O$, and the change of galena, PbS, to anglesite, $PbSO_4$, are examples of alteration

pseudomorphs. A core of the unaltered mineral may be found in such pseudomorphs.

STRUCTURAL COMPLEXITIES AND DEFECTS

In our prior discussion we assumed that atomic structures are perfect. That is, we assumed that a crystalline material consists of an ordered, three-dimensional, repetitive array of atoms and/or ions. In X-ray diffraction studies, on which almost all structure analyses of crystalline material are based, it is assumed that such repetitive, periodic order exists. However, specialized X-ray studies and applications of the transmission electron microscope (TEM) and the more recently developed high resolution transmission electron microscope technique (HRTEM) have shown that *structural defects* (or *imperfections*) on an atomic scale are common in three-dimensional structures. Such imperfections affect basic properties of crystalline materials such as strength, conductivity, mechanical deformation, and color. Before describing the various types of imperfections let us qualitatively illustrate some aspects of their occurrence. In Fig. 4.51 we have photographed a two-dimensional array of very small spheres. This represents a two-dimensional model of a crystal. In Fig. 4.51*a* we see an almost perfectly ordered array of spheres. In Fig. 4.51*b* we see structurally perfect domains with enclosed point defects ("holes") and traversed by linear defects. In Fig. 4.51*c* we see separate, well-ordered blocks of structure separated by highly defective boundaries. These regions would represent *domains* (of slightly differing atomic orientation) within a single crystal; as such the single crystal is said to consist of a mosaic of slightly misoriented blocks (domains). Imperfections in crystal structures are generally classified according to their geometry such as *point defects*, *line defects*, or *plane defects*. Several of these are illustrated schematically in Fig. 4.52.

Examples of point defects are *Schottky* and *Frenkel defects*. In a *Schottky defect* some cations (or anions) are absent from their normal sites in the crystal structure. If a crystal with such defects is to retain its electrical neutrality, the total charge on the cation vacancies must equal that of the anion vacancies. A *Frenkel defect* represents the absence of a cation (or anion) from its proper structural site location and a mislocation of this same cation (or anion) in an interstitial site. This is more common for cations than anions because anions tend to be larger. Yet another

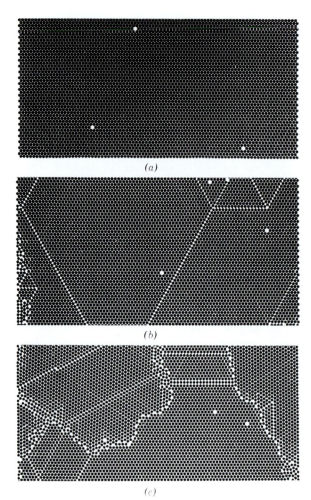

FIG. 4.51. A two-dimensional array of small spheres as a model for a crystal structure (6000 high precision stainless steel balls between two acrylic covers). *(a)* Regular, close packing of atoms with only three point defects in the pattern. *(b)* Point and line defects in the pattern. *(c)* A mosaic of domains separated by defective boundaries. (Photographs courtesy of Artorium, Inc., Montreal, Canada.)

type of point defect is known as an *impurity defect.* Such defective points are the result of the addition of a foreign ion (1) to an interstitial position in the crystal structure or (2) in place of one or several of the normally present ions. Impurity defects will change the chemical composition of the original "perfect" crystal, but the amounts of impurities are generally so small (in the range of parts per million or parts per billion) that they cannot be chemically detected. However, some property such as color may be strongly affected by such trace impurities.

Line defects involve, as the term implies, concentrations of defects along linear features in a crystal structure. Such line defects are commonly known as *dislocations* because they create an offset in the crystal structure. There are two types: *edge disloca-*

tions and *screw dislocations.* When a plane of atoms or ions in a crystal structure terminates in a line instead of continuing as would be required in a "perfect" crystal, it is said to contain an *edge dislocation* (see Fig. 4.52*d*). The presence of such line defects allows a crystal to deform under stress, by the atomic slippages of these linear defects throughout the structure. Such *slip planes* in the crystal structure represent planes of atomic misfit and therefore of lesser coherence. *Screw dislocations* are structural defects arranged along a screw axis that normally is not present in the structure. Figure 4.52*e* shows that the upper atomic surface of the schematic structure resembles a spiral ramp, centered about a vertical dislocation line, which is equivalent to a screw axis direction. Such spiral steps are of great importance during crystal growth because new ions or atoms, which are added to the outside surface of a growing crystal, are best housed along the ledge (see also Fig. 4.2).

Plane defects represent two-dimensional zones along which slightly misoriented blocks within a single crystal are joined. In ideally perfect crystals the whole internal structure is regarded as a rigorously continuous and symmetric repeat of unit cells. This implies short-range as well as long-range order. In less ideal (more real) crystals, blocks of structure may be slightly misoriented, as in a mosaic pattern. Each block in the mosaic has short-range order, but the whole crystal lacks long-range order. Figure 4.52*f* shows irregular zones (lines) of edge dislocations spaced at irregular intervals. In a three-dimensional structure these zones (or lines) will be irregular planar features along which ions (or atoms) have an irregular structural environment. These zones of irregularity are also known as *lineage structures.* The structures on either side of such lineages are slightly misoriented with respect to each other. When lineages are present, a continuous crystal structure must be viewed as made up of volumes of nearly perfect structure that occur in a mosaic of slight misorientation with each other.

Another planar defect is a *stacking fault* in which a regular sequence of layers (e.g., along the *c* axis of a structure) is interrupted by an improperly positioned layer, or when one of the ideally required layers is missing. Examples of this are sequences of ions (or atoms) in hexagonal closest packing (*AB, AB, AB,* . . .) interrupted by a layer from a cubic closest packing sequence (*ABC, ABC, ABC,* . . .) (see Chapter 3 for a discussion on closest packing).

Two other types of structural defects are exemplified by *omission solid solution* and *color centers.* Omission solid solution is discussed under "Compo-

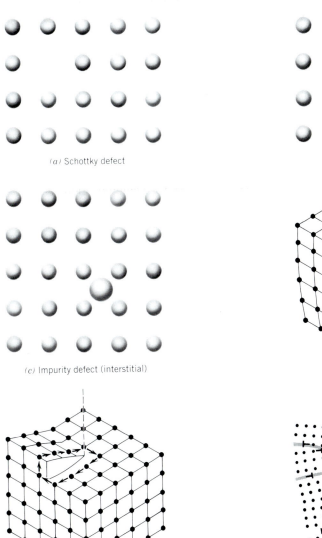

(a) Schottky defect

(c) Impurity defect (interstitial)

(e) Screw dislocation

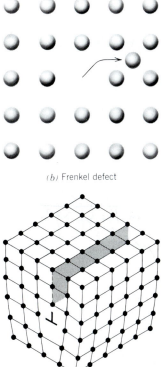

(b) Frenkel defect

(d) Edge dislocation

(f) Lineage structure

FIG. 4.52. Schematic representations of defects in crystal structures. *(a)* An ion (or atom) is missing from the structure. *(b)* An ion (or atom) is displaced from its normal site. *(c)* An interstitial impurity is randomly lodged in an otherwise regular structure. *(d)* A plane of atoms that stops along a dislocation line (the edge dislocation is shown by the usual graphic symbol—an upside-down T). *(e)* A screw dislocation line about which atomic planes wind in a helical form. *(f)* A crystal made of a mosaic of domains that differ only slightly in orientation. The irregular zones (which are planes in three dimensions) of defects are *lineage structures.*

sitional Variation in Minerals" (Chapter 3) and color centers under "Origin of Color" (later in this chapter).

Figure 4.53 shows several structural complexities and defects as resolved by high resolution transmission electron microscope (HRTEM) studies. Figure 4.53a is a structure image of crocidolite, a fibrous (asbestiform) variety of the amphibole riebeckite. It consists of a mosaic of domains that show slightly different crystallographic orientations, separated by grain boundaries (lineage structure). Figure 4.53b shows a structure image of grossular garnet in which a very small region (within the otherwise highly periodic and ordered garnet matrix) is dislo-

cated, and separated from the matrix by stacking faults. Figure 4.53c shows a deformed, curled region (C) in an otherwise well-layered matrix of lizardite (Lz), a variety of serpentine. Figure 4.53d depicts modulations in the structure of antigorite, a variety of serpentine. Modulations in the antigorite structure (see Fig. 11.37a) are wave-like features that result from a small dimensional misfit that occurs between the tetrahedral and octahedral layers. This misfit is accommodated by a curling of the structural layers, much like what happens when a telephone book is bent. A continuous curl results in the chrysotile structure (see Figs. 11.37 and 11.39).

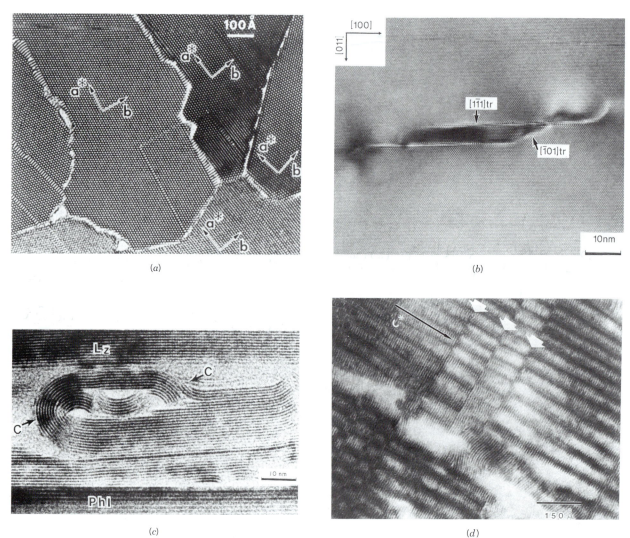

FIG. 4.53. Illustrations of structural complexities and defects. *(a)* HRTEM structure image of crocido-lite, a fibrous variety of riebeckite, viewed down the *c* axis. The locations of the *a** and *b* directions are shown (*a** = *a* sin β). Very slightly rotated domains (*subgrains*) are separated from each other along grain boundaries. The straight line features inside several of the domains are errors in the widths of the structural chains of amphiboles. (From J. H. Ahn and P. R. Buseck, 1991, Microstructures and fiber-formation mechanisms of crocidolite asbestos, *American Mineralogist* 76: 1467–78; see also Fig. 4.51.) *(b)* HRTEM structure image of a small dislocated region in an otherwise highly regular (ordered) matrix of grossular garnet. The central dislocated region is separated from the surrounding matrix by stacking faults. Crystallographic directions are shown as [011] and [100]. [1$\bar{1}$1]tr and [$\bar{1}$01]tr repre-sent the traces of the [1$\bar{1}$1] and [$\bar{1}$01] directions, respectively. Both traces represent stacking faults. Scale is in nanometers. (From F. M. Allen, B. K. Smith, and P. R. Buseck, 1987, Direct observation of dissociated dislocations in garnet, *Science* 238: 1695–97; photo by J. Berry.) *(c)* HRTEM structure image of locally deformed (curled) serpentine (C) inside a matrix of well-layered lizardite (Lz), also a variety of serpentine. The curled region occurs along a cleavage fracture. Phl = phlogopite mica. Scale is in nanometers. (From T. G. Sharp, M. T. Otten, and P. R. Buseck, 1990, Serpentinization of pholo-gopite phenocrysts from a micaceous kimberlite, *Contributions to Mineralogy and Petrology* 104: 530–39.) *(d)* HRTEM image of modulations in antigorite, a variety of serpentine. The modulations are about 50 Å in length. The *c** direction = *c* sin β. The large white arrows show defective boundaries, which may be out-of-phase or twin boundaries. (From P. R. Buseck and J. M. Cowley, 1983, Modu-lated and intergrowth structures in minerals and electron microscope methods for their study, *American Mineralogist* 68: 18–40; photo by G. Spinnler.)

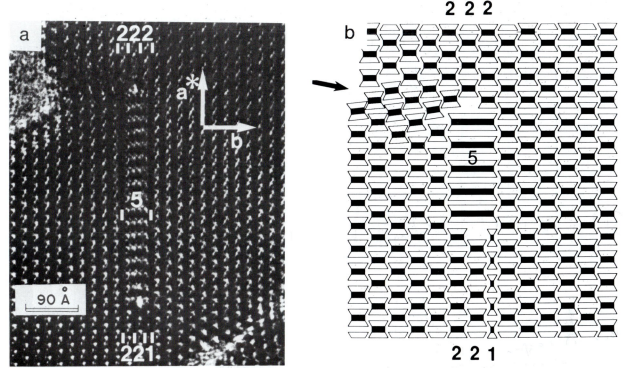

FIG. 4.54. *(a)* HRTEM structure image of crocidolite, a fibrous amphibole, viewed down the *c* axis of the structure. Note the scale bar of 90 Å (1 angstrom = 10^{-8} cm = 0.1 nanometer). Most of the field is composed of a regular pattern, which is the structure image of adjoining double (2) chains in the amphibole structure (see also page 461). However, the image also shows quintuple (5) and single (1) chain-width errors (these are marked by the appropriate numbers on the photograph). These defects appear as lines in the projected image, but represent continuous planar features in the three-dimensional structure. The a^* ($a^* = a \sin \beta$) and *b* directions are shown. *(b)* A schematic representation using I-beams to illustrate the complexities and defects observed in the structure image in *(a)*. The arrow at the left top of the figure points toward a disrupted region caused by structural mismatch. Similar mismatch occurs just "north" and "south" of the quintuple (5) chain. The mismatch to the "south" of the quintuple chain has been accommodated by an orderly arrangement of two double (2) chain widths and one single (1) chain width. (*a* and *b* from J. H. Ahn and P. R. Buseck, 1991, Microstructures and fiber-formation mechanisms of crocidolite asbestos, *American Mineralogist* 76: 1467–78.)

Figure 4.54*a* shows a variety of local defects and complexities in an amphibole structure image, and Fig. 4.54*b* is a schematic interpretation of these same features.

TWINNING

In the first part of Chapter 2 we introduced the morphological expression of twinned crystals. Now that we have a better understanding of crystal structures on which they are based, we will evaluate some of the internal aspects of twinning.

The more nearly perfect the periodic atomic arrangement of a crystal is, the lower its internal energy will be. Twinning is a deviation from perfection, and therefore the internal energy of a twin should be slightly greater than that of a "perfect" untwinned individual.

Under certain conditions of growth, two or more crystals may form a rational, symmetrical intergrowth. Such a *crystallographically controlled intergrowth* is called a *twin*. A twin thus comprises two (or more) *twinned crystals*. The lattice[2] directions of one crystal in a twin bear a definite crystallographic relation to the lattice directions of the other crystal. Figure 4.55 illustrates the twin relationship of two different orientations of the same orthorhombic lattice. The operations

[2]A lattice is defined as an imaginary pattern of points (or nodes) in which every point (or node) has an environment that is identical to that of any other point (node) in the pattern (for discussion see p. 214). A lattice is an abstraction of the original crystal structure.

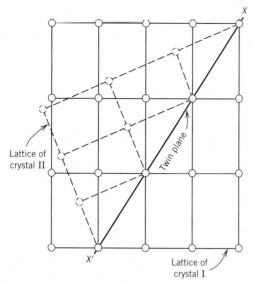

Lattice of
crystal II

Twin plane

X'

Lattice of
crystal I

FIG. 4.55. Cross-section through the orthorhombic lattice of a crystal (no. I) that is untwinned. A twinned counterpart of I is shown by II in the inclined position. Along the line XX' atoms are compatible with either of two crystal orientations, as shown by the solid and dashed circles. If the inclined position is chosen, a twinned crystal results. Such a twin may be regarded as the result of a rotation of 180° about the line XX' or of a mirror reflection along XX'.

that relate a crystal to its twinned counterpart can be: (1) reflection by a mirror (*twin plane*); (2) rotation about a line (*twin axis*) (although there are exceptions, the rotation is usually through an angle of 180°); and (3) inversion about a point (*twin center*). The twin operation is known as the *twin law*, in which the *twin elements* (twin mirror, twin axis) are parallel with lattice elements. Various twin laws are discussed on p. 208.

The description of a *twin law* consists of stating the crystallographic orientation of the twin element, axis, or plane. The two twinned orientations of an orthorhombic lattice (shown in Fig. 4.55) are related by a mirror reflection or a 180° rotation along the direction XX'. Along this direction there is a perfect coincidence of lattice nodes (or atomic positions) of crystals I and II. In the illustration the XX' direction (which would represent a planar surface in a three-dimensional lattice) is the contact between the two twinned crystals.[2]

Origin of Twinning

The various mechanisms by which twins are formed have been discussed by Buerger (1945) in terms of *growth twins*, *transformation twins*, and *gliding* (or *deformation*) *twins*.

Growth twins are the result of an emplacement of atoms, or ions (or groups of atoms or ions) on the

outside of a growing crystal in such a way that the regular arrangement of the original crystal structure (and, therefore, its lattice) is interrupted. For example, in Fig. 4.55 the line XX' may be considered the trace of an external face of a growing crystal. An ion or atom (or group of ions) would have the choice of attaching itself at structural sites that represent a continuation of the nodes in the lattice of crystal I or in nodes compatible with the lattice of crystal II. In the former case the original structure is continued without interruption, but in the latter a twinned relationship results. *Growth twinning* therefore reflects "accidents" during free growth ("nucleation errors") and can be considered as *primary twinning*. Growth twins can be recognized in mineral specimens by the interpenetration of two twinned crystals, or the presence of a single boundary between them. Most of the crystal drawings of twins in Chapter 5 represent growth twins.

Transformation twins occur in preexisting crystals and represent *secondary twinning*. Transformation twinning may result when a crystal that formed at high temperature is cooled and subsequently rearranges its structure to one whose symmetry is different from that of the high-temperature form (a polymorphic transition). For example high quartz, if cooled below 573°C, transforms to low quartz. Figure 4.56a illustrates the high quartz structure, whereas Fig. 4.56b shows the structure of low quartz in which part A is related to part B by a 180° rotation (see also Figs. 4.35 and 4.36). In the transition to the low quartz structure the original high quartz structure has a choice of two orientations, related by 180° rotation, for the trigonal structural arrangement of low quartz (see Fig. 4.56b). The relationship of these two orientations is known as a *Dauphiné twin* and is expressed as a 180° rotation about the vertical c axis or about the direction identified as [0001]. If such a twin of low quartz is heated above 573°C, the Dauphiné twinning will disappear and untwinned high quartz will form spontaneously. Figure 4.56c shows a basal section of low quartz with the 3-fold distribution of the twinned orientations of low quartz.

Another example of transformation twinning is provided by $KAlSi_3O_8$, which occurs in three different structural forms: high-temperature sanidine (with monoclinic symmetry), lower temperature orthoclase (also with monoclinic symmetry), and lowest temperature microcline (with triclinic symmetry); refer back to "Polymorphism." Microcline invariably shows the microscopic twin pattern shown in Fig. 5.41, which is known as *microcline* or *"tartan" twinning*. The cross-hatched pattern, which is usually vis-

High quartz

Low quartz

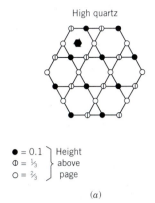

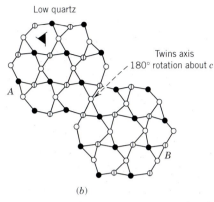

Twins axis
180° rotation about *c*

A

B

● = 0.1 ⎫ Height
◐ = ⅓ ⎬ above
○ = ⅔ ⎭ page

(*a*)

(*b*)

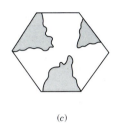

(*c*)

FIG. 4.56. *(a)* The distribution of silicon atoms in high quartz. The atomic positions are given in terms of thirds above the page. *(b)* A part of the low quartz structure *(A)* twinned with respect to another part *(B)* by a rotation axis of 180° perpendicular to the page. Note the presence of threefold screw axes instead of sixfold screw axes as in high quartz. *(c)* The threefold distribution of sectors of low quartz twinned according to the Dauphiné Law. The composition surfaces are generally irregular. This type of twin is visible only on sawn and etched basal sections of low quartz.

ible only between crossed polarizers, consists of two types of twins, twinned according to the *albite* and *pericline* laws. This combination of twinned relations is a key feature in the symmetry change from monoclinic (as in sanidine) to triclinic (as in microcline). In this transition from higher temperature, the mirror plane and the two-fold rotation axis are lost, leading to nucleation of triclinic "domains" that are related by twinning.

Gliding (or *deformation*) *twins* result when a crystal is deformed by the application of a mechanical stress. Deformation twinning is another type of *secondary twinning.* If the stress produces slippage of the atoms on a small scale, a twin may result (Fig. 4.57). If the movement of the atoms is large, slip or gliding may occur without twinning, which may finally lead to the rupture of the crystal. Deformation twinning is common in metals and is frequently present in metamorphosed limestones, as shown by the presence of polysynthetically twinned calcite (see Fig.

5.40*c*). Similarly, plagioclase feldspars from metamorphic terrains may show deformation twinning.

◢ ORIGIN OF COLOR

Minerals possess many properties, of which color is usually the first and most easily observed. For many it is characteristic and serves as a distinguishing criterion; indeed most people will recognize a number of the gem minerals and gemstones shown in Plates IX to XII (Chapter 13) on the basis of their colors alone. Yet in many minerals color is one of the most changeable and unreliable diagnostic properties.

Color is the response of the eye to the visible light range of the electromagnetic spectrum (see Fig. 4.58). Visible light represents a range of wavelengths from about 350 to 750 nanometers (nm, 1 nm = 10 angstroms). The energy of light, as of all electromagnetic radiation, can be expressed as follows:

$$E = h\nu = \frac{hc}{\lambda} = hc\bar{\nu}$$

FIG. 4.57. Deformation twinning in an oblique lattice due to the application of mechanical stress as indicated by the arrows. Note that the amount of movement of the first layer above and parallel with the twin plane in *(b)* is less than that of the successive layers further removed from the twin plane.

○ ○ ○ ○   ○ ○ ○ ○

○ ○ ○ ○   ○ ○ ○ ○

○ ○ ○ ○   − − ○ ○ ○ ○ ○ − − − −

○ ○ ○ ○   ○ ○ ○ ○ Composition
plane (= twin
○ ○ ○ ○   ○ ○ ○ ○   plane *m*)

(*a*)   (*b*)

where *E* denotes energy, *h* is Planck's constant, *c* is the speed of light (a constant), *ν* is frequency, λ is the wavelength, and $\bar{\nu}$ is the wavenumber. In Fig. 4.58 the spectral range of visible light is defined in terms of energy, wavelength, and wavenumber scales. Wavenumber, which is the reciprocal of the wavelength, is directly proportional to energy.

When white light strikes the surface of a mineral, it may be transmitted, scattered, reflected, refracted, or absorbed (see Fig. 4.59). The processes of scattering and reflection are part of the property perceived as the *luster* of a material. If the light suffers no

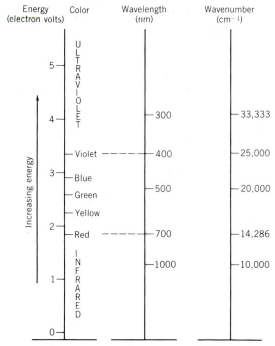

FIG. 4.58. The spectrum, with three ways of numerically specifying the colors. The wavelength scale is in nanometers (nm; 1 nm = 10 Å); the wavenumber scale expresses the number of wavelengths per unit length (cm).

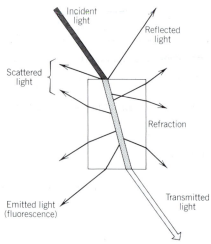

FIG. 4.59. Interaction of light with condensed matter causes reflection, refraction, scattering, and absorption. Some absorbed light can also be reemitted (usually at a longer wavelength) as fluorescence. (Redrawn after K. Nassau, 1980, The Causes of Color, *Scientific American* 243: 124–56.)

absorption, the mineral is colorless in reflected and transmitted light. Minerals are colored when certain wavelengths of light are absorbed, and the perceived color results from the combination of those remaining wavelengths that reach the eye. The wavelengths that are absorbed by minerals can be quantitatively measured by a spectrometer. An example of an *absorption spectrum* for the mineral beryl is given in Fig. 4.60. The peaks in this pattern represent absorption of specific wavelengths of light, and are the re-

sult of the interaction of such wavelengths with ions, molecules, and bonds in the irradiated structure. The absorptions between 0.4 and 0.7 μm are caused by chromophoric transition metal ions such as Fe^{3+} and Cr^{3+} (*chromophore* is derived from the Greek and means color-causing). In the infrared region between 1 and 4.5 μm there are absorptions due to molecules such as H_2O and CO_2, which are interstitial in the hexagonal channels of the beryl structure (see Fig. 3.60). Beyond 4.5 μm the absorptions are the result of vibrations of the crystal lattice, the so-called lattice modes.

As discussed in Chapter 3, the energies of electrons occur in discrete units, *quanta*. Furthermore, there are well-defined energy differences between

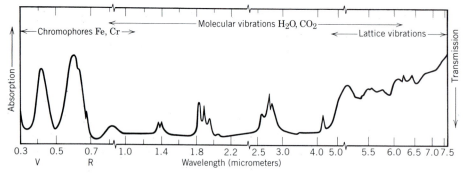

FIG. 4.60. The visible and infrared spectrum of beryl. Peaks correspond to absorption bands. Absorption in the visible region (0.1 to 0.8 μm) is caused by chromophores Fe and Cr; absorption in the 0.8 to 6.5 μm region results from molecular vibrations; and absorption above 4.5 μm is the result of lattice vibrations. 1 micrometer (μm) = 1000 nanometers (nm). V = violet, R = red. (After D. L. Wood and K. Nassau, 1968. The characterization of beryl and emerald by visible and infrared absorption and spectroscopy, *The American Mineralogist* 53: 777–801.)

these allowed energy levels. When electromagnetic radiation interacts with a material, those wavelengths whose energies correspond exactly to the energy differences between the electronic levels will be absorbed, resulting in electrons being excited from one level to another. In colored minerals, the energy differences between certain electron energy levels are in the range of energy of visible light. Thus, when white light shines on a mineral, certain wavelengths are absorbed (and therefore removed from the spectrum), causing excitation of electrons between these levels.

The electronic processes responsible for color in minerals can be classified as *crystal field transitions*, *molecular orbital (charge transfer) transitions*, and *color centers*. In our subsequent discussion of these phenomena we will draw extensively on the publications of Loeffler and Burns (1976) and Nassau (1978, 1980; see the end of this chapter for complete references).

Crystal Field Transitions

Crystal field transitions are electronic transitions between partially filled $3d$ orbitals of transition elements. Such transitions are most common in minerals containing the following transition elements: Ti, V, Cr, Mn, Fe, Co, Ni, and Cu. These elements belong to the first transition series with electronic configurations of the general form: $1s^2\ 2s^2\ 2p^6\ 3s^2\ 3p^6\ 3d^{10-n}\ 4s^{1-2}$, with partially filled $3d$ orbitals (see Tables 3.6 and 3.7). Of these elements Fe is most abundant in the Earth's crust, and for that reason is a dominant cause of color in minerals. The

electrons in the partially filled $3d$ orbitals can be excited by quanta of energy from the visible spectrum; these electronic transitions are the basis for the production of color. This is in contrast to ionic compounds made up of ions with a noble gas configuration, which are commonly colorless. This is because the energy gap between an occupied p orbital and the next available unoccupied orbital is considerably greater than the energy of visible light.

Crystal field theory accounts for these electronic transitions between partially filled d orbitals. The negative charges of the coordinating anions create an electrical field about the central transition metal ion. This is known as the *crystal field*, which has a specific symmetry and shape as a result of the number of anions, their distances from the cation, and their charges. The five $3d$ orbitals of a transition metal cation have the same energy in the absence of neighboring ions and they have distinctive probability distributions for the electrons (see Fig. 3.12c). Two of these orbitals, $d_{x^2-y^2}$ and d_{z^2}, with their maximum electron densities along the x, y, and z coordinate axes, are referred to as the e_g set. The other three orbitals, d_{xy}, d_{yz}, and d_{xz}, have their greatest electron density in directions between the coordinate axes. These are referred to as the t_{2g} set. When a transition element ion is surrounded by a spherically symmetrical cloud of negative charge, the electron orbitals are the same as those of the free ion (with all orbitals of the same energy), but their overall energy levels will exceed those of the free ion because the spherical negative field will repel all electrons equally in these orbitals, which adds to their potential energy (see Figs. 4.61a and b).

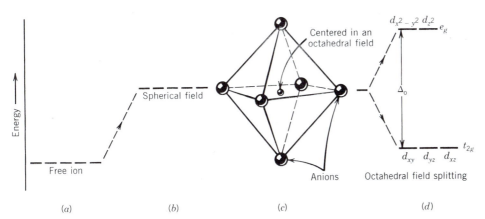

FIG. 4.61. Schematic representation of the energy levels of the five $3d$ orbitals in transition metals. *(a)* A free ion without surrounding neighbors, as in a gaseous state. *(b)* The ion surrounded by a uniform and spherically distributed negative charge. *(c)* The transition metal ion surrounded by an octahedral field of negative charges (anions) experiences the energy separation (or crystal field splitting, Δ_0) of the $3d$ orbitals, as shown in *(d)*. The five $3d$ orbitals are separated into a high-energy group (the e_g set) and a low-energy group (the t_{2g} set). (Adapted from B. M. Loeffler and R. G. Burns, 1976. Shedding light on the color of gems and minerals, *American Scientist* 64: 636–47.)

When a transition metal ion is placed in a coordination site in a mineral, there will be a nonuniform crystal field interaction on the various d orbitals by the neighboring anions. If the coordination polyhedron about the cation is octahedral (see Fig. 4.61c) the electrostatic repulsion between the anion orbitals and the centrally located cation orbitals raises the energy level of the $d_{x^2-y^2}$ and d_{z^2} orbitals (whose lobes of electron density are *along* the axes, see Fig. 3.12c) relative to the d_{xy}, d_{xz}, and d_{yz} orbitals (whose lobes of electron density are *between* the axes). This is known as *crystal field splitting*, meaning that the crystal field set up by the six surrounding anions splits the $3d$ energy levels of the central cation (Fig. 4.61c). In this illustration of crystal field splitting we have evaluated only what happens in an octahedral anion polyhedron. Minerals generally have several different coordination polyhedra; in these differing polyhedra, the energy level splitting of the transition metals' $3d$ orbitals will also be different. Furthermore, any distortion of the anionic coordination polyhedron about the central transition element will cause additional levels of splitting of $3d$ orbitals.

Let us now illustrate some of the above observations with respect to the color of three well-known minerals and their gem varieties: (1) peridot, a gem variety of olivine, with a yellow-green transmitted color (see Plate XI, no. 1, Chapter 13); (2) chrysoberyl, with a characteristic pale yellow-green color (see Plate X, no. 8); and (3) almandine, a member of the garnet group, with a dark red transmitted color (see Plate XI, no. 2).

Peridot is a member of the olivine series, $(Mg, Fe)_2SiO_4$. The olivine crystal structure consists of independent SiO_4 tetrahedra linked to Mg^{2+} and Fe^{2+} in octahedral coordination (see Fig. 11.4). The Fe^{2+} ions are distributed among two slightly different octahedral sites (designated as *M1* and *M2*). If white light shines on a peridot and one measures the amount of light absorption as a function of wavelength, the optical spectrum shown in Fig. 4.62 is obtained. Wavelengths of light that correspond in energy exactly to the energy differences caused in the $3d$ orbitals by crystal field splitting are absorbed. This absorbance in peridot takes place mainly in the infrared region, with some extension into the visible range. The absorption of the red component of white light is responsible for the yellow-green (transmitted) color of peridot. As such, Fe^{2+} in 6-coordination produces a characteristic green transmitted color in minerals.

The oxidation state of the transition element also affects the transmitted color. The mineral chrysoberyl, Al_2BeO_4, may contain some Fe^{3+} in substitution for Al^{3+}. The crystal structure of chrysoberyl is

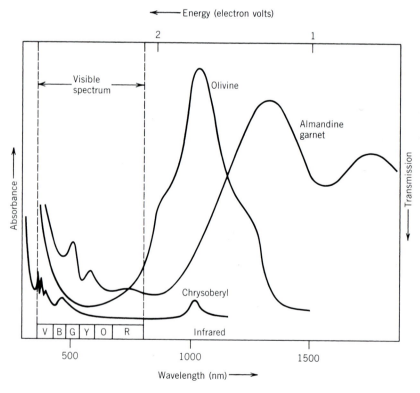

FIG. 4.62. Absorption spectra of two Fe^{2+}-bearing minerals, peridot, a gem variety of olivine, and almandine, a member of the garnet group. The spectrum of an Fe^{3+}-bearing mineral, chrysoberyl, is also shown. (From B. M. Loeffler and R. G. Burns, 1976, Shedding light on the color of gems and minerals, *American Scientist* 64: 636–47. Reprinted by permission of *American Scientist*, journal of Sigma Xi, The Scientific Research Society.)

very similar to that of olivine, with Al^{3+} (and small amounts of Fe^{3+}) in octahedral coordination. The optical absorption spectrum of chrysoberyl (see Fig. 4.62) is, however, very different from that of peridot. Chrysoberyl absorbs only weakly in the violet and blue region of the spectrum, giving the characteristic pale yellow color of Fe^{3+}-containing chrysoberyl. The differences in the absorption spectra (between peridot and chrysoberyl) are the result of the differences in the electronic structures of Fe^{2+} and Fe^{3+}.

A change in coordination polyhedron about a transition element also affects the absorption spectrum and resultant transmitted color. In the structure of garnet (see Fig. 11.5), for example, independent SiO_4 tetrahedra are linked to trivalent cations in octahedral coordination and divalent cations in 8-fold coordination (distorted cubic). In a common variety of garnet, almandine, $Fe_3Al_2Si_3O_{12}$, Fe^{2+} is housed in the 8-coordinated site. The absorption spectrum of such an almandine is given in Fig. 4.62. Now the main absorption peaks occur entirely outside the visible region, with lesser absorption peaks in the region of violet-blue-green-yellow. The transmission color resulting from this is a deep red.

Two other minerals that occur as highly prized gems illustrate yet another factor contributing to color. These are (1) emerald, an emerald-green, Cr-containing variety of beryl, $Be_3Al_2Si_6O_{18}$ (see Plate IX, nos. 2 and 5, Chapter 13), and (2) ruby, a ruby-red, Cr-containing variety of corundum, Al_2O_3 (see Plate IX, numbers 4 and 7).

In both minerals, small amounts of Cr^{3+} replace Al^{3+} in 6-coordinated sites (somewhat distorted octahedra). In beryl, a ring silicate (see Fig. 3.60), the six oxygens about Al^{3+} (or Cr^{3+}) are shared with SiO_4 and BeO_4 tetrahedra; in corundum, an oxide consisting of hexagonal close-packed layers of oxygen, Al^{3+} (or Cr^{3+}) occupies interstices between these layers. In the silicate structure of beryl there is a covalent component to the bonding (see Fig. 3.23), whereas in corundum the bonding is more truly ionic. This results in a weaker crystal field around the Cr^{3+} in beryl than in corundum. These differences are clearly reflected in Fig. 4.63, which compares the absorption spectra of the two gems. In emerald, the absorption peaks are at lower energy than in ruby. In emerald, the absorptions are in the violet and blue, in the yellow, and in the orange and red. This gives transmission in the green and hence the emerald-green color. In ruby, there is absorption in the violet, green, and yellow, with transmission in blue and red. The overall red color of ruby is furthermore intensified by a characteristic fluorescence in the red. That is, not only does ruby absorb most wavelengths such that red is transmitted, it also emits red light by fluorescence (see page 27).

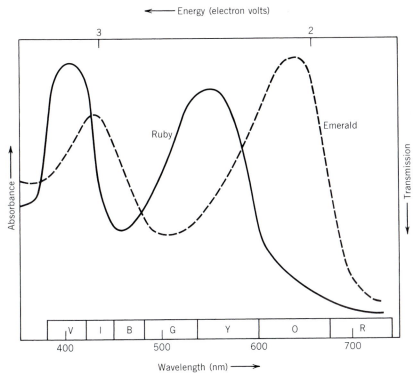

FIG. 4.63. Absorption spectra of emerald and ruby. In emerald, where the crystal field about Cr^{3+} is weaker, the absorption peaks are shifted to lower energy, producing transmission in green. In ruby, the absorption peaks are at higher energies, permitting transmission mainly in the blue and red regions. (From B. M. Loeffler and R. G. Burns, 1976, Shedding light on the color of gems and minerals, *American Scientist* 64: 636–47. Reprinted by permission of *American Scientist*, journal of Sigma Xi, The Scientific Research Society.)

TABLE 4.4 Examples of Common Minerals Whose Color Is Due to the Interaction of Transition Elements and Crystal Field Transitions*

Absorbing Ion	Mineral	Formula	Color
Cr^{3+}	Beryl (emerald)	$Be_3Al_2Si_6O_{18}$	Green
	Corundum (ruby)	Al_2O_3	Red
Mn^{3+}	Tourmaline (rubellite)	$Na(Li,Al)_3Al_6(BO_3)_3(Si_6O_{18})(OH)_4$	Pink
Mn^{2+}	Beryl (morganite)	$Be_3Al_2Si_6O_{18}$	Pink
	Spessartine garnet	$Mn_3Al_2(SiO_4)_3$	Yellow-orange
Fe^{3+}	Andradite garnet	$Ca_3Fe_2(SiO_4)_3$	Green
	Chrysoberyl	$BeAl_2O_4$	Yellow
Fe^{2+}	Olivine (peridot)	$(Mg,Fe)_2SiO_4$	Yellow-green
	Almandine garnet	$Fe_3Al_2(SiO_4)_3$	Dark red
Cu^{2+}	Turquoise	$CuAl_6(PO_4)_4(OH)_8 \cdot 5H_2O$	Light blue

*From Loeffler, B. M. and Burns, R. G., 1976, Shedding light on the color of gems and minerals, *American Scientist*, v. 64, pp. 636–647. Many of the minerals listed in the table are illustrated in Plates IX through XII, Chapter 13.

In summary, several factors that influence the transmitted color produced by crystal field interactions are:

1. the presence of a specific transition element,
2. its oxidation state (valence), which determines the number of electrons in $3d$ orbitals,
3. the geometry of the site in which the transition metal is housed, (octahedral, tetrahedral, etc.),
4. the strength of the crystal field (charges on anions, distortion of coordination polyhedra, etc.), and
5. the way in which the human eye interprets the pattern of transmitted wavelengths.

Table 4.4 is a summary of minerals whose color is the result of crystal field transitions.

Molecular Orbital Transitions

Molecular orbital transitions (also known as charge transfer transitions) occur in minerals when valence electrons transfer back and forth between adjacent ions. The electrons are contributed to *shared molecular orbitals* and as such are delocalized; that is, the valence electrons of a constituent atom are no longer in atomic orbitals centered on the atom. In such instances crystal field theory does not apply, but instead molecular orbital theory best describes the observed spectra.

Examples of molecular orbital transitions are found in many minerals. Of these, $Fe^{2+} \rightarrow Fe^{3+}$ and $Fe^{2+} \rightarrow Ti^{4+}$ are the most common metal-metal charge transfer transitions. In the $Fe^{2+} \rightarrow Fe^{3+}$ charge transfer transition, an electron is transferred from Fe^{2+} (in site A, making it Fe^{3+}) to Fe^{3+} (in site B, making it Fe^{2+}) such that $Fe^{2+}_{(A)} + Fe^{3+}_{(B)} \rightleftarrows Fe^{3+}_{(A)} + Fe^{2+}_{(B)}$. Energies of this reversible electron-hopping process generally correspond to wavelengths in visible light, and many minerals owe their intense blue color to such transitions. Examples are glaucophane (blue amphibole and crocidolite, blue amphibole asbestos), cordierite, kyanite (commonly blue), and sapphire (blue gem variety of corundum). The $Fe^{2+} \rightarrow Ti^{4+}$ charge transfer transition is also a large factor in the blue color of sapphire, Al_2O_3, which commonly contains small amounts of iron and titanium. Figure 4.64 shows the

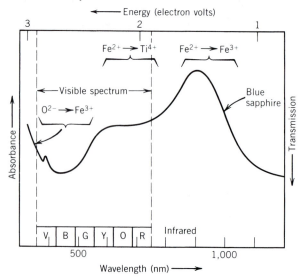

FIG. 4.64. Optical absorption spectrum of sapphire, blue gem corundum (see Plate IX, nos. 3 and 6, Chapter 13), and the molecular orbital transitions responsible for the absorption peaks ($Fe^{2+} \rightarrow Ti^{4+}$; $Fe^{2+} \rightarrow Fe^{3+}$; and $O^{2-} \rightarrow Fe^{3+}$ at the edge of the ultraviolet region). The only transmission in this spectrum is in the range of blue in the visible spectrum. (From B. M. Loeffler and R. G. Burns, 1976, Shedding light on the color of gems and minerals, *American Scientist* 64: 636–47. Reprinted by permission of *American Scientist*, journal of Sigma Xi, The Scientific Research Society.)

TABLE 4.5   Examples of Some Common Minerals Whose Color Is the Result of Charge–Transfer Transitions, Described by Molecular Orbital Theory

Ion Pair	Mineral	Formula	Color
$Fe^{2+} \rightarrow Fe^{3+}$	Beryl (aquamarine)	$Be_3Al_2Si_6O_{18}$	Blue-yellow
$Fe^{2+} \rightarrow Fe^{3+}$	Cordierite	$(Mg,Fe)_2Al_4Si_5O_{18}\cdot nH_2O$	Blue
$Fe^{2+} \rightarrow Ti^{4+}$	Corundum (sapphire)	Al_2O_3	Blue
$Fe^{2+} \rightarrow Ti^{4+}$	Kyanite	Al_2SiO_5	Blue
$O^{2-} \rightarrow Cr^{6+}$	Crocoite	$PbCrO_4$	Orange
$O^{2-} \rightarrow Fe^{3+}$	Beryl (heliodore)	$Be_3Al_2Si_6O_{18}$	Yellow

*From Loeffler, B. M. and Burns, R. G., 1976, Shedding light on the color of gems and minerals, *American Scientist*, v. 64, pp. 636–647.

optical absorption spectrum of blue sapphire, in which the main absorbance peaks are identified as due to $Fe^{2+} \rightarrow Ti^{4+}$, and $Fe^{2+} \rightarrow Fe^{3+}$ charge transfers. The main transmission of light is in the blue range of the visible spectrum (see Plate IX, nos. 3 and 6, Chapter 13 for color illustrations of sapphire). Table 4.5 is a summary of some minerals whose color is the result of molecular orbital transitions.

Color Centers

Coloration can also be caused by structural defects. This can be an excess electron that is unattached to any single atom and that is trapped at some structural defect, such as a missing ion or an interstitial impurity. A "hole," the absence of an electron, can have the same effect. These types are known as *color centers*, or *F* centers (from the German *Farbe*, color).

FIG. 4.65. Schematic illustration of the structure of fluorite, CaF_2, in which an electron fills a vacancy created by a fluorine ion that was removed. Here a color center is the result of the electron taking the place of the dislodged ion. (Adapted from K. Nassau, The causes of color, *Scientific American*, v. 243: 124–156. Copyright © 1980 by Scientific American, Inc. All rights reserved.)

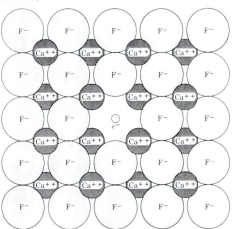

The coloring mechanism in purple fluorite, CaF_2, is known to be the result of Frenkel defects (see Fig. 4.52b) in the fluorite structure. Figure 4.65 is an illustration of the structure of fluorite in which an F^- ion is missing from its usual structural site. Such defects in the network of F^- ions can be the result of (1) high-energy radiation (e.g., X-rays) that displaced the F^- from its usual position to another one in the structure, (2) the growth of fluorite in a chemical environment with an excess of calcium, and (3) removal of some F from the crystal by the application of an electrical field. Because the overall structure must remain neutral, an electron usually occupies the empty position to produce an "electron color center," as in Fig. 4.65. Such an electron is not bound in place by a central nucleus, but by the electrical field (crystal field) of all of the surrounding ions. Within this field it can occupy a ground state and various excited states similar to those of the transition elements described above. The movement of electrons among these states can cause color and optical fluorescence. It should be noted that the original crystal structure of fluorite (without defects) can be restored by heating, whereupon the color fades.

The smoky color of some quartz crystals is attributed to the occurrence of a "hole color center." In such quartz some Al^{3+} substitutes for Si^{4+}, and this substitution is coupled with some interstitial Na^+ or H^+ ions in order to maintain electrical neutrality. When this type of quartz with some Al substituting for Si in tetrahedral sites, is exposed to an intense X-ray or gamma ray beam for a few minutes, or when it has been exposed to low levels of radiation over geological periods, "hole color centers" are produced. The radiation expels one electron from a pair of electrons in an oxygen atom adjacent to an Al^{3+} ion, thereby leaving a single, unpaired electron in the orbital. This is illustrated schematically in Fig. 4.66. The missing electron is called a "hole," and the remaining unpaired electron has a set of excited states

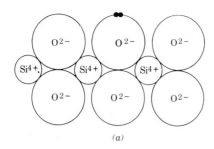

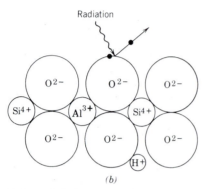

FIG. 4.66. Schematic illustration of the quartz structure. *(a)* The normal structure of pure SiO_2. *(b)* The structure with some ionic substitution of Al^{3+} for Si^{4+}, coupled with introduction of H^+ into the structure, in order to retain overall neutrality. Radiation ejects one of a pair of electrons from an O^{2-} and leaves a "hole" color center of smoky quartz. (After K. Nassau, 1978, The origins of color in minerals, *American Mineralogist* 63: 219–29.)

much like those of an excess electron, as described above. Table 4.6 lists some mineral examples in which coloration is due to color centers.

Other Causes of Color

Yet another coloring agent is the mechanical admixture of impurities, which can give a variety of colors to otherwise colorless minerals. Quartz may be

TABLE 4.6   Examples of Minerals in Which the Coloration Is Due to Color Centers*

Mineral	Color
Amethyst, fluorite	Purple
Smoky quartz	Brown to black
Irradiated diamond	Green, yellow, brown, blue, pink
Natural and irradiated topaz	Blue
Halite	Blue and yellow

*From Nassau, K., 1978, The origins of color in minerals, *American Mineralogist*, v. 63, pp. 219–229.

green because of the presence of finely dispersed chlorite; calcite may be black, colored by manganese oxide or carbon. Hematite, as the most common pigmenting impurity, imparts its red color to many minerals, including some feldspar and calcite and the fine-grained variety of quartz, jasper.

ORIGIN OF MAGNETIC PROPERTIES

Some minerals behave like magnets, whereas most do not. Such magnetic properties are the result of atomic properties that are specific to a number of elements. In Chapter 3 it was discussed that in order to specify the position of an electron in three-dimensional space, *three quantum numbers are needed: n*, the principal quantum number, *l*, the azimuthal (or orbital shape) quantum number, and *m*, the magnetic quantum number. In addition to these three quantum numbers, there is a fourth quantum number, the *spin quantum number, s*, which defines the spin of the electron in space (see Fig. 3.13). Because an electron can spin in only two directions, it has only two values, namely $+\frac{1}{2}$ and $-\frac{1}{2}$. A spinning electron behaves as a small magnet and will produce a magnetic field while moving around its orbit. This is analogous to the production of a magnetic field by an electrical current moving through a coiled wire. As such, each electron on an atom can be assigned values for four quantum numbers, *n*, *l*, *m*, and *s*, which determine the orbital in which the electron occurs and the direction of electron spin. There is, however, a restriction on the values these quantum numbers may have. This is known as the *Pauli exclusion principle*, which states that no two electrons in any one atom may have all four quantum numbers the same. This has the effect of limiting the number of electrons in any given orbital to two, and it also requires that the spins of the two electrons be in opposite directions (with *s* values of $+\frac{1}{2}$ and $-\frac{1}{2}$). This is summarized in Table 3.4.

The spin of the electron is mainly responsible for the magnetic properties of atoms and molecules. The spinning electron can be regarded as a minute magnet (or *magnetic dipole*) with a *magnetic moment* that is defined as a Bohr magneton, $\mu_\beta = 9.27 \times 10^{-24}$ Am^2, which is a product of the area about which the electron spins (in units of m^2) and the electron charge (in units of amperes, A). Because two electrons in the same orbital must have opposing spins, with one having an "up" and the other a "down" orientation of its pole, this produces a zero net magnetic moment. Such materials are known as *diamagnetic* because they experience no attraction for a magnet. They are, in fact, slightly repelled by a magnetic field, a result of the behavior of the electron cloud of the atom, not of the electron spin. In these substances there are the same number of electrons of each spin ("up" and "down"), so that their magnetic effects cancel. Many common minerals have no magnetic response; they are diamagnetic. They are composed of elements with core electron configurations of the rare gases, or with completely filled *d* orbitals (see Table 3.6). A few examples of the many common diamagnetic minerals are calcite, $CaCO_3$, albite, $NaAlSi_3O_8$, quartz, SiO_2, and apatite, $Ca_5(PO_4)_3(F, Cl, OH)$.

The most important elements that produce magnetic moments are those with unshared electrons (not involved in bonding) in 3*d* orbitals of the first transition series, including Ti, V, Cr, Mn, Fe, Co, Ni, and Cu (Z 22 through 29; see periodic table at the far end of this text or Table 3.6). This includes some very common mineral constituents, namely Fe, Mn, Ti, and Cr. The magnetic moments of these transition elements are the result of the spin of single, unpaired electrons and are proportional to the number of such electrons. The distribution of electrons in the five 3*d* orbitals is summarized in *Hund's rule*: *Electrons entering a subshell with more than one orbital will be distributed over the available orbitals with their spins in the same direction*. This means that in going from top to bottom in Table 4.7, electrons are first added to each orbital with their spins in the same direction and then doubled up until all orbitals are filled. This build-up continues until all electrons are paired, in all five 3*d* orbitals in Zn. The table shows that Fe^{3+}

and Mn^{2+}, each with five unpaired electrons, are among the most magnetic ions.

Although individual ions may be classified as more or less magnetic (as in Table 4.7), there is still the question of how such ions interact in crystal structures. If a structure has a *random arrangement of magnetic dipoles* (caused by specific constituent cations with unpaired spins, such as are listed in Table 4.7) it is said to be *paramagnetic*. When such a structure is placed into a magnetic field, the minute dipoles tend to align themselves with the external magnetic field. Thermal motion inside the structure, however, tends to randomize some of the dipole alignments. This results in only a small fraction of dipoles being aligned with the external magnetic field at any particular instant. As such, a paramagnetic material is drawn only weakly to an external magnetic field (shows low magnetic susceptibility). The magnetization imposed is also not permanent. Examples of two common minerals that show paramagnetic behavior are olivine $(Mg,Fe)_2SiO_4$, and augite $(Ca,Na)(Mg,Fe,Al)(Al,Si)_2O_6$.

Because of their different magnetic susceptibilities, minerals can be separated from each other by an electromagnet. Magnetic separation by a Franz Isodynamic Separator is a common procedure in the laboratory. Such a technique separates minerals that are paramagnetic from those that are diamagnetic. On a commercial scale, electromagnetic separation is used to separate ore minerals from gangue (waste).

Another property related to paramagnetism is called *ferromagnetism*, which is observed in metallic iron. In a paramagnetic material the magnetic dipoles are randomly oriented, but in a ferromagnetic

TABLE 4.7   Transition Elements (*Z* 21 Through *Z* 30), Their Common Ions, Number of 3*d* Electrons, Electron Spin Directions, and Magnetic Moment (Expressed in Terms of Bohr Magnetons, μ_β)

Elements	Ions	Spin Directions and Number of 3*d* Electrons for the Ions		Magnetic Moment
Sc	Ti^{3+}, V^{4+}	↑ _ _ _ _	1	$1\mu_\beta$
Ti	Ti^{2+}, V^{3+}	↑ ↑ _ _ _	2	$2\mu_\beta$
V	V^{2+}, Cr^{3+}, Mn^{4+}	↑ ↑ ↑ _ _	3	$3\mu_\beta$
Cr	Cr^{2+}, Mn^{3+}	↑ ↑ ↑ ↑ _	4	$4\mu_\beta$
Mn	Mn^{2+}, Fe^{3+}	↑ ↑ ↑ ↑ ↑	5	$5\mu_\beta$
Fe	Fe^{2+}, Co^{3+}	↑↓ ↑ ↑ ↑ ↑	6	$4\mu_\beta$
Co	Co^{2+}	↑↓ ↑↓ ↑ ↑ ↑	7	$3\mu_\beta$
Ni	Ni^{2+}	↑↓ ↑↓ ↑↓ ↑ ↑	8	$2\mu\beta$
Cu	Cu^{2+}	↑↓ ↑↓ ↑↓ ↑↓ ↑	9	$1\mu_\beta$
Zn	Zn^{2+}, Cu^+	↑↓ ↑↓ ↑↓ ↑↓ ↑↓	10	0

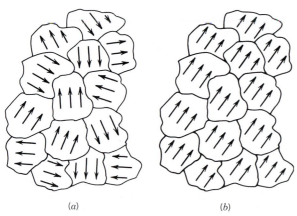

FIG. 4.67. Magnetic domains in a ferromagnetic solid. *(a)* Random domains when unmagnetized. *(b)* Parallel alignment of domains as a result of an external magnetic field.

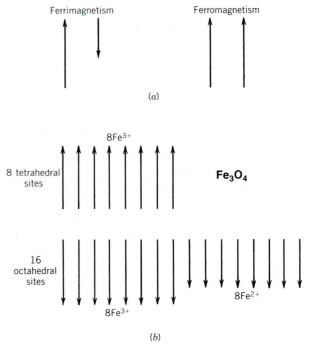

FIG. 4.68. *(a)* Schematic illustration of the spin alignments of dipoles in ferrimagnetic and ferromagnetic materials. In ferrimagnetic solids the ionic spins are antiparallel and their magnitudes are unequal. In ferromagnetic solids all spins are parallel and aligned in the same direction. *(b)* Schematic representation of the spin directions in the tetrahedral and octahedral sites of magnetite, Fe_3O_4. A net magnetic moment is due to the noncancellation of the dipole moments of Fe^{2+} ions.

substance these become aligned as a result of "exchange forces" that are the result of overlap of the orbitals of nearest neighbor atoms or ions (for further discussion of the exchange forces, see O'Reilly, 1984). In a substance such as metallic iron, *domains* exist containing large numbers of paramagnetic atoms with their dipole moments well aligned. Ordinarily these domains are randomly oriented (see Fig. 4.67*a*) so that their net magnetic effect is zero. When such a material is placed in a magnetic field, the domains become aligned with the external field (Fig. 4.67*b*), and it responds with strong magnetic attraction. This interaction is much stronger than normally experienced in paramagnetic materials. When the external magnetic field is removed from a paramagnetic substance, the magnetic domains randomize and no permanent magnetism remains. However, in a ferromagnetic material, the domains tend to remain in the orientation imposed by the external magnetic field even in the absence of the field. For example, a nail can be magnetized simply by running a permanent magnet over it; this process aligns ("poles") the magnetic domains in the nail fairly permanently. When a ferromagnetic material in which permanent magnetism has been induced is heated, the parallelism of the magnetic field is completely lost at the *Curie temperature*, above which it behaves paramagnetically. The Curie temperature for metallic iron is 770°C.

Yet another type of magnetism is known as *ferrimagnetism*, in which the ionic spin moments are antiparallel, instead of parallel as in ferromagnetism. In ferrimagnetic materials, the antiparallel spin moments are unequal, and as such there are permanent magnetic domains (see Fig. 4.68*a*). Examples of ferri-

magnetic minerals are members of the magnetite-ulvöspinel series, $Fe_3O_4 - Fe_2TiO_4$ (see spinel group, page 387), hematite-ilmenite solid solution members, $Fe_2O_3 - FeTiO_3$, and pyrrhotite, $Fe_{1-x}S$.

The distribution of magnetic dipoles in a ferrimagnetic material can be illustrated with reference to magnetite, a member of the spinel series. The formula of magnetite, Fe_3O_4, can be rewritten as $Fe^{3+}[Fe^{2+}Fe^{3+}]O_4$ in terms of a general XY_2O_4 formula for the spinel group. The spinel structure is based on cubic closest packing of oxygens with the cations in tetrahedral and octahedral interstices. The X cations occupy $\frac{1}{8}$ of the 64 (= 8) tetrahedral sites (per unit cell of spinel) and the Y cations occupy $\frac{1}{2}$ of the 32 (= 16) octahedral sites (per unit cell). The Fe^{3+} ions are therefore distributed in two different lattice sites, but with opposing magnetic spin directions. The Fe^{2+} ions (with lesser magnetic moment; see Table 4.7) are responsible for net unpaired spin and thus for the permanent magnetic domains in magnetite (see Fig. 4.68*b*). This net magnetization is considerably less than if the magnetic moments of all the cations were parallel, as in magnetized native iron, which is ferromagnetic. The Curie temperature of magnetite is 580°C, above which the magnetic ordering completely disappears.

The permanent magnetism of ferrimagnetic minerals in various rock types allows for the study of the ancient geomagnetic field of the Earth, known as *paleomagnetism*. The study of the natural remanent magnetization of rocks can yield a record of the Earth's magnetic field. For example, in the cooling of igneous rocks, from the temperature above to below the Curie temperature, the direction of the Earth's magnetic field is recorded in the orientation of the magnetic domains of ferrimagnetic minerals. *Lodestone*, a naturally occurring magnet of magnetite composition, is a ferrimagnetic substance in which all the net magnetic moments are strongly aligned ("poled"). The natural magnetism of lodestone is attributed to its having cooled from a melt (as part of an igneous rock) while under the influence of the Earth's magnetic field.

ORIGIN OF RADIOACTIVITY

The nuclei of some elements are unstable and can change (decay) spontaneously to different kinds of nuclei, with the release of radioactive energy in the process. As radioactivity is a statistically random process, the probability that a nucleus will decay in a given time interval is expressed in terms of a decay constant, λ, the fraction of the radioactive nuclei present that will decay in a unit of time. The equation that expresses the decay is:

$$\frac{dP}{dt} = -\lambda P$$

where P is the number of parent atoms at time t and λ is the *disintegration* or *decay constant*, whose units are reciprocal time. The decay constants are now well known for many unstable elements, from laboratory measurements. It has been found that these decay constants are truly constant in any terrestrial environment. This observation is basic to any radioactive "time clock," which provides radiometric ages of minerals and rocks.

Examples of geologically important unstable nuclei are the following isotopes: ^{40}K, ^{87}Rb, ^{232}Th, ^{238}U, and ^{235}U (see also Table 4.8). In the decay process of such unstable nuclei, nuclear particles (including alpha or beta particles) are emitted, as are gamma rays. An *alpha particle* consists of two protons and two neutrons that are tightly bonded together. The alpha particle, which is identical with the nucleus of

TABLE 4.8   Radioactive Isotopes Used in Radiometric Dating, Their Half-Lives, and Daughter Elements

Element	Isotope	Half-life* (in years)	Ultimate Daughter Elements[†]
Potassium	^{40}K	1.28×10^9 yr	^{40}Ca and ^{40}Ar
Rubidium	^{87}Rb	5×10^{11} yr	^{87}Sr
Thorium	^{232}Th	1.41×10^{10} yr	^{208}Pb and ^4He
Uranium	^{238}U	4.51×10^9 yr	^{206}Pb and ^4He
	^{235}U	7.1×10^8 yr	^{207}Pb and ^4He

*Half-life is the time required for one half of the original number of radioactive atoms to decay.
[†]The daughter elements are the new atoms formed at the expense of the disintegrated ones, the parent elements.

^4He, is ejected from the unstable nucleus during alpha decay. In such decay the atomic number Z of the nucleus decreases by two because of the removal of two protons, and the mass number decreases by four. An example of such decay is:

$$^{238}_{92}U \rightarrow ^{234}_{90}Th + \alpha + \gamma + \text{energy}$$

A *beta particle* is a negatively charged particle with the mass of an electron that is emitted from the unstable parent nucleus. Nuclei do not contain electrons, so we must regard the origin of this electron as the decay of a neutron into a proton, with the emission of the beta particle ($n = p + \beta^-$). By this process the charge of the nucleus is increased by 1, and the parent nucleus becomes the nucleus of the next higher element in the periodic table:

$$^{87}_{37}Rb \rightarrow ^{87}_{38}Sr + \beta^- + \text{energy}$$

In the decay series of $^{238}U \rightarrow ^{206}Pb$, $^{235}U \rightarrow ^{207}Pb$, and $^{232}Th \rightarrow ^{208}Pb$, which form the basis for three independent methods of age determination, several alpha and beta particles are produced in a sequence of intermediate radioactive products.

The process of disintegration can also occur by *electron capture*, in which an orbital electron is captured by the nucleus; this converts a proton to a neutron ($p + e^- = n$). The nuclear charge decreases by one without any significant change in mass. An example of this reaction is:

$$^{40}_{19}K + \text{orbital e}^- \rightarrow ^{40}_{18}Ar + \gamma + \text{energy}$$

In the above equations γ stands for high-energy electromagnetic radiation emitted by an excited nucleus as it drops into a less-excited state. Such emitted radiation is usually a by-product of alpha and beta decay and electron capture. Gamma radiation is on the short wavelength side of X-radiation (see Fig. 7.31).

FIG. 4.69. Autoradiograph of a dendritic aggregate of uraninite, UO_2. Locality: Ruggles Pegmatite, Grafton Center, New Hampshire (Harvard Mineralogical Collection). (From C. Frondel, 1958, *Systematic mineralogy of uranium and thorium*, U.S. Geological Survey Bulletin no. 1064, 400 pp.)

In minerals with considerable amounts of U and Th, this radiation effect is readily measured and is therefore a diagnostic property in the characterization of radioactive minerals. Examples of radioactive minerals are uraninite (UO_2), thorianite (ThO_2), and autunite ($Ca(UO_2)_2(PO_4)_2 \cdot 10–12H_2O$). The radiation is most easily measured, in the laboratory and in the field, by a Geiger counter or scintillation counter. The presence of radiation can also be shown by placing unexposed film in an opaque wrapper against a radioactive specimen. The radiation will expose the film; this is known as *autoradiography* (see Fig. 4.69).

As a result of the production of energetic alpha particles (and the associated recoil energy these particles impart to the daughter nucleus), beta particles, and gamma radiation, and the changes in ionic size in going from parent to daughter elements, the crystal structure in which these processes occur is generally profoundly affected. For this reason, most U- and Th-rich minerals have undergone partial or complete destruction of their structures, leading to various stages of *metamictization* (see "Metamict Minerals," page 148).

The production of gamma radiation, as a result of the various decay processes, is of major importance in the exploration for radioactive minerals. Not only can Geiger counters be used in field exploration programs, but airborne gamma-ray spectrometers can detect and distinguish the radiation emanating from ^{238}U, ^{232}Th, and ^{40}K decay. Such airborne surveys detect mainly radioactive materials that occur on the uppermost surface of the Earth, because even a thin cover will effectively block the radiation.

REFERENCES AND SELECTED READING

Banerjee, S. K. 1991. Magnetic properties of Fe-Ti oxides. In *Oxide minerals: Petrologic and magnetic significance. Reviews in Mineralogy* 25. Mineralogical Society of America, Washington, D. C.

Bohlen, S. R., A. Montana, and D. M. Kerrick. 1991. Precise determinations of the equilibria kyanite ⇆ sillimanite and kyanite ⇆ andalusite and a revised triple point for Al_2SiO_5 polymorphs. *American Mineralogist* 76: 677–80.

Bowen, N. L. 1913. The melting phenomena of the plagioclase feldspars. *American Journal of Science* 34: 577–99.

Brookins, D. G. 1988. *Eh-pH diagrams for geochemistry*. New York: Springer-Verlag.

Buerger, M. J. 1945. The genesis of twin crystals. *American Mineralogist* 30: 469–82.

Chopin, C. 1984. Coesite and pure pyrope of the Western Alps: A first record and some consequences. *Contributions to Mineralogy and Petrology* 86: 107–18.

Denbigh, K. 1981. *The principles of chemical equilibrium*. 4th ed. New York: Cambridge University Press.

Ehlers, E. G. 1972. *The interpretations of geological phase diagrams*. San Francisco: W. H. Freeman.

Ernst, W. G. 1976. *Petrologic phase equilibria*. San Francisco: W. H. Freeman.

Frondel, C. 1958. *Systematic mineralogy of uranium and thorium*. U. S. Geological Survey Bulletin no. 1064.

Garrels, R. M. and C. L. Christ. 1965. *Solutions, minerals, and equilibria*. San Francisco: Freeman, Cooper and Co.

Holdaway, M. J. 1971. Stability of andalusite and the aluminum silicate phase diagram. *American Journal of Science* 271: 97–131.

Keffer, F. 1967. The magnetic properties of materials. *Scientific American* 217: 222–38.

Kern, R., and A. Weisbrod. 1967. *Thermodynamics for geologists.* San Francisco: Freeman, Cooper, and Co.

Liu, L. and W. A. Bassett. 1986. *Elements, oxides, silicates: High-pressure phases with implications for the Earth's interior.* New York: Oxford University Press.

Loeffler, B. M., and R. G. Burns. 1976. Shedding light on the color of gems and minerals. *American Scientist* 64: 636–47.

Mitchell, R. H. 1986. *Kimberlites: Mineralogy, geochemistry and petrology.* New York: Plenum Press.

Nassau, K. 1978. The origins of color in minerals. *American Mineralogist* 63: 219–29.

——, 1980. *Gems made by man.* Radnor, Penn.: Chilton Book Co. (Specifically, see Chapter 26, "The origin of color in gemstones.")

——, 1980. The causes of color. *Scientific American* 243: 124–56.

——, 1983. *The physics and chemistry of color: The fifteen causes of color.* New York: Wiley.

Nordstrom, D. K., and J. L. Munoz. 1985. *Geochemical thermodynamics.* Menlo Park: Calif.: Benjamin/Cummings Publishing Co.

O'Reilly, W. 1984. *Rock and mineral magnetism.* London, England: Blackie; distributed by New York: Chapman and Hall.

Putnis, A. 1992. *Introduction to mineral sciences.* New York: Cambridge University Press.

Ribbe, P. H. 1975. The chemistry, structure, and nomenclature of feldspar. *Feldspar Mineralogy, Reviews in Mineralogy* 2. Mineralogical Society of America, Washington, D. C.

Shutong, X. A. I. Okay, J. Shouyuan, A. M. C. Sengör, S. Wen, L. Yican, and J. Laili. 1992. Diamond from the Dabie Shan metamorphic rocks and its implication for tectonic setting. *Science* 256:80–82.

Silica: Physical behavior, geochemistry, and materials applications. 1994. Edited by P. J. Keaney, C. T. Prewitt and G. V. Gibbs. Reviews in Mineralogy 29, Mineralogical Society of America, Washington, D. C.

Smith, J. V., and W. L. Brown. 1988. *Feldspar minerals.* Vol. 1, *Crystal structures, physical, chemical, and microtextural properties.* New York: Springer-Verlag.

Ultrahigh-pressure mineralogy: Physics and chemistry of the Earth's deep interior. 1988. Edited by R. J. Hemley. Reviews in Mineralogy 34, Mineralogical Society of America, Washington, D. C.

Wood, D. L., and K. Nassau. 1968. The characterization of beryl and emerald by visible and infrared absorption spectroscopy. *American Mineralogist* 54: 777–801.

CHAPTER 5

OVERVIEW OF CRYSTALLOGRAPHIC CONCEPTS

"A reasonable knowledge of crystal geometry should be part of the technical background of anyone who deals with the crystalline state, or who wishes to read intelligently any of the literature concerned with crystalline matter.... The main feature of crystal geometry is that it deals with orderly repetition. In fact, the geometry of crystals is the geometry of order."   Martin J. Buerger, 1971.

The definition of a mineral includes the phrase "a solid with a highly ordered atomic arrangement." This ordered internal arrangement may be expressed in the well-developed external form of crystals. External crystal form (also known as *morphology*) can be observed with the naked eye on macroscopic crystals. The inherent symmetry of such crystals can be visually evaluated, if the crystals are well-formed and undistorted. Using optical instrumentation, such as a reflecting goniometer, careful measurements of the angular positions of crystal faces can be made. This leads to better definition of the inherent symmetry. X-ray diffraction and electron diffraction techniques are used to study the internal orderly atomic arrangement of crystals. This is known as *crystal structure analysis*. The orderly repetition of atoms, ions, or molecules (as in organic crystal structures) is closely tied to symmetry.

This chapter serves as a broad overview of some important crystallographic concepts. Chapter 6 will address further aspects of morphological crystallography and space groups. In the present chapter, concepts based on external morphology precede those of internal structure; it is initially simpler to locate symmetry elements in physical objects (e.g., wooden crystal models) than in patterns of motifs or in models that represent the internal structures of minerals.

FROM SYMMETRY OPERATIONS TO SPACE GROUPS

After defining various symmetry operations, as expressed by the external form of crystals, and after having evaluated their various possible combinations, we will learn that there are only 32 nonidentical symmetry elements and combinations thereof. These 32 are known as the 32 *point groups*, or *crystal classes* (see Table 5.4), which can be distributed among six *crystal systems*. This grouping into six crystal systems leads to the recognition of appropriate *crystallographic axis* choices for each of the six crystal systems (see Figs. 5.26 and 5.27). All of the foregoing are treated in animated graphics in module II of the CD-ROM.

Having completed this overview of symmetry expressed by the morphology of crystals, we will evaluate the symmetry content and translational components of crystal structures. Because a crystal structure can be regarded as consisting of regularly repeated motifs (atoms, ions, or groups of atoms or ions) we will first evaluate the regular repeat of motifs in two dimensions. This leads to the conclusion that there are only five possible, nonidentical periodic arrays of points (also known as the five *plane lattices* or *nets*; these nets are constructed through animations in module III of the CD-ROM under the heading "Two-Dimensional Order"; see also Fig.

5.50). Considering the various symmetry contents of two-dimensional motifs (containing rotations or mirrors) we arrive at the conclusion that in ordered two-dimensional patterns there are only 10 nonidentical symmetry contents possible. These are the 10 *two-dimensional point groups* (see Fig. 5.55).

When one considers the interaction of the symmetry elements in the 10 two-dimensional point groups with the 5 plane lattices (or nets), as well as some possible *glide* reflections (in place of mirrors), one arrives at the conclusion that there are only 17 *two-dimensional plane groups* (see Fig. 5.59). This limited number is the result of the fact that there are constraints imposed in combining the 5 nets (plane lattice types) with the 10 motif symmetries, because not all motif symmetries are compatible with every net type. Seven of the 17 plane groups are illustrated in module III of the CD-ROM under the heading "Two-dimensional Order."

Finally, when we consider regular patterns in three dimensions instead of two, we conclude that instead of 5 plane nets we can construct 14 unique but different three-dimensional lattices (using the 5 plane nets as a base). These are known as the *14 Bravais lattices*. Animations constructing 10 of these 14 lattice types are given in module III of the CD-ROM under the heading "Three-dimensional order"; see also Figs. 5.62 and 5.63). These 14 three-dimensional lattice types can be combined with the symmetries inherent in the 32 point groups. Combining the symmetry inherent in the 32 point groups with the 14 Bravais lattices, as well as two additional symmetry operations that involve translation (*screw* and *glide* operations), one arrives at the 230 space groups. These 230 *space groups* represent the various ways in which motifs can be arranged in space in a homogeneous array (see Table 5.10). Various aspects of space groups are illustrated in module III of the CD-ROM under the heading "Three-dimensional Order."

MINERALS AS CRYSTALLINE SOLIDS

Minerals, with few exceptions, possess the internal, ordered arrangement that is characteristic of crystalline solids. When conditions are favorable, they may be bounded by smooth plane surfaces and assume regular geometric forms known as *crystals*. Today, most scientists use the term *crystal* to describe any solid with an ordered internal structure, regardless of whether it possesses external faces. Because bounding faces are mostly an accident of growth and their absence in no way changes the fundamental properties of a crystal, this usage is reason-

able. Thus, we may frame a broader definition of a crystal as a *homogeneous solid possessing long-range, three-dimensional internal order*. The study of crystalline solids and the principles that govern their growth, external shape, and internal structure is called *crystallography*. Although crystallography was originally developed as a branch of mineralogy, today it has become a separate science dealing not only with minerals but with all crystalline matter.

In this text, the general term *crystalline* is used to denote the ordered arrangement of atoms in the crystal structure. The term *crystal*, without a modifier, is generally used in the traditional sense of a regular geometric solid bounded by smooth plane surfaces. *Crystal* is also used in its broader sense with modifiers indicating perfection of development. Thus, a crystalline solid with well-formed faces is *euhedral*; if it has imperfectly developed faces, it is *subhedral*; and without faces, *anhedral*. These adjectives are derived from the Greek *hedron*, meaning face, and the Greek roots *eu*, meaning good, *an*, meaning without, and the Latin root *sub*, meaning somewhat, less than.

Crystalline substances may occur in such fine-grained aggregates that their crystalline nature (or crystallinity) can be determined only with the aid of a microscope. These are designated as *microcrystalline*. If the aggregates are so fine that the individual crystallites cannot be resolved with the microscope but can be detected by X-ray diffraction techniques, they are referred to as *cryptocrystalline*. Although most substances, both natural and synthetic, are crystalline, some lack any ordered internal atomic arrangement and are called *amorphous*, or noncrystalline. Naturally occurring amorphous substances are designated as *mineraloids* (see Chapter 4).

Internal Order

The internal order or crystal structure of a mineral can be thought of as a motif (or group of atoms) repeated on a lattice (which is a periodic array of points in space). The lattice expresses the *translation* component of the internal order, which will be discussed subsequently in this chapter. The motif, or group of atoms, has a certain symmetry (or lack thereof) that may be reflected in the crystal's external shape.

The first scientist to demonstrate that the external crystal form of a mineral (its morphology) is an expression of its internal order was René-Just Haüy (1743–1822; see Fig. 1.6). Figure 5.1 is an illustration of Haüy's concept of "integral molecules" that are regularly stacked together to achieve various

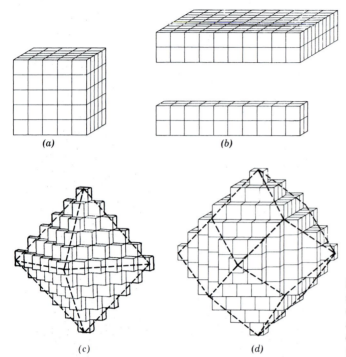

FIG. 5.1.   Different external shapes produced by systematic stacking of cubic unit cells. *(a)* Perfect cube, *(b)* distorted cubes, *(c)* octahedron, and *(d)* dodecahedron. The octahedral and dodecahedral forms are the result of systematic additions of units along directions of accelerated growth. Compare with Fig. 1.6.

commonly developed forms. Haüy coined the word *molécule*, by which he meant the modern concept of unit cells. A unit cell is the smallest unit of a structure (or pattern) that can be infinitely repeated to generate the whole structure (or to generate the complete print of a pattern).

The *three-dimensional internal order* of a crystal can be considered as the periodic repetition of a *motif* (a unit of pattern) in such a way that the environment of and around each repeated motif is identical. A simple and ordered arrangement of a motif in two dimensions is shown in Fig. 5.2 with a comma as the motif. (A comma was chosen as the motif because in a two-dimensional drawing it lacks all symmetry. Therefore, in such a representation the comma does not contribute symmetry elements to the overall ordered pattern; only translation is present.) In crystals the motifs may be molecules such as H_2O, anionic groups such as $(CO_3)^{2-}$, $(SiO_4)^{4-}$, or $(PO_4)^{3-}$, cations such as Ca^{2+}, Mg^{2+}, Fe^{2+}, atoms such as Cu, or combinations of anionic groups, ions, and/or atoms. Figure 5.3 illustrates the regular (ordered) arrangement of triangular $(CO_3)^{2-}$ groups and Ca^{2+} ions in the rhombohedral structural outline of a unit cell of calcite ($CaCO_3$).

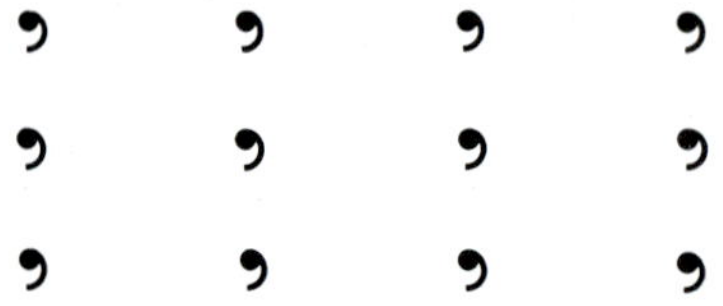

Symmetry

In prior discussions of coordination polyhedra and crystal structure (see Chapter 3), it became obvious that ordered geometrical arrangements contain various types of symmetry. Figure 5.4 illustrates how

FIG. 5.3.   The atomic structure of calcite, $CaCO_3$. The outline of the unit cell is rhombohedral in shape. The locations of the carbon, calcium, and oxygen are shown. The carbonate group, $(CO_3)^{2-}$, has the shape of an equilateral triangle with carbon in its center and oxygen at the three corners. The calcium ions and carbonate groups are the motif units of the structure.

FIG. 5.2.   Two-dimensional order. The comma is the motif.

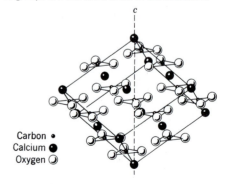

Carbon •
Calcium ●
Oxygen ◐

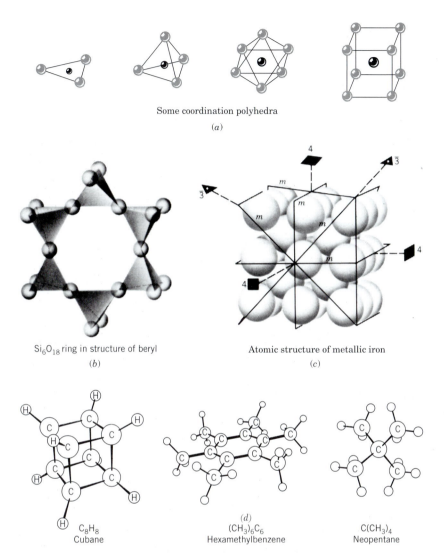

Some coordination polyhedra

(a)

Si_6O_{18} ring in structure of beryl

(b)

Atomic structure of metallic iron

(c)

C_8H_8
Cubane

(d)
$(CH_3)_6C_6$
Hexamethylbenzene

$C(CH_3)_4$
Neopentane

FIG. 5.4. Symmetry in chemistry. *(a)* Variable symmetry contents in some examples of regular coordination polyhedra (see also Fig. 3.36). *(b)* An (Si_6O_{18}) ring with sixfold rotational symmetry in the structure of beryl, $Be_3Al_2Si_6O_{18}$. *(c)* An illustration of the atomic structure of metallic iron. The iron atoms are arranged in what is known as body-centered cubic packing. The overall symmetry is clearly cubic (part of the isometric system) and a very few of the many symmetry elements present in this crystal structure are shown. *(d)* Three different organic molecules with variable inherent symmetry contents. (Taken with permission of the author, from Bernal, I., Hamilton, W. C., and Ricci, J. S., 1972, *Symmetry: A Stereoscopic Guide for Chemists*, New York: W. H. Freeman and Co. with permission).

fundamental and pervasive symmetry is in coordination polyhedra, in inorganic crystal structures (of minerals), and in organic molecules. This internal symmetry finds expression in the external symmetry of well-formed and undistorted crystals.

Although the main concern in this chapter is the evaluation of various symmetry elements (and their combinations) and the periodicity of crystal structures, one should realize that such concepts are part of many fields of endeavor and human thought. In the living world around us, plants and animals usually exhibit symmetry in their external forms. Man himself has bilateral symmetry, and the perfection of such symmetry is found to lead to "facial attractiveness" that has genetic significance (see Box 5.1). Examples of symmetry in architecture are commonplace, especially the bilateral symmetry in what is considered classical architecture (e.g., Greek, and Roman and the imitation thereof; see Box 5.1).

BOX 5.1
BILATERAL SYMMETRY IN HUMANS AND ARCHITECTURE

Leonardo da Vinci's famous drawing displays the proportions of the human body and stresses its bilateral symmetry. It now appears that human body symmetry is a parameter in sexual selection; the more symmetric the body the more attractive it is to a partner (R. Thornhill and S. W. Gangestad, 1994, Human fluctuating asymmetry and sexual behavior, *Psychological Science*, v. 5, pp. 297–302). The human face contains the same mirror symmetry as the overall body, and human facial symmetry is found to be linked to attractiveness as well. R. Thornhill and S. W. Gangestad, 1999, discuss this subject in their paper entitled "Facial Attractiveness" published in *Trends in Cognitive Sciences,* vol. 3, pp. 452–460. They note that "when members of a species discriminate between potential mates with regard to their physical appearance, as humans do, a reasonable working hypothesis is that the discrimination reflects special-purpose adap-

Leonardo da Vinci, 1452–1519, Schema delle proporzione del corpo umano.

tations responsive to cues that had mate value in evolutionary history". They review several studies that show that those with more symmetrical faces are generally rated as more attractive even though faces are not actually perfectly symmetrical.

Bilateral symmetry is common in buildings of many cultures. Examples shown here are: (*a*) The J. Paul Getty Museum in Malibu, California; this is a 1974 reconstruction and adaptation of the Roman Villa dei Papiri at Herculaneum which was buried by the eruption of mount Vesuvius in A.D. 79 (it is a highly symmetric building with everything planted around it enhancing strict bilaterial symmetry as well) and (*b*) The Taj Mahal, in Agra, India which was built between 1630 and 1648 by the Mogul emperor Shah Jahan in memory of his wife who died in 1629. This specific photograph shows more than bilateral symmetry on account of the additional reflection in the pool in front.

The J. Paul Getty Museum, Malibu, California; photograph by Jack Ross, used with permission;

The Taj Mahal, Agra, Uttar Pradesh, India; photograph by Wolfgang Kaehler, CORBIS, used with permission.

CRYSTALS AND SPECIFIC SYMMETRY ELEMENTS

The external shape of a well-formed crystal may reflect the presence or absence of certain symmetry elements. These are rotation axes, mirror (reflection) planes, center of symmetry, and rotoinversion axes. The presence of these *symmetry elements* can be detected, in a well-formed crystal, by the angular arrangement of the bounding faces and sometimes by their size and shape. (Examples of each of these symmetry elements as part of a well-formed crystal are shown in Fig. 5.5). In poorly developed or distorted crystals the symmetry is generally not obvious, but can be derived from careful measurement of the angular relations of the bounding faces. The recognition of symmetry elements in crystals, or wooden models that display the morphology of perfect crystals, is generally part of the laboratory sessions associated with the mineralogy course. Many such aspects of morphological symmetry are treated, with animations, in module II of the CD-ROM.

In the following discussion of symmetry operations we note that these operations are without translation. In our visual study of the morphology of crystals we can see the presence of rotation axes, mirrors, and centers of symmetry but we do not see atomic translations. Only inside the atomic structure of minerals can one evaluate translations on an angstrom level. Therefore, symmetry elements that involve translations will be introduced only later as part of a discussion of regular patterns.

Symmetry Elements (Without Translation)

The motif used in Fig. 5.2, the comma, contains *no symmetry*. However, many motifs, such as those used to create two-dimensional printed patterns, do contain symmetry. *Rotation axes, mirror planes*, and *centers of symmetry* are examples of such symmetry elements. The act of rotation about an axis, the act of reflection by a mirror, and the act of inversion about a central point are collectively referred to as *symmetry operations*.

Rotation

Rotation alone, through an angle (α), about an imaginary axis, generates another motif, or several other motifs. In Fig. 5.6a the angle α of 180° generates a pattern with two hands. A *rotation axis* is a line about which a motif unit may be rotated and repeat itself in appearance once or several times during a complete rotation (see Fig. 5.6a). Therefore, an *axis of rotation* is an imaginary line through a crystal about which the crystal may be rotated and repeat itself in appearance 1, 2, 3, 4, or 6 times during a complete rotation (see Fig. 5.10). Figure 5.5a depicts a 6-fold rotation axis. When rotated about this axis, the crystal repeats itself each 60° or six times in a 360° rotation.

Rotational symmetry is generally expressed by any whole number (n) from 1 to infinity. The number n expresses the number of times a motif unit is repeated during a complete (360°) rotation. A rotational symmetry of $n = 1$ means that after a complete rotation of 360° about an axis, all aspects of an object (or figure) come into coincidence with themselves just once. The other limiting case of rotational symmetry is with a rotation axis of infinite order ($n = \infty$). An object possessing this kind of an axis may be made to coincide with itself by any angle of rotation, because the amount of rotation necessary is infinitely small. Figure 5.7 shows how a

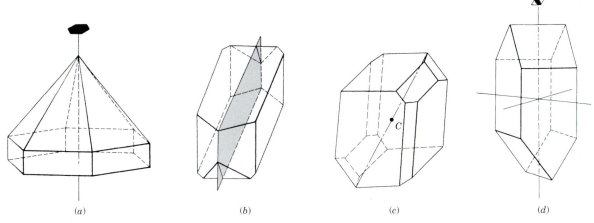

FIG. 5.5. Translation-free symmetry elements as expressed by the morphology of crystals. *(a)* sixfold axis of rotation (6); *(b)* mirror plane (*m*); *(c)* center of symmetry (*i*); *(d)* fourfold axis of rotoinversion ($\bar{4}$).

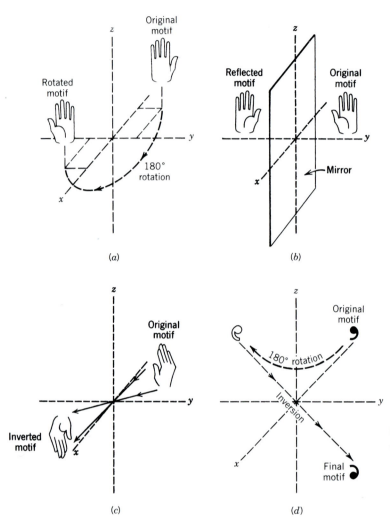

FIG. 5.6. *(a)* Generation of a pattern by rotation of a motif through an angle of 180°. *(b)* Right- and left-handed motifs related by a mirror reflection. *(c)* Motifs related by inversion through a center. *(d)* Motifs related to 180° rotation and subsequent inversion; this is known as rotoinversion (see also Fig. 5.12*a*).

hollow cylinder is compatible with a rotational axis, parallel to the length of the cylinder, of $n = \infty$. It also shows that this same hollow cylinder with a single notch has rotational symmetry of $n = 1$.

We can fabricate objects, or draw shapes, with rotational symmetries that lie between the two extremes of $n = \infty$ and $n = 1$ (where n is a whole number and an aliquot part of 360°). We might, for example, decide to construct a 36-sided cannister in which each side is offset from its adjoining sides by 10°. Such an object would have rotational symmetry of $n = 36$. The rotational symmetries of some shapes and objects are illustrated in Fig. 5.8.

It should be noted, however, that until now we have concerned ourselves with the rotational symmetry of independent objects. We have not considered the fact that we may wish to arrange such objects into an ordered pattern. When the symmetry of the motif is constrained by the translational sym-

metry of the lattice, only certain rotations are possible (see p. 221).

The types of rotation found in internally ordered crystals, and also expressed in their external shape (morphology) are one-fold ($\alpha = 360°$), two-fold ($\alpha = 180°$), three-fold ($\alpha = 120°$), four-fold ($\alpha = 90°$), and six-fold ($\alpha = 60°$). A five-fold axis and seven- and higher-fold axes are not possible. This will be proved geometrically on p. 221, after lattice translation has been discussed. Intuitively, this becomes clear when one tries to completely cover a plane surface with a five-sided motif such as a pentagon, without mismatches and gaps (see Fig. 5.9*a*). On the other hand, Fig. 5.9*b* shows how hexagons can completely cover a surface. The geometric picture in Fig. 5.9*a* shows mismatches and gaps, which tend *not* to occur in the crystal structures of minerals. On an atomic scale, such gaps could represent unsatisfied chemical bonds or overstretched bonds between ions (or

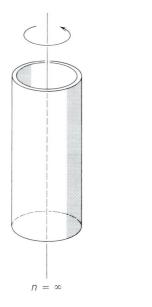

$n = \infty$ $n = 1$

Plan view:

(a)   (b)

FIG. 5.7. The two extremes of rotational symmetry ($n = \infty$ and $n = 1$) as shown by independent objects: *(a)* a perfect, hollow cylinder and *(b)* a hollow cylinder with a *V*-shaped notch at the top.

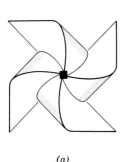
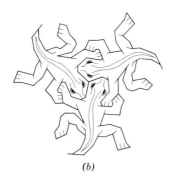
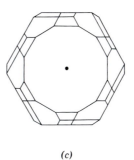

(a)   (b)   (c)

FIG. 5.8. Objects showing elements of rotational symmetry only. *(a)* A child's pinwheel with a fourfold axis (4) perpendicular to the face of the wheel. *(b)* A pattern of three lizards that are related to each other by a threefold rotation axis (3) perpendicular to the page. The threefold axis intersects the page at a point between the three heads. (Redrawn from plate 38 by M. C. Escher in *Fantasy and Symmetry; The Periodic Drawings of M. C. Escher*, 1965 by Caroline H. MacGillavry. New York: Harry N. Abrams, Inc., copyright ©1939, M. C. Escher/Cordon Art, Baarn, Holland.) *(c)* View of an apatite crystal along its *c*-axis. This drawing shows a large basal face (with dot marked in its center) and an array of modifying faces around it. The overall symmetry of this crystal requires a sixfold rotation axis (6) at the location of the dot. There are no mirror planes visible in this orientation of an apatite crystal.

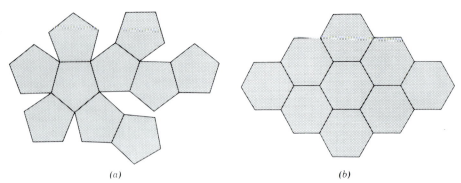

FIG. 5.9. *(a)* Arrangement of pentagons, which individually have fivefold symmetry axes perpendicular to the page, leads to gaps in the pattern. *(b)* Arrangement of hexagons with sixfold axes perpendicular to the page, as in honeycombs.

atoms), both of which are absent in ordered crystal structures.

The possible rotation axes are portrayed in Fig. 5.10 with the graphic symbols used to represent them. The number of duplications of the motif during a 360° rotation gives the rotation axis its name. For example, two equivalent units per 360° rotation are related by a *two-fold rotation axis*.

Rotation produces patterns in which the original motif and those generated from it are identical in orientation with respect to each other. In other words, the original motif and the newly generated one have the same "handedness." The original motif and those generated from it by rotation are, therefore, said to be *congruent*.

Reflection (Mirror)

A *reflection* produces a mirror image across a mirror plane, *m* (Fig. 5.6*b*). In this case the generated motif has the opposite handedness of the original motif, and the two make an *enantiomorphic pair* (meaning that the motifs are related by a mirror and that they cannot be superimposed on each other). This is the same as the relationship between your right and your

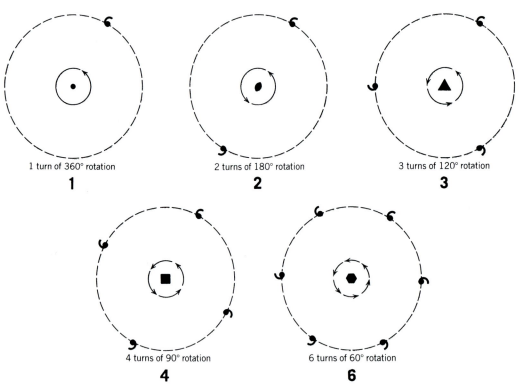

FIG. 5.10. Illustration of rotations that allow the motif to coincide with an identical unit for one-, two-, three-, four-, or sixfold rotation axes. Symbols in the center of the circles represent the graphic illustrations of these axes. The diagram for **2** represents a projection onto the *xy* plane of Fig. 5.6*a*.

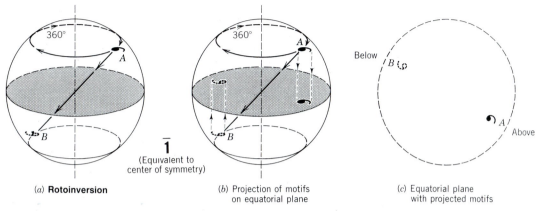

(a) **Rotoinversion** $\overline{1}$ (Equivalent to center of symmetry)   (b) Projection of motifs on equatorial plane   (c) Equatorial plane with projected motifs

FIG. 5.11.   *(a)* Illustration of an operation of rotoinversion, consisting of 360° rotation and subsequent inversion through the center of the globe. *(b)* Projection of the two motif units (*A* and *B*) from the outer skin of the globe onto the equatorial plane. *(c)* Location of the projected motifs on the equatorial plane (see also Fig. 5.12).

left hand. Therefore, a *mirror plane* is an imaginary plane that divides a crystal into halves, each of which, in a perfectly developed crystal, is the mirror image of the other. Figure 5.5*b* illustrates the nature and position of a single mirror in a crystal, also called a *symmetry plane.*

Center of Symmetry

A *center of symmetry*, also known as *inversion (i)*, produces an inverted object through an *inversion center.* An inversion involves drawing imaginary lines from every point on the object through the inversion center and out an equal distance on the other side of the inversion center. The inverted object is then "recreated" by connecting the points (Fig. 5.6*c*). Inversion, like reflection, produces an *enantiomorphic pair.*

A *center of symmetry* is present in a crystal if an imaginary line can be passed from any point on its surface through its center and a similar point is found on the line at an equal distance beyond the center. This is equivalent to $\overline{1}$, or inversion. Figure 5.5*c* illustrates a center of symmetry in a crystal.

Rotation with Inversion

In addition to the symmetrical order generated by operations of rotation axes, there are one-, two-, three-, four-, and sixfold rotations that can be combined with inversion and are known as *rotoinversion operations* (see Fig. 5.6*d*). The rotation axes have been illustrated by motifs that lie in the same plane, as in Fig. 5.10. In combining rotation with inversion it is best to observe the order of a pattern in three dimensions.

Figure 5.11 illustrates the combination of symmetry operations in a onefold rotoinversion. This is

known as a onefold rotoinversion axis, and is symbolized as $\overline{1}$ (read: bar one). The original motif is rotated 360°, so that it returns to its original position and is then inverted through a center. This combination of operations produces the same result as does the presence of a *center of symmetry.* The $\overline{1}$ operation is therefore also referred to as a *center of symmetry,* or *i* (for inversion). The right-hand portion of Fig. 5.11 illustrates how the three-dimensional arrangement of the motif commas appears when projected on the equatorial plane of the globe in Fig. 5.11*a*. The rotoinversion operations for $\overline{2}$, $\overline{3}$, $\overline{4}$, and $\overline{6}$ (read: bar two, bar three, etc.) are shown in Fig. 5.12. The $\overline{2}$ operation is equivalent to the operation of a mirror plane coincident with the equatorial plane of the globe (Fig. 5.12*a*). The $\overline{3}$ operation is the equivalent of a threefold rotation axis and inversion (*i*), which is the same as a threefold rotation and a center of symmetry. The $\overline{4}$ operation is not resolvable into other operations and as such is unique. The $\overline{6}$ operation is equivalent to a threefold axis of rotation with a mirror plane perpendicular to the rotation axis.

It should be noted that the original motif unit (denoted *A* in all illustrations of Figs. 5.11 and 5.12) has an *enantiomorphic*[1] relationship with the second motif unit (denoted *B*), because of the inversion. The third motif unit (denoted *C*), however, is congruent (similar) with the original motif unit (*A*). All of these symmetry operations (rotation, reflection, and rotoinversion) generate only a finite number of motifs. On

[1]Two enantiomorphic motifs are related by mirror reflection or inversion.

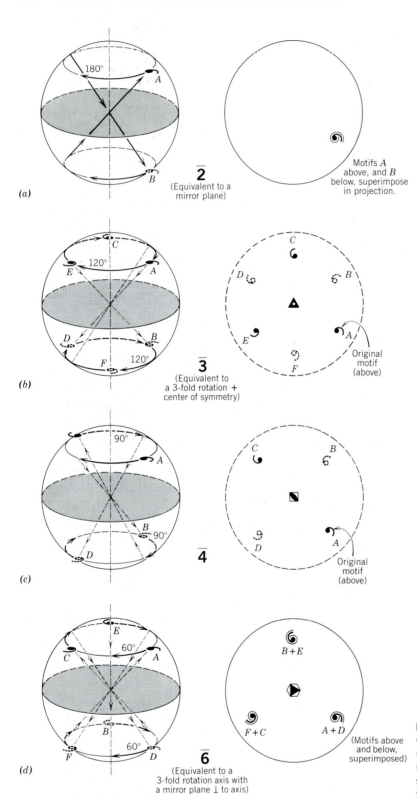

(a)

180°

A

B

$\overline{2}$
(Equivalent to a mirror plane)

Motifs A above, and B below, superimpose in projection.

(b)

120°

C

E   A

D   B

F   120°

$\overline{3}$
(Equivalent to a 3-fold rotation + center of symmetry)

C

D   B

E

F

Original motif (above)

(c)

90°

A

B   90°

D

$\overline{4}$

C   B

D   A

Original motif (above)

(d)

E

60°

C   A

B

F   60°   D

$\overline{6}$
(Equivalent to a 3-fold rotation axis with a mirror plane ⊥ to axis)

B + E

F + C   A + D

(Motifs above and below, superimposed)

FIG. 5.12. Illustration of operations of rotoinversion on motif units. To go from unit A to B (to C, etc.) involves rotation through an angle α (360°, 180°, 120°, 90°, or 60°) as well as inversion through the center (see Fig. 5.11 for illustration of projection scheme).

the other hand, translation, and translational symmetry operations (such as glide and screw operations; see p. 234) will repeat a motif infinitely. As such, rotation, reflection, and rotoinversion operations are classified as "without translation."

An *axis of rotoinversion* within a crystal is an imaginary line that relates rotation about an axis with inversion. Figure 5.5*d* illustrates the morphological expression of a fourfold rotoinversion axis.

These symmetry operations are illustrated, with animations, in module II of the CD-ROM.

Combinations of Rotations

Until now we have considered patterns generated by a *single* axis of rotation or rotoinversion. We can, however, combine various axes of rotation and generate regular three-dimensional patterns. Symmetry axes can only be put together in symmetrically consistent ways, such that an infinite set of axes is not generated. For instance, if one fourfold axis, *A*, is put at an acute angle to another fourfold axis, *B*, each will operate on the other, generating an infinite set of axes. To avoid this, axes must be put together at 90°, or at 54°44′, as in the special case of cubic symmetry. Furthermore, all symmetry operators must intersect at a single point.

For example, we might combine a fourfold rotation axis (4) perpendicular to the plane of the page with a twofold rotation axis (2) in the plane of the page. Another example would be a combination of a sixfold rotation axis (6) perpendicular to the plane of the page with a twofold axis (2) in the plane of the page. Both examples are illustrated in Fig. 5.13. The four- and sixfold axes in both symmetry combinations are at point *A*, the center of the circle, perpendicular to the page. The twofold axis is to the right of *A*, along the east–west direction. The presence of the four- and sixfold axes will generate three and five more twofold axes, respectively, shown as *dashed* lines (ignore, for the moment, the dotted axes). Although we have generated three and five twofold axis extensions, they constitute only two twofold axis directions at 90° to each other in Fig. 5.13*b* and three twofold axis directions at 120° to each other in Fig. 5.13*c*. Let us now enter a comma (marked by *B* in the drawings) above the page in a position slightly north of the original twofold axis. This comma is marked (+), indicating that it lies above the page, in the positive direction of the *z* axis (see Fig. 5.13*a*). The original twofold axis (in the east–west direction) will generate another comma from the one given at *B*, namely, on the south side of the twofold axis, and

below the page. This generated comma is accompanied by a minus (−) sign, indicating its position below the page. The four- and sixfold axes will generate three and five additional motif pairs, respectively, as shown by dashed commas in Figs. 5.13*b* and *c*. If now we carefully observe the arrangement of all of the commas, it becomes clear that we have generated yet another set of twofold axes. These axes are dotted and at 45° to the original twofold axes in Fig. 5.13*b* and at 30° to the original twofold axes in Fig. 5.13*c*. The total symmetry in Figs. 5.13*a* and *b*, therefore, consists of a fourfold rotation axis perpendicular to the page and two sets of twofold axes, the original set in the E–W, N–S directions, and the second at 45° thereto. The total symmetry in Fig. 5.13*c* consists of a sixfold rotation axis perpendicular to the page and two sets of three twofold axes in the plane of the page. The two sets of twofold axes are at 30° to each other. In each case, each *set* of twofold axes is symmetrically equivalent to the other set. That is, if one twofold axis is given, the other symmetry elements generate the other twofolds; in the example given, as two independent sets of twofolds. These types of combinations of axes can be represented by a sequence of digits for the types of rotation axes involved. In such symbols, each symmetrically equivalent set of symmetry elements is listed. For the examples in Fig. 5.13 this would result in 422 and 622, respectively. Three-dimensional representations of the locations of the rotation axes in the 422 and 622 combinations are given in Figs. 5.13*d* and *e*.

Other possible combinations of rotational symmetry elements are 222, 32, 23, and 432. Note that in 32 there are only two rotational symmetries indicated rather than three as in the other combinations. Figure 5.14 shows that combining a threefold axis with a twofold axis in a plane perpendicular to it generates no symmetry axes in addition to three twofolds. The 432 symmetry axis combination is one of high symmetry with the locations of the axes in specialized positions. Figure 5.15 shows the location of such axes with reference to a cubelike outline. The fourfold axes are perpendicular to the cube faces, the threefold axes are at the corners of the cube, and the twofold axes are located at the centers of the edges of the cube. 23 is also cubic symmetry, but now the twofolds are perpendicular to the cube faces. All cubic or isometric symmetry combinations have a set of four threefold axes (along the body diagonals of the cube). These threefolds meet the face normal axes (4 or 2) at 54°44′. (For a rigorous derivation of the limits on combinations of

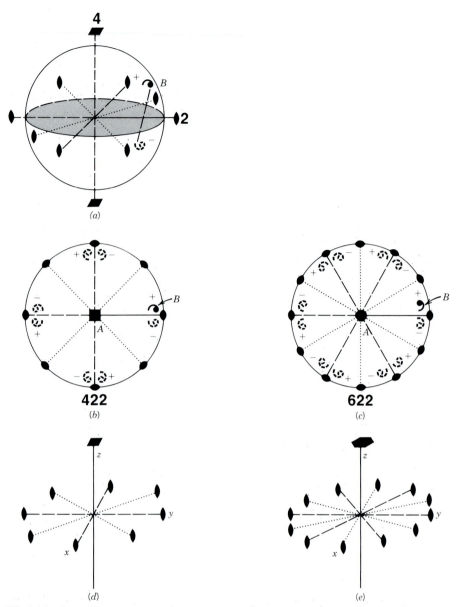

FIG. 5.13. Combination of rotation axes and the generation of ordered patterns from such combinations. *(a)* Perspective view of a vertical fourfold rotation axis combined with a twofold rotation axis E–W in the equatorial plane. *B* is the original motif above the equatorial plane to the back of the E–W twofold rotation axis. One motif generated by this E–W axis is also shown. *(b)* and *(c)* Plan views of the location of the symmetry axes and the motif units. *(d)* and *(e)* Three-dimensional sketches of the distribution of symmetry axes. See text for the development of these figures.

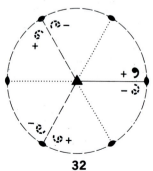

FIG. 5.14. Combination of a threefold rotation axis perpendicular to the page and a twofold rotation axis lying E–W in the page. The original motif units are on the right-hand side. Generated motif units and symmetry axes are indicated by dashes and dots. The resultant symmetry sequence is 32, not 322. Compare with Fig. 5.13.

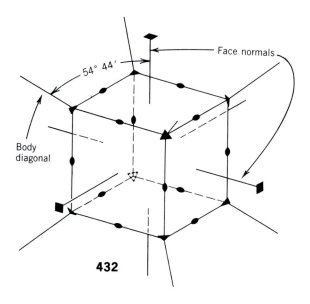

432

FIG. 5.15.   The location of symmetry axes in 432 with respect to a cubelike outline.

rotational operators, see Boisen and Gibbs, 1990; the complete reference is given at the end of this chapter.)

Combinations of Rotation Axes and Mirrors

In the previous section we discussed the possibility of combining rotational symmetries as in the sequences 622, 422, 32, and so on. Let us now consider some examples of combinations of rotation axes and mirror planes. As a general rule, mirror planes within crystals are either perpendicular to or parallel to any rotation axes that are present. In Fig. 5.16a a fourfold rotation axis is combined with a mirror plane perpendicular to the axis and in Fig. 5.16b a sixfold axis is combined with a mirror perpendicular to it. In Fig. 5.16a the arrangement of the motifs is compatible with the fourfold rotation axis and is shown by commas above the mirror plane. These are reflected by the mirror plane giving rise to another set of four commas below as shown by the dashed commas. Normally the symmetry elements and motif units are shown in a two-dimensional projection as on the right side of Fig. 5.16a. The motif units above the mirror as well as those below the mirror are projected onto the mirror itself. This causes the commas from above and below to coincide. In order to distinguish motif units that lie above the plane of projection (the mirror plane, in this case) from those below the plane, motif units above the plane are generally shown as solid dots and those below the plane as small open circles. When this convention is used in the projection of

Fig. 5.16a, it results in a fourfold rotation axis surrounded by four motif units (dots) above, and four identical motif units (circles) below the mirror plane. Note that the mirror plane is conventionally shown by a solid circle. This type of combination of symmetry elements is represented by $4/m$ (read: four over m). The symmetry combination in Fig. 5.16b is represented by $6/m$. Other similar combinations are $2/m$ and $3/m$.

In a previous section we derived several combinations of rotation axes such as 622, 422, 222. If we add mirror planes perpendicular to each of the rotation axes, the following symmetry combinations result: $6/m2/m2/m, 4/m2/m2/m$, and $2/m2/m2/m$. In Figs. 5.17a, b, and c illustrations are given for the combinations 422, $4/m2/m2/m$, and $4mm$ (compare with Fig. 5.13). The motif units are shown as solid dots and equivalent open circles indicating their position above and below the plane of projection, respectively. In 422, we have four motif units above the plane and four below, making a total of eight symmetrically related motifs. In $4/m2/m2/m$ we have a mirror plane perpendicular to the fourfold axis (shown in Fig. 5.17b as a solid circle), and we have mirror planes perpendicular to each of the four twofold rotation axes that lie in the equatorial plane. The trace of the mirror plane perpendicular to the E–W axis coincides with the N–S axis, and so forth. The $4/m2/m2/m$ symmetry combination, therefore, contains four vertical mirror planes in addition to the one horizontal mirror perpendicular to the fourfold axis. In Figs. 5.17b and c the traces of the mirrors are shown, by convention, as solid lines. In Fig. 5.17b, we generate eight motif

Perspective   Symbol   Projection

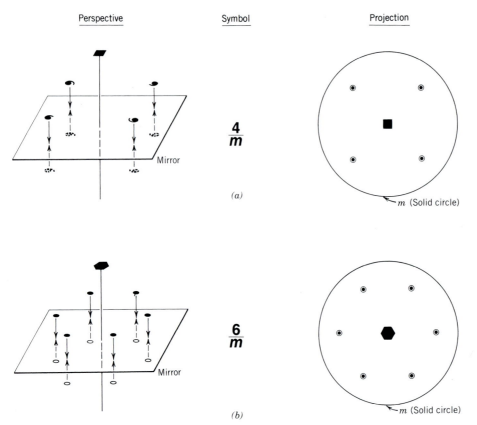

FIG. 5.16. *(a)* Combination of a 4-fold symmetry axis and a mirror plane perpendicular to it. The motif units that can be represented by commas are more conventionally shown by solid dots and small open circles in order to differentiate motif units above and below the mirror plane, respectively. *(b)* Combination of a 6-fold rotation axis and a mirror plane perpendicular to it.

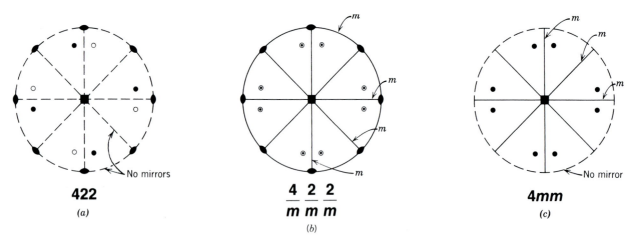

FIG. 5.17. *(a)* Combination of a 4-fold rotation axis and two sets of 2-fold rotation axes (see also Fig. 5.13). Motif units above the page are shown as solid dots, those below the page as open circles. *(b)* Combination of a 4-fold rotation axis, four 2-fold rotation axes, and mirror planes perpendicular to each of the axes. *(c)* Combination of a 4-fold rotation axis and two sets of mirror planes parallel to the 4-fold axis.

units above the equatorial mirror and eight below, making a total of 16 symmetrically related units. Yet another possible combination of fourfold rotation and mirror planes is expressed by 4mm. This symbolism denotes one fourfold rotation axis and four mirror planes that are parallel to (and intersect in) the fourfold axis and whose traces in the equatorial plane lie N–S, E–W, and 45° to these directions These mirror planes are called *vertical* mirrors because they are perpendicular to the equatorial plane. In 4/m, for instance, the mirror is a horizontal mirror *in* the equatorial plane and perpendicular to the fourfold axis. The total number of motif units related by the combined symmetry elements in

4mm is eight, all of which lie on one side of the projection (above the page, as shown in Fig. 5.17c). Other similar combinations are 6mm, 3m, and 2mm.

At this stage it is instructive to evaluate the interdependence of mirrors that intersect each other, and the rotational symmetry (along the line of intersection) that results therefrom. Several examples of such intersections are given in Fig. 5.18. When two vertical mirrors intersect each other at 90°, the vertical line of intersection is equivalent to a twofold axis of rotation (Fig. 5.18a). When three vertical mirrors intersect each other at 60°, the vertical line of intersection is one of threefold rotation.

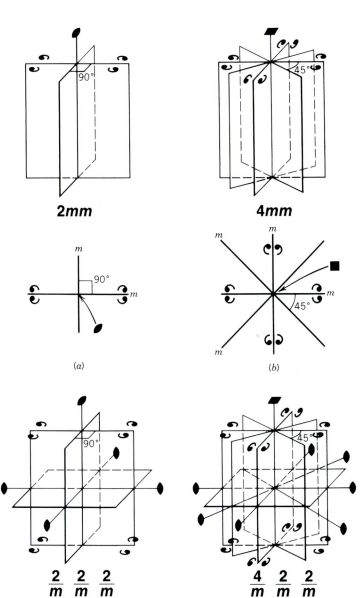

FIG. 5.18. Illustrations of intersecting mirrors and the resultant lines of intersection, equivalent to rotation axes. *(a)* and *(b)* Perspective and plan views of 2mm and 4mm. In *(a)* the vertical mirrors are at 90° to each other; in *(b)* the vertical mirrors are at 45° to each other. In *(c)* and *(d)* horizontal mirrors are added to the drawings in *(a)* and *(b)* respectively. The horizontal intersection lines become 2-fold rotations in both illustrations. Compare Fig. 5.18d with the plan view in Fig. 5.17b.

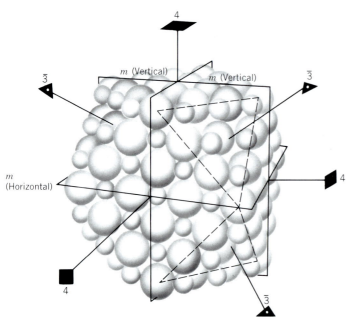

FIG. 5.19.   A structure model of the packing of Na^+ (small spheres) and Cl^- (large spheres) in NaCl (halite). The model has a cubo-octahedral outline showing square cube faces and triangular octahedral faces at the corners of the cube. This structure contains all of the symmetry elements that are also present in the morphology of a cube, that is, three 4-fold axes (all are shown), four 3-fold rotoinversion axes (three are shown at corners of the cube), six 2-fold axes (at the cube edge; none are shown), and nine mirror planes in various orientations (only three are shown that are perpendicular to each other).

When four vertical mirrors intersect each other at 45°, the line of intersection is one of fourfold rotation (Fig. 5.18b). And when six vertical mirrors intersect each other at 30°, the vertical intersection line is one of sixfold rotation. In Fig. 5.18c an additional horizontal mirror has been added to the configuration in Fig. 5.18a. This horizontal mirror reflects the motifs in a downward direction, and the horizontal intersection lines for all of the mirrors become twofold rotations. In Fig. 5.18d an additional horizontal mirror has been added to the configuration in Fig. 5.18b. Now the motifs at the top are reflected vertically downward and all

the horizontal intersections between mirrors are axes of twofold rotation.

Before proceeding with further aspects of the morphology of crystals, it should be noted that all of the nontranslational symmetry elements that have been introduced are the easily observed expression (in well-formed crystals or wooden models) of the internal atomic arrangement of the structure. This is shown in Fig. 5.19. This illustration depicts the regular packing of Na^+ and Cl^- ions in the structure of NaCl, halite. This specific ionic packing scheme is a function of relative ionic sizes (for Na^+ and Cl^-), their electrical charges, and

TABLE 5.1   Thirty-Two Possible Symmetry Elements and Combinations of Symmetry Elements

	Increasing Rotational Symmetry ⟶				
Rotation axis only	1	2	3	4	6
Rotoinversion axis only	$\bar{1}(=i)$	$\bar{2}(=m)$	$\bar{3}$	$\bar{4}$	$\bar{6}(=3/m)$
Combinations of rotation axes		222	32	422	622
One rotation with perpendicular mirror		2/m	$3/m(=\bar{6})$	4/m	6/m
One rotation with parallel mirrors		2mm	3m	4mm	6mm
Rotoinversion with rotation and mirror			$\bar{3}2/m$	$\bar{4}2m$	$\bar{6}2m$
Three rotation axes and perpendicular mirrors		2/m2/m2/m		4/m2/m2/m	6/m2/m2/m
Additional symmetry combinations present in isometric crystals		23 $2/m\bar{3}$		432 (see Fig. 5.15) $\bar{4}3m$	$4/m\bar{3}2/m$

the types of bonds between the ions. Although the NaCl structure (as shown in Fig. 5.19) also contains translational symmetry elements, it should be obvious that all of the geometric principles discussed thus far are basic to an understanding of the internal structures (not just the morphology) of crystalline materials.

Résumé of Symmetry Operations Without Translation

In the preceding discussion we introduced several symmetry elements: rotation axes (1, $\underline{2}$, 3, 4, and 6), rotoinversion axes ($\bar{1}$, $\bar{2}$, $\bar{3}$, $\bar{4}$, and $\bar{6}$), a center of symmetry (*i*), and mirror planes (*m*). We have also discussed some of the combinations of rotation axes such as 622, 422, 222, and of rotation axes and mirror planes such as 6/m2/m2/m, 4/m2/m2/m, and 4mm. The number of possible symmetry combinations is not unlimited. Indeed, the total number of nonidentical symmetry elements and combinations of symmetry elements is only 32. In Table 5.1 they are arranged in a sequence from the lowest rotational symmetry (1) to the highest rotational symmetry (6). In this table, as in the previous discussion, we have used symbols for the symmetry elements and the combinations of symmetry elements that are referred to as the *Hermann-Mauguin notation* after their inventors. Because of their universal acceptance, they are also called the *international symbols*. The 32 possible elements and combinations of elements are identical to the *32 possible crystal classes* to which crystals can be assigned on the basis of their morphology (see Table 5.4) or their internal atomic arrangement.

We have illustrated a number of combinations of symmetry elements, but we have not rigorously derived the possible 32 nonidentical symmetry elements or combinations of symmetry elements, which are also known as the *32 point groups*. The word *point* indicates that the symmetry operations leave one particular point, at least, of the pattern unmoved. The word *group* relates to the mathematical theory of groups that allows for a systematic derivation of all the possible and nonidentical symmetry combinations (see, for example, D. E. Sands 1975, or Boisen and Gibbs 1990; for complete references, see the end of this chapter).

A crystal, under favorable circumstances of growth, will develop smooth planes or "faces" that may assume regular geometric forms that express its internal, regular, atomic arrangement. In crystals with well-developed faces, one can recognize the ele-

ments of symmetry such as rotation axes, rotoinversion axes, a center of symmetry, and mirror planes. A systematic study of the external forms of crystals leads to 32 possible symmetries or symmetry combinations, which are the same 32 as the point groups already noted.

Certain of the 32 *crystal classes* have symmetry characteristics in common with others, permitting them to be grouped together in one of six *crystal systems*. Table 5.2 shows the conventional arrangement of crystal systems and classes. Compare this with Table 5.1 and note that the symmetry elements and combinations are the same in both tables, but that their groupings are somewhat different.

In Chapter 6 we will discuss the crystal morphology of 19 of the 32 crystal classes (point groups) as arranged within the six crystal systems. That treatment is based on the study and measurement of the external form of crystals. A graphic illustration of all 32 symmetries is given in Fig. 5.20. This figure precedes the systematic treatment of crystal form because it is based on the distribution of motif units, without reference to crystals or crystallographic axes. In Fig. 5.20 the presence or absence of a center of symmetry (or inversion, *i*) is not indicated by a specific symbol. Of the 32 crystal classes there are 21 without a center of symmetry and 11 with a center. Table 5.3 distinguishes the 32 crystal classes according to the absence or presence of a symmetry center. Visual inspection of the circles with the distribution of motif units in Fig. 5.20 allows one to determine whether a center of symmetry is present or absent. In such an evaluation one must keep in mind that a center of symmetry inverts the motif unit through the center of the circle; for example, a motif unit at the upper right (above the page) is balanced by a motif unit at the lower left (below the page).

TABLE 5.2 Thirty-Two Crystal Classes (see also Table 5.4)

Crystal System	Symmetry of Crystal Classes
Triclinic	1 and $\bar{1}$
Monoclinic	2, *m*, and 2/m
Orthorhombic	222, 2mm, and 2/m2/m2/m
Tetragonal	4, $\bar{4}$, 4/m, 422, 4mm, $\bar{4}$2m, and 4/m2/m2/m
Hexagonal	3, $\bar{3}$, 32, 3m, and $\bar{3}$ 2/m
	6, $\bar{6}$, 6/m, 622, 6mm, $\bar{6}$m2, and 6/m2/m2/m
Isometric	23, 2/m$\bar{3}$, 432, $\bar{4}$3/m, and 4/m$\bar{3}$2/m

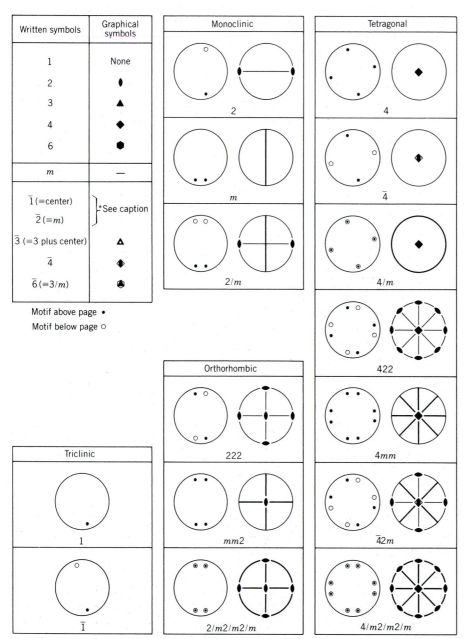

FIG. 5.20. Graphic representation of the distribution of motif units compatible with the symmetry elements of each of the 32 crystal classes (point groups). For all crystal classes, excepting triclinic, there are two circular diagrams, with the left-hand diagram showing the distribution of motif units and the right-hand diagram illustrating the symmetry elements consistent with these motif units. The motif units above the page are equivalent to those below the page, but they are differentiated by dots (above the page) and circles (below the page). The symbols for the symmetry elements are given at the top left corner of the diagram. The presence of a center of symmetry is not shown by any symbol; its presence can be deduced from the arrangement of motif units. Instead of $\bar{2}$ the symbol for a mirror (m) is used. The diagrams for the monoclinic system are shown in what crystallographers refer to as the "second setting," with m vertical (perpendicular to the page) and the 2-fold axis in an east-west orientation. Monoclinic symmetry can also be shown by setting the 2-fold rotation axis perpendicular to the page, and orienting the mirror parallel to the page; this is referred to as the "first setting."

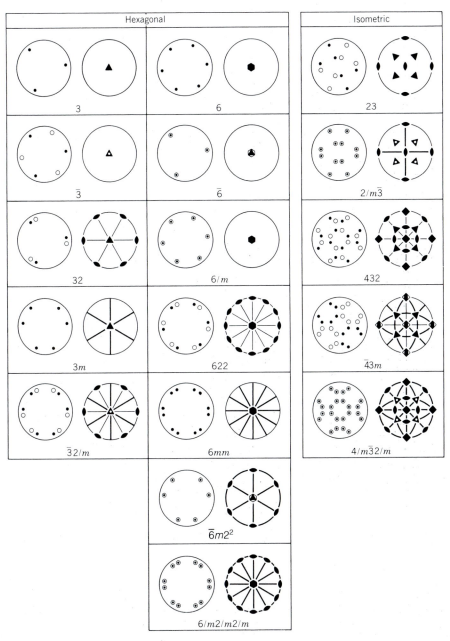

FIG. 5.20. (Continued)

TABLE 5.3 The 32 Crystal Classes Grouped According to the Presence or Absence of a Center of Symmetry

Crystal System	No Center	Center Present
Triclinic	1	$\bar{1}$
Monoclinic	$2, \bar{2}(= m)$	$2/m$
Orthorhombic	$222, mm2$	$2/m2/m2/m$
Tetragonal	$4, \bar{4}, 422$	$4/m, 4/m2/m2/m$
	$4mm, \bar{4}2m$	
Hexagonal	$3, 32, 3m$	$\bar{3}, \bar{3}2/m$
	$6, \bar{6}, 622$	$6/m, 6/m2/m2/m$
	$6mm, \bar{6}m2$	
Isometric	$23, 432, \bar{4}3m$	$2/m\bar{3}, 4/m\bar{3}2/m$

CRYSTAL MORPHOLOGY

Because crystals are formed by the repetition in three dimensions of a unit of structure, the limiting surfaces, which are known as the faces of a crystal, depend in part on the shape of the unit. They also depend on the conditions in which the crystal grows. These conditions include all the external influences such as temperature, pressure, nature of solution, direction of flow of the solution, and availability of open space for free growth. The angular relationships, size, and shape of faces on a crystal are aspects of *crystal morphology*.

If a cubic unit cell is repeated in three dimensions to build up a crystal with n units along each edge, a larger cube will result containing n^3 units. With a similar orderly repeat mechanism, different shapes may result, as shown in Fig. 5.1 for distorted cubes, octahedron, and dodecahedron. The octahedral and dodecahedral forms are common in many crystals, but because the unit cell dimensions are on the angstrom level the steps are invisible to the eye and the resulting faces appear as smooth, plane surfaces.

With a given internal structure, a limited number of planes bound a crystal, and only a comparatively few are common. In determining the types of crystal faces that may develop on a crystal we must also consider the internal lattice. Faces are most likely to form on crystals parallel to lattice planes that have a high density of *lattice points* (or *nodes*). The frequency with which a given face is observed is roughly proportional to the number of nodes it intersects in the lattice: the larger the number of nodes, the more common the face, as is illustrated in Fig. 5.21. This rule, known as the *law of Bravais*, is generally confirmed by observations. Although there are exceptions to the law, as pointed out by Donnay and Harker in 1937, it is usually possible to choose the lattice in such a way that the rule holds true.

Because crystal faces have a direct relationship to the internal structure, it follows that the faces have a definite relationship to each other. This fact was observed in 1669 by Nicolaus Steno, who pointed out that the angles between corresponding faces on crystals of quartz are always the same. This observation is generalized today as Steno's law of the constancy of interfacial angles, which states: *The angles*

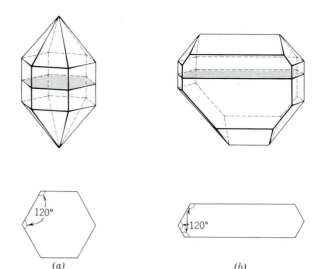

FIG. 5.22. Constancy of interfacial angles is shown in the comparison of a well-formed and highly symmetric quartz crystal (*a*) and a distorted quartz crystal (*b*). The sections across the direction of elongation show identical interfacial angles of 120° in both, irrespective of the asymmetric habit in (*b*).

between equivalent faces of crystals of the same substance, measured at the same temperature, are constant (see Fig. 1.5). For this reason crystal morphology is frequently a valuable tool in mineral identification. A mineral may be found in crystals of widely varying shapes and sizes, but the angles between pairs of corresponding faces are always the same. An illustration of such constancy of interfacial angles, in horizontal sections of two quite different-looking quartz crystals, is given in Fig. 5.22.

Although crystals possess a regular, orderly internal structure, different planes and directions within them have differences in atomic environments. Consider Fig. 5.23, which illustrates the packing of ions in sodium chloride (NaCl), the mineral halite. Any plane parallel to the front face of the cube is composed of half Na^+ and half Cl^- ions. On the other hand, cutting the corner of the cube shows planes containing only Na^+ ions alternating with planes containing only Cl^- ions.

These different atomic arrangements along different crystal planes or directions give rise to *vectorial properties*. Because the magnitude of the property is dependent on direction, it differs for different crystallographic directions. Some of the vectorial properties of crystals are hardness, conductivity for heat and electricity, thermal expansion, speed of light, growth rate, solution rate, and diffraction of X-rays.

Of these properties, some vary continuously with direction within the crystal. Hardness, electrical

FIG. 5.21. This figure represents one layer of *lattice points* in a cubic lattice. Several lines are possible through the network that include a greater or lesser number of *lattice points* (or nodes). These lines represent the traces of possible crystal planes. It is found that the planes with the highest density of lattice points are the most common, such as *AB* and *AC*.

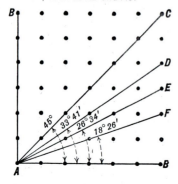

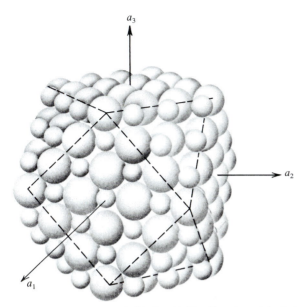

FIG. 5.23. Packing model of halite, NaCl, with cubo-octahedral outline, Na$^+$ small, Cl$^-$ large. Note that the cube faces have sheets of equal numbers of Na$^+$ and Cl$^-$ ions, whereas the octahedral planes (at the corners of the cube) consist of alternating sheets of Na$^+$ and sheets of Cl$^-$ ions. Na$^+$ and Cl$^-$ ions are both surrounded by six closest neighboring ions (six coordination) in a face-centered lattice. This type of structure is also found in PbS, galena, MgO, and many other *AX* compounds.

and heat conductivity, thermal expansion, and the speed of light in crystals are all examples of such *continuous vectorial properties.*

The *hardness* of some crystals varies so greatly with crystallographic direction that the difference may be detected by simple scratch tests. Thus, kyanite, a mineral that characteristically forms elongate bladed crystals, may be scratched with an ordinary pocket-knife in a direction parallel to the elongation, but it cannot be scratched by the knife perpendicular to the elongation. Some directions in a diamond crystal are much harder than others. When diamond dust is used for cutting or grinding, a certain percentage of the grains always presents the hardest surface and, hence, the dust is capable of cutting along planes in the crystal of lesser hardness. If a perfect sphere cut from a crystal is placed in a cylinder with abrasive and tumbled for a long time, the softer portions of the crystal wear away more rapidly. The nonspherical solid resulting serves as a hardness model for the substance being tested.

The directional character of *electrical conductivity* is of great importance in the manufacture of silicon and germanium diodes, tiny bits of silicon and germanium crystals used to rectify alternating current. In order to obtain the optimum rectifying effect, the small bit of semimetal must be oriented crystallo-

graphically, as the conduction of electricity through such crystals varies greatly with orientation.

Ball bearings of synthetic ruby sound very attractive, because the great hardness of ruby would cut down wear and give long life to the bearing. However, when heated, ruby expands differently along different crystallographic directions, and ruby ball bearings would rapidly become nonspherical with the rise of temperature from friction during operation. Because the *thermal expansion* figure of ruby is an ellipsoid of revolution with a circular cross section, however, cylindrical roller bearings (cut parallel to the crystal's threefold axis) are practical. Most minerals have unequal coefficients of thermal expansion in different directions, leading to poor resistance to thermal shock and easy cracking with heating or cooling. SiO$_2$ glass, which has an irregular internal structure as compared with quartz crystal, is more resistant to thermal shock than the mineral.

The *velocity of light* in all transparent crystals, except those that are isotropic (see Chapter 7), varies continuously with crystallographic direction. Of all the vectorial properties of crystals, the optical parameters are most easily determined quantitatively and are expressed as the index of refraction, the reciprocal of the velocity of light in the crystal relative to the velocity of light in air or vacuum.

Discontinuous vectorial properties, on the other hand, pertain only to certain definite planes or directions within the crystal. There are no intermediate values of such properties connected with intermediate crystallographic directions. An example of such a property is *rate of growth.* The rate of growth of a plane in a crystal is intimately connected with the density of lattice points in the plane. We saw that a plane such as *AB* in Fig. 5.21 has a much greater density of nodes than planes *AD, AE,* or *AF.* Calculations of the energy involved indicate that the energy of particles in a plane such as *AB,* in which there is a high density of nodes, is less than the energy of particles in less densely populated planes such as *AF.* Hence, the plane *AB* will be the most stable, because in the process of crystallization the configuration of lowest energy is that of maximum stability. Planes *AF, AD, AE,* and so on, will however grow faster than *AB* because fewer particles need to be added per unit area. In the growth of a crystal from a nucleus, the early forms that appear will be those of relatively high energy and rapid growth. Continued addition of material to these planes will build them out while the less rapidly growing planes lag behind (the rate of growth of a face is inversely proportional to the node density, because fewer nodes require less material to be added for growth, resulting in

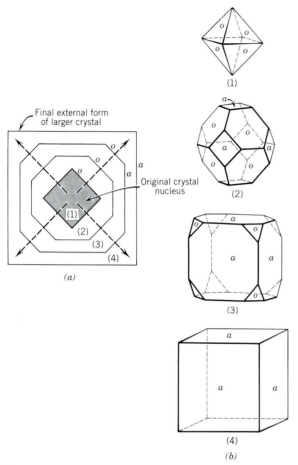

FIG. 5.24. *(a)* Schematic cross section of a crystal that grew from stage (1), a nucleus (with only *o* faces), via stages (2) and (3) to the final form of (4), with only *a* faces. The arrows are growth vectors representing the direction of fastest crystal growth. Note that the faces perpendicular to these growth vectors (*o* faces) are finally eliminated in stage (4). *(b)* Illustration of the complete crystals (at the various stages 1 to 4), for which *(a)* provides the schematic cross section. The form consisting of *o* faces only is an octahedron, the one with *a* faces only is a cube, and the two intermediate forms are combinations of the octahedron and the cube in different stages of development. (See also CD-ROM module II, under "Crystal Growth.")

more rapid growth). Figure 5.24 shows how a crystal, during its development from a nucleus to a larger crystal, may change its external form in various stages. (An animation of this process is given in module II of the CD-ROM under the heading "Crystal Growth.") Thus, the edges and corners of a cube may be built out by addition of material to the planes cutting off the corners and edges, whereas little material is added to the cube faces. As growth progresses, the rapidly growing faces disappear, literally growing themselves out of existence, building the slower-growing, more stable forms in the process. After this stage is complete, growth is much slower,

as addition is now entirely to the slower-growing, lowest-energy form. Thus, crystals themselves, if taken at various stages of their development, serve as models of the rate of growth for the compound being studied.

The *rate of solution* of a crystal in a chemical solvent is similarly a discontinuous vectorial process, and solution of a crystal or of any fragment of a single crystal may yield a more or less definite solution polyhedron. A clearer illustration of the vectorial nature of the rate of solution is afforded by etch pits. If a crystal is briefly treated with a chemical solvent that attacks it, the faces are etched or pitted. The shape of these pits is regular and depends on the structure of the crystal, the face being attacked, the presence of chemical impurities and inclusions, and the nature of the solvent. Valuable information about the internal geometry of arrangement of crystals may be obtained from a study of such etch pits.

Cleavage may be thought of as a discontinuous vectorial property and, like crystal form, reflects the internal structure, as cleavage always takes place along those planes across which there exist the weakest bonding forces. Those planes are generally the most widely spaced and the most densely populated.

CRYSTAL SYMMETRY

In our earlier discussion of symmetry elements (and their combinations), we noted the existence of 32 point groups (or crystal classes; see Tables 5.1 and 5.2). These point groups, uniquely defined by their symmetry content, are graphically illustrated in Fig. 5.25 and listed in Table 5.4. Each of these point groups has, in the past, been given a name, in accordance with the name of the general form in each class. These names are not given in Table 5.4 but are given in subsequent discussion of point groups in Chapter 6. The reason for omitting the generally accepted names of the point groups in Table 5.4 is that *each point group is uniquely defined by its Herman–Mauguin notation, but not by its name.*

As can be seen in Tables 5.2 and 5.4, certain groups of crystal classes have common symmetry characteristics. These groups of crystal classes are known as *crystal systems.* There are six such systems, with the hexagonal system having two subdivisions (hexagonal and rhombohedral) on the basis of the lattice symmetry being hexagonal or rhombohedral (see footnote to Table 5.4).

The 19 most important point groups are listed in bold type in Table 5.4. These are the point groups that contain the largest number of minerals (as well

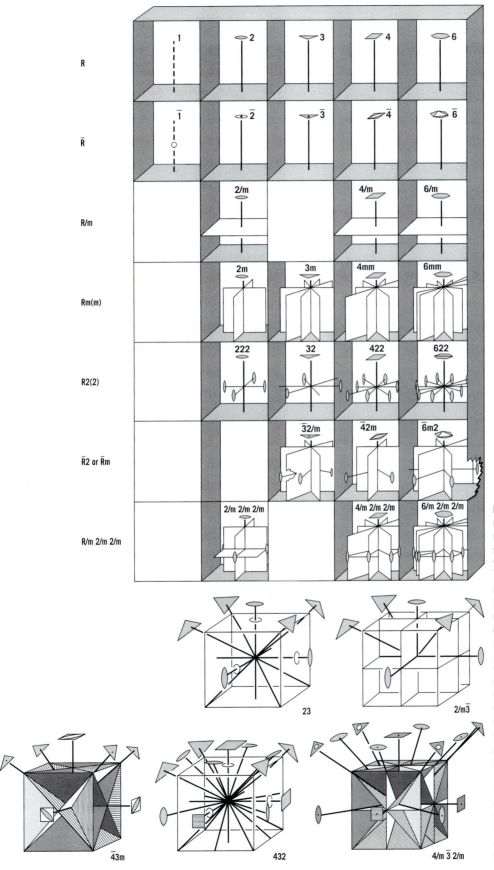

FIG. 5.25. Graphical representation of the geometric relationships between symmetry elements and the 32 point groups. *R* and $\overline{R}$ represent rotation and rotoinversion, respectively. R/*m*, R*m*(*m*), and R2(2) represent the combination of rotation with a mirror plane perpendicular to it, parallel to it, or with a twofold axis perpendicular to it. $\overline{R}$2 or $\overline{R}$*m* represents the combination of a rotoinversion axis with a twofold axis perpendicular to it or a mirror plane parallel to it (Modified from Fig. 1.15 in F. D. Bloss, 1994, *Crystallography and crystal chemistry*, reprinted original text of 1971; for complete reference see listing at the end of this chapter; with permission of the author).

TABLE 5.4   The Thirty-Two Point Groups (Crystal Classes) and Their Symmetry

Crystal System	Crystal Class	Symmetry Content	Crystal System	Crystal Class	Symmetry Content
Triclinic	1	none	Hexagonal*	3	$1A_3$
	$\overline{1}$	i		$\overline{3}$	$1\overline{A}_3 (= i + 1A_3)$
Monoclinic	2	$1A_2$		32	$1A_3, 3A_2$
	m	$1m$		3m	$1A_3, 3m$
	2/m	$i, 1A_2, 1m$		$\overline{3}\,2/m$	$1\overline{A}_3, 3A_2, 3m$
Orthorhombic	222	$3A_2$			$(1\overline{A}_3 = i + 1A_3)$
	$mm2$	$1A_2, 2m$		6	$1A_6$
	$2/m2/m2/m$	$i, 3A_2, 3m$		$\overline{6}$	$1\overline{A}_6 (= 1A_3 + m)$
Tetragonal	4	$1A_4$		6/m	$i, 1A_6, 1m$
	$\overline{4}$	$1\overline{A}_4$		622	$1A_6, 6A_2$
	4/m	$i, 1A_4, m$		6mm	$1A_6, 6m$
	422	$1A_4, 4A_2$		$\overline{6}m2$	$1\overline{A}_6, 3A_2, 3m$
	4mm	$1A_4, 4m$			$(1\overline{A}_6 = 1A_3 + m)$
	$\overline{4}\,2m$	$1\overline{A}_4, 2A_2, 2m$		6/$m2/m2/m$	$i, 1A_6, 6A_2, 7m$
	$4/m2/m2/m$	$i, 1A_4, 4A_2, 5m$	Isometric	23	$3A_2, 4A_3$
				$2/m\overline{3}$	$3A_2, 3m, 4\overline{A}_3$
					$(1\overline{A}_3 = 1A_3 + i)$
				432	$3A_4, 4A_3, 6A_2$
				$\overline{4}\,3m$	$3\overline{A}_4\, 4A_3, 6m$
				$4/m\overline{3}2/m$	$3A_4, 4\overline{A}_3, 6A_2, 9m$
					$(1\overline{A}_3 = 1A_3 + i)$

*In this table all point groups beginning with 6, $\overline{6}$, 3, and $\overline{3}$ are grouped in the hexagonal system. In earlier editions of the *Manual of Mineralogy* the hexagonal system was divided into the hexagonal and rhombohedral divisons. The use of these two subdivisions, as based on the presence of 6 or $\overline{6}$ versus 3 or $\overline{3}$ axes in the morphological symmetry of a crystal, results in confusion when subsequent X-ray investigations show a specific crystal with, for example, 32 symmetry to be based on a hexagonal lattice. This is the case in low quartz, which shows morphological symmetry 32 but is based on a primitive hexagonal lattice, resulting in space group $P3_12$ (or $P3_22$).

The hexagonal system can, however, be divided according to whether the lattice symmetry is hexagonal (6/$m2/m2/m$) or rhombohedral ($\overline{3}2/m$). This results in the following groupings:

Hexagonal (Hexagonal Lattice Division)		Rhombohedral (Rhombohedral Lattice Division)
6/$m2/m2/m$		
$\overline{3}\,2/m$		
$\overline{6}m2$	$\overline{3}2/m$	3m
6mm	3m	3 2
6 2 2 and	3 2	$\overline{3}$
6/m	$\overline{3}$	3
$\overline{6}$	3	
6		

as synthetic compounds) or minerals of widespread geologic occurrence. These same 19 point groups are shown in animations in module II of the CD-ROM under the heading "Crystal Classes"; they are also described in detail in Chapter 6.

Crystallographic Axes

In the description of crystals it is convenient to refer the external forms or internal symmetry to a set of three (or four) reference axes. These imaginary reference lines are known as the *crystallographic axes* and are generally taken parallel to the intersection edges of major crystal faces. Such axes are in most instances fixed by the symmetry and coincide with symmetry axes or with normals to symmetry planes.

For some crystals there may be more than one choice of crystallographic axes when the selection is made on morphology alone. Ideally the axes should be parallel to and their lengths proportional to the edges of the unit cell. A graphical illustration of the relationship of the choice of crystallographic axes to morphology is shown in Fig. 5.26.

All crystals, with the exception of those belonging to the hexagonal system, are referred to three crystallographic axes designated as *a*, *b*, and *c*; see Fig. 5.27. In the general case (triclinic system) all the axes are of different lengths and at oblique angles to each other. The ends of each axis are designated plus or minus; the front end of *a*, the right-hand end of *b*, and the upper end of *c* are positive; the opposite ends are negative. The angles between the posi-

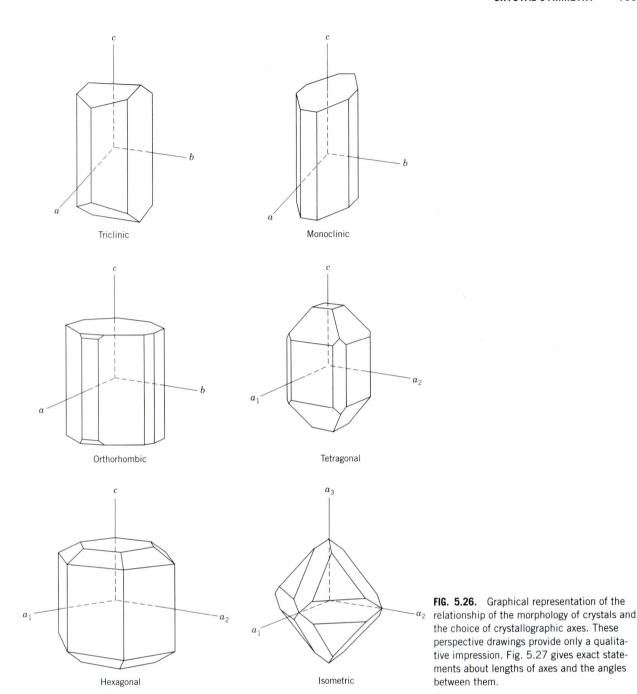

FIG. 5.26. Graphical representation of the relationship of the morphology of crystals and the choice of crystallographic axes. These perspective drawings provide only a qualitative impression. Fig. 5.27 gives exact statements about lengths of axes and the angles between them.

tive ends of the axes are conventionally designated by the Greek letters α, β, and γ. The α angle is enclosed between axial directions b and c, the β angle between a and c, and the γ angle between a and b. In summary the six crystal systems are referred to the following axial directions and axial angles (see also Fig. 5.27):

Triclinic. Three unequal axes all intersecting at oblique angles.

Monoclinic. Three unequal axes, two of which are inclined to each other at an oblique angle

and the third perpendicular to the plane of the other two.

Orthorhombic. Three mutually perpendicular axes all of different lengths.

Tetragonal. Three mutually perpendicular axes, two of which (the horizontal axes) are of equal length (a_1 and a_2), but the vertical axis is shorter or longer than the other two.

Hexagonal. Referred to four crystallographic axes; three equal horizontal axes (a_1, a_2, and a_3) intersect at angles of 120°, the fourth (vertical) is

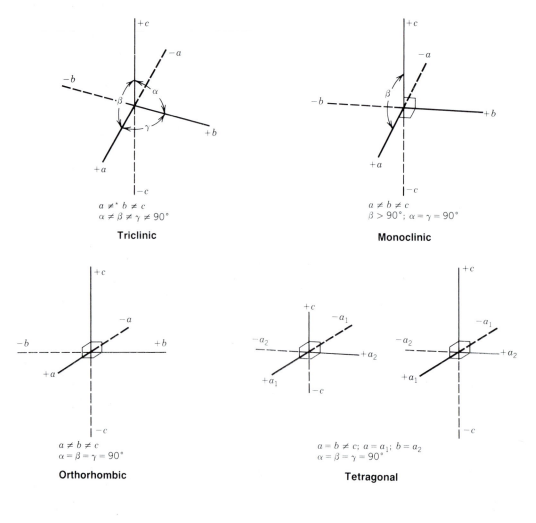

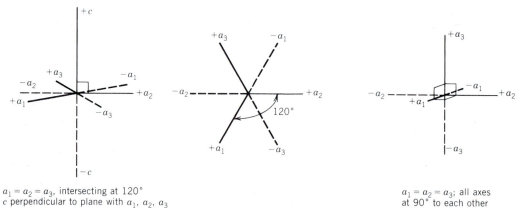

FIG. 5.27.   Illustrations of the conventional system of crystallographic axes adopted for each of the six crystal systems. The three directions are labeled a, b, and c, unless symmetry makes them equivalent. The vertical is labeled c, except in the isometric system where all axes are equal. The two horizontal axes are labeled a and b. If they are equal, they are labeled with the same letter, such as a_1 and a_2. In the hexagonal system there are three equal horizontal axes labeled a_1, a_2, and a_3. See Table 5.5 for the relationship of symmetry notation (as in the Hermann–Mauguin system) and crystallographic axes.

*The sign $\neq$ implies nonequality by reason of symmetry; accidental equality may occur.

of different length and perpendicular to the plane of the other three.

Isometric. Three mutually perpendicular axes of equal lengths (a_1, a_2, and a_3).

Table 5.5 provides a synopsis of the interrelationships of symmetry content, choice of crystallographic axes, and Hermann–Mauguin notation.

Crystallographic Notation for Planes

In this section we explain the system of notation that is used in crystallography for referring to planes and axes (or zones) as observed in the external forms of crystals.

Face Intercepts

Crystal faces can be defined by indicating their intercepts on the crystallographic axes. Thus, in describing a crystal face it is necessary to determine whether it is parallel to two axes and intersects the third, is parallel to one axis and intersects the other two, or intersects all three. In addition, one must determine at what *relative* distance the face intersects the different axes. In defining the position of a crystal face one must remember that crystal faces are parallel to a family of possible lattice planes. In Fig. 5.28a we have an a–b plane in an orthorhombic net based on the unit cell dimensions of the mineral sulfur.

The lattice plane AA is parallel to the b and c axes and intersects the a axis at one length (taken as unit length) along the a axis. The intercepts for this plane would be: $1a$, ∞b, ∞c. Similarly the plane $A'A'$, which is parallel to AA but intersects the a axis at two unit lengths, would have intercepts: $2a$, ∞b, ∞c. The plane BB, which is parallel to the a and c axes and intersects the b axis at unit distance, has intercepts: ∞a, $1b$, ∞c. The plane AB intersects both horizontal axes (a and b) at unit distance but is parallel to c, which leads to the intercepts: $1a$, $1b$, ∞c. A plane that intersects all three axes at unit distances would have intercepts $1a$, $1b$, and $1c$. Figure 5.28b shows the development of crystal faces some of which are parallel to the lattice planes shown in Fig. 5.28a. It must be remembered that the face intercepts shown on the faces are strictly relative values and do not indicate any actual cutting lengths.

When intercepts are assigned to the faces of a crystal, without any knowledge of its unit cell dimensions, one face that cuts all three axes is arbitrarily assigned the units 1a, 1b, and 1c. This face, which is

---

TABLE 5.5 Characteristic Symmetry, and Relationships Between Crystal Axes and Symmetry Notation of Crystal Systems

Crystal Class	System	Characteristic Symmetry	Hermann–Mauguin Notation
1, $\bar{1}$	Triclinic	onefold (inversion or identity) symmetry only	Because of low symmetry, no crystallographic constraints.
2, m, $2/m$	Monoclinic	one twofold rotation axis and/or mirror	The twofold axis is taken as the b axis, and the mirror (the a–c plane) is vertical (second setting).
222, $mm2$ $2/m2/m2/m$	Orthorhombic	three mutually perpendicular directions about which there is binary symmetry (2 or m)	The symbols refer to the symmetry elements in the order a, b, c; twofold axes coincide with the crystallographic axes.
4, $\bar{4}$, $4/m$ 422, $4mm$, $\bar{4}2m$, $4/m2/m2/m$	Tetragonal	one fourfold axis	The fourfold axis refers to the c axis; the second symbol (if present) refers to the axial directions (a_1 and a_2); the third symbol (if present) to directions at 45° to a_1 and a_2.
6, $\bar{6}$, $6/m$ 622, $6mm$ $\bar{6}m2$, $6/m2/m2/m$	Hexagonal*	one sixfold axis	The first number refers to the c axis; the second and third symbols (if present) refer respectively to symmetry elements parallel to and perpendicular to the crystallographic axes a_1, a_2, and a_3.
3, $\bar{3}$, 32, $3m$, $\bar{3}2/m$		one threefold axis	
23, $2/m\bar{3}$, 432, $\bar{4}3m$, $4/m\bar{3}2/m$	Isometric	four threefold axes each inclined at 54°44' to the crystallographic axes (see Fig. 5.15)	The first number refers to the three crystallographic axes a_1, a_2, and a_3; the second number refers to four diagonal directions of 3-fold symmetry (between corners of a cube); the third number or symbol (if present) refers to six directions between the edges of a cube.

*The accepted orientation of the symmetry elements in two crystal classes of the hexagonal system is not straightforward. These are $\bar{6}m2$ and $3m$. The location of the six- or threefold axis is unambiguous. However, the location of the next symmetry element is not obvious. In $\bar{6}m2$, the third entry (twofold rotation axes) coincides with the perpendiculars to a_1, a_2, and a_3; the m's are coincident with these same directions. In $3m$ the m's are located in directions perpendicular to a_1, a_2, and a_3.

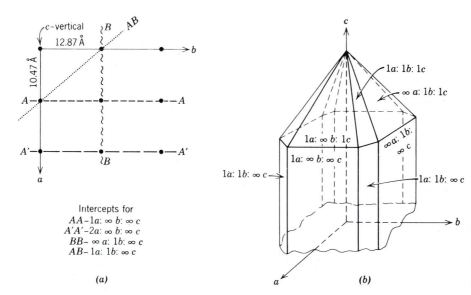

Intercepts for
AA– $1a$: $\infty\,b$: $\infty\,c$
$A'A'$– $2a$: $\infty\,b$: $\infty\,c$
BB– $\infty\,a$: $1b$: $\infty\,c$
AB– $1a$: $1b$: $\infty\,c$

(a)

(b)

FIG. 5.28. (a) Intercepts for some planes in an orthorhombic lattice. (b) Intercepts for some crystal faces on the upper half of an orthorhombic crystal.

referred to as the *unit face*, is generally the largest in case there are several faces that cut all three axes. Figure 5.29 shows an orthorhombic crystal that consists of faces, all of which cut all three crystallographic axes. The largest face (shaded) that intersects all three crystallographic axes at their positive ends is taken as the unit face. Its intercepts are $1a$, $1b$, and $1c$ as shown. The intercepts of the smaller face above it can now be estimated by extending the edges of this face in the directions of the a and b axes. The intercepts for the upper face become $2a$, $2b$, $\frac{2}{3}c$, with respect to the unit face. These intercepts

can be divided by the common factor 2, resulting in $1a$, $1b$, $\frac{1}{3}c$. This example illustrates that the units $1a$ and $1b$ do not represent actual cutting distances but express only relative values. The intercepts of a face have no relation to its size, for *a face may be moved parallel to itself for any distance without changing the relative values of its intersections with the crystallographic axes.*

Miller Indices

Various methods of notation have been devised to express the intercepts of crystal faces upon the crystal axes. The most universally employed is the system of indices proposed by W. H. Miller, which has many advantages over the system of intercepts discussed above.

The Miller indices of a face consist of a series of whole numbers that have been derived from the intercepts by their inversions and, if necessary, the subsequent clearing of fractions. The indices of a face are always given so that the three numbers (four in the hexagonal system) refer to the a, b, and c axes, respectively, and therefore the letters that indicate the different axes are omitted. Like the intercepts, the Miller indices express a ratio, but for the sake of brevity the ratio sign is also omitted. For the two upper faces in Fig. 5.29, which cut the positive segments of the crystallographic axes, the intercepts are $1a$, $1b$, $1c$, and $2a$, $2b$, $\frac{2}{3}c$, respectively. Inverting these intercepts leads to $\frac{1}{1}\,\frac{1}{1}\,\frac{1}{1}$ and $\frac{1}{2}\,\frac{1}{2}\,\frac{3}{2}$, respectively. For the unit-face this gives a Miller index of (111), and clearing the fractions for $\frac{1}{2}\,\frac{1}{2}\,\frac{3}{2}$ by multiplying all by 2 leads to a Miller index of (113) for the other face. The Miller index (111) is read as "one-one-one." Further examples of conversions of intercepts

FIG. 5.29. Relative intersections of faces on an orthorhombic crystal, all of which cut three crystallographic axes.

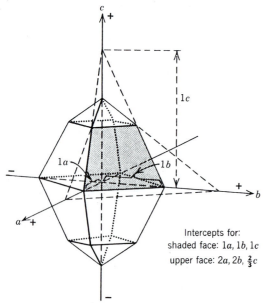

Intercepts for:
shaded face: $1a$, $1b$, $1c$
upper face: $2a$, $2b$, $\frac{2}{3}c$

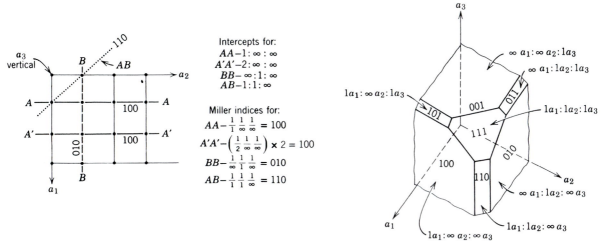

FIG. 5.30. *(a)* Intercepts and Miller indices for some planes in an isometric lattice. *(b)* Intercepts and Miller indices of some faces modifying the corner of a cube.

to Miller indices for an isometric lattice and crystal forms are given in Fig. 5.30. Commas are used in Miller indices only when two-digit numbers appear, as in (1, 14, 3). For faces that intersect negative ends of crystallographic axes, a line is placed over the appropriate number, as shown in Fig. 5.31. For example, $(1\ \bar{1}\ 1)$ reads "one, minus one, one" or "one, bar one, one." It should be noted that indices given above for specific faces are placed in parentheses. This is to distinguish them from similar symbols used to designate crystal forms (see below), zones, and axial directions. Furthermore, the $\bar{1}$ ("bar one") that is part of a Miller index representation is *not* the same as $\bar{1}$, a rotoinversion axis. An interactive animation on Miller index derivation is available in module II of the CD-ROM under the heading "Miller Indices."

FIG. 5.31. Miller indices with respect to positive and negative ends of crystallographic axes.

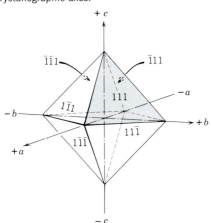

It is sometimes convenient when the exact intercepts are unknown to use a *general symbol* (*hkl*) for the Miller indices; here *h*, *k*, and *l* are, respectively, the reciprocals of rational but undefined intercepts along the *a*, *b*, and *c* axes. The symbol (*hkl*) would indicate that a face cuts all three crystallographic axes without implying relative units along the axes. If a face is parallel to one of the crystallographic axes and intersects the other two, the general symbols would be written as (0*kl*), (*h0l*), and (*hk*0). A face parallel to two of the axes may be considered to intersect the third at unity, and the indices would, therefore, be: (100), (010), and (001), as well as the negative equivalents such as ($\bar{1}$00), (0$\bar{1}$0), and (00$\bar{1}$).

Early in the study of crystals it was discovered that for a given face the indices could always be expressed by simple whole numbers or zero. This is known as the *law of rational indices*.

Crystals belonging to the hexagonal system are distinct from other systems because they possess one unique axis that is either six- or threefold in symmetry. This unique axis is perpendicular to the plane which contains three identical axes, labeled a_1, a_2, and a_3. Because of the presence of four crystallographic axes, instead of three, as in the other crystal systems, a four-number system for indexing crystal faces was developed. This is known as the Bravais–Miller system of indexing. The indices are derived from the intercepts on the axes in the same way as the three-number Miller indices. Figure 5.32 illustrates the conversion of intercepts on the four-axis system to Bravais–Miller indices for three differently oriented crystal faces. The general symbol in this index system is (*hkīl*), in which the first three letters refer to the a_1, a_2, and a_3 axes and the last letter

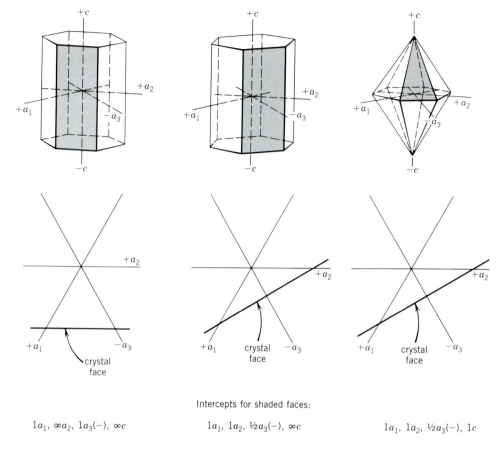

Intercepts for shaded faces:

$1a_1, \infty a_2, 1a_3(-), \infty c$ $1a_1, 1a_2, \tfrac{1}{2}a_3(-), \infty c$ $1a_1, 1a_2, \tfrac{1}{2}a_3(-), 1c$

Miller–Bravais indices for shaded faces:

$\dfrac{1}{1} \quad \dfrac{1}{\infty} \quad \dfrac{\bar{1}}{1} \quad \dfrac{1}{\infty}$ $\dfrac{1}{1} \quad \dfrac{1}{1} \quad \dfrac{\bar{1}}{\tfrac{1}{2}} \quad \dfrac{1}{\infty}$ $\dfrac{1}{1} \quad \dfrac{1}{1} \quad \dfrac{\bar{1}}{\tfrac{1}{2}} \quad \dfrac{1}{1}$

$1\,0\,\bar{1}\,0$ $1\,1\,\bar{2}\,0$ $1\,1\,\bar{2}\,1$

FIG. 5.32.   Derivation of the four-digit Bravais–Miller index from the intercepts of three different crystal faces in the hexagonal system.

to c. In this notation $h + k + i = 0$ holds invariably. For example, with respect to the three faces indexed in Fig. 5.32:

$10\bar{1}0, 1 + 0 + \bar{1} = 0$
$11\bar{2}0, 1 + 1 + \bar{2} = 0$
$11\bar{2}1, 1 + 1 + \bar{2} = 0$

Zones

One of the conspicuous features on many crystals is the arrangement of a group of faces with parallel intersection edges. Considered collectively, these faces form a *zone*. A line through the center of the crystal that is parallel to the lines of face intersections is called the *zone axis*. In Fig. 5.33 the faces m', a, m, and b are in one zone, and b, r, c, and r' are in another. The lines, designated [001] and [100], are the zone axes.

A zone is indicated by a symbol similar to that for Miller indices of faces, the generalized expression of which is $[uvw]$. Any two nonparallel faces determine a zone. Assume one wishes to determine the zone axis of two such faces, (hkl) and (pqr).

FIG. 5.33.   Crystal zones and zone axes.

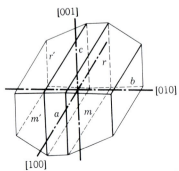

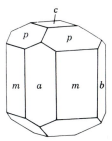

FIG. 5.34. Conventional lettering of forms on crystal drawings.

The method usually used is to write twice the symbol of one face and directly beneath, twice the symbol of the other face. The first and last digits of each line are disregarded and the remaining numbers, joined by sloping arrows, multiplied. In each set the product of 2 is subtracted from the product of 1:

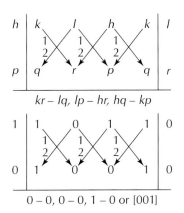

$$kr - lq, \ lp - hr, \ hq - kp$$

$$0 - 0, \ 0 - 0, \ 1 - 0 \ \text{or} \ [001]$$

As a specific example assume m, Fig. 5.34 is (hk0) with index (110) and b is (010). The zone axis is thus [001], as shown above. It should be noted that the

zone symbols are enclosed in square brackets, as [uvw], to distinguish them from face and form symbols.

Crystal Habit

The term *habit* is used to denote the general shapes of crystals as cubic, octahedral, prismatic. Because habit is controlled by the environment in which crystals grow, it may vary with locality. At one place it may be equant and at others tabular or fibrous. Only rarely do crystals present an ideal geometrical shape. But even in asymmetric crystals evidence of the symmetry is present in the physical appearance of the faces and in the constancy of the interfacial angles. Given in Fig. 5.35 are various crystal forms, first ideally developed and then distorted.

Form

As we discussed earlier in the text, the term *form* is often used to indicate general outward appearance. In crystallography, external shape is denoted by the word *habit*, whereas *form* is used in a special and restricted sense. Thus, *a form consists of a group of crystal faces, all of which have the same relation to the elements of symmetry* and display the same chemical and physical properties because all are underlain by like atoms in the same geometrical arrangement. The relationship between form and the elements of symmetry of a crystal is an important one to understand. For example, as in Fig. 5.36 where we have symmetry elements that belong to two crystal classes, $\overline{1}$ (in the triclinic system) and

FIG. 5.35. Examples of some crystal habits in the isometric system. *(a)* Cube and distorted cube. *(b)* Octahedron and asymmetric octahedron. *(c)* Dodecahedron and distorted dodecahedron.

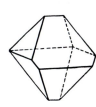

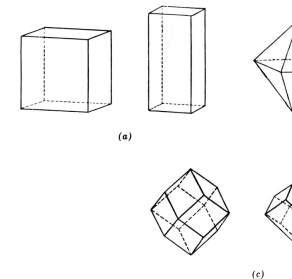

(a)

(b)

(c)

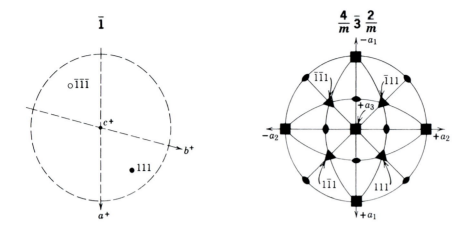

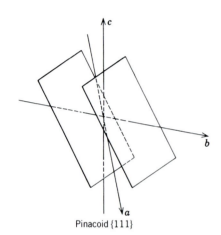

Pinacoid {111}

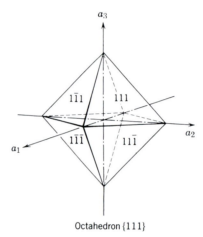

Octahedron {111}

FIG. 5.36. Development of a form from one face with Miller index (111) in the $\bar{1}$ and the $4/m\bar{3}\,2/m$ crystal classes. In the case of $\bar{1}$ symmetry only a two-faced form, a pinacoid, develops. In the illustration of the isometric class the pole to the (111) face coincides with the position of the threefold rotoinversion axis. Faces at the top of the crystal are generally shown by dots, those at the bottom by open circles. Because of the complexity of the isometric figure only the positions of the four top faces of the octahedron are shown by Miller indices; there exists a similar set of four faces underneath.

$4/m\,\bar{3}\,2/m$ (in the isometric system) we may wish to develop the full form for the unit face (111). In the case of the $\bar{1}$ symmetry (which is equivalent to a center of symmetry) we develop an additional face by inverting through the origin of the three crystallographic axes. This additional face will have indices $(\bar{1}\bar{1}\bar{1})$. The form in the case of $\bar{1}$ consists, therefore, of two parallel faces only, and is known as a pinacoid. In the case of $4/m\,\bar{3}\,2/m$ the symmetry elements will generate seven additional faces for (111) with indices $(\bar{1}11)$, $(1\bar{1}1)$, $(11\bar{1})$, $(\bar{1}\bar{1}1)$, $(1\bar{1}\bar{1})$, $(\bar{1}1\bar{1})$, and $(\bar{1}\bar{1}\bar{1})$. This form is known as an octahedron. It should be clear, therefore, that *the number of faces that belongs to a form is determined by the symmetry of the crystal class (or point group)*.

Although faces of a form may be of different sizes and shapes because of distortion of the crystal, the similarity is frequently evidenced by natural stri-

FIG. 5.37. Different appearances of different forms. *(a)* Quartz—hexagonal. *(b)* Apophyllite—tetragonal. *(c)* Pyrite—isometric.

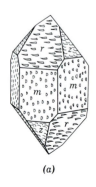

(a)

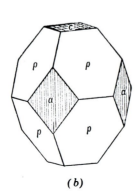

(b)

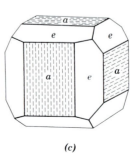

(c)

ations, etchings, or growths, as shown in Fig. 5.37. On some crystals the differences between faces of different forms can be seen only after etchings with acid. In Figs. 5.37a and b there are three forms, each of which has a different physical appearance from the others, and in Fig. 5.37c there are two forms.

In our discussion of Miller indices we said that a crystal face may be designated by a symbol enclosed in parentheses as (hkl), (010), or (111). Miller indices also may be used as form symbols and are then enclosed in braces as $\{hkl\}$, $\{010\}$. Thus in Fig. 5.36 (111) refers to a specific face, whereas $\{111\}$ embraces all faces of the form. In choosing a form symbol it is desirable to select, if possible, the face symbol with positive digits: $\{111\}$ rather than $\{1\bar{1}\bar{1}\}$, $\{010\}$ rather than $\{0\bar{1}0\}$.

In each crystal class there is a form, the faces of which intersect all of the crystallographic axes at different lengths; this is the general form, $\{hkl\}$. All other forms that may be present are *special forms*. In the orthorhombic, monoclinic, and triclinic crystal systems, $\{111\}$ is a general form, for the unit length along each of the axes is different. In the crystal systems of higher symmetry in which the unit distances along two or more of the crystallographic axes are the same, a general form must intersect the like axes at different multiples of the unit length. Thus, $\{121\}$ is a general form in the tetragonal system but a special form in the isometric system and $\{123\}$ is a general form in the isometric system. The concept of a *general form* can also be related to the symmetry elements of a specific crystal class. *An (hkl) face will not be parallel or perpendicular to a symmetry axis or plane, regardless of the crystal class.* A *special form*, however, consists of faces that are parallel or perpendicular to any of the symmetry elements in the crystal class. For most crystal classes the general form contains a larger number of faces than any of the special forms of that same class.

In Fig. 5.36 we developed a two-faced form and an eight-faced form. The two-faced form $\{111\}$ for crystal class $\bar{1}$ is referred to as an *open form*, because it consists of two parallel faces that do not enclose space (see also Fig. 5.38). The eight-faced form $\{111\}$ in crystal class $4/m\,\bar{3}\,2/m$ is a *closed form*, as the eight faces together enclose space (see Figs. 5.31 and 5.38).

A crystal will often display several forms in combination with one another but may have only one, provided that it is a closed form. Because any combination of forms must enclose space, a minimum of two open forms is necessary. The two may exist together or be in combination with other closed or open forms.

Names of Forms

In this, as in previous editions of this textbook, we will follow a scheme of form nomenclature originally proposed by Groth in 1895 and modified by A. F. Rogers in 1935. This scheme recognizes 48 kinds of crystal forms as distinguished by the angular relations of the crystal faces. Of these 48 forms, 32 are the general forms found in the 32 crystal classes (or point groups); 10 are special, closed forms of the isometric system; and 6 are special open forms (prisms) of the hexagonal and tetragonal systems. In this scheme of nomenclature the name of each of the 32 general forms, $\{hkl\}$ (or $\{hkil\}$ in the hexagonal system), becomes the descriptive name of each of the 32 crystal classes. For example, in $\bar{1}$ the general form $\{hkl\}$ is a two-faced form known as a pinacoid, and the descriptive name for $\bar{1}$ is pinacoidal class. Similarly, the name of the general form $\{hkl\}$ in $4/m2/m2/m$ (of the tetragonal system) is ditetragonal–dipyramid; accordingly, the class $4/m2/m2/m$ is commonly referred to as the ditetragonal–dipyramidal class. This nomenclature for general forms and crystal classes has been widely used in the English-speaking world.

Another, similar system of form nomenclature has been developed by crystallographers of the Fedorov Institute in Moscow, Russia. This system has been recommended for use internationally. However, for reasons of simplicity we will adopt only the Groth–Rogers scheme. It should be noted that, although form names are often useful, such names are not absolutely necessary because a form is uniquely defined by a combination of its Miller indices and the Hermann–Mauguin notation of its point group symmetry.

Illustration and Description of Forms

The forms listed in Tables 5.6 and 5.7 are illustrated in Fig. 5.38. In the case of prisms, pyramids, and dipyramids (nos. 5 through 25 in Fig. 5.38) the three-dimensional representations cannot properly convey the shapes of the cross-sections. For this reason the various prisms numbered 5 through 11 are accompanied by a cross-section perpendicular to the long axis of the prism; the shapes of these cross-sections represent sections perpendicular to the long axes of the pyramids (nos. 12 through 18) and dipyramids (nos. 19 through 25) as well. The total symmetry content of each of these forms is discussed in detail in Chapter 6 under the heading "Nineteen of the Thirty-Two Point Groups." Here follow only brief de-

Non-isometric forms

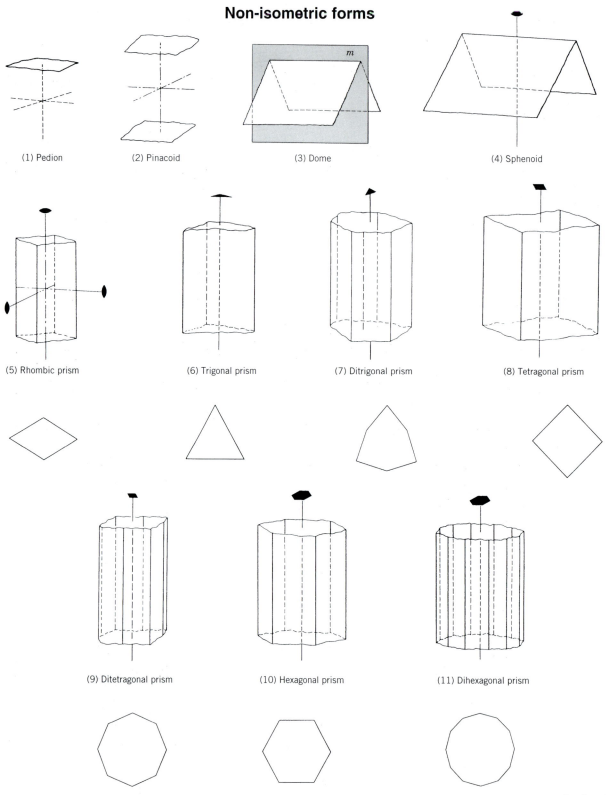

(1) Pedion   (2) Pinacoid   (3) Dome   (4) Sphenoid

(5) Rhombic prism   (6) Trigonal prism   (7) Ditrigonal prism   (8) Tetragonal prism

(9) Ditetragonal prism   (10) Hexagonal prism   (11) Dihexagonal prism

(continued)

FIG. 5.38. The 48 different crystal forms and some of their symmetry elements.

Non-isometric forms (cont'd)

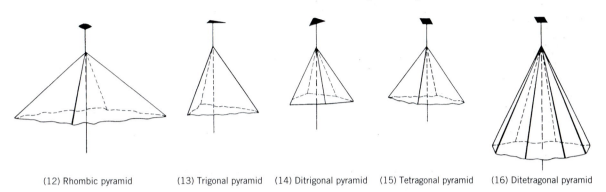

(12) Rhombic pyramid   (13) Trigonal pyramid   (14) Ditrigonal pyramid   (15) Tetragonal pyramid   (16) Ditetragonal pyramid

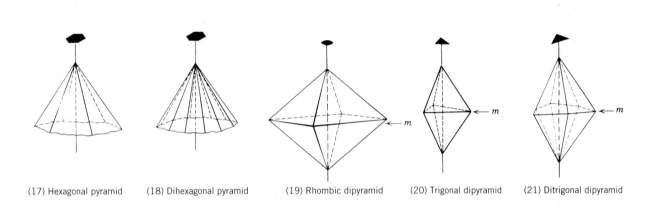

(17) Hexagonal pyramid   (18) Dihexagonal pyramid   (19) Rhombic dipyramid   (20) Trigonal dipyramid   (21) Ditrigonal dipyramid

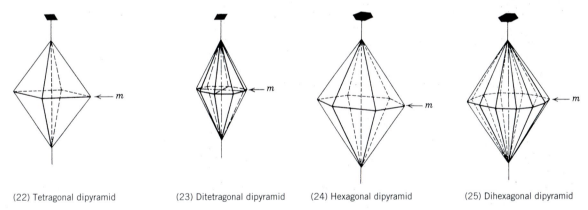

(22) Tetragonal dipyramid   (23) Ditetragonal dipyramid   (24) Hexagonal dipyramid   (25) Dihexagonal dipyramid

FIG. 5.38. (Continued)

Non-isometric forms (cont'd)

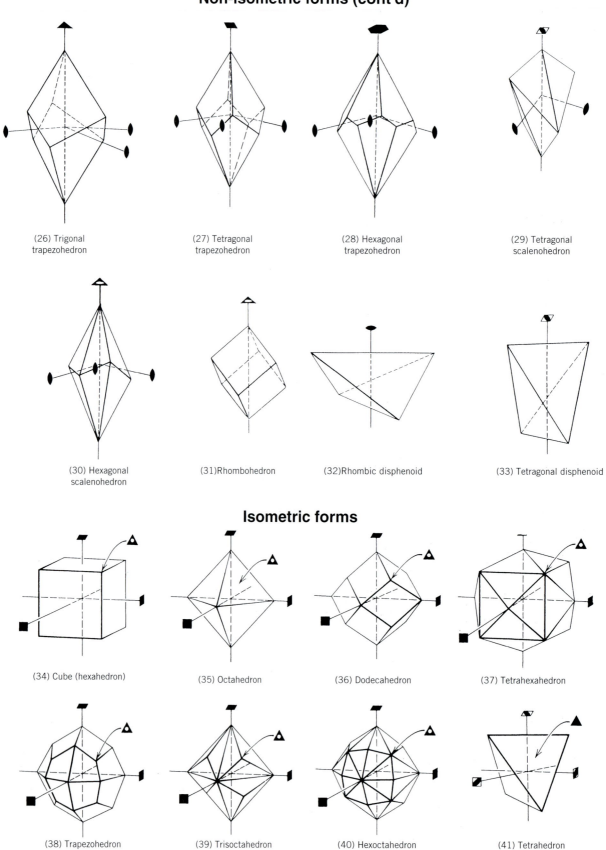

(26) Trigonal trapezohedron

(27) Tetragonal trapezohedron

(28) Hexagonal trapezohedron

(29) Tetragonal scalenohedron

(30) Hexagonal scalenohedron

(31) Rhombohedron

(32) Rhombic disphenoid

(33) Tetragonal disphenoid

Isometric forms

(34) Cube (hexahedron)

(35) Octahedron

(36) Dodecahedron

(37) Tetrahexahedron

(38) Trapezohedron

(39) Trisoctahedron

(40) Hexoctahedron

(41) Tetrahedron

FIG. 5.38.   (Continued)

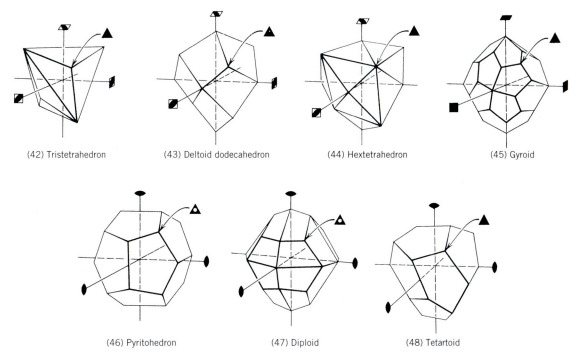

(42) Tristetrahedron   (43) Deltoid dodecahedron   (44) Hextetrahedron   (45) Gyroid

(46) Pyritohedron   (47) Diploid   (48) Tetartoid

FIG. 5.38.   (Continued)

scriptions of the forms shown in Fig. 5.38 (the numbers in parentheses in the following descriptions all relate to Fig. 5.38).

Pedion. A single face comprising a form (1).

Pinacoid. An open form made up of two parallel faces (2).

Dome. An open form consisting of two nonparallel faces symmetrical with respect to a mirror plane (*m*) (3).

Sphenoid. Two nonparallel faces symmetrical with respect to a twofold rotation axis (4).

Prism. An open form composed of 3, 4, 6, 8, or 12 faces, all of which are parallel to the same axis. Except for certain prisms in the monoclinic system, the axis is one of the principal crystallographic axes (5–11).

Pyramid. An open form composed of 3, 4, 6, 8, or 12 nonparallel faces that meet at a point (12–18).

Dipyramid. A closed form having 6, 8, 12, 16, or 24 faces (19–25). A dipyramid can be considered as formed from two pyramids by reflection of one of them into the other across a horizontal mirror plane.

Trapezohedron. A closed form that has 6, 8, or 12 faces in all, with 3, 4, or 6 upper faces offset from 3, 4, or 6 lower faces (26–28). These forms

are the result of a three-, four-, or sixfold axis combined with perpendicular twofold axes. In addition there is an isometric *trapezohedron* (no. 38) a 24-face form. In a well-developed single trapezohedron each face is a trapezium.

Scalenohedron. A closed form with 8 or 12 faces (29 and 30), grouped in symmetrical pairs. In the *tetragonal scalenohedron,* pairs of upper faces are related by an axis of fourfold rotoinversion to pairs of lower faces. The 12 faces of the *hexagonal scalenohedron* display three pairs of upper faces and three pairs of lower faces in alternating positions. The pairs are related by the center of symmetry, which coexists with a threefold axis in rotoinversion axis $\bar{3}$. In an ideally developed scalenohedron, each face is a scalene triangle.

Rhombohedron. A closed form composed of six faces of which three faces at the top alternate with three faces at the bottom, the two sets of faces being offset by 60° (no. 31). Rhombohedrons are found only in point groups $\bar{3}2/m$, 32, and $\bar{3}$.

Disphenoid. A closed form consisting of two upper faces that alternate with two lower faces, offset by 90° (32 and 33).

The specialized forms in the isometric system are numbered 34 through 48. Every one of these

TABLE 5.6   The Names of the 33 Different Types of Nonisometric Crystal Forms*

Name According to System of Groth-Rogers	No. of Faces
1. Pedion	1
2. Pinacoid	2
3. Dome	2
4. Sphenoid	2
5. Rhombic prism	4
6. Trigonal prism	3
7. Ditrigonal prism	6
8. Tetragonal prism	4
9. Ditetragonal prism	8
10. Hexagonal prism	6
11. Dihexagonal prism	12
12. Rhombic pyramid	4
13. Trigonal pyramid	3
14. Ditrigonal pyramid	6
15. Tetragonal pyramid	4
16. Ditetragonal pyramid	8
17. Hexagonal pyramid	6
18. Dihexagonal pyramid	12
19. Rhombic dipyramid	8
20. Trigonal dipyramid	6
21. Ditrigonal dipyramid	12
22. Tetragonal dipyramid	8
23. Ditetragonal dipyramid	16
24. Hexagonal dipyramid	12
25. Dihexagonal dipyramid	24
26. Trigonal trapezohedron	6
27. Tetragonal trapezohedron	8
28. Hexagonal trapezohedron	12
29. Tetragonal scalenohedron	8
30. Hexagonal scalenohedron	12
31. Rhombohedron	6
32. Rhombic disphenoid	4
33. Tetragonal disphenoid	4
Total no. of forms = 33	

*The number of faces of each form is given. The numbers on the left refer to Fig. 5.38.

TABLE 5.7   The Names of the 15 Different Types of Forms in the Isometric System*

Name According to System of Groth-Rogers	No. of Faces
34. Cube	6
35. Octahedron	8
36. Dodecahedron (rhombic)	12
37. Tetrahexahedron	24
38. Trapezohedron	24
39. Trisoctahedron	24
40. Hexoctahedron	48
41. Tetrahedron	4
42. Tristetrahedron	12
43. Deltoid dodecahedron	12
44. Hextetrahedron	24
45. Gyroid	24
46. Pyritohedron	12
47. Diploid	24
48. Tetartoid	12

*The number of faces of each form is given. The numbers on the left refer to Fig. 5.38.

forms contains four threefold rotation axes or $\bar{3}$ rotoinversion axes. One such direction (in the upper right, positive quadrant) is indicated in the illustrations.

In crystal drawings it is convenient to indicate the faces of a form by the same letter. The choice of which letter shall be assigned to a given form rests largely with the person who first describes the crystal. However, there are certain simple forms that, by convention, usually receive the same letter. Thus the three pinacoids that cut the a, b, c axes are lettered respectively a, b, and c, the letter m is usually given to {110} and p to {111} (see Fig. 5.34).

SOME COMMON TWINS

The concept of twinning was briefly introduced in Chapter 2 and the origin of twinning was discussed in Chapter 4. Here follows a brief overview of the morphological expression of twinning and of some of the most common types of twins.

A twin is a symmetrical intergrowth of two (or more) crystals of the same substance. Such crystallographically controlled intergrowths are called *twinned crystals.* The two or more individuals of the twinned aggregate are related by a symmetry element that is absent in a single (untwinned) crystal. The new symmetry element (*twin element*) brings one individual (crystal) into coincidence with another individual (crystal) in a twinned position. It is generally necessary to do careful morphological measurements (by reflecting goniometer) as well as X-ray diffraction studies (mainly by X-ray precession methods) to distinguish a twin from a random intergrowth of crystals.

The operations (*twin elements*) that may relate a crystal to its twinned counterpart are (1) reflection by a mirror plane (*twin plane*); (2) rotation about a crystal direction common to both (*twin axis*) with the angular rotation normally 180°; and (3) inversion about a point (*twin center*). Twinning is defined by a *twin law*, which states whether there is a center, an axis, or a plane of twinning, and gives the crystallographic orientation for the axis or plane. A twin plane is identified by its Miller index (e.g., (010)) and a twin axis direction is identified by a zone symbol (e.g., [001]).

Contact Twins

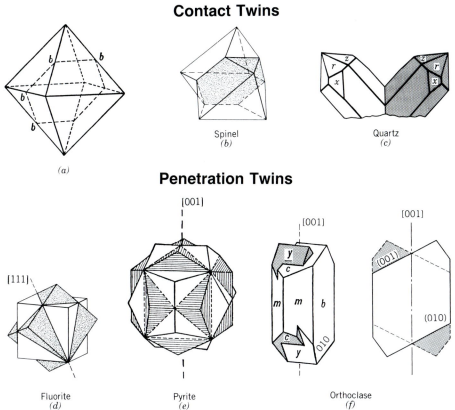

Spinel
(b)

Quartz
(c)

(a)

Penetration Twins

[001]

[111]

[001]

[001]

(010)

Fluorite
(d)

Pyrite
(e)

Orthoclase
(f)

FIG. 5.39. *(a)* Octahedron with possible twin plane $b–b(\bar{1}\bar{1}1)$. This is one of four octahedral directions in the form {111}. *(b)* Octahedral twinning {111} as shown by spinel. *(c)* Right- and left-handed quartz crystals twinned along $(11\bar{2}2)$, the *Japan twin law. (d)* Two interpenetrating cubes of fluorite twinned on [111] as the twin axis. *(e)* Two pyritohedral crystals (of pyrite) forming an *Iron Cross*, with twin axis [001]. *(f)* Orthoclase exhibiting the *Carlsbad twin law* in which two interpenetrating crystals are twinned by 180° rotation about the *c* axis, [001] direction. The schematic cross section, parallel to (010), reveals the presence of the twofold twin axis along [001].

The surface on which two individuals are united is known as the *composition surface.* If this surface is a plane, it is called the *composition plane.* The composition plane is commonly, but not invariably, the twin plane. If the twin law can be defined only by a twin plane, however, the twin plane is always parallel to a possible crystal face but never parallel to a plane of symmetry. The twin axis is a zone axis or a direction perpendicular to a rational lattice plane; but it can never be an axis of even rotation (two-, four-, sixfold) if the twin rotation involved is 180°. In some crystals a 90° rotation about a twofold axis can be considered a twin operation.

Types of Twins

Twinned crystals are usually designated as either *contact twins* or *penetration twins.* Contact twins have a definite composition surface separating the two individuals, and the twin is defined by a twin plane such as $(\bar{1}\bar{1}1)$ in Fig. 5.39*a.* This $(\bar{1}\bar{1}1)$ plane is one of four possible and crystallographically equivalent directions in the octahedron {111} of the isometric system. Therefore, if one wishes to describe all possible octahedral twin planes, one uses the {111} form symbol

instead of the notation (111) for a specific plane. Penetration twins are made up of interpenetrating individuals having an irregular composition surface, and the twin law is usually defined by a twin axis direction (e.g., [111] or [001]; see Figs. 5.39*d* to *f*).

Repeated or *multiple twins* are made up of three or more parts twinned according to the same law. If all the successive composition surfaces are parallel, the resulting group is a *polysynthetic twin* (Figs. 5.40*a*, *b*, and *c*). If successive composition planes are not parallel, a *cyclic twin* results (Figs. 5.40*d* and *e*). When a large number of individuals in a polysynthetic twin are closely spaced, crystal faces or cleavages crossing the composition planes show striations, owing to the reversed positions of adjacent individuals.

Twinning in the lower symmetry groups generally produces a resulting aggregate symmetry higher than that of each individual because the twin planes, or twin axes, are added symmetry elements.

Common Twin Laws

Triclinic System

The feldspars best illustrate the twinning in the triclinic system. They are almost universally twinned

Polysynthetic Twins

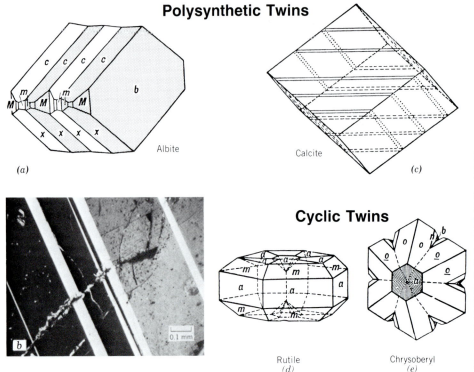

Albite

(a)

b

Calcite

(c)

Cyclic Twins

Rutile
(d)

Chrysoberyl
(e)

FIG. 5.40. *(a)* Albite polysynthetically twinned on {010}. *(b)* The same polysynthetic twinning as in *a* but as seen in a polarizing microscope. The dark and light lamellae in albite are related by reflection across (010). *(c)* Polysynthetic twinning in calcite on ($\overline{1}012$), which is one of the three directions of the negative rhombohedron. *(d)* Cyclic twin in rutile with the twin planes parallel to faces of the form {011}. *(e)* Cyclic twin in chrysoberyl with the twin planes parallel to faces of the form {031}.

according to the *albite law*, with the {010} twin plane, as shown in Figs. 5.40*a* and *b*. Another important type of twinning in triclinic feldspar is according to the *pericline law*, with [010] the twin axis. When, as frequently occurs in microcline, albite and pericline twins are closely interwoven, a typical cross-hatched or "tartan" pattern can be seen under a polarizing microscope (Fig. 5.41). In addition, triclinic feldspars twin according to the same laws as monoclinic feldspars (see below).

Monoclinic System

In the monoclinic system twinning on {100} and {001} is most common. Figure 5.42 illustrates gypsum with {100} the twin plane (*swallow-tail twin*). This same figure also shows three twin laws that occur in the mineral orthoclase. Two of these are contact twins: a *Manebach twin* with {001} as the twin plane, and a *Baveno twin* with {021} as the twin plane. The most common twin in orthoclase is the *Carlsbad twin*, an interpenetration twin in which the *c* axis, [001], is the twin element. In this case the two individuals are united on an irregular surface roughly parallel to (010).

Orthorhombic System

In the orthorhombic system the twin plane is most commonly parallel to a prism face. The contact twin of aragonite and the cyclic twins of aragonite and cerussite are all twinned on {110} (see Figs. 5.43*a* and *b*). The pseudohexagonal appearance of the cyclically twinned aragonite results from the fact that (110) ∧ ($1\overline{1}0$) is nearly 60°.

The mineral staurolite, which is monoclinic with a β angle of 90°, is pseudo-orthorhombic and morphologically appears orthorhombic. It is commonly found in two types of penetration twins. In one, with

FIG. 5.41. Photomicrograph of transformation twinning (see p. 156) in microcline. The specimen is viewed under a microscope with crossed polarizers. The section of the photograph is approximately parallel to (001). The twin laws represented are albite with twin and composition plane (010), and pericline with twin axis direction [010].

Monoclinic Twins

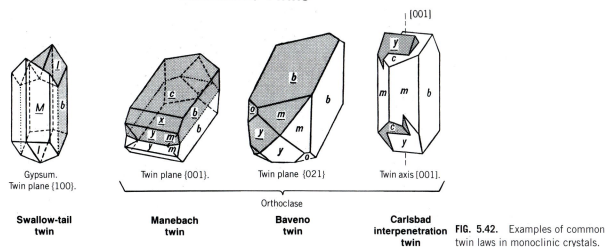

Gypsum.
Twin plane {100}.

Swallow-tail twin

Twin plane {001}.

Manebach twin

Twin plane {021}

Baveno twin

Orthoclase

Twin axis [001].

Carlsbad interpenetration twin

FIG. 5.42. Examples of common twin laws in monoclinic crystals.

{031} as twin plane, a right angle cross results; in the other, with twin plane {231}, a 60° cross is formed (Fig. 5.43c).

Tetragonal System
The most common type of twin in the tetragonal system has {011} as the twin plane. Crystals of cassi-

terite and rutile, twinned according to this law, are shown in Fig. 5.44.

Hexagonal System
In the hexagonal system carbonates, especially calcite, serve as excellent illustrations of three twin laws. The twin plane may be {0001}, with c the twin

Orthorhombic Twins

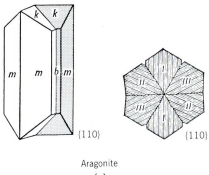

{110}

{110}

Aragonite

(a)

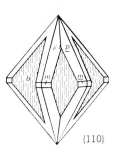

{110}

Cerussite

(b)

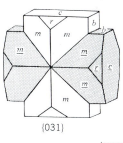

{031}

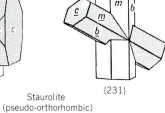

{231}

Staurolite
(pseudo-orthorhombic)

(c)

FIG. 5.43. Examples of common twins in orthorhombic crystals. (a) Contact and cyclic twinning on {110} in aragonite. (b) A cyclic twin on {110} in cerussite. (c) Staurolite twinned on {031} and {231}. The staurolite structure is actually monoclinic with β= 90°; it therefore appears pseudo-orthorhombic. It is illustrated here because of its orthorhombic-looking morphology.

Tetragonal Twins

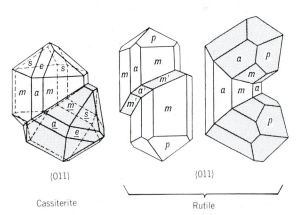

{011}

Cassiterite

{011}

Rutile

FIG. 5.44. Examples of common twin laws in tetragonal crystals.

axis (Fig. 5.45a), or it may be the positive rhombohedron {10$\bar{1}$1}. But twinning on the negative rhombohedron {01$\bar{1}$2} is most common and may yield contact twins or polysynthetic twins as the result of pressure (Fig. 5.45b). The ease of twinning according to this law can be demonstrated by the artificial twinning of a cleavage fragment of Iceland spar by the pressure of a knife blade.

In the class 622, quartz shows several types of twinning. Figure 5.45c illustrates the *Brazil law* with the twin plane parallel to {11$\bar{2}$0}. Here, right- and left-handed individuals have formed a penetration twin. Figure 5.45d shows a Dauphiné twin, a penetration twin with c the twin axis. Such twins are composed either of two right-hand or two left-hand individuals. Figure 5.45e illustrates the *Japan law* with the twin plane {11$\bar{2}$2}. The reentrant angles usu-

Hexagonal Twins

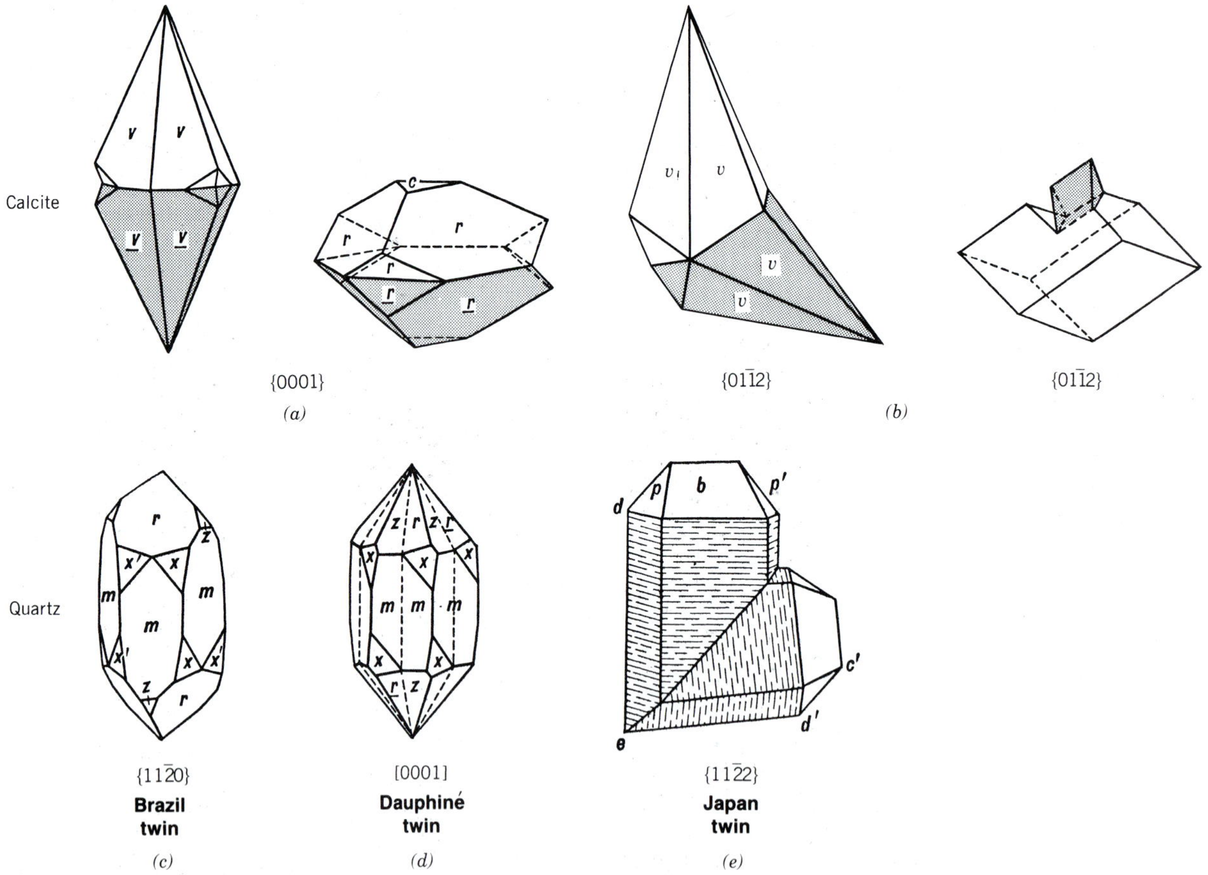

Calcite

{0001}

(a)

{01$\bar{1}$2}

{01$\bar{1}$2}

(b)

Quartz

{11$\bar{2}$0}

Brazil twin

(c)

[0001]

Dauphiné twin

(d)

{11$\bar{2}$2}

Japan twin

(e)

FIG. 5.45. Examples of twins in the hexagonal system. *(a)* and *(b)* Various twins in calcite. The calcite twin on the right is artificial and can be produced by pressure with a knife edge. *(c)* A *Brazil twin* in quartz. *(d)* A *Dauphiné twin* in quartz formed by rotation of 180° about the c-axis is [0001]; see also Fig. 4.56. *(e)* A *Japan twin* in quartz.

ally present on twinned crystals do not show on either Brazil or Dauphiné twins.

Isometric System

In the holohedral class of the isometric system ($4/m\bar{3}2/m$) the twin axis, with a few rare exceptions, is a 3-fold symmetry axis, and the twin plane is thus parallel to a face of the octahedron. Figures 5.39*a* and *b* show an octahedron with plane *bb* a possible twin plane, as well as an octahedron twinned according to this law, forming a contact twin. This type of twin is especially common in gem spinel and hence is called a *spinel twin*. Figure 5.39*d* shows two cubes forming a penetration twin with the 3-fold rotoinversion axis [1$\bar{1}$1] the twin axis.

In the class $2/m\bar{3}$, two pyritohedrons may form a penetration twin (Fig. 5.39*e*) with a 90° rotation about the twin axis [001]. This twin is known as the *iron cross*.

THE INTERNAL ORDER AND SYMMETRY OF MINERALS

In the earlier part of this chapter we discussed the external form and inherent symmetries of minerals. This led us to the formulation of the 32 crystal classes, or point groups. Before the discovery in 1912 of X-ray diffraction by minerals, it had long been suspected that the regular (and symmetrical) external form of euhedral crystals was a reflection of some type of internal order. Indeed, René J. Haüy had suggested as early as 1784 that crystals were built by stacking together tiny identical building blocks, which he referred to as "integral molecules" (see Fig. 1.6). The concept of *integral molecules* is essentially that of unit cells in modern X-ray crystallography.

Now that we are about to address the *three-dimensional periodicity* of mineral (atomic) structures, we should look once more at the statement that the internal structure of minerals is based on an *ordered atomic arrangement*. This statement implies that a certain atom (or ion) is present in exactly the same structural (atomic) site throughout an essentially infinite atomic array. An atom in the same atomic site means that it is surrounded by an identical arrangement of neighboring atoms, throughout the structure which consists of millions of unit cells with dimensions on the order of 5 to 20 angstroms (Å, equivalent to 0.5 to 2 nanometers, nm). Such complete order is present in "ideal" crystals, of which examples of the internal structure are shown

in Figs. 1.14 and 5.64. However, high-resolution transmission electron microscopy (HRTEM), with magnification levels on the order of 1,000,000×, has shown that atomic arrangements at and below the level of the unit cell may deviate considerably from perfect crystalline periodicity. Such enormous magnifications, which allow for the resolution of patterns and atoms on an angstrom level, are extremely far removed from those of visual observation of minerals (at 1× magnification) and those of optical microscopy (at magnification levels of about 1000×). This means that a mineral that appears to be homogeneous during optical microscopic observation may be found to be heterogeneous when studied by X-ray diffraction and, even more strikingly, with HRTEM techniques. Such inhomogeneities may be due to structural misfits (defects; see p. 151) or to chemical zonation. Furthermore, all structures show increased thermal vibrations of the atoms as a function of increased temperature. This may result in a random distribution of atoms (or ions) that at lower temperatures would be located in one or more specific structural (atomic) sites. Such randomization of atoms in a structure as a function of increased temperature is referred to as *disorder* (see also p. 138). The imperfections in crystals and the disorder of atoms in an otherwise periodic structure are considered as local phenomena in an overall well-crystallized structure with three-dimensional periodicities.

In this part of the chapter, we will first discuss ordered arrangements, beginning with examples of one- and two-dimensional order before we evaluate three-dimensional order.

A crystal structure may be thought of as a repetition of a motif or a group of atoms on a lattice, or periodic array of points. The ordered patterns that characterize crystalline materials represent a lower energy state than random patterns. Intuitively, a brick wall constructed with carefully positioned bricks in an ordered arrangement provides a more stable (and less energetic configuration) than a brick wall made with a random arrangement of identical bricks (see Fig. 5.46*a*). The brick in these walls can be considered as the motif and can be replaced by a comma, as in Fig. 5.46*b*. This analogy shows how an ordered pattern is generated by a motif repeated in a regular sequence of new locations. Any motion that brings the original motif into coincidence with the same motif elsewhere in the pattern is referred to as an *operation*. Many wallpaper patterns, for example, are based on a two-dimensional pattern in which the motif (sprays of flowers, dots, figurines) is arranged in

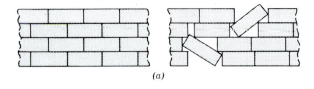

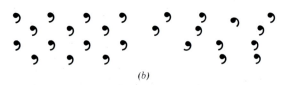

FIG. 5.46. *(a)* Two-dimensional ordered and more random arrangement of bricks in a brick wall. *(b)* The brick patterns are represented by commas as motifs.

a regular geometrical pattern. More abstract and less symmetrical arrangements, as are found in many contemporary wallpapers, may have a much less clearly defined pattern.

Translation Directions and Distances

As stated earlier, a crystal is a homogeneous solid possessing, long range, *three-dimensional internal order*. Such order is the result of the repeat of motif units (these are chemical units—for example, copper atoms) by regular translations in three dimensions. The three-dimensional pattern is said to be *homogeneous* if the angles and distances from one motif to surrounding motifs in one location of the pattern are the same in all parts of the pattern (see Fig. 5.47).

Figure 5.47*a* illustrates a two-dimensional array of motifs (commas). The order in such an array can be expressed in terms of two translations (t_1 and t_2) at 90° to each other. A somewhat less symmetrical pattern of motifs is shown in Fig. 5.47*b*, where the t_1 direction is the same as in Fig. 5.47*a* but the t_2 translation is at an angle <90° to the t_1 translation. These two illustrations can be thought of as infinite strings of units along the t_1 direction that have been repeated by parallel and identical infinite strings along a translation direction t_2. The translations, as marked by t_1 and t_2, are vectors.

A three-dimensional ordered pattern can be obtained by adding yet another translation component (t_3) that does not lie in the plane of t_1 and t_2 (see Fig. 5.47*c*). This results in a pattern that is infinite in three dimensions. In crystals, such a pattern is not exactly infinite, although it is generally considered so. The magnitude of the translations in inorganic crystals is

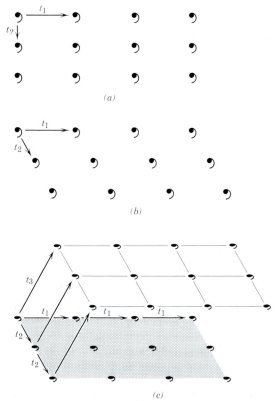

FIG. 5.47. *(a)* A two-dimensional pattern with translation components t_1 and t_2 at 90° to each other. *(b)* A two-dimensional pattern with translation components t_1 and t_2 at <90° to each other. *(c)* A three-dimensional pattern with translation components t_1, t_2, and t_3. None of the translation components make 90° with each other.

on the order of 1 to 10 angstroms[2] (Å, where Å = 10^{-8} cm), because that is the scale of ionic radii in crystals. This means that a dimension of 1 cm in a crystal would contain approximately 100 million translations; indeed this can be considered infinite!

It is often convenient to ignore the actual shape of the motif units in a pattern and to concentrate only on the geometry of the repetitions in space. If the motifs (commas in Fig. 5.47) are replaced by points, we have a regular pattern of points that is referred to as a lattice. *A lattice is*, therefore, *an imaginary pattern of points (or nodes) in which every point (node) has an environment that is identical to that of any other point (node) in the pattern. A lattice has no specific origin, as it can be shifted parallel to itself.*

A three-dimensional crystal structure can be viewed as the result of three-dimensional translations

[2]Named after the Swedish physicist Anders Jonas Angström, 1814–1874.

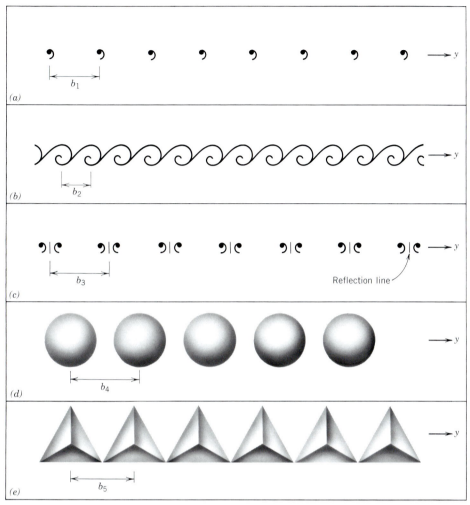

FIG. 5.48. Rows of objects at various spacings b along a translation direction y. (a) Regularly spaced asymmetric motifs, commas (asymmetric means "without any kind of symmetry"). (b) A scroll design of asymmetric motifs. (c) A row of equally spaced motifs in which one part of the motif is related to the other by a line of reflection (marked by a short line between pairs of motifs). (d) A row of spheres that may represent atoms in a structure. (e) A row of tetrahedra that may represent anionic complexes such as tetrahedral SiO_4 or GeO_4 groups.

acting upon motif units (the chemical units of the structure). The translations inside such a crystal structure are extremely small (on an atomic scale) and cannot be seen by the naked eye. The sizes of these translations are expressed in nanometers (10^{-7} cm) or angstroms. The only way in which such extremely small distances can be imaged is by transmission electron microscopy (see Fig. 1.14). *It is very important, therefore, to realize that the external form of a crystal, although an expression of its internal structure, is translation-free. The symmetry elements, observable in the external form development of crystals, are therefore also translation-free.*

Let us now develop lattices systematically, beginning with one-dimensional rows.

One-Dimensional Order (Rows)

A sequence of equally spaced equivalent points (or motifs) along a line represents order in one dimen-

sion, or a *row* (see Fig. 5.48). In such a row the magnitude of the *unit translation* (in this case b) determines the spacing. The motif, the unit of pattern or the atom printed at each lattice point, determines the ultimate pattern. Figure 5.48 illustrates several rows of objects with different spacings, b, along a direction defined as y, and with different motifs. Such rows can be found as borders along illustrations, in wallpaper, along friezes, and in the structures of crystalline materials. Interactive animations of one-dimensional order are given in module III of the CD-ROM.

Two-Dimensional Order (Plane Lattices)

Two-dimensional order is the result of regular translations in two different directions, designated x and y. Figure 5.49a shows a two-dimensional, ordered array of motifs (commas, in this instance) on which various choices of the x and y coordinate axes (the

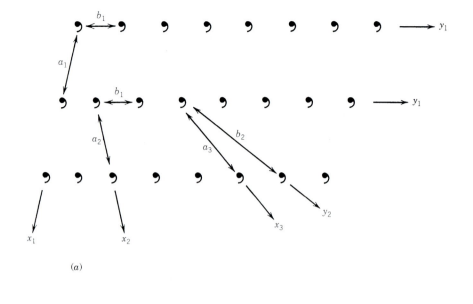

(a)

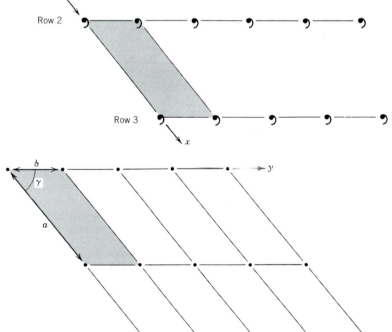

(b)

(c)

FIG. 5.49. *(a)* A two-dimensional, ordered array of motifs showing three different ways to generate the array by combining translation directions $(x_1, x_2, x_3; y_1, y_2)$ and distances $(a_1, a_2, a_3; b_1, b_2)$. These three different choices all generate the same pattern. *(b)* Yet another choice of translation directions and translation distances. The angle between the two directions of *x* and *y* is γ. This arrangement represents an oblique ($\gamma \neq 90°$), two-dimensional (planar), and ordered array of motifs. *(c)* A planar lattice based on the array of motifs in *(b)*. A lattice is, by definition, *infinitely extending*, but only finite portions can be shown in illustrations. The shaded parallelogram in *(b)* represents the smallest unit of pattern and a unit cell in the lattice of *(c)*.

axes along which the translations take place) have been superimposed. The units of the translation distance are marked as a (or a_1, a_2, etc.) and b (or b_1, b_2, etc.).

Such translational patterns can be described by translation vectors, where the vectors are noted as **a** and **b**, the magnitudes of the translations as a and b, and the coordinate axes along which the translations are repeated as x and y. The angle between x and y axes is denoted by γ.

Figure 5.49b shows how a regular two-dimensional pattern is produced with two different spacings (the unit of translation along the row = b; the unit of translation between the rows = a). This arrangement is an oblique array of commas because we chose an angle γ between the x and y directions that is neither 90° nor 60° nor 120°. In order to visualize the array of motifs (commas) without reference to the shape of the motif, it is standard practice to replace each motif with a point (thus eliminating any sense of shape or symmetry of the motif) and to connect such points (or nodes) by lines. This creates a two-dimensional *net* or *plane lattice*, as shown in Fig. 5.49c. The regularly spaced points (or nodes) represent the locations of motifs, which in chemical structures may be atoms, ions, molecules, or ionic complexes. The smallest building unit in the two-dimensional pattern of Fig. 5.49c is that of the shaded parallelogram; this is known as the unit cell. If this *unit cell* is repeated indefinitely by translations a and b along directions x and y, the array shown in Fig. 5.49b and the lattice shown in Fig. 5.49c will result.

There are only five possible and distinct plane lattices (also known as *nets*). The five choices are the result of repeating a row (with translation distance b along direction y) along direction x with repeat distance a. The five plane lattice types that result depend on the choice of the angle γ (between directions x and y; that is, whether γ = 90°, 60°, or some other angle) and on the size of a relative to b (i.e., whether $a = b$, or not). These five plane lattices (nets), illustrated in Fig. 5.50, represent the only possible ways to arrange points *periodically* in two dimensions. Interactive animations that build these five net types are found in module III of the CD-ROM under the heading "2-dimensional order."

In Fig. 5.50a, row no. 1 is infinitely repeated by translations along direction x, with translation distance a. Here a is unequal to b; the γ angle is not 90°, and an *oblique lattice* (or *clinonet*) results.

In Fig. 5.50b, row no. 1 is infinitely repeated by translations along direction x, with translation distance a. Here a is unequal to b; the angle γ = 90°,

and a *primitive rectangular lattice* (or *orthonet*) results (*primitive* implies nodes occur at the corners of the chosen unit cell only).

In Fig. 5.50c, row no. 1 is infinitely repeated by translations along direction x, with translation distance a, and with angle γ such that cos γ = $a/2b$. The resulting lattice (or net) is conventionally described in terms of two orthogonal directions (x and y'), resulting in centering of the net inside the rectangular unit cell choice. This is known as a *centered rectangular lattice* (or *centered orthonet*). The same array of nodes can be described by two vectors (**a**′ and **b**′, where **a**′ = **b**′ and γ' ≠ 90°, 60°, or 120°), resulting in a *primitive lattice* with a diamond shape. This alternate, primitive lattice choice is referred to as a *diamond lattice*. Either of these two unit cell choices (primitive or nonprimitive [= centered]) will, when repeated infinitely along two directions, produce the pattern of nodes in Fig. 5.50c.

In Fig. 5.50d, row no. 1 is infinitely repeated by translations in direction x, with a translation distance of a, such that $a = b$ (or $a_1 = a_2$) and γ = 60°. This results in a *hexagonal lattice* (or *hexanet*).

In Fig. 5.50e, row no. 1 is infinitely repeated by translations in direction x, with a translation distance of a, such that $a = b$ (or $a_1 = a_2$) and γ = 90°. This results in a *square lattice* (or *tetranet*).

The smallest units of repeat in these lattices outline the unit cells (shaded in Fig. 5.50). They range from a parallelogram (Fig. 5.50a), to *two types of rectangle* (Figs. 5.50b and c), to a *diamond shape* (Fig. 5.50c), to a *rhombus* (Fig. 5.50d), and a *square* (Fig. 5.50e). In Fig. 5.50c there is a choice of two differently shaped and sized unit cells. The preferable choice is the rectangular (centered, and larger) unit cell because of the orthogonality of its shape (orthogonal means "involving right angles or perpendiculars"), and because the mirror lines inherent in the symmetry of the pattern are parallel to the axial directions of the unit cell. Of the various choices of unit cells in Fig. 5.50, only one contains a central node, namely, the *centered rectangular lattice*. All other choices contain only corner nodes and are referred to as *primitive lattices* (e.g., primitive oblique lattice, etc.).

The five different plane lattices, or nets, in Fig. 5.50 should have been drawn as infinitely extending. However, limitations on illustration size restrict the number of points (or nodes) that can be shown. The shaded unit cell in the vertical column of nets, on the left side of the figure, is enlarged in the column on the right to illustrate the distribution of symmetry elements in one unit cell of each of the patterns. The locations of the various rotational symmetry axes (2,

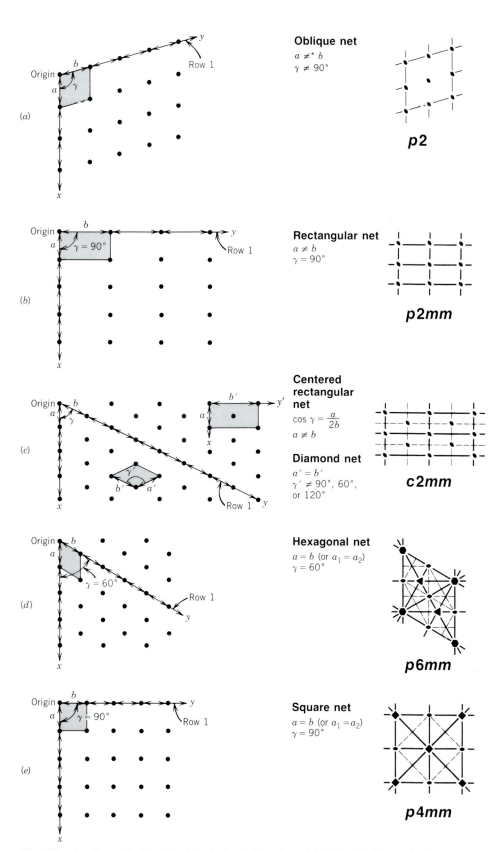

Oblique net
$a \neq^* b$
$\gamma \neq 90°$

p2

Rectangular net
$a \neq b$
$\gamma = 90°$

p2mm

Centered rectangular net
$\cos \gamma = \dfrac{a}{2b}$
$a \neq b$

Diamond net
$a' = b'$
$\gamma' \neq 90°, 60°,$ or $120°$

c2mm

Hexagonal net
$a = b$ (or $a_1 = a_2$)
$\gamma = 60°$

p6mm

Square net
$a = b$ (or $a_1 = a_2$)
$\gamma = 90°$

p4mm

FIG. 5.50. Development of the five distinct plane lattices (or nets) by the infinite repeat of a row (along direction *y*, with specified translation distance *b*), along direction *x* with repeat distance *a*; γ is the angle between *x* and *y*. The total symmetry content of each of the unit cell choices is given in the right-hand column. Rotational axes are shown by standard symbols, mirrors by broad lines, and glide lines by dashed lines.

*The sign $\neq$ implies nonequality by reason of symmetry; accidental equality may occur.

218

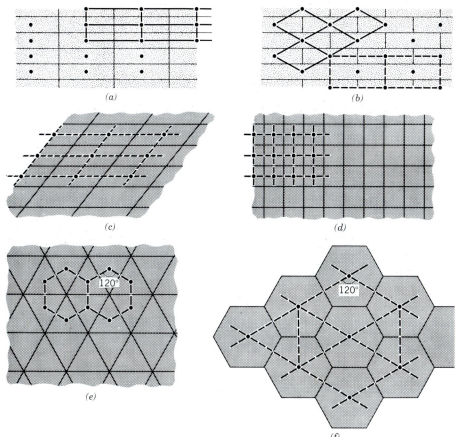

FIG. 5.51. (a) and (b) represent brick walls; (c) through (f) are possible arrangements of tiles. In each case the original motifs (bricks or tiles) are in part replaced by lattice points (nodes). When these are connected, the shape of the lattice becomes obvious. These shapes are (a) a rectangular lattice; (b) a centered rectangular lattice, or a diamond lattice; (c) an oblique lattice; (d) a square lattice; (e) a hexagonal lattice; and (f) a hexagonal lattice (with a centering node) or a noncentered rhombus lattice.

3, 4, and 6), perpendicular to the page, are shown by the standard symbols. The locations of mirrors perpendicular to the page are shown by broad black lines. But remember, there are no mirrors or symmetry axes in the plane of the page (i.e., parallel to the page). In the lattices in Figs. 5.50d and e, the dashed lines represent *glide lines*, which are a combination of translation and reflection (see p. 224).

The listing of the symmetry content for each lattice type is similar to that introduced in the prior part of this chapter. However, each symmetry listing is preceded by a small letter *p* (for primitive) or *c* (for centered) to indicate the lattice choice.

Various types of two-dimensional ordered patterns can be seen in our daily surroundings. For example, brick or tiled walls can be regarded as the result of repeating a motif (e.g., a brick or a ceramic tile) along two different translation directions (t_1 and t_2) parallel to the wall. Figure 5.51 shows some examples as well as their lattice types and unit cell choices. Such two-dimensional coverings of a wall by tiles are known as *tesselations*. Figure 5.51 represents a restatement of the fact that the five plane lattices (nets) in Fig. 5.50 are the only possible ways to arrange points periodically in two dimensions; namely, these five two-dimensional nets represent the only shapes that can "tile" a planar surface with no gaps (see also Box 5.2).

It is useful at this time to illustrate that an observer can have a number of choices of unit cell, once the lattice array of motifs has been established. Figure 5.52 shows a regular array of nodes that was originally generated by the two vectorial directions **a** and **b** and angle γ, as shown in the primitive, oblique unit cell marked A. Another unit cell choice, which through infinite repeat along two vectors in the plane of the page would have generated the same pattern, is unit cell B. The spacings and angle are the same as in A. It is an equivalent primitive, oblique unit cell. A third choice, designated as C, is also primitive and oblique but with $a_1 = a_2$ and with a different subtended angle. All the other choices, D, E, and F, are *nonprimitive* or *multiple* unit cells, because each contains in addition to parts of each of the four corner nodes (a total of one node), one more node per unit cell. In D the

BOX 5.2
PATTERNS IN OUR ENVIRONMENT

Mexican tiling, Santa Fe, New Mexico.

Sinuous rows of new tract housing in Palm City, Florida.

Regular, planar patterns are all around us. In print designs of dresses and ties, in brick patterns in sidewalks, in tiling arrangements along walls and in floors. All such designs are the result of a motif having been repeated infinitely along two (non-coincident) directions. Very interesting regular patterns can also be observed in the plantings and plowing of farmer's fields (especially when these are viewed from an aircraft at relatively low altitude).

Boats and moored docks off Lincoln Park in Chicago creating a daisy pattern on Lake Michigan (both aerial photographs by Alex S. MacLean/ Landslides, used with permission).

Examples of remarkable patterns, photographed by Alex MacLean, are available in the beautifully illustrated book *Look at the Land: Aerial Reflections on America,* 1993, by Alex MacLean, with text by Bill McKibben, Rizzoli International Publications, Inc., New York, N.Y. The jacket of this book reads: "Photographer and aviator Alex MacLean has spent twenty years traveling the length and breadth of America by air. With one hand navigating his plane and the other steadying his camera, he has skillfully captured in intensely colorful, unique photographs what can only be seen from above; the interconnectedness and extraordinary patterns of our natural and created environments." Here, two of these remarkable images are reproduced as black-and-white photographs.

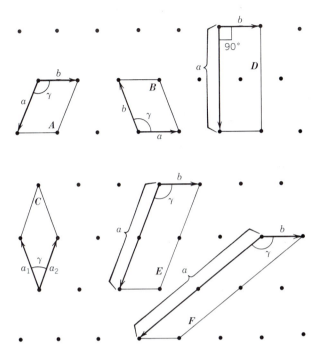

FIG. 5.52.   A regular array of nodes that can be generated by the infinite repeat of various unit cell choices (*A*, *B*, *C*, *D*, *E*, or *F*) along two vectors **a** and **b**.

additional node is centered (a *centered unit cell*), and in *E* and *F* it is made up of two half nodes on two of the sides. Generally, the smallest unit cell (e.g., *A*, *B*, or *C*) or an orthogonal and centered unit cell (e.g., *D*) would be the most appropriate choice.

FIG. 5.53.   Choice of alternate unit cells in a wallpaper design. The motif of the wallpaper contains symmetry 2*mm*.

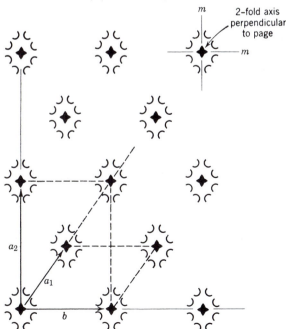

Figure 5.53 illustrates how a wallpaper designer has used the concepts of symmetry and translation in repetition of a motif. Each motif in this design contains a twofold rotation axis perpendicular to the page and two reflection lines (*m*) perpendicular to each other (planar point group symmetry 2*mm*). The pattern of the motifs can be described by a primitive, oblique unit cell with edges a_1 and *b*, but the most appropriate choice, in view of the 2*mm* symmetry content of the motifs, would be a rectangular centered unit cell with edges a_2 and *b*. For three-dimensional patterns several rules have been made to restrict the possible choices of unit cells (see page 231).

Rotation Angle Restrictions

Now that we have introduced the concepts of two-dimensional ordered arrays and plane lattices, we can evaluate geometrically why certain rotational axes are possible and others (e.g., fivefold rotation) are not. Figure 5.54 illustrates the geometric restrictions on rotation axes in ordered arrangements that also contain translation. If the motif units, represented by large nodes in Fig. 5.54, are part of an ordered arrangement, then the distances *AB* and *BC* must be equal. If the motif at *B* contains a rotation axis with the axis ⊥ to the plane of the figure, then the translations require similar axes at *A* and *C*. Furthermore, if points *D*, *E*, *F*, and *G* are related to *B* by a rotation, then *BC* = *BD* = *BE* = *BF* = *BG* = *t*. This also means that the distance *ED*, which lies on a line parallel to *AC*, must be equal to *AB* or a multiple thereof. In other words, *ED* = *u* = *mt* where *m* = integer. If the rotation by which *A*, *F*, *G*, *C*, *D*, and *E*

FIG. 5.54.   Motifs separated by translation (*t*) and a possible axis of rotation, perpendicular to the page, at each of the motif units. One axis of rotation at motif *B* is shown.

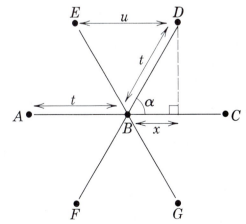

are related is through an angle α, the following geometric relations hold:

$$\cos \alpha = x/t \quad \text{and also} \quad x = \tfrac{1}{2}ED = \tfrac{1}{2}u$$
$$\therefore \cos \alpha = \tfrac{1}{2}u/t = u/2t$$
$$\therefore 2\,t \cos \alpha = u$$

Combining $u = mt$ and $u = 2t \cos \alpha$ gives

$$mt = 2t \cos \alpha \quad \text{or} \quad \cos \alpha = m/2$$

where m is an integer. This leads to restrictions on the solutions possible for the angle of rotation α. For $m = 2$

$$m/2 = 1, \alpha = 0° \text{ or } 360°$$

For $m = 1$,

$$m/2 = 1/2, \alpha = 60°$$

For $m = 0$,

$$m/2 = 0, \alpha = 90°$$

For $m = -1$,

$$m/2 = -1/2, \alpha = 120°$$

For $m = -2$,

$$m/2 = -1, \alpha = 180°$$

Any other integral values of m produce values of $\cos \alpha$ greater or less than ± 1, which is possible but mathematically meaningless. Other rotation angles produce noninteger values of m. For example, a five-fold rotation axis would require an angle of rotation of 72°. This leads to a value for $\cos 72° = 0.30902$. Such a number cannot equal $m/2$ in which m must be an integer. Therefore, a fivefold rotation axis is not possible in an ordered, crystalline structure. Five-fold symmetry is, however, not uncommon in objects in the biological world (e.g., in the distribution of petals in a geranium flower).

Symmetry Content of Planar Motifs

Two-dimensional motifs, as are often seen in wall-paper designs, in printed cloth, in ceramic tiles, and elsewhere, can display variable symmetry contents. However, because such motifs are printed on one side of a paper (with the other side generally blank), there are no symmetry elements that lie in the plane of the paper. That is, there is no mirror plane parallel to the paper, nor are there axes of rotational symmetry parallel to the paper. However, there may be a number of symmetry elements perpendicular to the plane of the drawing. These are mirror lines (m) (in three-dimensional patterns, m's are referred to as mirror planes; in two-dimensional patterns they are known as mirror lines; their reflec-

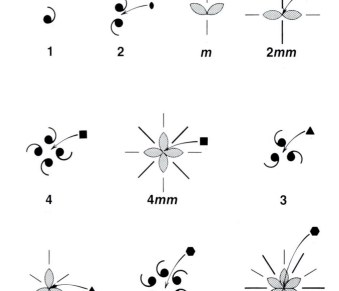

FIG. 5.55. The symmetry content of two-dimensional motifs. Locations of mirror lines (m) are shown by solid lines, and rotational axes by the standard symbols.

tion operations are equivalent) and rotation axes (1, 2, 3, 4, and 6). Although one can create an independent motif unit with 5-, or 11-, or more-fold rotational axes, the symmetry content of motif units that are part of a repetitive and ordered (crystalline) array can contain only one of five rotational axes (1, 2, 3, 4, and 6). There are only ten possible symmetry contents for two-dimensional motifs that, through regular translation, can become part of two-dimensional ordered patterns. These are shown in Fig. 5.55. The letters in this figure refer to the rotational symmetry (perpendicular to the page) inherent in the motif, and the m's note the location of mirror lines in various directions. These ten different symmetry contents represent the symmetry about a central (stationary) point and are referred to as the *ten planar point groups*. There are ten because each of the six symmetry elements, 1, 2, 3, 4, 6, and m, can occur individually, and the other four consist of possible combinations of rotational symmetry and mirrors, as in 2*mm*, 3*m*, 4*mm*, and 6*mm*. The significance of these symbols or groups of symbols is the same as discussed earlier in this chapter for the Hermann–Mauguin (international) notation of point groups (or crystal classes). The numerals refer to rotations about a point. The m's refer to reflection lines. The m's in 2*mm* and 3*m* and the first *m* in 4*mm* and 6*mm* refer to reflection lines that are coincident with the axial directions. These lines are at right angles in orthogonal arrays but at 120° to each other in sixfold and threefold patterns. Thus, in 4*mm* the first *m* refers to two reflection lines at right angles to each other (see Fig. 5.55). The second *m* in 4*mm* and 6*mm* refers to intermediate reflection lines, which in 4*mm* are at 45° to the first set of lines and in 6*mm* are at 30°. *The ten planar point groups shown in Fig. 5.55 are the two-dimensional analogues to the 32 three-*

dimensional point groups (crystal classes) discussed prior in this chapter.

Symmetry Content of Plane Lattices

The arrangements of nodes (or lattice points) in the five planar lattices reflect various inherent symmetry elements as shown in Fig. 5.50. Illustrations of the complete symmetry of some of the lattice types are given in Fig. 5.56. This illustration shows the motif distributions and symmetry elements compatible

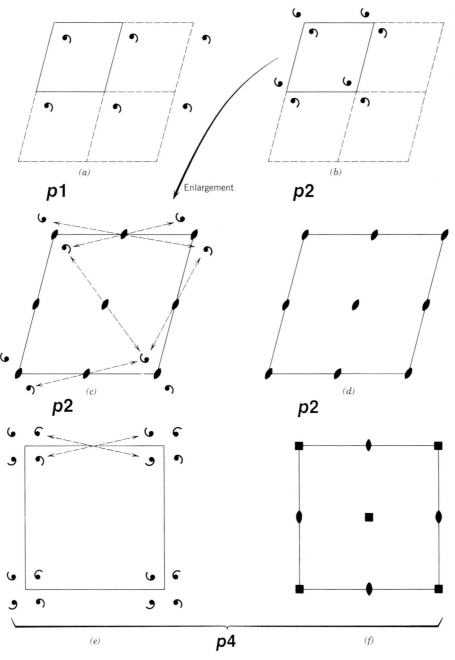

FIG. 5.56. Examples of rotational symmetry elements compatible with a primitive oblique lattice (*a* through *d*) and a primitive square lattice (*e* and *f*). See text for explanation.

with selected lattice types. In Fig. 5.56*a* we have a primitive oblique unit cell based on a regular distribution of commas. No rotation axis except 1 (which is equivalent to 0 or 360° rotation) is compatible with this pattern; mirror lines are also absent. Because the lattice is primitive (*p*) and contains only onefold rotational symmetry (1), it is referred to as *p*1. Figure 5.56*b* shows the same lattice as in *a*, but now it represents the repeat of two commas related by twofold rotation. It is still a primitive oblique unit cell, and it lacks mirror lines. It is described as *p*2. All of the possible locations of twofold rotations are shown in Fig. 5.56*c*, in an enlarged unit cell drawing of Fig. 5.56*b*. The twofold rotations at the corner nodes are probably obvious. However, there are others, halfway along each of the cell edges, as well as at a location in the very center of the cell. Figure 5.56*c* shows, by dashed lines, how some of the commas are related by such twofold rotations. It is standard procedure (e.g., in *International Tables for Crystallography*, vol. A) to represent the total symmetry content of a lattice such as *p*2 without reference to motif units. This is shown in Fig. 5.56*d*. A square lattice choice, as in Fig. 5.56*e*, contains points of fourfold rotation not only at the corner nodes, but there is an additional fourfold rotation point at the center if the distribution of commas (as motifs) is carefully evaluated. Furthermore, there are twofold rotations at the centers of the edges of the square. The total symmetry content of this primitive square lattice, referred to as *p*4, is shown in Fig. 5.56*f*.

It was noted in Fig. 5.50 that reflection lines (mirror symmetry, *m*) are compatible with four of the five planar lattice types (no reflections can be present in the oblique lattice). Because we are dealing here with the symmetry content of two-dimensional lattices that contain translational elements, we must also concern ourselves with the possible combination of reflection and translation. Such a combined operation (*m* + translation) is referred to as a *glide* operation, a *glide line*, or a *glide reflection*. A glide reflection causes a motif to be reflected across a reflection line and to be translated parallel to the reflection line. Figure 5.57 shows how motif units are related by a glide line (or glide plane) that has a translation component of *t*/2, where *t* is the shortest translation paralleling the glide plane.

Consideration of the 10 two-dimensional (planar) point groups (Fig. 5.55) in conjunction with the five plane lattices (Fig. 5.50), and the possibility of glide reflections (*g*) in addition to (or in place of) possible mirror reflections (*m*) leads to the so-called two-dimensional *plane groups*.

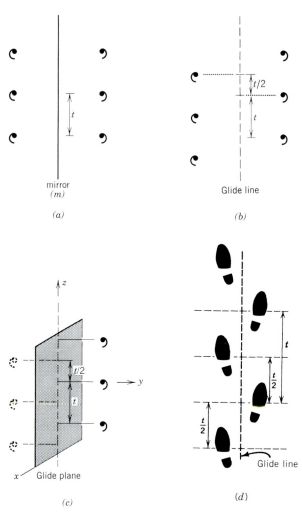

FIG. 5.57. Mirror symmetry and glide planes. *(a)* A two-dimensional array of an asymmetric motif with a spacing of *t* related by a reflection line, or a mirror perpendicular to the page. *(b)* A two-dimensional view of motifs that are related, across a glide line, by a glide component of *t*/2. *(c)* A three-dimensional illustration of a glide plane with glide component *t*/2. (In graphic illustrations, reflection lines and the traces of mirrors perpendicular to the page are shown by solid lines; glide lines and the traces of glide planes perpendicular to the page are shown by dashed lines.) *(d)* Human tracks showing relationship of motifs (footprints) by a glide line. Glide component *t*/2.

Two-Dimensional Plane Groups

Two-dimensional plane groups represent the infinite repetition of motifs on a plane. In the evaluation of such two-dimensional groups we must concern ourselves with the shapes (oblique, rectangular, hexagonal, or square) and the possible multiplicity (*p* for primitive and *c* for centered) of the plane lattice types. We must also take into account the translation-free point groups (1, 2, 3, 4, 6, *m*, 2*mm*, 3*m*, 4*mm*, and 6*mm*) and their compatibility with the lattice types (see Fig. 5.56) as well as the possible presence of glide reflections (g) in place of or in addition to possible mirror reflections (m).

Because we have already given examples of combinations of lattice type with planar point groups (e.g., p1, p2, and p4 in Fig. 5.56) we will now concern ourselves with combinations of lattice types (p or c) and m or g. Both of these symmetry elements are compati-

ble with four of the five plane lattices; an oblique lattice is the only one that *cannot* accommodate these operations. Figure 5.58 shows examples of a rectangular lattice (primitive as well as centered) containing m or g operations, or both, and an example of a square lattice containing both m and g operations.

In order to derive all possible two-dimensional plane groups one must consider systematically all

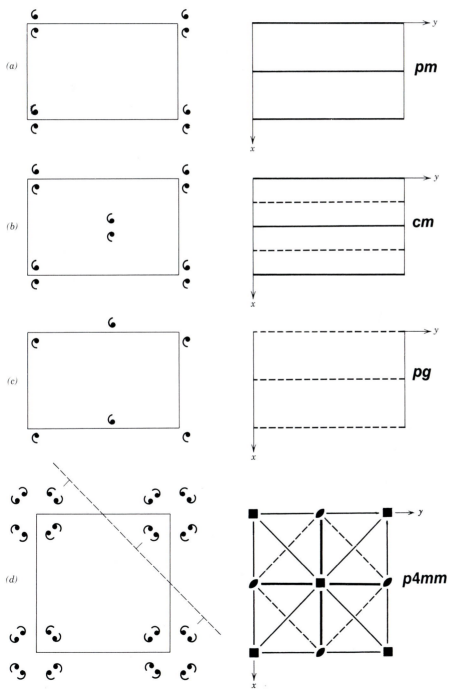

FIG. 5.58. Examples of mirror and glide symmetry compatible with a rectangular lattice type (a through c) and a square lattice (d). In the left-hand column are given the shape of the unit cell and the distribution of the motifs with respect to the outline of the unit cell. In the right-hand column only the distribution of the symmetry elements with respect to the unit cell outline is given. Crystallographic directions are indicated by axes x and y. Mirror lines (m) are shown by solid heavy lines and glide lines (g) by dashed lines. Rotation axes are shown by the standard symbols (see Tables 6.1 and 6.3).

(a) A primitive rectangular cell with mirror lines parallel to the y axis.

(b) A centered rectangular cell. The combination of centering and mirror lines produces glide lines parallel to and interleaved with the mirrors.

(c) A primitive rectangular cell with glides parallel to the y axis.

(d) In the left-hand illustration the location of one of four possible diagonal glide lines is shown on the basis of the distribution of motif clusters. In the right-hand column the total symmetry of this square planar array consists of fourfold rotations at corners and the center of the cell; twofold rotations at the centers of edges; mirror lines parallel to the two axes x and y; mirror lines in two diagonal positions; and glide lines interleaved with the diagonal mirror lines.

possible combinations of lattice types with permissible symmetry elements (or groupings of elements). As we have already seen (Figs. 5.56a and b), for an oblique cell only p1 and p2 are appropriate. For a rectangular cell we have other choices of symmetry

TABLE 5.8   Two-Dimensional Point Groups and Plane Groups*

Lattice	Point Group	Plane Group
Oblique p	1	p1
	2	p2
Rectangular p and c	m	pm
		pg
		cm
	2mm	p2mm
		p2mg
		p2gg
		c2mm
Square p	4	p4
	4mm	p4mm
		p4gm
Hexagonal p	3	p3
	3m	p3m1†
		p31m†
	6	p6
	6mm	p6mm

*From *International Tables for X-ray Crystallography*, 1969, v. 1, N. F. M. Henry and K. Lonsdale, eds.: Symmetry Groups. International Union of Crystallography, Kynoch Press, Birmingham, England.

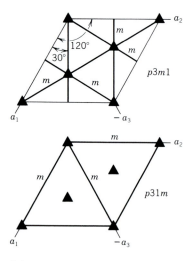

†There are two distinct groups for 3m—p3m1 and p31m. They have the same total symmetry content and shape. However, the conventional location of cell edges (as defined by three axes a₁, a₂, and a₃) differs by 30° in the two groups. In p3m1 the mirror lines bisect the 60° angle between cell edges; in p31m the reflection lines coincide with the cell edges.

elements (m and g) as well as the choice of a primitive (p) or a centered (c) cell. For example, for point group m we might expect pm, pg, cm, and cg as possible two-dimensional plane groups. Similarly, for 2mm we might consider pmm, cmm, pmg, pgg, cmg, and cgg as permissible two-dimensional plane group choices. When we look at a tabulation of all possible two-dimensional plane groups compatible with point group symmetries m and 2mm (see Table 5.8), we find that only 7 of the above 10 possibilities actually occur. Table 5.8 lists the 17 possible two-dimensional plane groups. The reason for this relatively small number is that not all combinations lead to new or different plane groups. Furthermore, *the interaction of the symmetry of the motif (planar point group) with the symmetry of the various plane lattices affects the overall resultant symmetry content of the planar pattern.* The final pattern displays the symmetry of the lattice when the symmetry elements of the motif are aligned with the corresponding symmetry elements of the lattice. If the motif has less symmetry than the lattice, the pattern will express the motif's lesser degree of symmetry, with the symmetry elements of the motif aligned with the corresponding symmetry elements of the lattice. Figure 5.59 illustrates the 17 possible plane patterns (known as *plane groups*). Interactive graphical developments of 7 of the 17 possible plane groups are given in module III of the CD-ROM under the heading "Two-dimensional Order: Generation of Seven Plane Groups." More detailed illustrations of each of these two-dimensional plane groups are given in *International Tables for X-ray Crystallography* 1: 57–72.

It has been mentioned (see also Fig. 5.51) that tesselations (two-dimensional coverings of a wall by tiles) can represent various patterns of the 17 two-dimensional plane groups. Artistic and often complicated drawings by the Dutch graphic artist M. C. Escher are commonly filled with designs of fish, horsemen, and birds. These drawings have been compiled by Caroline H. MacGillavry in a book titled *Fantasy and Symmetry: The Periodic Drawings of M. C. Escher* (see references). Two of the Escher drawings from this book have been redrawn in Fig. 5.60. Following is a brief passage from the introduction to this book (p. IX):

It occurred to several scientists attending this meeting [Fifth International Congress of the International Union of Crystallography, held in Cambridge, England, in 1960] that Escher's periodic drawings [in an exhibition arranged for that same meeting by crystallographers J. D. H. and Gabrielle Donnay] would make excellent material for teaching the principles of symmetry. These patterns are

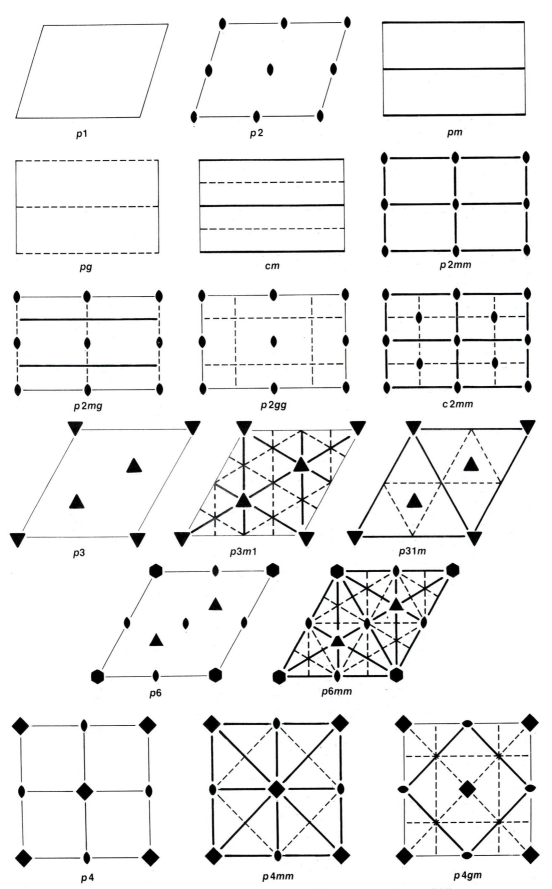

FIG. 5.59. Graphic representation of the symmetry content of the 17 plane groups. Heavy solid lines and dashed lines represent mirrors and glide lines, respectively, perpendicular to the page.

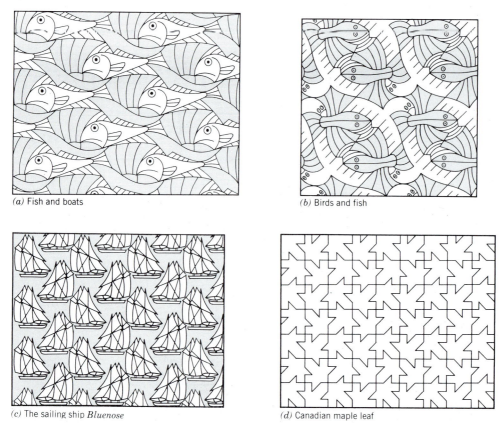

(a) Fish and boats

(b) Birds and fish

(c) The sailing ship *Bluenose*

(d) Canadian maple leaf

FIG. 5.60. Illustration of two-dimensional plane group symmetry in periodic drawings. *(a)* and *(b)* are redrawn from illustrations (plates 1a and 2, respectively) by M. C. Escher, as published by Caroline H. MacGillavry (see reference list); copyright © 1962 and 1963, respectively, M. C. Escher/Cordon Art, Baarn, Holland. *(c)* and *(d)* are redrawn from François Brisse in his publication entitled "La Symétrie bidimensionelle et le Canada" (see reference list at the end of this chapter). The two-dimensional plane groups represented by the illustrations are *(a)* p1, *(b)* p2, *(c)* pg, and *(d)* p4gm.

complicated enough to illustrate clearly the basic concepts of translation and other symmetry, which are so often obscured in the clumsy arrays of little circles, pretending to be atoms, drawn on blackboard by teachers of crystallography classes. On the other hand, most of the designs do not present too great difficulties for the beginner in the field.

The student who is interested in the scientific as well as esthetic aspects of two-dimensional periodic patterns should consult the book by MacGillavry (and other books on M. C. Escher listed in the reference section to this chapter). Any of these books are probably available in an art or architecture library. Figure 5.60 also contains redrawn illustrations of 2 of the 13 designs by François Brisse (1981), which he prepared especially for the 12th Congress of the International Union of Crystallography, held in Ottawa, Canada, in 1981.

The arrangements of motifs in the four illustrations of Fig. 5.60 represent 4 of the 17 possible two-dimensional plane groups listed in Table 5.8 and illustrated in Fig. 5.59. The best way for the student to evaluate the shape and size of the unit cell, as well as the symmetry content of an infinitely extending periodic drawing, is to place a sheet of transparent paper over it. On this transparent overlay one can substitute opaque circles (nodes) for the smallest motif, or a part of the motif (a motif unit). If symmetry is present, it is best to locate the nodes on the location of such symmetry elements (rotations, *m* or *g*). Once the nodes have been located, the lattice of the design can be chosen by drawing lines between the nodes. It is especially instructive to use the standard symbols for rotations perpendicular to the page (◗ ▲ ■ ⬣) and for mirrors (solid lines) and glides (dashed lines). This will reveal the similarity of the two-dimensional plane group (as deduced from these four artistic illustrations) with the representations in Fig. 5.59. (see also Klein, 1994; complete reference at end of chapter).

Three-Dimensional Order

Until now we have discussed aspects of order in one direction and in two-dimensional patterns. The concepts presented thus far are basic to an understanding of the regularity expressed by three-dimensional objects such as crystalline matter. In an assessment of three-dimensional order, a third direction (vector) will be needed to describe the distribution of nodes in a three-dimensional (space) lattice. Many of the symmetry elements introduced earlier will also be found in three-dimensional periodic structures. Rotational symmetry about a point will become rotational symmetry about a line (or axis); reflection (*m*) or glide reflection (*g*) across a line becomes a mirror plane or a glide plane, respectively. However, we will also need to introduce some new symmetry operations in three-dimensional patterns that are not present in planar periodicity. These new operations combine rotation and translation and are known as *screws* (see Figs. 5.66 and 5.67); the direction along which a screw operation takes place is known as a *screw axis*.

In order to fully describe the unit cell shape of the lattice (of a three-dimensionally ordered structure) and the total symmetry content compatible with this lattice, we will need to consider (as we did in planar patterns) combinations of lattice type with translation-free symmetry elements (e.g., *m*, rotation axes, and inversion) and translational symmetry elements (screw axes and glide planes). Such combinations will lead to 230 possible three-dimensional space groups, as compared with 17 two-dimensional plane groups. Each of these space groups belongs to one of the 32 point groups. Each space group will also be built upon a specific lattice type. In three-dimensional arrays there will be a choice of 14 different lattice types (these 14 include primitive and multiple lattices) as compared with only five lattice types for the two-dimensional patterns.

Three-Dimensional Lattices

Three-dimensional lattices can be constructed by adding one additional translation direction (vector) to the plane lattices of Fig. 5.50; this third vector must not lie in the plane of the two-dimensional nets. Vector space is referred to three non-coplanar axes, *x*, *y*, and *z*, which intersect at the origin. The unit cell vectors are denoted as **a**, **b**, and **c**, and the unit cell translations along *x*, *y*, and *z*, respectively, are noted as *a*, *b*, and *c* (see Fig. 5.61). The *x*, *y*, and *z* coordinate axes are commonly referred to as the *a*, *b*, and *c* axes. The unit cell dimensions are expressed in angstrom or nanometer units.

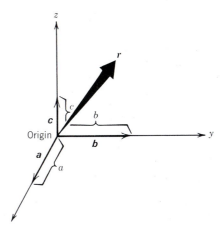

FIG. 5.61. *(a)* Representation of vectors **a**, **b**, and **c**, coordinate axis directions *x*, *y*, and *z*, and unit cell translations *a*, *b*, *c*, along directions *x*, *y*, and *z*, respectively. A general vector, **r**, in this three-dimensional space can be expressed as a linear combination of **a**, **b**, **c**, such that $\mathbf{r} = x\mathbf{a} + y\mathbf{b} + z\mathbf{c}$, where *x*, *y*, and *z* are real numbers. The *x*, *y*, and *z* coordinate axes (associated with cell edges *a*, *b*, and *c*, respectively) are commonly referred to as the *a*, *b*, and *c* axes.

Figure 5.62 illustrates the various three-dimensional (space) lattice types that result when the five planar nets (shown in Fig. 5.50) are stacked in various ways along a third direction (*z*). The space lattices that result may be primitive or nonprimitive. *A primitive space lattice is a parallelepiped with lattice points only at its corners.* If the unit cell is nonprimitive, the centering may occur on a pair of opposite faces of the unit cell and is called *A*-, *B*-, or *C*-centered, depending on whether the centering takes place along the direction of the *x*, *y*, or *z* axis. The centering may also occur on all faces of the unit cell and is then referred to as *F* (for face-centered), or it may be present in the center of the unit cell and is referred to as *I*, body-centered (*I* from the German word *innenzentriert*). These various types of unit cells are shown in Fig. 5.62, as well as the two choices of unit cells in space lattices that are derived from stacking a hexagonal net (see Fig. 5.50*d*). Space lattice no. 10, in Fig. 5.62, is based on a net with two equal translations (a_1 and a_2) that make an angle of 120° with each other. The unit cell in lattice no. 11, in Fig. 5.62, is known as a rhombohedral (*R*) unit cell in which the translation directions are a_R and the angles between the three equivalent edges of the unit cell are α_R.

In our prior discussion of unit cell choices in planar patterns, we noted that for a specific array of nodes (e.g., see Fig. 5.52), a considerable number of unit cell choices are possible. In order to reduce the number of choices (in three-dimensional arrays),

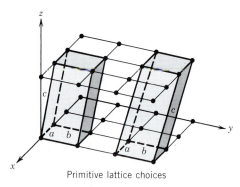

Primitive lattice choices

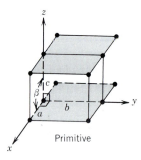

Primitive

(1) Stacking of an oblique net (or plane lattice) at an arbitrary angle results in *primitive triclinic lattices*.

(2) Stacking of a primitive rectangular net in a vertical direction (z), with $x \wedge z$ angle (β) $\neq 90°$, leads to a *primitive monoclinic lattice*.

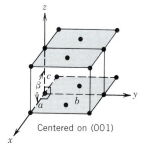

Centered on (001)

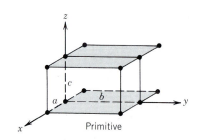

Primitive

(3) Stacking of a centered rectangular net in a vertical direction (z), with $x \wedge z$ angle (β) $\neq 90°$, results in a *centered monoclinic lattice*.

(4) Stacking of a primitive rectangular net in a vertical direction (z), with the $x \wedge z$ angle $= 90°$, leads to a *primitive orthorhombic lattice*.

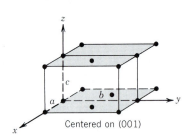

Centered on (001)

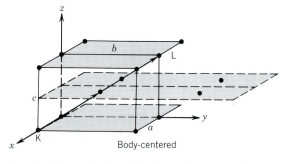

Body-centered

(5) Stacking of a centered rectangular net in a vertical direction (z), with the $x \wedge z$ angle $= 90°$, leads to a *centered orthorhombic lattice*.

(6) Stacking of a primitive rectangular net along the direction between nodes K and L results in an orthorhombic lattice with a central node. This is an *orthorhombic body-centered lattice*.

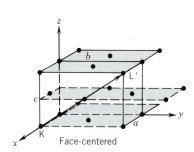

Face-centered

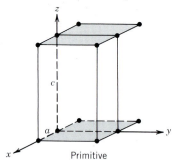

Primitive

(7) Stacking of a centered rectangular net along the direction between nodes K and L′ (on the front face) leads to centering on all faces of the three-dimensional lattice. This is a *face-centered orthorhombic lattice*.

(8) Stacking of a square net along the z direction, with angle $x \wedge z = 90°$, and with the c translation $\neq a_1$ or a_2, results in a *primitive tetragonal lattice*.

FIG. 5.62. Stacking of the five nets (plane lattices; see Fig. 5.50) in various ways (as specifically noted in this figure) leads to the 14 possible space lattices. These 14 lattice types are also known as the 14 Bravais lattices (see also Fig. 5.63).

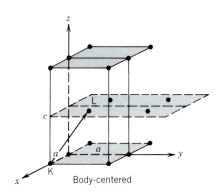

Body-centered

(9)   Stacking of the same net as in (8) but now in a direction between nodes K and L results in a *body-centered tetragonal lattice*.

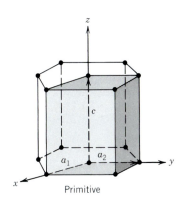

Primitive

(10)   Stacking of a hexagonal net in a *z* direction such that angle $x \wedge z = 90°$ leads to a primitive hexagonal lattice. If this lattice choice is rotated 3 times about *z*, it results in a *c-centered hexagonal lattice*.

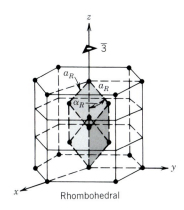

Rhombohedral

(11)   A hexagonal net can also be stacked along the edge directions of a rhombohedron (a_R). This results in a *rhombohedral lattice*, the edge directions of which are symmetrical with respect to the $\bar{3}$ axis along the *z* direction.

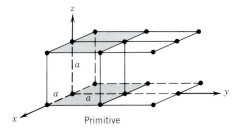

Primitive

(12)   Stacking of a square net along the *z* direction, with $x \wedge z$ angle = 90° and with *c* translation = a_1, and a_2, results in a *primitive isometric lattice*.

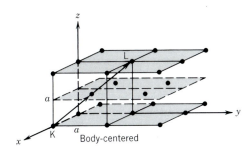

Body-centered

(13)   Stacking of a square net along the direction between nodes K and L (a body diagonal) results in a *body-centered isometric lattice*.

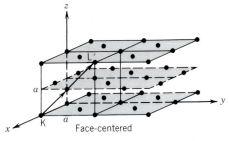

Face-centered

(14)   Stacking of a square net along the direction between nodes K and L′ (along the front face), results in a *face-centered isometric lattice*.

FIG. 5.62.   *(Continued)*

crystallographers have drawn up the following restrictions as to unit cell choice:

1. The edges of the unit cell should coincide, if possible, with symmetry axes of the lattice.
2. The edges should be related to each other by the symmetry of the lattice.
3. The smallest possible cell should be chosen in accordance with (1) and (2).

Clearly, any regular three-dimensional array of nodes can be outlined by a primitive lattice. However, it is frequently desirable and appropriate to choose a nonprimitive unit cell. In Table 5.4 we outlined the 32 nonidentical symmetry elements and combinations of symmetry elements in terms of the crystal classes and crystal systems. *The types of space lattices compatible with these 32 point groups*

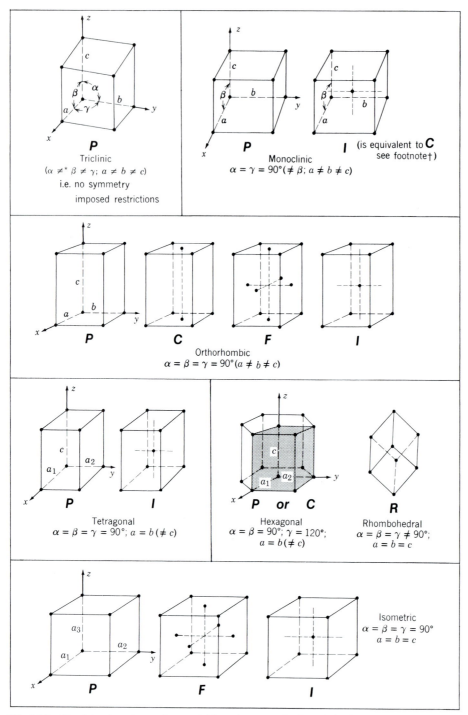

FIG. 5.63. The 14 unique types of space lattices, known as the Bravais lattices. Axial lengths are indicated by *a*, *b*, and *c* and axial angles by α, β, and γ. Each lattice type has its own symmetry constraints on lengths of edges *a*, *b*, and *c* and angles between edges, α, β, and γ. In the notations the nonequivalence of angles or edges that usually exist but are not mandatory are set off by parentheses.

*The sign $\neq$ implies nonequality by reason of symmetry; accidental equality may occur.

†In the monoclinic system the unit cell can be described by a body-centered or a *C*-face-centered cell by a change in choice of the length of the *a* axis and the angle β. Vectorially these relations are: $a_l = c_c + a_c$; $b_l = b_c$; $c_l = -c_c$; and $a_l \sin \beta_l = a_c \sin \beta_c$. Subscripts *l* and *c* refer to the unit cell types.

are known as the *14 Bravais lattices* and are shown in Fig. 5.62. These lattice types are unique, as was shown by Auguste Bravais (1811–1863), after whom they are named. That is, *they represent the only possible ways in which points can be arranged periodically in three dimensions.* Ten of the 14 Bravais lattice types are developed through animations in module III of the CD-ROM under the heading "Three-dimensional Order: Generation of 10 Bravais Lattices."

In Fig. 5.63 the 14 lattice types are arranged by crystal system. The names of the crystal systems reflect the characteristic symmetries of the lattice types. For example, in the triclinic system, which includes the symmetries 1 and $\bar{1}$, the unit cell compatible with these symmetries has no constraints, and thus its shape is one of low symmetry. The isometric system, however, contains very high symmetry ($4/m\bar{3}2/m$, $\bar{4}32$, $\bar{4}3m$, $2/m\bar{3}$, and 23), which is reflected in the unit cell having the highest symmetry constraints. There is a primitive lattice for each of the six crystal systems, and centered lattices occur in five of them. It should also be noted that only one face-centered lattice (namely, *C*) is shown in Fig. 5.63. If the lattice had been chosen in such a way as to be *A*-centered or *B*-centered rather than *C*-centered, this would not introduce a new category of lattice type. The *A*-, *B*-, and *C*-centered lattices are symmetrically identical and can be converted into each other by an appropriate exchange of the crystallographic axes. Table 5.9 provides a synopsis of the above discussion. The right-hand column in Table 5.9 has the heading "Multiplicity of Cell." This allows for a numerical distinction between *primitive* and *nonprimitive* lattice choices. A *primitive* lattice has nodes only at

the corners. Each corner node is shared between eight adjoining cells. For such a primitive lattice, there are eight corner nodes, of which $\frac{1}{8}$th of each node contributes to the cell. That is, it has a multiplicity of $8 \times \frac{1}{8} = 1$. In a face-centered cell, each node on a face is shared between two adjoining cells. Therefore, the total node content of a side-centered cell is $8 \times \frac{1}{8} = 1$ (for corner nodes) $+ 2 \times \frac{1}{2} = 1$ (for face-centered nodes), resulting in a multiplicity of $1 + 1 = 2$. Any node in the interior of a cell, as in a body-centered choice, belongs only to its own cell.

The shape and size of unit cells of minerals are most commonly determined by X-ray diffraction techniques (see Chapter 7). During the last decade, however, high-resolution transmission electron microscopy (HRTEM) has allowed the direct observation of projected images of crystal structures on photographic plates. Such a structure image is shown in Fig. 5.64 for the mineral cordierite. The dark parts of the photograph outline the projected image of the structure and the superimposed lines outline a rectangular unit cell for cordierite.

The internal (atomic) structure of crystalline materials is generally determined by a combination of X-ray, neutron, and electron diffraction techniques, and may be supplemented by a combination of spectroscopic methods. These methods, used singly or in combination, provide a quantitative three-dimensional reconstruction of the location of the atoms (or ions), the chemical bond types and their orientations, and the overall internal symmetry of the structure. From such structural information, the appropriate three-dimensional (space) lattice is derived. Figure 5.65 illustrates the derivation of the appropriate lattice type and unit cell from the structure of low (α)

TABLE 5.9 Description of Space Lattice Types and Distribution of the 14 Bravais Lattices Among the Six Crystal Systems

Name and Symbol	Location of Nonorigin Nodes	Multiplicity of Cell
Primitive (*P*)	. . . .	1
Side-centered (*A*)	Centered on A face (100)	2
(*B*)	Centered on B face (010)	2
(*C*)	Centered on *C* face (001)	2
Face-centered (*F*)	Centered on all faces	4
Body-centered (*I*)	An extra lattice point at center of cell	2
Rhombohedral (*R*)	A primitive rhombohedral cell	1
Primitive (*P*) in each of the 6 crystal systems		= 6
Body-centered (*I*) in monoclinic, orthorhombic, tetragonal, and isometric		= 4
Side-centered (*A* = *B* = *C*) in orthorhombic		= 1
Face-centered (*F*) in orthorhombic and isometric		= 2
Rhombohedral (*R*) in hexagonal		= 1
	Total	= 14

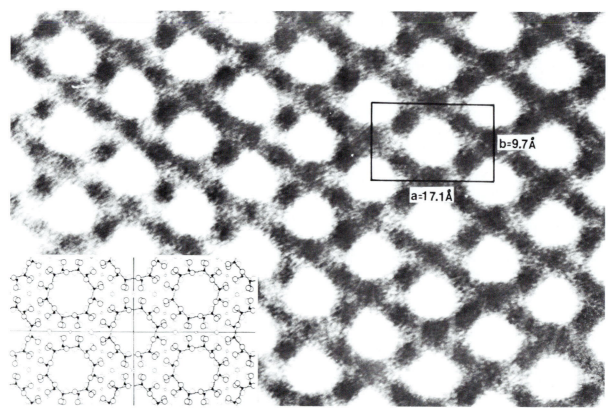

FIG. 5.64. High magnification structure image of an *a–b* section through the mineral cordierite. An orthorhombic unit cell is outlined and distances are given in angstroms. The insert shows the idealized structure of cordierite, as determined by X-ray diffraction techniques. The scales of the idealized structure and the electron transmission image are identical. (From Buseck and Iijima, 1974, *American Mineralogist* 59: 1–22).

quartz (SiO_2). This illustration also shows that the equivalent points (or *identipoints*) in a lattice are generally not atomic positions. Rather, as in Fig. 5.65, they are geometric points in the structure that have the same angle and distance relationships to the Si and O atoms (and hence to the threefold and twofold axes of the structure).

Screw Axes and Glide Planes

Earlier, we discussed the translation-free symmetry operations (rotation, rotoinversion, mirror, and center of symmetry), and now we have introduced pure translation (as in lattices), as well as the concept of a glide line in two-dimensional patterns. In three-dimensional periodic arrays we find two symmetry elements that combine a symmetry operation (rotation or mirror reflection) with a translation component. A rotational operation with translation (*t*) parallel to the axis of rotation is known as a *screw operation*, and a mirror reflection with a translation component (*t*/2 or *t*/4) parallel to the mirror is known as a *glide operation*. Animations of screw axis and

glide operations are given in module III of the CD-ROM under the headings "Three-dimensional Order: Screw Axes and Glide Planes."

The two-, three-, four-, and sixfold rotational operations can all be combined with a translation. (A onefold rotation axis combined with a translation is equivalent to a translation only.) Figure 5.66 compares the differences in motif distribution for a fourfold rotation and a fourfold screw operation. Rotation alone, through an angle (α), about an imaginary axis, generates a sequence of the motif along a circle. For a 90° rotation angle a pattern with four motifs is generated. In the fourfold screw operation the four motifs are generated from the original unit in a three-dimensional, helical path (as in a screw motion; see Fig. 5.66*b*). Screw axes are said to be *isogonal* (from the Greek meaning "same angle"), with the equivalent rotational axes. This means that fourfold screw operations rotate the motif through 90° angles while translating the motif parallel to the rotation axis. All possible screw axes, isogonal with rotational axes, are illustrated in Fig. 5.67.

THE INTERNAL ORDER AND SYMMETRY OF MINERALS

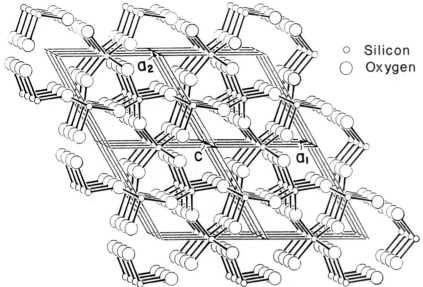

○ Silicon
◯ Oxygen

FIG. 5.65. A drawing of the structure of low (α) quartz (SiO_2), with the normally vertical z axis tilted at a small angle to better show the repeat distance c (of the unit cell) along the z direction. The primitive hexagonal space lattice, outlined by the various parallelepipeds, shows that each unit cell (with edges a_1, a_2, and c) contains a complete and representative unit of the repeating pattern of the structure. (From M. G. Boisen and G. V. Gibbs, 1990, *Mathematical Crystallography* rev. ed. *Reviews in Mineralogy* 15. Mineralogical Society of America, Washington, D.C.)

As with any screw motion, *screw axes are right- or left-handed.* A right-handed screw may be defined as one that advances away from the observer when rotated clockwise. The screw axis symbols consist of the symbols for rotation axes (2, 3, 4, and 6) followed by a subscript that represents the fraction of the translation (t) inherent in the operation. For example, 2_1 means that $\frac{1}{2}t$ (obtained by placing the subscript over the main axis symbol, as in a fraction) is the translation involved. For a threefold rotation there are two possible screw axes, namely, 3_1 and 3_2. The translation component in both screw axes is

$\frac{1}{3}t$, but a convention allows for the distinction between the directions of the screw. When the ratio of the subscript to the number of the rotation axes (as $\frac{1}{3}$ for 3_1) is less than $\frac{1}{2}$, the screw is right-handed. When this ratio is more than $\frac{1}{2}$, it is left-handed (as in 3_2), and when the ratio is $\frac{1}{2}$, the screw is considered neutral in direction (see Fig. 5.67). In other words, 3_1 and 3_2 are an *enantiomorphous pair of screw axes*, with 3_1 right-handed and 3_2 left-handed. Similarly, the following pairs are enantiomorphous: 4_1 and 4_3, 6_1 and 6_5, and 6_2 and 6_4. See Chapter 6 for further discussion of the representation of screw axes.

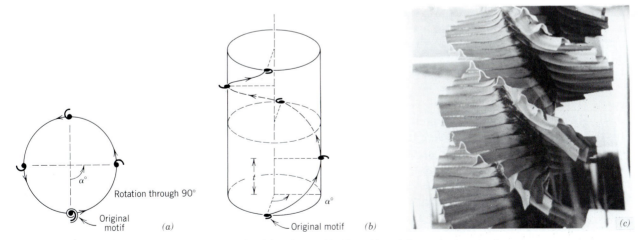

FIG. 5.66. Generation of patterns by fourfold rotation (a) and a combination of translation and rotation (b), which results in a screw motion (4_1). (c) Example of a many-fold screw axis in the vertical direction of a palm. The screw operation is shown by the trimmed remains of the leaves. The rotational symmetry is approximately 30. Photographed in Darwin, Northern Territory, Australia. As discussed on p. 176, independent objects can have unlimited rotational symmetry, whereas such symmetry is limited to 1, 2, 3, 4, and 6 in ordered arrays.

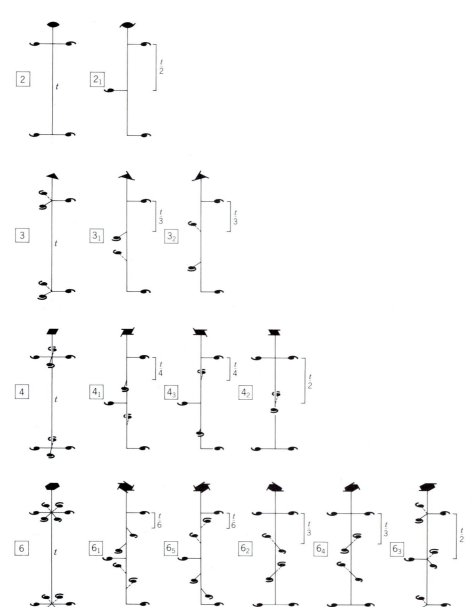

FIG. 5.67. Repetition of motif units by screw axes. The left column represents rotation axes and the columns to the right represent isogonal screw axes. The symbols at the top of the rotation and screw axes are internationally accepted. For projections of these screw axes, see Fig. 6.51.

In addition to the repetition of a motif by a mirror reflection, a regular pattern can be generated by a combination of a mirror reflection and a translation. This operation is referred to as a *glide plane* or *glide reflection.* Figure 5.57c shows how motifs are related to a glide plane that has a translation component of $t/2$. In our prior discussion of planar patterns (see Figs. 5.58b and c) we noted how a glide line repeats motifs on either side of the line with a periodicity of half the lattice translation. In three-dimensional patterns a wider variety of glide movements can occur.

Specific glide directions can be identified in two- and three-dimensional patterns and expressed in terms of a set of axes such as *x, y,* and *z.* However, many crystallographers refer the internal order

as well as external morphology of crystals to three axes, *a, b,* and *c.* The *c* axis is vertical and the *a* and *b* axes lie in a plane that does not contain *c.* If the glide component $(t/2)$ in a three-dimensional ordered arrangement is parallel to the *a* axis, it is referred to as an *a glide* and is represented by the symbol *a.* Similarly, if the glide component $(t/2)$ is parallel to the *b* or *c* axes, the glide is referred to as a *b* or *c glide,* respectively. If the glide component can be represented by $a/2 + b/2$, $a/2 + c/2$, $b/2 + c/2$, or $a/2 + b/2 + c/2$, it is referred to as a *diagonal glide* and is represented by the symbol *n.* If the glide component can be represented by $a/4 + b/4$, $b/4 + c/4$, $a/4 + c/4$, or $a/4 + b/4 + c/4$, it is known as a *diamond glide* and is symbolized by the letter *d.* It should be obvious that in the *diamond* (*d*) *glide* the simultaneous translations are one-fourth of the cell edges, whereas for a *diagonal* (*n*) *glide* the translation components are equal to one-half of the cell edges. See Chapter 6 (p. 280) for further discussion of the representation of various glide operations.

Space Groups

When we combine the 14 possible space lattice types (Bravais lattices) with the symmetry inherent in the 32 crystal classes (the translation-free point group symmetries), as well as the two symmetry operations that involve translation (screws and glides), we arrive at the concept of *space groups*. *Space groups, therefore, represent the various ways in which motifs* (such as atoms in crystals) *can be arranged in space in a homogeneous array* (homogeneous meaning that each motif is equivalent to every other motif in the pattern).

In the last decade of the nineteenth century three men of different nationalities and interests derived independently how many unique patterns could occur in three-dimensional periodic arrays. These were E. von Federov, a Russian crystallographer; Artur Schoenflies, a German mathematician; and William Barlow, a British amateur. Their unanimous conclusion was that there are 230 such unique patterns; they are known as the 230 space groups. It should be remembered that *the presence of periods of translation (as in lattices or glide planes or screw axes) cannot be detected morphologically because the translations involved are on the order of 1 to 10 Å.* The translation-free symmetry combinations are point groups, whereas space groups define the symmetry and translations in space. If we were to ignore the translation components in the 230 space groups, we would arrive at the 32 point groups.

Space groups have the following characteristics. (1) They are based on one of the 14 Bravais lattices that is compatible with a specific point group, and (2) they are isogonal with one of the 32 point groups. Isogonal implies that rotation and screw axes with the same rotational repeat have the same rotational angle (e.g., 60° in a sixfold rotation or sixfold screw axis). This means that the screw axes 6_1, 6_2, 6_3, 6_4, and 6_5 are isogonal with the rotation axis 6. In other words, *the point group is the translation-free residue of a family of possible isogonal space groups.*

A specific point group designation consists of a series of symmetry elements, as in $2/m2/m2/m$. For each of the specific point group symmetry elements there is a possible space group element. That is, instead of a mirror plane perpendicular to the first twofold rotation axis and a mirror plane perpendicu-

lar to the second twofold rotation axis, there may be glide planes in their places that would be symbolized as $2/b2/a2/m$. The space group symbol is additionally preceded by a symbol that designates the general lattice type (*P, A, B, C, I, F,* or *R*). The full symbol for a space group that is isogonal with $2/m2/m2/m$ might be $I2/b2/a2/m$. Another example of a space group isogonal with $2/m2/m2/m$ would be $P2_1/b2_1/c2_1/a$, and so on. Various aspects of space groups are illustrated with animations in module III of the CD-ROM under the heading "Three-dimensional Order: Space Group Elements in Structures."

Table 5.10 gives a listing of the 32 crystal classes (point groups) and the 230 space groups compatible therewith. When one compares the listing in the column entitled "Crystal Class" with the entries in the column "Space Groups," the isogonality of the relationships of these two groupings becomes obvious. The reason for including Table 5.10 is *not* for the beginning student to memorize all or many of the unique (230) space groups. It is included to illustrate the ease with which a space group symbol can be reduced to its isogonal point group (crystal class). All of the entries in the "Space Group" column contain as the first entry a symbol for the lattice type. Furthermore, many space group representations contain notations for screw axes and glide planes. Lattice type, screw, and glide operations all contain translational elements. If these translational elements are removed and the translation-free equivalent operations are substituted for them, the point group, which is isogonal with the space group, will result. For example, the last space group in Table 5.10 is listed as $I4_1/a\bar{3}2/d$. Because lattice types cannot be reflected in translation-free symbols for crystal classes (point groups), the lattice type must be dropped altogether. The remaining symbols must become translation-free as well. This means that *a* and *d* glides must be replaced by the translation-free mirror symbol *m*. This leads to the point group notation $4/m\bar{3}2/m$ in the isometric system.

If screw axes are present, a similar process allows easy conversion from space group to point group. For example, in space group $P6_122$, the screw axis is replaced by the isogonal sixfold rotation and the lattice type (*P*) is dropped. This results in the isogonal point group 622 in the hexagonal system. The ease in relating point and space group symmetries is one of the powerful aspects of the Hermann–Mauguin (international) notation. It allows the reader to "translate" a space group notation for a crystalline material into the isogonal and simpler point group notation.

One aspect of the Hermann–Mauguin notation that may need some clarification is the use by crystal-

TABLE 5.10   The 230 Space Groups, and the Isogonal 32 Crystal Classes (Point Groups). The Space Group Symbols Are, in General, Unabbreviated*

Crystal Class	Space Group
1	$P1$
$\bar{1}$	$P\bar{1}$
2	$P2$, $P2_1$, $C2$
m	Pm, Pc, Cm, Cc
$2/m$	$P2/m$, $P2_1/m$, $C2/m$, $P2/c$, $P2_1/c$, $C2/c$
222	$P222$, $P222_1$, $P2_12_12$, $P2_12_12_1$, $C222_1$, $C222$, $F222$, $I222$, $I2_12_12_1$
$mm2$	$Pmm2$, $Pmc2_1$, $Pcc2$, $Pma2$, $Pca2_1$, $Pnc2$, $Pmn2_1$, $Pba2$, $Pna2_1$, $Pnn2$, $Cmm2$, $Cmc2_1$, $Ccc2$, $Amm2$, $Abm2$, $Ama2$, $Aba2$, $Fmm2$, $Fdd2$, $Imm2$, $Iba2$, $Ima2$
$2/m2/m2/m$	$P2/m2/m2/m$, $P2/n2/n2/n$, $P2/c2/c2/m$, $P2/b2/a2/n$, $P2_1/m2/m2/a$, $P2/n2_1/n2/a$, $P2/m2/n2_1/a$, $P2_1/c2/c2/a$, $P2_1/b2_1/a2/m$, $P2_1/c2_1/c2/n$, $P2/b2_1/c2_1/m$, $P2_1/n2_1/n2/m$, $P2_1/m2_1/m2/n$, $P2_1/b2/c2_1/n$, $P2_1/b2_1/c2_1/a$, $p2_1/n2_1/m2_1/a$, $C2/m2/c2/m$, $C2/m2/c2_1/a$, $C2/m2/m2/m$, $C2/c2/c2/m$, $C2/m2/m2/a$, $C2/c2/c2/a$, $F2/m2/m2/m$, $F2/d2/d2/d$, $I2/m2/m2/m$, $I2/b2/a2/m$, $I2/b2/c2/a$, $I2/m2/m2/a$
4	$P4$, $P4_1$, $P4_2$, $P4_3$, $I4$, $I4_1$
$\bar{4}$	$P\bar{4}$, $I\bar{4}$
$4/m$	$P4/m$, $P4_2/m$, $P4/n$, $P4_2/n$, $I4/m$, $I4_1/a$
422	$P422$, $P42_12$, $P4_122$, $P4_12_12$, $P4_222$, $P4_22_12$, $P4_322$, $P4_32_12$, $I422$, $I4_122$
$4mm$	$P4mm$, $P4bm$, $P4_2cm$, $P4_2nm$, $P4cc$, $P4nc$, $P4_2mc$, $P4_2bc$, $I4mm$, $I4cm$, $I4_1md$, $I4_1cd$
$\bar{4}2m$	$P\bar{4}2m$, $P\bar{4}2c$, $P\bar{4}2_1m$, $P\bar{4}2_1c$, $P\bar{4}m2$, $P\bar{4}c2$, $P\bar{4}b2$, $P\bar{4}n2$, $I\bar{4}m2$, $I\bar{4}c2$, $I\bar{4}2m$, $I\bar{4}2d$
$4/m2/m2/m$	$P4/m2/m2/m$, $P4/m2/c2/c$, $P4/n2/b2/m$, $P4/n2/n2/c$, $P4/m2_1/b2/m$, $P4/m2_1/n2/c$, $P4/n2_1/m2/m$, $P4/n2_1/c2/c$, $P4_1/m2/m2/c$, $P4_2/m2/c2/m$, $P4_2/n2/b2/c$, $P4_2/n2/n2/m$, $P4_2/m2_1/b2/c$, $P4_2/m2_1/n2/m$, $P4_1/n2_1/m2/c$, $P4_2/n2_1/c2/m$, $I4/m2/m2/m$, $I4/m2/c2/m$, $I4_1/a2/m2/d$, $I4_1/a2/c2/d$
3	$P3$, $P3_1$, $P3_2$, $R3$
$\bar{3}$	$P\bar{3}$, $R\bar{3}$
32	$P312$, $P321$, $P3_112$, $P3_121$, $P3_212$, $P3_221$, $R32$
$3m$	$P3m1$, $P31m$, $P3c1$, $P31c$, $R3m$, $R3c$
$32/m$	$P31m$, $P31c$, $P3m1$, $P3c1$, $R3m$, $R3c$
6	$P6$, $P6_1$, $P6_5$, $P6_2$, $P6_4$, $P6_3$
$\bar{6}$	$P\bar{6}$
$6/m$	$P6/m$, $P6_3/m$
622	$P622$, $P6_122$, $P6_522$, $P6_222$, $P6_422$, $P6_322$
$6mm$	$P6mm$, $P6cc$, $P6_3cm$, $P6_3mc$
$\bar{6}m2$	$P\bar{6}m2$, $P\bar{6}c2$, $P\bar{6}2m$, $P\bar{6}2c$
$6/m2/m2/m$	$P6/m2/m2/m$, $P6/m2/c2/c$, $P6_3/m2/c2/m$, $P6_2/m2/m2/c$
23	$P23$, $F23$, $I23$, $P2_13$, $I2_13$
$2/m\bar{3}$	$P2/m\bar{3}$, $P2/n\bar{3}$, $F2/m\bar{3}$, $F2/d\bar{3}$, $I2/m\bar{3}$, $P2_1/a\bar{3}$, $I2_1/a\bar{3}$
432	$P432$, $P4_232$, $F432$, $F4_132$, $I432$, $P4_332$, $P4_132$, $I4_132$
$\bar{4}3m$	$P\bar{4}3m$, $F\bar{4}3m$, $I\bar{4}3m$, $P\bar{4}3n$, $F\bar{4}3c$, $I43d$
$4/m\bar{3}2/m$	$P4/m\bar{3}2/m$, $P4/n\bar{3}2/n$, $P4_2/m\bar{3}2/n$, $P4_2/n\bar{3}2/m$, $F4/m\bar{3}2/m$, $F4/m\bar{3}2/c$, $F4_1/d\bar{3}2/m$, $F4_1/d\bar{3}2/c$, $I4/m\bar{3}2/m$, $I4_1/a\bar{3}2/d$

*From *International Tables for Crystallography*, 1983, v. A, T. Hahn, ed: Space Group Symmetry. International Union of Crystallography, Reidel Publ. Co., Boston, USA.

lographers of what is known as *abbreviated symbols*. In our discussion of point groups we have always used the complete symbol such as 2/m2/m2/m and 4/m2/m2/m. The abbreviated symbolism for these two point groups would be *mmm* and 4/*mmm*, respectively (see also Table 6.2) The reason for the abbreviation for 2/m2/m2/m is the understanding that the three mutually perpendicular mirror planes intersect each other along twofold axes. Similar reasoning explains the abbreviation 4/*mmm*. In much of this text, for reasons of clarity, we will use only complete point group symbols. Abbreviated symbols are generally used in the literature for space group notation, however. For example, a mineral with point group 4/m$\bar{3}$2/m may have its space group reported as *Fm3m*, because the presence of fourfold and twofold rotation axes is implied. Such abbreviated symbols for space group notation will be used in the five chapters on systematic mineralogy.

Another system of space group notation may be encountered in the literature, especially in older texts. This is known as the *Schoenflies notation*. Because this notation does not logically follow from the Hermann–Mauguin point group symbols, it is not used here. References at the end of this chapter will enable you to find the Schoenflies system.

REFERENCES AND SELECTED READING

Bernal, I., W. C. Hamilton, and J. Ricci. 1972. *Symmetry: A stereoscopic guide for chemists.* San Francisco: W. H. Freeman & Company.

Bloss, F. D. 1994. *Crystallography and crystal chemistry: An introduction.* Reprint of the original text of 1971 by the Mineralogical Society of America, Washington, D. C.

Boisen, M. B., Jr., and G. V. Gibbs. 1990. Mathematical crystallography, *Review in Mineralogy* 15, Mineralogical Society of America, Washington, D. C.

Boldyrev, A. K. 1936. Are there 47 or 48 simple forms possible in crystals? *American Mineralogist* 21: 731–34.

Bragg, W. L., and G. F. Claringbull. 1965. *Crystal structures of minerals.* Ithaca, N. Y.: Cornell University Press.

Brisse, F. 1981. La Symétrie bidimensionelle et le Canada. *Canadian Mineralogist* 19: 217–24 (all illustrations are in color).

Buerger, M. J. 1978. *Elementary crystallography: An Introduction to the fundamental geometric features of crystals.* Rev. ed. Cambridge, Mass.: MIT Press.

———, 1971. *Introduction to crystal geometry.* New York: McGraw Hill.

Cotteril, R. 1985. *The Cambridge guide to the material world.* New York: Cambridge University Press.

Escher, M. C. 1968. *The graphic works of M. C. Escher.* New York: Hawthorn Books.

Goldschmidt, V. 1913–23. *Atlas der Kristallformen* (9 volumes and 9 atlases). Heidelberg, Universitätsbuchhandlung.

International Tables for Crystallography. 1983. Vol. A, *Space group symmetry.* Edited by T. Hahn. Boston: International Union of Crystallography, D. Reidel Publishing Company.

International Tables for X-ray Crystallography. 1969. Vol. 1, *Symmetry groups.* Edited by N. F. M. Henry and K. Lonsdale. Birmingham, England: International Union of Crystallography, Kynoch Press.

Hargittai, I., and Hargittai, M. 1994. *Symmetry: A unifying concept.* Bolinas, Calif.: Shelter Publications.

Klein, C. 1994. *Minerals and rocks: Exercises in crystallography, mineralogy, and hand specimen petrology.* Rev. ed. New York: Wiley.

MacGillavry, C. H. 1976. *Fantasy and Symmetry: The periodic drawings of M. C. Escher.* New York: Harry N. Abrams.

Phillips, F. C. 1971. *An introduction to crystallography.* New York: Wiley.

Rogers, A. F. 1935. A tabulation of crystal forms and discussion of form names, *American Mineralogist* 20: 838–51.

Sands, D. E. 1975. *Introduction to crystallography.* New York: W. A. Benjamin.

Schattschneider, D. 1990. *Visions of symmetry: Notebooks, periodic drawings, and related work of M. C. Escher.* New York: W. H. Freeman & Company.

Shubnikov, A. V., and V. A. Koptsik. 1974. *Symmetry in science and art.* New York: Plenum Press.

Stevens, P. S. 1991. *Handbook of regular patterns: An introduction to symmetry in two dimensions.* Cambridge, Mass.: MIT Press.

CHAPTER 6

SELECTED POINT GROUPS AND SPACE GROUPS

Having introduced the concepts of point group and space group in the preceding chapter, the question remains of how such concepts can be presented graphically for ease of visualization. A point group is the synopsis of the morphological symmetry content of a well-formed crystal, and this is expressed in Hermann–Mauguin notation. This chapter develops a useful projection technique (i.e., stereographic projection) that allows for the graphical representation of point group symmetry, together with information about the distribution of faces (specifically face poles) on a crystal. Subsequently, this chapter provides detailed discussions of nineteen of the thirty-two point groups (or crystal classes). Space group notation (as presented in space group symbols) contains symmetry elements (as in the Hermann–Mauguin notation of point groups), as well as translational elements (lattice type and symbols for screw axes and glide planes). This chapter ends with examples of space group projection onto a two-dimensional page using internationally accepted graphical symbols.

The prior chapter presented the 32 point groups (crystal classes) and their assignment to six crystal systems. The listing of point groups in Table 5.4 shows 19 of them in bold type. These are the point groups that contain the largest number of minerals (as well as synthetic compounds) or minerals of widespread geologic occurrence. These 19 point groups are treated systematically in this chapter after having introduced the graphical method for point group presentation known as *stereographic projection*. Such a projection allows for the superposition of information about the poles to crystal faces (face poles) and all symmetry elements.

As mentioned earlier, the word *point* (in point group) indicates that the symmetry operations leave one particular point, at least, of the projected pattern unmoved. This is the central point of the stereographic projection about which the symmetry elements and face poles are distributed. Before the discussion of the projection technique and the subsequent systematic treatment of 19 of the point groups, there is a brief description of the techniques of accu-

rate measurement of the angles between crystal faces.

MEASUREMENT OF CRYSTAL ANGLES

As stated by Steno's law, the angles between equivalent faces of the same substance are constant, regardless of whether the crystal is misshapen or ideal. The angles that are measured between the normals to the crystal faces characterize a crystal and should be measured carefully. A graphic representation of the distribution of the angles, and the normals to the crystal faces, may define the symmetry of the crystal and therefore its crystal class. Angles are measured with *goniometers*, and for accurate work a *reflecting goniometer* is used. A crystal mounted on this instrument can be rotated about a zone axis and will reflect a collimated beam of light from its faces through a telescope to the eye. The angle through which a crystal must be turned in order to throw successive beams of light from two adjacent faces into the telescope

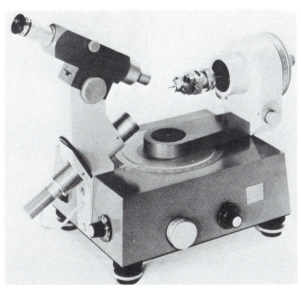

FIG. 6.1. Two-circle reflecting goniometer with crystal mounted at the right, a telescope at the upper left, and a light source at the lower left. (Courtesy of Huber Diffraktionstechnik, X-ray Crystallography Equipment, Rimstig, Germany.) Compare with Fig. 1.7d.

determines the angle between the faces. Figure 6.1 illustrates a modern, two-circle reflecting goniometer.

A simpler instrument used for approximate work and with larger crystals is known as a *contact goniometer* (Fig. 6.2a). In using a contact goniometer it is imperative that the plane determined by the two arms of the goniometer be exactly at right angles to the edge between the measured faces. It becomes clear from the construction in Fig. 6.2b that the angle between the poles to the crystal faces, the *internal*

angle, is measured. Thus, in Fig. 6.2a the angle should read as 40°, not as 140°.

CRYSTAL PROJECTIONS

A crystal projection is a means of representing the three-dimensional crystal on a two-dimensional plane surface. Different projections are used for different purposes, but each is made according to some definite rule so that the projection bears a known and reproducible relationship to the crystal. The crystal drawings in this book are known as *clinographic projections* and are a type of perspective drawing that yields a portrait-like picture of the crystal in two dimensions. This is the best means of conveying the appearance of a crystal and generally serves much better for the purpose than a photograph.

Spherical Projection

Because the actual size and shape of the different faces on a crystal are chiefly the result of accidents of growth, we wish to minimize this aspect of crystals in projecting, but at the same time emphasize the angular relation of the faces. We can do this using the *spherical projection*. We can envisage the construction of such a projection in the following way (see Fig. 6.3). Imagine a hollow model of a crystal containing a bright point source of light. Now let us place this model within a large hollow sphere of translucent material in such a way that the light source is at the center of the sphere. If we now make a pinhole in each of the faces so that the ray of light

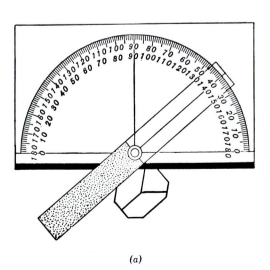

(a)

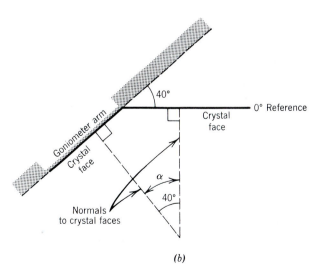

(b)

FIG. 6.2. Contact goniometer. (b) Schematic enlargement of (a) showing the measurement of the internal angle α.

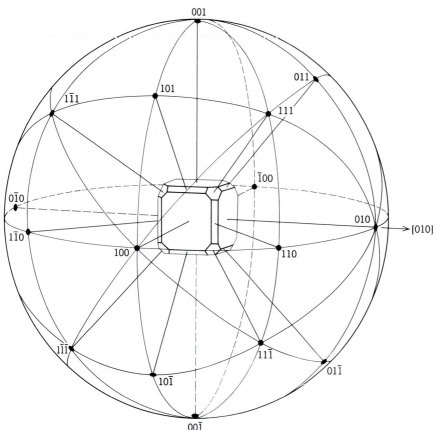

FIG. 6.3. Spherical projection of some isometric forms.

emerging from the hole is perpendicular to the face, these rays of light will fall on the inner surface of the sphere and make bright spots. The whole situation resembles a planetarium in which the crystal model with its internal light source and pinholes is the projector and the translucent sphere is the dome. If we now mark on the sphere the position of each spot of light, we may remove the model and have a permanent record of its faces. Each of the crystal faces is represented on the sphere by a point called the face *pole*. This is the spherical projection.

The position of each pole and thus its angular relationship to other poles can be described using angular coordinates on the sphere. This is done in a manner similar to the location of points on the Earth's surface by means of longitude and latitude. For example, the angular coordinates 74°00′ west longitude, 40°45′ north latitude, locate a point in New York City. This means that the angle, measured at the center of the Earth, between the plane of the equator and a line from the center of the Earth through that point in New York is exactly 40°45′; and the angle between the Greenwich meridian and the meridian passing through the point in New York, measured west in the plane of the equator, is exactly 74°00′. These relations are shown in Fig. 6.4.

A similar system may be used to locate the poles of faces on the spherical projection of a crystal. There is one major difference between locating points on a spherical projection and locating points on the Earth's surface. On the Earth, the latitude is measured in degrees north or south from the equator, whereas the angle used in the spherical projection is the colatitude, or *polar angle*, which is measured in degrees from the north pole. The north pole of a crystal projection is thus at a colatitude of 0°, the equator at 90°. The colatitude of New York City is 49°15′. This angle is designated in crystallography by the Greek letter ρ (rho).

The *crystal longitude* of the pole of a face on the spherical projection is measured, just as are longitudes on Earth, in degrees up to 180°, clockwise and counterclockwise from a starting meridian analogous to the Greenwich meridian of geography. To locate this starting meridian, the crystal is oriented in conventional manner with the (010) face to the right of the crystal. The meridian passing through the pole of this face is taken as zero. Thus, to determine the crystal longitude of any crystal face, a meridian is passed through the pole of that face, and the angle between it and the zero meridian is measured in the plane of the equator. This angle is designated by the Greek letter ϕ (phi).

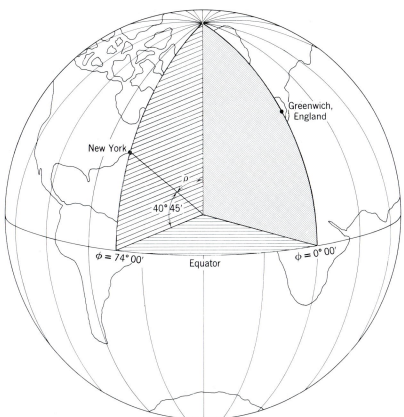

FIG. 6.4. Latitude and longitude of New York City.

If any plane is passed through a sphere, it will intersect the surface of the sphere in a circle. The circles of maximum diameter are those formed by planes passing through the center and having a diameter equal to the diameter of the sphere. These are called *great circles*. All other circles formed by passing planes through the sphere are *small circles*. The meridians on the Earth are great circles, as is the equator, whereas the parallels of latitude are small circles.

The spherical projection of a crystal brings out interesting zonal relationships, for the poles of all the faces in a zone lie along a great circle of the projection. In Fig. 6.3 (spherical projection) faces (001), (101), (100), (10$\bar{1}$), and (00$\bar{1}$) lie in a zone with the zone axis [010]. Because the great circle along which the poles of these faces lie passes through the north and south poles of the projection, it is called a *vertical great circle*. The zone axis is always perpendicular to the plane containing the face poles; thus all vertical circles have horizontal zone axes.

Stereographic Projection

In reducing the spherical projection of a crystal to a plane surface, it is important that the angular rela-

tions of the faces be preserved in such a way as to reveal the true symmetry. This can best be done with the *stereographic projection*.

The stereographic projection is a representation in a plane of half of the spherical projection, usually the northern hemisphere. The plane of the projection is the equatorial plane of the sphere, and the *primitive* circle (the circle outlining the projection) is the equator itself. If one were to view the poles of crystal faces located in the northern hemisphere of the spherical projection with the eye at the south pole, the intersection of the lines of sight with the equatorial plane would be corresponding points representing the poles on the stereographic projection. We can thus construct a stereographic projection by drawing lines from the south pole to the face poles in the northern hemisphere. The points corresponding to these poles on the stereographic projection are located where these lines intersect the equatorial plane. The relationship of the two projections is illustrated in Fig. 6.5. Various animations that illustrate the principles of stereographic projection are provided in module II of the CD-ROM.

In practice one plots poles directly on the stereographic projection. It is, therefore, necessary to determine stereographic distances in relation to angles

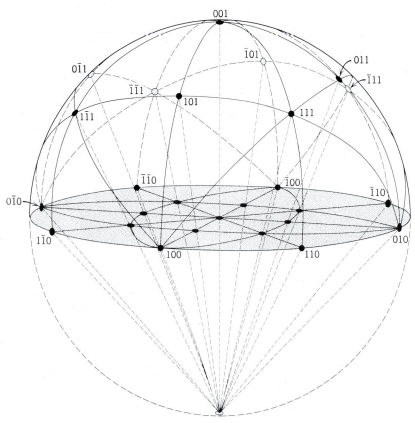

FIG. 6.5. Relation of spherical and stereo-graphic projections. (After E. E. Wahlstrom, 1951, *Optical crystallography*. New York: Wiley.)

of the spherical projection. Figure 6.6 shows a vertical section through the spherical projection of a crystal in the plane of the "zero meridian," that is, the plane containing the pole of (010). *N* and *S* are, respectively, the north and south poles of the sphere of

projection, *O* is the center of the projected crystal. Consider the face (011). *OD* is the perpendicular to the face (011), and *D* is the pole of this face on the spherical projection. The line from the south pole, *SD*, intersects the trace of the plane of the equator,

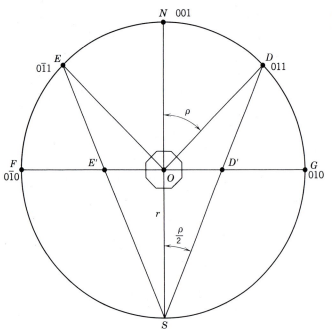

FIG. 6.6. Section through the sphere of projection showing relation of spherical to stereographic poles.

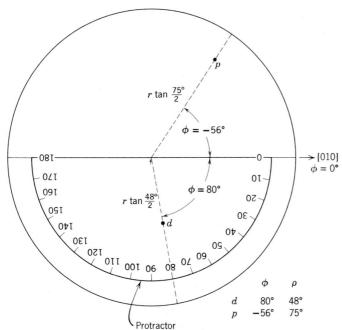

FIG. 6.7. Stereographic projection of crystal faces.

FG, in the point *D'*, the stereographic pole of (011). The angle *NOD* will be recognized as the angle ρ (rho). In order to plot *D'* directly on the stereographic projection, it is necessary to determine the distance *OD'* in terms of angle ρ. Because ΔSOD is isosceles, $\angle ODS = \angle OSD$.

$$\angle ODS + \angle OSD = \angle NOD = \rho$$

Therefore, $\angle OSD = \rho/2$. $OS = r$, the radius of the primitive of the projection.

$$\tan \rho/2 = OD'/r \quad \text{or} \quad OD' = r \tan \rho/2$$

To sum up, in order to find the stereographically projected distance from the center of the projection of the pole of any face, find the natural tangent of one-half of ρ of that face and multiply by the radius of the projection. The distance so obtained will be in whatever units are used to measure the radius of the primitive circle of the projection.

In addition to determining the distance a pole should lie from the center of the projection, it is also necessary to determine its *longitude*, or φ (phi) angle. Because the angle is measured in the plane of the equator, which is also the plane of the stereographic projection, it may be laid off directly on the primitive circle by means of a circular protractor. It is first necessary to fix the "zero meridian" by marking a point on the primitive circle to represent the pole of (010). A straight line drawn through this point and the center of the projection is the zero meridian. With the protractor edge along this line and the center point at the center of the projection, the φ angle

can be marked off. On a construction line from the center of the projection through this point lie all possible face poles having the specified φ angle. Positive φ angles are laid off clockwise from (010); negative φ angles are laid off counterclockwise, as shown in Fig. 6.7.

In order to plot the pole of the face having this given φ value, it is necessary to find the natural tangent of one-half of ρ, to multiply by the radius of the projection, and to lay off the resulting distance along the φ line. Although any projection radius may be chosen, one of 10 cm is usually used. This is large enough to give accuracy but not be unwieldy, and at the same time simplifies the calculation. With a 10 cm radius, it is only necessary to look up the natural tangent, move the decimal point one place to the right, and plot the result as centimeters from the center of the projection.

When the poles of crystal faces are plotted stereographically as explained above, their symmetry of arrangement should be apparent (see Fig. 6.8). We have seen that a great circle in the spherical projection is the locus of poles of faces lying in a crystal zone. When projected stereographically, vertical great circles become diameters of the projection; all other great circles project as circular arcs that subtend a diameter. The limiting case of such great circles is the primitive of the projection, which is a great circle common to both the spherical and stereographic projections. The poles of vertical crystal faces lie on the primitive and thus are projected without angular distortion.

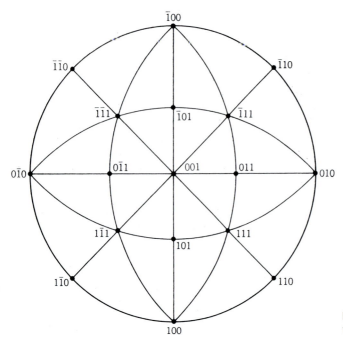

FIG. 6.8. Stereographic projection of some isometric crystal faces. Compare with Fig. 6.5.

Stereographic Net

Both the measurement and plotting of angles on the stereographic projection are greatly facilitated by means of a stereographic net. A net with radius of 5 cm is shown in Fig. 6.9 and a net with radius of 10 cm is given on an endpaper at the back of this text.

This type of net is also called the Wulff net, named after G. V. Wulff, Russian crystallographer (1863–1925). Both great and small circles are drawn on the net at intervals of 1° or 2°. In Fig. 6.9 the intervals are 2°. In practice the use of a net with a 10 cm radius, mounted on a stiff backing, is most convenient. If

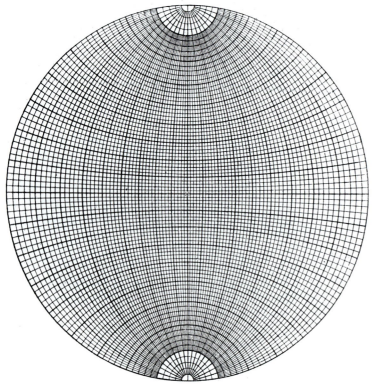

FIG. 6.9. Stereographic (Wulff) net. Radius equals 5 centimeters. See the back of this text for a stereographic net of 10 centimeter radius.

such a net is not available, it can be constructed by photocopying the net on the endpaper and mounting it on thin poster board. If a 5 cm net is considered adequate, Fig. 6.9 can be photocopied and backed by thin mounting board. The exact center of the net can now be pierced by a thumbtack from the back. The sharp point of the thumbtack, at the center of the stereographic net, will function as the pivot about which a sheet of tracing paper (or an $8\frac{1}{2} \times 11$ in. sheet of onionskin paper) can be rotated. Figure 6.10 shows this superposition of a semitransparent sheet on top of the stereographic net. The user should trace the outer (primitive) circle onto the overlay and should also mark the E–W and N–S directions. The eastern end of the E–W line should be marked $\varphi = 0°$, the southern end of the N–S line as $\varphi = 90°$, and the northern end of this same line as $\varphi = -90°$. Now the φ angles can be plotted directly along the primitive circle in clockwise (+) or counterclockwise (−) directions. The angle ρ can be located directly along the two vertical great circles (the N–S and E–W directions that intersect in the center of the projection); only along these two directions are graduations available for the direct plotting of angles.

The ρ angle of any face that projects at the center of the projection (at the location of the thumbtack) is equal to 0°. Any face that lies on the outside perimeter of the primitive circle has a ρ angle = 90°. Therefore, any ρ angle between 0° and 90° is plotted

outward from the center of the projection (away from the location of the thumbtack) along the E–W or N–S directions. If a combination of φ and ρ angles is such that the φ angle is neither 0° nor 90°, and the ρ is also neither 0° nor 90°, the transparent overlay must be rotated about the center until the φ direction of the specific plane coincides with either the N–S or E–W direction. Only then can the ρ angle of that plane be plotted. For example (see Fig. 6.10), if a plane has a $\varphi = 30°$ and a $\rho = 60°$, the location of the φ angle of the pole to the plane is marked by a short line at $\varphi = 30°$ on the circumference; subsequently this $\varphi = 30°$ location is rotated to be coincident with the E–W or N–S lines, and the $\rho = 60°$ is plotted directly using the graduations available on the underlying stereographic net. This is the location of the face pole, of a face with $\varphi = 30°$, $\rho = 60°$.

Instead of measuring φ and ρ angles the beginning student usually measures interfacial angles. Such interfacial angles can often be expressed as angles with respect to (010) or (001) of the crystal (if indeed the crystal shows the presence of these two forms). The measured interfacial angles, or the φ and ρ angles available in reference works in the literature, can be plotted easily with the aid of a stereographic net. We will illustrate the plotting of the face poles for two types of crystals: (1) for an orthorhombic crystal in which the three axial directions (a, b, and c) are at 90° to each other, and (2) for a mono-

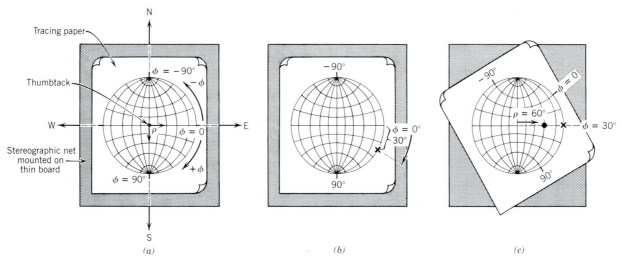

FIG. 6.10. *(a)* Illustration of the use of a stereographic net (mounted on thin board), pierced in its center by a thumbtack from the back, and overlain by a somewhat transparent paper (tracing paper or onionskin paper). The primitive circle, as well as the locations of $\varphi = 0°$, $\varphi = 90°$, and $\varphi = -90°$, must always be marked on the transparent overlay before any angles are plotted. *(b)* To project the pole of a plane with $\varphi = 30°$ and $\rho = 60°$, the angle $\varphi = 30°$ is plotted (*x*) on the primitive in a clockwise direction from $\varphi = 0°$. *(c)* The direction of $\varphi = 30°$ has been rotated to coincide with the E–W direction and the angle $\varphi = 60°$ can be measured directly along the vertical great circle. The black dot is the pole of the crystal face with $\varphi = 30°$, $\rho = 60°$.

clinic crystal where two of the three crystallographic directions are nonorthogonal (not 90°).

Projection of an Orthorhombic Crystal

Figure 6.11a is a crystal drawing of the orthorhombic mineral anglesite, $PbSO_4$, and Fig. 6.11b lists the φ and ρ angles of the crystal faces that lie in the positive quadrant. In Fig. 6.11c we illustrate the process of locating the poles of these faces on the stereographic projection. The starting point, as in all projections, is (010), face b. The pole of this face should be placed on the primitive at 0°. The interfacial angles $b \wedge n = 32\frac{1}{2}°$ and $b \wedge m = 52°$, can be measured and plotted as angles on the primitive. Face c is (001); it makes an angle of 90° with b and its pole should be placed at the center of the projection.

Face o is in a zone with faces c and b and thus has a φ angle of 0°. Its ρ angle, $c \wedge o = 52°$, can be measured directly and plotted along the vertical great circle. Face d lies in a vertical zone at 90° to the zone c, o, b. It thus has $\varphi = 90°$ and $c \wedge d = 39\frac{1}{4}°$ which can be plotted along the vertical circle of the net. The pole of face y cannot be plotted directly, but the angles $b \wedge y = 45°$ and $c \wedge y = 57°$ can be measured. To locate this pole the projection is rotated 90° (to the N–S direction) so that b lies along the radii of the small circles of the net, and a tracing of the 45° circle is made. This small circle is the locus of all poles 45° from b. The locus of $c \wedge y = 57°$ is a circle centered on the pole of 001, which is the center of the stereographic projection where the thumbtack pierces the paper. This circle is one of co-

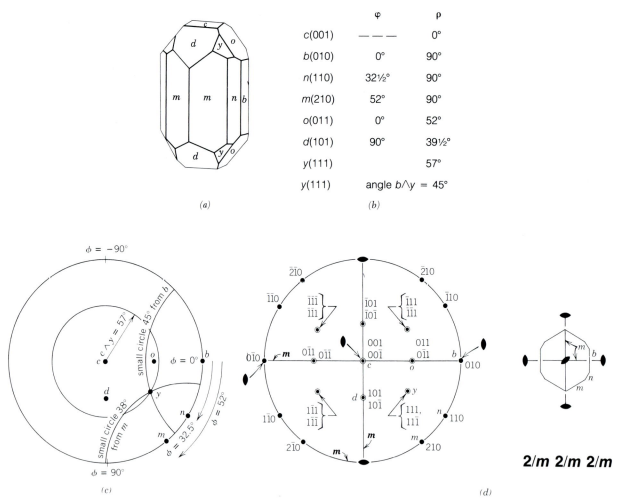

	φ	ρ
c(001)	– – –	0°
b(010)	0°	90°
n(110)	32½°	90°
m(210)	52°	90°
o(011)	0°	52°
d(101)	90°	39½°
y(111)		57°
y(111)	angle b∧y = 45°	

(a)

(b)

(c)

(d)

2/m 2/m 2/m

FIG. 6.11. *(a)* A crystal of the orthorhombic mineral anglesite with symmetry 2/m2/m2/m. *(b)* Listing of φ and ρ angles, as reported in the literature, and originally measured by reflecting goniometer for the faces of the crystal of anglesite shown in *(a)*. *(c)* A stereographic projection of all of the faces, in the positive quadrant, of the mineral anglesite. *(d)* The location of all of the face poles, with their Miller indices, as well as the symmetry elements consistent with the location of these face-poles. A horizontal cross section of the crystal in *(a)* is shown as well.

latitude (as explained earlier under spherical projection) and can be most easily drawn with a compass, after the spread of the compass has been measured along, for example, the E–W direction of the stereographic net at $\rho = 57°$. Where this colatitude circle intersects the small circle of $b \wedge y = 45°$ is the location of the pole of y.

To check this position, measure the angle $m \wedge y = 38°$. This is most easily done by locating the pole to m at the $\varphi = 90°$ position (southern end) of the stereographic net and reading the angular difference between m and y along one of the inclined great circles. Indeed, one can trace a small circle with m at $\varphi = 90°$, which will intersect the pole of y. Now that the pole y is located, its φ and ρ angles can be read directly from the stereographic net. The φ value will read $32\frac{1}{2}°$ and the ρ value will be $57°$.

In Fig. 6.11d we have plotted *all* the faces of the crystal of anglesite shown in Fig. 6.11b, not just those in the upper, positive quadrant of the stereographic projection as in Fig. 6.11c. The distribution of these face poles is consistent with the combination of symmetry elements $2/m2/m2/m$. There are three twofold axes at 90° to each other, as well as three major planes that intersect at the center of the stereographic projection. Two of these mirrors stand vertically as shown along the N–S and E–W directions. The third mirror is parallel to the page, in the plane of the primitive circle.

Projection of a Monoclinic Crystal

The location of the cystallographic axes in a monoclinic crystal is somewhat less straightforward than in the orthorhombic case, described above. In this text we locate the twofold rotation axis in monoclinic crystals in a horizontal, east–west direction. This orientation is referred to as the "second setting" (in the "first setting" the twofold axis is vertical). This means that a possible mirror plane (as in $2/m$), which is perpendicular to the twofold axis, stands vertically. If the monoclinic crystal shows an elongated habit, the direction of elongation is commonly chosen as parallel to the c axis. Further, if there is a prominent sloping plane, such as c in Fig. 6.12a, the a axis is taken as parallel to it and at right angles to b.

The monoclinic crystal shown in Fig. 6.12a is that of diopside, $CaMgSi_2O_6$. It is oriented as described above with the mirror plane vertical, the twofold axis parallel to the b axis, and horizontal (in an E–W direction). As shown in a cross-section in Fig. 6.12c, the c axis is parallel to the elongation of the crystal, and the a axis is parallel to the sloping c faces (001) and ($00\bar{1}$). In Fig. 6.12b there is a listing of the pertinent interfacial angles measured on the

crystal shown in Fig. 6.12a. The interfacial angle $c \wedge a$ ((001) $\wedge$ (100)) $= 74°$ is shown in the cross-section of the crystal, as is the one nonorthogonal angle (β) between the three crystallographic axes a, b, and c. In descriptions of monoclinic minerals in this book, the angle between the a and c crystallographic axes, known as the β angle, is taken as the angle that is not equal to 90°. In this crystal the β angle is 106°. Because of the presence of this nonorthogonal angle between the two axial directions of a and c, the location of one of these two axes will be different from what we have seen in the stereographic projection of, for example, the orthorhombic crystal in Fig. 6.11. In order to clarify this, we have drawn the upper part of the diopside crystal as surrounded by the sphere of the spherical projection in Fig. 6.12d. This figure shows that the pole to (001) is not coincident with the piercing point of the ($+$) end of the c axis. Similarly, the pole of (100) is not coincident with the piercing point of the ($+$) end of the a axis.

A cross-section parallel to the a–c plane of the crystal in the spherical projection is shown in Fig. 6.12e. The angular relations shown here are identical to those shown in Fig. 6.12c. In this cross section it is important to note the inclined position of the a axis. Only the ($-$) end of the a axis is above the trace of the stereographic projection (horizontal diameter of figure); however, the ($+$) end of the a axis lies below it. When the ends of the a axis are projected onto the stereographic projection, the ($-$) end will project as indicated in Fig. 6.12e. Its piercing point will be located 16° in from the primitive circle; its projected location is shown as an opaque dot in Fig. 6.12f. The ($+$) end of the a axis lies below the trace of the stereographic projection and must be projected upward. The point where it pierces the spherical projection is indicated as an open circle 16° in from the primitive. This is also shown in Fig. 6.12f. Once the a and c axes are located, the various face poles can be plotted. All those in the vertical zone (// c axis) can be located on the perimeter of the stereographic net; these are b, m, and a.

The interfacial angle $c \wedge a$ ((001) $\wedge$ (100) $= 74°$) implies that the pole of the (001) face is 16° away from the piercing point of the c axis, in a southward direction along the N–S vertical great circle. This is also shown in Fig. 6.12f. The only poles still to be located are for the two inclined faces u, which are symmetrically oriented with respect to the a–c plane (the mirror plane). The interfacial angle $c \wedge u = 34°$. This angle can be plotted with respect to c (001) after the c pole has been moved to the E–W vertical great circle. The 34° angle between c and u can now be plotted because these two faces lie in the same zone;

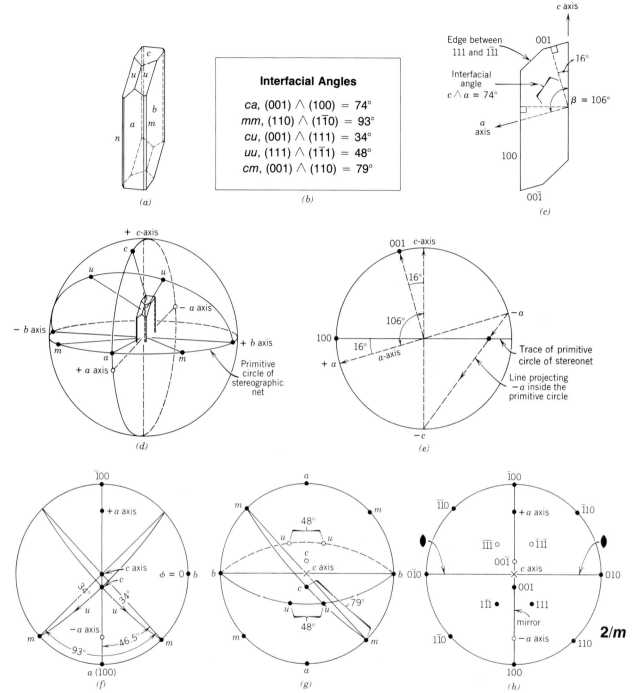

Interfacial Angles

ca, $(001) \wedge (100) = 74°$
mm, $(110) \wedge (1\bar{1}0) = 93°$
cu, $(001) \wedge (111) = 34°$
uu, $(111) \wedge (1\bar{1}1) = 48°$
cm, $(001) \wedge (110) = 79°$

(a)

(b)

(c)

(d)

(e)

(f)

(g)

2/m

(h)

FIG. 6.12. Plotting of face poles for a monoclinic crystal. *(a)* A crystal of the monoclinic mineral diopside. *(b)* Interfacial angles, as reported in the literature and originally measured by reflecting goniometer, for the faces of the monoclinic crystal of diopside, shown in *(a)*. *(c)* A vertical cross section through the diopside crystal showing the location of the *c* axis (parallel to the elongation of the crystal) and of the *a* axis (parallel to the (001) face). The β angle of 106° and the interfacial angle $c \wedge a = 74°$ (between (001) and (100)) are shown. *(d)* The upper half of the diopside crystal centered within the sphere of the spherical projection. Note the location of the piercing points of the +*c* and +*a* axes as well as the poles of *c* (001), *a* (100) and the two sets of faces marked *m* ((110) and (1$\bar{1}$0)) and *u* ((111) and 1$\bar{1}$1)). *(e)* A cross section through the sphere in *(d)*; this section con-

tains the *a* and *c* axes and the poles of (100) and (001). It should be noted that the (+) end of the *a* axis lies below the equatorial plane of the stereographic projection (the primitive circle) and that the (−) end of *a* axis lies above it. Therefore in projecting the piercing points of −*a* and +*a* from the sphere onto the stereographic projection, the "end points" of *a* move inward along the N–S direction (great circle) by 16° each. The −*a* end projects from above (as a solid dot) and the +*a* end projects from below (as an open circle). *(f)* Plotting of the poles of *b* ((010) at φ = 0°), *a* at φ = 90°, the two faces *m* with an interfacial angle = 93° (symmetrically on either side of *a*), and *c* (001) with an interfacial angle with *a* = 74°. The *c* axis is at the center of the projection and the − end of the *a* axis lies 16° inward from the pole to (100). *(Caption continued on next page)* that is, they both lie on the same great circle. In this case the great circle passing through *c* and *u* is inclined by 16° from the vertical. Two such inclined great circles are shown with solid lines in Fig. 6.12*f*. Once both poles of the two *u* faces in the upper hemisphere are plotted the *u* ∧ *u* interfacial angle (as given in Fig. 6.12*b*) can be verified. The faces *b*, *u*, *u*, *b* lie in the same zone. As such, the poles to these four faces lie on the same (inclined) great circle. On this circle (solid in Fig. 6.12*g*) the interfacial angle of 48° can be measured. The interfacial angle *c* ∧ *m* = 79° can be verified in a similar manner: the faces *m*, *c*, *m* lie in the same zone; as such their poles lie on the same inclined great circle.

In Fig. 6.12*h* we have plotted *all* of the face poles of the diopside crystal, and have assigned a Miller index to each pole. Furthermore, the symmetry elements compatible with the distribution of the face poles are also shown; this turns out to be 2/*m* with the twofold axis horizontal in an E–W direction and the mirror plane in a vertical position perpendicular to the twofold axis.

NINETEEN OF THE THIRTY-TWO POINT GROUPS

In the following section, 19 of the 32 point groups listed in Table 5.4 are described under the crystal systems in which they are grouped. These are the same 19 point groups for which animations are given in module II of the CD-ROM. The crystal systems will be treated in order of increasing symmetry. Within each system, however, the point groups will be discussed in order of decreasing symmetry. *The symmetry of each class is given by the Hermann–Mauguin notation but it is also shown by means of stereograms, giving projections of all the*

(Caption continued from previous page)
The interfacial angle *c* ∧ *u* ((001) ∧ (111)) is listed as 34° in *(b)*. With *c* and *u* on the same great circle, this 34° angle can be measured, symmetrically on either side of *c* (001). The projection shows that *m*, *u*, *c* lie on the same great circle and therefore in the same zone. *(g)* Here all the face poles of the crystal in *(a)* are plotted. Those marked *u* and shown as black dots are for face poles in the upper part of the sphere (now projected onto the primitive circle), and those shown as open circles are for poles projected upward from the lower part of the sphere. The interfacial angles *u* ∧ *u* ((111) ∧ (11$\bar{1}$)), and *c* ∧ *m* ((001) ∧ (110)) are shown along great circles as 48° and 79°, respectively (see listing in *b*). *(h)* The location of all of the face poles and their Miller indices and the symmetry consistent with the distribution of the face poles.

faces of the general forms. These are the forms from which the classes derive their names. In the stereograms it is necessary to show faces in the southern hemisphere as well as in the northern hemisphere in order to give completely the symmetry of the class. This is done by superimposing stereographic projections of the two hemispheres, with the poles in the northern hemisphere represented by solid points and those in the southern hemisphere by open circles. Thus, if two poles lie directly one above the other on the sphere, they will be represented by a solid point surrounded by a circle. A vertical face is represented by a single point on the primitive, for, although such a pole would appear on projections of both top and bottom of the crystal, it represents but one face.

Figure 6.13*a* is a drawing of a crystal with a horizontal symmetry plane. The stereogram of this crystal, Fig. 6.13*b*, consequently, shows all faces as solid points with circles to indicate corresponding faces at the top and bottom of the crystal. Figure 6.14*a* is a drawing of a crystal lacking a horizontal mirror plane. Its stereogram, Fig. 6.14*b*, has 12 solid points as poles of faces in the northern hemisphere and an additional 12, independent circles as poles of faces in the southern hemisphere.

The number of minerals (and synthetic crystalline compounds) falling within each of the six crystal systems is highly variable. A listing of 3837 crystalline compounds and their distribution among the 32 crystal classes is given in Bloss (1971, p. 28). For these compounds the distribution is as follows:

Triclinic	2%
Monoclinic	21
Orthorhombic	20
Tetragonal	12
Hexagonal	19
Isometric	26

Within these systems the largest number of mineral species is concentrated in the crystal class with the highest symmetry of each crystal system. Such classes, with the highest symmetry in a crystal system, are called the *holohedral* classes, from the Greek *holos* meaning "whole" or "complete." These are: 1 in triclinic, 2/*m* in monoclinic, 2/*m*2/*m*2/*m* in orthorhombic, 4/*m*2/*m*2/*m* in tetragonal, $\bar{3}$2/*m* in the hexagonal system (when based on a rhombohedral lattice), 6/*m*2/*m*2/*m* in the hexagonal system (when based on a hexagonal lattice), and 4/*m*$\bar{3}$2/*m* in the isometric system.

In the following systematic treatment of the 19 point groups, stereographic illustrations of the various crystal forms and their accompanying symmetry

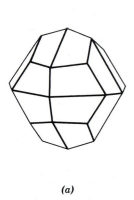

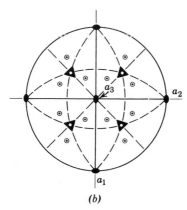

(a)   *(b)*

FIG. 6.13. Crystal with symmetry $2/m\bar{3}$. The stereogram shows a mirror plane at right angles to each of the twofold rotation axes, and four threefold axes of rotoinversion. The solid primitive circle denotes a horizontal mirror plane; faces at the bottom of the crystal lie directly below those at the top.

elements are extensively used. In these the same graphic symbols for symmetry elements and crystallographic directions are used throughout. A listing of these graphic symbols is given in Table 6.1. These are the standard symbols used in the *International Tables for X-ray Crystallography* (1969).

Throughout our discussion of the 19 point groups we will use the unabbreviated notation in describing their symmetry content. For example, we will use the symmetry representation $2/m2/m2/m$ for the orthorhombic dipyramidal class. Crystallographers commonly used abbreviated symbols, and would represent $2/m2/m2/m$ by simply *mmm*.

Similarly abbreviated Hermann–Mauguin notations are given in Table 6.2 for the six crystal classes in which it is possible to choose between a lengthier (and more descriptive) and an abbreviated notation. The reason for the use of the short forms is one of brevity. To the crystallographer *mmm* means three mutually perpendicular mirror planes that are perpendicular to the *a*, *b*, and *c* axes of the orthorhombic system, respectively. The lines of intersection between the three sets of intersecting

mirrors turn out to be axes of twofold rotation (see Figs. 5.18 and 6.23); in other words, *mmm* implies the same symmetry content as $2/m2/m2/m$. Similar reasoning holds for the elimination of the twofold rotational symbols in the short forms $4/mmm$ and $6/mmm$.

In the case of $\bar{3}2/m$ the combination of the threefold rotoinversion axis and three vertical mirror planes (see Fig. 6.38) generates three twofold axes that bisect the angles between the mirrors. As such the twofold axes in $\bar{3}2/m$ are the inherent result of the symmetry implied in the short form $\bar{3}m$. The short forms for the isometric symbols are less obvious. The short symbol *m3m* implies (1) a mirror plane parallel to the cube face (for the first *m* in *m3m*), (2) a threefold axis that runs from corner to corner in the cube (see Fig. 6.43), and (3) a symmetry plane that runs diagonally across a cube, from edge to edge (referring to the last *m* in *m3m*). The interaction of these three symmetry elements will produce all of the symmetry content of the more descriptive form $4/m\bar{3}2/m$. If the diagonal mirror plane were absent from *m3m*, the overall symmetry

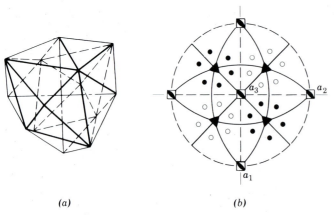

(a)   *(b)*

FIG. 6.14. Crystal with symmetry $\bar{4}3m$. Stereogram shows three fourfold axes of rotoinversion, six mirror planes, and four threefold rotation axes. The broken primitive circle indicates the lack of a horizontal mirror plane and faces at the top of the crystal do not lie above those at the bottom.

TABLE 6.1   Graphic Symbols Used in Stereographic Illustrations

Axes of rotation	Written symbol	Graphic symbol
1-fold	1	none
2-fold	2	⬬
3-fold	3	▲
4-fold	4	■
6-fold	6	⬢

Axes of rotoinversion		
1-fold	$\bar{1}$	(equivalent to a center of symmetry or i, inversion)*
2-fold	$\bar{2}$	(equivalent to a mirror, m)
3-fold	$\bar{3}$	◮
4-fold	$\bar{4}$	◈
6-fold	$\bar{6}$	⬣

Center of symmetry	i	none (see footnote)*

Mirror planes	m (horizontal, parallel to the plane of the page)	

solid line along primitive circle

m (vertical, perpendicular to the plane of the page)

m absent, in both horizontal and vertical positions

Both solid lines

Crystallographic directions	a, b, c	- - - - a, b, or c broken lines labeled with the appropriate letter

If m includes a crystallographic direction, a solid line is used—with a, b, or c

*A center of symmetry is shown in writing by the letter i for inversion, which is equivalent to $\bar{1}$. If a symmetry center occurs at the center of the sphere of projection, its presence is not shown by a symbol on the stereogram, but it can be detected from the arrangement of poles of equivalent faces (see also Fig. 5.20).

TABLE 6.2   Listing of the Equivalent Long and Abbreviated Forms of Six Crystal Classes

System	Long Form	Short Form
Orthorhombic	$2/m2/m2/m$	mmm
Tetragonal	$4/m2/m2/m$	$4/mmm$
Hexagonal	$\bar{3}2/m$	$\bar{3}m$
	$6/m2/m2/m$	$6/mmm$
Isometric	$2/m\bar{3}$	$m3$
	$4/m\bar{3}2/m$	$m3m$

would read only $m3$. These two symmetry elements combine to generate the symmetry of the nonabbreviated form $2/m\bar{3}$.

In the following systematic treatment of the six crystal systems and the 19 point groups, the introductory paragraphs to each crystal system will address (among other matters) the relationship of the symmetry elements (as expressed by the Herman–Mauguin notation) and the orientation of the crystallographic axes. Table 5.5 provides the student with a synopsis of the conventions in such orientations (see also Fig. 5.27).

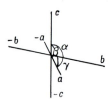

FIG. 6.15. Triclinic crystal axes.

The geometrically perfect model is considered in each of the form descriptions on the following pages. Keep in mind that in nature this ideal is rarely obtained, and that crystals not only are frequently malformed but also are usually bounded by a combination of forms. In the laboratory study of crystal forms wooden models may be used that similarly represent highly idealized crystals: their symmetry-related faces have been made equal in size and shape by the manufacturer of the models. Such perfection aids the student in the study of symmetry content and the recognition of forms.

Throughout this book the drawings of the external forms of crystals are in a standardized orientation with the c axis vertical, the b axis east-west, and the a axis toward the observer. Because of this standardized orientation the position of the axes has been omitted in most crystal drawings.

Triclinic System

Crystallographic Axes. In the triclinic system the crystal forms are referred to three crystallographic axes of unequal lengths that make oblique angles with each other (Fig. 6.15). The three rules to follow in orienting a triclinic crystal and thus in determining the position of the crystallographic axes are as follows:
1. The most pronounced zone should be vertical. The axis of this zone then becomes c.
2. {001} should slope forward and to the right.
3. Two forms in the vertical zone should be selected: one as {100} the other as {010}.
The directions of the a and b axes are determined by the intersections of {010} and {100}, respectively, with {001}. The b axis should be longer than the a axis. In reporting on

the crystallography of a new triclinic mineral or one that has not been recorded in the literature, the convention should be followed that $c < a < b$. The relative lengths of the three axes and the angles between them can be established only by X-ray diffraction techniques. The angles between the positive ends of b and c, c and a, and a and b are designated respectively as α, β, and γ (see Fig. 6.15).

$\bar{1}$

Symmetry—i. The symmetry consists of a onefold axis of rotoinversion, which is equivalent to the center of symmetry, or inversion (i). Figure 6.16 illustrates a triclinic pinacoid and its stereogram. This class is referred to as the *pinacoidal class* after its general {hkl} form.

Forms. All the forms are *pinacoids* and thus consist of two identical and parallel faces. Once a crystal is oriented, the Miller indices of the pinacoid establish its position.
1. {100}, {010}, and {001} *Pinacoids.* Each of these pinacoids intersects one crystallographic axis and is parallel to the other two. The front or a pinacoid, {100}, intersects the a axis and is parallel to the other two; the side or b pinacoid, {010}, intersects only the b axis; the basal or c pinacoid, {001}, intersects only the c axis.
2. {0kl}, {h0l}, and {hk0} *Pinacoids.* The {0kl} form is parallel to a and can be positive {0kl}, or negative {0k̄l}; the {h0l} form is parallel to b, {h0l} positive, and {h̄0l} negative; the {hk0} form is parallel to c, {hk0} positive, and {hk̄0} negative.
3. {hkl} *Pinacoids.* {hkl} positive right, {h̄kl} positive left, {hk̄l} negative right, {h̄k̄l} negative left. Each of these two-faced forms can exist independently of the others.

Various combinations of the above pinacoids are illustrated in Fig. 6.17.

Among the minerals that crystallize in $\bar{1}$ are:

amblygonite	polyhalite
chalcanthite	rhodonite
microcline	turquoise
pectolite	ulexite
plagioclase feldspars	wollastonite

Of the above-mentioned minerals microcline, rhodonite, and chalcanthite are fairly common as well-formed crystals (see Fig. 6.18).

$\bar{1}$

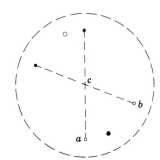

FIG. 6.16. Triclinic pinacoid and stereogram.

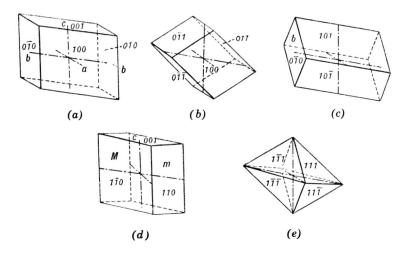

FIG. 6.17. Triclinic pinacoids. *(a)* Front {100}, side {010}, and base {001}. *(b)* {011} positive, {0$\bar{1}$1} negative. *(c)* {101} positive, {10$\bar{1}$} negative. *(d)* {110} positive, {1$\bar{1}$0} negative. *(e)* Four different forms.

Monoclinic System

Crystallographic Axes. Monoclinic crystals are referred to three axes of unequal lengths. The only restrictions in the angular relations are that $a \wedge b$ (γ) and $c \wedge b$ (α) = 90°. For most crystals the angle between $+a$ and $+c$ is greater than 90°, but in rare instances it too may equal 90° (e.g., staurolite). In such cases the monoclinic symmetry is not apparent from the morphology and the crystal is said to be pseudoorthorhombic. The twofold rotation axis or the direction perpendicular to the mirror plane is usually taken as the b axis; the a axis is inclined downward toward the front; and c is vertical. This orientation, known as the "second setting," is traditional for mineralogists.[1]

Figure 6.19 represents the crystallographic axes of the monoclinic mineral orthoclase, with β = 116°01'. Although the direction of the b axis is fixed by symmetry, the directions that serve as the a and c axes are matters of choice and depend on crystal habit and cleavage. If crystals show an elongated development (prismatic habit) parallel to a direction in the a–c plane, that direction often serves as the c axis. Further, if there is a prominent sloping plane (or planes), such as planes c or r in the drawings of

[1]Some crystallographers orient monoclinic crystals according to the "first setting" in which the twofold axis or the normal to the mirror plane is chosen as the c axis instead of the b axis. Here, all monoclinic crystals are referred to the "second setting."

Fig. 6.22 the a axis may be taken as parallel to these. It is quite possible that there may be two, or even more, choices that are equally good, but in the description of a new mineral it is conventional to orient the crystals such that $c < a$.

Cleavage is also an important factor in orienting a monoclinic crystal. If there is good pinacoidal cleavage parallel to the b axis, as in orthoclase, it is usually taken as the basal cleavage. If there are two equivalent cleavage directions, as in the amphiboles and pyroxenes, they are usually taken to be vertical prismatic cleavages.

2/m

Symmetry—i, 1A_2, 1m. The axis of twofold rotation is chosen as the b axis, and the a and c axes lie in the mirror plane, which is perpendicular to the b axis (Fig. 6.20a). The stereogram in Fig. 6.20b shows the symmetry of an {hkl} prism, the general form. Because the a axis slopes down and to the front, it does not lie in the equatorial plane; the positive end intersects the sphere of projection in the southern hemisphere (for further discussion see "Projection of a Monoclinic Crystal," page 249). This class is referred to as the *prismatic class*, because the general form {hkl} is a *prism*.

Forms. There are only two types of forms in this monoclinic class: *pinacoids* and *prisms*.
1. *Pinacoids.* (See Fig. 6.21). Front or a pinacoid, {100}, side or b pinacoid, {010} and basal or c pinacoid, {001}. There are also {$h0l$} and {$\bar{h}0l$} pinacoids; these

FIG. 6.18. Triclinic crystals. *(a)* Rhodonite. *(b)* Chalcanthite.

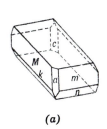

(a)

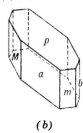

(b)

FIG. 6.19. Monoclinic crystal axes.

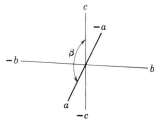

2/m

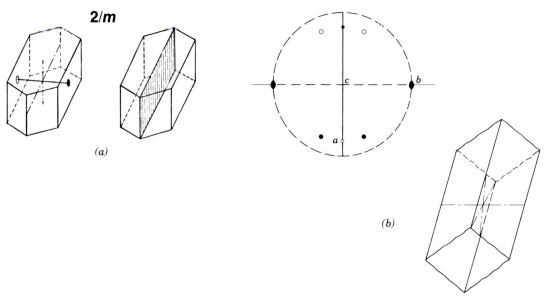

(a)

(b)

FIG. 6.20.   *(a)* Symmetry elements for 2/*m*. *(b)* Monoclinic prism {*hkl*} and its stereogram.

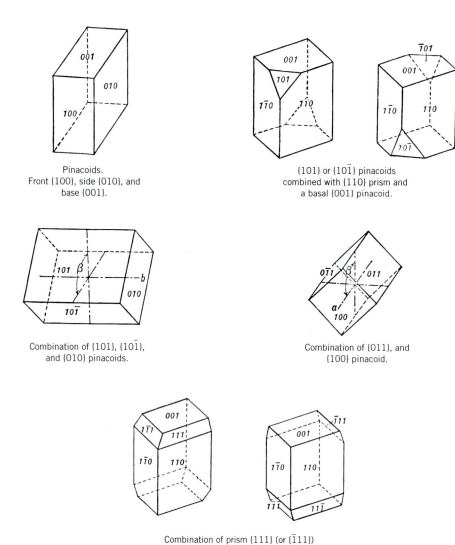

FIG. 6.21.   Commonly developed forms and form combinations in 2/*m*.

Pinacoids.
Front {100}, side {010}, and
base {001}.

{101} or {10$\bar{1}$} pinacoids
combined with {110} prism and
a basal {001} pinacoid.

Combination of {101}, {10$\bar{1}$},
and {010} pinacoids.

Combination of {011}, and
{100} pinacoid.

Combination of prism {111} (or {$\bar{1}$11})
and prism {110} and basal pinacoid {001}.

two pinacoids are independent of each other and the presence of one does not necessitate the presence of the other (see Figs. 6.21 and 6.22).

2. *Prisms.* The four-faced prism {*hkl*} is the general form, but {0*kl*} and {*hk*0} are prisms as well. The {0*kl*} prism intersects the *b* and *c* axes and is parallel to the *a* axis. The general form can occur as two independent prisms {*hkl*} and {*h̄kl*}. Prisms are illustrated in Figs. 6.21 and 6.22.

The only form in this crystal class that is fixed by the choice of the *b* axis as the twofold rotation axis is the {010} pinacoid. The other forms may vary with the choice of the *a* and *c* axes. For instance, the {100} pinacoid, {001} pinacoid, and {*h*0*l*} pinacoids may be converted into each other by a rotation about the *b* axis. In the same manner the prisms can be changed from one position to another.

Many minerals crystallize in the monoclinic, prismatic class. Some of the most common are:

azurite	kaolinite
borax	malachite
chlorite	mica (group)
clinoamphibole (group)	orpiment
clinopyroxene (group)	orthoclase
datolite	realgar
epidote	spodumene
gypsum	talc
heulandite	titanite

Orthorhombic System

Crystallographic Axes. The crystal forms in the orthorhombic system are referred to three crystallographic axes of unequal length that make angles of 90° with each other (see Fig. 6.23*a*). The relative lengths of the axes must be determined for each orthorhombic mineral. In orienting

an orthorhombic crystal, the convention is to set the crystal such that $c < a < b$. In the past, however, this convention has not necessarily been observed, and it is customary to confirm to the orientation given in the literature. One finds, therefore, that any one of the three axes may have been chosen as *c*. The longer of the other two is then taken as *b* and the shorter as *a*.

The decision as to which of the three axes should be chosen as the vertical axis rests largely on the crystal habit of the mineral. If its crystals commonly show an elongation in one direction, this direction is usually chosen as the *c* axis (see topaz crystals in Fig. 6.24). If, on the other hand, the crystals show a prominent pinacoid and therefore are tabular, this pinacoid is usually taken as {001} with *c* normal to it (see barite and celestite crystals in Fig. 6.24). Cleavage also aids in orienting orthorhombic crystals. If, as in topaz, there is one pinacoidal cleavage, it is taken as {001}. If, as in barite, there are two equivalent cleavage directions, they are set vertical and their intersection edges determine *c*. After the orientation has been determined, the length of the axis chosen as *b* is taken as unity, and the relative lengths of *a* and *c* are given in terms of it. Figure 6.23*a* represents the crystallographic axes for the orthorhombic system.

In the Hermann–Mauguin notation for the orthorhombic system, the symbols refer to the symmetry elements in the order *a*, *b*, *c*. For example, in the class *mm*2, the *a* and *b* axes lie in vertical mirror planes and *c* is an axis of twofold rotation.

2/*m*2/*m*2/*m*

Symmetry—*i*, 3A₂, 3*m*. The three axes of twofold rotation coincide with the three crystallographic axes; perpendicular to each of the axes is a mirror plane (Fig. 6.23*b*). The general form, rhombic dipyramid {*hkl*}, and its

FIG. 6.22. Monoclinic crystals with 2/*m* symmetry. Forms: *a* {100}, *b* {010}, *c* {001}, *m* {110}, *p* {111}, *o* {2̄21}, *r* {011}, *e* {120}, *x* {1̄01}, *y* {2̄01}, *z* {130}.

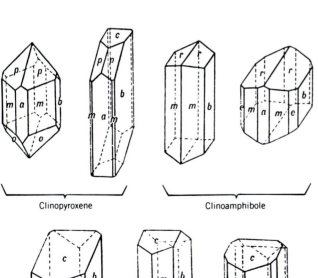

Clinopyroxene   Clinoamphibole

Orthoclase

2/m 2/m 2/m

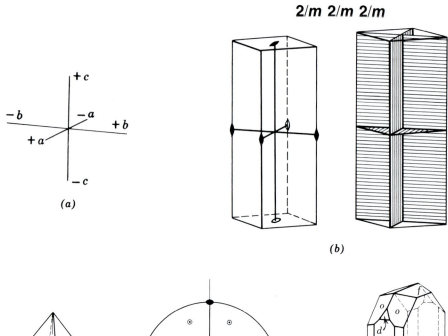

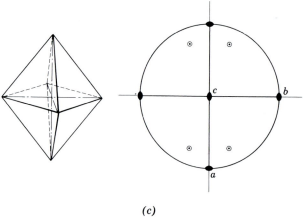

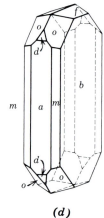

FIG. 6.23. *(a)* Orthorhombic crystal axes. *(b)* Rotational axes and mirror planes in 2/m2/m2/m. *(c)* The rhombic dipyramid {*hkl*} and its stereogram. *(d)* A crystal that is part of the orthopyroxene series showing the rhombic dipyramid, *o*.

stereogram are shown in Fig. 6.23c. This class is named the *rhombic-dipyramidal class.*

Forms. There are three types of forms in this class: pinacoids, prisms, and dipyramids.
1. *Pinacoid.* The pinacoid, consisting of two parallel faces, can occur in three different crystallographic orientations. These are: {100}, front or *a* pinacoid, which intersects *a* and is parallel to *b* and *c*; {010}, side or *b* pinacoid, which intersects *b* and is parallel to *a* and *c*; and {001}, basal or *c* pinacoid, which intersects *c* and is parallel to *a* and *b* (see Fig. 6.24).
2. *Rhombic Prisms.* Rhombic prisms consist of four faces that are parallel to one axis and intersect the other two. In the {0*kl*} prism the faces are parallel to *a* but intersect *b* and *c*; in the {*h0l*} prism the faces are parallel to *b* but intersect *a* and *c*; and in the {*hk*0} prism, the faces are parallel to *c* but intersect *a* and *b*. Examples of {011}, {101}, and {110} prisms are given in Fig. 6.24. Because all prisms intersect two axes and parallel the third, one prism will be transformed into another by a different choice of axes.
3. *Rhombic Dipyramid* {*hkl*}. A rhombic dipyramid has eight triangular faces, each of which intersects all three

crystallographic axes. In Fig. 6.24 is an illustration of the unit dipyramid {111}.

Combinations. Practically all orthorhombic crystals consist of combinations of two or more forms. Characteristic combinations for various minerals are given in Fig. 6.24.

There are many mineral representatives in this class. Among the more common are the following:

andalusite	goethite
anthophyllite (and other	marcasite
orthorhombic amphiboles)	olivine
aragonite (group)	sillimanite
barite (group)	stibnite
brookite	sulfur
chrysoberyl	topaz
columbite	
cordierite	
enstatite (and other	
orthopyroxenes)	

mm2

Symmetry—2m, 1A₂. The twofold rotation axis coincides with the *c* crystallographic axis. Two mirror planes at

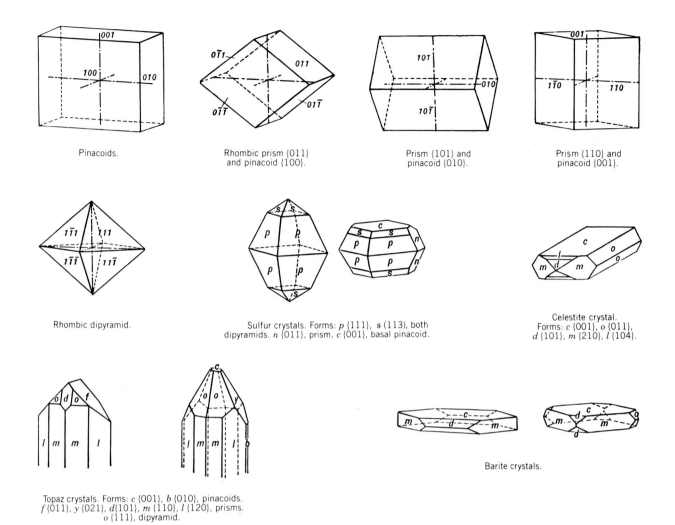

Pinacoids.

Rhombic prism {011} and pinacoid {100}.

Prism {101} and pinacoid {010}.

Prism {110} and pinacoid {001}.

Rhombic dipyramid.

Sulfur crystals. Forms: p {111}, s (113), both dipyramids. n {011}, prism, c {001}, basal pinacoid.

Celestite crystal. Forms: c {001}, o {011}, d {101}, m {210}, l {104}.

Barite crystals.

Topaz crystals. Forms: c {001}, b {010}, pinacoids. f {011}, y {021}, d{101}, m {110}, l {120}, prisms. o {111}, dipyramid.

FIG. 6.24.   Commonly developed forms and form combinations in 2/m2/m2/m.

right angles to each other intersect in this axis. A rhombic pyramid {hkl} and its stereogram are shown in Fig. 6.25. This class is referred to as the *rhombic-pyramidal class*.

Forms.   Because of the absence of a horizontal mirror, the forms at the top of the crystal are different from those at the bottom. The rhombic dipyramid thus becomes two rhombic pyramids, {hkl} at the top, and {hk$\bar{l}$} at the bottom. Similarly, {0kl} and {h0l} prisms do not exist, but each occurs as two domes. These domes have indices {0kl} and {0k$\bar{l}$} as well as {h0l} and {h0$\bar{l}$}. There are also pedions, {001} and {00$\bar{1}$}, and {hk0} prisms.

Only a few minerals crystallize in this class—the most common representatives are hemimorphite, $Zn_4Si_2O_7$-$(OH)_2 \cdot H_2O$ (Fig. 6.25) and bertrandite, $Be_4Si_2O_7(OH)_2$.

Tetragonal System

Crystallographic Axes.   The forms of the tetragonal system are referred to three crystallographic axes that make right angles with each other. The two horizontal axes, *a*, are equal in length and interchangeable, but the vertical

axis, *c*, is of a different length. Figure 6.26*a* represents the crystallographic axes for the tetragonal mineral zircon with *c* less than *a*. Figure 6.26*b* represents the crystallographic axes of a mineral with *c* greater than *a*. Figure 6.26 also shows the proper orientation of the crystallographic axes and the method of their notation.

In the Hermann–Mauguin notation of the symmetry elements in the tetragonal system, the first part of the symbol (consisting of 4 or $\bar{4}$ refers to the *c* axis, whereas the second or third parts refer to the axial (*a₁* and *a₂*) and diagonal symmetry elements, respectively.

4/m2/m2/m

Symmetry—*i*, 1A₄, 4A₂, 5*m*.   The vertical crystallographic axis is one of fourfold rotation. There are four horizontal axes of twofold symmetry, two of which coincide with the crystallographic axes (*a₁* and *a₂*) and the others at 45° to them. There are five mirror planes perpendicular to the symmetry axes. One of the horizontal symmetry axes lies in each of the vertical mirror planes. The position of the axes and mirrors is shown in Fig. 6.27. The general

mm2

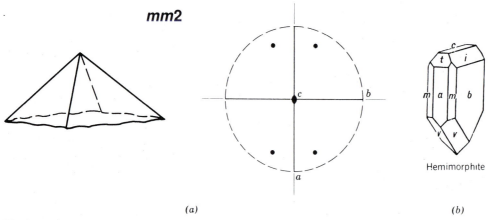

FIG. 6.25. *(a)* Rhombic pyramid {*hkl*} and its stereogram. *(b)* A hemimorphite crystal showing a rhombic pyramid {*hkl*}, *v*, at the lower end.

form {*hkl*}, the ditetragonal dipyramid, is illustrated in Fig. 6.27 with its stereogram. This class is known as the *ditetragonal-dipyramidal class*.

Forms

1. *Basal Pinacoid* {001}. A form composed of two parallel faces perpendicular to the fourfold axis and thus parallel with the horizontal *m*. It is shown in combination with various prisms in Fig. 6.28.
2. *Tetragonal Prisms* {010} and {110}. The {010} prism consists of four faces that are perpendicular to the twofold axes of the first kind and are therefore parallel with the mirrors in the first 2/*m* in the symbol. The {110} prism has its faces perpendicular to the twofold axes of the second kind and thus is parallel with the mirrors in the second 2/*m* in the symbol.
3. *Ditetragonal Prism* {*hk*0}. Consists of eight rectangular vertical faces, each of which intersects the two horizontal crystallographic axes unequally. There are various ditetragonal prisms, depending on their differing relations to the horizontal axes. A common form represented in Fig. 6.28 has indices {120}.
4. *Tetragonal Dipyramids* {*hhl*} and {0*kl*}. The {*hhl*} dipyramid has eight isosceles triangular faces, each of which intersects all three crystallographic axes, with

FIG. 6.26. Tetragonal crystal axes.

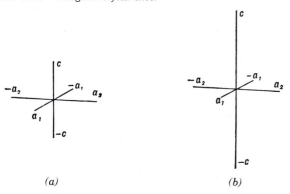

FIG. 6.27.   *(a)* Symmetry axes and planes for 4/*m*2/*m*2/*m*. *(b)* The ditetragonal dipyramid {*hkl*} and its stereogram.

4/m 2/m 2/m

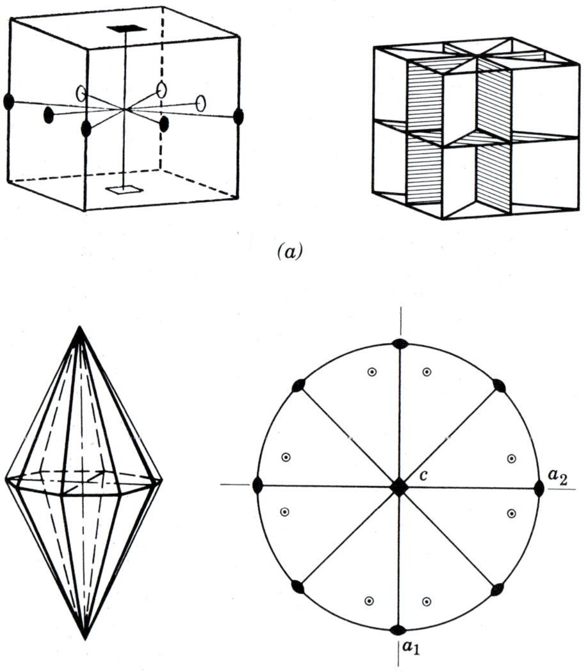

equal intercepts upon the two horizontal axes. There are various such dipyramids, depending on the inclination of their faces to *c*. The unit dipyramid {111} (Fig. 6.28), which intersects all the axes at their unit lengths, is most common. Indices of other similar dipyramids are {221}, {331}, {112}, {113}, and so on, or, in general, {*hhl*}. The {0*kl*} dipyramid is composed of eight isosceles triangular faces, each of which intersects one horizontal axis and the vertical axis and is parallel to the

Tetragonal Prisms

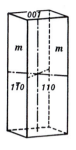

Prism {110} and
pinacoid {001}.

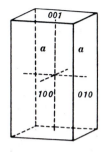

Prism {010} and
pinacoid {001}.

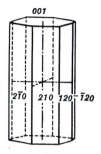

Prism {120} and
pinacoid {001}.

Tetragonal Dipyramids

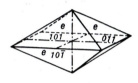

{111}

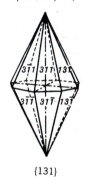

{011}

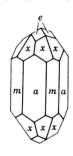

{131}

Tetragonal Crystals

Forms: e {011}, u {021}, c {001}, a {010}, m {110}, x {211}.
In this illustration the Miller indices for the forms are based on
the knowledge of the orientation of the unit cell. If the forms
were indexed on the basis of morphology e would be {111}, a
{110}, and u {221}.

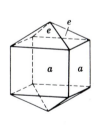

Zircon

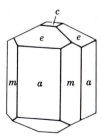

Vesuvianite

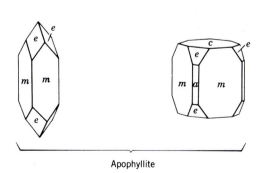

Apophyllite

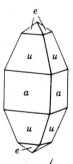

FIG. 6.28. Commonly developed
forms and form combinations in
4/m2/m2/m.

second horizontal axis. There are various such dipyramids with different intersections upon the vertical axis. The most common is the unit dipyramid {011} (Fig. 6.28). Miller indices for other similar dipyramids are {021}, {031}, {012}, {013}, or, in general {0kl}.

5. *Ditetragonal Dipyramid* {hkl}. Composed of 16 triangular faces; each face intersects all three of the crystallographic axes, cutting the two horizontal axes at different lengths. There are various ditetragonal dipyramids, depending on the different intersections on the crystallographic axes. One of the most common is the dipyramid {131}, shown in Fig. 6.28.

Several common minerals crystallize in 4/m2/m2/m. Major representatives are rutile (TiO_2), anatase (TiO_2), cassiterite (SnO_2), apophyllite ($KCa_4Si_8O_{20}(OH,F) \cdot 8H_2O$), zircon ($ZrSiO_4$), and vesuvianite ($Ca_{10}Mg_2Al_4(SiO_4)_5(Si_2O_7)_2(OH_4)$).

Tetragonal Combinations. Characteristic combinations of forms in this class, as found on crystals of different minerals, are shown in Fig. 6.28.

$\overline{4}2m$

Symmetry—$1\overline{A}_4$, $2A_2$, $2m$. The fourfold rotoinversion axis is chosen as the c axis and the axes of twofold rotation as the two a axes. At 45° to the a axes are two vertical mirror planes intersecting in the vertical axis (see Fig. 6.29a). Figure 6.29b illustrates a tetragonal scalenohedron {hkl} and its stereogram. This class is known as the *tetragonal-scalenohedral class.*

Forms

1. *Tetragonal Disphenoids* {hhl} *positive,* {h$\overline{h}$l} *negative* are the only important forms in this class. They consist of four isosceles triangular faces that intersect all three of the crystallographic axes, with equal intercepts on the two horizontal axes. There may be different disphenoids, depending on their varying intersections with the vertical axis. Two different disphenoids and a combination of a positive and a negative disphenoid are shown in Fig. 6.29c.

    The tetragonal disphenoid differs from the tetrahedron in the isometric system in that its vertical crystallographic axis is not of the same length as the horizontal axes. The only common mineral in this class is chalcopyrite, crystals of which ordinarily show only the disphenoid {112}. This disphenoid closely resembles a tetrahedron, and it requires accurate measurements to prove its tetragonal character.

2. *Tetragonal Scalenohedron* {hkl}. This form, Fig. 6.29b, if it were to occur by itself, is bounded by eight similar scalene triangles. It is a rare form and observed only in combination with others. Other forms that may be present are pinacoid, tetragonal prisms, ditetragonal prisms, and tetragonal dipyramids.

    Chalcopyrite ($CuFeS_2$) and stannite (Cu_2FeSnS_4) are the only common minerals that crystallize in this class.

4/m

Symmetry—i, $1A_4$, $1m$. There is only the vertical fourfold rotation axis with a symmetry plane perpendicular to it. Figure 6.30 illustrates a tetragonal dipyramid and its

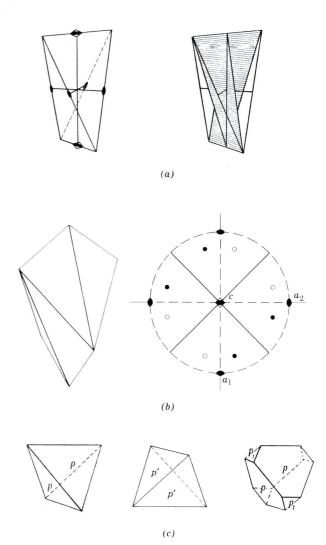

FIG. 6.29. (a) Symmetry axes and planes for $\overline{4}2m$. (b) The tetragonal scalenohedron {hkl} and its stereogram. (c) Tetragonal disphenoids {hhl} and {h$\overline{h}$l} and combination of the two types.

stereogram. This class is known as the *tetragonal-dipyramidal class* after the general form {hkl}.

Forms. The tetragonal dipyramid, {hkl}, is an eight-faced form having four upper faces directly above four lower faces. This form by itself appears to have higher symmetry, and it must be in combination with other forms to reveal the absence of vertical symmetry planes. The pinacoid {001} and tetragonal prisms {hk0} may be present. The tetragonal prism {hk0} is equivalent to the four alternate faces of the ditetragonal prism and is present in those classes of the tetragonal system that have no vertical mirror planes or horizontal twofold rotation axes.

Mineral representatives in this class are scheelite ($CaWO_4$), powellite ($CaMoO_4$), fergusonite ($YNbO_4$), and members of the scapolite series ($Na_4Al_3Si_9O_{24}Cl$ to $Ca_4Al_6Si_6O_{24}CO_3$). Figure 6.30 illustrates a crystal of fergusonite in which the tetragonal dipyramid z reveals the true symmetry of this class.

4/m

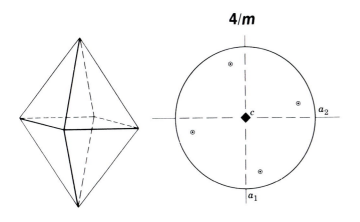

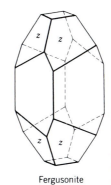

FIG. 6.30. Tetragonal dipyramid {*hkl*} and its stereogram. By itself this form appears to have higher symmetry. In the crystal of fergusonite the presence of this form (*z*) reveals the true symmetry, 4/*m*.

Fergusonite

Hexagonal System

All of the crystal classes in the hexagonal system can be based on a hexagonal lattice, whether their symbol begins with 6, $\bar{6}$, 3, or $\bar{3}$. The five beginning with 3 or $\bar{3}$ can also be based on a rhombohedral lattice.

Crystallographic Axes. The forms of the hexagonal system are referred to four crystallographic axes as proposed by Bravais. Three of these, designated a_1, a_2, and a_3, lie in the horizontal plane and are of equal length with angles of 120° between the positive ends; the fourth axis, *c*, is vertical. When properly oriented, one horizontal crystallographic axis, a_2, is left to right, and the other two make 120° angles on either side of it (Fig. 6.31*a*). The positive end of a_1 is to the front and left, the positive end of a_2 is to the right, and the positive end of a_3 is to the back and left. Figure 6.31*b* shows the four axes in clinographic projection. In stating the indices for any face of a hexagonal crystal, four numbers (the Bravais–Miller symbol) must be given. The numbers expressing the reciprocals of the intercepts of a face on the axes are given in the order a_1, a_2, a_3, *c*. Therefore, (11$\bar{2}$1), which represents the intercepts $3a_1$, $3a_2$, $-3/2a_3$, $3c$, refers to a face that cuts the positive ends of the a_1 and a_2 axes at twice the distance it cuts the negative end of the a_3 axis; it cuts the *c* axis at the same relative number of units (3) as it cuts the a_1 and a_2 axes. The general Bravais–Miller form symbol is {*hkil*}. The third digit of the index is the sum of the first two times −1; or, stated another way, $h + k + i = 0$ (see also Fig. 5.32 and page 199).

FIG. 6.31. Hexagonal crystal axes.

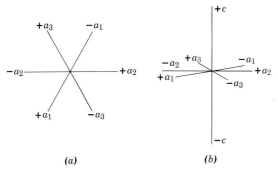

(a)   (b)

In the Hermann–Mauguin notation the first number refers to the principal axis of symmetry coincident with *c*. The second and third symbols, if present, refer respectively to symmetry elements parallel with and perpendicular to the crystallographic axes a_1, a_2, and a_3.

6/*m*2/*m*2/*m*

Symmetry—*i*, 1A_6, 6A_2, 7*m*. The vertical axis is an axis of sixfold rotation. There are six horizontal axes of twofold rotation, three of which coincide with the crystallographic axes (a_1, a_2, and a_3); the other three lie midway between them. There are seven mirror planes, each perpendicular to one of the symmetry axes. See Fig. 6.32 for

FIG. 6.32. (a) Symmetry axes and planes for 6/*m*2/*m*2/*m*. (b) The dihexagonal dipyramid {*hkīl*} and its stereogram.

6/*m* 2/*m* 2/*m*

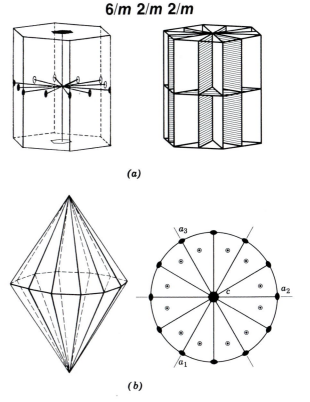

(a)

(b)

the location of the symmetry elements and a drawing of the general form, the dihexagonal dipyramid, as well as its stereogram. This class is known as the *dihexagonal-dipyramidal class*, after the general form.

Forms

1. *Pinacoid* {0001}. Composed of two parallel faces, perpendicular to the sixfold axis, and thus parallel with the horizontal *m*. It is commonly referred to as a basal pinacoid. It is shown in combination with various prisms in Fig. 6.33.

2. *Hexagonal Prisms* {$10\bar{1}0$} and {$11\bar{2}0$}. The {$10\bar{1}0$} prism consists of six vertical faces, each of which intersects two of the horizontal crystallographic axes equally and is parallel to the third. The faces of this prism are parallel to the twofold axes of the first kind. The {$11\bar{2}0$} prism also has six vertical faces, but each of these intersects two of the horizontal axes equally and the intermediate horizontal axis at one-half this distance. These two types of hexagonal prisms are geometrically identical forms; the distinction between them is only in orientation. See Fig. 6.33 for illustrations.

3. *Dihexagonal Prism* {$hki0$}. This form consists of 12 vertical faces, each of which intersects all three of the horizontal crystallographic axes at different lengths. There are various dihexagonal prisms, depending on their intercepts with the horizontal axes. A common dihexagonal prism with indices {$21\bar{3}0$} is shown in Fig. 6.33.

4. *Hexagonal Dipyramids* {$h0\bar{h}l$} and {$hh\overline{2h}l$}. The {$h0\bar{h}l$} hexagonal dipyramid consists of 12 isosceles triangular faces, each of which intersects two horizontal crystallographic axes equally, is parallel to the third, and intersects the vertical axis. Various hexagonal dipyramids are possible, depending on the inclination of the faces to the *c* axis. The unit form has the indices {$10\bar{1}1$} (see Fig. 6.33). The {$hh\overline{2h}l$} dipyramid is also composed of 12 isosceles triangular faces. Each face intersects two of the horizontal axes equally and the third (the intermediate horizontal axis) at one-half this distance; each face also intersects the vertical axis. Various such dipyramids are possible, depending on the inclination of the faces to *c*. A common form has the indices {$11\bar{2}2$} (see Fig. 6.33).

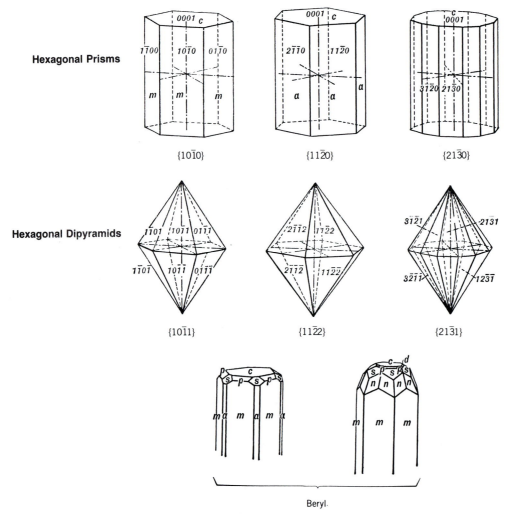

Hexagonal Prisms

{$10\bar{1}0$}   {$11\bar{2}0$}   {$21\bar{3}0$}

Hexagonal Dipyramids

{$10\bar{1}1$}   {$11\bar{2}2$}   {$21\bar{3}1$}

Beryl.

FIG. 6.33. Commonly developed forms and form combinations in 6/*m*2/*m*2/*m*.

6mm

 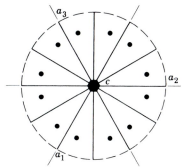 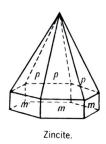

Zincite.

FIG. 6.34. Dihexagonal pyramid {hkīl} and its stereogram. The same pyramid (p) as shown in zincite crystals.

5. *Dihexagonal Dipyramid* {hkil}. This is composed of 24 triangular faces. Each face is a scalene triangle that intersects all three of the horizontal axes differently and also intersects the vertical axis. A common form, {21̄31̄}, as well as various combinations of forms in this class are shown in Fig. 6.33.

Beryl, $Be_3Al_2Si_6O_{18}$, affords the best example of a mineral representative in this class. Other minerals are molybdenite, MoS_2, pyrrhotite, $Fe_{1-x}S$, and nickeline (synonymous with niccolite), NiAs.

6mm

Symmetry—1A₆, 6m. The sixfold rotation axis is chosen as the c axis, and six vertical mirror planes intersect in this axis. Figure 6.34 shows a dihexagonal pyramid and its stereogram. This class is known as the *dihexagonal-pyramidal class* after the {hkil} general form.

Forms. The forms of this class are similar to those of class 6/m2/m2/m, but inasmuch as a horizontal mirror plane is lacking, different forms appear at the top and bottom of the crystal. The dihexagonal pyramid is thus two forms: {hkil} upper and {hkīl} lower. The hexagonal-pyramidal forms are {h0h̄l} upper and {h0h̄l̄} lower; and {hh2̄hl} upper and {hh2̄hl̄} lower. The pinacoid cannot exist here, but instead there are two pedions {0001} and {0001̄}. Hexagonal prisms and the dihexagonal prism may be present.

Wurtzite, ZnS, greenockite, CdS, and zincite, ZnO, are the most common mineral representatives in this class. Figure 6.34 shows a zincite crystal with a hexagonal prism terminated above by a hexagonal pyramid and below by a pedion.

622

Symmetry—1A₆, 6A₂. The symmetry axes are the same as those in class 6/m2/m2/m (see Fig. 6.32a), but mirror planes and the center of symmetry are lacking. This class is referred to as the *hexagonal-trapezohedral class*, after its general form {hkil}.

Forms. The hexagonal trapezohedrons {hkil} right and {ihkl} left are enantiomorphic forms, each with 12 trapezium-shaped faces (see Fig. 6.35). Other forms that may be present are the pinacoid, hexagonal prisms, dipyramids, and dihexagonal prisms.

High quartz, SiO_2, and kalsilite, $KAlSiO_4$, are the only mineral representatives in this class.

6/m

Symmetry—i, 1A₆, 1m. There is only the vertical sixfold rotation axis with a symmetry plane perpendicular to it. Figure 6.36 illustrates a hexagonal dipyramid {hkīl}, and its stereogram. This class is known as the *hexagonal-dipyramidal class*.

Forms. The general forms of this class are the *hexagonal dipyramids*, {hkil} positive, {khīl} negative. These forms consist of 12 faces, six above and six below, which correspond in position to one-half the faces of a dihexagonal dipyramid. Pinacoid and prisms may also be present.

This class has as its chief mineral representatives the minerals of the apatite group, $Ca_5(PO_4)_3(OH, F, Cl)$. The

622

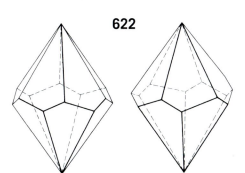 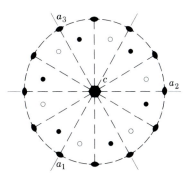

Left   Right

FIG. 6.35. Enantiomorphic, left- and right-handed hexagonal trapezohedrons, {ihk̄l} and {hkīl} respectively. A stereogram of the left-handed form.

6/m

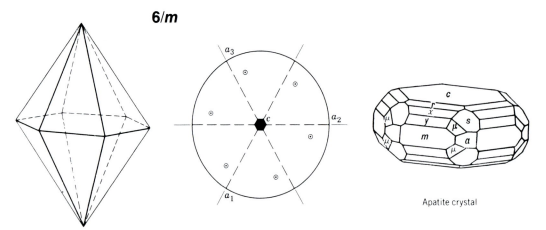

FIG. 6.36. Hexagonal dipyramid {$hk\bar{i}l$} and its stereogram. This form by itself appears to have higher symmetry, but in combination with other forms it reveals its low symmetry content. The form (μ) is the hexagonal dipyramid in the apatite crystal.

dipyramid revealing the symmetry of the class is rarely seen but is illustrated as face μ in Fig. 6.36.

$\bar{3}2/m$

Symmetry—$1\bar{A}_3$, $3A_2$, $3m$. The threefold rotoinversion axis is the vertical axis and the three twofold rotation

axes are the three horizontal crystallographic axes (a_1, a_2, and a_3). Three vertical mirror planes bisect the angles between the horizontal axes (see Fig. 6.37a). Figure 6.37c illustrates the general form {$hk\bar{i}l$}, a hexagonal scalenohedron and its stereogram. This class is known as the *hexagonal-scalenohedral class*.

FIG. 6.37. (a) Symmetry axes and planes for $\bar{3}2/m$. (b) Relationship between the rhombohedron {$h0\bar{h}l$} and a hexagonal dipyramid {$hk\bar{i}l$}. (c) Hexagonal scalenohedron {$hkil$} and its stereogram. (d) Relationship between the scalenohedron and a dihexagonal dipyramid.

$\bar{3}2/m$

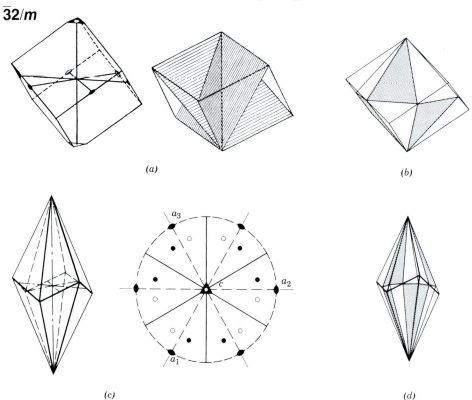

Forms

1. *Rhombohedron, {h0h̄l} positive, {0hh̄l} negative.* The rhombohedron is a form consisting of six rhomb-shaped faces, which correspond in their positions to the alternate faces of a hexagonal dipyramid {h0h̄l}. The relation of these two forms to each other is shown in Fig. 6.37*b*. The rhombohedron may also be thought of as a cube deformed in the direction of one of the axes of threefold rotoinversion. The deformation may appear either as an elongation along the rotoinversion axis, producing an acute solid angle, or compression along the rotoinversion axis, producing an obtuse solid angle. Depending on the angle, the rhombohedron is known as acute or obtuse.

Depending on the orientation, the rhombohedron may be positive or negative (Fig. 6.38). There are vari-

ous rhombohedrons that differ from each other in the inclination of their faces to the *c* axis. The index symbol of the unit positive rhombohedron is {101̄1} and of the unit negative rhombohedron {011̄1}.

2. *Scalenohedron, {hkı̄l} positive, {khı̄l} negative.* This form consists of 12 scalene triangular faces corresponding in position to alternate *pairs* of faces of a dihexagonal dipyramid (Fig. 6.37*d*). The scalenohedron is differentiated from the dipyramid by the zigzag appearance of the middle edges.

There are many different scalenohedrons; most frequently seen is {21̄31}, a common form on calcite; indicated by *v* in crystals in Fig. 6.38.

The rhombohedron and scalenohedron of this class may combine with forms found in classes of higher hexagonal

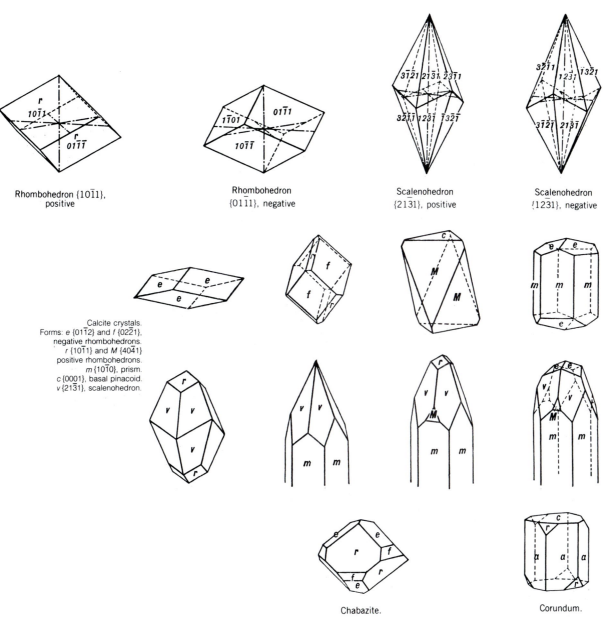

Rhombohedron {101̄1}, positive

Rhombohedron {011̄1}, negative

Scalenohedron {21̄31}, positive

Scalenohedron {1̄231}, negative

Calcite crystals.
Forms: *e* {011̄2} and *f* {022̄1}, negative rhombohedrons.
r {101̄1} and *M* {404̄1} positive rhombohedrons.
m {101̄0}, prism.
c {0001}, basal pinacoid.
v {21̄31}, scalenohedron.

Chabazite.

Corundum.

FIG. 6.38. Commonly developed forms and form combinations in 3̄2/*m*.

symmetry. Thus, they may be in combination with hexagonal prisms, dihexagonal prisms, hexagonal dipyramid, and pinacoid (see calcite, chabazite, and corundum crystals in Fig. 6.38).

Several common minerals crystallize in this class. Chief among them is calcite ($CaCO_3$) and the other members of the calcite group. Other minerals are corundum (Al_2O_3), hematite (Fe_2O_3), brucite ($Mg(OH)_2$), nitratite ($NaNO_3$, synonymous with soda niter), arsenic (As), millerite (NiS), antimony (Sb), and bismuth (Bi).

3*m*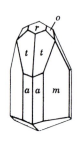

Symmetry—$1A_3$, 3*m*. The vertical axis is a threefold rotation axis, and three mirror planes intersect in this axis. In the Hermann–Mauguin notation of this class the 3 refers to the vertical *c* axis and the *m* refers to three planes per-

pendicular to the three horizontal axes a_1, a_2, and a_3. These three mirror planes intersect in the vertical threefold axis. A ditrigonal pyramid {$hki\bar{l}$} and its stereogram are shown in Fig. 6.39. This class is known as the *ditrigonal pyramidal class*, after the general form.

Forms. The forms are similar to those of the class $\bar{3}2/m$ but with only half the number of faces. Because of the lack of twofold rotation axes, the faces at the top of the crystals belong to different forms from those at the bottom. There are four possible *ditrigonal pyramids*, with indices {$hki\bar{l}$}, {$khi\bar{l}$}, {$hki\bar{l}$}, and {$khi\bar{l}$}. Other forms that may be present are pedions, hexagonal prisms and pyramids, trigonal pyramids, trigonal prisms, and ditrigonal prisms. There are four possible trigonal pyramids with indices {$h0\bar{h}l$}, {$0h\bar{h}l$}, {$0h\bar{h}l$}, and {$h0\bar{h}l$}.

Tourmaline (Fig. 6.39) is the most common mineral crystallizing in this class, but in addition there are members of the proustite (Ag_3AsS_3)–pyrargyrite (Ag_3SbS_3) series and alunite ($KAl_3(SO_4)_2(OH)_6$).

FIG. 6.39. *(a)* Ditrigonal pyramid {$hki\bar{l}$} and its stereogram. In the conventional orientation of the symmetry elements of this crystal class the mirror planes are perpendicular to a_1, a_2, and a_3 (see footnote to Table 5.5). *(b)* Tourmaline crystals showing 3*m* symmetry.

3*m*

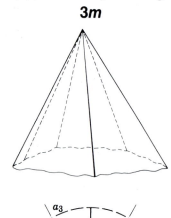

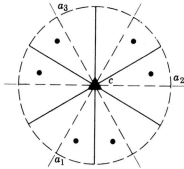

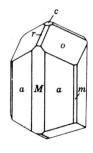

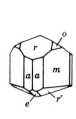

32

Symmetry—$1A_3$, $3A_2$. The four axial directions are occupied by the rotation axes. The vertical crystallographic axis is an axis of threefold rotation and the three horizontal crystallographic axes are axes of twofold rotation. The symmetry axes are similar to those in class $\bar{3}2/m$, but planes of symmetry are lacking. Figure 6.40*a* shows a positive left and a positive right trigonal trapezohedron and a stereogram for the positive right form. The class is referred to as the *trigonal-trapezohedral class*, after the general {$hki\bar{l}$} form.

Forms. There are four trigonal trapezohedrons, each made up of six trapezium-shaped faces. Their Miller indices are {$hki\bar{l}$}, {$\bar{i}kh l$}, {$khi\bar{l}$}, and {$ki h\bar{l}$}. These forms can be grouped into two enantiomorphic pairs, each with a right and left form (see Fig. 6.40*a*). Other forms that may be present are pinacoid, trigonal prisms, hexagonal prism, ditrigonal prisms, and rhombohedrons.

Low-temperature quartz is the most common mineral crystallizing in this class, but only rarely can faces of the trigonal trapezohedron be observed. When this form is present, the crystals can be distinguished as right-handed or left-handed (Fig. 6.40*b*), depending on whether, with a prism face fronting the observer, the trigonal trapezohedral faces, *x*, truncate the edges between prism and the top rhombohedron faces at the right or at the left. The faces marked *s* are trigonal dipyramids.

Cinnabar, HgS, and the rare mineral berlinite, $AlPO_4$, also crystallize in this class.

$\bar{3}$

Symmetry—$1\bar{A}_3$. The vertical axis is a threefold axis of rotoinversion. This is equivalent to a threefold rotation axis and a center of symmetry. Figure 6.41 illustrates a rhombohedron and its stereogram. This class is known as the *rhombohedral class* after the general {$hki\bar{l}$} form.

Forms. As general forms of this class there are four different *rhombohedrons*, each corresponding to six faces of the dihexagonal-dipyramid. If one of these appeared

32

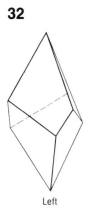
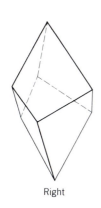
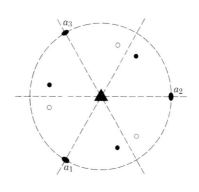

Left   Right

(a)

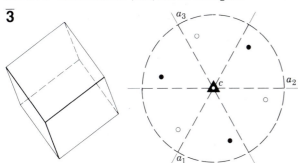

Left-handed   Right-handed

(b)

FIG. 6.40. (a) Trigonal trapezohedron {hkīl}, positive left- and positive right-handed forms, with a stereogram of the positive right form. (b) Left-handed and right-handed quartz crystals. The trigonal trapezohedral faces are marked by x.

alone on a crystal, it would have the morphological symmetry of class $\bar{3}2/m$. It is only in combination with other forms that its true symmetry becomes apparent. The {0001} pinacoid and hexagonal prisms may be present.

Dolomite, $CaMg(CO_3)_2$, is the most common mineral crystallizing in this class; other representatives are ilmenite, $FeTiO_3$, willemite, Zn_2SiO_4, and phenakite, Be_2SiO_4.

Isometric System

Crystallographic Axes. The crystal forms of classes of the isometric system are referred to three axes of equal length that make right angles with each other. Because the axes are identical, they are interchangeable, and all are

FIG. 6.41. Rhombohedron {hkīl} and its stereogram.

$\bar{3}$

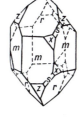

designated by the letter a. When properly oriented, one axis, a_1, is horizontal and oriented front to back, a_2 is horizontal and right to left, and a_3 is vertical (see Fig. 6.42).

In the Hermann–Mauguin notation the first number (4, $\bar{4}$, or 2) refers to the three crystallographic axes a_1, a_2, and a_3. If the number is 4 or $\bar{4}$, it means that there are three fourfold axes of rotation or inversion coincident with the three crystallographic axes. If it is 2, there are three twofold axes, coincident with the three crystallographic axes. The second number ($\bar{3}$ or 3) refers to the four diagonal directions of threefold symmetry, between the corners of a cube (see Fig. 6.43a). The third number or symbol (if present) refers to symmetry elements between the six pairs of opposing cube edges (see Fig. 6.43a). If it is 2 (as in 432), there are six twofold axes perpendicular to the edges; if it is m (as in $\bar{4}3m$), there are six mirror planes; if it is 2/m (as in $4/m\bar{3}2/m$), there are six twofold axes with mirrors perpendicular to them.

FIG. 6.42. Isometric crystal axes.

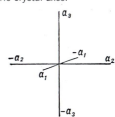

4/m32/m

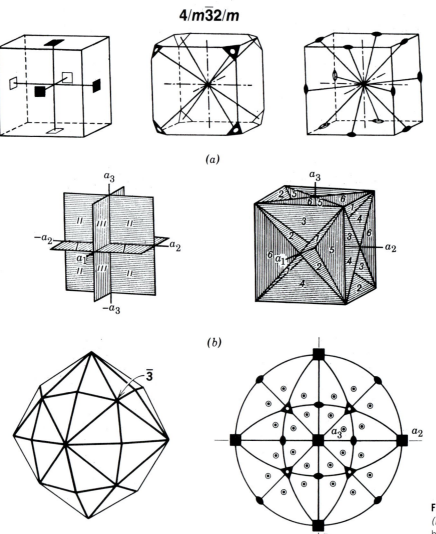

(a)

(b)

(c)

FIG. 6.43. *(a)* Symmetry axes and *(b)* planes for 4/m32/m, and *(c)* the hexoctahedron {*hkl*} and its stereogram. The location of one $\overline{3}$ is noted in the model.

Form Symbols. Although the symbol of any face of a crystal form might be used as the form symbol, it is conventional, when possible, to use one in which h, k, and l are all positive. In forms that have two or more faces with h, k, and l positive, the rule followed is to take the form symbol with $h < k < l$. For example, the form with a face symbol (123) also has faces with symbols (132), (213), (231), (312), and (321). Following the rule, {123} would be taken as the form symbol, for there $h < k < l$.

In giving the φ and ρ angles of a form, it is customary to give those for only one face; the others can be determined by knowing the symmetry. The face for which these coordinates are given is the one with the smallest φ and ρ values. This is the face of the form in which $h < k < l$.

4/m32/m

Symmetry—3A₄, 4A₃, 6A₂, 9m. The three crystallographic axes are axes of fourfold rotation. There are also four diagonal axes of threefold rotoinversion; these axes emerge in the middle of each of the octants formed by the intersection of the crystallographic axes. Further, there are six diagonal directions of twofold rotation, each of which bisects one of the angles between two of the crystallographic axes. There is also a center of symmetry because $\overline{3}$ is equivalent to $3 + i$. These symmetry elements are shown in Fig. 6.43a.

This class has nine mirror planes. Three of them are known as the axial planes, because each includes two crystallographic axes, and six are called diagonal planes, because each bisects the angle between two of the axial planes (see Fig. 6.43b). This combination of symmetry elements defines the highest symmetry possible in crystals. Every crystal form and every combination of forms that belongs to this class must show its complete symmetry. It is important to remember that in this class the three crystallographic axes are axes of fourfold rotation. Thus one can easily locate the crystallographic axes and properly orient the crystal.

The hexoctahedron, the general form from which this *hexoctahedral class* derives its name, is shown in Fig. 6.43c with a stereogram.

Forms. Illustrations of the most common forms and combinations of forms in this class are given in Figs. 6.44 and 6.45. Fig. 5.38 illustrates the 15 isometric forms.

1. *Cube* {001}. The cube is composed of six square faces that make 90° angles with each other. Each face intersects one of the crystallographic axes and is parallel to the other two.
2. *Octahedron* {111}. The octahedron is composed of eight equilateral triangular faces, each of which intersects all three of the crystallographic axes equally. When in combination with a cube, the octahedron can be recognized by its eight similar faces, each of which is equally inclined to the three crystallographic axes. It should be noted that the faces of an octahedron truncate symmetrically the corners of a cube.
3. *Dodecahedron* {011}. The dodecahedron is composed of 12 rhomb-shaped faces. Each face intersects two of the crystallographic axes equally and is parallel to the

third. Figure 6.44 shows a simple dodecahedron as well as combinations of a dodecahedron and a cube, of a dodecahedron and an octahedron, and of a cube, octahedron, and dodecahedron. Note that the faces of a dodecahedron truncate the edges of both the cube and the octahedron.

4. *Tetrahexahedron* {0kl}. The tetrahexahedron is composed of 24 isosceles triangular faces, each of which intersects one axis at unity and the second at some multiple and is parallel to the third. There are a number of tetrahexahedrons that differ from each other with respect to the inclination of their faces. The most common is {012}. The indices of other forms are {013}, {014}, {023}, and so on, or in general, {0kl}. It is helpful to note that the tetrahexahedron, as its name indicates, resembles a cube each of whose faces have been raised to accommodate four others. Figure 6.44 shows a simple tetrahexahedron as well as a cube with its edges beveled by the faces of a tetrahexahedron.
5. *Trapezohedron* {hhl}. The trapezohedron is composed of 24 trapezium-shaped faces, each of which intersects

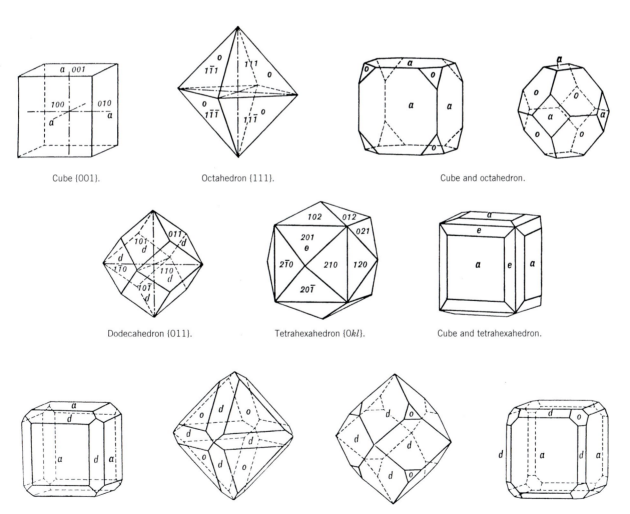

Cube {001}.   Octahedron {111}.   Cube and octahedron.

Dodecahedron {011}.   Tetrahexahedron {0kl}.   Cube and tetrahexahedron.

Combinations of cube and dodecahedron, octahedron and dodecahedron, and cube, octahedron, and dodecahedron.

FIG. 6.44. Some of the commonly developed forms and form combinations in 4/m$\overline{3}$2/m (see also Fig. 6.45).

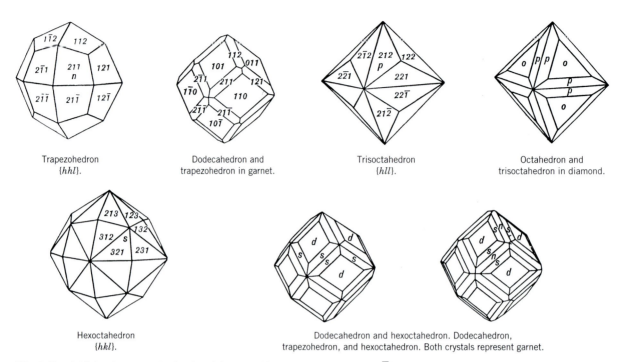

Trapezohedron {hhl}.

Dodecahedron and trapezohedron in garnet.

Trisoctahedron {hll}.

Octahedron and trisoctahedron in diamond.

Hexoctahedron {hkl}.

Dodecahedron and hexoctahedron. Dodecahedron, trapezohedron, and hexoctahedron. Both crystals represent garnet.

FIG. 6.45. Additional, commonly developed forms and form combinations in $4/m\bar{3}2/m$ (see also Fig. 6.44).

one of the crystallographic axes at unity and the other two at equal multiples. There are various trapezohedrons with their faces having different angles of inclination, but the most common is {112} (Fig. 6.45).

Figure 6.45 shows the common trapezohedron n {112} truncating the edges of the dodecahedron. Both forms by themselves and in combination are common on the mineral garnet.

6. *Trisoctahedron* {hll}. The trisoctahedron is composed of 24 isosceles triangular faces, each of which intersects two of the crystallographic axes at unity and the third axis at some multiple. There are various trisoctahedrons, the faces of which have different inclinations, but most common is {122} (Fig. 6.45). The trisoctahedron, like the trapezohedron, is a form that may be conceived as an octahedron, each face of which has been raised to accommodate three others. Figure 6.45 shows a combination of an octahedron and trisoctahedron.

7. *Hexoctahedron* {hkl}. The hexoctahedron is composed of 48 triangular faces, each of which intersects all three crystallographic axes at different lengths. There are several hexoctahedrons that have varying ratios of axial intercepts. A common hexoctahedron has indices {123}. Other hexoctahedrons have indices {124}, {135}, and so on, or, in general, {hkl}. Figure 6.45 shows a simple hexoctahedron as well as combinations with other isometric forms.

Determination of Indices of Forms. In determining the forms present on any crystal in this class it is first necessary to locate the crystallographic axes (axes of fourfold symmetry). Once the crystal has been oriented by these

axes, the faces of the cube, dodecahedron, and octahedron are easily recognized, because they intersect respectively one, two, and three axes at unit distances. The indices can be quickly obtained for faces of other forms that truncate symmetrically the edges between known faces. The algebraic sums of the h, k, and l indices of two faces give the indices of the face symmetrically truncating the edge between them. Thus, in Fig. 6.45 the algebraic sum of the two dodecahedron faces (101) and (011) is (112), or the indices of a face of a trapezohedron.

Occurrence of Isometric Forms in Class $4/m\bar{3}2/m$. The cube, octahedron, and dodecahedron are the most common isometric forms. The trapezohedron is also frequently observed as the only form on a few minerals. The other forms, the tetrahexahedron, trisoctahedron, and hexoctahedron, are rare and are ordinarily observed only as small truncations in combinations.

A large group of minerals crystallize in this class. Some of the most common are:

analcime	galena	silver
copper	garnet	spinel group
cuprite	gold	sylvite
diamond	halite	uraninite
fluorite	lazurite	

$\bar{4}3m$

Symmetry—$3\bar{A}_4$, $4A_3$, $6m$. The three crystallographic axes are axes of fourfold rotoinversion. The four diagonal axes are axes of threefold rotation, and there are six diagonal mirror planes, the same planes shown in Fig. 6.43b for class $4/m\bar{3}2/m$. The location of all of these symmetry ele-

$\overline{4}3m$

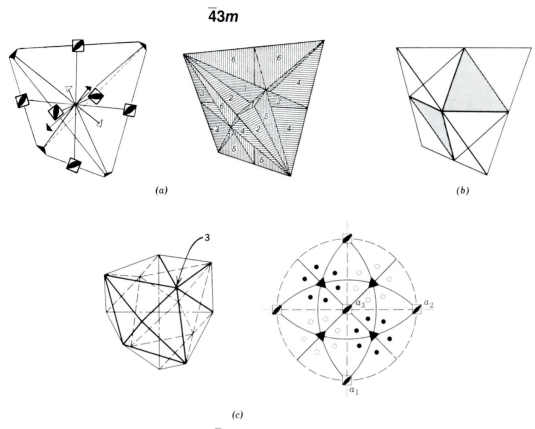

(a)

(b)

(c)

FIG. 6.46. *(a)* Symmetry axes and planes for $\overline{4}3m$. *(b)* Relation of octahedron and tetrahedron. *(c)* Hextetrahedron {*hkl*} and its stereogram. The location of one 3 is noted in the model.

ments is shown in Fig. 6.46a. The general form, the hextetrahedron, and its stereogram are illustrated in Fig. 6.46c. This class is known as the *hextetrahedral class.*

Forms

1. *Tetrahedron,* {111} *positive,* {1$\overline{1}$1} *negative.* The tetrahedron is composed of four equilateral triangular faces, each of which intersects all the crystallographic axes at equal lengths. The tetrahedral form can be considered as derived from the octahedron in class $4/m\overline{3}2/m$ by the omission of the alternate faces and the extension of the others, as shown in Fig. 6.46b. The positive tetrahedron {111} is shown in Fig. 6.46a. If the other four faces of the octahedron in Fig. 6.46b had been extended, the tetrahedron resulting would have had a different orientation, as shown in Fig. 6.47. This is the negative tetrahedron, {1$\overline{1}$1}. The positive and negative tetrahedrons are geometrically identical. The existence of both must be recognized for they may occur together as shown in Fig. 6.47. If the positive and negative tetrahedron are equally developed on the same crystal, the combination could not be distinguished from an octahedron unless, as often happens, the faces of the two forms showed different lusters, etchings, or striations that would serve to differentiate them.

Figure 6.47 illustrates the forms and many of the form combinations that are common in this class.

2. *Tristetrahedron,* {*hhl*} *positive,* {*h$\overline{h}$l*} *negative.* These forms with 12 faces can be conceived as a tetrahedron, each face of which has been raised to accommodate three others. The positive form may be made negative by a rotation of 90° about the vertical axis.

3. *Deltoid Dodecahedron* {*hll*} *positive,* {*h$\overline{l}$l*} *negative.* This is a 12-faced form in which three four-sided faces occur in place of one face of the tetrahedron.

4. *Hextetrahedron,* {*hkl*} *positive,* {*h$\overline{k}$l*} *negative.* The hextetrahedron has 24 faces that can be viewed as a tetrahedron, each face of which has been raised to accommodate six others.

Members of the tetrahedrite-tennantite series, (Cu, Fe, Zn, Ag)$_{12}$Sb$_4$S$_{13}$ to (Cu, Fe, Zn, Ag)$_{12}$As$_4$S$_{13}$, are the only common minerals that ordinarily show distinct hextetrahedral forms. Sphalerite, ZnS, occasionally exhibits them, but commonly its crystals are complex and distorted.

$2/m\overline{3}$

Symmetry—3A$_2$, 4$\overline{A}_3$, 3m. The three crystallographic axes are axes of twofold rotation; the four diagonal axes, each of which emerges in the middle of an octant, are axes of threefold rotoinversion; the three axial planes are mirror planes. This class has a center of symmetry because $\overline{3}$ is equivalent to $3 + i$. The combination of symmetry elements

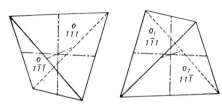

Tetrahedrons
positive {111} and negative {1$\bar{1}$1}.

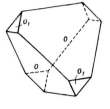

Combination (+) and (−).

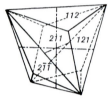

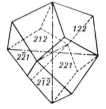

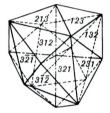

Tristetrahedron {hhl}.

Deltoid dodecahedron {hll}.

Positive hextetrahedron {hkl}.

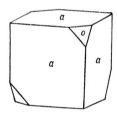

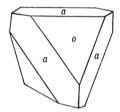

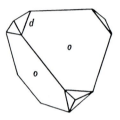

Combination of cube and tetrahedron.

Tetrahedron and dodecahedron.

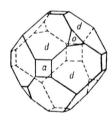

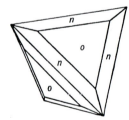

FIG. 6.47. Commonly developed forms and form combinations in $\bar{4}3m$.

Dodecahedron, cube and tetrahedron.

Tetrahedron and tristetrahedron.

2/m$\bar{3}$

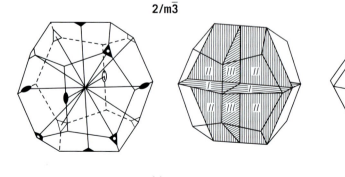

(a)

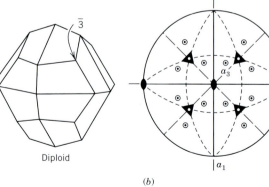

Diploid

(b)

FIG. 6.48. *(a)* Symmetry axes and planes for 2/m$\bar{3}$, and *(b)* the diploid {hkl} and its stereogram. The location of one $\bar{3}$ is noted on the model.

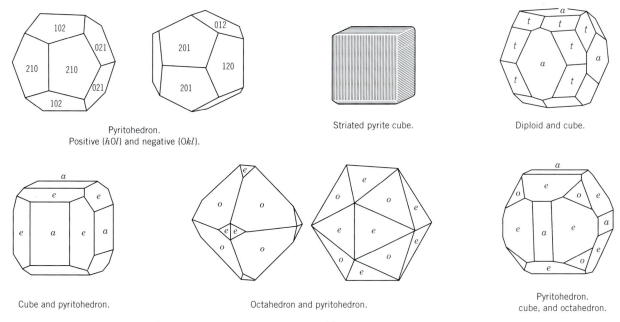

Pyritohedron.
Positive {h0l} and negative {0kl}.

Striated pyrite cube.

Diploid and cube.

Cube and pyritohedron.

Octahedron and pyritohedron.

Pyritohedron.
cube, and octahedron.

FIG. 6.49. Commonly developed forms and form combinations in 2/m$\bar{3}$.

and the positive diploid and its stereogram are shown in Fig. 6.48. This class is known as the *diploidal class* after the general form.

Forms

1. *Pyritohedron*, {h0l} *positive*, {0kl} *negative*. This form consists of 12 pentagonal faces, each of which intersects one crystallographic axis at unity, intersects the second axis at some multiple of unity, and is parallel to the third. A rotation of 90° about a crystallographic axis brings the positive pyritohedron into the negative position. There are a number of pyritohedrons that differ from each other with respect to the inclination of their faces. The most common positive pyritohedron has indices {102} (Fig. 6.49). This figure also shows the corresponding negative pyritohedron.

2. *Diploid*, {hkl} *positive*, {khl} *negative*. The diploid is a rare form composed of 24 faces that correspond to one-half of the faces of a hexoctahedron. The diploid may be pictured as having two faces built up on each face of the pyritohedron. As in the case of the pyritohedron, a rotation of 90° about one of the crystallographic axes brings the positive diploid into the negative position.

In addition to the pyritohedron and the diploid, the cube, dodecahedron, octahedron, trapezohedron, and trisoctahedron may be present. On some crystals these forms may appear alone and so perfectly developed that they cannot be distinguished from the forms of class 4/m$\bar{3}$2/m. This is often true of octahedrons and cubes of pyrite. Usually, however, they will show by the presence of striation lines or etch figures that they conform to the symmetry of class 2/m$\bar{3}$. This is shown in Fig. 6.49 by a cube of pyrite with characteristic striations showing the

lower symmetry. Figure 6.49 also shows combinations of the pyritohedron with forms of the hexoctahedral class, as well as a combination of cube and diploid {124}.

The chief mineral crystallizing in this class is pyrite (FeS$_2$); other rarer minerals are members of the skutterudite-nickel skutterudite series (CoAs$_{2-3}$ to NiAs$_{2-3}$), gersdorffite (NiAsS), and sperrylite (PtAS$_2$).

23

Symmetry—3A$_2$, 4A$_3$. The three crystallographic axes are axes of twofold rotation, and the four diagonal directions are axes of threefold symmetry. Figure 6.50 shows drawings of the positive left and positive right tetartoids and a stereogram of a positive right form. This class is known as the *tetartoidal class*.

Forms. There are four separate forms of the *tetartoid*: positive right {hkl}, positive left {khl}, negative right {k$\bar{h}$l}, and negative left {$\bar{h}$kl}. They comprise two enantiomorphic pairs, positive right and left, and negative right and left. Other forms that may be present are the cube, dodecahedron, pyritohedron, tetrahedron, and deltoid dodecahedron. Cobaltite, (Co,Fe)AsS, is the most common mineral representative crystallizing in this class.

Characteristics of Isometric Crystals

Four threefold symmetry axes are common to all isometric crystals. Symmetrically developed crystals are equidimensional in the three directions of the crystallographic axes. The crystals commonly show faces that are squares, equilateral triangles, or these shapes with truncated corners. All forms are *closed forms*. Thus, crystals are characterized by

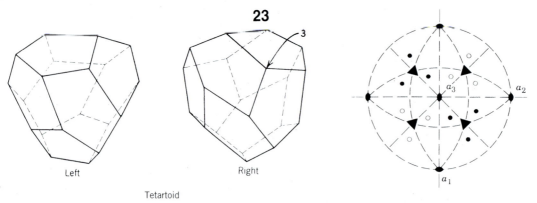

23

Left   Right

Tetartoid

FIG. 6.50. Enantiomorphic forms of the tetartoid. Positive left and positive right and a stereogram of the positive right {*hkl*} form. The location of one 3 is noted on the positive right model.

a large number of similar faces; the smallest number of any form of the hexoctahedral class is six.

Some important *interfacial angles* of the isometric system that may aid in the recognition of the most common forms are as follows:

Cube (100) $\wedge$ cube (010) = 90°00'
Octahedron (111) $\wedge$ octahedron ($\bar{1}$11) = 70°32'
Dodecahedron (011) $\wedge$ dodecahedron (101) = 60°00'
Cube (100) $\wedge$ octahedron (111) = 54°44'
Cube (100) $\wedge$ dodecahedron (110) = 45°00'
Octahedron (111) $\wedge$ dodecahedron (110) = 35°16'

REPRESENTATION OF SOME SPACE GROUPS

In the latter part of Chapter 5 (see p. 229) the concepts of three-dimensional lattices (the 14 Bravais lattice types, see Figs. 5.62 and 5.63), screw axes, and glide planes were introduced. Furthermore, the notation used in space group symbols was discussed. The following sections cover aspects of the graphical representation of space groups and the relationship of space group notation to morphology and structure.

TABLE 6.3   Symbols for Symmetry Axes (All Graphic Symbols Are for Axes Normal to the Page, Unless Otherwise Noted)

Symbol	Symmetry Axis	Graphic Symbol	Type of Translation (If Present)	Symbol	Symmetry Axis	Graphic Symbol	Type of Translation (If Present)
1	onefold rotation	None	None	4	fourfold rotation	■	None
$\bar{1}$	onefold rotoinversion	°	None	4_1	fourfold screw (right-handed)	◗	$\frac{1}{4}c$
2	twofold rotation	●	None	4_2	fourfold screw (neutral)	◧	$\frac{2}{4}c = \frac{1}{2}$
		→ (parallel to paper)		4_3	fourfold screw (left-handed)	◖	$\frac{3}{4}c$
2_1	twofold screw	◗	$\frac{1}{2}c$ $\frac{1}{2}a$ or $\frac{1}{2}b$	$\bar{4}$	fourfold rotoinversion	◨	None
				6	sixfold rotation	⬣	None
3	threefold rotation	▲	None	6_1	sixfold screw (right-handed)	⬢	$\frac{1}{6}c$
3_1	threefold screw (right-handed)	▲	$\frac{1}{3}c$	6_2	sixfold screw (right-handed)	⬢	$\frac{2}{6}c$
3_2	threefold screw (left-handed)	▲	$\frac{2}{3}c$	6_3	sixfold screw (neutral)	⬢	$\frac{3}{6}c = \frac{1}{2}$
$\bar{3}$	threefold rotoinversion	△	None	6_4	sixfold screw (left-handed)	⬢	$\frac{4}{6}c$
				6_5	sixfold screw (left-handed)	⬢	$\frac{5}{6}c$
				$\bar{6}$	sixfold rotoinversion	⬣	None

The relationship of screw axes to rotation axes was shown in Fig. 5.67, and the various types of glide components were explained in the text (p. 234). Here, we will introduce various graphical ways for the illustration of screw and glide operations on a two-dimensional page. The symbols used in the representations are those of the *International Tables for Crystallography*, 1983, volume A (a complete reference is given at the end of this chapter). Table 6.3 lists the conventional symbols for all types of rotational symmetry, and Figure 6.51 shows motif units related by symmetry axes (rotation and screw axes) and their projection onto the plane of the page (projected from above the page). This figure is very similar to the illustrations in Figs. 5.10 and 5.12 for rotation and rotoinversion axes, respectively. The fractions next to the motifs in Fig. 6.51 represent the distance the motif units lie above the surface of the page. The fractions that represent t/n are preceded by a plus sign (+) to indicate that the motif units lie above the page (in the positive direction of z in an x, y, z coordinate system). Table 6.4 gives the standard symbols used in graphic illustrations of glide planes and mirrors. Figure 6.52 illustrates the results of mirror and glide operations in a three-dimensional sketch. Figure 6.53 shows the results of these same operations in two-dimensional motif distributions.

The 230 space groups are listed in Table 5.10. As an introduction to the further understanding of space groups it is informative to ask what concepts are involved in the derivation of space groups. Because the derivation of all 230 is a lengthy and complex assignment, we will, as an example, tackle only a small number of lower symmetry space groups. This means that we will limit ourselves here to a few space groups in the triclinic and monoclinic systems. In the triclinic system only two possible space groups can occur, namely, $P1$ and $P\bar{1}$; these are the combinations of the two possible point groups in the triclinic system, 1 and $\bar{1}$ (see Table 5.4), and the only possible lattice type, P (see Fig. 5.63). In the monoclinic system, however, we have three possible point groups (2, m, and $2/m$) and two possible lattice types (P and I) to consider. For this illustration, let us restrict ourselves to a consideration of one of the point groups (2) and the two possible lattice types P and I (I in the monoclinic system can be transposed into A, B, or C by a different choice of coordinate axes; see footnote to Fig. 5.63). The four possible space group notations are $P2$, $P2_1$, and $I2$, and $I2_1$. Figure 6.54 illustrates the arrangement of motifs (commas) about the twofold rotation and twofold screw axes in relation to a monoclinic unit cell. Of the four possible space groups, only three turn out to be unique because $I2$ and $I2_1$ are equivalent in their arrangement

of symmetry elements, except for a change in the location of the origin chosen for the lattice. Therefore, the three unique space groups for the monoclinic point group 2 are $P2$, $P2_1$, and $I2$ (which is equivalent to $C2$). A more complete discussion of some additional space groups can be found in Buerger (1978).

Illustrations of Space Groups. In the prior section of this chapter we illustrated 19 of the 32 point groups. If our aim were analogous, in-depth coverage of three-dimensional periodic symmetry, we would have to likewise describe and illustrate a majority of the 230 space groups. This would be a formidable task, requiring a great deal of text and many illustrations. Instead, we will introduce the reader only to some general aspects of space group illustrations, now that all the necessary "operators" of space group symmetry have been defined. We will illustrate and discuss a few representative space groups that are found in common (rock-forming) minerals. After that we will discuss and illustrate the relationship of external morphology (point group symmetry) to internal structure and the space group link between these two concepts. We will do this for three well-known minerals. Readers interested in systematic coverage of space groups should consult some of the standard references on the subject, three of which are profusely illustrated: *International Tables for X-ray Crystallography*, vols. 1 and A; *Elementary Crystallography*; and *Mathematical Crystallography* (complete references are given at the end of this chapter).

In the illustrations of Fig. 6.55 the conventions of the *International Tables for X-ray Crystallography* (vol. A) are used. The coordinate axes are oriented as follows. The a axis is toward the reader, the b to the right (in an E–W position), and the c axis is perpendicular to the paper. The origins of the drawings are in the upper left corner. The motifs are represented by small circles instead of commas (as used extensively until this part of the book). The heights of motif locations are indicated by + or − next to them: + means a distance upward (along c) from the page, and − means an equivalent distance in a downward direction. Motifs and symmetry operators can be accompanied by fractions (e.g., $\frac{1}{4}$, $\frac{1}{2}$, $\frac{1}{3}$) indicating a fractional distance (upward or downward) within the unit cell. The open circle (○) is considered a right-hand motif, whereas an open circle with a small comma inside (◉) is the left-hand equivalent. These two symbols are thus enantiomorphic and can be related by a mirror, glide, or inversion operation. The two types of circles are analogous to the enantiomorphic commas used in prior illustrations. The symbols for the various symmetry operators are given in Tables 6.3 and 6.4. An inversion point (center of symmetry) is noted by a very small circle. If its height is not given, it is assumed to be zero (in the plane of the page). If a mirror exists parallel to the plane of projection (the plane of the page), the superimposed motifs are indicated by the small circle used for motifs, but now divided vertically down the middle (⊕). This symbol is accompanied by + and − signs at its side to indicate that one motif (above, +) superimposes with another below (−). The

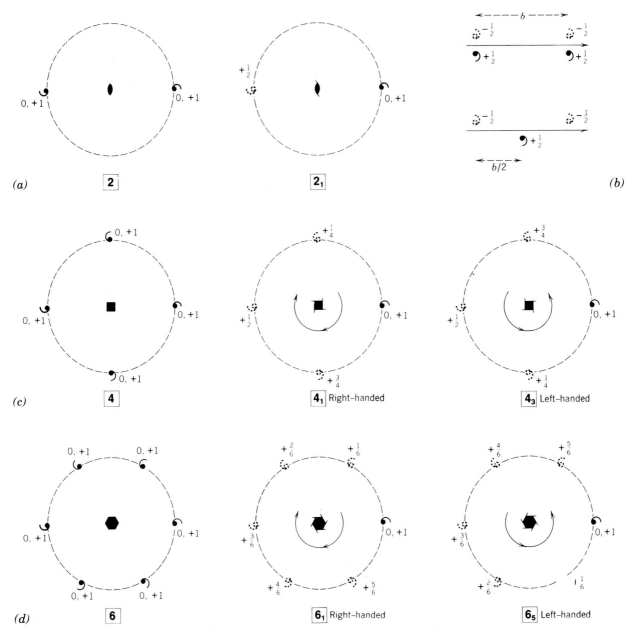

FIG. 6.51. Examples of several axes of rotation and some of their isogonal screw axes. The effect of these symmetry operations on motifs is shown as well. All diagrams are set such that the *a* and *b* axes (*x* and *y* directions) lie in the plane of the page. The height above the plane of the page, along the +*c* axis (in the +*z* direction), is indicated by a (+) sign. For example, 0 and +1 mean that the motif lies in the plane of the page (at zero height along the *c* axis) and is also repeated along the *c* axis by a unit lattice repeat to +1*c*. Fractions (e.g., $+\frac{1}{2}, +\frac{1}{3}, +\frac{1}{4}$) refer to heights above the page along the *c* axis (*z* direction). Motif units that do not lie in the plane of the page (but have been projected from above) are shown by a dashed comma design. *(a)* Twofold rotation axis and isogonal screw axis. The circles show the effect of these axes on motifs when the axes are oriented perpendicular to the page. *(b)* These same axes when oriented parallel to the plane of the page (or, when they lie within the plane of the page). *(c)* Fourfold rotation axis and two enantiomorphic fourfold screw axes. The rotational directions of the screws ("handedness") are shown by the arrows. *(d)* Sixfold rotation axis and two enantiomorphic sixfold screw axes.

TABLE 6.4   Symbols for Mirror and Glide Planes*

| | | Graphic Symbol | | |
Symbol	Symmetry Plane	Normal to Plane of Projection	Parallel to Plane of Projection†	Nature of Glide Translation
m	Mirror	———————	⌐ 120° ＼	None
a, b	Axial glide plane	– – – – – – – –	⌐↓ ←	a/2 along [100] or b/2 along [010]
c		····················	None	c/2 along the c axis
n	Diagonal glide plane	—·—·—·—·—	◿	a/2 + b/2; a/2 + c/2; b/2 + c/2; or a/2 + b/2 + c/2 (tetragonal and isometric)
d	Diamond glide plane	—·—←·—·— —·—·—→·—	◿	a/4 + b/4; b/4 + c/4; a/4 + c/4; or a/4 + b/4 + c/4 (tetragonal and isometric)

*From *International Tables for X-ray Crystallography*, 1969, v. 1, N. F. M. Henry and K. Lonsdale, eds.; Birmingham, England: Symmetry Groups. International Union of Crystallography, Kynoch Press.
†When planes are parallel to the paper, heights other than zero are indicated by writing the z coordinate next to the symbol (e.g., $\frac{1}{4}$ or $\frac{3}{8}$). The arrows indicate the direction of the glide component.

enantiomorphic relation in such an occurrence is shown by the small comma inside one half of the circle (e.g., ⊕).

The left-hand diagrams in Fig. 6.55 show the distribution of motifs in the unit cell of the three-dimensional periodic array, and the right-hand diagrams locate the various symmetry elements with respect to this unit cell.

Figure 6.55a shows the monoclinic space group C2/m, which is common in rock-forming minerals such as clinoamphiboles and sanidine. The space group illustrations are oriented such that the y axis (with the b cell edge) is the twofold axis in an east–west position. This means that we have adopted the second setting for this space group (in the first setting the c axis would be the unique twofold axis).[2]

This monoclinic space group, C2/m, illustrates an important aspect of space group notation. The notation shows only twofold rotation axes, but the right-hand side of Fig. 6.55a also shows twofold screw axes interleaved with the twofold rotation axes. Similarly, only m's are given in the space group symbol, but glide planes are interleaved with them. This is related to the choice of a c-centered lattice (C). In space groups with nonprimitive lattices, screw axes and glide planes are introduced because of centering. These new elements, however, are not noted in the sym-

bols. Therefore, particular attention should be paid to such occurrences in nonprimitive lattice types. The C2/m diagrams also show the presence of m's and twofold rotation axes at distances of one-half the appropriate cell lengths. These extra symmetry elements are oriented in accordance with the symmetry elements of the space group; such extra elements are present in all space groups. Centers of symmetry (inversion) are indicated by small open circles.

Figure 6.55b illustrates the orthorhombic space group $P2_1/n2_1/m2_1/a$, which is isogonal with point group 2/m2/m2/m. The abbreviated (short) symbol for this point group is mmm, and the analogous short symbol for the space group is Pnma. This space group notation is one of six possible choices; these reflect different ways of orienting an orthorhombic cell with respect to the a, b, and c axes. The six equivalent permutations are Pnma, Pbnm, Pmcn, Pnam, Pmnb, and Pcmn. The structure of the common rock-forming mineral olivine is based on this space group. The left-hand side of Fig. 6.55b shows the location of motifs, and the right-hand side illustrates the lattice type and symmetry elements compatible therewith. All twofold rotation axes in the point group are represented by twofold screw axes in the space group (each axial direction, a, b, and c, is coincident with 2_1 symmetry). The diagonal (n) glide perpendicular to the a axis is shown by the E–W dot-dash lines; the m's perpendicular to the b axis by the solid lines, and the a glide by the graphic symbol for a plane parallel to the plane of the page (but at one-fourth of the unit cell above the plane of projection).

The space group of the common form of quartz (low quartz) is illustrated in Fig. 6.55c. This is $P3_121$, which is one of a pair of enantiomorphic space groups, $P3_121$, and $P3_221$. These are isogonal with point group 321 (also referred to as

[2]Because the monoclinic system is referred to three axes a, b, and c and a β angle—which is not 90°—the projection of lattices in this system, with b to the right—in the plane of the page—and c perpendicular to the page, will not allow the a axis to lie in the plane of the page. Instead, an inclined a axis is projected onto the page and the north–south direction is one of asinβ instead of a, as in orthogonal systems.

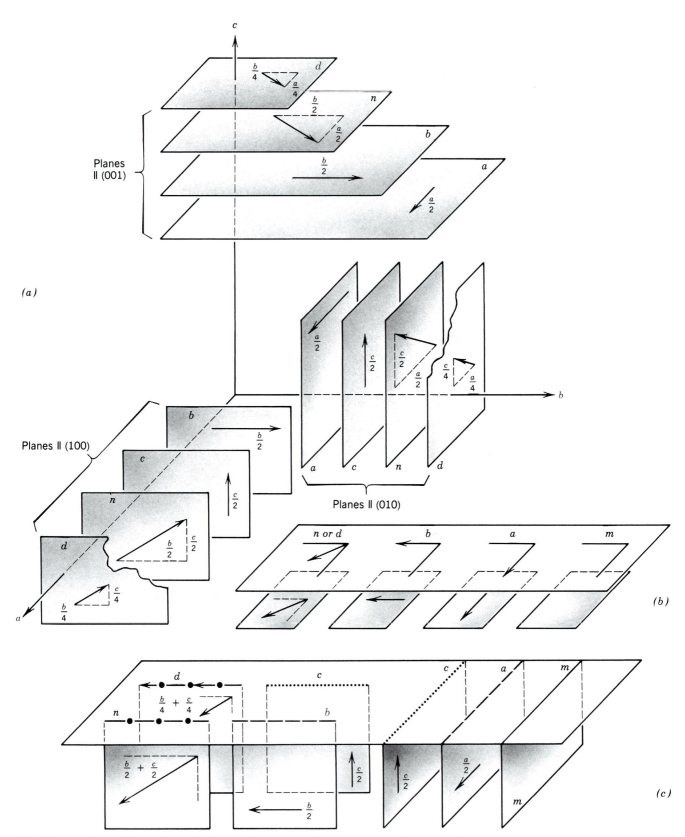

FIG. 6.52. *(a)* Sketch of the various glide planes and their translation components with reference to orthorhombic coordinate axes. *(b)* Symbols for glide and mirror planes when such planes are parallel to the plane of projection (001) or (0001). *(c)* Symbols for glide and mirror planes when such planes are perpendicular to the standard plane of projection. (Adapted from F. D. Bloss, 1971, *Crystallography and Crystal Chemistry: An Introduction*, Figs. 7.9 and 7.11, copyright © 1971 by Holt, Rinehart, and Winston, Inc., New York. Reprinted by permission of the author.)

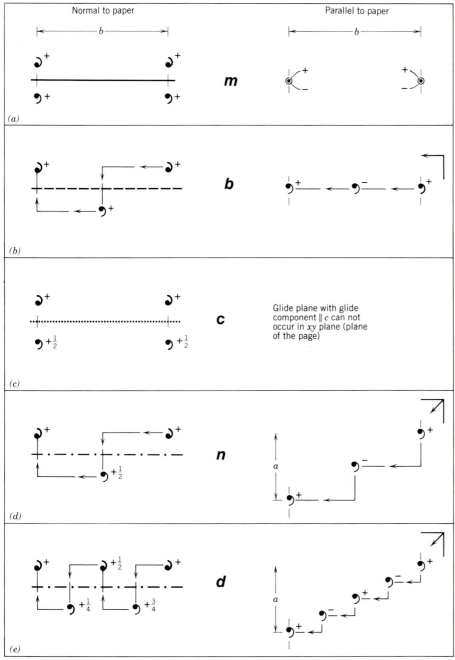

FIG. 6.53. Examples of symmetry planes and the distribution of motifs about them. In all drawings the plane of the paper is considered the *ab* plane. The location of a motif above the plane (in the (+) direction of *c*) is marked by +, or $+\frac{1}{4}$, $+\frac{1}{2}$, and so on. The (−) *c* direction is shown by (−) signs next to the motif. In the right-hand figure of *(a)* the motif units superimpose exactly; the upper motif is shown by •, the lower equivalent by a small circle about it. Note conventional symbols; see also Table 6.4. *(a)* A mirror plane. *(b)* A glide plane with a glide component parallel to the *b* axis (*b*/2). *(c)* A glide plane with a glide component parallel to the *c* axis (*c*/2). *(d)* A diagonal glide plane (*n*) with simultaneous glide components parallel to the *a* and *b* axes (*a*/2 + *b*/2). *(e)* A diamond glide plane *(d)* with simultaneous glide components parallel to the *a* and *b* axes (*a*/4 + *b*/4).

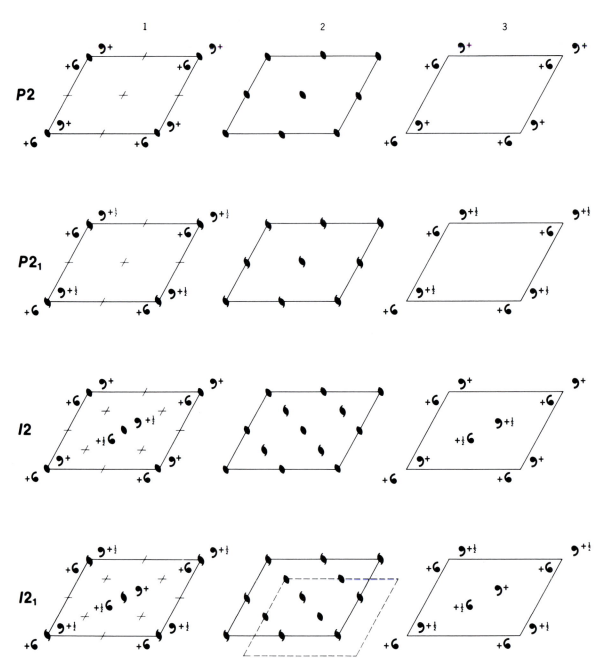

FIG. 6.54. Derivation of the monoclinic space groups $P2$, $P2_1$, $I2$, and $I2_1$. Column 1 depicts only the lattice type, the motif units, and the symmetry element specified for each lattice point (corners and centers). In the illustration in column 1 it should become clear that additional symmetry elements are inherent in the unit cell at the points indicated by $+$. Column 2 depicts all the symmetry elements inherent in the pattern. Column 3 shows the arrangement of motif units only. For space group $I2_1$ it becomes clear that the symmetry contained in the unit cell is equivalent to the symmetry in the unit cell for $I2$. Only a shift in origin of the unit cell is needed to obtain a symmetry pattern identical to that for $I2$ (column 2). The dashed unit cell indicates the actual shift necessary. In other words, $I2$ and $I2_1$ are identical and therefore only one of the two arrangements is unique. In Table 5.10 the space group $I2$ is listed as $C2$ because these two notations are equivalent (see footnote to Fig. 5.63).

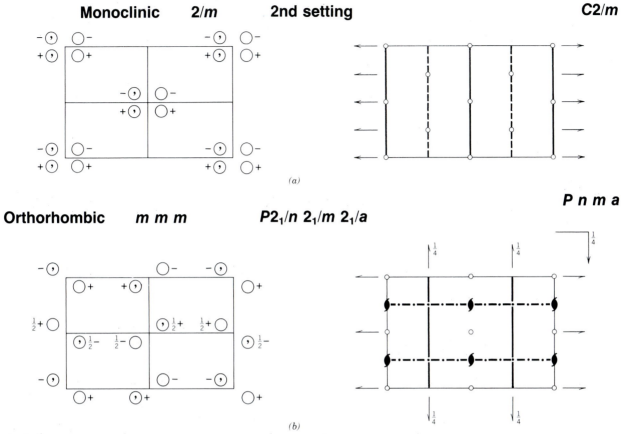

FIG. 6.55. Examples of illustrations of space groups. (Reproduced from *The International Tables for X-ray Crystallography*, vol. 1; see text for discussion.)

32). The unit cell has a primitive hexagonal outline. The location of motifs in the left-hand diagram immediately reveals the threefold screw axes at the four corners of the unit cell; two additional locations are present in the centers of the triangular halves of the unit cell. The locations of the twofold rotation axes are coincident with the a_1, a_2, and a_3 axis directions. The last numeral, 1, refers to directions at 30° to the axial directions of the unit cell (the diagonal directions) and indicates that there is no symmetry in these directions.[3] This space group also contains twofold screw axes that occur halfway along the cell translations and are parallel to the

twofold rotation axes. These are the result of combining a rotational axis with a nonparallel axial translation (e.g., the twofold rotation axes that are not at 90° to the edges of the unit cell).

The tetragonal space group $P4_2/nnm$ (that of zircon) is illustrated in Fig. 6.55d. This space group is isogonal with point group $4/mmm$. The full (unabbreviated) symbols for space and point groups are $P4_2/n2/n2/m$ and $4/m2/m2/m$, respectively. The square lattice and inherently high symmetry content are obvious in the illustrations. The fourfold rotation axes of the point group appear as alternating and parallel fourfold rotoinversion and fourfold screw axes (neutral as a result of 4_2). The n glide (in $4_2/n$) is parallel to the plane of the page at location $+\frac{1}{4}c$. The axial directions (a_1 and a_2) contain twofold rotation axes and have vertical n glides parallel and perpendicular to the a_1 and a_2 axes (these locations are shown by the "dot-dash" lines). The diagonal directions (at 45° to the a_1 and a_2 axes) contain mirror planes as well as twofold rotation axes with interleaved twofold screw axes. Centers of symmetry are located at $+\frac{1}{4}c$.

Space Groups as an Expression of Morphology and Structure. Space groups are an elegant and powerful "shorthand" for the characterization of the internal structure

---

[3]The numeral 1 is commonly used in space group notation in order to distinguish different orientations of the same overall symmetry content. For example, point group 32 has, among others, isogonal space groups $P312$ and $P321$. The symmetry content is the same in both representations. In $P312$, however, there is no symmetry along the a_1, a_2, and a_3 axes, and the twofold rotation axes lie in directions pependicular to a_1, a_2, and a_3. In $P321$ the a_1, a_2, and a_3 axes are the directions of twofold rotation and the directions perpendicular to a_1, a_2, and a_3 lack any symmetry. Other examples are: $P3m1$ and $P31m$; $P32/m1$ and $P312/m$ (see footnote to Table 5.8). Examples in the monoclinic system are: $P121$, $P12_11$, and $C121$ for space groups compatible with crystal class 2 (second setting). Here the b axis is one of twofold rotation symmetry, and the a and c axial directions lack symmetry (as noted by the two numerals 1 in the a and c locations of the symbol).

Hexagonal   321

P3₁21

Tetragonal   4/m m m

P 4₂/n 2/n 2/m   P 4₂/n n m

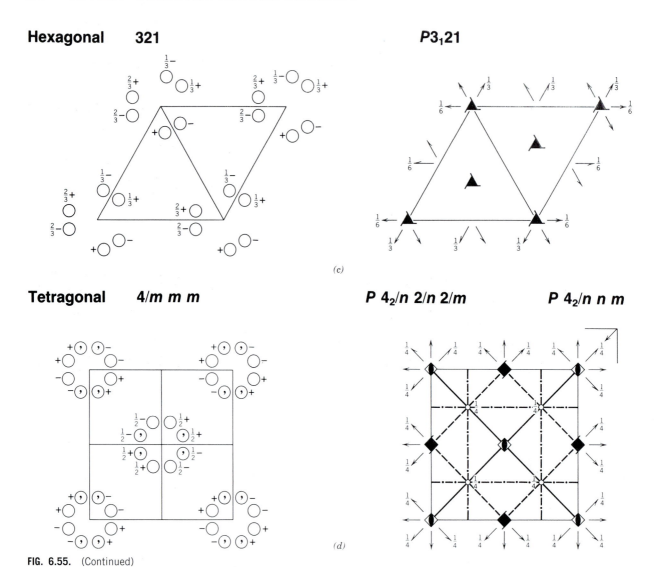

(c)

(d)

FIG. 6.55. (Continued)

of crystalline material. We know that a point group is the translation-free residue of an isogonal space group. In other words, an atomic pattern that might contain screw axes and/or glide planes will exhibit an external form that reflects only rotation axes and/or mirrors. The angular relations between the faces of a crystal will not be affected by the specific nature of the internally present axes and planes. Screw axes and glide planes cause displacements in the internal structure because of their translational components, but such displacements are so small they cannot be observed morphologically and are therefore absent in our evaluation of point group symmetry. Indeed, such translations are so small (2 to 5 Å) that only X-ray and electron diffraction techniques can resolve the distances.

A clear understanding of what space groups mean can be obtained from correlations between point group symmetry (as based on morphology), internal structure, and the resultant space group symbolism. The most informative way to evaluate an internal structure is by the vi-

sual study and inspection of three-dimensional models of the structure. (Various views of crystal structures and their space group elements are given in module III of the CD-ROM under the heading "Three-dimensional Order: Space Group Elements in Structures.") However, in our illustrations in Fig. 6.56, 6.57, and 6.58 (as in all other structure illustrations in this book) we are restricted to two-dimensional representations of three-dimensional arrangements. In Fig. 6.56 we will correlate the symmetry deduced from the morphology of diopside, $CaMgSi_2O_6$ (one of the members of the clinopyroxene series), with the internal structure and the space group derived therefrom. The diopside crystal shown in Fig. 6.56a has point group symmetry 2/m, as shown in Fig. 6.56b. In our orientation of the crystal we have chosen the b axis as the unique, twofold axis (this is referred to as the second setting). Figure 6.56c is a sketch of the simplest possible lattice (primitive) compatible with point group symmetry 2/m. In this sketch the locations of all twofold rotation axes and mirror planes are shown; locations of centers of

Monoclinic   2/m   2nd Setting

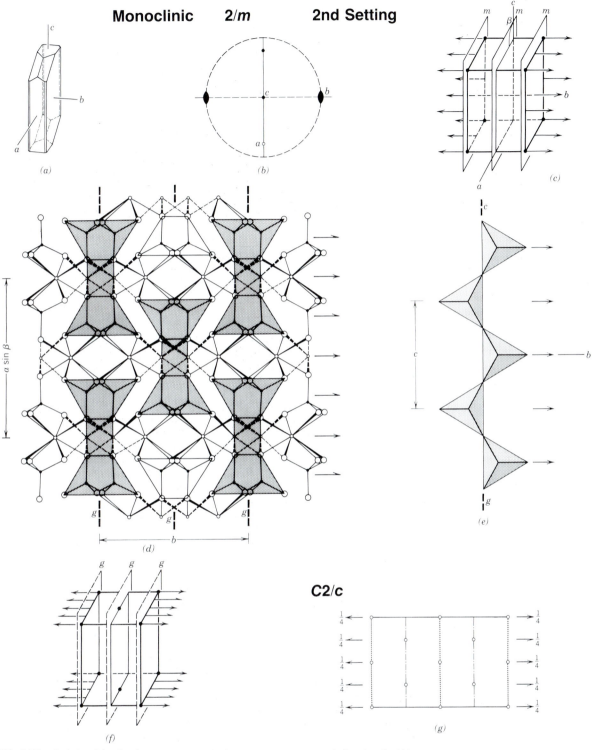

C2/c

FIG. 6.56. Relationship of point group symmetry to space group representation for diopside, a member of the clinopyroxene group. *(a)* Crystal of diopside. *(b)* Stereogram of 2/m symmetry. *(c)* Sketch of a possible monoclinic, primitive lattice compatible with 2/m. Locations of twofold axes and mirrors are noted. *(d)* A view of the clinopyroxene structure along the *c* axis. (Redrawn from M. Cameron and J. J. Papike, 1980, Crystal chemistry of pyroxenes, in *Reviews in Mineralogy,* 7.) Centering of the cell is outlined by the distribution of shaded units in the structure. Rotation and screw axes parallel to *b* are noted, as are glide planes (marked *g*). *(e)* The location of vertical glide planes, with a *c*/2 component, in an idealized tetrahedral pyroxene chain. *(f)* A sketch of the centered, monoclinic lattice, as based on the structural information of parts *(d)* and *(e)*. For simplicity, only some of the rotation axes, screw axes and glide planes are shown. *(g)* An illustration of the *C2/c* space group of diopside.

Hexagonal   6/m 2/m 2/m

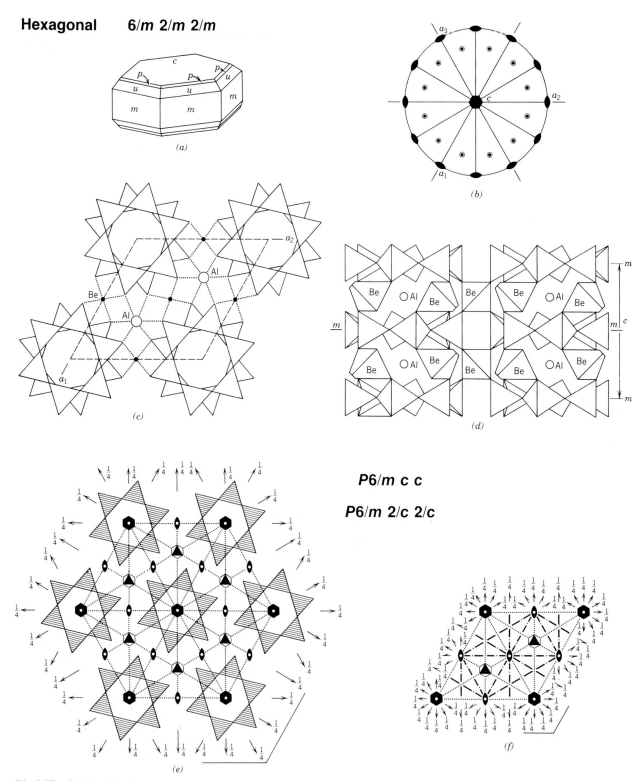

P6/m c c

P6/m 2/c 2/c

FIG. 6.57. Relationship of point group symmetry to space group representation for the mineral beryl. *(a)* A beryl crystal flattened on (0001), typical of cesium-rich beryl. *(b)* A stereogram of the point group symmetry 6/m2/m2/m as reflected by the morphology. *(c)* The structure of beryl as projected onto (0001). A unit cell is shown by dashed lines. *(d)* A vertical section through the beryl structure. *(e)* A projection of some of the structural elements of beryl and their relationships to each other as shown by the presence of some of the symmetry elements. (From A. V. Shubnikov and V. A. Koptsik, 1974, *Symmetry in science and art.* New York: Plenum Press.) *(f)* The conventional representation of the space group *P6/m2/c2/c*, which is compatible with the beryl structure.

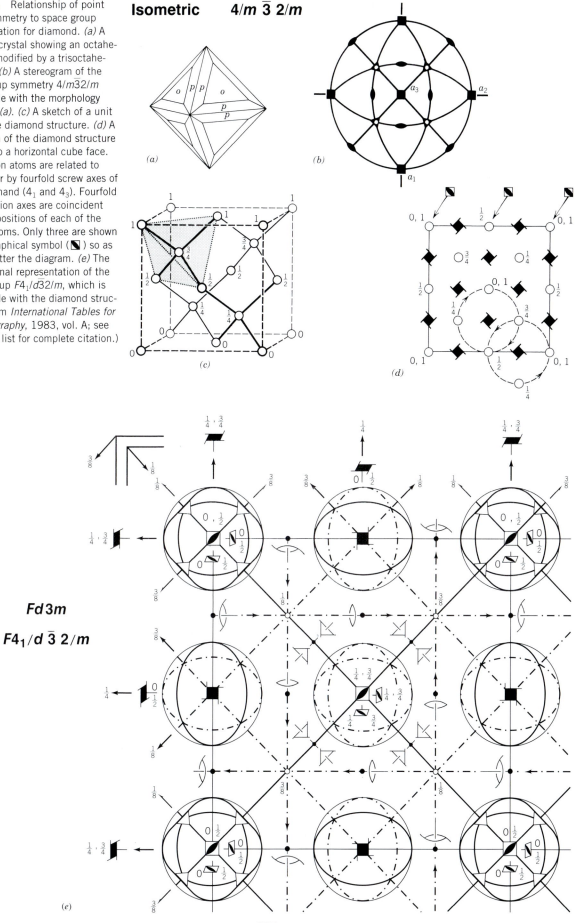

FIG. 6.58. Relationship of point group symmetry to space group representation for diamond. *(a)* A diamond crystal showing an octahedron (*o*) modified by a trisoctahedron (*p*). *(b)* A stereogram of the point group symmetry $4/m\bar{3}2/m$ compatible with the morphology shown in *(a)*. *(c)* A sketch of a unit cell of the diamond structure. *(d)* A projection of the diamond structure in *(c)* onto a horizontal cube face. The carbon atoms are related to each other by fourfold screw axes of opposite hand (4_1 and 4_3). Fourfold rotoinversion axes are coincident with the positions of each of the carbon atoms. Only three are shown by the graphical symbol (◨) so as not to clutter the diagram. *(e)* The conventional representation of the space group $F4_1/d\bar{3}2/m$, which is compatible with the diamond structure. (From *International Tables for Crystallography*, 1983, vol. A; see reference list for complete citation.)

Isometric $4/m\,\bar{3}\,2/m$

(a)

(b)

(c)

(d)

$Fd\bar{3}m$

$F4_1/d\,\bar{3}\,2/m$

(e)

symmetry (inversion) are not indicated. In order to arrive at the correct (not just the simplest) choice of lattice type (and unit cell), one must refer to the internal structure of diopside (and similar clinopyroxenes). Figures 6.56d and e show two sections through this type of structure. Figure 6.56d is a plan view (down the c axis), whereas Fig. 6.56e is a view of part of the b–c plane in the structure. The shaded parts of Fig. 6.56d identify repeats of some of the motifs in the structure; these are the cross-sections of tetrahedral–octahedral–tetrahedral (t–o–t) chains that are parallel to the c axis (see the section on the Pyroxene Group in Chapter 11 for further discussion). These structural units are repeated in a centered pattern in the plan view; in three dimensions this is compatible with a c-centered monoclinic lattice (C). The locations of twofold rotation axes (parallel to the b axis) and interleaved twofold screw axes (the result of the centering) are shown. The vertical section in Fig. 6.56e shows the presence of a glide component of c/2 in the tetrahedral chain. This is expressed by the c in the space group symbol C2/c, in place of m in the point group notation 2/m. The locations of these glide planes are shown in the plan view of the pyroxene structure in Fig. 6.56d and are marked g. Interleaved, halfway between these glides, will be additional glide planes that result from the centering of the unit cell choice. In Fig. 6.56f a sketch is shown of the three-dimensional lattice that is appropriate to the structure of diopside; it is monoclinic and C-centered. For purposes of clarity, the drawing shows only some of the symmetry elements present; additional glides, twofold rotation, and screw axes would be present in locations equivalent to those shown in Fig. 6.56c. In Fig. 6.56g the conventional space group representation is given for C2/c. In this discussion of the pyroxene structure we did not locate the various positions of inversion symmetry (centers of symmetry). It is conventional to choose the origin of a space group at the location of such a center of symmetry. It is thus that the twofold rotation and screw axes appear at $+\frac{1}{4}$ in Fig. 6.56g; the centers, at the positions noted, lie $\frac{1}{4}c$ below the symmetry axes.

Figure 6.57 illustrates the correlation between point group symmetry, structure, and resultant space group for the mineral beryl, $Be_3Al_2Si_6O_{18}$. Figure 6.57a shows the hexagonal crystal forms of beryl, and the resultant point group symmetry is shown in Fig. 6.57b. The choice of lattice type for this mineral must obviously be hexagonal. Two sections through the crystal structure of beryl are shown in Figs. 6.57c and d. The horizontal projection onto (0001) (Fig. 6.57c) shows the unique sixfold Si_6O_{18} rings. If the centers of four of these rings are chosen as the positions for possible lattice nodes, the rhombus shape of the lattice becomes obvious (a primitive hexagonal lattice choice, as in P of the space group symbol). The sixfold symmetry about the centers of the rings is also clear. In a vertical cross-section of the structure (Fig. 6.57d) the location of horizontal mirrors through the centers of the rings is shown (the hexagonal outline of and horizontal mirrors through these rings account for 6/m in the space group

symbol). Be and Al provide cross-links between the Si_6O_{18} rings; Be is in tetrahedral (fourfold) coordination and Al in octahedral (sixfold) coordination. The Be ions occupy positions where three directions of twofold rotation axes intersect (222) and the Al ions occupy locations where vertical threefold axes intersect with horizontal twofold axes (32). These symmetry locations are superimposed on the horizontal section through the beryl structure in Fig. 6.57e. For reasons of clarity not all possible symmetry elements have been shown in this representation. The location of the horizontal mirror (perpendicular to the c axis) is indicated by the 120° bracket at the bottom right side of the figure. The many horizontal twofold rotation axes are located at $\frac{1}{4}c$ above the plane of the diagram. Figure 6.57d shows that the structure contains sixfold rings that are not equivalent in orientation to the rings above and below. These rotations between interleaved rings (along the c direction) are the result of reflections by vertical glide planes (marked as c's in the space group symbol). Figure 6.57f shows the conventional representation of the space group P6/mcc, which reflects the beryl structure.

The last illustration of correlating point group, space group, and internal structure is that of the mineral diamond, C. Figs. 6.58a and b depict a possible morphology of diamond and the point group symmetry derived therefrom. Figure 6.58c shows a unit cell of diamond based on a cubic space lattice. The structure of diamond consists of carbon atoms in tetrahedral coordination with four nearest carbon neighbors. The lines joining carbon atoms in the upper left part of the unit cell of Fig. 6.58c show this coordination by the dotted outline of a tetrahedron. Upon careful inspection of the distribution of carbon atoms in the unit cell, it becomes clear that each of the faces of the cube contains (in addition to carbon atoms at all of the corners) one carbon atom at its center. This "all-face-centered" distribution of nodes in the lattice is reflected in F of the space group notation. When this same unit cell is projected onto a horizontal cube face at the bottom of the unit cell, the drawing in Fig. 6.58d results. This projection reveals the presence, at positions halfway between the carbon atoms (or projected carbon atoms), of two types of fourfold screw axes. These two enantiomorphic screw axis types (4_1 and 4_3) are located in alternate rows and are parallel to the fourfold rotation axes of the point group notation. The helical paths for two of the screw rotations are shown. The symmetry about each of the carbon atoms is $\bar{4}3m$, because of the tetrahedral coordination. A projection of the symmetry elements in the space group of diamond is shown in Fig. 6.58e. In addition to the 4_1 and 4_3 axes there are fourfold rotoinversion axes (◣) interspaced with the screw rotations. Centers of symmetry (small circles with fractions such as $\frac{1}{8}$ and $\frac{3}{8}$) are noted. Diamond glide planes (d) (shown by dot-dash lines with arrows) are parallel to the cube faces of the unit cell. These planes are inclined to the plane of the figure because they have translation components of the type b/4 + c/4, a/4 + c/4. Vertical mirror planes (and interleaved glide planes) occur in diagonal directions.

REFERENCES AND SELECTED READING

Bloss, F. D. 1994. *Crystallography and crystal chemistry: An introduction*. Reprint of the original text of 1971 by the Mineralogical Society of America, Washington, D. C.

Boisen, M. B., Jr., and G. V. Gibbs. 1990. Mathematical crystallography, *Reviews in Mineralogy* 15, Mineralogical Society of America, Washington, D. C.

Bragg, W. L., and G. F. Claringbull. 1965. *Crystal structures of minerals*. Ithaca, N. Y.: Cornell University Press.

Buerger, M. J. 1978. *Elementary crystallography: An Introduction to the fundamental geometric features of crystals*. Rev. ed. Cambridge, Mass.: MIT Press.

———, 1971. *Introduction to crystal geometry*. New York: McGraw Hill.

International Tables for Crystallography. 1983. Vol. A, *Space group symmetry*. Edited by T. Hahn. Boston: International Union of Crystallography, D. Reidel Publishing Company.

International Tables for X-ray Crystallography. 1969. Vol. 1, *Symmetry groups*. Edited by N. F. M. Henry and K. Lonsdale. Birmingham, England: International Union of Crystallography, Kynoch Press.

Klein, C. 1994. *Minerals and rocks: Exercises in crystallography, mineralogy, and hand specimen petrology*. Rev. ed. New York: Wiley.

O'Keffe, M., and B. G. Hyde. 1996. *Crystal structures I: Patterns and symmetry*. Monograph. Washington, D. C.: Mineralogical Society of America.

CHAPTER 7

ANALYTICAL METHODS
IN MINERAL SCIENCE

The physical and chemical properties of minerals can be quantitatively evaluated by a wide array of analytical (instrumental) techniques. The well-established methods include scanning electron microscopy (SEM), optical microscopy, X-ray diffraction (XRD), transmission electron microscopy (TEM) including electron diffraction and high-resolution transmission electron microscopy (HRTEM), X-ray fluorescence (XRF), flame atomic absorption spectroscopy (FAA), and electron microprobe analysis (EMPA). Two more instrumental techniques have become available only over the last fifteen years: secondary ion mass spectrometry (SIMS) and atomic force microscopy (AFM). This chapter provides the reader with a basic overview of these techniques and their applications to mineral studies. A student in mineralogy will probably have direct introduction to only two analytical techniques: optical microscopy and X-ray powder diffraction. These two techniques are thus treated in greater detail than the other methods. However, the overview of even these two techniques is not adequate for a thorough understanding of all of their underlying concepts; for this, additional references are given.

The various analytical techniques that will be presented in this chapter can be grouped in terms of the types of data they provide. For example, scanning electron microscopy is excellent at providing the investigator with high-magnification photographic images of crystal morphology, and optical microscopy is essential in rapid (commonly unambiguous) mineral identification and in the interpretation of mineral intergrowth (rock) textures. These two techniques, SEM and optical microscopy, are follow-up techniques to physical property tests (in hand specimens) discussed in Chapter 2. Whereas physical properties, as determined in hand specimens, aid in the tentative identification of, for example, a silicate that is part of a solid-solution series, optical microscopy (and refractive index measurements) will generally allow for unique identification.

X-ray diffraction studies (XRD) can be divided into two categories: (1) X-ray single-crystal diffraction and (2) X-ray powder diffraction. Single-crystal diffraction techniques are most commonly used in the determination of crystal structures, unit cell sizes, atomic parameters, and bond distributions. Powder diffraction is a rapid technique of mineral identification but can also provide information about unit cell sizes. Electron diffraction results (as part of TEM techniques) commonly compliment structural data obtained from X-ray single crystal experiments. Another application of the transmission electron microscope (TEM) is to the study of mineral reactions that occur on such a fine-grained scale that any evidence of reaction is not visible using lower resolution techniques such as that of high-power optical microscopy. The high resolution transmission electron microscope (HRTEM) mode allows for the projection of structures on a two-dimensional image with a resolution of about 1.5Å.

Several analytical techniques are specifically designed to obtain quantitative chemical data. These are flame atomic absorption (FAA), X-ray fluorescence (XRF), and electron microprobe analysis methods (EMPA). Secondary ion mass spectrometry (SIMS) uses a focused ion beam to remove ions from a mineral surface, and analyzes their mass and charge with a mass spectrometer.

Atomic force microscopy (AFM) allows for the characterization of surfaces of minerals on an atomic or near-atomic scale.

SCANNING ELECTRON MICROSCOPY (SEM)

The scanning electron microscope has an electron optical column in which a finely focused electron beam can be scanned over a specified (small) area of the specimen. The electron column in an SEM is similar to that shown in Fig. 1.11 for an electron microprobe. In an electron microprobe the electron beam is generally focused in a stationary mode, on a specific mineral grain, whereas in the SEM the beam is scanned across the specimen. The impact of the electron beam on the surface of a solid sample causes various types of radiation signals that can be selected by detectors above the specimen. These signals include secondary electrons, backscattered electrons, X-rays, cathodoluminescence radiation (this is the emission of electromagnetic radiation in the ultraviolet, visible, or infrared wavelengths during electron bombardment), and electrons absorbed by the specimen (also known as *specimen current*). An illustration of a backscatter electron detector with respect to the electron beam is given in Fig. 7.1. Figure 7.2 is an illustration of a typical SEM photo—with much depth of field—of a fibrous crystalline material (see also Fig. 2.1).

FIG. 7.2. Scanning electron microscope (SEM) photograph, using secondary electrons, of fibrous hollandite, $BaMn_8O_{16}$. (Photography courtesy of M. Spilde, Institute of Meteoritics, University of New Mexico.)

FIG. 7.1. Schematic drawing of a finely focused electron beam impinging on a material surface producing backscattered electrons (as well as other signals) inside the electron column of a scanning electron microscope (SEM). A backscatter electron detector (or secondary electron detector) is shown.

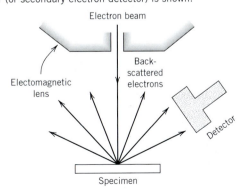

An important feature of the SEM is the three-dimensional appearance of the specimen image that results from the large depth of field. Resolution in an SEM ranges from about 50 to 25Å (5 to 2.5 nm).

Many SEM installations have an energy-dispersive X-ray detector system (EDS), which allows for the spectral analysis of X-rays generated from the specimen directly under the electron beam. Because most specimens studied by SEM have a rough (unpolished) surface, the chemical data obtained are generally not of the same high quality as those obtained by electron microprobe analysis (EMPA). Nonetheless, an SEM with EDX allows for topographic, crystallographic, and compositional information to be obtained rapidly and simultaneously from the same area.

OPTICAL MICROSCOPY

Optical studies of crystalline solids make use of a petrographic microscope, in which minerals can be studied with polarized light. Most of the discussion that follows will deal with translucent materials and transmitted-light optics although a brief section, at the end, will deal with reflected light optics for the study of opaque minerals.

The materials that are commonly studied with a petrographic microscope are small crystal fragments that were purposely broken off from a larger hand specimen that could not be identified with certainty just on the basis of properties assessed in hand specimen. Other samples are thin sections of crystalline materials (most commonly rocks) that are mounted on glass slides, with or without cover glasses, with a transparent cement. On both sample materials (small crystal fragments or thin sections) various optical measurements are made that may lead to their unique identification. The most quantitative optical identification technique involves the measurement of one or several refractive indices (RI) and such parameters as optic angle and sign, color, and orientation of optical directions with respect to crystallographic directions. Such optical data can be compared with reference data in tables of optical parameters, most of which are arranged in terms of refractive index (see Table 14.3 in Chapter 14, and references at the end of this chapter).

The optical technique is the quickest route to mineral identification, although such identification (in, e.g., a silicate that is part of a complex solid solution series) may well have to be supplemented with further chemical tests or information from X-ray diffraction. If the unknown mineral that is being identified (using optical parameters) is part of a relatively simple solid solution series (e.g., binary), considerable information about its chemical composition can be obtained if a variation diagram (relating optical parameters to changing chemical composition) is available (see Fig. 7.3). The optical technique of mineral identification is a quick and handy one for those who have mastered the various optical concepts that underly the measurements. However, students will find that gaining proficiency and self-confidence in optical measurements is a generally slow and a quite demanding learning experience.

Nature of Light

To account for all the properties of light, it is necessary to resort to two theories: the *wave theory* and the *corpuscular theory*. Here, we will concentrate on

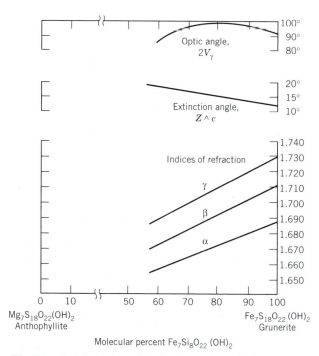

FIG. 7.3. Variation in several optical properties as a function of changing chemical composition in the monoclinic amphibole series known as the cummingtonite–grunerite series. The $Mg_7Si_8O_{22}(OH)_2$ composition is orthorhombic and is called anthophyllite (from C. Klein, 1964, Cummingtonite-grunerite Series: Chemical, Optical and X-ray study. *American Mineralogist* 49: 963–82).

the wave theory in explaining the optical behavior of crystals. This theory assumes that visible light, as part of the electromagnetic spectrum, travels in straight lines with a transverse wave motion; that is, it vibrates at right angles to the direction of propagation. The wave motion is similar to that generated by dropping a pebble into still water with waves moving out from the central point. The water merely rises and falls, it is only the wave front that moves forward. The *wavelength* (λ) of such a wave motion is the distance between successive crests (or troughs); the *amplitude* is the displacement on either side of the position of equilibrium; the *frequency* is the number of waves per second passing a fixed point; and the *velocity* is the frequency multiplied by the wavelength. Similarly light waves (Fig. 7.4) have length, amplitude, frequency, and velocity, but their transverse vibrations, perpendicular to the direction of propagation, take place in all possible directions.

Visible light occupies a very small portion of the electromagnetic spectrum (see Figs. 4.58 and 7.31). The wavelength determines the color and varies from slightly more than 7000 Å at the red end to about 4000 Å at the violet end. White light is composed of

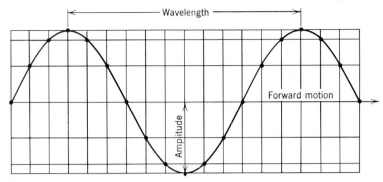

FIG. 7.4. Wave motion.

all wavelengths between these limits, whereas light of a single wavelength is called *monochromatic*.

Reflection and Refraction

When a light ray passes from a rare medium, such as air, into a denser medium, such as glass, part of it is reflected from the surface back into the air and part enters the glass (Fig. 7.5). The reflected ray obeys the laws of reflection:

1. The angle of incidence (*i*) equals the angle of reflection (*r'*), when both angles are measured from the surface normal.
2. The incident and reflected rays lie in the same plane.

The light that passes into the glass travels with a lesser velocity than in air and no longer follows the path of the incident ray but is bent or *refracted*. The amount of bending depends on the obliquity of the incident ray and the relative velocities of light in the two media; the greater the angle of incidence and the greater the velocity difference, the greater the refraction. In other words, *refraction* is the change in direction of wave propagation as light passes from one medium to another.

Refractive Index and Snell's Law

When light passes from one material into another it is either speeded up or slowed down as a function of the differing atomic structures of the two materials.

FIG. 7.5. Reflected and refracted light.

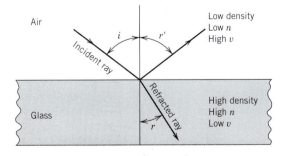

The refractive index (*n* or *R.I.*) of a substance is defined in terms of wave velocity:

$$n = \frac{\text{wave velocity in vacuum}}{\text{wave velocity in the material}} = \frac{c(\text{constant})}{v}$$

If follows from this that the *n* of a vacuum = 1.0, and for all other materials *n* is greater than 1.0. The velocity of light in vacuum is taken as unity and because, in comparison, the velocity of light in air = 0.9997 (almost the same as in vacuum), *n* air may also be considered 1.0. This allows for the following:

$$n = \frac{v(\text{air})}{v(\text{material})}$$

Because *n* air = 1.0, $n = \frac{1}{v(\text{material})}$. Stated another way, the index of refraction is the reciprocal of the velocity.

If the light velocity in two materials is different, the direction of propagation changes according to *Snell's law:*

$$\frac{\sin i}{\sin r} = \frac{n_2}{n_1} = \frac{v_1}{v_2},$$

Where *i* = angle of incidence, *r* = angle of refraction, n_1 and n_2 are refractive indices (see Fig. 7.5), and v_1 and v_2 are velocities of light.

A light ray passing from a medium of low *n* into one of higher *n* (see Fig. 7.5) is refracted toward the normal to the interface, and vice versa.

The velocity of light in glass is equal to frequency multiplied by wavelength; therefore, with fixed frequency, the longer the wavelength the greater the velocity. Red light with its longer wavelength has a greater velocity than violet light and because of the reciprocal relation between velocity and refractive index, *n* for red light is less than *n* for violet light (Fig. 7.6). A crystal thus has different refractive indices for different wavelengths of light. This phenomenon is known as *dispersion* and because of it,

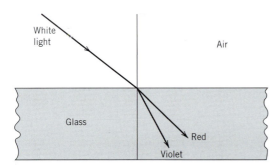

FIG. 7.6. Different refraction for different wavelengths of light.

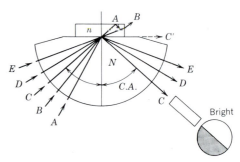

FIG. 7.8. Pulfrich refractometer and measurement of the critical angle *C.A.* $N = 1.90$, *C.A.* $= 50°$, $n = 1.455$.

monochromatic light is used for accurate determination of refractive index.

Total Reflection and the Critical Angle

We have seen (Fig. 7.5) that light is refracted toward the normal when it passes from a medium with a lower refractive index to a medium with a higher refractive index. When the conditions are reversed, as in Fig. 7.7, and the light moves from the higher to the lower index medium, it is refracted away from the normal. In Fig. 7.7, assume that lines *A*, *B*, *C*, and so forth represent light rays moving through glass, and into air at point *O*. The greater the obliquity of the incident ray, the greater the angle of refraction. Finally, an angle of incidence is reached, as at ray *D*, for which the angle of refraction is 90°, and the ray then grazes the surface. The angle of incidence at which this takes place is known as the *critical angle*. Rays such as *E* and *F*, striking the interface at a greater angle, are totally reflected back into the higher index medium.

The measurement of the critical angle is a quick and easy method of determining the refractive index of both liquids and solids. The instrument used is a refractometer, of which there are many types. A description of one of these, the Pulfrich refractometer, will suffice to illustrate the underlying principles. This instrument employs a polished hemisphere of high refractive index glass (Fig. 7.8). A crystal face or polished surface of the mineral is placed on the equatorial plane of the hemisphere and, depending on the

angle of incidence, light is either partly refracted through the unknown or totally reflected back through the hemisphere. If a telescope is placed in a position to receive the reflected rays, one can observe a sharp boundary between the portion of the field intensely illuminated by the totally reflected light and the remainder of the field. When the telescope is moved so that its cross hairs are precisely on the contact, the critical angle is read on a scale. Knowing this angle and the index of refraction of the hemisphere, *N*, one can calculate the index of refraction of the mineral: *n* (mineral) = sin critical angle × *N* (hemisphere).

Isotropic and Anisotropic Crystals

For optical considerations all transparent substances can be divided into two groups: *isotropic* and *anisotropic*. The isotropic group includes such non-crystalline substances as gases, liquids, and glass, but it also includes crystals that belong to the isometric crystal system. In them light moves in all directions with equal velocity. Hence, each isotropic substance has a single refractive index. In anisotropic substances, which include all crystals except those of the isometric system, the velocity of light varies with crystallographic direction, and thus, there is a range of refractive indices.

In general, light passing through an anisotropic crystal is broken into two polarized rays vibrating in mutually perpendicular planes. Thus, for a given orientation, a crystal has two indices of refraction, one associated with each polarized ray.

Polarized Light

We have seen that light can be considered a wave motion with vibrations taking place in all directions at right angles to the direction of propagation. When the wave motion is confined to vibrations in a single plane, the light is said to be *plane polarized* (Fig. 7.9). Two ways of polarizing light are by absorption and reflection.

FIG. 7.7. Light rays moving through glass and striking the glass–air interface at angle *C.A.*, and greater angles, are totally reflected.

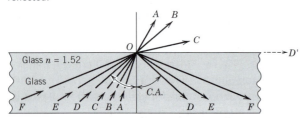

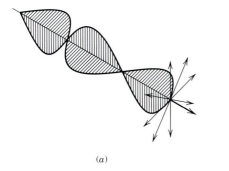

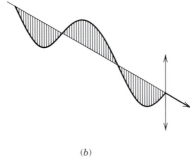

FIG. 7.9. Plane polarized light. *(a)* Unpolarized light with light vibrating in all directions at right angles to the direction of propagation. *(b)* Plane polarized light consisting of waves confined to vibration in only one plane through the line of propagation.

Polarized Light by Absorption

The polarized rays in which light is divided in anisotropic crystals may be differentially absorbed. If one ray suffers nearly complete absorption and the other very little, the emerging light will be plane polarized. This phenomenon is well illustrated by some tourmaline crystals (Fig. 7.10a). Light passed through the crystal at right angles to [0001] emerges essentially plane polarized, with vibrations parallel to the *c* axis. The other ray, vibrating perpendicularly to it, is almost completely absorbed. When two crystals are placed at right angles one above the other, the polarized ray emerging from one is absorbed by the other. Polarizing sheets, such as *Polaroid*, are made by aligning crystals on an acetate base. These crystals absorb very little light in one vibration direction, but are highly absorptive in the other (see Fig. 7.10b). The light transmitted by the sheet is thus plane polarized. Because they are thin and can be made in large sheets, manufactured polarizing plates are extensively used in optical equipment, including most polarizing microscopes.

Polarized Light by Reflection

Light reflected from a smooth, nonmetallic surface is partially polarized with the vibration directions parallel to the reflecting surface (see Fig. 7.11). The extent of polarization depends on the angle of incidence and the index of refraction of the reflecting surface. The fact that reflected light is polarized can

FIG. 7.10. Polarized light by absorption. *(a)* Tourmaline. *(b)* Polaroid. Arrows indicate directions of maximum transmission; directions of maximum absorption are at right angles.

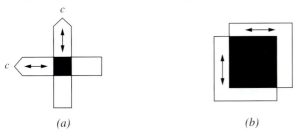

be easily demonstrated by viewing it through a polarizing filter. When the vibration direction of the filter is parallel to the reflecting surface, the light passes through the filter with only slight reduction in intensity (Fig. 7.11a); when the filter is turned 90° only a small percentage of the light reaches the eye (Fig. 7.11b). An example of this is the use of Polaroid lenses in sunglasses. Much light that is reflected from the ground surface, as well as objects, is horizontally polarized. Polaroid sunglasses are vertically polarized and therefore eliminate much reflected light (see Fig. 7.11c).

The Polarizing Microscope

The polarizing microscope is the most important instrument for determining the optical properties of minerals; with it more information can be obtained easily and quickly than with more specialized devices. Several manufacturers each make a number of models of polarizing microscopes that vary in complexity of design and sophistication and hence in price. A student model made by Nikon Inc. is illustrated in Fig. 7.12 with the essential parts named.

Although a polarizing microscope differs in detail from an ordinary compound microscope, its primary function is the same: to yield an enlarged image of an object placed on the stage. The magnification is produced by a combination of two sets of lenses: the objective and the ocular. The function of the objective lens, at the lower end of the microscope tube, is to produce an image that is sharp and clear. The ocular merely enlarges this image including any imperfection resulting from a poor quality objective. For mineralogical work it is desirable to have three objectives: low, medium, and high power. In Fig. 7.12 these are shown mounted on a revolving nosepiece and can be successively rotated into position. The magnification produced by an objective is usually indicated on its housing, such as 2 × (low), 10 × (medium), and 50 × (high). Oculars also have different magnifications such as 5×, 7×, 10×. The total magnification of the image can be de-

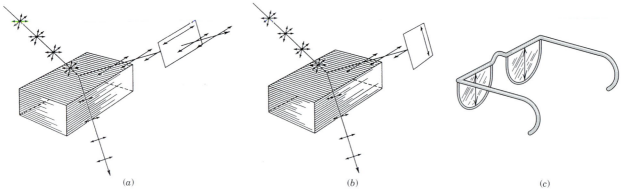

FIG. 7.11. Polarized light by reflection and refraction. *(a)* Polaroid filter transmits most reflected light. *(b)* Polaroid filter blocks reflected light. *(c)* Polaroid sunglasses that would block the reflected glare.

termined by multiplying the magnification of the objective by that of the ocular as: $50\times \cdot 10\times = 500\times$. Although in routine work the three objectives are frequently interchanged, a single ocular usually suffices. The ocular assembly, which slips into the upper end of the microscope tube, carries cross hairs—one N–S (front–back) the other E–W. These

FIG. 7.12. Polarizing microscope, Labophot-Pol, manufactured by Nikon Inc. 1. Oculars. 2. Analyzer. 3. Slot for accessory plate. 4. Revolving nosepiece for objectives. 5. Objectives. 6. Rotating stage. 7. Lever for swinging in and out condenser lens. 8. Rotatable polarizer. 9. Vertical adjustment of stage, for focusing (knobs on both sides). 10. Field diaphragm. 11. Substage illuminator. 12. Intensity adjustment knob for illuminator. 13. On/off switch. (Courtesy of S & M Microscopes Inc., Colorado Springs, CO.)

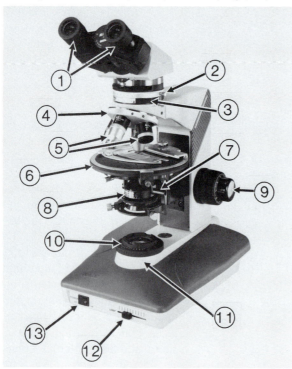

enable one to locate under high power a particular mineral grain that has been brought to the center of the field under low power. They are also essential in aligning cleavage fragments for making angular measurements. A condenser is located below the stage. The upper lens of the condenser, used with high-power objectives, makes the light strongly converging and can be rotated easily into or out of the optical system. The iris diaphragm, also located below the stage, can be opened or closed to control the depth of focus and to regulate the intensity of light striking the object.

In addition to the lenses, condenser, and diaphragm mentioned above which are common to all compound microscopes, the polarizing microscope has several other features. The *polarizer* below the stage is a polarizing plate that transmits plane polarized light vibrating in a N–S (front-back) direction. The *analyzer*, fitted in the tube above the stage, is a similar plate or prism that transmits light vibrating only in an E–W direction. The polarizer and analyzer are collectively called *polars*. When both polars are in position, they are said to be crossed and, if no anisotropic crystal is between them, no light reaches the eye. The polarizer remains fixed but the analyzer can be removed from the optical path at will. The Bertrand lens is an accessory that is used to observe interference figures (see page 300). In working with mineral grains it is frequently necessary to change their orientation. This is accomplished by means of a rotating stage, whose axis of rotation is the same as the microscope axis.

Microscopic Examination of Minerals and Rocks
The polarizing microscope is also called the *petrographic* microscope because it is used in the study of rocks. In examining thin sections of rocks, the textural relationships are brought out and certain optical

properties can be determined. It is equally effective in working with powdered mineral fragments. On such loose grains all the optical properties can be determined, and in most cases they characterize a mineral sufficiently to permit its identification.

The optimum size of mineral grains for examination with a polarizing microscope is minus 50 mesh to plus 100 mesh, but larger or smaller sizes may be used. To prepare a mount for examination (1) put a few mineral grains on an object glass, a slide 40 mm × 27 mm, (2) immerse the grains in a drop of liquid of known refractive index, and (3) place a cover glass on top of the liquid. When this type of mount is used the refractive indices of mineral grains are determined by the *immersion method.*

In using this method there should be available a series of calibrated liquids ranging in refractive index from 1.41 to 1.77, with a difference of 0.01 or less between adjacent liquids. These liquids cover the refractive index range of most of the common minerals. The immersion method is one of trial and error and involves comparing the refractive index of the unknown with that of a known liquid.

Isotropic Crystals and the Becke Line

Because light moves in all directions through an isotropic substance with equal velocities, there is no double refraction and only a single index of refraction. With a polarizing microscope, objects are always viewed in polarized light, which conventionally is vibrating N–S. If the object is an isotropic mineral, the light passes through it and continues to vibrate in the same plane. If the analyzer is inserted, darkness results because this polar permits light to pass only if it is vibrating in an E–W direction. Darkness remains as the position of the crystal is changed by rotating the microscope stage. This characteristic distinguishes isotropic from anisotropic crystals.

Let us first consider how the single refractive index of isotropic substances is determined. The first mount may be made by using any liquid, but if the

mineral is a complete unknown, it is wise to select a liquid near the middle of the range. When the grains are brought into sharp focus using a medium-power objective and plane polarized light, in all likelihood they will stand in relief, that is, they will be clearly discernible from the surrounding liquid. This is because the light is refracted as it passes from one medium to another of different refractive index. The farther apart the indices of refraction of mineral and liquid, the greater is the relief. But when the indices of the two are the same, there is no refraction at the interface, and the grains are essentially invisible. Relief shows that the index of refraction of the mineral is different from that of the liquid; but is it higher or lower? The answer to this important question can be found by means of the *Becke line* (Fig. 7.13). If the mineral grain is thrown slightly out of focus by raising the microscope tube (in most modern microscopes this is accomplished by lowering the stage), a narrow line of light will form at its edge and move toward the medium of higher refractive index. Thus if the Becke line moves into the mineral grain, a new mount must be made using a liquid of higher refractive index. After several tries it may be found that there is no Becke line and the mineral grains are invisible in a given liquid. The index of refraction of the mineral is then the same as that of the calibrated liquid. More frequently, however, the refractive index of the mineral is found to be greater than that of one liquid but less than that of its next higher neighbor. In such cases it is necessary to interpolate. If the Becke line moving into the mineral in the lower index liquid is more intense than the line moving out of the mineral into the next higher index liquid, it can be assumed that the refractive index of the mineral is closer to that of the higher liquid. In this way it is usually possible to report a refractive index to ±0.003.

Frequently the Becke line can be sharpened by restricting the light by means of the substage diaphragm. Most liquids have a greater dispersion than

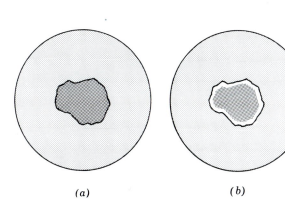

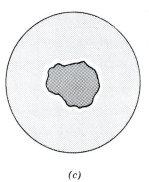

FIG. 7.13. The Becke Line. When thrown out of focus by raising of microscope tube (or lowering of stage), white line moves into medium of higher refractive index. *(a)* In focus. *(b)* n of grain > n of liquid. *(c)* n of grain < n of liquid.

(a)   *(b)*   *(c)*

minerals. Thus, if the refractive index of liquid and mineral are matched for a wavelength near the center of the spectrum, the mineral has a higher index than the liquid for red light but a lower index for violet light. This is evidenced, when observed in white light, by a reddish line moving into the grain while a bluish line moves out.

Aside from color, the single index of refraction is the only significant optical characteristic of isotropic minerals. It is, therefore, important in mineral identification to consider other properties such as cleavage, fracture, color, hardness, and specific gravity. Table 14.3 (chapter 14) gives a listing of increasing refractive index for many minerals.

Uniaxial Crystals

We have seen that light moves in all directions through an isotropic substance with equal velocity and vibrates in all directions at right angles to the directions of propagation. In hexagonal and tetragonal crystals, there is one and only one direction in which light moves in this way. This is parallel to the c axis, with vibrations in all directions in the basal plane (see Fig. 7.14a). For this reason the c axis is called the *optic axis*, and hexagonal and tetragonal crystals are called optically *uniaxial*. This distinguishes them from orthorhombic, monoclinic, and triclinic crystals, which have two optic axes and are called *biaxial*.

When light moves in uniaxial crystals in any direction other than parallel to the c axis, it is broken into two rays traveling with different velocities (see Fig. 7.14b). One, the *ordinary ray*, vibrates in the basal plane; the other, the *extraordinary ray*, vibrates at right angles to it and thus in a plane that includes the c axis. Such a plane, of which there is an infinite number, is referred to as the *principal section*. The nature of these two rays can be brought out in the following way. Assume that the direction of the incident beam is varied to make all possible angles with the crystal axes, and that the distance traveled by the resulting rays in any instant can be measured. We would find that:

1. One ray, with waves always vibrating in the basal plane, traveled the same distance in the same time. Its surface can be represented by a sphere, and because it acts much as ordinary light does it is the *ordinary ray (O ray)*.

2. The other ray, with waves vibrating in the plane that includes the c axis, traveled in the same time different distances depending on the orientation of the incident beam. If the varying distances of this, the *extraordinary ray (E ray)* were plotted, they would outline an ellipsoid of revolution, with the optic axis the axis of revolution.

Uniaxial crystals are divided into two optical groups: positive and negative. They are *positive* if the O ray has the greater velocity, and *negative* if the E

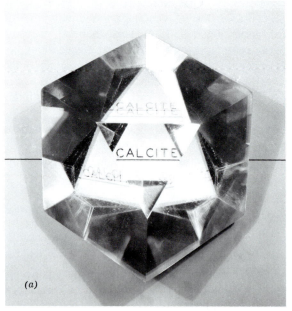

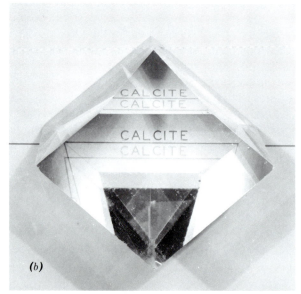

FIG. 7.14. *(a)* Calcite crystal with a triangular face cut perpendicular to the c axis. This means the viewer looks at the crystal parallel to the c axis (the optic axis). In this orientation calcite shows no double refraction. *(b)* Calcite showing double refraction when viewed normal to a rhombohedral face. The double repetition of "calcite" at the top of the photograph is seen obliquely through a face cut on the specimen parallel to the (0001) face. (Specimens from the Harvard Mineralogical Museum.)

ray has the greater velocity. Cross sections of the ray velocity surfaces are shown in Fig. 7.15. Note that in both positive and negative crystals the O and E rays have the same velocity when traveling along the optic c axis. But the difference in their velocities becomes progressively greater as the direction of light propagation moves away from the optic axis, reaching a maximum at 90°.

Because the two rays have different velocities, there are two indices of refraction in uniaxial crystals. Each index is associated with a vibration direction. The index related to vibration along the ordinary ray is designated ω (omega), whereas that associated with the extraordinary ray is ϵ (epsilon) or ϵ'. In positive crystals the O ray has a greater velocity than the E ray, and ω is less than ϵ. But in negative crystals with the E ray having the greater velocity, ω is greater than ϵ. The two principal indices of refraction of a uniaxial crystal are ω and ϵ and the difference between them is the *birefringence*.

The *uniaxial indicatrix* is a geometrical figure that is helpful in visualizing the relation of the refractive indices and their vibration directions that are perpendicular to the direction of propagation of light through a crystal. For positive crystals the indicatrix is a prolate spheroid of revolution; for negative crystals it is an oblate spheroid of revolution (Fig. 7.16). In their construction the direction of radial lines is proportional to the refractive indices. First, consider light moving parallel to the optic axis. It is not doubly refracted but moves through the crystal as the ordinary ray with waves vibrating in all directions in the basal plane. This is why light moving parallel to the c axis of calcite (Fig. 7.14a) produces a single image. There is a single refractive index for all these vibrations, proportional to the radius of the equatorial circle of the indicatrix. Now consider light traveling perpendicular to the optic axis. It is doubly refracted. The waves of the ordinary ray vibrate, as always, in the basal plane and the associated refractive index, ω, is again an equatorial radius of the indicatrix. The vibration direction of waves of the extraordinary ray must

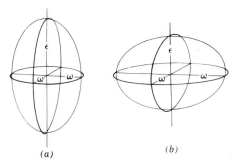

FIG. 7.16. Optical indicatrix, uniaxial crystals. *(a)* Positive. *(b)* Negative.

be at right angles to both the vibration direction of the ordinary waves and the direction of propagation. Thus, in this special case, it is parallel to the optic axis. The axis of revolution of the indicatrix is then proportional to ϵ, the greatest index in (+) crystals and the least in (−) crystals. It can be seen that light moving through a crystal in a random direction gives rise to two rays: (1) the O ray with waves vibrating in the basal section, the associated index, ω, and (2) the E ray with waves vibrating in the principal section in a direction at right angles to propagation. The length of the radial line along this vibration direction is ϵ', a refractive index lying between ω and ϵ.

A study of the indicatrix shows that (1) ω can be determined on *any* crystal grain; and only ω can be measured when light moves parallel to the optic axis, (2) ϵ can be measured only when light moves normal to the c axis, and (3) a randomly oriented grain yields, in addition to ω, an index intermediate to ω and ϵ, called ϵ'. The less the angle between the direction of light propagation and the normal to the optic axis, the closer is the value ϵ' to true ϵ.

Uniaxial Crystals Between Crossed Polars

Extinction. We have seen that because isotropic crystals remain dark in all positions between crossed polars, they can be distinguished from anisotropic crystals. However, there are special conditions under which uniaxial crystals present a dark field when viewed between crossed polars. One of these conditions is when light moves parallel to the optic axis. Moving in this direction, light from the polarizer passes through the crystal as through an isotropic substance and is completely cut out by the analyzer. The other special condition is when the vibration direction of light from the polarizer coincides exactly with one of the vibration directions of the crystal. In this situation, light passes through the crystal as either the O ray or the E ray to be completely eliminated by the analyzer, and the crystal is said to be at *extinc-*

FIG. 7.15. Ray velocity surfaces of uniaxial crystals.

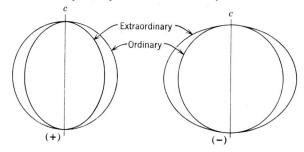

tion. As the crystal is rotated from this extinction position it becomes progressively lighter, reaching a maximum brightness at 45°. There are four extinction positions in a 360° rotation, one every 90°.

Interference. Let us consider how the crystal affects the behavior of polarized light as it is rotated from one extinction position to another. Figure 7.17 represents five positions of a tiny quartz crystal elongated on the *c* axis and lying on a prism face. In the diagrams it is assumed that light from the polarizer is moving upward, normal to the page, and vibrating in direction *P–P*. The vibration direction of the analyzer is *A–A*. The crystal in *a* is at an extinction position and light moves through it as the *E* ray vibrating parallel to the *c* axis. At 90°, position *b*, the crystal is also at extinction with light moving through it as the *O* ray. When the crystal is turned as in *c*, *d*, and *e*, polarized light entering it is resolved into two components. One moves through it as the *O* ray, vibrating in the basal plane; the other as the *E* ray, vibrating in the principal section. In *c* most light is transmitted as the *E* ray but in *e* it is transmitted mostly as the *O* ray. In *d*, the 45° position, the amounts of light transmitted by the two rays are equal.

When these rays from the crystal enter the analyzer, each is broken up into an *O* and *E* ray conforming in vibration directions to those of the analyzer. Only the components of the rays vibrating in an E–W direction are permitted to pass. During their passage through the crystal the two rays travel with different velocities and thus on emerging there is a phase difference because one is ahead of the other. The amount it is ahead depends both on the difference in velocities and on the thickness of crystal traversed. Because both rays vibrate in the same plane of the analyzer, they interfere. For monochro-

matic light, if one ray is an integral number ($n = 1$, 2, 3, etc.) of wavelengths ($n\lambda$) behind the other, it is said to be retarded (retardation (Δ) is expressed as $\Delta = n\lambda$, with $n = 1, 2, 3$, etc.). Such inference conditions result in darkness. On the other hand, if the path difference is $\lambda/2$, $3\lambda/2$, or in general retardation, $\Delta = n + \frac{1}{2}\lambda$, the waves reinforce one another to produce maximum brightness.

Each wavelength has its own set of critical conditions when interference produces darkness. Consequently, when white light is used, "darkness" for one wavelength means its elimination from the spectrum and its complementary color appears. The colors thus produced are called *interference colors*. There are different *orders* of interference, depending on whether the color results from a path difference of 1λ, 2λ, 3λ, . . . , $n\lambda$. These are called first-order, second-order, third-order, and so forth interference colors, as indicated on the interference color plate that appears on the endpaper at the back of this text.

Interference colors depend on three factors: optical orientation, thickness, and *birefringence*, which is the difference between the two refractive indices (i.e., $\omega - \epsilon$ for negative crystals and $\epsilon - \omega$ for positive crystals). Birefringence is an inherent characteristic of a given mineral, as are its refractive indices. The colored chart of interference colors (at the back of this text) is a graphical representation of the relationship between retardation (Δ), birefringence and crystal thickness. From left to right the chart displays the normal sequence of interference colors between crossed polars. Each color results from a specific retardation as expressed in nanometers (nm) along the bottom of the chart. Retardation, and the resulting color, are dependent on thickness. The thickness is given along the left, vertical scale, in micrometers (μm). Birefringence ($n_2 - n_1$) values are listed at the top of the chart with lines radiating from the lower left corner representing values of equal birefringence as a function of thickness. In summary, the interference color chart is a graphical representation of the formula $\Delta = d (n_2 - n_1)$, which expresses the fact that retardation (Δ) is a function of crystal thickness (d) and birefringence ($n_2 - n_1$).

With a continuous change in direction of light, from parallel to perpendicular to the optic axis, there is a continuous increase in interference colors. For a given orientation, the thicker the crystal and the greater its birefringence the higher the order of interference color. If a crystal plate is of uniform thickness, as a cleavage flake may be or a grain in a rock thin section, it will show a single interference color. As seen in immersion liquids, grains commonly vary in thickness and a variation in interference colors reflects this irregularity.

FIG. 7.17. Quartz crystal between crossed polars.

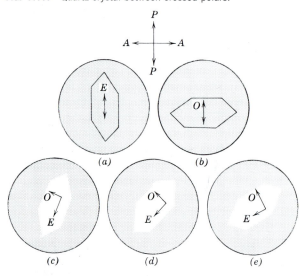

Accessory Plates

The gypsum plate, mica plate, and quartz wedge are accessory plates used with the polarizing microscope; their function is to produce interference of known amounts and thus predetermined colors. They are all constructed so that the fast ray (the vibration direction of the lesser refractive index) is parallel to the long dimension. The *gypsum plate* is made by cleaving a gypsum crystal to such a thickness that in white light it produces a uniform red interference color: *red of the first order*. The *mica plate* is made with a thin mica flake, cleaved to a thickness such that for yellow light it yields a path difference of a quarter of a wavelength. It is thus also called the *quarter wave plate*. The *quartz wedge* is an elongated, wedge-shaped piece of quartz with the vibration direction of the fast ray (ω) parallel to its length and the slow ray (ϵ) across its length. As thicker portions of the wedge are placed in the optical path, the path difference of the rays passing through it also increases, producing a succession of interference colors. The number of orders depends on the wedge angle: the greater the angle the more orders per unit of length.

Uniaxial Crystals in Convergent Polarized Light

What are known as *interference figures* are seen when properly oriented crystal sections are examined in convergent polarized light. To see them, the polarizing microscope (usually used as an orthoscope) is converted to a conoscope by swinging in the upper substage condensing lens, so that the section can be observed in strongly converging light, using a high-power objective. The interference figure then appears as an image just above the upper lens of the objective and can be seen between crossed polars by removing the ocular and looking down the microscope tube. If the Bertrand lens, an accessory lens located above the analyzer, is inserted, an enlarged image of the figure can be seen through the ocular.

The principal interference figure of a uniaxial crystal, the optic axis figure, Fig. 7.18, is seen when one views the crystal parallel to the *c* axis. Only for

the central rays from the converging lens is there now double refraction; the others, traversing the crystal in directions not parallel to the *c* axis, are resolved into *O* and *E* rays having increasing path difference as the obliquity to the *c* axis increases. The interference of these rays produces concentric circles of interference colors. The center is black with no interference, but moving outward there is a progression from first-order to second-order to third-order, and so forth, interference colors. If the crystal section is of uniform thickness no change will be noted as it is moved horizontally. If, however, the thickness varies, as in a wedge-shaped fragment, the positions of the colors change with horizontal movement. At the thin edge there may be only gray of the first-order, but as the crystal is moved so the light path through it becomes greater, all the first-order colors may appear. And with increasing crystal thickness, the path difference of the two rays may be great enough to yield second-, third-, and higher-order interference colors.

The reason for the black cross superimposed on the rings of interference colors is brought out in Fig. 7.18*a*. In this drawing the radial dashes indicate the vibration directions of the *E* ray and those at right angles indicate the vibration directions of the *O* ray. It will be seen that where these vibration directions are parallel or nearly parallel to the vibration directions of the polarizer and analyzer no light passes and thus the formation of the dark cross.

Figure 7.18 illustrates a centered optic axis figure as obtained on a crystal plate whose *c* axis coincides with the axis of the microscope; as the stage is rotated, no movement of the figure is seen. If the optic axis of the crystal makes an angle with the axis of the microscope, the black cross is no longer symmetrically located in the field of view (Fig. 7.19). When the stage is rotated, the center of the cross moves in a circular path, but the bars of the cross remain parallel to the vibration directions of the polarizer and analyzer. Even if the inclination of the optic axis is so great that the center of the cross does not appear, on rotation of the crystal the bars move

(a)

(b)

(c)

FIG. 7.18. Uniaxial optic axis interference figures. *(a)* Radial lines indicate vibration directions of the *E* ray; tangential lines indicate vibration directions of the *O* ray. *(b)* and *(c)* show isochromatic curves.

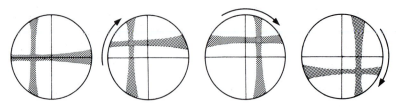

FIG. 7.19. Positions of an off-centered uniaxial optic axis figure on clockwise rotation of the microscope stage.

across the field, maintaining their parallelism to the vibration directions of the polars.

The *flash figure* is an interference figure produced by a uniaxial crystal when its optic axis is normal to the axis of the microscope; that is, a hexagonal or tetragonal crystal lying on a face in the prism zone. When the crystal is at an extinction position, the figure is an ill-defined cross occupying much of the field. On rotation of the stage, the cross breaks into two hyperbolas that rapidly leave the field in those quadrants containing the optic axis. The cross forms because the converging light is broken into O and E rays with vibration directions mostly parallel or nearly parallel to the vibration directions of the polarizer and analyzer. A centered flash figure not only indicates the vibration direction of the E ray, but assures one that in this direction a true value of ϵ can be obtained in plane polarized light.

Determination of Optic Sign

The mica plate, the gypsum plate, and the quartz wedge may be used with a uniaxial optic axis figure to determine the optic sign; that is, whether the crystal is positive or negative. They are inserted below the analyzer in a slot in the microscope tube so positioned that when the plates are in place, their vibration directions make angles of 45° with the vibration directions of the polars.

From the previous discussion we have learned that in the optic axis interference figure, the E ray vibrates radially and the O ray tangentially. By use of an accessory plate in which vibration directions of the slow and fast rays are known, one can tell whether the E ray of the crystal is slower (positive

crystals) or faster (negative crystals) than the O ray and thus determine the optic sign. For most American-made equipment the vibration of the slow ray is at right angles to the length of the plate and is so marked on the metal carrier. However, before using an accessory the vibration directions should be checked. The principle in the use of all the plates is the same: to add or subtract from the path difference of the O and E rays of the crystal.

If the *mica plate* is superimposed on a uniaxial optic axis figure in which the ordinary ray is slow (negative crystal), the interference of the plate reinforces the interference colors in the SE and NW quadrants, causing them to shift slightly toward the center. At the same time subtraction causes the colors in the NE and SW quadrants to shift slightly away from the center. The most marked effect produced by the mica plate is the formation of two black spots near the center of the black cross in the quadrants where subtraction occurs (Fig. 7.20).

The *gypsum plate* is usually used to determine the optic sign when low-order interference colors or no colors at all are seen in the optic axis figure. It has the effect of superimposing red of the first-order on the interference figure. If the figure shows several orders of interference colors, one should consider the color effect on the grays of the first-order near the center. In the quadrants where there is addition, the red plus the gray gives blue; in the alternate quadrants the red minus the gray gives yellow. The arrangement of colors in positive crystals is: yellow SE–NW; blue NE–SW; and in negative crystals, yellow NE–SW, blue SE–NW. It is suggested that the student insert the colors in Fig. 7.21.

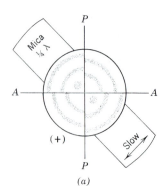

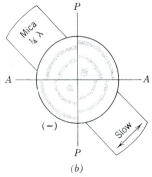

FIG. 7.20. Determination of optic sign with a mica plate.

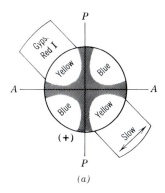
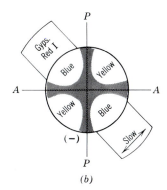

(a)   (b)

FIG. 7.21. Determination of optic sign with a gypsum plate.

The *quartz wedge* is most effective in determining optic sign when high-order interference colors are present in the optic axis figure. The wedge is usually inserted with the thin edge first. If its retardation is added to that of the crystal, the interference colors in two opposite quadrants will increase progressively as the wedge moves through the microscope tube. If the retardation is subtracted from that produced by the crystal, the order of colors will decrease. Thus as the quartz wedge is slowly inserted over an optic axis figure of a negative crystal, the color bands in the SE–NW quadrants move toward the center and disappear. At the same time in the NE–SW quadrants the colors move outward to the edge of the field. In a positive crystal similar phenomena are observed, but the colors move in the opposite directions, that is, away from the center in the SE–NW quadrants and toward the center in the NE–SW quadrants.

Sign of Elongation

Hexagonal and tetragonal crystals are frequently elongated on the *c* axis or have prismatic cleavage that permits them to break into splintery fragments also elongated parallel to *c*. If such an orientation is known, one can determine the optic sign by turning the elongated grain to the 45° position and inserting the gypsum plate. If the interference colors rise (gray of

FIG. 7.22. Determination of sign of elongation with a gypsum plate. In diagrams Y = yellow, B = blue. *(a)* Positive elongation, *(b)* negative elongation.

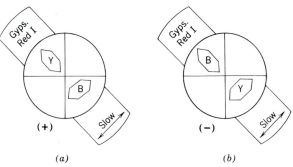

(a)   (b)

mineral plus first-order red equals blue), the slow ray of the gypsum has been superimposed on the slow ray of the mineral. If this is also the direction of elongation, it means that the *E* ray is slow (ϵ is the higher refractive index) and the mineral has positive elongation and is optically positive. When the slow ray of the gypsum plate is parallel to the elongation of the mineral grain and the interference colors fall (gray of mineral minus first-order red equals yellow), the mineral has negative elongation and is optically negative (Fig. 7.22).

Very commonly the interference colors of small grains are grays of the first order. Thus, on superimposing red of the first order, addition gives a blue color and subtraction a yellow color.

Absorption and Dichroism

In the discussion of polarized light (page 294) it was pointed out that in some tourmaline the absorption of one ray is nearly complete but for the other ray it is negligible. Although less striking, many crystals show a similar phenomenon; more light is absorbed in one vibration direction than in the other. In tourmaline where absorption of the *O* ray is greatest, it is expressed as, absorption: $O > E$ or $\omega > \epsilon$. In other crystals certain wavelengths may be absorbed in one direction and the complementary colors are transmitted. Thus, the crystal has different colors in different vibration directions and is said to be *dichroic*. Dichroism is expressed by giving the colors, for example, *O* or ω = yellow, *E* or ϵ = pink. Absorption is independent of other properties and is considered, as are refractive indices, a fundamental optical property of crystals.

Biaxial Crystals

Orthorhombic, monoclinic, and triclinic crystals are called optically biaxial because they have two directions in which light travels with zero birefringence. In uniaxial crystals there is only one such direction.

Light moving through a biaxial crystal, except along an optic axis, travels as two rays with mutually perpendicular vibrations. The velocities of the rays differ from each other and change with changing crystallographic direction. The vibration directions of the fastest ray, *X*, and the slowest ray, *Z*, are at right angles to each other. The direction perpendicular to the plane defined by *X* and *Z* is designated as *Y*. For biaxial crystals there are thus three indices of refraction resulting from rays vibrating in each of these principal optical directions. The numerical difference between the greatest and least refractive indices is the *birefringence*. Various letters and symbols have been used to designate the refractive indices, but the most generally accepted are the Greek letters as follows:

Index*		Direction	Ray Velocity
(alpha)	α lowest	*X*	highest
(beta)	β middle	*Y*	intermediate
(gamma)	γ highest	*Z*	lowest

*Other equivalent designations are: $\alpha = nX, n_x, N_x, N_p$; $\beta = nY, n_y, N_y, N_m$; $\gamma = nZ, n_z, N_z, N_g$.

The Biaxial Indicatrix

The biaxial indicatrix is a visualization of relative values of the three principal indices of refraction and is used in much the same manner as the uniaxial indicatrix. Its general shape (see Fig. 7.23*a*) is that of a triaxial ellipsoid with its three axes mutually perpendicular optical directions *X*, *Y*, and *Z*. The lengths of the semi-axes are proportional to the refractive indices: α along *X*, β along *Y*, and γ along *Z*. Sections through this triaxial ellipsoid are all ellipses (see Fig. 7.23*b* and *c*) except for two sections that are marked as *circular sections* (in Fig. 7.23*c*) with their radii noted as *S*.

The two directions normal to these sections are the *optic axes*; and the *XZ* plane in which they lie is called the *optic plane*. The *Y* direction perpendicular to this plane is the *optic normal*. Light moving along the optic axes and vibrating in the circular sections shows no birefringence and gives the constant refractive index, β. The optic axis of a uniaxial crystal is analogous to these directions, because light moving parallel to it also vibrates in a circular section with constant refractive index.

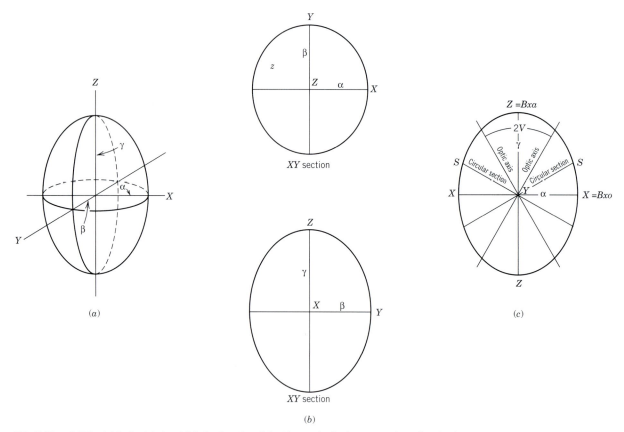

FIG. 7.23. *(a)* Biaxial indicatrix in which the lengths of the three principal axes are the refractive indices α, β, γ, measured along the optical directions *X*, *Y*, and *Z*. *(b)* and *(c)* Principal sections through the biaxial indicatrix of a positive crystal. The section in *(c)* shows the location of the two circular sections with the optic axes perpendicular to them.

With variation in refractive indices there is a corresponding variation in the axial lengths of the biaxial indicatrix. Some crystals are nearly uniaxial and in these the intermediate index, β, is very close to either α or γ. If β is close to α, the circular sections make only a small angle with the XY plane and the optic axes make the same angle with the Z direction. This is angle V and the angle between the two optic axes, known as the *optic angle*, is 2V. The optic angle is always acute and because, in this case, it is bisected by Z, Z is called the *acute bisectrix* (Bxa); X is the *obtuse bisectrix* (Bxo) because it bisects the obtuse angle between the optic axes. When Z is the Bxa, the crystal is optically positive.

If β is closer to γ than to α, the acute angle between the optic axes is bisected by X and the obtuse angle bisected by Z. In this case, with X the Bxa, the crystal is negative. When β lies exactly halfway between α and γ, the optic angle is 90°.

The relation between the optic angle and the indices of refraction is expressed by formula (a). A close approximation to the optic angle can be made using formula (b).

$$\text{(a) } \cos^2 V_x = \frac{\gamma^2(\beta^2 - \alpha^2)}{\beta^2(\gamma^2 - \alpha^2)}$$

$$\text{(b) } \cos^2 V_x = \frac{\beta - \alpha}{\gamma - \alpha}$$

The error using the simplified formula increases with increase of both birefringence and V and always yields values for V' less than true V. It should be noted that in using either formula *half* the optic angle is calculated, and that it is determined with X the bisectrix. Thus, when V < 45°, the crystal is negative but when V > 45°, the crystal is positive.

Optical Orientation in Biaxial Crystals

The orientation of the optical indicatrix is one of the fundamental optical properties. It is given by expressing the relationship of X, Y, and Z optical directions to the crystallographic axes, a, b, and c.

In *orthorhombic crystals*, each of the crystallographic axes is coincident with one of the principal optical directions. For example, the optical orientation of anhydrite is X = c, Y = b, Z = a. Usually the optical directions coinciding with only two axes are given, for this completely fixes the position of the indicatrix.

It is difficult or impossible to determine the axial directions on microscopic grains of some minerals. To do it, one must use a fragment oriented by X-ray study or broken from a faced crystal. However, even in small particles, orientation can be expressed relative to cleavages. Powdered fragments tend to lie on cleavages, which in orthorhombic crystals are commonly pinacoidal or prismatic. For example, barite has {001} and {210} cleavage and most grains lie on faces of these forms. Those lying on {001} will be diamond shaped (Fig. 7.24a) and have *symmetrical extinction*. That is, the extinction position makes equal angles with the bounding cleavage faces. Fragments lying on {210} will have *parallel extinction* (Fig. 7.24b). Symmetrical and parallel extinction are characteristic of orthorhombic crystals.

Barite is (+), X = c, Y = b, Z = a. Thus, grains lying on {001} yield a centered Bxo figure and one can determine β in the b direction and γ in the a direction. Grains showing parallel extinction do not give a centered interference figure, but α can be measured parallel to c.

In *monoclinic crystals* one of the principal optical directions (X, Y, or Z) of the indicatrix coincides with the b axis; the other two lie in the a–c plane of the crystal. The orientation is given by stating which optical direction equals b and indicating the *extinction angle*, the angle between the c axis and one of the other optical directions. If the extinction lies between the + ends of the a and c axes, the angle is positive; between +c and −a, the angle is negative. In gypsum Y = b and Z ∧ c = 53°. Thus a fragment lying on the {010} cleavage would yield an optic normal interference figure and α and γ could be determined; γ at the extinction position +53°; α at extinction position −37° (see Fig. 7.24c). A grain lying

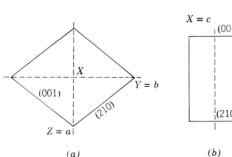 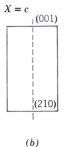 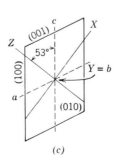

FIG. 7.24. Optical orientation. *(a)* Barite on {001} showing symmetrical extinction. *(b)* Barite on {210} showing parallel extinction. *(c)* Gypsum on {010} showing extinction angle.

(a)   (b)   (c)

on the {100} cleavage would show parallel extinction and β could be measured at right angles to the trace of {010}. Crystal fragments lying on any other face in the [001] zone, as on {110}, will show an extinction angle, but the angle to be recorded, usually that of *maximum extinction*, is observed on {010}. Parallel extinction indicates the grain is lying on a face in the [010] zone and the index of the ray vibrating parallel to *b* can be measured.

In *triclinic* crystals the optic indicatrix can occupy any position relative to the crystallographic axes. Thus, a complete optical orientation necessitates giving φ and ρ angles of the principal optical directions. But in most cases it suffices to give the extinction angles observed on grains lying on known cleavage faces.

Biaxial Crystals in Convergent Polarized Light

Biaxial interference figures are obtained and observed in the same manner as uniaxial figures, that is, with converging light, high-power objective, and Bertrand lens. Although interference figures can be observed on random sections of biaxial crystals, the most symmetrical and informative are obtained on sections normal to the optical directions *X*, *Y*, and *Z* and to an optic axis.

The *acute bisectrix figure* is observed on a crystal plate cut normal to the acute bisectrix. If 2*V* is very small, there are four positions during a 360° rotation at which the figure resembles the uniaxial optic axis figure. That is, a black cross is surrounded by circular bands of interference colors. However, as the stage is turned, the black cross breaks into two hyperbolas that have a slight but maximum separation at a 45° rotation; and the color bands, known as *isochromatic curves*, assume an oval shape. The hyperbolas are called *isogyres* and the dark spots, called *melatopes*, at their vertices in the 45° position, result from light rays that traveled along the optic

axes in the crystal. Thus with increasing optic angle the separation of the isogyres increases, and the isochromatic curves are arranged symmetrically about the melatopes as shown in Fig. 7.25. For most crystals when 2*V* exceeds 60° the isogyres leave the field at the 45° position; the larger the optic angle, the faster they leave.

The portion of the interference figure occupied by the isogyres is dark, for here, light as it emerges from the section has vibration directions parallel to those of the polarizer and analyzer. The dark cross is thus present when the obtuse bisectrix and optic normal coincide with the vibration directions of the polars. The bar of the cross parallel to the optic plane is narrower and better defined than the other bar (Fig. 7.25). Because light travels along the optic axes with no birefringence, their points of emergence are, of course, dark in all positions of the figure.

The Apparent Optic Angle

The distance between the points of emergence of the optic axes is dependent not only on 2*V* but on β as well. The refractive index of crystals is β for light rays moving along the optic axes. These rays are refracted on leaving the crystal, giving an apparent optic angle, 2*E*, greater than the real angle, 2*V* (Fig. 7.26). The higher the β refractive index, the greater the refraction. Thus if two crystals have the same 2*V*, the one with the higher β index has the larger apparent angle which results in the optic axes being farther apart as they emerge in the interference figure.

The *optic axis figure* is observed on mineral grains cut normal to an optic axis. Such grains are easy to select, for they remain essentially dark between crossed polars on complete rotation. The figure consists of a single isogyre, at the center of which is the emergence of the optic axis. When the optic plane is parallel to the vibration direction of either polar, the isogyre crosses the center of the field as a

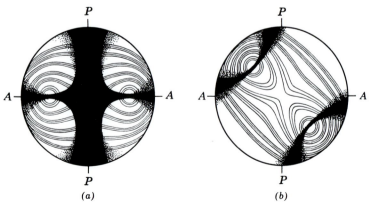

(a) (b)

FIG. 7.25. Acute bisectrix interference figure. *(a)* Parallel position. *(b)* 45° position.

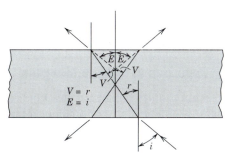

FIG. 7.26. Relation of 2V to 2E.

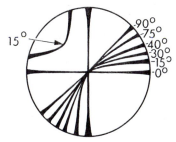

FIG. 7.27. Curvature of isogyre in optic axis figure from 0° to 90° 2V.

straight bar. On rotation of the stage it swings across the field forming a hyperbola in the 45° position. In this position the figure can be pictured as half an acute bisectrix figure with the convex side of the isogyre pointing toward the acute bisectrix. As 2V increases, the curvature of the isogyre decreases and when 2V = 90°, the isogyre is straight (Fig. 7.27).

The *obtuse bisectrix figure* is obtained on a crystal section cut normal to the obtuse bisectrix. When the plane of the optic axes is parallel to the vibration direction of either polar, there is a black cross. On rotation of the stage the cross breaks into two isogyres that move rapidly out of the field in the direction of the acute bisectrix. Although not as informative as an acute bisectrix figure, a centered obtuse bisectrix figure indicates that an accurate determination of β and of either α or γ can be made on the mineral section producing it.

The *optic normal figure* is obtained on sections cut parallel to the plane of the optic axes and resembles the flash figure of a uniaxial crystal. When the X

and Z optical directions are parallel to the vibration directions of the polars, the figure is a poorly defined cross. On slight rotation of the stage it splits into hyperbolas that move rapidly out of the field in the quadrants containing the acute bisectrix. An optic normal figure is obtained on sections with maximum birefringence and indicates that α and γ can be determined on this section.

Determination of Optic Sign of a Biaxial Crystal

The optic sign of biaxial crystals can best be determined on acute bisectrix or optic axis figures with the aid of accessory plates. Let us assume that a in Fig. 7.28 represents an acute bisectrix figure of a negative crystal in the 45° position. By definition, X is the acute bisectrix and Z the obtuse bisectrix. OP is the trace of the optic plane and Y the vibration direction of β, at right angles. The velocity is constant for all rays moving along the optic axes, and for them the refractive index of the crystal is β, including those vibrating in the optic plane. Consider the velocities of other rays,

FIG. 7.28. Optic sign determination of negative crystal with gypsum plate. *(a)* Acute bisectrix figure. *(b)* Optic axis figure.

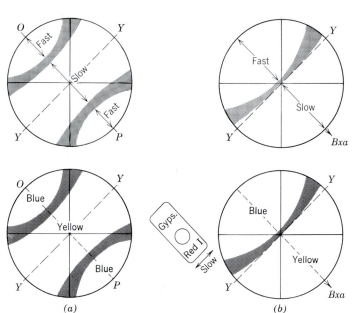

(a)   (b)

vibrating in the optic plane. In a negative crystal, those emerging between the isogyres of an acute bisectrix figure have a lesser velocity, those emerging outside the isogyres have a greater velocity. If the gypsum plate is superimposed over such a figure, the slow ray of the plate combines with the fast ray of the crystal and subtraction of interference colors produces a yellow color on the convex side of the isogyres. On the concave sides of the isogyres a blue color is produced by addition, the slow ray of the plate over the slow ray of the crystal. In a positive crystal the reverse color effect is seen, for here Z is the acute bisectrix.

An optic axis figure in the 45° position can be used in a manner similar to the acute bisectrix figure in determining optic sign. Insertion of the gypsum plate yields for $(-)$ crystal: convex side yellow, concave side blue (Fig. 7.28b), for $(+)$ crystal: convex side blue, concave side yellow.

The quartz wedge may be used to determine optic sign if several isochromatic bands are present. As the wedge is inserted, the colors move out in those portions of the figure where there is subtraction and move in where there is addition. In other words, in the areas that a gypsum plate would render blue, the colors move in; in those that it would render yellow, they move out.

There is a tendency for students learning optical techniques to search for interference figures in the initial immersion when, in all probability, the indices of the mineral are far removed from the refractive index of the liquid. It is time saving, particularly in working with biaxial minerals, to first match as closely as possible the refractive indices of the unknown with n of the liquid. An interference figure, then, in addition to the optic sign, will yield other useful information. Using a grain giving an optic axis figure, β can be compared with n of the liquid. On a grain yielding an acute bisectrix figure, β and either α or γ can be compared with the refractive index of the liquid. The indices of a grain showing highest interference colors are most likely to be close to α and γ. Therefore, in the same mount, one should check such a grain to estimate how far α and γ are from the refractive index of the liquid.

Absorption and Pleochroism

The absorption of light in biaxial crystals may differ in the X, Y, and Z optical directions. If the difference is only in intensity and X has the greatest absorption and Z the least, it is expressed as $X > Y > Z$. If different wavelengths are absorbed in different directions, the mineral is said to be pleochroic and the color of the transmitted light is given. For example, for a member of the orthopyroxene series, the *pleochroism* may be stated as X = brownish red, Y = reddish yellow, Z = green. The term *pleochroism* is com-

monly used to denote all differential absorption in both uniaxial and biaxial crystals.

Optical Properties of Opaque Minerals

The discussion in this chapter has dealt entirely with the optical properties of nonopaque minerals. It should be mentioned, however, that opaque minerals also possess optical properties by which they are characterized. Specially built microscopes are used for viewing polished sections of these minerals in reflected light. Because most of the ore minerals are opaque, the techniques for polished section study with the *ore microscope* (or *reflecting light microscope*) have been largely developed and used by the student of ores (see Fig. 7.29 for an illustration of the light path in a reflecting light microscope). In a polished section, color is the first and frequently the most important property observed, and the skilled microscopist is able to distinguish one mineral from another by a subtle color difference. Figure 7.30 illustrates the appearance of some opaque minerals in the ore microscope. Using polarized light many nonisometric minerals show a *bireflectance*, a property analogous to pleochroism in nonopaque minerals.

FIG. 7.29. Schematic cross section through a reflecting light microscope showing the light path and some of the components of the microscope.

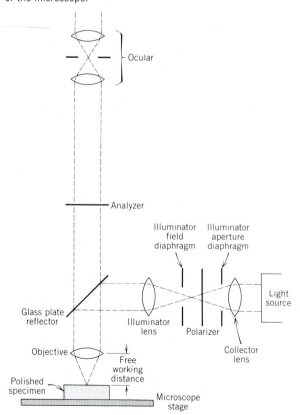

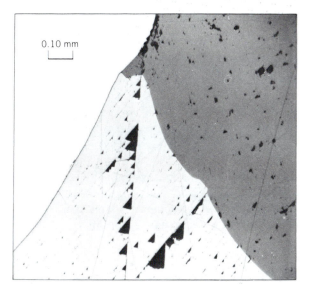

FIG. 7.30. Photomicrograph of a polished section of opaque minerals from Balmat, New York, as obtained with a reflecting light microscope. The minerals are: white, pyrite; light gray with triangular pits, galena; dark gray, sphalerite. (Courtesy of Charles Grocetti, Harvard University.)

That is, there is a change in brightness or color as the mineral is rotated on the microscope stage. Different minerals show this property in varying degrees. When viewed between crossed polars, nonisometric opaque minerals will show, as do nonopaque minerals, four positions of extinction in a 360° rotation. Isotropic minerals show neither bireflectance nor extinction positions and thus can be distinguished from anisotropic minerals.

The microscopic study of polished sections of opaque minerals gives important clues to the paragenesis and origin of mineral deposits. From the textural relationship of the minerals one can determine their order of deposition and their subsequent replacement and exsolution phenomena. There is a large literature on ore microscopy, but even a brief discussion of the methods and techniques is beyond the scope of this book.

X-RAY DIFFRACTION TECHNIQUES (XRD)

Ever since the discovery of X-rays by Wilhelm Conrad Roentgen in 1895, and the first application of an X-ray experiment to the study of crystalline material in 1912 by Max von Laue (see p. 7), X-ray diffraction techniques have been fundamental to crystal structure analysis. All that is now known about the location of atoms, their sizes, and their bonding in crystal structures—as well as our knowledge of a structure's space group symmetry and chemical composition of its unit cell—has been derived largely from X-ray diffraction studies. X-ray crystallography is a science that is pursued by researchers with a broad knowledge base in mathematics, physics, chemistry and computational skills.

This section of the text briefly introduces X-rays and their diffraction effects, and then there is a short overview of crystal structure analysis, using *single-crystal techniques* and an introduction to the *X-ray powder diffraction method*, which is routinely used in mineral identification. Powder diffraction equipment is generally available in most Geology departments and this identification method is especially powerful in the study of minerals (and other materials) that are too fine-grained to be evaluated by optical microscope techniques. Examples of such materials are members of the clay mineral and zeolite groups.

X-Ray Spectra

Electromagnetic waves form a continuous series varying in wavelength from long radio waves with wavelengths of the order of thousands of meters to cosmic radiation whose wavelengths are of the order of 10^{-12} meters (a millionth of a millionth of a meter!). All forms of electromagnetic radiation have certain properties in common, such as propagation along straight lines at a speed of 300,000 km per second in a vacuum, reflection, refraction according to Snell's law, diffraction at edges and by slits or gratings and a relation between energy and wavelength given by the Einstein equation:

$$E = h\nu = hc/\lambda$$

where E is energy, ν frequency, c velocity of propagation, λ wavelength, and h Planck's constant. Thus, the shorter the wavelength the greater its energy and the greater its powers of penetration. X-rays occupy only a small portion of the spectrum, with wavelengths varying between slightly more than 100 Å and 0.02 Å (see Fig. 7.31). X-rays used in the investigation of crystals have wavelengths of the order of

FIG. 7.31. The electromagnetic spectrum.

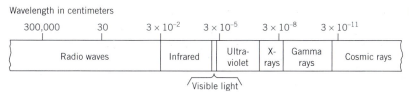

Wavelength in centimeters

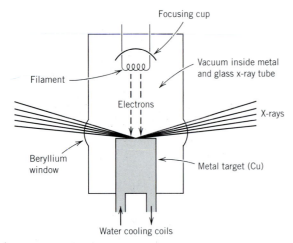

FIG. 7.32. Schematic representation of a sealed vacuum X-ray tube. The tungsten filament is heated to very high temperature, making electrons boil off. The differential voltage between the filament and the target metal accelerates the electrons toward the target. When the electrons strike the target, X-rays are produced that are able to leave the X-ray tube housing through beryllium windows.

1 Å. Visible light has wavelengths between 7200 and 4000 Å, more than 1000 times as great, and hence is less penetrating and energetic than X-radiation.

When electrons moving at high velocity strike the atoms of any element, as is the case in an X-ray tube where electrons bombard a target material, X-rays are produced (see Fig. 7.32). These X-rays result in two types of X-ray spectra—*continuous* and *characteristic* (Fig. 7.33).

In a modern X-ray tube there is nearly a complete vacuum. The tube is fitted with a tungsten filament as a cathode that provides the source of electrons. The anode consists of one of a number of metals such as Mo, Cu, or Fe, and acts as the target for the electrons. When the filament is heated by passage of a current, electrons are emitted which are accelerated toward the target anode by a high voltage applied across the tube. X-rays are generated when the electrons impact on the target (anode). The nature of the X-rays depends on the metal of the target and the applied voltage. No X-rays are produced until the voltage reaches a minimum value dependent on the target material. At that point a *continuous spectrum* is generated. With increasing potential, the intensity of all wavelengths increases, and the value of the minimum wavelength becomes progressively less (Fig. 7.33*a*). The *continuous spectrum*, also referred to as *white radiation*, is caused by the stepwise loss of energy of bombarding electrons in series of encounters with atoms of the target material. When an electron is instantaneously stopped, it loses all of its energy at once and the X-ray radiation emitted is that of the shortest wavelength (see Fig. 7.33*a*). The stepwise energy losses from a stream of electrons produce a continuous range of wavelengths that can be plotted as a smooth function of intensity against wavelength (Fig. 7.33*a*). The curve begins at the short wavelength limit, rises to a maximum, and extends toward infinity at very low intensity levels.

If the voltage across the X-ray tube is increased to a critical level, which is dependent on the element

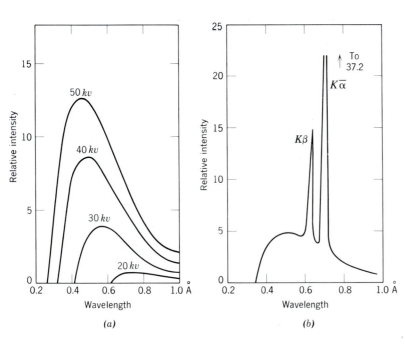

FIG. 7.33. X-ray spectra. *(a)* Distribution of intensity with wavelength in the continuous X-ray spectrum of tungsten at various voltages. *(b)* Intensity curve showing characteristic wavelengths superimposed on the continuous X-ray spectrum of molybdenum. (After C. T. Ulrey, 1918, An experimental investigation of the energy in the continuous X-ray spectra of certain elements. *Phys. Reviews* 11: 401–10.)

of the target, there becomes superimposed on the continuous spectrum a *line spectrum of characteristic radiation* peculiar to the target material. This characteristic radiation, many times more intense than the continuous spectrum, consists of several isolated wavelengths, as shown in Fig. 7.33b by β and α peaks.

The characteristic X-ray spectrum is produced when the bombarding electrons have sufficient energy to dislodge electrons from the inner electron shells in the atoms of the target material. When these inner electrons are expelled, they leave vacancies that are filled by electrons from surrounding electron shells. The electron transitions, from outer to inner shells, are accompanied by the emission of X-radiation with specific wavelengths. Electron transitions from the *L*- to the *K*-shell produce $K\overline{\alpha}$ radiation, and those from the *M*- to the *K*-shell cause *K*β radiation (see Fig. 7.54). The *K*β peak can be eliminated by an appropriate filter yielding essentially a single wavelength, which by analogy to monochromatic light is called *monochromatic X-radiation*. The $K\overline{\alpha}$ radiation consists of two peaks, $K\alpha_1$ and $K\alpha_2$, which are very close together in wavelength.

The wavelengths of the characteristic X-radiation emitted by various metals have been accurately determined. The $K\overline{\alpha}$ wavelengths (weighted averages of $K\alpha_1$ and $K\alpha_2$) of the most commonly used metals are:

	Å		Å
Molybdenum	0.7107	Cobalt	1.7902
Copper	1.5418	Iron	1.9373
		Chromium	2.2909

Diffraction Effects and the Bragg Equation

Crystals consist of an ordered three-dimensional structure with characteristic periodicities, or *identity periods*, along the crystallographic axes. When an X-ray beam strikes such a three-dimensional arrangement, it causes electrons in its path to vibrate with a frequency of the incident X-radiation. These vibrating electrons absorb some of the X-ray energy and, acting as a source of new wave fronts, emit (scatter) this energy as X-radiation of the same frequency and wavelength. In general, the scattered waves interfere destructively, but in some specific directions they reinforce one another, producing a cooperative scattering effect known as *diffraction*.

In a row of regularly spaced atoms that is bombarded by X-rays every atom can be considered the center of radiating, spherical wave shells (Fig. 7.34). When the scattered waves interfere constructively, they produce wave fronts that are *in phase*, and dif-

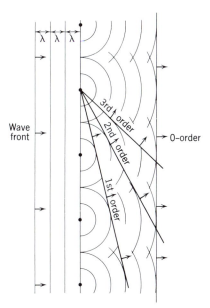

FIG. 7.34. Scattering of X-rays by a row of equally spaced, identical atoms.

fraction will occur. Figure 7.35 illustrates that rays 1 and 2 will be in phase only when distance *AB* represents an integral number of wavelengths—in other words, when $AB = n\lambda = c \cos \phi$ (where *n* represents whole numbers such as 0, 1, 2, 3, . . . , *n*). For a specific value of *n*λ, φ is constant, and the locus of all possible diffracted rays will be represented by a cone with the row of scattering points as the central axis. Because the scattered rays will be in phase for the same angle φ on the other side of the incident beam, there will be another similar but inverted cone on that side (see Fig. 7.36). The two cones with *n* = 1 would have φ (as in Fig. 7.36) as the angle between the cone axis and the outer surface of the cone. When *n* = 0, the cone becomes a plane that includes the incident beam. The greater the value of *n*

FIG. 7.35. Conditions for X-ray diffraction from a row of atoms.

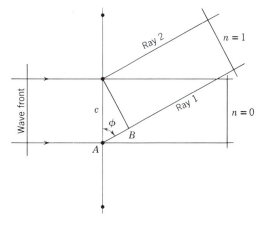

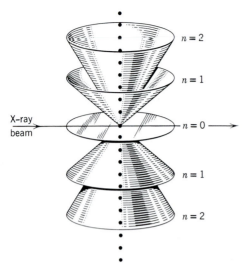

FIG. 7.36. Diffraction cones from a row of atoms.

the larger the value of cos φ, and hence the smaller the angle φ and the narrower the cone. All have the same axis, however, and all have their vertices at the same point, the intersection of the incident beam and the row of atoms.

In a three-dimensional lattice there are three axial directions, each with its characteristic periodicity of scattering points and each capable of generating its own series of nested cones with characteristic apical angles. Diffraction cones from any three non-coplanar rows of scattering atoms may or may not intersect each other, but only when all three intersect in a common line is there a diffracted beam (Fig. 7.37). This direction (shown by the arrow in Fig. 7.37) represents the direction of the diffracted beam, which can be recorded on a film or registered electronically. The geometry of the three intersecting

FIG. 7.37. Diffraction cones from three noncoplanar rows of scattering atoms, intersecting in a common line.

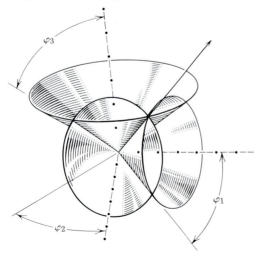

cones in Fig. 7.6 can be expressed by three independent equations (the Laue equations) in which the three cone angles (ϕ_1, ϕ_2, and ϕ_3) define a common direction along the path of the arrow (the common intersection of the three cones). In order to produce a diffraction effect (a spot on a photographic plate or film) these three geometric equations must be simultaneously satisfied. The three equations are named after Max von Laue, who originally formulated them.

Shortly after the publication of these equations, W. L. Bragg, working on X-ray diffraction in England, pointed out that although X-rays are indeed diffracted by crystals, the diffracted X-rays act as though they were reflected from planes within the crystal. Unlike the reflection of light, however, X-rays are not "reflected" continuously from a given crystal plane. Using a given wavelength, λ, Bragg showed that a "reflection" takes place from a family of parallel planes only under certain conditions. These conditions must satisfy the equation: $n\lambda = 2d \sin \theta$, where n is an integer (1, 2, 3, ..., n), λ the wavelength, d the distance between successive parallel planes, and θ the angle of incidence and "reflection" of the X-ray beam from the given atomic plane. This equation, known as the *Bragg law*, expresses in a simpler manner the simultaneous fulfillment of the three Laue equations.

We have seen that the faces most likely to appear on crystals are those parallel to atomic planes having the greatest density of lattice nodes. Parallel to each face is a family of equispaced identical planes. When an X-ray beam strikes a crystal, it penetrates it, and the resulting diffraction effect is not from a single plane but from an almost infinite number of parallel planes, each contributing a small bit to the total diffraction maximum. In order that the diffraction effect ("reflection") be of sufficient intensity to be recorded, the individual "reflections" must be *in phase* with one another. The following conditions necessary for reinforcement were demonstrated by W. L. Bragg.

In Fig. 7.38 the lines p, p_1, and p_2 represent the traces of a family of atomic planes with spacing d. X-rays striking the outer plane pp would be reflected at the incident angle θ, whatever the value of θ. However, to reinforce one another in order to give a "reflection" that can be recorded, all "reflected" rays must be *in phase*. The path of the waves along *DEF* "reflected" at *E* is longer than the path of the waves along *ABC* "reflected" at *B*. If the two sets of waves are to be *in phase*, the path difference of *ABC* and *DEF* must be a whole number of wavelengths (*n*λ). In Fig. 7.38 *BG* and *BH* are drawn perpendicular to *AB* and *BC*, respectively, so that *AB* = *DG* and *BC* = *HF*. To satisfy the condition that the two waves be in phase, *GE* + *EH* must be equal to an integral number of wavelengths. *BE* is perpendicular to the lines p and

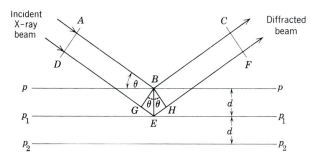

FIG. 7.38. Geometry of X-ray "reflection."

p_1 and is equal to the interplanar spacing d. In ΔGBE, $d \sin \theta = GE$; and, in ΔHBE, $d \sin \theta = EH$. Thus for in phase "reflection" $GE + EH = 2d \sin \theta = n\lambda$.

This equation,

$$n\lambda = 2d \sin \theta,$$

is the *Bragg equation*. For a given interplanar spacing (d) and given λ, "reflections" (diffraction maxima) occur only at those angles of θ that satisfy the equation. Suppose, for example, a monochromatic X-ray beam is parallel to a cleavage plate of halite and the plate is supported in such a way that it can be rotated about an axis at right angles to the X-ray beam. As the halite is slowly rotated there is no "reflection" until

the incident beam makes an angle θ that satisfies the Bragg equation, with $n = 1$. On continued rotation there are further "reflections" only when the equation is satisfied at certain θ angles with $n = 2, 3$, and so on. These are known are the first-, second-, third-order, etc., "reflections." These "reflections" are in actuality the diffraction effects that occur when the three diffraction cones about three noncoplanar rows of atoms intersect in a common direction (see Fig. 7.37).

Single-Crystal X-ray Diffraction and Structure Analysis

Single-crystal X-ray diffraction, as the name implies, concerns itself with the interaction of an X-ray beam with a very small single-crystal (about 1 mm in size or smaller) of a mineral (or other crystalline substance). The X-ray beam and the crystalline structure produce X-ray diffraction effects that can be recorded on film or measured by an electronic device, an X-ray counter (or detector). Until about 1970 almost all X-ray diffraction studies involved film methods which consist of a crystal positioned in an X-ray beam at a fixed distance from a film wrapped in a light-tight envelope. Figure 7.39*a* shows this arrangement for the *Laue method* in which the crystal is stationary and

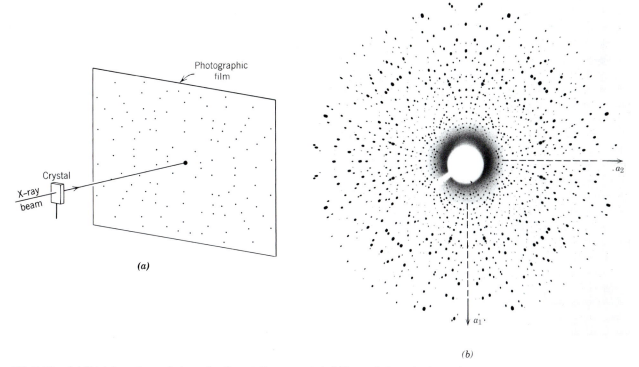

FIG. 7.39. *(a)* Obtaining a Laue photograph with a stationary crystal. *(b)* Laue photograph of vesuvianite with point group symmetry $4/m2/m2/m$. The photograph was taken along the fourfold rotation axis (*c* axis) of vesuvianite, thus revealing fourfold symmetry and mirrors in the arrangement of diffraction spots. The axial directions, a_1 and a_2, were inked onto the photograph after it had been developed.

the X-ray beam is unfiltered (meaning that it contains X-rays represented by $K\beta$ radiation as well as the continuum in addition to the $K\alpha$ radiation). Although the *Laue method* is now mainly of historic interest, it is mentioned here because the resulting diffraction effects, as recorded on a flat film, display the symmetry of the crystal as long as one of the crystallographic axes of the crystal is aligned parallel with the X-ray beam. Figure 7.39b illustrates the fourfold symmetry arrangement of X-ray diffraction spots (about the center of the film) for the mineral vesuvianite. Here the X-ray beam was parallel to the *c* axis, the fourfold axis in vesuvianite, which is tetragonal in symmetry.

In other single-crystal film methods the film is generally not stationary and furthermore, the film may not be contained in a flat but instead in a cylindrical housing. The most commonly used single-crystal method in which a flat film and a single crystal move on a complex gyratory motion, is known as the *precession method*. An example of an X-ray single-crystal precession photograph of vesuvianite is shown in Fig. 7.40.

Although an enormous number of crystal structures have been solved on the basis of data obtained on film using single-crystal camera techniques, the modern approach to data acquisition is not by film technique, but by *single-crystal diffractormeter*. In such instrumentation (see also "X-ray Powder Diffraction and Mineral Identification," later in this chapter)

the intensity of X-ray reflections is not evaluated from the intensity of a spot on an X-ray film, but instead is measured by an electronic device, an X-ray counter (or detector). Such detectors greatly improve the accuracy of X-ray intensity measurements over those obtained by film techniques. Furthermore, in automated instrumentation, such detectors can measure large numbers of X-ray reflections with high accuracy. The most commonly used automated technique in X-ray structure analysis is the *four-circle diffractometer*. The name four-circle arises from its possession of four arcs that are used to orient the single crystal so as to bring desired (atomic) planes into reflecting positions. An automated four-circle diffractometer is shown in Fig. 7.41.

The Determination of Crystal Structures

The orderly arrangement of atoms in a crystal is known as the *crystal structure*. It provides information on the location of all the atoms, bond positions and bond types, space group symmetry, and the chemical content of the unit cell.

The first steps in determining the atomic structure of a crystal are the measurement of its unit cell size and the evaluation of its space group. The systematic measurement of the geometric distribution of diffracted X-ray beams on single crystal X-ray photographs in one or more crystallographic orientations (precession photographs are most convenient) yields

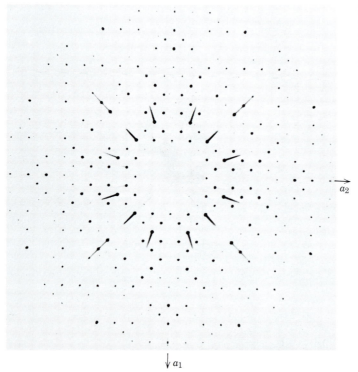

FIG. 7.40. Precession photograph of vesuvianite, with point group symmetry $4/m2/m2/m$. The photo was taken along the fourfold rotation *c* axis, thus revealing the fourfold symmetry about the axis, as well as mirror planes. Compare with Fig. 7.39b.

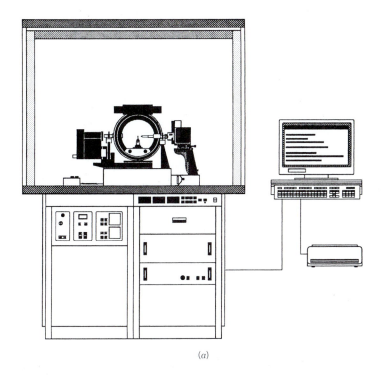

(a)

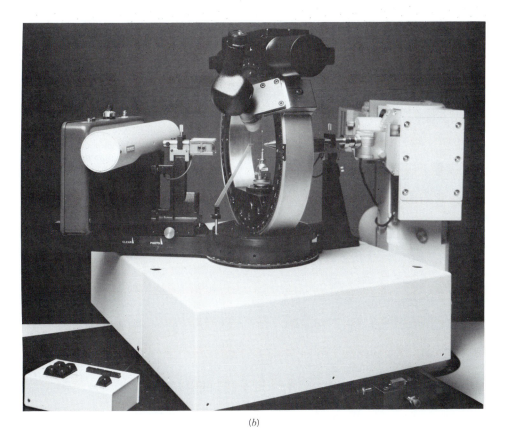

(b)

FIG. 7.41. *(a)* Schematic illustration of the P4 single-crystal X-ray diffractometer manufactured by Siemens Industrial Automation, Inc. *(b)* Close-up of the four-circle goniometer for controlling the orientation of the single crystal (in center of photograph). To the right is the X-ray tube, and to the left is the X-ray detector, a scintillation counter. (Courtesy of Siemens Industrial Automation, Inc., Madison, Wis.)

information about the geometry of that crystal's unit cell, specifically the lengths of the unit cell edges and the angles between them. Information about the space group and the crystal structure (i.e., the symmetry of the atomic arrangement and the coordinates of atoms within the unit cell) is contained in the intensities of the diffracted beams, which are measured either from X-ray single crystal photographs, or more commonly with quantum counting detectors on single-crystal diffractometers. The diffracted beams are identified by Bragg indices *hkl* associated with the lattice planes of the same indices.

The relationship between the intensity (*I*) of the diffracted beam associated with lattice planes having Bragg diffraction indices *hkl* and coordinates $x_j y_j z_j$ of atom *j* in the unit cell is

$$I = k\left[\sum_j f_j e^{i2\pi(hx_j + ky_j + lz_j)}\right]^2 = kF_{hkl}^2$$

where *k* is a term containing several physical constants and experimental factors, including a scale factor; f_j is the scattering factor of atom *j* (which depends on atomic number and scattering angle and includes corrections for atomic thermal motion); and the summation is over all atoms in the unit cell. The summation is termed the *structure factor*, F_{hkl}, and its value clearly depends on the kinds of atoms in the cell and their positions.

Because the intensities of diffracted beams are related to F_{hkl}^2 rather than F_{hkl}, the atomic coordinates cannot be extracted straightforwardly from measured intensities. Over the past 70 years a great deal of effort has been devoted to methods by which these intensities can be made to yield atomic positions, with remarkable success. Modern methods are based either on analysis using Fourier techniques, which are very powerful, particularly when the positions of a small number of heavy atoms are known, or on so-called direct methods, which are essentially statistical in nature and have become extremely successful as computing power has increased. An illustration of a now-classical electron density map (and the structure derived therefrom) by W. L. Bragg in 1929 is given in Fig. 7.42. Once the atom positions are known crudely, they can be refined to high precision, using least-squares analysis on many measured F_{hkl}^2, a procedure that can yield amplitudes of atomic

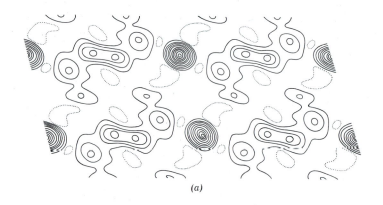

(a)

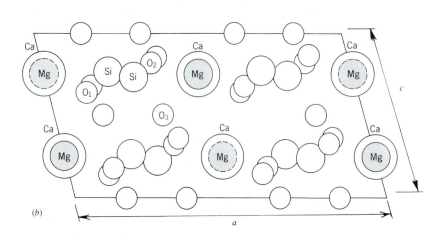

(b)

FIG. 7.42. (a) The summation of Fourier series for diopside, $CaMgSi_2O_6$, projected on (010). The distribution of scattering matter is indicated by contour lines drawn through points of equal density in the projection. (b) The atomic positions of diopside (with space group *C 2/c*) projected on (010), as derived from the distribution in (a). (Redrawn after W. L. Bragg, 1929, *Zeitschrift für Kristallographie* 70:488.)

thermal vibration as well as site occupancies of substituting atomic species in minerals that are members of solid solution series. Accurate interatomic distances (bond lengths) can then be calculated from refined atomic coordinates. In recent years highly accurate electron density maps, calculated as a Fourier series in which the structure factors are the Fourier coefficients, have yielded information about bonding and the spatial distribution of valence electrons. Examples of such detailed electron density maps are given in "Molecules as models for bonding in silicates," by G. V. Gibbs (1982); complete reference is given at the end of this chapter.

Supplementing structure analyses based on diffraction phenomena (these are X-ray, neutron, and electron diffraction techniques), a variety of spectroscopic methods, including infrared, optical, Mössbauer, and resonance techniques such as nuclear magnetic resonance (NMR) yield information about the local environments or ionization states of certain atomic species. Thus a complete structural picture of a mineral at the atomic level usually requires information obtained by both diffraction and spectroscopy.

X-Ray Powder Diffraction and Mineral Identification

The relative rarity of well-formed crystals and the difficulty of achieving the precise orientation required by single-crystal methods led to the discovery of the *powder method* of X-ray investigation. In X-ray diffraction studies of powders, the original specimen is ground as finely as possible and bonded with an amorphous material into a small spindle (for powder film methods) or mounted on a glass slide, or in a special rectangular sample holder (for powder diffractometer techniques). The *powder mount* consists ideally of crystalline particles in completely random orientation. To ensure randomness of orientation of these tiny particles with respect to the impinging X-ray beam, the spindle mount (used in film cameras) is generally rotated in the path of the beam during exposure.

When a beam of monochromatic X-ray strikes the mount, all possible diffractions take place simultaneously. If the orientation of the crystalline particles in the mount is truly random, for each family of atomic planes with its characteristic interplanar spacing (d), there are many particles whose orientation is such that they make the proper θ angle with the incident beam to satisfy the *Bragg law*: $n\lambda = 2d \sin \theta$. The diffraction maxima from a given set of planes form cones with the incident beam as axis

and the internal angle 4θ. Any set of atomic planes yields a series of nested cones corresponding to "reflections" of the first, second, third, and higher orders ($n = 1, 2, 3, \ldots$). Different families of planes with different interplanar spacings will satisfy the Bragg law at appropriate values of θ for different integral values of n, thus giving rise to separate sets of nested cones of "reflected" rays.

If the rays forming these cones are permitted to fall on a flat photographic plate at right angles to the incident beam, a series of concentric circles will result (Fig. 7.43). However, only "reflections" with small values of the angle 2θ can be recorded in this manner.

In order to record all the possible diffraction cones that may occur in three dimensions (see Figs. 7.36 and 7.37) a film method is used in which the film is wrapped around the inside of a cylindrical camera. This camera is known as a *powder camera* in which the film fits snugly to the inner curve of the camera (see Fig. 7.44a). This type of mounting is known as the *Straumanis method*. Figure 7.44b shows the circular film strip with two holes cut into it, one to allow the X-ray beam to enter the camera and the other for a lead-lined beam catcher. Although this powder camera method has been used extensively for mineral identification, the X-ray *powder diffractometer* is the instrumental method currently used most often. This powerful analytical tool uses essentially monochromatic X-radiation and a finely powdered sample, as does the powder film method, but records the information about the "reflections" present as a graphical recording, or as electronic counts (X-ray counts) that can be stored and printed out by computer.

The sample for powder diffractometer analysis is prepared by grinding it to a fine powder, which is then spread uniformly over the surface of a glass slide, using a small amount of adhesive binder. (If enough powder is available it can be compressed

FIG. 7.43. X-ray diffraction from a powder mount recorded on a flat plate.

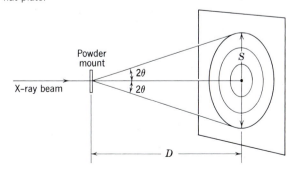

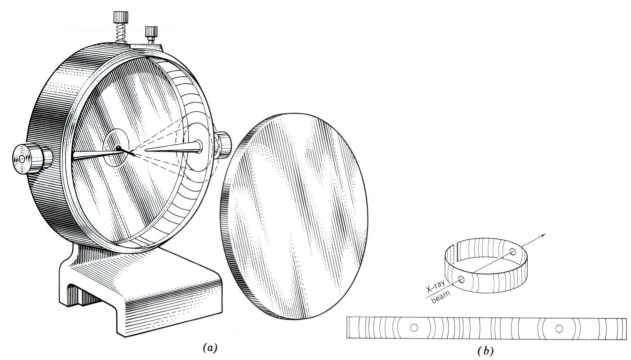

FIG. 7.44. *(a)* Metal powder diffraction camera with a powder spindle in the center and film strip against the inner cylindrical wall of the camera. *(b)* Circular film strip with curved lines that represent the conical "reflections" produced inside the camera.

into a rectangular sample holder.) The instrument is so constructed that the sample, when clamped in place, rotates in the path of a collimated X-ray beam while an X-ray detector, mounted on an arm, rotates about it to pick up the diffracted X-ray signals (see Fig. 7.45). When the instrument is set at zero position, the X-ray beam is parallel to the base of the sample holder and passes directly into the X-ray detector. The slide mount and the counter are driven by a motor through separate gear trains so that, while the sample rotates through the angle θ, the detector rotates through 2θ.

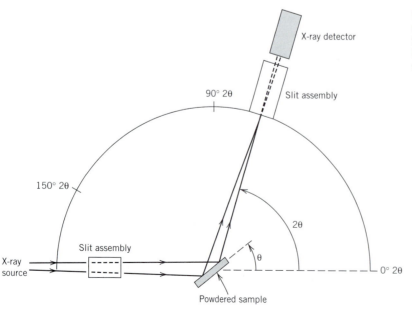

FIG. 7.45. Schematic illustration of the essential components of a powder X-ray diffractometer. In such an instrument the sample holder rotates (θ)° while the detector arm rotates 2(θ)°.

If the specimen has been properly prepared, there will be thousands of tiny crystalline particles in random orientation. As in powder photography, all possible "reflections" from atomic planes take place simultaneously. Instead of recording all of them on a film at one time, however, the X-ray detector maintains the appropriate geometrical relationship to receive each diffraction maximum separately.

In operation, the sample, the X-ray detector, and the recording device are activated simultaneously. If an atomic plane has an interplanar spacing (d) such that a reflection occurs at $\theta = 20°$, there is no evidence of this reflection until the counting tube has been rotated through 2θ, or $40°$. At this point the diffracted beam enters the X-ray detector, causing it to respond. The pulse thus generated is amplified and causes an electronic response on a vertical scale that represents peak height. The angle 2θ at which the diffraction occurs is read on a horizontal scale. The heights of the peaks are directly proportional to the intensities of the diffraction effects. An example of a diffractometer tracing for low quartz is given in Figure 7.46. The 2θ positions of the diffraction peaks in such a tracing can be read off directly or they can be tabulated as 2θ positions by an online computer. The interplanar spacings giving rise to them are calculated using the equation $n\lambda = 2d \sin \theta$.

Once a diffractometer tracing has been obtained and the various diffraction peaks have been tabulated in a sequence of decreasing interplanar spacings (d), together with their relative intensities (I, with the strongest peak represented by 100 and all

the other peaks scaled with respect to 100), the investigator is ready to begin the mineral identification process. Through a computer search technique (doing searches for comparable or identical diffraction patterns on the basis of the strongest lines or the largest interplanar spacings), the diffraction pattern of an unknown can be compared with records stored in the Powder Diffraction File (PDF) published by the International Center for Diffraction Data (ICDD). This file is primarily available on CD-ROM and is compatible with PC, VAX/VMS, and Macintosh platforms.

The Powder Diffraction File is the world's largest and most complete collection of X-ray powder diffraction patterns. The PDF contains more than 70,000 experimental patterns compiled by the ICDD since 1941, as well as more than 42,000 calculated patterns. Each pattern includes a table of interplanar (d) spacings, relative intensities (I), and Miller indices, as well as additional information such as chemical formula, compound name, mineral name, structural formula, crystal system, physical data, experimental parameters, and references. An example printout for low quartz from this file is shown in Fig. 7.47. Entries are indexed to allow for subfile searches of inorganics, organics, minerals, metals and alloys, common phases, pharmaceuticals, zeolites, and many others.

By this outlined comparative search procedure a completely unknown substance may generally be identified in a short time using a very small volume of sample.

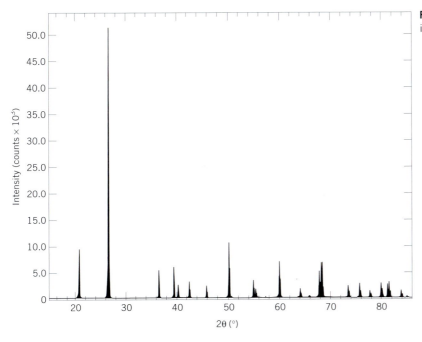

FIG. 7.46. X-ray powder diffractometer tracing for a finely powdered sample of low quartz.

PDF#33-1161 (Deleted Card): QM = Star (+); d = Diffractometer, I = Diffractometer		PDF Card

Quartz, syn
SiO2

Radiation = CuKa1	Lambda = 1.540598	Filter =
Calibration = Internal (Si)	d-Cutoff =	I/Ic (RIR) = 3.6
Ref = Natl. Bur. Stand. (U.S.) Monogr. 25, 18 61 (1981)		

Hexagonal—(Unknown), P3221(154)   Z = 3   mp =
Cell = 4.9134 × 5.4053   Pearson = hP9 (O2 Si)
Density (c) = 2.649   Density (m) = 2.656   Mwt = 60.08   Vol = 113.01   F(30) = 76.8 (.0126,31)
Ref = Ibid.

NOTE: Sample from the Glass Section at NBS, Gaithersburg, MD, USA, ground single-crystals of optical quality. To replace 5-490 and validated by calculated pattern. Plus 6 additional reflections to 0.9089. Pattern taken at 25 C. Pattern reviewed by Holzer, J., McCarthy, G., North Dakota State Univ., Fargo, ND, USA, ICDD Grant-in-Aid (1990). Agrees well with experimental and calculated patterns. Deleted by 46-1045, higher F#N, more complete, LRB 1/95.

Color: Colorless

Strong Line: 3.34/X 4.26/2 1.82/1 1.54/1 2.46/1 2.28/1 1.37/1 1.38/1 2.13/1 2.24/1

39 Lines, Wavelength to Compute Theta = 1.54056A (Cu), 1%-Type = (Unknown)

#	d(A)	I(f)	h	k	l	2-Theta	Theta	1/(2d)	#	d(A)	I(f)	h	k	l	2-Theta	Theta	1/(2d)
1	4.2570	22.0	1	0	0	20.850	10.425	0.1175	21	1.2285	1.0	2	2	0	77.660	38.830	0.4070
2	3.3420	100.0	1	0	1	26.651	13.326	0.1495	22	1.1999	2.0	2	1	3	79.875	39.938	0.4167
3	2.4570	8.0	1	1	0	36.541	18.271	0.2035	23	1.1978	1.0	2	2	1	80.044	40.022	0.4174
4	2.2820	8.0	1	0	2	39.455	19.727	0.2191	24	1.1843	3.0	1	1	4	81.145	40.572	0.4222
5	2.2370	4.0	1	1	1	40.283	20.141	0.2235	25	1.1804	3.0	3	1	0	81.470	40.735	0.4236
6	2.1270	6.0	2	0	0	42.464	21.232	0.2351	26	1.1532	1.0	3	1	1	83.818	41.909	0.4336
7	1.9792	4.0	2	0	1	45.808	22.904	0.2526	27	1.1405	1.0	2	0	4	84.969	42.484	0.4384
8	1.8179	14.0	1	1	2	50.139	25.070	0.2750	28	1.1143	1.0	3	0	3	87.461	43.731	0.4487
9	1.8021	1.0	0	0	3	50.610	25.305	0.2775	29	1.0813	2.0	3	1	2	90.855	45.428	0.4624
10	1.6719	4.0	2	0	2	54.867	27.434	0.2991	30	1.0635	1.0	4	0	0	92.819	46.410	0.4701
11	1.6591	2.0	1	0	3	55.327	27.663	0.3014	31	1.0476	1.0	1	0	5	94.662	47.331	0.4773
12	1.6082	1.0	2	1	0	57.236	28.618	0.3109	32	1.0438	1.0	4	0	1	95.115	47.558	0.4790
13	1.5418	9.0	2	1	1	59.947	29.973	0.3243	33	1.0347	1.0	2	1	4	96.223	48.112	0.4832
14	1.4536	1.0	1	1	3	63.999	32.000	0.3440	34	1.0150	1.0	2	2	3	98.734	49.367	0.4926
15	1.4189	1.0	3	0	0	65.759	32.879	0.3524	35	0.9898	1.0	4	0	2	102.195	51.098	0.5052
16	1.3820	6.0	2	1	2	67.748	33.874	0.3618	36	0.9873	1.0	3	1	3	102.556	51.278	0.5064
17	1.3752	7.0	2	0	3	68.128	34.064	0.3636	37	0.9783	1.0	3	0	4	103.880	51.940	0.5111
18	1.3718	8.0	3	0	1	68.321	34.160	0.3645	38	0.9762	1.0	3	2	0	104.195	52.098	0.5122
19	1.2880	2.0	1	0	4	73.460	36.730	0.3882	39	0.9636	1.0	2	0	5	106.141	53.071	0.5189
20	1.2558	2.0	3	0	2	75.668	37.834	0.3982									

FIG. 7.47.   Example of the printout for SiO$_2$, low quartz, as obtained from the powder diffraction file (PDF-2) licensed by the International Center for Diffraction Data (JCPDS), 12 Campus Boulevard, Newton Square, PA., 19073-3273; copyright © JCDPS-ICDD, 1999. The printout was obtained using Jade 5.0 from Materials Data Inc. (MDI).

The powder method is of wider usefulness, however, and there are several other applications in which it is of great value. Variations in chemical composition of a known substance involve the substitution of ions, generally of a somewhat different size, in specific sites in a crystal structure. As a result of this substitution the unit cell dimensions and hence the interplanar spacings are slightly changed, and the positions of the lines in the powder diffractogram corresponding to these interplanar spacings are shifted accordingly. By measuring these small shifts in position of the lines in powder patterns of substances of known structure, changes in chemical composition may often be accurately detected. Figure 7.48 illustrates a variation diagram that correlates unit cell dimensions (b and unit cell volume, V) and changes in the position of a specific diffraction maximum (1, 11, 0) with composition in the cummingtonite-grunerite series.

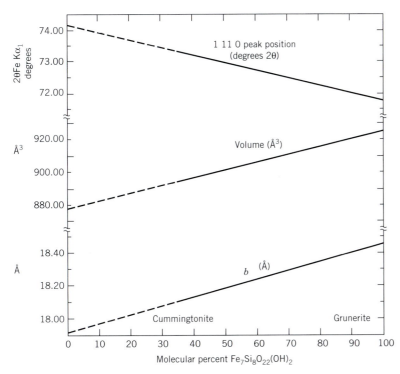

FIG. 7.48. Variation of *b* and *V* (volume) of the unit cell and the position of the 1,11,0 peak as a function of composition in the monoclinic cummingtonite–grunerite series with composition ranging from $Fe_2Mg_5Si_8O_{22}(OH)_2$ to $Fe_7Si_8O_{22}(OH)_2$. (After C. Klein and D. R. Waldbaum, 1967, X-ray crystallographic properties of the cummingtonite-grunerite series. *Jour. Geology* 75:379–92).

Further, the relative proportions of two or more known minerals in a mixture are often conveniently determined by the comparison of the intensities of the same peaks in diffractometer patterns of prepared control samples of known composition.

Several student exercises using X-ray powder diffraction film and diffractometer techniques are given in Klein (1994; see reference list for complete reference).

Until recent years, most of the structural studies of crystalline materials were based on single-crystal X-ray and neutron diffraction experiments, but a newly developed X-ray powder diffraction technique, the *Rietveld refinement method*, allows the extraction of structural information from powdered instead of single-crystal specimens. This is an especially important development for the determination of the crystal structures of minerals that are typically finely crystalline and are not found in well-developed single crystals. Examples of such finely crystalline or poorly ordered minerals are clay minerals, manganese and iron oxides and hydroxides, and some zeolites. There are three basic requirements of the Rietveld refinement:

1. Accurate powder diffraction intensity data measured at specific intervals of 2θ
2. A basic understanding ("starting model") of the actual crystal structure of the material that is being studied

3. A quantitative understanding of shapes, width, and any systematic errors in the positions of X-ray peaks in the powder pattern

For further discussion of the Rietveld method, see Post and Bish (1989; see reference list for complete reference).

TRANSMISSION ELECTRON MICROSCOPY (TEM)

A transmission electron microscope consists of a finely focused electron beam that impinges on a very thin foil of an object (the unknown material under investigation) and that, upon transmission through the object, can be used to display electron diffraction patterns and high resolution transmission electron microscope (HRTEM) images. A schematic drawing of the electron column in a TEM with its electromagnetic lenses was shown in Fig. 1.13 and a photograph of a modern instrument was shown in Fig. 1.12. The thin foil of the material to be investigated is produced by ion bombardment or sputter-etching methods for nonconducting materials or by ultramicrotomy. For metals and conducting materials, foil preparation is done by electrochemical methods. The thin foil, which is on the order of a few nanometers thick, is held in place with a sample holder centered in the electron beam (see Fig. 1.12).

Modern instruments allow for textural, crystallographic, and chemical evaluation from areas as small as a square nanometer (10Å × 10Å) and from features that are as little as 0.14 nanometer (1.4Å) apart. By comparison, most minerals have crystal structures whose basic unit cells are greater than 0.4 nm² (4Å²).

As discussed under X-ray diffraction, the structures of almost all minerals have been determined by single-crystal X-ray diffraction techniques. Such structure analyses, based on the averaged information obtained from many thousands of unit cells (all of which are assumed to be identical), provide accurately the positions of atoms in the unit cell on a scale of 1 to 100 Å. The TEM technique is especially powerful in elucidating structural features that range in size from 100 to 10,000 Å, which cannot be directly evaluated by X-ray diffraction techniques. Furthermore, TEM studies allow for phase identification of extremely small particles or intergrowths, of minerals with exsolution features, of polytype stackings, and defect structures.

An example of an electron diffraction pattern is given in Fig. 7.49. Such patterns which can be in-

FIG. 7.50. High resolution transmission electron microscopy (HRTEM) image of a biotite mica, with the *c* crystallographic axis almost vertical. Bright (light colored) regions represent relatively low electron density within the structure. Perfect {0001} cleavage is shown by the whitest, linear, horizontal features. The dark horizontal lines represent Fe and Mg located in octahedral sheets (high electron density means darkness). Magnification approximately 10 million times. (After H. Xu and D. R. Veblen, 1995, Periodic and nonperiodic stacking in biotite from the Bingham Canyon Porphyry Copper Deposit, Utah, *Clays and Clay Minerals* 43: 159–73; photograph courtesy of Huifang Xu.)

FIG. 7.49. Electron diffraction pattern from an isometric spinel grain with the [111] axis of the grain perpendicular to the page (i.e., parallel to the electron beam). The [111] axis is the $\bar{3}$ axis at the corner of a cube. Compare this electron diffraction photo with a very similar appearing single-crystal X-ray diffraction photo in Fig. 7.40. (After A. J. Brearley, D. C. Rubie, and E. Ito, 1992, Mechanisms of transformations between the α, β, and γ polymorphs of Mg_2SiO_4 at 15 GPa, *Phys. Chem. Minerals* 18: 343–58; photograph courtesy of A. J. Brearley.)

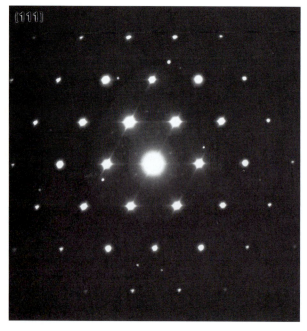

dexed with Miller indices (similar to the indexing of X-ray diffraction patterns) provide information about symmetry; distances among indexed diffraction maxima provide information about unit cell size. This, in turn, allows for phase identification. An example of an image obtained by TEM in the high resolution transmission electron microscope (HRTEM) mode is given in Fig. 7.50. This shows the image of structural features of mica at a resolution of about 0.1 nm (1 Å). Similar HRTEM images are used throughout this text to illustrate defect structures and exsolution features (see e.g., Chapter 4).

CHEMICAL ANALYTICAL TECHNIQUES

Chemical analyses of minerals (and rocks) are obtained by a variety of analytical techniques. Prior to about 1947, quantitative mineral analyses were obtained mainly by "wet" analytical techniques, in which the mineral is dissolved by some appropriate means. Since 1960 the majority of analyses have been made by instrumental techniques such as flame atomic absorption analysis (FAA), X-ray fluorescence analysis (XRF), and electron microprobe analysis (EMPA). Each of these techniques has its own specific sample preparation requirements and fairly well-established detection limits and error ranges. The results of any analysis are generally presented in a table of weight percentages of the elements or oxide components in the mineral analyzed.

Minerals submitted for chemical analysis ideally consist of one mineral species only (the one being analyzed) and must be free of weathering or other alteration products and inclusions. Because it is often very difficult, if not impossible, to separate as much as 0.1 gram to 1 gram of clean material for analysis, analysis results not infrequently must be accompanied by a statement concerning the amount of impurities. An instrumental method, such as electron probe microanalysis, allows for quantitative analysis, *in situ* (in a polished section), of mineral grains as small as 1 micron in diameter (1 μm = 10^{-3} mm).

It may be helpful at this point to distinguish a *qualitative* from a *quantitative* chemical analysis. A *qualitative* analysis involves the detection and identification of all the constituents of a compound ("What is present?"). A *quantitative analysis* involves the determination of the weight percentages (or parts per million composition) of elements in a compound ("How much of each is present?"). A preliminary qualitative analysis is commonly helpful in deciding on the methods to be followed in a subsequent quantitative analysis.

Flame Atomic Absorption Spectroscopy (FAA)

This analytical technique is commonly considered a "wet" analytical procedure because the original sample must be completely dissolved in a solution before it can be analyzed. It is impossible to adequately describe the impact atomic absorption spectroscopy has had on the chemical determination of elements since the technique was first introduced by Alan Walsh in 1955. A classical wet chemist of that time instantly appreciated the speed, accuracy, and elimination of the need for most chemical separations that came with FAA. While it is true that much FAA work is now being supplemented by inductively coupled plasma (ICP) and ICP–mass spectrometric (ICP–MS) methods, the low cost, accuracy, and ease of operation of FAA will ensure it a prominent place in the analytical laboratory for years to come.

The basic concepts of FAA can be explained with reference to Fig. 7.51. The energy source in this technique is a light source (a hollow cathode lamp), the energy of which ranges from the visible to the ultraviolet portion of the electromagnetic spectrum, with wavelengths down to about 190 nm (see Fig. 7.52). The energy of a quantum (see also Chapter 3) is proportional to the wavelength of the radiation, in accordance with the Einstein equation

$$E = \frac{hc}{\lambda}$$

where E is the energy of the quantum, c is the speed of light, λ is the wavelength of the radiation, and h is Planck's constant. The cloud in Fig. 7.51 ideally represents atoms that are free from any molecular bonding forces. When the light energy is equivalent to the energy required to raise the atom from its low energy levels to higher energy levels, it is absorbed and causes excitation of the atom. In Fig. 7.51 the light beam enters the sample cloud and the absorption of the light beam in the cloud is monitored by a spectrometer, across the atomic cloud. In order to determine element concentrations by this analytical

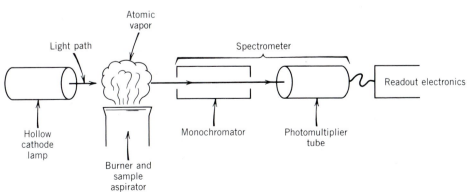

FIG. 7.51. Schematic representation of the major components in an atomic absorption spectrometer.

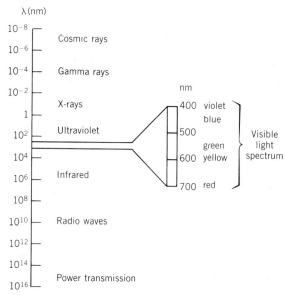

FIG. 7.52. The electromagnetic spectrum with the visible region expanded. Wavelength (λ) values are expressed in nm (nanometers), where 1 nm = 10 angstroms and 1 nm = 10^{-9} meters.

technique, the atoms must be completely free of any of the bonding that exists in the solid or liquid state, because the electrons will not be free to absorb specific wavelengths from the lamp if they are bound to surrounding atoms. To achieve this, the sample solution is aspirated as a fine mist into a flame, where it is converted to an atomic vapor. The amount of light radiation that is measured by the spectrometer (after absorption in the atomic cloud) is expressed as

$$A - \log I_0/I$$

where A is the *absorbance*, I_0 is the incident light intensity, and I the transmitted light intensity.

The incident light intensity is supplied by a hollow cathode lamp. This lamp, at high lamp current, causes emission of a line spectrum that is characteristic of the element being excited in the cathode tube. From this light spectrum a single high-intensity line of specific wavelength (λ) is selected as the light source (Fig. 7.53a). λ is selected so that the spectral line emitted by the cathode lamp is the same as that of the element being analyzed in the atomic vapor.

Prior to the vaporization and analysis procedure, a spectrometer (consisting of a photomultiplier tube, a monochromator, and associated electronics) is preset on the wavelength (λ) of the emission spectral line (of the hollow cathode tube).

This is achieved with the aid of the monochromator, which allows the isolation of a specific spectral line. The sample being vaporized absorbs energy at the λ value of the cathode tube (Fig. 7.53b), and the final reduction in intensity (due to absorption) is measured by the photomultiplier tube (Fig. 7.53c). The detectibility for some elements, such as magnesium and sodium, is extremely high, with quantitative determinations possible in the ppm range.

FIG. 7.53. (a) Line spectrum emitted by the hollow-cathode lamp. This is the same spectrum as that of the element to be analyzed. (b) The element in the atomic vapor (above the flame of the burner) absorbs the energy of the specific spectral line selected for the analysis. (c) The photomultiplier tube records the absorption of the specific spectral line.

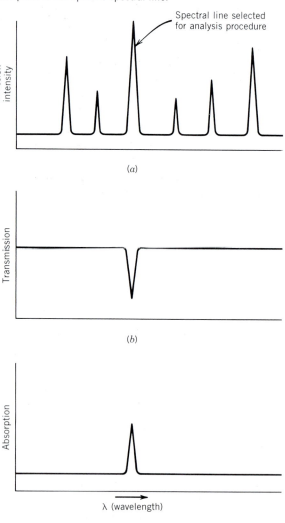

X-Ray Fluorescence Analysis (XRF)

This analytical technique, also known as *X-ray emission spectrography*, is used in most research laboratories that study the chemistry of inorganic substances, but it is also routinely used in a wide range of industrial applications. Examples of such applications are in the mining industry (for quality control of the product shipped to the consumer), in the glass and ceramics industry, in the manufacture of metals and alloys, and in environmental protection and pollution control applications.

The analysis sample in this technique is ground to a fine powder and subsequently compressed into a circular pellet, or into a disc with the admixture of a binder. This sample preparation is distinctly different from the procedure used in the "wet" technique discussed earlier. The pellet or disc of sample is irradiated (for a short period of time) with X-rays generated in a high-intensity X-ray tube (see Fig. 7.52 for the λ range of X-rays). These incident X-rays from the X-ray tube are, to a considerable extent, absorbed according to Beer's Law:

$$\log \frac{I_0}{I} = K_d \Delta d$$

where I_0 is the incident X-ray intensity, I is the intensity of the X-ray beam that was not absorbed in the sample, K_d is a proportionality constant, and Δd is the thickness of the sample. The X-ray energy that is absorbed in the sample results in the generation of an *X-ray emission spectrum that is characteristic* for each element in the sample. In the process of the absorption of X-ray energy in the sample, electrons are dislodged from the innermost shells (known as K, L, M; see Chapter 3). An expelled electron (from, e.g., the K shell) must be replaced, and it is highly probable that the vacancy will be filled from the next outer shell (the L shell) rather than from a more remote shell. This creates a new vacancy, which is filled from the next shell, and so on. Electrons that "fall into" inner electron shells move from higher to lower energy levels and as a result emit energy in the form of characteristic X-radiation (see Fig. 7.54). The groupings of spectral lines are classified as K-, L-, or M-spectra in terms of the patterns in which the outer electrons fall into the lower energy states. These generated characteristic X-rays are known as *secondary X-rays* and the emission phenomenon is called *X-ray fluorescence*. Each element has characteristic spectral lines with specific wavelengths superimposed on a low-intensity continuous background spectrum. Examples of two characteristic K-spectra for two different elements are given in Fig. 7.55.

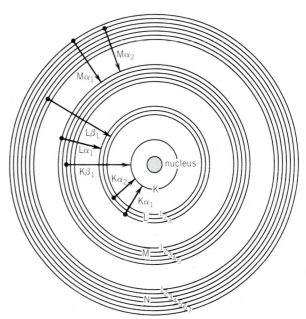

FIG. 7.54. Schematic illustration of the production if K and L characteristic spectra as a result of electrons cascading from upper to lower energy levels in the atomic structure. X-ray emission lines are labeled with a capital letter representing the shell whose vacancy is being filled. The Greek letter is α if the electron that fills the vacancy originates in the next highest shell, β if the electron comes from two shells up, and so on. The Arabic numeral subscripts indicate specific subshells (1 for *s*, 2 for *p*) for the origin of the electron that fills the vacancy.

FIG. 7.55. Characteristic K-spectra for the elements Mo and Cu, superimposed on a continuous spectrum. This continuous spectrum is considered part of the background X-ray intensity in analytical techniques.

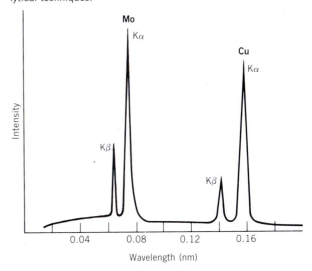

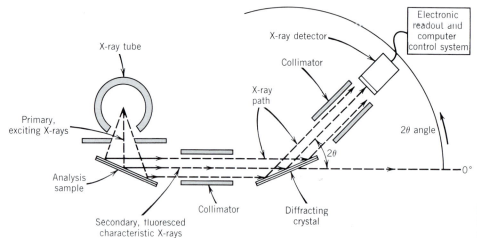

FIG. 7.56. Schematic illustration of the major components in an X-ray fluorescence analyzer.

The generated (fluoresced or secondary) X-ray spectrum may consist of a very large number of spectral lines in a sample that consists of more than one or two elements. Such a spectrum must be resolved into its spectral line components, such that the lines can be identified by wavelengths (in nm or Å) that are specific to the elements responsible for their production. This is achieved with an X-ray spectrometer consisting of a diffracting crystal (with known spacing between adjoining atomic planes) and an X-ray detector. The crystal will diffract the various λ values of the impinging X-rays according to the Bragg equation

$$n\lambda = 2d \sin \theta$$

where n is generally a small number, ranging from 1 to 2 or 3 (known as the "order of diffraction"), λ is the wavelength of a specific spectral line, d is the distance between a specific set of atomic planes in the diffracting crystal, and θ is the angle over which the X-ray is "reflected" by the crystal.

The intensity and position of each spectral line diffracted by the crystal in the spectrometer is recorded by an electronic X-ray counting device (generally a scintillation counter or flow proportional counter). This is traced on a recorder or presented on a high-resolution visual display terminal. A schematic diagram of the major components that constitute an X-ray fluorescence analyzer is given in Fig. 7.56.

Qualitative X-ray fluorescence analysis involves identification of the various spectral lines with the elements responsible for them (see Fig. 7.57). Quantitative analysis is more involved because each X-ray intensity must be quantitatively compared with that of a standard (of known composition) of the same elemental makeup. Both peak and background intensities near the peak are counted to permit estimation of peak heights. Online computers handle quantitative correction programs in extremely short time.

X-ray fluorescence analysis can be used for the determination of major elements (those in the one to many tens of percent range). However, it is also very sensitive to accurate determinations of some trace element components (e.g., Y, Zr, Sr, Rb, in the ppm range) because of near-zero background.

Electron Microprobe Analysis (EMPA)

Electron microprobe analysis methods are based on the same principles as those outlined for X-ray fluorescence, except that the original energy source is not an X-ray tube but instead is a finely focused beam of electrons. Electrons are charged particles that can be focused using electrical fields. (X-rays, by contrast, cannot be focused, they can only be collimated.) As the name implies, this is a *microanalytical* technique, because it allows the qualitative and/or quantitative analysis of a minute volume of material.

The energy source used for generating electrons that will dislodge inner-shell electrons in the sample to be analyzed is a tungsten filament that at very high temperatures and high voltage becomes a source of free electrons (inside the high vacuum of the electron gun and column of the instrument; see Fig. 1.11). These electrons can be focused into a very fine beam through a set of electromagnetic lenses between the electron source (the W filament) and the sample to be analyzed (see Fig. 1.11). Under fairly optimal operating conditions, this electron beam can be focused to a smaller than 1 μm diameter. Because

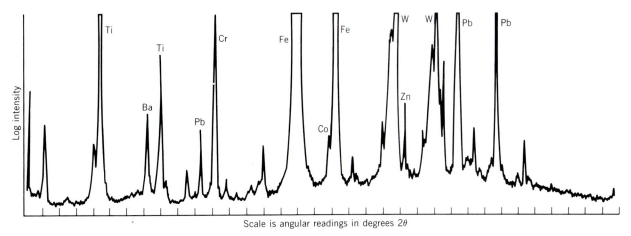

Scale is angular readings in degrees 2θ

FIG. 7.57. Chart recording of the X-ray fluorescence spectrum obtained of elements present in a genuine bank note. The elements responsible for specific peaks are identified. The W peaks are due to the X-ray tube used with a W target; they are not part of the chemical composition of the bank note. The horizontal scale is in angles of 2θ, expressed as degrees. (From H. A. Liebhafsky, H. G. Pfeiffer, E. H. Winslow, and P. D. Zemany, 1960, *X-ray Absorption and Emission in Analytical Chemistry.* New York: Wiley.)

electrons in the beam impinge upon the sample at high velocity, they will penetrate a distance that may be about three times larger than the diameter of the beam (see Fig. 7.58). The minimal analysis volume that can be obtained will range from about 10 to 20 μm^3, which in weight is approximately 10^{-11} grams (for a silicate material).

FIG. 7.58. Schematic representation of the very small volume irradiated by electrons incident upon the polished surface of a sample. The characteristic X-rays generated are dispersed and analyzed by a spectrometer (with a diffracting crystal, as in X-ray fluorescence) or an energy dispersive analysis system (see Fig. 1.10 for a photograph of an electron microprobe and Fig. 1.11 for a schematic cross section of the instrument).

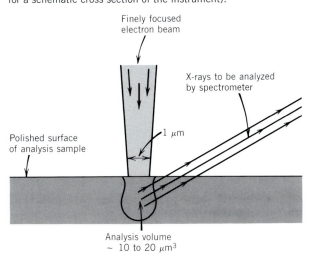

Finely focused electron beam

X-rays to be analyzed by spectrometer

Polished surface of analysis sample

1 μm

Analysis volume ~ 10 to 20 μm^3

In this minute analysis volume, the incident higher energy electrons displace inner-shell electrons of the constituent atoms of the sample. Outer-shell electrons fill these inner-shell vacancies, losing energy by the emission of characteristic X-rays (just as in X-ray fluorescence, discussed above; see Figs. 7.54 and 7.55). These characteristic X-ray spectra are analyzed by a crystal spectrometer (as in the case of X-ray fluorescence) or by an energy dispersive X-ray analysis system. A qualitative electron microprobe analysis is relatively quick and easy to perform, once the sample is in the proper condition for analysis (see below). A quantitative analysis may involve some complex correction procedures that address various interactions between the generated X-rays and their travel path through the sample before detection by the spectrometer. Such corrections, known as matrix corrections, are generally performed by a high-speed computer that is online with the instrument. See Fig. 1.10 for a photograph of a computer-automated electron microprobe.

Because the analytical technique involves a finely focused electron beam that can cause considerable heating of the sample analysis area, and because a minute analysis volume is involved, the sample preparation for electron microprobe techniques is very different from that for X-ray fluorescence. Most commonly, the analysis sample is a polished section (or polished thin section) of a mineral, or rock, or other solid material to be analyzed. The polished surface can be accurately located under the finely focused electron beam by a precise and generally computer-controlled stage mechanism. This allows for the location of grains, or analysis area, as small as 1 μm in diameter, in the field of a high-powered optical (transmitted and reflected light) microscope, which is built into the electron beam column.

The absolute detection limit for most elements analyzed by this technique is not as good as that for X-ray fluorescence because of the presence of a continuum (background) spectrum. However, the ability to obtain a quantitative chemical analysis on a minute volume of material (or a specific mineral grain) is the main reason for the great popularity of this technique in studies of minerals, rocks, ceramics, alloys, and so on.

As in a scanning electron microscope (SEM), the electron beam in an electron microprobe can be moved at high speed across a small area of the specimen. Such a raster scan generates X-rays, cathodoluminescence radiation (i.e., the emission of

electromagnetic radiation in the ultraviolet, visible, or infrared wavelengths during electron bombardment), secondary and backscattered electrons and specimen current. An energy-dispersive detector system (EDS) or a wavelength spectrometer can record the X-ray radiation generated over the small area scanned. Figure 7.59 is an example of Mn zonation in calcite as qualitatively expressed by cathodoluminescence. A quantitative line traverse across this zoned calcite is shown in the left corner of the photography. The central graph consists of a horizontal distance bar with quantitative weight percentage data for Mn and Fe along the vertical axis. These quantitative data were produced with a stationary electron beam positioned at specific spots along the traverse; at each analysis point X-ray signals resulted that were converted to weight percentage values of Mn and Fe.

Secondary Ion Mass Spectrometry (SIMS)

This technique offers, *in situ*, very high-sensitivity quantitative elemental analysis with detection limits in the parts per million (ppm) to parts per billion range (ppb). It can analyze most chemical elements from H to U, and it can also determine the isotopic composition of a sample. The lateral resolution of this surface analytical technique is approximately 5 μm (micrometer).

The instrumentation employs a focused beam of ions that impinges on the solid surface of the sample. These atoms of the surface of the sample are extracted as secondary ions for analysis by a mass spectrometer. The collision process (see Fig. 7.60)

FIG. 7.59. A polished section of calcite that shows clearly defined zones as outlined by cathodoluminescence under an electron beam that was rastered over the surface of the specimen. The total area scanned and photographed is 1000 × 1000 μm (micrometers). The zones with high luminescence (bright regions) are high in manganese content as compared to the darker regions that are low. A quantitative line traverse, along the inclined line in the left corner, shows the variation in Fe and Mn in weight percentage. (After R. F. Denniston, C. K. Shearer, G. D. Layne, and D. T. Vaniman, 1977, SIMS analyses of minor and trace element distributions in fracture calcite from Yucca Mountain, Nevada, USA, *Geochimica et Cosmochimica Acta* 61: 1803–18.)

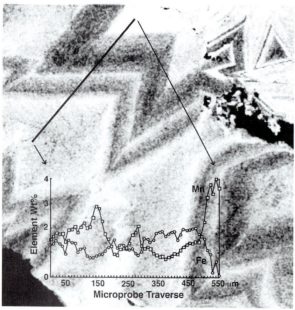

FIG. 7.60. Schematic illustration of the sputtering process showing emission of secondary particles induced by a collision cascade of primary particles. (After N. MacRae, 1995, Secondary-Ion Mass Spectrometry. *The Canadian Mineralogist* 33: 219–36.)

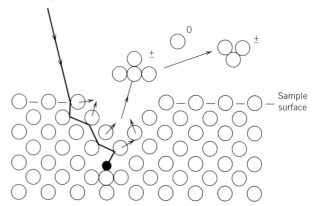

causes atoms at or near the surface of the sample to escape due to the energy received from the ion beam. In this process particles are ejected as atoms or molecules in a neutral, excited, or ionized state, but only the ionized species can be used in SIMS analysis. The instrument consists of a source of bombarding primary ions (commonly O^- or Cs^+) that are focused onto the specimen sample by lens systems. In most of the instruments used for mineralogical analysis the secondary ions (produced by sputtering at the sample surface) are extracted, and focused into a double focusing mass spectrometer which separates ions on the basis of energy and mass. Figure 7.61 shows the general layout of the instrumentation.

Although the instrument can be used to determine the concentration of any chemical element (especially those that occur at very low levels) it is commonly applied to the determination of the abundance and distribution of the rare earth elements (REE). These are commonly used as petrogenetic indicators (petrogenesis is that part of petrology that deals with the origin and formation of rocks). REE abundances are often so low in geologic samples that they are generally below the detection levels of the other microanalysis technique, electron microprobe analysis (EMPA). The SIMS technique also allows collecting isotopic compositions for age dating and diffusion studies as well as production of high-resolution isotopic-distribution images of geologic samples.

From the design of earlier SIMS instruments came the development of a new ion microprobe known as the Sensitive High-Resolution Microprobe (SHRIMP), which has the unique capability of accurate microprobe analysis for U and Pb on a microscale (area of analysis is about 30 micrometers). This technique has been especially successful in the U and Pb analyses of individual zircon grains for age dating many of the oldest rock types in the Precambrian.

ATOMIC FORCE MICROSCOPY (AFM)

This technique allows for the study of surface morphology and structure at the atomic scale. Atomic force microscopy was developed as an outgrowth of work on the scanning tunneling microscope (STM), which is applied to electrically conductive materials such as metals and semiconductors such as sulfides. In the STM analytical method, an atomically sharp, conducting voltage-biased tip is brought close to the surface of the conducting material. Electrons tunnel from the material to the tip, or from the tip to the material. The spatial variation in current is a reflection of the surface at an atomic or near-atomic scale. Atomic force microscopy produces similar-scale images for insulators.

The atomic force microscope makes use of the interatomic forces between atoms forming the surface of the sample under study and the atoms forming the end of a sharp probe tip attached to a cantilever-type spring (see Fig. 7.62). The application

FIG. 7.61. Schematic diagram of an ion microprobe. (Illustration courtesy of C. Shearer, Institute of Meteoritics, University of New Mexico, Albuquerque, New Mexico.)

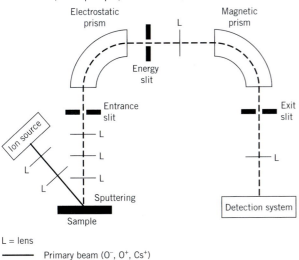

L = lens

——— Primary beam (O^-, O^+, Cs^+)

- - - - Secondary ion beam

FIG. 7.62. Schematic illustration of some of the components in an atomic force microscope (AFM). (Diagram courtesy of Huifang Xu, Department of Earth and Planetary Sciences, University of New Mexico.)

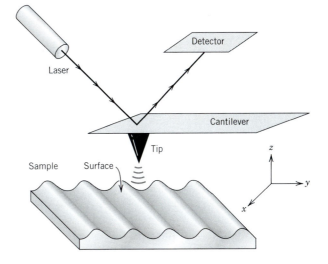

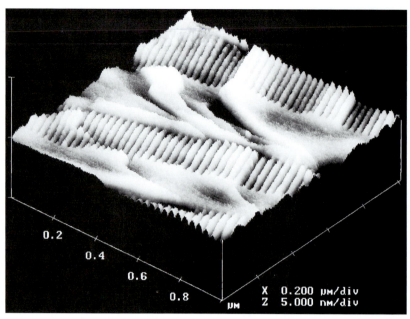

FIG. 7.63. An AFM image of the complex texture of the (001) cleavage surface in cryptoperthite (an extremely fine exsolution intergrowth of Na- and K-feldspar). The Na-rich (albite) lamellae show closely spaced albite twins as expressed by the regularly spaced ridges. The K-rich lamellae (close to orthoclase in composition) have an essentially flat surface. The horizontal scales are expressed in increments of 0.2 μm (micrometers); the vertical scale is in 5 nanometers per division. (After H. Xu, 1997, Surface characters of twin domains and exsolution lamellae in a feldspar crystal: Atomic force microscopy study. *Materials Research Bulletin* 32: 1121–27; photograph courtesy of Huifang Xu.)

of a small tracking force keeps the tip in contact with the surface. Any force acting on the probe tip deflects the cantilever, and such deflection reflects the atomic topography of the sample surface. The tip displacement is proportional to the force between the surface and the tip. The resultant bending of the cantilever can be measured optically by laser interferometry or beam deflection. One important advantage of AFM is that material can commonly be imaged "as is" with essentially no sample preparation. Examples of images obtained by AFM are shown in Figures 7.63 and 7.64. Figure 7.63 shows a

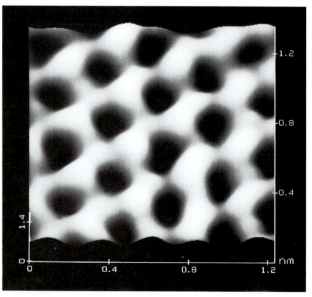

FIG. 7.64. An AFM image of the atomic structure of an octahedral sheet that forms the surface of the sheet silicate lizardite, one of three polymorphs of serpentine, $Mg_3Si_2O_5(OH)_4$. The hydroxyl groups (pale gray) are arranged in triangles and are bonded to magnesium ions (medium gray). The black areas are empty sites. The horizontal scales are shown in increments of 0.4 nanometers (4 Å). (After F. J. Wicks, K. Kjoller, R. K. Eby, F. C. Hawthorne, G. S. Henderson, and G. A. Vrdoljak, 1993, Imaging the internal atomic structure of layer silicates using the atomic force microscope. *The Canadian Mineralogist* 31: 541–50; photograph courtesy of F. J. Wicks.)

(001) cleavage surface of cryptoperthite and its exsolution lamellae, at a horizontal resolution of about 0.02 micrometers and a vertical resolution of about 0.1 nanometer (1Å). Figure 7.64 is an AFM image of an octahedral sheet in lizardite, one of three polymorphs of serpentine. The horizontal resolution (as shown by the rectangular axes) is on the order of 0.1 nanometer (1Å) and the vertical resolution is about 0.1 nanometer (1Å). Such high-resolution images allow scientists to evaluate processes such as crystallization, dissolution, adsorption, and alteration at the atomic scale.

REFERENCES AND SELECTED READING

Optical Microscopy: Nonopaque Minerals

Ehlers, E. G. 1987. *Optical mineralogy.* Vol. 1, *Theory and techniques*; Vol. 2, *Mineral descriptions.* Palo Alto, Calif.: Blackwell Scientific Publications.

Heinrich, E. W. 1965. *Microscopic identification of minerals.* New York: McGraw-Hill.

Larsen, E. S., and H. Berman. 1934. *The microscopic determination of the nonopaque minerals.* 2nd ed. U. S. Geological Survey Bulletin no. 848. Washington, D. C.: U. S. Government Printing Office.

MacKenzie, W. S., and C. Guilford. 1980. *Atlas of rock-forming minerals in thin section.* New York: Wiley.

Nesse, W. D. 1991. *Introduction to optical mineralogy.* 2nd ed. New York: Oxford University Press.

Phillips, W. R. 1971. *Minerals optics: Principles and techniques.* San Francisco: W. H. Freeman & Co.

Phillips, W. R., and D. T. Griffen. 1981. *Optical mineralogy: The nonopaque minerals.* San Francisco: W. H. Freeman & Co.

Shelley, D. 1985. *Optical mineralogy.* 2nd ed. New York: Elsevier Science Publishers.

Stoiber, R. E., and Morse, S. A. 1981. *Microscopic identification of crystals.* Rev. ed. Robert E. Krieger Publishing Co.

Wahlstrom, E. E. 1979. *Optical crystallography.* 5th ed. New York: Wiley.

Optical Microscopy: Opaque Minerals

Craig, J. R., and D. J. Vaughan. 1981. *Ore microscopy and ore petrology.* New York: Wiley.

Galopin, R., and N. F. M. Henry. 1972. *Microscopic study of opaque minerals.* London: McCrone Research Associates Ltd.

Ixer, R. A. 1990. *Atlas of opaque and ore minerals in their associations.* New York: Van Nostrand Reinhold.

Ramdohr, P. 1969. *The ore minerals and their intergrowths.* New York: Pergamon Press.

Stanton, R. L. 1972. *Ore petrology.* New York: McGraw-Hill.

Uytenbogaardt, W., and E. A. J. Burke. 1971. *Tables for microscopic identification of ore minerals.* 2nd ed. New York: Elsevier-North Holland.

X-ray Diffraction

Azaroff, L. V. 1968. *Elements of X-ray Crystallography.* New York: McGraw-Hill.

Azaroff, L. V., and M. J. Buerger. 1958. *The powder method in X-ray crystallography.* New York: McGraw-Hill.

Bragg, W. L. 1949. *The crystalline state: A general survey.* London: G. Bell and Sons, Ltd.

Buerger, M. J. 1964. *The precession method in X-ray crystallography.* New York: Wiley.

Cullity, B. D. 1978. *Elements of X-ray diffraction.* 2nd ed. Reading, Mass.: Addison-Wesley.

Gibbs, G. V. 1982. Molecules as models for bonding in silicates. *American Mineralogist* 67: 421–451.

Klein, C. 1994. *Minerals and rocks: Exercises in crystallography, mineralogy, and hand specimen petrology.* Rev. ed. New York: Wiley.

Klug, H. P., and L. E. Alexander. 1974. *X-ray diffraction procedures for polycrystalline and amorphous materials.* 2nd ed. New York: Wiley.

Nuffield, E. W. 1966. *X-ray diffraction methods.* New York: Wiley.

Post, J. E., and D. L. Bish. 1989. Rietveld refinement of crystal structures using X-ray powder diffraction data. In *Modern powder diffraction,* edited by D. L. Bish and J. E. Post. *Reviews in Mineralogy* 20: 277–308.

Scanning Electron Microscopy and Electron Microprobe Analysis

Goldstein, J. I., D. E. Newbury, P. Echlin, D. C. Joy, A. D. Romig Jr., C. E. Lyman, C. Fiori, and E. Lifshin. 1992. *Scanning electron microscopy and X-ray microanalysis, a text for biologists, material scientists, and geologists.* 2nd ed. New York: Plenum Press.

Electron Microscopy

Buseck, P. R., and S. Iijima. 1974. High resolution electron microscopy of silicates. *American Mineralogist* 59: 1–24.

Electron microscopy in mineralogy. 1976. Coordinating ed. H. R. Wenk. New York: Springer-Verlag.

Hirsch, P., A. Howie, R. B. Nicholson, D. W. Pashley, and M. J. Whelan. 1977. *Electron microscopy of thin crystals.* Malabar, Fla.: Robert E. Krieger Publishing Company.

Veblen, D. R. 1985. Direct imaging of complex structures and defects in silicates. *Annual Review of Earth and Planetary Sciences* 13: 119–46.

Chemical and X-ray Fluorescence Analysis

Laitenen, H. A., and S. Harris. 1975. *Chemical Analysis.* 2nd ed. New York: McGraw-Hill.

Liebhafsky, H. A., H. G. Pfeiffer, E. H. Winslow, and P. D. Zemany. 1960. *X-ray absorption and emission in analytical chemistry.* New York: Wiley.

Skoog, D. A., and J. L. Leary. 1992. *Principles of instrumental analysis.* 4th ed. New York: Saunders College Publishing.

Strobel, H. A., and W. R. Heineman. 1989. *Chemical instrumentation: A systematic approach.* 3rd ed. New York: Wiley.

Willard, H. H., L. M. Merritt, J. A. Dean, and F. A. Settle. 1981. *Instrumental methods of analysis.* 6th ed. New York: D. Van Nostrand Company.

Ion Microprobe Analysis

Reed, S. J. B. 1989. Ion microprobe analysis—A review of geological applications. *Mineralogical Magazine* 53: 3–24.

Sykes, D. E. 1989. Dynamic secondary ion mass spectrometry. In *Methods of surface analysis.* Edited by J. M. Walls. New York: Cambridge University Press.

Atomic Force Microscopy

Binnig, G., C. F. Quate, and C. Gerber. 1986. Atomic force microscopy. *Physics Review Letters* 56: 930–33.

Hochella, M. F., Jr., C. M. Eggleston, V. B. Elings, and M. S. Thompson. 1990. Atomic structure and morphology of the albite {010} surface: An atomic force microprobe and diffraction study. *American Mineralogist* 75: 723–30.

Wicks, F. J., K. Kjoller, R. K. Eby, F. C. Hawthorne, G. S. Henderson, and G. A. Vrdoljak. 1993. Imagining the internal atomic structure of layer silicates using the atomic force microscope. *The Canadian Mineralogist* 31: 541–50.

Additional References

Microbeam Techniques in the earth sciences. 1995. Edited by F. C. Hawthorne and R. F. Martin. *The Canadian Mineralogist* 33, part 2: 201–508.

Spectroscopic methods in mineralogy and geology. 1988. Edited by F. C. Hawthorne. *Reviews in Mineralogy* 18, Mineralogical Society of America, Washington, D. C.

Zolensky, M. E., C. Pieters, B. Clark, and J. J. Papike. 2000. Small is beautiful: The analysis of nanogram-sized astromaterials. *Meteorites and Planetary Science* 35: 9–29.

CHAPTER 8

CRYSTAL CHEMISTRY AND SYSTEMATIC DESCRIPTIONS OF NATIVE ELEMENTS, SULFIDES, AND SULFOSALTS

This chapter and four subsequent chapters will discuss the crystal chemistry, followed by systematic descriptions, of the common minerals and those rarer ones of most economic importance. These five chapters together will treat in detail about 200 mineral species, which is a relatively small number compared with the 3814 minerals that are recognized.[1] The names and chemical compositions of some other, related minerals are also given. Of the 3814 recognized mineral species, 105 are native elements and 579 are sulfides and sulfosalts combined. This first chapter provides information on a considerable number of ore minerals. The term ore mineral *means that it is part of an ore, usually metallic, which is economically desirable, as contrasted with* gangue, *which represents the valueless rock or mineral aggregate in an ore. An ore is a naturally occurring material from which a mineral or minerals of economic value can be extracted at a reasonable profit.*

The systematic description of all minerals[2] follows a common scheme of presentation. The headings used and the data given under each are as follows:

Crystallography. Under this heading is given the following crystallographic information:

The crystal system and symbol of the crystal class.

Crystallographic features usually observed by inspection, such as habit, twinning, and crystal forms.

Other pertinent data are listed without subheadings. For example, consider the following data for cerussite:

Pmcn (space group), $a = 5.15$, $b = 8.47$, $c = 6.11$ Å (unit cell dimensions), $Z = 4$ (formula units per unit cell).

*d*s: 3.59(10), 3.50(4), 3.07(2), 2.49(3), 2.08(3) (strongest X-ray lines in angstrom units with relative intensities in parentheses).

Physical Properties. The physical properties *cleavage*, **H** (hardness), **G** (specific gravity), *luster*, *color*, and *optics* (brief summary of optical data) are listed.

Composition and Structure. The chemical composition is given. Commonly, it includes the percentage of elements or oxides, as well as the elements that may substitute for those given in the chemical formula. The most important as-

---

[1]All numerical data regarding mineral species are courtesy of J. A. Mandarino, Toronto, Ontario, Canada.

[2]Mineral names are given in boldface capitals (e.g., **GOLD**) or lowercase (e.g., **Acanthite**). Those in boldface capitals are considered most common or important.

pects of the crystal structure are briefly described.

Diagnostic Features. The outstanding properties and tests that aid one in recognizing the mineral and distinguishing it from others are listed.

Occurrence. A brief statement of the mode of occurrence and characteristic mineral associations is given. The localities where a mineral is or has been found in notable amount or quality are mentioned. For minerals found abundantly the world over, emphasis is on North American localities.

Use. For a mineral of economic value, there is a brief statement of its uses.

Similar Species. The similarity of the species listed to the mineral whose description precedes it may be on the basis of either chemical composition or crystal structure.

MINERAL CLASSIFICATION

Chemical composition has been the basis for the classification of minerals since the middle of the nineteenth century. According to this scheme, minerals are divided into classes depending on the dominant anion or anionic group (e.g., oxides, halides, sulfides, silicates, etc.). There are a number of reasons why this criterion is a valid basis for the broad framework of mineral classification. First, minerals having the same anion or anionic group dominant in their composition have unmistakable family resemblances, in general stronger and more clearly marked than those shared by minerals containing the same dominant cation. Thus, the carbonates resemble each other more closely than do the minerals of copper. Second, minerals related by dominance of the same anion tend to occur together or in the same or similar geological environment. Thus, the sulfides occur in close mutual association in deposits of vein or replacement type, whereas the silicates make up the great bulk of the rocks of the Earth's crust. Third, such a scheme of mineral classification agrees well with the current chemical practice in naming and classifying inorganic compounds.

However, it was early recognized that chemistry alone does not adequately characterize a mineral. A full appreciation of the nature of minerals was to wait until X-rays were used to determine internal structures. It is now clear that mineral classification must be based on chemical composition *and* internal structure, because these together represent the essence of a mineral and determine its physical properties. Crystallochemical principles

were first used by W. L. Bragg and V. M. Goldschmidt for silicate minerals. This large mineral group was divided into subclasses partially on the basis of chemical composition but principally in terms of internal structure. Within the silicate class, therefore, framework, chain and sheet silicate (etc.) subclasses exist on the basis of the structural linkage of SiO_4 tetrahedra. Such structural principles in combination with chemical composition provide a logical classification. It is this scheme of classification that is used in the subsequent sections on systematic mineralogy.

The broadest divisions of the classification used in this book (as based on C. Palache, H. Berman, and C. Frondel, *Dana's System of Mineralogy*, 7th ed., and H. Strunz, *Mineralogische Tabellen*, 5th ed.; see references) are as follows:

1. Native elements
2. Sulfides   } Chapter 8
3. Sulfosalts
4. Oxides
    (a) Simple and multiple } Chapter 9
    (b) Hydroxides
5. Halides
6. Carbonates
7. Nitrates
8. Borates   } Chapter 10
9. Phosphates
10. Sulfates
11. Tungstates
12. Silicates   Chapters 11 and 12

These classes are subdivided into *families* on the basis of chemical types, and the family in turn may be divided into *groups* on the basis of structural similarity. A group is made up of *species*, which may form *series* with each other. Species have the same structure but different chemistries. A *species* may be subdivided into chemical *varieties* by adjectival modifiers that reflect the presence of unusual amounts of chemical constituents. The following are some examples of such modifiers:

aluminian	: Al-rich
calcian	: Ca-rich
chromian	: Cr-rich
ferroan	: Fe^{2+}-rich
ferrian	: Fe^{3+}-rich
magnesian	: Mg-rich
manganoan	: Mn-rich

These are used in the minerals manganoan aegirine, ferrian diopside, and magnesian augite, for example.

In each of the classes, the mineral with the highest ratio of metal to nonmetal is given first, followed by those containing progressively less metal. Because a relatively small number of minerals is described in this book, often only one member of a group or family is represented, and thus a rigorous adherence to division and subdivision is impractical.

CRYSTAL CHEMISTRY OF NATIVE ELEMENTS, SULFIDES, AND SULFOSALTS

Native Elements

With the exception of the free gases of the atmosphere, only about 20 elements are found in the native state. These elements can be divided into (1) metals; (2) semimetals; (3) nonmetals. The more common *native metals*, which display very simple structures, constitute three groups: the *gold group* (space group: $Fm3m$), gold, silver, copper, and lead, all of which are isostructural; the *platinum group* (space group: $Fm3m$), platinum, palladium, iridium and osmium, all of which are isostructural; and the *iron group*, iron and nickel-iron, of which pure Fe as well as kamacite have space group $Im3m$ and the more Ni-rich variety of nickel-iron (taenite) $Fm3m$.

Metals		Semimetals	
Gold group		Arsenic group	
Gold	Au	(Arsenic	As)
Silver	Ag	(Bismuth	Bi)
Copper	Cu	Nonmetals	
Platinum group		Sulfur	S
Platinum	Pt	Diamond	C
Iron group		Graphite	C
Iron	Fe		
(Kamacite	Fe, Ni)		
(Taenite	Fe, Ni)		

In addition, mercury, tantalum, tin, and zinc have been found. The *native semimetals* form two isostructural groups: arsenic, antimony, and bismuth (space group $R3m$), and the less common selenium and tellurium (space group $P3_121$). The important *nonmetals* are sulfur and carbon in the form of diamond and graphite.

Native Metals

It is fitting that this section begin with a discussion of the gold group, for our knowledge of the properties and usefulness of metals arose from the chance discovery of nuggets and masses of these minerals. Many relatively advanced early cultures were restricted in their use of metal to that found in the native state.

The elements of the *gold group* belong to the same group in the periodic table of elements and, hence, their atoms have somewhat similar chemical properties; all are sufficiently inert to occur in an elemental state in nature. When uncombined with other elements, the atoms of these metals are united into crystal structures by the rather weak metallic bond. The minerals are isostructural and are built on the face-centered cubic lattice with atoms in 12-coordination (see Fig. 8.1a). Complete solid solution takes place between gold and silver, as these two elements have the same atomic radii (1.44 Å). Copper, because of its smaller atomic radius (1.28 Å), exhibits only limited solid solution in gold and silver. Conversely, native copper carries only traces of gold and silver in solid solution.

The similar properties of the members of this group arise from their common structure. All are rather soft, malleable, ductile, and sectile. All are excellent conductors of heat and electricity, display metallic luster and hackly fracture, and have rather low melting points. These are properties conferred by the metallic bond. All are isometric hexoctahedral and have high densities resulting from the cubic closest packing of their structures.

Those properties with respect to which the minerals of this group differ arise from the properties of the atoms of the individual elements. Thus, the yellow of gold, the red of copper, and the white of silver are atomic properties. The specific gravities likewise depend on atomic properties and show a rough proportionality to the atomic weights.

Although only platinum is discussed here, the *platinum group* also includes the rarer minerals palladium, platiniridium, and iridosmine. The last two are respectively alloys of iridium and platinum and iridium and osmium, with hexagonally close-packed structures and space group $P6_3/mmc$. Platinum and iridium, however, have cubic closest packed structures, similar to metals of the gold group, with space group $Fm3m$. The platinum metals are harder and have higher melting points than metals of the gold group.

Members of the *iron group* metals are isometric and include pure iron (Fe), which occurs only rarely on the Earth's surface, and two species of nickel-iron (*kamacite* and *taenite*), which are common in meteorites. Iron and nickel have almost identical atomic radii (1.26 Å and 1.25 Å, respectively), and thus nickel can and usually does substitute for some of the iron. Pure iron and kamacite, which contains up to about 5.5 weight percent Ni, show cubic close packing with space group $Im3m$. Taenite, which shows a range in Ni content from 27 to 65 weight percent, is cubic closest packed with space group $Fm3m$ (see Figs. 8.1a and 8.1c). These two minerals

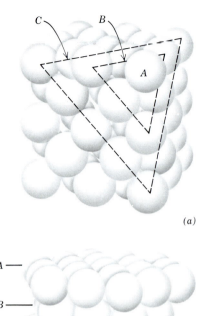

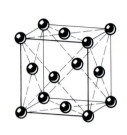

(a)

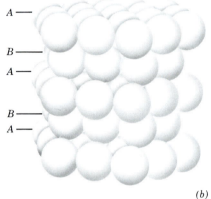

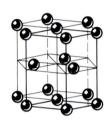

(b)

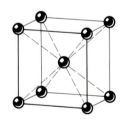

(c)

FIG. 8.1. *(a)* Close-packed model of cubic closest packing (*ABCABC* . . .) of equal spheres as shown by Cu, Au, Pt, and many other metals. Each metal atom is surrounded by 12 closest neighbors (see also Figs. 3.37 and 3.38). Close-packed layers are parallel to {111}. The face-centered cubic lattice (*F*) compatible with this packing sequence is illustrated on the right. *(b)* Close-packed model of hexagonal closest packing (*ABAB* . . .) of equal spheres as shown by Mg, Zn, and Cd. Each metal atom is surrounded by 12 closest neighbors (see also Figs. 3.37 and 3.38). This type of stacking leads to a hexagonal (*H*) lattice, as illustrated on the right, which can be reinterpreted as a rhombohedral (*R*) lattice (see Fig. 5.62; no. 11). *(c)* Close-packed model of body-centered cubic packing of equal spheres, as shown by Fe. Each sphere is surrounded by 8 closest neighbors. This packing is not as close as that exhibited by CCP and HCP. The body-centered (*I*) lattice compatible with this packing model is illustrated on the right.

are characteristic of iron meteorites and it is believed that Fe–Ni alloys of this type constitute a large part of the Earth's core.

Native Semimetals

The semimetals arsenic, antimony, and bismuth belong to an isostructural group with space group $R\bar{3}m$. Their structures, unlike those of the metals, cannot be represented as a simple packing of spheres, because each atom is somewhat closer to three of its neighbors than to the remainder of the surrounding atoms. In Fig. 8.2*a*, which represents the structure of arsenic (or antimony), the spheres intersect each other along the flat circular areas. The bonding between the four closest atoms, forming pyramidal groups (see Fig. 8.2*b*), is due to the covalent nature of these bonds. This covalency is related to the position of these semimetals in group V of the periodic table toward the electronegative end of a row. The relatively strong bonding to four closest neighbors results in a layered structure, as shown in Fig. 8.2*b*. These layers are parallel to {0001}, and the weak bonding between them gives rise to a good cleavage. Members of this group have similar physical properties. They are rather brittle and much poorer conductors of heat and electricity than the native metals.

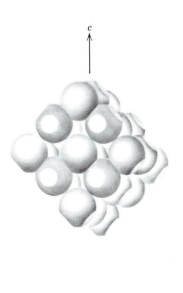

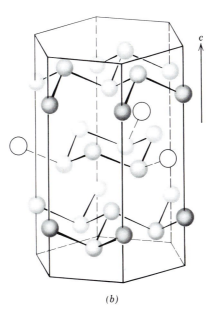

(a)   (b)

FIG. 8.2. *(a)* Close-packed model of the structures of As and Sb. Flat areas represent overlap between adjoining atoms. *(b)* Expanded view of the structure of As or Sb, showing corrugated layers of covalently bonded groups, parallel to {0001}.

These properties reflect a bond type intermediate between metallic and covalent. Hence, it is stronger and more directional than pure metallic in its properties, leading to lower symmetry.

Because native arsenic and bismuth as minerals are uncommon, there will be no detailed descriptions of these species.

Native Nonmetals

The structure of the nonmetals, sulfur, diamond, and graphite are very different from those of the metals and semimetals. *Sulfur* ordinarily occurs in or-

thorhombic crystal form (with space group *Fddd*) in nature; two monoclinic polymorphs are very rare in nature but are commonly produced synthetically. The orthorhombic structure of S is stable at atmospheric pressure below 95.5°C, and a monoclinic polymorph is stable from 95.5 to 119°C. This monoclinic polymorph melts above 119°C. The unit cell of the orthorhombic form of sulfur contains a very large number of S atoms, namely 128. The sulfur atoms are arranged in puckered rings of eight atoms that form closely bonded S_8 molecules (see Fig. 8.3*a*). These rings are bonded to each other by van der

FIG. 8.3. *(a)* S_8 rings in orthorhombic sulfur as seen parallel (below) and perpendicular to the rings (above). *(b)* Unit cell of the structure of sulfur, showing the arrangement of S_8 rings.

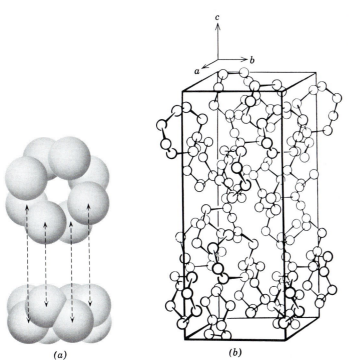

(a)   (b)

Waals forces, with a relatively large spacing between the rings (see Fig. 8.3*b*); 16 of such rings are present in the unit cell. Small amounts of selenium (atomic radius 1.16 Å) may substitute for S (atomic radius 1.04 Å) in the structure.

The structures of the two carbon polymorphs, *diamond* and *graphite*, are shown in Figs. 8.4*a* and *b*. Diamond has an exceptionally close-knit and strongly bonded structure in which each carbon atom is bound by powerful and highly directive covalent bonds to four carbon neighbors at the apices of a regular tetrahedron. The resulting structure, al-

though strongly bonded, is not closed packed, and only 34% of the available space is filled. The presence of rather widely spaced sheets of carbon atoms parallel to {111} (see Fig. 8.4*a*) accounts for prominent octahedral cleavage in diamond. The {111} sheets are planes of maximum atomic population.

The structure of graphite, illustrated in Fig. 8.4*b*, consists of six-membered rings in which each carbon atom has three near neighbors arranged at the apices of an equilateral triangle. Three of the four valence electrons in each carbon atom may be considered to be locked up in tight covalent bonds with three close

FIG. 8.4. *(a)* Partial representation of the structure of diamond. The horizontal plane is (111). *(b)* The structure of graphite with sheets // {0001}. Dashed vertical lines link atoms in successive sheets; these lines do not represent bonds.

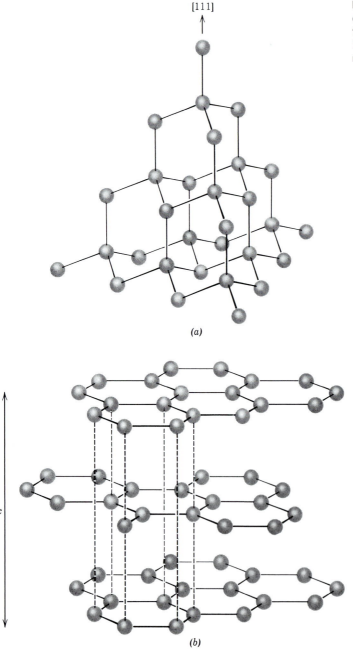

[111]

(a)

c

(b)

carbon neighbors in the plane of the sheet. The fourth is free to wander over the surface of the sheet, creating a dispersed electrical charge that bestows on graphite its relatively high electrical conductivity. In contrast, diamond, in which all four valence electrons are locked up in covalent bonds, is among the best electrical insulators.

The sheets composing the graphite crystal are stacked in such a way that alternate sheets are in identical positions, with the intervening sheet translated a distance of one-half the identity period in the plane of the sheets (see Fig. 8.4b). The distance between sheets is much greater than one atomic diameter, and van der Waals bonding forces perpendicular to the sheets are very weak. The wide separation and weak binding give rise to the perfect basal cleavage and easy gliding parallel to the sheets. Because of this open structure, only about 21% of the available space in graphite is filled, and the specific gravity is proportionately less than that of diamond. See modules III and IV on the CD-ROM for various illustrations of the structures of diamond and graphite.

Sulfides

The sulfides form an important class of minerals that includes a majority of the ore minerals. With them are classed the similar but rarer sulfarsenides, arsenides, and tellurides.

Most of the sulfide minerals are opaque with distinctive colors and characteristically colored streaks. Those that are nonopaque, such as cinnabar, realgar, and orpiment, have extremely high refractive indices and transmit light only on thin edges.

The general formula for the sulfides is given as X_mZ_n in which X represents the metallic elements and Z the nonmetallic element. The general order of listing of the various minerals is in a decreasing ratio of $X : Z$.

The sulfides can be divided into small groups of similar structures but it is difficult to make broad generalizations about their structure. Regular octahedral or tetrahedral coordination about sulfur is found in many simple sulfides such as in galena, PbS, (with an NaCl-type structure), and in sphalerite, ZnS (see Fig. 8.5a). In more complex sulfides, as well as sulfosalts, distorted coordination polyhedra may be found (see tetrahedrite, Fig. 8.5c). Many of the sulfides have ionic and covalent bonding, whereas others, displaying most of the properties of metals, have metallic bonding characteristics. The structures and some aspects of the crystal chemistry of a few of the most common sulfides (e.g., sphalerite, chalcopyrite, pyrite, marcasite, and covellite) will be discussed.

ZnS occurs in two polymorphic forms: the sphalerite-type (Fig. 8.5a) and the wurtzite-type structures (Fig. 8.6). In both the sphalerite and wurtzite structures Zn is surrounded by four sulfurs in tetrahedral

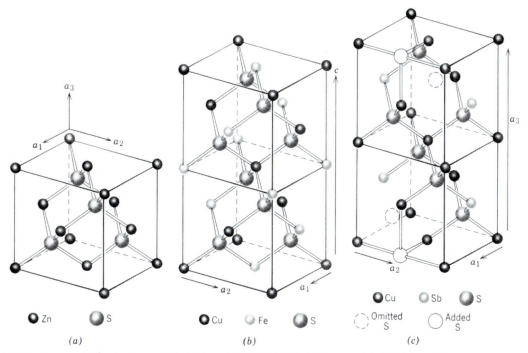

FIG. 8.5. The sphalerite structure and derivatives. (a) Sphalerite, ZnS. Compare this illustration with Fig. 3.53a. (b) Chalcopyrite, $CuFeS_2$. (c) Tetrahedrite, $Cu_{12}Sb_4S_{13}$. (After B. J. Wuensch, 1974, in *Sulfide mineralogy, Reviews in Mineralogy* 1, Mineralogical Society of America, Washington, D. C.)

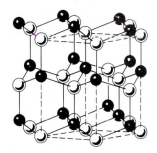

FIG. 8.6. The wurtzite type polymorph of ZnS. See Fig. 8.5*a* for the sphalerite polymorph.

coordination, but in sphalerite the Zn atoms are arranged in a face-centered cubic lattice, whereas in wurtzite they are approximately in positions of hexagonal closest packing. The atomic arrangement in sphalerite is like that in diamond (see Fig. 8.4*a*) in which the carbon has been replaced by equal amounts of Zn and S. Greenockite, CdS, a relatively rare sulfide, is isostructural with ZnS and occurs in both the sphalerite and wurtzite type structures. Natural sphalerite shows extensive Fe^{2+} and very limited Cd^{2+} substitution for Zn (see Table 3.15).

Sulfides, Sulfarsenides, and Arsenides

Acanthite (Argentite)	Ag_2S	Cinnabar	HgS
Chalcocite	Cu_2S	Realgar	AsS
Bornite	Cu_5FeS_4	Orpiment	As_2S_3
Galena	PbS	Stibnite	Sb_2S_3
Sphalerite	ZnS	Pyrite	FeS_2
Chalcopyrite	$CuFeS_2$	Marcasite	FeS_2
Pyrrhotite	$Fe_{1-x}S$	Molybdenite	MoS_2
Nickeline	NiAs	Cobaltite	(Co, Fe)AsS
Millerite	NiS	Arsenopyrite	FeAsS
Pentlandite	$(Fe, Ni)_9S_8$	Skutterudite	$(Co, Ni)As_3$
Covellite	CuS		

Chalcopyrite, $CuFeS_2$, has a structure (Fig. 8.5*b*) that can be derived from the sphalerite structure by regularly substituting Cu and Fe ions for Zn in sphalerite; this leads to a doubling of the unit cell. Stannite, Cu_2FeSnS_4, has a structure based on sphalerite in which layers of ordered Fe and Sn alternate with layers of Cu (chalcopyrite can be rewritten as $Cu_2Fe_2S_4$; this illustrates chemically the close similarity to Cu_2FeSnS_4).

Covellite, CuS, is an example of a chemically very simple substance with a rather complex structure (see Fig. 8.7) in which part of the Cu is tetrahedrally coordinated by four S, and part coordinated by three S in the form of a triangle. The structure can, therefore, be viewed as made of sheets of CuS_3 triangles, between double layers of CuS_4 tetrahedra; covalent sulfur-sulfur bonds link the layers.

Pyrite, FeS_2, has a cubic structure as shown in Fig. 8.8*a*. The structure contains covalently bonded S_2 pairs, which occupy the position of Cl in the NaCl

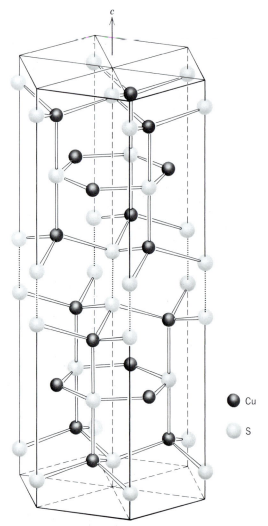

FIG. 8.7. Structure of covellite, CuS. The dotted lines indicate covalent S-S bonds. (After B. J. Wuensch, 1974, in *Sulfide Mineralogy, Reviews in Mineralogy* 1, Mineralogical Society of America, Washington, D. C.)

structure type. The pyrite structure may be considered as derived from the NaCl structure in which Fe is found in the Na positions of NaCl. Another polymorphic form of FeS_2 is marcasite (see Fig. 8.8*b*) with an orthorhombic structure. This structure, as pyrite, contains closely spaced S_2 pairs. It is still unclear what the stability fields of the two polymorphs, pyrite and marcasite, are. From geological occurrences one would conclude that marcasite occurs over a range of low to medium temperature in sedimentary rocks and metalliferous veins. Pyrite, however, occurs also as a magmatic (high *T*) constituent.

The structure of arsenopyrite, FeAsS, can be derived from the marcasite (FeS_2 or FeSS) structure (see Fig. 8.8*c*) in which one S per unit formula is replaced by As. The general coordination of the atoms in marcasite and FeAsS is approximately the same in both structures.

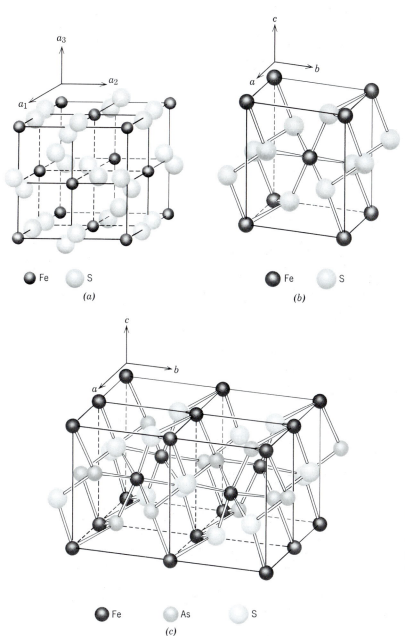

FIG. 8.8. (a) The structure of pyrite, as based on NaCl type structure (compare with Fig. 3.44). Note the closely spaced sulfurs in S_2 groups. (b) The structure of marcasite, showing coordination of ions to closest neighbors. (c) The structure of arsenopyrite, FeAsS, a derivative of the marcasite structure type. (b and c after B. J. Wuensch, 1974 in *Sulfide Mineralogy*, *Reviews in Mineralogy* 1, Mineralogical Society of America, Washington, D. C.)

Sulfosalts

The term *sulfosalt* was originally proposed to indicate that a compound was a salt of one of a series of acids in which sulfur had replaced the oxygen of an ordinary acid. Because such acids are purely hypothetical, it is somewhat misleading to endeavor to thus explain this class of minerals. Nevertheless, the term sulfosalt has been retained for indicating a certain type of unoxidized sulfur mineral that is structurally distinct from a sulfide.

The sulfosalts comprise a diverse and relatively large group of minerals with about 100 species. They differ from sulfides, sulfarsenides, and arsenides in that As, Sb, and rarely Bi play a role more or less like that of metals in the structure; in sulfarsenides and arsenides the semimetals take the place of sulfur in the structure. For example, in arsenopyrite, FeAsS (Fig. 8.8c), the As is substituted for S in a marcasite type of structure. In enargite, Cu_3AsS_4, on the other hand, As enters the metal position of the wurtzite type of structure and is coordinated to four neighboring S ions (see Fig. 8.9). Sulfosalts may be considered double sulfides. As such enargite, Cu_3AsS_4, may be considered $3Cu_2S \cdot As_2S_5$.

The sulfosalts usually occur as minor minerals in hydrothermal veins associated with the more com-

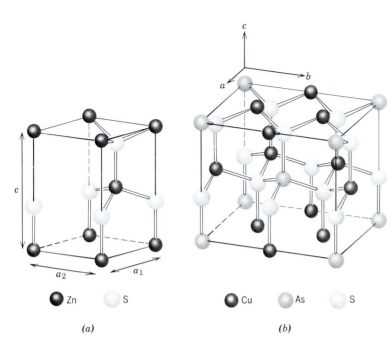

FIG. 8.9. *(a)* Wurtzite structure type of ZnS. *(b)* Enargite, Cu_3AsS_4, an orthorhombic derivative of the wurtzite structure. (After B. J. Wuensch, 1974, in *Sulfide Mineralogy, Reviews in Mineralogy* 1, Mineralogical Society of America, Washington, D. C.)

mon sulfides. With only rare exceptions they are compounds containing silver, copper, or lead but only a very few are abundant enough to have served as ores of these metals. Mention is made below of only five of the more important sulfosalts.

SYSTEMATIC DESCRIPTIONS

Native Metals

GOLD—Au

Cystallography. Isometric; $4/m\overline{3}2/m$. Crystals are commonly octahedral (Fig. 8.10*a*), rarely showing the faces of the dodecahedron, cube, and trapezohedron {113}. Often in arborescent crystal groups with crystals elongated in the direction of a threefold symmetry axis, or flattened parallel to an octahedron

FIG. 8.10. *(a)* Distorted gold octahedron. *(b)* Dendritic gold.

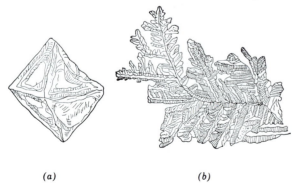

(a)   *(b)*

face. Crystals are irregularly formed, passing into filiform, reticulated, and dendritic shapes (Fig. 8.10*b*). Seldom shows crystal forms; usually in irregular plates, scales, or masses.

Fm3m; a = 4.079 Å; *Z* = 4. *d*s: 2.36(10), 2.04(7), 1.443(6), 1.229(8), 0.785(5).

Physical Properties. H $2\frac{1}{2}$–3, **G** 19.3 when pure. The presence of other metals decreases the specific gravity, which may be as low as 15. *Fracture* is hackly. Very malleable and ductile. Opaque. *Color* is various shades of yellow, depending on the purity becoming paler with increase of silver (see Plate I, no. 1).

Composition and Structure. A complete solid solution series exists between Au and Ag, and most gold contains some Ag. California gold carries 10 to 15% Ag. When Ag is present in amounts of 20% or greater, the alloy is known as *electrum*. Small amounts of Cu and Fe may be present, as well as traces of Bi, Pb, Sn, Zn, and the platinum metals. The purity or *fineness* of gold is expressed in parts per 1000. Most gold contains about 10% of other metals and thus has a fineness of 900. The structure of gold is based on cubic closest packing of Au atoms (see Fig. 8.1*a*).

Diagnostic Features. Gold is distinguished from the yellow sulfides pyrite and chalcopyrite and from yellow flakes of altered micas by its sectility and its high specific gravity.

Occurrence. Gold has an average abundance of 0.004 ppm in the Earth's crust, and is therefore a rare element. It occurs in nature widely distributed in small amounts. It is found most commonly in veins that bear a genetic relation to silicic types of igneous rocks. Most gold occurs as the native metal. For a listing of some other, rare gold minerals, see "Similar Species," below.

The chief sources of gold are hydrothermal gold-quartz veins where, together with pyrite and other sulfides, gold was deposited from ascending mineral-bearing solutions. Gold is also recovered as a byproduct from sulfide deposits mined essentially for the base metals. The gold is merely mechanically mixed with the sulfides and is not in chemical substitution. In most of the veins, gold is so finely divided and uniformly distributed that its presence in the ore can only be detected by microscopic techniques.

When gold-bearing veins are weathered, the gold liberated either remains in the soil mantle as an eluvial deposit, or is washed into the neighboring streams to form a placer deposit. Because of its high specific gravity, gold works its way through the lighter sands and gravels to lodge behind irregularities or to be caught in crevices in the bedrock. The rounded or flattened nuggets of placer gold can be removed by panning, a process of washing away all but the heavy concentrate from which the gold can be easily separated. On a larger scale, gold-bearing sand is washed through sluices where the gold collects behind crossbars or riffles and amalgamates with mercury placed behind the riffles. Most placer mining is today carried on with dredges, some of which are gigantic and can extract the gold from thousands of cubic yards or gravel a day.

Of the estimated world production of gold in 1999 about 20% came from the Republic of South Africa. The principal sources of South African gold are the Precambrian Witwatersrand conglomerate "the Rand," in the Transvaal and similar conglomerates in the Orange Free State. The "reefs" in these conglomerates in which the gold is concentrated are thought to be fossil placers. Other major gold producers are the United States, Australia, Russia, and Canada.

The most important gold-producing states in the United States, in decreasing importance are Nevada, South Dakota, Utah, Arizona, and Montana. The gold from Utah and Arizona is essentially a byproduct of copper mining. For 100 years following the discovery of gold in the streams of California in 1848, that state was the leading producer. The most important gold-producing districts were those of the Mother Lode, gold-quartz veins along the western slope of the Sierra Nevada.

Use. Most of the existing gold is owned by various countries as bullion and is used for international settlements. An increasing amount in the form of medallions and small bars is used for investment purposes. Other uses are in jewelry, scientific instruments, electroplating, gold leaf, and dental appliances.

Similar Species. In addition to native gold there are 19 relatively rare to extremely uncommon minerals in which gold combines with other elements. The majority of these are tellurides. The most prominent examples are: *calaverite*, $AuTe_2$, associated with other tellurides at Cripple Creek, Colorado, and at Kalgoorlie, Western Australia; and *sylvanite*, $(Au,Ag)Te_2$, also found at Cripple Creek, Colorado, and at Kalgoorlie, Western Australia. Others are *krennerite*, $(Au,Ag)Te_2$; *petzite*, Ag_3AuTe_2; *kostovite*, $CuAuTe_4$; *montbrayite*, $(Au,Sb)_2Te_3$; and *muthmannite*, $(Ag,Au)Te$. Other compounds are: *uytenbogaardtite*, Ag_3AuS_3; *fischesserite*, Ag_3AuSe_2; and *nagyagite*, $Pb_5Au(Te,Sb)_4S_{5-8}$. For a complete listing of gold-containing minerals see Wilson (1982).

SILVER—Ag

Crystallography. Isometric; $4/m\bar{3}2/m$. Crystals are commonly malformed and in branching, arborescent, or reticulated groups. Found usually in irregular masses, plates, and scales; in places as coarse or fine wire (Fig. 8.11).

$Fm3m$; $a = 4.09$ Å, $Z = 4$. ds: 2.34(10), 1.439(6), 1.228(8), 0.936(7), 0.934(8).

Physical Properties. H $2\frac{1}{2}$–3. **G** 10.5 when pure, 10 to 12 when impure. *Fracture* is hackly. Malleable and ductile. *Luster* metallic. *Color* and *streak* are silver-white, often tarnished to brown or gray-black (see Plate I, no. 2).

FIG. 8.11. Silver, Kongsberg, Norway.

Composition and Structure. Native silver commonly contains alloyed Au, Hg, and Cu, more rarely traces of Pt, Sb, and Bi. *Amalgam* is a solid solution of Ag and Hg. The structure of silver is based on cubic closest packing of Ag atoms (see Fig. 8.1*a*).

Diagnostic Features. Silver can be distinguished by its malleability, color on a fresh surface, and high specific gravity.

Occurrence. Native silver is widely distributed in small amounts in the oxidized zone of ore deposits. However, native silver in larger deposits is the result of deposition from primary hydrothermal solutions. There are three types of primary deposits: (1) Associated with sulfides, zeolites, calcite, barite, fluorite, and quartz as typified by the occurrence in Kongsberg, Norway. Mines at this locality were worked for several hundred years and produced magnificent specimens of wire silver. (2) With arsenides and sulfides of cobalt, nickel, and silver and with native bismuth. Such was the type at the old silver mines at Freiberg and Schneeberg in Saxony, Germany, and at Cobalt, Ontario. (3) With uraninite and cobalt-nickel minerals. Deposits at Joachimsthal, Bohemia, Czechoslovakia, and at Great Bear Lake, Northwest Territories, Canada, are of this type.

In the United States small amounts of primary native silver are associated with native copper on the Keweenaw Peninsula, Michigan. This has been the best U.S. source of crystallized silver. Elsewhere in the United States native silver is largely secondary. Most of the world's supply comes from silver sulfides and sulfosalts. Large producers of silver are Canada, Peru, Mexico, and the United States, with Mexico a very important producer historically.

Use. An ore of silver, although most of the world's supply comes from other minerals such as *acanthite*, Ag_2S, *proustite*, Ag_3AsS_3 and *pyrargyite*, Ag_3SbS_3. Silver has many uses, chief of which are in photographic film emulsions, plating, brazing alloys, tableware, and electronic equipment. Because of declining production and increased price, silver in coinage has been largely replaced by other metals such as nickel and copper.

COPPER—Cu

Crystallography. Isometric; $4/m\bar{3}2/m$. Tetrahexahedron faces are common, as well as the cube, dodecahedron, and octahedron. Crystals are usually malformed and in branching and arborescent groups (Figs. 8.12*a* and *b*). Usually occurs in irregular masses, plates, and scales, and in twisted and wire-like forms.

$Fm3m$; $a = 3.615$ Å; $Z = 4$. *d*s: 2.09(10), 1.81(10), 1.28(10), 1.09(10), 1.05(8).

Physical Properties. **H** $2\frac{1}{2}$–3. **G** 8.9. Highly ductile and malleable. *Fracture* hackly. *Luster* metallic. *Color* copper-red on fresh surface, usually dark with dull luster because of tarnish (see Plate I, no. 3).

Composition and Structure. Native copper often contains small amounts of Ag, Bi, Hg, As, and Sb. The structure of copper is based on cubic closest packing of Cu atoms (see Fig. 8.1*a*).

Diagnostic Features. Native copper can be recognized by its red color on fresh surfaces, hackly fracture, high specific gravity, and malleability.

Occurrence. Small amounts of native copper have been found at many localities in the oxidized zones of copper deposits associated with cuprite, malachite, and azurite.

Most primary deposits of native copper are associated with basaltic lavas, where deposition of copper resulted from the reaction of hydrothermal solutions with iron oxide minerals. The only major deposit of this type is on the Keneenaw Peninsula,

FIG. 8.12. *(a)* Dendritic copper, Broken Hill, New South Wales, Australia. *(b)* Native copper, Keweenaw Peninsula, Michigan. (Harvard Mineralogical Museum).

Michigan, on the southern shore of Lake Superior. The rocks of the area are composed of Precambrian basic lava flows interbedded with conglomerates. The series, exposed for 100 miles along the length of the peninsula, dips to the northwest beneath Lake Superior to emerge on Isle Royale 50 miles away. Some copper is found in veins cutting these rocks, but it occurs principally either in the lava flows filling cavities or in the conglomerates filling interstices or replacing pebbles. Associated minerals are prehnite, datolite, epidote, calcite, and zeolites. Small amounts of native silver are present. Michigan copper had long been used by the American Indians, but it was not until 1840 that its source was located. Exploitation began shortly thereafter, and during the next 75 years there was active mining through the length of the Keweenaw Peninsula. Most of the copper in the district was in small irregular grains, but notable large masses were found; one found in 1857 weighed 520 tons.

Sporadic occurrences of copper similar to the Lake Superior district have been found in the sandstone areas of the eastern United States, notably in New Jersey, and in the glacial drift overlying a similar area in Connecticut. In Bolivia at Corocoro, southwest of La Paz, there is a noted occurrence in sandstone. Native copper, sometimes as fine crystal groups, occurs in small amounts associated with the oxidized copper ores of Ajo, Bisbee, and Ray, Arizona, and the Chino Mine, Santa Rita, New Mexico.

Use. A minor ore of copper; copper sulfides are today the principal ores of the metal. The greatest use of copper is for electrical purposes, mostly as wire. It is also extensively used in alloys, such as brass (copper and zinc), bronze (copper and tin with some zinc), and German silver (copper, zinc, and nickel). These and many other minor uses make copper second only to iron as a metal essential to modern civilization.

PLATINUM—Pt

Crystallography. Isometric; $4/m\bar{3}2/m$. Cubic crystals are rare and commonly malformed. Usually found in small grains and scales. In some places it occurs in irregular masses and nuggets of larger size.

$Fm3m$; $a = 3.923$ Å; $Z = 4$. ds: 2.27(9), 1.180(10), 1.956(8), 1.384(8).

Physical Properties. H $4-4\frac{1}{2}$ (unusually high for a metal). G 21.45 when pure, 14 to 19 when native. Malleable and ductile. *Color* steel-gray, with bright luster. Magnetic when rich in iron.

Composition and Structure. Platinum is usually alloyed with several percent Fe and with smaller amounts of Ir, Os, Rh, Pd; also Cu, Au, Ni. The structure of platinum is based on cubic closest packing of atoms (see Fig. 8.1a).

Diagnostic Features. Determined by its high specific gravity, malleability, and steel-gray color.

Occurrence. Most platinum occurs as the native metal in ultrabasic rocks, especially dunites, associated with olivine, chromite, pyroxene, and magnetite. It has been mined extensively in placers, which are usually close to the platinum-bearing igneous parent rock.

Platinum was first discovered in Colombia, South America. It was taken to Europe in 1735, where it received the name *platina* from the word *plata* (Spanish for silver) because of its resemblance to silver. A small amount of platinum is still produced in Colombia from placers in two districts near the Pacific Coast. In 1822 platinum was discovered in placers on the Upper Tura River on the eastern slope of the Ural Mountains in a large district surrounding Nizhniy Tagil, Russia. From that time until 1934, most of the world's supply of platinum came from placers of that district, which centered around the town of Nizhniy Tagil. In 1934, because of the large amount of platinum recovered from the copper-nickel ore of Sudbury, Ontario, Canada, that area became the leading producer. In 1954 the Republic of South Africa moved into first place. Part of the South African production is a byproduct from gold mining on the Rand, but the chief source is the Merensky Reef in the ultrabasic rocks of the Bushveld igneous complex. The Merensky Reef is a horizon of this layered intrusive about 12 inches thick and extending for many miles with a uniform platinum content of about one-half ounce per ton of ore. The chief mining operations are near Rustenburg in the Transvaal. It is estimated that today about half of the world's platinum-group metals comes from Russia. In the United States small amounts are produced as a byproduct of copper mining. In Canada most of the platinum comes from *sperrylite*, $PtAs_2$, which occurs in the copper-nickel ore in Sudbury, Ontario.

Use. Many of the uses of platinum depend on its high melting point (1755°C), resistance to chemical attack, and superior hardness. But its major use is as catalysts to control automobile exhaust emissions and in the chemical and petroleum industries. It is also used in dentistry, surgical instruments, jewelry, and electrical equipment.

Similar Species. *Iridium*; *iridosmine*, an alloy of iridium and osmium; and *palladium* are rare minerals of the platinum group associated with platinum.

Iron—Fe

Crystallography. Isometric; $4/m\bar{3}2/m$. Crystals are rare. Terrestrial: in blebs and large masses; meteoritic (*kamacite*): in plates and lamellar masses, and in regular intergrowth with nickel-iron (*taenite*).

Ordinary iron (α-Fe and kamacite) body-centered cubic, $Im3m$, $a = 2.86$ Å; $Z = 2$. ds: 2.02(9), 1.168(10), 1.430(7), 1.012(7). Nickel-iron (taenite), face-centered cubic, $Fm3m$, $a = 3.56$ Å, $Z = 4$. ds: 2.06(10), 1.073(4), 1.78(3), 1.27(2).

Physical Properties. (α-Fe). *Cleavage* {001} poor. **H** $4\frac{1}{2}$. **G** 7.3–7.9. *Fracture* hackly. Malleable. Opaque. *Luster* metallic. *Color* steel-gray to black. Strongly magnetic.

Composition and Structure. α-Fe always contains some Ni and frequently small amounts of Co, Cu, Mn, S, C. *Kamacite* contains approximately 5.5 weight percent Ni. *Taenite* is highly variable in Ni content and ranges from about 27 to 65 weight percent Ni. The structures of α-Fe and kamacite are based on body-centered cubic packing of atoms (see Fig. 8.1c), whereas the structure of taenite is based on face-centered cubic packing (see Fig. 8.1a).

Diagnostic Features. Iron can be recognized by its strong magnetism, its malleability, and the oxide coating usually on its surface.

Occurrence. Occurs sparingly as terrestrial iron but commonly in meteorites. Iron in the elemental (native) state is highly unstable in the oxidizing conditions in rocks of the upper crust and in the Earth's atmosphere. Iron is normally present as Fe^{2+} or Fe^{3+} in oxides such as magnetite, Fe_3O_4, or hematite, Fe_2O_3, or as the hydroxide, goethite, FeO·OH. Terrestrial iron is regarded as a primary magmatic constituent or a secondary product formed by the reduction of iron compounds by assimilated carbonaceous material. The most important locality is Disko Island, Greenland, where fragments ranging from small grains to masses of many tons are included in basalts. Masses of nickel-iron have been found in Josephine and Jackson counties, Oregon.

Iron meteorites are composed largely of a regular intergrowth of *kamacite* and *taenite*. Evidence of the intergrowth, the Widmanstätten pattern (Fig. 3.3), is seen on polished and etched surfaces of such meteorites. Stony-iron meteorites consist of mixtures of silicates, kamacite, taenite, and troilite (FeS).

Native Nonmetals

SULFUR—S

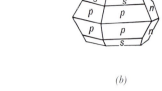

Crystallography. Orthorhombic; $2/m2/m2/m$. Pyramidal habit common (Fig. 8.13a), often with two dipyramids, prism {011}, and base in combination (Figs. 8.13b and 8.14). Commonly found in irregular masses imperfectly crystallized. Also massive, reniform, stalactitic, as incrustations, earthy.

$Fddd$; $a = 10.47$, $b = 12.87$, $c = 24.49$ Å; $Z = 128$. ds: 7.76(4), 5.75(5), 3.90(10), 3.48(4), 3.24(6).

Physical Properties. *Fracture* conchoidal to uneven. Brittle. **H** $1\frac{1}{2}$–$2\frac{1}{2}$. **G** 2.05–2.09. *Luster* resinous. *Color* sulfur-yellow, varying with impurities to yellow shades of green, gray, and red. Transparent to translucent (see Plate I, no. 4). *Optics*: (+); $\alpha = 1.957$, $\beta = 2.037$, $\gamma = 2.245$, $2V = 69°$.

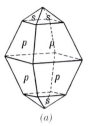

FIG. 8.13. Sulfur crystals.

Sulfur is a poor conductor of heat. When a crystal is held in the hand close to the ear, it will be heard to crack. This is due to the expansion of the surface layers because of the heat from the hand, while the interior, because of the slow heat conductivity, is unaffected. Crystals of sulfur should therefore be handled with care.

Composition and Structure. Native sulfur may contain small amounts of selenium in substitution for S. The structure of the orthorhombic polymorph is given in Fig. 8.3. This structure consists of covalently bonded S_8 groups in the shape of puckered rings. These discrete and compact S_8 rings are relatively loosely bound to each other by van der Waals bonds. The unit cell of sulfur contains 128 atoms of sulfur distributed among 16 S_8 rings. Sulfur melts at 119°C to a liquid in which the covalently bonded S_8 rings retain their identity up to 160°C. The monoclinic polymorphs (α and β sulfur) are very rare in nature.

Diagnostic Features. Sulfur can be identified by its yellow color and the ease with which it burns; it ignites in a candle flame. The absence of a good cleavage distinguishes it from orpiment.

Occurrence. Sulfur often occurs at or near the crater rims of active or extinct volcanoes, where it has been derived from the gases given off in fumaroles.

FIG. 8.14. Sulfur crystals, Cinciana, Sicily, Italy (Harvard Mineralogical Museum).

These may furnish sulfur as a direct sublimation product or by the incomplete oxidation of hydrogen sulfide gas. It is also formed from sulfates, by the action of sulfur-forming bacteria. Sulfur may be found in veins associated with metallic sulfides and formed by the oxidation of the sulfides. It is most commonly found in Tertiary sedimentary rocks associated with anhydrite, gypsum, and limestone; often in clay rocks; frequently with bituminous deposits. The large deposits near Girgenti, Sicily, are world famous for the fine crystals associated with celestite, gypsum, calcite, aragonite. Sulfur is also found associated with the volcanoes of Mexico, Hawaii, Japan, Argentina, and at Ollague, Chile, where it is mined at an elevation of 19,000 feet.

In the United States the most productive deposits are in Texas and Louisiana, where sulfur is associated with anhydrite, gypsum, and calcite in the cap rock of salt domes. There are 10 to 12 producing localities, but the largest are Boling Dome in Texas and Grand Ecaille, Plaquemines Parish, Louisiana. Sulfur is also recovered from salt domes in Mexico and offshore in the shallow waters of the Gulf of Mexico. Sulfur is obtained from these deposits by the Frasch method. Superheated water is pumped down to the sulfur horizon, where it melts the sulfur; compressed air then forces the sulfur to the surface.

About half of the world's sulfur is produced as the native element; the remainder is recovered as a byproduct in smelting sulfide ores, from sour natural gas, and from pyrite.

Use. Sulfur is used primarily for the manufacture of sulfur compounds, such as sulfuric acid (H_2SO_4) and hydrogen sulfide (H_2S). Large quantities of elemental sulfur are used in insecticides, synthetic fertilizers, and the vulcanization of rubber. Sulfur compounds are used in the manufacture of soaps, textiles, leather, paper, paints, dyes, and in the refining of petroleum.

DIAMOND—C

Crystallography. Isometric; $4/m\bar{3}2/m$. Crystals usually octahedral but may be cubic or dodecahedral. Curved faces, especially of the octahedron (see Fig. 8.15) and hexoctahedron are frequently observed (Figs. 8.16a and b; see Plate I, no. 5 and also Plate IX, no. 1, in Chapter 13). Crystals may be flattened on {111}. Twins on {111} (spinel law) are common, usually flattened parallel to the twin plane (Fig. 8.16c). *Bort*, a variety of diamond, has rounded forms and rough exterior resulting from a radial or cryptocrystalline aggregate. The term is also applied to badly colored or flawed diamonds without gem value.

$Fd\bar{3}m$; $a = 3.567$ Å; $Z = 8$. ds: 2.06(10) 1.26(8), 1.072(7), 0.813(6), 0.721(9).

Physical Properties. *Cleavage* {111} perfect. **H** 10 (hardest known mineral). **G** 3.52. *Luster* adamantine; uncut crystals have a characteristic greasy appearance. The very high refractive index, 2.42, and the strong dispersion of light account for the bril-

FIG. 8.15. Diamond crystals all showing a pronounced octahedral habit. (Courtesy of the Diamond Information Center, New York, N.Y.)

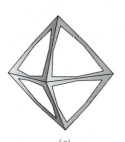   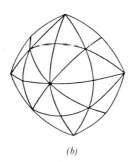   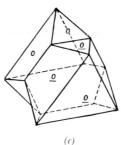

(a)   (b)   (c)   **FIG. 8.16.** Diamond crystals.

liancy and "fire" of the cut diamond. *Color* usually pale yellow or colorless; also pale shades of red, orange, green, blue, and brown. Deeper shades are rare. *Carbonado* or *carbon* is black or grayish-black bort. It is noncleavable, opaque, and less brittle than crystals.

Composition and Structure. The composition is pure carbon. The structure of diamond is illustrated in Figs. 8.4a and 6.58. (Many aspects of the diamond structure are illustrated with animations on the CD-ROM, module III, under the heading "Three-dimensional Order; Space Group Elements in Structures.") Figure 8.4a shows that each carbon is surrounded by four neighboring carbon atoms in tetrahedral coordination. This is the result of covalent bonding in which four valence electrons in each carbon fill the bonding orbitals of the four neighboring carbon atoms by electron sharing. Every carbon atom is linked in this way to four others, forming a continuous network. Refer back to Fig. 6.58c to see the face-centered arrangement of the carbon atoms in a cubic unit cell. Figure 6.58d shows the location of some of the symmetry elements in the unit cell of diamond. Figure 6.58e illustrates the many complexities of the diamond space group, $F4_1/d\bar{3}2/m$ (which is abbreviated to $Fd3m$).

Diamond, as is to be expected from its high specific gravity and fairly close packing, is the high-pressure polymorph (see Box 8.1). At low pressures or temperatures it is unstable with respect to graphite and may be converted to graphite at moderate temperatures. The reason that diamond and graphite can coexist at room temperatures and pressures is because the reconstructive polymorphic reaction between the two minerals is very sluggish. In order to permit the change from graphite to diamond, extremely high temperature, that is, high activation energy, is needed to cause the carbon atoms in the graphite structure to break loose by thermal agitation, and to make them available for building the diamond structure. Such temperatures also increase the pressure required to bring about the polymorphic reaction (see Box 8.1).

Diagnostic Features. Diamond is distinguished from minerals that it resembles by its great hardness, adamantine luster, and cleavage.

Occurrence. Diamonds have been discovered in many different localities but in only a few in notable amounts. Most commonly, diamond is found in alluvial deposits, where it accumulates because of its inert chemical nature, its great hardness, and its fairly high specific gravity. In several countries in Africa and North America, and in China, Venezuela, Australia, India, and Siberia, diamonds have been found *in situ*. The rock in which they occur is called *kimberlite*, after the type locality of Kimberley, South Africa. Kimberlites are a group of volatile-rich (mainly CO_2), potassic ultrabasic rocks with a range of megacryst compositions set in a fine-grained groundmass (a megacryst is a megascopically visible crystal that stands out clearly from the finer groundmass, but whether it formed in the same kimberlite magma is uncertain). The megacryst assemblage may consist of ilmenite, pyrope garnet, olivine, clinopyroxene, phlogopite, enstatite, and chromite. The matrix mineralogy is complex and may include olivine, phlogopite, perovskite, spinel, and diopside. Although these intrusive bodies vary in size and shape, many are roughly circular with a pipelike shape and are referred to as "*kimberlite*" or "*diamond pipes.*" In the deepest mines, in e.g., Kimberley, South Africa, the diameter of the pipes decreased with depth and mining stopped at a depth of 3500 feet, although the pipes continued. At the surface the kimberlite is weathered to a soft yellow rock, "yellow ground," that in depth gives way to a harder "blue ground." The ratio of diamonds to barren rock varies from one pipe to another. In the Kimberley mine it was 1 : 8,000,000, but in some it may be as high as 1 : 30,000,000.

Diamonds were first found in India, which remained virtually their only source until they were discovered in Brazil in 1725. The early diamonds came from stream gravels in southern and central India, and it is estimated that 12 million carats were produced from this area (one carat weighs 0.2g).

BOX 8.1
DIAMOND SYNTHESIS

The synthesis of diamond is based on the stability fields of the various carbon polymorphs as shown in the accompanying pressure-temperature phase diagram for carbon (based on experimental data from various sources). There are three major problems associated with the synthesis of diamond. First, in order to achieve the dense, strongly bonded structure of diamond (it is the hardest material known to man), extremely high pressure must be achieved in the laboratory. Second, even when such high pressure is achieved, a very high temperature is necessary to permit the carbon atoms to diffuse for the conversion from other forms of carbon to diamond (the *P-T* region used for diamond growth is shaded in the diagram). Third, most of the synthetic diamonds produced are in small grains and growth of large single crystals suitable for gemstones presents yet further problems.

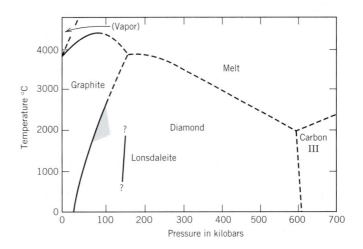

The first synthetic diamonds were made in 1953 by Dr. Erik Lundblad in the research laboratories of ASEA in Sweden. These resulted from experimental conditions of about 2700°C at about 76 kilobars using an apparatus that simultaneously exerted very high pressures and withstood extremely high temperatures. The ASEA technique was not patented at that time because the synthesis mechanism was not yet fully understood. In 1955 General Electric began the synthetic manufacture of small diamonds. World production of synthetic diamonds is now on the order of 300 million carats (one carat = 200 milligrams; a unit weight in gemstones), or about three times the production of diamond mined from the Earth. The synthetic material, which is identical to natural diamonds, offers the prospect of an assured and unlimited supply of diamonds. In recent years experimental techniques have produced diamond as large as 11.1 ct. of good quality; however, the synthetic process is still too costly to be competitive with high-quality larger diamonds of natural origin.

The success of diamond synthesis encouraged experimentation with boron nitride, BN, whose structure is similar to that of graphite. At very high pressures and temperatures (on the order of 85 kilobars and 2400°C) it transforms, similar to the graphite-diamond transition to cubic boron nitride (with the trade name of *Borazon*). This is extremely strong and hard, second only to diamond. It is manufactured in considerable quantities and is used mainly as an abrasive. This synthesis illustrates the application of crystal chemical concepts first evolved and tested in minerals.

After the initial synthesis of diamond, scientists at the General Electric Company continued experimentation. In 1967 they produced a hexagonal transparent polymorph of carbon with specific gravity and refractive indices close to those of diamond. Almost simultaneously a hexagonal diamond was found in meteoritic material from the Cañon Diablo, Arizona, meteorite crater. The name *lonsdaleite* is given to this naturally occurring hexagonal polymorph of diamond. Lonsdaleite is also found closely intergrown with polycrystalline diamond, *carbonado* in *yakutiite*, named after its occurrence in alluvial deposits of northern Yakutia, Siberia. It is likely that yakutiite is the result of a meteorite impact.

Much recent research involving the synthesis of diamond is centered on the deposition of diamond from a carbonaceous gas at less than atmospheric pressure in order to obtain a thin film of polycrystalline diamond on substrates such as glass, silicon, tungsten, and molybdenum. These thin diamond films can provide protective coatings on other materials, and may improve the performance of cutting tools and semiconductor materials.

Today the Indian production is only a few hundred carats a year.

Following their discovery in stream gravels in the state of Minas Gerais, Brazil, diamonds were also found in the states of Bahia, Goias, and Mato Grosso. Brazil's production today is chiefly from stream gravels. Extensive upland deposits of diamond-bearing gravels and clays are also worked. The black carbonado comes only from Bahia, Brazil.

In 1866 diamonds were discovered in the gravels of the Vaal River, South Africa, and in 1871 in the "yellow ground" of several pipes located near the present city of Kimberley. Although some diamonds are still recovered from gravels, the principal South African production is from kimberlite pipes. The deposits were originally worked as open pits, but as they deepened, underground methods were adopted. The world's largest and most productive diamond mine is the Premier, 24 miles east of Pretoria, South Africa. Since mining began there in 1903, nearly 30 million carats or six tons of diamonds have been produced. It was at the Premier Mine in 1905 that the world's largest diamond, the Cullinan, weighing 3106 carats, was found. Prospecting for diamonds has located in South Africa over 700 kimberlite pipes, dikes, and sills, most of which are barren. As recently as 1966 the Finsch Mine, 80 miles west of Kimberley, came into production. Even more recently diamonds have been found in kimberlite in Botswana and in northern South Africa, close to the border with Zimbabwe. The most productive pipe of all was found in about 1979 in Western Australia, and is now known as the Argyle Mine. This pipe consists of *lamproite*. Lamproite is a rock name that describes highly potassic and somewhat aluminum-poor igneous rocks with the following range of minerals: phlogopite, K-rich soda tremolite (also known as richterite), leucite, sanidine, diopside, and a variety of rare K-rich, Ba-rich, Ti-rich, and Zr-rich oxides and silicates.

In Cape Province, South Africa, on the desert coast just south of the mouth of the Orange River, terrace deposits containing high-quality stones were discovered in 1927; previously, in 1908, diamonds were found near Luderitz on the coast of Namibia. Elsewhere in Africa diamonds, mostly alluvial, have been found in a dozen different countries. The principal producers are Zaire, Botswana, Angola, Ghana, Sierra Leone, the Central African Republic, and Tanzania. Zaire is by far the largest producer and furnishes from kimberlites and minor placer deposits about 20% of the world's supply. These Zaire diamonds are mostly of industrial grade, with only 10% of gem quality. Russia is also a major producer of diamonds from both pipes and placers. Important kimberlite localities in Russia are Mir kimberlite, Mirny, and Udachnaya kimberlite, Udachnaya, both in Yakutia; and Pomorskaya kimberlite, Arkhangelsk, Onega Peninsula, White Sea. Important alluvial deposits with diamonds in rivers and associated placers are found in the Urals and in Yakutia.

The world's main diamond producers, listed for 1995, with their natural rough diamond production recorded in millions of carats are

Australia	40.8
Zaire	20.0
Botswana	16.8
Russia	12.5
South Africa	9.1
Angola	1.9
Namibia	1.3

(Data *Metals and Minerals Annual*, 1996, p. 27, published by Mining Journal Ltd., London.) For a discussion of diamond synthesis see Box 8.1.

Diamonds have been found sparingly in various parts of the United States. Small stones have occasionally been discovered in the stream sands along the eastern slope of the Appalachian Mountains in Virginia south to Georgia. Diamonds have also been reported from the gold sands of northern California and southern Oregon, and sporadic occurrences have been noted in the glacial drift in Wisconsin, Michigan, and Ohio. In 1906 diamonds were found in a lamproite pipe near Murfreesboro, Pike County, Arkansas. This locality, resembling the diamond pipes of South Africa, has yielded about 40,000 stones but is at present unproductive. In 1951 the old mine workings were opened to tourists who, for a fee, were permitted to look for diamonds. This area is now a state park.

Use. *In industry*, fragments of diamond crystals are used to cut glass. The fine powder is employed in grinding and polishing diamonds and other gemstones. Wheels are impregnated with diamond powder for cutting rocks and other hard materials. Steel bits are set with diamonds, especially the cryptocrystalline variety, *carbonado*, for diamond drilling in exploratory mining work. Diamond is also used in wire drawing and in tools for the truing of grinding wheels.

The diamond is also one of the most highly prized *gemstones*. Its value depends on its hardness, its brilliancy, which results from its high index of refraction, and its "fire," resulting from its strong dispersion. In general, the most valuable are those flawless stones that are colorless or possess a "blue-white" color. A faint straw-yellow color, which diamond often shows, detracts from its value. Diamonds colored deep shades of yellow, red, green, or blue, known as *fancy stones*, are greatly prized and bring very high prices. Diamonds can be colored deep shades of green by irradiation with high-energy nuclear particles, neutrons, deuterons, and alpha particles, and blue by exposing them to fast-moving electrons. A stone colored green by irradiation can be made a deep yellow by proper heat treatment. These artificially colored stones are difficult to distinguish from those of natural color. See Chapter 13 for further discussion and illustrations of gem diamonds.

Name. The name diamond is a corruption of the Greek word *adamas*, meaning *invincible*. The luster term, *adamantine*, which describes the very high luster of minerals with a high index of refraction, such as of diamond and cerussite, is derived from *adamas*.

Similar Species. For a discussion of materials that are marketed as gem diamond imitations see Chapter 13 and specifically Table 13.2.

GRAPHITE—C

Crystallography. Hexagonal; $6/m2/m2m$. In tabular crystals of hexagonal outline with prominent basal plane. Distinct faces of other forms very rare. Triangular markings on the base are the result of gliding along an $\{hh\overline{2}hl\}$ pyramid. Usually in foliated or scaly masses, but may be radiated or granular. A photograph of two polymorphs of carbon, diamond, and graphite, is given in Fig. 8.17.

$P6_3/mmc$; $a = 2.46$, $c = 6.74$ Å; $Z = 4$. *ds*: 3.36(10), 2.03(5), 1.675(8), 1.232(3), 1.158(5).

Physical Properties. *Cleavage* {0001} perfect. **H** 1–2 (readily marks paper and soils the fingers). **G** 2.23. *Luster* metallic, sometimes dull earthy. *Color* and streak, black. Greasy feel. Folia flexible but not elastic (see Plate I, no. 6).

Composition and Structure. Carbon. Some graphite impure with iron oxide, clay, or other minerals. The structure is illustrated in Fig. 8.4*b*.

Diagnostic Features. Graphite is recognized by its color, foliated nature, and greasy feel. Distinguished

FIG. 8.17. The two polymorphs of carbon: fibrous graphite from Buckingham, Quebec, Canada, and an octahedral diamond from the Republic of South Africa (Harvard Mineralogical Museum).

from molybdenite by its black color (molybdenite has a blue tone), and black streak on glazed porcelain.

Occurrence. Graphite most commonly occurs in metamorphic rocks such as crystalline limestones, schists, and gneisses. It may be found as large, crystalline plates or disseminated in small flakes in sufficient amount to form a considerable proportion of the rock. In these cases, it has probably been derived from carbonaceous material of organic origin that has been converted into graphite during metamorphism. Metamorphosed coal beds may be partially converted into graphite, as in the graphite coals of Rhode Island and in the coal fields of Sonora, Mexico. Graphite also occurs in hydrothermal veins associated with quartz, biotite, orthoclase, tourmaline, apatite, pyrite, and titanite, as in the deposits of Ticonderoga, New York. The graphite in these veins may have been formed from hydrocarbons introduced into them during the metamorphism of the region and derived from the surrounding carbon-bearing rocks. Graphite occurs occasionally as an original constituent of igneous rocks as in the basalts of Ovifak, Greenland, in a nepheline syenite in India, and in a graphite pegmatite in Maine. It is also found in some iron meteorites as graphite nodules.

The principal countries producing natural graphite are: China, Russia, North and South Korea, India, and Mexico.

Synthesis. Graphite is manufactured on a large scale in electrical furnaces using anthracite coal or petroleum coke as the raw materials. The use of artificial graphite in the United States is considerably in excess of that of the natural mineral.

Use. Used in the manufacture of refractory crucibles for the steel, brass, and bronze industries. Flake graphite for crucibles comes mostly from Sri Lanka and the Malagasy Republic. Mixed with oil, graphite is used as a lubricant, and mixed with fine clay, it forms the "lead" of pencils. It is employed in the manufacture of protective paint for structural steel and is used in foundry facings, batteries, electrodes, generator brushes, and in electrotyping.

Name. Derived from the Greek word meaning *to write*, in allusion to its use in pencils.

Sulfides, Sulfarsenides, and Arsenides

Acanthite—Ag₂S

Crystallography. Monoclinic, $2/m$ below 173°C; isometric, $4/m\overline{3}2/m$ above 173°C. Crystals, twinned polymorphs of the high-temperature form, commonly show the cube, octahedron, and dodecahedron and frequently are arranged in branching or reticulated groups. Most commonly massive or as a coating. Historically acanthite has been referred to as argentite.

$P2_1/n$; $a = 4.23$, $b = 6.93$, $c = 7.86$ Å, $\alpha = 99°35'$; $Z = 4$. ds: 2.60(10), 2.45(8), 2.38(5), 2.22(3), 2.09(4).

Physical Properties. H $2-2\frac{1}{2}$. G 7.3. Very sectile; can be cut with a knife like lead. *Luster* metallic. *Color* black. *Streak* black, shining. Opaque. Bright on fresh surface but on exposure becomes dull black.

Composition and Structure. Ag 87.1, S 12.9%. Ag_2S has space group $P2_1/n$ below 173°C, and space group $Im3m$ above 173°C. On cooling Ag_2S from above 173°C the structure twins pervasively to produce apparently cubic crystals of twinned acanthite, known to mineralogists as *argentite*. However, it seems that acanthite is the only stable form of Ag_2S at ordinary temperatures.

Diagnostic Features. Acanthite can be distinguished by its color, sectility, and high specific gravity.

Occurrence. Acanthite is an important primary silver mineral found in veins associated with native silver, the ruby silvers, polybasite, stephanite, galena, and sphalerite. It may also be of secondary origin. It is found in microscopic inclusions in argentiferous galena. Acanthite is an important ore in the silver mines of Guanajuato and elsewhere in Mexico. Formerly important European localities are Freiberg, Saxony, Germany; Joachimsthal, Schemnitz, and Kremnitz, Czechoslovakia, and Kongsberg, Norway. In the United States it has been an important ore mineral in Nevada, notably at the Comstock Lode and at Tonopah. It is also found in the silver districts of Colorado, and in Montana at Butte associated with copper ores.

Use. An important ore of silver.

Name. The name acanthite comes from the Greek word meaning *thorn*, in allusion to the shape of the crystals. The name argentite comes from the Latin *argentum*, meaning silver.

CHALCOCITE—Cu₂S

Crystallography. Monoclinic, pseudo-orthorhombic; $2/m$ or m, (below 105°C); above 105°C, hexagonal. Crystals are uncommon, usually small and tabular with hexagonal outline; striated parallel to the a axis (Fig. 8.18). Commonly fine-grained and massive.

$P2_1/c$ or Pc; $a = 11.82$, $b = 27.05$, $c = 13.43$Å, $\beta = 90°$; $Z = 96$. ds: 3.39(3), 2.40(7), 1.969(8), 1.870(10), 1.695(4).

Physical Properties. *Cleavage* {110} poor. *Fracture* conchoidal. H $2\frac{1}{2}-3$. G 5.5–5.8. *Luster* metallic.

FIG. 8.18. Chalcocite crystals.

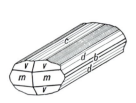
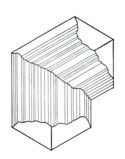

Imperfectly sectile. *Color* shining lead-gray, tarnishing to dull black on exposure. *Streak* grayish-black. Some chalcocite is soft and sooty.

Composition and Structure. Cu 79.8, S 20.2%. May contain small amounts of Ag and Fe. Below 105°C the structure is based on hexagonal closest packing of sulfur atoms with a monoclinic space group. Above 105°C it inverts to high chalcocite with space group $P6_3/mmc$.

Diagnostic Features. Chalcocite is distinguished by its lead-gray color and sectility.

Occurrence. Chalcocite is one of the most important copper-ore minerals. Fine crystals are rare but have been found in Cornwall, England, and Bristol, Connecticut. Chalcocite may occur as a primary mineral in veins with bornite, chalcopyrite, enargite, and pyrite. But its principal occurrence is as a supergene mineral in enriched zones of sulfide deposits. Under surface conditions the primary copper sulfides are oxidized; the soluble sulfates formed move downward, reacting with the primary minerals to form chalcocite and thus enriching the ore in copper. The water table is the lower limit of the zone of oxidation, and here a "chalcocite blanket" may form (See Box 8.2). Many famous copper mines owe their greatness to this process of secondary enrichment, as: Rio Tinto, Spain; Ely, Nevada; Morenci, Miami, and Bisbee, Arizona; and Butte, Montana.

Much of the world's copper is today produced from what is called "porphyry copper" ore. In these deposits primary copper minerals disseminated through the rock, usually a porphyry, have been altered, at least in part, to chalcocite and thus enriched to form a workable ore body. The amount of copper in such deposits is small, rarely greater than 1 or 2% and may be as low as 0.40%.

Use. An important copper ore.

Name. From the Greek word *chalkos* meaning *copper*.

Similar Species. *Digenite*, Cu_9S_5, is blue to black, associated with chalcocite. *Djurleite*, $Cu_{31}S_6$, a black to dark gray mineral, occurs in many of the large porphyry copper deposits of the Western Cordillera.

BORNITE—Cu₅FeS₄

Crystallography. Tetragonal. $\bar{4}2m$, below 228°C; isometric, $4/m\bar{3}2/m$, above 228°C. The most common form in ores is tetragonal. Rarely in rough pseudocubic and less commonly dodecahedral and octahedral crystals. Usually massive.

$P\bar{4}2_1c$; $a = 10.94$, $c = 21.88$ Å; $Z = 16$. ds: 3.31(4), 3.18(6), 2.74(5), 2.50(4), 1.94(10).

$Fm3m$; $a = 5.50$ Å; $Z = 1$. ds: 3.17(5), 2.75(5), 1.94(10), 1.66(1).

BOX 8.2
VEINS AND VEIN MINERALIZATION

Although ore and mineral deposits can be of igneous, metamorphic, or sedimentary origin a large number of mineral deposits exist as tabular or lenticular bodies known as *veins*. The veins have been formed by the filling with minerals of preexisting fractures or fissures.

Vein mineralization is generally the result of what is known as deposition from *hydrothermal solutions*. This term refers to heated or hot magmatic emanations rich in water as well as heated aqueous solutions that have no demonstrable magmatic affiliations. An example of relatively high-temperature vein mineralization in and adjacent to granitic intrusions occurs in Cornwall, England. The mineral assemblage in such veins consists of quartz-mica-tourmaline-wolframite-stannite-cassiterite-molybdenite. In many mining districts, however, there is no apparent relationship between the ore veins and possible igneous activity.

Associated with the economically useful minerals (*ore minerals*) are minerals of no commercial value (*gangue*). Careful assemblage studies, involving both ore and gangue mineralizations, together with the study of very small inclusions of remaining hydrothermal fluid (fluid inclusions) in mineral grains allow for the division of hydrothermal ore deposits in terms of temperature of origin: low (50°–150°C), intermediate

(150°–400°C), and high temperature (400°–600°C).

The *primary* or hypogene ore minerals may alter near the surface to secondary or *supergene* minerals. The primary sulfide minerals such as pyrite, chalcopyrite, galena, and sphalerite are particularly subject to such alteration. Under the influence of low-temperature and oxygenated meteoric waters, sphalerite (ZnS) alters to hemimorphite, $Zn_4(Si_2O_7)(OH)_2 \cdot H_2O$, and smithsonite, $ZnCO_3$; galena (PbS) to anglesite, $PbSO_4$, and cerussite, $PbCO_3$; copper sulfides such as chalcopyrite and bornite to covellite, chalcocite, native copper, cuprite, malachite, and azurite. The circulating meteoric waters dissolve primary minerals at near-surface conditions, producing a barren and leached zone (see the schematic cross-section of a vein that was originally rich in primary Cu-sulfides; the leached zone is the result of original pyrite, as well as other sulfides, having been oxidized

producing goethite and/or limonite), and frequently redeposit part of the soluble species below and close to the groundwater table, producing *secondary enrichment zones*. This process of *supergene enrichment* is especially important in ore formation because metals leached from the oxidized upper parts of mineral deposits may be redeposited at depth. Because all such chemical processes take place at essentially atmospheric conditions (about 25°C and 1 atmosphere pressure) Eh-pH stability diagrams for appropriate chemical systems can be applied to the assemblages in supergene deposits. Such a diagram is shown at the right for the system Cu-H_2O-O_2-S-CO_2 at 25° and 1 atmosphere total pressure. (Adapted from Anderson, J., in Garrels, R. M. and Christ, C. L., 1965, *Solutions, Minerals, and Equilibria*, Freeman, Cooper, and Company, San Francisco, Cal., Fig. 7.27b on p. 240; see same book for construction of such diagrams.)

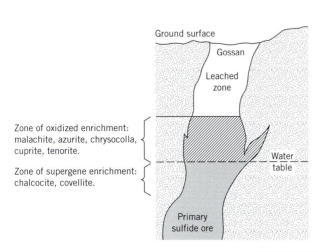

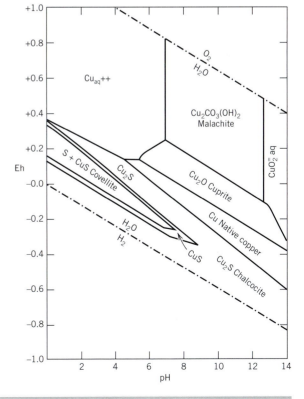

BOX 8.3
SULFIDE MINERALS AS ORES AND AS MINING-RELATED CONTAMINANTS

In sulfide ore-mineral assemblages pyrite is commonly the most abundant sulfide. Chalcopyrite is generally the next most common sulfide species and represents the most important copper mineral. Bornite is commonly associated with pyrite and chalcopyrite as well. Chalcocite, and to a lesser degree, covellite may occur as primary constituents in some ore deposits (Most commonly these two Cu-sulfides occur in supergene enriched sulfide deposits; see Box 8.2.) Other sulfides that may be present are pyrrhotite, molybdenite (MoS_2), galena (PbS), and sphalerite (ZnS). The triangular diagram in this box shows some of the most common sulfides that can be represented in the Cu-Fe-S system. Many of these sulfides (e.g., bornite and chalcopyrite) naturally exhibit some solid solution of especially Cu and Fe; this is not shown in the diagram. *Tielines* (the bold lines in the diagram) connect commonly occurring pairs of minerals. Triangles indicate coexistences of three sulfides. Tielines are a graphical representation for the common occurrence of two or more minerals in a mineral assemblage. The Fe-FeS (troilite) coexistence is common in iron meteorites.

In the mining process the sulfide ore minerals are extracted from the various rock types in which they occur. This produces commonly very large accumulations of waste rock that become essentially permanent deposits in the area surrounding the mining operation. This also happens in the coal mining industry where pyrite is a common minor constituent of the coal beds. These waste dumps are exposed to natural precipitation and oxygen from the atmosphere which together cause the slow dissolution of the sulfides and the subsequent production of acidic sulfate-and metal-rich waters; this is known as acid rock drainage (ARD). A key reaction in this process is:

$$FeS_2 + {}^7/_2 O_2 + H_2O \rightarrow Fe^{2+} + 2SO_4^{2-} + 2H^+$$
pyrite   aqueous   liquid   aqueous aqueous aqueous

The dissolved Fe^{2+}, SO_4^{2-}, and H^+ represent an increase in the total dissolved solids and acidity of the water. The increasing acidity of the water results in a decrease in pH.

This same process is intensified if the mining company applies methods of waste rock leaching in which the waste dumps are naturally and artificially leached by meteoric as well as irrigation water so as to recover any metals that are still in the waste rock. Depending on the local hydrologic conditions these acidic, metal-rich solutions may enter nearby streams, rivers, and lakes as well as the groundwater system. Such pollution is common in many sulfide and coal mining areas and may be contained with very expensive and complex remediation procedures.

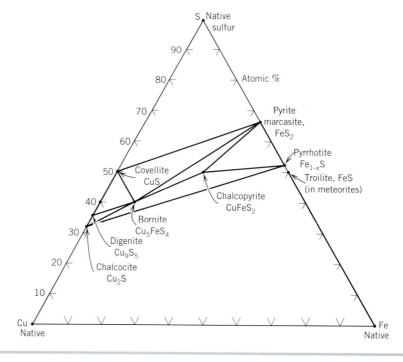

Physical Properties. H 3. G 5.06–5.08. *Luster* metallic. *Color* brownish-bronze on fresh fracture but quickly tarnishing to variegated purple and blue (hence called *peacock ore*; see Plate I, no. 7) and finally to almost black on exposure. *Streak* grayish-black.

Composition and Structure. Cu 63.3, Fe 11.2, S 25.5% for stoichiometric Cu_5FeS_4. Shows extensive solid solution within the Cu-Fe-S system. Its stoichiometric composition in terms of Cu-Fe-S is shown in the diagram in Box 8.3, with the compositions of other common sulfides in this system. The structure of the high T polymorph of bornite is relatively complex with sulfur atoms in a face-centered cubic lattice and the Cu and Fe atoms in tetrahedral coordination to S. The structure of the low temperature form is derived from the high *T* form with the addition of defects. This defect structure gives rise to large variations in Cu, Fe, and S contents.

Diagnostic Features. Bornite is distinguished by its characteristic bronze color on the fresh fracture and by the purple tarnish.

Alteration. Bornite alters readily to chalcocite and covellite.

Occurrence. Bornite is a widely occurring copper ore usually found associated with other sulfides (chalcocite, chalcopyrite, covellite, pyrrhotite and pyrite; see diagram in Box 8.3) in hypogene deposits. It is less frequently found as a supergene mineral, in the upper, enriched zone of copper veins. It occurs disseminated in basic rocks, in contact metamorphic deposits, in replacement deposits, and in pegmatites. It is not as important an ore of copper as chalcocite and chalcopyrite.

Good crystals of bornite have been found associated with crystals of chalcocite at Bristol, Connecticut, and in Cornwall, England. Found in large masses in Chile, Peru, Bolivia, and Mexico. In the United States it has been an ore mineral at the Magma Mine, Arizona; Butte, Montana; Engels Mine, Plumas County, California; Halifax County, Virginia; and Superior, Arizona.

Use. An ore of copper.

Name. Bornite was named after the Austrian mineralogist Ignatius von Born (1742–1791).

GALENA—PbS

Crystallography. Isometric; $4/m\bar{3}2/m$. The most common form is the cube, sometimes truncated by the octahedron (Figs. 8.19 and 8.20). Dodecahedron and triscotahedron rare.

Fm3m; $a = 5.936$Å; $Z = 4$. *d*s: 3.44(9), 2.97(10), 2.10(10), 1.780(9), 1.324(10).

Physical Properties. *Cleavage* perfect {001}. **H** $2\frac{1}{2}$. **G** 7.4–7.6. *Luster* bright metallic. *Color* and *streak* lead-gray (see Plate I, no. 8).

Composition and Structure. Pb 86.6, S 13.4%. Silver is usually present as admixtures of silver minerals such as acanthite or tetrahedrite but also in solid solution. Inclusions probably also account for the small amounts of Zn, Cd, Sb, As, and Bi that may be present. Selenium may substitute for sulfur and a complete series of PbS–PbSe has been reported. Galena has an NaCl type of structure with Pb in place of Na and S in place of Cl.

Diagnostic Features. Galena can be easily recognized by its good cleavage, high specific gravity, softness, and lead-gray streak.

Alteration. By oxidation galena is converted into anglesite, $PbSO_4$, and cerussite, $PbCO_3$.

Occurrence. Galena is a very common metallic sulfide, found in veins associated with sphalerite, pyrite, marcasite, chalcopyrite, cerussite, anglesite, dolomite, calcite, quartz, barite, and fluorite. When found in hydrothermal veins galena is frequently associated with silver minerals; it often contains silver itself and so becomes an important silver ore. A large part of the supply of lead comes as a secondary product from ores mined chiefly for their silver. In a second type of deposit typified by the lead–zinc ores of the Mississippi Valley, galena, associated with sphalerite, is found in veins, open space filling, or replacement bodies in limestones. These are low-temperature deposits, located at shallow depths, and usually contain little silver. Galena is also found in contact metamorphic deposits, in pegmatites, and as disseminations in sedimentary rocks.

Famous world localities are Freiberg, Saxony, and the Harz Mountains, Germany; Pribram, Bohemia, Czechoslovakia; Derbyshire and Cumbria, England; Sullivan Mine, British Colombia; and Broken Hill, Australia.

In the past, many districts in the United States were significant lead producers, but today only three are of importance. Chief of these is in southeast Missouri, which yields nearly 90% of the domestic production. Here galena is disseminated through limestone. A small production comes from the Coeur d'Alene district, Idaho, where galena is the chief ore mineral in the lead–silver veins, and from Colorado, where lead is recovered mostly as a byproduct from the mining of silver and gold ores. Formerly, the tristate district of Missouri, Kansas, and Oklahoma was an important lead producer. In this region galena, associated with sphalerite, occurred in irregular veins and pockets in limestone and chert. Because the minerals could grow into open spaces, the tristate district is famous for its many beautifully crystallized specimens of galena, sphalerite, marcasite, chalcopyrite, calcite, and dolomite.

Use. Practically the only source of lead and an important ore of silver. The largest use of lead is in storage batteries, but nearly as much is consumed in making metal products such as pipe, sheets, and

FIG. 8.19. Galena crystals.

FIG. 8.20. *(a)* Galena crystals, Wellington Mine, Breckenridge, Colorado. *(b)* Galena on sphalerite and dolomite crystals, Jasper County, Missouri (both specimens from the Harvard Mineralogical Museum).

shot. Lead is converted into the oxides (*litharge*, PbO, and *minium*, Pb_3O_4) used in making glass and in giving a glaze to earthenware, and into white lead (the basic carbonate), the principal ingredient of many paints. However, this latter use is diminished because of the poisonous nature of lead-based paints. Diminished also is its use in gasoline anti-knock additives because of environmental restrictions. Lead is a principal metal of several alloys as solder (lead and tin), type metal (lead and antimony), and low-melting alloys (lead, bismuth, and tin). Lead is used as shielding around radioactive materials.

Name. The name galena is derived from the Latin *galena*, a name originally given to lead ore.

Similar Species. *Altaite*, PbTe, and *alabandite*, MnS, like galena, have an NaCl type of structure.

SPHALERITE—ZnS

Crystallography. Isometric; $\bar{4}3m$. Tetrahedron, dodecahedron, and cube common forms (Fig. 8.21), but the crystals, frequently highly complex and usually distorted or in rounded aggregates, often show polysynthetic twinning on {111}. Usually found in cleavable masses, coarse to fine granular. Compact, botryoidal, cryptocrystalline.

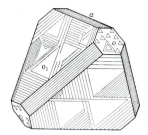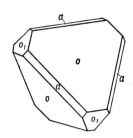

FIG. 8.21.   Sphalerite crystals.

$F\bar{4}3m$; a = 5.41 Å; Z = 4. ds: 3.12(10), 1.910(8), 1.631(7), 1.240(4), 1.106(5).

Physical Properties. *Cleavage*, {011} perfect but some sphalerite is too fine grained to show cleavage. **H** $3\frac{1}{2}$–4. **G** 3.9–4.1. *Luster* nonmetallic and resinous to submetallic; also adamantine. *Color* colorless when pure, and green when nearly so (see Plate I, no. 9). Commonly yellow, brown to black, darkening with increase in iron. Also red (ruby zinc). Transparent to translucent. *Streak* white to yellow and brown. *Optics*: n = 2.37.

Composition and Structure. Zn 67, S 33% when pure. Nearly always contains some Fe, with the amount of Fe dependent on the temperature and chemistry of the environment. If Fe is in excess, as indicated by association with pyrrhotite, the amount of FeS in sphalerite can reach 50 mole percent (see also Table 3.15). If sphalerite and pyrrhotite crystallize together, the amount of iron is an indication of the temperature of formation, and sphalerite can be used as a geological thermometer. Mn and Cd are usually present in small amounts in solid solution (see Table 3.15).

The structure of sphalerite is similar to that of diamond with one half of the carbon atoms replaced by Zn and the other half by S (see Figs. 3.53, 8.4*a*, and 8.5*a*). Sphalerite is considered the low-temperature cubic polymorph of ZnS, and wurtzite the high-temperature polymorph stable above 1020°C at 1 atmosphere pressure. The *wurtzite* polymorph (space group $P6_3mc$), with Zn atoms in hexagonal closest packing (see Fig. 8.6), shows a very large number of stacking sequences, referred to as wurtzite *polytypes*, which differ in the length of their *c* axes. For example the 4*H* polytype has a *c* axis of 12.46 Å and the 8*H* polytype a *c* axis of 24.96 Å (*H* = hexagonal).

Diagnostic Features. Sphalerite can be recognized by its striking resinous luster and perfect cleavage. The dark varieties (black jack) can be told by the reddish-brown streak, always lighter than the massive mineral.

Occurrence. Sphalerite, the most important ore mineral of zinc, is extremely common. Its occurrence and mode of origin are similar to those of galena, with which it is commonly found. In the shallow seated lead–zinc deposits of the tristate district of Missouri, Kansas, and Oklahoma (now largely exhausted), these minerals are associated with marcasite, chalcopyrite, calcite, and dolomite. Sphalerite with only minor galena occurs in hydrothermal veins and replacement deposits associated with phyrrhotite, pyrite, and magnetite. Sphalerite is also found in veins of igneous rocks and in contact metamorphic deposits.

Zinc is mined in significant amounts in more than 40 countries. Although in a few places the ore minerals are hemimorphite and smithsonite and, at Franklin and Sterling Hill, New Jersey, willemite, zincite, and franklinite, most of the world's zinc comes from sphalerite. The principal producing countries are: Canada, Russia, United States, Australia, Poland, Mexico, and Japan. In the United States nearly 60% of the zinc is produced east of the Mississippi River with Tennessee, New York, Pennsylvania, and New Jersey the principal producing states. In the western United States, Idaho, Colorado, and Utah are the chief producers. Elmwood, Tennessee, produces some outstanding specimen material.

Use. The most important ore of zinc. The chief uses for metallic zinc, or *spelter*, are in galvanizing iron; making brass, an alloy of copper and zinc; in electric batteries; and as sheet zinc. Zinc oxide, or zinc white, is used extensively for making paint. Zinc chloride is used as a preservative for wood. Zinc sulfate is used in dyeing and in medicine. Sphalerite also serves as the most important source of cadmium, indium, gallium, and germanium.

Name. Sphalerite comes from the Greek meaning *treacherous*. It was called blende because, although often resembling galena, it yielded no lead; from the German word meaning *blind* or *deceiving*.

Similar Species. *Greenockite*, CdS, a rare mineral mined as a source of cadmium, is isostructural in two polymorphic forms with sphalerite and wurtzite. Cadmium is recovered from greenockite associated with zinc minerals, especially sphalerite.

CHALCOPYRITE—CuFeS$_2$

Crystallography. Tetragonal; $\bar{4}2m$. Commonly tetrahedral in aspect with the disphenoid p {112} dominant (see Figs. 8.22 and 8.23). Other forms shown in Fig. 8.23 are rare. Usually massive.

$I\bar{4}2d$; a = 5.25, c = 10.32 Å; Z = 4. ds: 3.03(10), 1.855(10), 1.586(10), 1.205(8), 1.074(8).

Physical Properties. H $3\frac{1}{2}$–4. **G** 4.1–4.3. *Luster* metallic. *Color* brass-yellow; often tarnished to bronze or iridescent. *Streak* greenish-black. Brittle.

FIG. 8.22. Chalcopyrite and quartz crystals, St. Agnes, Cornwall, England (Harvard Mineralogical Museum).

Composition and Structure. Cu 34.6, Fe 30.4, S 35.0%. It deviates very little from ideal $CuFeS_2$. See diagram in Box 8.3 for chalcopyrite composition in the Cu–Fe–S system. Its structure can be regarded as a derivative of the sphalerite structure in which half the Zn is replaced by Cu and the other half by Fe. This leads to a doubling of the unit cell (see Fig. 8.5b).

Diagnostic Features. Recognized by its brass-yellow color (see Plate II, no. 1) and greenish-black streak. Distinguished from pyrite by being softer than steel and from gold by being brittle. Known as *fool's gold*, a term also applied to pyrite.

Occurrence. Chalcopyrite is the most widely occurring copper mineral and one of the most impor-

tant sources of that metal. Most sulfide ores contain some chalcopyrite, but the most important economically are the hydrothermal vein (see Box 8.3) and replacement deposits. In the low-temperature deposits as in the tristate district, it occurs as small crystals associated with galena, sphalerite, and dolomite. Associated with pyrrhotite and pentlandite, it is the chief copper mineral in the ores of Sudbury, Ontario, and similar high-temperature deposits. Chalcopyrite is the principal primary copper mineral in the "porphyry-copper" deposits. Also occurs as an original constituent of igneous rocks; in pegmatite dikes; in contact metamorphic deposits; and disseminated in schistose rocks. It may carry gold or silver and become an ore of those metals—often in subordinate amount with large bodies of pyrite, making them serve as low-grade copper ores.

A few of the localities and countries in which chalcopyrite is the chief ore of copper are Cornwall, England; Falun, Sweden; Schemnitz and Schlaggenwald, Czechoslovakia; Freiberg, Saxony, Germany; Rio Tinto, Spain; South Africa; Zambia; and Chile. Found widely in the United States but usually with other copper minerals in equal or greater amount; found at Butte, Montana; Bingham, Utah; Jerome, Arizona; Ducktown, Tennessee; and various districts in California, Colorado, and New Mexico. In Canada the most important occurrences of chalcopyrite are at

FIG. 8.23. Chalcopyrite crystals.

Sudbury, Ontario, and in the Rouyn-Noranda district, Quebec.

Alteration. Chalcopyrite is the principal source of copper for the secondary minerals malachite, azurite, covellite, chalcocite, and cuprite. Concentrations of copper in the zone of supergene enrichment are often the result of such alteration and removal of copper in solution with its subsequent deposition (see Box 8.2).

Use. Important ore of copper.

Name. Derived from Greek word *chalkos*, meaning *copper*, and from *pyrites*.

Similar Species. *Stannite*, Cu_2FeSnS_4, tetragonal, is a rare mineral and a minor ore of tin. The crystal structure of stannite can be derived from that of chalcopyrite by substitution of Fe and Sn in place of part of the Cu in chalcopyrite, see page 340.

PYRRHOTITE—$Fe_{1-x}S$

Crystallography. Monoclinic; $2/m$, for low-temperature form, stable below about 250°C; hexagonal, $6/m2/m2/m$, for high-temperature forms. Hexagonal crystals, usually tabular but in some cases pyramidal (Fig. 8.24), indicate formation as the high-temperature polymorph.

$A2/a$; $a = 12.78$, $b = 6.86$, $c = 11.90$, $\beta = 117°17'$; $Z = 4$. No specific ds are given here because there are several monoclinic forms.

$C6/mmc$; $a = 3.44$, $c = 5.73$ Å; $Z = 2$. ds: 2.97(6), 2.63(8), 1.06(10), 1.718(6), 1.045(8).

Physical Properties. H 4. G 4.58–4.65. *Luster* metallic. *Color* brownish-bronze (see Plate II, no. 2). *Streak* black. Magnetic, but varying in intensity; the greater the amount of iron, the lesser the magnetism. Opaque.

Composition and Structure. Most pyrrhotites have a deficiency of iron with respect to sulfur, as indicated by the formula $Fe_{1-x}S$, with x between 0 and 0.2. This is referred to as omission solid solution (see page 93 and Fig. 3.61). A complete solid solution series from FeS (with 50 atomic percent Fe) to pyrrhotite, with 44.9 atomic percent Fe, exists in the high-temperature part of the pyrrhotite stability field (from its melting point, at 1190°C, to about 400°C). This phase field has hexagonal symmetry. At lower temperatures (see Fig. 8.25), the pyrrhotite phase

FIG. 8.24. Pyrrhotite crystal.

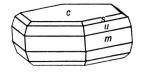

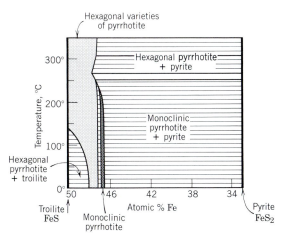

FIG. 8.25. The stability field of pyrrhotite, $Fe_{1-x}S$, in a temperature–composition section for the join FeS–FeS$_2$. The stability field of monoclinic pyrrhotite is fairly narrow and centers on Fe_7S_8. Its maximum temperature limit is 254°C. The hexagonal pyrrhotite field is narrow at 0° but widens all the way to the FeS axis at 140°C. The extent of solid solution is shown by shading. (Simplified after S. A. Kissin, 1974, in *Sulfide Mineralogy, Reviews in Mineralogy* 1, Mineralogical Society of America, Washington, D.C., p. CS–25.)

field narrows and the single, hexagonal solid solution field of pyrrhotite gives way to several hexagonal and monoclinic types. The monoclinic pyrrhotite compositions center at about Fe_7S_8. The monoclinic variety is stable from 0° to 254°C, where it inverts to the hexagonal type (see Fig. 8.25).

The structure of pyrrhotite is a complex derivative of the NiAs structure type. The sulfur atoms are arranged in approximate hexagonal closest packing. A large number of ordered structures exist for this nonstoichiometric sulfide. Several hexagonal structure types are stable over various temperature ranges in the field labeled "hexagonal varieties of pyrrhotite" in Fig. 8.25. The monoclinic structure is stable only below 254°C. The structural relations within the various polymorphs are complex. The mineral *troilite* is essentially FeS in composition.

Diagnostic Features. Recognized usually by its massive nature, bronze color, and magnetism.

Occurrence. Pyrrhotite is commonly associated with basic igneous rocks, particularly norites. It occurs in them as disseminated grains or, as at Sudbury, Ontario, as large masses associated with pentlandite, chalcopyrite, or other sulfides. At Sudbury vast tonnages of pyrrhotite are mined principally for the copper, nickel, and platinum that are extracted from associated minerals. Pyrrhotite is also found in contact metamorphic deposits, in vein deposits, and in pegmatites.

Large quantities are known in Finland, Norway, Sweden, and Russia. Fine crystals are found in Santa Eulalia, Chihuahua, Mexico; in the Morro Velho gold mine, Nova Lima, Minas Gerais, Brazil; and in Trepča, Yugoslavia. In the United States considerable amounts are found at Ducktown, Tennessee.

Use. It is mined for its associated nickel, copper, and platinum. At Sudbury, Ontario, it is also a source of sulfur and an ore of iron.

Name. The name pyrrhotite comes from the Greek meaning *reddish*.

Similar Species. *Troilite*, FeS, common in iron meteorites.

Nickeline—NiAs

Crystallography. Hexagonal; $6/m2/m2/m$. Rarely in tabular crystals. Usually massive, reniform with columnar structure.

$C6/mmc$; $a = 3.61$, $c = 5.02$ Å; $Z = 2$. ds: 2.66(10), 1.961(9), 1.811(8), 1.328(3), 1.071(4).

Physical Properties. H $5–5\frac{1}{2}$. G 7.78. *Luster* metallic. *Color* pale copper-red (hence called *copper nickel*), with gray to blackish tarnish. *Streak* brownish-black. Opaque.

Composition and Structure. Ni 43.9, As 56.1%. Usually a little Fe, Co, and S. As frequently replaced in part by Sb. The structure of NiAs is based on hexagonal closest packing of arsenic atoms giving a sequence *ABABAB.* . . . The nickel atoms are coordinated to six closest As atoms.

Diagnostic Features. Characterized by its copper-red color. Gives nickel test with dimethylglyoxime.

Alteration. Quickly alters to *annabergite* (green nickel bloom, $Ni_3(AsO_4)_2 \cdot 8H_2O$) in moist atmosphere.

Occurrence. Nickeline with other nickel arsenides and sulfides, pyrrhotite, and chalcopyrite, frequently occurs in, or is associated with, norites. Also found in vein deposits with cobalt and silver minerals.

Found in Germany in the silver mines of Saxony, the Harz Mountains, in Hessen–Nassau; and at Cobalt, Ontario.

Use. A minor ore of nickel.

Name. The first name of this mineral, *kupfernickel*, gave the name *nickel* to the metal. It is now called nickeline, after its nickel content.

Similar Species. *Breithauptite*, NiSb, is isostructural with nickeline with similar occurrence and association. Breithauptite has a distinctive light copper-red color on fresh fracture.

Millerite—NiS

Crystallography. Hexagonal; $\overline{3}2/m$ (low-temperature polymorph stable below 379°C). Usually in hairlike tufts and radiating groups of slender to capillary crystals, Fig. 8.26. In velvety incrustations. Rarely in coarse, cleavable masses.

$R3m$; $a = 9.62$, $c = 3.16$ Å; $Z = 9$. ds: 4.77(8), 2.75(10), 2.50(6), 2.22(6), 1.859(10).

FIG. 8.26.   Millerite in a calcite-lined vug, Keokuk, Iowa (Harvard Mineralogical Museum).

Physical Properties. *Cleavage* $\{10\overline{1}1\}$, $\{01\overline{1}2\}$ good. **H** $3–3\frac{1}{2}$. **G** 5.5 ± 0.2. *Luster* metallic. *Color* pale brass-yellow; with a greenish tinge when in fine hairlike masses. *Streak* black, somewhat greenish.

Composition and Structure. Ni 64.7, S 35.3%. The low-temperature form, stable below 379°C, shows little if any solid solution. Ni and S are both in 5-coordination in the low-temperature form. The high-temperature form (stable above 379°C) has an NiAs type of structure and exhibits considerable metal deficiency, $Ni_{1-x}S$, similar to pyrrhotite.

Diagnostic Features. Characterized by its capillary crystals and distinguished from minerals of similar color by nickel tests.

Occurrence. Millerite forms as a low-temperature mineral often in cavities and as an alteration of other nickel minerals, or as crystal inclusions in other minerals.

In the United States, it is found with hematite and ankerite at Antwerp, New York; in vugs and geodes in limestone in Harrodsburg, Indiana; Keokuk, Iowa; and Milwaukee, Wisconsin. In coarse, cleavable masses it is a major ore mineral at the Marbridge Mine, Lamotte Township, Quebec.

Use. A subordinate ore of nickel.

Name. In honor of the mineralogist, W. H. Miller (1801–1880), who first studied the crystals.

Pentlandite—(Fe,Ni)$_9$S$_8$

Crystallography. Isometric; $4/m\overline{3}2/m$. Massive, usually in granular aggregates with octahedral parting.

$Fm3m$; $a = 10.07$ Å; $Z = 4$. ds: 5.84(2), 3.04(6), 2.92(2), 2.31(3), 1.78(10).

Physical Properties. Parting on $\{111\}$. **H** $3\frac{1}{2}$–4. **G** 4.6–5.0. Brittle. *Luster* metallic. *Color* yellowish-bronze. *Streak* light bronze-brown. Opaque. Nonmagnetic.

Composition and Structure. (Fe,Ni)$_9$S$_8$. Usually the ratio of Fe : Ni is close to 1 : 1. Commonly contains small amounts of Co. A rather complicated face-centered cubic structure with the metal atoms in octahedral as well as tetrahedral coordination with sulfur, see Fig. 8.27. Pure (Fe,Ni)$_9$S$_8$, without Co, is stable up to 610°C in the Fe–Ni–S system. Pentlandite with as much as 40.8 weight percent Co is stable up to 746°C. Pentlandite commonly occurs as exsolution lamellae within pyrrhotite.

Diagnostic Features. Pentlandite closely resembles pyrrhotite in appearance but can be distinguished from it by octahedral parting and lack of magnetism. Gives nickel test with dimethyglyoxime.

Occurrence. Pentlandite usually occurs in basic igneous rocks where it is commonly associated with other nickel minerals and pyrrhotite, and chalcopyrite, and has probably accumulated by magmatic segregation.

Found at widely separated localities in small amounts but its chief occurrences are in Canada where, associated with pyrrhotite, it is the principal source of nickel at Sudbury, Ontario, and the Lynn Lake area, Manitoba. It is also an important ore mineral in similar deposits in the Petsamo district of Karelia in Russia and in the Kimbalda area of Western Australia.

Use. The principal ore of nickel. The chief use of nickel is in steel. Nickel steel contains $2\frac{1}{2}$–$3\frac{1}{2}$% nickel, which greatly increases the strength and toughness of the alloy, so that lighter machines can

be made without loss of strength. Nickel is also an essential constituent of stainless steel (see Box 9.1). The manufacture of Monel metal (68% Ni, 32% Cu) and Nichrome (38–85% Ni) consumes a large amount of the nickel produced. Other alloys are German silver (Ni, Zn, and Cu); metal for coinage—the 5-cent coin of the United States is 25% Ni and 75% Cu; low-expansion metals for watch springs; and other instruments. Nickel is used in plating; although chromium now largely replaces it for the surface layer, nickel is used for a thicker underlayer.

Name. After J. B. Pentland, who first noted the mineral.

Covellite—CuS

Crystallography. Hexagonal; 6/m2/m2/m. Rarely in tabular hexagonal crystals. Usually massive as coatings or disseminations through other copper minerals.

*P*6$_3$/*mmc*; $a = 3.80$, $c = 16.36$ Å; $Z = 6$. *d*s: 3.06(4), 2.83(6), 2.73(10), 1.899(8), 1.740(5).

Physical Properties. *Cleavage* {0001} perfect giving flexible plates. **H** $1\frac{1}{2}$–2. **G** 4.6–4.76. *Luster* metallic. *Color* indigo-blue or darker. *Streak* lead-gray to black. Often iridescent. Opaque.

Composition and Structure. Cu 66.4 S 33.6%. A small amount of Fe may be present. Covellite has a rather complex structure (see Fig. 8.7). One type of Cu atom is in tetrahedral coordination with S, with the tetrahedra sharing corners to form layers. A second type of Cu is in trigonal coordination with S to build planar layers. The excellent {0001} cleavage is parallel to this layer

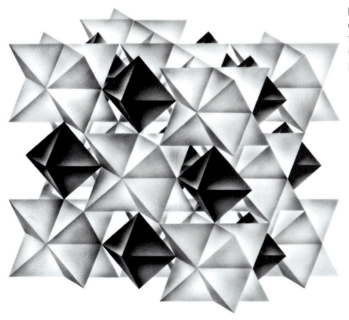

FIG. 8.27. The structure of pentlandite, (Fe,Ni)$_9$S$_8$, which consists of Fe and Ni in octahedral and tetrahedral coordination with S. The starlike clusters consist of eight linked (Fe,Ni)S$_4$ tetrahedra. The sulfur atoms are in cubic closest packing.

structure. Covellite is stable up to 507°C, its temperature of decomposition.

Diagnostic Features. Characterized by the indigo-blue color, micaceous cleavage yielding flexible plates, and association with other copper sulfides.

Occurrence. Covellite is not an abundant mineral but is found in most copper deposits as a supergene mineral, usually as a coating, in the zone of sulfide enrichment (see Box 8.2). It is associated with other copper minerals, principally chalcocite, chalcopyrite, bornite, and enargite, and is derived from them by alteration. Primary covellite is known but uncommon.

Found at Bor, Serbia, Yugoslavia; and Leogang, Austria. In large iridescent crystals from the Calabona Mine, Alghero, Sardinia, Italy. In the United States covellite is found in appreciable amounts and excellent crystals at Butte, Montana, and Summitville, Colorado. Formerly found at Kennecott, Alaska.

Use. A minor ore of copper.

Name. In honor of N. Covelli (1790–1829), the discoverer of the Vesuvian covellite.

CINNABAR—HgS

Crystallography. Hexagonal; 32 (low-temperature polymorph stable below approximately 344°C). High-temperature form, known as *metacinnabar*, isometric; $\bar{4}3m$. Crystals of cinnabar usually rhombohedral, often in penetration twins. Trapezohedral faces are rare. Usually fine granular, massive; also earthy, as incrustations and disseminations through the rock. Metacinnabar crystals are tetrahedral, with rough faces; usually massive.

$P3_121$ or $P3_221$; $a = 4.146$, $c = 9.497$ Å; $Z = 3$. ds: 3.37(10), 3.16(8), 2.87(10), 2.07(8), 1.980(8).

$F\bar{4}3m$; $a = 5.852$; $Z = 4$. ds: 3.38(10), 2.93(3), 2.07(5), 1.764(4).

Physical Properties. *Cleavage* $\{10\bar{1}0\}$ perfect. **H** $2\frac{1}{2}$. **G** 8.10. *Luster* adamantine when pure to dull earthy when impure. *Color* vermilion-red when pure to brownish-red when impure. *Streak* scarlet. Transparent to translucent. *Metacinnabar* has a metallic luster and grayish-black color. *Hepatic cinnabar* is an inflammable liver-brown variety of cinnabar with bituminous impurities; usually granular or compact, sometimes with a brownish streak.

Composition and Structure. Hg 86.2, S 13.8%, with small variations in Hg content. Traces of Se and Te may replace S. Frequently impure from admixture of clay, iron oxide, bitumen. The structure of cinnabar, which is different from that of any other sulfide, is based on infinite spiral Hg–S–Hg chains which extend along the c axis; it can be represented by one of two enantiomorphic space groups (see above). *Metacinnabar*, isometric, is stable above about 344°C. Some Hg deficiency is found in this polymorph and its composition may be represented as $Hg_{1-x}S$, with x ranging from 0 to 0.08.

Diagnostic Features. Recognized by its red color and scarlet streak, high specific gravity, and cleavage.

Occurrence. Cinnabar is the most important ore of mercury but is found in quantity at comparatively few localities. Occurs as impregnations and as vein fillings near recent volcanic rocks and hot springs and evidently deposited near the surface from solutions which were probably alkaline. Associated with pyrite, marcasite, stibnite, and sulfides or copper in a gangue of opal, chalcedony, quartz, barite, calcite, and fluorite.

The important localities for the occurrence of cinnabar are at Almaden, Spain; Idria, Yugoslavia; and in superb twinned crystals in the provinces of Kweichow and Hunan, China. In the United States deposits are in California at New Idria, and New Almaden.

Use. The only important source of mercury. The principal uses of mercury are in electrical apparatus, industrial control instruments, electrolytic preparation of chlorine and caustic soda, and in mildew proofing of paint. Other lesser but important uses are in dental preparations, scientific instruments, drugs, catalysts, and in agriculture. Formerly a major use of mercury was in the amalgamation process for recovering gold and silver from their ores, but amalgamation has been essentially abandoned in favor of other methods of extraction.

Name. The name cinnabar is supposed to have come from India, where it is applied to a red resin.

Realgar—AsS

Crystallography. Monoclinic; $2/m$. Found in short, vertically striated, prismatic crystals (Fig. 8.28). Frequently coarse to fine granular and often earthy and as an incrustation.

FIG. 8.28. Realgar.

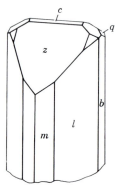

$P2_1/n$; $a = 9.29$, $b = 13.53$, $c = 6.57$ Å; $\beta = 106°33'$, $Z = 16$. ds: 5.40(10), 3.19(9), 2.94(8), 2.73(8), 2.49(5).

Physical Properties. *Cleavage* {010} good. **H** $1\frac{1}{2}$–2. **G** 3.48. Sectile. *Luster* resinous. *Color* and *streak* red to orange. Translucent to transparent.

Composition and Structure. As 70.1, S 29.9%. The realgar structure contains ringlike groups of As_4S_4, somewhat similar to the rings of S_8 in native sulfur. Each As is bonded covalently to another arsenic as well as to two sulfur atoms.

Diagnostic Features. Realgar is distinguished by its red color, resinous luster, orange-red streak, and almost invariable association with orpiment.

Alteration. On long exposure to light disintegrates to a reddish-yellow powder.

Occurrence. Realgar is found in veins of lead, silver, and gold ores associated with orpiment, other arsenic minerals, and stibnite. It also occurs as a volcanic sublimation product and as a deposit from hot springs.

Found in good crystals at Nagyág, Romania; Binenthal, Switzerland; and Allchar, Macedonia. In the United States realgar is found at Mercur, Utah; at Manhattan, Nevada; and deposited from the geyser waters in the Norris Geyser Basin, Yellowstone National Park.

Use. Realgar was used in fireworks to give a brilliant white light when mixed with saltpeter and ignited. Today, artificial arsenic sulfide is used for this purpose. It was formerly used as a pigment.

Name. The name is derived from the Arabic, Rahj al ghar, *powder of the mine*.

Orpiment—As_2S_3

Crystallography. Monoclinic; $2/m$. Crystals small tabular or short prismatic (Fig. 8.29), and rarely distinct; many pseudo-orthorhombic. Usually in foliated or columnar masses.

$P2_1/n$; $a = 11.49$, $b = 9.59$, $c = 4.25$ Å; $\beta = 90°27'$; $Z = 4$. ds: 4.78(10), 2.785(4), 2.707(6), 2.446(6), 2.085(4).

Physical Properties. *Cleavage* {010} perfect; cleavage laminae flexible but not elastic. Sectile. **H** $1\frac{1}{2}$–2. **G** 3.49. *Luster* resinous, pearly on cleavage face. *Color* lemon-yellow. *Streak* pale yellow. Translucent.

Composition and Structure. As 61, S 39%. Contains up to 2.7% Sb. In the structure of orpiment trigonal pyramids of AsS_3 can be recognized that share corners to form six-membered rings. These rings are linked to produce a corrugated As_2S_3 layer structure. The bonds within the layers are essentially covalent, whereas those between the layers are van der Waals in nature. The perfect {010} cleavage is parallel to these layers.

Diagnostic Features. Characterized by its yellow color and foliated structure. Distinguished from sulfur by its perfect cleavage.

Occurrence. Orpiment is a rare mineral, associated usually with realgar and formed under similar conditions. Found in good crystals at several localities in Romania and Czechoslovakia. In the United States it occurs at Mercur, Utah, and at Manhattan, Nevada. Deposited with realgar from geyser waters in the Norris Geyser Basin, Yellowstone National Park.

Use. Used in dyeing and in a preparation for the removal of hair from skins. Artificial arsenic sulfide is largely used in place of the mineral. Both realgar and orpiment were formerly used as pigments but this use has been discontinued because of their poisonous nature.

Name. Derived from the Latin, *auripigmentum*, "golden paint," in allusion to its color and because the substance was supposed to contain gold.

STIBNITE—Sb_2S_3

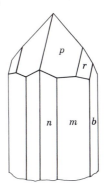

Crystallography. Orthorhombic; $2/m2/m2/m$. Slender prismatic habit, prism zone vertically striated. Crystals often steeply terminated (Figs. 8.30 and 8.31) and sometimes curved or bent (Fig. 8.31). Often in radiating crystal groups or in bladed forms with prominent cleavage. Massive, coarse to fine granular.

Pbnm; $a = 11.22$, $b = 11.30$, $c = 3.84$ Å; $Z = 4$. ds: 5.07(4), 3.58(10), 2.76(3), 2.52(4), 1.933(5).

Physical Properties. *Cleavage* {010} perfect, showing striations parallel to [100]. **H** 2. **G** 4.52–4.62. *Luster* metallic, splendent on cleavage surfaces. *Color* and *streak* lead-gray to black. Opaque (see Plate II, no. 3).

Composition and Structure. Sb 71.4, S 28.6%. May carry small amounts of Au, Ag, Fe, Pb, and Cu. The structure of stibnite is composed if zigzag chains of closely bonded Sb and S atoms that are parallel to the c axis. Sb–S distances in the chains range from

FIG. 8.29. Orpiment.

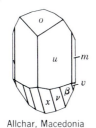

Allchar, Macedonia

FIG. 8.30. Stibnite.

FIG. 8.31. Cluster of stibnite crystals, and a single, curved crystal, Ischinokowa, Japan (Harvard Mineralogical Museum).

2.5Å to 3.1Å; these represent covalent linking. Distances between adjoining bands are larger, ranging from 3.2 to 3.6 Å (see Fig. 8.32). The long, striated prisms (// c) of stibnite are parallel to these structural chains. The excellent {010} cleavage occurs between Sb–S chains.

Diagnostic Features. Characterized by its easy fusibility, bladed habit, perfect cleavage in one direction, lead-gray color, and soft black streak. Fusible in a candle flame.

Occurrence. Stibnite is found in low-temperature hydrothermal veins or replacement deposits and in hot spring deposits. It is associated with other antimony minerals that have formed as the product of its decomposition, and with galena, cinnabar, sphalerite, barite, realgar, orpiment, and gold.

There are numerous localities in Romania for well-crystallized stibnite, but the finest crystals have come from the province of Iyo, Island of Shikoku, Japan. The world's most important producing district is in the province of Hunan, China. Found in quantity at only a few localities in the United States, notably Manhattan, Nevada.

Use. The chief ore of antimony but much of the metal is produced as a byproduct from smelting lead ores. Antimony trioxide is used as a pigment and for making glass.

Name. The name stibnite comes from an old Greek word that was applied to the mineral.

Similar Species. *Bismuthinite*, Bi_2S_3, is a rare mineral isostructural with stibnite and with similar physical properties.

PYRITE—FeS₂

Crystallography. Isometric; $2/m\bar{3}$. Frequently in crystals (Figs. 8.33, 8.34, and 8.35). The most common forms are the cube, the faces of which are usually striated, the pyritohedron, and the octahedron. Figure 8.33*f* shows a penetration twin, known as the *iron cross* with [001] the twin axis. Also massive, granular, reniform, globular, and stalactitic.

*Pa*3; $a = 5.42$ Å; $Z = 4$. *d*s: 2.70(7), 2.42(6), 2.21(5), 1.917(4), 1.632(10).

Physical Properties. *Fracture* conchoidal. Brittle. **H** $6–6\frac{1}{2}$ (unusually hard for a sulfide). **G** 5.02. *Luster* metallic, splendent. *Color* pale brass-yellow, see Plate

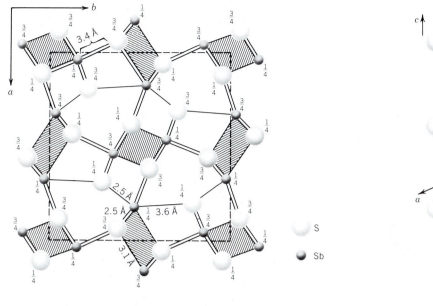

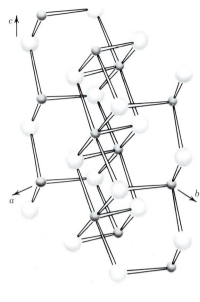

(a)   *(b)*

FIG. 8.32.   *(a)* The structure of stibnite, Sb_2S_3, projected on (001). The heights of the atoms are indicated by fractions along the *c* axis. The regions with the lined patterns represent the projections of zigzag chains parallel to the *c* axis. The interatomic distances within the chains are shown as ranging between 2.5 and 3.1 Å; those between chains range from 3.4 to 3.6 Å. (Redrawn after E. Hellner and G. Leineweber, 1956, *Über komplex zusammengesetzte sulfidische Erze. Zeitschrift für Kristallographie* 107: 105–154.) *(b)* Representation of one of the zigzag chains in the structure of stibnite.

II, no. 4; may be darker because of tarnish. *Streak* greenish or brownish-black. Opaque. Paramagnetic.

Composition and Structure. Fe 46.6, S 53.4%. May contain small amounts of Ni and Co. Some analyses show considerable Ni, and a complete solid solution series exists between pyrite and *bravoite*, $(Fe,Ni)S_2$. Frequently carries minute quantities of Au

and Cu as microscopic impurities. The structure of pyrite can be considered as a modified NaCl type of structure with Fe in the Na and S_2 in the Cl positions (Fig. 8.8*a*). FeS_2 occurs in two polymorphs, pyrite and *marcasite*.

Diagnostic Features. Distinguished from chalcopyrite by its paler color and greater hardness, from

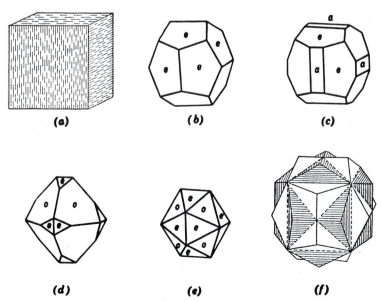

FIG. 8.33.   Pyrite crystals. *(a)* Striated cube. *(b)* Pyritohedron {210}; *(c)* Cube and pyritohedron. *(d)* and *(e)* Octahedron and pyritohedron. *(f)* Twinned pyritohedrons, *iron cross*.

FIG. 8.34. Pyrite, Bingham, Utah (Harvard Mineralogical Museum).

gold by its brittleness and hardness, and from marcasite by its deeper color and crystal form.

Alteration. Pyrite is easily altered to oxides of iron, usually limonite. In general, however, it is much more stable than marcasite. Pseudomorphic crystals of limonite after pyrite are common. Pyrite veins are usually capped by a cellular deposit of limonite, termed *gossan* (see Box 8.2). Rocks that

FIG. 8.35. Pyrite (Harvard Mineralogical Museum).

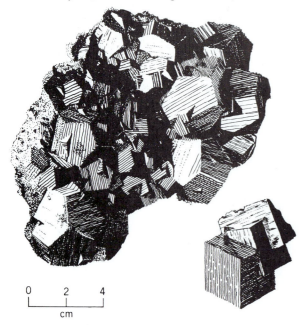

contain pyrite are unsuitable for structural purposes because the ready oxidation of pyrite would serve both to disintegrate the rock and to stain it with iron oxide.

Occurrence. Pyrite is the most common and widespread of the sulfide minerals and the world-wide occurrences of fine crystals are too many to enumerate. A few of the notable localities of crystals in the United States are: Chester, Vermont; Leadville and Central City, Colorado; and Bingham, Utah (Fig. 8.34). Fine crystals are found in La Libertad and Huanzala, Peru. Sculptural groups of crystals occur at Ambasaguas and Navajun, Spain.

Pyrite has formed at both high and low temperatures, but the largest masses probably at high temperature. It occurs as magmatic segregations, as an accessory mineral in igneous rocks, and in contact metamorphic deposits and hydrothermal veins. Pyrite is a common mineral in sedimentary rocks, being both primary and secondary. It is associated with many minerals but found most frequently with chalcopyrite, sphalerite, and galena.

Large and extensively developed deposits occur at Rio Tinto and elsewhere in Spain and also in Portugal. Important deposits of pyrite in the United States are in Virginia, where it occurs in large lenticular masses that conform in position to the foliation of the enclosing schists; in St. Lawrence County, New York; at the David Mine, near Charlemont, Massachusetts; and in various places in California, Colorado, and Arizona.

Use. Pyrite is often mined for the gold or copper associated with it. Because of the large amount of sulfur present in the mineral, it is used as an iron ore only in those countries where oxide ores are not available. Its chief use is a source of sulfur for sulfuric acid and *copperas* (ferrous sulfate). Copperas is used in dyeing, in the manufacture of inks, as a preservative of wood, and as a disinfectant. Pyrite may be cut as a gemstone which is sold under the name of its polymorph, marcasite.

Name. The name *pyrite* is from a Greek word meaning *fire*, in allusion to the brilliant sparks emitted when struck by steel.

MARCASITE—FeS$_2$

Crystallography. Orthorhombic; $2/m2/m2/m$. Crystals commonly tabular {010}; less commonly prismatic [001] (Fig. 8.36). Often twinned, giving cockscomb and spear-shaped groups (Figs. 8.37 and 8.38). Usually in radiating forms. Often stalactitic, having an inner core with radiating structure and covered with irregular crystal groups. Also globular and reniform.

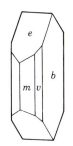
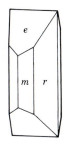

FIG. 8.36. Marcasite crystals.

Pmnn; *a* = 4.45, *b* = 5.42, *c* = 3.39 Å; *Z* = 2. *d*s: 2.70(10), 2.41(6), 2.32(6), 1.911(5), 1.755(9).

Physical Properties. **H** 6–6$\frac{1}{2}$. **G** 4.89. *Luster* metallic. *Color* pale bronze-yellow to almost white on fresh fracture; hence, called *white iron pyrites*. Yellow to brown tarnish. *Streak* grayish-black. Opaque.

Composition and Structure. Of constant composition, FeS_2, dimorphous with pyrite. The marcasite structure is shown in Fig. 8.8*b*. The configuration of nearest neighbor atoms is the same as in pyrite. The stability relationships of pyrite and marcasite are still unclear. Experimental evidence indicates that marcasite is metastable relative to pyrite and pyrrhotite above about 157°C. Geological occurrences of marcasite indicate a lower temperature stability range than for pyrite which may occur in magmatic segregations.

Diagnostic Features. Usually recognized and distinguished from pyrite by its pale yellow color, its crystals, or its fibrous habit.

Alteration. Marcasite usually disintegrates more easily than pyrite with the formation of ferrous sulfate and sulfuric acid. The white powder that forms on marcasite is *melanterite*, $FeSO_4 \cdot 7H_2O$.

Occurrence. Marcasite is found in metalliferous veins, frequently with lead and zinc ores. It is less stable than pyrite, being easily decomposed, and is much less common. It is deposited at low temperatures from acid solutions and is commonly formed under surface conditions as a supergene mineral. Marcasite most frequently occurs as replacement de-

posits in limestone, and often in concretions embedded in clays, marls, and shales.

Found abundantly in clay near Carlsbad and elsewhere in Czechoslovakia; in the chalk marl of Folkestone and Dover, England. In the United States marcasite is found with zinc and lead deposits of the Joplin, Missouri, district; at Mineral Point, Wisconsin; and at Galena, Illinois.

Use. Marcasite is used to a slight extent as a source of sulfur.

Name. Derived from an Arabic word, at one time applied generally to pyrite.

MOLYBDENITE—MoS$_2$

Crystallography. Hexagonal; 6/*m*2/*m*2/*m*. Crystals in hexagonal plates or short, slightly tapering prisms. Commonly foliated, massive, or in scales.

*P*6$_3$/*mmc*; *a* = 3.16, *c* = 12.32 Å; *Z* = 2. *d*s: 6.28(10), 2.28(9), 1.824(6), 1.578(4), 1.530(4).

Physical Properties. *Cleavage* {0001} perfect, laminae flexible but not elastic. Sectile. **H** 1–1$\frac{1}{2}$. **G** 4.62–4.73. Greasy feel. *Luster* metallic. *Color* lead-gray (see Plate II, no. 5). *Streak* grayish-black. Opaque.

Composition and Structure. Mo 59.9, S 40.1%, of essentially constant composition. In the structure of molybdenite a sheet of Mo atoms is sandwiched between two sheets of S atoms, the three sheets together forming a layered structure. The bond strengths within the layer are much stronger than between the layers, giving rise to the excellent {0001} cleavage. MoS_2 occurs as two polytypes of which one is hexagonal (2*H*) and the other rhombohedral (3*R*).

Diagnostic Features. Resembles graphite but is distinguished from it by higher specific gravity; by a blue tone to its color, whereas graphite has a brown tinge. On glazed porcelain, it gives a greenish streak, graphite a black streak.

Alteration. Molybdenite alters to yellow *ferrimolybdite*, $Fe_2(MoO_4)_3 \cdot 8H_2O$.

Occurrence. Molybdenite forms as an accessory mineral in certain granites; in pegmatites and aplites; also associated with porphyry copper deposits. Com-

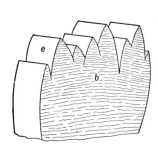

FIG. 8.37. "Cockscomb" marcasite.

FIG. 8.38. Marcasite, Ottawa County, Oklahoma (Harvard Mineralogical Museum).

monly in high-temperature vein deposits associated with cassiterite, scheelite, wolframite, and fluorite. Also in contact metamorphic deposits with lime silicates, scheelite, and chalopyrite.

Exceptionally good crystals are found at Kingsgate, New South Wales, Australia. In the United States molybdenite is found in many localities; examples are Okanogan County, Washington, and Questa, New Mexico. From various pegmatites in Ontario, Canada. The bulk of the world's supply comes from Climax, Colorado, where molybdenite occurs in quartz veinlets in silicified granite with fluorite and topaz. Much molybdenum is produced at Bingham Canyon, Utah, as a byproduct of the copper mining.

Use. The principal ore of molybdenum.

Name. The name molybdenite comes from the Greek word *molybdos* meaning *lead*.

COBALTITE—(Co,Fe)AsS

Crystallography. Orthorhombic; *mm2*. Pseudoisometric with forms that appear isometric, that is, cubes and pyritohedrons with the faces striated as in pyrite. Also granular.

$Pca2_1$; $a = 5.58$, $b = 5.58$, $c = 5.58$ Å; $Z = 4$. *d*s: 2.77(8), 2.49(10), 2.27(9), 1.680(10), 1.490(8).

Physical Properties. *Cleavage* pseudocubic, perfect. Brittle. **H** $5\frac{1}{2}$. **G** 6.33. *Luster* metallic. *Color* silver-white, inclined to red. *Streak* grayish-black.

Composition and Structure. Usually contains considerable Fe (maximum about 10%) and lesser amounts of Ni. *Gersdorffite*, NiAsS, and cobaltite form a complete solid solution series, but intermediate compositions are rare. The structure of cobaltite is closely related to that of pyrite (Fig. 8.8*a*) in which

half the S_2 pairs are replaced by As. Such a substitution causes a lowering of the symmetry of the structure of cobaltite as compared to that of pyrite. Natural cobaltites have a somewhat disordered distribution of As and S in the structure.

Diagnostic Features. Although in crystal form cobaltite resembles pyrite, it can be distinguished by its silver color and cleavage.

Occurrence. Cobaltite is usually found in high-temperature deposits, as disseminations in metamorphosed rocks, or in vein deposits with other cobalt and nickel minerals. Notable occurrences of cobaltite are at Tunaberg, Sweden, and Cobalt, Ontario. The largest producer of cobalt today is Zaire, where oxidized cobalt and copper ores are associated.

Use. An ore of cobalt.

Name. In allusion to its chemistry.

ARSENOPYRITE—FeAsS

Crystallography. Monoclinic; 2/*m*, pseudoorthorhombic. Crystals are commonly prismatic elongated on *c* and less commonly on *b* (Fig. 8.39). Twinning on {100} and {001} produces pseudo-orthorhombic crystals; on {110} generates contact or penetration twins; may be repeated, as in marcasite.

$P2_1$; $a = 5.74$, $b = 5.67$, $c = 5.78$ Å, $\beta = 112°17'$; $Z = 4$. *d*s: 2.68(10), 2.44(9), 2.418(9), 2.412(9), 1.814(9).

Physical Properties. *Cleavage* {101} poor. **H** $5\frac{1}{2}$ –6. **G** 6.07. *Luster* metallic. *Color* silver-white. *Streak* black. Opaque.

Composition and Structure. Close to FeAsS with some variation in As and S contents ranging from $FeAs_{0.9}S_{1.1}$ to $FeAs_{1.1}S_{0.9}$. Cobalt may replace part of the Fe and a series extends to *glaucodot*, (Co,Fe)AsS.

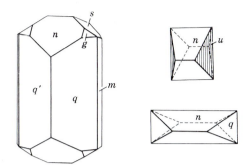

FIG. 8.39. Aresnopyrite crystals.

The structure of arsenopyrite is a derivative of the marcasite structure type (see Fig. 8.8b) in which one-half of the S is replaced by As.

Diagnostic Features. Distinguished from marcasite by its silver-white color. Its crystal form distinguishes it from skutterudite.

Occurrence. Arsenopyrite is the most common mineral containing arsenic. It occurs with tin and tungsten ores in high-temperature hydrothermal deposits, associated with silver and copper ores, galena, sphalerite, pyrite, chalcopyrite. Frequently associated with gold. Often found sparingly in pegmatites, in contact metamorphic deposits, disseminated in crystalline limestones.

Arsenopyrite is a widespread mineral and is found in considerable abundance in many localities, as at Freiberg and Munzig, Saxony, Germany; with tin ores in Cornwall, England; from Tavistock, Devonshire; in various places in Bolivia. In the United States it is associated with gold at Lead, South Dakota. Large quantities occur at Deloro, Ontario.

Use. The principal source of arsenic. Most of the arsenic produced is recovered in the form of the oxide, as a byproduct in the smelting of arsenical ores for copper, gold, lead, and silver. Metallic arsenic is used in some alloys, particularly with lead in shot metal. Arsenic is used chiefly, however, in the form of white arsenic or arsenious oxide in medicine, insecticides, preservatives, pigments, and glass. Arsenic sulfides are used in paints and fireworks.

Name. Arsenopyrite is a contraction of the older term *arsenical pyrites*.

SKUTTERUDITE—(Co,Ni)As₃

Crystallography. Isometric; $2/m\bar{3}$. Common crystal forms are cube and octahedron, more rarely dodecahedron and pyritohedron. Usually massive, dense to granular.

$Im3$; $a = 8.21 - 8.29$ Å; $Z = 8$; ds: 2.61(10), 2.20(8), 1.841(9), 1.681(7), 1.616(9).

Physical Properties. H $5\frac{1}{2}$–6. **G** 6.5 ± 0.4. Brittle. *Luster* metallic. *Color* tin-white to silver-gray. *Streak* black. Opaque.

Composition and Structure. Essentially $(Co,Ni)As_3$ but Fe usually substitutes for some Ni or Co. The high nickel varieties are called *nickel skutterudite*, $(Ni,Co)As_3$. A remarkable feature of the structure of skutterudite is the square grouping of As as As_4. The unit cell contains $8(Co + Ni)$ atoms and six As_4 groups which leads to the formula $(Co,Ni)As_3$. *Smaltite*, $(Co,Ni)As_{3-x}$ and *chloanthite*, $(Co,Ni)As_{3-x}$ (with $x = 0.5 - 1.0$) have skutterudite-like structures but a deficiency of As.

Diagnostic Features. Tin-white to silver-gray color. Chemical tests may be necessary for identification.

Occurrence. Skutterudite is usually found with cobaltite and nickeline in veins formed at moderate temperature. Native silver, bismuth, arsenopyrite, and calcite are also commonly associated with it.

Notable localities are Skutterud, Norway; Annaberg and Schneeberg, in Saxony, Germany; and Cobalt, Ontario, where skutterudite is associated with silver ores. Well-crystallized material is found near Bou Azzer, Morocco.

Use. An ore of cobalt and nickel. Cobalt is chiefly used in alloys for making permanent magnets and high-speed tool steel. Cobalt oxide is used as blue pigment in pottery and glassware.

Name. Skutterudite from the locality, Skutterud, Norway.

Similar Species. *Linnaeite*, Co_3S_4, associated with cobalt and nickel minerals.

Sulfosalts

Enargite—Cu₃AsS₄

Crystallography. Orthorhombic; *mm*2. Crystals elongated parallel to *c* and vertically striated, also tabular parallel to {001}. Columnar, bladed, massive.

$Pnm2_1$; $a = 6.41$, $b = 7.42$, $c = 6.15$ Å; $Z = 2$. ds: 3.22(10), 2.87(8), 1.859(9), 1.731(6), 1.590(5).

Physical Properties. *Cleavage* {110} perfect, {100} and {010} distinct. **H** 3. **G** 4.45. *Luster* metallic. *Color* and *streak* grayish-black to iron-black. Opaque.

Composition and Structure. In Cu_3AsS_4: Cu 48.3, As 19.1, S 32.6%. Sb substitutes for As up to 6% by weight, and some Fe and Zn are usually present. The structure of enargite can be regarded as a derivative of the wurtzite structure (Fig. 8.9), in which $\frac{3}{4}$ of Zn is replaced by Cu and $\frac{1}{4}$ of Zn by As. In chemical terms Zn_4S_4 becomes Cu_3AsS_4. The low-temperature polymorph of Cu_3AsS_4 (stable below 320°C) is *luzonite* with a tetragonal structure ($I\bar{4}m$). *Famatinite*, Cu_3SbS_4, the antimony analogue of enargite is isostructural with luzonite. Extensive solid solution exists in famatinite toward luzonite, and famatinite toward enargite.

Diagnostic Features. Characterized by its color and cleavage. Distinguished from stibnite by a test for Cu.

Occurrence. Enargite is a comparatively rare mineral, found in vein and replacement deposits formed at moderate temperatures associated with pyrite, sphalerite, bornite, galena, tetrahedrite, covellite, and chalcocite.

Notable localities: Morococha, Quiruvilca, and Cerro de Pasco, Peru; also from Chile and Argentina; Island of Luzon, Phillippines. In the United States it was an important ore mineral at Butte, Montana, and to a lesser extent in the Tintic district, Utah. Occurs in the silver mines of the San Juan Mountains, Colorado.

Use. An ore of copper.

Name. From the Greek word *enarges* meaning *distinct*, in allusion to the cleavage.

Pyrargyrite—Ag_3SbS_3, Proustite—Ag_3AsS_3

These minerals are known as the *ruby silvers* and in places have been important ores. They are isostructural with similar crystals forms, physical properties, and occurrences but there is little solid solution between them. Ruby red, hexagonal (3*m*) prismatic crystals make handsome mineral specimens.

Tetrahedrite—$Cu_{12}Sb_4S_{13}$, Tennantite—$Cu_{12}As_4S_{13}$

These two isostructural minerals form a complete solid solution series. They are so similar in crystallographic and physical properties that it is impossible to distinguish them by inspection. They are isometric, $\overline{4}3m$, and frequently occur in tetrahedral crystals (Fig. 8.40). Fe, Zn, and less commonly Ag, Pb, and Hg may substitute for Cu. The argentiferous variety, *freibergite*, may contain 18% Ag and thus become a silver ore.

REFERENCES AND SELECTED READING

Anthony, J. W., R. A. Bideaux, K. W. Bladh, and M. C. Nichols. 1990. *Handbook of mineralogy.* Vol. 1, *Elements, sulfides, sulfosalts.* Tucson, Ariz.: Mineral Data Publications.

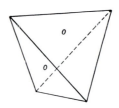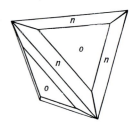

FIG. 8.40. Tetrahedrite crystals.

Blackburn, W. H., and W. H. Dennen. 1997. Encyclopedia of mineral names. *The Canadian Mineralogist,* Special publication 1.

Bragg, L., and G. F. Claringbull. 1965. *Crystal structures of minerals.* London: G. Bell and Sons, Ltd.

Cabri, L. J. 1972. The mineralogy of the platinum-group elements. *Minerals Science and Engineering* 4: 3–29.

Deer, W. A., R. A. Howie, and J. Zussman. 1962. *Rock forming minerals.* Vol. 5: *Non-Silicates.* New York: Wiley.

———. 1992. *An introduction to the rock forming minerals.* 2nd ed. New York: Wiley.

Fleischer, M., and J. A. Mandarino. 1991. *Glossary of mineral species* 1991. Tucson, Ariz.: Mineralogical Record Inc.

Gaines, R. B., H. W. C. Skinner, E. E. Foord, B. Mason, and A. Rosenzweig. 1997. *Dana's new mineralogy.* New York: Wiley.

Nickel, E. H., and M. C. Nichols. 1991. *Mineral reference manual.* New York: Van Nostrand Reinhold.

Palache, C., H. Berman, and C. Frondel. 1944. *The system of mineralogy.* 7th ed. Vol. 1, *Elements, sulfides, sulfosalts, oxides.* New York: Wiley.

Strunz, H. 1970. *Mineralogische tabellen.* 5th ed. Akademische Verlagsgesellschaft, Geest und Portig K. G., Leipzig.

Sulfide mineralogy. 1974 *Reviews in Mineralogy.* Vol. 1. Mineralogical Society of America. Washington, D.C.

Vaughan, D. J., and J. R. Craig. 1978. *Mineral chemistry of metal sulfides.* New York: Cambridge University Press.

Wilks, J., and E. Wilks. 1991. *Properties and applications of diamond.* Oxford: Butterworth-Heinemann Ltd.

Wilson, W. E. 1982. The gold-containing minerals: A review. *Mineralogical Record* 13: 389–400.

CHAPTER 9

CRYSTAL CHEMISTRY AND SYSTEMATIC DESCRIPTIONS OF OXIDES, HYDROXIDES, AND HALIDES

As in the prior chapter, the first part of this chapter concerns itself with an overview of the crystal chemistry. This is followed by systematic descriptions of the most common (and/or important) oxides, hydroxides, and halides. There are 395 known oxides and hydroxides, and 140 known halides. In this chapter only 16 oxides, 6 hydroxides, and 6 halides are discussed. Each group includes some important ore minerals.

The *oxides* are a group of minerals that are relatively hard, dense, and refractory and generally occur as accessory minerals in igneous and metamorphic rocks and as resistant detrital grains in sediments. The *hydroxides*, on the other hand, tend to be lower in hardness and density and are found mainly as secondary alteration or weathering products. The *halides* comprise about 140 chemically related minerals with diverse structure types and of diverse geological origins. Here we will discuss only six of the most common halides.

is treated in Chapters 11 and 12 with the silicates because the structure of quartz (and its polymorphs) is most closely related to that of other Si–O compounds. The multiple oxides, XY_2O_4, have two nonequivalent metal atom sites (*A* and *B*). Within the oxide class are several minerals of great economic importance. These include the chief ores of iron (hematite and magnetite), chromium (chromite), manganese (pyrolusite, as well as the hydroxides, manganite, and romanechite), tin (cassiterite), and uranium (uraninite).

CRYSTAL CHEMISTRY OF OXIDES

The oxide minerals include those natural compounds in which oxygen is combined with one or more metals. They are here grouped as simple oxides and multiple oxides. The simple oxides, compounds of one metal and oxygen, are of several types with different $X : O$ ratios (the ratio of metal to oxygen) as X_2O, XO, X_2O_3. Although not described on the following pages, ice, H_2O (see Fig. 4.13), is a simple oxide of the X_2O type in which hydrogen is the cation. The most common of all oxides, SiO_2, quartz and its polymorphs, is not considered in this chapter, but instead

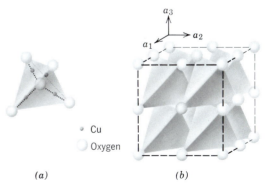

FIG. 9.1. Structure of cuprite. *(a)* Oxygen atoms at the corners and the center of a tetrahedral group; Cu halfway between two oxygens. *(b)* Oxygens in a cubic lattice, as shown in a polyhedral representation of the cuprite structure.

The bond type in oxide structures is generally strongly ionic, in contrast to the sulfide structures with ionic, covalent, as well as metallic bonding. One of the first structure determinations reported by Sir Lawrence Bragg in 1916 was that of cuprite, Cu_2O, in which the oxygen atoms are arranged at the corners and center of tetrahedral groups. The Cu atoms lie halfway between the oxygens (see Fig. 9.1) within the tetrahedral groups. The oxide mineral structures that will be discussed in this section are listed in the table on p. 373.

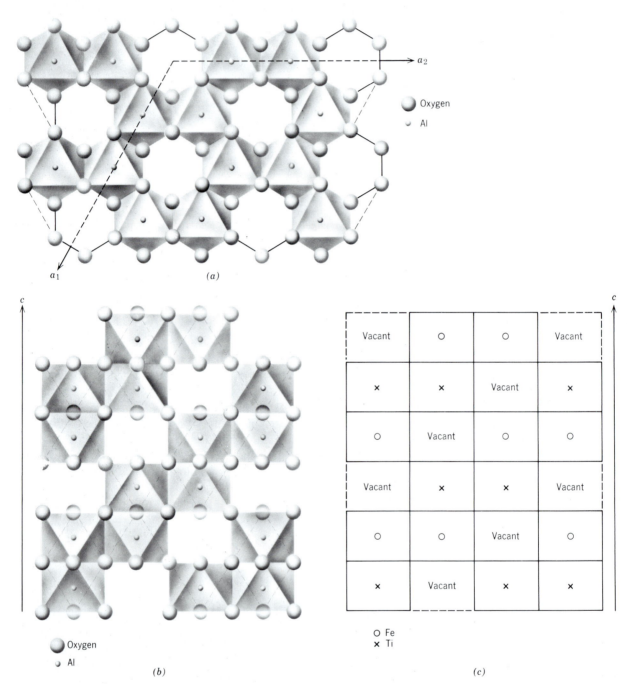

FIG. 9.2. Structures of the hematite group. *(a)* Basal sheet of octahedra in corundum, Al_2O_3, or hematite, Fe_2O_3, with one octahedron vacant for every two octahedra with Al or Fe^{3+} in center. *(b)* Vertical section through the corundum structure showing the locations of filled and empty octahedra. Note locations of Al^{3+} within octahedra. *(c)* Schematic vertical cross section through the ilmenite structure. This representation is the same as in Fig. 9.2b with oxygen positions and octahedral outlines eliminated; only cation locations are shown.

X_2O and XO types				
Cuprite	Cu_2O			
Periclase	MgO			
(Zincite	ZnO)			
X_2O_3 type			**XY_2O_4 type**	
Hematite Group			**Spinel Group**	
Corundum	Al_2O_3		Spinel	$MgAl_2O_4$
Hematite	Fe_2O_3		Gahnite	$ZnAl_2O_4$
Ilmenite	$FeTiO_3$		Magnetite	Fe_3O_4
			Franklinite	(Zn,Fe,Mn)-
				$(Fe,Mn)_2O_4$
			Chromite	$FeCr_2O_4$
XO_2 type				
(excluding SiO_2)				
Rutile Group				
Rutile	TiO_2	Chrysoberyl	$BeAl_2O_4$	
Pyrolusite	MnO_2	Columbite	(Fe,Mn)-	
Cassiterite	SnO_2		$(Nb,Ta)_2O_6$	
Uraninite	UO_2			

The structures of the *hematite group* are based on hexagonal closest packing of oxygens with the cations in octahedral coordination between them (see Fig. 9.2a as well as various animations of the corundum, Al_2O_3, structure in module III of the CD-ROM under the title of "Three-dimensional Order: Space Group Elements in Structures"). As can be seen in a basal projection of this structure only $\frac{2}{3}$ of the octahedral spaces are occupied by Fe^{3+} or Al^{3+}. The presence of $\frac{1}{3}$ oxygen octahedra without central Al^{3+} or Fe^{3+} ions is related to the electrostatic valency (e.v.) or bond strength of the $Al^{3+}-O^{2-}$ and $Fe^{3+}-O^{2-}$ bonds. Because Al^{3+} is surrounded by six oxygens, the e.v. of each of the six Al–O bonds $= \frac{1}{2}$. Each oxygen is shared between four octahedra, which means that four bonds of e.v. $= \frac{1}{2}$ can radiate from an oxygen position. In the (0001) plane, as in Fig. 9.2a, this allows for only *two* Al–O bonds from each oxygen, as shown by the geometry of two octahedra sharing one oxygen corner. A vertical sec-

tion through the corundum (or hematite) structure (Fig. 9.2b) shows the arrangement of the Al^{3+} (or Fe^{3+}) ions and of the omitted cations as well. In the vertical stacking of octahedra, each octahedron shares a face between two adjoining layers. The cations, Al^{3+} or Fe^{3+}, within the octahedra that share faces will tend to move away from the shared face on account of the repulsive forces between them (see Fig. 9.2b). The corundum and hematite structures have space group $R\bar{3}m$. The ilmenite structure, in which Fe and Ti are arranged in alternating Fe–O and Ti–O layers (Fig. 9.2c), has a lower symmetry (space group $R\bar{3}$) because of the ordered Fe and Ti substitution for Al^{3+} (as in corundum) or Fe^{3+} (as in hematite).

The structure of periclase, MgO, is identical to that of NaCl with cubic space group $Fm3m$. In a section parallel to {111} this structure appears very similar (see Fig. 9.3) to that of members of the *hematite group*. In such a section a "layer" of octahedra is seen without any cation omissions in the octahedral sites. With Mg in 6-coordination, the e.v. of each of the Mg–O bonds is $\frac{1}{3}$. This, in turn, allows for six bonds, each with e.v. $= \frac{1}{3}$, to radiate from every oxygen ion in the structure. In other words, each oxygen is shared between six Mg–O octahedra, and not between just four Al–O (or Fe–O) octahedra as in the structures of the *hematite group*. The MgO type of structure, therefore, shows no cation vacancies.

The structures of XO_2 type oxides that we will consider fall into two structure types. One is the structure typified by *rutile*, in which the cation is in 6-coordination with oxygen. The radius ratio, $R_X : R_O$, in this structure lies approximately between the limits of 0.732 and 0.414 (see Table 9.1). The other has the *fluorite* structure (see Fig. 9.11), in which each oxygen has four cation neighbors arranged about it at the apices of a more or less regular tetrahedron, whereas each cation has eight oxygens surrounding

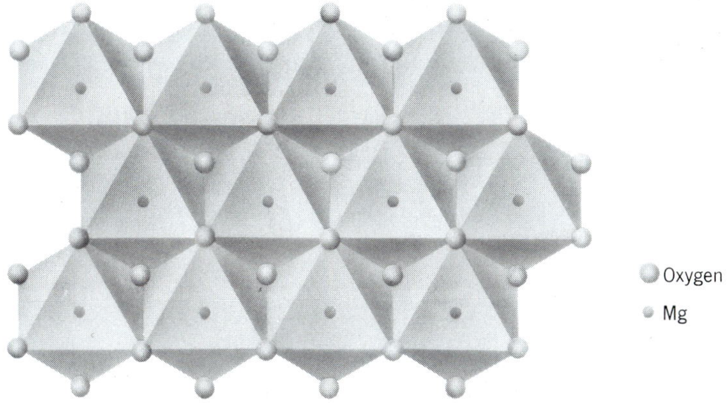

FIG. 9.3. Structure of periclase, MgO, as seen parallel to (111). Compare this with Fig. 9.2a as well as with Fig. 9.7, of brucite.

Oxygen

Mg

TABLE 9.1   Radius Ratios in XO_2 Type Oxides ($R_{oxygen} = 1.36$ Å)

R_x	[C.N.]	R_x/R_O	Ion	Mineral	Structure Type
0.53	[6]	0.39	Mn^{4+}	Pyrolusite	Rutile
0.61	[6]	0.45	Ti^{4+}	Rutile	Rutile
0.69	[6]	0.51	Sn^{4+}	Cassiterite	Rutile
0.97	[8]	0.71	Ce^{4+}	Cerianite	Fluorite
1.00	[8]	0.74	U^{4+}	Uraninite	Fluorite
1.05	[8]	0.77	Th^{4+}	Thorianite	Fluorite

it at the corners of a cube. The oxides of 4-valent uranium, thorium, and cerium, of great interest because of their connection with nuclear chemistry, have this structure. In general, all dioxides in which the radius ratio of cation to oxygen lies within and close to the limits for 8-coordination (0.732–1) may be expected to have this structure and to be isometric hexoctahedral (see Table 9.1).

The members of the *rutile group* are isostructural with space group $P4_2/mnm$. (Rutile itself has two additional polymorphs, anatase and brookite, with slightly different structures.) In the rutile structure (see Figs. 9.4 and 3.55) Ti^{4+} is located at the center of oxygen octahedra, the edges of which are shared forming chains parallel to the c axis. These chains are cross-linked by the sharing of octahedral corners (see Fig. 9.4b). Each oxygen is linked to three titanium ions and the e.v. of each of the Ti–O bonds $= \frac{2}{3}$. The prismatic habit of minerals of the rutile group is a reflection of the chainlike arrangement of octahedra.

The minerals of the *spinel group*, with the general formula XY_2O_4, are based on an arrangement of oxygens in approximate cubic closest packing along (111) planes in the structure (see Fig. 9.5). Several end members are listed in Table 9.2. The cations that are interstitial to the oxygen framework are in octahedral and tetrahedral coordination polyhedra with oxygen. In a unit cell of spinel, with edge length of approximately 8 Å, there are 32 possible octahedral sites and 64 possible tetrahedral sites; of these, 16 octahedral and 8 tetrahedral sites are occupied by cations. Occupied octahedra are joined along edges to form rows and planes parallel to {111} of the structure and occupied tetrahedra, with their apices along $\overline{3}$ axes, provide cross-links between layers of octahedra (see Fig. 9.5). A plan view of an oxygen layer parallel to (111), and its coordination with cations, is given in Fig. 9.6.

The general chemical formula of the spinel group is XY_2O_4 (or $X_8Y_{16}O_{32}$ per unit cell), where X and Y are various cations with variable valence. In magnetite, $X = Fe^{2+}$ and $Y = Fe^{3+}$, whereas in ulvöspinel, $X = Ti^{4+}$, and $Y = Fe^{2+}$. There are two types of spinel structures, and these are referred to as the "normal spinel" and the "inverse spinel" structures. In the normal spinel structure, the eight X cations occupy the 8 tetrahedral sites and the Y cations occupy the 16 octahedral sites, giving the formula $X_8Y_{16}O_{32}$ (which is equivalent to XY_2O_4). In the inverse spinel structure 8 of the 16 Y cations occupy the eight tetrahedral sites, resulting in the formula $Y(YX)O_4$ (see Table 9.2). Most naturally occurring spinels have cation distributions between the normal and inverse structure types.

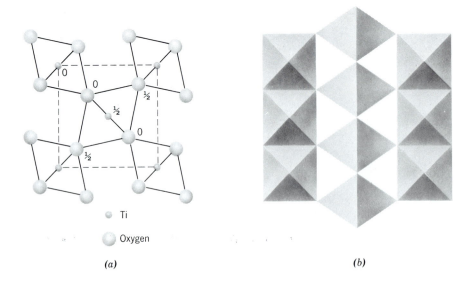

(a)

(b)

○ Ti

◯ Oxygen

FIG. 9.4. Structure of rutile. (a) Projection on (001) showing location of Ti and O atoms and the outline of the unit cell (dashed lines). (b) Projection on (110) showing chains of octahedra parallel to c. Compare these illustrations with Fig. 3.55.

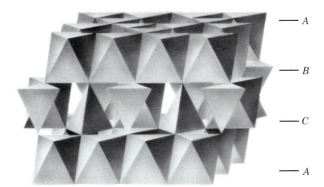

— A

— B

— C

— A

FIG. 9.5. The spinel (AB_2O_4) structure ($F4_1/d\bar{3}2/m = Fd3m$) with alternating layers parallel to {111} of octahedral and octahedral-tetrahedral polyhedra, as based upon approximate cubic closest packing. (From G. A. Waychunas, 1991, Crystal chemistry of oxides and hydroxides, in *Oxide Minerals, Reviews in Mineralogy* 25. Mineralogical Society of America, Washington, D.C.)

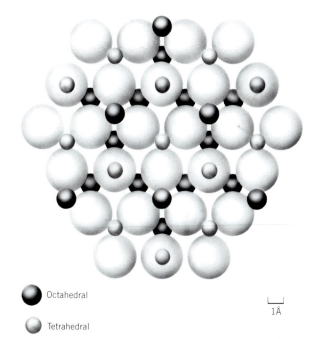

● Octahedral

○ Tetrahedral

$\overset{1Å}{\underset{\rule{1cm}{0.4pt}}{}}$

FIG. 9.6. An oxygen layer in the spinel structure, projected onto the (111) plane. See Fig. 9.5 for the position of such planes in the structure. The large circles are oxygen in approximate cubic closest packing. The cation layers on either side of the oxygen layer are shown as well. (Redrawn after D. H. Lindsley, 1976, The crystal chemistry and structure of oxide minerals as exemplified by the Fe-Ti oxides, in *Oxide Minerals, Reviews in Mineralogy* 3. Mineralogical Society of America, Washington, D.C.)

The coordination polyhedra about the various cations in spinel are not what might be predicted on the basis of the ionic sizes of the cations. Because Mg^{2+} is larger than Al^{3+}, one would expect Mg to occur in octahedral and Al in tetrahedral coordination with oxygen. In the normal spinel structure, however, the general concepts of radius ratio do not apply; indeed the larger cation is in the smaller polyhedron, and vice versa. When crystal field stabilization energies are considered instead of the geometric aspects of the ions, it becomes clear that the larger cation may occupy tetrahedral sites.

The spinel structure has a coordination scheme similar to that of the silicates of the olivine series. The composition of members of the forsterite-fayalite series with a total solid solution series from Mg_2SiO_4 to Fe_2SiO_4 can be represented as $X_2^{2+}Y^{4+}O_4$. Although this is not the same as $X^{2+}Y_2^{3+}O_4$ as in spinel, in both cases the total cation charge is identical. If one compares the structure of an Fe–Mg olivine with that of a possible Fe–Mg spinel, one finds that the spinel structure is about 12% denser than the olivine

structure for the same composition. This observation has led petrologists to suggest that olivine, which is thought to be abundant in the mantle, is converted to spinel-type structures at great depth. Two of the high-pressure olivine polymorphs are known as *ringwoodite* and *wadsleyite*, respectively (see page 111). In 1969 an olivine, *ringwoodite*, from a stony meteorite, was found to have a spinel structure. This same olivine composition $(Fe_{0.7}Mg_{0.3})_2SiO_4$ had been previously synthesized as a spinel in the laboratory of A. E. Ringwood, Australian National University, Canberra, Australia.

TABLE 9.2   End Members of the Spinel Group (XY_2O_4)

Normal Spinel Structure		Inverse Spinel Structure	
Spinel	$MgAl_2^{3+}O_4$	Magnetite	$Fe^{3+}(Fe^{2+}Fe^{3+})O_4$
Hercynite	$FeAl_2^{3+}O_4$	Magnesioferrite	$Fe^{3+}(Mg^{2+}Fe^{3+})O_4$
Gahnite	$ZnAl_2^{3+}O_4$	Jacobsite	$Fe^{3+}(Mn^{2+}Fe^{3+})O_4$
Galaxite	$MnAl_2^{3+}O_4$	Ulvöspinel	$Fe^{2+}(Fe^{2+}Ti^{4+})O_4$
Franklinite	$ZnFe_2^{3+}O_4$		
Chromite	$Fe^{2+}Cr_2^{3+}O_4$		
Magnesiochromite	$Mg^{2+}Cr_2^{3+}O_4$		

CRYSTAL CHEMISTRY OF HYDROXIDES

The hydroxide minerals to be considered in this section are as follows:

Brucite	$Mg(OH)_2$
Manganite	$MnO(OH)$
Romanechite	$BaMn^{2+} Mn_8^{4+}O_{16}(OH)_4$

Goethite Group

Diaspore	$\alpha\ AlO(OH)$
Goethite	$\alpha\ FeO(OH)$
Bauxite—mixture of diaspore, gibbsite, and boehmite	

All structures in this group are characterized by the presence of the hydroxyl $(OH)^-$ group, or H_2O molecules. The presence of $(OH)^-$ groups causes the bond strengths in these structures generally to be much weaker than in the oxides.

The structure of *brucite* (see Fig. 9.7) consists of Mg^{2+} octahedrally coordinated to $(OH)^-$, with the octahedra sharing edges to form a layer. Because each $(OH)^-$ group is shared between three adjoining octahedra, the Mg^{2+} to $(OH)^-$ bond strengths have an e.v. $= \frac{1}{3}$. With three such bonds $(3 \times \frac{1}{3} = 1)$ the $(OH)^-$ group is neutralized. For this reason, the layers in the brucite structure are held together by only weak bonds (compare Fig. 9.7 with Fig. 9.3). The structure of *gibbsite*, $Al(OH)_3$, is in principle identical to that of brucite except that, because of charge requirements, $\frac{1}{3}$ of the octahedrally coordinated cation positions are vacant (see Fig. 9.2a). A basal structural layer in corundum (as in Fig. 9.2) is structurally equivalent to the Al–OH sheets in gibbsite. The brucite structure type is referred to as *trioctahedral* (each $(OH)^-$ group is surrounded by three occupied octahedral positions) and the gibbsite structure type as *dioctahedral* (only two out of three octahedrally coordinated cation sites are filled). Such trioc-

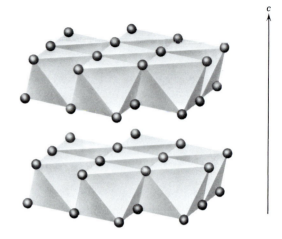

OH

FIG. 9.7. Structure of brucite, composed of parallel layers of Mg^{2+} in octahedral coordination with $(OH)^-$. Large spacing along the c axis is caused by weak bonding between adjacent layers.

tahedral and dioctahedral layers are essential building units of the phyllosilicates (see page 462).

The structure of *diaspore* ($\alpha AlO(OH)$; space group *Pbnm*) is shown in Fig. 9.8. Oxygen and $(OH)^-$ groups are arranged in hexagonal closest packing with Al^{3+} in octahedral coordination between them. A chainlike pattern is produced by $Al(O,OH)_6$ octahedra extending along the c axis. The octahedra in each chain share edges, and the chains are joined to each other by adjoining apical oxygens. *Goethite*, $\alpha FeO(OH)$, is isostructural with diaspore. Both compositions, $AlO(OH)$ and $FeO(OH)$, occur in nature in two crystal forms, $\gamma AlO(OH)$, *boehmite*, and $\gamma FeO(OH)$, *lepidochrocite*, both with space group *Amam*, and show a somewhat different linkage of the octahedrally coordinated cations from that found in diaspore or goethite as seen in Fig. 9.9. The octahedra are linked by their

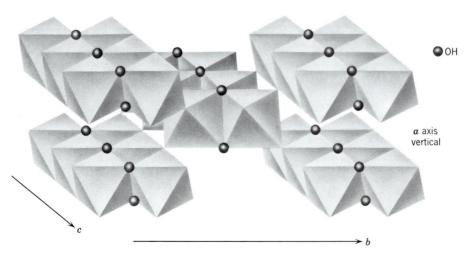

a axis
vertical

OH

FIG. 9.8. The structure of diaspore, $\alpha AlO(OH)$, and goethite, $\alpha FeO(OH)$. The double chains of $AlO_3(OH)_3$ or $FeO_3(OH)_3$ octahedra run parallel to the c axis. Only (OH) groups are indicated; all unmarked apices of octahedra represent oxygen.

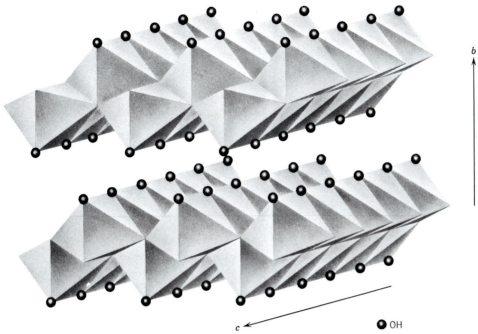

FIG. 9.9. The structure of boehmite, γAlO(OH) and lepidocrocite, γFeO(OH), showing the arrangement of $AlO_4(OH)_2$ or $FeO_4(OH)_2$ octahedra. The corrugated sheets are parallel to (010). Only (OH) groups are indicated; all unmarked apices of octahedra represent oxygen.

apices to form chains; the chains, in turn, are joined by the sharing of octahedral edges, which results in corrugated sheets parallel to {010}. The sheets are weakly held together by hydrogen bonds between pairs of oxygens. Such bonding is represented as (OH)⁻ groups in Fig. 9.9.

CRYSTAL CHEMISTRY OF HALIDES

The chemical class of halides is characterized by the dominance of the electronegative halogen ions, Cl^-, Br^-, F^-, and I^-. These ions are large, have a charge of only −1, and are easily polarized. When they combine with relatively large, weakly polarized cations of low valence, both cations and anions behave as almost perfectly spherical bodies. The packing of these spherical units leads to structures of the highest possible symmetry.

The structure of NaCl, shown in Figs. 9.10, 3.44, 3.48, and 3.51, was the first to be determined by X-ray diffraction techniques by W. H. and W. L. Bragg in 1913. The arrangement of the ions in the structure showed unambiguously that no molecules exist in the NaCl structure. Each cation and each anion is surrounded by six closest neighbors in octahedral coordination. Many halides of the XZ type crystallize with the NaCl structure (see Fig. 3.51); some mineral examples are, *sylvite*, KCl, *carobbiite*, KF and *chlorargyrite*, AgCl. Some XZ sulfides and oxides that

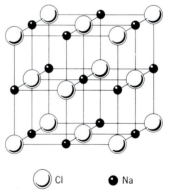

FIG. 9.10. The structure of halite. Compare this illustration with Fig. 3.51.

have the NaCl-type structure are *galena*, PbS, *alabandite*, MnS, and *periclase*, MgO.

Several alkali halides such as CsCl, CsBr, and CsI, none of which is naturally occurring, do not have the NaCl type structure but crystallize with a geometric arrangement of eight closest neighbors (cubic coordination) around the cation as well as the anion. This is known as the CsCl-structure type (see Fig. 3.52). The radius ratio ($R_X : R_Z$) is the primary factor in determining which of the two structure types is adopted by a given alkali halide of the XZ type.

The structures of several of the XZ_2 halides are identical to that of fluorite, CaF_2 (see Fig. 9.11), in which the Ca^{2+} ions are arranged at the corners and

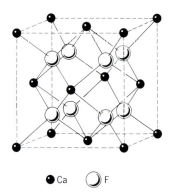

● Ca   ◐ F

FIG. 9.11. The structure of fluorite. Compare this illustration with Fig. 3.54.

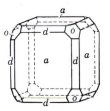 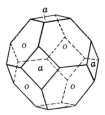

FIG. 9.12. Cuprite crystals.

face centers of a cubic unit cell. F^- ions are tetrahedrally coordinated to four Ca^{2+}. Each Ca^{2+} is coordinated to eight F^-, which surround it at the corners of a cube. The radius ratio ($R_{Ca} : R_F = 0.752$) leads to 8-coordination for Ca. Several oxides such as *uraninite*, UO_2, and *thorianite*, ThO_2, have a fluorite type structure.

Because the weak electrostatic charges are spread over the entire surface of the nearly spherical ions, the halides are the most perfect examples of pure ionic bonding. The isometric halides all have relatively low hardness and moderate to high melting points and are poor conductors of heat and electricity in the solid state. Any conduction of electricity that takes place does so by electrolysis—that is, by transport of charges by ions rather than by electrons. As the temperature increases and ions are liberated by thermal disorder, electrical conductivity increases rapidly, becoming excellent in the molten state. Commercial methods take advantage of this conductivity in the preparation of sodium and chlorine by electrolysis of molten sodium chloride.

When the halogen ions are combined with smaller and more strongly polarizing cations than those of the alkali metals, structures of lower symmetry result, and the bond has somewhat more covalent properties. In such structures, water or hydroxyl may enter as essential constituents, as in atacamite. Although there are 140 halide species, we will consider only 6 in detail.

SYSTEMATIC DESCRIPTIONS

Oxides

Cuprite—Cu_2O
Crystallography. Isometric; $4/m\bar{3}2/m$. Commonly occurs in crystals showing the cube, octahedron, and dodecahedron, frequently in combination (Fig. 9.12).

Sometimes in elongated capillary crystals, known as "plush copper" or *chalcotrichite*.

$Pn3m$; $a = 4.27$ Å; $Z = 2$. ds: 2.46(10), 2.13(6), 1.506(5), 1.284(4), 0.978(3).

Physical Properties. H $3\frac{1}{2}$–4. **G** 6.1. *Luster* metallic-adamantine in clear crystallized varieties. *Color* red of various shades; ruby-red in transparent crystals, called "ruby copper." *Streak* brownish-red.

Composition and Structure. Cu 88.8, O 11.2%. Usually pure, but FeO may be present as an impurity. The structure, as based on O atoms at the corners and centers of tetrahedral groups, is shown in Fig. 9.1.

Diagnostic Features. Usually distinguished from other red minerals by its crystal form, high luster, streak, and association with limonite.

Occurrence. Cuprite is a supergene copper mineral and in places an important copper ore. It is found in the upper oxidized portions of copper veins, associated with limonite and secondary copper minerals such as native copper, malachite, azurite, and chrysocolla (see Box 8.2).

Noteworthy foreign countries where cuprite is an ore are Chile, Bolivia, Australia, and Zaire. Fine crystals have been found at Cornwall, England; Chessy, France; and the Onganja Mine and Tsumeb, Namibia. Found in the United States in excellent crystals in the copper deposits at Bisbee, Arizona, and Chino at Santa Rita, New Mexico. Also found at Clifton and Morenci, Arizona.

Use. A minor ore of copper.

Name. Derived from the Latin *cuprum*, copper.

Similar Species. *Tenorite*, CuO, is a black supergene mineral.

Zincite—ZnO
Crystallography. Hexagonal; 6mm. Crystals are rare, terminated at one end by faces of a steep pyramid and the other by a pedion. Usually massive with platy or granular appearance.

$P6_3mc$; $a = 3.25$, $c = 5.19$ Å; $Z = 2$. ds: 2.83(7), 2.62(5), 2.49(10), 1.634(6), 1.486(6).

Physical Properties. *Cleavage* {10$\bar{1}$0} perfect; {0001} parting. **H** 4. **G** 5.68. *Luster* subadamantine. *Color* deep red to orange-yellow. *Streak* orange-yellow. Translucent. *Optics*: (+); $\omega = 2.013$, $\epsilon = 2.029$.

Composition and Structure. Zn 80.3, O 19.7%. Mn^{2+} often present and probably colors the mineral; pure ZnO is white. The structure of zincite is similar to that of wurtzite (see Fig. 8.6). Zn is in hexagonal closest packing and each oxygen lies within a tetrahedral group of four Zn ions.

Diagnostic Features. Distinguished chiefly by its red color, orange-yellow streak, and the association with franklinite and willemite.

Occurrence. Zincite is confined almost exclusively to the zinc deposits at Franklin and Sterling Hill, New Jersey, where it is associated with franklinite and willemite in calcite, often in an intimate mixture. Reported only in small amounts from other localities.

Use. An ore of zinc, particularly used for the production of zinc white (zinc oxide).

Similar Species. *Periclase*, MgO, is another oxide of XO composition. The structure of MgO is of the NaCl type with space group *Fm3m*. Continuous octahedral sheets can be identified in the structure parallel to {111}, see Fig. 9.3. It is found in contact metamorphosed Mg-rich limestones by the reaction: $CaMg(CO_3)_2 \rightarrow CaCO_3 + MgO + CO_2$.

HEMATITE GROUP, X_2O_3[1]
CORUNDUM—Al_2O_3

Crystallography. Hexagonal; $\overline{3}2/m$. Crystals are commonly tabular on {0001} or prismatic {11$\overline{2}$0}. Often tapering hexagonal dipyramids (Fig. 9.13), rounded into barrel shapes with deep horizontal striations. May show rhombohedral faces. Usually rudely crystallized or massive; coarse or fine granular. Polysynthetic twinning is common on {10$\overline{1}$1} and {0001}.

$R\overline{3}c$; $a = 4.76$, $c = 12.98$ Å; $Z = 6$. ds: 2.54(6), 2.08(9), 1.738(5), 1.599(10), 1.374(7).

Physical Properties. Parting on {0001} and {10$\overline{1}$1}, the latter giving nearly cubic angles; more rarely prismatic parting. **H** 9. Corundum may alter to mica, and care should be exercised in obtaining a

[1]The long-established morphological unit with *c* one-half the length of *c* of the unit cell is retained here for members of this group; the indices of forms of these minerals are expressed accordingly.

fresh surface for a hardness test. **G** 4.02. *Luster* adamantine to vitreous. Transparent to translucent. *Color* usually some shade of brown, pink, or blue, but may be colorless or any color. *Ruby* is red gem corundum, see Plate II, no. 6; *sapphire* is gem corundum of any other color. Rubies and sapphires having a stellate opalescence when viewed in the direction of the *c* crystal axis are termed *star ruby* or *star sapphire* (see Fig. 2.13). *Emery* is a black granular corundum intimately mixed with magnetite, hematite, or hercynite. *Optics*: $(-)$, $\omega = 1.769$, $\epsilon = 1.760$.

Composition and Structure. Al 52.9, O 47.1%. Rubies contain trace amounts of Cr (ppm) as a coloring agent and sapphires are blue due to trace amounts of Fe and Ti (see pages 161 and 162). The structure of corundum is illustrated in Fig. 9.2 and consists of oxygen in hexagonal closest packing and Al in octahedral coordination. Two-thirds of the octahedra are occupied by Al and $\frac{1}{3}$ are vacant (see also discussion on page 373).

Diagnostic Features. Characterized chiefly by its great hardness, high luster, specific gravity, and parting. Infusible.

Occurrence. Corundum is common as an accessory mineral in some metamorphic rocks, such as crystalline limestone, mica-schist, gneiss. Found also as an original constituent of silica deficient (undersaturated) igneous rocks such as syenites and nepheline syenites. May be found in large masses in the zone separating peridotites from adjacent country rocks. It is disseminated in small crystals through certain lamprophyric dikes and is found in large crystals in pegmatites. Found frequently in crystals and rolled pebbles in detrital soil and stream sands, where it has been preserved through its hardness and chemical inertness. Associated minerals are commonly chlorite, micas, olivine, serpentine, magnetite, spinel, kyanite, and diaspore.

The finest rubies have come from Burma; the most important locality is near Mogok, 90 miles north of Mandala. Some stones are found here in the metamorphosed limestone that underlies the

FIG. 9.13. Corundum crystals.

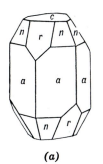

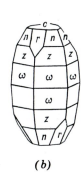

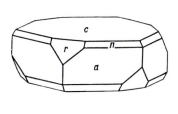

(a)   *(b)*   *(c)*

area, but most of them have been recovered from the overlying soil and associated stream gravels. Recently rubies of fine quality occurring in situ have been mined in southern Kenya. Darker, poorer quality rubies have been found in alluvial deposits near Bangkok, Thailand, and at Battambang, Cambodia. In Sri Lanka rubies of relatively inferior grade, associated with more abundant sapphires and other gemstones, are recovered from stream gravels. In the United States a few rubies have been found associated with the large corundum deposits of North Carolina.

Sapphires are found in the alluvial deposits of Thailand, Sri Lanka, and Cambodia associated with rubies. The stones from Cambodia of a cornflower-blue color are most highly prized. Sapphires occur in Kashmir, India, and are found over an extensive area in central Queensland, Australia. In the United States small sapphires of fine color are found in various localities in Montana. They were first discovered in the river sands east of Helena during placer operations for gold, and later found embedded in the rock of a lamprophyre dike at Yogo Gulch.

Common abrasive corundum has in the past been mined in many countries, but today Zimbabwe, Russia, and India are the only significant producers. At one time it was extensively mined in North Carolina, where it occurs in large masses at the edges of intruded olivine rock (dunite), and in Craigmont, Ontario, where, as a primary constituent of nepheline syenite, corundum makes up more than 10% of the rock mass.

Emery is found in large quantities on Cape Emeri on the island of Naxos, Greece, where mining continues today as it has for many centuries. In the United States emery was mined at Chester, Massachusetts, and at Peekskill, New York.

Synthesis. Synthetic corundum is manufactured from bauxite on a large scale. This material, together with other manufactured abrasives, notably silicon carbide, has largely taken the place of natural corundum as an abrasive.

Synthetic rubies and sapphires, colored with small amounts of Cr and Ti, are synthesized by fusing alumina powder in an oxy-hydrogen flame, which on cooling forms single crystal *boules*. This,

the Verneuil process, has been in use since 1902. In 1947 the Linde Air Products of the United States succeeded in synthesizing star rubies and star sapphires. This was accomplished by introducing titanium, which, during proper heat treatment, exsolves as oriented rutile needles to produce the star. The synthetic gems rival the natural stones in beauty, and it is difficult for the untrained person to distinguish them (see also Chapter 13).

Use. As a gemstone and abrasive. The deep red ruby is one of the most valuable of gems, second only to emerald. The blue sapphire is also valuable, and stones of other colors may command good prices (for color illustrations see Plate IX in Chapter 13). Stones of gem quality are used as watch jewels and as bearings in scientific instruments. Corundum is used as an abrasive, either ground from the pure massive mineral or in its impure form as emery.

Name. Probably from Kauruntaka, the Indian name for the mineral.

HEMATITE—Fe_2O_3

Crystallography. Hexagonal; $\overline{3}2/m$. Crystals are usually thick to thin tabular on {0001}, basal planes often show triangular markings, and the edges of the plates may be beveled with rhombohedral forms (Fig. 9.14b). Thin plates may be grouped in rosettes (*iron roses*) (Fig. 9.15). More rarely crystals are distinctly rhombohedral, often with nearly cubic angles, and may be polysynthetically twinned on {0001} and {10$\overline{1}$1}. Also in botryoidal to reniform shapes with radiating structure, *kidney ore* (see Fig. 2.9). May also be micaceous and foliated, *specular*, see Plate II, no. 7. Usually earthy; called *martite* when in octahedral pseudomorphs after magnetite.

$R\overline{3}c$; $a = 5.04$, $c = 13.76$ Å; $Z = 6$. ds: 2.69(10), 2.52(8), 2.21(4), 1.843(6), 1.697(7).

Physical Properties. *Parting* on {10$\overline{1}$1} with nearly cubic angles, also on {0001}. **H** $5\frac{1}{2}$–$6\frac{1}{2}$, **G** 5.26 for crystals. *Luster* metallic in crystals and dull in earthy varieties. *Color* reddish-brown to black. Red earthy variety is known as *red ocher*, platy and metallic variety as *specularite*. *Streak* light to dark red, which becomes black on heating. Translucent.

Composition and Structure. Fe 70, O 30%; essentially a pure substance at ordinary temperatures, al-

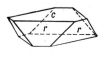

(a)

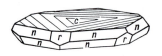

(b)

(c)

FIG. 9.14. Hematite crystals.

FIG. 9.15. Hematite, iron rose, St. Gotthard, Switzerland. (Harvard Mineralogical Museum.)

though small amounts of Mn and Ti have been reported. Forms a complete solid solution series with ilmenite above 950°C (see Figs. 9.16 and 9.26). The structure is similar to that of corundum (see Fig. 9.2 and page 373).

Diagnostic Features. Distinguished chiefly by its characteristic red streak.

Occurrence. Hematite is widely distributed in rocks of all ages and forms the most abundant and important ore of iron. It may occur as a sublimation product in connection with volcanic activities. Oc-

curs in contact metamorphic deposits and as an accessory mineral in feldspathic igneous rocks such as granite. Found from microscopic scales to enormous masses in regionally metamorphosed rocks where it may have originated by the oxidation of limonite, siderite, or magnetite. It is found in red sandstones as the cementing material that binds the quartz grains together. The quantity of hematite in a deposit that can be mined economically must be measured in tens of millions of tons. Such major accumulations are largely sedimentary in origin and many of them, through leaching of associated silica by meteoric waters, have been enriched to high-grade (direct shipping) ores (over 50% Fe). Like limonite, it may be formed in irregular masses and beds as the result of the weathering and oxidation of iron-bearing rocks. The oölitic ores are of sedimentary origin and may occur in beds of considerable size.

Noteworthy localities for hematite crystals are on the island of Elba; St. Gotthard, Switzerland, in "iron roses"; in the lavas of Vesuvius; at Cleator Moor, Cumbria, England; and in Minas Gerais, Brazil.

In the United States the columnar and earthy varieties are found in enormous bedded, Precambrian deposits that have furnished a large proportion of the iron ore of the world. The chief iron ore districts of the United States are grouped around the southern and northwestern shores of Lake Superior in Michigan, Wisconsin, and Minnesota. The chief districts, which are spoken of as iron ranges, are, from east to west, the Marquette in northern Michigan; the Menominee in Michigan to the southwest of the Marquette; the Penokee-Gogebic in northern Wisconsin. In Minnesota the Mesabi, northwest of Duluth; the Vermilion, near the Canadian boundary; and the Cuyuna, southwest of the Mesabi. The iron ore of these different ranges varies from the hard specular variety to the soft, red earthy type. Of the several ranges the Mesabi is the largest, and since 1892 has yielded over 2.5 billion tons of high-grade ore, over twice the total production of all the other ranges.

Oölitic hematite is found in the United States in the rocks of the Clinton formation, which extends from central New York south along the line of the Appalachian Mountains to central Alabama. The most important deposits lie in eastern Tennessee and northern Alabama, near Birmingham. Hematite has been found at Iron Mountain and Pilot Knob in southeastern Missouri. Deposits of considerable importance are located in Wyoming in Laramie and Carbon counties.

Although the production of iron ore within the United States remains considerable, the rich deposits have been largely worked out. In the future much of

FIG. 9.16. Compositions of naturally occurring oxide minerals in the system FeO-Fe₂O₃-TiO₂. Solid lines (tielines) indicate common geological coexistences at relatively low temperatures. Dashed lines indicate complete solid solution between end members. Hematite-ilmenite is a complete series above 950°C. Magnetite-ulvöspinel is a complete series above about 600°C. Magnetite commonly contains ilmenite lamellae. It is possible that such intergrowths result from the oxidation of members of the magnetite-ulvöspinel series. Compare with Fig. 9.26.

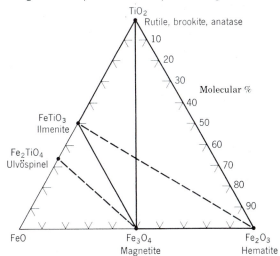

BOX 9.1
ORE MINERALS FOR THE STEEL INDUSTRY

Iron is the mainstay of modern civilization. It is used in an enormous range of applications—among them, in transportation (automobiles, trains, and ships), in construction, in industrial machines, in containers (e.g., oil drums), and in household appliances. The major iron ore minerals are **hematite, magnetite, and goethite**. They are found in large quantities in iron ore deposits that have formed as a result of the leaching of silica (SiO_2) that is normally a major component of the banded iron-formations (BIF) in which such ore deposits are hosted. Supergene enrichment and oxidation processes (see Box 8.2) concentrate the iron minerals into richer ore types. Some of the main iron-producing countries are Russia, Australia, and Brazil.

The BIFs in which the iron ore deposits are hosted are all Precambrian, ranging mainly from 3.8 billion to about 1.8 billion years in age. Their abundance peaks at about 2.5 billion years ago with a much smaller reoccurrence between 0.8 and 0.6 billion years ago (see diagram). Most BIFs have similar bulk chemistries (large amounts of Fe and Si) and consist of magnetite, hematite, Fe-silicates, Fe-carbonates (such as siderite), and chert. These are chemical sedimentary sequences that are the result of precipitation from ocean systems in which deep ocean hydrothermal input (of Fe^{2+} and Si) mixed with seawater. Such upwelling, Fe-enriched waters appear to have been reducing, or anaerobic, or even anoxic (so as to allow for the transport of Fe^{2+}

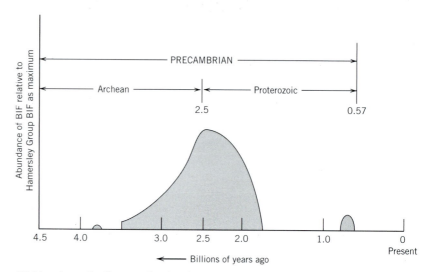

Highly schematic diagram showing the relative abundance of Precambrian BIFs versus geologic time. The maximum at 2.5 billion years ago is represented by the BIFs in the Hamersley Basin of Western Australia. (From C. Klein and N. J. Beukes, Time distribution, stratigraphy, and sedimentologic setting, and geochemistry of Precambrian iron-formations. The Proterozoic Biosphere, *J. W. Schopf and C. Klein, editors, New York: Cambridge University Press 1992, pp. 139–47).*

in solution), with the atmosphere gradually changing from originally reducing to slightly oxidizing in the early Proterozoic.

In addition to iron, other chemical elements are commonly used in alloying to yield special steels with specific properties. These include manganese, chromium, nickel, niobium, molybdenum, tungsten, vanadium, and cobalt.

Common manganese ore minerals are **pyrolusite, manganite**, and **romanechite** (formerly known as **psilomelane**). The largest reserves of manganese ore are in the Kalahari manganese field of South Africa, of early Proterozoic age. This area holds 81 percent of the world's known land-based reserves.

The only chromium ore mineral is **chromite**. Chromite is a rock-forming mineral that occurs as magmatic segregations in ultrabasic rocks such as peridotites. Two major chromite-producing regions are the Precambrian Bushveld igneous complex of South Africa and the Great Dike of Zimbabwe.

The most common nickel ore mineral is **pentlandite**, $(Fe,Ni)_9S_8$, which is always associated with pyrrhotite and chalcopyrite. Major nickel-producing districts are in Sudbury, Ontario, Canada, and in the southwestern part of Western Australia.

the iron must come from low-grade deposits or must be imported. The low-grade, silica-rich iron-formation from which the high-grade deposits have been derived is known as *taconite*, which contains about 25% iron. The iron reserves in taconite are far greater than were the original reserves of high-grade ore.

Exploration outside the United States has been successful in locating several ore bodies with many hundreds of millions of tons of high-grade ore. These are notably in Venezuela, Brazil, Canada, and Australia. Brazil's iron mountain, Itabira, is estimated to have 15 billion tons of very pure hematite. In 1947 Cerro Bolivar in Venezuela was discovered as an ex-

tremely rich deposit of hematite, and by 1954 ore was being shipped from there to the United States. In Canada major Precambrian iron ore deposits have been located in the Labrador Trough close to the boundary between Quebec and Labrador. A 350-mile railroad built into this previously inaccessible area began in 1954 to deliver iron ore to Seven Islands, a major port on the St. Lawrence River. Large Australian deposits are located in the Precambrian rocks of the Hamersley Range, Western Australia. The main producers of iron ore are Australia, Russia, the United States, Brazil, Canada, China, and Sweden (see also Box 9.1).

Use. Most important ore of iron for steel manufacture. Also used in pigments, red ocher, and as polishing powder. Black crystals may be cut as gems.

Name. Derived from a Greek word *haimatos* meaning *blood*, in allusion to the color of the powdered mineral.

Similar Species. *Maghemite*, γFe_2O_3, dimorphous with hematite. Formed by weathering or low-temperature oxidation of spinels containing ferrous iron, commonly magnetite.

ILMENITE—FeTiO₃

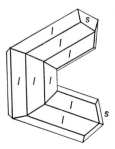

Crystallography. Hexagonal; $\bar{3}$. Crystals are usually thick tabular with prominent basal planes and small rhombohedral truncations. Often in thin plates. Usually massive, compact; also in grains or as sand.

$R\bar{3}$; $a = 5.09$, $c = 14.06$ Å; $Z = 6$. ds: 2.75(10), 2.54(7), 1.867(5), 1.726(8), 1.507(4).

Physical Properties. **H** $5\frac{1}{2}$–6. **G** 4.7. *Luster* metallic to submetallic. *Color* iron-black. *Streak* black to brownish-red. May be magnetic without heating. Opaque.

Composition and Structure. Fe 36.8, Ti 31.6, O 31.6% for stoichiometric FeTiO₃. The formula can be more realistically expressed as $(Fe, Mg, Mn)TiO_3$ with limited Mg and Mn substitution. Ilmenite may contain limited amounts of Fe_2O_3 (less than 6 weight percent) at ordinary temperatures. However, a complete solid solution exists between Fe_2O_3 and $FeTiO_3$ above 950°C (see Figs. 9.16 and 9.26). Related species are *geikielite*, $MgTiO_3$, and *pyrophanite*, $MnTiO_3$. The structure of ilmenite is similar to that of corundum (Fig. 9.2*b*) with Fe and Ti ordered in alternate octahedrally coordinated layers perpendicular to the *c* axis (see Fig. 9.2*c*).

Diagnostic Features. Ilmenite can be distinguished from hematite by its streak and from magnetite by its lack of strong magnetism.

Occurrence. Ilmenite is a common accessory mineral in igneous rocks. It may be present in large masses in gabbros, diorites, and anorthosites as a product of magmatic segregation intimately associated with magnetite. It is also found in some pegmatites and vein deposits. As a constituent of black sands it is associated with magnetite, rutile, zircon, and monazite.

Found in large quantities at Krägerö and other localities in Norway; in Finland; and in crystals at Miask in the Ilmen Mountains, Russia. A large percentage of the production of ilmenite is recovered from beach sands, notably in Australia, Republic of South Africa, India, Brazil, and, in the United States from Florida. Until recently ilmenite was actively mined at Tahawas, New York. A large ilmenite-hematite deposit is mined at Allard Lake, Quebec, Canada. The world production of ilmenite is about 4 million tons.

Use. The major source of titanium. It is used principally in the manufacture of titanium dioxide for paint pigments, replacing older pigments, notably lead compounds. As the metal and in alloys, because of its high strength-to-weight ratio and high resistance to corrosion, titanium is used for aircraft and space vehicle construction in both frames and engines.

Name. From the Ilmen Mountains, Russia.

Similar Species. *Perovskite*, $CaTiO_3$, a pseudocubic titanium mineral found usually in nepheline syenites and carbonatites. *Pseudobrookite*, Fe_2TiO_5, and solid solutions toward *ferropseudobrookite*, $FeTi_2O_5$ (see Fig. 9.26), occur in igneous rocks, in kimberlites, and in meteorite-impacted basalts.

RUTILE GROUP, XO₂

RUTILE–TiO₂

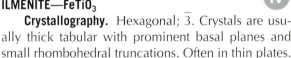

Crystallography. Tetragonal; $4/m2/m2/m$. Usually in prismatic crystals with dipyramid terminations common and vertically striated prism faces. Frequently in elbow twins, often repeated (Figs. 9.17 and 9.18) with twin plane {011}. Crystals frequently slender acicular. Also compact, massive.

$P4_2/mnm$; $a = 4.59$, $c = 2.96$Å; $Z = 2$. ds: 3.24(10), 2.49(5), 2.18(3), 1.687(7), 1.354(4).

Physical Properties. *Cleavage* {110} distinct. **H** 6–6$\frac{1}{2}$. **G** 4.18–4.25. *Luster* adamantine to submetallic. *Color* red, reddish-brown to black, see Plate II, no. 8. *Streak* pale brown. Usually subtranslucent, may be transparent. *Optics*: (+) $\omega = 2.612$, $\epsilon = 2.899$.

Composition and Structure. Ti 60, O 40%. Although rutile is essentially TiO_2, some analyses report considerable amounts of Fe^{2+}, Fe^{3+}, Nb, and Ta. When Fe^{2+} substitutes for Ti^{4+}, electrical neutrality is

FIG. 9.17. Rutile crystals.

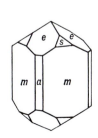

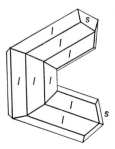

FIG. 9.18. Rutile. Elbow twin, Pfitch, Austria. Cyclic twin, Parksburg, Chester County, Pennsylvania (Harvard Mineralogical Museum).

maintained by twice as much Nb^{5+} and Ta^{5+} entering the structure. This leads to the formulation: $Fe_x(Nb,Ta)_{2x}Ti_{1-3x}O_2$. The structure of rutile is given in Figs. 9.4 and 3.55 and described on page 374. TiO_2 occurs in two additional, relatively rare polymorphs, tetragonal *anatase* (space group $I4_1/amd$) and orthorhombic *brookite* (space group $Pbca$). In anatase TiO_6 octahedra share four edges, in brookite the octahedra share three edges, whereas in rutile they share only two edges. The stability fields of the three polymorphs are still not well defined. Rutile is generally considered to be a high T polymorph of TiO_2 and occurs in rocks which formed over a wide range of P and T. Its large stability range may suggest that the polymorph in which only two edges are shared between adjoining octahedra is indeed the most stable structure (see page 79).

Diagnostic Features. Characterized by its peculiar adamantine luster and red color. Lower specific gravity distinguishes it from cassiterite.

Occurrence. Rutile is found in granites, granite pegmatites, gneisses, mica schists, metamorphic limestone, and dolomite. It may be present as an accessory mineral in the rock or in quartz veins traversing it. Often occurs as slender crystals inside quartz and micas. Is found in considerable quantities in black sands associated with ilmenite, magnetite, zircon, and monazite.

Notable European localities are: Krägerö, Norway; Yrieix, near Limoges, France; in Switzerland; and the Tyrol. Rutile from beach sands of northern New South Wales and southern Queensland makes Australia the largest producer of rutile. In the United States remarkable crystals have come from Graves Mountain, Lincoln County, Georgia. Rutile is found in Alexander County, North Carolina, and at Magnet Cove, Arkansas. It has been mined in Amherst and Nelson counties, Virginia, and derived in commercial quantities from the black sands of northeastern Florida.

Synthesis. Single crystals of rutile have been manufactured by the Verneuil process (see page 578). With the proper heat treatment they can be made transparent and nearly colorless, quite different from the natural mineral. Because of its high refractive index and dispersion, this synthetic material makes a beautiful cut gemstone with only a slight yellow tinge. It is sold under a variety of names, some of the better known are *titania*, *kenya gem*, and *miridis*.

Use. Most of the rutile produced is used as a coating of welding rods. Some titanium derived from rutile is used in alloys; for electrodes in arc lights; to give a yellow color to porcelain and false teeth. Manufactured oxide is used as a paint pigment (see ilmenite).

Name. From the Latin *rutilus*, red, an allusion to the color.

Similar Species. *Stishovite*, a very high-pressure polymorph of SiO_2, isostructural with rutile, is found rarely in meteorite impact craters. This dense form of SiO_2 is the only example of Si in 6-coordination rather than its common tetrahedral coordination. *Leucoxene*, a fine-grained, yellow to brown alteration product of minerals high in Ti (ilmenite, perovskite, titanite), consists mainly of rutile, less commonly anatase.

PYROLUSITE—MnO_2

Crystallography. Tetragonal; $4/m2/m2/m$. Rarely in well-developed crystals. Usually in radiating fibers or columns. Also granular massive; often in reniform coatings and dendritic shapes (Fig. 9.19) finely intergrown with other Mn-oxides and hydroxides. Frequently pseudomorphous after manganite.

$P4_2/mnm$; $a = 4.39$, $c = 2.86$ Å; $Z = 2$. ds: 3.11(10), 2.40(5), 2.11(4), 1.623(7), 1.303(3).

Physical Properties. *Cleavage* {110} perfect. **H** 1–2 (often soiling the fingers). For coarsely crystalline *polianite* the hardness is $6–6\frac{1}{2}$. **G** 4.75. *Luster* metallic. *Color* and *streak* iron-black, see Plate II, no. 9. *Fracture* splintery. Opaque.

Composition and Structure. Mn 63.2, O 36.8%. Commonly contains some H_2O. The structure is the same as that of rutile (see Fig. 9.4) with Mn in 6-coordination with oxygen.

Diagnostic Features. Characterized by and distinguished from other manganese minerals by its black streak and low hardness.

Occurrence. Manganese is present in small amounts in most crystalline rocks. When dissolved

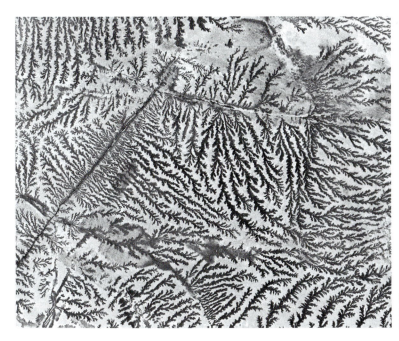

FIG. 9.19. Dendrites of manganese oxide minerals on limestone, Sardinia, Italy (Harvard Mineralogical Museum).

from these rocks, it may be redeposited as various minerals, but chiefly as pyrolusite. Nodular deposits of pyrolusite are found in bogs, on lake bottoms, and on the floors of seas and oceans. Nests and beds of manganese ores are found enclosed in residual clays, derived from the decay of manganiferous limestones. Also found in veins with quartz and various metallic minerals.

Pyrolusite is the most common manganese ore and is widespread in its occurrence. The chief manganese-producing countries are Russia, Republic of South Africa, Brazil, China, and India. Current production in the United States is negligible.

Use. Most important manganese ore. Manganese is used with iron in the manufacture of *spiegeleisen* and *ferromanganese*, employed in making steel (see also Box 9.1). This is its principal use, for about 13.2 pounds of manganese are consumed in the production of one ton of steel. Also used in various alloys with copper, zinc, aluminum, tin, and lead. Pyrolusite is used as an oxidizer in the manufacture of chlorine, bromine, and oxygen; as a disinfectant in potassium permanganate; as a drier in paints; as a decolorizer of glass; and in electric dry-cells and batteries. Manganese is also used as a coloring material in bricks, pottery, and glass.

Name. *Pyrolusite* is derived from two Greek words, *pyros* meaning *fire* and *louo to wash*, because it is used to free glass through its oxidizing effect of the colors due to iron.

Similar Species. *Alabandite*, MnS, is comparatively rare, associated with other sulfides in veins.

Wad is the name given to manganese ore composed of an impure mixture of hydrous manganese oxides.

CASSITERITE—SnO_2

Crystallography. Tetragonal; $4/m2/m2/m$. The common forms are the prisms {110} and {010} and the dipyramids {111} are {011} (Fig. 9.20*a*). Frequently in elbow-shaped twins with a characteristic notch, giving rise to the miner's term *visor tin* (Figs. 9.20*b* and 9.21); the twin plane is {011}. Usually massive granular; often in reniform shapes with radiating fibrous appearance, *wood tin*.

$P4_2/mnm$; $a = 4.73$, $c = 3.18$ Å; $Z = 2$. *d*s: 2.36(8), 2.64(7), 1.762(10), 1.672(4), 1.212(5).

Physical Properties. *Cleavage* {010} imperfect. **H** 6–7. **G** 6.8–7.1 (unusually high for a nonmetallic mineral). *Luster* adamantine to submetallic and dull. *Color* usually brown or black; rarely yellow or white. *Streak* white. Translucent, rarely transparent. *Optics:* (+); $\omega = 1.997$, $\epsilon = 2.093$.

FIG. 9.20. Cassiterite crystals.

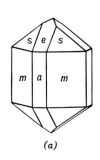

(a)

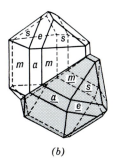

(b)

FIG. 9.21. Twinned crystal of cassiterite, Schlaggenwald, Bohemia, Czechoslovakia (Harvard Mineralogical Museum).

Composition and Structure. Close to SnO_2 with Sn 78.6, O 21.4%. Small amounts of Fe^{3+} may be present and lesser amounts of Nb and Ta substitute for Sn. The structure is that of rutile (see Fig. 9.4 and page 374).

Diagnostic Features. Recognized by high specific gravity, adamantine luster, and light streak.

Occurrence. Cassiterite is widely distributed in small amounts but is produced on a commercial scale in only a few localities. It has been noted as an original constituent of igneous rocks and pegmatites, but it is more commonly found in high-temperature hydrothermal veins in or near granitic rocks. Tin veins usually have minerals that contain fluorine or boron, such as tourmaline, topaz, fluorite, and apatite, and the minerals of the wall rocks are commonly much altered. Frequently associated with wolframite, molybdenite and arsenopyrite. Cassiterite is commonly found in the form of rolled pebbles in placer deposits, *stream tin.*

Most of the world's supply of tin comes from alluvial deposits chiefly in Malaysia, Russia, Indonesia, and Thailand. Bolivia is the only country where a significant production comes from vein deposits; but in the past the mines of Cornwall, England, were major producers. Fine specimens originate from

Schlaggenwald, Czechoslovakia, from Cornwall, England, and from Araca, Bolivia. In the United States cassiterite is not found in sufficient quantities to warrant mining but is present in small amounts in numerous pegmatites. *Wood tin* is found in rhyolites in New Mexico and Mexico.

Use. Principal ore of tin. The chief use of tin was in the manufacture of *tin plate* and *tern plate* for food containers. Tern plate is made by applying a coating of tin and lead instead of pure tin. At present aluminum, glass, paper, plastic, and tin-free steel are replacing tin-plated containers. Tin is also used with lead in solders, in babbit metal with antimony and copper, and in bronze and bell-metal with copper. "Phosphor bronze" contains 89% Cu, 10% Sn, and 1% P. Synthetic tin oxide is a polishing powder.

Name. From the Greek word *kassiteros* meaning *tin.*

Similar Species. *Stannite*, Cu_2FeSnS_4, is structurally similar to chalcopyrite and sphalerite. It is a minor ore of tin.

URANINITE—UO_2

Crystallography. Isometric; $4/m\bar{3}2/m$. The rare crystals are usually octahedral with subordinate cube and dodecahedron faces. More commonly as the variety *pitchblende*: massive or botryoidal with a banded structure.

$Fm3m$; $a = 5.46$ Å; $Z = 4$. ds: 3.15(7), 1.926(6), 1.647(10), 1.255(5), 1.114(5).

Physical Properties. H $5\frac{1}{2}$. G 7.5–9.7 for crystals; 6.5–9 for pitchblende. The specific gravity decreases with oxidation of U^{4+} to U^{6+}. *Luster* submetallic to pitchlike, dull. *Color* black. *Streak* brownish-black.

Composition and Structure. Uraninite is always partially oxidized, and thus the actual composition lies between UO_2 and U_3O_8 (= $U^{4+}O_2$ + $2U^{6+}O_3$). Th can substitute for U and a complete series between uraninite and *thorianite*, ThO_2, has been prepared synthetically. In addition to Th, analyses usually show the presence of small amounts of Pb, Ra, Ce, Y, N, He, and A. The lead occurs as one of two stable isotopes (^{206}Pb and ^{207}Pb) that result from the radioactive decay of uranium. For example: ^{238}U $\rightarrow$ ^{206}Pb + 8^{4}He and ^{235}U $\rightarrow$ ^{207}Pb + 7^{4}He. Thorium decays as follows: ^{232}Th $\rightarrow$ ^{208}Pb + 6^{4}He. In addition to ionized helium atoms (α particles), electrons (β particles) are emitted during the decay process. Because the radioactive disintegration proceeds at a uniform and known rate, the accumulation of both helium and lead can be used as a measure of the time elapsed since the mineral crystallized. For example, for the decay of ^{238}U, the *half-life*, T, which is the time needed to reduce the number of ^{238}U atoms by one-half, is 4.51×10^9 years (see Table 4.8). Both lead-uranium and helium-uranium ratios have been used by geologists to determine the age of rocks (see also page 167).

It was in uraninite that helium was first discovered on Earth, having been previously noted in the sun's spectrum. Also radium was discovered in uraninite. The structures of uraninite and thorianite are like that of fluorite (Fig. 9.11) in which the metal atom is at the center of eight oxygens at the corner of a cube, with each oxygen at the center of a tetrahedral group of metal atoms.

Diagnostic Features. Characterized chiefly by its pitchy luster, high specific gravity, color, and streak. Because of its radioactivity, uraninite, as well as other uranium compounds, can be detected in small amounts by Geiger and scintillation counters.

Occurrence. Uraninite occurs as a primary constituent of granitic rocks and pegmatites. It also is found in high-temperature hydrothermal veins associated with cassiterite, chalcopyrite, pyrite, and arsenopyrite as at Cornwall, England; or in medium-temperature veins with native silver and Co-Ni-As minerals as at Joachimsthal, Czechoslovakia, and Great Bear Lake, Canada. The world's most productive uranium mine during and immediately following World War II was the Shinkolobwe Mine in Zaire. Here uraninite and secondary uranium minerals are in vein deposits associated with cobalt and copper minerals. The ores also carry significant amounts of Mo, W, Au, and Pt.

Wilberforce, Ontario, has been a famous Canadian locality, but the present major Canadian sources are the mines at Great Bear Lake, the Beaverlodge region, Saskatchewan; and the Blind River area, Ontario. At Blind River uraninite occurs as detrital grains in a Precambrian quartz conglomerate. In the same type of deposit it is found in the gold-bearing Witwatersrand conglomerates, Republic of South Africa.

In the United States uraninite was found long ago in isolated crystals in pegmatites at Middletown,

Glastonbury, and Branchville, Connecticut, and the mica mines of Mitchell County, North Carolina. Following World War II, exploration has located many workable deposits of uraninite and associated uranium minerals on the Colorado Plateau in Arizona, Colorado, New Mexico, and Utah.

Use. Uraninite is the chief ore of uranium although other minerals are important sources of the element such as *carnotite* (page 439), *tyuyamunite* (page 440), *torbernite* (page 439), and *autunite* (page 439).

Uranium has assumed an important place among the elements because of its susceptibility to nuclear fission, a process by which the nuclei of uranium atoms are split apart with the generation of tremendous amounts of energy. This energy, first demonstrated in the atomic bomb, is now produced by nuclear-power reactors for generating electricity.

Uraninite is also the source of radium but contains it in very small amounts. Roughly 750 tons of ore must be mined in order to furnish 12 tons of concentrates; chemical treatment of these concentrates yields about 1 gram of a radium salt. In the form of various compounds uranium has a limited use in coloring glass and porcelain, in photography, and as a chemical reagent.

Name. Uraninite in allusion to the composition.

Similar Species. *Thorianite*, ThO_2, is dark gray to black with submetallic luster. Found chiefly in pegmatites and as water-worn crystals in stream gravels. *Cerianite*, $(Ce,Th)O_2$, an extremely rare mineral, has, as does thorianite, the fluorite structure.

SPINEL GROUP, XY_2O_4

The minerals of the *spinel group* show extensive solid solution between the various end-member compositions listed in Table 9.2. There is, for example, extensive solid solution between magnetite, Fe_3O_4, ulvöspinel, Fe_2TiO_4, and a synthetic end-member composition, Mg_2TiO_4 (a magnesian titanate with inverse spinel structure). Furthermore, there are substitutions between chromite, $FeCr_2O_4$,

---

TABLE 9.2   End Members of the Spinel Group (XY_2O_4)

Normal Spinel Structure		Inverse Spinel Structure	
Spinel	$MgAl_2^{3+}O_4$	Magnetite	$Fe^{3+}(Fe^{2+}Fe^{3+})O_4$
Hercynite	$FeAl_2^{3+}O_4$	Magnesioferrite	$Fe^{3+}(Mg^{2+}Fe^{3+})O_4$
Gahnite	$ZnAl_2^{3+}O_4$	Jacobsite	$Fe^{3+}(Mn^{2+}Fe^{3+})O_4$
Galaxite	$MnAl_2^{3+}O_4$	Ulvöspinel	$Fe^{2+}(Fe^{2+}Ti^{4+})O_4$
Franklinite	$ZnFe_2^{3+}O_4$		
Chromite	$Fe^{2+}Cr_2^{3+}O_4$		
Magnesiochromite	$Mg^{2+}Cr_2^{3+}O_4$		

and magnesiochromite, $MgCr_2O_4$; between spinel, $MgAl_2O_4$, and hercynite, $FeAl_2O_4$, and so on. The complexity of the chemical substitutions in this group makes it very difficult to use triangular composition diagrams for expression of the various solid solutions' extents; instead, a "spinel prism" is used for such chemical representations. Figure 9.22 shows three such prisms: (a) for the compositional space between normal spinels at the base of the prism and inverse spinels, with (Fe^{2+} + Ti^{4+}) substitutions, at the top edge; (b) for the compositional space between normal spinels at the base of the prism and inverse spinels, with Fe^{3+} substitutions, along the upper edge; and (c) the general nomenclature for compositions in this prism.

Spinel—$MgAl_2O_4$

Crystallography. Isometric; $4/m\bar{3}2/m$. Usually in octahedral crystals or in twinned octahedrons (spinel twins) (Figs. 9.23*a* and *b*). Dodecahedron may be present as small truncations (Fig. 9.23*c*), but other forms rare. Also massive and as irregular grains.

$Fd3m$; a = 8.10 Å; Z = 8. *d*s: 2.44(9), 2.02(9), 1.552(9), 1.427(10), 1.053(10).

Physical Properties. **H** 8. **G** 3.5–4.1. **G** 3.55 for the composition as given. Nonmetallic. *Luster* vitreous. *Color* various: white, red, lavender, blue, green, brown, black. *Streak* white. Usually translucent, may be clear and transparent. *Optics:* n = 1.718.

Composition and Structure. MgO 28.2, Al_2O_3 71.8%. Fe^{2+}, Zn, and less commonly Mn^{2+} substitute for Mg in all proportions. Fe^{3+} and Cr may substitute in part for Al. The clear red, nearly pure magnesium spinel is known as *ruby spinel*. *Ferroan spinel*, intermediate between spinel and hercynite, $FeAl_2O_4$, is dark green to black, and *chromian spinel*, intermediate between hercynite and *chromite*, $FeCr_2O_4$, is yellowish to greenish-brown (see Fig. 9.22 for compositional definitions). The structure of spinel is illustrated in Figs. 9.5 and 9.6 and discussed on page 374.

Diagnostic Features. Recognized by its hardness (8), its octahedral crystals, and its vitreous luster. The iron spinel can be distinguished from magnetite by its nonmagnetic character and white streak.

Occurrence. Spinel is a common high-temperature mineral occurring in contact metamorphosed limestones and metamorphic argillaceous rocks poor

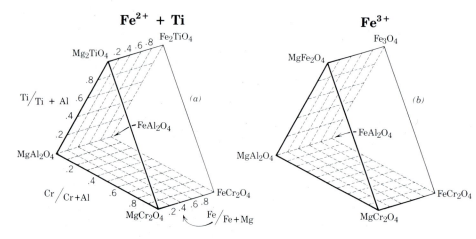

FIG. 9.22. End member compositions in the spinel group as represented in a spinel prism. *(a)* Compositions in this space range from those of normal spinels (in the base of the prism) to those with [Fe^{2+} (or Mg^{2+}) + Ti] end members along the upper edge (these are inverse spinels). *(b)* Compositions in this space range from those of the normal spinels (in the base) to those with Fe^{3+}-rich end members along the upper edge (these are inverse spinels). *(c)* Nomenclature for members of the spinel group as based on the chemical compositions represented in *(a)* and *(b)*.

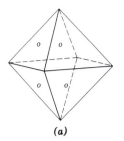 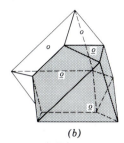 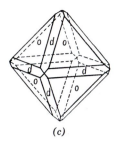

FIG. 9.23. Spinel crystals.

in SiO_2. Occurs also as an accessory mineral in many dark igneous rocks. In contact metamorphic rocks it is associated with phlogopite, pyrrhotite, chondrodite, and graphite. Found frequently as rolled pebbles in stream sands, where it has been preserved because of its resistant physical and chemical properties. The ruby spinels are found in this way, often associated with gem corundum, in the sands of Sri Lanka, Thailand, Upper Burma, and Malagasy Republic. Ordinary spinel is found in various localities in New York and New Jersey.

Use. When transparent and finely colored, spinel is used as a gem (see Plate X, no. 10, Chapter 13). Usually red and known as *ruby spinel* or *balas ruby*. Some stones are blue. The largest cut stone known weighs about 80 carats. The stones are usually comparatively inexpensive.

Synthesis. Synthetic spinel has been made by the Verneuil process (see corundum) in various colors rivaling the natural stones in beauty. Synthetic spinel is also used as a refractory.

Similar Species. *Hercynite*, $FeAl_2O_4$, is associated with corundum in some emery; also found with andalusite, sillimanite, and garnet. *Ferroan galaxite*, $(Fe^{2+},Mn)Al_2O_4$, an Fe^{2+}-rich variety of *galaxite*, $MnAl_2O_4$, occurs sporadically in nature.

Gahnite—$ZnAl_2O_4$
Crystallography. Isometric; $4/m\bar{3}2/m$. Commonly octahedral with faces striated parallel to the edge between the dodecahedron and octahedron. Less frequently showing well-developed dodecahedrons and cubes.

$Fd3m$; $a = 8.12$ Å; $Z = 8$. ds: 2.85(7), 2.44(10), 1.65(4), 1.48(6), 1.232 (8).

Physical Properties. H $7\frac{1}{2}$–8. **G** 4.55. *Luster* vitreous. *Color* dark green. *Streak* grayish. Translucent. *Optics*: $n = 1.80$.

Composition and Structure. Fe^{2+} and Mn^{2+} may substitute for Zn; and Fe^{3+} for Al. Structure is that of spinel (see Figs. 9.5 and 9.6 and page 374).

Diagnostic Features. Characterized by crystal form (striated octahedrons) and hardness.

Occurrence. Gahnite is a rare mineral. It occurs in granite pegmatites, in zinc deposits, and also as a metamorphic mineral in crystalline limestones. Found in large crystals at Bodenmais, Bavaria, Germany; in a talc schist near Falun, Sweden. In the United States it is found at Charlemont, Massachusetts, and Franklin, New Jersey.

Name. After the Swedish chemist J. G. Gahn, the discoverer of manganese.

MAGNETITE—Fe_3O_4
Crystallography. Isometric; $4/m\bar{3}2/m$. Magnetite is frequently in octahedral crystals (Figs. 9.24a and 9.25), more rarely in dodecahedrons (Fig. 9.24b). Dodecahedrons may be striated parallel to the intersection with the octahedron (Fig. 9.24c). Other forms rare. Usually granular massive, coarse-, or fine-grained.

$Fd3m$; $a = 8.40$ Å; $Z = 8$. ds: 2.96(6), 2.53(10), 1.611(8), 1.481(9), 1.094(8).

Physical Properties. Octahedral parting on some specimens. **H** 6. **G** 5.18. *Luster* metallic. *Color* iron-black, see Plate III, no 1. *Streak* black. Strongly magnetic; may act as a natural magnet, known as *lodestone*. Opaque.

Composition and Structure. Fe 72.4, O 27.6%. The composition of much magnetite corresponds closely to Fe_3O_4. However, analyses may show con-

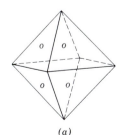 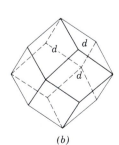 

FIG. 9.24. Magnetite crystals.

FIG. 9.25. Octahedral magnetite crystal. Binnenthal, Switzerland (Harvard Mineralogical Museum).

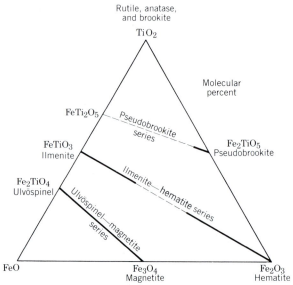

FIG. 9.26. Solid solution series and their extent among oxides in the system TiO_2-FeO-Fe_2O_3. The extent of solid solution between end member compositions is shown by solid bars, for mineral assemblages synthesized in the laboratory at about 600°C. At higher temperatures, the extent of solid solution will increase considerably in, for example, the ilmenite-hematite and pseudo-brookite series. Lack of solid solution at about 600°C is shown by dashed lines. Compare with Fig. 9.16. (Compiled from S. E. Haggerty, 1976, Oxidation of opaque mineral oxides in basalts, and opaque mineral oxides in terrestrial igneous rocks; articles in *Oxide Minerals, Reviews in Mineralogy.* Mineralogical Society of America, Washington, D.C.)

siderable percentages of Mg and Mn^{2+} substituting for Fe^{2+} and Al, Cr, Mn^{3+}, and Ti^{4+} substituting for Fe^{3+}. Above 600°C a complete solid solution series is possible between magnetite and *ulvöspinel*, Fe_2TiO_4 (see Figs. 9.16 and 9.26). The structure of magnetite is that of an *inverse spinel* (see page 374), which can be expressed by rewriting the formula as $Fe^{3+}(Fe^{2+},Fe^{3+})_2O_4$ (see Table 9.2, p. 387).

Diagnostic Features. Characterized chiefly by its strong magnetism, its black color, and its hardness (6). Can be distinguished from magnetic franklinite by streak.

Occurrence. Magnetite is a common mineral found disseminated as an accessory through most igneous rocks. In certain types of rocks, through magmatic segregation, magnetite may become one of the chief constituents of the rock and may thus form large ore bodies. Such bodies are often highly titaniferous. Commonly associated with crystalline metamorphic rocks and may occur as large beds and lenses. It is a common constituent of sedimentary and metamorphic banded Precambrian iron-formations. In such occurrences it is of chemical, sedimentary origin. Found in the black sands of the seashore. Occurs as thin plates and dendritic growths between plates of mica. Often intimately associated with corundum, forming *emery*.

The largest magnetite deposits in the world are in northern Sweden at Kiruna and Gellivare associated with apatite, and are believed to have formed by magmatic segregation. Other important deposits are in Norway, Romania, and the Ukraine. The most powerful natural magnets are found in Siberia, in the Harz Mountains, on the Island of Elba, and in the

Bushveld igneous complex, Republic of South Africa. The extensive Precambrian iron deposits in the Lake Superior region and in the Labrador Trough, Canada, carry considerable magnetite as an ore mineral.

The Precambrian iron-formations of the Lake Superior region, containing about 25% iron, mostly in the form of magnetite, were a major source of iron in the United States. This low-grade ore, *taconite*, has been actively mined and the magnetite separated magnetically from the waste material. Found as lodestone and in crystals at Magnet Cove, Arkansas.

Use. An important iron ore (see also Box 9.1).

Name. Probably derived from the locality Magnesia, bordering on Macedonia. A fable told by Pliny ascribes its name to a shepherd named Magnes, who first discovered the mineral on Mount Ida by noting that the nails of his shoes and the iron ferrule of his staff adhered to the ground.

Similar Species. *Magnesioferrite*, $Fe^{3+}(Mg,Fe^{3+})_2O_4$ (or $MgFe_2O_4$) with an inverse spinel structure, is a rare mineral found chiefly in fumaroles. *Jacobsite*, $Fe^{3+}(Mn,Fe^{3+})_2O_4$ (or $MnFe_2O_4$), an inverse spinel and a rare mineral found at Långban, Sweden. *Ulvöspinel*, $Fe^{3+}(Ti,Fe^{2+})_2O_4$ (or Fe_2TiO_4), also with an inverse spinel structure, is not uncommonly present as exsolution blebs and lamellae within magnetite (see also Table 9.2). *Maghemite*, γFe_2O_3, dimorphous with hematite. Formed by weathering or low-temperature oxidation of spinels containing ferrous iron, commonly magnetite. *Martite*, hematite pseudomorphous after magnetite.

Franklinite—(Zn,Fe,Mn)(Fe,Mn)$_2$O$_4$

Crystallography. Isometric; $4/m\bar{3}2/m$. Crystals are octahedral (see Fig. 9.27) with dodecahedral truncations; commonly rounded. Also massive, coarse, or fine granular, in rounded grains.

$Fd3m$; $a = 8.42$ Å; $Z = 8$. ds: 2.51(10), 1.610(10), 1.480(9), 1.278(6), 1.091(6).

Physical Properties. H 6. G 5.15. *Luster* metallic. *Color* iron-black. *Streak* reddish-brown to dark brown. Slightly magnetic.

Composition and Structure. Dominantly $ZnFe_2O_4$ but always with some substitution of Fe^{2+} and Mn^{2+} for Zn, and Mn^{3+} for Fe^{3+}. Analyses show a wide range in the proportions of the various elements. The structure of franklinite is that of a normal spinel (see page 374 and Figs. 9.5 and 9.6).

Diagnostic Features. Resembles magnetite but is only slightly magnetic and has a dark brown streak. Usually identified by its characteristic association with willemite and zincite.

Occurrence. Franklinite, with only minor exceptions, is confined to the zinc deposits at Franklin, New Jersey, where, enclosed in a granular limestone, it is associated with zincite and willemite.

Use. As an ore of zinc and manganese. The zinc is converted into zinc white, ZnO, and the residue is smelted to form an alloy of iron and manganese, *spiegeleisen*, used in the manufacture of steel.

Name. From Franklin, New Jersey.

CHROMITE—FeCr$_2$O$_4$

Crystallography. Isometric; $4/m\bar{3}2/m$. Habit octahedral, but crystals are small and rare. Commonly massive, granular to compact.

$Fd3m$; $a = 8.36$ Å; $Z = 8$. ds: 4.83(4), 2.51(10), 2.08(5), 1.602(6), 1.473(8).

Physical Properties. H $5\frac{1}{2}$. G 4.6. *Luster* metallic to submetallic; frequently pitchy. *Color* iron-black to brownish-black. *Streak* dark brown. Subtranslucent. *Optics:* n = 2.16.

Composition and Structure. For $FeCr_2O_4$, FeO 32.0, Cr_2O_3 68.0%. Some Mg is always present substituting for Fe^{2+} and some Al and Fe^{3+} may substitute for chromium. There is extensive solid solution between chromite and magnesiochromite, $MgCr_2O_4$. The structure of chromite is that of a normal spinel (see page 374 and Figs. 9.5 and 9.6).

Diagnostic Features. The submetallic luster usually distinguishes chromite.

Occurrence. Chromite is a common constituent of peridotites and other ultrabasic rocks and of serpentines derived from them. One of the first minerals to separate from a cooling magma; large chromite ore deposits are thought to have been derived by such magmatic differentiation. For example, the Bushveld igneous complex of South Africa and the Great Dike of Zimbabwe contain many seams of

FIG. 9.27. Franklinite, Franklin, New Jersey (Harvard Mineralogical Museum).

chromite enclosed in pyroxenites. Associated with olivine, serpentine, and corundum.

The important countries for its production are Republic of South Africa, Russia, Albania, and Zimbabwe. Chromite is found only sparingly in the United States, but in the past, Pennsylvania, Maryland, North Carolina, Wyoming, California, Alaska, and Oregon were small producers. During World War II, bands of low-grade chromite were mined in the Stillwater igneous complex, Montana.

Use. The only ore of chromium. Chromite ores are grouped into three categories—metallurgical, refractory, and chemical—on the basis of their chrome content and their Cr/Fe ratio. As a metal, chromium is used as a ferroalloy to give steel the combined properties of high hardness, great toughness, and resistance to chemical attack. Chromium is a major constituent in stainless steel (see Box 9.1). *Nichrome*, an alloy of Ni and Cr, is used for resistance in electrical heating equipment. Chromium is widely used in plating plumbing fixtures, automobile accessories, and so forth.

Because of its refractory character, chromite is made into bricks for the linings of metallurgical furnaces. The bricks are usually made of crude chromite and coal tar but sometimes of chromite with kaolin, bauxite, or other minerals. Chromium is a constituent of certain green, yellow, orange, and red pigments and in $K_2Cr_2O_7$ and $Na_2Cr_2O_7$, which are used as mordants to fix dyes.

Similar Species. *Magnesiochromite*, $MgCr_2O_4$, is similar to chromite in both occurrence and appearance.

Name. Named in allusion to the composition.

Chrysoberyl—BeAl$_2$O$_4$

Crystallography. Orthorhombic; $2/m2/m2/m$. Usually in crystals tabular on {001}, the faces of which are striated parallel to [100]. Commonly twinned on {130}, giving pseudohexagonal appearance (Fig. 9.28).

Pmnb; $a = 5.47$, $b = 9.39$, $c = 4.42$ Å; $Z = 4$. *d*s: 3.24(8). 2.33(8), 2.57(8), 2.08(10), 1.61(10).

Physical Properties. *Cleavage* {110}. **H** $8\frac{1}{2}$. **G** 3.65–3.8 *Luster* vitreous. *Color* various shades of green,

FIG. 9.28. Twinned chrysoberyl.

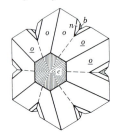

brown, yellow; may be red by transmitted light. *Optics:* (+), $\alpha = 1.746$, $\beta = 1.748$, $\gamma = 1.756$; 2V = 45°; X = c, Y = b.

Alexandrite is a gem variety, emerald-green in daylight, but red by transmitted light and usually red by artificial light. *Cat's eye*, or *cymophane*, is a chatoyant variety that, when cut as an oval or round *cabochon gem*, shows a narrow band of light on its surface. The effect results from minute tubelike cavities or needlelike inclusions parallel to the c axis.

Composition and Structure. BeO 19.8, Al_2O_3 80.2%. Be 7.1%. From its formula, $BeAl_2O_4$, it would appear that chrysoberyl is a member of the spinel group. However, because of the small size of Be^{2+}, chrysoberyl has a structure of lower symmetry. It consists of oxygen atoms in hexagonal closest packing with Be in 4-coordination with O and Al in 6-coordination with O. Morphologically and structurally chrysoberyl is very similar to olivine $(Mg,Fe)_2SiO_4$. Be is in the same coordination as Si, and Al in the same coordination as Fe^{2+} or Mg. The hexagonal closest packed arrangement of O in chrysoberyl leads to a pseudohexagonal lattice, angles, and twinning.

Diagnostic Features. Characterized by its high hardness, its yellowish to emerald-green color, and its twin crystals.

Occurrence. Chrysoberyl is a rare mineral. It occurs in granitic rocks, pegmatites, and mica schists. Frequently in river sands and gravels. The outstanding alluvial gem deposits are found in Brazil and Sri Lanka; the *alexandrite* variety comes from the Ural Mountains and Brazil. In the United States chrysoberyl of gem quality is rare, but it has been found in Oxford County and elsewhere in Maine; Haddam, Connecticut; and Greenfield, New York. Recently found in Colorado.

Use. As a gemstone. The ordinary yellowish-green stones are inexpensive; the varieties alexandrite and cat's eye are highly prized gems (see Chapter 13).

Name. *Chrysoberyl* means *golden beryl*. Cymophane is derived from two Greek words meaning *wave* and *to appear*, in allusion to the chatoyant effect. *Alexandrite* was named in honor of Alexander II of Russia.

Columbite–Tantalite—(Fe,Mn)Nb$_2$O$_6$–(Fe,Mn)Ta$_2$O$_6$

Crystallography. Orthorhombic; $2/m2/m2/m$. Commonly in crystals. The habit is short prismatic or thin tabular on {010}; often in square prisms because of prominent development of {100} and {010} (Fig. 9.29). Also in heart-shaped twins, twinned on {201}.

Pcan; $a = 5.74$, $b = 14.27$, $c = 5.09$ Å; $Z = 4$. *d*s for columbite: 3.66(7), 2.97(10), 1.767(6), 1.735(7), 1.712(8).

Physical Properties. *Cleavage* {010} good. **H** 6. **G** 5.2–7.9, varying with the composition, increasing with rise in percentage of Ta_2O_5 (Fig. 9.30). *Luster* submetallic. *Color* iron-black, frequently iridescent. *Streak* dark red to black. Subtranslucent.

Composition and Structure. A complete solid solution exists from *columbite*, $(Fe,Mn)Nb_2O_6$, to *tantalite*, $(Fe,Mn)Ta_2O_6$. Often contains small amounts of Sn and W. A variety known as *manganotantalite* is essentially tantalite

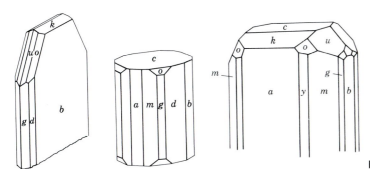

FIG. 9.29. Columbite-tantalite crystals.

with Mn^{2+} substituting for most of the Fe^{2+}. In the structure octahedral chains of $(Mn,Fe)O_6$ and $(Ta,Nb)O_6$ exist in which octahedra join through sharing of edges. The chains are linked to each other by common apices.

Diagnostic Features. Recognized usually by its black color with lighter-colored streak, and high specific gravity. Distinguished from wolframite by lower specific gravity and less distinct cleavage.

Occurrence. The minerals occur in granitic rocks and pegmatites, associated with quartz, feldspar, mica, tourmaline, beryl, spodumene, cassiterite, wolframite, microlite, and monazite.

Notable localities for their occurrence are Bernie Lake, Manitoba, Canada; Zaire; Nigeria; Brazil; near Moss, Norway; Bodenmais, Bavaria, Germany; Ilmen Mountains, Russia. Western Australia (*manganotantalite*); and Malagasy Republic. In the United States it is found at Standish, Maine; Haddam, Middletown, and Branchville, Connecticut; in Amelia County, Virginia; Mitchell County, North Carolina; Black Hills, South Dakota; and near Canon City, Colorado.

Use. Source of tantalum and niobium. Because of its resistance to acid corrosion, tantalum is employed in chemical equipment, in surgery for skull plates and sutures, also in some tool steels and in electronic tubes. Niobium has its chief use in alloys in weldable high-speed steels, stainless steels, and alloys resistant to high temperatures, such as used in the gas turbine of the aircraft industry.

Name. *Columbite* from Columbia, a name for America, whence the original specimen was obtained. *Tantalite* from the mythical Tantalus in allusion to the difficulty in dissolving in acid.

Similar Species. *Microlite*, $Ca_2Ta_2O_6(O,OH,F)$, is found in pegmatites; *pyrochlore*, $(Ca,Na)_2(Nb,Ta)_2O_6(O,OH,F)$, and *fergusonite*, $YNbO_4$, an oxide of niobium, tantalum, and rare earths (represented by Y in the fergusonite formula) are found associated with alkalic rocks.

Hydroxides

Brucite—Mg(OH)₂

Crystallography. Hexagonal; $\overline{3}2/m$. Crystals are usually tabular on $\{0001\}$ and may show small rhombohedral truncations. Commonly foliated (Fig. 9.31), massive.

$C\overline{3}m$; $a = 3.13$, $c = 4.74$ Å; $Z = 1$. ds: 4.74(8), 2.37(10), 1.793(10), 1.372(7), 1.189(9).

Physical Properties. *Cleavage* $\{0001\}$ perfect. Folia flexible but not elastic. Sectile. **H** $2\frac{1}{2}$. **G** 2.39. *Luster* on base pearly, elsewhere vitreous to waxy. *Color* white, gray, light green. Transparent to translucent. *Optics*: (+), $\omega = 1.566$, $\epsilon = 1.581$.

Composition and Structure. For $Mg(OH)_2$: MgO 69.0, H_2O 31.0%. Fe^{2+} and Mn^{2+} may substitute for Mg. The structure is illustrated in Fig. 9.7. The perfect $\{0001\}$ cleavage is parallel to the octahedral sheets. Upon heating, brucite transforms into periclase (MgO).

Diagnostic Features. Recognized by its foliated nature, light color, and pearly luster on cleavage face. Distinguished from talc by its greater hardness and lack of greasy feel, and from mica by being inelastic.

Occurrence. Brucite is found associated with serpentine, dolomite, magnesite, and chromite; as an alteration product of periclase and magnesium silicates, especially serpentine. It is also found in crystalline limestone.

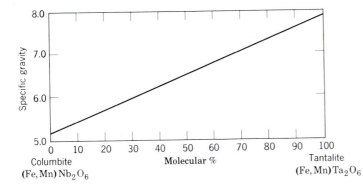

FIG. 9.30. Columbite-tantalite. Variation of specific gravity with composition.

FIG. 9.31. Brucite crystal, Wood's Mine, Texas, Pennsylvania (Harvard Mineralogical Museum).

Notable foreign localities for its occurrence are at Unst, one of the Shetland Islands, Scotland, and Aosta, Italy. In the United States found at Tilly Foster Iron Mine, Brewster, New York; at Wood's Mine, Texas, Pennsylvania; and in Gabbs, Nevada.

Use. Brucite is used as a raw material for magnesia refractories and is a minor source of metallic magnesium.

Name. In honor of the early American mineralogist Archibald Bruce.

Similar Species. *Gibbsite*, $Al(OH)_3$, one of the three hydroxides of Al that are the main constituents of bauxite. The structure of gibbsite is like that of brucite, with one-third of the octahedral cation positions vacant (compare with Fig. 9.2a).

MANGANITE—MnO(OH)

Crystallography. Monoclinic; $2/m$ (pseudo-orthorhombic). Crystals are usually prismatic parallel to c and vertically striated (Figs. 9.32 and 9.33). Often columnar to coarse fibrous. Twinned on {011} as both contact and penetration twins.

$B2_1/d$; $a = 8.84$, $b = 5.23$, $c = 5.74$ Å, $\beta = 90°$; $Z = 8$. ds: 3.38(10), 2.62(9), 2.41(6), 2.26(7), 1.661(9).

Physical Properties. *Cleavage* {010} perfect, {110} and {001} good. **H** 4. **G** 4.3. *Luster* metallic. *Color* steel-gray to iron-black, see Plate III, no. 2. *Streak* dark brown. Opaque.

Composition and Structure. Mn 62.4, O 27.3, H_2O 10.3%. The structure consists of hexagonal closest packing of oxygen and $(OH)^-$ groups with Mn^{3+} in octahedral coordination with O^{2-} and $(OH)^-$. In this respect the structure is similar to that of diaspore (see Fig. 9.8); however the distribution of the cations is quite different.

Diagnostic Features. Recognized chiefly by its black color and prismatic crystals. Hardness (4) and brown streak distinguish it from pyrolusite.

Occurrence. Manganite is found associated with other manganese oxides in deposits formed by meteoric waters. Found often in low-temperature hydrothermal veins associated with barite, siderite, and calcite. It frequently alters to pyrolusite.

Occurs at Ilfeld, Harz Mountains, Germany, in fine crystals; also at Ilmenau, Thuringia, Germany,

FIG. 9.32. Manganite.

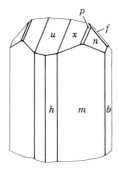

FIG. 9.33. Manganite, Ilfeld, Germany (Harvard Mineralogical Museum).

and Cornwall, England. In the United States at Negaunee, Michigan. In Nova Scotia, Canada.

Use. A minor ore of manganese.

Name. Named in allusion to the composition.

Romanechite—BaMn^{2+}Mn$_8^{4+}$O$_{16}$(OH)$_4$

Crystallography. Orthorhombic, 222. Massive, botryoidal, stalactitic (Fig. 9.34). Appears amorphous.

$P222$; $a = 9.45$, $b = 13.90$, $c = 5.72$; $Z = 2$. ds: 3.46(7), 2.87(7), 2.41(7), 2.190(10), 1.820(7).

Physical Properties. H 5–6. G 3.7–4.7. *Luster* submetallic. *Color* black. *Streak* brownish-black. Opaque.

Composition and Structure. Small amounts of Mg, Ca, Ni, Co, Cu, and Si may be present. The structure is somewhat similar to that of rutile (Fig. 9.4) with complex chains of MnO,OH octahedra and large channels between adjoining chains. Both Ba and adsorbed H$_2$O are located in

FIG. 9.34. Romanechite, Ironwood, Michigan (Harvard Mineralogical Museum).

these channels. Upon heating to about 600°C, romanechite transforms into *hollandite*, Ba$_2$Mn$_8$O$_{16}$ (see Fig. 7.2).

Diagnostic Features. Distinguished from the other manganese oxides by its greater hardness and botryoidal form, and from limonite by its black streak.

Occurrence. Romanechite is a secondary mineral; it occurs usually with pyrolusite, and its origin and associations are similar to those of that mineral.

Use. An ore of manganese. (See pyrolusite.)

Similar Species. Many of the hard botryoidal masses formerly called *psilomelane* are now known to be a mixture of several manganese oxides of which romanechite is a major constituent. Some of the other minerals commonly present in the mixture are *cryptomelane*, KMn$_8$O$_{16}$, *manjiroite*, (Na,K)Mn$_8$O$_{16}$·nH$_2$O, and *todorokite* (Mn,Ca,Mg)Mn$_3$O$_7$·H$_2$O. The presence of these and other minerals accounts for the oxides of Na, K, Ca, Co, Cu, Al, and Fe reported in chemical analyses of psilomelane.

Name. From the locality at Romanèche, France.

Diaspore—αAlO(OH)

Crystallography. Orthorhombic; $2/m2/m2/m$. Usually in thin crystals, tabular parallel to {010}; sometimes elongated on [001]. Bladed, foliated, massive, disseminated.

Pbnm; $a = 4.41$, $b = 9.40$, $c = 2.84$ Å; $Z = 4$. ds: 3.98(10), 2.31(8), 2.12(7), 2.07(7), 1.629(8).

Physical Properties. *Cleavage* {010} perfect. H 6$\frac{1}{2}$–7. G 3.35–3.45. *Luster* vitreous except on cleavage face, where it is pearly. *Color* white, gray, yellowish, greenish. Transparent to translucent. *Optics:* (+), α = 1.702, β = 1.722, γ = 1.750; 2V = 85°; X = c, Y = b; r < v.

Composition and Structure. Al$_2$O$_3$ 85, H$_2$O 15%. The structure is illustrated in Fig. 9.8 and consists of Al in 6-coordination with oxygen and OH$^-$ forming AlO$_3$(OH)$_3$ octahedra. The structure of *boehmite*, γAlO(OH), is shown in Fig. 9.9.

Diagnostic Features. Characterized by its good cleavage, its bladed habit, and its high hardness.

Occurrence. Diaspore is commonly associated with corundum in emery rock, in dolomite, and in chlorite schist. In a fine-grained massive form it is a major constituent of much bauxite.

Notable localities are Mugla, Turkey; the island of Naxos, Greece; and Postmasburg, South Africa. In the United States it is found in Chester County, Pennsylvania; at Chester, Massachusetts; and with alunite, forming rock masses at Mt. Robinson, Rosite Hills, Colorado. It is found abundantly in the bauxite and aluminous clays of Arkansas, Missouri, and elsewhere in the United States.

Use. As a refractory.

Name. Derived from a Greek word meaning *to scatter*, in allusion to its decrepitation when heated.

GOETHITE—αFeO(OH)

Crystallography. Orthorhombic; $2/m2/m2/m$. Rarely in distinct prismatic, vertically striated crystals. Often flattened parallel to {010}. In acicular crystals. Also massive, reniform, stalactitic, in radiating fibrous aggregates (Fig. 9.35). Foliated. The so-called *bog ore* is generally loose and porous.

Pbnm; $a = 4.65$, $b = 10.02$, $c = 3.04$ Å; $Z = 4$. ds: 4.21(10), 2.69(8), 2.44(7), 2.18(4), 1.719(5).

Physical Properties. *Cleavage* {010} perfect. **H** $5–5\frac{1}{2}$. **G** 4.37; may be as low as 3.3 for impure material. *Luster* adamantine to dull, silky in certain fine, scaly, or fibrous varieties. *Color* yellowish-brown to dark brown, see Plate III, no. 3. *Streak* yellowish-brown. Subtranslucent.

Composition and Structure. Fe 62.9, O 27.0, H_2O 10.1%. Mn is often present in amounts up to 5%. The massive varieties often contain adsorbed or capillary H_2O. Goethite is isostructural with diaspore (see Fig. 9.8). *Lepidocrosite*, $\gamma FeO(OH)$, a polymorph of goethite, is a platy mineral and is often associated with goethite; it is isostructural with boehmite (see Fig. 9.9).

Diagnostic Features. Distinguished from hematite by its streak.

Occurrence. Goethite is one of the most common minerals and is typically formed under oxidizing conditions as a weathering product of iron-bearing minerals. It also forms as a direct inorganic or biogenic precipitate from water and is widespread as a deposit in bogs and springs. Goethite forms the gossan or "iron hat" over metalliferous veins (see Box 8.2). Large quantities of goethite have been found as residual lateritic mantles resulting from the weathering of serpentine.

Goethite in some localities constitutes an important ore of iron. It is the principal constituent of the valuable minette ores of Alsace-Lorraine, France. Other notable European localities are: Eiserfeld in Westphalia, Germany; Příbram, Bohemia, Czechoslovakia; and Cornwall, England. Large deposits of iron-rich laterites composed essentially of goethite are found in the Mayari and Moa districts of Cuba.

In the United States goethite is common in the Lake Superior hematite deposits and has been obtained in fine specimens at Negaunee, near Marquette, Michigan. Goethite is found in iron-bearing limestones along the Appalachian Mountains, from western Massachusetts as far south as Alabama. Such deposits are particularly important in Alabama, Georgia, Virginia, and Tennessee. Finely crystallized material occurs with smoky quartz and microcline in Colorado at Florissant and in the Pikes Peak region.

Use. An ore of iron (see also Box 9.1).

Name. In honor of Goethe, the German poet.

Similar Species. *Limonite*, $FeO \cdot OH \cdot nH_2O$, is used mainly as a field term to refer to natural hydrous iron oxides of uncertain identity.

FIG. 9.35. Goethite from an iron mine near Negaunee, Michigan (Harvard Mineralogical Museum).

BAUXITE[2]—A Mixture of Diaspore, Gibbsite, and Boehmite

Crystallography. A mixture. Pisolitic, in round concretionary grains (Fig. 9.36); also massive, earthy, claylike.

Physical Properties. H 1–3. **G** 2–2.55. *Luster* dull to earthy. *Color* white, gray, yellow, red. Translucent.

Composition. A mixture of hydrous aluminum oxides in varying proportions. Some bauxites closely approach the composition of *gibbsite*, $Al(OH)_3$ (see brucite, "Similar Species"), but most are a mixture and usually contain some Fe. As a result, bauxite is not a mineral and should be used only as a rock name. The principal constituents of the rock bauxite are *gibbsite*, see Fig. 9.37; *boehmite*, $\gamma AlO(OH)$ (see Fig. 9.9), and *diaspore*, $\alpha AlO(OH)$ (see Fig. 9.8), any one of which may be dominant.

Diagnostic Features. Can usually be recognized by its pisolitic character.

Occurrence. Bauxite is of supergene origin, commonly produced under subtropical to tropical climatic conditions by prolonged weathering and leaching of silica from aluminum-bearing rocks. Also may be derived from the weathering of clay-bearing limestones. It has apparently originated as a colloidal precipitate. It may occur in place as a direct derivative of the original rock, or it may have been transported and deposited in a sedimentary formation. In the tropics deposits known as *laterites*, consisting largely of hydrous aluminum and ferric oxides, are

FIG. 9.37. Scanning electron micrograph of very small prismatic crystals of gibbsite (monoclinic) in bauxite from Surinam, South America. (Courtesy of H. H. Murray, Indiana University.)

found in the residual soils. These vary widely in composition and purity, but many are valuable as sources of aluminum and iron.

Bauxite occurs over a large area in the south of France, an important district being at Baux, near Arles, France. The principal world producers are Surinam, Jamaica, and Guiana. Other major producing countries are Indonesia, Russia, Australia, and Hungary. In the United States, the chief deposits are found in Arkansas, Georgia, and Alabama. In Arkansas bauxite has formed by the alteration of a nepheline syenite.

Use. The ore of aluminum. Eighty-five percent of the bauxite produced is consumed as aluminum ore. Because of its low density and great strength, aluminum has been adapted to many uses. Sheets, tubes, and castings of aluminum are used in automobiles, airplanes, and railway cars, where light weight is desirable. It is manufactured into cooking utensils, food containers, household appliances, and furniture. Aluminum is replacing copper to some extent in electrical transmission lines. Aluminum is alloyed with copper, magnesium, zinc, nickel, silicon, silver, and tin. Other uses are in paint, aluminum foil, and numerous salts.

The second largest use of bauxite is in the manufacture of Al_2O_3, which is used as an abrasive. It is also manufactured into aluminous refractories. Synthetic alumina is also used as the principal ingredient in heat-resistant porcelain such as spark plugs.

Name. From its occurrence at Baux, France.

FIG. 9.36. Pisolitic bauxite, Bauxite, Arkansas.

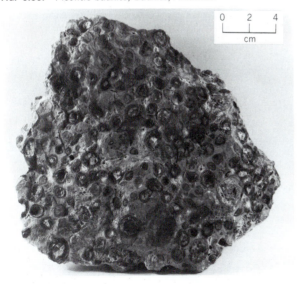

0 2 4
cm

[2]Although bauxite is not a mineral species, it is described here because of its importance as the ore of aluminum.

BOX 9.2
EVAPORITE MINERALS

When a restricted body of seawater or the waters of saline lakes evaporate, the elements in solution (see table) are precipitated in what are known as *evaporites*. More than 80 minerals (excluding clastic material) have been recorded in evaporites, and most of these are chlorides, sulfates, carbonates, and borates. Only about 11 rank as major constituents (see table). Upon evaporation, the general sequence of precipitation is this: some *calcite* (when the original volume of seawater is reduced by evaporation to about one-half), *gypsum* or *anhydrite* (with the volume reduced to one-fifth of the original), *halite* (with the volume reduced to one-tenth of the original), and finally *sulfates* and *chlorides* of Mg and K. If all the salt in a 1000-foot (305 m) column of seawater were pre-

cipitated, it would form 0.5 feet (0.15 m) of calcium sulfate, 11.8 feet (3.6 m) of NaCl, and 2.6 feet (0.8 m) of K- and Mg-bearing salts, producing a total salt column of 15 feet (4.6 m) thick.

The photograph shows the fine and regular banding of calcite (darker bands in the photograph) alternating with gypsum (lighter bands) in the upper Permian Castile Evaporite sequence of West Texas and New Mexico. Seasonal changes in temperature and/or ionic concentration in the evaporating body of water result in regular cycles of mineral precipitation, and each couplet is concluded to represent a *varve*, which is an annual layer of sedimentation resulting from evaporation (R. Y. Anderson, W. E. Dean, Jr., D. W. Kirkland, and H. Snider, 1972, Permian Castile varved

evaporite sequence. *Geological Society of America Bulletin* 83: 59–86).

In natural deposits, the less-soluble minerals that precipitate early in the evaporation sequence tend to show increased abundance. Hence, gypsum and anhydrite are by far the most abundant evaporate minerals and commonly form massive beds. The deposition of calcium sulfate as gypsum or anhydrite depends on the temperature and salinity of the brine; anhydrite is formed at higher salt concentrations and at higher temperatures than gypsum. Halite forms about 95% of the chloride minerals in an evaporite sequence, but the more soluble salts are commonly missing in natural deposits. When present, thick and extensive beds of rock salt overlie the gypsum-anhydrite zone of marine evaporites.

Commercially, the amount of salt consumed as sodium chloride is relatively small compared to the amount used to produce chemicals such as sodium hydroxide, chlorine, and chlorides. Common salt is also extracted from seawater in shallow impoundments where the modern climate allows evaporation, as in Western Australia, Mexico, Bahamas, India, Brazil, and Spain. Deposits of the more soluble salts such as sylvite, carnallite, and polyhalite are rare, because the deposition from brine may be interrupted or the salts may be redissolved. Nevertheless, large deposits (particularly of sylvite) have formed in places and are mined extensively as the chief source of potassium. Potassium is the third most common component of most chemical fertilizers (NPK: nitrogen, phosphorus, and potassium; see also Box 10.1).

Major Ionic Constituents of Seawater and Most Common Minerals in Evaporite Sequences*

	Normal Seawater (Ion Concentration as Parts per million)	Common Minerals in Marine Evaporites
K^+	380	Halite—$NaCl$
Na^+	10,556	Sylvite—KCl
Ca^{2+}	400	Carnallite—$KMgCl_3 \cdot 6H_2O$
Mg^{2+}	1,272	Anhydrite—$CaSO_4$
Cl^{-1}	18,980	Gypsum—$CaSO_4 \cdot 2H_2O$
SO_4^{-2}	2,649	Langbeinite—$K_2Mg_2(SO_4)_3$
HCO_3^{-1}	140	Polyhalite—$K_2Ca_2Mg(SO_4)_4 \cdot 2H_2O$
Total	34,387	Kieserite—$MgSO_4 \cdot H_2O$
		Calcite—$CaCO_3$
		Magnesite—$MgCO_3$
		Dolomite—$CaMg(CO_3)_2$

*After F. H. Stewart, 1963, *Marine Evaporites*. U.S. Geological Survey Professional Paper no. 440-Y.

Finely varved outcrop specimen showing alternating bands of calcite (dark on account of enclosed organic matter) and gypsum (light). Diamond drill cores show that several hundred feet below the outcrop surface the original anhydrite has not yet been transformed (hydrated) to gypsum.

Halides

HALITE—NaCl

Crystallography. Isometric; $4/m\overline{3}2/m$. Habit cubic, see Plate III, no. 4; other forms are very rare. Some crystals hopper-shaped (Fig. 9.38). Found in crystals or granular crystalline masses showing cubic cleavage, known as *rock salt*. Also massive, granular to compact.

$Fm3m$; $a = 5.640$ Å; $Z = 4$. ds: 2.82(10), 1.99(4), 1.628(2), 1.261(2), 0.892(1).

Physical Properties. *Cleavage* {001} perfect. **H** $2\frac{1}{2}$. **G** 2.16. *Luster* transparent to translucent. *Color* colorless or white, or when impure may have shades of yellow, red, blue, purple. Salty taste. *Optics*: $n = 1.544$.

Composition and Structure. Na 39.3, Cl 60.7%. Commonly contains impurities, such as calcium and magnesium sulfates and calcium and magnesium chlorides. The structure of halite is illustrated in Figs. 9.10 and 3.51. This structure exists in a large number of XZ compounds with a radius ratio of between 0.41 and 0.73.

Diagnostic Features. Characterized by its cubic cleavage and taste, and distinguished from sylvite by less bitter taste.

Occurrence. Halite is a common mineral, occurring often in extensive beds and irregular masses, precipitated by evaporation with gypsum, sylvite, anhydrite, and calcite. Halite is dissolved in the waters of salt springs, salt lakes, and the ocean. It is a major salt in playa deposits of enclosed basins.

The deposits of salt have been formed by the gradual evaporation and ultimate drying up of en-closed bodies of salt water (see Box 9.2). The salt beds formed in this way may have subsequently been covered by other sedimentary deposits and gradually buried beneath the rock strata formed on them. Salt beds range between a few feet to over 200 feet in thickness and it is estimated that some are buried beneath as much as 35,000 feet of overlying strata.

Extensive bedded deposits of salt are widely distributed throughout the world and are mined in many countries. Important production comes from China, Russia, Great Britain, Germany, Canada, and Mexico.

The United States is the world's largest producer; salt in commercial amounts is, or has been, produced in every state, either from rock-salt deposits or by evaporation of saline waters. Thick beds of rock salt extend from New York State through Ontario, Canada, and into Michigan. Salt is recovered from these beds at many localities. Notable deposits are also found in Ohio, Kansas, and New Mexico, and in Canada in Nova Scotia and Saskatchewan. Salt is obtained by the evaporation of sea waters in California and Texas and from the waters of the Great Salt Lake in Utah.

Salt is also produced from *salt domes*, nearly vertical pipelike masses of salt that appear to have punched their way upward to the surface from an underlying salt bed. Anhydrite, gypsum, and native sulfur are commonly associated with salt domes. Geophysical prospecting for frequently associated petroleum has located several hundred salt domes along the Gulf coast of Louisiana and Texas and far out into the Gulf itself. Salt domes are also found in Germany, Romania, Spain, and Iran. In Iran, in an

FIG. 9.38. Halite, hopper-shaped crystals.

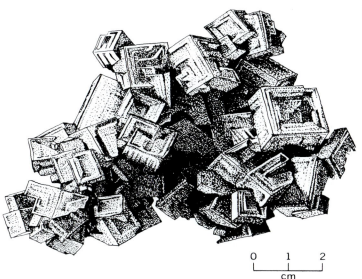

0   1   2
cm

area of complete aridity, salt that has punched its way to the surface is not dissolved but moves down slope as a salt glacier.

Use. Halite finds its greatest use in the chemical industry, where it is the source of sodium and chlorine for the manufacture of hydrochloric acid and a large number of sodium compounds.

Salt is used extensively in the natural state in tanning hides, in fertilizers, in stock feeds, in salting icy highways, and as a weed killer. In addition to its familiar functions in the home, salt enters into the preparation of foods of many kinds, such as the preservation of butter, cheese, fish, and meat.

Name. Halite comes from the Greek word *halos* meaning *salt*.

SYLVITE—KCl

Crystallography. Isometric; $4/m\bar{3}2/m$. Cube and octahedron frequently in combination. Usually in granular crystalline masses showing cubic cleavage; compact.

$Fm3m$; $a = 6.293$ Å; $Z = 4$. ds: 3.15(10), 2.22(6), 1.816(2), 1.407(2), 1.282(4).

Physical Properties. *Cleavage* {001} perfect. **H** 2. **G** 1.99. Transparent when pure. *Color* colorless or white; also shades of blue, yellow, or red from impurities. Readily soluble in water. Salty taste but more bitter than halite. *Optics:* $n = 1.490$.

Composition and Structure. K 52.4, Cl 47.6%. May contain admixed NaCl. Sylvite has the NaCl structure (see Fig. 9.10) but because of the difference in the ionic radii of Na^+ (1.02 Å) and K^+ (1.38 Å) there is little solid solution between KCl and NaCl.

Diagnostic Features. Distinguished from halite by its more bitter taste.

Occurrence. Sylvite has the same origin, mode of occurrence, and associations as halite but is much rarer. It remains in the mother liquor after precipitation of halite and is one of the last salts to be precipitated (see Box 9.2).

It is found in quantity and frequently well crystallized, associated with the salt deposits at Stassfurt, Germany. In the United States it is found in large amount in the Permian salt deposits near Carlsbad, New Mexico, and in western Texas. More recently, deposits have been located in Utah. The most important world reserves are in Saskatchewan, Canada, where extensive bedded deposits have been found at depths greater than 3000 feet.

Use. It is the chief source of potassium compounds, which are principally used as fertilizers. (see Box 10.1).

Name. Potassium chloride is the *sal digestivus Sylvii* of early chemistry, whence the name for the species.

Other Potassium Salts. Several other potassium minerals are commonly associated with sylvite and are found in Germany and Texas in sufficient amount to make them valuable as sources of potassium salts. These are *carnallite*, $KMgCl_3 \cdot 6H_2O$, a usually massive to granular, generally light-colored mineral; *kainite*, $KMg(Cl,SO_4) \cdot 2\frac{3}{4}H_2O$; and *polyhalite*, $K_2Ca_2Mg(SO_4)_4 \cdot 2H_2O$.

Chlorargyrite—AgCl

Crystallography. Isometric; $4/m\bar{3}2/m$. Habit cubic, but crystals are rare. Usually massive, resembling wax; often in plates and crusts.

$Fm3m$; $a = 5.55$ Å; $Z = 4$. ds: 3.20(5), 2.80(10), 1.97(5), 1.67(2), 1.61(2).

Physical Properties. **H** 2–3. **G** 5.5 ±. Sectile, can be cut with a knife; hornlike appearance, hence the name *horn silver*. Transparent to translucent. *Color* pearl-gray to colorless. Rapidly darkens to violet-brown on exposure to light. *Optics:* $n = 2.07$.

Composition and Structure. Ag 75.3, Cl 24.7%. A complete solid solution series exists between AgCl and *bromargyrite*, AgBr. Small amounts of F may be present in substitution for Cl or Br. Some specimens contain Hg. Chlorargyrite is isostructural with NaCl (see Fig. 9.10).

Diagnostic Features. Distinguished chiefly by its waxlike appearance and its sectility.

Occurrence. Chlorargyite is an important supergene ore of silver found in the upper, enriched zone of silver deposits. It is found associated with native silver, cerussite, and secondary minerals in general.

Notable amounts have been found at Broken Hill, Australia; and in Peru, Chile, Bolivia, and Mexico. In the United States chlorargyrite was an important mineral in the mines at Leadville and elsewhere in Colorado, at the Comstock Lode in Nevada, and in crystals at the Poorman's Lode in Idaho.

Use. A silver ore.

Name. Chlorargyrite, from its composition.

Similar Species. Other closely related minerals that are less common but form under similar conditions, are *bromargyrite* AgBr, and *iodian bromargyrite*, Ag(Cl,Br,I), isostructural with chlorargyrite; and *iodargyrite*, AgI, which is hexagonal.

Cryolite—Na₃AlF₆

Crystallography. Monoclinic; $2/m$. Prominent forms are {001} and {110}. Crystals are rare, usually cubic in aspect, and in parallel groupings growing out of massive material. Usually massive.

$P2_1/n$; $a = 5.47$, $b = 5.62$, $c = 7.82$ Å, $\beta = 90°11'$; $Z = 2$. ds: 4.47(2), 3.87(2), 2.75(7), 2.33(4), 1.939(10).

Physical Properties. Parting on {110} and {001} produces cubical forms. **H** $2\frac{1}{2}$. **G** 2.95–3.0. *Luster* vitreous to greasy. *Color* colorless to snow-white. Transparent to translucent. *Optics*: (+), $\alpha = 1.338$, $\beta = 1.338$, $\gamma = 1.339$; $2V = 43°$; $X = b$, $Z \wedge c = -44°$, $r < v$. The low refractive index, near that of water, gives the mineral the appearance of watery snow or paraffin, and causes the powdered mineral to almost disappear when immersed in water.

Composition and Structure. Na 32.8, Al 12.8, F 54.4%. In the structure of cryolite Al is octahedrally coordinated to six F^-. The Na^+ ions are also surrounded by six F^- ions, but in a somewhat less regular pattern. At high temperature (above 550°C) cryolite transforms to an isometric form with space group $Fm3m$.

Diagnostic Features. Characterized by pseudocubic parting, white color, and peculiar luster; and for the Greenland cryolite, the association of siderite, galena, and chalcopyrite.

Occurrence. The only important deposit of cryolite is at Ivigtut, on the west coast of Greenland. Here, in a large mass in granite, it is associated with siderite, galena, sphalerite, and chalcopyrite; and less commonly quartz, wolframite, fluorite, cassiterite, molybdenite, arsenopyrite, columbite. It is found at Miask, Russia; in the United States, at the foot of Pikes Peak, Colorado; and in crystals in Montreal, Quebec, Canada.

Use. Cryolite is used for the manufacture of sodium salts, of certain kinds of glass and porcelain, and as a flux for cleansing metal surfaces. It was early used as a source of aluminum. When bauxite became the ore of aluminum, cryolite was used as a flux in the electrolytic process. Today, with the essential exhaustion of the Ivigtut deposit, the sodium aluminum fluoride used in the aluminum industry is manufactured from fluorite.

Name. Name is derived from two Greek words, *kryos* meaning *frost* and *lithos* meaning *stone*, in allusion to its icy appearance.

FLUORITE—CaF$_2$

Crystallography. Isometric; $4/m\bar{3}2/m$. Usually in cubes, often as penetration twins twinned on [111] (Fig. 9.39a). Other forms are rare, but examples of all forms of the hexoctahedral class have been observed; the tetrahexahedron (Fig. 9.39b) and hexoctahedron (Fig. 9.39c) are characteristic. Usually in crystals or in cleavable masses. Also massive; coarse or fine granular; columnar.

$Fm3m$; $a = 5.46$ Å; $Z = 4$. *ds*: 3.15(9), 1.931(10), 1.647(4), 1.366(1), 1.115(2).

Physical Properties. *Cleavage* {111} perfect, see Plate III, no. 5. **H** 4. **G** 3.18. Transparent to translucent. *Luster* vitreous. *Color* varies widely; most commonly light green (see Plate XII, no. 1, Chapter 13), yellow, bluish-green, or purple; also colorless, white, rose, blue, brown. The color in some fluorite results from the presence of a hydrocarbon. A single crystal may show bands of varying colors; the massive variety is also often banded in color. The phenomenon of fluorescence (see page 27) received its name because it was early observed in some varieties of fluorite. *Optics*: $n = 1.433$.

Composition and Structure. Ca 51.3, F 48.7%. The rare earths, particularly Y and Ce, may substitute for Ca. The fluorite structure is shown in Figs. 9.11 and 3.54.

Diagnostic Features. Determined usually by its cubic crystals and octahedral cleavage; also vitreous luster and usually fine coloring, and by the fact that it can be scratched with a knife.

Occurrence. Fluorite is a common and widely distributed mineral. Usually found in hydrothermal veins in which it may be the chief mineral or as a gangue mineral with metallic ores, especially those of lead and silver. Common in vugs in dolomites and limestone and has been observed also as a minor accessory mineral in various igneous rocks and pegmatites. Associated with many different minerals, as calcite, dolomite, gypsum, celestite, barite, quartz, galena, sphalerite, cassiterite, topaz, tourmaline, and apatite.

Fluorite is found in quantity in England, chiefly from Cumbria, Derbyshire, and Durham; the first two

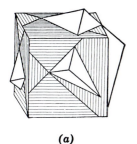

(a)

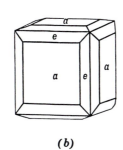

(b)

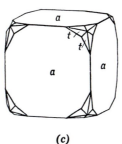

(c)

FIG. 9.39. Fluorite. *(a)* Penetration twin. *(b)* Cube and tetrahexahedron. *(c)* Cube and hexoctahedron.

localities are famous for their magnificent crystallized specimens. Found commonly in the mines of Saxony, Germany. Fine specimens come from the Alps. The large producers of commercial fluorite (fluorspar) are: Mongolia, Russia, Republic of South Africa, China, Spain, and Thailand. Compared with these countries, the production in the United States is small, but there are significant amounts mined in southern Illinois near Rosiclare and Cave-in-Rock. Much of the fluorite at Cave-in-Rock is in coarse crystalline aggregates lining flat, open spaces, and thus the locality is the source of many beautifully crystallized specimens. (Fig. 9.40). Fluorite is also mined in small amounts in Nevada, Texas, and Utah.

Use. The bulk of the fluorite produced is used in the chemical industry (over 50%), mainly in the preparation of hydrofluoric acid, and as a flux in the making of steel (over 40%). Other uses are in the manufacture of glass, fiberglass, pottery, and enamel. Formerly used extensively as an ornamental material and for carving vases and dishes. Small amounts of fluorite are used for lenses and prisms in various optical systems, but most of the optical material is now made synthetically.

Name. From the Latin *fluere*, meaning to flow, because it melts more easily than other minerals with which it was (in the form of cut stones) confused.

Atacamite—$Cu_2Cl(OH)_3$

Crystallography. Orthorhombic; $2/m2/m2/m$. Commonly found in slender prismatic crystals with vertical striations. Also tabular parallel to {010}. Usually in confused crystalline aggregates; fibrous; granular.

Pnam; $a = 6.02$, $b = 9.15$, $c = 6.85$ Å; $Z = 4$. *d*s: 5.48(10), 5.03(7), 2.84(5), 2.78(5), 2.76(6).

Physical Properties. *Cleavage* {010} perfect. **H** $3-3\frac{1}{2}$. **G** 3.75–3.77. *Luster* adamantine to vitreous. *Color* various shades of green. Transparent to translucent. *Optics:* $(-)$, $\alpha = 1.831$, $\beta = 1.861$, $\gamma = 1.880$; $2V = 75°$; $r < v$. $X = b$, $Y = a$.

Composition and Structure. Cu 14.88, CuO 55.87, Cl 16.60, H_2O 12.65%. In the structure of atacamite part of the Cu atoms is in 6-coordination with five (OH) groups and one Cl. The remaining Cu is in 6-coordination with four (OH) groups and two Cl.

Diagnostic Features. Characterized by its green color and granular crystalline aggregates. Distinguished from malachite by its lack of effervescence in acid.

Occurrence. Atacamite is a comparatively rare copper mineral. Found originally as sand in the province of Atacama in Chile. Occurs in arid regions as a supergene mineral in the oxidized zone of copper deposits. It is associated with other secondary minerals in various localities in Chile, especially Chuquicamata, and in some of the copper districts of South Australia. In the United States occurs sparingly in the copper districts of Arizona.

Use. A minor ore of copper.

Name. From the province of Atacama, Chile.

FIG. 9.40. Yellow fluorite cubes coated by white quartz crystals, Cave-in-Rock, Harding County, Illinois (Harvard Mineralogical Museum).

REFERENCES AND SELECTED READING

Anthony, J. W., R. A. Bideaux, K. W. Bladh, and M. C. Nichols. 1997. *Handbook of mineralogy.* Vol. 3, *Halides, hydroxides, oxides.* Tucson, Ariz.: Mineral Data Publications.

Bragg, L., and G. F. Claringbull. 1965. *Crystal structures of minerals.* London: G. Bell and Sons, Ltd.

Deer, W. A., R. A. Howie, and J. Zussman. 1962. *Rock forming minerals.* Vol. 5: *Non-Silicates.* New York: Wiley.

Evans, R. C. 1966. *An introduction to crystal chemistry. 2nd ed.* Cambridge, England: Cambridge University Press.

Oxide minerals. *1976. Reviews in Mineralogy 3.* Mineralogical Society of America, Washington, D. C.

Oxide minerals: Petrologic and magnetic significance. *1991. Reviews in Mineralogy 25.* Edited by D. H. Lindsley. Mineralogical Society of America, Washington, D. C.

Palache, C., H. Berman, and C. Frondel. 1944 and 1951. *The system of mineralogy. 7th ed.* Vols. 1 and 2. New York: Wiley.

CHAPTER 10

CRYSTAL CHEMISTRY AND SYSTEMATIC DESCRIPTIONS OF CARBONATES, NITRATES, BORATES, SULFATES, CHROMATES, TUNGSTATES, MOLYBDATES, PHOSPHATES, ARSENATES, AND VANADATES

Carbonates, nitrates, borates, sulfates, chromates, tungstates, molybdates, phosphates, arsenates, and vanadates together number 1395 species, but most of them are uncommon minerals and only a small number will be considered here. The five largest groups are as follows: phosphates with 409 species, sulfates with 286 species, arsenates with 228 species, carbonates with 213 species, and borates with 125 species. In the prior two chapters it was noted that among the minerals discussed in those two chapters there are a large number of important ore minerals. This chapter, as well as the two subsequent chapters on silicates, contains large numbers of minerals that are described as industrial minerals. Industrial minerals (or materials) are defined as any rock, mineral, or other naturally occurring substance of economic value, exclusive of metallic ores, mineral fuels, and gemstones; they are also referred to as the nonmetallics.

These chemically diverse minerals are treated together because most of them contain *anionic complexes*, which can be recognized as strongly bonded units in their structures. Examples of such anionic complexes are: $(CO_3)^{2-}$ in carbonates, $(NO_3)^{1-}$ in nitrates, $(PO_4)^{3-}$ in phosphates, $(SO_4)^{2-}$ in sulfates, $(CrO_4)^{2-}$ in chromates, $(WO_4)^{2-}$ in tungstates, and $(AsO_4)^{3-}$ in arsenates. The bond strengths within such anionic complexes are always stronger than those between the anionic complex and other ions of the structure; these compounds

are therefore referred to as *anisodesmic* (see page 78). For example, in the carbonates, radius ratio considerations predict three closest oxygen neighbors about carbon. This arrangement is triangular with carbon at the center and oxygen at each of the corners of the triangle (see Fig. 10.1); the (NO_3) group is also triangular. The e.v. of the bonds between carbon and each of the three closest oxygens is $\frac{1}{3} \times 4 = 1\frac{1}{3}$. This means that each oxygen has a residual charge of e.v. $= \frac{2}{3}$ for bonding other ions in the carbonate structure. Figure 10.1 illustrates the

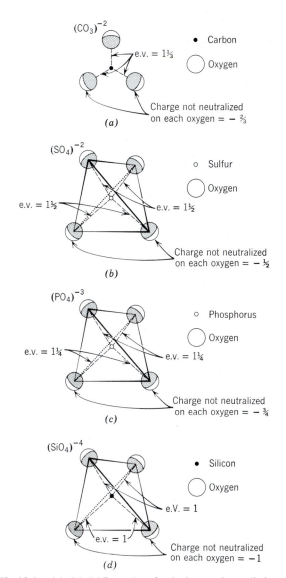

FIG. 10.1. *(a), (b), (c)* Examples of anionic complexes, their bond strengths between the central cation and oxygen, and the residual charges on the oxygens. *(d)* The tetrahedral (SiO_4) group in which the e.v.'s between oxygen and the central cation are the same as the residual charge on the oxygen ($= 1$).

anisodesmic character of structures with tetrahedral (PO_4) and (SO_4) anionic groups as well. Borates with triangular (BO_3) groups and silicates and tetrahedral (SiO_4) groups are examples of *mesodesmic* bonding (see page 79).

CRYSTAL CHEMISTRY OF CARBONATES

The anionic $(CO_3)^{2-}$ complexes of carbonates are strongly bonded units and do not share oxygens with each other (as noted above, the residual e.v. of $\frac{2}{3}$ does not allow this). The triangular carbonate groups

are the basic building units of all carbonate minerals and are largely responsible for the properties peculiar to the group.

Although the bond between the central carbon and its coordinated oxygens in the (CO_3) group is strong, it is not as strong as the covalent bond in CO_2. In the presence of hydrogen ion, the carbonate group becomes unstable and breaks down to yield CO_2 and water, according to $2H^+ + CO_3 \rightarrow H_2O + CO_2$. This reaction is the cause of the familiar "fizz" test with acid, which is widely used in the identification of carbonates.

The important anhydrous carbonates fall into three structurally different groups: the *calcite group*, the *aragonite group*, and the *dolomite group*. Aside from the minerals in these groups, the hydrous copper carbonates, azurite and malachite, are the only important carbonates.

Calcite Group		Aragonite Group	
(Hexagonal; $R\bar{3}c$)		(Orthorhombic; *Pmcn*)	
Calcite	$CaCO_3$	Aragonite	$CaCO_3$
Magnesite	$MgCO_3$	Witherite	$BaCO_3$
Siderite	$FeCO_3$	Strontianite	$SrCO_3$
Rhodochrosite	$MnCO_3$	Cerussite	$PbCO_3$
Smithsonite	$ZnCO_3$		

Dolomite Group	
(Hexagonal; $R\bar{3}$)	
Dolomite	$CaMg(CO_3)_2$
Ankerite	$CaFe(CO_3)_2$

Monoclinic Carbonates with (OH)	
Malachite	$Cu_2CO_3(OH)_2$
Azurite	$Cu_3(CO_3)_2(OH)_2$

Calcite Group

The above five members of the calcite group are isostructural with space groups $R\bar{3}c$. The structure of calcite, one of the earliest to be analyzed by X-rays by W. L. Bragg in 1914 (see Fig. 10.2*a*), can be thought of as a derivative of the NaCl structure in which triangular (CO_3) groups replace the spherical Cl, and Ca is in place of Na. The triangular shape of the (CO_3) groups causes the resulting structure to be rhombohedral instead of isometric as in NaCl. The (CO_3) groups lie in planes at right angles to the 3-fold (*c*) axis (Fig. 10.2*a*) and the Ca ions, in alternate planes, are in 6-coordination with oxygens of the (CO_3) groups. Each oxygen is coordinated to two Ca ions as well as to a carbon ion at the center of the (CO_3) group.

Calcite shows perfect rhombohedral cleavage to which, traditionally, the indices of $\{10\bar{1}1\}$ have been assigned. In the morphological descriptions and indexing of forms of the calcite and dolomite group minerals that follow, this convention has been preserved. However, X-ray structural determinations have shown that this rhombohedron does not correspond to the correct unit cell and that the simplest unit cell is a much steeper rhombohedron (Fig. 10.2*b*). Therefore, the structural axial ratios differ from the morphological. It should be noted that the radius ratio of Ca : O (= 0.714) in $CaCO_3$ is so close to the limiting value between 6- and 8-coordination (0.732) that $CaCO_3$ can occur in two structure types: *calcite*, with 6-coordination of Ca to O and *aragonite*, with 9-coordination of Ca to O.

Aragonite Group

When the (CO_3) group is combined with large divalent cations (ionic radii greater than 1.0 Å), the radius ratios generally do not permit stable 6-coordination and orthorhombic structures result. This is the *aragonite structure type* (Fig. 10.3 and Table 3.13, p. 84) with space group *Pmcn*. $CaCO_3$ occurs in both the *calcite* and *aragonite structure types* because although Ca is somewhat large for 6-coordination (calcite), it is relatively small, at room temperature, for 9-coordination (aragonite); calcite is the stable form of $CaCO_3$ at room temperature (see Fig. 10.4). Carbonates with larger cations such as $BaCO_3$, $SrCO_3$, and $PbCO_3$, however, have the aragonite structure, stable at room temperature. In the aragonite structure the (CO_3) groups as in calcite, lie perpendicular to the *c* axis, but in two structural planes, with the (CO_3) triangular groups of one plane pointing in opposite directions to those of the other. In calcite all (CO_3) groups lie in a single structural plane and point in the same direction (Fig. 10.2*a*). Each Ca is surrounded by nine closest oxygens. The cations have an arrangement in the structure approximating hexagonal closest packing, which gives rise to marked pseudohexagonal symmetry. This is reflected in both the crystal angles and in the pseudohexagonal twinning which is characteristic of all members of the group.

Solid solution within the aragonite group is somewhat more limited than in the calcite group, and it is interesting to note that Ca and Ba, respectively the smallest and largest ions in the group, form an ordered compound *barytocalcite*, $BaCa(CO_3)_2$, analogous to the occurrence of dolomite, $CaMg(CO_3)_2$, in the system $CaCO_3$–$MgCO_3$. The differences in physical properties of the minerals of the

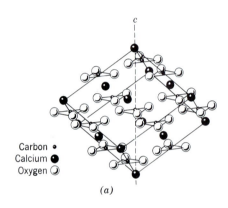

Carbon ●
Calcium ●
Oxygen ◐

(a)

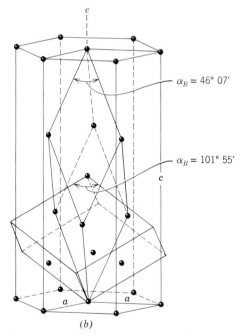

$\alpha_R = 46°\ 07'$

$\alpha_R = 101°\ 55'$

(b)

FIG. 10.2. *(a)* Structure of calcite, $CaCO_3$. *(b)* The relation of the steep, true unit cell to the cleavage rhombohedron, which is face-centered. A hexagonal cell (rhomb-based prism) is also shown.

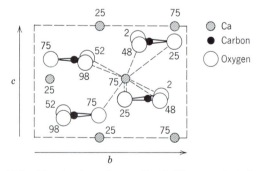

Ca ◍
Carbon ●
Oxygen ○

FIG. 10.3. The structure of aragonite, $CaCO_3$, as projected on (100). Oxygens that would normally superimpose have been made visible by some displacement. Numbers represent heights of atomic positions above the plane of origin, marked with respect to *a*. Dashed rectangle outlines the unit cell. Ca–O bonds are also shown.

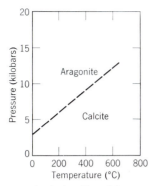

FIG. 10.4. The approximate location of the experimentally determined stability fields of calcite and aragonite.

aragonite group are conferred largely by the cations. Thus the specific gravity is roughly proportional to the atomic weight of the metal ions (Table 2.2).

Dolomite Group

The dolomite group includes *dolomite*, $CaMg(CO_3)_2$, *ankerite*, $CaFe(CO_3)_2$, and *kutnahorite*, $CaMn(CO_3)_2$. These three carbonates are isostructural with space group $R\bar{3}$. The structure of dolomite is similar to that of calcite but with Ca and Mg layers alternating along the *c* axis. The large difference in size of the

Ca^{2+} and Mg^{2+} ions (33%) causes *cation ordering* with the two cations in specific, and separate levels in the structure. With the nonequivalence of Ca and Mg layers, the 2-fold rotation axes of calcite do not exist and the symmetry is reduced to that of the rhombohedral class, $\bar{3}$. The composition of dolomite is intermediate between $CaCO_3$ and $MgCO_3$ with Ca : Mg = 1 : 1. The occurrence of this ordered compound, however, does not imply that solid solution exists between $CaCO_3$ and $MgCO_3$ (see Fig. 10.5). In the dolomite structure, especially at low temperatures, each of the two divalent cations occupies a structurally distinct position. At higher temperatures (above 700°C) dolomite shows small deviations from the composition with Ca : Mg = 1 : 1 as shown in Fig. 10.6. This diagram also shows that at elevated temperatures calcite coexisting with dolomite becomes more magnesian. Similarly dolomite in dolomite–calcite pairs becomes somewhat more calcic. At temperatures above about 1000° to 1100°C a complete solid solution exists between calcite and dolomite, but not between dolomite and magnesite. The compositions of coexisting dolomite and calcite have been used for an estimate of temperatures of crystallization of rocks containing both carbonates, using the temperature scale in Fig. 10.6.

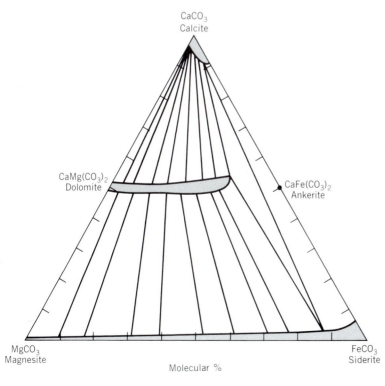

CaCO₃
Calcite

CaMg(CO₃)₂
Dolomite

CaFe(CO₃)₂
Ankerite

MgCO₃
Magnesite

FeCO₃
Siderite

Molecular %

FIG. 10.5. Carbonates and the extent of their solid solution in the system $CaO–MgO–FeO–CO_2$. The extent of solid solution series in this diagram is based upon chemical analyses of carbonates in metamorphic rocks that have been metamorphosed to about 400°C (biotite zone of the greenschist facies). Tielines connect commonly coexisting carbonate species. Calcite-dolomite coexistences are common in Mg-containing limestones; ankerite-siderite coexistences are found in banded iron-formations. (Adapted from L. M. Anovitz, and E. J. Essence, 1987, Phase equilibria in the system $CaCO_3$-$MgCO_3$-$FeCO_3$. *Journal of Petrology*, v. 28, pp. 389–415.)

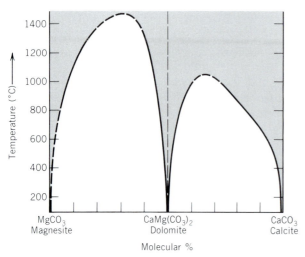

FIG. 10.6. $CaCO_3$–$MgCO_3$ system at CO_2 pressures sufficient to prevent decomposition of the carbonates. Vertical dashed line is ideal dolomite composition. Solid solution is shown by shading. (Adapted from L. M. Anovitz, and E. J. Essene, 1987, Phase equilibria in the system $CaCO_3$–$MgCO_3$–$FeCO_3$. *Journal of Petrology* 28:389–415.)

CRYSTAL CHEMISTRY OF NITRATES

The minerals in this group are structurally similar to the carbonates with planar, triangular $(NO_3)^{1-}$ groups, much like the $(CO_3)^{2-}$ group. Like C in the (CO_3) group, the highly charged and highly polarizing N^{5+} ion binds its three coordinated oxygens into a close-knit group in which the strength of the oxygen-nitrogen bond (e.v. = $1\frac{2}{3}$) is greater than any other possible bond in the structure. Because of the greater strength of this N–O bond as compared with the C–O bond, nitrates are less readily decomposed by acids than carbonates. There are eight nitrate minerals but, with the exception of nitratite and niter, they are very rare.

CRYSTAL CHEMISTRY OF BORATES

Within the borate group of minerals BO_3 units are capable of polymerization (similar to the polymerization of SiO_4 tetrahedral groups in the silicates) in the form of chains, sheets, and isolated multiple groups (Fig. 10.7). This is possible because the small B^{3+} ion, which generally coordinates three oxygens in a triangular group, has bond strengths to each O with an e.v. = 1; this is exactly half the bonding energy of the oxygen ion. This permits a single oxygen to be shared between two boron ions linking the BO_3 triangles into expanded structural units (double triangles, triple rings, sheets, and chains). Because the triangular coordination of BO_3 is close to the upper stability limit of

3-coordination, boron is found also in 4-coordination in tetrahedral groups. In addition to BO_3 and BO_4 groups, natural borates may contain complex ionic groups such as $[B_3O_3(OH)_5]^{2-}$ (Fig. 10.7e) that consist of one triangle and two tetrahedra. In the structure of *colemanite*, $CaB_3O_4(OH)_3 \cdot H_2O$, complex infinite chains of tetrahedra and triangles occur, and in *borax*, $Na_2B_4O_5(OH)_4 \cdot 8H_2O$, a complex ion, $[B_4O_5(OH)_4]^{2-}$ consisting of two tetrahedra and two triangles is found (Fig. 10.7f). Borates can be classified on the basis of the structural anionic linking (or lack thereof) as insular (independent single or double BO_3 or BO_4 groups), chain, sheet, and framework structures.

Although it is possible to prepare a three-dimensional boxwork made up of BO_3 triangles only and having the composition B_2O_3, such a configuration has a very low stability and disorders readily, yielding a glass. Because of the tendency to form somewhat disordered networks of BO_3 triangles, boron is regarded as a "network-former" in glass manufacture and is used in the preparation of special glasses of light weight and high transparency to energetic radiation.

CRYSTAL CHEMISTRY OF SULFATES

In the discussion of the structures of sulfide minerals we have seen that sulfur occurs as the large, divalent sulfide anion. This ion results from the filling by captured electrons of the two vacancies in the outer, or valence, electron shell. The six electrons normally present in this shell may be lost, giving rise to a small, highly charged and highly polarizing positive ion (radius = 0.12 Å). It occurs in tetrahedral coordination with surrounding oxygens. The sulfur to oxygen bond in such an ionic group is very strong (e.v. = $1\frac{1}{2}$; see Fig. 10.1b) and covalent in its properties and produces tightly bound groups that are not capable of sharing oxygens. These anionic $(SO_4)^{2-}$ groups are the fundamental structure units of the sulfate minerals.

The most important and common of the anhydrous sulfates are members of the *barite group* (space group *Pnma*), with large divalent cations coordinated with the sulfate ion. Part of the structure of *barite*, $BaSO_4$, is illustrated in Fig. 10.8. Each barium ion is coordinated to 12 oxygen ions belonging to seven different (SO_4) groups. The barite type structure is also found in manganates (with MnO_4 tetrahedral groups) and chromates (with CrO_4 tetrahedral groups) which contain large cations.

Anhydrite, $CaSO_4$, because of the smaller size of Ca^{2+} compared to Ba^{2+}, has a very different structure from that of barite (see Fig. 10.9). Each Ca^{2+} is coor-

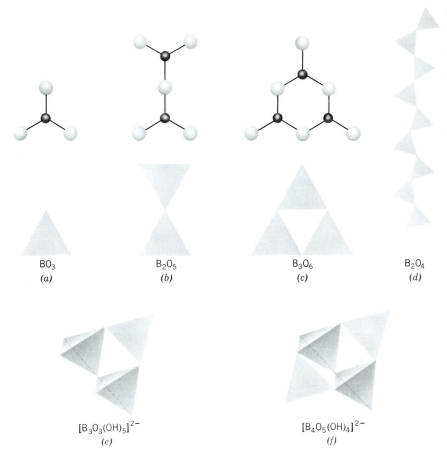

BO₃
(a)

B₂O₅
(b)

B₃O₆
(c)

B₂O₄
(d)

$[B_3O_3(OH)_5]^{2-}$
(e)

$[B_4O_5(OH)_4]^{2-}$
(f)

FIG. 10.7. Independent BO_3 triangles (a), multiple groups (b) and (c), and chains (d) in borates. Also complex triple and quadruple groups (e) and (f). The triple group in (e) consists of two $BO_2(OH)_2$ tetrahedra and a BO_2OH triangle. The group in (f) contains two BO_3OH tetrahedra and two BO_2OH triangles; this type of group is present in the structure of borax (see Fig. 10.32.)

dinated to eight nearest neighbor oxygens from the tetrahedral (SO_4) groups.

Of the hydrous sulfates, *gypsum*, $CaSO_4 \cdot 2H_2O$, is the most important and abundant. The structure of gypsum is illustrated in Fig. 10.10. Gympsum is monoclinic with space group $C2/c$. The structure consists of layers parallel to {010} of $(SO_4)^{2-}$ groups strongly bonded to Ca^{2+}. Successive layers of this type are separated by sheets of H_2O molecules. The bonds between H_2O molecules in neighboring sheets are weak, which explains the excellent {010} cleavage in gypsum. Loss of the water molecules causes collapse of the structure and transformation into a metastable polymorph of anhydrite ($\gamma CaSO_4$) with a large decrease in specific volume and loss of the perfect cleavage.

FIG. 10.8. The structure of barite, $BaSO_4$, as projected on (010). The dashed lines outline the unit cell. The oxygen atoms, which would normally superimpose, have been displaced from their positions in this projection.

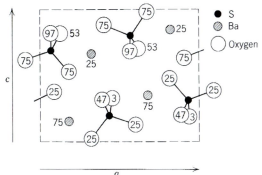

FIG. 10.9. The structure of anhydrite, $CaSO_4$.

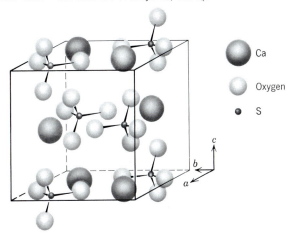

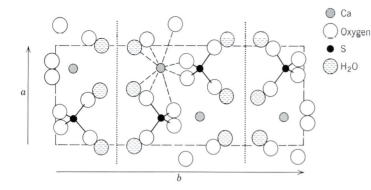

Ca
Oxygen
S
H_2O

FIG. 10.10. The structure of gypsum, $CaSO_4 \cdot 2H_2O$, projected on (001). Perfect (010) cleavage is indicated by dotted lines. Dashed lines outline unit cell. The 8-coordination (six oxygens and two water molecules) is indicated for one Ca^{2+}.

CRYSTAL CHEMISTRY OF TUNGSTATES AND MOLYBDATES

W^{6+} and Mo^{6+} are considerably larger than S^{6+} and P^{5+}. Hence, when these ions enter into anisodesmic ionic groups with oxygen, the four coordinated oxygen ions do not occupy the apices of a regular tetrahedron, as is the case in the sulfates and phosphates, but form a somewhat flattened grouping of square outline. Although W (at. wt. 184) has a much greater atomic weight than Mo(96), both belong to the same family of the periodic table and, because of the lanthanide contraction, have the same ionic radii. As a result, each may freely substitute for the other as the coordinating cation in the warped tetrahedral oxygen groupings. In nature, however, the operation of the processes of geochemical differentiation often separates these elements, and it is not uncommon to find primary tungstates almost wholly free of Mo and vice

versa. In secondary minerals, the two elements are more commonly in solid solution with each other.

The minerals of this chemical class fall mainly into two isostructural groups. The *wolframite group* contains compounds with fairly small divalent cations such as Fe^{2+}, Mn^{2+}, Mg, Ni, and Co in 6-coordination with (MoO_4). Complete solid solution occurs between Fe^{2+} and Mn^{2+} in the minerals of this group.

The *scheelite group* contains compounds of larger ions such as Ca^{2+} and Pb^{2+} in 8-coordination with $(WO_4)^{2-}$ and $(MoO_4)^{2-}$ groups. The structure of scheelite (see Fig. 10.11) is close to that of anhydrite and zircon, $ZrSiO_4$, but differs from these in the manner of linking of the CaO_8 polyhedra. The (WO_4) tetrahedra are somewhat flattened along the *c* axis and join edges with CaO_8. W and Mo may substitute for each other, forming partial series between *scheelite*, $CaWO_4$, and *powellite*, $CaMoO_4$; and *stolzite*, $PbWO_4$, and *wulfenite*, $PbMoO_4$. The substitution of Ca and Pb for one another forms partial series between scheelite and stolzite and between powellite and wulfenite.

FIG. 10.11. Structure of scheelite, $CaWO_4$.

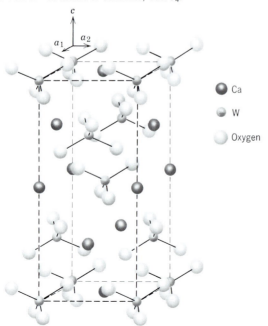

Ca
W
Oxygen

CRYSTAL CHEMISTRY OF PHOSPHATES, ARSENATES, AND VANADATES

P^{5+} is only slightly larger than S^{6+} and, hence, like sulfur, forms a tetrahedral anionic $(PO_4)^{3-}$ group with oxygen (see Fig. 10.1). All phosphates contain this phosphate anionic complex as the fundamental building unit. Similar tetrahedral units, $(AsO_4)^{3-}$ and $(VO_4)^{3-}$ occur in arsenates and vanadates. P^{5+}, As^{5+}, and V^{5+} may substitute for each other in the anionic groups. This type of substitution is best shown in the pyromorphite series of the apatite group. *Pyromorphite*, $Pb_5(PO_4)_3Cl$, *mimetite*, $Pb_4(AsO_4)_3Cl$, and *vanadinite*, $Pb_5(VO_4)_3Cl$, are isostructural, and all gradations of composition between the end members exist.

The structure of *apatite*, $Ca_5(PO_4)_3(OH,F,Cl)$, which is the most important and abundant phosphate,

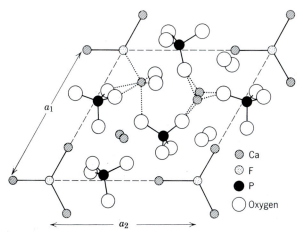

FIG. 10.12. Structure of fluorapatite, $Ca_5(PO_4)_3F$, projected on the (0001) plane. The dashed parallelogram outlines the base of the unit cell. The tetrahedral (PO_4) groups, triangular coordination of F to Ca, and examples of the two types of coordination about Ca are shown.

is illustrated in Fig. 10.12. The oxygens of the (PO_4) groups are linked to Ca in two different structural sites. In one site Ca is in irregular 9-coordination, and in the other in irregular 8-coordination. Each fluorine (or Cl or OH) lies in a triangle with three calciums. Apatite shows extensive solid solution with respect to anions as well as cations. (PO_4) may be substituted for

by (AsO_4) or (VO_4) as noted above, but also in part by tetrahedral (CO_3OH) groups, giving rise to *carbonate–apatite*, $Ca_5F(PO_4,CO_3OH)_3$. Small amounts of (SiO_4) and (SO_4) may also be present in substitution for (PO_4); these types of substitution must be coupled with other cation substitutions in apatite in order to retain the electrical neutrality of the structure. F may be replaced by (OH) or Cl producing *hydroxylapatite*, $Ca_5(PO_4)_3(OH)$ and *chlorapatite*, $Ca_5(PO_4)_3Cl$. Mn^{2+} and Sr^{2+} may substitute for Ca. These varied ionic substitutions are typical of the phosphates which generally have rather complicated structures.

SYSTEMATIC DESCRIPTIONS

Carbonates

Of the 213 carbonates known only 13 are described on the following pages.

Calcite—CaCO₃
Crystallography. Hexagonal; $\bar{3}2/m$. Crystals are extremely varied in habit and often highly complex. More than 300 different forms have been described (Fig. 10.13). Three important habits exist: (1) prismatic, in long or short prisms, in which the prism faces are prominent, with base or rhombohedral ter-

FIG. 10.13. Calcite crystals. Forms. $c\{0001\}$, $m\{10\bar{1}0\}$, $e\{01\bar{1}2\}$, $r\{10\bar{1}1\}$, $f\{02\bar{2}1\}$, $v\{2\bar{1}\bar{3}1\}$.

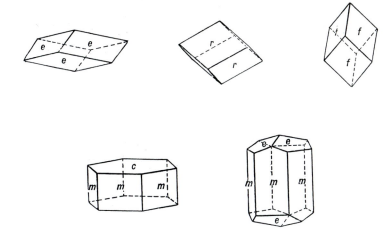

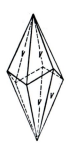

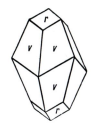

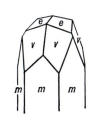

minations; (2) rhombohedral, in which rhombohedral forms predominate; the unit (cleavage) form *r* is not common; (3) scalenohedral, in which scalenohedrons predominate, often with prism faces and rhombohedral truncations. The most common scalenohedron is {$21\bar{3}1$}. All possible combinations and variations of these types are found.

Twinning with the twin plane, {$01\bar{1}2$}, very common (Fig. 10.14*a*); often produces twinning lamellae that may, as in crystalline limestones, be of secondary origin. This twinning may be produced artificially (see page 157). Twins with {0001}, the twin plane common (Fig. 10.14*b*). Calcite is usually in crystals or in coarse- to fine-grained aggregates. Also fine-grained to compact, earthy, and stalactitic.

R$\bar{3}$c. Hexagonal cell, *a* = 4.99, *c* = 17.06 Å; *Z* = 6; rhombohedral cell, *a* = 6.37 Å, α (rhombohedral angle) = 46°05′; *Z* = 2. *d*s: 3.04(10), 2.29(2), 2.10(2), 1.913(2), 1.875(2).

Physical Properties. *Cleavage* {$10\bar{1}1$} perfect (cleavage angle = 74°55′). *Parting* along twin lamellae on {$01\bar{1}2$}. **H** 3 on cleavage, $2\frac{1}{2}$ on base. **G** 2.71. *Luster* vitreous to earthy. *Color* usually white to colorless, but may be variously tinted, gray, red, green, blue, yellow; also when impure, brown to black. Transparent to translucent. The chemically pure and optically clear, colorless variety is known as *Iceland spar* because of its occurrence in Iceland, see Plate III, no. 6. *Optics:* (−); ω = 1.658, ε = 1.486.

Composition and Structure. Most calcites tend to be relatively close to pure $CaCO_3$ with CaO 56.0 and CO_2 44.0%. Mn^{2+}, Fe^{2+}, and Mg may substitute for Ca and a complete solid solution series extends to rhodochrosite, $MnCO_3$, above 550°C; a very partial series, with up to 5 weight percent FeO in calcite, exists between calcite and siderite, $FeCO_3$.

Some inorganic calcites may contain from 0 to about 2 weight percent MgO. Calcites in the hard parts of living organisms, however, may show a range of $MgCO_3$ of 2 to 16 molecular percent. See Fig. 10.5 for solid solution series in the system $CaO–MgO–FeO–CO_2$. The structure of calcite is shown in Fig. 10.2*a* and discussed on page 405.

Diagnostic Features. Fragments effervesce readily in cold dilute HCl. Characterized by its hardness (3), rhombohedral cleavage, light color, vitreous luster. Distinguished from dolomite by the fact that coarse fragments of calcite effervesce freely in cold HCl and distinguished from aragonite by lower specific gravity and rhombohedral cleavage.

Occurrence. As a *rock-forming mineral:* Calcite is one of the most common widespread minerals. It occurs in extensive sedimentary rock masses in which it is the predominant mineral; in *limestones*, it is essentially the only mineral present. Crystalline, metamorphosed limestones are *marbles. Chalk* is a fine-grained pulverulent deposit of calcium carbonate. Calcite is an important constituent of calcareous marls and calcareous sandstones. Limestone has, in great part, been formed by the deposition on a sea bottom of great thicknesses of calcareous material in the form of shells and skeletons of sea animals. A smaller proportion of these rocks has been formed directly by precipitation of calcium carbonate.

As cave deposits, and so on: Waters carrying calcium carbonate in solution and evaporating in limestone caves often deposit calcite as stalactites, stalagmites, and incrustations. Such deposits, usually semitranslucent and of light yellow colors, are often beautiful and spectacular. An example is Carlsbad Caverns, New Mexico. Both hot and cold calcareous spring water may form cellular deposits of calcite known as *travertine*, or *tufa*, around their mouths. The deposit at Mammoth Hot Springs, Yellowstone Park, is more spectacular than most, but of similar origin. *Onyx marble* is banded calcite and/or aragonite used for decorative purposes. Because much of this material comes from Baja California, Mexico, it is also called Mexican onyx.

Siliceous calcites: Calcite crystals may enclose considerable amounts of quartz sand (up to 60%) and form what are known as sandstone crystals. Such occurrences are found at Fontainebleau, France (Fontainebleau limestone), and in the Bad Lands, South Dakota.

Calcite occurs as a primary mineral in some igneous rocks such as carbonatites and nepheline syenites. It is a late crystallization product in the cavities in lavas. It is also a common mineral in hydrothermal veins associated with sulfide ores.

FIG. 10.14. Twinned calcite crystals. Twin planes: *(a)* {$01\bar{1}2$}. *(b)* {0001}.

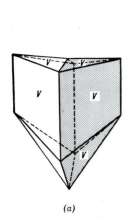
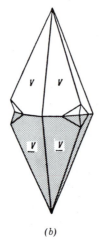

(a)   *(b)*

It is impossible to specify all the important districts for the occurrence of calcite in its various forms. Some of the more notable classic localities in which finely crystallized calcite is found are as follows: Andreasberg in the Harz Mountains, Germany; in Cumbria (Fig. 10.15), and Lancashire, England; Iceland; and Guanajuato, Mexico. In the United States at Joplin, Missouri, and the Lake Superior copper district.

Use. The most important use for calcite is for the manufacture of cements and lime for mortars. Limestone is the chief raw material, which when heated to about 900°C forms *quicklime*, CaO, by the reaction: $CaCO_3 \rightarrow CaO + CO_2^{\nearrow}$. The CaO, when mixed with water, forms one or several CaO-hydrates (slaked lime), swells, gives off much heat, and hardens or, as commonly termed, *sets*. Quicklime when mixed with sand forms common mortar.

The greatest consumption of limestone is in the manufacture of cements. The type known as Portland cement is most widely produced. It is composed of about 75% calcium carbonate (limestone) with the remainder essentially silica and alumina. Small amounts of magnesium carbonate and iron oxide are also present. In some limestone, known as *cement rocks*, the correct proportions of silica and alumina are present as impurities. In others these oxides are contributed by clay or shale mixed with the limestone before "burning." When water is mixed with cement, hydrous calcium silicates and calcium aluminates are formed.

Limestone is a raw material for the chemical industry, and finely crushed is used as a soil conditioner, for whiting and whitewash. Great quantities are quarried each year as a flux for smelting various metallic ores, as an aggregate in concrete, and as road material. A fine-grained limestone is used in lithography.

Calcite in several forms is used in the building industry. Limestone and marble as dimension stone are used both for construction purposes and for decorative exterior facings. Polished slabs of travertine and Mexican onyx are commonly used as ornamental stone for interiors. Indiana is the chief source of building limestone (the Salem limestone) in the United States, with the most productive quarries in Lawrence and Monroe counties. Many of the federal buildings in Washington, D.C., have been constructed from this limestone. The most important marble quarries are in Vermont, New York, Georgia, and Tennessee.

Iceland spar, named for its occurrence in Iceland, is valuable for various optical instruments; its best known use was in the form of the Nicol prism to produce polarized light, prior to the use of Polaroid plates.

Name. From the Latin word *calx*, meaning burnt lime.

Magnesite—$MgCO_3$

Crystallography. Hexagonal; $\overline{3}2/m$. Crystals are rhombohedral, $\{10\overline{1}1\}$, but rare. Usually cryptocrystalline in white, compact, earthy masses, less frequently in cleavable granular masses, coarse to fine.

$R\overline{3}c$. Hexagonal cell, $a = 4.63$, $c = 15.02$ Å; $Z = 6$; rhombohedral cell, $a = 5.62$ Å, α (rhombohedral angle) $= 48°10'$, $Z = 2$. ds: 2.74(10), 2.50(2), 2.10(4), 1.939(1), 1.700(3).

Physical Properties. *Cleavage* $\{10\overline{1}1\}$ perfect. **H** $3\frac{1}{2}$–5. **G** 3.0–3.2. *Luster* vitreous. *Color* white, gray, yellow, brown. Transparent to translucent. *Optics:* $(-)$; $\omega = 1.700$, $\epsilon = 1.509$.

FIG. 10.15. Group of calcite crystals, Cumbria, England (Harvard Mineralogical Museum).

Composition and Structure. MgO 47.8, CO_2 52.2%. Fe^{2+} substitutes for Mg and a complete series extends to siderite (see Figs. 10.5 and 10.6). Small amounts of Ca and Mn may be present. Magnesite is isostructural with calcite (see Fig. 10.2).

Diagnostic Features. Cleavable varieties are distinguished from dolomite by higher specific gravity and absence of abundant calcium. The white massive variety resembles chert and is distinguished from it by inferior hardness. Scarcely acted upon by cold HCl, but dissolves with effervescence in hot HCl.

Occurrence. Magnesite commonly occurs in veins and irregular masses derived from the alteration of Mg-rich metamorphic and igneous rocks (serpentinites and peridotites) through the action of waters containing carbonic acid. Such magnesites are compact, cryptocrystalline, and often contain opaline silica. Beds of crystalline cleavable magnesite are (1) of metamorphic origin associated with talc schists, chlorite schists, and mica schists, and (2) of sedimentary origin, formed as a primary precipitate or as a replacement of limestones by Mg-containing solutions, dolomite being formed as an intermediate product.

Notable deposits of the sedimentary type of magnesite are in China; at Satka in the Ural Mountains, Russia; and at Styria, Austria. The most famous deposit of the cryptocrystalline type is on the Island of Euboea, Greece. The best crystals originate from veins in Oberdorf, Austria, and Bahia, Brazil.

In the United States the compact variety is found in irregular masses in serpentine in the Coast Range, California. The sedimentary type is mined at Chewelah in Stevens County, Washington, and in the Paradise Range, Nye County, Nevada. There are numerous minor localities in the eastern United States in which magnesite is associated with serpentine, talc, or dolomite rocks.

Use. Dead-burned magnesite, MgO, that is, magnesite that has been calcined at a high temperature and contains less than 1% CO_2, is used in manufacturing bricks for furnace linings. Magnesite is the source of magnesia for industrial chemicals. It has also been used as an ore of metallic Mg, but at present the entire production of Mg comes from brines and seawater.

Name. Magnesite is named in allusion to the composition.

Siderite—FeCO$_3$

Crystallography. Hexagonal; $\overline{3}2/m$. Crystals are usually unit rhombohedrons, frequently with curved faces. In globular concretions. Usually cleavable granular. May be botryoidal, compact, and earthy.

$R\overline{3}c$. Hexagonal cell, $a = 4.72$, $c = 15.45$ Å; $Z = 6$; rhombohedral cell, $a = 5.83$ Å, $\alpha = 47°45'$, $Z = 2$. ds: 3.59(6), 2.79(10), 2.13(6), 1.963(6), 1.73(8).

Physical Properties. *Cleavage* $\{10\overline{1}1\}$ perfect. **H** $3\frac{1}{2}$–4. **G** 3.96 for pure $FeCO_3$, but decreases with presence of Mn^{2+} and Mg. *Luster* vitreous. *Color* usually light to dark brown. Transparent to translucent. *Optics:* (−); $\omega = 1.875$, $\epsilon = 1.633$.

Composition and Structure. For pure $FeCO_3$, FeO 62.1, CO_2 37.9%. Fe 48.2%. Mn^{2+} and Mg substitute for Fe^{2+} and complete series extend to rhodochrosite and magnesite (see Figs. 10.5 and 10.6). The substitution of Ca for Fe^{2+} is limited due to the large difference in size of the two ions. Siderite is isostructural with calcite (see Fig. 10.2 and page 405).

Diagnostic Features. Distinguished from other carbonates by its color and high specific gravity, and from sphalerite by its rhombohedral cleavage. Soluble in hot HCl with effervescence.

Alteration. Pseudomorphs of limonite after siderite are common.

Occurrence. Siderite is frequently found as *clay ironstone*, impure by admixture with clay materials, in concretions with concentric layers. As *black-band ore* it is found, contaminated by carbonaceous material, in extensive stratified formations lying in shales and commonly associated with coal measures. These ores have been mined extensively in Great Britain in the past, but at present are mined only in North Staffordshire and Scotland. Clay ironstone is also abundant in the coal measures of western Pennsylvania and eastern Ohio, but it is not used to any great extent as an ore. Siderite is also formed by the replacement action of Fe-rich solutions upon limestones, and if such occurrences are extensive, they may be of economic value. The most notable deposit of this type is in Styria, Austria, where siderite is mined on a large scale. Siderite, in its crystallized form, is a common vein mineral associated with various metallic ores containing silver minerals, pyrite, chalcopyrite, tetrahedrite, and galena. When siderite predominates in such veins, it may be mined, as in southern Westphalia, Germany. Siderite is also a common constituent of banded Precambrian iron deposits, as in the Lake Superior region. A famous classic locality is Cornwall, England. Modern localities are the Morro Velho Gold Mine, Nova Lima, Brazil, and Llallagua, Bolivia.

Use. An ore of iron. Important in Great Britain and Austria, but unimportant elsewhere.

Name. From the Greek word meaning *iron*. The name *spherosiderite* of the concretionary variety was shortened to siderite to apply to the entire species.

Chalybite, used by some mineralogists, was derived from the Chalybes, ancient iron workers, who lived by the Black Sea.

Rhodochrosite—MnCO₃

Crystallography. Hexagonal; $\bar{3}2/m$. Only rarely in crystals of the unit rhombohedron; frequently with curved faces. Usually cleavable, massive; granular to compact.

$R\bar{3}c$. Hexagonal cell, $a = 4.78$, $c = 15.67$ Å; $Z = 6$; rhombohedral cell, $a = 5.85$ Å, $\alpha = 47°46'$, $Z = 2$. ds: 3.66(4), 2.84(10), 2.17(3), 1.770(3), 1.763(3).

Physical Properties. *Cleavage* {10$\bar{1}$1} perfect. **H** $3\frac{1}{2}$–4. **G** 3.5–3.7. *Luster* vitreous. *Color* usually some shade of rose-red, see Plate III, no. 7; may be light pink to dark brown. *Streak* white. Transparent to translucent. *Optics:* (−); $\omega = 1.816$, $\epsilon = 1.597$.

Composition and Structure. For pure MnCO₃, MnO 61.7, CO₂ 38.3%. Fe^{2+} substitutes for Mn^{2+} forming a complete solid solution series between rhodochrosite and siderite (see Fig. 10.16). Ca^{2+} shows some substitution for Mn^{2+}. The occurrence of *kutnahorite*, $CaMn(CO_3)_2$, with an ordered structure of the dolomite type, suggests that only limited solid solution occurs, at ordinary temperatures, between $CaCO_3$ and $MnCO_3$. Mg may also substitute for Mn but the $MnCO_3$–$MgCO_3$ series is incomplete. Considerable amounts of Zn may substitute for Mn (see smithsonite). Rhodochrosite is isostructural with calcite (see Fig. 10.2 and page 405).

Diagnostic Features. Characterized by its pink color and rhombohedral cleavage; the hardness (4) distinguishes it from rhodonite, $MnSiO_3$, with hardness of 6. Infusible. Soluble in hot HCl with effervescence.

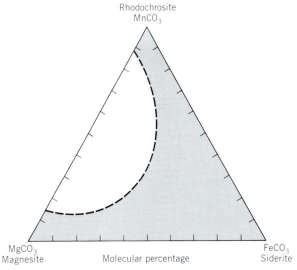

FIG. 10.16. The approximate extent of solid solution for carbonates in the three-component system MgCO₃–FeCO₃–MnCO₃. The lack of solid solution beteen MnCO₃ and MgCO₃ may be the result of a considerable difference in the size of the cationic radii, or it may reflect a lack of carbonate-rich rock types that bridge the gap between MnCO₃ and MgCO₃. (From E. J. Essene, 1983, Solid solutions and solvi among metamorphic carbonates with applications to geologic thermobarometry, in *Carbonates: Mineralogy and Chemistry. Reviews in Mineralogy*, v. 11, pp. 77–96.)

Occurrence. Rhodochrosite is a comparatively rare mineral, occurring in hydrothermal veins with ores of silver, lead, and copper, and with other manganese minerals. Beautiful banded rhodochrosite is mined for ornamental and decorative purposes at Capillitas, Catamarca, Argentina (see Fig. 10.17). Excellent crystals are found in the Kalahari manganese region of Cape Province, Republic of South Africa,

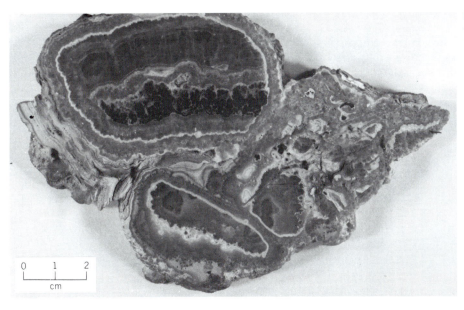

FIG. 10.17. Rhodochrosite, Capillitas, Catamarca, Argentina (Harvard Mineralogical Museum).

and at Pasto Bueno, Peru. In the United States it is found at Butte, Montana, where it has been mined as a manganese ore. In good crystals at Alicante, Lake County; Alma Park County; and elsewhere in Colorado.

Use. A minor ore of manganese. Small amounts used for ornamental purposes.

Name. Derived from two Greek words meaning *rose* and *color*, in allusion to its rose-pink color.

Smithsonite—ZnCO₃

Crystallography. Hexagonal; $\bar{3}2/m$. Rarely in small rhombohedral or scalenohedral crystals. Usually reniform, botryoidal (Fig. 10.18), or stalactitic, and in crystaline incrustations or in honeycombed masses known as *dry-bone ore*. Also granular to earthy.

$R\bar{3}c$. Hexagonal cell, $a = 4.66$, $c = 15.02$ Å; $Z = 6$; rhombohedral cell, $a = 5.63$ Å, $\alpha = 48°20'$, $Z = 2$. ds: 2.75(10), 3.55(5), 2.33(3), 1.946(3), 1.703(4).

Physical Properties. *Cleavage* {10$\bar{1}$1} perfect. **H** 4–4½. **G** 4.30–4.45. *Luster* vitreous. *Color* usually dirty brown. May be colorless, white, green, blue, or pink. The yellow variety contains Cd and is known as *turkey-fat ore*. Streak white. Translucent. *Optics:* (−); $\omega = 1.850$, $\epsilon = 1.623$.

Composition and Structure. For pure $ZnCO_3$, ZnO 64.8, CO_2 35.2%. Considerable Fe^{2+} may substitute for Zn, but there appears to be a gap in the $ZnCO_3$-$FeCO_3$ series. Mn^{2+} is generally present in only a few percent, but the occurrence of a zincian rhodochrosite with Zn : Mn = 1 : 1.2 suggests there may be a complete series between $ZnCO_3$ and $MnCO_3$. Ca and Mg are present in amounts of only a few weight percent. Small amounts of Co are found in a pink, and small amounts of Cu in a blue-green variety of smithsonite. Smithsonite is isostructural with calcite (see Fig. 10.2 and page 405).

Diagnostic Features. Soluble in cold HCl with effervescence. Distinguished by its effervescence in acids, tests for zinc, its hardness, and its high specific gravity.

Occurrence. Smithsonite is a zinc ore of supergene origin, usually found with zinc deposits in limestones. Associated with sphalerite, galena, hemimorphite, cerussite, calcite, and limonite. Often found in pseudomorphs after calcite. Smithsonite is found in places in translucent green or greenish-blue material which is used for ornamental

FIG. 10.18. Smithsonite, Kelly Mine near Socorro, New Mexico (Harvard Mineralogical Museum).

purposes. Laurium, Greece, is noted for this ornamental smithsonite, and Sardinia, Italy, for yellow stalactites with concentric banding. Fine crystallized specimens have come from the Broken Hill Mine, Zambia, and from Tsumeb, Namibia. In the United States smithsonite occurs as an ore in the zinc deposits of Leadville, Colorado; Arkansas and Wisconsin. Fine greenish-blue material has been found at the Kelly mine, Magdalena district, New Mexico.

Use. An ore of zinc. A minor use is for ornamental purposes.

Name. Named in honor of James Smithson (1754–1829), who founded the Smithsonian Institution in Washington, D.C. English mineralogists formerly called the mineral *calamine*.

Similar Species. *Hydrozincite,* $Zn_5(CO_3)_2(OH)_6$, occurs as a secondary mineral in zinc deposits.

Aragonite—CaCO₃

Crystallography. Orthorhombic; $2/m2/m2/m$. Three habits of crystallization are common. (1) Acicular pyramidal; consisting of a vertical prism terminated by a combination of a very steep dipyramid and {110} prism (Fig. 10.19a). Usually in radiating

FIG. 10.19. *(a)* and *(b)* Aragonite crystals. *(c)* and *(d)* Aragonite twins on {110} with a repeated twin producing a pseudohexagonal outline as in *(d)*.

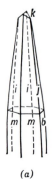

(a)

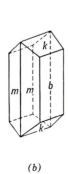

(b)

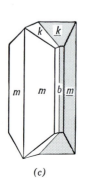

(c)

(d)

groups of large to very small crystals (see Fig. 10.20 and Plate III, no. 8). (2) Tabular; consisting of prominent {010} modified by {110} and a low prism, k{011} (Fig. 10.19b); often twinned on {110} (Fig. 10.19c). (3) In pseudohexagonal twins (Fig. 10.19d) showing a hexagonal-like prism terminated by a basal plane. These are formed by an intergrowth of three individuals twinned on {110} with {001} planes in common. The cyclic twins are distinguished from true hexagonal forms by noting that the basal surface is striated in three different directions and that, because the prism angle of the simple crystals is not exactly 60°, the composite prism faces for the twin will often show slight reentrant angles. Also found in reniform, columnar, and stalactitic aggregates.

Pmcn; *a* = 4.96, *b* = 7.97, *c* = 5.74 Å; *Z* = 4. *d*s: 3.04(9), 2.71(6), 2.36(7), 1.975(10), 1.880(8).

Physical Properties. *Cleavage* {010} distinct, {110} poor. *Luster* vitreous. *Color* colorless, white, pale yellow, and variously tinted. Transparent to translucent. **H** $3\frac{1}{2}$–4. **G** 2.94 (harder and higher specific gravity than calcite). *Optics:* (−); α = 1.530, β = 1.680, γ = 1.685; 2V = 18°; X = c, Y = a. r < v weak.

Composition, Structure, and Synthesis. Most aragonite is relatively pure $CaCO_3$. Small amounts of Sr and Pb substitute for Ca. The structure of aragonite is given in Fig. 10.3 and discussed on page 406. Calcite can be transformed into the aragonite structure by extensive grinding with a mortar and pestle. This structural transformation can be determined only by X-ray diffraction techniques. The phase diagram for the polymorphs of $CaCO_3$ is given in Fig. 10.4. Aragonite, with a somewhat denser structure than calcite, is stable on the high *P*, low *T* side of the curve relating the two polymorphs. If the temperature of metamorphism in a metamorphic terrane can be arrived at independently, this curve allows for the estimation of a maximum or minimum pressure of metamorphism in calcite- or aragonite-bearing rocks.

Diagnostic Features. Effervesces in cold HCl. Distinguished from calcite by its higher specific gravity and lack of rhombohedral cleavage. Cleavage fragments of columnar calcite are terminated by a cross cleavage that is lacking in aragonite. Distinguished from witherite and strontianite by lower specific gravity.

Alteration. Pseudomorphs of calcite after aragonite are common. $CaCO_3$ secreted by mollusks as aragonite is usually changed to calcite on the outside of the shell.

Occurrence. Aragonite is less stable than calcite under atmospheric conditions and much less common. It is precipitated in a narrow range of physicochemical conditions represented by low temperature, near-surface deposits. Experiments have shown that carbonated waters containing calcium more often deposit aragonite when they are warm and calcite when they are cold. The pearly layer of many shells and the pearl itself is aragonite. Aragonite is deposited by hot springs; found associated with beds of gypsum and deposits of iron ore where it may occur in forms resembling coral, and is called *flos ferri*. It is found as fibrous crusts on serpentine and in amygdaloidal cavities in basalt. The occurrence of aragonite in some

FIG. 10.20. Aragonite, Cleator Moor, Cumbria, England (Harvard Mineralogical Museum).

metamorphic rocks as in the Franciscan Formation in California and in blue schists of New Zealand is the result of recrystallization at high pressures and relatively low temperature.

Notable localities for the various crystalline types are as follows: pseudohexagonal twinned crystals are found in Aragon, Spain; and at Girgenti, Sicily, Italy, associated with native sulfur. Prismatic crystals were found near Bílina, Czechoslovakia. The acicular type is found at Alston Moor and Cleator Moor, Cumbria, England. Flos ferri is found in the iron mines of Styria, Austria. Some *onyx marble* from Baja California, Mexico, is aragonite. In the United States pseudohexagonal twins are found at Lake Arthur, New Mexico. Flos ferri occurs in the Organ Mountains, New Mexico, and at Bisbee, Arizona.

Name. From Aragon, Spain, where the pseudohexagonal twins were first recognized.

Witherite—$BaCO_3$

Crystallography. Orthorhombic; $2/m2/m2/m$. Crystals are always twinned on {110} forming pseudohexagonal dipyramids by the intergrowth of three individuals (Fig. 10.21). Crystals often deeply striated horizontally and by a series of reentrant angles have the appearance of one pyramid capping another. Also botryoidal to globular; columnar or granular.

Pmcn; $a = 5.31$, $b = 8.90$, $c = 6.43$ Å; $Z = 4$. *d*s: 3.72(10), 2.63(6), 2.14(5), 2.03(5), 1.94(5).

Physical Properties. *Cleavage* {010} distinct, {110} poor. **H** $3\frac{1}{2}$. **G** 4.31. *Luster* vitreous. *Color* colorless, white, gray. Translucent. *Optics:* $(-)$, $\alpha = 1.529$, $\beta = 1.676$, $\gamma = 1.677$; $2V = 16°$; $X = c$, $Y = b$; $r > v$ weak.

Composition and Structure. BaO 77.7, CO_2 22.3%. Small amounts of Sr and Ca may substitute for Ba. Witherite is isostructural with aragonite (see Fig. 10.3 and page 406).

Diagnostic Features. Soluble in cold HCl with effervescence. Witherite is characterized by high specific gravity. It is distinguished from barite by its effervescence in acid.

Occurrence. Witherite is a comparatively rare mineral, most frequently found in veins associated with galena. Found in England in fine crystals near Hexham in Northumberland and Alston Moor in Cumbria. In the

United States found with fluorite at the Minerva Mine, Cave-in-Rock, Illinois.

Use. A minor source of barium.

Name. In honor of D. W. Withering (1741–1799), who discovered and first analyzed the mineral.

Strontianite—$SrCO_3$

Crystallography. Orthorhombic; $2/m2/m2/m$. Crystals are usually acicular, radiating like type 1 under aragonite. Twinning on {110} frequent, giving pseudohexagonal appearance. Also columnar, fibrous, and granular.

Pmcn; $a = 5.11$, $b = 8.41$, $c = 6.03$ Å; $Z = 4$. *d*s: 3.47(10), 2.42(5), 2.02(7), 1.876(5), 1.794(7).

Physical Properties. *Cleavage* {110} good. **H** $3\frac{1}{2}$–4. **G** 3.78. *Luster* vitreous. *Color* white, gray, yellow, green. Transparent to translucent. *Optics:* $(-)$, $\alpha = 1.520$, $\beta = 1.667$, $\gamma = 1.669$; $2V = 7°$; $X = c$, $Y = b$. $r < v$ weak.

Composition and Structure. SrO 70.2, CO_2 29.8% for pure $SrCO_3$. Ca may be present in substitution for Sr to a maximum of about 25 atomic percent. Strontianite is isostructural with aragonite (see Fig. 10.3 and page 406).

Diagnostic Features. Characterized by high specific gravity and effervescence in HCl. Can be distinguished from celestite by poorer cleavage and effervescence in acid.

Occurrence. Strontianite is a low-temperature hydrothermal mineral associated with barite, celestite, and calcite in veins in limestone or marl, and less frequently in igneous rocks and as a gangue mineral in sulfide veins. Occurs in commercial deposits in Westphalia, Germany; Spain; Mexico; and England. Fine specimens are found in Oberdorf, Austria. Uncommon in the United States; found with fluorite at Cave-in-Rock, Illinois.

Use. Source of strontium. Strontium has no great commercial application; used in fireworks, red flares, military rockets, in the separation of sugar from molasses, and in various strontium compounds.

Name. From Strontian in Argyllshire, Scotland, where it was originally found.

Cerussite—$PbCO_3$

Crystallography. Orthorhombic; $2/m2/m2/m$. Crystals of varied habit are common and show many forms. Often tabular on {010} (Fig. 10.22a). May form reticulated groups with the plates crossing each other at 60° angles, see Fig. 10.23; frequently in

FIG. 10.21. Witherite.

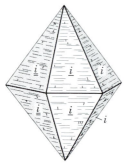

FIG. 10.22. Cerussite.

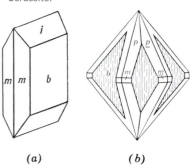

(a)   (b)

FIG. 10.23. Cerussite, Tsumeb, Namibia (Harvard Mineralogical Museum).

pseudohexagonal twins with deep reentrant angles in the vertical zone (Fig. 10.22b). Also in granular crystalline aggregates; fibrous; compact; earthy.

Pmcn; $a = 5.19$, $b = 8.44$, $c = 6.15$ Å; $Z = 4$. *d*s: 3.59(10), 3.50(4), 3.07(2), 2.49(3), 2.08(3).

Physical Properties. *Cleavage* {110} good, {021} fair. **H** $3-3\frac{1}{2}$. **G** 6.58. *Luster* adamantine. *Color* colorless, white, or gray. Transparent to subtranslucent. *Optics:* (−); $\alpha = 1.804$, $\beta = 2.077$, $\gamma = 2.079$; $2V = 9°$, $X = c$, $Y = b$, $r > v$.

Composition and Structure. Most cerussite is very close in composition to $PbCO_3$, with PbO 83.5 and CO_2 16.5%. It is isostructural with aragonite (see Fig. 10.3 and page 406).

Diagnostic Features. Recognized by its high specific gravity, white color, and adamantine luster. Crystal form and effervescence in nitric acid serve to distinguish it from anglesite.

Occurrence. Cerussite is an important and widely distributed supergene lead ore formed by the action of carbonated waters on galena. Associated with the primary minerals galena and sphalerite, and various secondary minerals such as anglesite, pyromorphite, smithsonite, and limonite.

Notable localities for its occurrence are Mies, Bohemia, Czechoslovakia; on the island of Sardinia, Italy; at Tsumeb, Namibia; Touissit, Morocco; and Broken Hill, New South Wales, Australia. In the United States found in various districts in Arizona; from the Organ Mountains, New Mexico; and in the Coeur d'Alene district in Idaho.

Use. An important ore of lead.

Name. From the Latin word meaning *white lead*.

Similar Species. *Phosgenite*, $Pb_2CO_3Cl_2$ (tetragonal), is a rare carbonate.

Dolomite—CaMg(CO$_3$)$_2$; Ankerite—CaFe(CO$_3$)$_2$

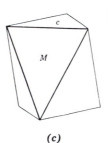

Crystallography. For *dolomite*. Hexagonal; $\bar{3}$. Crystals are usually the unit rhombohedron (Fig. 10.24a), more rarely a steep rhombohedron and base (Fig. 10.24c). Faces are often curved, some so acutely as to form "saddle-shaped" crystals (Figs. 10.24b, 10.25, and Plate III, no. 9). Other forms rare. In coarse, granular, cleavable masses to fine-grained

FIG. 10.24. Dolomite.

(*a*)

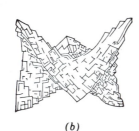

(*b*)

(*c*)

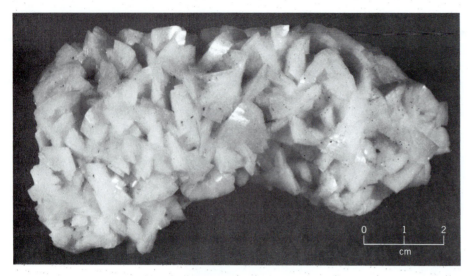

FIG. 10.25. Dolomite, St. Eustace, Quebec, Canada (Harvard Mineralogical Museum).

and compact. Twinning on {0001} common; lamellar twinning on {02$\bar{2}$1}. *Ankerite* generally does not occur in well-formed crystals. When it does, the crystals are similar to those of dolomite.

$R\bar{3}$; a = 4.84, c = 15.95 Å; Z = 3. ds: 2.88(10), 2.19(4), 2.01(3), 1.800(1), 1.780(1).

Physical Properties. *Cleavage* {10$\bar{1}$1} perfect. **H** $3\frac{1}{2}$–4. **G** 2.85. *Luster* vitreous; pearly in some varieties, *pearl spar*. *Color* commonly some shade of pink, flesh color; may be colorless, white, gray, green, brown, or black. Transparent to translucent. *Optics:* (−); ω = 1.681, ϵ = 1.500. With increasing substitution of Fe (toward ankerite composition) **G** and indices of refraction increase. Ankerite is typically yellowish-white but, due to oxidation of the iron, may appear yellowish-brown.

Composition and Structure. *Dolomite:* CaO 30.4, MgO 21.7, CO$_2$ 47.9%. *Ankerite:* CaO 25.9, FeO 33.3, CO$_2$ 40.8%. Naturally occurring dolomite deviates somewhat from Ca : Mg = 1 : 1 (see Fig. 10.6) with the Ca : Mg ratio ranging from 58 : 42 to 47$\frac{1}{2}$: 52$\frac{1}{2}$. A complete solid solution series extends to *ankerite* and there is also a complete solution toward *kutnahorite*, CaMn(CO$_3$)$_2$ (see Fig. 10.26). Members of the dolomite group are isostructural. The structure of dolomite is discussed on page 407.

Diagnostic Features. For *dolomite.* In cold, dilute HCl large fragments are slowly attacked but are soluble with effervescence only in hot HCl; the powdered mineral is readily soluble in cold acid. Crystallized variety told by its curved rhombohedral crystals and usually by its flesh-pink color. The massive rock variety is distinguished from limestone by the less vigorous reaction with HCl. *Ankerite* is similar to dolomite in its physical and chemical proper-

ties except for its color which ranges from yellow-brown to brown.

Occurrence. *Dolomite* is found in many parts of the world chiefly as extensive sedimentary strata, and the crystallized equivalent, dolomitic marble. Dolomite as a rock mass is generally thought to be secondary in origin, formed from ordinary limestone by the replacement of some of the Ca by Mg. The replacement may be only partial and thus most dolomite rocks (*dolostones*) are mixtures of dolomite and calcite. The mineral occurs also as a hydrothermal vein mineral, chiefly in the lead and zinc veins

FIG. 10.26. The approximate extent of solid solution between CaMn(CO$_3$)$_2$-CaMg(CO$_3$)$_2$-CaFe(CO$_3$)$_2$. (From E. J. Essene, 1983, Solid solutions and solvi among metamorphic carbonates with applications to geologic thermobarometry, in *Carbonates: Mineralogy and Chemistry. Reviews in Mineralogy*, v. 11, pp. 77–96.)

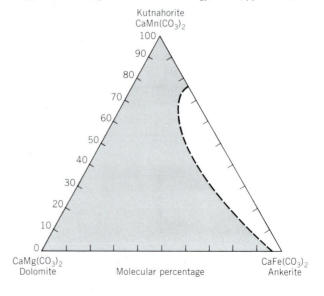

SYSTEMATIC DESCRIPTIONS   421

that traverse limestone, associated with fluorite, calcite, barite, and siderite. *Ankerite* is a common carbonate in Precambrian iron-formations.

Dolomite is abundant in the Dolomite region of southern Tyrol; in crystals from the Binnenthal, Switzerland; Traversella in Piedmont, Italy; northern England; and Guanajuato, Mexico. In the United States it is found as masses of sedimentary rock and gray to pink saddle-shaped crystals in vugs of rocks of many of the middle-western states, especially in the Joplin, Missouri, district.

Use of Dolomite. As a building and ornamental stone. For the manufacture of certain cements. For the manufacture of magnesia used in the preparation of refractory linings of the converters in the basic steel process. Dolomite is a potential ore of metallic Mg.

Name. *Dolomite* in honor of the French chemist, Dolomieu (1750–1801). *Ankerite* after M. J. Anker (1772–1843), Austrian mineralogist.

Similar Species. *Kutnahorite*, $CaMn(CO_3)_2$, is isostructural with dolomite.

Malachite—$Cu_2CO_3(OH)_2$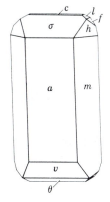

Crystallography. Monoclinic; $2/m$. Crystals are slender prismatic but seldom distinct. Crystals may be pseudomorphous after azurite. Usually in radiating fibers forming botryoidal or stalactitic masses (Fig. 10.27). Often granular or earthy.

$P2_1/a$; $a = 9.48$, $b = 12.03$, $c = 3.21$ Å, $\beta = 98°44$; $Z = 4$. ds: 6.00(6), 5.06(8), 3.69(9), 2.86(10), 2.52(6).

Physical Properties. *Cleavage* $\{\overline{2}01\}$ perfect but rarely seen. **H** $3\frac{1}{2}$–4. **G** 3.9–4.03. *Luster* adamantine to vitreous in crystals; often silky in fibrous varieties; dull in earthy type. *Color* bright green, see Plate IV, no. 1. *Streak* pale green. Translucent. *Optics:* (−), α

= 1.655, β = 1.875, γ = 1.909; $2V = 43°$; $Y = b$, $Z \wedge c = 24°$; pleochroism: *X* colorless, *Y* yellow green, *Z* deep green.

Composition and Structure. CuO 71.9, CO_2 19.9, H_2O 8.2%. Cu 57.4%. Cu^{2+} is octahedrally coordinated by O^{2-} and $(OH)^{1-}$ in $CuO_2(OH)_4$ and $CuO_4(OH)_2$ octahedra. These octahedra are linked along the edges forming chains that run parallel to the *c* axis. The chains are cross-linked by triangular $(CO_3)^{2-}$ groups.

Diagnostic Features. Soluble in HCl with effervescence, yielding a green solution. Recognized by its bright green color and botryoidal forms, and distinguished from other green copper minerals by its effervescence in acid.

Occurrence. Malachite is a widely distributed supergene copper mineral found in the oxidized portions of copper veins associated with azurite, cuprite, native copper, iron oxides. It usually occurs in copper deposits associated with limestone.

Notable localities for its occurrence are at Nizhniy Tagil in the Ural Mountains, Russia; pseudomorphous after cuprite at Chessy, near Lyons, France, and associated with azurite; from Tsumeb, Namibia; Katanga, Zaire; and Broken Hill, New South Wales, Australia. In the United States, formerly an important copper ore in the southwestern copper districts; at Bisbee, Morenci, and other localities in Arizona.

Use. A minor ore of copper. It is used extensively as an ornamental and gem mineral. In the nineteenth century Russia was the principal source, but today most of the material comes from Zaire.

Name. Derived from the Greek word *malache* for *mallows*, in allusion to its green color.

Azurite—$Cu_3(CO_3)_2(OH)_2$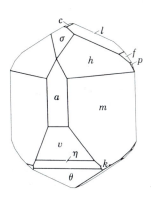

Crystallography. Monoclinic; $2/m$. Habit varies (Fig. 10.28); crystals are frequently complex and malformed. Also in radiating spherical groups.

FIG. 10.27. Malachite, cut and polished.

FIG. 10.28. Azurite crystals.

$P2_1c$; $a = 4.97$, $b = 5.84$, $c = 10.29$ Å, $\beta = 92°24'$; $Z = 2$. ds: 5.15(7), 3.66(4), 3.53(10), 2.52(6), 2.34(3).

Physical Properties. *Cleavage* {011} perfect, {100} fair. **H** $3\frac{1}{2}$–4. **G** 3.77. *Luster* vitreous. *Color* intense azure-blue, see Plate IV, no. 1. Transparent to translucent. *Optics:* (+), $\alpha = 1.730$, $\beta = 1.758$, $\gamma = 1.838$; $2V = 67°$; $X = b$, $Z \wedge c = -13°$; pleochroic in blue $Z > Y > X$; $r > V$.

Composition and Structure. CuO 69.2, CO_2 25.6, H_2O 5.2%. Cu 55.3%. The structure of azurite contains Cu^{2+} ions in square, coplanar groups with 2 O^{2-} and $2(OH)^{1-}$. These square groups are linked into chains parallel to the b axis. Each $(OH)^{1-}$ group is shared by 3 Cu^{2+} and each oxygen of the triangular (CO_3) group is bonded to one copper.

Diagnostic Features. Characterized chiefly by its azure-blue color and effervescence in HCl.

Alteration. Pseudomorphs of malachite after azurite commonly observed; less common after cuprite.

Occurrence. Azurite is less common than malachite but has the same origin and associations. It was found in fine crystals at Chessy, near Lyons, France; Tsumeb, Namibia; Touissit, Morocco; and Broken Hill, New South Wales, Australia. In the United States at the Copper Queen and other mines, Bisbee and Morenci, Arizona. Widely distributed with copper ores.

Use. A minor ore of copper.

Name. Named in allusion to its color.

Rare Hydrous Carbonates. *Aurichalcite*, $(Zn,Cu)_5$-$(CO_3)_2(OH)_3$ pale green to blue in orthorhombic acicular crystals. *Gaylussite*, $Na_2Ca(CO_3)_2 \cdot 5H_2O$, monoclinic and *trona*, $Na_3(CO_3)(HCO_3) \cdot 2H_2O$, monoclinic, are both found in saline-lake deposits.

Nitrates

Nitratite—NaNO₃. In nitratite $(NO_3)^{1-}$ groups combine in one-to-one proportion with monovalent Na^+ in 6-coordination forming a structure analogous to that of calcite (see Fig. 10.2). Nitratite and calcite are thus isostructural with the same crystallography and cleavage. Because of the lesser cationic charge, nitratite is softer (**H** 1–2) and melts at a low temperature and, because of the lower atomic weight of sodium, has a lower specific gravity (**G** 2.29). Nitratite is soluble in water and is thus found only in arid regions. The most notable occurrence is in northern Chile where it is recovered as a source of nitrogen.

Niter (Saltpeter)—KNO₃. Niter is isostructural with aragonite (see Fig. 10.3) with similar crystallography and analogous pseudohexagonal twinning on {110}. Like nitratite it fuses easily and is soluble in water. Niter is less common than nitratite but has been recovered from the soils in several countries as a source of nitrogen for fertilizer.

Borates

Of the 125 borate minerals known, only the four most common are considered here:

Kernite $Na_2B_4O_6(OH)_2 \cdot 3H_2O$
Borax $Na_2B_4O_5(OH)_4 \cdot 8H_2O$
Ulexite $NaCaB_5O_6(OH)_6 \cdot 5H_2O$
Colemanite $CaB_3O_4(OH)_3 \cdot H_2O$

Kernite—Na₂B₄O₆(OH)₂·3H₂O

Crystallography. Monoclinic; $2/m$. Rarely in crystals; usually in coarse, cleavable aggregates.

$P2/a$; $a = 15.68$, $b = 9.09$, $c = 7.02$ Å; $\beta = 108°52'$; $Z = 4$. ds: 7.41(10), 6.63(8), 3.70(3), 3.25(2), 2.88(2).

Physical Properties. *Cleavage* {001} and {100} perfect. Cleavage fragments are thus elongated parallel to the b crystallographic axis, see Fig. 10.29. The

FIG. 10.29. Kernite showing typically excellent {100} and {001} cleavage. Scale = 1 cm. From Boron, Kern County, California.

angle between the cleavages: $(001) \wedge (100) = 71°8'$. **H** 3. **G** 1.95. *Luster* vitreous to pearly. *Color* colorless to white; colorless specimens become chalky white on long exposure to the air owing to a surface film of *tincalconite*, $Na_2B_4O_5(OH)_4 \cdot 3H_2O$. *Optics:* $(-)$; $\alpha = 1.454$, $\beta = 1.472$, $\gamma = 1.488$; $2V = 70°$. $Z = b$, $X \wedge c = 71°$. $r > v$.

Composition and Structure. Na_2O 22.7, B_2O_3 51.0, H_2O 26.3%. The structure of kernite contains complex chains, parallel to the *b* axis, of composition $[B_4O_6(OH)_2]^{2-}$. These chains consist of BO_4 tetrahedra, joined at the vertices (as in the pyroxenes of the silicates) with remaining vertices connected to BO_3 triangles. The chains are linked to each other by Na^+ (in 5-coordination) and hydroxyl–oxygen bonds.

Diagnostic Features. Characterized by long, splintery cleavage fragments and low specific gravity. Slowly soluble in cold water.

Occurrence. The original locality and principal occurrence of kernite is in the Mohave Desert at Boron, California. Here, associated with borax, colemanite, and ulexite in a bedded series of Tertiary clays, it is present by the million tons. This deposit of sodium borates is 4 miles long, 1 mile wide, and as much as 250 feet thick, and lies from 150 to 1000 feet beneath the surface. The kernite occurs near the bottom of the deposit and is believed to have formed from borax by recrystallization caused by increased temperature and pressure. Kernite is also found associated with borax at Tincalayu, Argentina.

Use. Boron compounds are used in the manufacture of glass, especially in glass wool for insulation purposes. They are also used in soap, in porcelain enamels for coating metal surfaces, and in the preparation of fertilizers and herbicides.

Name. From Kern County, California, where the mineral is found.

Borax—$Na_2B_4O_5(OH)_4 \cdot 8H_2O$

Crystallography. Monoclinic; $2/m$. Commonly in prismatic crystals (Fig. 10.30). Also as massive, cellular material or encrustations.

FIG. 10.30. Borax.

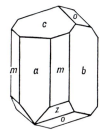

$C2/c$; $a = 11.86$, $b = 10.67$, $c = 12.20$ Å; $\beta = 106°$; $Z = 4$. *d*s: 5.7(2), 4.86(5), 3.96(4), 2.84(5), 2.57(10).

Physical Properties. *Cleavage* {100} perfect. **H** $2-2\frac{1}{2}$. **G** 1.7 $\pm$. *Luster* vitreous. *Color* colorless or white. Translucent. Sweetish-alkaline taste. Clear crystals effloresce and turn white with the formation of *tincalconite*, $Na_2B_4O_5(OH)_4 \cdot 3H_2O$, see Fig. 10.31. *Optics:* $(-)$; $\alpha = 1.447$, $\beta = 1.469$, $\gamma = 1.472$; $2V = 40°$. $X = b$, $Z \wedge c = -56°$. $r > v$.

Composition and Structure. Na_2O 16.2, B_2O_3 36.6, H_2O 47.2%. The structure of borax is shown in Fig. 10.32. It consists of $[BO_3OH]$ tetrahedra and $[BO_2OH]$ triangles that are joined parallel to the *c* crystallographic axis. The Na^+ is in 6-coordination with H_2O molecules, producing $Na(H_2O)_6$ octahedra that have common edges, forming ribbons parallel to the *c* crystallographic axis. The boron groups are linked to the $Na-H_2O$ ribbons by van der Waals and hydrogen bonding.

Diagnostic Features. Characterized by its crystals and tests for boron. Readily soluble in water.

Occurrence. Borax is the most widespread of the borate minerals. It is formed on the evaporation of enclosed lakes and as an efflorescence on the surface in arid regions. Deposits in Tibet furnished the first borax, under the name of *tincal*, to reach western civilization. In the United States it was first found in Lake County, California, and later in the desert region of southeastern California, in Death Valley, Inyo County, and in San Bernardino County. Associated with kernite, it is the principal mineral mined from the bedded deposits at Boron, California. It is also obtained commercially from the brines of Searles Lake at Trona, California.

Although the United States is the largest producer of borax, the world's most extensive deposits of borate minerals are in Turkey, notably in the Kirka area. Associated with other borates, borax is mined in the Andes Mountains in contiguous parts of Argentina, Bolivia, and Chile. Minerals commonly associated with borax are halite, gypsum, ulexite, colemanite, and various rare borates.

Use. Although boron is obtained from several minerals, it is usually converted to borax, the principal commercial product. There are many uses of borax. Its greatest use is in the manufacture of glass fibers for insulation and textiles. It is also used in soaps and detergents; as an antiseptic and preservative; in medicine; as a solvent for metallic oxides in soldering and welding; and as a flux in various smelting and laboratory operations. Elemental boron is used as a deoxidizer and alloy in nonferrous metals; and as a neutron absorber in shields for atomic

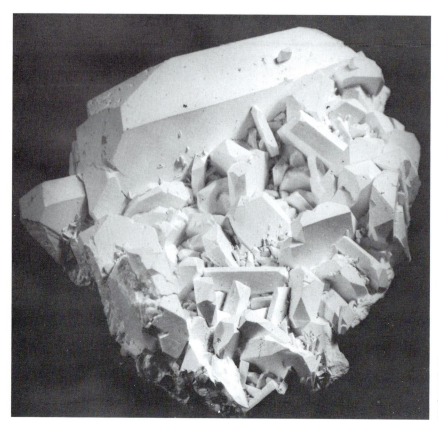

FIG. 10.31. Borax crystals, altered to white, chalky tincalconite. Baker Mine, Kramer borate district, California. (Courtesy of Richard C. Erd, U.S. Geological Survey.)

FIG. 10.32. The structure of borax as seen on (100).

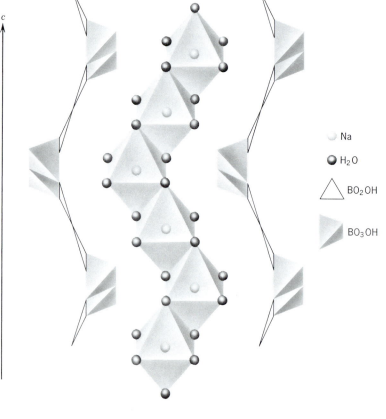

reactors. Boron is used in rocket fuels and as an additive in motor fuel. Boron carbide, harder than corundum, is used as an abrasive.

Name. Borax comes from an Arabic name for the substance.

Ulexite—NaCaB$_5$O$_6$(OH)$_6$·5H$_2$O

Crystallography. Triclinic; $\bar{1}$. Usually in rounded masses of loose texture, consisting of fine fibers that are acicular or capillary crystals, "cottonballs." Rarely in closely packed parallel fibers showing fiber-optical properties and called "television rock."

$P\bar{1}$; $a = 8.73$, $b = 12.75$, $c = 6.70$ Å; $\alpha = 90°16'$, $\beta = 109°8'$, $\gamma = 105°7'$; $Z = 2$. ds: 12.3(10), 7.89(7), 6.60(7), 4.19(7), 2.67(7).

Physical Properties. *Cleavage* {010} perfect. H $2\frac{1}{2}$; the aggregate has an apparent hardness of 1. G 1.96. *Luster* silky. *Color* white. Tasteless. *Optics:* (+); $\alpha = 1.491$, $\beta = 1.504$, $\gamma = 1.520$; 2V = 73°.

Composition and Structure. Na$_2$O 7.7, CaO 13.8, B$_2$O$_3$ 43.0, H$_2$O 35.5%. The structure contains large anionic groups of composition [B$_5$O$_6$(OH)$_6$]$^{3-}$. Each Ca^{2+} cation is surrounded by an irregular polyhedron of five oxygens, three (OH) groups, and two H$_2$O molecules. Each Na$^+$ cation is octahedrally coordinated by (OH) groups and H$_2$O molecules. Neighboring octahedra share edges to make continuous chains parallel to c.

Diagnostic Features. The soft "cottonballs" with silky luster are characteristic of ulexite.

Occurrence. Ulexite crystallizes in arid regions from brines that have concentrated in enclosed basins, as in playa lakes. Usually associated with borax. It occurs abundantly in the dry basins of northern Chile and in Argentina. In the United States it has been found abundantly in certain of the enclosed basins of Nevada and California and with colemanite in bedded Tertiary deposits.

Use. A source of borax.

Name. In honor of the German chemist, G. L. Ulex (1811–1883), who discovered the mineral.

Colemanite—CaB$_3$O$_4$(OH)$_3$·H$_2$O

Crystallography. Monoclinic; $2/m$. Commonly in short prismatic crystals, highly modified (see Fig. 10.33). Cleavable massive to granular and compact.

$P2_1/a$; $a = 8.74$, $b = 11.26$, $c = 6.10$ Å, $\beta = 110°7'$; $Z = 4$. ds: 5.64(5), 4.00(4), 3.85(5), 3.13(10), 2.55(5).

Physical Properties. *Cleavage* {010} perfect. H $4-4\frac{1}{2}$. G 2.42. *Luster* vitreous. *Color* colorless to white. Transparent to translucent. *Optics:* (+); $\alpha = 1.586$, $\beta = 1.592$, $\gamma = 1.614$; 2V = 55°; $X = b$, $Z \wedge c = 84°$. $r > v$.

Composition and Structure. CaO 27.2, B$_2$O$_3$ 50.9, H$_2$O 21.9%. The structure contains infinite chains with composition [B$_3$O$_4$(OH)$_2$]$^{2-}$ that are parallel to the a axis. The chains contain BO$_3$ triangles and BO$_2$(OH)$_2$ and BO$_3$(OH) tetrahedra. Interspersed between the chains are Ca ions and H$_2$O molecules.

Diagnostic Features. Characterized by one direction of highly perfect cleavage and exfoliation on heating.

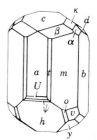

FIG. 10.33. Colemanite.

Occurrence. Colemanite deposits are interstratified with lake bed deposits of Tertiary age. Ulexite and borax are usually associated, and colemanite is believed to have originated by their alteration. Found in California in Los Angeles, Ventura, San Bernardino, and Inyo counties, and in Nevada in the Muddy Mountains and White Basin, Clark County. Also occurs extensively in Argentina and Turkey.

Use. A source of borax that, at the time of the discovery of kernite, yielded over half of the world's supply.

Name. In honor of William T. Coleman, merchant of San Francisco, who marketed the product of the colemanite mines.

Similar Species. Other borates, locally abundant, are *boracite*, Mg$_3$ClB$_7$O$_{13}$; *hydroboracite*, CaMgB$_6$O$_8$(OH)$_6$·3H$_2$O and *inyoite*, CaB$_3$O$_3$(OH)$_5$·4H$_2$O.

Sinhalite, Mg(Al,Fe)BO$_4$ is a rare gem mineral found in the gravels of Sri Lanka and Burma. Before its description as a new mineral in 1952, cut stones had been misidentified as brown peridot.

Sulfates and Chromates

Although 303 minerals belong to these two classes only a few of them are common.

Anhydrous Sulfates and Chromate	
Barite group	
Barite	BaSO$_4$
Celestite	SrSO$_4$
Anglesite	PbSO$_4$
Anhydrite	CaSO$_4$
Crocoite	PbCrO$_4$

Hydrous and Basic Sulfates	
Gypsum	CaSO$_4$·2H$_2$O
Antlerite	Cu$_3$SO$_4$(OH)$_4$
Alunite	KAl$_3$(SO$_4$)$_2$(OH)$_6$

BARITE GROUP

The sulfates of Ba, Sr, and Pb form an isostructural group with space group *Pnma*. They have closely related crystal constants and similar habits. The members of the group are: *barite*, *celestite*, and *anglesite*.

Barite—BaSO$_4$

Crystallography. Orthorhombic; $2/m2/m2/m$. Crystals are usually tabular on {001}; often diamond-shaped (Fig. 10.34). Both {0kl} and {h0l} prisms are

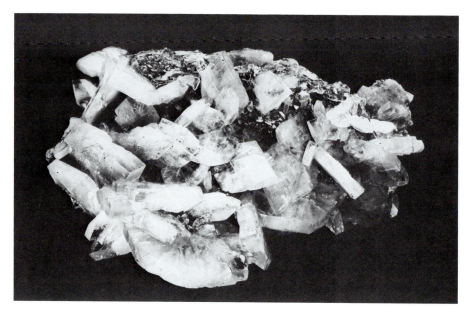

FIG. 10.34. Barite, Cumbria, England (Harvard Mineralogical Museum).

usually present, either beveling the corners of the diamond-shaped crystals or the edges of the tables and forming rectangular prismatic crystals elongated parallel to either *a* or *b* (Fig. 10.35). Crystals may be very complex. Frequently in divergent groups of tabular crystals forming "crested barite" or "barite roses," see Fig. 10.36 and Plate IV, no. 2. Also coarsely laminated; granular, earthy.

Pnma; $a = 8.87$, $b = 5.45$, $c = 7.14$ Å; $Z = 4$. *d*s: 3.90(6), 3.44(10), 3.32(7), 3.10(10), 2.12(8).

Physical Properties. *Cleavage* {001} perfect, {210} less perfect. **H** $3-3\frac{1}{2}$. **G** 4.5 (heavy for a nonmetallic mineral). *Luster* vitreous; on some specimens pearly on base. *Color* colorless, white, and light shades of blue, yellow, red (see Plate XII, no. 11, Chapter 13). Transparent to translucent. *Optics:* (+); $\alpha = 1.636$, $\beta = 1.637$, $\gamma = 1.648$; 2V = 37°. $X = c$, $Y = b$. $r < v$.

Composition and Structure. BaO 65.7, SO$_3$ 34.3% for pure barite. Sr substitutes for Ba and a complete solid solution series extends to celestite, but most material is near one end or the other of the series. A small amount of Pb may substitute for Ba.

The structure of barite is illustrated in Fig. 10.8 and discussed on page 408.

Diagnostic Features. Recognized by its high specific gravity and characteristic cleavage and crystals.

Occurrence. Barite is a common mineral of wide distribution. It occurs usually as a gangue mineral in hydrothermal veins, associated with ores of silver, lead, copper, cobalt, manganese, and antimony. It is found in veins in limestone with calcite, or as residual masses in clay overlying limestone. Also in sandstone with copper ores. In places acts as a cement in sandstone. Deposited occasionally as a sinter by waters from hot springs.

Notable localities for the occurrence of barite crystals are in Westmoreland, and Cumbria, England; Felsobanya and other localities, Romania; and

FIG. 10.36. Crested barite, Cumbria, England (Harvard Mineralogical Museum).

FIG. 10.35. Barite crystals.

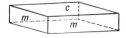
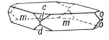
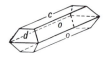

Saxony, Germany. Massive barite, occurring usually as veins, nests, and irregular bodies in limestones, has been quarried in the United States in Georgia, Tennessee, Missouri, and Arkansas.

Use. More than 80% of the barite produced is used in oil- and gas-well drilling as "heavy mud" to aid in the support of drill rods, and to help prevent blowing out of gas. Barite is the chief source of Ba for chemicals. A major use of barium is in *lithopone*, a combination of barium sulfide and zinc sulfate that forms an intimate mixture of zinc sulfide and barium sulfate. Lithopone is used in the paint industry and to a lesser extent in floor coverings and textiles. Precipitated barium sulfate, "blanc fixe," is employed as a filler in paper and cloth, in cosmetics, as a paint pigment, and for barium meals in radiology.

Name. From the Greek word *barys* meaning *heavy*, an allusion to its high specific gravity.

Celestite—SrSO$_4$

Crystallography. Orthorhombic; $2/m2/m2/m$. Crystals closely resemble those of barite. Commonly tabular parallel to {001} or prismatic parallel to *a* or *b* with prominent development of {0kl} and {h0l} prisms. Crystals elongated parallel to *a* are frequently terminated by nearly equal development of faces of *d*{101} and *m*{210} (Fig. 10.37). Also radiating fibrous; granular.

Pnma; $a = 8.38$, $b = 5.37$, $c = 6.85$ Å; $Z = 4$. *d*s: 3.30(10), 3.18(6), 2.97(10), 2.73(6), 2.04(6).

Physical Properties. *Cleavage* {001} perfect, {210} good. **H** $3–3\frac{1}{2}$. **G** 3.95–3.97. *Luster* vitreous to pearly. *Color* colorless, white, often faintly blue or red. Transparent to translucent. *Optics:* (+); $\alpha = 1.622$, $\beta = 1.624$, $\gamma = 1.631$; $2V = 50°$; $X = c$, $Y = b$.

Composition and Structure. SrO 56.4, SO$_3$ 43.6% for pure celestite. Ba substitutes for Sr and a complete solid solution series exists between celestite and barite. At ordinary temperatures only a limited series exists between anhydrite, CaSO$_4$, and SrSO$_4$. Celestite is isostructural with barite (see Fig. 10.8).

Diagnostic Features. Closely resembles barite but is differentiated by lower specific gravity and by chemical test for strontium.

Occurrence. Celestite is found usually disseminated through limestone or sandstone, or in nests

and lining cavities in such rocks. Associated with calcite, dolomite, gypsum, halite, sulfur, fluorite. Also found as a gangue mineral in lead veins.

Notable localities for its occurrence are with the sulfur deposits of Silicy; and Yate, Gloucestershire, England. Found in the United States at Clay Center, Ohio, and elsewhere in northwestern and southeastern Michigan; Lampasas, Texas; and with colemanite in Inyo County, California.

Use. Used in the preparation of strontium nitrate for fireworks and tracer bullets, and with other strontium salts used in the refining of beet sugar.

Name. Derived from the Latin *caelestis* meaning *heavenly*, in allusion to the faint blue color of the first specimens described.

Anglesite—PbSO$_4$

Crystallography. Orthorhombic; $2/m2/m2/m$. Frequently in crystals with habit often similar to that of barite but more varied. Crystals may be prismatic parallel to any one of the crystal axes and frequently show many forms, with a complex development (Fig. 10.38). Also massive, granular to compact. Frequently earthy, in concentric layers that may have an unaltered core of galena.

Pnma; $a = 8.47$, $b = 5.39$, $c = 6.94$ Å; $Z = 4$. *d*s: 4.26(9), 3.81(6), 3.33(9), 3.22(7), 3.00(10).

Physical Properties. *Cleavage* {001} good, {210} imperfect. Fracture conchoidal. **H** 3.0. **G** 6.2–6.4 (unusually high). *Luster* adamantine when crystalline, dull when earthy. *Color* colorless, white, gray, pale shades of yellow. May be colored dark gray by impurities. Transparent to translucent. *Optics:* (+); $\alpha = 1.877$, $\beta = 1.883$, $\gamma = 1.894$; $2V = 75°$; $X = c$, $Y = b$. $r < v$.

Composition and Structure. PbO 73.6, SO$_3$ 26.4%. Anglesite is isostructural with barite (see Fig. 10.8 and page 408).

Diagnostic Features. Recognized by its high specific gravity, its adamantine luster, and frequently by its association with galena. Distinguished from cerussite in that it will not effervesce in nitric acid.

Occurrence. Anglesite is a common supergene mineral found in the oxidized portions of lead de-

FIG. 10.37. Celestite crystals.

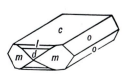 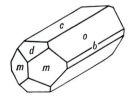

FIG. 10.38. Anglesite.

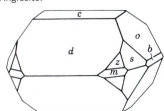

posits. It is formed through the oxidation of galena, either directly, as is shown by concentric layers of anglesite surrounding a core of galena, or by solution and subsequent deposition and recrystallization. Anglesite is commonly associated with galena, cerussite, sphalerite, smithsonite, hemimorphite, and iron oxides.

Notable localities for its occurrence are Monte Poni, Sardinia, Italy; Island of Anglesey, Wales; Derbyshire, England; and Leadhills, Scotland. It can also be found in Sidi-Amorben-Salem, Tunisia; Tsumeb, Namibia; Broken Hill, New South Wales, and Dundas, Tasmania, Australia; and Touissit, Morocco. In addition, it is found in crystals embedded in sulfur at Los Lamentos, Chihuahua, Mexico. In the United States it is in Phoenixville, Pennsylvania; Tintic district, Utah; and Coeur d'Alene district, Idaho.

Use. A minor ore of lead.

Name. Named from the original locality on the Island of Anglesey.

Anhydrite—CaSO₄

Crystallography. Orthorhombic. $2/m2/m2/m$. Crystals are rare; when observed are thick tabular on {010}, {100}, or {001} (see Fig. 10.39), also prismatic parallel to b. Usually massive or in crystalline masses resembling an isometric mineral with cubic cleavage. Also fibrous, granular, massive.

Amma; $a = 6.95$, $b = 6.96$, $c = 6.21$ Å; $Z = 4$. *d*s: 3.50(10), 2.85(3), 2.33(2), 2.08(2), 1.869(2).

Physical Properties. *Cleavage* {010} perfect, {100} nearly perfect, {001} good. **H** $3-3\frac{1}{2}$. **G** 2.89–2.98. *Luster* vitreous to pearly on cleavage. *Color* colorless to bluish or violet. Also may be white or tinged with rose, brown, or red. *Optics:* (+); $\alpha = 1.570$, $\beta = 1.575$, $\gamma = 1.614$; $2V = 44°$; $X = b$, $Y = a$; $r < v$.

Composition and Structure. CaO 41.2, SO₃ 58.8%. The structure illustrated in Fig. 10.9 is very different from the barite type. In anhydrite Ca is in 8-coordination with oxygen from SO₄ groups, whereas in barite Ba is in 12-coordination with oxygen. A metastable polymorph of anhydrite (γCaSO₄) is hexagonal and is formed as the result of slow dehydration of gypsum (see Fig. 10.42).

FIG. 10.39. Anhydrite.

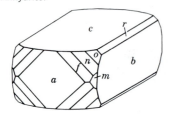

Diagnostic Features. Anhydrite is characterized by its three cleavages at right angles. It is distinguished from calcite by its higher specific gravity and from gypsum by its greater hardness. Some massive varieties are difficult to recognize, and one should test for sulfate.

Alteration. Anhydrite by hydration changes to gypsum with an increase in volume, and in places large masses have been thus altered.

Occurrence. Anhydrite occurs in much the same manner as gypsum and is often associated with that mineral but is not nearly as common. Found in beds associated with salt deposits in the cap rock of salt domes, and in limestones. Found in some amygdaloidal cavities in basalt.

Notable foreign localities are: Wieliczka, Poland; Aussee, Styria and Hall near Innsbruck, Tyrol, Austria; and Bex, Switzerland. In the United States found in Lockport, New York; West Paterson, New Jersey; New Mexico; and Texas. Found in large beds in Nova Scotia.

Use. Ground anhydrite is used as a soil conditioner and to a minor extent as a setting retardant in Portland cement. In Great Britain and Germany it has been used as a source of sulfur for the production of sulfuric acid.

Name. Anhydrite is from the Greek meaning *without water*, in contrast to the more common hydrous calcium sulfate, gypsum.

Crocoite—PbCrO₄

Crystallography. Monoclinic; $2/m$. Commonly in slender prismatic crystals, vertically striated, and columnar aggregates. Also granular.

$P2_1/n$; $a = 7.11$, $b = 7.41$, $c = 6.81$ Å, $\beta = 102°33'$; $Z = 4$. *d*s: 4.96(3), 2.85(3), 2.33(2), 2.08(2), 1.869(2).

Physical Properties. *Cleavage* {110} imperfect. **H** $2\frac{1}{2}-3$. **G** 5.9–6.1. *Luster* adamantine. *Color* bright hyacinth-red. *Streak* orange-yellow. Translucent.

Composition and Structure. PbO 68.9, CrO₃ 31.1%. Crocoite is isostructural with *monazite*, CePO₄, with Pb in 9-coordination with oxygen, linking six CrO₄ tetrahedra.

Diagnostic Features. Crocoite is characterized by its color, high luster, and high specific gravity. It may be confused with wulfenite, PbMoO₄, but can be distinguished from it by its redder color, lower specific gravity, and crystal form.

Occurrence. Crocoite is a rare mineral found in the oxidized zones of lead deposits in those regions where lead veins have traversed rocks containing chromite. Associated with pyromorphite, cerussite, and wulfenite. Notable localities are: Dundas, Tasmania and Beresovsk near Sverdlovsk, Ural Mountains, Russia. In the United States it is found in small quantities in the Vulture district, Arizona.

Use. Not abundant enough to be of commercial value, but of historic interest, because the element chromium was first discovered in crocoite.

Name. From the Greek meaning *saffron*, in allusion to the color.

Gypsum—CaSO₄·2H₂O

Crystallography. Monoclinic; $2/m$. Crystals are of simple habit (Figs. 10.40a and 10.41); tabular on {010}; diamond-shaped, with edges beveled by {120} and {$\overline{1}$11}. Other forms are rare. Twinning is common, on {100} (Fig. 10.40b), often resulting in swallowtail twins.

$C2/c$; $a = 6.28$, $b = 15.15$, $c = 5.67$ Å, $\beta = 114°12'$; $Z = 4$. ds: 7.56(10), 4.27(5), 3.06(6), 2.87(2), 2.68(3).

Physical Properties. *Cleavage* {010} perfect yielding thin folia; {100}, with conchoidal surface; {011}, with fibrous fracture. **H** 2. **G** 2.32. *Luster* usually vitreous; also pearly and silky. *Color* colorless, white, gray; various shades of yellow, red, brown, from impurities. Transparent to translucent, see Plate IV, no. 3.

Satin spar is a fibrous gypsum with silky luster. *Alabaster* is the fine-grained massive variety. *Selenite* is a variety that yields broad colorless and transparent cleavage folia. *Optics:* (+); $\alpha = 1.520$, $\beta = 1.523$, $\gamma = 1.530$; $2V = 58°$. $Y = b$, $X \wedge c = -37°$; $r > v$.

Composition and Structure. CaO 32.6, SO₃ 46.5, H₂O 20.9%. As the result of dehydration of gypsum, CaSO₄·2H₂O, several phases may form. These are: γ CaSO₄, when all H₂O is lost and a metastable phase CaSO₄·½H₂O. During the dehydration process the first 1½ molecules of H₂O in gypsum are lost rel-

atively continuously between 0° and about 65°C (see Fig. 10.42) perhaps with only slight changes in the gypsum structure. At about 70°C the remaining ½H₂O molecule in CaSO₄·½H₂O is still retained relatively strongly, but at about 95°C this is lost and the structure transforms to that of a polymorph of anhydrite. The structure of gypsum is given in Fig. 10.10 and discussed on page 409.

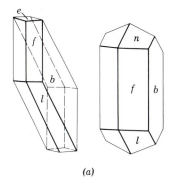

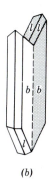

FIG. 10.40. *(a)* Gypsum crystals. *(b)* Gypsum twin.

FIG. 10.41. Gypsum crystals, Ellsworth, Ohio (Harvard Mineralogical Museum).

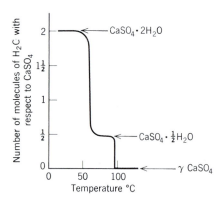

FIG. 10.42. Dehydration curve of gypsum showing formation of metastable $CaSO_4 \cdot \frac{1}{2}H_2O$ at about 65°C. At about 100°C the γ polymorph of $CaSO_4$ is formed.

Diagnostic Features. Characterized by its softness and its three unequal cleavages.

Occurrence. Gypsum is a common mineral widely distributed in sedimentary rocks, often as thick beds. If frequently occurs interstratified with limestones and shales and is usually found as a layer underlying beds of rock salt, having been deposited there as one of the first minerals to crystallize on the evaporation of saltwaters (see Box 9.2). May recrystallize in veins, forming *satin spar*. Occurs also as lenticular bodies or scattered crystals in clays and shales. Frequently formed by the alteration of anhydrite, and under these circumstances may show folding because of increased volume. Found in volcanic regions, especially where limestones have been acted upon by sulfur vapors. Also common as a gangue mineral in metallic veins. Associated with many different minerals, the more common being halite, anhydrite, dolomite, calcite, sulfur, pyrite, and quartz.

Gypsum is the most common sulfate, and extensive deposits are found in many localities through the world. The principal world producers are the United States, Canada, France, Japan, and Iran. In the United States commercial deposits are found in many states, but the chief producers are located in California, Iowa, New York, Texas, and Oklahoma. Gypsum is found in large deposits in Arizona and New Mexico in the form of windblown sand. Spectacular crystals originate from several mines in Mexico, especially Naica and Chihuahua.

Use. Gypsum is used chiefly for the production of plaster of Paris. In the manufacture of the material, the gypsum is ground and then heated until about 75% of the water has been driven off, producing the substance $CaSO_4 \cdot \frac{1}{2}H_2O$. This material, when mixed with water, slowly absorbs the water, crystallizes and thus hardens, or *sets*. Plaster of Paris is used extensively for *staff*, the material from which temporary exposition buildings are built, for gypsum lath, wallboard, and for molds and casts of all kinds. Gypsum is employed in making *adamant plaster* for interior use. Serves as a soil conditioner, *land plaster*, for fertilizer. Uncalcined gypsum is used as a retarder in Portland cement. Satin spar and alabaster are cut and polished for various ornamental purposes but are restricted in their use because of their softness.

Name. From the Greek name for the mineral, applied especially to the calcined mineral.

Antlerite—$Cu_3SO_4(OH)_4$

Crystallography. Orthorhombic; $2/m2/m2/m$. Crystals are usually tabular on {010} (Fig. 10.43b). May be in slender prismatic crystals vertically striated. Also in parallel aggregates, reniform, massive.

Pnam; $a = 8.24$, $b = 11.99$, $c = 6.03$ Å; $Z = 4$. ds: 4.86(10), 3.60(8), 3.40(3), 2.68(6), 2.57(6).

Physical Properties. *Cleavage* {010} perfect. H $3\frac{1}{2}$–4. G $3.9\pm$. *Luster* vitreous. *Color* emerald to blackish-green. *Streak* pale green. Transparent to translucent. *Optics:* (+); $\alpha = 1.726$, $\beta = 1.738$, $\gamma = 1.789$; $2V = 53°$; $X = b$, $Y = a$; pleochroism X yellow-green, Y blue-green, Z green.

Composition and Structure. CuO 67.3, SO_3 22.5, H_2O 10.2%. The structure of antlerite contains Cu in two types of octahedral coordination: $CuO(OH)_5$ and $CuO_3(OH)_3$. These octahedra are linked with the SO_4 tetrahedra.

Diagnostic Features. Antlerite is characterized by its green color, {010} cleavage, and association. Lack of effervescence in HCl distinguishes it from malachite. It is commonly associated with atacamite and brochantite, and one must use optical or chemical properties to distinguish it from these minerals.

Occurrence. Antlerite is found in the oxidized portions of copper veins, especially in arid regions. It was formerly considered a rare mineral, but in 1925 it was recognized as the chief ore mineral at Chuquicamata, Chile, the world's largest copper mine. It may form directly as a secondary mineral on chalcocite, or the copper may go into solution and later be deposited as antlerite, filling cracks. In the United States it is found at Bisbee, Arizona.

FIG. 10.43. Antlerite crystals.

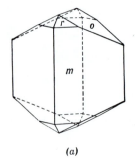

(a)

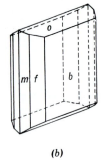

(b)

Use. An ore of copper.

Name. From the Antler Mine, Arizona, from which locality it was originally described.

Similar Species. *Brochantite*, $Cu_4SO_4(OH)_6$, is similar in all its properties to antlerite, but, although more widespread, it is nowhere abundant. Until the antlerite find in 1925, it was considered to be the chief ore mineral at Chuquicamata, Chile. Another rare hydrous copper sulfate is *chalcanthite*, $CuSO_4 \cdot 5H_2O$, which occurs abundantly at Chuquicamata and other arid areas in Chile. *Epsomite*, $MgSO_4 \cdot 7H_2O$, occurs as an efflorescence on the rocks in mine workings and on the walls of caves. *Melanterite*, $FeSO_4 \cdot 7H_2O$, is also a secondary mineral found as an efflorescence on mine timbers.

Alunite—$KAl_3(SO_4)_2(OH)_6$

Crystallography. Hexagonal; $3m$. Crystals are usually a combination of positive and negative trigonal pyramids resembling rhombohedrons with nearly cubic angles (90°50′). May be tabular on {0001}. Commonly massive or disseminated.

$R3m$; $a = 6.97$, $c = 17.38$ Å; $Z = 3$. ds: 4.94(5), 2.98(10), 2.29(5), 1.89(6), 1.74(5).

Phyical Properties. *Cleavage* {0001} imperfect. **H** 4. **G** 2.6–2.8. *Luster* vitreous to pearly in crytals, earthy in massive material. *Color* white, gray, or reddish. Transparent to translucent. *Optics:* (+); $\omega = 1.572$, $\epsilon = 1.592$.

Composition and Structure. K_2O 11.4, Al_2O_3 37.0, SO_3 38.6, H_2O 13.0%. Na may replace K at least up to Na : K = 7 : 4. When Na exceeds K in the mineral it is called *natroalunite*, $(Na,K)Al_3(SO_4)_2(OH)_6$. In the structure of alunite K and Al are in 6-coordination with respect to O and (OH). The SO_4 groups share some of their oxygens with the K-containing octahedra.

Diagnostic Features. Alunite is usually massive and difficult to distinguish from rocks such as limestone and dolomite, and other massive minerals such as anhydrite and granular magnesite.

Occurrence. Alunite, also called *alumstone*, is usually formed by sulfuric acid solutions acting on rocks rich in potash feldspar, and in some places large masses have thus been formed. It is found in smaller amounts about volcanic fumaroles. In the United States it is found at Red Mountain in the San Juan district, Colorado; Goldfield, Nevada; and Marysvale, Utah.

Use. In the production of alum. At Marysvale, Utah, alunite has been mined and treated in such a way as to recover potassium and aluminum.

Name. From the Latin meaning alum.

Similar Species. *Jarosite*, $KFe_3(SO_4)_2(OH)_6$, the Fe^{3+} analogue of alunite, is a secondary mineral found as crusts and coatings on ferruginous ores.

Tungstates and Molybdates

The tungstates and molybdates comprise a group of 42 minerals, of which we will discuss only 3: wolframite, scheelite, and wulfenite.

Wolframite—$(Fe,Mn)WO_4$

Crystallography. Monoclinic; $2/m$. Crystals are commonly tabular parallel to {100} (Fig. 10.44) giving bladed habit, with faces striated parallel to c. In bladed, lamellar, or columnar aggregates. Massive granular.

$P2/c$; $a = 4.79$, $b = 5.74$, $c = 4.99$ Å; $\beta = 90°28′$; $Z = 2$. ds: 4.76(5), 3.74(5), 2.95(10), 2.48(6), 1.716(5). Cell parameters are slightly less for *ferberite*, $FeWO_4$, and greater for *huebnerite*, $MnWO_4$.

Physical Properties. *Cleavage* {010} perfect. **H** $4-4\frac{1}{2}$. **G** 7.0–7.5, higher with higher Fe content. *Luster* submetallic to resinous. *Color* black in ferberite to brown in huebnerite. *Streak* from nearly black to brown.

Composition and Structure. Fe^{2+} and Mn^{2+} substitute for each other in all proportions and a complete solid solution series exists between *ferberite*, $FeWO_4$, and *huebnerite*, $MnWO_4$. The percentage of WO_3 is 76.3 in ferberite and 76.6 in huebnerite. The structure of wolframite consists of distorted tetrahedral (WO_4) groups and octahedral $(Fe,Mn)O_6$ groups. From the interatomic distances around W, however, one may also conclude that W is in distorted octahedral coordination.

Diagnostic Features. The dark color, one direction of perfect cleavage, and high specific gravity serve to distinguish wolframite from other minerals.

Occurrence. Wolframite is a comparatively rare mineral found usually in pegmatites and high-temperature quartz veins associated with granites. More rarely in sulfide veins. Minerals commonly associated include cassiterite, scheelite, bismuth, quartz, pyrite, galena, sphalerite, and arsenopyrite. In some veins wolframite may be the only metallic mineral present.

Found in fine crystals from Schlaggenwald and Zinnwald, Bohemia, Czechoslovakia, and in the various tin districts of Saxony, Germany, and Cornwall, England. Important producing countries are China, Russia, Korea, Thailand, Bolivia, and Australia. Wolframite occurs in the United States in the Black Hills, South Dakota. Ferberite has been mined extensively

FIG. 10.44. Wolframite.

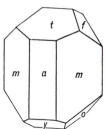

in Boulder County, Colorado. Huebernite is found near Silverton, Colorado; Mammoth district, Nevada; and Black Hills, South Dakota. More than 90% of the world's estimated tungsten resources are outside the United States, with almost 60% in southeastern China.

Use. Chief ore of tungsten. Tungsten is used as hardening metal in the manufacture of high-speed tool steel, valves, springs, chisels, files, and so on. Its high melting point (3410°C) requires a special chemical process for reduction of the metal, which is produced in the form of a powder. By powder-metallurgy, pure metal products such as lamp filaments are fabricated. A large use for tungsten is in the manufacture of carbides, harder than any natural abrasives (other than diamond), used for cutting tools, rock bits, and hard facings. Sodium tungstate is used in fireproofing cloth and as a mordant in dyeing.

Name. Wolframite is derived from an old word of German origin.

Scheelite—CaWO$_4$

Crystallography. Tetragonal; $4/m$. Crystals are usually simple dipyramids, {011} (Fig. 10.45). The dipyramid, {112}, closely resembles the octahedron in angles. Also massive granular.

$I4/a$; $a = 5.25$, $c = 11.40$ Å; $Z = 4$. ds: 4.77(7), 3.11(10), 1.94(8), 1.596(9), 1.558(7).

Physical Properties. *Cleavage* {101}, distinct. **H** $4\frac{1}{2}$–5. **G** 5.9–6.1 (unusually high for a nonmetallic mineral). *Luster* vitreous to adamantine. *Color* white, yellow, green, brown. Translucent; some specimens transparent (see Plate XII, no. 14, Chapter 13). Most scheelite will fluoresce with bluish-white color in short ultraviolet radiation. *Optics:* (+); $\omega = 1.920$, $\epsilon = 1.934$.

Composition and Structure. CaO 19.4, WO$_3$ 80.6%. Mo can substitute for W and a partial series extends to *powellite*, CaMoO$_4$. The structure of scheelite (Fig. 10.11) consits of flattened (WO$_4$) tetrahedra and CaO$_8$ polyhedra.

Diagnostic Features. Recognized by its high specific gravity, crystal form, and fluorescence in short ultraviolet light. The test for tungsten may be necessary for positive identification.

Occurrence. Scheelite is found in granite pegmatites, contact metamorphic deposits, and high-temperature hydrothermal veins associated with granitic rocks. Associated with cassiterite, topaz, fluorite, apatite, molybdenite, and wolframite. Occurs with tin in deposits of Bohemia, Czechoslovakia, Saxony, Germany, and Cornwall, England; and in quantity in New South Wales and Queensland, Australia. In the United States scheelite is mined near Mill City and Mina, Nevada; near Atolia, San Bernardino County, California; and in lesser amounts in Arizona, Utah, and Colorado.

Use. An ore of tungsten. Wolframite furnishes most of the world's supply of tungsten, but scheelite is more important in the United States. Transparent crystals may be cut as faceted gems.

Name. After K. W. Scheele (1742–1786), the discoverer of tungsten.

Wulfenite—PbMoO$_4$

Crystallography. Tetragonal; 4 or $4/m$. Crystals are usually square, tabular in habit with prominent {001}; Figs. 10.46a and 10.47. Some crystals are very thin. More rarely pyramidal (Fig. 10.46b). Some crystals may be twinned on {001} giving them a dipyramidal habit.

$I4_1/a$; $a = 5.42$, $c = 12.10$ Å; $Z = 4$. ds: 3.24(10), 3.03(2), 2.72(2), 2.02(3), 1.653(3).

Physical Properties. *Cleavage* {011} distinct. **H** 3. **G** 6.8 ±. *Luster* vitreous to adamantine. *Color* yellow, orange, red, gray, white. *Streak* white. Transparent to subtranslucent. Piezoelectricity suggets symmetry 4 rather than $4/m$ as determined from structure. *Optics:* (−); $\omega = 2.404$, $\epsilon = 2.283$.

Composition and Structure. PbO 60.8, MoO$_3$ 39.2%. Ca may substitute for Pb, indicating at least a partial series to *powellite*, Ca(Mo,W)O$_4$. Wulfenite is isostructural with scheelite (see Fig. 10.11).

Diagnostic Features. Wulfenite is characterized by its square tabular crystals, orange to yellow color, high luster, and association with other lead minerals. Distinguished from crocoite by test for Mo.

FIG. 10.46. Wulfenite crystals.

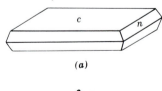

(a)

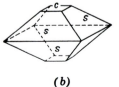

(b)

FIG. 10.45. Scheelite.

FIG. 10.47. Wulfenite, Stephenson-Bennett Mine, Dona Ana County, New Mexico (Harvard Mineralogical Museum).

Occurrence. Wulfenite is found in the oxidized portions of lead veins with other secondary lead minerals, especially cerussite, vanadinite, and pyromorphite. Found in the United States in several places in the Southwest. Found in beautiful crystals at the Red Cloud, Glove, and Mammoth mines in Arizona; also at the Bennett mine in New Mexico; and in Schwarzenbach, Yugoslavia.

Use. A minor source of molybdenum. Molybdenite, MoS_2, is the chief ore.

Name. After X. F. Wulfen, Austrian mineralogist.

Phosphates, Arsenates and Vanadates

This mineral group, composed mostly of phosphates, is very large (with 698 known species) but most of its members are so rare that they need not be mentioned here. Of the list of phosphates, arsenates, and vanadates given here, only apatite can be considered as common.

Triphylite—	$Li(Fe,Mn)PO_4$
Lithiophilite	$Li(Mn,Fe)PO_4$
Monazite	$(Ce,La,Y,Th)PO_4$
Apatite Group	
Apatite	$Ca_5(PO_4)_3(F,Cl,OH)$
Pyromorphite	$Pb_5(PO_4)_3Cl$
Vanadinite	$Pb_5(VO_4)_3Cl$
Erythrite	$Co_3(AsO_4)_2 \cdot 8H_2O$
Amblygonite	$LiAlPO_4F$
Lazulite—	$(Mg,Fe)Al_2(PO_4)_2(OH)_2$
Scorzalite	$(Fe,Mg)Al_2(PO_4)_2(OH)_2$
Wavellite	$Al_3(PO_4)_2(OH)_3 \cdot 5H_2O$
Turquoise	$CuAl_6(PO_4)_4(OH)_8 \cdot 4H_2O$
Autunite	$Ca(UO_2)_2(PO_4)_2 \cdot 10-12H_2O$
Carnotite	$K_2(UO_2)_2(VO_4)_2 \cdot 3H_2O$

Triphylite—$Li(Fe,Mn)PO_4$
Lithiophilite—$Li(Mn,Fe)PO_4$

Crystallography. Orthorhombic; $2/m2/m2/m$. Crystals are rare. Commonly in cleavable masses. Also compact.

Pmcn. For triphylite: $a = 6.01$, $b = 4.86$, $c = 10.36$ Å; $Z = 4$. *d*s: 4.29(8), 3.51(9), 3.03(9), 2.54(10), 1.75(5).

Physical Properties. *Cleavage* {001} nearly perfect, {010} imperfect. **H** $4\frac{1}{2}$–5. **G** 3.42–3.56 increasing with Fe content. *Luster* vitreous to resinous. *Color* bluish-gray in triphylite to salmon-pink or clove-brown in lithiophilite. May be stained black by manganese oxide. Translucent. *Optics:* (+); $\alpha = 1.669-1.694$, $\beta = 1.673-1.695$, $\gamma = 1.682-1.700$; $2V = 0°-55°$, $X = c$, $Y = a$, $Z = b$. Indices increase with Fe content.

Composition and Structure. A complete Fe^{2+}-Mn^{2+} series exists between two essentially pure end members. In the structure of members of this series Li and (Mn,Fe) are in 6-coordination. These octahedra are linked along their edges into zigzag chains that are connected by (PO_4) tetrahedra.

Diagnostic Features. Characterized by two cleavages at right angles, resinous luster, and association.

Occurrence. Triphylite and lithiophilite are pegmatite minerals associated with other phosphates, spodumene, and beryl. Notable localities are in Bavaria, Germany, and Finland. Lithiophilite is found at several localities in Argentina. In the United States triphylite is found in large crystals at the Palermo Mine, North Grothon, New Hampshire, and elsewhere in New England, and in the Black Hills, South Dakota. Lithiophilite is found at Branchville, Connecticut.

Name. Triphylite from the Greek words meaning *three* and *family* in allusion to containing three cations. Lithiophilite from the Greek words meaning *lithium* and *friend*.

Monazite—(Ce,La,Y,Th)PO₄

Crystallography. Monoclinic; $2/m$. Crystals are rare and usually small, often flattened on {100}, or elongated on b. Usually in granular masses, frequently as sand.

$P2_1/n$; $a = 6.79$, $b = 7.01$, $c = 6.46$ Å; $\beta = 103°38'$; $Z = 4$. ds: 4.17(3), 3.30(5), 3.09(10), 2.99(2), 2.87(7).

Physical Properties. *Cleavage* {100} poor. *Parting* {001}. **H** 5–5$\frac{1}{2}$. **G** 4.6–5.4. *Luster* resinous. *Color* yellowish to reddish-brown. Translucent. *Optics:* (+); $\alpha = 1.785$–1.800, $\beta = 1.787$–1.801, $\gamma = 1.840$–1.850; 2V = 10°–20°; $X = b$, $Z \wedge c = 2°$–6°.

Composition and Structure. A phosphate of the rare-earth metals essentially (Ce,La,Y,Th)PO₄. Th content ranges from a few to 20% ThO₂. Si is often present up to several percent SiO₂. The Si has been ascribed to admixture of *thorite*, ThSiO₄, but may be in part due to substitution of (SiO₄) for (PO₄). In the structure of monazite the rare earths are in 9-coordination with oxygen, linking six PO₄ tetrahedra. It is isostructural with *crocoite*, PbCrO₄ (see page 428).

Diagnostic Features. Radioactive. Large specimens may be distinguished from zircon by crystal form and inferior hardness, and from titanite by crystal form and higher specific gravity. On doubtful specimens it is usually wise to make a chemical phosphate test.

Occurrence. Monazite is a comparatively rare mineral occurring as an accessory in granites, gneisses, aplites, and pegmatites, and as rolled grains in the sands derived from the decomposition of such rocks. It is concentrated in sands because of its resistance to chemical attack and its high specific gravity, and is thus associated with other resistant and heavy minerals such as magnetite, ilmenite, rutile, and zircon.

The bulk of the world's supply of monazite comes from beach sands in Brazil, India, and Australia. A dikelike body of massive granular monazite was mined near Van Rhynsdorp, Cape Province, South Africa. Found in the United States in North Carolina, both in gneisses and in the stream sands; and in the beach sands of Florida.

Use. Monazite is the chief source of thorium oxide, which it contains in amounts varying between 1 and 20%; commercial monazite usually contains between 3 and 9%. Thorium oxide is used in the manufacture of mantles for incandescent gas lights.

Thorium is a radioactive element and is receiving considerable attention as a source of atomic energy. A commercial thorium-fueled reactor is in operation at Fort St. Vrain, Colorado.

Name. The name *monazite* is derived from the Greek word *monachos* meaning *solitary*, in allusion to the rarity of the mineral.

APATITE GROUP
Apatite—Ca₅(PO₄)₃(F,Cl,OH)

Crystallography. Hexagonal; $6/m$. Commonly occurs in crystals of long prismatic habit, see Fig. 10.48; some short prismatic or tabular. Usually terminated by prominent dipyramid, {10$\bar{1}$1}, and fre-

FIG. 10.48. Apatite in marble, Renfrew County, Ontario, Canada.

quently a basal plane. Some crystals show faces of a hexagonal dipyramid (μ, Fig. 10.49c) which reveals the true symmetry. Also in massive granular to compact masses.

$P6_3/m$; $a = 9.39$, $c = 6.89$ Å; $Z = 2$. ds: 2.80(10), 2.77(4), 2.70(6), 1.84(6), 1.745(3).

Physical Properties. *Cleavage* {0001} poor. **H** 5 (can just be scratched by a knife). **G** 3.15–3.20. *Luster* vitreous to subresinous. *Color* usually some shade of green or brown; also blue, violet, colorless, see Plate IV, no. 4. Transparent to translucent. *Optics:* (−); $\omega = 1.633$, $\epsilon = 1.630$ (fluorapatite).

Composition and Structure. Ca₅(PO₄)₃F, *fluorapatite* is most common; more rarely Ca₅(PO₄)₃Cl, *chlorapatite*, and Ca₅(PO₄)₃(OH), *hydroxylapatite*. F, Cl, and OH can substitute for each other, giving complete series. (CO₃, OH) may substitute for (PO₄) giving *carbonate-apatite*. The (PO₄) group can be partially replaced by (SO₄) as well as (SiO₄). The P^{5+} to S^{6+} replacement is compensated by the coupled substitution of Ca^{2+} by Na^+. Furthermore the P^{5+} to S^{6+} replacement may be balanced by substitution of Si^{4+} for P^{5+}. Mn and Sr can substitute in part for Ca. The structure of apatite is illustrated in Fig. 10.12.

Collophane. The name collophane has been given to the massive, cryptocrystalline types of apatite that constitute the bulk of phosphate rock and fossil bone. X-ray study shows that collophane is essentially apatite and does not warrant designation as a separate species. In its physical appearance, collophane is usually dense and massive with a concretionary or colloform structure. It is usually impure and contains small amounts of calcium carbonate.

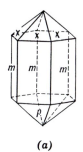 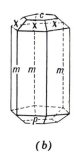 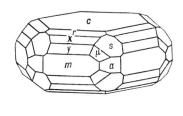

(a)   *(b)*   *(c)*   **FIG. 10.49.** Apatite crystals.

Diagnostic Features. Apatite is usually recognized by its crystals, color, and hardness. Distinguished from beryl by the prominent pyramidal terminations of its crystals and by its being softer than a knife blade.

Occurrence. Apatite is widely disseminated as an accessory constituent in all classes of rocks—igneous, sedimentary, and metamorphic. It is also found in pegmatites and other veins, probably of hydrothermal origin. Found in titaniferous magnetite bodies. Occasionally concentrated into large deposits or veins associated with alkalic rocks. Phosphate materials of bones and teeth are members of the apatite group.

Apatite occurs in large amounts along the southern coast of Norway, between Langesund and Arendal, where it is found in veins and pockets associated with gabbro. It is distributed through the magnetite iron ore at Kiruna, Sweden, and Tahawas, New York. In Ontario and Quebec, Canada, large apatite crystals were formerly mined. The world's largest deposit of apatite is located on the Kola Peninsula, near Kirovsk, Karelia, Russia. Here apatite in granular aggregates intimately associated with nepheline and titanite, is found in a great lens between two types of alkalic rocks.

Finely crystallized apatite in pegmatites and veins occurs at various localities in the Tyrol, Austria; in Switzerland; in Panasquiera, Portugal; and Jumilla, Spain. In the United States at Auburn, Maine; St. Lawrence County, New York; Alexander County, North Carolina; and San Diego County, California.

The variety collophane is an important constituent of the rock *phosphorite* or *phosphate rock.* Bone is calcium phosphate, and large bodies of phosphorite are derived from the accumulation of animal remains as well as from chemical precipitation from seawater. Commercial deposits of phosphorite are found in northern France, Belgium, Spain, and especially in northern Africa in Tunisia, Algeria, and Morocco. In the United States, high-grade phosphate deposits are found in Tennessee

and in Wyoming and Idaho. Deposits of "pebble" phosphate are found at intervals all along the Atlantic coast from North Carolina to Florida. The most productive deposits in the United States are in Florida.

Use. Crystallized apatite has been used extensively as a source of phosphate for fertilizer, but today only the deposits on the Kola Peninsula are of importance and phosphorite deposits supply most of the phosphate for fertilizer (see Box 10.1). The calcium phosphate is treated with sulfuric acid and changed to superphosphate to render it more soluble in the dilute acids that exist in the soil. Transparent varieties of apatite of fine color are occasionally used for gems (see Plate XII, no. 8 Chapter 13). The mineral is too soft, however, to allow its extensive use for this purpose.

Name. From the Greek word *apate* meaning *deceipt,* because the gem varieties were confused with other minerals.

Pyromorphite—Pb$_5$(PO$_4$)$_3$Cl

Crystallography. Hexagonal; 6/m. Crystals are usually prismatic with basal plane (Fig. 10.50). Rarely shows pyramid truncations. Often in rounded barrel-shaped forms. Sometimes cavernous, the crystals being hollow prisms. Also in parallel groups. Frequently globular, reniform, fibrous, and granular.

FIG. 10.50. Pyromorphite.

BOX 10.1
THE SOURCE OF CHEMICALS IN FERTILIZERS

Containers for chemical fertilizers prominently display three numbers that represent the weight percentages of three major components: nitrogen, phosphorus, and potassium–**(NPK)**. Other components such as S, Mg, and Fe are generally present in smaller amounts and are listed in a table that is in smaller type.

The nitrogen component in fertilizers is obtained from ammonia, NH_3, and nitrate (NO_3^-) salts. Most of the ammonia is produced by the fractional distillation of liquid air. The four major producers of ammonia are Russia, China, the United States, and the European Community.

The phosphorus component in fertilizers is derived from sedimentary phosphate rock (also known as phosphorite). About 90% of the world's production of phosphate is consumed in chemical fertilizers. The mineral that is mined is **apatite**, $Ca_5(PO_4)_3(F,$ Cl, OH). Apatite occurs in igneous as well as sedimentary rocks. Apatite of igneous origin is commonly close to $Ca_5(PO_4)_3(OH, F)$ in composition, whereas sedimentary apatite, also known as carbonate-apatite, is closer to $Ca_5(PO_4, CO_3, OH)_3(F, OH)$. The sedimentary deposits contain apatite in nodular or compact masses. The apatite is derived from a variety of sources, including marine invertebrates that secrete shells of calcium phosphate and the bones and excre-

ment of vertebrates. Sedimentary deposition of phosphate continues today. Typical phosphorite beds contain about 30% P_2O_5. The United States is the world's largest exporter of phosphate rock, whereas Morocco may have the largest reserves of phosphate raw materials. Major U.S.-producing regions are along the eastern coastline (South Carolina, Georgia, and Florida) and in the Phosphoria Formation in Idaho.

Potash is the third component in most chemical fertilizers. This is mined from potash-bearing salt deposits that have resulted from the evaporation of seawater (see Box 9.2). The most important potassium minerals are **sylvite**, KCl, **carnallite**, and **langbeinite** (see Box 9.2 for compositions). A mixture of KCl and NaCl, known as *sylvinite*, is the highest-grade potash ore. The largest high-grade sylvinite ore reserves are found in Saskatchewan, Canada. Russia ranks second with about 32% of the world's potash reserves in several sedimentary basins.

Sulfur is a common additional component of chemical fertilizers. Only about 10% of sulfur production is

shipped in international markets as elemental sulfur. Most is recovered or

FERTILIZER 16-8-8	
	%
Total nitrogen (N)	16.00
Available phosphoric acid (P_2O_5)	8.00
Soluble potash (K_2O)	8.00
Sulfur	14.00

converted to sulfuric acid, H_2SO_4. Sulfuric acid is not only an important basic industrial chemical, but the largest use of sulfuric acid is in the chemical process of converting phosphate rock into components of chemical fertilizers. The four largest world producers of sulfur are the United States, Russia, Canada, and the European Community. Most of the world's production is from *salt domes*. These are largely subsurface geologic structures that consist of a vertical cylinder of salt embedded in horizontal or inclined strata. Salt domes along the coast of the Gulf of Mexico contain abundant sulfur, as well as NaCl and potash. Sulfur is also recovered (ultimately as sulfuric acid) as a byproduct from natural gas and crude oil, and from sulfide ores in which a common constituent is pyrite, FeS_2.

$P6_3m$; $a = 9.97$, $c = 7.32$ Å; $Z = 2$. ds: 4.31(6), 4.09(9), 2.95(10), 2.05(8), 1.94(7).

Physical Properties. H $3\frac{1}{2}$–4. G 7.04. *Luster* resinous to adamantine. *Color* usually various shades of green, brown, yellow; rarely orange-yellow, gray, white. Subtransparent to translucent. *Optics:* (−), $\omega = 2.058$, $\epsilon = 2.048$.

Composition and Structure. For pure $Pb_5(PO_4)_3Cl$, PbO 82.2, P_2O_5 15.7, Cl 2.6%. (AsO_4) substitutes for (PO_4) and a complete series extends to *mimetite*, $Pb_5(AsO_4)_3Cl$. Ca may substitute in part for Pb. Isostructural with apatite (see Fig. 10.12).

Diagnostic Features. Pyromorphite is characterized by its crystal form, high luster, and high specific gravity.

Occurrence. Pyromorphite is a supergene mineral found in the oxidized portions of lead veins, associated with other oxidized lead and zinc minerals.

Notable localities for its occurrence are the lead mines at Ems in Nassau and at Zschopau, Saxony, Germany; Příbram, Bohemia, Czechoslovakia; Beresovsk, Ural Mountains; in Cumbria, England and at Leadhills, Scotland. In the United States it is found at Phoenixville, Pennsylvania, and the Coeur d'Alene district, Idaho.

Use. A subordinate ore of lead.

Name. Derived from two Greek words meaning *fire* and *form*, in allusion to the apparent crystalline form it assumes on cooling from fusion.

Similar Species. *Mimetite*, $Pb_5(AsO_4)_3Cl$, is isostructural with pyromorphite and is similar in appearance, occurrence, and most of its physical and chemical properties.

Vanadinite—$Pb_5(VO_4)_3Cl$

Crystallography. Hexagonal; $6/m$. Most commonly occurs in prismatic crystals with $\{10\bar{1}0\}$ and $\{0001\}$. May have small pyramidal faces, rarely the hexagonal dipyramid. In rounded crystals; in some cases cavernous. Also in globular forms. As incrustations.

$P6_3/m$; $a = 10.33$, $c = 7.35$ Å; $Z = 2$. ds: 4.47(3), 4.22(4), 3.38(6), 3.07(9), 2.99(10).

Physical Properties. **H** 3. **G** 6.9. *Luster* resinous to adamantine. *Color* ruby-red, orange-red, brown, yellow. Transparent to translucent. *Optics:* ($-$); $\omega = 2.25–2.42$, $\epsilon = 2.20–2.35$. Indices lowered by substitution of As or P for V.

Composition. PbO 78.7, V_2O_5 19.4, Cl 2.5%. PO_4 and AsO_4 may substitute in small amounts for VO_4. In the variety *endlichite*, intermediate between vanadinite and mimetite, the proportion of V_2O_5 to As_2O_5 is nearly 1 : 1. Small amounts of Ca, Zn, and Cu substitute for Pb. Vanadinite is isostructural with apatite (see Fig. 10.12).

Diagnostic Features. Characterized by crystal form, high luster, and high specific gravity; distinguished from pyromorphite and mimetite by color.

Occurrence. Vanadinite is a rare secondary mineral found in the oxidized portion of lead veins associated with other secondary lead minerals. Found in fine crystals near Oudjda, Morocco, and Grootfontein, Namibia. In the United States it occurs in various districts in Arizona and New Mexico.

Use. Source of vanadium and minor ore of lead. Vanadium is obtained chiefly from other ores, such as *patronite*, VS_4; the vanadate *carnotite*, $K_2(UO_2)_2(VO_4)_2 \cdot 3H_2O$; and a vanadium mica, *roscoelite*, $KV_2(AlSi_3O_{10})(OH)_2$. Vanadium is used chiefly as a steel-hardening metal. Metavanadic acid, HVO_3, is a yellow pigment, known as vanadium bronze. Vanadium oxide is a mordant in dyeing.

Name. In allusion to the composition.

Erythrite—$Co_3(AsO_4)_2 \cdot 8H_2O$

Crystallography. Monoclinic; $2/m$. Crystals are prismatic and vertically striated. Usually as crusts in globular and reniform shapes. Also pulverulent and earthy.

$C2/m$; $a = 10.20$, $b = 13.37$, $c = 4.74$ Å; $\beta = 105°$; $Z = 2$. ds: 6.65(10), 3.34(1), 3.22(1), 2.70(1), 2.32(1).

Physical Properties. *Cleavage* $\{010\}$ perfect. **H** $1\frac{1}{2}–2\frac{1}{2}$. **G** 3.06. *Luster* adamantine to vitreous, pearly on cleavage. *Color* crimson to pink. Translucent. *Optics:* ($-$); $\alpha = 1.626$, $\beta = 1.661$, $\gamma = 1.699$; $2V = 90°\pm$; $X = b$, $Z \wedge c = 31°$; $r > v$. Pleochroism X pink, Y violet, Z red.

Composition and Structure. CoO 37.5, As_2O_5 38.4, H_2O 24.1%. Ni substitutes for Co to form a complete series to *annabergite*, $Ni_3(AsO_4)_2 \cdot 8H_2O$. *Annabergite*, or *nickel bloom*, is light green in color. The structure of erythrite is of a layer type with strong bonds between (AsO_4) tetrahedra and $Co(O,H_2O)$ octahedra that are linked by common

vertices. The layers parallel to (010) are held together by weak residual bonds.

Diagnostic Features. The association of erythrite with other cobalt minerals and its pink color are usually sufficient to distinguish it from all other minerals.

Occurrence. Erythrite is a rare secondary mineral. In pink crusts known as *cobalt bloom* it occurs as an alteration product of cobalt arsenides. It is rarely present in large amounts and usually forms as crusts or fine aggregates filling cracks. Notable localities are at Schneeberg, Saxony, Germany, and Bon Azzer, Morocco.

Use. Although erythrite has no economic importance, it is used by the prospector as a guide to other cobalt minerals and associated native silver.

Name. From the Greek word *erythros*, meaning *red*.

Similar Species. *Vivianite*, $Fe_3(PO_4)_2 \cdot 8H_2O$, a rare mineral which is a weathering product of primary Fe–Mn phosphates in pegmatites.

Amblygonite—$LiAlFPO_4$

Crystallography. Triclinic: $\bar{1}$. Usually occurs in coarse, cleavable masses. Crystals are rare, equant, and usually rough when large. Frequently twinned on $\{1\bar{1}1\}$.

$P\bar{1}$; $a = 5.19$, $b = 7.12$, $c = 5.04$ Å; $\alpha = 112°02'$, $\beta = 97°50'$, $\gamma = 68°8'$; $Z = 2$. ds: 4.64(10), 3.15(10), 2.93(10), 2.39(5), 2.11(4).

Physical Properties. *Cleavage* $\{100\}$ perfect, $\{110\}$ good, $\{0\bar{1}1\}$ distinct. **H** 6. **G** 3.0–3.1. *Luster* vitreous, pearly on $\{100\}$ cleavage. *Color* white to pale green or blue, rarely yellow (see Plate XII, no. 4, Chapter 13). Translucent. *Optics:* usually ($-$); $\alpha = 1.58–1.60$, $\beta = 1.50–1.62$, $\gamma = 1.60–1.63$; $2V = 50–90°$; $r > v$. Indices increase with increase in (OH) in substitution for F.

Composition and Structure. Li_2O 10.1, Al_2O_3 34.4, F 12.9, P_2O_5 47.9%. Na substitutes for Li; (OH) substitutes for F and probably forms a complete series. When OH > F, the mineral is known as *montebrasite*. In the structure of amblygonite AlO_6 octahedra and PO_4 tetrahedra are linked by vertices; Li is in 5-coordination and lies between PO_4 tetrahedra and the nearest Al octahedra. The structure is fairly compact as reflected in the relatively high density.

Diagnostic Features. Cleavage fragments may be confused with feldspar, but are distinguished by cleavage angles.

Occurrence. Amblygonite is a rare mineral found in granite pegmatites with spodumene, tourmaline, lepidolite, and apatite. Found at Montebras, France. In the United States it occurs at Hebron, Paris, Auburn, and Peru, Maine; Portland, Connecticut, and the Black Hills, South Dakota.

Use. A source of lithium.

Name. From the two Greek words *amblys,* meaning *blunt,* and *gonia,* meaning *angle,* in allusion to the angle between the cleavages.

Similar Species. *Beryllonite,* $NaBePO_4$, colorless and *brazilianite,* $NaAl_3(PO_4)_2(OH)_4$, colorless to yellow, are rare gem minerals found in pegmatites.

Lazulite—$(Mg,Fe)Al_2(PO_4)_2(OH)_2$; Scorzalite—$(Fe,Mg)Al_2(PO_4)_2(OH)_2$

Crystallography. Monoclinic; $2/m$. Crystals showing steep $\{hkl\}$ prisms are rare. Usually massive, granular to compact.

$P2_1/c$; $a = 7.12$, $b = 7.26$, $c = 7.24$ Å; $\beta = 118°55'$; $Z = 2$. *d*s: 6.15(8), 3.23(8), 3.20(7), 3.14(10), 3.07(10).

Physical Properties. *Cleavage* $\{110\}$ indistinct. **H** $5-5\frac{1}{2}$. **G** 3.0–3.1. *Luster* vitreous. *Color* azure-blue. Translucent. *Optics:* (−); $\alpha = 1.604-1.639$, $\beta = 1.626-1.670$, $\gamma = 1.637-1.680$, 2V = 60°; $Y = b$, $X \wedge c = 10°$; $r < v$. Absorption $X < Y < Z$. Indices increase with increasing Fe^{2+} content.

Composition and Structure. Probably a complete solid solution series exists from lazulite to scorzalite with the substitution of Fe^{2+} for Mg. In the structure $(Mg,Fe)(O,OH)_6$ octahedra are linked by edges and faces with $Al(O,OH)_6$ octahedra to form groups. These groups are joined to each other and to PO_4 tetrahedra.

Diagnostic Features. If crystals are lacking, lazulite is difficult to distinguish from other blue minerals without optical or chemical tests.

Occurrence. The members of the lazulite-scorzalite series are rare minerals found in high-grade metamorphic rocks and in pegmatites. They are usually associated with kyanite, andalusite, corundum, rutile, sillimanite, and garnet. Notable localities are Salzburg, Austria; Krieglach, Styria; and Horrsjoberg, Sweden. In the United States they are found with corundum on Crowder's Mountain, Gaston County, North Carolina; with rutile on Graves Mountain, Lincoln County, Georgia; and with andalusite in the White Mountains, Inyo County, California.

Use. A minor gemstone.

Name. Lazulite derived from an Arabic word meaning *heaven,* in allusion to the color of the mineral. Scorzalite after E. P. Scorza, Brazilian mineralogist.

Wavellite—$Al_3(PO_4)_2(OH)_3 \cdot 5H_2O$

Crystallography. Orthorhombic; $2/m2/m2/m$. Single crystals rare. Usually in radiating, spherulitic (Fig. 10.51) and globular aggregates.

Pcmn; $a = 9.62$, $b = 17.34$, $c = 6.99$ Å; $Z = 4$. *d*s: 8.39(10), 5.64(6), 3.44(8), 3.20(8), 2.56(8).

Physical Properties. *Cleavage* $\{110\}$ and $\{101\}$ good. **H** $3\frac{1}{2}-4$. **G** 2.36. *Luster* vitreous. *Color* white, yellow, green, and brown. Translucent. *Optics:* (+); $\alpha = 1.525$, $\beta = 1.535$, $\gamma = 1.550$; 2V = 70°. $X = b$, $Y = a$; $r > v$.

Composition and Structure. Al_2O_3 38.0, P_2O_5 35.2, H_2O 26.8%. F may substitute for OH. Details of the structure are uncertain. $Al(O,OH)_6$ octahedra are linked by common vertices to PO_4 tetrahedra.

FIG. 10.51. Wavellite, Hot Springs, Arkansas (Harvard Mineralogical Museum).

Diagnostic Features. Almost invariably found in radiating globular aggregates.

Occurrence. Wavellite is a secondary mineral found in small amounts in crevices in aluminous, low-grade metamorphic rocks, and in limonite and phosphorite deposits. Although it occurs in many localities, only rarely is it found in quantity. It is abundant in the tin veins of Llallagua, Bolivia. In the United States wavellite occurs in a number of localities in Arkansas.

Name. After Dr. William Wavel, who discovered the mineral.

Turquoise—$CuAl_6(PO_4)_4(OH)_8 \cdot 5H_2O$

Crystallography. Triclinic; $\bar{1}$. Rarely in minute crystals, usually cryptocrystalline. Massive compact, reniform, stalactitic. In thin seams, incrustations, and disseminated grains.

$P\bar{1}$; $a = 7.48$, $b = 9.95$, $c = 7.69$; $\alpha = 111°39'$, $\beta = 115°23'$, $\gamma = 69°26'$; $Z = 1$. *d*s: 6.16(7), 4.80(6), 3.68(10), 3.44(7), 3.28(7).

Physical Properties. *Cleavage* $\{001\}$ perfect, $\{010\}$ good (rarely seen). **H** 6. **G** 2.6–2.8. *Luster* wax-like. *Color* blue, bluish-green, green (see Plates IV, no. 5 and XI, no. 7). Transmits light on thin edges. *Optics:* (+); $\alpha = 1.61$, $\beta = 1.62$, $\gamma = 1.65$; 2V = 40°; $r < v$ strong.

Composition and Structure. Fe^{3+} substitutes for Al and a complete series exists between turquoise and *chalcosiderite,* $CuFe_6^{3+}(PO_4)_4(OH)_8 \cdot 4H_2O$. The structure consists of PO_4 tetrahedra and (Al,Fe^{3+}) octahedra linked by common vertices. Fairly large holes in the structure contain Cu which is coordinated to 4(OH) and 2 H_2O molecules.

Diagnostic Features. Turquoise can be recognized by its color. It is harder than chrysocolla, the only common mineral which it resembles. There is much imitation material on the market (see Chapter 13).

Occurrence. Turquoise is a secondary mineral usually found in the form of small veins and stringers traversing more or less decomposed volcanic rocks in arid regions. The famous Persian deposits are found in trachyte near Nishapur in the province of Khorasan, Iran. In the United States it is found in a much altered trachytic rock in the Los Cerillos Mountains, near Santa Fe, and elsewhere in New Mexico. Turquoise is also found in Arizona, Nevada, and California. Small crystals have been found at Lunch Station, Virginia.

Use. As a gemstone. It is always cut in round or oval forms. Much cut turquoise is veined with the various gangue materials, and such stones are sold under the name of *turquoise matrix*.

Name. Turquoise is French and means *Turkish*, the original stones having come into Europe from the Persian locality through Turkey.

Similar Species. *Variscite*, $Al(PO_4) \cdot 2H_2O$, is a massive, bluish-green mineral somewhat resembling turquoise and used as a gem material. It has been found in nodules in a large deposit at Fairfield, Utah.

Autunite—Ca(UO₂)₂(PO₄)₂·10–12H₂O

Crystallography. Tetragonal; $4/m2/m2/m$. Crystals are tabular on {001}; subparallel growths are common; also foliated and scaly aggregates.

$I4/mmm$; $a = 7.00$, $c = 20.67$ Å; $Z = 2$. *d*s: 10.33(10), 4.96(8), 3.59(7), 3.49(7), 3.33(7).

Physical Properties. *Cleavage* {001} perfect. **H** $2-2\frac{1}{2}$. **G** 3.1–3.2. *Luster* vitreous, pearly on {001}. *Color* lemon yellow to pale green. *Streak* yellow. In ultraviolet light fluoresces strongly yellow-green. *Optics:* (−); $\omega = 1.577$, $\epsilon = 1.553$. Pleochroism E pale yellow, O dark yellow.

Composition and Structure. Small amounts of Ba and Mg may substitute for Ca. The H_2O content apparently ranges from 10 to 12 H_2O. On drying and slight heating autunite passes reversibly to *meta-autunite* I (a tetragonal phase with $6\frac{1}{2}-2\frac{1}{2}$ H_2O). On continued heating to about 80°C this passes irreversibly to *meta-autunite* II (an orthorhombic phase with 0–6 H_2O). Neither meta-I nor the meta-II hydrate occurs as a primary phase in nature. The structure of autunite consists of (PO_4) tetrahedra and $(UO_2)O_4$ polyhedra, which are joined into tetragonal corrugated layers of composition $UO_2(PO_4)$ parallel to {001}. These layers are held together by weak hydrogen bonds to H_2O molecules.

Diagnostic Features. Autunite is characterized by yellow-green tetragonal plates and strong fluorescence in ultraviolet light. Radioactive.

Occurence. Autunite is a secondary mineral found chiefly in the zone of oxidation and weathering derived from the alteration of uraninite or other uranium minerals. Notable localities are near Autun, France; Sabugal and Vizeu, Portugal; Johanngeorgenstadt district and Falkenstein, Germany; Cornwall, England; and the Katanga district of Zaire. In the United States autunite is found in many pegmatites, notably at the Ruggles Mine, Grafton Center, New Hampshire; Black Hills, South Dakota; and Spruce Pine, Mitchell County, North Carolina. The finest specimens have come from the Daybreak Mine near Spokane, Washington.

Use. An ore of uranium (see uraninite, page 386).

Name. From Autun, France.

Similar Species. *Torbernite*, $Cu(UO_2)_2(PO_4)_2.8-12H_2O$, is isostructural with autunite and has similar properties, but there is no evidence of a solid solution series. Color green, nonfluorescent. Associated with autunite.

Carnotite—K₂(UO₂)₂(VO₄)₂·3H₂O

Crystallography. Monoclinic; $2/m$. Only rarely in imperfect microscopic crystals flattened on {001} or elongated on b. Usually found as a powder or as loosely coherent aggregates; disseminated.

$P2_1a$; $a = 10.47$, $b = 8.41$, $c = 6.91$ Å; $\beta = 103°40'$; $Z = 2$. *d*s: 6.56(10), 4.25(3), 3.53(5), 3.25(3), 3.12(7).

Physical Properties. *Cleavage* {001} perfect. *Hardness* unknown, but soft. **G** 4.7–5. *Luster* dull or earthy. *Color* bright yellow to greenish-yellow. *Optics:* (−); $\alpha = 1.75$, $\beta = 1.93$, $\gamma = 1.95$; $2V = 40°\pm$, $Y = b$, $X = c$. $r < v$. Indices increase with loss of water.

Composition and Structure. The water content of carnotite is partly zeolitic and varies with the humidity at ordinary temperatures; the $3H_2O$ is for fully hydrated material. Small amounts of Ca, Ba, Mg, Fe, and Na have been reported. The structure of carnotite consists of a layer pattern that is the result of strong bonds between the (VO_4) groups and the $UO_2(O_5)$ polyhedra. These layers, of composition $(UO_2)_2(VO_4)_2$, are held together by hydroxyl–hydrogen bonds to water molecules and also by K, Ca, and Ba between the layers.

Diagnostic Features. Carnotite is characterized by its yellow color, pulverulent nature, radioactivity, and occurrence. Unlike many secondary uranium minerals, carnotite will not fluoresce in ultraviolet light.

Occurrence. Carnotite is of secondary origin, and its formation is usually ascribed to the action of meteoric waters on preexisting uranium and vanadium minerals. It has a strong pigmenting power and when present in a sandstone in amounts even less than 1% will color the rock yellow. It is found principally in the plateau region of southwestern Colorado and in adjoining districts of Utah, where it occurs disseminated in a cross-bedded sandstone. Concentrations of relatively pure carnotite are found around petrified tree trunks.

Use. Carnotite is an ore of vanadium and, in the United States, a principal ore of uranium.

Name. After Marie-Adolphe Carnot (1839–1920), French mining engineer and chemist.

Similar Species. *Tyuyamunite*, $Ca(UO_2)_2(VO_4)_2 \cdot 5-8\frac{1}{2}H_2O$, is the calcium analogue of carnotite and similar in physical properties except for a slightly more greenish color and yellow-green fluorescence. It is found in almost all carnotite deposits. Named for Tyuya Muyum, southeastern Turkistan, where it is mined as a uranium ore.

REFERENCES AND SELECTED READING

Anthony, J. W., R. A. Bideaux, K. W. Bladh, and M. C. Nichols. *Handbook of Mineralogy,* Vol. IV, *Arsenates, Phosphates, Vanadates.* Tucson, Ariz.: Mineral Data Publications.

Aristarain, L. F., and C. S. Hurlbut Jr. 1972. Boron minerals and deposits. *Mineralogical Record* 3:165–72, 213–20.

Boron mineralogy, petrology and geochemistry, 1996. *Reviews in Mineralogy,* 33. Mineralogical Society of America, Washington, D. C.

Carbonates. 1983. *Reviews in Mineralogy* 11. Mineralogical Society of America, Washington, D. C.

Carr, D. C., ed. 1994. *Industrial minerals and rocks.* Littleton, Colorado: Society for Mining, Metallurgy, and Exploration.

Chang, L. L. Y., R. A. Howie, and J. Zussman. 1996. *Rock-forming minerals* 5B. Nonsilicates. Longman Group Limited.

Gaines, R. B., H. C. W. Skinner, E. E. Foord, B. Mason, and A. Rosenzweig. 1997. *Dana's new mineralogy.* New York: Wiley.

Morgan, V., and R. C. Erd. 1969. Minerals of the Kramer borate district, California. *Mineral Information Service* (California Division of Mines and Geology) 22: 143–53, 165–72.

Palache, C., H. Berman, and C. Frondel. 1951. *The system of mineralogy.* 7th ed. Vol. 2. New York: Wiley.

Phosphate, potash, and sulfur—A special issue. 1979. *Economic Geology* 74: 191–496.

Strunz, H. 1970. *Mineralogische Tabellen.* 5th ed. Akademische Verlagsgesellschaft, Leipzig.

CHAPTER 11

CRYSTAL CHEMISTRY OF ROCK-FORMING SILICATES

The silicate mineral class is of great importance because about 27% of the known minerals and nearly 40% of the common minerals are silicates. With few exceptions all the igneous rock-forming minerals are silicates, and thus they constitute well over 90% of the Earth's crust (see Fig. 11.1). Rock-forming is said of those minerals that are part of the mineral makeup of rocks and are used in their classification. The more important rock-forming silicate minerals include olivine, garnet, pyroxenes, amphiboles, micas, clay minerals, feldspars, and quartz. Other rock-forming minerals, among the carbonates, are calcite and dolomite.

When the average weight percentages of the eight most common elements in the Earth's crust are recalculated on the basis of atomic percent (see Fig. 3.2), we find that out of every 100 atoms 62.5 are O, 21.2 are Si, and 6.5 are Al. Fe, Mg, Ca, Na, and K each account for about two to three more atoms. With the possible exception of Ti, all other elements are present in insignificant amounts in the upper levels of the Earth's crust (see Fig. 3.2). When we recalculate the atomic percentages of the eight most abundant elements in terms of volume percentages (see Fig. 3.2, last column) we find that the Earth's crust can be regarded as a packing of oxygen ions, with interstitial metal ions, such as Si^{4+}, Al^{3+}, Fe^{2+}, Ca^{2+}, Na^+, K^+, and so forth.

The dominant minerals of the crust are thus shown to be silicates, with oxides and other oxygen compounds such as carbonates in subordinate amounts. Of the different assemblages of silicate minerals that characterize igneous, sedimentary, and metamorphic rocks, ore veins, pegmatites, weathered rocks, and soils, each has the potential to tell us something of the environment in which it was formed.

We have a further deep and compelling reason to study the silicates. The soil from which our food is ultimately drawn is made up in large part of silicates. The brick, stone, concrete, and glass used in the construction of our buildings either are silicates or are largely derived from silicates. Even with the coming of the space age, we need not fear obsolescence of our studies of the silicates, but rather an enlargement of their scope, because we now know that the moon and the four terrestrial planets of our solar system

FIG. 11.1. Estimated volume percentages for the common minerals in the Earth's crust, inclusive of continental and oceanic crust. Ninety-two percent are silicates. (From A. B. Ronov, and A. A. Yaroshevsky, 1969, Chemical composition of the Earth's crust. American Geophysical Union Monograph no. 13, p. 50.)

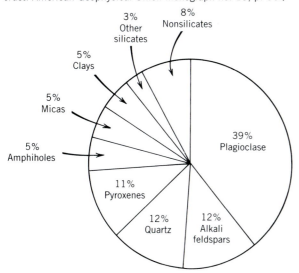

have rocky crusts made of silicates and oxides much like those of Earth.

The fundamental unit on which the structure of all silicates is based consists of four O^{2-} at the apices of a regular tetrahedron surrounding and coordinated by one Si^{4+} at the center (Figs. 10.1d and 11.2). The powerful bond that unites the oxygen and silicon ions is the cement that holds the Earth's crust together. This bond may be estimated by use of Pauling's electronegativity concept (see Fig. 3.23) as 50% ionic and 50% covalent. That is, although the bond arises in part from the attraction of oppositely charged ions, it also involves sharing of electrons and interpenetration of the electronic clouds of the ions involved. The bond is strongly localized in the vicinity of these shared electrons.

Although electron sharing is present in the Si–O bond, the total bonding energy of Si^{4+} is still distributed equally among its four closest oxygen neighbors. Hence, the strength of any single Si–O bond is equal to just one-half the total bonding energy available in the oxygen ion. Each O^{2-} has, therefore, the potentiality of bonding to another silicon ion and entering into another tetrahedral grouping, thus uniting the tetrahedral groups through the shared (or bridging) oxygen. Such linking of tetrahedra is often referred to as polymerization, a term borrowed from organic chemistry, and the capacity for polymerization is the origin of the great variety of silicate structures. In no case, however, are three or even two oxygens shared between two adjacent tetrahedra in nature. Such sharing would place two highly charged Si^{4+} ions close together, and the repulsion between them would render the structure unstable.

The sharing of oxygens may involve one, two, three, or all four of the oxygen ions in the tetrahedron, giving rise to a diversity of structural configurations. Figure 11.3 illustrates the various ways in which SiO_4 tetrahedra can be combined. Silicates with independent tetrahedral SiO_4 groups (in which the tetrahedra are not linked to each other) are known as nesosilicates (from the Greek nesos, meaning island) or orthosilicates (from the Greek orthos, meaning normal). Silicates in which two SiO_4 groups

are linked, giving rise to Si_2O_7 groups are classed as sorosilicates (from the Greek soros, meaning heap) or disilicates (in reference to the double tetrahedral groupings). If more than two tetrahedra are linked, closed ringlike structures are formed of a general composition Si_xO_{3x}. Fourfold tetrahedral rings have composition Si_4O_{12}. This group is known as the ring silicates, or the cyclosilicates (from the Greek kyklos, meaning circle). Tetrahedra may also be joined to form infinite single chains with a unit composition Si_2O_6 (or SiO_3). Infinite double chains give a ratio of Si : O = 4 : 11, resulting in Si_4O_{11} (or Si_8O_{22}). Both of these types of chain silicates are also known as inosilicates (from the Greek inos, meaning thread). When three of the oxygens of a tetrahedron are shared between adjoining tetrahedra, infinitely extending flat sheets are formed of unit composition Si_2O_5. Such sheet silicates are also referred to as phyllosilicates (from the Greek phyllon, meaning leaf). When all four oxygens of a SiO_4 tetrahedron are shared by adjoining tetrahedra, a three-dimensional network of unit composition SiO_2 results. These framework silicates are also known as tectosilicates (from the Greek word tecton, meaning builder).

In the subsequent treatment of silicates in this book, we will use the above structural classification (as illustrated in Fig. 11.3) of silicates. There are, however, alternate classifications such as those proposed by Liebau (1985) and Zoltai (1960).

Next to O and Si, the most important constituent of the crust is Al. Al^{3+} has a radius of 0.39 Å and thus the radius ratio Al : O = 0.286, corresponding to 4-coordination with oxygen. However, the radius ratio is sufficiently close to the upper limit for 4-coordination so that 6-coordination is also possible. It is this capacity for playing a double role in silicate minerals that gives Al^{3+} its outstanding significance in the crystal chemistry of the silicates. When Al coordinates four O's arranged at the apices of a regular tetrahedron, the resultant grouping occupies approximately the same space as a silicon–oxygen tetrahedron and may link with silicon tetrahedra in polymerized groupings. On the other hand, Al^{3+} in 6-coordination may serve to link the tetrahedral groupings through simple ionic bonds, weaker than those that unite the ions in the tetrahedra. It is thus possible to have Al in silicate structures both in the tetrahedral sites, substituting for Si, and in the octahedral sites with 6-coordination, involved in solid solution relations with elements such as Mg and Fe^{2+}.

Mg, Fe^{2+}, Fe^{3+}, Mn^{2+}, Al^{3+}, and Ti^{4+} all tend to occur in silicate structures in 6-coordination with oxygen (see Table 11.1). Although divalent, trivalent, and tetravalant ions are included here, all have

FIG. 11.2.   Close packing representation of an SiO_4 tetrahedron.

Class	Arrangement of SiO$_4$ tetrahedra (central Si^{4+} not shown)	Unit composition	Mineral example
Nesosilicates	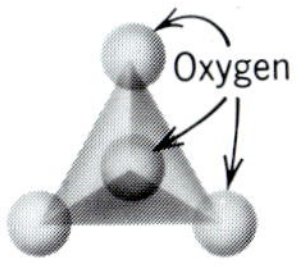 Oxygen	$(SiO_4)^{4-}$	Olivine, $(Mg, Fe)_2 SiO_4$
Sorosilicates	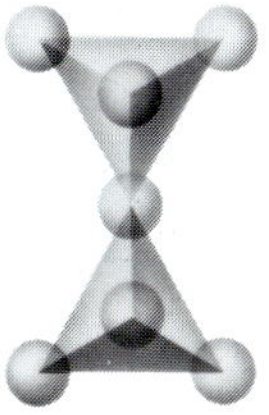	$(Si_2O_7)^{6-}$	Hemimorphite, $Zn_4Si_2O_7(OH) \cdot H_2O$
Cyclosilicates	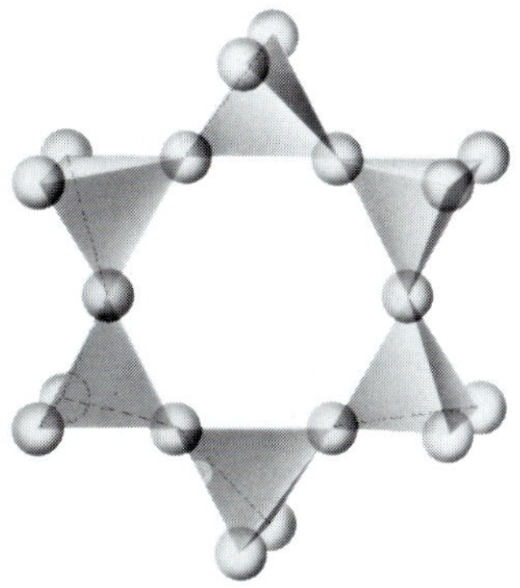	$(Si_6O_{18})^{12-}$	Beryl, $Be_3Al_2Si_6O_{18}$
Inosilicates (single chain)	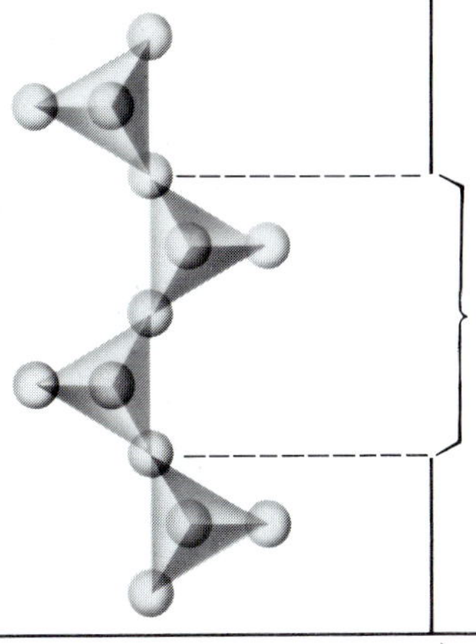	$(Si_2O_6)^{4-}$	Pyroxene e.g. Enstatite, $MgSiO_3$

FIG. 11.3. Silicate classification.

about the same space requirements and about the same radius ratio relations with oxygen, and, hence, tend to occupy the same type of atomic site. Solid solution relations between ions of different charge are possible through the mechanism of coupled substitution (see page 91).

The larger and more weakly charged cations, Ca^{2+} and Na^+, of ionic radii 1.12 Å and 1.16 Å, respectively, generally enter sites having 8-coordination with respect to oxygen. Although the sizes of these two ions are very similar, their charges are different. This charge difference is compensated for

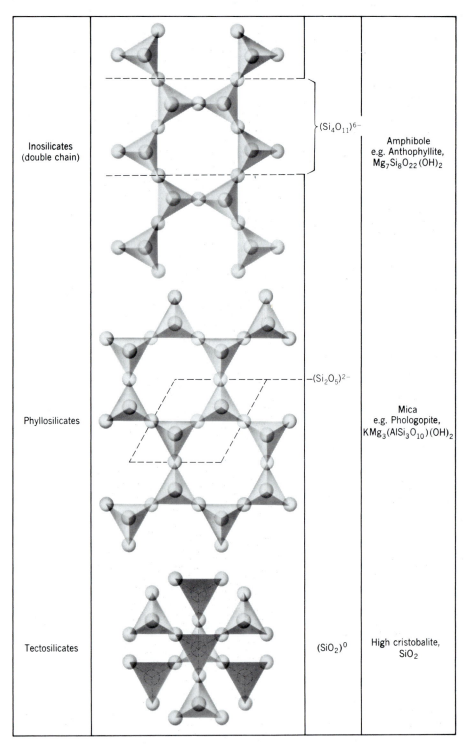

Inosilicates (double chain) — $(Si_4O_{11})^{6-}$ — Amphibole e.g. Anthophyllite, $Mg_7Si_8O_{22}(OH)_2$

Phyllosilicates — $(Si_2O_5)^{2-}$ — Mica e.g. Phologopite, $KMg_3(AlSi_3O_{10})(OH)_2$

Tectosilicates — $(SiO_2)^0$ — High cristobalite, SiO_2

FIG. 11.3. (continued)

by coupled substitution such as in the plagioclase feldspars where $(Na^+ + Si^{4+})$ substitutes for $(Ca^{2+} + Al^{3+})$ in order to keep the overall structure neutral.

The largest ions common in silicate structure are those of K, Rb, Ba, and the rarer alkalis and alkali earths. These ions generally do not enter readily into

Na or Ca sites and are found in high-coordination-number sites of unique type. Hence, solid solution between these ions and the other common ions is limited.

Ionic substitution is generally common and extensive between elements that are grouped together in Table 11.1. The usual role played by the com-

TABLE 11.1   Coodination of Common Elements in Silicates, Arranged in Decreasing Ionic Size*

	Ion	Coordination Number with Oxygen		IonicRadius Å
X	K^+	8–12		1.51[8]–1.64[12]
X	Na^+	8–6	cubic to octahedral	1.18[8]–1.02[6]
	Ca^{2+}	8–6		1.12[8]–1.00[6]
Y	Mn^{2+}	6		0.83[6]
	Fe^{2+}	6		0.78[6]
	Mg^{2+}	6	octahedral	0.72[6]
	Fe^{3+}	6		0.65[6]
	Ti^{4+}	6		0.61[6]
	Al^{3+}	6		0.54[6]
Z	Al^{3+}	4	tetrahedral	0.39[4]
	Si^{4+}	4		0.26[4]

*See Table 3.11 for a listing of ionic radii.

monest elements permits us to write a general formula for all silicates:

$$X_mY_n(Z_pO_q)W_r,$$

where X represents large, weakly charged cations in 8- or higher coordination with oxygen; Y represents medium-sized, two to four valent ions in 6-coordination; Z represents small, highly charged ions in tetrahedral coordination; O is oxygen; and W represents additional anionic groups such as $(OH)^-$ or anions

such as Cl^- or F^-. The ratio $p : q$ depends on the degree of polymerization of the silicate framework, and the other subscript variables, m, n, and r, depend on the need for electrical neutrality. Any common silicate may be expressed by suitable substitution in this general formula.

The subsequent treatment of individual silicates will be on the basis of subclasses that reflect their internal structure (neso-, soro-, cyclosilicates, etc.) and their chemical composition. Such a scheme is the basis of *Mineralogische Tabellen* (1970).

NESOSILICATES

In the nesosilicates the SiO_4 tetrahedra are isolated (Fig. 11.3) and bound to each other only by ionic bonds from interstitial cations. Their structures depend chiefly on the size and charge of the interstitial cations. The atomic packing of the nesosilicate structure is generally dense, causing the minerals of this group to have relatively high specific gravity and hardness. Because the SiO_4 tetrahedra are independent and not linked into chains or sheets, for example, the crystal habit of the nesosilicates is generally equidimensional and pronounced cleavage directions are absent. Although Al^{3+} substitutes commonly and easily in the Si position of silicates, the

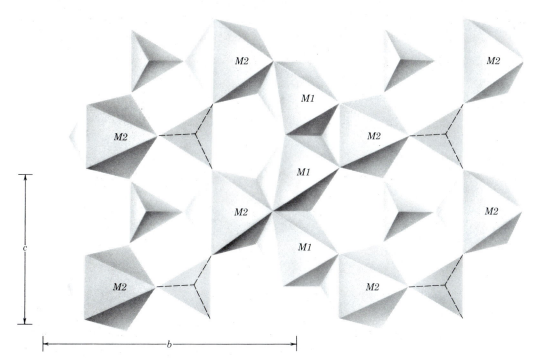

FIG. 11.4. Portion of the structure of olivine projected on (100). *M1* and *M2* are octahedral sites. The *M1* site is most distorted and the *M2* site is somewhat more regular. Extensive edge sharing among the polyhedra causes these distortions because shared edges are shortened as cations repel each other; see also page 79. (Redrawn after J. J. Papike and M. Cameron, 1976, Crystal chemistry of silicate minerals of geophysical interest. *Reviews of Geophysics and Space Physics* 14: 37–80).

amount of Al substitution in SiO_4 tetrahedra in nesosilicates is generally low.

Very common members, especially in high-temperature igneous rocks, of the nesosilicate group are *forsterite*, Mg_2SiO_4, and *fayalite*, Fe_2SiO_4, end members of the $(Mg,Fe)_2SiO_4$ olivine series. The structure of olivine, which is shown in Fig. 11.4, can be viewed as consisting of layers parallel to {100}. These layers consist of octahedra cross-linked by independent SiO_4 tetrahedra. The octahedrally coordinated sites are known as *M1* and *M2* with *M1* distorted and *M2* somewhat more regular. In the $(Mg,Fe)_2SiO_4$ olivines, Mg and Fe^{2+} occupy the *M1* and *M2* sites without any specific preference for either site. In the calcic olivines, however (e.g., *monticellite*, $CaMgSiO_4$), Ca enters into the *M2* site and Mg into *M1*.

Garnets are another group of very common nesosilicate minerals, especially in metamorphic rocks. Their structural formula may be represented as $A_3B_2(SiO_4)_3$, where *A* and *B* refer respectively to 8- and 6-coordinated cationic sites. The *A* sites are occupied by rather large divalent cations, whereas the *B* sites house smaller trivalent cations. Because of these size considerations in the filling of the *A* sites we may expect a fairly well-defined division of the garnets into those with Ca and those with easily interchangeable divalent ions such as Mg^{2+}, Fe^{2+},

Mn^{2+}. Similarly, because of the limited substitution possible in the *B* sites, we may expect a separation of garnets into Al^{3+}, Fe^{3+}, and Cr^{3+} bearing. These two trends are well marked and have given rise to a grouping of garnets into two series: *pyralspite* (Ca absent in *A*; *B* = Al) and *ugrandite* (*A* = Ca).

Pyralspite		**Ugrandite**	
Pyrope	$Mg_3Al_2Si_3O_{12}$	*Uvarovite*	$Ca_3Cr_2Si_3O_{12}$
Almandine	$Fe_3Al_2Si_3O_{12}$	*Grossular*	$Ca_3Al_2Si_3O_{12}$
Spessartine	$Mn_3Al_2Si_3O_{12}$	*Andradite*	$Ca_3Fe_2^{3+}Si_3O_{12}$

This classification serves as an excellent mnemonic aid for the names and formulas. Another grouping on the basis of the ions in the *B* site yields three unequal groups:

Aluminum Garnets	**Ferri-Garnet**	**Chrome-Garnet**
Pyrope	Andradite	Uvarovite
Almandine		
Spessartine		
Grossular		

Hydroxyl, as tetrahedral $(OH)_4$ groups, may substitute to a limited extent for SiO_4 tetrahedra in hydrogarnets such as *hydrogrossular*, $Ca_3Al_2Si_2O_8(SiO_4)_{1-m}(OH)_{4m}$, with *m* ranging from 0 to 1. Ti^{4+} may enter into the *B* sites concomitant with replacement of Ca by Na in the *A* sites, producing the black *melanite*.

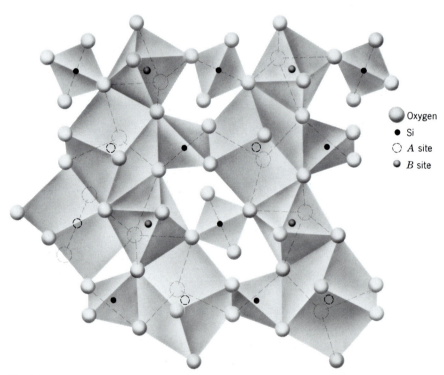

FIG. 11.5. Portion of the garnet structure projected on (001). Tetrahedra, octahedra and 8 coordination polyhedra (triangular dodecahedra, drawn as distorted cubes) are shown. (After G. A. Novak and G. V. Gibbs, 1971. The crystal chemistry of the silicate garnets, *American Mineralogist* 56: 791–825.)

Oxygen
Si
A site
B site

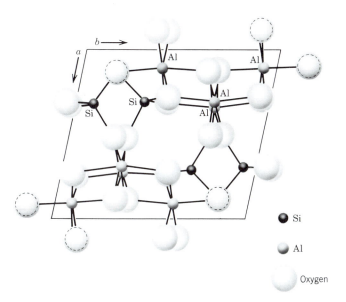

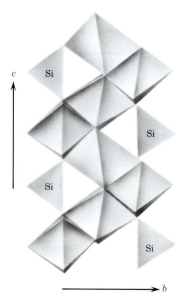

● Si

● Al

○ Oxygen

FIG. 11.6. The structure of kyanite. *(a)* Projected on (001) and *(b)* showing zigzag chains of edge-sharing octahedra parallel to the *c* axis. (Redrawn after C. W. Burnham, 1963, Refinement of the crystal structure of kyanite, *Zeitschrift für Kristallographie* 118: 337–360; and J. K. Winter, and S. Ghose, 1979, Thermal expansion and high-temperature crystal chemistry of the Al₂SiO₅ polymorphs, *American Mineralogist* 64: 573–86; see also *Orthosilicates, Reviews in Mineralogy* 1980, Mineralogical Society of America, Washington, D.C.)

The structure of garnet, which is illustrated in Fig. 11.5, consists of alternating SiO_4 tetrahedra and BO_6 octahedra that share corners to form a continuous three-dimensional network. The *A* sites are surrounded by 8 oxygens in irregular coordination polyhedra.

The aluminosilicates of the nesosilicate group, kyanite, sillimanite, and andalusite, are commonly found in medium- to high-grade metamorphic rocks of Al-rich bulk composition. All three minerals are poly-morphs of Al_2SiO_5, which may be stated structurally as $Al^{[4-6]}Al^{[6]}SiO_5$ (digits in square brackets indicate coordination number). Chains of edge-sharing octahedra, parallel to the *c* axis, are characteristic of all three structures. In *kyanite*, $Al^{[6]}Al^{[6]}SiO_5$, the triclinic polymorph, with space group $P\bar{1}$, all of the Al is octahedrally coordinated. It occurs as octahedral chains parallel to *c*, and as isolated Al octahedra (see Fig. 11.6). In *sillimanite*, $Al^{[4]}Al^{[6]}SiO_5$, an orthorhombic

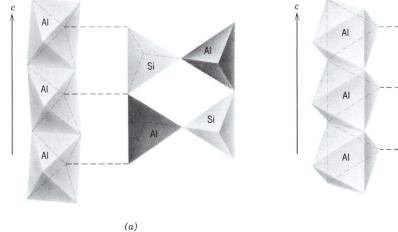

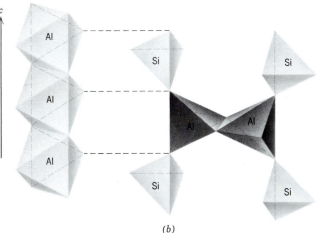

(a)   (b)

FIG. 11.7. *(a)* Projection of the sillimanite structure showing octahedral chains parallel to the *c* axis. (Redrawn after C. W. Burnham, 1963, Refinement of the crystal structure of sillimanite, *Zeitschrift für Kristallographie:* 127–148.) *(b)* Projection of the andalusite structure showing octahedral chains parallel to the *c* axis, and the presence of AlO₅ polyhedra between SiO₄ tetrahedra. (Redrawn after C. W. Burnham and M. J. Buerger, 1961, Refinement of the crystal structure of andalusite, *Zeitschrift für Kristallographie*, pp. 269–290.)

polymorph, with space group *Pbnm*, the octahedrally coordinated Al is found in octahedral chains (see Fig. 11.7*a*) and adjacent tetrahedral chains consist of alternating tetrahedral AlO_4 and SiO_4 groups. In *andalusite*, $Al^{[5]}Al^{[6]}SiO_5$, another orthorhombic polymorph, with space group *Pbnm*, half the Al is found in octahedral chains and the other half occurs in 5-coordinated polyhedra (see Fig. 11.7*b*) which are linked by SiO_4 tetrahedra. The stability fields for these three polymorphs are shown in Fig. 12.12.

SOROSILICATES

The sorosilicates are characterized by isolated, double tetrahedral groups formed by two SiO_4 tetrahedra sharing a single apical oxygen (Fig. 11.8). The resulting ratio of silicon to oxygen is 2 : 7. More than 70 minerals are known in this group, but most of them are rare. The table below lists six species, of which members of the *epidote group* and *vesuvianite* are most important.

Hemimorphite	$Zn_4(Si_2O_7)(OH)_2 \cdot H_2O$
Lawsonite	$CaAl_2(Si_2O_7)(OH)_2 \cdot H_2O$
Epidote group	
Clinozoisite	$Ca_2Al_3O(SiO_4)(Si_2O_7)(OH)$
Epidote	$Ca_2(Fe^{3+},Al)Al_2O(SiO_4)(Si_2O_7)(OH)$
Allanite	$X_2Y_3O(SiO_4)(Si_2O_7)(OH)$
Vesuvianite	$Ca_{10}(Mg,Fe)_2Al_4(SiO_4)_5(Si_2O_7)_2(OH)_4$

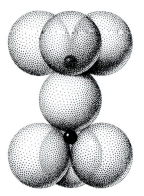

FIG. 11.8.   Close-packed representation of an Si_2O_7 group.

The structure of epidote (Fig. 11.9) contains both independent SiO_4 tetrahedra as well as Si_2O_7 groups. Chains of AlO_6 and $AlO_4(OH)_2$ octahedra, which share edges, run parallel to the *b* axis. (Similar octahedral chains are also present in the three polymorphs of Al_2SiO_5.) An additional octahedral position occurs outside the chains; this site is occupied by Al in clinozoisite and by Fe^{3+} and Al in epidote. The chains are linked by independent SiO_4 and Si_2O_7 groups. Ca is in irregular 8-coordination with oxygen. The site in which Ca^{2+} is housed may in part be filled by Na^+. The octahedral site outside the chains may house, in addition to Al, Fe^{3+}, Mn^{3+} and more rarely Mn^{2+}.

FIG. 11.9.   Schematic representation of the structure of epidote, projected on (010). Dashed lines outline unit cell.

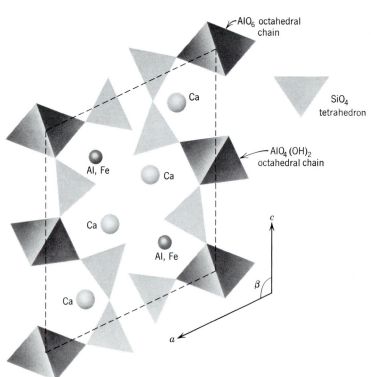

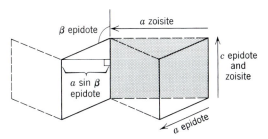

FIG. 11.10. Relation between unit cells of monoclinic epidote and orthorhombic zoisite, as projected on (010). The monoclinic unit cell outlined in solid can be related to its dashed equivalent by a mirror reflection (twinning). The orthorhombic unit cell is shown by shading.

All members of the epidote group are isostructural and form monoclinic crystals characteristically elongate on *b*. Orthorhombic *zoisite* has a structure that may be derived from that of its monoclinic polymorph, *clinozoisite*, by a twinlike doubling of the cell along the *a* axis, such that *a* of zoisite = 2 *a* sin β of clinozoisite (or epidote) (see Fig. 11.10). The structure of clinozoisite is the same as that of epidote, with all octahedral positions occupied by Al.

Allanite may be derived from the epidote structure by replacing some of the Ca^{2+} by rare earths and adjusting the electrostatic charge balance by replacing some of the Fe^{3+} by Fe^{2+}. As a result of the very similar structures of the members of the epidote group, the various ionic substitutions provide the major variables. These are:

	Ions in Ca Site	Ions in Al Site Outside Chains
Clinozoisite	Ca^{2+}	Al^{3+}
Epidote	Ca^{2+}	Fe^{3+}, Al^{3+}
Piemontite	Ca^{2+}	$Mn^{3+}, Fe^{3+}, Al^{3+}$
Allanite	$Ca^{2+}, Ce^{3+}, La^{3+}, Na^+$	$Fe^{3+}, Fe^{2+}, Mg^{2+}, Al^{3+}$

CYCLOSILICATES

The cyclosilicates contain rings of linked SiO_4 tetrahedra having a ratio of Si : O = 1 : 3. Three possible closed cyclic configurations of this kind may exist, as shown in Fig. 11.11. The simplest is the Si_3O_9 ring, represented among minerals only by the rare ti-

FIG. 11.11. Close-packed representation of ring structures in the cyclosilicates.

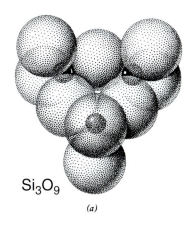

Si_3O_9

(a)

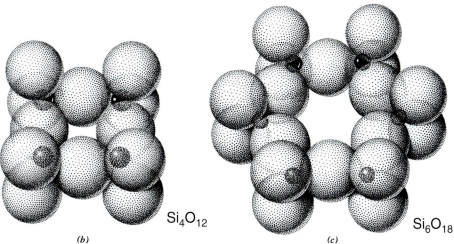

Si_4O_{12}

(b)

Si_6O_{18}

(c)

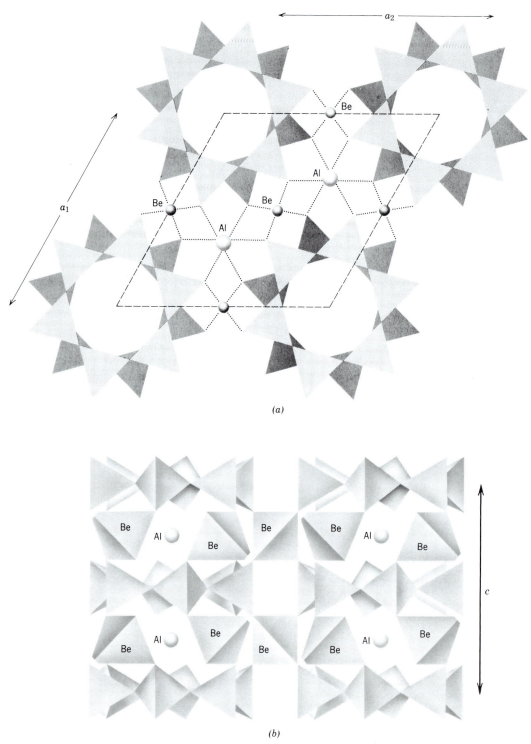

(a)

(b)

FIG. 11.12.   The structure of beryl. *(a)* Projected onto (0001). Dashed lines outline the unit cell. *(b)* View of the beryl structure with *c* axis vertical.

tanosilicate *benitoite*, BaTiSi$_3$O$_9$. The Si$_4$O$_{12}$ ring occurs only in a few very rare silicates. An example is *papagoite*, Ca$_2$Cu$_2$Al$_2$Si$_4$O$_{12}$(OH)$_6$. *Axinite* was originally classified on the basis of Si$_4$O$_{12}$ rings in its structure. More recent data, however, have shown it to be made of more complex B$_2$Si$_8$O$_{30}$ groups. The Si$_6$O$_{18}$ ring, however, is the basic framework of the structure of the common and important minerals, *beryl*, Be$_3$Al$_2$Si$_6$O$_{18}$, and *tourmaline*. In the structure of beryl, Si$_6$O$_{18}$ rings are arranged in layers parallel to {0001}. See Figs. 11.12a and b as well as several interactive illustrations of the beryl structure in module III of the CD-ROM under the heading "3-dimensional Order: Space Group Elements in Structures." Sheets of Be and Al ions lie between the layers of rings. Be in 4-coordination and Al in 6-coordination tie the layers together horizontally and vertically. The silicon–oxygen rings are so arranged as to be nonpolar; that is, a mirror plane can be pictured passing through the tetrahedra in the plane of the ring (Fig. 11.12b). Rings are positioned one above the other in the basal sheets so that the central holes correspond, forming prominent channels parallel to the c axis. In these channels a wide variety of ions, neutral atoms, and molecules, can be housed. (OH)$^-$, H$_2$O, F, He, and ions of Rb, Cs, Na, and K

are housed in beryl in this way. With monovalent alkalis such as Na$^+$ in the channels, the overall charge of the structure is neutralized by the substitution 2 alkalis^{+1} ⇆ Be^{2+}, or by 3 alkalis^{+1} + 1 Al^{3+} ⇆ 3 Be^{2+}. Although beryl is here included as a member of the cyclosilicates on the basis of the Si$_6$O$_{18}$ rings, consideration of the presence of the BeO$_4$ tetrahedra (see Fig. 11.12b) as an equally essential part of the structure reveals an overall three-dimensional network for this structure. For this reason, beryl can also be classified as a member of the framework (or tecto) silicates. For further discussion of the beryl structure and space group, see Fig. 6.57 and related text.

Cordierite, (Mg,Fe)$_2$Al$_4$Si$_5$O$_{18}$·nH$_2$O, has a high-temperature polymorph, *indialite*, which is isostructural with beryl in which the Al is randomly distributed in the (Si,Al)$_6$O$_{18}$ rings. The common, lower-temperature form, *cordierite*, is orthorhombic (pseudohexagonal) in which two of the tetrahedra in the 6-fold ring are occupied by Al, producing an ordered structure (see Fig. 11.13). Al occupies the Be sites and Mg and Fe^{2+} the octahedral Al sites of the beryl structure. H$_2$O molecules may reside in the channels of the structure. Gibbs (1966) points out that cordierite should be classified with the tectosilicates rather than the cyclosilicates, because Al- and Si-

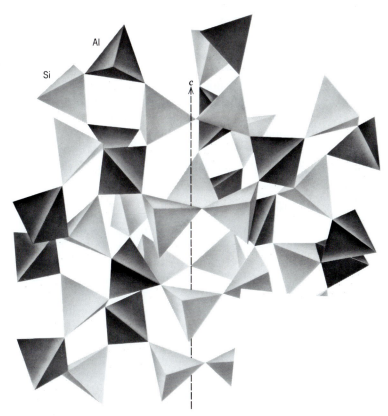

FIG. 11.13. Idealized drawing of the (Al$_4$Si$_5$O$_{18}$) framework in low cordierite. The octahedral coordination of Mg and Fe^{2+} is not shown. (After G. V. Gibbs, 1966, The polymorphism of cordierite I: The crystal structure of low cordierite. *American Mineralogist* 51: 1068–1087.)

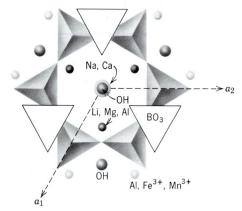

FIG. 11.14.   Part of the structure of tourmaline projected on (0001).

tetrahedra are in perfect alternation in all directions of the structure except for two SiO_4 tetrahedra that share a common oxygen in the six-membered ring.

The structure of *tourmaline*, with a complex chemical composition, had long remained a mystery. It has been shown to be built on Si_6O_{18} rings along the center of which Na^+ and $(OH)^-$ alternate (see Fig. 11.14). The Si_6O_{18} rings in tourmaline are polar; that is, the net strength of bonds to one side of the ring is not the same as the strength of bonds extending to the other, looking first in one direction, then in the other along the *c* axis. Interlayered with the rings are sheets of triangular BO_3 groups. $(Li,Mg,Al)O_4(OH)_2$ octahedral groups link the Si_6O_{18} rings and BO_3 groups together. The columns of Si_6O_{18} rings are linked to each other by $(Al,Fe,Mn)O_5(OH)$ groups. The composition of tourmaline is complex, but the following generalizations apply: Ca may substitute for Na; Al may be replaced by Fe^{3+}, Mn^{3+}; and Mg may be replaced by Fe^{2+}, Mn^{2+}, Al^{3+}, and Li^+. The varieties are determined by the relative proportions of the cations, and ionic substitution follows the usual pattern, with extensive mutual substitution of Mg by Fe^{2+} and Mn^{2+}, and Na^+ by Ca^{2+}, with concomitant coupled substitution to maintain electrical neutrality.

INOSILICATES

SiO_4 tetrahedra may link into chains by sharing oxygens (Fig. 11.3). Such simple chains may be joined side by side by further sharing of oxygens in alternate tetrahedra to form bands or double chains (Fig. 11.3). In the simple chain structure, two of the four oxygens in each SiO_4 tetrahedron are shared, giving a ratio of Si : O = 1 : 3. In the band structure half of the tetra-

hedra share three oxygens and the other half share two oxygens, yielding a ratio of Si : O = 4 : 11.

Included in the inosilicates are two important rock-forming groups of minerals: the *pyroxenes* as single chain members and the *amphiboles* as double chain members. Many similarities exist between the two groups in crystallographic, physical and chemical properties. Although most pyroxenes and amphiboles are monoclinic, both groups have orthorhombic members. Also in both the repeat distance along the chains, that is, the *c* dimension of the unit cell, is approximately 5.2 Å. The *a* cell dimensions are also analogous but, because of the double chain, the *b* dimension of amphiboles is roughly twice that of the corresponding pyroxenes.

The same cations are present in both groups, but the amphiboles are characterized by the presence of (OH), which is lacking in pyroxenes. Although the color, luster, and hardness of analogous species are similar, the (OH) in amphiboles gives them, in general, slightly lower specific gravity and refractive indices than their pyroxene counterparts. Furthermore, the crystals have somewhat different habits. Pyroxenes commonly occur in stout prisms, whereas amphiboles tend to form more elongated crystals, often acicular. Their cleavages are distinctly different and can be directly related to the underlying chain structure (see Figs. 11.18 and 11.24).

Pyroxenes crystallize at higher temperatures than their amphibole analogues and hence are generally formed early in a cooling igneous melt and occur also in high-temperature metamorphic rocks rich in Mg and Fe. If water is present in the melt or as a metamorphic fluid, the early-formed pyroxene may react with the liquid at lower temperatures to form amphibole. Under prograde metamorphic conditions amphiboles commonly react to form pyroxenes, and under retrograde metamorphic conditions pyroxenes commonly give way to aphiboles (see Fig. 12.35).

Pyroxene Group

The chemical composition of pyroxenes can be expressed by a general formula as XYZ_2O_6, where *X* represents Na^+, Ca^{2+}, Mn^{2+}, Fe^{2+}, Mg^{2+}, and Li^+ in the *M2* crystallographic site; *Y* represents Mn^{2+}, Fe^{2+}, Mg^{2+}, Fe^{3+}, Al^{3+}, Cr^{3+}, and Ti^{4+} in the *M1* site; and *Z* represents Si^{4+} and Al^{3+} in the tetrahedral sites of the chain. (It should be noted that the *X* cations in general are larger than the *Y* cations, in accordance with the cation size requirements of the sites *M2* and *M1*.) The pyroxenes can be divided into several groups, the most common of which can be represented as part of the chemical system $CaSiO_3$ (wollastonite, a pyroxenoid)–$MgSiO_3$(enstatite)–FeSiO$_3$ (ferrosilite). The trapezium part of this system includes members of the common series *diopside*, CaMgSi$_2$O$_6$, *hedenbergite*, CaFeSi$_2$O$_6$, and of the series *enstatite-ferrosilite* (see Fig. 11.15*a*). Only the names for end member composi-

tions of the common pyroxenes are shown in Fig. 11.15*a*. In prior years, intermediate compositions have been given various species names, such as bronzite and hypersthene (for members of the orthopyroxene series) and salite and ferrosalite (for

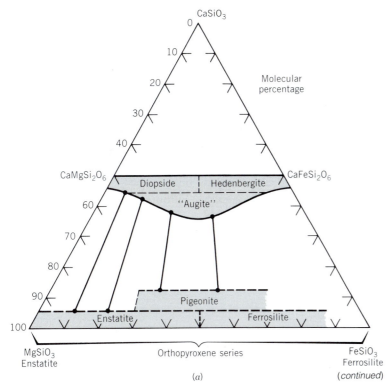

(*a*)   (*continued*)

FIG. 11.15. *(a)* Pyroxene compositions in the system CaSiO$_3$–MgSiO$_3$–FeSiO$_3$. General compositional fields are outlined. Representative tielines across the miscibility gap between augite and more Mg–Fe-rich pyroxenes are shown. The "augite" field is labeled with quotation marks because all augite compositions contain considerable Al, which is not considered in this triangular composition diagram. *(b)* Nomenclature of pyroxenes in the system Wo (wollastonite), En (enstatite), and Fs (ferrosilite) as derived from chemical compositional information. Members of the orthopyroxene series range from En$_{100}$ to En$_0$ (which is equivalent to Fs$_0$ to Fs$_{100}$). Any other, more general compositions are expressed in molecular percentages of Wo, En, and Fs (e.g., Wo$_{50}$En$_{25}$Fs$_{25}$). The space groups of the various solid solution series are shown.

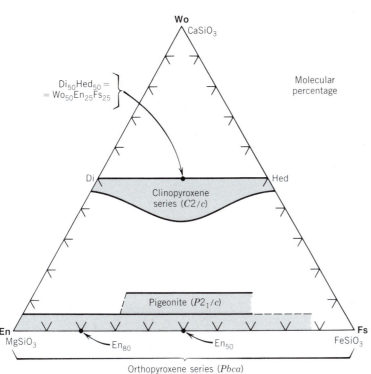

(*b*)

members of the diopside-hedenbergite series). Modern nomenclature rulings (Morimoto et al., 1988; for complete reference see end of this chapter) disallow such species designations. The end member names, as shown in Fig. 11.15a, apply from the end member composition to the 50 molecular percentage point (the halfway point) in the diopside-hedenbergite and enstatite-ferrosilite series. If chemical analyses are available for intermediate compositions they can be expressed in terms of molecular percentages of the end members (see Fig. 11.15b). For example, pure enstatite can be stated as En_{100}. A member of the two-component orthopyroxene series composed of 80 molecular percent enstatite and 20 molecular percent ferrosilite can be listed as En_{80}. Similarly, the compositions of the two-component diopside-hedenbergite series can be expressed by molecular percentages that appear as subscripts to Di and Hed (e.g., $Di_{50}Hed_{50}$). Any more general chemical composition in the trapezium part of Fig. 11.15b can be expressed in terms of molecular percentages of the three end members, Wo (for wollastonite), En (for enstatite), and Fs (for ferrosilite). An example,

shown in Fig. 11.15b, would be $Wo_{50}En_{25}Fs_{25}$. Table 3.17 illustrates the recalculation of a pyroxene analysis in terms of Wo-En-Fs end member components, and Fig. 3.64 (and related text) explains the procedure for plotting chemical compositions on triangular diagrams.

Compositionally *augite* is closely related to members of the diopside–hedenbergite series but with some substitution of, for example, Na for Ca in *M2*, Al for Mg (or Fe^{2+}) in *M1*, and Al for Si. *Pigeonite* represents a field of Mg–Fe solid solutions with a Ca-content somewhat larger than in the *enstatite–ferrosilite* series, which represents the composition field of the orthopyroxenes. Sodium-containing pyroxenes are *aegirine*, $NaFe^{3+}Si_2O_6$, and *jadeite*, $NaAlSi_2O_6$. Aegirine and augite represent a complete solid solution series as shown by members of intermediate composition, *aegirine-augite*. *Omphacite* represents a solid solution series between augite and jadeite. *Spodumene*, $LiAlSi_2O_6$, is a relatively rare pyroxene found in Li-rich pegmatites.

The pyroxene structure is based on single SiO_3 chains that run parallel to the *c* axis. Figure 11.16 illustrates this tetrahedral chain as well as the double

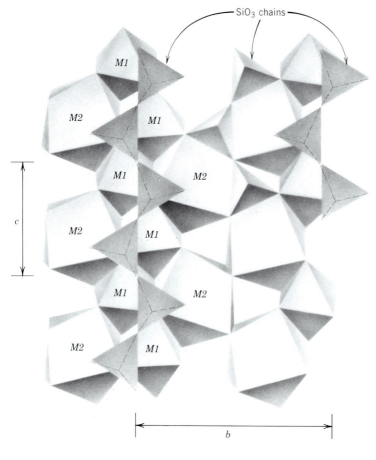

- SiO₃ chains -

FIG. 11.16. The structure of jadeite, $NaAlSi_2O_6$, a monoclinic pyroxene in an approximate projection onto (100). The *M2* site is occupied by Na^+, the *M1* site by Al^{3+}. (After C. T. Prewitt and C. W. Burnham, 1966, The crystal structure of jadeite, $NaAlSi_2O_6$. *American Mineralogist* 51: 956–75.)

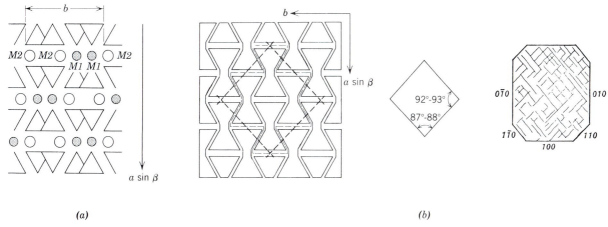

FIG. 11.17.   *(a)* Schematic projection of the monoclinic pyroxene structure on a plane perpendicular to the *c* axis. *(b)* Control of cleavage angles by *t-o-t* strips (also referred to as "I-beams") in the pyroxene structure, as compared with naturally occurring pyroxene cleavage.

octahedral chain to which it is bonded. See also Fig. 11.3 and several interactive illustrations of the clinopyroxene structure in module III of the CD-ROM under the heading: "Three-dimensional Order: Space Group Elements in Structures." The structure contains two types of cation sites, labeled *M1* and *M2*. The *M1* site is a relatively regular octahedron, but, especially in the monoclinic pyroxenes, the *M2* site is an irregular polyhedron of 8-coordination (in orthorhombic pyroxenes with Mg in the *M2* site this polyhedron is closer to a regular octahedron). Figure 11.17*a* shows the pyroxene structure and the distribution of cation sites as seen in a direction parallel to the *c* axis. The cations in the *M1* sites are all coordi-

nated by oxygens of two opposing SiO_3 chains, and as such produce a tetrahedral–octahedral–tetrahedral (*t-o-t*) strip. The coordination of cations in the *M2* position, however, is such that several of these *t-o-t* strips are cross-linked. These *t-o-t* strips are often schematically represented as in Fig. 11.17*b*; this in turn shows the relationship of the *t-o-t* strips to the cleavage angles in pyroxenes. Figure 11.18 is a direct structure image of the features shown in Fig. 11.17.

The majority of pyroxenes can be assigned to one of three space groups, two monoclinic (*C2/c* and *P2₁/c*) and one orthorhombic (*Pbca*) (see Fig. 11.15*b*). The *C2/c* structure is found in most of the common clinopyroxenes such as diopside, $CaMgSi_2O_6$, jadeite,

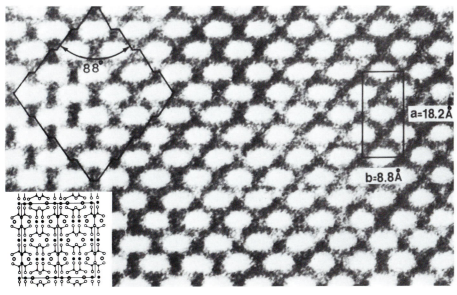

FIG. 11.18.   High-resolution transmission electron microscope (HRTEM) image of an *a-b* section of enstatite. The white regions correspond to areas between *M2* sites in the structure. A unit cell and cleavage surfaces are outlined. The insert shows the enstatite structure at the same scale as the structure image. (From P. R. Buseck, and S. Iijima, 1974, High resolution electron microscopy of silicates. *American Mineralogist* 59: 1–21.)

$NaAlSi_2O_6$, and augite. This structure is illustrated in Fig. 11.16. See Fig. 6.56 for an illustration of the $C2/c$ space group. The $M1$ site is generally occupied by cations that are smaller than those of the $M2$ site. For example, in the diopside–hedenbergite series, $CaMgSi_2O_6$–$CaFeSi_2O_6$, the $M1$ site is occupied by Mg and Fe^{2+} in random distribution, whereas the $M2$ is occupied solely by the larger Ca^{2+} ion, in 8-coordination. The $M2$ site, however, can also house Mn^{2+}, Fe^{2+}, Mg, or Li, in which case the coordination is sixfold. The $P2_1/c$ space group is found in pigeonite (see Fig. 11.15), which may be represented as $Ca_{0.25}(Mg,Fe)_{1.75}Si_2O_6$ in which the $M2$ site is occupied by all of the Ca in the formula as well as additional Mg and Fe so as to make the composition of the site $[Ca_{0.25}(Fe + Mg)_{0.75}]$. The $M2$ site, as present in the clinopyroxenes with space group $C2/c$, is too large to accommodate these smaller cations, so that $M2$ is somewhat smaller in pigeonite (hence, this site has irregular 7-coordination) due to a slight shift in the pigeonite structure as compared with the more regular diopside type structure. The orthorhombic $Pbca$ structure is that found in the orthopyroxene series which contains virtually no Ca. The Mg and Fe^{2+} ions are distributed among $M1$ and $M2$ with the larger cation (Fe^{2+}) showing strong preference for the somewhat larger and distorted $M2$ site. The coordination of both the $M1$ and $M2$ sites in orthopyroxenes is sixfold.

The unit cells of the orthorhombic pyroxenes are related to the monoclinic unit cells by a twinlike mirror across {100} accompanied by an approximate doubling of the a cell dimension (e.g., a of enstatite $\approx 2\,a \sin\beta$ of diopside; see Fig. 11.19a). Figures 11.19b and c show schematically the development of monoclinic and orthorhombic pyroxene structures built up from t-o-t strips. The compositions represented by the orthopyroxene series may in rare occurrences be found in a monoclinic form, known as the *clinoenstatite-clinoferrosilite series*.

Pyroxenoid Group

There are a number of silicate minerals that have, as do the pyroxenes, a ratio of Si : O = 1 : 3, but with structures that are not identical to those of the pyroxenes. Both pyroxene and pyroxenoid structures contain octahedrally coordinated cations between SiO_3 chains but in pyroxenoids the geometry of the chains is not of the simple, infinitely extending type with a repeat distance of about 5.2 Å along the direction of the chain (see Fig. 11.20). In wollastonite, $CaSiO_3$, the smallest repeat of the chain consists of three twisted tetrahedra with a repeat distance of 7.1 Å (see also Fig. 12.40) and in rhodonite, $MnSiO_3$, the unit repeat is built of five twisted tetrahedra with a repeat distance of 12.5 Å. Because of the lower symmetry of the chains (as compared with the pyroxene chain) the structures of pyroxenoids are triclinic. The chain structure in the pyroxenoids is expressed by their generally splintery cleavages and sometime fibrous habit.

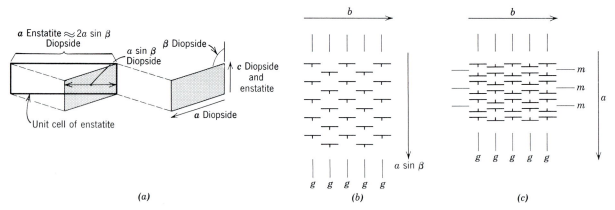

FIG. 11.19. *(a)* Relationship between unit cells of clinopyroxene (e.g., diopside) and orthopyroxene (e.g., enstatite) as projected on (010). The monoclinic unit cell outlined by shading can be related to the larger orthorhombic unit cell by a mirror reflection (compare with Fig. 11.25). *(b)* Schematic representation of a possible monoclinic pyroxene (looking down the c axis) with the t-o-t strips (or "I-beams") represented by strike and dip symbols; g's indicate glide planes (with a translation component parallel to c). *(c)* Schematic representation of a possible orthorhombic pyroxene; m's represent mirrors. (*b* and *c* after J. B. Thompson, Jr., Harvard University, pers. comm.)

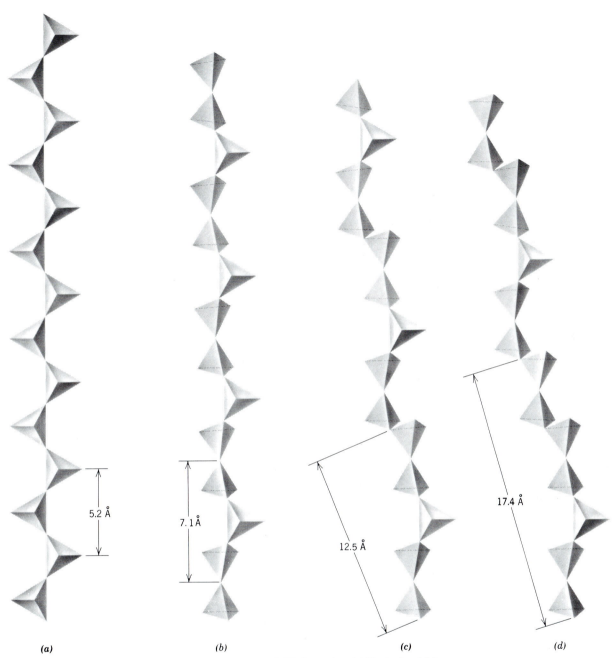

FIG. 11.20. SiO$_3$ chains in pyroxene *(a)* and pyroxenoids. *(b)* Wollastonite. *(c)* Rhodonite. *(d)* Pyrox-mangite. (After F. Liebau, 1959, Über die Kristallstruktur des Pyroxmangits (Mn, Fe, Ca, Mg) SiO$_3$, *Acta Crystallographica* 12: 177–181.)

Amphibole Group

The chemical composition of members of the amphibole group can be expressed by the general formula $W_{0-1}X_2Y_5Z_8O_{22}(OH,F)_2$, where W represents Na$^+$ and K$^+$ in the A site, X denotes Ca^{2+}, Na$^+$, Mn^{2+}, Fe^{2+}, Mg^{2+}, and Li$^+$ in the $M4$ sites, Y represents Mn^{2+}, Fe^{2+}, Mg^{2+}, Fe^{3+}, Al^{3+}, and Ti^{4+} in the $M1$, $M2$, and $M3$ sites, and Z refers to Si^{4+} and Al^{3+} in the tetrahedral sites. Essentially complete ionic substitution may take place between Na and Ca and among Mg, Fe^{2+}, and Mn^{2+}. There is limited substitution between Fe^{3+} and Al and between Ti and other Y-type ions; and partial substitution of Al for Si in the tetrahedral sites of the double chains. Partial substitution of F and O for OH in the hydroxyl sites is also common. Several common series of the amphibole group can be represented compositionally in the chemical system $Mg_7Si_8O_{22}(OH)_2$ (anthophyllite)–$Fe_7Si_8O_{22}(OH)_2$ (grunerite)–"$Ca_7Si_8O_{22}(OH)_2$," a hypothetical end member composition, as in Fig. 11.21. This illustration is very similar to the Wo–En–Fs diagram for the pyroxene group (Fig. 11.15). A complete series exists from *tremolite*, $Ca_2Mg_5Si_8O_{22}(OH)_2$, to *ferroactiolite*, $Ca_2Fe_5Si_8O_{22}(OH)_2$. The commonly occurring *actinolite* composition is a generally Mg-rich member of the tremolite–ferroactinolite series. The compositional range from $Mg_7Si_8O_{22}(OH)_2$ to about $Fe_2Mg_5Si_8O_{22}(OH)_2$ is represented by the orthorhombic species, *anthophyllite*. The *cummingtonite-grunerite* series is monoclinic and extends from about $Fe_2Mg_5Si_8O_{22}(OH)_2$ to $Fe_7Si_8O_{22}(OH)_2$. Miscibility gaps are present between anthophyllite and the tremolite-actinolite series, as well as between the members of the cummingtonite–grunerite series and the calcic amphiboles. These gaps are reflected in the common occurrence of anthophyllite–tremolite and grunerite–actinolite pairs, as well as exsolution textures between members of the Mg–Fe and the Ca-containing amphiboles of Fig. 11.21. No Ca-amphibole compositions are possible above the 2/7th line (representing two Ca out of a total of seven $X + Y$ cations) because Ca can be housed only in the two *M4* sites of the amphibole structure. *Hornblende* may be regarded as a tremolite–ferroactinolite type composition with additional partial substitution of Na in the *A* and *M4* sites, Mn, Fe^{3+}, and Ti^{4+} for *Y* cations, and Al for Si in the tetrahedral sites, leading to a very complex general formula. Sodium-containing amphiboles are represented by members of the *glaucophane*, $Na_2Mg_3Al_2Si_8O_{22}(OH)_2$,-*riebeckite*, $Na_2Fe_3^{2+}Fe_2^{3+}Si_8O_{22}(OH)_2$, series. *Arfvedsonite*,

$NaNa_2Fe_4^{2+}Fe^{3+}Si_8O_{22}(OH)_2$, contains additional Na in the *A* site of the structure. Table 11.2 compares the composition of some of the common members of the pyroxene and amphibole groups, and Table 3.18 illustrates the recalculation of an amphibole analysis.

The amphibole structure is based on double Si_4O_{11} chains that run parallel to the *c* axis. Figure 11.22 illustrates this chain as well as the octahedral strip to which it is bonded. See also Fig. 11.3 and an interactive animation of the amphibole structure in module I of the CD-ROM under the heading "Solid Solution Mechanisms: Substitutional Solid Solution in Amphiboles." The structure contains several cation sites, labeled *A*, *M4*, *M3*, *M2*, *M1*, as well as the tetrahedral sites in the chains. The *A* site has 10- to 12-coordination with oxygen and (OH) and houses mainly Na, and at times small amounts of K. The *M4* site has 6- to 8-coordination and houses *X* type cations (see Table 11.2). The *M1*, *M2*, and *M3* octahedra accommodate *Y* type cations and share edges to form octahedral bands parallel to *c*. *M1* and *M3* are coordinated by four oxygens and two (OH, F) groups, whereas *M2* is coordinated by six oxygens. Figure 11.23*a* shows the monoclinic amphibole structure and the distribution of cation sites as seen in a direction parallel to the *c* axis. The *t-o-t* strips (Fig. 11.23*b*) are approximately twice as wide (in the *b* direction) as the equivalent *t-o-t* strips in pyroxenes (see Fig. 11.17*b*) because of the doubling of the chain width in amphiboles. This wider geometry causes the typical 56° and 124° cleavage angles as shown in Fig. 11.23. Figure 11.24 is a direct structure image of the features shown in Fig. 11.23.

The majority of amphiboles can be assigned to one of three space groups, two monoclinic (*C2/m* and *P2₁/m*) and one orthorhombic (*Pnma*). The *C2/m* structure is found in all of the common clinoamphiboles such as tremolite, $Ca_2Mg_5Si_8O_{22}(OH)_2$, and

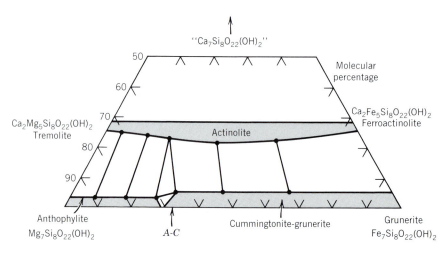

FIG. 11.21. Amphibole compositions in the system $Mg_7Si_8O_{22}(OH)_2$–$Fe_7Si_8O_{22}(OH)_2$–"$Ca_7Si_8O_{22}(OH)_2$". General compositional fields are outlined. Representative tielines across miscibility gaps are shown. Compare with Fig. 11.15. The composition marked *A–C* refers to an anthophyllite-cummingtonite intergrowth illustrated in Fig. 11.25*b*.

TABLE 11.2   Ions in Common Pyroxenes and Amphiboles

Pyroxenes			Amphiboles			
Atomic Sites		Name		Atomic Sites		Name
M2	*M1*		*A*	*M4*	(*M1* + *M2* + *M3*)	
Mg	Mg	Enstatite	□*	Mg	Mg	Anthophyllite
		other members of	□	Fe	Mg	Cummingtonite
Fe	Mg	the orthopyroxene				
		series	□	Fe	Fe	Grunerite
Ca	Mg	Diopside	□	Ca	Mg	Tremolite
Ca	Fe	Hedenbergite	□	Ca	Fe	Ferroactinolite
Ca	Mn	Johannsenite				
	Mg, Fe, ⎫		□	Ca, Na	Mg, Fe²⁺, Mn, ⎫	
Ca,	Mn, Al, ⎬	Augite			Al, Fe³⁺,Ti ⎭	Hornblende
Na	Fe³⁺, Ti ⎭					
Na	Al	Jadeite	□	Na	Mg, Al	Glaucophane
Na	Fe³⁺	Aegirine	□	Na	Fe²⁺, Fe³⁺	Riebeckite
			Na	Na	Fe²⁺, Fe³⁺	Afrvedsonite
Li	Al	Spodumene	□	Li	Mg, Fe³⁺ ⎫	Holmquistite
					Al, Fe²⁺ ⎭	

*□ represents a vacant atomic site.

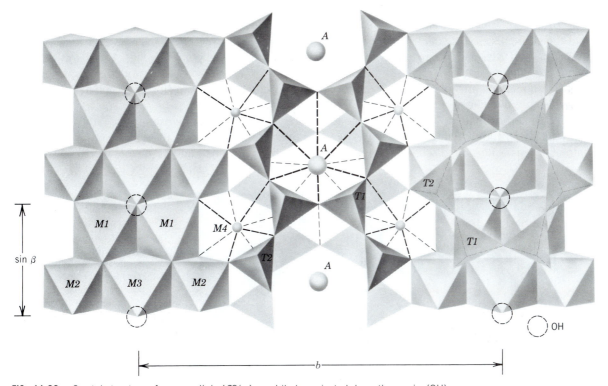

FIG. 11.22. Crystal structure of a monoclinic (*C2/m*) amphibole projected down the *a* axis. (OH) groups are located in the center of the large holes of the rings in the chains. *M1*, *M2*, and *M3* sites house *Y* cations and are 6-coordinated. The *M4* site houses larger *X* cations in 6- to 8-coordination. The ion in the *A* site, between the backs of double chains is 10- to 12-coordinated. (After J. J. Papike, et al., 1969, Mineralogical Society of America Special Paper no. 2, p. 120.)

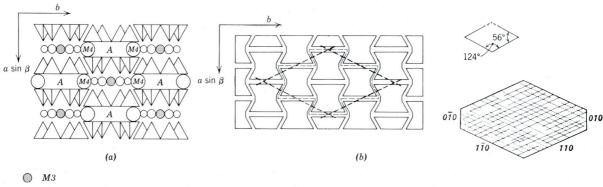

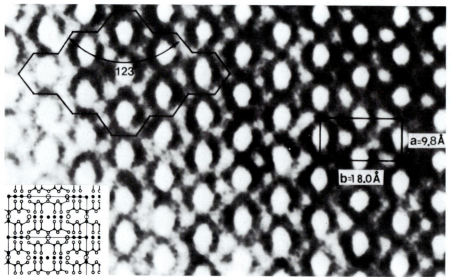

FIG. 11.23. *(a)* Schematic projection of the monoclinic amphibole structure on a plane perpendicular to the *c* axis (after Colville et al., 1966, *American Mineralogist* 51: 1739). Compare with Fig. 11.17*a*. *(b)* Control of cleavage angles by *t-o-t* strips (also referred to as "I-beams") in the amphibole structure, as compared with naturally occurring cleavage angles.

hornblende. This type of structure is illustrated in Fig. 11.22. The $P2_1/m$ space group is found in some Mg-rich cummingtonites, close to $(Mg, Fe)_2Mg_5Si_8O_{22}(OH)_2$ in composition, and results because the *M4* site is somewhat collapsed to house the relatively small Mg and Fe^{2+} ions. The orthorhombic *Pnma* structure is found in members of the anthophyllite, $Mg_7Si_8O_{22}(OH)_2$ to $Fe_2Mg_5Si_8O_{22}(OH)_2$, series and in gedrite, an Al and Na-containing anthophyllite, and in holmquistite. The presence of cations of small size in *M4*, *M3*, *M2*, and *M1* leads to an orthorhombic rather than a monoclinic structure. The unit cells of orthorhombic amphiboles are related to the monoclinic unit cells by a twinlike mirror across (100) accompanied by an approximate doubling of the *a* cell dimension (e.g., *a* of anthophyllite ≈ 2 *a* sin β of cummingtonite). This is illustrated schematically and

also with a high-resolution transmission electron microscope (HRTEM) image in Fig. 11.25. This relationship between the structures of clinoamphiboles and orthoamphiboles is identical to that for clinopyroxenes and orthopyroxenes (see Fig. 11.19).

The presence of (OH) groups in the structure of amphiboles causes a decrease in their thermal stabilities as compared with the more refractory pyroxenes. This causes amphiboles to decompose to anhydrous minerals (often pyroxenes) at elevated temperatures below the melting point (see Figs. 12.35 and 12.48).

In 1977, Veblen et al. reported several new, ordered structures that are closely related to amphiboles. These are referred to as *biopyriboles*, a term derived from *bio*tite (mica), *pyrox*ene, and amph*ibole*. This collective term reflects the close architectural connection between numbers of the pyroxene, amphi-

FIG. 11.24. High-resolution transmission electron microscope (HRTEM) image of an *a-b* section of hornblende. The white regions correspond to the *A* sites. A unit cell and cleavage surfaces are outlined. The insert shows the hornblende structure at the same scale as the structure image. (From P. R. Buseck and S. Iijima, 1974, High resolution electron microscopy of silicates. *American Mineralogist* 59: 1–21.)

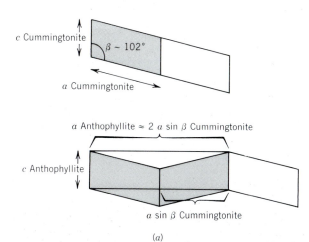

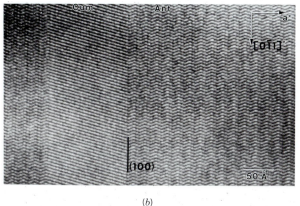

(a)

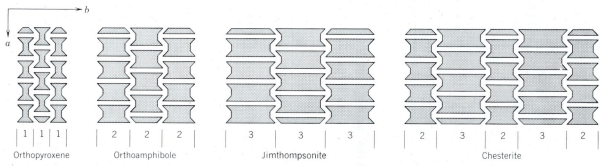

(b)

FIG. 11.25. *(a)* Relationship between unit cells of clinoamphiboles (e.g., cummingtonite or tremolite) and orthoamphibole (anthophyllite). The monoclinic unit cell (outlined by shading) can be repeated by a mirror reflection, which results in a zigzag pattern. This repeat pattern is described by the larger orthorhombic unit cell outline. This relationship is identical in the pyroxenes (see Fig. 11.19). *(b)* High-resolution transmission electron microscope (HRTEM) image provides a direct picture of the relationships shown in *(a)*. A cummingtonite (Cum) lamella (about 140 Å thick) is set in a matrix of anthophyllite (Ant). The stacking of the unit cells in cummingtonite is consistently in the same direction. However, the stacking of the unit cells in anthophyllite is the result of a mirror reflection parallel to (100), resulting in the zigzag pattern. Both amphiboles have approximately the same composition: $Na_{0.1}Fe_{2.2}Mg_{4.5}Al_{0.1}Si_8O_{22}(OH)_2$. In other words, the structure image is one of two polymorphs. The value of (Mg/Fe × 100)% for this formula is 67%, which plots it at the 67% molecular percentage point along the bottom edge of the triangle in Fig. 11.21, marked *A-C*. (Micrograph courtesy of Eugene A. Smelnik, Princeton University.)

bole, and layer silicate groups. Indeed, two of the new structures discovered, *jimthompsonite*, $(Mg,Fe)_{10}Si_{12}O_{32}(OH)_4$, and *chesterite*, $(Mg,Fe)_{17}Si_{20}O_{54}(OH)_6$, have chain widths and chain repeat sequences that can be interpreted in terms of single chain repeats (SiO_3 chains as in pyroxenes), double chain repeats ($Si_4O_{11}(OH)$ chains as in amphiboles), and triple chain repeats. These triple chain repeats are wider than those known in pyroxenes and amphiboles, and their increased width is suggestive of mica structures with infinitely extending sheets. Schematic illustrations of the pyroxene, amphibole, and new biopyribole structures are given in Fig. 11.26 in terms of I-beams. A direct image of some of these same structures, by high-resolution transmission electron microscopy (HRTEM), is given in Fig. 11.27.

FIG. 11.26. Schematic illustration of "I-beams" projected onto the (001) plane in orthopyroxene, orthoamphibole, jimthompsonite, and chesterite. The digits mean: 1—single chain width as in pyroxene, 2—double chain width as in amphibole, and 3—triple chain width. (Redrawn after D. R. Veblen, P. R. Buseck, and C. W. Burnham, 1977, Asbestiform chain silicates: New minerals and structural groups. *Science* 198: 359–65.)

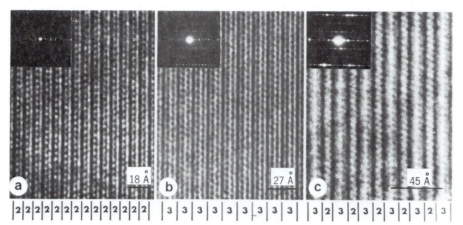

FIG. 11.27. High-resolution transmission electron microscope (HRTEM) images of *(a)* anthophyllite, *(b)* jimthompsonite, and *(c)* chesterite. In each image the *c* axis of the mineral is coincident with the vertical striping. In anthophyllite the 2s locate double Si_4O_{11} chains. In jimthompsonite the 3s refer to triple chains, and in chesterite the alternating 2s and 3s refer to alternating double and triple chains. The inset in each image is an electron diffraction pattern; this is used mainly by the microscopist for crystallographic orientation. (From D. R. Veblen, P. R. Buseck, and C. W. Burnham, 1977, Asbestiform chain silicates: New minerals and structural groups. *Science* 198: 359–65; copyright © 1977 by the AAAS. See also D. R. Veblen, 1981, Non-classical pyriboles and polysomatic reactions, in *Amphiboles and other hydrous pyriboles*, *Reviews in Mineralogy* 9A, Mineralogical Soc. of America, pp. 189–236.)

FIG. 11.28. High-resolution transmission electron microscope (HRTEM) image of a disordered chain silicate, with chains of variable width aligned perpendicular to the plane of the photograph. The black regions represent the chains, and the white spots represent regions of low electron density between chains. The 2, 3, and 4 refer to double, triple, and quadruple chains, respectively. The precursor of this silicate material was an amphibole with double chains only. (From P. R. Buseck and D. R. Veblen, 1978, Trace elements, crystal defects, and high resolution microscopy. *Geochimica et Cosmochimica Acta* 42: 669–78; see also P. R. Buseck, 1983. Electron microscopy of minerals. *American Scientist* 71: 175–85.) Compare this illustration with Fig. 4.54.

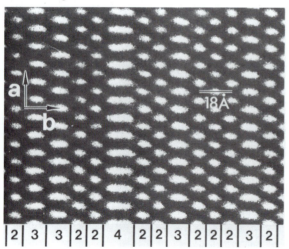

The new ordered structure types were found in a fine intergrowth with anthophyllite, and are considered low-temperature alteration (and hydration) products of original enstatite and anthophyllite. They appear to be intermediate stages of structural development between pyroxenes (anhydrous, high-temperature minerals) and phyllosilicates (hydrous, lower-temperature minerals; see also Fig. 11.43). Many of the intermediate reaction products, however, turn out to be disordered rather than ordered structures. Figure 11.28 is an example of a random intergrowth of double, triple, and quadruple chain widths in a material whose precursor was an amphibole, with only double chain widths, prior to alteration. Solid state reactions, possibly as a result of low-temperature metamorphism, facilitated the wide chain construction.

PHYLLOSILICATES

As the derivation of the name of this important group implies (Greek: *phyllon*, leaf), most of its many members have a platy or flaky habit and one prominent cleavage. They are generally soft, of relatively low specific gravity, and may show flexibility or even elasticity of the cleavage lamellae. All these characterizing peculiarities arise from the dominance in the structure of the infinitely extended sheet of SiO_4 tetrahedra. In this sheet (see Fig. 11.3) three of the four oxygens in each SiO_4 tetrahedron are shared with neighboring tetrahedra, leading to a ratio of $Si : O = 2 : 5$. Each sheet, if undistorted, has sixfold symmetry.

Most of the members of the phyllosilicates are hydroxyl bearing, with the (OH) group located in the center of the sixfold rings of tetrahedra, at the same height as the unshared apical oxygens in the SiO_4 tetrahedra (see Fig. 11.29). When ions, external to the Si_2O_5 sheet, are bonded to the sheets, they are coordinated to two oxygens and one OH, as shown in Fig. 11.29. The size of the triangle between two oxygens and one (OH) is just about the same as (but not identical to, see page 469) the triangular face of an XO_6 octahedron (with X commonly Mg or Al). This means that it is possible to bond to a regular net of apical oxygens and (OH) groups of $(Si_2O_5OH)^{3-}$ composition a sheet of regular octahedra, in which each octahedron is tilted onto one of its triangular sides (Fig. 11.30). When such tetrahedral and octahedral sheets are joined, one obtains the general geometry of the layered *lizardite* or *kaolinite* structures (Fig. 11.31a).

The cations in the octahedral sheet may be divalent or trivalent. When the cations are divalent, for example, Mg or Fe^{2+}, the sheet will have the geometry of that of brucite (Fig. 9.7) in which each cation site is occupied. In such a sheet six bonds, each with e.v. $= \frac{2}{6} = \frac{1}{3}$, originate from the Mg^{2+} ion. Three such bonds radiate from each oxygen, or (OH) group, thus neutralizing one-half the charge of the oxygen but all

of the OH. A sheet in which each oxygen or (OH) group is surrounded by three cations, as in the brucite, $Mg(OH)_2$, structure, is known as *trioctahedral*. When the cations in the octahedral sheet are trivalent, charge balance is maintained when one out of every three cations sites is unoccupied (see Fig. 11.31b). This is the case in gibbsite, $Al(OH)_3$, and corundum structures (Fig. 9.2; see page 372). Such a sheet structure in which each oxygen or (OH) group is surrounded only by two cations is known as *dioctahedral*. On the basis of the chemistry and geometry of the octahedral sheets, the phyllosilicates are divided into two major groups: *trioctahedral* and *dioctahedral*. (See the various interactive stacking animations in module I of the CD-ROM under the heading "Architecture of Layer Silicates.")

It was pointed out above that the geometries of an average Si_2O_5 sheet and a sheet of XO_6 octahedra are generally compatible such that one can be bonded to the other. Let us look at this in terms of the chemistries of the octahedral and tetrahedral sheets. Brucite, $Mg(OH)_2$, consists of two (OH) planes between which Mg is coordinated in octahedra. We can state this symbolically as $Mg_3 \frac{(OH)_3}{(OH)_3}$. If we replace two of the (OH) groups on one side of a brucite sheet by two apical oxygens of an Si_2O_5 sheet, we obtain $Mg_3 \frac{Si_2O_5(OH)}{(OH)_3}$. This means that the other side of the brucite structure is not attached to an Si_2O_5 sheet as in Fig. 11.31a. This is known as the *lizardite*, $Mg_3Si_2O_5(OH)_4$, structure. The equivalent structure with a dioctahedral sheet is *kaolinite*, $Al_2Si_2O_5(OH)_4$. In short, the antigorite and kaolinite structures are built of one tetrahedral ("t") and one octahedral ("o") sheet, giving *t-o* layers (see Fig. 11.34). Such *t-o* layers are electrically neutral and are bonded to one another by weak van der Waals bonds. As noted above, in the lizardite and kaolinite structures only one side of the octahedral sheet is coordinated to a tetrahedral sheet. However, we can derive further members of the phyllosilicate group by

FIG. 11.29. Undistorted, hexagonal ring in an Si_2O_5 sheet showing the location of apical oxygens and (OH) group. In an ideal sheet structure the size of the triangles (outlined by shading) is the same as the size of triangular faces of XO_6 octahedra.

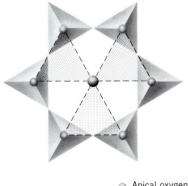

◔ Apical oxygens
● OH group

FIG. 11.30. Infinitely extending sheet of XO_6 octahedra. All octahedra lie on triangular faces.

◔ Oxygen
● Most commonly Mg or Al

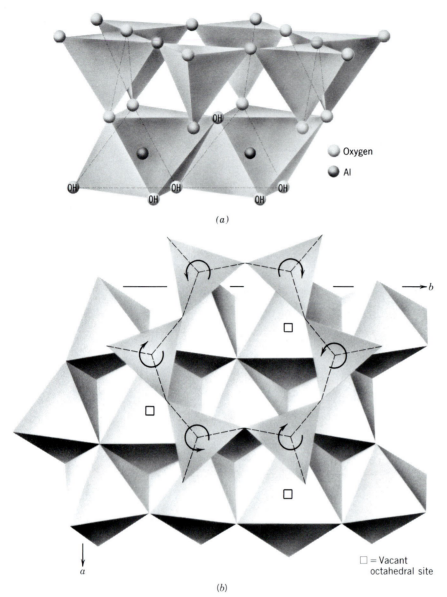

FIG. 11.31. *(a)* Diagrammatic sketch of the structure of kaolinite with a tetrahedral sheet bonded on one side by an octahedral sheet. (From R. E. Grim, 1968, *Clay Mineralogy*. New York: McGraw-Hill Book Co). *(b)* Plan view of the structure of muscovite in which the overall dimensions of an octahedral sheet are somewhat smaller than those of an idealized tetrahedral sheet. To accommodate this slight misfit, the SiO_4 tetrahedra in the tetrahedral sheet are rotated (in opposite directions as shown by the arrows) about axes normal to the sheet. This changes the shape of the six-membered tetrahedral rings from hexagonal to trigonal. The octahedra marked with a square are vacant (see discussion of dioctahedral micas). (From S. W. Bailey, 1984, Crystal structure of the true micas, in *Micas, Reviews in Mineralogy* 13: 13–60, Mineralogical Society of America, Washington, D.C.)

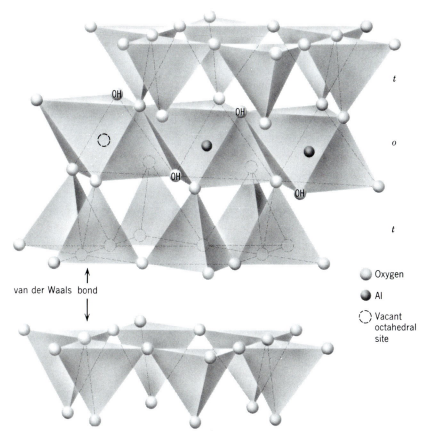

Oxygen

Al

Vacant
octahedral
site

van der Waals bond

FIG. 11.32. Diagrammatic sketch of a *t-o-t* type layer as in pyrophyllite (after Grim 1968).

joining tetrahedral sheets on both sides of the *o* sheet. This leads to *t-o-t* types layers (Fig. 11.32) as in *talc*, $Mg_3Si_4O_{10}(OH)_2$, and *pyrophyllite*, $Al_2Si_4O_{10}(OH)_2$. We may again begin with brucite, $Mg_3 \dfrac{(OH)_3}{(OH)_3}$, and replace two (OH) groups in both the upper and lower (OH) planes by two oxygens of Si_2O_5 sheets, giving rise to $Mg_3 \dfrac{Si_2O_5(OH)}{Si_2O_5(OH)}$ or $Mg_3Si_4O_{10}(OH)_2$, talc. Similarly, gibbsite, $Al_2 \dfrac{(OH)_3}{(OH)_3}$, would result in $Al_2 \dfrac{Si_2O_5(OH)}{Si_2O_5(OH)}$ by replacement of two out of three (OH) groups in the (OH) planes by oxygens of Si_2O_5 sheets; this leads to pyrophyllite, $Al_2Si_4O_{10}(OH)_2$ (see Figs. 11.32 and 11.34). The *t-o-t* layers are electrically neutral and form stable structural units that are joined to each other by van der Waals bonds. Because this is a very weak bond, we

may expect such structures to have excellent cleavage, easy gliding, and greasy feel, as indeed found in the minerals talc and pyrophyllite.

We can carry the process of evolution of phyllosilicate structures one step farther by substituting Al for some of the Si in the tetrahedral sites of the Si_2O_5 sheets. Because Al is trivalent, whereas Si is tetravalent, each substitution of this kind causes a free electrical charge to appear on the surface of the *t-o-t* layer. If Al substitutes for every fourth Si, a charge of significant magnitude is produced to bind univalent cations (*interlayer cations*), in regular 12-coordination to *t-o-t* layers (see Fig. 11.34). By virtue of these *t-o-t*-interlayer cation bonds, the structure is more firmly held together, the ease of gliding is diminished, the hardness is increased, and slippery feel is lost. The resulting mineral structures are those of the true *micas* (see Fig. 11.31*b*, 11.33, and 11.34). In the trioctahedral group of micas the univalent interlayer cation is K^+; in the dioctahedral

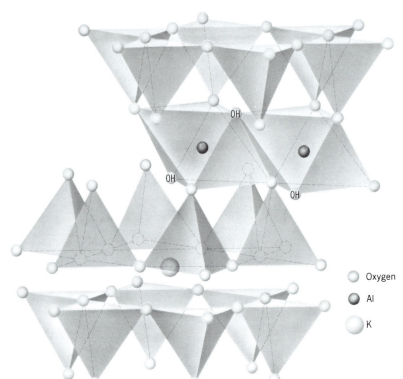

Oxygen

Al

K

FIG. 11.33. Diagrammatic sketch of the muscovite structure (after Grim 1968).

group the cation can be K^+, as in muscovite, or Na^+, as in paragonite. It is easy to remember the formulas of the micas if it is recalled that *one* of the aluminums belongs in the tetrahedral sheet and the formulas are written accordingly. Thus,

Trioctahedral	$KMg_3(AlSi_3O_{10})(OH)_2$	Phlogopite
Dioctahedral	$KAl_2(AlSi_3O_{10})(OH)_2$	Muscovite

If half the Si in the tetrahedral sites of the Si_2O_5 sheets is replaced by Al, two charges per *t-o-t* layer become available for binding an interlayer cation. Such ions as Ca^{2+}, and, to a smaller extent, Ba^{2+}, may enter between two layers of the mica structure (see Fig. 12.62). The interlayer cations are held by ionic bonds so strong that the quality of the cleavage is diminished, the hardness increased, the flexibility of the layers is almost wholly lost, and the density is increased. The resulting minerals are the *brittle micas*, typified by

Trioctahedral	$CaMg_3(Al_2Si_2O_{10})(OH)_2$	Clintonite
Dioctahedral	$CaAl_2(Al_2Si_2O_{10})(OH)_2$	Margarite

There is little solid solution between members of the dioctahedral and trioctahedral groups, although there may be extensive and substantially complete ionic substitution of Fe^{2+} for Mg, of Fe^{3+} for Al, and

of Ca for Na in appropriate sites. Ba may replace K, Cr may substitute for Al, and F may replace OH to a limited extent. Mn, Ti, and Cs are rarer constituents of some micas. The lithium micas are structurally distinct from muscovite and biotite because of the small lithium ion.

Additional members of the phyllosilicate group can be developed. The important group of *chlorites* may be viewed as consisting of two sheets of talc (or pyrophyllite) separated by a brucite- (or gibbsite-) like octahedral sheet (see Figs. 11.35 and 12.62). This leads to a formula such as $Mg_3Si_4O_{10}(OH)_2 \cdot Mg_3(OH)_6$. However, in most chlorites Al, Fe^{2+}, and Fe^{3+} substitute for Mg in octahedral sites both in the talc and brucite-like sheets, and Al substitutes for Si in the tetrahedral sites. A more general chlorite formula would be:

$$(Mg,Fe^{2+},Fe^{3+},Al)_3(Al,Si)_4O_{10}(OH)_2-(Mg,Fe^{2+},Fe^{3+},Al)_3(OH)_6$$

The various members of the chorite group differ from one another in amounts of substitution and the manner in which the successive octahedral and tetrahedral layers are stacked along the c axis.

The *vermiculite* minerals may be derived from the talc structure by interlamination of water mole-

Trioctahedral   **Dioctahedral**

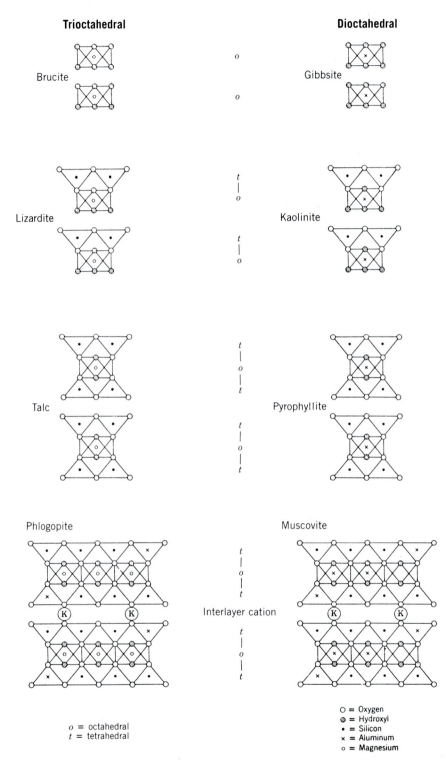

Brucite o Gibbsite

                                           o

Lizardite $\begin{array}{c} t \\ | \\ o \end{array}$ Kaolinite

                                       $\begin{array}{c} t \\ | \\ o \end{array}$

Talc $\begin{array}{c} t \\ | \\ o \\ | \\ t \end{array}$ Pyrophyllite

                                   $\begin{array}{c} t \\ | \\ o \\ | \\ t \end{array}$

Phlogopite   Muscovite

                                   $\begin{array}{c} t \\ | \\ o \\ | \\ t \end{array}$

                    Interlayer cation

                                   $\begin{array}{c} t \\ | \\ o \\ | \\ t \end{array}$

O = Oxygen
⊘ = Hydroxyl
• = Silicon
× = Aluminum
o = Magnesium

o = octahedral
t = tetrahedral

FIG. 11.34. Schematic development of some of the phyllosilicate structures (compare with Fig. 12.62).

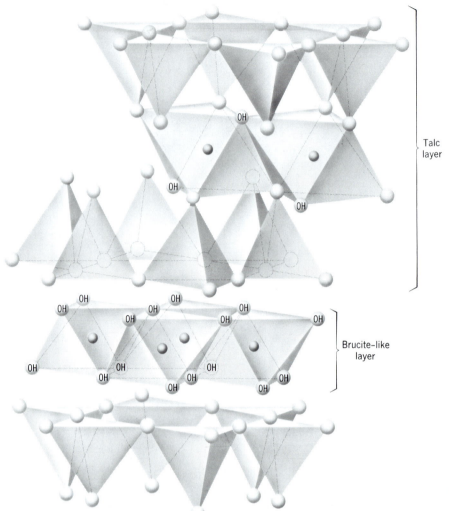

FIG. 11.35. Diagrammatic sketch of the structure of chlorite (after Grim 1968).

cules in definite sheets 4.98 Å thick, which is about the thickness of two water molecules (see Fig. 11.36). An example of a specific vermiculite formula would be $Mg_3(Si,Al)_4O_{10}(OH)_2 \cdot 4.5H_2O[Mg]_{0.35}$ in which [Mg] represents exchangeable ions in the structure. The presence of exchangeable ions located between layers of H_2O molecules, and the ability of the structure to retain variable amounts of H_2O, are of great importance in agriculture. When the vermiculite structure is saturated with H_2O, the basal spacing is approximately 14.8 Å. This water can be gradually extracted as shown by a discontinuous collapse sequence along the c axis, leading to a basal spacing of about 9.0 Å for a vermiculite without interlayer water. The structure of the *smectite* group may be derived from the pyrophyllite structure by insertion of sheets of molecular water containing exchangeable cations between the pyrophyllite t-o-t layers, leading to a structure that is essentially identical to that of vermiculite. The members of the vermiculite and montmorillonite groups exhibit an unrivaled capacity for swelling when wetted because of their ability to incorporate large amounts of interlayer water (see also Box 12.2).

If occasional random substitution of aluminum for silicon in the tetrahedral sites of pyrophyllite sheets takes place, there may not be enough of an aggregate charge on the t-o-t layers to produce an ordered mica structure with all possible interlayer cation sites filled. Locally, however, occasional cation sites may be occupied, leading to properties intermediate between those of clays and micas. Introduction of some molecular water may further complicate this picture. K-rich minerals of this type,

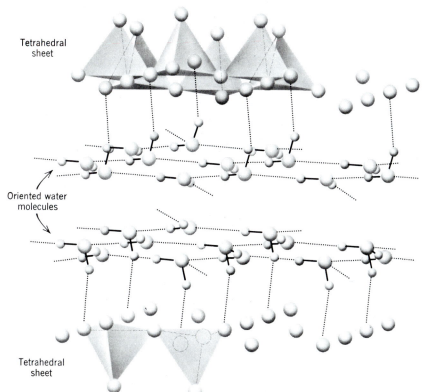

Tetrahedral sheet

Oriented water molecules

Tetrahedral sheet

FIG. 11.36. Diagrammatic sketch of the vermiculite structure, showing layers of water (after Grim 1968).

intermediate between montmorillonite clays and the true micas, are referred to as *alkali-deficient* micas, of which *illite* is an example.

In our previous discussion of the derivation of various types of structures among the phyllosilicates (see Figs. 11.34 and 12.62) we suggested that there is a good geometric fit between an octahedral brucite sheet and a tetrahedral Si_2O_5 sheet. In detail, however, one finds that there is considerable mismatch between a brucite sheet and an undistorted Si_2O_5 sheet with hexagonal rings. The misfit is due to the fact that the edges of an $Mg(OH)_6$ octahedron in the brucite sheet are somewhat greater than the distances between apical oxygens in the Si_2O_5 or $(Si,Al)_2O_5$ sheet; this means that the geometry shown in Fig. 11.29 is an oversimplification. In the case of the *serpentine minerals*, *antigorite* and *chrysotile*, both $Mg_3Si_2O_5(OH)_4$, this misfit is compensated for by a bending (stretching of the distance between apical oxygens) of the tetrahedral sheet so as to provide a better fit with the adjoining octahedral brucite sheet (see Fig. 11.37). In the variety *antigorite*, the bending is not continuous but occurs as corrugations. In the fibrous variety, *chrysotile*, the mismatch is resolved by a continuous bending of the structure into cylindrical tubes (see Figs. 11.38 and 11.39). These high-

magnification electron micrographs reveal the existence of tubelike structures for the layers that occur in essentially parallel sheets in the kaolinite group.

In our discussion of the phyllosilicates we have pointed out the existence of hexagonal rings in the undistorted Si_2O_5 sheet. In order to join the *t* sheet to a brucite *o* layer, or to an *o-t* sequence so as to pro-

FIG. 11.37. *(a)* Diagrammatic sketch of the antigorite structure as viewed along the *b* axis, showing curved *t* and *o* sheets. *(b)* Highly schematic representation of the possible curvature in the chrysotile structure.

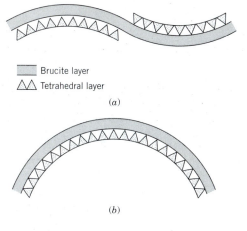

▨ Brucite layer
◿◺ Tetrahedral layer

(a)

(b)

FIG. 11.38. Electron micrograph showing tubular fibers of chrysotile from Globe, Arizona. Magnification 35,000 ×. (After B. Hagy and G. T. Gaust, 1956, Serpentines: Natural mixtures of chrysotile and antigorite, *American Mineralogist* 41: 817–38.)

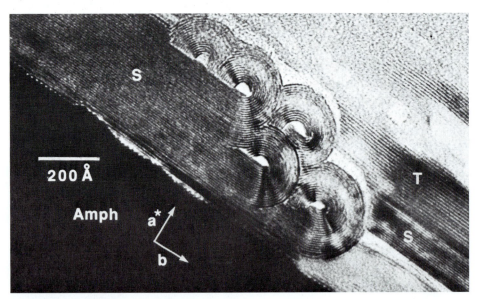

FIG. 11.39. High-resolution transmission electron microscope (HRTEM) image of an intergrowth of planar and coiled structures in serpentine. The planar serpentine (S) is seen ending in rolls of *chrysotile*. Talc (T) is interlayered with the planar serpentine, *lizardite*. The two types of serpentine are formed against grains of amphibole (Amph). (From D. R. Veblen and P. R. Buseck, 1979. Serpentine minerals: Intergrowths and new combination structures. *Science* 206: 1398–1400; copyright ©1979 by the AAAS. See also P. R. Buseck, 1983. Electron microscopy of minerals. *American Scientist* 71: 175–85.)

duce *t-o* or *t-o-t* type layers, the octahedral sheet is staggered with respect to the tetrahedral sheets (see any of the phyllosilicate illustrations). This stagger reduces the symmetry of the overall structure to monoclinic (see Fig. 11.40), even though the tetrahedral sheets contain hexagonal holes. Most of the phyllosilicates, therefore, have monoclinic structures, some are triclinic, and a few orthorhombic or trigonal.

Because of the hexagonal to trigonal symmetry about the (OH) group in the $Si_2O_5(OH)$ tetrahedral sheets, there are three alternate directions (see Fig. 11.41) in which $Si_2O_5(OH)$ sheets can be stacked. These three directions can be represented by three vectors (1, 2, and 3; and negative directions $\bar{1}$, $\bar{2}$, and $\bar{3}$) at 120° to each other. If the stacking of $Si_2O_5(OH)$ sheets is always in the same direction (see Fig. 11.41*b* and module I of the CD-ROM for various stacking polytype sequences under the heading "Polytypism"), the resulting structure will have monoclinic symmetry, with space group *C2/m*; this is referred to as the 1*M* (*M* = monoclinic) *polytype* with stacking sequence [1]. If the stacking sequence of

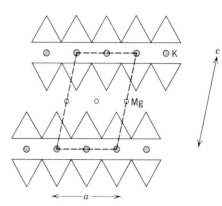

FIG. 11.40. Schematic projection along the *b* axis of the mica structure. Note stagger of *t* and *o* sheets with respect to each other. Unit cell is outlined by dashed lines.

$Si_2O_5(OH)$ sheets consists of alternating 1 and $\bar{1}$ directions (in opposing directions along the same vector), a structure results which is best described as orthorhombic with space group $Ccm2_1$. This stacking sequence can be expressed as [$1\bar{1}$] and the polytype

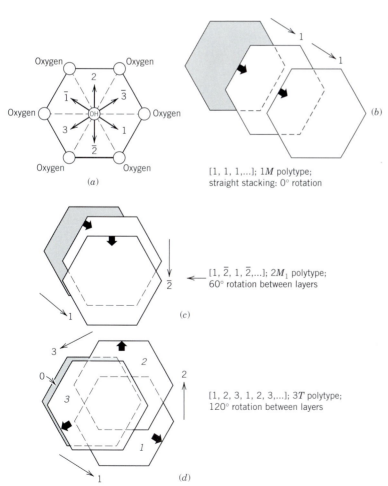

[1, 1, 1,...]; 1*M* polytype; straight stacking: 0° rotation

(*a*)

(*b*)

[1, $\bar{2}$, 1, $\bar{2}$,...]; 2*M*₁ polytype; 60° rotation between layers

(*c*)

[1, 2, 3, 1, 2, 3,...]; 3*T* polytype; 120° rotation between layers

(*d*)

FIG. 11.41. Schematic illustration of some possible stacking polymorphs (*polytypes*) in mica. (*a*) Three vectorial directions for possible location of the (OH) group in an $Si_2O_5(OH)$ sheet, which is stacked above or below the hexagonal ring shown. (*b*) Stacking of $Si_2O_5(OH)$ sheets in the same direction. (*c*) Stacking in two directions at 60° to each other. (*d*) Stacking in three directions at 120° to each other. Here, sheet *3* should lie directly above sheet *O* but is offset slightly for illustrative purposes.

is known as 2*O* (*O* = orthorhombic). If the stacking sequence consists of two vectors at 60° to each other, that is, [12], another monoclinic polytype results; this is known as 2*M*₁ with space group *C*2/*c* (see Fig. 11.41*c*). When all three vectors come into play, such as in [123], a trigonal polytype, 3*T* (*T* = trigonal), results which is compatible with either of two enantiomorphic space groups: *P*3₁12 or *P*3₂12. Additional *polytype* stacking sequences are possible, but most micas that commonly show *polytypism* belong to 1*M*(*C*2/*m*), 2*M*₁(*C*2/*c*), or 3*T*(*P*3₁12) polytypes. Polytypism (see also page 141) is found in many layer silicates, among them serpentine, mica,

and chlorite. Figure 11.42*a* illustrates two different polytypes for planar serpentine structures, and Fig. 11.42*b* shows two different polytypes of mica.

The great importance of the phyllosilicates lies in part in the fact that the products of rock weathering, and hence the constituents of soils, are mostly of this structural type. The release and retention of plant foods, the stockpiling of water in the soil from wet to dry seasons, and the accessibility of the soil to atmospheric gases and organisms depend in large part on the properties of the sheet silicates.

Geologically, the phyllosilicates are of great significance. The micas are the chief minerals of schists

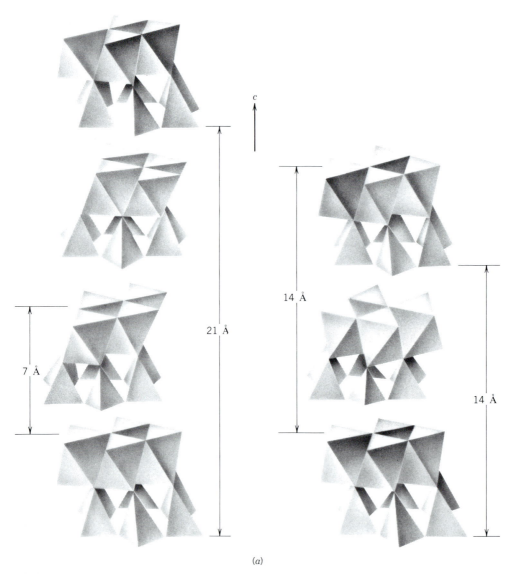

(*a*)

FIG. 11.42. *(a)* Two different polytypes (different stackings of *t-o* sheets along the *c* crystallographic axis) for planar serpentine.

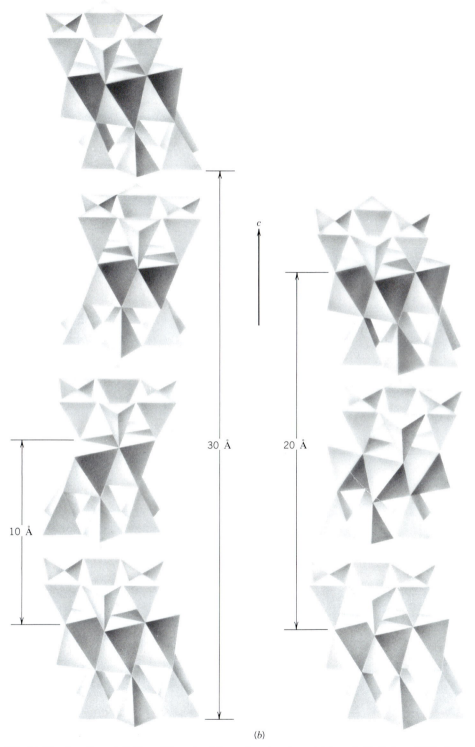

FIG. 11.42. (*continued*) (*b*) Two different polytypes for mica. (*a* and *b* from S. W. Bailey, 1988, X-ray diffraction identification of the polytypes of mica, serpentine, and chlorite. *Clays and Clay Minerals* 36: 193–213.)

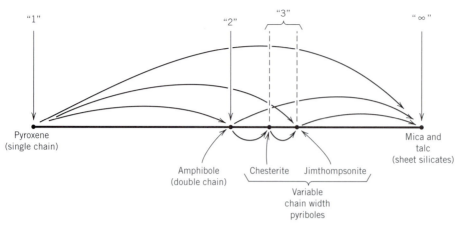

FIG. 11.43. The many possible reaction paths that lead from a high temperature pyroxene to lower temperature sheet silicates. Ch and Jt are pyriboles, chesterite, and jimthompsonite, respectively, "1," "2," and "3" refer to chain width: "1" in pyroxenes, "2" in amphiboles, and "3" in biopyriboles with variable chain widths; "∞" means infinite chain width, resulting in sheets (see also Figs. 11.26 and 11.27). (From D. R. Veblen and P. R. Buseck, 1981, Hydrous pyriboles and sheet silicates in pyroxenes and uralites: Intergrowth microstructures and reaction mechanisms. *American Mineralogist* 66: 1107–1134.)

and are widespread in igneous rocks. They form at lower temperatures than amphiboles or pyroxenes and frequently are formed as replacements of earlier minerals as a result of hydrothermal alteration. The chemical changes and structural transformations that take place in such alteration reactions are complex. Figure 11.43 is a schematic representation of the various reaction paths an originally high-temperature py-

roxene (of igneous origin) might take during lower-temperature reactions (such as alteration by hydrothermal solutions). The lower-temperature end products, marked on the far right of the diagram, are layer silicates. The intermediate reaction products are mainly amphiboles, but also various pyriboles (see Figs. 11.26 and 11.27 and related text). Figure 11.44 gives two high-resolution transmission electron mi-

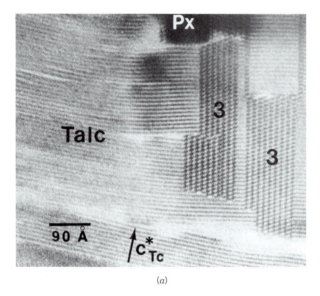
(a)

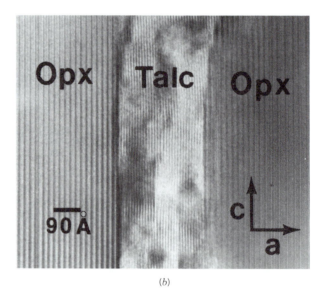
(b)

FIG. 11.44. High-resolution transmission electron microscope (HRTEM) images of altered pyroxenes. (a) A complex intergrowth of high-temperature pyroxene and alteration products of talc and triple-chain silicate ("3" = pyribole). (b) Direct replacement of original orthopyroxene by later talc along a fracture. (From D. R. Veblen and P. R. Buseck, 1981, Hydrous pyriboles and sheet silicates in pyroxenes and uralites: Intergrowth microstructures and reaction mechanisms. *American Mineralogist* 66: 1107–1134.)

croscope (HRTEM) structure images for igneous pyroxenes from the Palisades Sill, New Jersey. Figure 11.44*a* shows the presence of a triple chain pyribole between a relict (unaltered) augite and sheet silicate which is mainly talc. The pyribole material appears to be an intermediate product in the reaction sequence: pyroxene $\rightarrow$ triple chain silicate $\rightarrow$ talc (see Fig. 11.43). In Figure 11.44*b* talc appears to have directly replaced orthopyroxene along a fracture.

TECTOSILICATES

Approximately 64% of the rocky crust of the Earth is made up of minerals built about a three-dimensional framework of linked SiO_4 tetrahedra (see Fig. 11.1). These minerals belong to the *tectosilicate* class in which all the oxygen ions in each SiO_4 tetrahedron are shared with neighboring tetrahedra. This results in a stable, strongly bonded structure in which the ratio of Si : O is 1 : 2 (Fig. 11.3).

SiO₂ Group

An SiO_2 framework that does not contain other structural units is electrically neutral. There are at least nine different ways in which such a framework can be built. These modes of geometrical arrangement correspond to nine known polymorphs of SiO_2, one of which is synthetic (see Table 11.3). Each of these polymorphs has its own space group, cell dimensions, characteristic morphology, and lattice energy. Which polymorph is stable is determined chiefly by energy considerations; the higher-temperature forms with great lattice energy possess the more expanded structures that are reflected in lower specific gravity

and refractive index. In addition to the nine polymorphs of SiO_2 there are two related and essentially amorphous substances, *lechatelierite*, a high silica glass of variable composition, and *opal*, $SiO_2 \cdot nH_2O$, with a locally ordered structure of silica spheres and highly variable H_2O content.

The principal naturally occurring SiO_2 polymorphs fall into three structural categories: *low quartz*, with the lowest symmetry and the most compact structure; *low tridymite*, with higher symmetry and a more open structure; and *low cristobalite*, with the highest symmetry and the most expanded structure of the three polymorphs. These polymorphs are related to each other by reconstructive transformations, a process that requires considerable energy. The sluggishness and energy requirements of reconstructive transformations allow the phases to exist metastably for long periods of time. The temperatures of reconstructive inversions vary widely, depending chiefly on the rate and direction of temperature change. Each of the three above structure types has also a high–low inversion, as shown by the existence of high and low quartz, high and low tridymite, and high and low cristobalite (see Table 11.3). These transformations are displacive; they take place quickly and reversibly at a fairly constant and sharply defined temperature of inversion and may be repeated over and over again without physical disintegration of the crystal. The nearly instant high $\rightarrow$ low inversions take place with the release of a fairly constant amount of energy; very near the same temperature the low $\rightarrow$ high inversions take place with the absorption of energy (see Table 11.4). The structures of several SiO_2 polymorphs are illustrated in Fig. 11.45 (see Fig. 4.36 for high and low quartz structures).

TABLE 11.3   Polymorphs of SiO₂

Name	Symmetry	Space Group	Specific Gravity	Refractive Index (Mean)
Stishovite*	Tetragonal	$P4_2/mnm$	4.35	1.81
Coesite	Monoclinic	$C2/c$	3.01	1.59
Low (α) quartz	Hexagonal	$P3_221$ (or $P3_121$)	2.65	1.55
High (β) quartz	Hexagonal	$P6_222$ (or $P6_422$)	2.53	1.54
Keatite (synth.)	Tetragonal	$P4_12_12$ (or $P4_32_12$)	2.50	1.52
Low (α) tridymite	Monoclinic or	$C2/c$ (or Cc)	2.26	1.47
	Orthorhombic	$C222_1$		
High (β) tridymite	Hexagonal	$P6_3/mmc$	2.22	1.47
Low (α) cristobalite	Tetragonal	$P4_12_12$ (or $P4_32_12$)	2.32	1.48
High (β) cristobalite	Isometric	$Fd3m$	2.20	1.48

*Only polymorph with Si in octahedral coordination with oxygen.

TABLE 11.4   Inversion Temperatures for Displacive Transformations in Some SiO$_2$ Polymorphs

High T Polymorph	Minimum Crystallization T for Stable Form at 1 Atmosphere P	Inversion to Low T Form at 1 Atmosphere P
High cristobalite	1470°C	~268°C
High tridymite	870°C	~120°–140°C
High quartz	574°C	573°C

The low-temperature form of each of the displacive polymorphic pairs has a lower symmetry than the higher-temperature form (see Table 11.3), but this symmetry change is less than in the reconstructive transformations. The effect of increased pressure is to raise all inversion temperatures and for any temperature to favor the crystallization of the polymorph most economical of space (Fig. 11.46).

The most dense of the silica polymorphs are *coesite* and *stishovite*. *Coesite* was synthesized in 1953 and *stishovite*, which is isostructural with rutile, TiO$_2$, in 1961 and it was not until later that they were found in nature. They have both been identified in very small amounts at Meteor Crater, Arizona. Their formation is attributed to the high pressure and high temperature resulting from the impact of a meteorite. *Coesite* has been found in a sanidine-coesite eclogite from a kimberlite pipe in South Africa and in ultrahigh pressure metamorphic rocks (see Fig. 4.10; also Fig. 4.16 and related text). *Keatite* has not been found in nature.

Feldspar Group

The compositions of the majority of common feldspars can be expressed in terms of the system KAlSi$_3$O$_8$ (orthoclase; Or)–NaAlSi$_3$O$_8$ (albite; Ab)–CaAl$_2$Si$_2$O$_8$ (anorthite; An). The members of the series between KAlSi$_3$O$_8$ and NaAlSi$_3$O$_8$ are known as the *alkali feldspars* and the members in the series between NaAlSi$_3$O$_8$ and CaAl$_2$Si$_2$O$_8$ as the *plagioclase feldspars*. Members of both of these feldspar groups are given specific names as shown in Fig. 11.47. The chemical compositions of feldspars in this ternary system are generally expressed in terms of molecular percentages or Or, Ab, and An; for example, Or$_{20}$Ab$_{75}$An$_5$. Barium feldspars such as *celsian*, BaAl$_2$Si$_2$O$_8$, and *hyalophane*, (K,Ba)(Al,Si)$_2$Si$_2$O$_8$, are relatively rare. All feldspars show good cleavages in two directions which make an angle of 90°, or close to 90°, with each other. Their hardness is about 6,

and specific gravity ranges from 2.55 to 2.76 (excluding the Ba feldspars).

The unambiguous characterization of a feldspar requires a knowledge not only of the chemical composition but also of the structural state of the species. The structural state, which refers to the Al and Si distribution in tetrahedral sites of the framework structure, is a function of the crystallization temperature and subsequent thermal history of a feldspar. In general, feldspars that cooled rapidly

FIG. 11.45. Structures of some polymorphs of SiO$_2$ (see Fig. 4.36 for illustration of low and high quartz structures). *(a)* Tetrahedral layers in high tridymite projected onto (0001). *(b)* Portion of the high cristobalite structure projected onto (111).

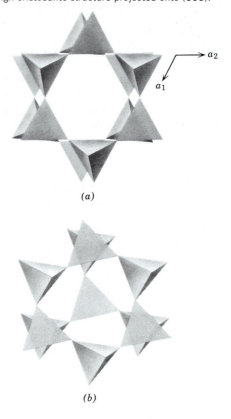

(a)

(b)

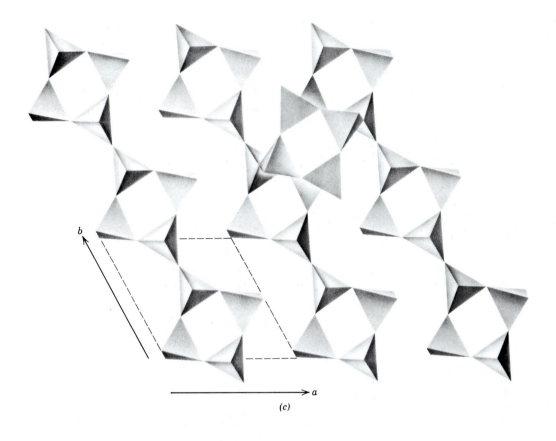

(c)

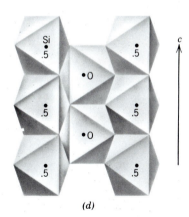

(d)

FIG. 11.45. (*continued*) (*c*) Coesite structure showing four-membered tetrahedral rings that lie parallel to (001). (*d*) Structure of stishovite, with Si in octahedral coordination with oxygen, projected on (100) (*a, b, c,* and *d* after J. J. Papike, and M. Cameron, 1976, Crystal chemistry of silicate minerals of geophysical interest. *Reviews of Geophysics and Space Physics* 14: 37–80.)

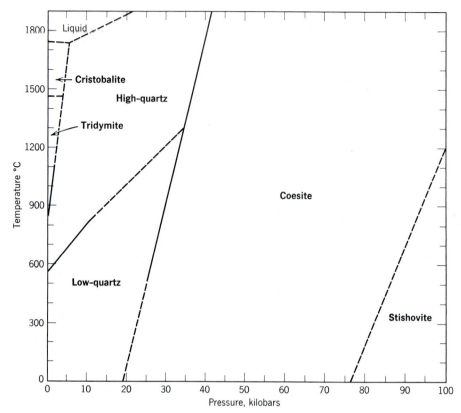

FIG. 11.46. Stability relations of the SiO_2 polymorphs.

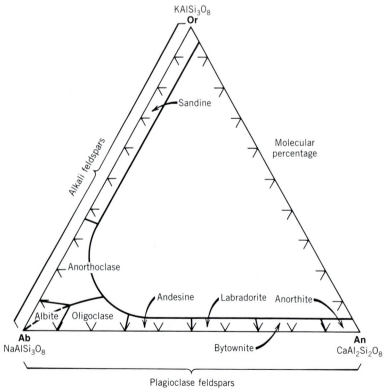

FIG. 11.47. Nomenclature for the plagioclase feldspar series and the high-temperature alkali feldspars. (After W. A. Deer, R. A. Howie, and J. Zussman, 1963, *Rock-Forming Minerals* 4. New York: Wiley, p. 2.)

after crystallization at high temperature show a disordered Al–Si distribution (high structural state). Those that cooled very slowly from high temperatures or those that crystallized at low temperatures generally show an ordered Al–Si distribution (low structural state).

Structure

The feldspar structure, similar to the structures of the various polymorphs of SiO_2, consists of an infinite network of SiO_4 as well as AlO_4 tetrahedra. The feldspar structure can be considered a "stuffed" derivative of the SiO_2 structures, by incorporation of Al into the tetrahedral network, and concomitant housing of Na^+ (or K^+ or Ca^{2+}) in available voids. When only one Si^{4+} (per feldspar formula unit) is substituted by Al^{3+}, the structure can be neutralized by incorporating one K^+ or one Na^+. Similarly when two Si^{4+} (per feldspar formula unit) are substituted for by Al^{3+}, the electrostatic charge of the network can be balanced by a divalent cation such as Ca^{2+}. We may state this as follows:

$$Si_4O_8 \text{ framework} \rightarrow Na(AlSi_3O_8) \rightarrow Ca(Al_2Si_2O_8)$$

In the plagioclase structures the amount of tetrahedral Al varies in proportion to the relative amounts of Ca^{2+} and Na^+ so as to maintain electrical neutrality; the more Ca^{2+}, the greater the amount of Al^{3+}.

The general architecture of the feldspar structure can be illustrated with the aid of a drawing of the high-temperature polymorph of $KAlSi_3O_8$, *sanidine*, with space group $C2/m$ (see Fig. 11.48). In this structure the Al–Si distribution is completely disordered, meaning that the Al and Si ions are randomly distributed among the two crystallographically distinct tetrahedral sites, *T1* and *T2*. The K^+ ions, bonded to nine nearest oxygens in large interstices, occupy special positions on mirror planes perpendicular to the b axis. The Si–Al tetrahedral framework consists of four-membered rings of tetrahedra that are linked into chains (of a double crankshaft type) parallel to the a axis (see Fig. 11.49).

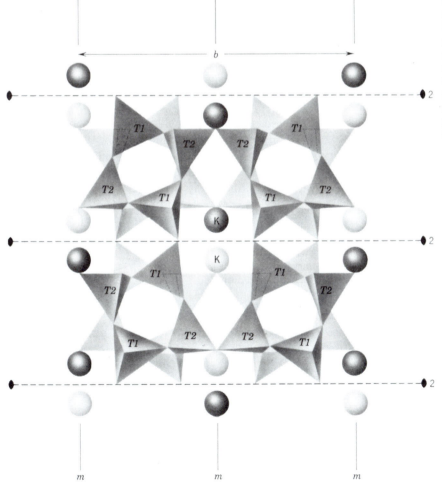

FIG. 11.48. The structure of high sanidine, $KAlSi_3O_8$, projected on ($\overline{2}01$). Mirror planes (*m*) and twofold rotation axes (2) are shown. Other symmetry elements such as glide planes and twofold screw axes are also present but their location is not shown here. (After J. J. Papike and M. Cameron, 1976, Crystal chemistry of silicate minerals of geophysical interest. *Reviews of Geophysics and Space Physics* 14: 66.)

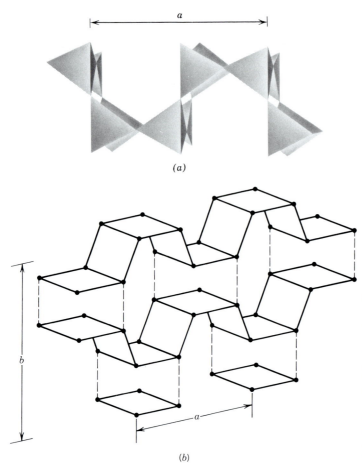

FIG. 11.49. *(a)* The four-membered rings in Fig. 11.48 are linked to form crankshaft-like chains that run parallel to the *a* axis. (After J. J. Papike and M. Cameron, 1976, Crystal chemistry of silicate minerals of geophysical interest. *Reviews of Geophysics and Space Physics* 14: 67.) *(b)* Schematic representation of the location and orientation of the four-membered crankshaft-like chains in the feldspar structure. Black dots are locations of Si. Dashed lines locate bonds between adjoining $(Si, Al)O_4$ tetrahedra. (Modified with permission after P. Ribbe, 1987, Feldspar. *McGraw-Hill Encyclopedia of Science and Technology*, 6th ed., v. 7, pp. 38–47, McGraw-Hill Book Co., New York; reprinted with permission of McGraw-Hill.)

The square, blocky outline of these chains, imparted by the four-membered rings, finds outward expression in the right-angled cleavage and pseudotetragonal habit characteristic of feldspar. The structure of *microcline,* a low-temperature polymorph of $KAlSi_3O_8$, has triclinic symmetry (space group $C\bar{1}$) and lacks the mirror planes and rotation axes of sanidine as shown in Fig. 11.48. In other words, its structure is less symmetric and the K^+ ions are no longer located in special positions. The Al–Si distribution is completely ordered in what is known as low-temperature or maximum microcline (maximum refers to maximum *triclinicity,* which results from the complete order). The tetrahedra that contain Al in this structure can be unambiguously located, whereas in sanidine the Al–Si distribution is completely random. *Orthoclase* represents a polymorph of $KAlSi_3O_8$ in which the Al–Si distribution is between the total randomness found in sanidine and the total order of microcline. Orthoclase, with space group $C2/m$, crystallizes at intermediate temperatures (see Fig. 11.51). Although unambiguous distinction between the three structurally different K-feldspars, sanidine, orthoclase, and microcline, is based on careful

measurements of unit cell dimensions and/or optical parameters, such as 2V and/or extinction angle $b \wedge Z$, we will use the terms orthoclase and microcline more broadly. Their descriptions (in Chapter 12) will be based on generally recognizable characteristics in hand specimens. It must be remembered, however, that the definitions of the three types of K-feldspar as used in the literature are based on parameters that can be obtained only by X-ray and optical techniques (see Smith and Brown 1988 or Smith 1974). Such measurements allow for the definition of maximum microcline, and high, intermediate, and low orthoclase, and provide the investigator with information about the state of order or disorder of the Al–Si distribution in the feldspar structure.

The question of whether a specific, originally high-temperature feldspar retains its high-temperature (disordered) structure type or whether it will transform (upon cooling) to a lower temperature (more ordered) structural state is much influenced by the cooling rate of the process. Figure 11.50 is a schematic illustration of the various cooling paths for a potassium feldspar, as a function of temperature and cooling rate.

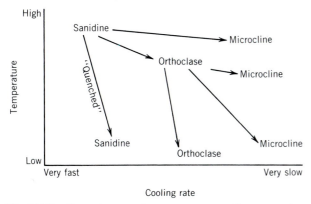

FIG. 11.50. The various possible temperature-cooling rate paths that an originally high-temperature K-feldspar (sanidine) can follow. The path marked "quenched" depicts a path in which the original high-temperature sanidine is cooled so rapidly that all characteristics of the high-temperature state are preserved in the final cooled product. (Modified with permission from A. Putnis, and J. D. C. McConnell, 1980, *Principles of mineral behavior*, Fig. 7.8, Blackwell Scientific Publications, Oxford, England, 257 pp; reprinted with permission.)

Microcline is particularly characteristic of deepseated rocks and pegmatites, orthoclase of intrusive rocks formed at intermediate temperatures and sanidine of extrusive, high-temperature lavas.

The general structure of members of the plagioclase series is very similar to that of microcline. The Na end member, *albite*, is generally triclinic (space

group $C\bar{1}$) with a low albite form that shows a highly ordered Al-Si distribution and a high albite form with a highly disordered Al-Si distribution. A monoclinic variety of albite occurs at very high temperature and is known as *monalbite*. The Ca end member, *anorthite*, is also triclinic with space group $P\bar{1}$ at room temperature, and perfect Al-Si ordering in the structure. At elevated temperatures the structure of anorthite becomes body-centered with space group $I\bar{1}$. The general stability fields of the various forms of feldspar are shown in Figs. 11.51, 11.54, and 11.56.

Composition

The alkali feldspar series ($NaAlSi_3O_8$ to $KAlSi_3O_8$) shows complete solid solution only at high temperatures (see Fig. 11.51). For example, members of the sanidine–high albite series are stable at elevated temperatures but at lower temperature two separate phases, low albite and microcline, become stable. As can be seen from Fig. 11.51, the compositional ranges of low albite and microcline are very small. If a homogeneous feldspar, of composition $Or_{50}Ab_{50}$, in which the Na^+ and K^+ ions are randomly distributed, is allowed to cool slowly, segregation of Na^+ and K^+ ions will result because the size requirements of the surrounding structure become more stringent. The Na^+ will diffuse to form Na-rich regions and the K^+ will segregate into K-rich regions in the structure, causing the originally homogeneous feldspar to be-

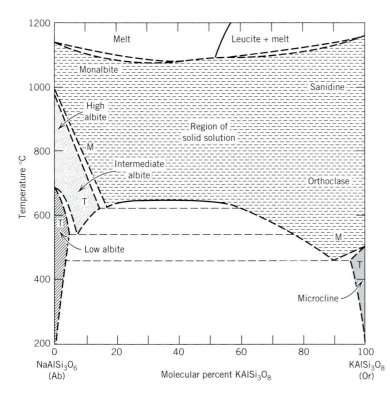

FIG. 11.51. Schematic phase diagram for the system $NaAlSi_3O_8$ (Ab)–$KAlSi_3O_8$ (K-spar) showing a large miscibility gap at temperatures below approximately 650°C. M and T mean monoclinic and triclinic, respectively. Compare with Fig. 4.19. (Modified with permission after J. V. Smith and W. L. Brown, 1988, *Feldspar minerals* 1, Fig. 1.2, New York: Springer Verlag, 828 pp.)

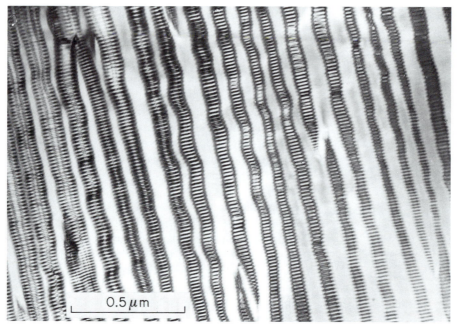

FIG. 11.52. Microstructure in an alkali feldspar of composition 57.3 weight percent Or. The Na-rich lamellae are twinned according to the albite law. This very high magnification photograph was taken with a transmission electron microscope. (From P. E. Champness and G. W. Lorimer, 1976, Exsolution in Silicates, chapter 4.1 in *Electron microscopy in mineralogy*, H. R. Wenk, ed., New York: Springer-Verlag.)

come a heterogeneous intergrowth. The separation most commonly results in thin layers of albite in a host crystal of K feldspar. Such intergrowths are known as *perthites,* and are the result of *exsolution* (see Fig. 4.44 and discussion on page 143). In the alkali feldspar series the orientation of the exsolution lamellae is roughly parallel to {100}. When these intergrowths are visible to the naked eye they are known as *macroperthite,* when visible only by optical microscope they are referred to as *microperthite,* and when detectable only by X-ray or electron microscope techniques, they are called *cryptoperthite* (see Fig. 11.52). More rarely the host mineral is a plagioclase feldspar and the lamellae are of $KAlSi_3O_8$ composition; this is called *antiperthite.*

Only very limited solid solution occurs between $KAlSi_3O_8$ and $CaAl_2Si_2O_8$ (see Fig. 11.53). Essentially complete solid solution, however, exists at elevated temperatures in the plagioclase series ($NaAlSi_3O_8$ to $CaAl_2Si_2O_8$) (see Fig. 11.54). The general formula of a feldspar in this series may be written as: $Na_{1-x}Ca_x(Si_{3-x}Al_{1+x})O_8$, where x ranges from 0 to 1. The structural interpretation of the region of essentially complete solid solution is complicated because of the varying ratio of Al/Si from albite, $NaAlSi_3O_8$, to anorthite, $CaAl_2Si_2O_8$. Three types of exsolution textures found in the plagioclase series are not visible to the naked eye, but may be detected because of iridescence. *Peristerite* intergrowths occur in the

range An_2 to An_{15} (see Fig. 11.55). *Bøggild intergrowths* occur in some plagioclase with composition between An_{47} and An_{58}; their presence is indicated by the play of colors in labradorite. A third intergrowth occurs in the An_{60} to An_{85} region and is known as *Huttenlocher intergrowths.* These disconti-

FIG. 11.53. Experimentally determined extent of solid solution in the system Or-Ab-An at P_{H_2O} = 1 kilobar. (After P. H. Ribbe, 1975, *Feldspar mineralogy, Reviews in Mineralogy* 2, Fig. R-1, Mineralogical Soc. of America, Washington, D.C.)

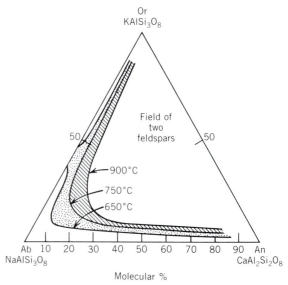

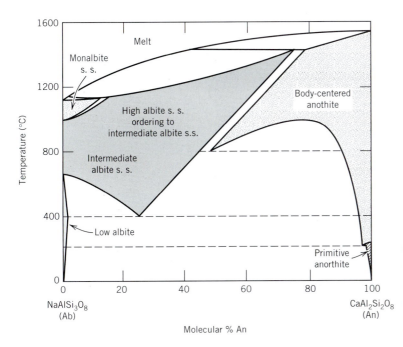

FIG. 11.54. Schematic phase diagram for the plagioclase feldspar series showing a range of almost complete solid solution at high temperatures, and three miscibility regions at lower temperatures. (Simplified with permission after J. V. Smith and W. L. Brown, 1988, *Feldspar minerals* 1, Fig. 1.4, New York: Springer Verlag, 828 pp.)

nuities in the solid solution series between Ab and An are on a very fine scale, however, and most properties, such as specific gravity or refractive index, show a generally linear change with chemical composition. Thus, determination of a suitable property with sufficient precision permits a close approximation of the chemical composition in the plagioclase series. Figure 11.56 is a simplified representation of the three-component feldspar system ($KAlSi_3O_8$—K-spar; $NaAlSi_3O_8$—albite; $CaAl_2Si_2O_8$—anorthite) show-

FIG. 11.55. Microstructure in the peristerite region of the plagioclase series (between An_2 and An_{15}), showing sharply defined, very fine lamellae approximately parallel to $(0\bar{4}1)$. These lamellae act as a diffraction grating for white light, producing the delicate bluish sheen of moonstone. This photograph was taken with a transmission electron microscope. (From P. E. Champness and G. W. Lorimer, 1976, Exsolution in silicates, chapter 4.1, in *Electron microscopy in mineralogy*, H. R. Wenk, ed., New York: Springer-Verlag.)

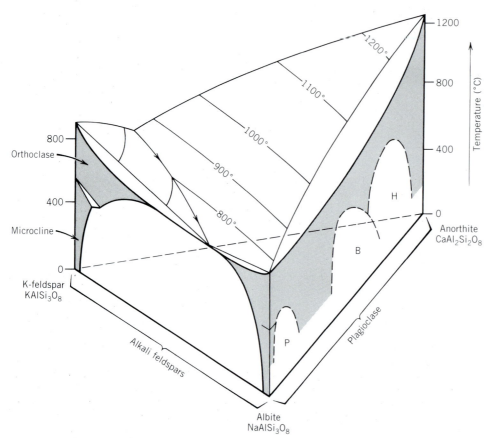

FIG. 11.56. Highly schematic temperature-composition diagram for the three-component system KAlSi$_3$O$_8$ (K-feldspar)–NaAlSi$_3$O$_8$ (albite)–CaAl$_2$Si$_2$O$_8$ (anorthite), at water pressure of about 5 kilobars. Details of the interior are complex and have been omitted. The upper, contoured surface of the diagram is the liquidus surface. The three intergrowth regions, as a result of miscibility gaps, in the lower temperature range of the plagioclase series are: P = peristerite, B = Bøggild intergrowth, H = Huttenlocher intergrowth. Solid solution is shown by shading. Compare the two vertical sides of this diagram with more detailed information given in Figs. 11.51 and 11.54. (Adapted with permission from P. H. Ribbe, 1987, Feldspars, in *McGraw-Hill Encyclopedia of Science and Technology*, 6th ed., v. 7, p. 45. McGraw-Hill Book Co., New York; reprinted with permission of McGraw-Hill.)

ing the major miscibility gaps, and melting temperatures at water pressure of about five kilobars (compare with Fig. 4.19).

Feldspathoid Group

The feldspathoids are anhydrous framework silicates that are chemically similar to the feldspars. The chief chemical difference between feldspathoids and feldspars lies in the SiO$_2$ content. The feldspathoids contain only about two-thirds as much silica as alkali feldspar and hence tend to form from melts rich in alkalis (Na and K) and poor in SiO$_2$ (see Fig. 11.57). The structures of the feldspathoids are closely related to those of the feldspars and silica minerals. However, several of the feldspathoids show the development of somewhat larger structural cavities (than

FIG. 11.57. Feldspathoid compositions, as compared to those of the feldspars, in the system SiO$_2$ (quartz)–NaAlSiO$_4$ (nepheline)–KAlSiO$_4$ (kalsilite)–CaAl$_2$O$_4$. Shaded areas represent average extents of solid solution (compare with Figs. 4.21 and 4.24).

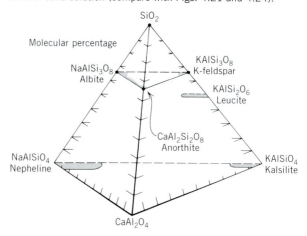

feldspar or silica minerals) as a result of four- and six-membered tetrahedral linkages. This general increase in openness of the feldspathoid structures, as compared with those of the feldspars, is expressed in their ranges of specific gravity:

G range for feldspars: 2.54–2.75
G range for feldspathoids: 2.15–2.5

Examples of the aluminosilicate frameworks in the feldspathoids are given in Figs. 11.58 to 11.60. The structure of *leucite*, $KAlSi_2O_6$, has tetragonal symmetry (space group $I4_1/a$) at low to intermediate temperatures; at approximately 605°C it inverts to a cubic structure with space group $Ia3d$. In the low-temperature form the Al and Si tetrahedra share corners to form four- and six-membered rings (see Fig. 11.58). The K^+ ions are in 12-coordination with oxygen in large cavities in the structure. The structure of *nepheline*, $(Na,K)AlSiO_4$ (space group $P6_3$), can be considered as a derivative of the high tridymite structure with Al replacing Si in half the tetrahedra and

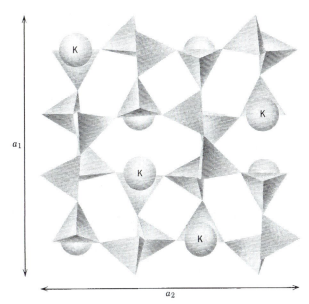

FIG. 11.58. Portion of the leucite structure projected down the *c* axis. (After J. J. Papike and M. Cameron, 1976, Crystal chemistry of silicate minerals of geophysical interest. *Reviews of Geophysics and Space Physics* 14: 74.)

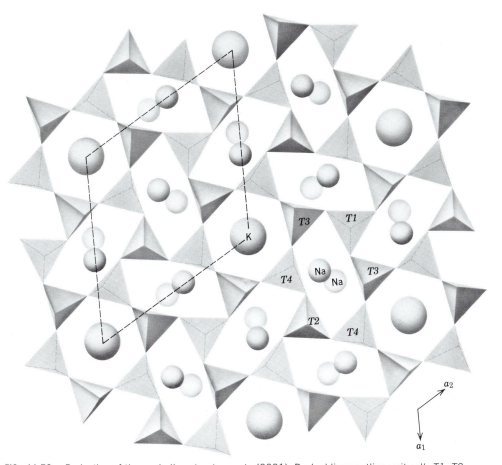

FIG. 11.59. Projection of the nepheline structure onto (0001). Dashed lines outline unit cell. *T1*, *T2*, *T3*, and *T4* refer to crystallographically distinct tetrahedral sites. (After J. J. Papike and M. Cameron, 1976, Crystal chemistry of silicate minerals of geophysical interest. *Reviews of Geophysics and Space Physics* 14: 56.)

Na and K in interstitial voids. The Si and Al are ordered in specific tetrahedral sites of the structure; the *T1* and *T4* sites are Al-rich, the *T2* and *T3* sites Si-rich. One-fourth of the interstitial sites, which are filled by K, are nearly hexagonal in geometry (see Fig. 11.59). The other three-fourths of the sites are irregular in configuration and are occupied by Na. The structure of *sodalite*, $Na_8(AlSiO_4)_6$ (see Fig. 11.60), has large cavities occupied by Na^+ and Cl^-. The structural framework consists of corner-sharing, alternating SiO_4 and AlO_4 tetrahedra. The cagelike cavities result from the linking of six four-membered and eight six-membered tetrahedral rings. The six-membered rings form channelways parallel to the cube diagonals of the structure. The large central cavities are occupied by Cl^-, which are tetrahedrally coordinated by Na^+.

Some of the members of the feldspathoid group contain unusual anions. Sodalite contains Cl and cancrinite incorporates CO_3. Noselite houses SO_4 and lazurite SO_4, S, and Cl ions. These large anionic groups and anions are located in large interstices of the structure.

Zeolite Group

The zeolites form a large group of hydrous silicates that show close similarities in composition, association, and mode of occurrence. They are framework aluminosilicates with Na, Ca, and K, and highly variable amounts of H_2O in the generally large voids of the framework. Many of them fuse readily with marked intumescence (a swelling up which results when water is expelled during heating), hence the name *zeolite*, from two Greek words meaning *to boil* and *stone*. Traditionally, the zeolites have been thought of as well-crystallized minerals found in cavities and veins in basic igneous rocks. More recently, large deposits of zeolites have been found in the western United States and in Tanzania as alterations of volcanic tuff and volcanic glass. Such zeolites are the diagenetic alteration products of silicic tuffs in Cenozoic lacustrine deposits, especially those that were deposited in highly saline and alkaline waters. The occurrence of zeolites in various rock types is used to define the low-grade regional metamorphic zone known as the *zeolite facies*.

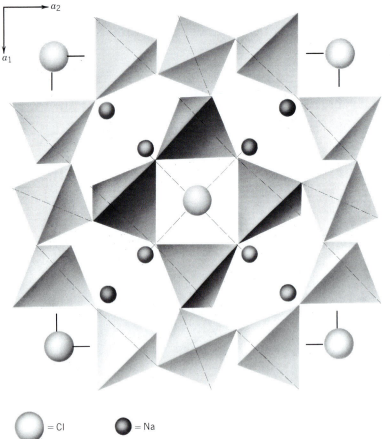

FIG. 11.60. The structure of sodalite, $Na_8(AlSiO_4)_6Cl_2$. This can be visualized as a cubic unit cell with a four-membered tetrahedral ring on each cube face, and a six-membered tetrahedral ring around each corner of the cube. (From J. J. Papike, 1988, Chemistry of the rock-forming silicates: Multiple chain, sheet, and framework structures. *Reviews of Geophysics* 26: 407–44.)

a_2

a_1

○ = Cl   ● = Na

Physical properties, such as hardness (**H**) and specific gravity (**G**) of members of the zeolite group show a pronounced decrease, when compared to those of the silica minerals, feldspars, and feldspathoids:

	H	**G**
Quartz	7	2.65
Feldspars	6	2.54–2.75
Feldspathoids	5–6	2.15–2.5
Zeolites	$3\frac{1}{2}$–$5\frac{1}{2}$	2.0–2.4

The zeolites, like the feldspars and feldspathoids, are built of frameworks of AlO_4 and SiO_4 tetrahedra; the zeolite frameworks, however, are very open with large interconnecting spaces or channels. The zeolites can be divided into those with a fibrous habit and an underlying chain structure (Fig. 11.61), those with a platy habit and an underlying sheet structure, and those with an equant habit and an underlying framework structure (Fig. 11.62).

Much of the interest in zeolites derives from the presence of the spacious channels and the water molecules and variable amounts of Na, Ca, and K that are housed in the channels and intraframe cavities. The water molecules are weakly tied by hydrogen bonding to anionic framework atoms. When a zeolite is heated, the water in the channelways is given off easily and continuously as the temperature rises, leaving the structure intact. This phenomenon of continuous water loss as a function of increasing temperature is illustrated in Fig. 11.63 for several common, natural zeolites. The formulas in Fig. 11.63

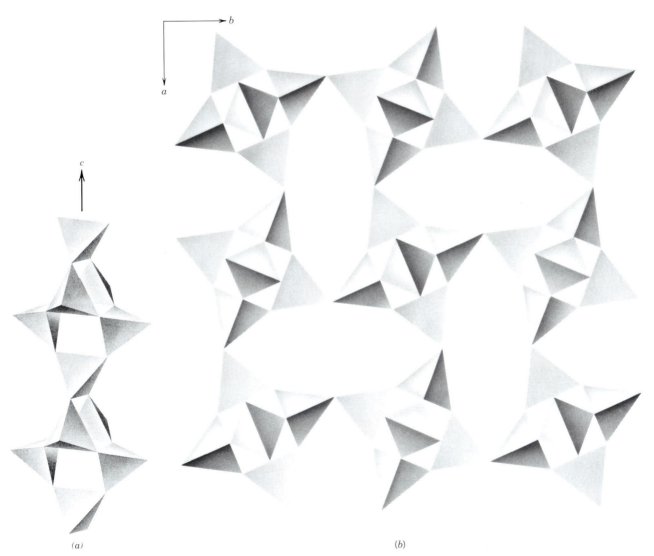

(a)

(b)

FIG. 11.61. *(a)* Chains of AlO_4 and SiO_4 tetrahedra as found in natrolite and other fibrous zeolites. *(b)* The structure of natrolite projected on (001).

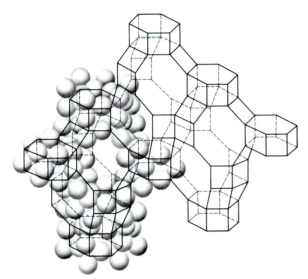

FIG. 11.62. Schematic representation of the structure of chabazite. Si and Al (not shown) occupy the corners of the framework outlined by the lines. The balls represent oxygen in close packing. Each framework unit contains a cavity which is connected to adjacent cavities by channels. The channel-diameter in chabazite is 3.9 Å. (From D. W. Breck and J. V. Smith, 1959, Molecular sieves. *Scientific American* 200, no. 1, p. 88. All rights reserved.)

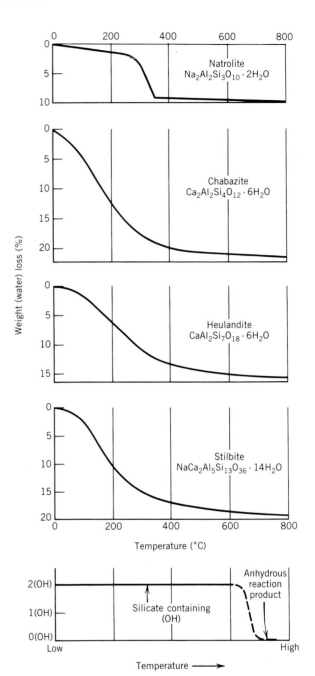

FIG. 11.63. Thermal gravimetric analysis (TGA) curves for four selected zeolites. In each of these four curves there is a continuous water loss with increasing temperature. In every case about 80% to 90% of the water in these structures is lost by about 350°C. (From G. Gottardi, and E. Galli, 1985, *Natural Zeolites.* New York: Springer-Verlag, with permission.) The schematic curve at the bottom illustrates the behavior of silicates with (OH) groups (e.g., amphiboles, micas) and the abrupt loss of this (OH) at some high temperature, at which the original (OH)-containing structure is destroyed and converts to an anhydrous reaction product + H_2O.

show the large and variable numbers of water molecules per formula unit. The thermal gravimetric analysis (TGA) curves show a continuous loss of water with increasing temperature and also illustrate the fact that 80% to 90% of all the water is lost from the structure below about 350°C. This relatively low-temperature water loss occurs via the channels of the structures without collapse of the framework. Collapse and overall destruction of the framework (known as decrepitation) commonly occurs at temperatures above 600°C. Such water loss behavior is in sharp contrast to that observed in silicates with (OH) groups in which (OH) is an integral part of their structures (e.g., amphiboles and micas). Such (OH) groups are generally lost over a very small temperature range at high temperatures. As the result of such loss, new, commonly anhydrous minerals form (see the schematic curve at the bottom of Fig. 11.63). Dehydration behavior similar to that of zeolites, but at much lower temperature than for (OH)-containing silicates, is shown by gypsum, $CaSO_4 \cdot 2H_2O$ (see Fig. 10.42), which converts initially (at about 65°C) to $CaSO_4 \cdot \frac{1}{2}H_2O$ and at about 100°C to the γ polymorph of $CaSO_4$.

The dehydrated zeolite structure can be completely rehydrated when it is immersed in water. In industrial applications, the property of rehydration allows zeolites to be used as desiccants, such as in the removal of water from gaseous hydrocarbons and petroleum. Zeolites in their dehydrated state can absorb other molecules as well, as long as the overall size of the zeolite channels is large enough to accommodate such molecules. Molecules too large to pass through the channels are excluded, giving rise to the "molecular sieving" property of most zeolites. For example, a specific zeolite with channel diameters of about 4.5 Å is able to absorb normal hydrocarbons, such as octane and pentane (with effective cross-sectional diameters of about 4.3 Å), but will exclude branch chain hydrocarbons, such as isooctane and isopentane (with effective cross-sectional diameters of 5.0 Å or larger). This sieving property is illustrated in Fig. 11.64.

Zeolites, in addition to their commercial use as molecular sieves, are also much exploited on account of their cation exchange properties (see Box 12.5). The cations (such as Na^+, K^+, and Ca^{2+}) are only loosely bonded to the tetrahedral framework and can be removed or exchanged easily by washing with a strong solution of another ion. This ability for cation exchange is the basis for water softeners, in which "hard" water (with high Ca^{2+} concentrations) is made "soft" by exchanging the Ca^{2+} (in the water) with Na^{2+} (supplied by a natural or synthetic zeolite, e.g., $Na_2Al_2Si_3O_{10} \cdot 2H_2O$). "Hard" water, that is, water containing many calcium ions in solution, is passed through a tank filled with zeolite grains. The Ca^{2+} ions replace the Na^+ ions in the zeolite, forming $CaAl_2Si_3O_{10} \cdot 2H_2O$, contributing Na^+ ions to the solution. Water containing sodium does not form

scum and is said to be "soft." When the zeolite in the tank has become saturated with calcium, a strong NaCl brine is passed through the tank. The high concentration of sodium ions forces the reaction to go in the reverse direction, and the $Na_2Al_2Si_3O_{10} \cdot 2H_2O$ is reconstituted, calcium going into solution. By base exchange many ions, including silver, may be substituted for the alkali-metal cations in the zeolite structure. These cation exchange properties are used in removing harmful ions from radioactive waste, or ammonia from sewage and agricultural waste.

Because of the great demand for various zeolites used in commercial applications such as in catalytic cracking (cracking is the molecular weight reduction process by which the heavier components of crude oil are converted to lighter, more volatile materials such as those used in gasoline), there is a large industrial production (through laboratory synthesis) of zeolites. The first synthetic zeolites were produced in the 1950s by the Linde Division of the Union Carbide Corporation. Such syntheses have produced many structures and compositions that have no natural counterpart. The number of naturally occurring zeolites is about 46, and only a few of these are found in sufficient quantity and purity to be used as a commercial product. Synthetic zeolite production now occurs on a large scale with a production of more than 12,000 tons per year. Very specific zeolite structures (with unique channel diameters and/or cation exchange properties) are being produced for unique industrial applications (see Box 12.5).

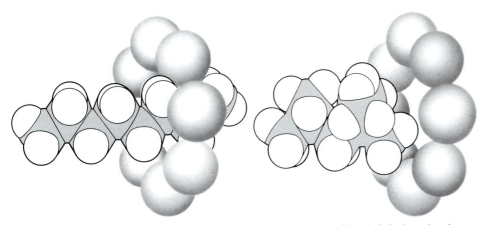

FIG. 11.64. Schematic illustration of how a zeolite sieve separates a straight chain hydrocarbon (e.g., octane) from a branching chain hydrocarbon (e.g., isooctane). These two organic compounds have almost identical properties and thus are very hard to separate by other methods. Isooctane has the higher "antiknock" rating. (From E. M. Flanigen, and F. A. Mumpton, 1977, Commercial properties of natural zeolites, in *Mineralogy and Geology of Natural Zeolites*, *Reviews in Mineralogy* 4, Mineralogical Soc. America, pp. 165–75.)

REFERENCES AND SELECTED READING

Anthony, J. W., R. A. Bideaux, K. W. Bladh, and M. C. Nichols. 1995. *Handbook of mineralogy.* Vol. 2, *Silica, silicates.* Tucson, Ariz.: Mineral Data Publications.

The Al₂SiO₅ polymorphs. 1990. *Reviews in Mineralogy* 22. Mineralogical Society of America, Washington, D. C.

Amphiboles. 1981. *Reviews in Mineralogy* 9A. Mineralogical Society of America, Washington, D. C.

Breck, D. W. 1974. *Zeolite molecular sieves.* New York: Wiley.

Buseck, P. R. 1983. Electron microscopy of minerals. *American Scientist* 71: 175–85.

Buseck, P. R., and S. Iijima. 1974. High resolution electron microscopy of silicates. *American Mineralogist* 59: 1–22.

Darragh, P. J., A. J. Gaskin, and J. V. Sanders. 1976. Opals. *Scientific American* 234, 4: 84–95.

Deer, W. A., R. A. Howie, and J. Zussman. 1962 and 1963. *Rock-forming minerals,* v. 1–4. New York: Wiley.

———. 1978. *Single-chain silicates,* v. 2A. 1982, *Orthosilicates,* v. 1A; 1986; *Disilicates and ring silicates* v. 1B; *Double-chain silicates,* v. 2B. For vols. 2A, 1A, and 1B. New York: Wiley. For vol. 2B, London: The Geological Society.

———. 1992. *An introduction to the rock-forming minerals.* 2nd ed. New York: Wiley.

Feldspar mineralogy. 1983. *Reviews in Mineralogy* 2. 2nd ed. Mineralogical Society of America, Washington, D. C.

Frondel, C. 1962. *The system of mineralogy 3. Silica Minerals.* New York: Wiley.

Gibbs, G. V. 1966. The polymorphism of cordierite I: The crystal structure of low cordierite. *American Mineralogist:* 1068–88.

Gottardi, G., and E. Galli. *Natural zeolites.* New York: Springer-Verlag.

Grim, R. E. 1968. *Clay mineralogy.* 2nd ed. New York: McGraw-Hill.

Hawthorne, F. C. 1983. The crystal chemistry of amphiboles. *Canadian Mineralogist* 21: 173–480.

Hydrous phyllosilicates. 1988. *Reviews in Mineralogy* 19. Mineralogical Society of America, Washington, D. C.

Liebau, F. 1985. *Structural chemistry of silicates: Structure, bonding, and classification.* New York: Springer-Verlag.

Micas. 1984. *Reviews in Mineralogy* 13. Mineralogical Society of America, Washington, D. C.

Mineralogy and geology of natural zeolites. 1977. *Reviews in Mineralogy* 4. Mineralogical Society of America, Washington, D. C.

Orthosilicates. 1980. *Reviews in Mineralogy* 5. Mineralogical Society of America, Washington, D. C.

Papike, J. J., and M. Cameron. 1976. Crystal chemistry of silicate minerals of geophysical interest. *Reviews of Geophysics and Space Physics* 14: 37–80.

Putnis, A. 1992. *Introduction to mineral sciences.* New York: Cambridge University Press.

Pyroxenes. 1980. *Reviews in Mineralogy* 7. Mineralogical Society of America, Washington, D. C.

Smith, J. V. 1974. *Feldspar minerals,* vols. 1, 2. New York: Springer-Verlag.

Smith, J. V., and W. L. Brown. 1988. *Feldspar minerals.* Vol. 1, *Crystal structures, physical, chemical, and microtextural properties.* New York: Springer-Verlag.

Smyth, J. R., and D. L. Bish. 1988. *Crystal structures and cation sites for the rock-forming minerals.* Boston: Allen and Unwin.

Strunz, H. 1970. *Mineralogische Tabellen,* 5th ed. Leipzig: Akademische Verlagsgesellschaft, Geest und Portig, K. G.

Veblen, D. R., P. R. Buseck, and C. W. Burnham. 1977. Asbestiform chain silicates: new minerals and structural groups. *Science* 198: 359:65.

Wenk, H. R., ed. 1976. *Electron microscopy in mineralogy.* New York: Springer-Verlag.

Zoltai, T. 1960. Classification of silicates and other minerals, with tetrahedral structures. *American Mineralogist* 45: 960–73.

CHAPTER 12

SYSTEMATIC DESCRIPTIONS OF ROCK-FORMING SILICATES

This chapter contains 75 systematic silicate mineral entries. This number is very small compared with that of 1002 which represents the total number of silicates that are known. The small number is a reflection of the fact that only relatively few silicates are common as rock-forming minerals. The remaining 900 or so silicates range from uncommon to very rare. Another reason for the small number of systematic entries in this chapter is that several major silicate groups are treated under one mineral heading, such as olivine and garnet. In such instances a number of separately named mineral species are described under one mineral group name. Such a group may show considerable to extensive solid solution with two or more end member compositions (with specific species names given to such end-member compositions). Of the 75 entries, 28 of the mineral names are printed in lowercase (instead of boldface capitals) because these are considered to be less common or important. The silicate mineral group includes a considerable number of industrial minerals.

NESOSILICATES

The following minerals in the nesosilicate group will be discussed in detail:

Phenacite Group			
Phenacite	Be_2SiO_4		
Willemite	Zn_2SiO_4		
Olivine Group			
Forsterite	Mg_2SiO_4		
Fayalite	Fe_2SiO_4		
Garnet Group $A_3B_2(SiO_4)_3$			
Pyrope	Mg_3Al_2	Uvarovite	Ca_3Cr_2
Almandine	Fe_3Al_2	Grossular	Ca_3Al_2
Spessartine	Mn_3Al_2	Andradite	$Ca_3Fe_2^{3+}$
Zircon Group			
Zircon	$ZrSiO_4$		
Al_2SiO_5 Group			
Andalusite			
Sillimanite	Al_2SiO_5		
Kyanite			
Topaz	$Al_2SiO_4(F,OH)_2$		
Staurolite	$Fe_2Al_9O_6(SiO_4)_4(O,OH)_2$		

Humite Group	
Chondrodite	$Mg_5(SiO_4)_2(OH,F)_2$
Datolite	$CaB(SiO_4)(OH)$
Titanite	$CaTiO(SiO_4)$
Chloritoid	$(Fe,Mg)_2Al_4O_2(SiO_4)_2(OH)_4$

Phenacite Group

Phenacite—Be_2SiO_4

Crystallography. Hexagonal; $\bar{3}$. Crystals are usually flat rhombohedral or short prismatic. Often with complex development. Frequently twinned on $\{10\bar{1}0\}$.

$R\bar{3}$; $a = 12.45$, $c = 8.23$ Å; $Z = 18$. ds: 3.58(6), 2.51(8), 2.35(6), 2.18(8), 1.258(10).

Physical Properties. *Cleavage* $\{11\bar{2}0\}$ imperfect. **H** $7\frac{1}{2}$– 8. **G** 2.97–3.00. *Luster* vitreous. *Color* colorless, white. Transparent to translucent. *Optics*: (+); $\omega = 1.654$, $\epsilon = 1.670$.

Composition and Structure. BeO 45.6, SiO_2 54.4%. The structure consists of SiO_4 and BeO_4 tetrahedra with each oxygen linked to two Be and one Si at the corners of an equilateral triangle.

Diagnostic Features. Characterized by its crystal form and great hardness.

Occurrence. Phenacite is a rare pegmatite mineral associated with topaz, chrysoberyl, beryl, and apatite. Fine crystals are found at the emerald mines in the Ural Mountains, Russia, and in Minas Gerais, Brazil. In the United States found at Mount Antero, Colorado.

Use. Occasionally cut as a gemstone.

Name. From the Greek *phenakos* meaning a *deceiver*, in allusion to its resemblance to quartz.

Willemite—Zn_2SiO_4

Crystallography. Hexagonal; $\bar{3}$. In hexagonal prisms with rhombohedral terminations. Usually massive to granular. Rarely in crystals.

$R\bar{3}$; $a = 13.96$, $c = 9.34$ Å; $Z = 18$. *ds*: 2.84(8), 2.63(9), 2.32(8), 1.849(8), 1.423(10).

Physical Properties. *Cleavage* {0001} good. **H** $5\frac{1}{2}$. **G** 3.9–4.2. *Luster* vitreous to resinous. *Color* yellow-green, flesh-red, and brown; white when pure. Transparent to translucent. Most willemite from Franklin, New Jersey, fluoresces under ultraviolet light. *Optics*: (+); $\omega = 1.691$, $\epsilon = 1.719$.

Composition and Structure. ZnO 73.0, SiO_2 27.0%. Mn^{2+} often replaces a considerable part of the Zn (manganiferous variety called *troostite*); Fe^{2+} may also be present in small amount. Willemite is isostructural with phenacite, with SiO_4 and ZnO_4 tetrahedra. Because Zn^{2+} (radius = 0.60 Å) is much larger than Be^{2+} (radius = 0.27 Å) the structure of willemite is much expanded over that of phenacite.

Diagnostic Features. Willemite from Franklin, New Jersey, can usually be recognized by its association with franklinite and zincite.

Occurrence. Willemite is found in crystalline limestone and may be the result of metamorphism of earlier hemimorphite or smithsonite. It is also found sparingly as a secondary mineral in the oxidized zone of zinc deposits.

Found at Altenberg, near Moresnet, Belgium; Tsumeb, Namibia. The most important locality is in the United States at Franklin, New Jersey, where willemite occurs associated with franklinite and zincite and as grains embedded in calcite. It has also been found at the Merritt mine, New Mexico, and Tiger, Arizona.

Use. A valuable zinc ore at Franklin, New Jersey. Occasionally used as a gem.

Name. In honor of the King of the Netherlands, Willem I (1553–1584).

Olivine Group

The composition of the majority of olivines can be represented in the system $CaO-MgO-FeO-SiO_2$ (Fig. 12.1). The most common series in this system is from *forsterite*, Mg_2SiO_4, to *fayalite*, Fe_2SiO_4. Relatively rare olivines occur also along the *monticellite*, $CaMgSiO_4$, to *kirschteinite*, $CaFe^{2+}SiO_4$, join. Very little, if any, solid solution exists between these two

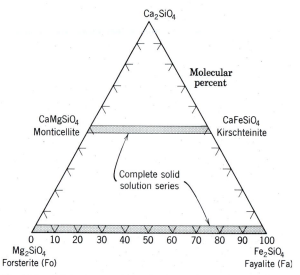

FIG. 12.1. Olivine compositions in the system Ca_2SiO_4-Mg_2SiO_4-Fe_2SiO_4.

series. Mn^{2+} may substitute for Fe^{2+}, forming a relatively rare series between fayalite and *tephroite*, Mn_2SiO_4. The structure of olivine, with fairly similar *M1* and *M2* octahedral sites, and independent SiO_4 tetrahedra, is shown in Fig. 11.4.

Members of the forsterite–fayalite series are common as primary crystallization products in Fe- and Mg-rich, silica-poor melts (magmas). *Dunites* and *peridotites* are pure olivine and olivine plus pyroxene rocks, respectively. Olivine concentrations in igneous rocks may result from the accumulation of olivine crystals, under the influence of gravity, during the cooling stages of a magma. Members of the forsterite–fayalite series are highly refractory, as can be seen from Fig. 12.2 (forsterite melting point = 1890°C and fayalite melting point = 1205°C). This diagram represents a complete solid solution series

FIG. 12.2. Temperature-composition diagram for the system Mg_2SiO_4-Fe_2SiO_4 at atmospheric pressure (see also Fig. 4.18 and related discussion).

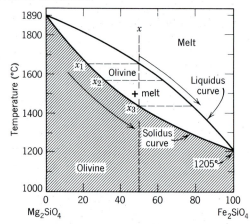

without a maximum or minimum (see Fig. 4.18 and related discussion). When a melt with composition x (50 weight percent Fe_2SiO_4) is cooled to the liquidus curve, olivine crystals of composition x_1 will start to form. The liquid, as a result of the Mg-rich olivine crystallization, will become more Fe-rich, as shown by the upper arrow. This, in turn, will cause a more Fe-rich olivine to crystallize, as shown by x_2, until finally all liquid is used up. At this point, under equilibrium conditions, the final composition to crystallize is of composition x_3, which is the same as the original x. Chemical zoning in olivines with Mg-rich cores and more Fe-rich rims may be found in high-temperature igneous rocks. Such rimming is the result of nonequilibrium crystallization in the direction of the arrows in Fig. 12.2.

OLIVINE—$(Mg,Fe)_2SiO_4$

Crystallography. Orthorhombic; $2/m2/m2/m$. Crystals are usually a combination of the three prisms, the

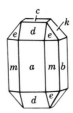

FIG. 12.3. Olivine.

three pinacoids, and a dipyramid. Often flattened parallel to either {100} or {010} (see Fig. 12.3). Usually appears as embedded grains or in granular masses.

Pbnm; $Z = 4$. Mg_2SiO_4: $a = 4.75$, $b = 10.20$, $c = 5.98$ Å. Fe_2SiO_4: $a = 4.82$, $b = 10.48$, $c = 6.09$ Å. ds for common olivine: 4.29(10), 2.41(8), 2.24(7), 1.734(8), 1.498(7). d_{130} can be used to determine compositions in this series (see Fig. 12.4a).

Physical Properties. *Fracture* conchoidal. **H** $6\frac{1}{2}$–7. **G** 3.27–4.37, increasing with increase in Fe content (see Fig. 12.4b). *Luster* vitreous. *Color* pale

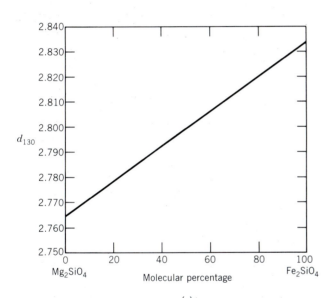

(a)

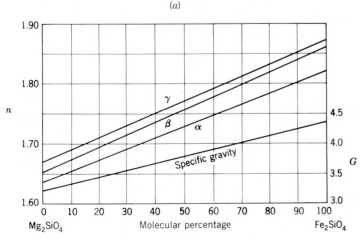

(b)

FIG. 12.4. (a) Relationship of the interplanar spacing of 130 (= d_{130}) and composition for the forsterite-fayalite series. (From W. A. Deer, R. A. Howie, and J. Zussman, 1982. *Orthosilicates*, vol. 1A, New York: Wiley; and London: Longman.) (b) Variations of refractive indices and specific gravity with composition.

yellow-green to olive-green in forsterite, see Plate IV, no. 6; darker, brownish-green with increasing Fe^{2+}. Transparent to translucent. *Optics*: forsterite (+), others (−). For $Mg_{1.6}Fe_{0.4}SiO_4$: $\alpha = 1.674$, $\beta = 1.692$, $\gamma = 1.712$, $2V = 87°$; $X = b$, $Z = a$ (see Fig. 12.4*b*).

Composition and Structure. A complete solid solution exists from *forsterite*, Mg_2SiO_4, to *fayalite*, Fe_2SiO_4 (see Fig. 12.1). The more common olivines are richer in Mg than in Fe^{2+}. An example of the recalculation of an olivine analysis is given in Table 3.16. Compositions intermediate to the end members, forsterite (Fo) and fayalite (Fa), are commonly expressed as Fo_xFa_y, for example, $Fo_{60}Fa_{40}$, which is shortened to Fo_{60}. The structure of olivine is given in Fig. 11.4 and discussed on page 446. An HRTEM structure image is given in Fig. 12.5. Under very high pressures the olivine structure transforms to one of two different high-pressure polymorphs known as wadsleyite and ringwoodite, respectively (see Fig. 12.6). Olivine is thought to be abundant in the mantle and occurs as ringwoodite at great depths (see Fig. 4.9 and related text).

Diagnostic Features. Distinguished usually by its glassy luster, conchoidal fracture, green color, and granular nature.

Occurrence. Olivine is a rather common rock-forming mineral, varying in amount from that of an accessory to that of a main constituent. It is found principally in the dark-colored igneous rocks such as gabbro, peridotite, and basalt. In these rock types it coexists with plagioclase and pyroxenes. The rock, *dunite*, is made up almost wholly of olivine. Forsterite is not stable in the presence of free SiO_2

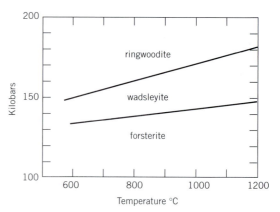

FIG. 12.6. The stability fields of various polymorphs of Mg_2SiO_4: forsterite, wadsleyite, and ringwoodite. (Adapted from K. Suito, 1972, Phase transformations of pure Mg_2SiO_4 into a spinel structure under high pressures and temperatures. *Journal of Physics of the Earth* 20: 225–43.)

and will react with it to form pyroxene, according to this reaction:

$$Mg_2SiO_4 + SiO_2 \rightarrow 2MgSiO_3$$
forsterite   quartz   enstatite

Found as glassy grains in stony and stony-iron meteorites. Occasionally found in crystalline dolomitic limestones, where it is formed by the reaction:

$$2CaMg(CO_3)_2 + SiO_2 \rightarrow$$
dolomite   quartz

$$Mg_2SiO_4 + 2CaCO_3 + 2CO_2$$
forsterite   calcite

Associated often with pyroxene, plagioclase, magnetite, corundum, chromite, and serpentine.

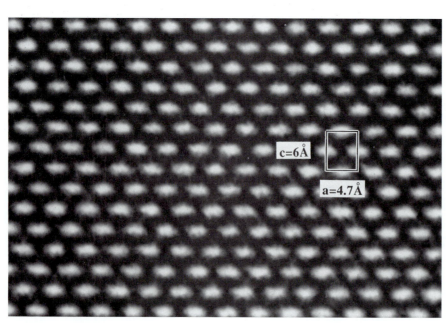

c=6Å

a=4.7Å

FIG. 12.5. High-resolution transmission electron microscope (HRTEM) image of an *a–c* section of olivine, of composition $(Mg_{1.76}Fe_{0.24})SiO_4$. A unit cell is outlined. This image shows the highly regular and homogeneous nature of the structure of this olivine crystal. Compare with the radiation-damaged olivine in Fig. 4.49 and the olivine to clay alteration in Fig. 12.54 (courtesy of L. M. Wang, University of New Mexico).

The transparent gem variety is known as *peridot* (see Plate XI, no. 1, Chapter 13). It was used as a gem in ancient times in the East, but the exact locality for the stones is not known. Peridot is found in Burma, and in rounded grains associated with pyrope garnet in the surface gravels of Arizona and New Mexico, but the best quality material comes from Zebirget, an island in the Red Sea. Crystals of olivine are found in the lavas of Vesuvius. Larger crystals, altered to serpentine, come from Snarum, Norway. Olivine occurs in granular masses in volcanic bombs in the Eifel district, Germany, and in Arizona. Dunite rocks are found at Dun Mountain, New Zealand, and with the corundum deposits of North Carolina.

Alteration. Very readily altered to *serpentine* minerals such as *antigorite*, $Mg_3Si_2O_5(OH)_4$. Magnesite, $MgCO_3$ and iron oxides form at the same time as a result of the alteration. Olivines in metamorphosed basic igneous rocks commonly show *coronas*, which are concentric rims consisting of pyroxene and amphibole. Such coronas are the result of the instability of the high temperature olivine in a lower temperature, H_2O-containing environment.

Use. As the clear green variety, *peridot*, it is used as a gem. Olivine is mined as refractory sand for the casting industry and for the manufacture of refractory bricks.

Name. Olivine derives its name from the usual olive-green color. *Chrysolite* is a synonym for olivine. Peridot is an old name for the species.

Similar Species. Other rarer members of the olivine group are *monticellite*, $CaMgSiO_4$, a high-temperature contact metamorphic mineral in siliceous dolomitic limestones; and *tephroite*, Mn_2SiO_4.

Garnet Group
The garnet group includes a series of isostructural species with space group *Ia3d*; they crystallize in the hexoctahedral class of the isometric system and are similar in crystal habit. The structural arrangement (see Fig. 11.5) is such that the atomic population of the {100} and {111} families of planes is much depleted. As a result the cube and octahedron, common on most isometric hexoctahedral crystals, are rarely found on garnets.

Crystallography. Isometric; $4/m\bar{3}2/m$. Common forms are dodecahedron *d* (Fig. 12.7*a*) and trapezohedron, *n* (Fig. 12.7*b*), often in combination (Figs. 12.7*c*, *d*, and 12.8). Hexoctahedrons are observed occasionally (Fig. 12.7*e*). Other forms are rare. Usually distinctly crystallized; also appears in rounded grains; massive granular, coarse or fine.

Ia3d; cell edge (see Table 12.1); *Z* = 8. *d*s for pyrope: 2.89(8), 2.58(9), 1.598(9), 1.542(10), 1.070(8).

Physical Properties. **H** $6\frac{1}{2}-7\frac{1}{2}$. **G** 3.5–4.3, varying with composition (see Table 12.1). *Luster* vitreous to resinous. *Color* varying with composition (see below); most commonly red, also brown, yellow, white, green, black. *Streak* white. Transparent to translucent.

FIG. 12.7. Garnet crystals.

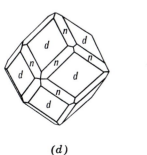

(a)

(b)

(c)

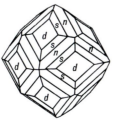

(d)

(e)

FIG. 12.8. Garnet crystals in chlorite schist.

Composition and Structure. Garnet compositions can be expressed by the general structural formula $A_3B_2(SiO_4)_3$ where the A site houses Ca, Mg, Fe^{2+}, or Mn^{2+} and the B site incorporates Al, Fe^{3+}, and Cr^{3+} (see Fig. 11.5 and page 446). The formulas of the main species are given in Table 12.1, with the refractive indices, specific gravity, and unit cell edges for the endmember compositions. There is extensive substitution among the *pyralspite* group and also among the *ugrandite* group but relatively little solid solution between these two major categories (see Fig. 12.9 and garnet groupings on p. 446). Rarely are the pure endmembers found. Hydrous garnets such as *hydrogrossular*, may contain up to 8.5% H_2O. This water, in the form of $(OH)_4^{4-}$ groups, substitutes probably for (SiO_4) tetrahedra in the structure, assuming the replacement $Si^{4+} \rightleftarrows 4H^+$.

Experimental studies at high pressures and temperatures show that the garnet structure is stable in the conditions of the Earth's mantle. It appears that pyrope type compositions are most likely in the P and T regime of the Earth's mantle (see p. 111).

Pyrope, Mg_3Al_2. Some Ca and Fe^{2+} usually present. Color is deep red to nearly black. Often transparent and then used as a gem. Name derived from Greek meaning *firelike*. *Rhodolite* is the name given to a pale rose-red or purple garnet, corresponding in composition to two parts of pyrope and one almandine.

Almandine, Fe_3Al_2. Fe^{3+} may replace Al, and Mg may replace Fe^{2+}. Color is fine deep red, transparent in precious garnet; brownish-red, translucent in common garnet, see Plate IV, no. 7. Name derived from Alabanda in Asia Minor, where in ancient times garnets were cut and polished.

Spessartine, Mn_3Al_2. Fe^{2+} usually replaces some Mn^{2+} and Fe^{3+} some Al. Color is brownish to red. Name derived from the locality Spessart, Germany.

Grossular, Ca_3Al_2 (*essonite, cinnamon stone*). Often contains Fe^{2+} replacing Ca, and Fe^{3+} replacing Al. Color white, green, yellow, cinnamon-brown, pale red. Name derived from the botanical name for gooseberry, in allusion to the light green color of the original grossular.

TABLE 12.1   Chemical Compositions and Physical Properties of Garnets

Species	Composition	n	G	Unit Cell Edge (Å)
Pyrope	$Mg_3Al_2Si_3O_{12}$	1.714	3.58	11.46
Almandine	$Fe_3Al_2Si_3O_{12}$	1.830	4.32	11.53
Spessartine	$Mn_3Al_2Si_3O_{12}$	1.800	4.19	11.62
Grossular	$Ca_3Al_2Si_3O_{12}$	1.734	3.59	11.85
Andradite	$Ca_3Fe_2Si_3O_{12}$	1.887	3.86	12.05
Uvarovite	$Ca_3Cr_2Si_3O_{12}$	1.868	3.90	12.00
Hydrogrossular	$Ca_3Al_2Si_2O_8(SiO_4)_{1-m}(OH)_{4m}$;	1.734 to	3.59 to	11.85 to
	$m = 0-1$	1.675	3.13	12.16

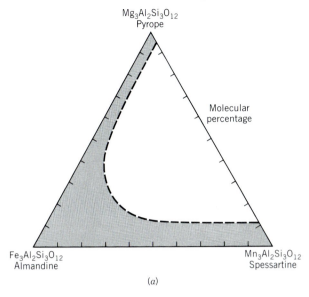

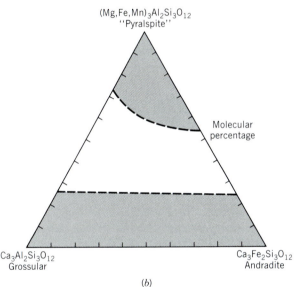

FIG. 12.9. *(a)* The extent of solid solution in garnets of the *pyralspite* group: pyrope-almandine-spessartine. *(b)* There is only limited solid solution between members of the *pyralspite* and *ugrandite* groups; ugrandite represents uvavorite-grossular-andradite.

Andradite, $Ca_3Fe_2^{3+}$. Common garnet in part. Al may replace Fe^{3+}; Fe^{2+}, Mn^{2+}, and Mg may replace Ca. Color various shades of yellow, green, brown to black. *Demantoid* is a green variety with a brilliant luster, used as a gem. Named after the Portuguese mineralogist, d'Andrada.

Uvarovite, $Ca_3Cr_2^{3+}$. Color emerald-green. Named after Count Uvarov.

Diagnostic Features. Garnets are usually recognized by their characteristic isometric crystals, their hardness, and their color. Specific gravity, refractive

index, and unit cell dimension taken together serve to distinguish members of the group.

Occurrence. Garnet is a common and widely distributed mineral; occurring abundantly in some metamorphic rocks (see Fig. 12.8) and as an accessory constituent in some igneous rocks. Its most characteristic occurrence is in mica schists, hornblende schists, and gneisses. It is frequently used as an index mineral in the delineation of isograds in metamorphic rocks. Found in pegmatite dikes, more rarely in granitic rocks. *Pyrope* occurs in ultrabasic rocks such a peridotites or kimberlites and in serpentines derived from them. The garnets, which in eclogites coexist with pyroxenes and kyanite, vary in composition from pyrope to almandine. *Almandine* is the common garnet in metamorphic rocks, resulting from the regional metamorphism of argillaceous sediments. It is also a widespread detrital garnet in sedimentary rocks. *Spessartine* occurs in skarn deposits and in Mn-rich assemblages containing rhodonite, Mn-oxides, and so forth. *Grossular* is found chiefly as a product of contact or regional metamorphism of impure limestones. *Andradite* occurs in geological environments similar to that of grossularite. It may be the result of metamorphism of impure siliceous limestone, by this reaction:

$$3CaCO_3 + 3SiO_2 + Fe_2O_3 \rightarrow Ca_3Fe_2Si_3O_{12} + 3\ CO_2^{\prime}$$
calcite   quartz   hematite   andradite

Melanite, a black variety of andradite, occurs in alkaline igneous rocks. *Uvarovite* is the rarest of this group of garnets and is found in serpentine associated with chromite. The best known locality is Outokumpu, Finland.

Pyrope of gem quality is found with clear grains of olivine (peridot) in the surface sands near Fort Defiance, close to the New Mexico–Arizona line. A locality near Meronitz, Bohemia, Czechoslovakia, is famous for pyrope gems. Almandine of gem quality is found in northern India, Sri Lanka, and Brazil. Fine crystals, although for the most part too opaque for cutting, are found in a mica schist on the Stikine River, Alaska. Grossular is used only a little in jewelry, but essonite or cinnamon stones of good size and color are found in Sri Lanka.

Alteration. Garnet often alters to other minerals, particularly talc, serpentine, and chlorite.

Use. All species, except uvarovite, are cut as gemstones (see Plate XI, no. 2, Chapter 13). The most valued is a green andradite, known as *demantoid*, which comes from the Ural Mountains, Russia. At Gore Mountain, New York, large crystals of almandine in an amphibolite are mined. The unusual angular fractures and high hardness of the garnets make them desirable for a variety of abrasive purposes, including garnet paper.

Name. *Garnet* is derived from the Latin *granatus*, meaning like a grain.

ZIRCON—ZrSiO$_4$

Crystallography. Tetragonal; $4/m2/m2/m$. Crystals usually show a simple combination of $a\{010\}$ and $e\{011\}$, but $m\{110\}$ and a ditetragonal dipyramid is also observed (Fig. 12.10). Usually in crystals; also in irregular grains.

$I4_1/amd$; $a = 6.60$, $c\ 5.98$ Å; $Z = 4$. ds: 4.41(7), 3.29(10), 2.52(8), 1.710(9).

Physical Properties. *Cleavage* $\{010\}$ poor. **H** $7\frac{1}{2}$. **G** 4.68. *Luster* adamantine. *Color* commonly some shade of brown, see Plate IV, no. 8; also colorless, gray, green, red. *Streak* uncolored. Usually translucent; in some cases transparent. *Optics*: (+), $\omega = 1.923–1.960$, $\epsilon = 1.968–2.015$. Metamict zircon, $n = 1.78$.

Composition and Structure. For ZrSiO$_4$, ZrO$_2$ 67.2, SiO$_2$ 32.8%. Zircon always contains some hafnium. Although the amount is usually small (1 to 4%), analyses have reported up to 24% HfO$_2$. The structure of zircon is shown in Fig. 12.11. Zr is in 8-coordination with oxygen in the form of distorted cubelike polyhedra. The eight oxygens surrounding each Zr belong to six different SiO$_4$ tetrahedra. Although the structure of zircon is resistant to normal chemical attack, it is often in a metamict state (see page 148). This is caused by the structural damage from Th and U, which are present in small amounts in many zircons. The "self-irradiation" from the decay of U and Th produces as a final result an isotropic glass with a reduction in density of about 16% and a lowering of refractive index (see Figs. 4.48 and 4.49).

Diagnostic Features. Usually recognized by its characteristic crystals, color, luster, hardness, and high specific gravity.

Occurrences. Zircon is a common and widely distributed accessory mineral in all types of igneous

FIG. 12.10. Zircon crystals.

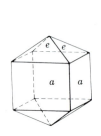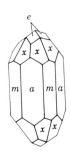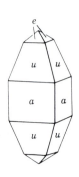

FIG. 12.11. Structure of zircon, ZrSiO$_4$, consisting of independent SiO$_4$ tetrahedra and distorted ZrO$_8$ cubes. (Redrawn from T. Zoltai, *Polyhedral Structure Models.* University of Minnesota, Minneapolis.)

rocks. It is especially common in the more silicic types such as granite, granodiorite, syenite, and monzonite, and very common in nepheline syenite. Found also in crystalline limestone, gneisses, and schists. Because zircon is a stable chemical compound, it is a common accessory mineral in many sediments. Its characteristics in the heavy mineral fraction of sandstones are often useful in the evaluation of provenance. Zircon is frequently found as rounded grains in stream and beach sands, often with gold. Zircon has been produced from beach sands in Australia, Brazil, and Florida.

Gem zircons are found in the stream sands at Matura, Sri Lanka, and in the gold gravels in the Ural Mountains, Russia, and Australia. In large crystals from the Malagasy Republic. Found in the nepheline syenites of Norway. Found in the United States in Orange and St. Lawrence counties, New York; in considerable quantity in the sands of Henderson and Buncombe counties, North Carolina. Large crystals have been found in Renfrew County, Ontario, Canada.

Use. When transparent it serves as a gemstone. It is colorless in some specimens, but more often of a brownish or red-orange color. Blue is not a natural color for zircon but is produced by heat treatment. The colorless, yellowish, or smoky stones are called *jargon*, because although resembling diamond they have little value. Serves as the source of zirconium oxide, which is one of the most refractory substances known. Platinum, which fuses at 1755°C, can be melted in crucibles of zirconium oxide.

Overshadowing its other uses since 1945 is the use of zircon as the source of metallic zirconium.

Pure zirconium metal is used in the construction of nuclear reactors. Its low neutron-absorption cross section, coupled with retention of strength at high temperatures and excellent corrosion resistance, makes it a most desirable metal for this purpose.

Mineral grains of zircon in igneous, metamorphic, or sedimentary rocks, are commonly used for radioactive age determinations.

Name. The name is very old and is believed to be derived from Persian words *zar*, gold, and *gun*, color.

Similar Species. *Thorite*, $ThSiO_4$, is isostructural with zircon. It is usually reddish-brown to black, hydrated, and radioactive.

Al_2SiO_5 Group

The three polymorphs of Al_2SiO_5 are *andalusite*, $Al^{[6]}Al^{[5]}SiO_5$ (space group *Pnnm*), *sillimanite*, $Al^{[6]}Al^{[4]}SiO_5$ (space group *Pbnm*), and *kyanite*, $Al^{[6]}Al^{[6]}SiO_5$ (space group $P\bar{1}$). The structures of the three polymorphs have been discussed on pages 447–448 and their structures are shown in Figs. 11.6 and 11.7. All three minerals may be found in metamorphosed aluminous rocks, such as pelitic schists. The stability relations of the three polymorphs have been determined experimentally, as shown in Fig. 12.12. A knowledge of these stability fields is of great value in the study of regionally and contact metamorphosed terranes. Rocks of appropriate compositions tend to form sillimanite in high-tempera-

FIG. 12.12. Experimentally determined stability fields for the polymorphs of Al_2SiO_5. (From M. J. Holdaway, 1971, Stability of andalusite and the aluminum silicate stability diagram. *Amer. Jour. of Science* 271: 97–131; see also *Orthosilicates, Reviews in Mineralogy*, 1980, Mineralogical Society of America. Washington, D.C., p. 190.)

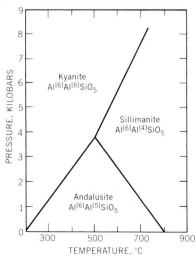

ture, regionally metamorphosed areas. Andalusite is frequently found in contact metamorphosed aureoles, and kyanite is located in metamorphic areas that have undergone considerable pressure. All three polymorphs are fairly easily recognized, in hand specimen, in medium- to coarse-grained rocks. The ease of recognition of these minerals and their presence in pelitic schists have led to their use as index minerals in the defining of metamorphic zones, as a function of temperature and pressure. Sillimanite can be used to define metamorphic zones that have been subjected to temperature of about 500°C and up (see Fig. 12.12). The metamorphic zones are named after the index mineral present, for example, sillimanite zone, kyanite zone. The boundaries between different zones are referred to as *isograds*. It should be noted that kyanite, which has **G** = 3.55 to 3.66 (as compared with **G** = 3.23 for sillimanite and **G** = 3.16–3.20 for andalusite) has the densest structure of the three polymorphs and its stability field is located in the direction of increasing pressure (Fig. 12.12). Andalusite has the largest specific volume and is, therefore, the polymorph stable at the lowest pressure conditions.

ANDALUSITE—Al_2SiO_5

Crystallography. Orthorhombic; $2/m2/m2/m$. Usually occurs in coarse, nearly square prisms terminated by {001}.

Pnnm; $a = 7.78$, $b = 7.92$, $c = 5.57$ Å; $Z = 4$. ds: 4.53(10), 3.96(8), 2.76(9), 2.17(10), 1.46(10).

Physical Properties. *Cleavage* {110} good. **H** $7\frac{1}{2}$. **G** 3.16–3.20. *Luster* vitreous. *Color* flesh-red, reddish-brown, olive-green. The variety *chiastolite* has dark-colored carbonaceous inclusions arranged in a regular manner forming a cruciform design (see Fig. 12.13 and Plate IV, no. 9). Transparent to translucent. *Optics*: (−), $\alpha = 1.632$, $\beta = 1.638$, $\gamma = 1.643$; $2V = 85°$; $X = c$, $Z = a$. In some crystals strong pleochroism; X red, Y and Z green to colorless; $r > v$.

Composition and Structure. Al_2O_3 63.2, SiO_2 36.8%. The structure consists of chains of AlO_6 octahedra parallel to c, cross-linked by SiO_4 tetrahedra and AlO_5 polyhedra (see Fig. 11.7b).

Diagnostic Features. Characterized by the nearly square prism and hardness. Chiastolite is readily recognized by the symmetrically arranged inclusions.

FIG. 12.13. Successive sections through a chiastolite crystal.

Alteration. Pseudomorphs of fine-grained muscovite (sericite) after andalusite are common.

Occurrence. Andalusite is formed typically in the contact aureoles of igneous intrusions in argillaceous rocks. Here it commonly coexists with cordierite. Andalusite can be found in association with kyanite, or sillimanite, or both in regionally metamorphosed terrane. Such occurrences may reflect variations in P and T during metamorphism, as well as the sluggishness of the reactions in the system Al_2SiO_5 (see page 121).

Notable localities are in Andalusia, Spain; the Austrian Tyrol; in gem crystals and water-worn pebbles from Minas Gerais, Brazil. Crystals of chiastolite are found at Bimbowrie, South Australia. In the United States found in the White Mountains near Laws, California and in Delaware County, Pennsylvania. Chiastolite is found at Westford, Lancaster, and Sterling, Massachusetts, and also in California.

Use. Andalusite has been mined in large quantities in California for use in the manufacture of spark plugs and other porcelains of a highly refractory nature. When transparent it may serve as a gemstone.

Name. From Andalusia, a province of Spain.

SILLIMANITE—Al_2SiO_5

Crystallography. Orthorhombic; $2/m2/m2/m$. Occurs in long, slender crystals without distinct terminations, see Plate V, no. 1; often in parallel groups; frequently fibrous and called *fibrolite*.

$Pbnm$; $a = 7.44$, $b = 7.60$, $c = 5.75$ Å; $Z = 4$. ds: 3.32(10), 2.49(7), 2.16(8), 1.677(7), 1.579(7).

Physical Properties. *Cleavage* {010} perfect. **H** 6–7. **G** 3.23. *Luster* vitreous. *Color* brown, pale green, white. Transparent to translucent. *Optics*: (+), $\alpha = 1.657$, $\beta = 1.658$, $\gamma = 1.677$; 2V = 20°, X = b, Z = c; r > v.

Composition and Structure. Oxide components as in andalusite. The structure consist of AlO_6 chains parallel to the c axis and linking SiO_4 and AlO_4 tetrahedra, which alternate along the c direction (see Fig. 11.7a).

Diagnostic Features. Characterized by slender crystals with one direction of cleavage.

Occurrence. Sillimanite occurs as a constituent of high-temperature metamorphosed argillaceous rocks. In contact metamorphosed rocks it may occur in sillimanite-cordierite gneisses or sillimanite-biotite hornfels. In regionally metamorphosed rocks it is found, for example, in quartz-muscovite-biotite-oligoclase-almandine-sillimanite schists. In silica-poor rocks it may be associated with corundum.

Notable localities for its occurrence are Maldau, Bohemia, Czechoslovakia; Fassa, Austrian Tyrol;

Bodenmais, Bavaria; and Freiberg, Saxony, Germany; and waterworn masses in diamantiferous sands from Minas Gerais, Brazil. In the United States found at Worcester, Massachusetts; at Norwich and Willimantic, Connecticut; and New Hampshire.

Name. In honor of Benjamin Silliman (1779–1864), professor of chemistry at Yale University.

Similar Species. *Mullite*, a nonstoichiometric compound of approximate composition $Al_6Si_3O_{15}$, is rare as a mineral but common in synthetic Al_2O_3-SiO_2 systems at high temperature. *Dumortierite*, $Al_7O_3(BO_3)(SiO_4)_3$, has been used in the manufacture of high-grade porcelain.

KYANITE—Al_2SiO_5

Crystallography. Triclinic; $\bar{1}$. Usually in long, tabular crystals, rarely terminated; see Plate V, no. 2. In bladed aggregates.

$P\bar{1}$; $a = 7.10$, $b = 7.74$, $c = 5.57$ Å; $\alpha = 90°6'$, $\beta = 101°2'$, $\gamma = 105°45'$; $Z = 4$. ds: 3.18(10), 2.52(4), 2.35(4), 1.93(5), 1.37(8).

Physical Properties. *Cleavage* {100} perfect. **H** 5 parallel to length of crystals, 7 at right angles to this direction. **G** 3.55–3.66. *Luster* vitreous to pearly. *Color* usually blue, often of darker shade toward the center of the crystal. Also, in some cases, white, gray, or green. Color may be in irregular streaks and patches. *Optics*: (−); $\alpha = 1.712$, $\beta = 1.720$, $\gamma = 1.728$; 2V = 82°; r > v.

Composition and Structure. Oxide components as in andalusite. The structure consists of AlO_6 octahedral chains parallel to c and AlO_6 octahedra and SiO_4 tetrahedra between the chains (see Fig. 11.6).

Diagnostic Features. Characterized by its bladed crystals, good cleavage, blue color, and different hardness in different directions.

Occurrence. Kyanite is typically a result of regional metamorphism of pelitic rocks and is often associated with garnet, staurolite, and corundum. It also occurs in some eclogites (rocks with omphacite-type pyroxenes associated with pyrope-rich garnets) and in garnet-omphacite-kyanite occurrences in kimberlite pipes. Both of these rock types reflect high to very high pressures of origin. Crystals of exceptional quality are found at St. Gotthard, Switzerland (Fig. 12.14); in the Austrian Tyrol; and at Pontivy and Morbihan, France. Commercial deposits are located in India, Kenya, and in the United States in North Carolina and Georgia.

Use. Kyanite is used, as is andalusite, in the manufacture of spark plugs and other high refractory porcelains. Transparent crystals may be cut as gemstones.

Name. Derived from the Greek word *kyanos*, meaning *blue*.

Plate I

1. Gold in quartz

2. Silver

3. Copper

4. Sulfur crystals

5. Diamond (rounded octahedral crystal) in kimberlite

6. Graphite

7. Bornite; broken surface with slight purplish tarnish

8. Galena (cubic crystals)

9. Sphalerite crystals

*All of the mineral specimens photographed for these Plates are from the Harvard University Mineralogical Museum. Carl A. Francis and William C. Metropolis provided access to the collections and assisted in selection. Photographs were taken by John Savickas, Albuquerque, New Mexico.

Plate II

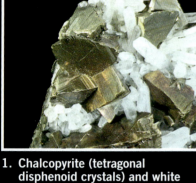

1. Chalcopyrite (tetragonal disphenoid crystals) and white quartz

2. Pyrrhotite (curved hexagonal crystals) with white quartz and black sphalerite

3. Stibnite crystals

4. Pyrite (cubic crystals)

5. Molybdenite

6. Corundum: common corundum in feldspar-mica gneiss and ruby in chromian zoisite

7. Hematite; variety, specularite

8. Rutile (twinned crystal)

9. Pyrolusite

Plate III

1. Magnetite (octahedral crystals)

2. Manganite

3. Goethite

4. Halite (cubic crystals)

5. Fluorite (cubic crystals and octahedral cleavage fragments)

6. Calcite (rhombohedral cleavage fragment); variety, Iceland spar

7. Rhodochrosite (rhombohedral crystals)

8. Aragonite (prismatic crystals)

9. Dolomite ("saddle-shaped" crystals)

Plate IV

1. Azurite and malachite

2. Barite crystals on botryoidal hematite

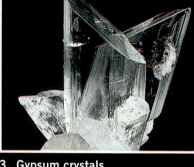

3. Gypsum crystals

4. Apatite crystals (blue) in coarse-grained calcite

5. Turquoise

6. Olivine

7. Garnet (almandine) crystals in mica schist

8. Zircon crystals

9. Andalusite; variety, chiastolite

Plate V

1. Sillimanite in quartz matrix

2. Kyanite and staurolite crystals in paragonite schist

3. Topaz crystal

4. Vesuvianite in calcite matrix

5. Beryl; common beryl and aquamarine

6. Tourmaline crystals; black ("schorl"), pink and green varieties

7. Enstatite cleavage fragment

8. Diopside with calcite

9. Augite with potassium feldspar

Plate VI

1. Augite crystal

2. Jadeite

3. Wollastonite and some green diopside grains

4. Rhodonite with calcite and brown willemite

5. Anthophyllite

6. Grunerite; asbestiform variety "amosite"

7. Actinolite in chlorite matrix

8. Hornblende crystals in cross-section in an albitic matrix

9. Chrysotile asbestos bands within serpentine matrix (a mixture of lizardite and antigorite)

Plate VII

1. Kaolinite, pseudomorphous after Carlsbad-twinned orthoclase

2. Talc

3. Muscovite crystal group

4. Phlogopite

5. Biotite

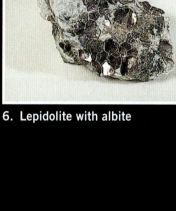

6. Lepidolite with albite

7. Chlorite

8. Quartz crystals

9. Chalcedony; banded, in agate

Plate VIII

1. Microcline cleavage fragment with perthitic texture

2. Close-up of perthitic texture in microcline of photo no. 1.

3. Microcline crystals; green variety, amazonite

4. Orthoclase; Carlsbad-twinned crystals in trachyte matrix

5. Plagioclase; variety, labradorite showing albite twin lamellae and iridescence (also known as labradorescence)

6. Leucite; trapezohedral phenocrysts in phonolite

7. Sodalite with nepheline and muscovite

8. Lazurite with pyrite in marble

9. Stilbite

FIG. 12.14. Bladed kyanite crystals and prismatic staurolite (dark) in mica schist, St. Gotthard, Switzerland (Harvard Mineralogical Museum).

TOPAZ—Al₂SiO₄(F,OH)₂ $\mathbf{IV}$

Crystallography. Orthorhombic, $2/m2/m2/m$. Commonly in prismatic crystals terminated by dipyramids, {0kl} and {h0l} prisms and basal pinacoid (see Fig. 12.15 and Plate V, no. 3). Vertical prism faces are frequently striated. Usually in crystals (Fig. 12.16) but also in crystalline masses; granular, coarse or fine.

Pbnm; $a = 4.65$, $b = 8.80$, $c = 8.40$ Å; $Z = 4$. ds: 3.20(9), 2.96(10), 2.07(9), 1.65(9), 1.403(10).

Physical Properties. *Cleavage* {001} perfect. **H** 8. **G** 3.4–3.6. *Luster* vitreous. *Color* colorless, yellow, pink, wine-yellow, bluish, greenish. Transparent to translucent. *Optics*: (+); $\alpha = 1.606$–1.629, $\beta = 1.609$–1.631, $\gamma = 1.616$–1.638; $2V = 48°$–68°; $X = a$, $Y = b$; $r > v$.

Composition and Structure. Al₂SiO₄(OH)₂ contains Al₂O₃ 56.6, SiO₂ 33.4 and H₂O 10.0%. Most of the (OH)⁻ is generally replaced by F⁻, the maximum fluorine content being 20.7%. The structure of topaz consists of chains parallel to the c axis of AlO₄F₂ octahedra, which are cross-linked by independent SiO₄ tetrahedra. This arrangement is morphologically expressed by the prismatic [001] habit. The perfect {001} cleavage of topaz passes through the structure without breaking Si–O bonds; only Al–O and Al–F bonds are broken. The structure is relatively dense because it is based on closest packing of oxygens and fluorine; this packing scheme is neither cubic nor hexagonal, but more complex, *ABAC*.

Diagnostic Features. Recognized chiefly by its crystals, basal cleavage, hardness (8), and high specific gravity.

Occurrence. Topaz is a mineral formed by fluorine-bearing vapors given off during the last stages of the solidification of siliceous igneous rocks. Found in cavities in rhyolite lavas and granite; a characteristic mineral in pegmatites, especially in those carrying tin. Associated with tourmaline, cassiterite, apatite, and fluorite; also with beryl, quartz, mica, and feldspar. Found in some localities as rolled pebbles in stream sands.

Notable localities for its occurrence are in Russia in the Nerchinsk district of Siberia in large wine-yellow crystals, and in Mursinsk, Ural Mountains, in pale blue crystals; in Saxony, Germany, from various tin localities; Omi and Mino provinces, Japan; and San Luis Potosi, Mexico. Minas Gerais, Brazil, has

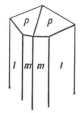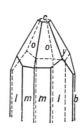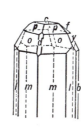

FIG. 12.15. Topaz crystals.

0   1   2

cm

FIG. 12.16. Topaz crystals on feldspar, Siberia, Russia (Harvard Mineralogical Museum).

long been the principal source of yellow gem-quality topaz. In the 1940s several well-formed, large crystals of colorless topaz were found there; the largest weighs 596 pounds. In the United States it has been found at Pikes Peak, near Florissant and Nathrop, Colorado; Thomas Range, Utah; Streeter, Texas; San Diego County, California; Stoneham and Topsham, Maine; Amelia, Virginia; and Jefferson, South Carolina.

Use. As a gemstone. Frequently sold as "precious topaz" to distinguish it from citrine quartz, commonly called topaz. The color of the stones varies, being colorless, wine-yellow, golden brown, pale blue, and pink (see Plate XI, no. 3, Chapter 13). The color of most deep blue topaz has been induced by irradiation plus heat treatment of colorless material (see Plate XI, no. 4, Chapter 13).

Name. Derived from Topazion, the name of an island in the Red Sea, but originally probably applied to some other species.

Similar Species. *Danburite*, $Ca(B_2Si_2O_8)$, a member of the tectosilicates, shows crystal form and physical properties that are similar to those of topaz.

It can be distinguished from topaz by testing for boron.

STAUROLITE—$Fe_2^{2+}Al_9O_6(SiO_4)_4(O,OH)_2$ **Ⅳ**

Crystallography. Monoclinic; $2/m$ (pseudoorthorhombic). Prismatic crystals with common forms $\{110\}$, $\{010\}$, $\{001\}$, and $\{101\}$ (Fig. 12.17a). Cruciform twins very common, of two types: (1) with twin plane $\{031\}$ in which the two individuals cross at nearly 90° (Fig. 12.17b); (2) with twin plane $\{231\}$ in which they cross at nearly 60° (Fig. 12.17c). In some cases both types are combined in one twin group. Usually in crystals; rarely massive.

$C2/m$; $a = 7.83$, $b = 16.62$, $c = 5.65$ Å; $\beta = 90°$, $Z = 2$. ds: 3.01(8), 2.38(10), 1.974(9), 1.516(5), 1.396(10).

Physical Properties. **H** $7-7\frac{1}{2}$. **G** 3.65–3.75. *Luster* resinous to vitreous when fresh; dull to earthy when altered or impure. *Color* red-brown to brownish-black, see Plate V, no. 2. Translucent. *Optics*: (+), $\alpha = 1.739–1.747$, $\beta = 1.745–1.753$, $\gamma = 1.752–1.761$; $2V = 82°–88°$; $X = b$ (colorless), $Y = a$ (pale yellow), $Z = c$ (deep yellow); $r > v$.

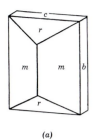

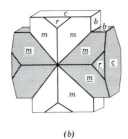

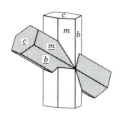

(a)   (b)   (c)   **FIG. 12.17.**   Staurolite.

Composition and Structure. FeO 16.7, Al$_2$O$_3$ 53.3, SiO$_2$ 27.9, H$_2$O 2.0% for the pure Fe end member; however, Mg, replacing Fe^{2+}, and Fe^{3+}, replacing Al, are generally both present in amounts of a few percent. The structure closely resembles that of kyanite with layers of 4Al$_2$SiO$_5$ composition (AlO$_6$ octahedra in chains parallel to the c axis) alternating with layers of Fe$_2$AlO$_3$(OH)$_2$ composition along [010]. The ideal formula, on the basis of 24(O,OH), should contain one OH; however, this OH content is generally exceeded.

Diagnostic Features. Recognized by its characteristic crystals and twins. Distinguished from andalusite by its obtuse prism.

Occurrence. Staurolite is formed during regional metamorphism of aluminum-rich rocks and is found in schists and gneisses. Often associated with almandine garnet and kyanite in high-grade metamorphic rocks. May grow on kyanite in parallel orientation (see Fig. 4.3a). In somewhat lower-grade metamorphic rocks it may occur with chloritoid. It is commonly used as an index mineral in medium-grade metamorphic rocks.

Notable localities are Monte Campione, Switzerland, and in large twin crystals in Brittany, France, and Scotland. In the United States found at Windham, Maine; Franconia and Lisbon, New Hampshire; also in North Carolina, Georgia, Tennessee, Virginia, New Mexico, and Montana.

Use. In North Carolina the right angle twins are sold as amulets under the name "fairy stone," but most of the crosses offered for sale are imitations carved from a fine-grained rock and dyed.

Name. Derived from the Greek word *stauros* meaning *cross*, in allusion to its cruciform twins.

Humite Group

The humite group includes the following four members:

Norbergite	Mg$_3$(SiO$_4$)(F,OH)$_2$
Chondrodite	Mg$_5$(SiO$_4$)$_2$(F,OH)$_2$
Humite	Mg$_7$(SiO$_4$)$_3$(F,OH)$_2$
Clinohumite	Mg$_9$(SiO$_4$)$_4$(F,OH)$_2$

All four species have similar structures, chemistry, and physical properties. The structure of the members of the humite group is closely related to that of olivine with layers parallel to {100}, which have the atomic arrangement of olivine, and alternating layers of Mg(OH,F)$_2$ composition. For example, the norbergite composition can be rewritten to express this layering as Mg$_2$SiO$_4$·Mg(OH,F)$_2$. The replacement of F by OH is extensive, but hydroxyl end members are unknown. The members of this group are rather restricted in their occurrence and are found mainly in metamorphosed and metasomatised limestones and dolomites, and skarns associated with ore deposits. Here we will discuss in detail only chondrodite, the most common member of the humite group.

Chondrodite—Mg$_5$(SiO$_4$)$_2$(F,OH)$_2$

Crystallography. Monoclinic; 2/m. Crystals are frequently complex with many forms. Usually in isolated grains. Also massive.

$P2_1/c$; a = 7.89, b = 4.74, c = 10.29, β = 109°; Z = 2. ds: 3.02(5), 2.76(4), 2.51(5), 2.26(10), 1.740(7).

Physical Properties. H 6–6$\frac{1}{2}$. G 3.1–3.2. *Luster* vitreous to resinous. *Color* light yellow to red. Translucent. *Optics*: (+); α = 1.592–1.615, β = 1.602–1.627, γ = 1.621–1.646; 2V = 71°–85°; Z = b, $X \wedge c$ = 25°; $r > v$.

Composition and Structure. The composition of chondrodite can be rewritten as 2Mg$_2$SiO$_4$·Mg(OH,F)$_2$ to reflect the olivine-type and brucite-like layering in the structure. The two main substitutions are Mg by Fe^{2+} and F by OH. The Fe substitution is limited to about 6 weight percent FeO.

Diagnostic Features. Characterized by its light yellow to red color and its mineral associations in crystalline limestone. The members of the humite group cannot be distinguished from one another without optical tests.

Occurrence. Chondrodite occurs most commonly in metamorphosed dolomitic limestones. The mineral association, including phlogopite, spinel, pyrrhotite, and graphite, is highly characteristic. In skarn deposits it is found with wollastonite, forsterite, and monticellite. Noteworthy localities of chondrodite are Monte Somma, Italy; Paragas, Finland; and Kafveltorp, Sweden. In the United States it was common at the Tilly Foster magnetite deposit near Brewster, New York.

Name. Chondrodite is from the Greek *chondros*, meaning *a grain*; alluding to its occurrence as isolated grains. Humite is named in honor of Sir Abraham Hume.

Datolite—CaB(SiO$_4$)(OH)

Crystallography. Monoclinic; $2/m$. Crystals are usually nearly equidimensional in the three axial directions and often complex in development (Fig. 12.18). Usually in crystals. Also coarse to finely granular. Compact and massive, resembling unglazed porcelain.

$P2_1/a$; $a = 4.83$, $b = 7.64$, $c = 9.66$, $\beta = 90°9'$; $Z = 4$. ds: 3.76(5), 3.40(3), 3.11(10), 2.86(7), 2.19(6).

Physical Properties. H $5-5\frac{1}{2}$. G 2.8–3.0. *Luster* vitreous. *Color* white, often with faint greenish tinge. Transparent to translucent. *Optics*: (−); $\alpha = 1.624$, $\beta = 1.652$, $\gamma = 1.668$; 2V = 74°; $Y = b$, $Z = c$; $r > v$.

Composition and Structure. CaO 35.0, B$_2$O$_3$ 21.8, SiO$_2$ 37.6, H$_2$O 5.6%. The structure can be regarded as consisting of layers parallel to {100} of independent SiO$_4$ and B(O,OH)$_4$ tetrahedra. The layers are bonded by Ca in 8-coordination (6 oxygens and 2 OH groups). Despite the layerlike structure datolite has no conspicuous cleavage.

Diagnostic Features. Characterized by its glassy luster, pale green color, and its crystals with many faces.

Occurrence. Datolite is a secondary mineral found usually in cavities in basaltic lavas and similar rocks. Associated with zeolites, prehnite, apophyllite, and calcite. Notable foreign localities are Andreasberg, Harz Mountains, Germany; in Italy near Bologna, and from Seiser Alpe and Theiso, Trentino; and Arendal, Norway. In the United States it occurs in the trap rocks of Massachusetts, Connecticut, and New Jersey, particularly at Westfield, Massachusetts, and Bergen Hill, New Jersey. Masses of porcelain-like datolite, some white and some red from disseminated native copper, were found in the Lake Superior copper deposits.

Name. Derived from a Greek word meaning *to divide*, in allusion to the granular character of a massive variety.

TITANITE (*sphene*)—CaTiO(SiO$_4$)   **Ⅳ**

Crystallography. Monoclinic; $2/m$. Wedge-shaped crystals common resulting from a combination of {001}, {110}, and {111} (Fig. 12.19). May be lamellar or massive.

$C2/c$, $a = 6.56$, $b = 8.72$, $c = 7.44$ Å, $\beta = 119°43'$; $Z = 4$. ds: 3.23(10), 2.99(9), 2.60(9), 2.27(3), 2.06(4).

Physical Properties. *Cleavage* {110} distinct. Parting on {100} may be present. **H** $5-5\frac{1}{2}$. **G**

FIG. 12.18. Datolite.

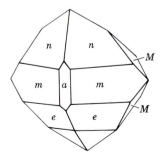

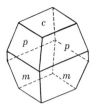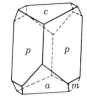

FIG. 12.19. Titanite crystals.

3.4–3.55. *Luster* resinous to adamantine. *Color* gray, brown, green, yellow, black. Transparent to translucent. *Optics*: (+); $\alpha = 1.900$, $\beta = 1.907$, $\gamma = 2.034$; 2V = 27°; $Y = b$, $Z \wedge c = 51°$; $r > v$.

Composition and Structure. CaO 28.6, TiO$_2$ 40.8, SiO$_2$ 30.6%. Small amounts of rare earths, Fe, Al, Mn, Mg, and Zr may be present. The structure of titanite is shown in Fig. 12.20. It consists of corner-sharing TiO$_6$ octahedra that form kinked chains parallel to the a axis. These chains are cross-linked by isolated SiO$_4$ tetrahedra. This tetrahedral-octahedral framework produces cavities that house Ca in 7-coordination.

Diagnostic Features. Characterized by its wedge-shaped crystals and high luster. Hardness is less than that of staurolite and greater than that of sphalerite.

Occurrence. Titanite in small crystals is a common accessory mineral in granites, granodiorites, diorites, syenites, and nepheline syenites. It is found in crystals of considerable size in metamorphic gneisses, chlorite schists, and crystalline limestone. Also found with iron ores, pyroxene, amphibole, scapolite, zircon, apatite, feldspar, and quartz.

The most notable occurrence of titanite is on the Kola Peninsula, Russia, where it is associated with apatite and nepheline syenite. It is mined there as a source of titanium. It is found in crystals in Tavetsch, Binnental, and St. Gotthard, Switzerland; Zillertal, Tyrol; Ala, Piemonte, and Vesuvius, Italy; and Arendal, Norway. In the United States in Diana, Rossie, Fine, Pitcairn, and Edenville, New York; and Riverside, California. Also in various places in Ontario and Quebec, Canada.

Use. As a source of titanium oxide for use as a paint pigment. A minor gemstone.

Name. In reference to its titanium content. The element titanium was named after the Titans, the mythical first sons of the Earth.

Similar Species. *Benitoite*, BaTiSi$_3$O$_9$, a cyclosilicate, occurs in association with *neptunite*, KNa$_2$Li(Fe,Mn)$_2$Ti$_2$O(Si$_4$O$_{11}$)$_2$, a complex inosilicate, in San Benito, California. *Astrophyllite, aenigmatite, lamprophyllite, ramsayite*, and *fersmannite* are rare Ti-bearing silicates found associated with alkalic rocks. *Benitoite* is a minor gemstone.

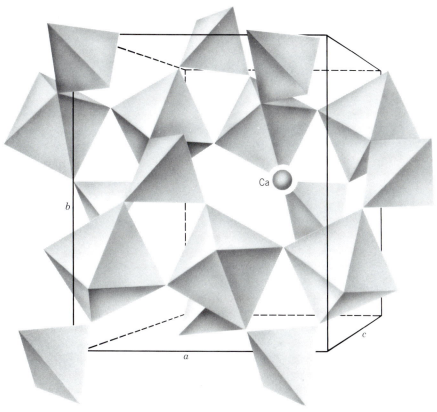

FIG. 12.20. Crystal structure of titanite (sphene) showing Si tetrahedra, Ti octahedra, and a 7-coordinated Ca site. The axes outline a unit cell. (From J. J. Papike, 1987, Chemistry of the rock-forming silicates: Ortho, ring, and single-chain structures. *Reviews of Geophysics* 25: 1483–1526.)

Chloritoid—$(Fe,Mg)_2Al_4O_2(SiO_4)_2(OH)_4$

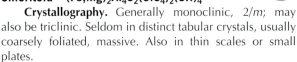

Crystallography. Generally monoclinic, 2/*m*; may also be triclinic. Seldom in distinct tabular crystals, usually coarsely foliated, massive. Also in thin scales or small plates.

$C2/c$; $a = 9.52$, $b = 5.47$, $c = 18.19$; $\beta = 101°39'$; $Z = 4$. *d*s: 4.50(10), 4.45(10), 3.25(6), 2.97(8), 2.70(7), 2.46(9).

Physical Properties. *Cleavage* {001} good (less so than in micas), producing brittle flakes. **H** $6\frac{1}{2}$ (much harder than chlorite). **G** 3.5–3.8. Luster pearly. *Color* dark green, greenish-gray, often grass-green in very thin plates. *Streak* colorless. *Optics*: (+); $\alpha = 1.713–1.730$, $\beta = 1.719–1.734$, $\gamma = 1.723–1.740$; 2V = 36°–60°; $X = b$, $Z \wedge c = 2°–30°$. Pleochroism strong: Z yellow-green, Y indigo-blue, X olive-green.

Composition and Structure. For the Fe^{2+} end member, SiO_2 23.8, Al_2O_3 40.5, FeO 28.5, and H_2O 7.2%. All chloritoids show some Mg replacing Fe^{2+} but this substitution is limited. Fe^{3+} substitutes to some extent for Al. The structure of chloritoid is illustrated in Fig. 12.21. It can be described as consisting of two close-packed octahedral layers; one of a bucite-like sheet with composition $[(Fe,Mg)_2AlO_2(OH)_4]^{1-}$ and another of a corundum-type composition, $[Al_3O_8]^{7-}$. These sheets alternate in the direction of the c axis. Independent SiO_4 tetrahedra link the brucite- and corundum-like sheets together. The SiO_4 tetrahedra, therefore, do not occur in continuous sheets, as in the micas and chlorite. Synthetic studies show that chloritoid gives way to staurolite at elevated temperature, according to the reaction: chloritoid + andalusite → staurolite + quartz + fluid.

Diagnostic Features. Chloritoid is often difficult to distinguish from chlorite, with which it is generally associated. Optical study is necessary for unambiguous identification.

Occurrence. Chloritoid is a relatively common constituent of low- to medium-grade regionally metamorphosed iron-rich pelitic rocks. It generally occurs as porphyroblasts in association with muscovite, chlorite, staurolite, garnet, and kyanite. The original chloritoid was described from Kosoibrod, in the Ural Mountains, Russia. It is most commonly found in greenschist facies rocks; it is not uncommon in Vermont, for example, and northern Michigan.

Name. Chloritoid is named for its superficial resemblance to chlorite.

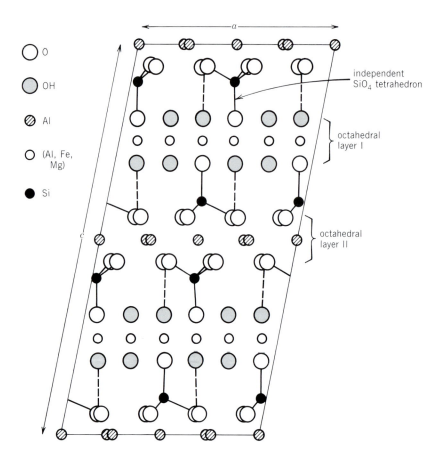

O   O

O   OH

⊘   Al

O   (Al, Fe, Mg)

● Si

independent SiO$_4$ tetrahedron

octahedral layer I

octahedral layer II

FIG. 12.21.   The structure of monoclinic chloritoid projected on (010). Octahedral layer (I) has approximate composition $[Fe_2^{2+}AlO_2(OH)_4]^{1-}$ with some Mg^{2+} and Mn^{2+} substituting for Fe^{2+}; octahedral layer (II) has the composition of $[Al_3O_8]^{7-}$. Isolated tetrahedra link these layers. (From F. W. Harrison and G. W. Brindley, 1957, The crystal structure of chloritoid. *Acta Crystallographica* 10: 77–82.)

SOROSILICATES

More than 70 minerals have been identified in this group, but most of them are rare. This section describes six species, of which members of the *epidote group* and *vesuvianite* are most important.

Hemimorphite	$Zn_4(Si_2O_7)(OH)_2 \cdot H_2O$
Lawsonite	$CaAl_2(Si_2O_7)(OH)_2 \cdot H_2O$
Epidote group	
Clinozoisite	$Ca_2Al_3O(SiO_4)(Si_2O_7)(OH)$
Epidote	$Ca_2(Fe^{3+},Al)Al_2O(SiO_4)(Si_2O_7)(OH)$
Allanite	$X_2Y_3O(SiO_4)(Si_2O_7)(OH)$
Vesuvianite	$Ca_{10}(Mg,Fe)_2Al_4(SiO_4)_5(Si_2O_7)_2(OH)_4$

HEMIMORPHITE—$Zn_4(Si_2O_7)(OH)_2 \cdot H_2O$

Crystallography. Orthorhombic; *mm2*. Crystals usually tabular parallel to {010}. They show prism faces and are terminated above usually by a combination of domes and pedion, and below by a pyramid (Fig. 12.22), forming polar crystals. Crystals often divergent, giving rounded groups with slight reentrant notches between the individual crystals,

forming knuckle or coxcomb masses. Also mammillary, stalactitic, massive, and granular.

Imm2, $a = 8.38$, $b = 10.72$, $c = 5.12$ Å; $Z = 2$. ds: 6.60(9), 5.36(6), 3.30(8), 3.10(10), 2.56(5).

Physical Properties. *Cleavage* {110} perfect. **H** $4\frac{1}{2}$–5. **G** 3.4–3.5. *Luster* vitreous. *Color* white, in some cases with faint bluish or greenish shade; also yellow to brown. Transparent to translucent. Strongly pyroelectric and piezoelectric. *Optics:* (+); $\alpha = 1.614$, $\beta = 1.617$, $\gamma = 1.636$; $2V = 46°$; $X = b$, $Z = c$; $r > v$.

Composition and Structure. ZnO 67.5, SiO$_2$ 25.0, H$_2$O 7.5%. Small amounts of Fe and Al may be pre-

FIG. 12.22. Hemimorphite.

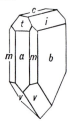

sent. The structure contains Si_2O_7 groups linked by $ZnO_3(OH)$ tetrahedra. The tetrahedra in the Si_2O_7 groups have their bases parallel to {001} and their apices all point the same way along the c direction. This orientation causes the polar character of the structure. Each (OH) group is bound to two Zn^{2+} ions. H_2O molecules lie in holes between the tetrahedra. The H_2O molecules are lost continuously by heating hemimorphite up to 500°C. At this temperature all H_2O molecules are driven off and only (OH) groups are retained. These can be driven off only at much higher temperatures with destruction of the crystal structure.

Diagnostic Features. Characterized by the grouping of crystals. Resembles prehnite but has a higher specific gravity.

Occurrence. Hemimorphite is a secondary mineral found in the oxidized portions of zinc deposits, associated with smithsonite, sphalerite, cerussite, anglesite, and galena.

A notable locality for its occurrence is in Chihuahua, Mexico. In the United States it is found at Sterling Hill, Ogdensburg, New Jersey; Friedensville, Pennsylvania; Wythe County, Virginia; with the zinc deposits of southwestern Missouri; Leadville, Colorado; Organ Mountains, New Mexico; and Elkhorn Mountains, Montana.

Use. An ore of zinc.

Name. From the hemimorphic character of the crystals. The mineral was formly called *calamine*.

Lawsonite—CaAl₂(Si₂O₇)(OH)₂·H₂O

Crystallography. Orthorhombic; 222. Usually in tabular or prismatic crystals. Frequently twinned polysynthetically on {110}.

$C222_1$; $a = 8.79$, $b = 5.84$, $c = 13.12$ Å; $Z = 4$. ds: 4.17(5), 3.65(6), 2.72(10), 2.62(7), 2.13(7).

Physical Properties. *Cleavage* {010} and {110} good. **H** 8. **G** 3.09. *Color* colorless, pale blue to bluish-gray. *Luster* vitreous to greasy. Translucent. *Optics*: (+); α = 1.665, β = 1.674, γ = 1.684; 2V = 84°; X = a, Z = c; r > v.

Composition and Structure. The composition of lawsonite is the same as that of anorthite, $CaAl_2Si_2O_8$, + H_2O. The structure consists of (AlO,OH) octahedra linked by Si_2O_7 groups. Ca^{2+} and H_2O molecules are located between these polyhedra.

Diagnostic Features. Lawsonite is characterized by its high hardness.

Occurrence. Lawsonite is a typical mineral of the glaucophane schist facies associated with chlorite, epidote, titanite, glaucophane, garnet, and quartz. The type locality is on the Tiburon Peninsula, San Francisco Bay, California. It is a common constituent of gneisses and schists formed under low temperature and high pressure.

Name. In honor of Professor Andrew Lawson of the University of California.

Similar Species. *Ilvaite*, $CaFe_2^{2+}Fe^{3+}O(Si_2O_7)(OH)$, is related to lawsonite with a similar although not identical structure.

Epidote Group

The structure and crystal chemistry of the epidote group are discussed on page 448 (see also Figs. 11.9 and 11.10).

CLINOZOISITE—Ca₂Al₃O(SiO₄)(Si₂O₇)(OH)
EPIDOTE—Ca₂(Al,Fe)Al₂O(SiO₄)(Si₂O₇)(OH)

Crystallography. Monoclinic; 2/m. Crystals are usually elongated parallel to b with a prominent development of the faces of the [010] zone, giving them a prismatic aspect (Figs. 12.23 and 12.24). Striated parallel to b. Twinning on {100} common. Usually coarse to fine granular; also fibrous.

$P2_1/m$; $a = 8.98$, $b = 5.64$, $c = 10.22$ Å, β = 115°24'; $Z = 2$. ds: 5.02(4), 2.90(10), 2.86(6), 2.53(6), 2.40(7).

Physical Properties. *Cleavage* {001} perfect and {100} imperfect. **H** 6–7. **G** 3.25–3.45. *Luster* vitreous. *Color* epidote: pistachio-green to yellowish-green to black; clinozoisite: pale green to gray. Transparent to translucent. *Optics*: Refractive indices and birefringence increase with iron content. Clinozoisite: (+): α = 1.670–1.715, β = 1.674–1.725, γ = 1.690–1.734; 2V = 14°–90°; Y = b, X ∧ c = −2° to −7°; r > v. Epidote: (−): α = 1.715–1.751, β = 1.725–1.784, γ = 1.734–1.797; 2V = 64°–90°; Y = b, X ∧ c = 1° to −5°; r > v. Absorption Y > Z > X. Transparent crystals may show strong absorption in ordinary light.

Composition and Structure. A complete solid solution series extends from clinozoisite (Al : Fe^{3+} = 3 : 0) to epidote (Al : Fe^{3+} = 2 : 1). *Piemontite*, $Ca_2Mn^{3+}Al_2O(SiO_4)(Si_2O_7)(OH)$, is isostructural with epidote and clinozoisite but contains mainly Mn^{3+} instead of Fe^{3+} or Al^{3+} in the Al site outside the chains of the epidote structure. The structure of epidote has been discussed on page 448 and is illustrated in Fig. 11.9.

Diagnostic Features. Epidote is characterized by its peculiar green color and one perfect cleavage.

FIG. 12.23. Epidote crystals.

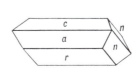
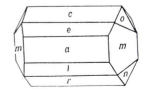

0   1   2
cm

(a)

(b)

FIG. 12.24.   *(a)* Epidote, Knappenwand, Austria (Harvard Mineralogical Museum). *(b)* Epidote with quartz crystals. Prince of Wales Island, Alaska.

Occurrence. Epidote forms under conditions of regional metamorphism of the epidote-amphibolite facies. Characteristic associations of actinolite–albite–epidote–chlorite occur in the upper part of the greenschist facies. Epidote forms also during retrograde metamorphism and forms as a reaction product of plagioclase, pyroxene, and amphibole. Epidote is common in metamorphosed limestones with calcium-rich garnets, diopside, vesuvianite, and calcite. *Epidotization* is a low-temperature metasomatism and is found in veins and joint fillings in some granitic rocks.

Epidote is a widespread mineral. Notable localities for its occurrence in fine crystals are Knappenwand, Untersulzbachthal, Salzburg, Austria (Fig. 12.24a); Bourg d'Oisans, Isère, France; and the Ala Valley and Traversella, Piemonte, Italy. In the United States found at Riverside, California, and on Prince of Wales Island, Alaska (Fig. 12.24b).

Use.   Sometimes cut as a gem.

Name. Epidote from the Greek meaning *increase*, because the base of the vertical prism has one side longer than the other. Zoisite was named after Baron von Zois and piemontite after the locality, Piemonte, Italy.

Similar Species. *Zoisite* is an orthorhombic polymorph (space group *Pnmc*) of clinozoisite. It is similar in appearance and occurrence to clinozoisite but is less common. In 1967 gem quality, blue-colored crystals were found in Tanzania. This variety is known as *tanzanite*.

Allanite—$(Ca,Ce)_2(Fe^{2+},Fe^{3+})Al_2O(SiO_4)(Si_2O_7)(OH)$

Crystallography. Monoclinic; $2/m$. Habit of crystals similar to epidote. Commonly massive and in embedded grains.

$P2_1/m$; $a = 8.98$, $b = 5.75$, $c = 10.23$ Å, $\beta = 115°0'$; $Z = 2$. ds: 3.57(6), 2.94(10), 2.74(8), 2.65(6), 2.14(4).

Physical Properties. H $5\frac{1}{2}$–6. G 3.5–4.2. *Luster* submetallic to pitchy and resinous. *Color* brown to pitch-black. Often coated with a yellow-brown alteration product. Subtranslucent, will transmit light on thin edges. Slightly radioactive. *Optics:* (−), usually, with $2V = 40°$–$90°$; when (+), $2V = 60°$–$90°$, $\alpha = 1.690$–1.791, $\beta = 1.700$–1.815, $\gamma = 1.706$–1.828; $Y = b$, $X \wedge c$ 1°–40°. Metamict allanites are isotropic with $n = 1.54$–1.72.

Composition and Structure. Of variable composition with Ce, La, Th, and Na in partial substitution for Ca, and Fe^{2+}, Fe^{3+}, Mn^{3+}, and Mg in partial substitution for some of the Al. The structure of well-crystallized allanite is the same as that of epidote (see Fig. 11.9). Allanite is commonly found in a metamict state as the result of "self-irradiation" by radioactive constituents in the original mineral. Total destruction of the structure leads to a glassy product that adsorbs considerable H_2O (see Fig. 4.48 and related discussion).

Diagnostic Features. Characterized by its black color, pitchy luster, and association with granitic rocks.

Occurrence. Allanite occurs as a minor accessory constituent in many igneous rocks, such as granite, syenite, diorite, and pegmatites. Frequently associated with epidote.

Notable localities are at Miask, Ural Mountains, Russia; Greenland; Falun and Ytterby, Sweden; and the Malagasy Republic. In the United States allanite is found at Franklin, New Jersey, and Barringer Hill, Texas.

Name. In honor of Thomas Allan, who first observed the mineral. *Orthite* is sometimes used as a synonym.

VESUVIANITE (*idocrase*)— $Ca_{10}(Mg,Fe)_2Al_4(SiO_4)_5(Si_2O_7)_2(OH)_4$

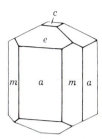

Crystallography. Tetragonal; $4/m2/m2/m$. Crystals prismatic often vertically striated. Common forms are {110}, {010}, and {001} (Fig. 12.25). Frequently found in crystals, but striated columnar aggregates more common. Also granular, massive.

$P4/nnc$; $a = 15.66$, $c = 11.85$ Å; $Z = 4$. ds: 2.95(4), 2.75(10), 2.59(8), 2.45(5), 1.621(6).

FIG. 12.25. Vesuvianite.

Physical Properties. *Cleavage* {010} poor. H $6\frac{1}{2}$. G 3.35–3.45. *Luster* vitreous to resinous. *Color* usually green or brown, see Plate V, no. 4; also yellow, blue, red. Subtransparent to translucent. *Streak* white. *Optics:* (−); $\omega = 1.703$–1.752, $\epsilon = 1.700$–1.746.

Composition and Structure. There is some substitution of Na for Ca; Mn^{2+} for Mg; Fe^{3+} and Ti for Al; and F for (OH). B and Be have been reported in some varieties. The structure of vesuvianite appears to be closely related to that of grossular garnet. Some parts of the structure are common to both minerals. Isolated SiO_4 tetrahedra as well as Si_2O_7 groups occur. Three-fourths of the Ca is in 8-coordination and one-fourth in 6-coordination with oxygen. The Al and Fe (and Mg) are in octahedral coordination with oxygen. A projection of a unit cell of the structure is given in Fig. 12.26. A structure image (obtained by high-resolution transmission electron microscopy) of this rather complicated silicate is given in Fig. 12.27. In this photograph a unit cell is outlined, which is equivalent to the structure shown in Fig. 12.26. Furthermore, part of the photograph contains a computer-simulated image of the structure. Such calculated images are very helpful to the investigator, because they aid in the understanding of features of the experimentally obtained structure image.

Diagnostic Features. Brown tetragonal prisms and striated columnar masses are characteristic of vesuvianite.

Occurrence. Vesuvianite is usually formed as the result of contact metamorphism of impure limestones. Associated with other contact minerals, such as grossular and andradite garnet, wollastonite, and diopside. Originally discovered in the ancient ejections of Vesuvius and in the dolomitic blocks of Monte Somma.

Important localities are Zermatt, Switzerland; in Italy, at Ala, Piemonte, at Monzoni, Trentino, and at Vesuvius; Achmatovsk, Ural Mountains, and River Vilui, Siberia, Russia; and Morelos, Mexico. In the United States it is found at Sanford, Maine; near

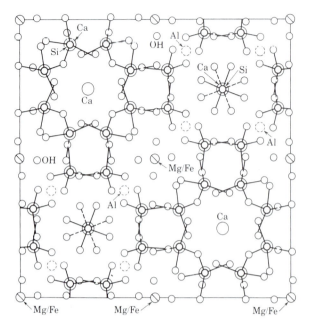

FIG. 12.26. The structure of a unit cell of vesuvianite, projected in (001). In order to compare this structural projection with the projected structure image in Fig. 12.27, one must note that regions of high atomic potential are dark areas in Fig. 12.27, whereas white regions represent parts of the structure with low electron density. The two vertical channels, with Ca at their centers, correspond with the darkest regions in the structure image of Fig. 12.27.

Olmsteadville, New York; Franklin, New Jersey; Magnet Cove, Arkansas; and Crestmore, California. Found in many contact metamorphic deposits in the western United States. A compact green variety resembling jade found in Siskiyou, Fresno, and Tulare counties, California, is known as *californite*. In Quebec, Canada, found at Litchfield, Pontiac County; at Templeton, Ottawa County; and at Asbestos.

Use. The green, massive variety californite is used as a jade substitute. Transparent crystals may be cut as faceted gems.

Name. From the locality Mount Vesuvius.

CYCLOSILICATES

We will describe in detail the following cyclosilicates:

Axinite	$(Ca,Fe^{2+},Mn)_3Al_2BSi_4O_{15}(OH)$
Beryl	$Be_3Al_2(Si_6O_{18})$
Cordierite	$(Mg,Fe)_2Al_4Si_5O_{18}\cdot nH_2O$
Tourmaline	$(Na,Ca)(Li,Mg,Al)_3(Al,Fe,Mn)_6$
	$(BO_3)_3(Si_6O_{18})(OH)_4$

Axinite—$(Ca,Fe^{2+},Mn)_3Al_2BSi_4O_{15}(OH)$

Crystallography. Triclinic; $\bar{1}$. Crystals are usually thin with sharp edges but varied in habit (Fig. 12.28). Frequently in crystals and crystalline aggregates; also massive, lamellar to granular.

$P\bar{1}$; $a = 7.15$, $b = 9.16$, $c = 8.96$ Å, $\alpha = 88°04'$, $\beta = 81°36'$, $\gamma = 77°42'$; $Z = 2$. ds: 6.30(7), 3.46(8), 3.28(6), 3.16(9), 2.81(10).

Physical Properties. *Cleavage* {100} distinct. **H** $6\frac{1}{2}$–7. **G** 3.27–3.35. *Luster* vitreous. *Color* clove-brown, violet, gray, green, yellow. Transparent to translucent. *Optics*: (−); $\alpha = 1.674$–1.693, $\beta = 1.681$–1.701, $\gamma = 1.684$–1.704; $2V = 63°80'$; $r < v$.

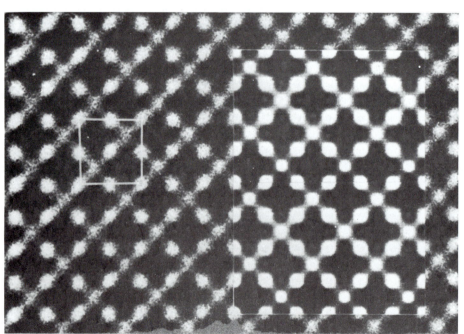

FIG. 12.27. High-resolution transmission electron microscope (HRTEM) image of vesuvianite. The edge of the unit cell, as outlined by a white square, is 15 Å. A computer-calculated image, which simulates the experimental image, is outlined by a rectangular area in the right-hand part of the illustration. The white areas are regions of low electron density in the vesuvianite structure. (From P. R. Buseck, 1978, Computed crystal structure images for high resolution electron microscopy. *Nature* 274: 322–24; reprinted by permission from *Nature*. Copyright © 1978, Macmillan Journals Limited.)

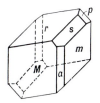

FIG. 12.28. Axinite.

Composition and Structure. A considerable range exists in composition with varying amounts of Ca, Mn, and Fe. Some Mg may be present. The complex structure of axinite was originally regarded as made up of Si_4O_{12} rings and BO_3 triangles and (OH) groups. Such Si_4O_{12} rings would classify axinite as one of the cyclosilicates. More recent structure analysis, however, has shown that the axinite structure is best viewed as being made up of $B_2Si_8O_{30}$ groups, in which BO_4 tetrahedra share three corners each, linking together four $S_{12}O_7$ groups. Zoltai (1960) would classify this structure under "complex silicate groups."

Diagnostic Features. Characterized by the triclinic crystals with very acute angles.

Occurrence. Axinite occurs in cavities in granite, and in the contact zones surrounding granitic intrusions. Notable localities for its occurrence are Bourg d'Oisans, Isère, France; various points in Switzerland; St. Just, Cornwall, England; and Obira, Japan. In the United States it occurs at Luning, Nevada, and a yellow manganous species at Franklin, New Jersey.

Use. A minor gem.

Name. Derived from a Greek word meaning *axe*, in allusion to the wedgelike shape of the crystals.

BERYL—$Be_3Al_2(Si_6O_{18})$ IV

Crystallography. Hexagonal; $6/m2/m2/m$. Strong prismatic habit. Frequently vertically striated and grooved. Cesium beryl frequently flattened on {0001}. Forms usually present consist only of {10$\bar{1}$0} and {0001} (Fig. 12.29a). Pyramidal forms are rare (Figs. 12.29b and 12.30). Crystals frequently of considerable size with rough faces. At Albany, Maine, a tapering crystal 27 feet long weighed over 25 tons.

*P*6/*mcc*; $a = 9.23$, $c = 9.19$ Å; $Z = 2$. *d*s: 7.98(9), 4.60(5), 3.99(5), 3.25(10), 2.87(10).

FIG. 12.29. Beryl crystals.

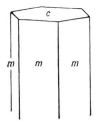

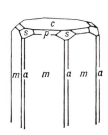

FIG. 12.30. Transparent, light green beryl crystal (length 12.5 cm), from Perm, Ural Mountains, Russia (Harvard Mineralogical Museum).

Physical Properties. *Cleavage* {0001} imperfect. **H** $7\frac{1}{2}$–8. **G** 2.65–2.8. *Luster* vitreous. *Color* commonly bluish-green or light yellow, may be deep emerald-green, gold-yellow, pink, white, or colorless (see Plate IX, nos. 2 and 5, and Plate XI, no. 8, Chapter 13). Transparent to translucent. Frequently the larger, coarser crystals show a mottled appearance due to the alternation of clear transparent spots with cloudy portions.

Color serves as the basis for several variety names of gem beryl. *Aquamarine* is the pale greenish-blue transparent variety; see Plate V, no. 5. *Morganite*, or rose beryl, is pale pink to deep rose. *Emerald* is the deep green, transparent beryl. *Golden beryl* is a clear golden-yellow variety. *Optics*: (−); $\omega = 1.560$–1.602, $\epsilon = 1.557$–1.599.

Composition and Structure. BeO 14.0, Al_2O_3 19.0, SiO_2 67.0 are the theoretical percentages of the oxides in the formula. However, the presence of alkalis (Na and Rb) and Li may considerably reduce the percentage of BeO. Small and variable amounts of H_2O and CO_2 are housed interstitially in the large, hexagonal, vertical channels (see page 92). The structure of beryl is illustrated in Fig. 11.12 and discussed on page 451. See also Fig. 6.57 and related text as well as module III on the CD-ROM under the heading "3-dimensional Order: Space Group Elements in Structures."

Diagnostic Features. Recognized usually by its hexagonal crystal form and color. Distinguished from apatite by greater hardness and from quartz by higher specific gravity.

Occurrence. Beryl, although containing the rare element Be, is rather common and widely distributed. It occurs usually in granitic rocks, or in pegmatites. It is also found in mica schists and associated with tin ores. The world's finest emeralds are found in Colombia in a dark bituminous limestone; the most notable localities are Muzo, Cosquez, and Chivor. Emeralds of good quality are found in mica schists in the Transvaal, South Africa; Sandawana, Zimbabwe; and near Sverdlovsk, Russia. Rather pale emeralds have been found in small amount in Alexander County, North Carolina, associated with the green variety of spodumene, *hiddenite*. Beryl of the lighter aquamarine color is much more common and is found in gem quality in many countries.

The world's major source of gem beryl, other than emerald, is Brazil. Fine specimens come from several localities, but the most important are in the state of Minas Gerais. Other important localities are in the Ural Mountains, Russia, Malagasy Republic, and Namibia. In the United States gem beryl, chiefly *aquamarine*, has been found in various places in New England, Idaho, North Carolina, and Colorado. The most important localities for *morganite* are in the Pala and Mesa Grande districts, San Diego County, California. Common beryl, as an ore of beryllium, is produced in small amounts in many countries but the principal producers are Russia, Brazil, and the United States.

Use. As a gemstone the emerald (see Plate IX, nos. 2 and 5, Chapter 13) ranks as one of the most valuable of stones, and may have a greater value than the diamond. Beryl is also the principal source of beryllium, a lightweight metal similar to aluminum in many of its properties. A major use of beryllium is as an alloy with copper. Beryllium greatly increases the hardness, tensile strength, and fatigue resistance of copper.

Name. The name *beryl* is of ancient origin, derived from the Greek word which was applied to green gemstones.

Similar Species. *Euclase,* $BeAl(SiO_4)(OH)$, and *gadolinite,* $YFe^{2+}Be_2(SiO_4)_2O_2$, are rare beryllium silicates. Blue *euclase* is sometimes cut as a gemstone.

Cordierite—$(Mg,Fe)_2Al_4Si_5O_{18}\cdot nH_2O$

Crystallography. Orthorhombic: $2/m2/m2/m$. Crystals are usually short prismatic, pseudohexagonal twins twinned on {110}. Also found as imbedded grains and massive.

Cccm; $a = 17.13$, $b = 9.80$, $c = 9.35$ Å; $Z = 4$. ds: 8.54(10), 4.06(8), 3.43(8), 3.13(7), 3.03(8).

Physical Properties. *Cleavage* {010} poor. **H** $7–7\frac{1}{2}$. **G** 2.60–2.66. *Luster* vitreous. *Color* various shades of blue to bluish-gray. Transparent to translucent. *Optics:* usually $(-)$, may be $(+)$. Indices increasing with Fe content. $\alpha = 1.522–1.558$, $\beta = 1.524–1.574$, $\gamma = 1.527–1.578$. $2V = 0°–90°$. $X = c$, $Y = a$; $r < v$. Pleochroism: cordierite is sometimes called *dichroite* because of pleochroism. Fe rich varieties: X colorless, Y and Z violet.

Composition and Structure. Although some substitution of Mg by Fe^{2+} occurs, most cordierites are magnesium-rich. Mn may replace part of the Mg. The Al content of cordierite shows little variation. Most analyses show appreciable but variable H_2O, which is probably located in the large channels parallel to c. Small amounts of Na and K may be similarly housed. The structure of the low-temperature form, also known as low cordierite, is shown in Fig. 11.13 and is discussed on page 451. A high-temperature polymorph, *indialite*, with random distribution of Al in the $(AlSi)_6O_{18}$ rings, is isostructural with beryl and has space group $P6/mcc$.

Diagnostic Features. Cordierite resembles quartz and is distinguished from it with difficulty. Distinguished from corundum by lower hardness. Pleochroism is characteristic if observed.

Alteration. Commonly altered to some form of mica, chlorite, or talc and is then various shades of grayish-green.

Occurrence. Cordierite is a common constituent of contact and regionally metamorphosed argillaceous rocks. It is especially common in hornfels produced by contact metamorphism of pelitic rocks. Common assemblages are sillimanite-cordierite-spinel, and cordierite-spinel-plagioclase-orthopyroxene. Cordierite is also found in regionally metamorphosed cordierite-garnet-sillimanite gneisses. Cordierite-anthophyllite coexistences have been described from several localities. It occurs also in norites resulting from the incorporation of argillaceous material in gabbroic magmas. It is found in some granites and pegmatites. Gem material has come from Sri Lanka.

Use. Transparent cordierite of good color is used as a gem known by jewelers as *iolite* or *dichroite*.

Name. After the French geologist P. L. A. Cordier (1777–1861). *Iolite* is sometimes used as a synonym.

TOURMALINE—(Na,Ca)(Li,Mg,Al)$_3$-(Al,Fe,Mn)$_6$(BO$_3$)$_3$(Si$_6$O$_{18}$)(OH)$_4$

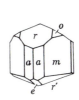
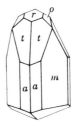
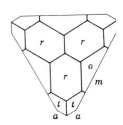

FIG. 12.31. Tourmaline.

Crystallography. Hexagonal; 3*m*. Usually in prismatic crystals with a prominent trigonal prism and subordinate hexagonal prism, {11$\bar{2}$0}, vertically striated. The prism faces may round into each other giving the crystals a cross section like a spherical triangle. When doubly terminated, crystals usually show different forms at the opposite ends of the vertical axis (Fig. 12.31). May be massive, compact; also coarse to fine columnar, either radiating or parallel.

*R*3*m*; *a* = 15.95, *c* = 7.24 Å; *Z* = 3. *d*s: 4.24(7), 4.00(7), 3.51(7), 2.98(9), 2.58(10).

Physical Properties. **H** 7–7$\frac{1}{2}$. **G** 3.0–3.25. *Luster* vitreous to resinous. *Color* varied, depending on the composition. *Fracture* conchoidal.

Black, Fe-bearing tourmaline (*schorl*) is most common; brown tourmaline (*dravite*) contains Mg. The rarer Li-bearing varieties (*elbaite*, containing Na and *liddicoatite*, containing Ca) are light colored in fine shades of green (*verdelite*), yellow-red-pink (*rubellite*), and blue (*indicolite*); see Plate V, no. 6. Rarely white or colorless *achroite*. A single crystal may show several different colors arranged either in concentric envelopes about the *c* axis or in layers transverse to the length. Strongly pyroelectric and piezoelectric. *Optics*: (−); ω = 1.635–1.675, ε = 1.610–1.650. Some varieties are strongly pleochroic, *O* > *E*.

Composition and Structure. A complex silicate of B and Al (see Fig. 11.14) with the following substitutions: Ca for Na along the centers of the ring channels; Mg and Al for Li in 6-coordination between Si$_6$O$_{18}$ rings and BO$_3$ groups; Fe^{3+} and Mn^{3+} for Al in polyhedra that link the Si$_6$O$_{18}$ rings. The structure is discussed on page 452.

Diagnostic Features. Usually recognized by the characteristic rounded triangular cross section of the crystals and conchoidal fracture. Distinguished from hornblende by absence of cleavage.

Occurrence. The most common and characteristic occurrence of tourmaline is in granite pegmatites (see Box 12.4) and in the rocks immediately surrounding them. It is found also as an accessory mineral in igneous rocks and metamorphic rocks. Most pegmatitic tourmaline is black and is associated with the common pegmatite minerals, microcline, albite,

FIG. 12.32. Tourmaline crystals with quartz and cleavelandite, Pala, California (Harvard Mineralogical Museum).

quartz, and muscovite. Pegmatites are also the home of the light-colored lithium-bearing tourmalines frequently associated with lepidolite, beryl, apatite, fluorite, and rarer minerals. The brown Mg-rich tourmaline is found in crystalline limestones.

Famous localities for the occurrence of the gem tourmalines are the Island of Elba, Italy; the state of Minas Gerais, Brazil; Ural Mountains near Sverdlovsk, Russia; and the Malagasy Republic. In the United States found at Paris and Auburn, Maine; and Mesa Grande, Pala (Fig. 12.32), Rincon, and Ramona in San Diego County, California. Brown crystals are found near Gouverneur, New York, and fine black crystals at Pierrepont, New York.

Use. Tourmaline forms one of the most beautiful of the gemstones (see Plate X, no. 11, Chapter 13). The color of the stones varies, the principal shades being olive-green, pink to red, and blue. Sometimes a stone is so cut as to show different colors in different parts. The green-colored stones are usually known by the mineral name, tourmaline, or as *Brazilian emeralds*. The red or pink stones are known as *rubellite*, and the rarer dark blue stones are called *indicolite*.

Because of its strong piezoelectric property, tourmaline is used in the manufacture of pressure gauges to measure transient blast pressures (for piezoelectricity, see page 35).

Name. *Tourmaline* comes from *turamali*, a name given to the early gems from Sri Lanka.

INOSILICATES

Pyroxene Group

We will describe in detail the following common pyroxenes:

Enstatite-Ferrosilite Series

Enstatite	$MgSiO_3$
Ferrosilite	$FeSiO_3$
Pigeonite	$Ca_{0.25}(Mg,Fe)_{1.75}Si_2O_6$

Diopside-Hedenbergite Series

Diopside	$CaMgSi_2O_6$
Hedenbergite	$CaFeSi_2O_6$
Augite	$XY(Z_2O_6)$

Sodium Pyroxene Group

Jadeite	$NaAlSi_2O_6$
Aegirine	$NaFe^{3+}Si_2O_6$
Spodumene	$LiAlSi_2O_6$

ENSTATITE—$MgSiO_3$-$(Mg,Fe)SiO_3$
FERROSILITE—$(Fe,Mg)SiO_3$

Crystallography. Orthorhombic; $2/m2/m2/m$. Prismatic habit, crystals rare. Usually massive, fibrous, or lamellar.

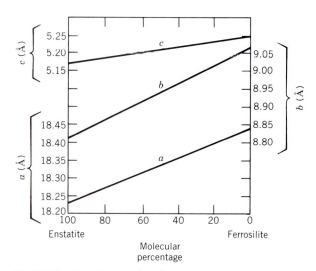

FIG. 12.33. Variation in unit cell parameters as a function of composition in the orthopyroxene series. (From W. A. Deer, R. A. Howie, and J. Zussman, 1978, *Single-Chain Silicates*, 2nd ed., v. 2A. New York: Wiley; and London: Longman.)

Pbca; $a = 18.22$, $b = 8.81$, $c = 5.17$ Å, for pure enstatite; $Z = 8$. *d*s: 3.17(10), 2.94(4), 2.87(9), 2.53(4), 2.49(5). Unit cell parameters increase with Fe content (see Fig. 12.33).

Physical Properties. *Cleavage* {210} good. Because of the doubling of the a dimension in orthorhombic pyroxenes, the cleavage form is {210} rather than {110} as in monoclinic pyroxenes. Frequently good parting on {100}, less common on {001}. **H** $5\frac{1}{2}$–6 and **G** 3.2–3.6 for pure enstatite; increasing with Fe content. *Luster* vitreous to pearly on cleavage surfaces; En_{80} has submetallic, bronzelike luster. *Color* grayish, yellowish, or greenish-white to olive-green and brown; see Plate V, no. 7. Translucent. *Optics:* enstatite (+); En_{90} to En_{10} (−). $\alpha = 1.653$–1.710, $\beta = 1.653$–1.728, $\gamma = 1.663$–1.725, for En_{100} to En_{50}. $X = b$, $Z = c$. Indices increase with Fe content (see Fig. 12.34); in ferrosilite $\beta = 1.765$.

Composition and Structure. Fe^{2+} may substitute for Mg in all proportions up to nearly 90% $FeSiO_3$. However, in the more common orthopyroxenes the ratio of Fe : Mg rarely exceeds 1 : 1. Pure enstatite contains MgO 40.0, SiO_2 60.0%. The maximum amount of CaO in orthopyroxenes generally does not exceed 1.5 weight percent. The nomenclature for the orthopyroxenes is shown in Fig. 11.15; chemical compositions are generally expressed in terms of molecular percentages, for example, $En_{40}Fs_{60}$. The pure end member $FeSiO_3$, *ferrosilite*, is rarely found in nature because in most geologically observed pressure and temperature ranges the compositionally equivalent assemblage Fe_2SiO_4 (fayalite) + SiO_2 is more stable; all other varieties of the orthopyroxene

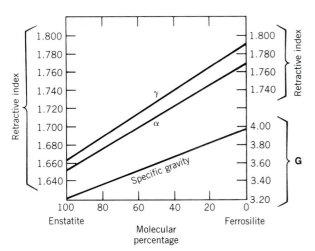

FIG. 12.34. Variation in two refractive indices (α and γ) and specific gravity as a function of composition in the orthopyroxene series. (From W. A. Deer, R. A. Howie, and J. Zussman, 1978, *Single-Chain Silicates*, 2nd ed., v. 2A. New York: Wiley; and London: Longman.)

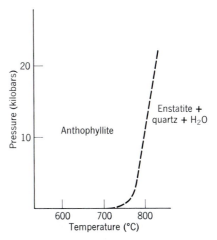

FIG. 12.35. Schematic *P-T* diagram for the stability fields of anthophyllite, $Mg_7Si_8O_{22}(OH)_2$, and reaction products, enstatite, $MgSiO_3$, + quartz + H_2O. (After H. J. Greenwood, 1963, *Journal of Petrology* 4: 325.)

series are found. The structure of members of the orthopyroxene series can be considered to consist of the monoclinic *t–o–t* strips twinned along {100} so as to essentially double the *a* dimension of orthopyroxenes as compared to the *a* of clinopyroxenes (see Fig. 11.19*a*). In the *Pbca* structure of orthopyroxenes Fe^{2+} shows a strong preference for the *M2* crystallographic site. Compositions between $MgSiO_3$–$FeSiO_3$ may also occur as members of the monoclinic series, *clinoenstatite-clinoferrosilite*, with space group $P2_1/c$. Experimental results of the stability fields of enstatite versus clinoenstatite are controversial. The common occurrence of orthopyroxenes versus the rare occurrence of clinopyroxenes in the series $MgSiO_3$–$FeSiO_3$ may indicate that the orthorhombic series is generally more stable, and at lower temperatures, than the monoclinic series.

Diagnostic Features. Usually recognized by their color, cleavage, and unusual luster. Varieties high in iron are black and difficult to distinguish from augite without optical tests.

Occurrence. Mg-rich orthopyroxene is a common constituent of peridotites, gabbros, norites, and basalts and is commonly associated with Ca-clinopyroxenes (e.g., augite; see also Box 12.1), olivine, and plagioclase. It may be the major constituent of pyroxenites. Orthopyroxenes may also be found in metamorphic rocks, some types of which are of high *T* and high *P* origin, such as in the granulite facies. Iron-rich members of the orthopyroxene series are common in metamorphosed iron-formation in association with grunerite. In all such occurrences orthopyroxenes commonly coexist with clinopyroxenes because of a large miscibility gap between the two groups (see Fig. 11.15*a*). Orthopyroxenes frequently show exsolved lamellae of a Ca-rich clinopyroxene. Enstatite as well as clinoenstatite occur in both iron and stony meteorites. In prograde metamorphic rocks orthopyroxenes form commonly at the expense of Mg–Fe amphibole (e.g., anthophyllite) and in retrograde metamorphic rocks orthopyroxenes may give way to Mg–Fe amphiboles (see Fig. 12.35).

Enstatite is found in the United States at the Tilly Foster Mine, Brewster, New York, and at Edwards, St. Lawrence County, New York; at Texas, Pennsylvania; Bare Hills, near Baltimore, Maryland; and Webster, North Carolina. Ferroan enstatite occurs in New York in the norites of the Cortland region, on the Hudson River, and in the Adirondack region. Fe-rich orthopyroxenes are common constituents of the metamorphosed Lake Superior and Labrador Trough iron-formations.

Use. A minor gemstone.

Name. Enstatite is named from the Greek word *enstates*, meaning *opponent*, because of its refractory nature.

Pigeonite ~ $Ca_{0.25}(Mg,Fe)_{1.75}Si_2O_6$

Crystallography. Monoclinic; $2/m$. Very rarely as well-formed phenocrysts with a prismatic habit parallel to *c*.

$P2_1/c$; $a = 9.71$, $b = 8.96$, $c = 5.25$, $\beta = 108°33'$; $Z = 8$. *d*s: 3.21(8), 3.02(10), 2.908(8), 2.904(10), 2.578(6).

Physical Properties. *Cleavage* {110} good; parting on {100} may be present. **H** 6. **G** 3.30–3.46. *Color* brown, greenish-brown, to black. *Optics*: (+); α = 1.682–1.722, β = 1.684–1.722, γ = 1.704–1.752, increasing with Fe^{2+}. Two orientations occur: (1) with $Y = b$, and (2) more common, $X = b$. $Z \wedge c = 37°–44°$, $2V = 0°–30°$.

BOX 12.1
THE TWO MOST COMMON CRUSTAL ROCK TYPES: BASALT AND GRANITE

There are two major types of igneous rocks, *extrusive* and *intrusive*. The first group includes those igneous rocks

Vesicular, dark gray basalt (composed of two pyroxene types and plagioclase feldspar)

that reached the Earth's surface in a molten or partly molten state. Modern volcanoes produce lava flows that pour from a vent or fracture in the Earth's

crust. Such *extrusive* or *volcanic* rocks tend to cool and crystallize rapidly, with the result that their grain size is generally small. **Basalt** is a very dark, green to black, fine-grained volcanic rock that is common on many volcanic islands, such as the Hawaiian Islands and Iceland. Furthermore, the upper layers of the oceanic crust consist of basalt (see Fig. 3.1*b*). The photograph illustrates a dark gray *vesicular basalt* (vesicular means that it contains cavities of variable shape that are the result of gas entrapment during solidification of the magma). Basalt and its coarser-grained equivalent known as **gabbro** consists commonly of two types of pyroxene (ortho- and clinopyroxene) and calcium-rich plagioclase, but may also contain olivine.

Intrusive or *plutonic* rocks are the result of crystallization from a magma that did not reach the Earth's surface. A magma that is deeply buried in the Earth's crust generally cools slowly, and the mineral constituents crystallizing from it have time to grow to considerable size, giving the rock a medium- to coarse-grained texture. The mineral grains in such rock can generally be identified

with the naked eye. **Granite** is generally a coarse-grained rock of light color and consists mainly of quartz and potassium feldspar, as well as sodium-rich plagioclase. Muscovite is commonly present, and so are small amounts of dark silicates such as amphiboles or biotite. The photograph illustrates a coarse-grained quartz–feldspar–hornblende granite. The extrusive equivalent of granite is known as **rhyolite**. The average composition of the continental crust is similar to that of granite because the total amount of sedimentary and metamorphic rocks is insignificant in comparison to the bulk of plutonic igneous rocks (see Fig. 3.1*b*).

The average chemical composition of basalt (and its plutonic equivalent, gabbro) and granite (and its volcanic equivalent, rhyolite) are listed in the table. Basalt is relatively low in SiO_2

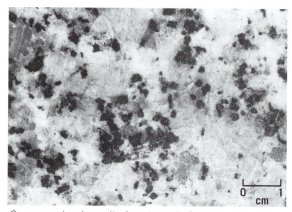

Coarse-grained granite (composed of quartz, K-feldspar, sodic plagioclase, and hornblende)

content, as compared with granite. However, basalt is much richer in FeO, MgO, and CaO than granite. The Na_2O and K_2O contents of granite are higher than those of basalt.

Average Chemical Compositions of Basalt and Granite*

Oxides (Weight %)	Basalt	Granite
SiO_2	48.36	72.08
TiO_2	1.32	0.37
Al_2O_3	16.84	13.86
Fe_2O_3	2.55	0.86
FeO	7.92	1.67
MnO	0.18	0.06
MgO	8.06	0.52
CaO	11.07	1.33
Na_2O	2.26	3.08
K_2O	0.56	5.46
H_2O	0.64	0.53
P_2O_5	0.24	0.18
Total	100.00	100.00

*Analyses from S. R. Nockolds, 1954, *Geological Society of America Bulletin* 65: 1007–1032.

Composition and Structure. Pigeonites are calcium-poor monoclinic pyroxenes that contain between 5 to 15 molecular percent of the $CaSiO_3$ component (see Fig. 11.15; field just above orthopyroxene base). The crystal structure of pigeonite is similar to that of diopside with all

of the Ca and additional Fe and Mg in the *M2* site, and the remaining Mg and Fe in *M1*. Fe shows a strong preference for the *M2* sites. Pigeonite is stable at high temperatures in igneous rocks and inverts commonly at lower temperatures to orthopyroxene with augite-type exsolution lamellae. A

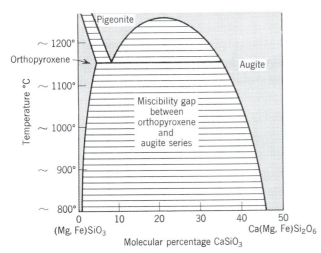

FIG. 12.36. Schematic *T*-composition section across the Wo-En-Fs diagram shown in Fig. 11.15*a*. The section is at about En$_{65}$Fs$_{35}$.

temperature-composition stability diagram is given in Fig. 12.36.

Diagnostic Features. Can be distinguished from other pyroxenes only by optical or X-ray techniques. Augite 2V > 39°; pigeonite 2V < 32°.

Occurrence. Pigeonite is common in high-temperature, rapidly cooled lavas and in some intrusives such as diabases. It is present as phenocrysts in some volcanic rocks, but is not known from metamorphic rocks. If pigeonite formed in an igneous rock that cooled slowly, it may have exsolved augite lamellae on {001} and may subsequently have inverted to orthopyroxene (see Fig. 12.36) through a reconstructive transformation. At even lower temperatures the orthopyroxene may have developed augite exsolution along {100} planes as well.

Name. After the locality, Pigeon Cove, Minnesota.

DIOPSIDE—CaMgSi$_2$O$_6$
HEDENBERGITE—Ca,FeSi$_2$O$_6$
AUGITE—(Ca,Na)(Mg,Fe,Al)(Si,Al)$_2$O$_6$

Diopside and hedenbergite form a complete solid solution series with physical and optical properties varying linearly with composition. Augite is a clinopyroxene in which some Na substitutes for Ca, some Al substitutes for both Mg (or Fe) and Si, and in which Fe and Mg contents are higher than in diopside or hedenbergite (see Fig. 11.15*a*). Although the crystal constants vary slightly from one member to another, a single description suffices for all.

Crystallography. Monoclinic; 2/*m*. In prismatic crystals showing square or eight-sided cross section (Fig. 12.37). Also granular massive, columnar, and lamellar. Frequently twinned polysynthetically on {001}; less commonly twinned on {100}.

*C*2/*c*; *a* = 9.73, *b* = 8.91, *c* = 5.25 Å; β = 105°50′; *Z* = 4. *d*s: 3.23(8), 2.98(10), 2.94(7),

2.53(4), 1.748(4). See Fig. 6.56 for an illustration of the dinopyroxene structure and related *C*2/*c* space group, as well as Module III on the CD-ROM under the heading "3-dimensional Order: Space Group Elements in Structures."

Physical Properties. *Cleavage* {110}, at 87° and 93°, imperfect. Frequently parting on {001}, and less commonly on {100} in the variety *diallage*. **H** 5–6. **G** 3.2–3.3. *Luster* vitreous. *Color* white to light green in diopside, see Plate V, no. 8; deepens with increase of Fe. Augite is black; see Plates V, no. 9 and VI, no. 1. Transparent to translucent. *Optics*: (+); α = 1.67–1.74, β = 1.67–1.74, γ = 1.70–1.76 (see Fig. 12.38). 2V = 55°–65°; *Y* = *b*, *Z* ∧ *c* = 39°–48°; *r* > *v*. Darker members show pleochroism; *X* pale green, *Y* yellow-green, *Z* dark green.

Composition and Structure. In the diopside–hedenbergite series Mg and Fe^{2+} substitute for each other in all proportions. In the majority of analyses of members of this series the Al$_2$O$_3$ content varies between 1 and 3 weight percent. In augite, in addition to varying Mg and Fe^{2+} contents, Al substitutes for both Mg (or Fe^{2+}) and Si. Mn, Fe^{3+}, Ti, and Na may also be present. A complete series exists toward aegirine-augite, (Na,Ca)(Fe^{3+},Fe^{2+},Mg,Al)Si$_2$O$_6$, by the replacement Ca(Mg,Fe^{2+}) ⇆ NaFe^{3+}. Compositional zoning is commonly found in igneous augites with the cores rich in the augite component and the rims tending toward aegirine-augite. Chemical analyses are generally recalculated in terms of molecular percentages of wollastonite (Wo), enstatite (En), and ferrosilite (Fs) and are expressed as Wo$_x$En$_y$Fs$_z$, where *x*, *y*, and *z* are molecular percentages (see Table 3.17 and Fig. 11.15). The structures of diopside, hedenbergite, and augite are all based on space group *C*2/*c*. Their structure is discussed on page 454, and a monoclinic structure is illustrated in Fig. 11.16. The Ca ions in the *M2* site are in 8-coordination, whereas the ions in *M1* are in 6-coordination.

Diagnostic Features. Characterized by crystal form and imperfect prismatic cleavage at 87° and 93°.

FIG. 12.37. Augite crystals.

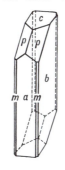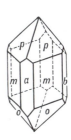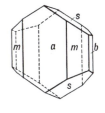

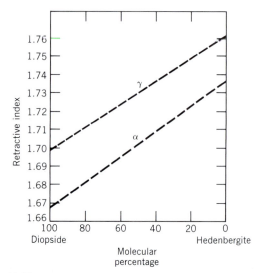

FIG. 12.38. Approximate trends of two refractive indices (α and γ) as a function of composition in the diopside-hedenbergite series. There is considerable scatter in the original data from which these lines were obtained. (From W. A. Deer, R. A. Howie, and J. Zussman, 1978, *Single-Chain Silicates*, v. 2A. New York: Wiley; and London: Longman.)

Occurrence. *Diopside* and *hedenbergite* are common in metamorphic rocks. Diopside, in association with forsterite and calcite, infrequently with monticellite, is the result of thermal metamorphism of siliceous, Mg-rich limestones, or dolomites. For example:

$$CaMg(CO_3)_2 + 2SiO_2 \rightarrow CaMgSi_2O_6 + 2CO_2\!\!\uparrow$$
$$\text{dolomite}\qquad \text{quartz}\qquad \text{diopside}$$

Other associations include tremolite, scapolite, vesuvianite, garnet, and titanite. Hedenbergite occurs in more Fe-rich metamorphic rocks. Diopside and hedenbergite are also known as products of igneous crystallization. Early formed Ca-rich clinopyroxenes may be very close to diopside in composition, whereas the latest stages of crystallization may be represented by hedenbergite compositions, due to enrichment of the residual magma in Fe. The Skaergaard intrusion, East Greenland, contains late, fine-grained hedenbergite interstitial to earlier formed, coarser grained diopside, and augite grains.

Fine crystals have been found in the Ural Mountains, Russia; Austrian Tyrol; Binnenthal, Switzerland; and Piemonte, Italy. At Nordmark, Sweden, fine crystals range between diopside and hedenbergite. In the United States they are found at Canaan, Connecticut; and DeKalb Junction and Gouverneur, New York.

Augite is the most common pyroxene and an important rock-forming mineral. It is found chiefly in the dark-colored igneous rocks, such as basaltic lavas and intrusives, gabbros, peridotites, and andesites (see Box 12.1). Chemically zoned augites are common in quickly cooled rocks such as the lunar basalts. The clinopyroxene crystallization history is very well documented for many lunar basalts and gabbros as well as for the Skaergaard intrusion, East Greenland, for example; early formed crystals are more magnesian than later pyroxene grains. Fine crystals of augite have been found in the lavas of Vesuvius and at Val di Fassa, Trentino, Italy; and at Bilin, Czechoslovakia.

Use. Transparent varieties of diopside have been cut and used as gemstones.

Name. Diopside is from two Greek words meaning *double* and *appearance*, because the vertical prism zone can apparently be oriented in two ways. Hedenbergite is named after M. A. Ludwig Hedenberg, the Swedish chemist who discovered and described it. Augite comes from a Greek word meaning *luster*. The name pyroxene, *stranger to fire*, is a misnomer and was given to the mineral because it was thought that it did not occur in igneous rocks.

Jadeite—NaAlSi$_2$O$_6$

Crystallography. Monoclinic; $2/m$. Rarely in isolated crystals. Usually granular in compact, massive aggregates.

$C2/c$; $a = 9.50$, $b = 8.61$, $c = 5.24$ Å; $\beta = 107°26'$; $Z = 4$. *d*s: 3.27(3), 3.10(3), 2.92(8), 2.83(10), 2.42(9).

Physical Properties. *Cleavage* {110} at angles of 87° and 93°. Extremely tough and difficult to break. **H** $6\frac{1}{2}$–7. **G** 3.3–3.5. *Color* apple-green (see Plate VI, no. 2) to emerald-green, white. May be white with spots of green. *Luster* vitreous, pearly on cleavage surfaces. *Optics*: (+); $\alpha = 1.654$, $\beta = 1.659$, $\gamma = 1.667$; $2V = 70°$; $X = b$, $Z \wedge c = 34°$; $r > v$.

Composition and Structure. Na$_2$O 15.4, Al$_2$O$_3$ 25.2, SiO$_2$ 59.4 for pure end member. There is no replacement of Si by Al in jadeite and Fe^{3+} substitution for Al is very limited. Jadeite has a composition that is intermediate between that of nepheline, NaAlSiO$_4$, and albite, NaAlSi$_3$O$_8$, but does not form under the normal crystallization conditions of these two minerals. However, under high pressures (10 to 25 kilobars) and elevated temperatures (between 600° and 1000°C), jadeite forms by:

$$NaAlSiO_4 + NaAlSi_3O_8 \rightleftarrows NaAlSi_2O_6$$
$$\text{nepheline}\qquad \text{albite}\qquad \text{jadeite}$$

Similarly, jadeite forms at high pressures at the expense of albite alone according to the following reaction:

$$NaAlSi_3O_8 \rightleftarrows NaAlSi_2O_6 + SiO_2$$

(see Fig. 12.39). The structure of jadeite is shown in Fig. 11.16.

Diagnostic Features. Characterized by its green color and tough aggregates of compact fibers. On polished surfaces nephrite has an oily luster, jadeite is vitreous.

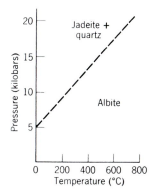

FIG. 12.39. Experimentally determined stability fields of albite and jadeite + quartz. Reaction curve is only approximately located as shown by dashes.

Occurrence. Jadeite is found only in metamorphic rocks. Laboratory experiments have shown that high pressure and only relatively low temperatures are necessary for the formation of jadeite. Such occurrences are found near the margins of the continental crust as in the Alps, California, and Japan. In the Fransiscan Formation of California, jadeitic pyroxene is associated with glaucophane, aragonite, muscovite, lawsonite, and quartz.

Use. Jadeite has long been highly prized in the Orient, especially in China, where it is worked into ornaments and utensils of great variety and beauty. It was also used by primitive people for various weapons and implements.

Name. The term *jade* includes both jadeite and the amphibole, nephrite.

Similar Species. *Omphacite* is a bright green variety of augite-type composition rich in $NaAlSi_2O_6$, with space group $C2/c$ or $P2/n$. It occurs in eclogites, which contain omphacite and pyrope-rich garnet as main constituents. These rocks are generally considered to be the result of high pressures and high temperatures of metamorphism. Eclogites are chemically equivalent to basalts, but contain denser minerals (see Fig. 4.27). Omphacite is also found in kimberlites. The name *omphacite* is from the Greek, *omphax*, an unripe grape, in allusion to its characteristic color.

AEGIRINE—$NaFe^{3+}Si_2O_6$

Crystallography. Monoclinic; $2/m$. Crystals are slender prismatic with steep terminations. Often in fibrous aggregates. Faces are often imperfect.

$C2/c$; $a = 9.66$, $b = 8.79$, $c = 5.26$ Å; $\beta = 107°20'$; $Z = 4$. ds: 6.5(4), 4.43(4), 2.99(10), 2.91(4), 2.54(6).

Physical Properties. *Cleavage* {110} imperfect at angles of 87° and 93°. **H** $6-6\frac{1}{2}$. **G** 3.40–3.55. *Luster* vitreous. *Color* brown or green. Translucent. *Optics* for aegirine: $(-)$; $\alpha = 1.776$, $\beta = 1.819$, $\gamma = 1.836$; $2V = 60°$; $Y = b$, $Z \wedge c = 8°$; $r > v$. Aegirine-augite: $(+)$ with lower indices and pleochroism in greens and brown.

Composition and Structure. Although the term *aegirine* is the recommended name for $NaFe^{3+}Si_2O_6$, both *aegirine* and *acmite* have been used for pyroxenes of this composition. Aegirines show a wide range of composition, but in most species the replacement is according to $NaFe^{3+} \rightleftarrows Ca(Mg,Fe^{2+})$, which causes a complete series to augite, with an intermediate member known as *aegirine-augite*. Compositional zoning is very common in aegirine with compositions trending toward augite. Earlier crystallized material is richer in augite (in cores), and rims tend to be enriched in the aegirine component. The structure of aegirine is similar to that of other $C2/c$ pyroxenes (see Fig. 11.16).

Diagnostic Features. The slender prismatic crystals, brown to green color, and association are characteristic. However, it is not easily distinguished without optical tests.

Occurrence. Aegirine is a comparatively rare rock-forming mineral found chiefly in igneous rocks rich in Na and poor in SiO_2 such as nepheline syenite and phonolite. Associated with orthoclase, feldspathoids, augite, and soda-rich amphiboles. It is also found in some metamorphic rocks associated with glaucophane or riebeckite. It occurs in the nepheline syenites and related rocks of Norway, southern Greenland, and Kola Peninsula, Russia. In the United States it is found in fine crystals at Magnet Cove, Arkansas. In Canada at Mount St. Hilaire, Quebec.

Name. After Aegir, the Scandinavian sea god, the mineral being first reported from Norway.

Spodumene—$LiAlSi_2O_6$

Crystallography. Monoclinic; $2/m$. Crystals prismatic, frequently flattened on {100}. Deeply striated vertically. Crystals usually coarse with roughened faces; some very large. Occurs also in cleavable masses. Twinning on {100} common.

$C2/c$; $a = 9.52$, $b = 8.32$, $c = 5.25$ Å; $\beta = 110°28'$; $Z = 4$. ds: 4.38(5), 4.21(6), 2.93(10), 2.80(8), 2.45(6).

Physical Properties. *Cleavage* {110} perfect at angles of 87° and 93°. Usually a well-developed parting on {100}. **H** $6\frac{1}{2}-7$. **G** 3.15–3.20. *Luster* vitreous. *Color* white, gray, pink, yellow, green. Transparent to translucent. *Optics*: $(+)$; $\alpha = 1.660$, $\beta = 1.666$, $\gamma = 1.676$; $2V = 58°$; $Y = b$, $Z \wedge c = 24°$. $r < v$. Absorption: $X > Y > Z$. The clear lilac-colored variety is called *kunzite*, and the clear emerald-green variety *hiddenite*.

Composition and Structure. Li_2O 8.0, Al_2O_3 27.4, SiO_2 64.6%. A small amount of Na usually substitutes for Li. The structure of spodumene is the same as that of other $C2/c$ pyroxenes. The cell volume of spodumene is smaller than that of diopside, for example, because the larger Ca and Mg ions are substituted for by smaller Li and Al. This reduction in ionic sizes causes a somewhat closer packing of the SiO_3 chains.

Diagnostic Features. Characterized by its prismatic cleavage and {100} parting. The angle formed by one cleavage direction and the {100} parting resembles the cleavage angle of tremolite.

Alteration. Spodumene easily alters to other species, becoming dull. The alteration products include clay minerals, albite, *eucryptite*, $LiAlSiO_4$, muscovite, and microcline.

Occurrence. Spodumene is found almost exclusively in lithium-rich pegmatites (see Box 12.4). Although it is a comparatively rare mineral, it occasionally occurs in very large crystals. At the Etta Mine, Black Hills, South Dakota, crystals measuring as much as 40 feet in length and weighing many tons have been found. It formerly was mined as the chief source of lithium, but today other minerals such as petalite, lepidolite, and amblygonite are of equal or greater importance. Also much of the lithium of commerce is extracted as Li_2CO_3 from brines. The major producers of lithium are Russia, China, and Zimbabwe.

The principal countries for the production of gem spodumene are Brazil and Afghanistan. In the United States, beautiful crystals of kunzite have come from California, notably the Pala district in San Diego County. Hiddenite comes from Stony Point, Alexander County, North Carolina.

Use. As a gemstone and as a source of lithium. A major use of lithium is in grease to help it retain its lubricating properties over a wide range of temperatures. It is also used in aluminum, ceramics, storage batteries, air conditioning, and as a welding flux.

Names. *Spodumene* comes from a Greek word meaning *ash-colored*. *Hiddenite* is named for W. E. Hidden; *kunzite*, for G. F. Kunz.

Similar Species. *Eucryptite*, $LiAlSiO_4$, is a major source of lithium at Bikita, Zimbabwe.

Pyroxenoid Group

We will discuss in detail the following three members of the pyroxenoid group:

Wollastonite	$CaSiO_3$
Rhodonite	$MnSiO_3$
Pectolite	$Ca_2NaH(SiO_3)_3$

Wollastonite—CaSiO₃

Crystallography. Triclinic; $\bar{1}$. Rarely in tabular crystals with either {001} or {100} prominent. Commonly massive, cleavable to fibrous; also compact. *Pseudowollastonite*, $CaSiO_3$, is a polymorphic form stable above 1120°C; it is triclinic, pseudohexagonal, with different properties than wollastonite.

$P\bar{1}$; $a = 7.94$, $b = 7.32$, $c = 7.07$; $\alpha = 90°2'$, $\beta = 95°22'$, $\gamma = 103°26'$; $Z = 6$. ds: 3.83(8), 3.52(8), 3.31(8), 2.97(10), 2.47(6).

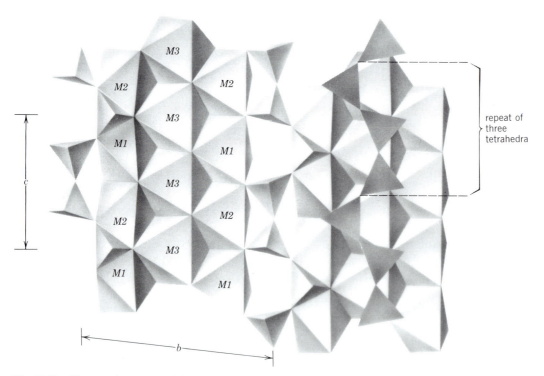

FIG. 12.40. The crystal structure of triclinic wollastonite. The tetrahedral silicate chain has a repeat of three tetrahedra. The Ca^{2+} ions are located in octahedral bands that are parallel to *c* and are three octahedra wide. Octahedral sites are identified as *M1*, *M2*, and *M3*. (Adopted from J. J. Papike, 1987, Chemistry of the rock-forming silicates: Ortho, ring, and single-chain structures. *Reviews of Geophysics and Space Physics* 25: 1483–1526.)

Physical Properties. *Cleavage* {100} and {001} perfect, {$\overline{1}$01} good giving splintery fragments elongated on *b*. **H** 5–5½. **G** 2.8–2.9. *Luster* vitreous, pearly on cleavage surfaces. May be silky when fibrous. *Color* colorless, white, or gray; see Plate VI, no. 3. Translucent. *Optics*: (−); α = 1.620, β = 1.632, γ = 1.634; 2V = 40°; *Y* near *b*, *X* ∧ *c* = 32°.

Composition and Structure. CaO 48.3, SiO$_2$ 51.7% for pure CaSiO$_3$. Most analyses are very close to the pure end member in composition, although considerable amounts of Fe and Mn, and lesser Mg, may replace Ca. The structure of wollastonite consists of infinite chains, parallel to the *c* axis, with a unit repeat of three twisted tetrahedra (see Figs. 11.20*b* and 12.40). Ca is in irregular octahedral coordination and links the SiO$_3$ chains. *Pseudowollastonite*, stable above 1120°C, has space group *P*$\overline{1}$ but a much larger unit cell (*Z* = 24 as compared to *Z* = 6 for wollastonite). The basic structure of pseudowollastonite is very similar to that of wollastonite.

Diagnostic Features. Characterized by its two perfect cleavages of about 84°. It resembles tremolite but is distinguished from it by the cleavage angle.

Occurrence. Wollastonite occurs chiefly as a contact metamorphic mineral in crystalline limestones, and forms by the reaction:

$$CaCO_3 + SiO_2 \rightarrow CaSiO_3 + CO_2\nearrow$$
calcite   quartz   wollastonite

It is associated with calcite, diopside, andradite, grossular, tremolite, plagioclase feldspar, vesuvianite, and epidote. During progressive metamorphism of siliceous dolomites the following approximate sequence of mineral formation is often found, beginning with the lowest-temperature product: talc→tremolite→diopside→forsterite→wollastonite→periclase→monticellite.

In places it may be so plentiful as to constitute the chief mineral of the rock mass. Such wollastonite rocks are found in the Black Forest, Germany; in Brittany, France; in Willsboro, New York; in California; and in Mexico. Crystals of the mineral are found at Csiklova in Romania; the Harz Mountains, Germany; and Chiapas, Mexico. In the United States found in New York at Diana, Lewis County, and St. Lawrence County. In California at Crestmore, Riverside County.

Use. Wollastonite is mined in those places where it constitutes a major portion of the rock mass and is used in the manufacture of tile.

Name. In honor of the English chemist, W. H. Wollaston (1766–1828).

RHODONITE—MnSiO$_3$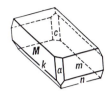

Crystallography. Triclinic; $\overline{1}$. Crystals commonly tabular parallel to {001} (Fig. 12.41); often rough with rounded edges. Commonly massive, cleavable to compact; in embedded grains.

P$\overline{1}$; *a* = 7.79, *b* = 12.47, *c* = 6.75 Å; α = 85°10′, β = 94°4′, γ = 111°29′; *Z* = 10. *d*s: 4.78(4), 3.15(5), 3.09(3), 2.98(8), 2.93(9), 2.76(10).

Physical Properties. *Cleavage* {110} and {1$\overline{1}$0} perfect. **H** 5½–6. **G** 3.4–3.7. *Luster* vitreous. *Color*

FIG. 12.41. Rhodonite.

rose-red, pink, brown, see Plate VI, no. 4; frequently with black exterior of manganese oxide. Transparent to translucent. *Optics*: (+); α = 1.716–1.733, β = 1.720–1.737, γ = 1.728–1.747; 2V = 60°–75°, *r* < *v*.

Composition and Structure. Rhodonite is never pure MnSiO$_3$ but always contains some Ca, with a maximum CaSiO$_3$ content of about 20 molecular percent. Fe^{2+} may replace Mn up to as much as 14 weight percent FeO (see Fig. 12.42). Zn may be present, and Zn-rich varieties are known as *fowlerite*. The structure of rhodonite consists of SiO$_3$ chains, parallel to the *c* axis, with a unit repeat of five twisted tetrahedra (see Fig. 11.20*c*). Layers of cations alternate with the chains. The structure is similar to that of wollastonite (see Fig. 12.40*b*), and pyroxmangite, (Mn,Fe)SiO$_3$.

Diagnostic Features. Characterized by its pink color and near 90° cleavages. Distinguished from rhodochrosite by its greater hardness.

Occurrence. Rhodonite occurs in manganese deposits and manganese-rich iron-formations, as a re-

FIG. 12.42. Extent of compositional fields for pyroxenoid and pyroxene compositions in the system CaSiO$_3$–FeSiO$_3$–MnSiO$_3$. The temperature and pressure conditions applicable to this diagram are about 600°C and 6 kilobars, respectively. Straight line segments (tielines) join naturally occurring mineral pairs. (Redrawn from P. E. Brown, E. J. Essene, and D. R. Peacor, 1980, Phase relations inferred from field data for Mn pyroxenes and pyroxenoids. *Contributions to Mineralogy and Petrology* 74: 417–25; see also *Pyroxenes, Reviews in Mineralogy*, complete reference at end of chapter.)

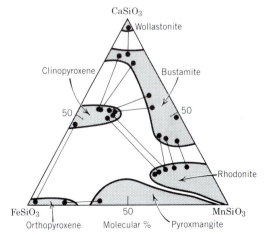

sult of metamorphic and commonly associated meta-somatic activity. It may form from rhodochrosite by the reaction:

$$\underset{\text{rhodochrosite}}{MnCO_3} + \underset{\text{quartz}}{SiO_2} \rightarrow \underset{\text{rhodonite}}{MnSiO_3} + CO_2 \nearrow$$

Rhodonite is found at Långban, Sweden, with other manganese minerals and iron ore; in large masses near Sverdlovsk in the Ural Mountains, Russia; and at Broken Hill, New South Wales, Australia. In the United States rhodonite occurs in good-sized crystals in crystalline limestone with franklinite, willemite, zincite, and so forth, at Franklin, New Jersey (Fig. 12.43).

Use. Some rhodonite is polished for use as an ornamental stone. This material is obtained chiefly from the Ural Mountains, Russia, and Australia.

Name. Derived from the Greek word for *a rose*, in allusion to the color.

Similar Species. *Pyroxmangite*, (Mn,Fe)SiO_3, is structurally very similar to rhodonite, but with a unit repeat of seven tetrahedra in the SiO_3 chain (see Fig. 11.20*d*). *Pyroxferroite*, Ca_{0.15}Fe_{0.85}SiO_3, isostructural with pyroxmangite, is a relatively common mineral in lunar lavas. *Bustamite*, (Mn,Ca,Fe)SiO_3, is very

FIG. 12.43.   Rhodonite, Franklin, New Jersey (Harvard Mineralogical Museum).

similar in structure to wollastonite. All of these pyroxenoids have extensive fields of solid solution, as shown in Fig. 12.42. *Tephroite*, Mn_2SiO_4, a red to gray mineral associated with rhodonite, is isostructural with olivine.

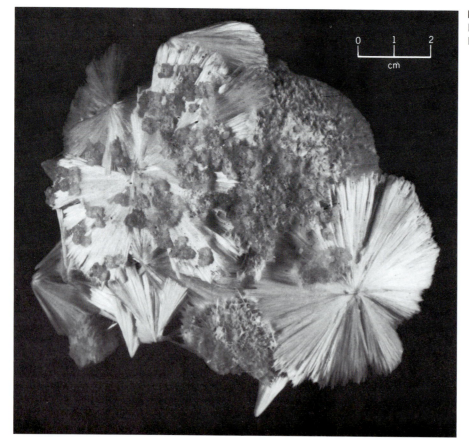

FIG. 12.44.   Pectolite, Paterson, New Jersey (Harvard Mineralogical Museum).

Pectolite—Ca$_2$NaH(SiO$_3$)$_3$

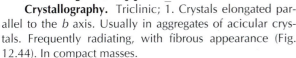

Crystallography. Triclinic; $\bar{1}$. Crystals elongated parallel to the *b* axis. Usually in aggregates of acicular crystals. Frequently radiating, with fibrous appearance (Fig. 12.44). In compact masses.

$P\bar{1}$; $a = 7.99$, $b = 7.04$, $c = 7.02$ Å; $\alpha = 90°31'$, $\beta = 95°11'$, $\gamma = 102°28'$; $Z = 2$. *d*s: 3.28(7), 3.08(9), 2.89(10), 2.31(7), 2.28(7).

Physical Properties. *Cleavage* {001} and {100} perfect. **H** 5. **G** 2.8±. *Luster* vitreous to silky. *Color* colorless, white, or gray. Transparent. *Optics:* (+); $\alpha = 1.595$, $\beta = 1.604$, $\gamma = 1.633$; 2V = 60°; $Z \approx b$, $X \wedge c = 19°$.

Composition and Structure. CaO 33.8, Na$_2$O 9.3, SiO$_2$ 54.2, H$_2$O 2.7%. In some pectolite Mn^{2+} substitutes for Ca. The structure of pectolite contains SiO$_3$ chains, parallel to the *b* axis, the unit repeat of which consists of three twisted tetrahedra (see Fig. 11.20*b*) similar to that found in wollastonite. The Ca ions are in octahedral coordination, and Na is present in very distorted octahedral coordination.

Diagnostic Features. Characterized by two directions of perfect cleavage, yielding sharp, acicular fragments that will puncture the skin if not handled carefully. Resembles wollastonite.

Occurrence. Pectolite is a secondary mineral similar in its occurrence to the zeolites. Found lining cavities in basalt, associated with various zeolites, prehnite, calcite, and so forth. Found at Bergen Hill and West Paterson, New Jersey. In Canada, at Asbestos and Mount St. Hilaire, Quebec.

Name. From the Greek word meaning *compact*, in allusion to its habit.

◢ AMPHIBOLE GROUP

We will describe in detail the following common amphiboles:

Anthophyllite	(Mg,Fe)$_7$Si$_8$O$_{22}$(OH)$_2$
Cummingtonite Series	
Cummingtonite	Fe$_2$Mg$_5$Si$_8$O$_{22}$(OH)$_2$
Grunerite	Fe$_7$Si$_8$O$_{22}$(OH)$_2$
Tremolite Series	
Tremolite	Ca$_2$Mg$_5$Si$_8$O$_{22}$(OH)$_2$
Actinolite	Ca$_2$(Mg,Fe)$_5$Si$_8$O$_{22}$(OH)$_2$
Hornblende	$X_{2-3}Y_5Z_8$O$_{22}$(OH)$_2$
Sodium Amphibole Group	
Glaucophane	Na$_2$Mg$_3$Al$_2$Si$_8$O$_{22}$(OH)$_2$
Riebeckite	Na$_2$Fe$_3^{2+}$Fe$_2^{3+}$Si$_8$O$_{22}$(OH)$_2$

Anthophyllite—(Mg,Fe)$_7$Si$_8$O$_{22}$(OH)$_2$

Crystallography. Orthorhombic; $2/m2/m2/m$. Rarely in distinct crystals. Commonly lamellar or fibrous.

Pnma; $a = 18.56$, $b = 18.08$, $c = 5.28$ Å; $Z = 4$. *d*s: 8.26(6), 3.65(4), 3.24(6), 3.05(10), 2.84(4).

Physical Properties. *Cleavage* {210} perfect; (210) $\wedge$ ($2\bar{1}0$) = 55°. **H** $5\frac{1}{2}$–6. **G** 2.85–3.2. *Luster* vitreous. *Color* gray to various shades of green and brown and beige; see Plate VI, no. 5. *Optics:* (−); $\alpha = 1.60$–1.69, $\beta = 1.61$–1.71, $\gamma = 1.62$–1.72. 2V = 70°–100°; $X = a$, $Y = b$. Absorption $Z > Y$ and X. Indices increase with Fe content.

Composition and Structure. Anthophyllite is part of a solid solution series from Mg$_7$Si$_8$O$_{22}$(OH)$_{22}$ to approximately Fe$_2$Mg$_5$Si$_8$O$_{22}$(OH)$_2$; at higher Fe contents the monoclinic cummingtonite structure results. *Gedrite* is an Al and Na-containing variety of anthophyllite, with an end member composition approximating Na$_{0.5}$(Mg,Fe^{2+})$_2$(Mg,Fe^{2+})$_{3.5}$(Al,Fe^{3+})$_{1.5}$Si$_6$Al$_2$O$_{22}$(OH)$_2$, with Na in the *A* site of the structure. At moderate temperatures a miscibility gap exists between anthophyllite and gedrite as shown by coexisting anthophyllite and gedrite grains. The structures of anthophyllite and gedrite are similar, both with orthorhombic space group *Pnma*. The relationship of the unit cell of orthoamphibole to that of clinoamphibole is shown in Fig. 11.25*a*.

Diagnostic Features. Characterized by its clove-brown color but unless in crystals cannot be distinguished from other amphiboles such as cummingtonite or grunerite without optical or X-ray tests.

Occurrence. Anthophyllite is a metamorphic product of Mg-rich rocks such as ultrabasic igneous rocks and impure dolomitic shales. It is common in cordierite-bearing gneisses and schists. It may also form as a retrograde product rimming relict orthopyroxenes and olivine (see Fig. 12.35). It occurs at Kongsberg, Norway, and in many localities in southern Greenland. In the United States it is found at several localities in Pennsylvania, in southwestern New Hampshire and central Massachusetts, in the Gravelly Range and Tobacco Root Mountains of southwestern Montana, and at Franklin, North Carolina.

Name. From the Latin *anthophyllum*, meaning *clove*, in allusion to the clove-brown color.

CUMMINGTONITE—(Mg,Fe)$_7$Si$_8$O$_{22}$(OH)$_2$
GRUNERITE—Fe$_7$Si$_8$O$_{22}$(OH)$_2$

Crystallography. Monoclinic; $2/m$. Rarely in distinct crystals. Commonly fibrous or lamellar, often radiated.

$C2/m$; for grunerite, $a = 9.59$, $b = 18.44$, $c = 5.34$, $\beta = 102°0'$; cell lengths decrease with increasing Mg (see Fig. 12.45); $Z = 2$. *d*s: 9.21(5), 8.33(10), 3.07(8), 2.76(9), 2.51(6).

Physical Properties. *Cleavage* {110} perfect. **H** $5\frac{1}{2}$–6. **G** 3.1–3.6. *Luster* silky; fibrous; see Plate VI,

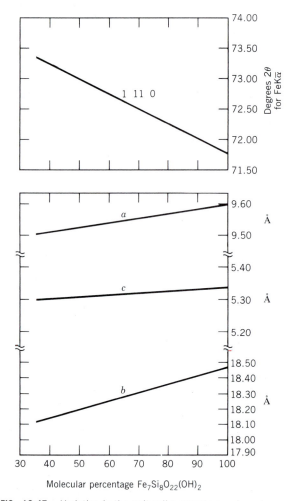

FIG. 12.45. Variation in the unit cell parameters *a*, *b*, and *c* as a function of composition in the cummingtonite-grunerite series. The topmost line relates the angular (2θ) position of the 1, 11, 0 X-ray diffraction peak (as recorded on a diffractometer pattern, or X-ray photograph; see Chapter 7) as a function of composition in the cummingtonite-grunerite series (using FeKα radiation). (From C. Klein and D. R. Waldbaum, 1967, X-ray crystallographic properties of the cummingtonite-grunerite series. *Journal of Geology* 75: 379–92.)

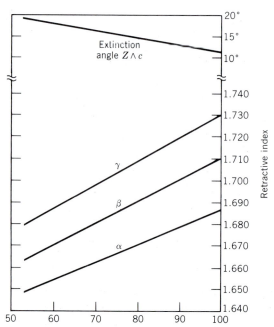

FIG. 12.46. Variation in refractive indices and extinction angle for members of the monoclinic cummingtonite-grunerite series. (From C. Klein, 1964, Cummingtonite-grunerite series: A chemical optical and X-ray study. *American Mineralogist* 49: 963–82.)

range up to a maximum of 0.4 and 0.9 weight percent, respectively. The structure (see Fig. 11.22) of the majority of members of the cummingtonite–grunerite series is $C2/m$, like that of tremolite; however, some Mg-rich cummingtonites have $P2_1/m$ symmetry.

Diagnostic Features. Characterized by light brown color and needlelike, often radiating habit. May be impossible to distinguish from anthophyllite or gedrite without optical or X-ray tests.

Occurrence. Cummingtonite is a constituent of regionally metamorphosed rocks and occurs in amphibolites. It commonly coexists with hornblende or actinolite (see tielines in Fig. 11.21). Mg-rich cummingtonite can also coexist with anthophyllite, because of a small miscibility gap between anthophyllite and the Mg-rich part of the cummingtonite–grunerite series (see Fig. 11.25*b*). Cummingtonite phenocrysts have been reported in some igneous rocks such as dacites. Mn-rich varieties occur in metamorphosed manganese-rich units. Grunerites are characteristic of metamorphosed iron-formations in the Lake Superior region and the Labrador Trough. Upon prograde metamorphism cummingtonite and grunerite give way to members of the orthopyroxene or olivine series (see Fig. 12.35).

Use. *Amosite*, a very rare asbestiform variety of grunerite, was mined only in the eastern part of the Transvaal Province of South Africa (see Fig. 12.47). The varietal name is derived from the word *Amosa*,

no. 6. *Color* various shades of light brown. Translucent; will transmit light on thin edges. *Optics*: (−) for grunerite; (+) for cummingtonite; α = 1.65–1.69; β = 1.67–1.71; γ = 1.69–1.73; 2V large; Y = *b*, Z ∧ *c* = 13°–20° (see Fig. 12.46). *r* < *v* for cummingtonite; *r* > *v* for grunerite. Essentially nonpleochroic.

Composition and Structure. The cummingtonite-grunerite series extends from approximately $Fe_2Mg_5Si_8O_{22}(OH)_2$ to the end member $Fe_7Si_8O_{22}(OH)_2$. Members with Mg < Fe (atomic percentage) are referred to as cummingtonite; those with Fe > Mg as grunerite (in the literature this division is often taken at 30 atomic percent Fe). Mn^{2+} can be present to as much as $Mn_2Mg_5Si_8O_{22}(OH)_2$, with probably most of the Mn in the M4 structure site. Al_2O_3 and CaO

FIG. 12.47. Amosite, an asbestiform variety of the amphibole grunerite (in the trade this is also known as "brown asbestos"), Penge, Transvaal Province, Republic of South Africa (compare with Figs. 12.50 and 12.52).

an acronym for the company "Asbestos Mines of South Africa." Few data are as yet available on the possible health hazards of this asbestos type (see Ross 1981 and 1982; complete references at end of chapter). For comparison and more extensive discussion of asbestos, see *chrysotile* and *crocidolite*.

Names. Cummingtonite is named after Cummington, Massachusetts, and grunerite after Grüner, nineteenth century French chemist.

TREMOLITE—Ca₂Mg₅Si₈O₂₂(OH)₂

TREMOLITE—$Ca_2Mg_5Si_8O_{22}(OH)_2$
ACTINOLITE—$Ca_2(Mg,Fe)_5Si_8O_{22}(OH)_2$

Crystallography. Monoclinic; 2/m. Crystals usually prismatic. The termination of the crystals is almost always formed by the two faces of a low {0kl} prism (similar to hornblende, see Fig. 12.49). Tremolite is often bladed and frequently in radiating columnar aggregates. In some cases in silky fibers. Coarse to fine granular. Compact.

$C2/m$; $a = 9.84$, $b = 18.05$, $c = 5.28$; $\beta = 104°42'$; $Z = 2$. ds: 8.38(10), 3.27(8), 3.12(10), 2.81(5), 2.71(9).

Physical Properties. *Cleavage* {110} perfect; (110) $\wedge$ ($1\bar{1}0$) = 56° often yielding a splintery surface. **H** 5–6. **G** 3.0–3.3 *Luster* vitreous; often with silky sheen in the prism zone. *Color* varying from white to green in actinolite. The color deepens and the specific gravity increases with increase in Fe content, see Plate VI, no. 7. Transparent to translucent. A felted aggregate of tremolite fibers goes under the name of *mountain leather* or *mountain cork*. A tough, compact variety that supplies much of the material known as *jade* is called *nephrite* (see jadeite). *Optics*: (−); α = 1.60–1.68, β = 1.61–1.69, γ = 1.63–1.70; 2V = 80°; Y = b, Z $\wedge$ c = 15°. Indices increase with iron content.

Composition and Structure. A complete solid solution exists from tremolite to ferroactinolite, $Ca_2Fe_5Si_8O_{22}(OH)_2$ (see Fig. 11.21). The name *actinolite* is used for intermediate members. In high-temperature metamorphic and in igneous occurrences a complete series exists from the tremolite-ferroactinolite series to hornblende as a result of a wide range of Al substitution for Si, and concomitant cation replacements elsewhere in the structure. The names *tremolite* and *ferroactinolite* are generally used to refer to the Al-poor members of this large composi-

FIG. 12.48. Comparison of the thermal stability limits of tremolite and ferractinolite. (After W. G. Ernst, 1968, *Amphiboles.* New York: Springer-Verlag, p. 58.)

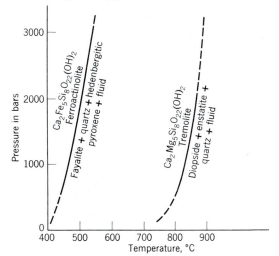

FIG. 12.49. Hornblende crystals.

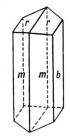
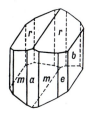

tional range. In lower-temperature occurrences tremolite (or actinolite) may coexist with hornblende as the result of a miscibility gap between tremolite (or actinolite) and hornblende. The structures of the members of the tremolite-ferroactinolite series are of the $C2/m$ type (see Figs. 11.22, 11.23, and page 458).

Diagnostic Features. Characterized by slender prisms and good prismatic cleavage. Distinguished from pyroxenes by the cleavage angle and from hornblende by lighter color.

Occurrence. Tremolite is most frequently found in metamorphosed dolomitic limestones where it forms according to this reaction:

$$5CaMg(CO_3)_2 + 8SiO_2 + H_2O$$

dolomite   quartz

$$\rightarrow Ca_2Mg_5Si_8O_{22}(OH)_2 + 3CaCO_3 + 7CO_2\nearrow$$

tremolite   calcite

At higher temperatures tremolite is unstable and yields to diopside (see Fig. 12.48). Actinolite is a characteristic mineral of the greenschist facies of metamorphism. It occurs also in glaucophane schists in coexistence with quartz, epidote, glaucophane, pumpellyite, and lawsonite.

Notable localities for crystals of tremolite are Ticino, Switzerland; in the Tyrol; and in Piemonte, Italy. In the United States from Russell, Gouverneur, Amity, Pierrepont, DeKalb, and Edwards, New York. Crystals of actinolite are found at Greiner, Zillerthal, Tyrol, Austria. The original jade was probably nephrite that came from the Khotan and Yarkand Mountains of Chinese Turkestan. Nephrite has been found in New Zealand, Mexico, and Central America; and, in the United States, near Lander, Wyoming.

Use. The compact variety, *nephrite*, is used as an ornamental and gem material.

Name. Tremolite is derived from the Tremola Valley near St. Gotthard, Switzerland. Actinolite comes from two Greek words meaning *a ray* and *stone*, in allusion to its frequently somewhat radiated habit.

HORNBLENDE—
(Ca,Na)$_{2-3}$(Mg,Fe,Al)$_5$Si$_6$(Si,Al)$_2$O$_{22}$(OH)$_2$

Crystallography. Monoclinic; $2/m$. Crystals prismatic, usually terminated by {011}. The vertical prism zone shows, in addition to the prism faces, usually {010}, and more rarely {100} (see Fig. 12.49). May be columnar or fibrous; coarse- to fine-grained.

$C2/m$; $a = 9.87$, $b = 18.01$, $c = 5.33$ Å; $\beta = 105°44'$; $Z = 2$. ds: 3.38(9), 3.29(7), 3.09(10), 2.70(10), 2.59(7).

Physical Properties. *Cleavage* {110} perfect. **H** 5–6. **G** 3.0–3.4. *Luster* vitreous; fibrous; fibrous varieties often silky. *Color* various shades of dark green to black; see Plate VI, no. 8. Translucent; will transmit light on thin edges. *Optics*: (−); $\alpha = 1.61–1.71$, $\beta = 1.62–1.72$, $\gamma = 1.63–1.73$; $2V = 30°–90°$; $Y = b$, $Z \wedge c = 15°–34°$. $r > v$ usually; may be $r < v$. Pleochroism X yellow-green, Y olive-green, Z deep green.

Composition and Structure. What passes under the name of hornblende is in reality a complex and large compositional range with variations in the ratios Ca/Na, Mg/Fe^{2+}, Al/Fe^{3+}, Al/Si, and OH/F. A generalized formula for a "common" hornblende is given above, and an example of the recalculation of an amphibole analysis is given in Table 3.18. At elevated temperatures hornblende shows a complete series toward tremolite-ferroactinolite, but at lower temperatures a miscibility gap exists. In some volcanic rocks oxyhornblendes are found with considerable Fe^{3+}/Fe^{2+} ratios and low OH contents; an example composition is NaCa$_2$Fe$_4^{2+}$Fe^{3+}(AlSi$_7$)O$_{23}$(OH). The structure of hornblende is similar to that of $C2/m$ tremolite (see Figs. 11.22 and 11.23). Because of the miscibility gap between Ca-amphiboles and Mg Fe amphiboles, hornblende often contains grunerite exsolution lamellae (see Fig. 4.42).

Diagnostic Features. Crystal form and cleavage angles serve to distinguish hornblende from dark pyroxenes. Usually distinguished from other amphiboles by its dark color.

Occurrence. Hornblende is an important and widely distributed rock-forming mineral, occurring in both igneous and metamorphic rocks; it is particularly characteristic of some medium-grade metamorphic rocks known as *amphibolites* in which hornblende and associated plagioclase are the major constituents. It characteristically alters from pyroxene both during the late magmatic stages of crystallization of igneous rocks and during metamorphism. Hornblende is a common constituent of syenites and diorites.

Name. From an old German word for any dark prismatic mineral occurring in ores but containing no recoverable metal.

Similar Species. Because of the large chemical variation in the hornblende series, many names have been proposed for members on the basis of chemical composition and physical and optical properties. The most common names on the basis of chemical composition are *edenite*, *pargasite*, *hastingsite*, and *tschermakite*.

Glaucophane—Na$_2$Mg$_3$Al$_2$Si$_8$O$_{22}$(OH)$_2$
Riebeckite—Na$_2$Fe$_3^{2+}$Fe$_2^{3+}$Si$_8$O$_{22}$(OH)$_2$

Crystallography. Monoclinic; $2/m$. In slender acicular crystals; frequently aggregated; riebeckite sometimes asbestiform.

FIG. 12.50. Crocidolite, an asbestiform variety of the amphibole riebeckite (in the trade this is also known as "blue asbestos"), Kuruman, Cape Province, Republic of South Africa (compare with Figs. 12.47 and 12.52).

$C2/m$; $a = 9.58$, $b = 17.80$, $c = 5.30$ Å; $\beta = 103°48'$; $Z = 2$. ds: 8.42(10); 4.52(5), 3.43(6), 3.09(8), 2.72(10).

Physical Properties. *Cleavage* {110} perfect. **H** 6. **G** 3.1–3.4. *Luster* vitreous. *Color* blue to lavender-blue to black, darker with increasing iron content. *Streak* white to light blue. Translucent. *Optics*: (−); $\alpha = 1.61–1.70$, $\beta = 1.62–1.71$, $\gamma = 1.63–1.72$; $2V = 40°–90°$; $Y = b$, $Z \wedge c = 8°$. Pleochroism in blue $X < Y < Z$.

Composition and Structure. Very few glaucophanes are close to the end member composition, $Na_2Mg_3Al_2Si_8O_{22}(OH)_2$, because of some substitution by Fe^{2+} for Mg and Fe^{3+} for Al. Riebeckite analyses do not conform well with the end member formulation because the sum of $X + Y$ cations is frequently larger than 7, with Na entering the normally vacant A site. A partial series exists between glaucophane and riebeckite, with intermediate compositions known as *crossite*. The structures of both sodic amphiboles are similar to that of $C2/m$ tremolite (see Fig. 11.22).

Diagnostic Features. Glaucophane and riebeckite are characterized by their generally fibrous habit and blue color.

Occurrence. *Glaucophane* is found only in metamorphic rocks such as schists, eclogite, and marble. The occurrences of glaucophane reflect low-temperature, relatively high-pressure metamorphic conditions, in association with jadeite, lawsonite, and aragonite. It is a major constituent of glaucophane schists in the Franciscan Formation of California. It is also found in metamorphic rocks from Mexico, Japan, Taiwan, Indonesia, and eastern Australia. *Riebeckite* occurs most commonly in igneous rocks such as granites, syenites, nepheline syenites, and related pegmatites. It is a conspicuous mineral in the granite of Quincy, Massachusetts. It is present in some schists of regional metamorphic origin. The asbestiform variety of riebeckite is known as *crocidolite* (see Fig. 12.50).

Use. Crocidolite made up about 4% of the total world production of asbestos. All production from mines in the Cape Province, South Africa, was stopped in 1995. There are extensive reserves of crocidolite in the Hamersley Range of western Australia, but crocidolite has not been mined there since 1966. The crocidolite is closely interbanded with Precambrian banded iron-formation sequences, both in South Africa and western Australia. Medical studies have shown (see Ross 1981 and 1982; also Mossman and Gee 1989, and Mossman et al. 1990) that crocidolite is a much greater health hazard than chrysotile asbestos. For comparison, see discussions of *chrysotile* and *amosite*.

In many places in South Africa there has been a pseudomorphous replacement of oxidized crocidolite by quartz. The preservation of the fibrous texture makes it an attractive ornamental material with a chatoyancy. This is widely used for jewelry under the name of *tiger's eye*.

Name. Glaucophane is from the two Greek words meaning *bluish* and *to appear*. Riebeckite is in honor of E. Riebeck.

Similar Species. *Arfvedsonite*, $Na_3Fe_4^{2+}Fe^{3+}Si_8O_{22}(OH)_2$, is a deep green amphibole similar to riebeckite in occurrence. It is commonly associated with aegirine, or aegirine-augite. In the structure of arfvedsonite the A site is completely or almost completely filled by Na (see Table 11.2).

PHYLLOSILICATES

We will describe in detail the following common phyllosilicates:

Serpentine Group	
Antigorite	
Chrysotile }	$Mg_3Si_2O_5(OH)_4$
Lizardite	
Clay Mineral Group	
Kaolinite	$Al_2Si_2O_5(OH)_4$
Talc	$Mg_3Si_4O_{10}(OH)_2$
Pyrophyllite	$Al_2Si_4O_{10}(OH)_2$
Mica Group	
Muscovite	$KAl_2(AlSi_3O_{10})(OH)_2$
Phlogopite	$KMg_3(AlSi_3O_{10})(OH)_2$
Biotite	$K(Mg,Fe)_3(AlSi_3O_{10})(OH)_2$
Lepidolite	$K(Li,Al)_{2-3}(AlSi_3O_{10})(OH)_2$
Margarite	$CaAl_2(Al_2Si_2O_{10})(OH)_2$
Chlorite Group	
Chlorite	$(Mg,Fe)_3(Si,Al)_4O_{10}(OH)_2\cdot(Mg,Fe)_3(OH)_6$

We will also cover three minerals closely related to the phyllosilicate groups:

Apophyllite	$KCa_4(Si_4O_{10})_2F \cdot 8H_2O$
Prehnite	$Ca_2Al(AlSi_3O_{10})(OH)_2$
Chrysocolla	$Cu_4H_4Si_4O_{10}(OH)_8$

Serpentine Group

ANTIGORITE, LIZARDITE AND CHRYSOTILE—$Mg_3Si_2O_5(OH)_4$

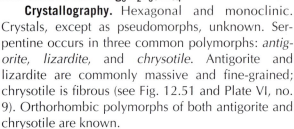

Crystallography. Hexagonal and monoclinic. Crystals, except as pseudomorphs, unknown. Serpentine occurs in three common polymorphs: *antigorite*, *lizardite*, and *chrysotile*. Antigorite and lizardite are commonly massive and fine-grained; chrysotile is fibrous (see Fig. 12.51 and Plate VI, no. 9). Orthorhombic polymorphs of both antigorite and chrysotile are known.

Antigorite either Pm or $P2/m$; lizardite-1T, $P31m$; lizardite-2H_1, $P6_3cm$; chrysotile, orthorhombic or monoclinic, depending on polytype. For monoclinic antigorite: $a = 5.30$, $b = 9.20$, $c = 7.46$ Å; $\beta = 91°24'$; $Z = 2$. ds: 7.30(10), 3.63(8), 2.52(2), 2.42(2), 2.19(1). For monoclinic chrysotile: $a = 5.34$, $b = 9.25$, $c = 14.65$ Å; $\beta = 93°16'$; $Z = 4$.

Physical Properties. **H** 3–5, usually 4. **G** 2.5–2.6. *Luster* greasy, waxlike in the massive varieties, silky when fibrous. *Color* often variegated, showing mottling in lighter and darker shades of green. Translu-

cent. *Optics*: (−); *Chrysotile*: $\alpha = 1.532$–1.549 (across fiber), $\gamma = 1.545$–1.556 (parallel to fiber length). *Antigorite*: nearly isotropic, $n = 1.55$–1.56.

The characteristic morphology of all asbestos minerals, in their natural form, is a parallel-sided fiber with a length-to-width ratio (referred to as an *aspect ratio*) in excess of 100 : 1 (Champness, Cliff, and Lorimer 1976). Chrysotile asbestos occurs in very narrow tubular fibers that are often hollow (see Figs. 11.38 and 11.39). Other asbestos minerals are *crocidolite* (Fig. 12.50), an asbestiform variety of the amphibole riebeckite, and *amosite* (Fig. 12.47), an asbestiform variety of the amphibole grunerite. The term *asbestiform* refers to minerals that are mined as asbestos and possess fibrosity typical of asbestos— that is, with small fiber thickness, flexibility, and separability.

Composition and Structure. Close to $Mg_3Si_2O_5(OH)_4$ with MgO 43.0, SiO_2 44.1, and H_2O 12.9%. Fe and Ni may substitute to some extent for Mg and Al for some Si. The idealized structure of lizardite is shown in Fig. 11.34 in terms of trioctahedral *t–o* layers, in analogy with the dioctahedral kaolinite structure. Because of a misfit between the *t* and *o* layers the antigorite structure consists of finite and corrugated layers parallel to {001} (see Fig. 11.37a and page 469). In chrysotile, this same mismatch is responsible for the curvature of the *t–o* layers forming cylindrical tubes (see Fig. 11.37b and page 469). Much of the matrix material of serpentine, known as *lizardite*, is extremely

FIG. 12.51. Veins of chrysotile asbestos in serpentine. Thetford, Quebec, Canada.

fine-grained and has a platy morphology as seen under the high magnification of the electron microscope (see Fig. 11.39).

Diagnostic Features. Recognized by its variegated green color and its greasy luster or by its fibrous nature.

Occurrence. Serpentine is a common mineral and is widely distributed: usually as an alteration of magnesium silicates, especially olivine, pyroxene, and amphibole. From forsterite, it may form by this reaction:

$$2Mg_2SiO_4 + 3H_2O \rightarrow Mg_3Si_2O_5(OH)_4 + Mg(OH)_2$$
forsterite   serpentine   brucite

It is frequently associated with magnesite, chromite, and magnetite. Found in both igneous and metamorphic rocks, frequently in disseminated particles, in places in such quantity as to make up practically the entire rock mass. Serpentine, as a rock name, is applied to such rock masses made up mostly of a mixture of fine-grained antigorite and lizardite. The asbestiform variety, *chrysotile*, as the principal asbestos mineral, is mined extensively. More than half of the world's production comes from deposits in Russia, just east of the Ural Mountains. Although for many years asbestos has been mined from large deposits in southeastern Quebec, making Canada the leading world producer, it is now second to Russia; South Africa was ranked third in world production of asbestos minerals. In the United States there is a small production of chrysotile from the Coalinga area, San Benito Mountains, California.

Use. Massive serpentine, which is translucent and of a light to dark green color, is often used as an ornamental stone and may be valuable as building material. Mixed with white marble and showing beautiful variegated coloring, it is called *verde antique* marble. A transparent yellow-green variety, *bowenite*, is used as a jade substitute.

The use of chrysotile, as asbestos, depends on its fibrous, flexible nature, which allows it to be made into felt and woven into cloth and other fabrics, and upon its incombustibility and slow conductivity of heat. Asbestos products, therefore, were used for fireproofing and as an insulation material against heat and electricity.

About 90% of the past and 95% of the present world production of asbestos was or is the chrysotile form (Fig. 12.52). This is known as "white asbestos." The remaining 5% of world production consisted of crocidolite (the asbestiform variety of the amphibole riebeckite), known as "blue asbestos," and amosite (the asbestiform variety of the amphibole grunerite), known as "brown asbestos." Other types of asbestos are of little economic importance.

Over the last many years there has been much attention given, in newspapers and on television, to the health hazards posed by asbestos. In such reports, no distinction is made between the various asbestos types, and furthermore, the health hazards that have been assessed in the dusty workplaces of earlier years (occupational settings such as sites of asbestos mining and asbestos board manufacturing) are linearly extrapolated to those in nonoccupational settings (e.g.,

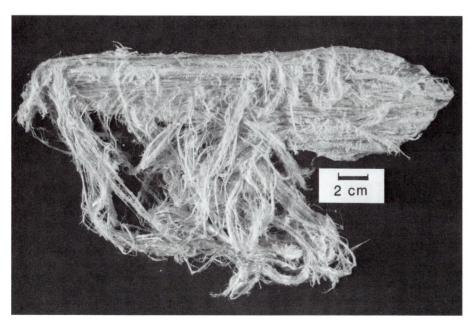

FIG. 12.52. Chrysotile, asbestiform variety of serpentine, Thetford, Quebec, Canada (compare with varieties of amphibole asbestos, Figs. 12.47 and 12.50).

schoolrooms and office buildings with asbestos ceiling and floor tiles). Medical studies on the pathogenicity of the different forms of asbestos (Mossman et al., 1990; Mossman and Gee, 1989; Ross, 1981, 1984; for complete references see the end of this chapter) show that "blue asbestos" (crocidolite, the asbestiform variety of the amphibole riebeckite) poses a much greater health hazard (in *occupational settings*) than chrysotile. Occupational exposure to asbestos can cause the following types of disorders; asbestosis, lung cancer, mesothelioma (cancer of the plural and peritoneal membranes), and benign changes in the pleura (Mossman et al. 1990). Crocidolite fibers appear to be most pathogenic, especially with respect to mesothelioma. Smoking is a strong contributor to the incidence of lung cancer (Ross 1984).

The main asbestos type used in U.S. buildings is chrysotile. Available data do not support the concept that low-level (*nonoccupational*) exposure to asbestos is a health hazard in buildings and schools (Mossman et al. 1990; see also Klein, C., 1993, Rocks, Minerals and a Dusty World in *Health Effects of Mineral Dusts. Reviews in Mineralogy*, vol. 28, Mineralogical Society of America, Washington, D.C., pp. 7–59). Furthermore, a low-level background of fibers is present globally, in the air and water, as a result of natural weathering processes of rocks with fibrous minerals. Abelson (1990) stated: "We live on a planet on which there is an abundance of serpentine- and amphibole-containing rocks. Natural processes have been releasing fibers throughout Earth history, we breathe about 1 million fibers per year."

Name. Serpentine refers to the green, serpentlike cloudings of the massive variety. Antigorite after Antigorio in Italy and chrysotile from the Greek words for *golden* and *fiber*.

Similar Species. *Greenalite*, $(Fe,Mg)_3Si_2O_5(OH)_4$, may be considered as the Fe-rich analogue of the serpentine composition. It has a modulated structure that is very different from that of antigorite, for example; it is fairly common in unmetamorphosed Precambrian iron-formations. *Garnierite*, $(Ni,Mg)_3Si_2O_5(OH)_4$, is an apple-green Ni-bearing serpentine formed as an alteration product of Ni-rich peridotites. It is mined as a nickel ore in New Caledonia, Russia, and Australia. In the United States it is found at Riddle, Oregon.

Clay Mineral Group

Clay is a rock term, and like most rocks it is made up of a number of different minerals in varying proportions. The term *clay* refers to a naturally occurring material composed primarily of fine-grained miner-als, which is generally plastic at appropriate water contents and will harden when dried or fired. With the use of X-ray techniques, clays have been shown to consist mainly of a group of crystalline substances known as the *clay minerals*. They are all essentially hydrous aluminum layer silicates. In some, Mg or Fe substitute in part for Al and alkalies or alkaline earths may be present as essential constituents. Although a clay may be made up of a single clay mineral, there are usually several mixed with other minerals such as feldspar, quartz, carbonates, and micas.

KAOLINITE—$Al_2Si_2O_5(OH)_4$

Crystallography. Triclinic; 1. In very minute, thin, rhombic or hexagonal plates (Fig. 12.53). Usually in claylike masses, either compact or friable.

$C\bar{1}$; $a = 5.14$. $b = 8.93$, $c = 7.37$ Å; $\alpha = 91°48'$; $\beta = 104°30'$, $\gamma = 90°$; $Z = 2$. ds: 7.15(10), 3.57(10), 2.55(8), 2.49(9), 2.33(10).

Physical Properties. *Cleavage* {001} perfect. **H** 2. **G** 2.6. *Luster* usually dull earthy; crystal plates pearly. *Color* white, see Plate VII, no. 1. Often variously colored by impurities. Usually unctuous and plastic. *Optics*: (−); $\alpha = 1.553–1.565$, $\beta = 1.559–1.569$, $\gamma = 1.560–1.570$; $2V = 24°–50°$. $r > v$.

Composition and Structure. Kaolinite shows little compositional variation; for $Al_2Si_2O_5(OH)_4$, Al_2O_3 39.5, SiO_2 46.5, H_2O 14.0%. The dioctahedral structure of kaolinite is shown schematically in Figs. 11.31a and 11.34. It consists of an Si_2O_5 sheet bonded to a gibbsite sheet. The clay minerals *dickite* and *nacrite* are chemically similar to kaolinite but show a *t–o* stacking different from that in kaolinite.

Diagnostic Features. Recognized usually by its claylike character, but without X-ray tests it is impossible to distinguish from the other clay minerals of similar composition which collectively make up *kaolin*.

Occurrence. Kaolinite is a common mineral, the chief constituent of kaolin or clay. It is always a secondary mineral formed by weathering or hydrothermal alteration of aluminum silicates, particularly feldspar. It is found mixed with feldspar in rocks that are undergoing alteration; in places it forms entire deposits where such alteration has been carried to completion. As one of the common products of the decomposition of rocks it is found in soils, and transported by water it is deposited, mixed with quartz and other materials, in lakes, and so on, in the form of beds of clay. Good pseudomorphs after potassium feldspar occur in Cornwall, England.

Use. Clay is one of the most important of the natural industrial substances (see also Box 12.2). It is available in every country of the world and is com-

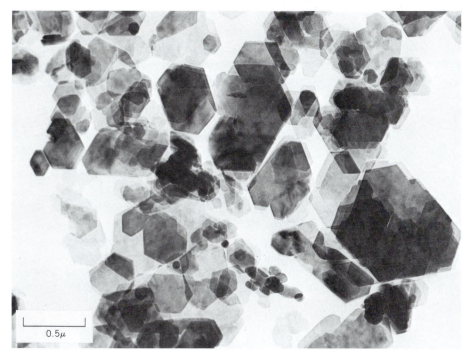

FIG. 12.53. Transmission electron micrograph of well-crystallized kaolinite from Georgia (courtesy of Dr. Kenneth M. Towe, Smithsonian Institution, Washington, D.C.).

mercially produced in nearly every state in the United States. Many and varied products are made from it which include common brick, paving brick, drain tile, and sewer pipe. The commercial users of clay recognize many different kinds having slightly different properties, each of which is best suited for a particular purpose. High-grade clay, which is known as *china clay* or *kaolin*, has many uses in addition to the manufacture of china and pottery. Its largest is as a filler in paper, but it is also used in the rubber industry and in the manufacture of refractories.

The chief value of clay for ceramic products lies in the fact that when wet it can be easily molded into any desired shape, and when it is heated, part of the combined water is driven off, producing a hard, durable substance.

Name. Kaolinite is derived from *kaolin*, which is a corruption of the Chinese *kauling*, meaning high ridge, the name of a hill near Jauchu Fa, where the material is obtained.

Similar Species. *Dickite* and *nacrite* are of the same composition as kaolinite but differ somewhat in their structure; they are less important constituents of clay deposits.

Halloysite has two forms: one with kaolinite composition, $Al_2Si_2O_5(OH)_4$, the other with composition $Al_2Si_2O_5(OH)_4 \cdot 2H_2O$. The second type dehydrates to the first with loss of interlayer water molecules.

The *smectite* group comprises a number of clay minerals composed of *t–o–t* layers of both dioctahe-dral and trioctahedral type. The outstanding characteristic of members of this group is their capacity to absorb water molecules between the sheets, thus producing marked expansion of the structure (see Fig. 11.36 for the similar structure of vermiculite; see also Box 12.2). The dioctahedral members are *montmorillonite*, *beidellite*, and *nontronite*; the trioctahedral members are *hectorite* and *saponite*.

Montmorillonite is the dominant clay mineral in *bentonite*, altered volcanic ash (see also Fig. 12.54). Bentonite has the unusual property of expanding several times its original volume when placed in water. This property gives rise to interesting industrial uses. Most important is as a drilling mud in which the montmorillonite is used to give the fluid a viscosity several times that of water. It is also used for stopping leakage in soil, rocks, and dams.

Illite is an alkali-deficient mica near the muscovite composition. The illites differ from the micas in having less substitution of Al for Si, in containing more water, and in having K partly replaced by Ca and Mg. Illite is the chief constituent in many shales.

TALC—$Mg_3Si_4O_{10}(OH)_2$

Crystallography. Triclinic; $\bar{1}$. Crystals rare. Usually tabular with rhombic or hexagonal outline. Foliated and in radiating foliated groups. When compact and massive, known as *steatite* or *soapstone*.

$C\bar{1}$; $a = 5.29$, $b = 9.17$, $c = 9.46$ Å; $\alpha = 90°28'$; $\beta = 98°41'$, $\gamma = 90°5'$; $Z = 4$. ds: 9.34(10), 4.66(9), 3.12(10), 2.48(7), 1.870(4).

BOX 12.2
CLAY MINERALS AND SOME OF THEIR APPLICATIONS

Commercial clays, or clays utilized as raw materials in manufacturing, are among the most important nonmetallic mineral resources. The various applications of **kaolinite** are described under "use" as part of the systematic descriptions of kaolionite.

Two other clay minerals that have many commercial applications are **montmorillonite** and **vermiculite**. Montmorillonite is the major component of bentonite which is a soft, plastic, light-colored rock that also contains some colloidal silica. It is the result of devitrification and accompanying chemical alteration of glassy igneous material, usually tuff or volcanic ash. See Fig. 12.54 for an illustration that shows this type of reaction. Montmorillonite has the following idealized formula: $(Na,Ca)_{0.3}(Al,Mg)_2Si_4O_{10} \cdot nH_2O$. This clay has the ability to swell as a result of the incorporation of water molecules between the $t–o–t$ sheets, in association with the interlayer cations, Na^+ and Ca^{2+}. When montmorillonite is heated in air, the interlayer water is driven off, thereby causing the collapse of the structure around the remaining interlayer cations. This process happens spontaneously in nature during burial metamorphism, leading to water expulsion and layer collapse, thereby transforming the montmorillonite structure to an illite-like structure.

The montmorillonite structure is always somewhat unbalanced, resulting

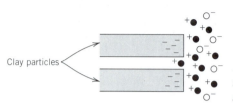

Clay particles

Schematic illustration of cations in solution, or in a water-rich soil, near a negatively charged clay particle

in an overall somewhat negative charge that is balanced by exchangeable cations that are absorbed around the edges of the fine clay particles (see diagram). Such cation absorbance, and the associated cation-exchange capacity, is an important inherent property of many clays and is used in commercial applications such as soil remediation, and in the fertilization of soils.

The above properties are the basis for the use of bentonite in many commercial applications: as drilling muds and catalysts in the petroleum industry, as bonding clays in foundries, as bonding agents in hematite-rich iron ore pellets, and as absorbents. Bentonite clays are also being used as backfill (a buffer material) around high-level radioactive waste (HLW) cannisters that are buried underground (Graver, R. 1994. Bentonite as a backfill material in a high-level waste repository. *Materials Research Society Bulletin*: 43–46.) Such clays are selected for sealing the space around HLW cannisters because (1) they have low hydraulic conductivity, as compared with the host rock in which the repository is located; (2) they have

good swelling properties to seal construction-caused joints and rock fractures; (3) their plasticity absorbs rock movements; (4) they show good retention of radionuclides; and (5) they are stable over a period of at least 10^6 years.

Vermiculite with an idealized formula of $Mg_3(Si,Al)_4O_{10}(OH)_2 \cdot 4 \cdot 5H_2O[Mg]_{0.35}$ (see discussion on p. 468 and Fig. 11.36 for a structure illustration) is similar in its properties to those of montmorillonite. The vermiculite structure can expand, as a result of the absorption of H_2O in the interlayer position, but to a lesser degree than observed in montmorillonite. Much of commercial vermiculite is used in agriculture for soil conditioning as a plant-growth medium, chemical fertilizer carrier, and carrier for pesticides and herbicides. When rapidly heated, vermiculite produces a lightweight expanded product that is widely used for thermal insulation and in potting soil. Mixed with plaster and cement, vermiculite is used to make lightweight versions of these materials. Vermiculite is also used as an absorbent for some environmentally hazardous liquids.

Physical Properties. *Cleavage* {001} perfect. Thin folia somewhat flexible but not elastic. Sectile. **H** 1 (will make a mark on cloth). **G** 2.7–2.8. *Luster* pearly to greasy. *Color* apple-green (see Plate VII, no. 2), gray, white, or silver-white; in soapstone often dark gray or green. Translucent. Greasy feel. *Optics*: (−); $\alpha = 1.539$, $\beta = 1.589$, $\gamma = 1.589$; $2V = 6°–30°$. $Z = b$, $X \perp \{001\}$; $r > v$.

Composition and Structure. There is little variation in the chemical composition of most talc; pure talc contains MgO 31.7, SiO_2 63.5, H_2O 4.8%. Small amounts of Al or Ti may substitute for Si and Fe may replace some of the Mg. *Minnesotaite*, $Fe_3Si_4O_{10}(OH)_2$, is common in low-grade metamor-

phic Precambrian iron deposits. Minnesotaite has a modulated structure that is different from that of talc; it is probable that an almost complete solid solution exists between talc and minnesotaite. The trioctahedral structure of talc is similar to that of dioctahedral pyrophyllite and consists of essentially neutral $t–o–t$ layers held together by weak residual bonds (see Fig. 11.34).

Diagnostic Features. Characterized by its micaceous habit, cleavage, softness, and greasy feel.

Occurrence. Talc is a secondary mineral formed by the alteration of magnesium silicates, such as olivine, pyroxenes (see Figs. 11.43 and 11.44), and amphiboles, and may be found as pseudomorphs

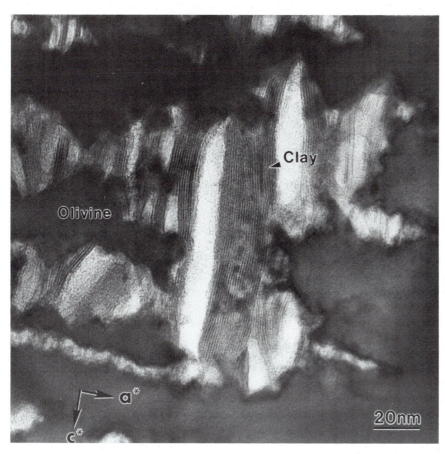

FIG. 12.54. Transmission electron micrograph of olivine that (through low-temperature hydrothermal alteration and weathering) has developed etch pits and channels lined with clay minerals (of the montmorillonite type). The white areas are very small void spaces that would have been completely filled if the montmorillonite had been completely hydrated. (From J. F. Banfield, D. R. Veblen, and B. F. Jones, 1990, Transmission electron microscopy of subsolidus oxidation and weathering of olivine. *Contributions to Mineralogy and Petrology* 106: 110–23.)

after these minerals. Characteristically in low-grade metamorphic rocks, where, in massive form, *soapstone*, it may make up nearly the entire rock mass. It may also occur as a prominent constituent in schistose rocks, as in the talc schist. Figure 12.55 gives a

P–T stability diagram for talc and related minerals in the system MgO–SiO$_2$–H$_2$O.

In the United States many talc or soapstone quarries are located along the line of the Appalachian Mountains from Vermont to Georgia. The

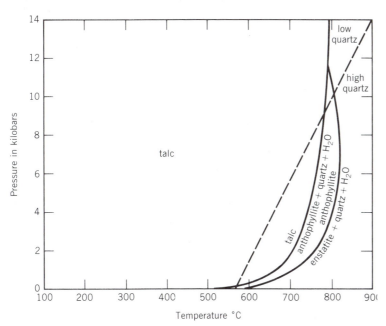

FIG. 12.55. Pressure-temperature diagram for the stability field of talc in the system MgO–SiO$_2$–H$_2$O. Talc is stable over a wide temperature range, from 100°C to between 500° and 700°C, depending on the pressure. At high temperature talc reacts to form anthophyllite, and at even higher temperature anthophyllite reacts to form enstatite. Compare with Fig. 11.43. The low–high quartz polymorphic transition is also shown. (Adapted from B. W. Evans and S. Guggenheim, 1988, Talc, pyrophyllite, and related minerals, in *Hydrous Phyllosilicates, Reviews in Mineralogy* 19, Mineralogical Soc. of America, Washington, D.C., pp. 225–94.)

major producing states are California, North Carolina, Texas, and Georgia.

Use. Most of the talc and soapstone produced is used in powdered form as an ingredient in paint, ceramics, rubber, insecticides, roofing, paper, and foundry facings. The most familiar use is in talcum powder. Talc is also used as an ornamental material for carving small objects.

Name. The name *talc* is of ancient and doubtful origin, probably derived from the Arabic, *talk*.

Similar Species. *Minnesotaite*, $Fe_3Si_4O_{10}(OH)_2$, the Fe-rich equivalent of talc. Occurs in Precambrian chem-formations.

Pyrophyllite—$Al_2Si_4O_{10}(OH)_2$

Crystallography. Triclinic; $\bar{1}$. Not in distinct crystals. Foliated, in some cases in radiating lamellar aggregates. Also granular to compact. Identical with talc in appearance.

$C\bar{1}$; $a = 5.16$, $b = 8.97$, $c = 9.35$ Å; $\alpha = 91°11'$; $\beta = 100°28'$, $\gamma = 89°38'$; $Z = 4$. ds: 9.21(6), 4.58(5), 4.40(2), 3.08(10), 2.44(2).

Physical Properties. *Cleavage* {001} perfect. Folia somewhat flexible but not elastic. **H** 1–2 (will make a mark on cloth). **G** 2.8 *Luster* pearly to greasy. *Color* white, apple-green, gray, brown. Translucent, will transmit light on thin edges. *Optics:* (−); $\alpha = 1.552$, $\beta = 1.588$, $\gamma = 1.600$; $2V = 57°$; $X \perp$ {001}; $r > v$.

Composition and Structure. Pyrophyllite shows little deviation from the ideal formula; Al_2O_3 28.3, SiO_2 66.7, H_2O 5.0%. The dioctahedral structure of pyrophyllite consists of essentially neutral *t–o–t* layers (Figs. 11.32 and 11.34) held together by weak van der Waals bonds.

Diagnostic Features. Characterized chiefly by its micaceous habit, cleavage, and greasy feel. X-ray diffraction techniques are needed for positive identification.

Occurrence. Pyrophyllite is a comparatively rare mineral. Found in metamorphic rocks; frequently with kyanite. Occurs in considerable amount in Guilford and Orange counties, North Carolina.

Use. Quarried in North Carolina and used for the same purpose as talc but does not command as high a price as the best grades of talc. A considerable part of the so-called *agalmatolite*, from which the Chinese carve small images, is this species.

Name. From the Greek meaning *fire* and a *leaf*, because it exfoliates on heating.

Mica Group

The micas, composed of *t–o–t* layers with interlayer cations and little or no exchangeable water, crystallize in the monoclinic system but with their crystallographic angle β close to 90°, so that their monoclinic symmetry is not easily seen. The crystals are usually tabular with prominent basal planes and have either a diamond-shaped or hexagonal outline with angles of approximately 60° and 120°. Crystals, as a rule,

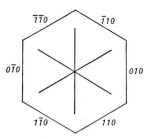

FIG. 12.56. Percussion figure in mica.

therefore, appear to be either orthorhombic or hexagonal. They are characterized by a highly perfect {001} cleavage. A blow with a somewhat dull-pointed instrument on a cleavage plate develops in all the species a six-rayed *percussion figure* (Fig. 12.56), two lines of which are nearly parallel to the prism edges and the third, which is most strongly developed, parallel to the mirror plane {010}. There is limited ionic substitution between various members of the group and two members may crystallize together in parallel position in the same crystal plate with the cleavage extending through both.

MUSCOVITE—$KAl_2(AlSi_3O_{10})(OH)_2$

Crystallography. Monoclinic; $2/m$. Distinct crystals rare; usually tabular with prominent {001}, see Plate VII, no. 3. The presence of prism faces {110} at angles of nearly 60° gives some plates a diamond-shaped outline, making them simulate orthorhombic symmetry (Fig. 12.57). If {010} is also present, the crystals have a hexagonal appearance. The prism faces are roughened by horizontal striations and frequently taper. Penetration twins with [310] the twin axis (see Fig. 12.58). Foliated in large to small sheets; in scales that are, in some cases, aggregated into

FIG. 12.57. Muscovite. Diamond-shaped crystals (Harvard Mineralogical Museum).

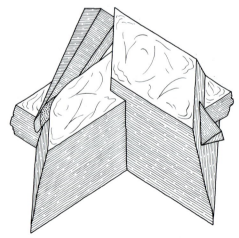

FIG. 12.58. Muscovite twins, twin axis [310], producing a penetration twin, with (001) of the individual crystals coplanar. Methuen Township, Ontario, Canada (Harvard Mineralogical Museum).

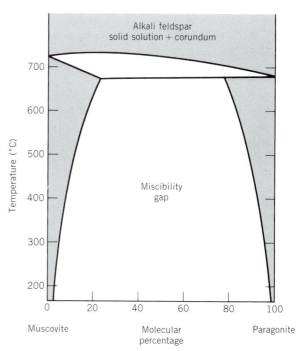

FIG. 12.59. Temperature-composition diagram for the system $KAl_2(AlSi_3O_{10})(OH)_2$-$NaAl_2(AlSi_3O_{10})(OH)_2$. Below about 700°C this shows a large miscibility gap, and above about 700°C it shows the reaction relations of (1) muscovite to K-feldspar and (2) paragonite to albite. At that temperature there is complete solid solution between K-feldspar and Na-feldspar (see also Fig. 11.51). Corundum is a reaction product in this reaction:

$$KAl_2(AlSi_3O_{10})(OH)_2 \rightarrow KAlSi_3O_8 + Al_2O_3 + H_2O\nearrow$$
$$\text{muscovite} \qquad \text{K-feldspar} \quad \text{corundum}$$

(From H. P. Eugster and H. S. Yoder, 1955. The join muscovite-paragonite. Carnegie Institute, Washington, Annual Report of the Director of the Geophysical Laboratory, 1954–55, p. 124.)

plumose or globular forms. Also cryptocrystalline and compact massive.

There are several polytypes such as $2M_1$, $1M$, and $3T$. $2M_1$ is most common with space group $C2/c$; $a = 5.19$, $b = 9.04$, $c = 20.08$ Å; $\beta = 95°30'$; $Z = 4$. ds: 9.95(10), 3.37(10), 2.66(8), 2.45(8), 2.18(8).

Physical Properties. *Cleavage* {001} perfect, allowing the mineral to be split into very thin sheets; folia flexible and elastic. **H** $2-2\frac{1}{2}$. **G** 2.76–2.88. *Luster* vitreous to silky or pearly. *Color* colorless and transparent in thin sheets. In thicker blocks translucent, with light shades of yellow, brown, green, red. Some crystals allow more light to pass parallel to the cleavage than perpendicular to it. *Optics:* (−); $\alpha = 1.560–1.572$, $\beta = 1.593–1.611$, $\gamma = 1.599–1.615$; $2V = 30°–47°$; $Z = Y$, $X \wedge c = 0°–5°$; $r > v$.

Composition and Structure. Essentially $KAl_2(AlSi_3O_{10})(OH)_2$. Little solid solution occurs among the dioctahedral mica group or between members of the dioctahedral and trioctahedral groups. Minor substitutions may be Na, Rb, Cs for K; Mg, Fe^{2+}, Fe^{3+}, Li, Mn, Ti, Cr for Al; F for OH. *Paragonite*, $NaAl_2(AlSi_3O_{10})(OH)_2$, which is isostructural with muscovite, is practically indistinguishable from it without X-ray methods. At low to moderate temperatures a large miscibility gap exists between muscovite and paragonite (Fig. 12.59). The structure of muscovite is illustrated in Figs. 11.33 and 11.34. It consists of t sheets of composition $(Si,Al)_2O_5$ linked to octahedral gibbsite-like sheets to form t-o-t layers. These layers have a net negative charge which is balanced by K or Na ions (in paragonite) between them (see discussion on page 465). Although the muscovite structure may show several polytypic forms due to various ways of stacking succeeding Si_2O_5 sheets (see page 471 and Figs. 11.41 and 11.42b),

the most commonly observed polytype is $2M_1$ with space group $C2/c$.

Diagnostic Features. Characterized by its highly perfect cleavage and light color.

Occurrence. Muscovite is a widespread and common rock-forming mineral. Characteristic of granites and granite pegmatites. In pegmatites, muscovite associated with quartz and feldspar may be in large crystals, called *books*, which in some localities are several feet across. It is also very common in metamorphic rocks, forming the chief constituent of certain mica schists. In the chlorite zone of metamorphism, muscovite is a characteristic constituent of albite–chlorite-muscovite schists (see Fig. 12.60 for a *P–T* stability diagram). Coexistences between muscovite and other micas are shown in Fig. 12.61. In some schistose rocks it occurs as fibrous aggregates of minute scales with a silky luster and is known as *sericite*. The development of sericite from feldspar

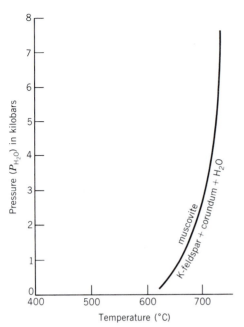

FIG. 12.60. Pressure-temperature stability diagram for muscovite, according to the reaction:

$$KAl(Al_2Si_3O_{10})(OH)_2 \rightarrow KAlSi_3O_8 + Al_2O_3 + H_2O \nearrow$$
muscovite   K-feldspar   corundum

(From D. A. Hewitt and D. R. Jones, 1984, Experimental phase relations of the micas, in *Micas, Reviews in Mineralogy* 13. Mineralogical Soc. America, Washington, D.C., pp. 201–56.)

and other minerals, such as topaz, kyanite, spodumene, and andalusite, is a common feature of retrograde metamorphism. Sericite also forms as an alteration of the wall rock of hydrothermal ore veins. As *illite* it is a constituent of some shales, soils, and recent sediments.

Large and important deposits of muscovite occur in Russia and India. The United States produces only a small amount of sheet mica, but is the largest producer of scrap (flake) mica with major production coming from North and South Carolina, Connecticut, Georgia, New Mexico, and South Dakota. Crystals measuring 7 to 9 feet across have been mined in Mattawan Township, Ontario, Canada. Good specimen material is obtained from pegmatites worldwide.

Use. Because of its high dielectric and heat-resisting properties, *sheet mica*, single cleavage plates, is used as an insulating material in the manufacture of electrical apparatus. The *isinglass* used in furnace and stove doors is sheet mica. Many small parts used for electrical insulation are built up of thin sheets of mica cemented together. They may thus be pressed into shape before the cement hardens. India is the largest supplier of mica used in this way. Ground mica is used in many ways: in the manufacture of wallpapers to give them a shiny luster; as a lubricant when mixed with oils; a filler; and as a fireproofing material.

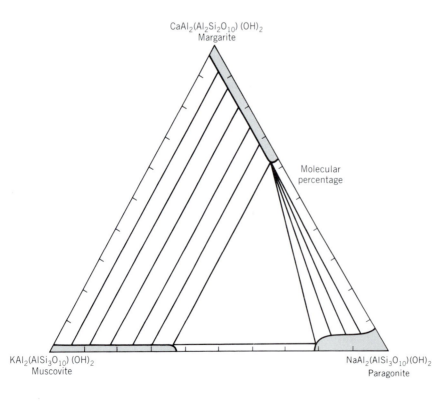

FIG. 12.61. The extent of solid solution for several micas in rocks that have been metamorphosed to a temperature of about 450°C (garnet zone metamorphism). Schematic tielines are drawn between the various solid solution fields, to indicate possible mica pairs, as well as a three-mica triangle (muscovite–paragonite–margarite). (From C. V. Guidotti, 1984, Micas in metamorphic rocks, in *Micas, Reviews in Mineralogy* 13, pp. 357–468, Mineralogical Society America, Washington, D.C.)

Name. *Muscovite* was so called from the popular name of the mineral, *Muscovy-glass*, because of its use as a substitute for glass in Old Russia (Muscovy). *Mica* was probably derived from the Latin *micare*, meaning to *shine*.

Phlogopite—$KMg_3(AlSi_3O_{10})(OH)_2$

Crystallography. Monoclinic; $2/m$. Usually in six-sided plates or in tapering prismatic crystals. Crystals frequently large and coarse. Found also in foliated masses.

$C2/m$; $a = 5.33$, $b = 9.23$, $c = 10.26$ Å; $\beta = 100°12'$; $Z = 2$. ds: 10.13(10), 3.53(4), 3.36(10), 3.28(4), 2.62(10).

Physical Properties. *Cleavage* {001} perfect. Folia flexible and elastic. **H** $2\frac{1}{2}$–3. **G** 2.86. *Luster* vitreous to pearly. *Color* yellowish-brown, green, white, often with copperlike reflections from the cleavage surface; see Plate VII, no. 4. Transparent in thin sheets. When viewed in transmitted light, some phlogopite shows asterism because of tiny oriented inclusions of rutile. *Optics*: (−); $\alpha = 1.53$–1.59, $\beta = 1.56$–1.64, $\gamma = 1.56$–1.64; 2V = 0°–15°; $Y = b$, $X \wedge c = 0°$–15'; $r < v$.

Composition and Structure. Small amount of Na and lesser amounts of Rb, Cs, and Ba may substitute for K. Fe^{2+} substitutes for Mg, forming a series toward biotite. F may substitute in part for OH. The trioctahedral structure of phlogopite is schematically illustrated in Fig. 11.34. It consists of *t–o–t* layers bonded by interlayer K^+ ions; it is very similar to the muscovite structure except for the difference in the octahedral layer. It occurs most commonly in the $1M$ (space group $C2/m$) polytype, but $2M_1$ and $3T$ polytypes occur occasionally (see Figs. 11.41 and 11.42*b*).

Diagnostic Features. Characterized by its micaceous cleavage and yellowish-brown color. It is, however, impossible to draw a sharp distinction between biotite and phlogopite.

Occurrence. Phlogopite is found in metamorphosed Mg-rich limestones, dolomites, and ultrabasic rocks. It is a common mineral in kimberlite. In the United States found chiefly in Jefferson and St. Lawrence counties, New York. Found abundantly in Canada in Ontario at North and South Burgess, and in various other localities in Ontario and Quebec.

Use. Same as for muscovite; chiefly as electrical insulator.

Name. Name from the Greek word *phlogos* meaning *fire*, an allusion to its color.

BIOTITE—$K(Mg,Fe)_3(AlSi_3O_{10})(OH)_2$

Crystallography. Monoclinic; $2/m$. Rarely in tabular or short prismatic crystals with prominent {001}

and pseudohexagonal outline. Usually in irregular foliated masses; often in disseminated scales or in scale aggregates.

$C2/m$; $a = 5.31$, $b = 9.23$, $c = 10.18$ Å; $\beta = 99°18'$; $Z = 2$. ds: 10.1(10), 3.37(10), 2.66(8), 2.54(8), 2.18(8).

Physical Properties. *Cleavage* {001} perfect. Folia flexible and elastic. **H** $2\frac{1}{2}$–3. **G** 2.8–3.2. *Luster* splendent. *Color* usually dark green, brown to black. More rarely light yellow. Thin sheets usually have a smoky color (differing from the almost colorless muscovite). *Optics*: (−); $\alpha = 1.57$–1.63, $\beta = 1.61$–1.70; $\gamma = 1.61$–1.70; 2V = 0°–25°; $Y = b$, $Z \wedge a = 0°$–9°; $r < v$.

Composition and Structure. The composition is similar to phlogopite but with considerable substitution of Fe^{2+} for Mg. There is also substitution by Fe^{3+} and Al for Mg and by Al for Si. In addition Na, Ca, Ba, Rb, and Cs may substitute for K. A complete solid solution series exists between phlogopite and biotite. The trioctahedral biotite structure is the same as that of phlogopite (see Fig. 11.34). The $1M$ polytype with space group $C2/m$ is most common but $2M_1$ and $3T$ also occur (see Figs. 11.41 and 11.42*b*).

Diagnostic Features. Characterized by its micaceous cleavage and dark color; see Plate VII, no. 5.

Occurrence. Biotite forms under a very great variety of geological environments. It is found in igneous rocks varying from granite pegmatites, to granites, to diorites, to gabbros and peridotites. It occurs also in felsite lavas and porphyries. In metamorphic rocks it is formed over a wide range of temperature and pressure conditions, and it occurs in regionally as well as contact metamorphosed rocks. The crystallization of biotite in argillaceous rocks is taken as the onset of the temperature and pressure conditions of the *biotite zone*. Typical associations at this metamorphic grade are biotite–chlorite and biotite–muscovite. Occurs in fine crystals in blocks included in the lavas of Vesuvius.

Name. In honor of French physicist J. B. Biot.

Similar Species. *Glauconite*, similar in composition to biotite, is an authigenic mineral found in green pellets in marine sedimentary rocks. *Vermiculite* is a platy mineral formed chiefly as an alteration of biotite. The structure is essentially that of talc with interlayered water molecules (see Fig. 11.36 and Box 12.2). The name comes from the Latin word meaning *to breed worms*, for when heated, the mineral expands into wormlike forms. In the expanded condition it is used extensively in heat and sound insulating. Vermiculite is mined at Libby, Montana, and Macon, North Carolina.

Stilpnomelane, with approximate composition $K_{0.6}(Mg,Fe^{2+},Fe^{3+})_6Si_8Al(O,OH)_{27} \cdot 2$–$4H_2O$, can be an important constituent of low-grade, regionally meta-morphosed schists, of glaucophane metamorphic facies rocks, and of essentially unmetamorphosed Precambrian banded iron-formations. It is impossible to distinguish stilpnomelane in hand specimen from biotite mica. Stilpnomelane has a modulated layer structure that is very different from that of biotite. Optical microscopic or chemical techniques are necessary to distinguish stilpnomelane from biotite.

LEPIDOLITE—$K(Li,Al)_{2-3}(AlSi_3O_{10})(O,OH,F)_2$

Crystallography. Monoclinic and hexagonal, depending on the polytype. Crystals usually in small plates or prisms with hexagonal outline. Commonly in coarse- to fine-grained scaly aggregates.

For $1M$ polytype, $C2/m$; $a = 5.21$, $b = 8.97$, $c = 20.16$ Å; $\beta = 100°48'$; $Z = 4$. ds: 10.0(6), 5.00(5), 4.50(5), 2.58(10), 1.989(8).

Physical Properties. *Cleavage* {001} perfect. **H** $2\frac{1}{2}$–4. **G** 2.8–2.9. *Luster* pearly. *Color* pink and lilac to grayish white; see Plate VII, no. 6. Translucent. *Optics*: (−); $\alpha = 1.53$–1.55, $\beta = 1.55$–1.59, $\gamma = 1.55$–1.59; 2V = 0°–60°; $Y = b$, $Z \wedge a = 0°$–7°; $r > v$.

Composition and Structure. The composition of lepidolite varies depending chiefly on the relative amounts of Al and Li in octahedral coordination. Analyses show a range of Li_2O from 3.3 to 7 weight percent. In addition Na, Rb, and Cs may substitute for K. It is possible that there is a continuous chemical series from dioctahedral muscovite to trioctahedral lepidolite, in which all of the octahedral sites are occupied. Lepidolite can occur as one of three polytypes ($1M$, $2M_2$, and $3T$) (see Figs. 11.41 and 11.42b).

Diagnostic Features. Characterized chiefly by its micaceous cleavage and usually by its lilac to pink color. Muscovite may be pink and lepidolite white, and X-ray powder diffraction techniques may be necessary to distinguish them.

Occurrence. Lepidolite is a comparatively rare mineral, found in pegmatites, usually associated with other lithium-bearing minerals such as pink and green tourmaline, amblygonite, and spodumene (see Box 12.4). Often intergrown with muscovite in parallel position. Notable foreign localities for its occurrence are Rozna, Moravia, Czechoslovakia; Bikita, Zimbabwe; and the Malagasy Republic. In the United States it is found in several localities in Maine; near Middletown, Connecticut; Pala, California; Dixon, New Mexico; and Black Hills, South Dakota.

Use. A source of lithium. Used in the manufacture of heat-resistant glass.

Name. Derived from the Greek word *lepidos* meaning *scale*.

Margarite—$CaAl_2(Al_2Si_2O_{10})(OH)_2$

Crystallography. Monoclinic; *m*. Seldom in distinct crystals. Usually in foliated aggregates with micaceous habit.

Cc; $a = 5.13$, $b = 8.92$, $c = 19.50$ Å; $\beta = 100°48'$; $Z = 4$. ds: 4.40(8), 3.39(8), 3.20(9), 2.51(10), 2.42(8).

Physical Properties. *Cleavage* {001} perfect. **H** $3\frac{1}{2}$–5 (harder than the true micas). **G** 3.0–3.1. *Luster* vitreous to pearly. *Color* pink, white, and gray. Translucent. Folia somewhat brittle; because of this brittleness margarite is known as a *brittle mica*. *Optics*: (−); $\alpha = 1.632$–1.638; $\beta = 1.643$–1.648; $\gamma = 1.645$–1.650; 2V = 40°–67°; $Z = b$, $Y \wedge a = 7°$.

Composition and Structure. Most analyses are close to the above end member composition with CaO 14.0, Al_2O_3 51.3, SiO_2 30.2 and H_2O 4.5%. A small amount of Na may replace Ca. The dioctahedral structure of margarite is very similar to that of muscovite (see Fig. 12.62). In margarite, however, the tetrahedral layer has the composition $(Si_2Al_2)O_{10}$ instead of $(Si_3Al)O_{10}$ as in muscovite. Because of the greater electrical charge on the $(Si_2Al_2)O_{10}$ sheet the structure can be balanced by incorporated divalent Ca^{2+} ions instead of monovalent K^+. The bond strength between the layers is therefore greater; this is expressed in the brittle nature of margarite.

Diagnostic Features. Characterized by its micaceous cleavage, brittleness, and association with corundum.

Occurrence. Margarite occurs usually with corundum and diaspore and apparently as an alteration product. It is found in this way with the emery deposits of Asia Minor and on the islands of Naxos and Nicaria, of the Grecian Archipelago. In the United States associated with emery at Chester, Massachusetts; Chester County, Pennsylvania; and with corundum deposits in North Carolina.

Name. From the Greek *margarites* meaning *pearl*.

Similar Species. Other *brittle micas* are *clintonite* and *xanthophyllite*, which may be regarded as the Ca analogues of phlogopite.

Chlorite Group

A number of minerals are included in the chlorite group, all of which have similar chemical, crystallographic, and physical properties. Without quantitative chemical analyses or careful study of the optical and X-ray properties, it is extremely difficult to distinguish between the members. The following is a composite description of the principal members of the group.

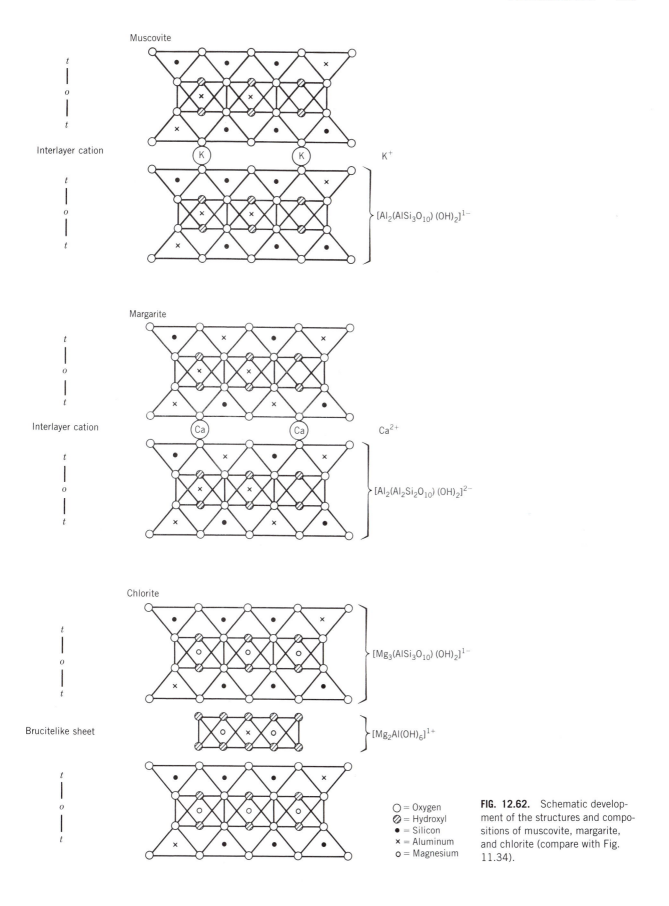

FIG. 12.62. Schematic development of the structures and compositions of muscovite, margarite, and chlorite (compare with Fig. 11.34).

Muscovite

Interlayer cation

K^+

$[Al_2(AlSi_3O_{10})(OH)_2]^{1-}$

Margarite

Interlayer cation

Ca^{2+}

$[Al_2(Al_2Si_2O_{10})(OH)_2]^{2-}$

Chlorite

$[Mg_3(AlSi_3O_{10})(OH)_2]^{1-}$

Brucitelike sheet

$[Mg_2Al(OH)_6]^{1+}$

○ = Oxygen
⊘ = Hydroxyl
● = Silicon
× = Aluminum
○ = Magnesium

CHLORITE—$(Mg,Fe)_3(Si,Al)_4O_{10}(OH)_2 \cdot (Mg,Fe)_3(OH)_8$ IV

Crystallography. Triclinic or monoclinic depending on the polytype. In pseudohexagonal tabular crystals, with prominent {001}. Similar in habit to crystals of the mica group, but distinct crystals rare. Usually foliated massive or in aggregates of minute scales; also in finely disseminated particles.

Cell parameters vary with composition. $C1$, $C\bar{1}$, Cm and $C2/m$, depending on polytype. For clinochlore: $C2/m$; $a = 5.2–5.3$, $b = 9.2–9.3$, $c = 28.6$ Å; $\beta = 96°50'$; $Z = 4$. ds: 3.54(10), 2.53(6), 2.00(6), 1.562(4), 1.534(7).

Physical Properties. *Cleavage* {001} perfect. Folia flexible but not elastic. **H** $2–2\frac{1}{2}$. **G** 2.6–3.3. *Luster* vitreous to pearly. *Color* green of various shades; see Plate VII, no. 7. Rarely yellow, white, rose-red. Transparent to translucent. *Optics:* Most (+), some (−); in all Bxa nearly ⊥ {001}. $\alpha = 1.57–1.66$, $\beta = 1.57–1.67$, $\gamma = 1.57–1.67$; $2V = 20°–60°$. Pleochro-

ism in green (+), $X, Y > Z$; (−) $X < Y, Z$. The indices increase with increasing iron content.

Composition and Structure. The composition of chlorite can be regarded as made up of t–o–t layers of composition $[Mg_3(AlSi_3O_{10})(OH)_2]^{1-}$ interleaved with a "brucitelike" sheet in which one out of three Mg^{2+} ions is replaced by one Al^{3+}, resulting in a composition of $[Mg_2Al(OH)_6]^{1+}$. This is illustrated schematically in Fig. 12.62. When the composition of the t–o–t layer and the "brucitelike" sheet are added, the composition $(Mg,Al)_6(Si,Al)_4O_{10}(OH)_8$ results, which is a general formula for magnesian chlorite. In most chlorites there is considerable deviation from this formula because of Fe^{2+}, Fe^{3+}, and additional Al substitution. A diagrammatic sketch of the chlorite structure is given in Fig. 11.35.

The general formula of chlorite may be represented as follows: $A_{5-6}Z_4O_{10}(OH)_8$, where $A =$ Al, Fe^{2+}, Fe^{3+}, Li, Mg, Mn, Ni, and $Z =$ Al, Si, Fe^{3+}. Because of the extensive solid solution, many vari-

Pyrophyllite—$Al_2Si_4O_{10}(OH)_2$
Kaolinite—$Al_2Si_2O_5(OH)_4$
Kyanite ⎤
Sillimanite ⎬ —Al_2SiO_5
Andalusite ⎦
Talc—$Mg_3Si_4O_{10}(OH)_2$
Anthophyllite—$Mg_7Si_8O_{22}(OH)_2$
Enstatite—$MgSiO_3$
Serpentine—$Mg_3Si_2O_5(OH)_4$
Forsterite—Mg_2SiO_4
Spinel—$MgAl_2O_4$
Cordierite—$Mg_2Al_4Si_5O_{18} \cdot nH_2O$
Pyrope—$Mg_3Al_2Si_3O_{12}$
Chlorite—$(Mg_5Al)(Si_3Al)O_{10}(OH)_8$

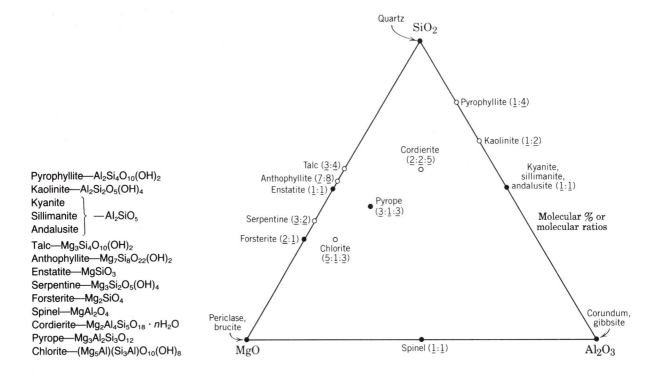

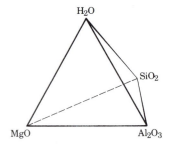

FIG. 12.63. The composition of an idealized chlorite, and those of other end member compositions of rock-forming minerals in the system MgO-Al_2O_3-SiO_2-H_2O. In this diagram, which is water-free, and which may be considered as a projection from an H_2O corner (in a tetrahedron, see insert), the hydrous formulas are plotted on the basis of their molecular (MgO, Al_2O_3, and SiO_2) ratios, without reference to the OH groups or H_2O content. The opaque dots are anhydrous compositions, and the open circles hydrous compositions. The molecular ratios ($MgO:SiO_2$; $Al_2O_3:SiO_2$; $MgO:Al_2O_3:SiO_2$, and $MgO:Al_2O_3$) are noted next to the mineral names.

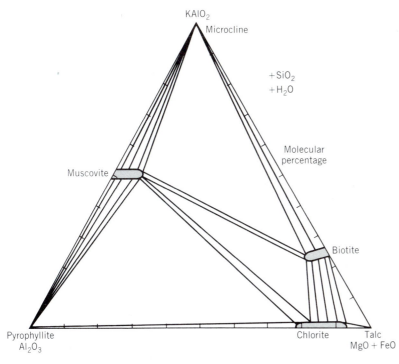

KAlO₂
Microcline

+SiO₂
+H₂O

Molecular
percentage

Muscovite

Biotite

Pyrophyllite
Al₂O₃

Chlorite   Talc
MgO + FeO

FIG. 12.64. Common assemblages involving chlorite, and other layer silicates, as well as microcline in metamorphic rocks of the biotite zone of the greenschist facies. The chemistry of this system is: $KAlO_2$–Al_2O_3–(Mg + Fe)O–(SiO_2)–(H_2O). The right-hand corner of the triangle shows two components, allowing for the representation of some solid solution extent in natural minerals. SiO_2 and H_2O are assumed to be present for the representation of the composition of anhydrous and hydrous silicates. Tielines connect coexisting mineral pairs. Triangles represent three-mineral assemblages. (Adapted from *Petrologic Phase Equilibria*, 21st edition, by W. G. Ernst. Copyright © 1976 by W. H. Freeman and Company. Reprinted by permission.)

etal names have been given to members of the chlorite group. Examples are *chamosite, clinochlore, pennantite,* and *sudoite*. The composition of an idealized Mg end member chlorite, $(Mg_5Al)(Si_3Al)O_{10}(OH)_8$, is plotted on an anhydrous basis in Fig. 12.63. Such a diagram, which is frequently used in petrologic discussions, allows for the comparison of, for example, the composition of chlorite with that of other rock-forming minerals. Figure 12.64 illustrates the common occurrence of chlorite, in association with other layer silicates, as well as microcline, in metamorphic rocks of the biotite zone of the greenschist facies.

The great range in composition of chlorite is reflected in variations in physical and optical properties as well as X-ray parameters (e.g., interplanar spacings and variation in unit cell size).

Diagnostic Features. Characterized by its green color, micaceous habit and cleavage, and by the fact that the folia are not elastic.

Occurrence. Chlorite is a common mineral in metamorphic rocks and it is the diagnostic mineral of the *greenschist facies*. In pelitic schists it occurs in quartz-albite-chlorite-sericite-garnet assemblages. It is also commonly found with actinolite and epidote. Chlorite is also a common constituent of igneous rocks where it has formed as an alteration of Mg–Fe silicates such as pyroxenes, amphiboles, biotite, and garnet. The green color of many igneous rocks is due to the chlorite to which the ferromagnesian silicates

have altered; and the green color of many schists and slates is due to finely disseminated particles of the mineral.

Name. Chlorite is derived from the Greek word *chloros* meaning *green*, in allusion to the common color of the mineral.

Related Species

Apophyllite—KCa₄(Si₄O₁₀)₂F·8H₂O

Crystallography. Tetragonal; $4/m2/m2/m$. Usually in crystals showing a combination of {110}, {011}, and {001} (Fig. 12.65). Crystals may resemble a combination of cube and octahedron (Fig. 12.66), but are shown to be tetragonal by difference in luster between faces of prism and base.

$P4/mnc$; a = 9.02, c = 15.8 Å; Z = 2. ds: 4.52(2), 3.94(10), 3.57(1), 2.98(7), 2.48(3).

Physical Properties. *Cleavage* {001} perfect. H $4\frac{1}{2}$–5. G 2.3–2.4. *Luster* of base pearly, other faces vitreous. *Color*

FIG. 12.65. Apophyllite.

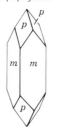
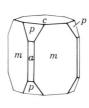

FIG. 12.66. Apophyllite crystals with stilbite, India.

FIG. 12.67. Prehnite, Paterson, New Jersey.

colorless, white, or grayish; may show pale shades of green, yellow, rose. Transparent to translucent. *Optics*: (+); $\omega = 1.537$, $\epsilon = 1.535$. May be optically (−).

Composition and Structure. The structure of apophyllite differs from that of the other phyllosilicates in that the sheets are composed of fourfold and eightfold rings. These sheets are linked to each other by Ca, K, and F ions. Water, which is bonded to the Si_2O_5 sheet by hydrogen bonding, is lost at about 250°C. This indicates that the water is more strongly held than adsorbed water, but less strongly than structural (OH) groups.

Diagnostic Features. Recognized usually by its crystals, color, luster, and basal cleavage.

Occurrence. Apophyllite occurs as a secondary mineral lining cavities in basalt and related rocks, associated with zeolites, calcite, datolite, and pectolite.

It is found in fine crystals at Andreasberg, Harz Mountains, Germany; near Bombay, India; Iceland; and Guanajuato, Mexico. In the United States at Bergen Hill and Paterson, New Jersey; and Lake Superior copper district. Found in fine crystals in Nova Scotia.

Name. Apophyllite is derived from two Greek words meaning *from* and *a leaf*, because of its tendency to exfoliate when ignited.

Prehnite—$Ca_2Al(AlSi_3O_{10})(OH)_2$

Crystallography. Orthorhombic; $2mm$. Distinct crystals are rare, commonly tabular parallel to {001}. Usually reniform, stalactitic (Fig. 12.67), and in rounded groups of tabular crystals.

$P2cm$; $a = 4.65$, $b = 5.48$, $c = 18.49$ Å; $Z = 2$. *ds*: 3.48(9), 3.28(6), 3.08(10), 2.55(10), 1.77(7).

Physical Properties. H $6-6\frac{1}{2}$. G 2.8–2.95. *Luster* vitreous. *Color* usually light green, passing into white. Translucent. *Optics*: (+); $\alpha = 1.616$, $\beta = 1.626$, $\gamma = 1.649$; $2V = 66°$; $X = a$, $Z = c$; $r > v$.

Composition and Structure. CaO 27.1, Al_2O_3 24.8, SiO_2 43.7, H_2O 4.4%. Some Fe^{3+} may replace Al. The structure of prehnite contains sheets of Al and Si tetrahedra parallel to {001}. Ca is in 7-coordination and lies between Si tetrahedra in adjoining sheets.

Diagnostic Features. Characterized by its green color and crystalline aggregates forming reniform surfaces. Resembles hemimorphite but is of lower specific gravity.

Occurrence. Prehnite occurs as a secondary mineral lining cavities in basalt and related rocks. Associated with zeolites, datolite, pectolite, and calcite. In the United States, it occurs notably at Paterson and Bergen Hill, New Jersey. In Canada, it occurs at Asbestos, Quebec.

Use. Sometimes used as an ornamental and gem material.

Name. In honor of Colonel Prehn, who brought the mineral from the Cape of Good Hope.

Chrysocolla ≈ $Cu_4H_4Si_4O_{10}(OH)_8$

Crystallography and Structure. Generally amorphous and thus in the strict sense cannot be considered a mineral. Partially crystalline material shows the presence of Si_4O_{10} layers in a very much defect structure. Massive, compact; in some cases earthy. Individual specimens inhomogeneous.

Physical Properties. Fracture conchoidal. H 2–4. G 2.0–2.4. *Luster* vitreous to earthy. *Color* green to greenish-blue; brown to black when impure. Refractive index variable, usually about 1.50.

Composition. Chrysocolla is a hydrogel or gelatinous precipitate and shows a wide range in composition. Chemical analyses report: CuO 32.4–42.2, SiO_2 37.9–42.5, H_2O 12.2–18.8%. In addition Al_2O_3 and Fe_2O_3 are usually present in small amounts.

Diagnostic Features. Characterized by its green or blue color and conchoidal fracture. Distinguished from turquoise by inferior hardness.

Occurrence. Chrysocolla forms in the oxidized zones of copper deposits (see Box 8.2) associated with malachite, azurite, cuprite, or native copper, for example. It is commonly found in the oxidized portions of porphyry copper ores. Found in the copper districts of Arizona and New Mexico; at Chuquicamata, Chile; and Zaire.

Use. A minor ore of copper. Sometimes cut as a gemstone, but most chrysocolla used in jewelry is intergrown with chalcedony and has the refractive index of chalcedony.

Name. Chrysocolla is derived from two Greek words meaning *gold* and *glue*, which was the name of a similar-appearing material used to solder gold.

Similar Species. *Dioptase*, $Cu_6(Si_6O_{18})\cdot 6H_2O$, is a rhombohedral cyclosilicate occurring in well-defined green rhombohedral crystals. It is a minor gem mineral. *Plancheite*, $Cu_8(Si_4O_{11})(OH)_2\cdot H_2O$, and *shattuckite*, $Cu_5(SiO_3)_4(OH)_2$, have inosilicate structures.

TECTOSILICATES

We will describe in detail the following tectosilicate groups and species:

SiO$_2$ Group

Quartz	
Tridymite	SiO_2
Cristobalite	
Opal	$SiO_2 \cdot nH_2O$

Feldspar Group
K-Feldspars

Microcline	
Orthoclase	$KAlSi_3O_8$
Sanidine	

Plagioclase Feldspars

Albite	$NaAlSi_3O_8$
Anorthite	$CaAl_2Si_2O_8$

Feldspathoid Group

Leucite	$KAlSi_2O_6$
Nepheline	$(Na,K)AlSiO_4$
Sodalite	$Na_8(AlSiO_4)_6Cl_2$
Lazurite	$(Na,Ca)_8(AlSiO_4)_6(SO_4,S,Cl)_2$
Petalite	$LiAlSi_4O_{10}$

Scapolite Series

Marialite	$Na_4(AlSi_3O_8)_3(Cl_2,CO_3,SO_4)$
Meionite	$Ca_4(Al_2Si_2O_8)_3(Cl_2,CO_3,SO_4)$
Analcime	$NaAlSi_2O_6\cdot H_2O$

Zeolite Group

Natrolite	$Na_2Al_2Si_3O_{10}\cdot 2H_2O$
Chabazite	$CaAl_2Si_4O_{12}\cdot 6H_2O$
Heulandite	$CaAl_2Si_7O_{18}\cdot 6H_2O$
Stilbite	$NaCa_2Al_5Si_{13}O_{36}\cdot 14H_2O$

SiO$_2$ Group

Quartz—SiO$_2$ (IV)

Crystallography. Quartz, hexagonal; 32. High-quartz, hexagonal; 622. Crystals commonly prismatic, with prism faces horizontally striated.

Terminated usually by a combination of positive and negative rhombohedrons, which often are so equally developed as to give the effect of a hexagonal dipyramid (see Fig. 12.68a and Plate VII, no. 8). In some crystals one rhombohedron predominates or occurs alone (Fig. 12.68b). It is chosen as r {10$\overline{1}$1}. The prism faces may be wanting, and the combination of the two rhombohedrons gives what appears to be a hexagonal dipyramid (a *quartzoid*) (Fig. 12.68c). Some crystals are malformed, but the recognition of the prism faces by their horizontal striations assists in the orientation. The trigonal trapezohedral faces x are occasionally observed and reveal the true symmetry. They occur at the upper right of alternate prism faces in right-hand quartz and to the upper left of alternate prism faces in left-hand quartz (Figs. 12.68d and e). The right- and left-hand trigonal trapezohedrons are enantimorphous forms and reflect the arrangement of the SiO_4 tetrahedra, either in the form of a right- or left-hand screw (see space groups). In the absence of x faces, the "hand" can be recognized by observing whether plane polarized light passing parallel to c is rotated to the left or right.

Crystals may be elongated in tapering and sharply pointed forms, and some appear twisted or bent. Equant crystals with apparent sixfold symmetry (Figs. 12.68a and c) are characteristic of high-temperature quartz, but the same habit is found in quartz that crystallized as the low-temperature form. Most quartz is twinned according to one or both of two laws (see Figs. 5.45c and d). These are *Dauphiné*, c the twin axis; and *Brazil*, {11$\overline{2}$0} the twin plane. Both

FIG. 12.68. Quartz crystals.

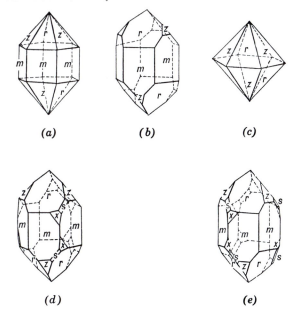

(a)   (b)   (c)

(d)   (e)

types are penetration twins and external evidence of them is rarely seen.

The size of crystals varies from individuals weighing several tons to finely crystalline coatings, forming "drusy" surfaces. Also common in massive forms of great variety. From coarse- to fine-grained crystalline to flintlike or cryptocrystalline, giving rise to many variety names, discussed in the next section. May form in concretionary masses.

$P3_221$ or $P3_121$, $a = 4.91$, $c = 5.41$ Å; $Z = 3$. ds: 4.26(8), 3.34(10), 1.818(6), 1.541(4), 1.081(5) (see Fig. 7.47).

Physical Properties. H 7. G 2.65. Fracture conchoidal. *Luster* vitreous, in some specimens greasy, splendent. *Color* usually colorless or white, but frequently colored by impurities and may then be any color. Transparent to translucent. Strongly piezoelectric and pyroelectric. *Optics*: (+); $\omega = 1.544$, $\epsilon = 1.553$.

Composition and Structure. Of all the minerals, quartz is most nearly a pure chemical compound with constant physical properties. Si 46.7, O 53.3%. The structure of low (α) quartz with one of two enantiomorphic space groups, $P3_121$ or $P3_221$, is discussed on page 137 and illustrated in Fig. 4.36b. At 573°C, and at atmospheric pressure, this structure transforms instantaneously to high (β) quartz, with hexagonal symmetry and one of two enantiomorphic space groups, $P6_222$ or $P6_422$. (This structure is shown in Fig. 4.36a). The displacive transformation from low to high quartz involves only minor atomic adjustments without breaking of Si-O bonds. Upon cooling high quartz, through the inversion point at 573°C, Dauphiné twinning may be produced (see page 137).

Diagnostic Features. Characterized by its glassy luster, conchoidal fracture, and crystal form. Distinguished from calcite by its hardness and from white varieties of beryl by its inferior hardness.

Varieties. A great many different forms of quartz exist, to which varietal names have been given. The more important varieties, with a brief description of each, follow.

Coarsely Crystalline Varieties

Rock Crystal. Colorless quartz, commonly in distinct crystals.

Amethyst. Quartz colored various shades of violet, often in crystals. The color results from the presence of trace amounts of iron as $[FeO_4]^{4-}$ color centers (see Table 4.6).

Rose Quartz. Coarsely crystalline but usually without crystal form, color a rose-red or pink. Often fades on exposure to light. Small amounts of Ti^{4+} appear to be the coloring agent.

Smoky Quartz; Cairngorm Stone. Frequently in crystals; smoky yellow to brown to almost black. Named *cairngorm* for the locality of Cairngorm in Scotland. The dark color is attributed to the presence of trace amounts of Al^{3+} ions, which produce $[AlO_4]^{4-}$ color centers upon irradiation of originally colorless quartz (see Table 4.6 and related discussion).

Citrine. Light yellow resembing topaz in color (see Plate X, nos. 1 to 4, Chapter 13 for illustrations of colored varieties of quartz).

Milky Quartz. Milky white owing to minute fluid inclusions. Some specimens have a greasy luster.

Quartz may contain parallel fibrous inclusions that give the mineral a chatoyancy. When stones are cut *en cabochon* they are called *quartz cat's eye* (see Fig. 2.13). *Tiger's eye* is a yellow fibrous quartz pseudomorphous after the fibrous amphibole crocidolite. It is also chatoyant.

With Inclusions. Many other minerals occur as inclusions in quartz and thus give rise to variety names. *Rutilated quartz* has fine needles of rutile penetrating it (Fig. 12.69). Tourmaline and other minerals are found in quartz in the same way. *Aventurine quartz* includes brilliant scales of colored minerals such as hematite (red) or chromium mica (green) and is used as a gem material. Liquids and gases may occur as inclusions; both liquid and gaseous carbon dioxide exist in some quartz.

FIG. 12.69. Rutilated quartz, Brazil (Harvard Mineralogical Museum).

Microcrystalline Varieties

IV

The microcrystalline varieties of quartz are divided into two types, depending on their structure: *fibrous* and *granular*. It is difficult to distinguish between them macroscopically.

A. Fibrous Varieties

Chalcedony is the general term applied to fibrous varieties. More specifically, it is a brown to gray, translucent variety, with a waxy luster, often mammillary and in other imitative shapes. Chalcedony has been deposited from aqueous solutions and is frequently found lining or filling cavities in rocks. Color and banding give rise to the following varieties:

Carnelian, a red chalcedony, grades into brown sard.

Chrysoprase is an apple-green chalcedony colored by nickel oxide.

Agate is a variety with alternating layers of chalcedony having different colors and porosity. The colors are usually in delicate, fine parallel bands. The bands are commonly curved, in some specimens concentric (see Fig. 12.70 and Plate VII, no. 9). Most agate used for commercial purposes is colored by artificial means. Some agates have the different colors not arranged in bands but irregularly distributed. *Moss agate* is a variety in which the variation in color is caused by visible impurities, commonly manganese oxide in mosslike patterns (see Plate X, nos. 5 and 6, Chapter 13).

FIG. 12.70.   Agate cut and polished, Brazil (Harvard Mineralogical Museum).

Wood that has been petrified by replacement by clouded agate is known as *silicified* or *agatized wood*.

Onyx, like agate, is a layered chalcedony, with layers arranged in parallel planes. *Sardonyx* is an onyx with sard alternating with white or black layers.

Heliotrope or *bloodstone* is a green chalcedony with small red spots of jasper in it.

B. Granular Varieties

Flint and *chert* resemble each other and there is no sharp distinction between them. Dark, siliceous nodules, usually found in chalk, are called flint, whereas lighter-colored bedded deposits are called chert.

Jasper is a granular microcrystalline quartz with dull luster usually colored red by included hematite.

Prase has a dull green color; otherwise it is similar to jasper, and occurs with it.

Occurrence. Quartz is a common and abundant mineral occurring in a great variety of geological environments. It is present in many igneous and metamorphic rocks and is a major constituent of granite pegmatites (see Boxes 12.1 and 12.4). It is the most common gangue mineral in hydrothermal and metal-bearing veins and in many veins is essentially the only mineral present. In the form of flint and chert, quartz is deposited on the sea floor contemporaneously with the enclosing rock; or solutions carrying silica may replace limestone to form chert horizons. On the breakdown of quartz-bearing rocks, the quartz, because of its mechanical and chemical stability, persists as detrital grains to accumulate as sand (see Box 12.3). Quartz-rich sandstone and its metamorphic equivalent, quartzite, may be composed mainly of quartz.

Rock crystal is found widely distributed, some of the more notable localities being the Alps; Minas Gerais, Brazil; the Malagasy Republic; and Japan. The best quartz crystals from the United States are found near Hot Springs, Arkansas, and Little Falls, Herkimer County, New York. Important occurrences of amethyst are in the Ural Mountains, Russia, Czechoslovakia, Uruguay, Zambia, and Brazil. Found at Thunder Bay on the north shore of Lake Superior. In the United States rock crystal is found in Delaware and Chester counties, Pennsylvania; Oxford County, Maine. Smoky quartz is found in large and fine crystals in Switzerland; and in the United States at Pikes Peak, Colorado; Alexander County, North Carolina; and Oxford County, Maine.

The chief source of agates at present is in southern Brazil and northern Uruguay. Most of these agates are cut at Idar-Oberstein, Germany, itself a famous agate locality. In the United States agate is

BOX 12.3
MINERAL DUST IN THE ENVIRONMENT

The issue of adverse human health effects caused by mineral dusts, especially asbestos and crystalline silica (SiO_2) dusts, has been a major concern. The incidence of cancer and asbestosis among professional asbestos workers and silicosis among hard-rock miners and sand blasters was the basis for subsequent evaluations of the interaction of mineral dust and health. Aspects of the asbestos question are discussed under the headings of crocidolite ("blue asbestos") use on p. 527 and chrysotile ("white asbestos") use on p. 529 (see also Klein 1993, Rocks, Minerals, and a Dusty World*; see complete reference in footnote).

Here we will address the question of whether there is good reason to be concerned about silica (SiO_2) dust in the air. Although there are several polymorphs of SiO_2 (see Table 11.3), the three that are of concern are quartz, cristobalite, and tridymite, with quartz by far the most common of the three. Quartz is a common constituent of many rock types and may constitute 100% of some beach sands. The photograph shows well-sorted, rounded sand size grains, in the size range of grains in a sandbox.

Inhalation of large quantities of fine quartz dust, over extended periods of time, may result in silicosis, in which the lung develops scar tissue and loses its air exchange capacity. On the basis of some animal experiments in which rats, mice, and hamsters were exposed to intense airborne quartz dusts, the International Agency for Research on Cancer (IARC) declared quartz a carcinogen in 1987. The results in these animal experiments were extremely inconsistent (see Saffiotti et al. 1993, Biological Studies on the carcinogenic mechanisms of quartz*), and there is always the fundamental question of how well such animal studies relate to human beings. As a result of this IARC decision the U.S. Occupational Safety and Health Administration (OSHA) now requires that any U.S. product that contains more than 0.1 percent "free silica" display a warning label. For example, bags of sand

| THIS PRODUCT MAY CONTAIN SILICA. SILICA DUST IF INHALED MAY CAUSE RESPIRATORY OR OTHER HEALTH PROBLEMS. PROLONGED INHALATION MAY CAUSE DELAYED LUNG INJURY, INCLUDING SILICOSIS AND POSSIBLY CANCER. |

and other quartz-bearing products now carry labels such as the one printed here labeling quartz as a carcinogen. What happens if quartz-rich beaches, sandboxes, dirt roads, farmlands, all kinds of commercial products, and the desert regions of Arizona, Nevada, California, and New Mexico (with commonly fierce dust storms) become identified, in the public mind, as carcinogenic to humans? Ross et al. 1993 (Health effects of mineral dusts other than asbestos*) reviews many epidemiological studies of workers employed since 1950 and exposed to mineral dusts, but finds no evidence for nonmalignant chronic disease (such as silicosis) and no justification for the prediction of increased cancer risk due to silica dust exposure. As such, declaring quartz, in the natural environment, a carcinogen is not only confusing but highly alarming to the public at large.

*All above three articles are part of Reviews in Mineralogy, vol. 28, entitled *Health Effects of Mineral Dusts*, G. D. Guthrie, Jr., and Brooke T. Mossman, editors. Mineralogical Society of America, Washington, D.C. Articles are on pp. 7–59 for C. Klein; on pp. 361–401 for Ross et al.; pp. 523–543 for Saffiotti et al.

Well-sorted, rounded mineral grains described as sand, on the basis of the size of the grains ranging between $\frac{1}{16}$ mm and 2 mm in diameter. Some of the sand grains in this photograph are made of quartz, others are feldspar. Sand from the Great Sand Dunes National Monument, Colorado.

found in numerous places, notably in Oregon and Wyoming. The chalk cliffs at Dover, England, are famous for the flint nodules that weather from them. Similar nodules are found on the French coast of the English Channel and on islands off the coast of Denmark. Massive quartz, occurring in veins or with feldspar in pegmatite dikes, is mined in Connecticut, New York, Maryland, and Wisconsin for its various commercial uses.

Use. Quartz has many and varied uses. It is widely used as gemstones or ornamental material, as amethyst, rose quartz, smoky quartz, tiger's eye, aventurine, carnelian, agate, and onyx. As sand, quartz is used in mortar, in concrete, as a flux, as an abrasive, and in the manufacture of glass and silica brick. In powdered form it is used in porcelain, paints, sandpaper, scouring soaps, and as a wood filler. In the form of quartzite and sandstone it is used as a building stone and for paving purposes.

Quartz has many uses in scientific equipment. Because of its transparency in both the infrared and ultraviolet portions of the spectrum, quartz is made into lenses and prisms for optical instruments. The *optical activity* of quartz (the ability to rotate the plane of polarization of light) is utilized in the manufacture of an instrument to produce monochromatic light of differing wavelengths. Quartz wedges, cut from transparent crystals, are used as an accessory to the polarizing microscope. Because of its piezoelectric property, quartz has specialized uses. It is cut into small oriented plates and used as radio oscillators to permit both transmission and reception on a fixed frequency. The tiny quartz plate used in digital quartz watches serves the same function as quartz oscillators used to control radio frequencies (see page 35). This property also renders it useful in the measurement of instantaneous high pressures such as result from firing a gun or an atomic explosion.

Synthesis. Since 1947 much of the quartz used for optical and piezoelectrical purposes has been manufactured by hydrothermal methods. More recently yellow, brown, blue, and violet quartz has been synthesized for use as gem material.

Name. The name *quartz* is a German word of ancient derivation.

Similar Species. *Lechatelierite*, SiO_2, is fused silica or silica glass. Found in fulgurites, tubes of fused sand formed by lightning, and in cavities in some lavas. Lechatelierite is also found at Meteor Crater, Arizona, where sandstone has been fused by the heat generated by the impact of a meteorite.

Tridymite—SiO$_2$

Crystallography. Low (α) tridymite: monoclinic or orthorhombic; $2/m$, m, or 222. High (β) tridymite: hexagonal; $6/m2/m2/m$. Crystals are small and commonly twinned and at room temperature are pseudomorphs after high tridymite.

Low tridymite, for $C2/c$ or Cc; $a = 18.54$, $b = 5.01$, $c = 25.79$ Å; $\beta = 117°40'$; $Z = 48$; for $C222_1$; $a = 8.74$, $b = 5.04$, $c = 8.24$ Å; $Z = 8$. ds: (low tridymite): 4.30(10), 4.08(9), 3.81(9), 2.96(6), 2.47(6). High tridymite, $P6_3/mmc$; $a = 5.04$, $c = 8.24$ Å; $Z = 4$.

Physical Properties. H 7. **G** 2.26. *Luster* vitreous. *Color* colorless to white. Transparent to translucent. *Optics*: (+); $\alpha = 1.468–1.479$; $\beta = 1.470–1.480$, $\gamma = 1.475–1.483$; 2V = 40°–90°.

Composition and Structure. Ideally SiO_2. However, small amounts of Na and Al may be in solid solution. The crystal structure of low (α) tridymite has not been investigated in detail but it is undoubtedly closely related to that of high tridymite. The high (β) tridymite structure consists of sheets of tetrahedra that lie parallel to {0001}; the tetrahedra within each sheet share corners to form six-membered rings (see Fig. 11.45a) and in these rings tetrahedra alternately point up or down, providing linkage between the sheets. Tridymite is the stable form of SiO_2 at temperatures between 870° and 1470°C, at atmospheric pressure (see Fig. 11.46). At higher temperatures it transforms to cristobalite, at lower temperatures to high quartz. These transformations are reconstructive and extremely sluggish.

Diagnostic Features. It is impossible to identify tridymite by macroscopic means, but under the microscope its crystalline outline and refractive index distinguish it from the other silica minerals.

Occurrence. Tridymite occurs commonly in certain siliceous volcanic rocks such as rhyolite, obsidian, and andesite, and for this reason may be considered an abundant mineral. It is most commonly the result of devitrification of volcanic glass, such as obsidian, and is commonly associated with sanidine and cristobalite. It is found in large amounts in the lavas of the San Juan district of Colorado. It is also found in stony meteorites and the lunar basalts.

Name. From the Greek meaning *threefold*, in allusion to its common occurrence in trillings.

Cristobalite—SiO$_2$

Crystallography. Low (α) cristobalite: tetragonal; 422. High (β) cristobalite: isometric; $4/m\bar{3}2/m$. Crystals small octahedrons; this form is retained on inversion from high to low cristobalite. Also in spherical aggregates.

Low cristobalite, $P4_12_12$ (or $P4_32_12$): $a = 4.97$, $c = 6.93$ Å; $Z = 4$. ds (low cristobalite): 4.05(10), 2.84(1), 2.48(2), 1.929(1), 1.870(1). High cristobalite, $Fd3m$: $a = 7.13$ Å; $Z = 8$.

Physical Properties. H $6\frac{1}{2}$. **G** 2.32. *Luster* vitreous. Colorless. Translucent. *Optics*: (+); ω 1.484, $\epsilon = 1.487$.

Composition and Structure. Ideally SiO_2, but most natural material contains some Na and Al in solid solution. The low cristobalite structure is tetragonal, whereas high cristobalite is isometric. In high cristobalite six-membered tetrahedral rings are stacked parallel to {111}; see Fig. 11.45b. High cristobalite is stable from 1470°C to the melting point, 1728°C, at atmospheric pressure (see Fig. 11.46). The transformation at 1470°C to tridymite is of the reconstructive type.

Diagnostic Features. The occurrence in small lava cavities as spherical aggregates is characteristic, but it cannot be determined with certainty without optical or X-ray measurements.

Occurrence. Cristobalite is present in many siliceous volcanic rocks, both as the lining of cavities and as an important constituent in the fine-grained groundmass. It is, therefore, an abundant mineral. It is most commonly the result of devitrification of volcanic glass, such as obsidian.

Associated with tridymite in the lavas of the San Juan district, Colorado.

Name. From the Cerro San Cristobal near Pachuca, Mexico.

OPAL—SiO₂·nH₂O

Crystallography. Generally amorphous. Massive; often botryoidal, stalactitic. Although X-ray studies indicate that much opal is essentially amorphous, precious opals contain silica spheres in an ordered packing (see Structure, below, and also Figs. 2.10 and 2.11).

Physical Properties. *Fracture* conchoidal. **H** 5–6. **G** 2.0–2.25. *Luster* vitreous; often somewhat resinous. *Color* colorless, white, pale shades of yellow, red, brown, green, gray, and blue. The darker colors result from impurities. Often has a milky or "opalescent" effect and may show a fine play of colors. Transparent to translucent. Some opal, especially *hyalite*, shows a greenish-yellow fluorescence in ultraviolet light. *Optics*: Refractive index 1.44–1.46.

Composition and Structure. SiO₂·nH₂O. The water content, usually between 4 and 9%, may be as high as 20%. The specific gravity and refractive index decrease with increasing water content.

Although opal is essentially amorphous, it has been shown to have an ordered structure. It is not a crystal structure with atoms in a regular three-dimensional array but is made up of closely packed spheres of silica in hexagonal and/or cubic closest packing (see Figs. 2.10 and 2.11). Air or water occupies the voids between the spheres. In common opal the domains of equal-size spheres with uniform packing are small or nonexistent, but in precious opal large domains are made up of regularly packed spheres of the same size. The sphere diameters vary from one opal to another and range from 1500 Å to 3000 Å. When white light passes through the essentially colorless opal and strikes planes of voids between spheres, certain wavelengths are diffracted and flash out of the stone as nearly pure spectral colors. This phenomenon has been described as analogous to the diffraction of X-rays by crystals. In X-ray diffraction the interplanar spacings (*d*) are of the same order of magnitude as the wavelengths of X-rays. In precious opal the spacings, determined by sphere diameters, are far greater but so are the wavelengths of visible light (4000–7000 Å). The different wavelengths that satisfy the Bragg equation are diffracted with change in the angle of incident light (θ). Because light is refracted when it enters precious opal, the equation must include the refractive index, μ (= 1.45), and the equation is written $n\lambda = \mu 2d \sin \theta$ (see Fig. 2.11).

Diagnostic Features. Distinguished from microcrystalline varieties of quartz by lesser hardness and specific gravity and by the presence of water.

Varieties. Precious opal is characterized by a brilliant internal play of colors that may be red, orange, green, or blue (see Plate X, no. 7, Chapter 13). The body color is white, milky-blue, yellow, or black (*black opal*). *Fire opal* is a variety with intense orange to red reflectins.

Common Opal. Milk-white, yellow, green, red, etc., without internal reflections.

Hyalite. Clear and colorless opal with a globular or botryoidal surface.

Geyserite or *Siliceous Sinter.* Opal deposited by hot springs and geysers. Found about the geysers in Yellowstone National Park.

Wood Opal. Fossil wood with opal as the petrifying material.

Diatomite. Fine-grained deposits, resembling chalk in appearance. Formed by sinking from near the surface and the accumulation on the sea floor of the siliceous tests of diatoms. Also known as *diatomaceous earth* or *infusorial earth.*

Occurrence. Opal may be deposited by hot springs at shallow depths, by meteoric waters, or by low-temperature hypogene solutions. It is found lining and filling cavities in rocks and may replace wood buried in volcanic tuff. The largest accumulations of opal are as siliceous tests of silica-secreting organisms.

Precious opals are found at Caernowitza, Hungary; in Querétaro, Mexico; Queensland and New South Wales, Australia; and Brazil. Black opal has been found in the United States in Virgin Valley, Nevada. Diatomite is mined in several western states, principally at Lompoc, California.

Synthesis. Pierre Gilson in Switzerland has synthesized precious opal that is identical to natural material in chemical and physical properties, including a beautiful play of colors.

Use. As a gem. Opal is usually cut in round shapes, *en cabochon*. Stones of large size and exceptional quality are very highly prized. Diatomite is used extensively as an abrasive, filler, filtration powder, and in insulation products.

Name. The name *opal* originated in the Sanskrit, *upala*, meaning stone or precious stone.

Feldspar Group

K-Feldspars
MICROCLINE—KAlSi₃O₈

Crystallography. Triclinic; $\bar{1}$. The habit and crystal forms are similar to those of orthoclase, and microcline may be twinned according to the same laws

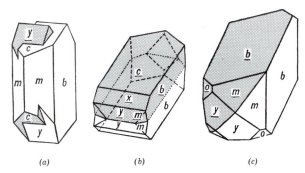

FIG. 12.71. Twinning in feldspar. *(a)* Carlsbad twin. *(b)* Manebach twin. *(c)* Baveno twin.

as orthoclase; *Carlsbad* twins are common, but *Baveno* and *Manebach* twins are rare (see Fig. 12.71). It is also twinned according to the *albite law*, with {010} the twin plane, and the *pericline law*, with {010} the twin axis. These two types of twinning are usually present in microcline and on {001} the lamellae cross at nearly 90° giving a characteristic "tartan" structure (see Fig. 5.41). Microcline is found in cleavable masses, in crystals, and as a rock constituent in irregular grains. Microcline probably forms some of the largest known crystals. In a pegmatite in Karelia, Russia, a mass weighing more than 2000 tons showed the continuity of a single crystal. Also in pegmatites microcline may be intimately intergrown with quartz, forming *graphic granite* (Fig. 12.72).

FIG. 12.72. Graphic granite, Hybla, Ontario. Quartz dark, microcline light.

Microcline frequently has irregular and discontinuous bands crossing {001} and {010} that result from the exsolution of albite. The intergrowth as a whole is called *perthite*, see Plate VIII, nos. 1 and 2, or, if fine, *microperthite* (see page 482 and Fig. 12.73).

$C\bar{1}$; for low microcline $a = 8.59$, $b = 12.97$, $c = 7.22$ Å; $\alpha = 90°41'$, $\beta = 115°59'$, $\gamma = 87°39'$; $Z = 4$. ds: 4.21(5), 3.83(3), 3.48(3), 3.37(5), 3.29(5), 3.25(10).

Physical Properties. *Cleavage* {001} perfect, {010} good at an angle of 89°30'. **H** 6. **G** 2.54–2.57. *Luster* vitreous. *Color* white to pale yellow, more rarely red or green. Green microcline is known as *amazonite*. Translucent to transparent. *Optics:* (−); $\alpha = 1.522$, $\beta = 1.526$, $\gamma = 1.530$; 2V = 83°; $r > v$.

Composition and Structure. For $KAlSi_3O_8$, K_2O 16.9, Al_2O_3 18.4, SiO_2 64.7%; there is some substitution of Na for K (see Fig. 11.51). The structure of microcline is triclinic and therefore less symmetrical than that of sanidine, which is shown in Fig. 11.48. Maximum low microcline shows perfect Al-Si order in the tetrahedral framework. With increasing disorder (increasing T) the terms *low*, *intermediate*, and *high* microcline are applied. At considerably higher T the microcline structure transforms to orthoclase or sanidine (see Fig. 11.51 and also Fig. 4.39 and related text).

Diagnostic Features. Distinguished from orthoclase only by observing the presence of microcline ("tartan") twinning, which can rarely be determined without the aid of the microscope. With only minor exceptions, deep green feldspar is microcline.

Occurrence. Microcline is a prominent constituent of igneous rocks such as granites and syen-

FIG. 12.73. Microperthite as seen in polarized light under the microscope.

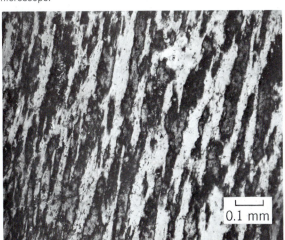

0.1 mm

BOX 12.4
MINERALS IN PEGMATITES

Pegmatites are extremely coarse-grained bodies that are commonly closely related genetically and in space to large masses of plutonic rocks. They may be found as veins or dikes traversing the granular igneous rock, but more commonly they extend out from it into the surrounding country rock. Granites more commonly than any other rock have pegmatites genetically associated with them; consequently, unless modified by other terms, pegmatite refers to *granite pegmatite*. The minerals in most pegmatites, therefore, are the common minerals in granite (quartz, feldspar, mica) but of extremely large size. Crystals of these minerals measuring a foot (approximately 30 cm) across are common, and in some localities they reach gigantic sizes. Quartz crystals weighing thousands of pounds and mica crystals more than 10 feet across have been found. A common characteristic of pegmatites is the simultaneous and interpenetrating crystallization of quartz and K-feldspar (usually microcline) to form *graphic granite*.

Although most pegmatites are composed entirely of the minerals found abundantly in granite, those of greatest interest contain other, rarer minerals. In these pegmatites there has apparently been a definite sequence in deposition. The earliest minerals are microcline and quartz, with smaller amounts of garnet and black tourmaline. These are followed and partly replaced by albite, lepidolite, gem tourmaline, beryl, spodumene, amblygonite, topaz, apatite, and fluorite. A host of rarer minerals such as triphylite, columbite, monazite, molybdenite, and uranium minerals may be present. In

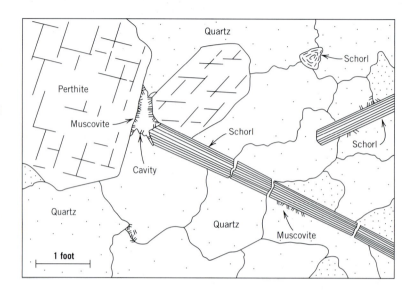

some places the above minerals are abundant and form large crystals that are mined for their rare constituent elements. Thus, spodumene crystals more than 40 feet long have been found in the Black Hills of South Dakota, and beryl crystals from Albany, Maine, have measured as much as 27 feet long and 6 feet in diameter.

The drawing, with a scale bar = one foot (1 foot = 30.4 cm), is of the giant pegmatite texture in a quartz–perthite–schorl (black variety of tourmaline) occurrence in a pegmatite dike near Sage, Riverside County, California. (Jahns, R. H. 1953. The genesis of pegmatites. I. Occurrence and origin of giant crystals. *American Mineralogist* 38: 563–98.)

The formation of pegmatite dikes is believed to be directly connected with the crystallization of the larger masses of associated plutonic rock. The process of crystallization brings about a

concentration of the volatile constituents in the residual melt of the magma. The presence of these volatiles (H_2O, B, F, Cl, and P) decreases the viscosity and thus facilitates crystallization. Such an end product of magmatic crystallization is also enriched in the rare elements originally disseminated through the magma. When this residual liquid is injected into the cooler surrounding rock, it crystallizes from the borders inward, frequently giving a zonal distribution of minerals with massive quartz at the center.

The world's most important gemstone-producing pegmatites are in the northeastern part of the state of Minas Gerais, Brazil. These pegmatites have produced large quantities of gem tourmaline, topaz, kunzite (a pink gem spodumene), morganite (a rose-colored gem beryl), aquamarine (a green-blue gem beryl), and chrysoberyl (see also Chapter 13).

ites that cooled slowly and at considerable depth. In sedimentary rocks it is present in arkose and conglomerate. In metamorphic rocks in gneisses. Microcline is the common K-feldspar of pegmatites and was quarried extensively in North Carolina, North and South Dakota, Colorado, Virginia, Wyoming,

Maine, and Connecticut (see Box 12.4). *Amazonite*, a green variety of microcline, is found in the Ural Mountains, Russia, and in various places in Norway and the Malagasy Republic. In the United States it is found at Pikes Peak, Colorado (see Fig. 12.74), and Amelia Court House, Virginia; see Plate VIII, no. 3.

FIG. 12.74. Microcline and smoky quartz, Crystal Peak, Colorado (Harvard Mineralogical Museum).

Use. Feldspar is used chiefly in the manufacture of porcelain. It is ground very fine and mixed with kaolin or clay, and quartz. When heated to high temperature the feldspar fuses and acts as a cement to bind the material together. Fused feldspar also furnishes the major part of the glaze on porcelain ware. A small amount of feldspar is used in the manufacture of glass to contribute alumina to the batch. Amazonite is polished and used as an ornamental material.

Name. *Microcline* is derived from two Greek words meaning *little* and *inclined*, referring to the slight variation of the cleavage angle from 90°. *Feldspar* is derived from the German word *feld*, field.

ORTHOCLASE—KAISi₃O₈

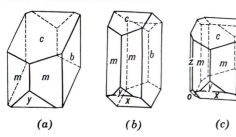

Crystallography. Monoclinic; 2/m. Crystals are usually short prismatic elongated parallel to *a*, or elongated parallel to *c* and flattened on {010}. The prominent forms are *b*{010}, *c*{001}, and *m*{110}, often with smaller {0kl} and {hkl} prisms (Fig. 12.75). Frequently twinned according to the following laws:

FIG. 12.75. Orthoclase and sanidine.

(a)   (b)   (c)

Carlsbad penetration twin with *c* as twin axis, see Plate VIII, no. 4; *Baveno* with {021} as twin and composition plane; *Manebach* with {001} as twin and composition plane (see Fig. 12.71*b*). Commonly in crystals or in coarsely cleavable to granular masses; more rarely fine-grained, massive, and cryptocrystalline. Most abundantly in rocks as formless grains.

$C2/m$; $a = 8.56$, $b = 12.96$, $c = 7.21$ Å; $\beta = 115°50'$; $Z = 4$. *d*s: 4.22(6), 3.77(7), 3.46(5), 3.31(10), 3.23(8).

Physical Properties. *Cleavage* {001} perfect, {010} good, {110} imperfect. **H** 6. **G** 2.57. *Luster* vitreous. *Color* colorless, white, gray, flesh-red, rarely yellow or green. *Streak* white. *Optics*: (−); $\alpha = 1.518$, $\beta = 1.524$, $\gamma = 1.526$; $2V = 10°–70°$; $Z = b$, $X \wedge a = 5°$; $r > v$.

Composition and Structure. Most analyses contain small amounts of Na, but a complete solid solution series is possible at high temperature (see Fig. 11.51). Orthoclase represents a K-feldspar structure that crystallizes at intermediate temperature and has a partially ordered Al-Si distribution. At higher temperatures the Al-Si distribution in the structure becomes less ordered and is known as sanidine (see Fig. 4.39 and related text).

Diagnostic Features. Is usually recognized by its color, hardness, and cleavage. Distinguished from the other feldspars by its right-angle cleavage and the lack of twin striations on the best cleavage surface.

Occurrence. Orthoclase is a major constituent of granites, granodiorites, and syenites, which have cooled at moderate depth and at reasonably fast

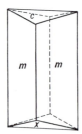

FIG. 12.76. Adularia.

FIG. 12.77. Polished slab of basalt with sanidine phenocrysts.

rates. In more slowly cooled granites and syenites microcline will be the characteristic K-feldspar.

Name. The name *orthoclase* refers to the right-angle cleavage possessed by the mineral.

Similar Species. *Adularia*, $KAlSi_3O_8$, is a color-less, translucent to transparent variety of K-feldspar which is commonly found in pseudo-orthorhombic crystals (Fig. 12.76). It occurs mainly in low-temperature veins in gneisses and schists. Some adularia shows an opalescent play of colors and is called *moonstone.*

SANIDINE—$(K,Na)AlSi_3O_8$

Crystallography. Monoclinic; $2/m$. Crystals are often tabular parallel to {010}; also elongated on *a* with square cross section as in Fig. 12.75*a*. Carlsbad twins common.

$C2/m$; for high sanidine, $a = 8.56$, $b = 13.03$, $c = 7.17$ Å; $\beta = 115°59'$; $Z = 4$. *d*s: 4.22(6), 3.78(8), 3.31(10), 3.278(6), 3.225(8).

Physical Properties. *Cleavage* {001} perfect, {010} good. **H** = 6. **G** = 2.56–2.62. *Luster* vitreous. *Color* colorless and commonly transparent. *Streak* white. *Optics*: (−); $\alpha = 1.518$–1.525; $\beta = 1.523$–1.530, $\gamma = 1.525$–1.531. Occurs in two orientations with optic plane parallel with {010} and $2V = 0°$–60°, and with optic plane normal to {010} and $2V = 0°$–25°.

Composition and Structure. A complete solid solution exists at high temperature between sanidine and high albite; part of the intermediate region is known as *anorthoclase* (see Fig. 11.47). The structure of sanidine shows a disordered (random) distribution of Al and Si in the tetrahedral framework (see Fig. 11.48). The Al and Si distribution in orthoclase is more ordered (see Fig. 4.39 and related text).

Diagnostic Features. Can be characterized with confidence only by optical or X-ray techniques. Optic orientation and 2V differ from orthoclase. Microcline and plagioclase show different types of twinning.

Occurrence. As phenocrysts (see Fig. 12.77) in extrusive igneous rocks such as rhyolites and tra-

chytes. Sanidine is characteristic of rocks that cooled quickly from an initial high temperature of eruption. Most sanidines are cryptoperthitic.

Name. Sanidine is derived from the Greek *sanis*, tablet, and *idos*, appearance, in allusion to its typical tabular habit.

Plagioclase Feldspar Series

The plagioclase feldspars form, at elevated temperatures, an essentially complete solid solution series from pure *albite* (Ab), $NaAlSi_3O_8$, to pure *anorthite* (An), $CaAl_2Si_2O_8$ (see Fig. 11.54). Discontinuities in this series, as shown by the existence of peristerite and other intergrowths (see page 482), can be detected only by careful electron optical and X-ray studies. The nomenclature for the series, as based on six arbitrary divisions, is given in Fig. 11.47. If species names are not given, the composition is expressed by $Ab_{20}An_{80}$, for example, which may be shortened to An_{80}. Most of the properties of the various species in this series vary in a uniform manner with the change in chemical composition (see Fig. 12.80). For this reason the series can be more easily understood if one comprehensive description is given, rather than six individual descriptions, and the dissimilarities between members are indicated. The distinction between high- and low-temperature forms of the series can be made only by X-ray and optical means.

ALBITE—NaAlSi₃O₈
ANORTHITE—CaAl₂Si₂O₈

Crystallography. Triclinic; $\bar{1}$. Crystals commonly tabular parallel to {010}; occasionally elongated on *b* (Fig. 12.78). In anorthite crystals may be prismatic elongated on *c*.

Crystals frequently twinned according to the various laws governing the twins of orthoclase, that is *Carlsbad, Baveno,* and *Manebach.* In addition, they are nearly always twinned according to the *pericline law.* Albite twinning with twin plane {010} is commonly polysynthetic and because (010) ∧ (001) ≈ 86°, {001} either as a crystal face or cleavage is crossed by parallel groovings or striations (Figs. 12.79 and 5.40*a*). Often these striations are so fine as to be invisible to the unaided eye, but on some specimens they are coarse and easily seen. The presence of the striations on the basal cleavage is one of the best proofs that a feldspar belongs to the plagioclase series. Pericline twinning with *b* as the twin axis is also polysynthetic. Striations resulting from it can be seen on {010}. Their direction on {010} is not constant but varies with the composition.

Distinct crystals are rare. Usually in twinned, cleavable masses; as irregular grains in igneous rocks.

For albite, $C\bar{1}$: *a* = 8.14, *b* = 12.8, *c* = 7.16 Å; α = 94°20′, β = 116°34′, γ = 87°39′; *Z* = 4. For anorthite, $I\bar{1}$ or $P\bar{1}$: *a* = 8.18, *b* = 12.88, *c* = 14.17 Å; α = 93°10′, β = 115°51′, γ = 91°13′; *Z* = 8. *d*s (Ab-An): 4.02–4.03(7), 3.77–3.74(5), 3.66–3.61(6), 3.21(7)–3.20(10), 3.18(10)–3.16(7).

Physical Properties. *Cleavage* {001} perfect, {010} good. **H** 6. **G** 2.62 in albite to 2.76 in anorthite (Fig. 12.80). *Color* colorless, white, gray; less frequently greenish, yellowish, flesh-red. *Luster* vitreous to pearly. Transparent to translucent. A beautiful play of colors is frequently seen, especially in labradorite (see Plate VIII, no. 5) and andesine. *Optics*: Albite: (+); α = 1.527, β = 1.532, γ = 1.538; 2V = 74°. Anorthite: (−); α = 1.577, β = 1.585, γ = 1.590, 2V = 77° (see Fig. 12.80).

FIG. 12.78. Albite crystals.

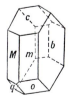 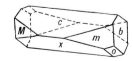

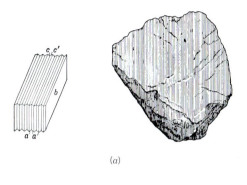

(a)

(b)

FIG. 12.79. *(a)* Albite twinning (see also Fig. 5.40). *(b)* Labradorite feldspar showing albite twinning striations on the basal pinacoid (001) (Harvard Mineralogical Museum).

Composition and Structure. An essentially complete solid solution series extends from NaAlSi₃O₈ to CaAl₂Si₂O₈ at elevated temperatures (see Fig. 11.54). Considerable K may be present, especially toward the albite end of the series (see Figs. 11.53 and

FIG. 12.80. Variation of refractive indices and specific gravity with composition in the plagioclase feldspar series.

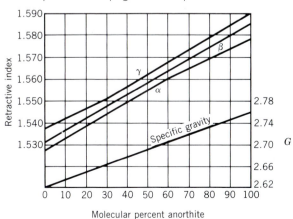

11.54). Very fine scale discontinuities in the series, which can be detected only by electron optical and X-ray means, are reflected in the occurrence of very fine lamellar intergrowths: peristerite, Bøggild and Huttenlocher intergrowths (see page 482). The structure of albite is triclinic ($C\overline{1}$) at low to moderate temperatures, and shows a highly ordered Al-Si distribution; low albite is structurally analogous to low microcline. High albite is also triclinic and has a highly disordered Al-Si distribution. At temperatures above 980°C a monoclinic variety of albite, known as monalbite, with a totally disordered Al-Si distribution exists. Anorthite has space group $P\overline{1}$ at low temperatures. At elevated temperatures the structure of anorthite inverts to $I\overline{1}$ (see Fig. 11.54).

Diagnostic Features. The plagioclase feldspars can be distinguished from other feldspars by the presence of albite twin striations on {001}. They can be placed accurately in their proper places in the series only by quantitative chemical analysis or optical tests, but they can be roughly distinguished from one another by specific gravity (see Fig. 12.80).

Occurrence. The plagioclase feldspars, as rock-forming minerals, are even more widely distributed and more abundant than the potash feldspars. They are found in igneous, metamorphic, and, more rarely, sedimentary rocks.

The classification of igneous rocks is based largely on the kind and amount of feldspar present. As a rule, the greater the percentage of SiO_2 in a rock the fewer the dark minerals, the greater amount of potash feldspar, and the more sodic the plagioclase; and conversely, the lower the percentage of SiO_2 the greater the percentage of dark minerals and the more calcic the plagioclase.

In an igneous crystallization sequence the more refractory minerals crystallize before the less refractory ones. The An end member of the plagioclase series has a much higher melting point than the Ab end member (see Fig. 11.54). For this reason plagioclase feldspars that crystallize early from a magma are generally more Ca-rich than those that crystallize later. Continuous chemical zoning occurs commonly in phenocrysts with Ca-rich centers and more Na-rich rims.

Albite is included with orthoclase and microcline in what are known as the *alkali* feldspars, all of which have a similar occurrence. They are commonly found in granites, syenites, rhyolites, and trachytes. Albite is common in pegmatites, where it may replace earlier microcline; as the platy variety, *cleavelandite*. Some albite and oligoclase shows an opalescent play of colors and is known as *moonstone*. The name *albite* is derived from the Latin *albus*, meaning *white*, in allusion to the color.

Oligoclase is characteristic of granodiorites and monzonites. In some localities, notably at Tvedestrand, Norway, it contains inclusions of hematite, which give the mineral a golden shimmer and sparkle. Such feldspar is called *aventurine* oligoclase, or *sunstone*. The name is derived from two Greek words meaning *little* and *fracture*, because it was believed to have less perfect cleavage than albite.

Andesine is rarely found except as grains in andesites and diorites. It is named for the Andes Mountains, where it is the chief feldspar in the andesite lavas.

Labradorite is the common feldspar in gabbros and basalts and in anorthosite it is the only important constituent. Found on the coast of Labrador in large cleavable masses that show a fine iridescent play of colors. The name is derived from this locality.

Bytownite is rarely found except as grains in gabbros. Named for Bytown, Canada (now the city of Ottawa).

Anorthite is rarer than the more sodic plagioclases. Found in rocks rich in dark minerals and in druses of ejected volcanic blocks and in granular limestones of contact metamorphic deposits. Name derived from the Greek word meaning *oblique*, because its crystals are triclinic.

Use. Plagioclase feldspars are less widely used than potash feldspars. Albite, or *soda spar*, as it is called commercially, is used in ceramics in a manner similar to microcline. Labradorite that shows a play of colors is polished and used as an ornamental stone. Those varieties that show opalescence are cut and sold under the name of *moonstone*.

Name. The name *plagioclase* is derived from the Greek word *plagios*, meaning *oblique*, an allusion to the oblique angle between the cleavages. (See under "Occurrence" for names of specific species.)

Feldspathoid Group

LEUCITE—KAlSi$_2$O$_6$

Crystallography. Tetragonal; $4/m$ below 605°C. Isometric, $4/m\overline{3}2/m$ above 605°C. Usually in trapezohedral crystals that at low temperature are pseudomorphs of the high-temperature form (Fig. 12.81).

Below 605°C: $I4_1/a$; $a = 13.04$, $c = 13.85$ Å; $Z = 16$. Above 605°C: $Ia3d$; $a = 13.43$ Å; $Z = 16$. ds (low T form): 5.39(8), 3.44(9), 2.27(10), 2.92(7), 2.84(7).

Physical Properties. H $5\frac{1}{2}$–6. G 2.47. *Luster* vitreous to dull. *Color* white to gray. Translucent. *Optics*: (+); $\omega = 1.508$, $\epsilon = 1.509$.

Composition and Structure. Most leucites are close to $KAlSi_2O_6$ with K_2O 21.5, Al_2O_3 23.5, SiO_2 55.0%. Small amounts of Na may replace K. The structure of the low-temperature polymorph is shown in Fig. 11.58.

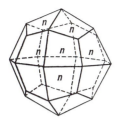

FIG. 12.81. Leucite.

Diagnostic Features. Characterized by its trapezohedral form. Leucite is usually embedded in a fine-grained matrix (see Fig. 12.82 and Plate VIII, no. 6), whereas analcime is usually in cavities in free-growing crystals.

Occurrence. Although leucite is a rather rare mineral it is abundant in certain recent lavas; rarely observed in deep-seated rocks. It is found only in silica-deficient rocks and thus never in rocks containing quartz. Its most notable occurrence is as phenocrysts in the lavas of Mount Vesuvius. In the United States it is found in rocks of the Leucite Hills, Wyoming, and in certain of the rocks in the Highwood and Bear Paw Mountains, Montana. Pseudomorphs of a mixture of nepheline, orthoclase, and analcime after leucite, *pseudoleucite*, are found in syenites of Arkansas, Montana, and Brazil.

Name. From the Greek word *leukos* meaning *white*.

Similar Species. *Pollucite*, $CsAlSi_2O_6 \cdot H_2O$, isostructural with leucite, is a rare isometric mineral usually occurring in pegmatites, notably at Bernie Lake, Manitoba.

NEPHELINE—(Na,K)AlSiO₄

Crystallography. Hexagonal; 6. Rarely in small prismatic crystals with base. Almost invariably massive, compact, and in embedded grains.

$P6_3$; $a = 10.01$, $c = 8.41$ Å; $Z = 8$. ds: 4.18(7), 3.27(7), 3.00(10), 2.88(7), 2.34(6).

Physical Properties. *Cleavage* $\{10\bar{1}0\}$ distinct. **H** $5\frac{1}{2}$–6. **G** 2.60–2.65. *Luster* vitreous in clear crystals; greasy in the massive variety. *Color* colorless, white, or yellowish. In the massive variety gray, greenish, and reddish. Transparent to translucent. *Optics*: (−); $\omega = 1.529$–1.546, $\epsilon = 1.526$–1.542.

Composition and Structure. The end member composition $NaAlSiO_4$ contains Na_2O 21.8, Al_2O_3 35.9, SiO_2 42.3%. The amount of K in natural nephelines may range from 3 to 12 weight percent K_2O. A large miscibility gap exists at low to moderate temperatures between $NaAlSiO_4$ and *kalsilite*, $KAlSiO_4$. However, at temperatures above about 1000°C a complete solid solution series exists between Na and K end members. Figure 12.83 illustrates the large miscibility gap in this chemical system. As a function of decreasing temperature, the Na^+ and K^+ ions in an originally homogeneous solid solution be-

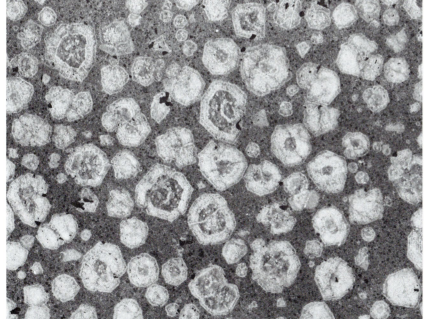

FIG. 12.82. Polished slab of basalt showing large leucite phenocrysts.

come segregated, forming finely intergrown exsolution lamellae of differing compositions. The process of exsolution in this chemical system is analogous to that discussed for the alkali feldspars (see Fig. 11.51 and page 481). The structure of nepheline is illustrated in Fig. 11.59 and discussed on page 485.

Diagnostic Features. Characterized in massive varieties by its greasy luster. Distinguished from quartz by inferior hardness.

Occurrence. Nepheline is a rock-forming mineral found in silica-deficient intrusive and extrusive rocks. Crystals are present in the lavas of Mount Vesuvius. The largest known mass of intrusive nepheline rocks is on the Kola Peninsula, Russia, where, locally, nepheline is associated with apatite. Extensive masses of nepheline rocks are found in Norway and South Africa. In the United States nepheline, both massive and in crystals, is found at Litchfield, Maine, associated with cancrinite. Found near Magnet Cove, Arkansas, and Beemerville, New Jersey. Common in the syenites of the Bancroft region of Ontario, Canada, where the associated pegmatites contain large masses of nearly pure nepheline.

Use. Iron-free nepheline, because of its high Al_2O_3 content, has been used in place of feldspar in the glass industry. Most nepheline for this purpose comes from Ontario, Canada. Nepheline produced as a by-product of apatite mining on the Kola Peninsula, Russia, is used by the Russians in several industries including ceramics, leather, textile, wood, rubber, and oil.

FIG. 12.83. Temperature-composition section for the join nepheline, $Na_3K(AlSiO_4)_4$–kalsilite, $K_4(AlSiO_4)_4$. Compare with Fig. 11.51, which is a similar diagram for the alkali feldspar system. (From J. Ferry and J. G. Blencoe, 1978, Subsolidus phase relations in the nepheline-kalsilite system at 0.5, 2.0, and 5.0 Kbar, *American Mineralogist* 63: 1225–40.)

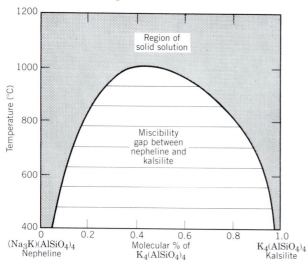

Name. *Nepheline* is derived from the Greek word *nephele* meaning a *cloud*, because when immersed in acid the mineral becomes cloudy.

Similar Species. *Cancrinite*, $Na_6Ca(CO_3)(AlSiO_4)_6\cdot2H_2O$, is a rare mineral similar to nepheline in occurrence and associations.

Sodalite—$Na_8(AlSiO_4)_6Cl_2$

Crystallography. Isometric; $\bar{4}3m$. Crystals rare, usually dodecahedrons. Commonly massive, in embedded grains.

$P\bar{4}3n$; $a = 8.83$–8.91 Å; $Z = 2$. ds: 6.33(8), 3.64(10), 2.58(5), 2.10(2), 1.75(3).

Physical Properties. *Cleavage* {011} poor. **H** $5\frac{1}{2}$–6. **G** 2.15–2.3. *Luster* vitreous. *Color* usually blue, also white, gray, green. Transparent to translucent. *Optics*: refractive index 1.483.

Composition and Structure. Na_2O 25.6, Al_2O_3 31.6, SiO_2 37.2, Cl 7.3%. Only small substitution of K for Na. In the structure of sodalite the $AlSiO_4$ framework is linked such as to produce cagelike, cubooctahedral cavities in which the Cl is housed (see Fig. 11.60).

Diagnostic Features. Usually identified by its blue color, and distinguished from lazurite by the different occurrence and absence of associated pyrite; see Plate VIII, no. 7. If color is not blue, a test for chlorine is necessary to distinguish it from analcime, leucite, and haüynite.

Occurrence. Sodalite is a comparatively rare rock-forming mineral associated with nepheline, cancrinite, and other feldspathoids in nepheline syenites, trachytes, phonolites, and so forth. Found in transparent crystals in the lavas of Vesuvius. The massive blue variety is found at Bancroft, Ontario; Ice River, British Columbia; and Namibia.

Use. May be carved as an ornamental material.

Name. Named in allusion to its sodium content.

Similar Species. Other rare feldspathoids are *haüynite*, $(Na,Ca)_{4-8}(AlSiO_4)_6(SO_4)_{1-2}$, and *noselite*, $Na_8(AlSiO_4)_6SO_4$.

Lazurite—$(Na,Ca)_8(AlSiO_4)_6(SO_4,S,Cl)_2$

Crystallography. Isometric; $\bar{4}3m$. Crystals rare, usually dodecahedral. Commonly massive, compact.

$P\bar{4}3m$; $a = 9.08$ Å; $Z = 1$. ds: 3.74(10), 2.99(10), 2.53(9), 1.545(9), 1.422(7).

Physical Properties. Cleavage {011} imperfect. **H** 5–$5\frac{1}{2}$. **G** 2.4–2.45. *Luster* vitreous. *Color* deep azure-blue, greenish-blue. Translucent. *Optics*: refractive index 1.50±.

Composition and Structure. Small amounts of Rb, Cs, Sr, and Ba may substitute for Na. There is considerable variation in the amounts of SO_4, S, and Cl. Lazurite appears isostructural with sodalite (see Fig. 11.60 and discussion on page 486). The large cagelike cavities in the structure house the Cl and S anions and the (SO_4) anionic groups.

Diagnostic Features. Characterized by its blue color and the presence of associated pyrite; see Plate VIII, no. 8.

Occurrence. Lazurite is a rare mineral, occurring usually in crystalline limestones as a product of contact metamorphism. *Lapis lazuli* is a mixture of lazurite with small amounts of calcite, pyroxene, and other silicates, and commonly contains small disseminated particles of pyrite. The best quality of lapis lazuli comes from north-eastern Afghanistan. Also found at Lake Baikal, Siberia, Russia, and in Chile.

Use. Lapis lazuli is highly prized as an ornamental stone, for carvings, and so on (see Plate XI, no. 5, Chapter 13). As a powder it was formerly used as the paint pigment ultramarine. Now ultramarine is produced synthetically.

Name. Lazurite is an obsolete synonym for azurite, and hence the mineral is named because of its color resemblance to azurite.

Petalite—Li(AlSi$_4$O$_{10}$)

Crystallography. Monoclinic; $2/m$. Crystals rare, flattened on {010} or elongated on [100]. Usually massive or in foliated cleavable masses.

$P2/a$; $a = 11.76$, $b = 5.14$, $c = 7.62$ Å; $\beta = 112°24'$; $Z = 2$. ds: 3.73(10), 3.67(9), 3.52(3), 2.99(1), 2.57(2).

Physical Properties. *Cleavage* {001} perfect, {201} good. Brittle. **H** $6-6\frac{1}{2}$. **G** 2.4. *Luster* vitreous, pearly on {001}. *Color* colorless, white, gray. Transparent to translucent. *Optics*: (+); $\alpha = 1.505$, $\beta = 1.511$, $\gamma = 1.518$; $2V = 83°$; $Z = b$. $X \wedge a = 2°-8°$. $r > v$.

Composition and Structure. LiO 4.9, Al$_2$O$_3$ 16.7, SiO$_2$ 78.4%. The structure of petalite consists of a framework of AlO$_4$ and SiO$_4$ tetrahedra that contains Si$_4$O$_{10}$ layers linked by AlO$_4$ tetrahedra. The Li is in tetrahedral coordination with oxygen.

Diagnostic Features. Characterized by its platy habit. Distinguished from spodumene by its cleavage and lesser specific gravity.

Occurrence. Petalite is found in pegmatites where it is associated with other lithium-bearing minerals such as spodumene, tourmaline, and lepidolite. Petalite was considered a rare mineral until it was found to be abundant at the Varutrask pegmatite, Sweden. More recently it has been found abundantly in Bikita, Zimbabwe, and in Namibia. At these localities, associated with lepidolite and eucryptite, it is mined extensively.

Use. An important ore of lithium. (See spodumene, page 519.) Colorless petalite may be faceted as a gem.

Name. From the Greek word *petalos* meaning *leaf*, alluding to the cleavage.

Scapolite Series

The *scapolites* are metamorphic minerals with compositions suggestive of the feldspars. There is a complete solid solution series between *marialite*, 3NaAlSi$_3$O$_8$·NaCl, and *meionite*, 3CaAl$_2$Si$_2$O$_8$·CaSO$_4$ (or CaCO$_3$). When the formulas are written in this way, it is clear that the composition of marialite is the equivalent of three formula weights of albite, NaAlSi$_3$O$_8$, plus one formula weight NaCl, and that the meionite composition is equivalent to three formula weights of anorthite, CaAl$_2$Si$_2$O$_8$, plus one formula weight of CaCO$_3$ or CaSO$_4$. In this series there is complete substitution of Ca for Na with charge compensation effected as in the feldspars by concomitant substitution of Al for Si. There is also complete substi-

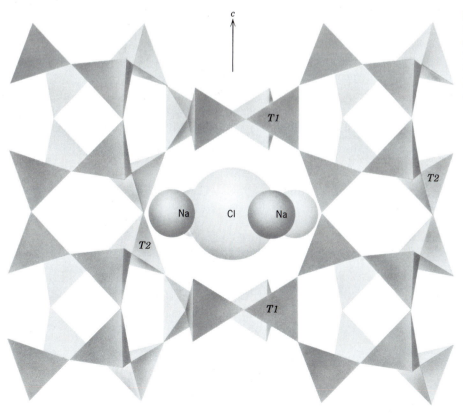

FIG. 12.84. The structure of scapolite projected on (100). (After J. J. Papike and M. Cameron, 1976, Crystal chemistry of silicate minerals of geophysical interest. *Reviews of Geophysics and Space Physics* 14: 75.)

tution of CO_3, SO_4, and Cl_2 for each other. The name *wernerite* has been used for members intermediate in composition between marialite and meionite.

The structure of scapolite consists of a framework of SiO_4 and AlO_4 tetrahedra with large cavities containing the (Ca,Na) ions and (CO_3,Cl_2,SO_4) anionic groups (see Fig. 12.84).

Marialite—Meionite

Crystallography. Tetragonal; $4/m$. Crystals usually prismatic. Prominent forms are prisms {010} and {110} and dipyramid {011}; rarely shows the faces of the tetragonal dipyramid z (Fig. 12.85). Crystals are usually coarse, or with faint fibrous appearance.

$I4/m$; $a = 12.2$, $c = 7.6$ Å; $Z = 2$. a varies slightly with composition. *d*s: (marialite), 4.24(7), 3.78(9), 3.44(10), 3.03(10), 2.68(9).

Physical Properties. *Cleavage* {100} and {110} imperfect but distinct. **H** 5–6. **G** 2.55–2.74. The specific gravity and refractive indices increase with increasing Ca content. *Luster* vitreous when fresh. *Color* white, gray, pale green; more rarely yellow, bluish, or reddish. Transparent to translucent. *Optics*: (−); ω 1.55–1.60, $\epsilon = 1.54$–1.56.

Composition and Structure. The extent of solid solution between *marialite*, $Na_4(AlSi_3O_8)_3(Cl_2,CO_3,SO_4)$, and *meionite*, $Ca_4(Al_2Si_2O_8)_3(Cl_2,CO_3,SO_4)$, is discussed on page 557. The structure, illustrated in Fig. 12.84, contains two crystallographically distinct tetrahedra, T1 and T2. In marialite compositions Al is confined to the T2 tetrahedra (46% Al, 54% Si) and only Si occupies the T1 sites. In meionite compositions Al is housed in both tetrahedral sites.

Diagnostic Features. Characterized by its crystals with square cross section and four cleavage directions at 45°. When massive, resembles feldspar but has a characteristic fibrous appearance on the cleavage surfaces.

Occurrence. Scapolite occurs in crystalline schists, gneisses, amphibolites, and granulite facies rocks. In many cases it is derived by alteration from plagioclase feldspars. It is also characteristic in crystalline limestones as a contact metamorphic mineral. Associated with diopside, amphibole, garnet, apatite, titanite, and zircon.

Crystals of gem quality with a yellow color occur in the Malagasy Republic. In the United States found in various places in Massachusetts, notably at Bolton; and in Orange, Lewis, and St. Lawrence counties, New York. Also found at various points in Ontario, Canada.

Use. Colorless crystals may be cut as faceted gemstones.

FIG. 12.85. Scapolite.

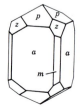

Name. From the Greek word *skapos* meaning a *shaft*, in allusion to the prismatic habit of the crystals.

Analcime—$NaAlSi_2O_6\cdot H_2O$[1]

Crystallography. Isometric; $4/m\bar{3}2/m$. Usually in trapezohedrons (Figs. 12.86a and 12.87). Cubes with trapezohedral truncations also known (Fig. 12.86b). Usually in crystals, also massive granular.

$Ia3d$; $a = 13.71$ Å; $Z = 16$. *d*s: 5.61(8), 4.85(4), 3.43(10), 2.93(7), 2.51(5).

Physical Properties. **H** 5–5½. **G** 2.27. *Luster* vitreous. *Color* colorless or white. Transparent to translucent. *Optics*: refractive index 1.48–1.49.

Composition and Structure. The chemical composition of most analcimes is fairly constant with minor amounts of K or Ca substituting for Na; also some Al substitution for Si. $NaAlSi_2O_6\cdot H_2O$ contains Na_2O 14.1, Al_2O_3 23.2, SiO_2 54.5, H_2O 8.2%. The structure is made of a framework of AlO_4 and SiO_4 tetrahedra with four- and six-membered tetrahedral rings. This framework contains continuous channels along the threefold axes of the structure which are occupied by H_2O molecules. The octahedrally coordinated Na is housed in somewhat smaller voids in the structure.

Diagnostic Features. Usually recognized by its free-growing crystals and its vitreous luster. Crystals resemble leucite, but leucite is always embedded in rock matrix.

Occurrence. Analcime occurs as a primary mineral in some igneous rocks and is also the product of hydrothermal action in the filling of basaltic cavities. It is an original constituent of analcime basalts and may occur in alkaline rocks such as aegirine-analcime-nepheline syenites. It is also found in vesicles of igneous rocks in association with prehnite, calcite, and zeolites. Fine crystals are found in the Cyclopean Islands near Sicily; in the Val di Fassa and on the Seiser Alpe, Trentino, Italy; in Victoria, Australia;

FIG. 12.86. Analcime crystals.

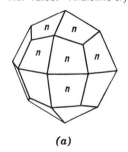

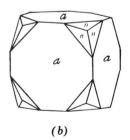

(a)   (b)

[1]Analcime may be classified as a member of the zeolite group; however, its structure, chemistry, and occurrence are very similar to those of the feldspathoids.

FIG. 12.87. Analcime from Ischia, Italy. Scanning electron microscope (SEM) photograph (From G. Gottardi and E. Galli, 1985, *Natural zeolites*. New York: Springer-Verlag; with permission.)

and Kerguelen Island in the Indian Ocean. In the United States found at Bergen Hill, New Jersey; in the Lake Superior copper district; and at Table Mountain, near Golden, Colorado. Also found at Cape Blomidon, Nova Scotia.

Name. Derived from the Greek word *analkidos* meaning *weak*, in allusion to its weak electrical property when heated or rubbed.

Zeolite Group

The number of naturally occurring zeolites is about 46 and here we will describe only four species in detail. These are: *natrolite* with a fibrous habit; *chabazite* with an equant habit; and *heulandite* and *stilbite* with platy habit and cleavage.

Natrolite—$Na_2Al_2Si_3O_{10} \cdot 2H_2O$

Crystallography. Orthorhombic; *mm2*. Prismatic, often acicular with prism zone vertically striated. Usually in radiating crystal groups (Fig. 12.88); also fibrous, massive, granular, or compact.

Fdd2; $a = 18.29$, $b = 18.64$, $c = 6.59$ Å; $Z = 8$. *d*s: 6.6(9), 5.89(8), 3.19(9), 2.86(10), 2.42(6).

Physical Properties. *Cleavage* {110} perfect. **H** $5-5\frac{1}{2}$. **G** 2.25. *Luster* vitreous. *Color* colorless or white, rarely tinted yellow to red. Transparent to translucent. *Optics:* (+); $\alpha = 1.480$, $\beta = 1.482$, $\gamma = 1.493$; $2V = 63°$; $X = a$; $Y = b$; $r < v$.

Composition and Structure. Na_2O 16.3, Al_2O_3 26.8, SiO_2 47.4, H_2O 9.5%. Some K and Ca may replace Na. Natrolite is one of the group of fibrous zeolites. Its structure consists of an Si-Al-O framework in which chains parallel to the *c* axis are prominent (see Fig. 11.61). The chains are linked laterally by the sharing of some oxygens of opposing tetrahedra. Na is coordinated between the chains to six oxygens, four of which are tetrahedral oxygens and two of which are from water molecules. The perfect {110} cleavage in natrolite is due to the relatively small number of bonds between chains as compared to the large number of bonds within the chainlike structure.

Diagnostic Features. Light color and commonly fibrous habit.

Occurrence. Natrolite is characteristically found lining cavities in basalt associated with other zeolites and calcite. Notable localities are Aussig and Salesel, Bohemia, Czechoslovakia; Puy-de-Dôme, France; and Val di Fassa, Trentino, Italy. In the United States found at Bergen Hill, New Jersey. Also found in various places in Nova Scotia.

Name. From the Latin *natrium*, meaning *sodium*, in allusion to its composition.

Similar Species. *Scolecite*, $CaAl_2Si_3O_{10} \cdot 3H_2O$, is another fibrous zeolite, similar in structure to natrolite but monoclinic in symmetry.

FIG. 12.88. Natrolite from Altavilla, Vicenza, Italy. (From G. Gottardi and E. Galli, 1985, *Natural zeolites*. New York: Springer-Verlag, with permission).

BOX 12.5
ZEOLITES AND THEIR MANY UNIQUE PROPERTIES

There is considerable discussion of the various behavioral properties of zeolites in Chapter 11, pp. 486 to 489. The focus here is on some of their additional commercial applications. About 46 natural zeolite minerals have been identified, and more than 100 zeolites have been synthesized (see Holmes, D. A. 1994. Zeolites. In *Industrial minerals and rocks*, 6th ed. Edited by D. C. Carr (senior editor). Littleton, Colo.: Society for Mining, Metallurgy, and Exploration, Inc.)

Zeolites are characterized by the following properties (from Holmes):

- High degree of hydration
- Low density and large void volume when dehydrated
- Stability of the crystal structure of many of them when dehydrated
- Cation exchange properties
- Uniform molecular-sized channels in the dehydrated crystals
- Ability to absorb gases and vapors
- Catalytic properties

Potential commercial utilization includes hundreds of possible applications. The principal uses are as follows:

- Ammonium ion removal in waste stream treatment, sewage treatment, pet litter and aquaculture
- Odor control
- Heavy metal ion removal from nuclear, mine, and industrial wastes
- Agricultural use, such as soil conditioner and animal feed supplement

Synthetic zeolites can be tailored in physical and chemical properties to serve many applications more closely, and they are more uniform in quality than their natural counterparts. However, natural zeolites are generally much lower in cost than synthetic zeolite products. Much ongoing research

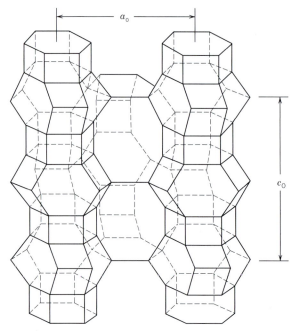

View of the framework structure of erionite. The lines connect Si and Al positions only; oxygen atoms and cations are now shown. (From Kokotailo, G. T., K. S. Sawru, and S. L. Lawton. 1972. Direct observation of stacking faults in the zeolite erionite. American Mineralogist 57: 439–44.)

in zeolite synthesis is focused on the production of zeolite structures with larger pore diameters and cage sizes. Pore diameters in zeolites are typically less than 10 Å. Synthetic zeolites with very large ring and cage sizes have recently been produced (examples of various such cages are shown in the diagram of the erionite structure).

An unusually fibrous natural zeolite, erionite, with formula $(Na_2K_2Ca_2)(Al_8Si_{28}O_{72}) \cdot 28H_2O$, has been identified as an especially serious health hazard in the region of the town of Karain, Turkey. The local geol-

ogy in that area consists of recent volcanic ash (tuff) deposits that are easily worked with tools to produce dimension stone for housing. These building materials represent the alteration of the original volcanic tuff by groundwater to produce relatively unconsolidated material consisting of montmorillonite clay and erionite. The erionite needles in the local geology and building stones have been identified as the cause of many cases of malignant mesothelioma (a fatal tumor arising from mesothelial cells in the pleura) in the region.

Chabazite—$Ca_2Al_2Si_4O_{12} \cdot 6H_2O$

Crystallography. Hexagonal, $\bar{3}2/m$ or monoclinic, $\bar{1}$. Usually in rhombohedral crystals, {$10\bar{1}1$}, with nearly cubic angles. May show several different rhombohedrons(Fig. 12.89). Often in penetration twins.

$R\bar{3}m$ or $P\bar{1}$; $a = 9.41$, $b = 9.42$, $c = 9.42$ Å; $\alpha = 94°11'$, $\beta = 94°16'$, $\gamma = 94°21'$; $Z = 6$. ds: 9.35(5), 5.03(3), 4.33(7), 3.87(3), 2.925(10).

Physical Properties. *Cleavage* {$10\bar{1}1$} poor. **H** 4–5. **G** 2.05–2.15. *Luster* vitreous. *Color* white, yellow, pink, red. Transparent to translucent. *Optics*: (+) or (−); $\alpha = 1.4848$, $\beta = 1.4852$, $\gamma = 1.4858$, $2V \sim 70°$.

Composition and Structure. The ideal composition is $Ca_2Al_2Si_4O_{12} \cdot 6H_2O$, but there is considerable replacement of Ca by Na and K as well as (Na,K)Si for CaAl. The structure of chabazite consists of an Al-Si-O framework with

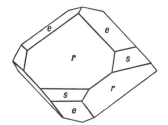

FIG. 12.89. Chabazite.

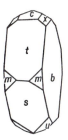

FIG. 12.91. Heulandite.

large cagelike openings bounded by rings of tetrahedra (see Fig. 11.62). The cages are connected to each other by channels that allow for the diffusion of molecules through the structure of a size comparable to the diameter (3.9 Å) of the channels. Argon (3.84 Å in diameter) is rapidly absorbed by the chabazite structure, but *iso*-butane (5.6 Å in diameter) cannot enter the structure. In this way chabazite can act as a molecular sieve (see Fig. 11.64).

Diagnostic Features. Recognized usually by its rhombohedral-appearing crystals (Fig. 12.90), and distinguished from calcite by its poorer cleavage and lack of effervescence in HCl.

Occurrence. Chabazite is found, usually with other zeolites, lining cavities in basalt. Notable localities are the Faeroe Islands; the Giant's Causeway, Ireland; Aussig, Bohemia, Czechoslovakia; Seiser Alpe, Trentino, Italy; Oberstein, Germany; and India. In the United States found at West Paterson, New Jersey, and Goble Station, Oregon. Also found in Nova Scotia and there known as *acadialite*.

Name. Chabazite is derived from a Greek word that was an ancient name for a *stone*.

Heulandite—$CaAl_2Si_7O_{18}\cdot6H_2O$

Crystallography. Monoclinic; $2/m$ but crystals often simulate orthorhombic symmetry (Figs. 12.91 and 12.92); often diamond-shaped with {010} prominent.

$C2/m$; $a = 17.71$, $b = 17.94$, $c = 7.46$ Å; $\beta = 116°20'$; $Z = 4$. ds: 8.9(7), 5.10(4), 3.97(10), 3.42(5), 2.97(7).

Physical Properties. *Cleavage* {010} perfect. **H** $3\frac{1}{2}$–4. **G** 2.18–2.2. *Luster* vitreous, pearly on {010}. *Color* colorless, white, yellow, red. Transparent to translucent. *Optics*: (+); $\alpha = 1.482$, $\beta = 1.485$, $\gamma = 1.489$; $2V = 50°$; $X = b$. $Y \wedge c = 35°$. $r > v$.

Composition and Structure. There is considerable variation in the Si/Al ratio with concomitant variation in the proportions of Ca and Na. The structure of heulandite consists of a very open Si-Al-O framework in which two-thirds of the $(Si,Al)O_4$ tetrahedra are linked to form networks of six-membered rings parallel to {010}; this is responsible for the perfect {010} cleavage. The framework contains several sets of channels that house the water molecules as well as Ca.

Diagnostic Features. Characterized by its crystal form and one direction of perfect cleavage with pearly luster.

Occurrence. Heulandite is usually found in cavities of basic igneous rocks associated with other zeolites and calcite. Found in notable quality in Iceland; the Faeroe Islands; Andreasberg, Harz Mountains, Germany; Tyrol,

FIG. 12.92. Heulandite from Val di Fassa, Italy. (From G. Gottardi and E. Galli, 1985, *Natural zeolites*. New York: Springer-Verlag, with permission.)

FIG. 12.90. Chabazite from Heinekeim, Germany. (From G. Gottardi and E. Galli, 1985, *Natural zeolites*. New York: Springer-Verlag, with permission.)

Austria; and India, near Bombay. In the United States found at West Paterson, New Jersey. Also found in Nova Scotia.

Name. In honor of the English mineral collector, H. Heuland.

Stilbite—NaCa$_2$Al$_5$Si$_{13}$O$_{36}$·14H$_2$O

Crystallography. Monoclinic; 2/m. Crystals usually tabular on {010} or in sheaflike aggregates (Figs. 12.93, 12.94 and Plate VIII, no. 9). They may also form cruciform penetration twins.

C2/m; a = 13.61, b = 18.24, c = 11.27 Å; β = 127°51'; Z = 4. ds: 9.1(9), 4.68(7), 4.08(10), 3.41(5), 3.03(7).

Physical Properties. *Cleavage* {010} perfect. H 3$\frac{1}{2}$–4. G 2.1–2.2. *Luster* vitreous; pearly on {010}. *Color* white, more rarely yellow, brown, red. Translucent. *Optics:* (−); α = 1.494, β = 1.498, γ = 1.500; 2V = 33°; Y = b, X ∧ a = 5°.

Composition and Structure. Na and K are usually present, substituting for Ca by the coupled substitution mechanisms Na$^+$Si^{4+} ⇆ Ca^{2+}Al^{3+} and (Na,K)$^+$Al^{3+} ⇆ Si^{4+}. The structure of stilbite is similar to that of heulandite with sheets of six-membered (Si,Al)O$_4$ tetrahedra parallel to {010}. This accounts for the tabular habit and excellent {010} cleavage.

Diagnostic Features. Characterized chiefly by its cleavage, pearly luster on the cleavage face, and common sheaflike groups of crystals.

Occurrence. Stilbite is found in cavities in basalts and related rocks associated with other zeolites and calcite. Notable localities are Poona, India; Isle of Skye; Faeroe Islands; Kilpatrick, Scotland; and Iceland. In the United States it is found in northeastern New Jersey. Also found in Nova Scotia.

Name. Derived from the Greek word *stilbo* meaning *luster*, in allusion to the pearly luster.

Similar Species. Other zeolites of lesser importance than those described are:

Phillipsite, KCa(Al$_3$Si$_5$O$_{16}$)·6H$_2$O, monoclinic
Harmotome, Ba(Al$_2$Si$_6$O$_{16}$)·6H$_2$O, monoclinic
Gmelinite, (Na$_2$,Ca)(Al$_2$Si$_4$O$_{12}$)·6H$_2$O, hexagonal
Laumontite, Ca(Al$_2$Si$_4$O$_{12}$)]4H$_2$O, monoclinic
Scolecite, Ca(Al$_2$Si$_3$O$_{10}$)·3H$_2$O, monoclinic

FIG. 12.93. Stilbite.

FIG. 12.94.   Stilbite, Jewel Tunnel, Bombay, India.

Thomsonite, NaCa$_2$(Al$_5$Si$_5$O$_{20}$)·6H$_2$O, orthorhombic. A compact, massive type is used as a gem. When polished, red, yellow, or green "eyes" are seen on a rounded surface.

REFERENCES AND SELECTED READING

In addition to these references, see references from Chapter 11.

Abelson, P. H. 1990. The asbestos removal fiasco (editorial). *Science* 247: 1017.

Carr, D. (ed.). *Industrial minerals and rocks.* 1994. 6th ed. Littleton, Colo.: Society for Mining, Metallurgy, and Exploration, Inc.

Champness, P. E., G. Cliff, and G. W. Lorimer. 1976. The identification of asbestos. *Journal of Microscopy* 108: 231–49.

Darragh, P. J., A. J. Gaskin, and J. V. Sanders. 1976. Opals. *Scientific American* 234, 4: 84–95.

Deer, W. A., R. A. Howie, and J. Zussman. 1962 and 1963. *Rock-forming minerals*, v. 1–4. New York: Wiley.

———. 1978. *Single-chain silicates*, v. 2A. 1982, *Orthosilicates*, v. 1A; 1986; *Disilicates and ring silicates* v. 1B; *Double-chain silicates*, v. 2B. For vols. 2A, 1A, and 1B. New York: Wiley. For vol. 2B, London: The Geological Society.

———. 1992. *An introduction to the rock-forming minerals.* 2nd ed. New York: Wiley.

Feldspar mineralogy. 1983. *Reviews in Mineralogy* 2. 2nd ed. Mineralogical Society of America, Washington, D. C.

Frondel, C. 1962. *The system of mineralogy 3. Silica Minerals.* New York: Wiley.

Gottardi, G., and E. Galli. *Natural zeolites.* New York: Springer-Verlag.

Grim, R. E. 1968. *Clay mineralogy.* 2nd ed. New York: McGraw-Hill.

Morimoto, N., J. Fabries, A. K. Ferguson, I. V. Ginzburg, M. Ross, F. A. Siefert, J. Zussman, K. Aoki, and G. Gottardi. 1988. Nomenclature of pyroxenes. *American Mineralogist*: 73: 1123–1133.

Mossman, B. T., J. Bignon, J. Corn, A. Seaton, and J. B. L. Gee. 1990. Asbestos: Scientific developments and implications for public policy. *Science*: 247: 294–301.

Mossman, B. T., and J. B. L. Gee. 1989. Asbestos-related diseases. *New England Journal of Medicine*: 320: 1721–29.

Nesse, W. D. 1991. *Introduction to optical mineralogy.* 2nd ed. New York: Oxford University Press.

Pyroxenes. 1980. *Reviews in Mineralogy 7.* Mineralogical Society of America, Washington, D. C.

Ross, M. 1981. The geologic occurrences and health hazards of amphiboles and serpentine asbestos, in *Amphiboles, Reviews in Mineralogy 9*: 279–325.

———. 1984. A survey of asbestos related disease in trades and mining occupations and in factory and mining communities as a means of predicting health risks of nonoccupational exposure to fibrous minerals. *American Society for Testing and Materials,* Special Publication no. 834: 51–104.

CHAPTER 13

GEM MINERALS

Because most gems are minerals, it is fitting to devote a brief section to gemology, the science of gemstones, in a textbook of mineral science. In fact, the use of the term gem *is legally restricted in the American gem trade to refer to stones of natural origin.*

Gemstones have always been a fascinating subject. In earlier centuries gems and jewelry were mainly reserved for the ruling classes, but today nearly everyone owns or has purchased gemstones as part of jewelry items.

This brief chapter introduces very few new concepts or techniques but draws mainly on subject matter already covered in prior chapters. Although the instructor may not cover this specific material in lectures, students who have mastered the subject matter in a mineral science course can easily read this chapter. After reading this chapter, you will be a well-informed customer in jewelry stores.

A gem is defined as follows:

a gem is a mineral which, by cutting and polishing, possesses sufficient beauty to be used in jewelry or for personal adornment.

Although beauty is the prime requirement, the more desirable and more valuable gems are *both rare and durable.* Included in the term *gem* are several organic gem materials, pearl, amber, coral, and jet, which, although products of nature, are not strictly minerals. They will, therefore, be omitted from our discussion. The reason for the restricted use of the term *gem* is to exclude all manufactured simulants; but in spite of the legislation, they are frequently referred to as *synthetic gems.* In a study of gems these synthetics cannot be ignored, for one of the major tasks of the gemologist, and frequently a difficult one, is to determine whether a stone is natural or synthetic. Such determinations are important because the rare natural gem may have a value several hundred times greater than a similar-appearing synthetic substitute. It is also important to know whether the color of a gemstone is natural or has been artificially induced. Thus, in the latter part of this chapter we will discuss synthetics briefly, how they are manufactured, how they can be distinguished from the

gems they simulate, and the manner in which the color of gems can be enhanced by treatment.

GEM MINERALS

Because a mineral must have certain qualifications to be placed in the special category of gem minerals, the number is limited. Of the approximately 3800 known mineral species, approximately 70 meet the requirements, and of these about 15 can be considered as important gem minerals.

Chapters 8 to 12 on "Systematic Mineralogy" include all the minerals commonly used as gems as well as many others less frequently encountered as cut stones. In the following list these minerals are arranged by chemical groups, the order in which they are described on earlier pages, with the names of the more important gem minerals in boldface type. However, it should be pointed out that the list is not all-inclusive and that other rarer minerals occasionally appear as gemstones. The gemologist is frequently called upon to identify the minerals from which ornamental objects have been carved. Therefore, included in Table 13.1 are some more commonly used for carvings than for cut stones.

Plate IX

1. Diamond crystals with 5 carat cut stone.

2. The "Patricia Emerald"; 6.6 cm high, 126 grams in weight. (The American Museum of Natural History.)

3. Natural sapphire crystals, 3.5 cm long.

4. Ruby crystal on calcite, Jagadalaic, Afghanistan. Crystal 1.3 cm wide.

5. The "Denise Emerald," 83.1 carats.

6. Sapphire, 6.5 carats.

7. Ruby 7.21 carats.

Plate X

1. Smoky quartz.
2. Amethyst.
3. Citrine.
4. Rose quartz.
5. Quartz (with the exception of aventurine) micro-crystalline. Clockwise from upper left: aventurine quartz, agate, carnelian, chrysoprase, tiger's eye.
6. Agate snuff bottle and moss agate.
7. Opal.
8. Chrysoberyl.
9. Zircon.
10. Spinel.
11. Tourmaline.

(Harvard Mineralogical Museum.)

Plate XI

1. Olivine (peridot).
2. Garnet.
3. Topaz, natural colors.
4. Topaz. From left to right: colorless stream pebble; colored by irradiation; heat treated after irradiation; below: gem cut from treated material.
5. Lapis lazuli.
6. Jadeite.
7. Turquoise.
8. Beryl.
9. Synthetic emerald crystals.
(Harvard Mineralogical Museum.)

Plate XII

1. Fluorite.
2. Sphalerite.
3. Petalite.
4. Amblygonite.
5. Sinhalite.
6. Diopside.
7. Dioptase.
8. Apatite.
9. Euclase.
10. Enstatite.
11. Barite.
12. Phenakite.
13. Brazilianite.
14. Scheelite.

(Courtesy of A. Rupenthal,
Idar-Oberstein, Germany.)

TABLE 13.1   Gem Minerals*

NATIVE	TUNGSTATES	SILICATES
ELEMENTS	Scheelite	(continued)
Diamond	PHOSPHATES	Axinite
SULFIDES	Beryllonite	**Beryl**
Sphalerite	Apatite	Cordierite
Pyrite	Amblygonite	**Tourmaline**
OXIDES	Brazilianite	Enstatite-
Zincite	**Turquoise**	hypersthene
Corundum	Variscite	Diopside
Hematite	SILICATES	**Jadeite** (jade)
Rutile	Phenacite	Spodumene
Anatase	Willemite	Rhodonite
Cassiterite	**Olivine**	**Tremolite-actinolite**
Spinel	**Garnet**	(nephrite jade)
Gahnite	**Zircon**	Serpentine
Chrysoberyl	Euclase	Talc
HALIDES	Andalusite	Prehnite
Fluorite	Sillimanite	Chrysocolla
CARBONATES	Kyanite	Dioptase
Calcite	**Topaz**	**Quartz**
Rhodochrosite	Staurolite	**Opal**
Smithsonite	Datolite	Feldspar
Aragonite	Titanite	Danburite
Malachite	Benitoite	Sodalite
Azurite	Zoisite	Lazurite
SULFATES	Epidote	Petalite
Gypsum	Vesuvianite	Scapolite
		Thomsonite

*Colored photographs of many of these minerals and gems cut from them are given in Plates IX–XII.

GEM QUALIFICATIONS

Of the several attributes of gemstones, beauty is most important. Factors contributing to beauty are color, luster, transparency, and, through skillful cutting, brilliance and fire. Most gems possess two or more of these qualities, but in some nontransparent stones, such as turquoise, beauty lies in color alone. Opal, one of the most beautiful gems, owes its attractiveness to flashes of spectral colors diffracted from the stone's interior, a phenomenon known as play of color.

Because most gemstones are used for personal adornment, they should be able to resist scratching and abrasion, which would dull their luster and mar their beauty. Thus durability, which depends on hardness and toughness, is the second requisite of a gem mineral. It is generally considered that a gemstone hard enough to resist abrasion should have a minimum hardness of 7, that of quartz. Only 10 or 12 gems satisfy this requirement, but others fashioned from softer minerals will retain their luster for many years if worn with care. This does not mean that care should not be exercised in wearing hard gemstones, for some are brittle and a sharp blow

may cause them to fracture or cleave. A few minerals with a hardness less than 7 may be extremely tough. This is true of nephrite jade. A blow that would shatter a diamond would have little effect on the nephrite.

In addition to the two intrinsic properties of beauty and durability, other factors influence the desirability of a gem. Chief of these is *rarity*, long an attribute of the highly prized gems—diamond, emerald, ruby, and sapphire. As the supply diminishes and they become rarer, their value increases. If, on the other hand, a gem through new discoveries becomes abundant, it decreases in value and loses its appeal to the few that formerly could afford it. The desirability of gems is also subject to the vagaries of *fashion*. For example, the dark red pyrope garnet was much in vogue during the nineteenth century but is in little demand today. A final factor that can be added to the desirable properties of gems is *portability*. Gemstones are unique in combining high value with small volume and weight and can be transported easily from one country to another during periods of political unrest and economic instability.

TYPES OF GEM CUTS

Space does not permit an account of the manner in which gem rough is fashioned into gemstones. Yet it is important to point out the two basic types of cuts because mention is frequently made of them on the following pages. There are the *cabochon* cut and the *faceted* cut. The cabochon has a smooth-domed top and most commonly a flat base (Fig. 13.1). Faceted gemstones are bounded by polished plane surfaces (facets) to which different names are given depending on their position. When viewed from above, the outline of faceted stones may be round, oval, rectangular, square, or other shapes. The brilliant cut, most common for diamond, is shown in Fig. 13.2. The upper part of the stone is called the *bezel*, *crown*, or *top*; the lower part is called the *pavilion*, *base*, or *back*. The central top facet is the *table* and the small facet parallel to it is the *culet*. The edge between bezel and pavilion is the *girdle*.

FIG. 13.1. Cross sections of cabochon cut stones. *(a)* Simple cabochon with domed top and flat base. *(b)* Double cabochon. *(c)* Hollow cabochon.

(a)   (b)   (c)

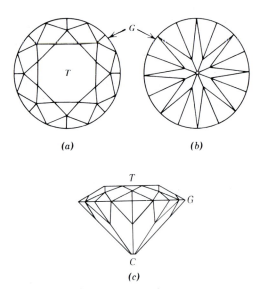

FIG. 13.2. A faceted gemstone (brilliant cut) with 58 facets. *(a)* The bezel showing the top facets. *(b)* The pavilion showing the bottom facets. *(c)* Front view showing the table (*T*), culet (*C*), and girdle (*G*).

THE EARLY USES OF GEMS

Although gemology as a science is modern, interest in gems is ancient and extends back beyond the dawn of history. Archaeological finds present clear evidence that our remote ancestors collected and treasured gem minerals as objects of beauty. By the beginning of history, numerous gems were not only known but were shaped to enhance their beauty and even pierced, permitting them to be strung together and worn as necklaces or bracelets. Four thousand years before the Christian Era, gems of lapis lazuli, chalcedony, amazonite, and jasper were cut and carved in Babylonia. The presence of lapis lazuli, undoubtedly from Afghanistan, shows that gem trade routes existed at that time. Later trade extended as far as Egypt, for lapis lazuli has been found in Egyptian tombs of the Twelfth Dynasty erected 4400 years ago. But far earlier in the predynastic period (5000–3000 B.C.) ornaments were fashioned in Egypt from a large number of minerals. They included several varieties of quartz—rock crystal, chalcedony, agate, carnelian, chrysoprase, and jasper—as well as turquoise, chrysocolla, amazonite, olivine, fluorite, and malachite. Emerald was also a gem well known to the early Egyptians. But unlike many of the other gems found in Egyptian tombs, whose source is unknown, emeralds were obtained from the Zabara Mountains in upper Egypt near the shore of the Red Sea. Even today, one can see abundant evidence of these ancient mining activities.

By the time of the ancient Greek and Roman civilizations, many of the gem minerals were known and the art of carving them was well advanced. The people of these civilizations valued gems as things of beauty and used them, as they are used today, for personal adornment. But of equal or perhaps greater importance were the presumed supernatural powers that they bestowed on the wearer. The belief in these mystical properties of gems was already old by Roman times, and each successive culture or civilization has added to the lore or independently created its own. Wearing gems as talismans or amulets was widespread, with different virtues attributed to different stones. There was virtually no disaster or human ailment that some gem would not guard against. They protected the wearer from disease, poison, lightning, fire, and intoxication; others rendered him invisible or invulnerable or endowed him with strength and wisdom.

Astrology played a part in the superstitions surrounding gems, remnants of which persist today. A particular stone was believed to possess special virtues if worn under zodiacal control, that is, during a given month. A definite gem was assigned to each sign of the zodiac and thus, for most effective influence, a different stone should be worn each month. Because this required 12 gems, which few could possess, and because the most potent stone corresponded to the sign in which the birthday of the wearer fell, the use of month stones or birthstones gradually evolved.

The astrologers who assigned gems to zodiacal signs were obviously influenced by earlier religious symbolism of gems, because the early lists of birthstones bear a striking resemblance to gems mentioned in the Bible. In Exodus 28 instructions are given for preparing the breastplate of the high priest of the 12 tribes of Israel as follows: "And thou shall set it in settings of stones, even four rows of stones; the first row shall be a sardius, a topaz, and a carbuncle. And the second row shall be an emerald, a sapphire, and a diamond; and the third row a ligure, an agate, and an amethyst; and the fourth row a beryl, and an onyx, and a jasper." With the exception of *ligure* (possibly zircon), all the names of the various stones are still in use, but certainly many do not refer to stones so known today. In ancient times minerals were classified chiefly by color, with a single name given to what we now recognize as several species. Carbuncle was red and could refer to ruby or spinel but most likely to garnet. Topaz was a green stone, possibly olivine, and a sapphire a blue one, perhaps lapis lazuli. What was called diamond could not have been today's gem, for one large enough for the breastplate was rare, if indeed known, and was too hard to engrave with the name of one of the tribes. It is possible that emerald, amethyst, beryl, sardius (sard), and onyx refer to gems as we know them.

Although the relation of some early described gems to modern gem and mineral names is uncertain, it seems probable that the gem minerals listed as "important" (Table 13.1) were, with few exceptions, known 2000 years ago. The esteem in which individual gems are held, and hence their value, has changed from time to time through the centuries and is still subject to change. Even at the current time there is no unanimity among gemologists as to the ordering of gems according to value.

IMPORTANT GEMS—YESTERDAY AND TODAY

Diamond

For most of historic time diamond has held the preeminent position as the most coveted gem, but it was rare among the ancients. Pliny, writing in 100 A.D., said it is "the most valuable of gemstones, but known only to kings." Its great value results from its high hardness, brilliant luster, and high dispersion, giving rise to flashes of spectral colors. Moreover, it is quite insoluble in acids or alkalis. Because of these resistant properties it was called *adamas*, from the Greek meaning "the invincible," from which the name *diamond* is derived.

India was the chief source of early diamonds. Diamonds were won from three separate alluvial diggings, but the most important area was Golconda. This was not a mine but a trading center whose name was given to a district in the southern part of the country that yielded many famous stones. Some of these gems are known today; the whereabouts of others are unknown. But each in its time has left a trail of treachery, intrigue, murder, and wars. The most celebrated of these ancient diamonds is the Koh-i-nor, which, after recutting weighs 108.83 carats,[1] and is included in the British crown jewels. The Great Mogul (180 carats), described in detail by the French gem merchant Tavernier in the middle of the seventeenth century, vanished after the sack of Delhi in 1739. The blue Hope diamond (45.52 carats), with a long history of tragedy and disaster, now occupies a central position in the gem display of the Smithsonian Institution, Washington, D.C.

With the discovery of diamonds in river gravels in Brazil in 1725, that country became and remained for 140 years the world's leading diamond producer. In 1866 diamonds were discovered in South Africa, first in stream gravels but shortly thereafter in diamond "pipes," the rock in which they formed. During the following 100 years, numerous finds of diamonds, both alluvial and in situ, were made in a dozen African countries, but South Africa remains the largest producer of African gem diamonds. It was at the Premier Mine in that country that the Cullinan (3106 carats), the world's largest gem diamond, was found in 1905. In recent years diamonds have been coming from Russia. Although the figures are unreliable, it is estimated that Russia accounts for about 13% of the world production and is expanding its capacity. In 1995 the world diamond production totaled about 108 million carats, but the supply is increasing. Each year brings reports of new finds and expanded production from old mines. For example, Botswana produces about 17% and Australia about 41% of total world diamond production, which includes gem as well as industrial diamonds.

About 75% of the diamonds produced are of industrial grade, not suitable for gems. Stones cut from the remaining 25% have great variation in value, which depends on what is sometimes called the four Cs; these are color, clarity, cut, and carat weight. Diamonds are graded into many categories on the basis of color and clarity (imperfections). At the top of the scale, the flawless stone of best color has a value at least ten times that of a stone of equal weight at the bottom of the scale. To give the optimum brilliance and fire, a diamond must be properly proportioned. A well-cut stone may be worth 50% more than one poorly cut but of equal weight, color, and clarity. For this reason many old diamonds are recut to increase their value even though they lose weight in the process. The price of a diamond is obviously related to its weight but not in proportion to it. If a 1-carat stone is worth $6500, a 2-carat stone of like quality may be worth $8400 per carat, and one of 5 carats $12,000 per carat. Examples of several diamond cuts are shown in Fig. 13.3.

Beryl

As a gem mineral, beryl occurs in transparent crystals in several color varieties to which special names have been given. These are *goshenite*, colorless; *morganite*, pink; *aquamarine*, blue-green; and *golden beryl*, yellow. These beryls make handsome gems but it is only the deep green variety, *emerald*, that has ranked as a "precious" stone. Since ancient

---

[1]One carat equals 0.2 gram.

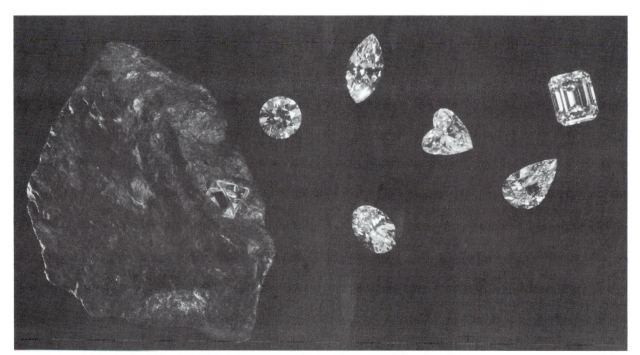

FIG. 13.3. Diamond, octahedral crystal in kimberlite matrix with faceted gems. The types of cuts, clockwise from upper left, are brilliant, marquise, heart-shaped, emerald cut, pendoloque (pear-shaped), oval. The brilliant cut stone (upper left) is approximately 4 carats. (Courtesy of the Diamond Information Center, New York, N.Y.)

times a high value has been placed on emerald, and today its value may equal, or even exceed, that of diamond and ruby.

Although emeralds were mined in ancient Egypt, nearly 4000 years ago, it was not until the Spanish conquest in the sixteenth century that large, high-quality stones reached Europe and the rest of the world. When the Spanish arrived in what is now Colombia, Ecuador, and Peru, they found the Indians there had large stones of emerald. These they seized, but it was not until later that they learned that the source was Colombia. Other emerald deposits have since been found in Russia (1830), Australia (about 1900), South Africa (1927), and Zimbabwe (1956). But Colombia remains the unquestioned source of fine stones, as it has been for over 400 years. There are two separate mining districts: Chivor, 75 km northeast of Bogota, and Muzo, 100 km north of Bogota. The mines at both localities have been at various times under private and government ownership. But whoever the operator, stealing, bribery, and even murder have accompanied the mining operation. Today at nationalized mines, the work is carried out under armed guard. Still the same problems exist and a

very high percentage of Colombian emerald production is mined and sold illegally.

Ruby and Sapphire

These gem varieties of corundum have since ancient times been held in high esteem, and with diamond and emerald they are regarded as the most valuable of gemstones. Ruby receives its name from the Latin *ruber*, meaning red. Sapphire comes from the Latin *sapphirus*, meaning blue, and was first used to denote any blue gemstone. Today all gem corundum, except red, is called sapphire but the color is specified as yellow sapphire, purple sapphire, and so on. In the gem trade, however, *sapphire* is assumed to be blue.

Rubies may be of various shades of red, but the deep red, known as pigeon-blood, is of greatest value. Flawless stones of this color are seldom larger than three carats, and those that exceed ten carats are extremely rare. In the past, large rubies were reported but the whereabouts of most of them are unknown. Travernier in the seventeenth century described a ruby owned by the King of Bijapur as weighing more than 50 carats. The German Emperor Rudolph II is reported to have possessed a ruby the size of a hen's egg. Because rubies have been confused with other red stones, particularly spinel and tourmaline, it may be that some of the earlier reported large rubies "disappeared" on the discovery that they were other gems. Blue sapphires are far more abundant than rubies of similar quality and thus are of less value. Also, large sapphire crystals are not uncommon and numerous stones exceeding 100 carats have been described, many of which are known today. Transparent rubies and sapphires are commonly faceted, but nontransparent stones of fine color may be cut cabochon. This is particularly true if there is indication of asterism in the rough. Care must be taken in cutting such a stone to have the c axis perpendicular to the base of the cabochon so that the star appearing on the upper curved surface will be centered. Since the advent of synthetic star rubies and star sapphires, the popularity of natural "stars" has decreased, and they are of no greater value than cabochons of similar size and color that lack asterism.

Since the fifteenth century, Burma has been the source of the finest rubies. They come from three separate areas, but the most important is near Mogok. The rubies originated in a metamorphic limestone, and mining of the parent rock has yielded some fine stones. However, most of the gems are recovered, as they have been since ancient times, from the soil of the mountain slopes and from the gravels of the stream valleys. Stones of poorer quality than the Burmese have long come from alluvial deposits in Cambodia, Thailand, and Sri Lanka. The most recent major ruby discovery is in the Tsavo National Park, Kenya. Mining, which began there in 1980, has produced stones of high quality said to rival those from Burma.

Because ruby and sapphire are both varieties of corundum, it is not surprising that they form under similar conditions and have similar occurrences. Excellent sapphires of a cornflower blue—the most desirable color—come from alluvial deposits near Batambang in Thailand and adjacent areas of Cambodia. In Sri Lanka sapphires of good quality are recovered from the gem gravels. In 1881 in Kashmir in the Himalaya Mountains, a landslide revealed the presence of sapphires. These "Kashmir" blue stones are highly prized but are somewhat lighter in color than those from Cambodia and Thailand.

Opal

Precious opal lacks the high hardness, clarity, and brilliance of most of the other important gems. But even without these properties, it is highly prized. The great value of precious opal lies in a subtle beauty resulting from an internal display of flashing colors. Various names have been given to precious opal, such as *black opal*, from which a brilliant play of color is emitted from a dark body color; and *fire opal*, in which a translucent background gives forth flashes of red to orange colors. Handsome carvings have been executed from opal but as a gem it is always cut cabochon. Many opal gemstones are doublets; that is, a thin slice of precious opal is cemented to a slice of nongem opal, or other material, and the assembly is cut and polished. Because most opals are nontransparent and the color flashes a surface phenomenon, the doublet resembles a stone cut from a single piece and the deception, if such is intended, is difficult to detect. Some doublets are capped with colorless quartz, which protects the softer opal from damage, yet does not detract from the beauty of the stone.

Opal was well known and greatly valued in Roman times. The high esteem in which it was held is indicated by Pliny in the story of Nonius, a Roman senator who possessed an opal the size of a hazelnut. The emperor Marcus Antonius demanded the stone, but rather than part with it, Nonius abandoned his other possessions and fled the country. With only minor fluctuations, opals maintained their high value until early in the nineteenth century. At that time a decrease in popularity and a consequent lessening of value may have resulted from the superstition that opal was an unlucky stone. Today, with the superstition either forgotten or ignored, opal commands a high place among important gems.

Ancient opals of fine quality are believed to have come from India and those of poorer quality from Egypt. From Roman times until late in the nineteenth century, most precious opal came from mines near Czerevenitza, Hungary (now in Czechoslovakia). Today Australia is the principal source, with production from several widely separated localities. Opal occurs in several locations in Mexico, making that country the second major producer.

Jade

The name *jade* is given to two different minerals: jadeite, a pyroxene, and tremolite-actinolite, an amphibole (called nephrite). The two types have in common an extreme toughness and generally a green color. Although nephrite is composed of a felted mass of fibers and jadeite is an aggregate of interlocking grains, the textural differences are obscured on polished pieces, making identification difficult. Distinction is best made by specific gravity: nephrite $G = 3.0\pm$; jadeite $G = 3.3$.

When the Spanish conquistadores arrived in Mexico and Central America, they found carved jadeite that was greatly valued by the natives for the important role it played in their social and religious life. It is reported that Montezuma considered a gift of two pieces of jade to Cortez equal to two cartloads of gold. Recent finds of jadeite in Central America have been made, but it is uncertain as to whether they are the source of the ancient jade of the Aztecs and Mayas.

On the discovery of New Zealand in 1769, Captain Cook found that the native Maoris also placed a high value on jade and carved it into ornaments, utensils, and weapons. This was nephrite found in stream boulders and along the west coast of the south island. The source of this jade has been located in the mountains to the east and is actively worked today.

Many centuries before the West knew of jade in America and New Zealand, it was valued in China above all other substances, and the art of working it into elaborate carvings was ancient. The early Chinese jade was nephrite from Chinese Turkestan; it was much later that jadeite, the rarer and more valued jade, was brought from Burma to be carved by Chinese artisans. Jadeite is still being recovered as stream boulders from the Uru River in Upper Burma.

Chrysoberyl

The most common chrysoberyl occurs in shades of yellow, green, or brown. Its high hardness $(8\frac{1}{2})$, transparency, and pleasing color make it a desirable gemstone. However, it is for the far less common varieties of *cat's eye* and *alexandrite* that the gem is best known and most valued. A cabochon cut cat's eye shows a marked chatoyancy, a sharp band of light crossing a yellow, green, or brown background. Numerous other gems show a chatoyancy, but the term *cat's eye* should be reserved for chatoyant chrysoberyl. These rare and highly prized stones come mostly from Sri Lanka.

Alexandrite is a dark-colored chrysoberyl with the remarkable property of appearing green in daylight and red in artificial (incandescent) light. Because of this color change, it has been described as "an emerald by day and a ruby by night." Alexandrite was discovered in the Ural Mountains, Russia, in 1830 on the day the Czarevitch, later Czar Alexander II of Russia, reached his majority, and was named for him. Alexandrites are also found in the gem gravels of Sri Lanka and, although their color change is less dramatic than in the Russian stones, they command a high price.

Topaz

For centuries the name *topaz* has been subject to much confusion, a situation that still exists with both jeweler and layperson. The ancients undoubtedly used the word for gem olivine, *peridot*, that came from Zebirget, an island in the Red Sea. The first use of the name for the mineral we know today as topaz was by Henckel in 1737 for yellow to sherry-colored gems from Saxony. Since then it has been used for stones of this color, sometimes called *precious topaz*, to set it apart from other similar-appearing gems that may be called topaz. The most common confusion is with yellow quartz, citrine, for which various names have been used to imply the stone is topaz or a close relative.

The confusion regarding topaz is compounded by the fact that gem-quality crystals of the mineral are not always wine-yellow. In fact, they are most commonly colorless, but may be pink, red, orange, greenish, pale blue, or brown. Furthermore, certain colorless topaz can be turned a deep aquamarine-blue by irradiation followed by heat treatment. Most of this treated topaz comes from Brazil, which is also the source of most of the precious topaz.

Tourmaline

Tourmaline is unique among gem minerals in occurring in shades of all colors as well as in multicolored crystals. Names have been given to the various colored varieties, but it is more meaningful to call them tourmaline with the appropriate color modifier. Tourmaline is a widespread mineral and undoubtedly was early used as a gem, but it was not until early in the eighteenth century that its identity became known. At that time it received its name from the Singhalese word, *turamali*, used in Ceylon (now Sri Lanka) for yellow zircon. It is reported that when a parcel of turamali sent to Holland turned out to be yellow tourmaline, the name was given to that mineral, and, although a misnomer, it has persisted.

There are numerous worldwide occurrences of gem tourmaline. Some of the more important are in Minas Gerais, Brazil; Malagasy Republic; Mozambique; and Namibia. In the United States the first significant gem find was of tourmaline at Mount Mica, Maine. In 1972 another major deposit was discovered not far away in Newry, Maine. Other important localities are in California, mostly in San Diego County. Mining of gem tourmaline began there 100 years ago and, with reopening of old mines and the discovery of new deposits, is still producing superb specimens.

Quartz

Quartz is listed with the "important" gem minerals not for its rarity or high value but because of its many common and abundant gem varieties. In all likelihood, quartz was the first mineral used for personal adornment and has occupied a prominent place among gems ever since. Of the 12 stones in the breastplate of the high priest, at least half appear to have been a quartz variety. Theophrastus in his treatise *On Stones*, written about 300 B.C., mentions quartz and its varieties more than any other mineral, and states "among the ancients, there was no precious stone in more common use." It is still in common use, and all but two or three of its many coarsely crystalline and fine-grained varieties are used as gem materials. One should refer to "Quartz" (Chapter 12) for the names and properties of the gem varieties.

Turquoise

Turquoise, valued chiefly for its color, is an ancient gem material. The oldest mines are at Sarabit Elkhadem on the Sinai Peninsula and date as far back as 4000 B.C. It appears certain that carved turquoise in bracelets of Queen Zer of Egypt's First Dynasty came from this locality. Nishapur, Iran (Persia), is the source of the finest turquoise today just as it has been since mining began there more than 2000 years ago. The only other deposits of importance today are in China, Tibet, and in southwestern United States.

Turquoise came from many localities in the American Southwest, but the most famous is in the Los Cerillos Mountains, New Mexico, where the American Indians may have begun mining as early as 1000 years ago. Indian jewelry set with turquoise has become very popular in recent years. The result has been not only a large increase in price but the use of substitutes and the manufacture of synthetic turquoise.

Garnet

The great range in the chemistry of garnet, greater than in any other gem material, is reflected in the diversity of color and other physical properties of its several varieties. For a description of the various species and the extensive solid solution between them, one should refer to the discussion of the garnet group in Chapters 11 and 12. Since biblical times the garnet used as gems is dark red and thus to many is the only color associated with the mineral. Yet garnet occurs in all colors except blue. Mention is made here of only the more unusual colors that may be encountered in garnet gemstones.

Pyrope and *almandine,* and solid solutions between them, are various shades of red and violet. These are the garnets used in early times as gems. *Grossular* may also be red, but it is the orange-yellow (hessonite) to orange-brown (essonite) material that has been most widely used as gemstones. Recently a transparent, emerald-green variety, called *tsavorite,* has been found in the Tsavo National Park, Kenya. Stones of tsavorite are small but highly valued. A jade-green, massive variety of grossular from South Africa has been used as a jade substitute. *Andradite* may be yellow-green to black, but it is only the transparent green variety, *demantoid,* that is important as a gem. It is the garnet of greatest value and occupies a high position among all gems. *Spessartine* of gem quality is rare, but yellow, yellow-brown, and orange-brown crystals are cut into lovely stones. *Uvarovite,* the chromium garnet, occurs in brilliant, deep green crystals. The crystals are too small for cutting but were they larger, they would make striking gems.

Zircon

Zircon is an Eastern gem that for centuries has come from the gem gravels of Sri Lanka and Indochina. It is recovered as water-worn crystals, usually in shades of red or brown, but it may be green, gray, or colorless. For a long time, the reddish-brown crystals have been heat treated to produce more attractive colors. Heating in air usually yields a yellow stone, but blue or colorless stones result when heated in a reducing environment. The most popular colored stone is blue, which is sold under the name of *starlite.*

Zircon has a high refractive index and a high dispersion, giving a brilliance and fire to a cut stone approaching those of a diamond. Because of the resemblance, colorless zircons from Matura, Sri Lanka, were called "Matura diamonds."

There are two types of zircon used as gems: a crystalline variety and a metamict type. The chemical composition is given as $ZrSiO_4$, but hafnium is always present, and sometimes small amounts of thorium and uranium are present as well. Originally all zircon was crystalline, but irradiation by these radioactive elements destroys the structure, making the crystals amorphous. Metamict zircons are usually green and have lower refractive indices and density than the crystalline variety (see p. 148 for a discussion of the metamict state).

Olivine

The olive-green gem variety of olivine is today known as *peridot,* but until recently was called chrysolite (a synonym for olivine) and "evening emerald." The gem was known to the ancients as topazian, a name presumably derived from Topazias, an island in the Red Sea. It was called topaz until the eighteenth century, when the name was transferred to another gem, our present-day topaz. Although the early stones undoubtedly came from the Red Sea island, for centuries their source remained a mystery until it was rediscovered early in the twentieth century. Since rediscovery, the island, known as Zebirget (formerly Saint John's Island), has yielded a large quantity of fine peridot crystals. In 1958, the deposit was nationalized by the Egyptian government. Although peridot has come from Burma, Australia, Norway, and the United States, Zebirget remains the outstanding locality for large stones of high quality.

GEM PROPERTIES AND INSTRUMENTS FOR THEIR DETERMINATION

Gems are basically minerals, so the student who has digested the material presented in the first twelve chapters of this book has in essence learned of the properties of gems and how to determine them. This section briefly reviews physical properties as they pertain to gem identification and then describes identification tools.

Physical Properties

Because the characterizing properties of gems are obviously the same as for the minerals from which they were fashioned, they are not given here; for them, one should refer to descriptions of the individual minerals. Chapter 14, however, has two tables helpful in gem identification: (1) a list of minerals in order of increasing specific gravity (Table 14.2), and (2) a list of nonopaque minerals in order of increasing refractive index (Table 14.3). In most cases an unknown gem can be identified by determination of these two properties. Specific gravity—along with cleavage, fracture, hardness, and fluorescence—are covered in the next few paragraphs. Refractive indices are described in the "Instruments" section.

Cleavage and Fracture

These properties that frequently enable one to determine a mineral specimen are far less important in gem identification. Evidence of them is usually obliterated in a well-cut and polished gemstone. However, careful inspection of a cut stone with a hand lens or microscope may reveal interior reflecting surfaces, indicating incipient cleavage cracks. A mounted stone

should also be examined at the points of contact with the prongs where pressure of mounting may have produced tiny fractures or cleavages.

Hardness

A scratch test for hardness of a gemstone should be made only when other identifying tests have failed, and then with extreme caution. To determine hardness the gemologist uses a set of *hardness points*. These are metal tubes set with sharp-pointed mineral fragments usually having hardnesses of 10, 9, $8\frac{1}{2}$, 8, $7\frac{1}{2}$, 7, and 6. When it is necessary to check hardness on a faceted stone, the scratch, as short as possible, should be attempted on a back facet near the girdle or on the girdle itself. To avoid scratching an unset stone, the gem girdle can be used as the "scratcher" with attempts to scratch the scale minerals. Begin with a mineral low in the scale and work toward those of higher hardness until no scratch is produced. A short scratch usually can be made on the back of a cabochon without injury to the stone.

Specific Gravity

The specific gravity is as characteristic a property of gems as it is of minerals. Its determination, coupled with refractive index, in many cases suffices for identification. But unlike the refractive index, discussed later, it cannot be determined for a mounted stone. The most reliable methods of determining specific gravity are by hydrostatic weighing or by suspension in a heavy liquid (page 33). The definite values obtained by these methods are more helpful in pinpointing a gem than approximate specific gravities obtained in the manner outlined below.

It is common practice for gemologists to maintain a number of bottles containing liquids of known densities. Densities found most convenient are 2.67 (bromoform + acetone), in which quartz floats and beryl sinks; 2.89 (bromoform), in which beryl floats and tourmaline sinks; 3.10 (methylene iodide + acetone), in which gem tourmaline just floats and fluorite sinks; 3.33 (methylene iodide), in which fluorite sinks and jadeite is suspended or sinks slowly. By making successive tests, the specific gravity of a gem can be bracketed or determined as greater than 3.33 or less than 2.67. For example, a gemstone that sinks in the 3.10 liquid but floats in the 3.33 liquid has a specific gravity between these limits. To avoid contamination, the stone should be well cleaned before transferring from one liquid to another.

Fluorescence

Some gemstones fluoresce under ultraviolet light but for only a few, as for their parent minerals, is the fluorescent color constant and therefore diagnostic. Of this small number that are occasionally cut as gems, willemite fluoresces a yellow-green, scheelite a pale blue, and benitoite a bright blue in shortwave ultraviolet. Gems cut from some minerals are always nonfluorescent, but those cut from other minerals may or may not fluoresce. Thus, in most cases fluorescence cannot be used as a definitive test for gem identification.

Instruments for Studying Gems

Many of the methods developed by the mineralogist for mineral determination are applied equally effectively to gemstone identification. In the study of rough gem material, the procedures are identical. In general, however, there is a fundamental difference in approach. Whereas the mineralogist can scratch, powder, or dissolve the mineral under investigation, the gemologist is restricted to nondestructive tests when working with cut and polished gemstones. As a result, specialized instruments and techniques have been developed for their study. The following pages describe the importance of these methods used in determining the properties of gems.

Observation

The first step in identification of a gem is to look at it with the unaided eye, for several properties are seen as surface phenomena. These include luster, color, and, in a faceted stone, fire caused by dispersion of the spectral colors. Other surface features, observed best in cabochon stones, are (with examples) play of color, *opal*; opalescence, *moonstone*; iridescence or labradorescence, *labradorite*; asterism, *star ruby*; and chatoyancy, *cat's eye*, (page 26). Such observations may yield sufficient information for the experienced gemologist to identify a gem.

Hand Lens

Perhaps the most important gemological instrument is a simple hand lens; one with a 10× magnification is most commonly used. With it one can determine the quality of the cut. Are the facets well polished, are they symmetrically disposed, do they meet at common points, and are the edges between them sharp or rounded? The cut of an imitation is usually poorer than that of the gem it simulates. The hand lens will also reveal major flaws or inclusions that detract from the value of a stone.

The Microscope

For a critical examination, one that will reveal imperfections not seen with a hand lens, it is necessary to use a microscope. The type most useful in gemol-

ogy is a low-power binocular microscope with magnification ranging from 10× to 60×. This instrument differs from the polarizing microscope (page 296) in that it gives true stereoscopic vision with an erect rather than an inverted image. Provision is made to view the stone in both incident and transmitted light. With the light source above the microscope stage, the external features are observed in light reflected from the surface. The interior of the stone is best seen in transmitted light, that is, in light passing through it from a substage illuminator. However, it is difficult to inspect the complete interior of a gemstone in normal transmitted light because of reflection and refraction by the facets. This problem is largely overcome in the gemological microscope (Fig. 13.4) by a substage adapter, giving what is called dark field illumination. Using this lighting technique, the gemstone is illuminated by a hollow cone of light that does not fall directly on the microscope objective. The gemstone is held at the apex of this cone and any flaws or inclusions within it ap-

FIG. 13.4. A low-power stereoscopic wide-field microscope, designed to provide several different methods for illuminating gemstones. (Courtesy of Gemological Institute of America, GEM Instruments Corp., Santa Monica, Calif.)

pear bright against a dark background. Study of these internal imperfections aids both in identification and in quality determination. Also, it frequently provides a means of determining whether a stone is of natural origin or is a synthetic imitation.

The Polariscope

The most informative tests are based on the optical properties and thus are usually the first made in gemstone identification. A determination as to whether a stone is isotropic or anisotropic can be made quickly using a polarizing microscope (Fig. 7.12). Equally effective for the purpose is a simple polariscope used by most gemologists. It is composed of two polarizing plates, one held above the other with a light source below, see Fig. 13.5. The plates are adjusted to the crossed position by turning the upper polar until a minimum of light passes through it. A transparent gemstone placed on the lower plate and slowly turned will remain dark if it is isotropic (isometric or noncrystalline), but will become alternately light and dark if anisotropic (page 294).

FIG. 13.5. Polariscope used for distinguishing isotropic from anisotropic materials, determining pleochroism, and obtaining interference figures. (Courtesy of Gemological Institute of America, GEM Instruments Corp., Santa Monica, Calif.)

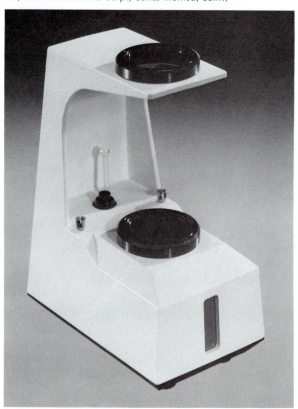

Refractive Index and the Refractometer

Of the several measurable properties of gems, refractive index is most informative and is determined most easily. The **RI** (the abbreviation used by gemologists for refractive index) can be determined on both set and unset gemstones, using especially designed refractometers. Several makes of refractometers are available, but all operate on the same principle: total reflection and the critical angle (page 294). But they are so constructed that, instead of measuring the angle, the refractive index is read directly from a scale. Central to all the instruments is a lead glass of high refractive index cut in the form of a hemisphere, hemicylinder, or prism with a flat, polished upper surface. The refractometer developed by the Gemological Institute of America (Fig. 13.6) uses a hemicylinder.

To make an **RI** determination, a facet of the gemstone, usually the table, is placed on the polished surface but separated from it by a film of liquid. Light entering through a ground glass at the back of the instrument passes through the hemicylinder, striking the gemstone at varying angles of incidence. When the light rays are at angles greater than the critical angle, they are totally reflected back through the hemicylinder to fall on a transparent scale. The image of the scale is reflected by a mirror and is observed through a telescope (Fig. 13.7a). The position of the boundary separating light from dark portions (the shadow edge) is determined by the refractive index, which can be read from the scale (Fig. 13.7b).

The contact liquid usually used with a standard jeweler's refractometer is a mixture of methelyne iodide, sulfur, and tetraiodoethylene with a **RI** of 1.81. Because the liquid must have an **RI** greater than that of the gemstone, the maximum refractive index determination is about 1.80. *Caution:* Refractive index liquids available in most mineralogical laboratories with **RI** greater than 1.78 should not be used. They are corrosive and will etch the lead glass of the refractometer. However, they may be used in a refractometer in which the glass is replaced by chemically inert cubic zirconia (**RI** 2.16). With this instrument, using a liquid of **RI** 2.10, measurements of refractive index as high as 2.08 can be made.

Anisotropic Gems. Although, basically, the method of refractive index determination is applicable to all gemstones, it strictly applies only to those that are isotropic with a single refractive index. Anisotropic crystals have two refractive indices if uniaxial and three if biaxial.

Uniaxial stones, in general, give rise to two shadow edges on the refractometer, one resulting from vibrations of the O ray, the other from vibra-

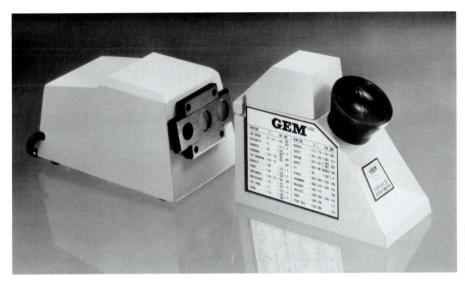

FIG. 13.6. Jeweler's refractometer. (Courtesy of Gemological Institute of America, GEM Instruments Corp., Santa Monica, Calif.)

tions of the E ray. If the stone is turned on the hemicylinder, one edge remains constant; this is ω, the **RI** of the O ray. The other edge, resulting from the E ray, varies as the stone is turned, giving values of ε'. The true value of ε is read when the two edges are farthest apart. The difference between ω and ε is the *birefringence*. The stone is optically positive if ω < ε and negative if ω > ε. If the polished surface of the gem in contact with the refractometer is perpendicular to the optic axis (c crystal axis), both shadow edges remain constant as the stone is turned. In this situation, measurements should be made with the stone lying on another facet. One can then determine whether the high or low index is variable and thus the optic sign.

Biaxial gemstones normally show two shadow edges on the refractometer but, unlike uniaxial, both edges move as the stone is turned. The maximum reading of the higher edge is γ, the minimum reading of the lower edge is α, and their difference is the birefringence. To determine the optic sign, it is necessary to know the third index, β. For (γ − β) > (β − α)

FIG. 13.7. (a) Schematic diagram of light rays entering a refractometer and striking a gemstone of spinel at varying angles of incidence. Rays at angles less than the critical angle (CA) pass upward through the stone. Rays at angles greater than CA are totally reflected through the hemicylinder. (b) Image of scale as seen through the optical system with reading at 1.720, the **RI** of spinel. The faint edge at 1.81 is the **RI** of the contact liquid.

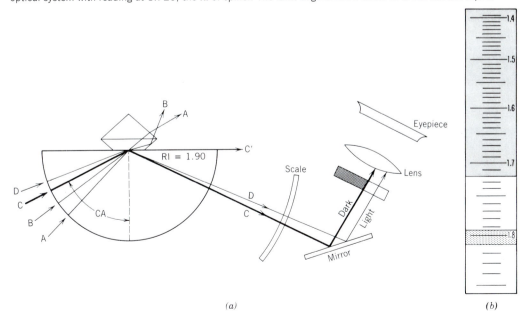

(a)   (b)

is positive and $(\gamma - \beta) < (\beta - \alpha)$ is negative. As the stone is rotated, the value of β is observed as *either* (1) the lowest reading of the higher edge or (2) the highest reading of the lower edge. If (1) is less than the midpoint between α and γ, β is nearer to α than to γ and the stone is positive; if (2) is greater than this midpoint, β nearer to γ than to α and the stone is negative. If neither shadow edge passes the midpoint the gemstone should be placed on a different face and readings (1) and (2), the two possibilities for β, again determined. The value common to both sets of readings is β.

Dispersion

Refractive index varies with the wavelength of light and is less for red than for violet light, a phenomenon called *dispersion*. Dispersion is a characterizing feature of gemstones and is given as a number representing the difference of **RI** determined in red light and in violet light. For diamond it is 0.044 and for quartz 0.013. The high dispersion of diamond gives it its fire, the flashes of color from a cut stone. If white light is used in determining the **RI** of a gemstone, the shadow edge on the scale is not a line but, because of dispersion, is a band of spectral colors. For precise measurements it is necessary to use a monochromatic source such as sodium light. The refractive index usually reported is that determined in sodium light.

The Dichroscope

Light passing through an anisotropic crystal may be absorbed differently in different vibration directions. The phenomenon is called *dichroism* in uniaxial crystals (page 303) and *pleochroism* in biaxial crystals (page 308) although the term *pleochroism* is commonly used for both. It may be evidenced only in a difference in intensity of light of a given wavelength, but frequently different wavelengths are absorbed in each vibration direction, resulting in color variations characteristic of a given gem.

Uniaxial gemstones with only two rays, O and E, can show only two pleochroic colors, which vary as do the refractive indices. When light moves parallel to the optic axis, only one color is seen, that of the O ray. The color of the E ray is most pronounced when light moves perpendicular to the optic axis. In some

uniaxial gems the color difference is only in intensity. For example, in green tourmaline the absorption of the O ray is much greater than that of the E ray and is expressed as O > E or $\omega > \epsilon$. When absorption results in different pleochroic colors, the colors are given. For example in emerald: O (ω) yellow-green, E (ϵ) blue-green. In biaxial crystals light absorption may be different for light vibrating in each of the principal crystallographic directions, X, Y, Z. For example, the variety of chrysoberyl, *alexandrite*, has strong pleochroism expressed as X (α) red, Y (β) orange, and Z (γ) green.

Because pleochroic colors result from rays vibrating at right angles to each other, it is possible to observe one by eliminating the other. This can be done with a polariscope by rotating the stone to an extinction position. The upper polar is then removed and the color noted in plane polarized light. If the stone is pleochroic, a different color or color intensity will be observed on rotating it 90°. A gemstone either uniaxial or biaxial may be pleochroic, yet no color difference will be seen if light passes through it parallel to an optic axis. Thus, in testing for pleochroism, it is advisable to make observations with light passing through the stone in several directions.

The dichroscope is an instrument long used by gemologists for detecting pleochroism. The instrument is a metal tube with a square or rectangular opening at one end and a lens at the other. Within the tube is an elongated cleavage piece of optical calcite (Iceland spar). Because of the strong double refraction of calcite, two images of the opening are seen through the instrument (Fig. 13.8). If a pleochroic stone is held over a bright light source and viewed through the dichroscope, the two images have different colors. Although the same colors are seen with a polariscope, they are observed simultaneously with the dichroscope and are compared more easily.

Color Filters

A colored gemstone absorbs certain wavelengths of white light, and the color results from a combination of the unabsorbed wavelengths. To the eye two gems may have nearly identical colors but the color of one may result from quite a different mixture of wavelengths than the color of the other. Color filters offer

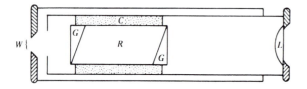

FIG. 13.8. Dichroscope, diagrammatic section. *R* is a calcite cleavage rhombohedron held in a metal tube by a cork setting, *C*. Glass prisms, *G*, are cemented to the calcite to aid in light transmission. To the right of the drawing is a double image of the window. *W*, as seen through lens, *L*.

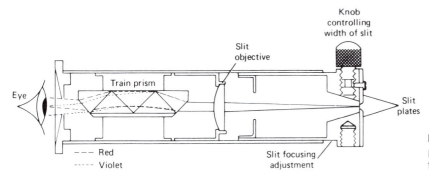

FIG. 13.9. Section through a direct vision prism spectroscope of the type manufactured by R. and J. Beck, London.

a means of resolving the basic difference and thus aid in gem identification.

The most commonly used filter is the *Chelsea*, sometimes called the "emerald filter" because its principal use is to distinguish emerald from other green gems and imitations. The filter transmits only the long red wavelengths and a band in the yellow-green portion of the spectrum. Emerald transmits red light but absorbs some of the yellow-green. It therefore appears red when illuminated by a bright incandescent light and viewed through the filter. Most other green gems as well as glass imitations absorb red light and appear green. Although this is a helpful test in emerald identification, it must be used with caution, because a few other green gems as well as synthetic emerald may appear red and the color of emerald from some localities remains unchanged.

The Spectroscope

Gemologists use a small, direct vision spectroscope. It is essentially a metal tube with an adjustable slit at one end and an eyecap window at the other end. Between the ends are arranged a lens to focus on the slit and a train of prisms, as shown in Fig. 13.9. When the slit is directed toward a source of white light, one observes a ribbon of the complete visible spectrum, red at one end, violet at the other. Differ-

ent portions of the spectrum are brought into sharp focus by pulling out or pushing in the drawtube carrying the slit.

When a gemstone is interposed between the bright light and the spectroscope, the continuous spectrum is visible but it may be crossed by dark bands or lines where certain wavelengths have been eliminated (Fig. 13.10). This is an absorption spectrum. The dark bands or lines fall in positions identical to the bright bands or lines of the emission spectrum but are not as sharp or as well defined. Nevertheless, their distribution depends on the elements present and yields information that aids in the identification of certain gems. Bands or lines resulting from both major elements and minor elements, acting as coloring agents, may be present. In some cases with data obtained with the spectroscope, one can differentiate between a natural or synthetic gem or tell whether the color is natural or has been artificially induced.

X-ray Diffraction

Of the several methods used today for mineral identification, the one yielding the most definitive information is an X-ray powder photograph or powder diffractogram (p. 317). Thus, in studying a mineral, this is frequently the first, and perhaps the only, test

FIG. 13.10. Absorption spectrum of a pale yellow diamond. Seen through the spectroscope the background would be a continuous spectrum of violet through red, crossed by dark lines. (Courtesy of Gemological Institute of America, Santa Monica, Calif.)

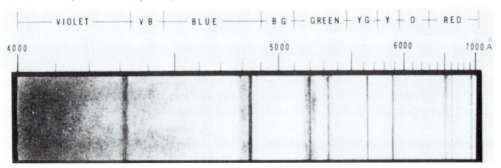

made, even though other, simpler tests might have been adequate.

In studying a gem, the reverse is true. Only when all other tests have failed to identify a gem is resort made to an X-ray powder photograph. This is due in part to the fact that X-ray equipment is unavailable and in part to the difficulty in obtaining the requisite sample. However, with care and without injury to the stone, sufficient material can be ground from the back of a cabochon or scraped from the girdle of a faceted gemstone to make a powder mount.

SYNTHESIS OF GEM MATERIALS

During the twentieth century mineralogical research resulted in the laboratory synthesis of many minerals. Most of these are fine-grained aggregates or in small crystals, but a few have been grown in large, single crystals. These latter are our immediate concern—particularly those that duplicate gem minerals. Because duplication is not only in appearance but in all respects—chemistry, crystal structure, and physical properties—detection is difficult.

The method of synthesis varies, depending on the desired product. Some techniques have been developed to produce only a single gem (e.g., diamond and opal), whereas others are used to synthesize several. The methods of more general application, outlined below, are followed by a consideration of the most important synthetic gem materials. Included are not only synthetic minerals but also synthetic products without natural equivalents.

Verneuil Process

We can consider that gem synthesis on a commercial scale had its beginning in 1902. In that year Auguste Verneuil announced that by a process of flame fusion he had synthesized ruby in sizes large enough and clear enough to be cut as gemstones. An industry based on the Verneuil technique grew rapidly, and within a decade millions of carats, not only of ruby, but of sapphires both colorless and in a variety of colors, were being produced annually. Later (ca. 1920) spinel and still later (ca. 1947) rutile were manufactured by the same process. The Verneuil method with only minor changes remains today the principal means of gem synthesis.

In the Verneuil process a powder with the chemical composition of the desired crystal (for ruby, Al_2O_3 plus a coloring agent) is melted as it passes through an oxygen–hydrogen flame. Droplets of the fused powder fall on a ceramic plate and, on cooling, form a carrot-shaped single crystal of corundum

called a *boule*. Most crystals grown by the Verneuil method contain small gas bubbles, usually spherical but sometimes elongated. Also present are curved striae resulting from successive layers of molten material spreading over the curved surface of the growing boule. Detection in a gemstone of these bubbles and growth striations signifies a synthetic origin.

Czochralski Process

By this technique, also called *crystal pulling*, large, high-quality crystals of several compounds including corundum can be grown. A melt of the composition of the desired crystal is contained in an iridium crucible. A "seed" crystal, held on the end of a rotating rod, is touched to the surface of the melt and then slowly withdrawn. On withdrawal, material continually crystallizes, forming a rod-shaped crystal. Ruby crystals 40 cm long and 5 cm in diameter are grown in this manner.

Flux Growth

In this method a powder of the composition of the desired crystal is mixed with a flux, a material having a relatively low melting point. As the mixture is heated in an inert crucible, the flux melts and, in the molten state, dissolves the other material. After the melt is thoroughly mixed, it is allowed to cool slowly. On reaching a critical temperature, crystal nuclei of the desired material form and gradually increase in size as the temperature is further lowered. When cool, the crystals are recovered by dissolving the flux. Several gem materials, including emerald and ruby, are grown by this method.

Hydrothermal Growth

In nature many minerals crystallize from hot, water-rich solutions. The constituent elements, held in solution at high temperatures, are precipitated as the solutions move toward the surface into regions of lower temperature. Using equipment designed to withstand high temperature and high pressure, hydrothermal growth in the laboratory closely duplicates the natural process. The apparatus, called an autoclave or "bomb," is a heavy-walled steel cylinder closed at one end. To prepare an autoclave for crystal growth of quartz, for example, pure quartz fragments as source material are placed in the bottom and seed plates of thin slices of single quartz crystals are held in the upper part. The bomb is then 85% filled with water to which a crystallization catalyst has been added to increase the solubility of the quartz fragments. After sealing, the bomb is placed in a furnace and heated to about 400°C in the lower part and to about 340°C in the upper part. As the temperature is raised, the water expands to fill the bomb and the pressure reaches 20,000 to 25,000 psi. The convection current generated by the temperature gradient causes the solution to rise, carrying silica dissolved from the feed material. When it reaches the zone of lower temperature, the silica is deposited on the seed plates. Although ruby and emerald are grown hydrothermally, quartz is by far the most important product grown by this method.

TREATMENT OF GEMSTONES

The color of a number of gemstones can be changed or enhanced by "treatment." The change in some is temporary, lasting only a few days or weeks, but in others it is permanent, resulting in a color identical to that of a naturally colored stone. The principal methods of treatment are dyeing, heating, and irradiation, or, for some gems, a combination of heating and irradiation.

Dyeing

An old and an obvious way to color a gemstone is to stain it with an appropriate dye. Because the material must be sufficiently porous to permit penetration of the pigmenting solution, the method is restricted to relatively few gems. Dyeing has been used most effectively in coloring chalcedony, jade, and turquoise.

Heat Treatment

It has long been known that heating certain gemstones will result in a color change or will render an off-color stone colorless. It is very likely that most, if not all, gem materials have been subjected to heat treatment in both oxidizing and reducing environments with the hope that they too could be similarly enhanced. However, only a few have responded favorably, and some have been downgraded by heating. The gems for which heat treatment is most successful are zircon, quartz, beryl, topaz, and zoisite (tanzanite).

Irradiation

Compared with dyeing and heat treatment, irradiation of gemstones is a new process, a twentieth-century development. Irradiation has produced de-sirable color changes in numerous gems but in only a few is the color permanent. There are several irradiation sources: X-rays, alpha particles from a cyclotron, neutrons from atomic piles, and gamma rays from cobalt 60. Irradiation is most effective in producing desirable color changes in diamond, topaz, and quartz.

SYNTHETIC AND TREATED GEMS

Gems that have been synthesized and those that have responded most favorably to treatment are listed below, designated s for synthesis and t for treatment.

Beryl (s, t)	Quartz (s, t)
Chrysoberyl (s)	Rutile (s)
Corundum (s, t)	Spinel (s)
Diamond (s, t)	Topaz (t)
Jade (t)	Turquoise (s, t)
Opal (s, t)	Zoisite (t)

These gems are discussed briefly, describing the method by which each is synthesized and/or treated. Also mentioned, under "Imitation," are the materials, both natural and synthetic, that are sometimes used to imitate the natural gem. Glass is not included with the materials used as imitations, but it could be in each case. Therefore, the gemologist should be ever alert to the possibility that a stone is glass or "paste," as it is sometimes called. Although glass may correspond in color to the gem it simulates, usually the deception can be quickly detected by determination of other properties.

Beryl

Synthesis. Beryl synthesis has been directed chiefly to emerald because of its high value compared with other gem beryls. Efforts have been highly successful and emerald has been synthesized by both flux-growth and hydrothermal methods. Also, a special hydrothermal method (the Lechtleitner process) has been developed by which an overgrowth of emerald can be produced on gemstones of colorless beryl.

Most natural emeralds are nonfluorescent but synthetics usually fluoresce a dull red under long-wave ultraviolet. Also, natural emeralds usually appear red when viewed through the Chelsea filter, whereas most synthetics and imitations appear green. Unfortunately neither test is unequivocal. Flux-grown emeralds contain wisplike inclusions that signify a synthetic origin. The refractive indices and specific gravity of synthetics made by the different processes differ slightly from one another as well as from natural emerald, as shown below.

Emerald Type	ω	ϵ	G
Natural Colombian	1.583	1.577	2.71–2.77
Hydrothermal	1.573	1.568	2.66–2.70
Flux grown (some)	1.579	1.571	2.65–2.68
Flux grown (normal)	1.564	1.561	2.65–2.66

Treatment. Gems of green, yellow, and pink beryl are usually of natural colors, but the deep blue of many aquamarines has been produced by heat treatment of greenish-yellow beryl. Emeralds with surface cracks may be "oiled," that is, immersed in an oil of the same refractive index as emerald. The cracks, filled with oil, are thus obscured and the color of the stone deepened.

Imitation. A composite stone simulating emerald, called a triplet, is made with a crown and pavilion of colorless beryl (or colorless synthetic spinel), between which there is a layer of green-colored cement.

Chrysoberyl

Synthesis. Alexandrite, the variety of chrysoberyl that appears green in daylight and red in incandescent light, has been synthesized by flux growth. The properties of these synthetics, including the color change, are very similar to those of natural crystals.

Imitation. Synthetic corundum and, to a lesser extent, synthetic spinel, are made to resemble alexandrite. However, the color change in these synthetics is less marked than in alexandrite, being more violet in artificial light and a grayish-green in daylight.

Corundum (Ruby and Sapphire)

Synthesis. Synthetic corundum is produced most extensively by the Verneuil process. It is synthesized as ruby (red) and as sapphire both colorless and in a wide variety of colors, some of which do not occur in nature. Synthetic sapphire in colors usually displayed by other gems may be sold erroneously under the names of those gems; for example, purple as "amethyst" and yellow as "topaz." Since 1947 star rubies and blue star sapphires have been manufactured by the Verneuil method. As cabochon stones they rival in beauty the natural stones they imitate. Tiny bubbles and curved growth striae are characteristic of corundum grown by the Verneuil process.

Corundum is also synthesized hydrothermally and by crystal pulling and flux-growth methods. Rubies made by Kashan, Inc. of Austin, Texas, by flux growth lack the characterizing features of Verneuil-grown crystals and detection as a synthetic is difficult, and in some cases impossible. Moreover, the color of "Kashan rubies" is strikingly similar to Burmese rubies and, for this reason they command a relatively high price.

Treatment. The color of some rubies and blue sapphires can be enhanced by heat treatment. Gray sapphires heated to a high temperature in the proper gaseous environment may become a deep blue. Also a surface color can be induced by diffusion of impurities (e.g., TiO_2) into a stone during heat treatment.

Diamond

Synthesis. Using apparatus designed to maintain both high temperature and high pressure, diamonds were synthesized in 1955 by the General Electric Company (see Box 8.1). Although these were small crystals, not suitable for gems, they signaled the beginning of an industry to manufacture abrasive diamond. By 1980 diamond grit was being made in several countries, with a total annual production of more than 100 million carats. In 1970 General Electric announced the synthesis of gem quality diamonds large enough (about 1 carat each) to be cut into gemstones. However, the process is so expensive that synthetic diamonds do not compete in the gem market with natural stones.

Treatment. Color can be produced or changed in diamonds by irradiation and heat treatment. Early in the twentieth century diamonds were colored green by exposure to radiation from radium compounds. This treatment induced not only color but a radioactivity that remains dangerously high. Today diamonds are colored green, without induced radioactivity, in a nuclear reactor and with a cyclotron. Heat treatment of these green stones changes the color to yellow or brown. Diamonds have been colored blue by high-energy electron bombardment. Distinguishing between treated and naturally colored stones presents a difficult problem for the gemologist.

Imitation. For a long time various colorless gems have been used to imitate diamond, but the properties of most of them are so different from those of diamonds as to permit easy recognition. Zircon, with **RI** beyond the range of the jeweler's refractometer and relatively high dispersion, is most deceptive, but unlike diamond, it is birefringent. With the manufacture of gem materials that have no natural counterparts, detection of diamond simulants became more difficult. These substances are briefly described at the end of this chapter under "Manufactured Gem Materials Without Natural Counterparts." They and

TABLE 13.2   Properties of Diamond and Diamond Imitations

Material	RI*	Disp†	Opt.‡	G§	H
Diamond	2.42	0.044	I	3.52	10
Rutile (syn.)	2.75±	0.33	A	4.26	6
Strontium titanate ‖	2.41	0.19	I	5.13	$5\frac{1}{2}$
Cubic zirconia ‖	2.16	0.06	I	5.9±	8
GGG ‖	1.97	0.038	I	7.02	7
Zircon	1.95±	0.039	A	4.7	7+
YAG ‖	1.83	0.028	I	4.55	8+
Sapphire (nat., syn.)	1.77	0.018	A	4.02	9
Spinel (syn.)	1.73±	0.020	I	3.64	8
Topaz	1.62±	0.014	A	3.5±	8
Beryl	1.58±	0.014	A	2.7	8
Quartz	1.55	0.013	A	2.65	7

*RI is average refractive index; ± indicates a variation greater than 0.01.
†Disp. is dispersion
‡Opt. I, isotropic; A, anisotropic.
§G is average specific gravity; ± indicates a variation of 0.1 or greater.
‖ Manufactured materials without natural counterparts.

the minerals used to imitate diamond are listed in Table 13.2 with several diagnostic properties, in order of decreasing refractive index. For comparison, diamond and its properties are given first.

Jade

Jade is not a single mineral, but includes the pyroxene, jadeite, and the amphibole, tremolite-actinolite, known as nephrite. Although the two types of jade are quite distinct mineralogically, they are frequently difficult to distinguish from one another.

Treatment. Jadeite occurs in all colors but the most highly prized has a deep emerald-green color. White jadeite has been dyed the color of this "imperial" jade. Under magnification, the color can be seen distributed along veinlets and grain boundaries.

Imitation. Several minerals have been used as jade substitutes, some of which closely resemble true jade. These include massive green vesuvianite, *californite*; massive grossular garnet; and particularly *bowenite*, a translucent yellow-green serpentine. Other substitutes, more easily distinguished because of low hardness, are *pseudophite*, a massive form of chlorite; *agalmatolite*, a compact talc or pyrophyllite; and *verdite*, a green rock containing the chrome mica, *fuchsite*.

Opal

Synthesis. The determination in 1964 that opal is made up of close-packed silica spheres (page 25) pointed the way to its eventual synthesis. By developing a process that would closely duplicate this "structure," Pierre Gilson produced synthetic opal in 1974. These synthetics have chemical and physical properties, including a play of color, that are essentially identical to those of natural precious opal. Detection of the synthetic product is thus difficult. Under magnification in transmitted light, however, synthetic opal shows a mosaic structure not seen in natural opal.

Treatment. A play of color can be induced in some opals by immersing them in water, which apparently fills voids between the spheres. But as the water slowly leaves, so does the color effect. In such an opal the play of color is retained if it is impregnated with paraffin or plastic.

Imitation. A material simulating opal, called Opal-Essence, is manufactured by John S. Slocum of Rochester, Michigan. It is basically a glass with included platelets from which emanate color flashes. It can be distinguished from opal by its greater refractive index (1.49–1.51) and specific gravity (2.41–2.50). A realistic opal imitation is made of plastic spheres embedded in a plastic matrix, but it has much lower hardness and specific gravity than does opal.

Quartz

Synthesis. The many efforts to synthesize quartz during and immediately following World War II met with success in 1947, when optical grade crystals large enough for scientific uses were grown by the hydrothermal method. Since then, increasingly large amounts have been produced each year, with an estimated world production today of more than 700 tons! Most of this quartz is colorless and used for technological purposes. But colored quartz is also synthesized, duplicating the natural colors of citrine, amethyst, and smoky quartz. Other synthetic crystals are grown in colors in which natural quartz does not occur, for example, dark green and deep blue. If not of an unusual color, it is difficult if not impossible to distinguish synthetic from natural quartz in a cut stone.

Because it is general practice in quartz synthesis to use seed plates cut from single, untwinned crystals, the resulting synthetic product is also in single, untwinned crystals. Although natural citrine and amethyst gems may be single crystals, they are commonly twinned according to the Brazil law (optical twinning). Thus, the presence of this twinning in a gemstone points to a natural origin.

Treatment. Most colorless quartz, natural and synthetic, will turn smoky on irradiation but, on heating, again becomes colorless. With heat treatment, amethyst usually becomes an orange-brown citrine but some turns green. In either case the violet amethyst color returns on irradiation.

Dyed Microcrystalline Quartz. Although agate and other types of chalcedony occur in several natural colors, much of this material used for ornamental purposes has been dyed. Depending on the dye, a gray to white agate may be rendered black or colored red, yellow, blue, or brown. Chert dyed the blue color of lapis lazuli has been sold under the name of "Swiss lapis."

Rutile

Although natural rutile is too dark to be used as a gem, synthetic rutile of gem quality has been produced by the Verneuil process since 1947. The boules as grown are black, but on annealing in oxygen, become transparent and pale yellow. When first made, synthetic rutile was a popular diamond substitute, but it is little used today. With its high refractive index (average 2.75) and extremely high dispersion (0.33) it outshines diamond and is thus easily detected. Furthermore, its high birefringence (0.287) causes a pronounced doubling of the back facets, giving a "fuzzy" appearance to cut stones.

Spinel

Synthetic spinel is made by the Verneuil process in a variety of colors not found in the mineral and, as such, is used more to imitate other gems than to simulate natural spinel. Most commonly it is colorless or blue to imitate diamond, blue sapphire, and aquamarine. Different shades of green imitate tourmaline,

peridot, emerald, and chrysoberyl. Alexandrite-colored synthetic spinel is used to simulate alexandrite, and a cloudy form makes a rather convincing imitation of moonstone.

Synthetic spinel has the curved striations and gas bubbles characteristic of Verneuil-grown crystals. Its refractive index (1.73) and specific gravity (3.64) are both slightly greater than of natural spinel, which are respectively 1.72 and 3.60.

Turquoise

Synthesis. Synthetic turquoise was introduced to the gem market in 1972 by Pierre Gilson as Gilson-created Turquoise. It is of a fine color and made both with and without "matrix," but the method of manufacture has not been revealed.

Treatment. Natural turquoise may be impregnated with oil, paraffin, or plastic to enhance its color and increase its hardness. Such treatment can usually be detected by touching the material with a red-hot needle.

Imitation. Glass, dyed chalcedony or chert, enamel, and plastic are used to imitate turquoise. Powdered turquoise, when compacted and bonded with plastic, forms a realistic imitation of the naturally massive material. When scratched with a knife, such impregnated turquoise tends to cut rather than to powder.

MANUFACTURED GEM MATERIALS WITHOUT NATURAL COUNTERPARTS

These manufactured materials that do not duplicate gem minerals are used primarily as diamond substitutes, although colored varieties of some make attractive cut stones in their own right. Included with them is, paradoxically, the mineral name *garnet*. It seems appropriate to do this because manufactured "garnets" differ both chemically and physically from natural garnets.

Garnet

The chemical composition of garnet is expressed by the general structural formula, $A_3B_2(SiO_4)_3$. Several different ions may occupy both the A and B sites, giving rise to a wide range in chemical composition. In synthetic garnets the general structure is preserved, but unusual elements substitute in the A and B sites, and the silicon site may be occupied by aluminum or gallium. The synthetic "garnets" used as diamond imitations are rare earth compounds and usually are designated by the abbreviations "YAG" and "GGG," reflecting the elements present.

YAG is yttrium aluminum "garnet," $Y_3Al_2(AlO_4)_3$. It is isometric with the following properties: **H** 8+, **G** 4.55, **RI** 1.833, disp. 0.028. In addition to colorless material, YAG is produced in red, yellow, green, and blue colors.

GGG is gadolinium gallium "garnet," $Gd_3Ga_2(GaO_4)_3$. It is isometric with the following properties; **H** $6\frac{1}{2}$, **G** 7.05, **RI** 1.97, disp. 0.038. Because of its higher refractive index and dispersion, GGG is a more impressive diamond substitute than YAG. Both YAG and GGG are produced by flux-growth and Czochralski techniques.

Strontium Titanate, SrTiO₃

Strontium titanate has been produced since 1955 by the Verneuil process. The boules as they come from the furnace are black due to a deficiency of oxygen, but on annealing in an oxygen atmosphere, they become quite colorless. Strontium titanate is isometric, with **H** $5\frac{1}{2}$ and **G** 5.13. The **RI** is 2.41, only slightly below that of diamond (2.42); but the dispersion, 0.19, over four times that of diamond (0.044), gives the faceted stone an excessive fire. Furthermore, because of the low hardness, polishing scratches and rounded facet junctions are usually present.

Cubic Zirconia

Since its introduction to the gem market in 1976, cubic zirconia, sometimes abbreviated CZ, has become the most important diamond simulant. It is isometric and has the following properties: **H** 8, **G** 5.9±, **RI** 2.16±, disp. 0.06. On consideration of these properties, it becomes immediately apparent why CZ is the most convincing diamond substitute. Because of its high hardness, it takes a good polish and resists scratching. The high refractive index produces a brilliance approaching that of diamond, and the dispersion, although high, is not excessive.

A unique type of melt growth called "skull melting" has been devised for the manufacture of cubic zirconia; the melting point of ZrO_2(2750°C) is so high that conventional crucibles cannot be used. The "skull" is a container that is open at the top with a cylindrical wall composed of closely spaced, vertical fingers of water-cooled copper tubes. The material to be melted is placed in the "skull," which remains relatively cool while, by use of a radio frequency generator, the central material is brought to the necessary temperature and melts within a shell of its own powder. On cooling, crystals grow upward from the bottom as CZ, the isometric polymorph of zirconia. To prevent the crystals from inverting to the low-temperature monoclinic form on further cooling, a stabilizer of yttrium oxide or calcium oxide is added to the ZrO_2 powder to be melted. The refractive index and specific gravity vary with the kind and amount of stabilizer.

REFERENCES AND SELECTED READING

Anderson, B. W. 1990. *Gem testing.* 10th ed. London: Butterworth & Co.

Arem, J. E. 1987. *Color encyclopedia of gemstones,* 2nd ed. New York: Van Nostrand Reinhold.

Branson, O. T. 1976. *Turquoise, the gem of the centuries.* Tuscon, Ariz.: Treasure Chest Publications, Inc.

Gübelin, E. J., and J. I. Koivula. 1986. *Photoatlas of inclusions in gemstones.* Zurich: ABC Edition.

Hurlbut, C. S., Jr., and R. C. Kammerling. 1991. *Gemology.* New York: Wiley.

Keller, P. 1990. *Gemstones and their origins.* New York: Van Nostrand Reinhold.

Liddicoat, R. T. 1989. *Handbook of gem identification.* 12th ed. Santa Monica, Calif.: Gemological Institute of America.

Nassau, K. 1980. *Gems made by man.* Santa Monica, Calif.: Gemological Institute of America.

Schumann, W. 1997. *Gemstones of the world.* Rev. and expanded edition. New York: Sterling Publishing Co.

Sinkankas, J. 1959, 1976. *Gemstones of North America,* v. 1, 2. New York: Van Nostrand Reinhold.

Webster, R. 1983. *Gems.* 4th ed. London: Butterworth & Co.

CHAPTER 14
DETERMINATIVE TABLES

This chapter contains three determinative tables. Table 14.1 is based on those physical properties of minerals that can be easily and quickly determined in hand specimen. Table 14.2 lists increasing specific gravity by which minerals can be identified. Table 14.3 lists increasing refractive indices (for nonopaque minerals) by which a mineral's identity can be established.

Table 14.1 allows a mineral to be identified on the basis of easily assessed physical properties. These properties should be used with the understanding that for many minerals there is a variation in physical properties from specimen to specimen. Color for some minerals is constant, but for others it is extremely variable. Hardness, although more definite, may vary slightly and, by change in the state of aggregation of a mineral, may appear to vary more widely. Cleavage may also be obscured in a fine-grained aggregate of a mineral. In using the tables it is often impossible to differentiate between two or three similar species. However, by consulting the descriptions of these possible minerals in the sections "Systematic Descriptions" and the specific tests given there, a definite decision can usually be made.

In the determinative tables only the common minerals or those that, though rarer, are of economic importance have been included. The chances of having to determine a mineral that is not included in these tables are small, but it must be borne in mind that there is such a possibility. The names of the minerals have been printed in three different styles of type, as **PYRITE**, CHALCOCITE, and Covellite, in order to indicate their relative importance and frequency of occurrence. Whenever there is ambiguity in placing a mineral, it has been included in the two or more possible divisions.

The general scheme of classification is outlined on page 334. The proper division in which to look for a mineral can be determined by means of the tests indicated there. The tables are divided into two main sections on the basis of luster: (1) metallic and submetallic and (2) nonmetallic. The metallic miner-als are opaque and give black or dark-colored streaks, whereas nonmetallic minerals are nonopaque and give either colorless or light-colored streaks. The tables are next subdivided according to hardness. These divisions are easily determined. For nonmetallic minerals they are as follows:

- $\leq 2\frac{1}{2}$, can be scratched by the fingernail
- $>2\frac{1}{2}-\leq 3$, cannot be scratched by fingernail, can be scratched by copper cent
- $>3-\leq 5\frac{1}{2}$, cannot be scratched by copper cent, can be scratched by a knife
- $>5\frac{1}{2}-\leq 7$, cannot be scratched by a knife, can be scratched by quartz
- >7 cannot be scratched by quartz.

Metallic minerals have fewer divisions of hardness; they are $\leq 2\frac{1}{2}$; $>2\frac{1}{2}-\leq 5\frac{1}{2}$; $>5\frac{1}{2}$.

Nonmetallic minerals are further subdivided according to whether they show a *prominent* cleavage. If the cleavage is not prominent and imperfect or obscure, the mineral is included with those that have no cleavage. Minerals in which cleavage may not be observed easily, because of certain conditions in the state of aggregation, have been included in both divisions.

The minerals that fall in a given division of the tables have been arranged according to various methods. In some cases, those that possess similar cleavages have been grouped together; frequently color determines the order, and so on. The column farthest to the left will indicate the method of arrangement.

The figures given in the column headed **G** are specific gravity. This is an important diagnostic physical property but less easily determined than luster, hardness, color, and cleavage. For a discussion of specific gravity and methods for its accurate determination, see page 33. However, if a specimen is pure and of sufficient size, its approximate specific gravity can be determined by simply weighing it in the hand. Below is a list of common minerals over a wide range of specific gravity. By experimenting with specimens of these one can become expert in the approximate determination of specific gravity. Following the determinative tables (Table 14.1) is a list of the common minerals arranged according to increasing specific gravity (Table 14.2).

Gypsum	2.32	Pyrite	5.02
Orthoclase	2.57	Arsenopyrite	6.07
Quartz	2.65	Cassiterite	6.95
Calcite	2.71	Galena	7.50
Fluorite	3.18	Cinnabar	8.10
Topaz	3.53	Copper	8.9
Corundum	4.02	Silver	10.5
Barite	4.45		

Many minerals are polymorphic. The crystal system given in the tables is that of the most commonly observed form.

GENERAL CLASSIFICATION OF THE TABLES
Luster—Metallic or Submetallic

I. Hardness: $\leq 2\frac{1}{2}$ (will leave a mark on the paper). Page 586

II. Hardness: $>2\frac{1}{2}$, $<5\frac{1}{2}$ (can be scratched by knife; will not readily leave a mark on paper). Page 587
III. Hardness: $>5\frac{1}{2}$ (cannot be scratched by knife). Page 590

Luster—Nonmetallic

I. Streak definitely colored. Page 592
II. Streak colorless.
   A. Hardness: $\leq 2\frac{1}{2}$ (can be scratched by fingernail). Page 594
   B. Hardness: $>2\frac{1}{2}$, ≤ 3 (cannot be scratched by fingernail; can be scratched by cent).
      1. Cleavage prominent. Page 596
      2. Cleavage not prominent.
         a. A small splinter is fusible in the candle flame. Page 597
         b. Infusible in candle flame. Page 598
   C. Hardness: >3, $\leq 5\frac{1}{2}$ (cannot be scratched by cent; can be scratched by knife).
      1. Cleavage prominent. Page 599
      2. Cleavage not prominent. Page 603
   D. Hardness: $>5\frac{1}{2}$, ≤ 7 (cannot be scratched by knife; can be scratched by quartz).
      1. Cleavage prominent. Page 606
      2. Cleavage not prominent. Page 609
   E. Hardness: >7 (cannot be scratched by quartz).
      1. Cleavage prominent. Page 612
      2. Cleavage not prominent. Page 613

TABLE 14.1   Minerals Arranged by Several Physical Properties

LUSTER: METALLIC OR SUBMETALLIC
1. Hardness: $\leq 2\frac{1}{2}$,
(Will leave a mark on paper)

Streak	Color	G	H	Remarks	Name, Composition, Crystal System
Black	Iron-black	4.7	1–2	Usually splintery or in radiating fibrous aggregates.	**PYROLUSITE** p. 384 MnO_2 Tetragonal
	Steel-gray to iron-black	2.23	$1-1\frac{1}{2}$	Cleavage perfect {0001}. May be in hexagonal-shaped plates. Greasy feel.	**GRAPHITE** p. 351 C Hexagonal
Black to greenish black	Blue-black	4.7	$1-1\frac{1}{2}$	Cleavage perfect {0001}. May be in hexagonal-shaped leaves. Greenish streak on glazed porcelain (graphite, black). Greasy feel.	MOLYBDENITE p. 367 MoS_2 Hexagonal
Gray-black	Blue-black to lead-gray	7.6	$2\frac{1}{2}$	Cleavage perfect cubic {100}. In cubic crystals. Massive granular.	**GALENA** p. 354 PbS Isometric
	Blue-black	4.5	2	Cleavage perfect {010}. Bladed with cross striations. Fuses in candle flame.	**STIBNITE** p. 363 Sb_2S_3 Orthorhombic
Bright red	Red to vermillion	8.1	$2-2\frac{1}{2}$	Cleavage perfect {10$\bar{1}$0}. Luster adamantine. Usually granular massive.	**CINNABAR** p. 362 HgS Hexagonal
	Ruby-red	5.55	$2-2\frac{1}{2}$	Cleavage {10$\bar{1}$1}. Fusible in candle flame. Associated with pyrargyrite.	Prousite p. 370 Ag_3AsS_3 Hexagonal
Red-brown	Red to vermillion	5.2	1+	Earthy. Frequently as pigment in rocks. Crystalline hematite is harder and black.	**HEMATITE** p. 380 Fe_2O_3 Hexagonal
Red-brown to brown-red	Deep red to black	5.85	$2\frac{1}{2}$	Cleavage {10$\bar{1}$1}. Fusible in candle flame. Shows dark ruby-red color in thin splinters. Associated with other silver minerals.	Pyrargyrite p. 370 Ag_3SbS_3 Hexagonal
Black; may mark paper	Steel-gray on fresh surface. Tarnishes to dull gray	7.3	$2-2\frac{1}{2}$	Usually massive or earthy. Easily sectile. Bright steel-gray on fresh surfaces; darkens on exposure.	ACANTHITE p. 351 Ag_2S Morph.: isometric
	Blue; may tarnish to blue-black	4.6	$1\frac{1}{2}-2$	Usually in platy masses or in thin six-sided platy crystals. Moistened with water turns purple.	Covellite p. 361 CuS Hexagonal

TABLE 14.1 (continued)

LUSTER: METALLIC OR SUBMETALLIC
II. Hardness: $>2\frac{1}{2}-\leq5\frac{1}{2}$,
(Can be scratched by a knife; will not readily leave a mark on paper)

Streak	Color	G	H	Remarks	Name, Composition, Crystal System
Black	Gray-black	4.4	3	Cleavage {110}. Usually in bladed masses showing cleavage. Associated with other copper minerals.	ENARGITE p. 369 Cu_3AsS_4 Orthorhombic
	Pale copper-red; may be silver-white, pinkish	7.8	$5-5\frac{1}{2}$	Usually massive. May be coated with green nickel bloom. Associated with cobalt and nickel minerals.	NICKELINE p. 360 NiAs Hexagonal
	Fresh surface brownish-bronze; purple tarnish	5.1	3	Usually massive. Associated with other copper minerals; chiefly chalcocite and chalcopyrite.	**BORNITE** p. 354 Cu_5FeS_4 Morph.: isometric
	Brownish-bronze	4.6	4	Small fragments magnetic. Usually massive. Often associated with chalcopyrite and pyrite.	**PYRRHOTITE** p. 359 $Fe_{1-x}S$ Morph.: hexagonal
		4.6 to 5.0	$3\frac{1}{2}-4$	Octahedral parting. Resembles pyrrhotite with which it usually is associated but nonmagnetic.	Pentlandite p. 360 $(Fe,Ni)_9S_8$ Isometric
	Brass-yellow	4.1 to 4.3	$3\frac{1}{2}-4$	Usually massive, but may be in crystals resembling tetrahedrons. Associated with other copper minerals and pyrite.	**CHALCOPYRITE** p. 359 $CuFeS_2$ Tetragonal
	Brass-yellow; slender crystals greenish	5.5	$3-3\frac{1}{2}$	Cleavage {10$\bar{1}$1}, rarely seen. Usually in radiating groups of hairlike crystals	MILLERITE p. 360 NiS Hexagonal
Gray-black; will mark paper	Lead-gray	4.6	2	Cleavage perfect {010}. Fuses easily in candle flame. In bladed crystal aggregates with cross striations.	**STIBNITE** p. 363 Sb_2S_3 Orthorhombic
		7.5	$2\frac{1}{2}$	Cleavage perfect {100}. In cubic crystals and granular masses. If held in the candle flame does not fuse but small globules of metallic lead collect on the surface.	**GALENA** p. 354 PbS Isometric
Black; may mark paper	Steel-gray on fresh surface. Tarnishes to dull gray	7.3	$2-2\frac{1}{2}$	Easily sectile. Usually massive or earthy. Rarely in cubic crystals.	Acanthite p. 351 Ag_2S Morph.: isometric

TABLE 14.1 (continued)

LUSTER: METALLIC OR SUBMETALLIC
II. Hardness: $>2\frac{1}{2} - \leq 5\frac{1}{2}$,
(Can be scratched by a knife; will not readily leave a mark on paper)

Streak	Color	G	H	Remarks	Name, Composition, Crystal System
Usually black; may be brownish	Black	3.7 to 4.7	5–6	Massive botryoidal and stalactitic. Usually associated with pyrolusite	ROMANECHITE p. 395 $BaMn^{2+}Mn^{4+}O_{16}(OH)_4$ Appears amorphous
Black; may have brown tinge	Steel-gray. May tarnish to dead black on exposure	4.7 to 5.0	$3-4\frac{1}{2}$	Massive or in tetrahedral crystals. Often associated with silver ores.	TETRAHEDRITE p. 370 $(Cu,Fe,Zn,Ag)_{12}Sb_4S_{13}$ Isometric
Gray-black		5.7	$2\frac{1}{2}-3$	Somewhat sectile. Usually compact massive. Associated with other copper minerals.	CHALCOCITE p. 352 Cu_2S Morph.: orthorhombic
Dark brown to black	Iron-black to brownish-black	4.6	$5\frac{1}{2}$	Luster pitchy. May be accompanied by yellow or green oxidation products. Usually in masses in peridotites.	**CHROMITE** p. 391 $FeCr_2O_4$ Isometric
	Brown to black	7.0 to 7.5	$5-5\frac{1}{2}$	Cleavage perfect {010}. With greater amounts of Mn streak and color are darker.	WOLFRAMITE p. 431 $(Fe,Mn)WO_4$ Monoclinic
Dark brown to black	Steel-gray to iron-black	4.3	4	In radiating fibrous or crystalline masses. Distinct prismatic crystals often grouped in bundles. Associated with pyrolusite.	MANGANITE p. 394 $MnO(OH)$ Morph.: orthorhombic
Light to dark brown	Dark brown to coal-black More rarely yellow or red	3.9 to 4.1	$3\frac{1}{2}-4$	Cleavage perfect {110} (6 directions). Usually cleavable, granular; may be in tetrahedral crystals. The streak is always of a lighter color than the specimen.	**SPHALERITE** p. 357 ZnS Isometric
Red-brown to brown-red	Dark brown to steel-gray to black	4.8 to 5.3	$5\frac{1}{2}-6\frac{1}{2}$	Usually harder than knife. Massive, radiating, reniform, micaceous.	**HEMATITE** p. 380 Fe_2O_3 Hexagonal
	Red-brown to deep red. Ruby-red if transparent	6.0	$3\frac{1}{2}-4$	Massive or in cubes or octahedrons. May be in very slender crystals. Associated with malachite, azurite, native copper.	CUPRITE p. 378 Cu_2O Isometric
Yellow-brown; yellow ocher	Dark brown to black	4.37	$5-5\frac{1}{2}$	Cleavage {010}. In radiating fibers, mammillary and stalactitic forms. Rarely in crystals.	**GOETHITE** p. 395 $\alpha FeO(OH)$ Orthorhombic
Dark red (some varieties mark paper)	Dark red to vermillion	8.10	$2\frac{1}{2}$	Cleavage {10$\bar{1}$1}. Usually granular or earthy. Commonly impure and dark red or brown. When pure, translucent or transparent and bright red.	**CINNABAR** p. 362 HgS Hexagonal

TABLE 14.1   (continued)

LUSTER: METALLIC OR SUBMETALLIC

II. Hardness: $>2\frac{1}{2}-\leq 5\frac{1}{2}$,

(Can be scratched by a knife; will not readily leave a mark on paper)

Streak	Color	G	H	Remarks	Name, Composition, Crystal System
Copper-red, shiny	Copper-red on fresh surface. Black tarnish	8.9	$2\frac{1}{2}-3$	Malleable. Usually in irregular grains. May be in branching crystal groups or in rude isometric crystals.	COPPER p. 344 Cu Isometric
Silver-white, shiny	Silver-white on fresh surface. Black tarnish	10.5	$2\frac{1}{2}-3$	Malleable. Usually in irregular grains. May be in wire, plates, branching crystal groups.	SILVER p. 343 Ag Isometric
Gray, shiny	White or steel-gray	14 to 19	$4-4\frac{1}{2}$	Malleable. Irregular grains or nuggets. Unusually hard for a metal. Rare.	Platinum p. 345 Pt Isometric
Gold-yellow, shiny	Gold-yellow	15.0 to 19.3	$2\frac{1}{2}-3$	Malleable. Irregular grains, nuggets, leaves. Very heavy; specific gravity varies with silver content.	GOLD p. 342 Au Isometric

TABLE 14.1   (continued)

LUSTER: METALLIC OR SUBMETALLIC
III. Hardness: $>5\frac{1}{2}$
(Cannot be scratched by a knife)

Streak	Color	G	H	Remarks	Name, Composition, Crystal System
Black	Silver- or tin-white	6.0 to 6.2	$5\frac{1}{2}$–6	Usually massive. Crystals pseudo-orthorhombic.	**ARSENOPYRITE** p. 368 FeAsS Monoclinic
		6.1 to 6.9	$5\frac{1}{2}$	Usually massive. Crystals pyritohedral. May be coated with pink nickel bloom.	Skutterudite-Nickel skutterudite p. 369 $(Co,Ni,Fe)As_3$-$(Ni,Co,Fe)As_3$ Isometric
		6.33	$5\frac{1}{2}$	Commonly in pyritohedral crystals with pinkish cast. Also massive.	Cobaltite-Gersdorffite p. 368 $(Co,Fe)AsS$-$NiAsS$ Pseudo-isometric
	Copper-red to silver-white with pink tone	7.5	5–$5\frac{1}{2}$	Usually massive. May be coated with green nickel bloom.	NICKELINE p. 360 NiAs Hexagonal
	Pale brass-yellow	5.0	6–$6\frac{1}{2}$	Often in pyritohedrons or striated cubes. Massive granular. Most common sulfide.	**PYRITE** p. 364 FeS_2 Isometric
	Pale yellow to almost white	4.9	6–$6\frac{1}{2}$	Frequently in "cock's comb" crystal groups and radiating fibrous masses.	**MARCASITE** p. 366 FeS_2 Orthorhombic
	Black	5.18	6	Strongly magnetic. Crystals octahedral. May show octahedral parting.	**MAGNETITE** p. 389 Fe_3O_4 Isometric
Dark brown to black	Black	9.0 to 9.7	$5\frac{1}{2}$	Luster pitchy. Massive granular, botryoidal crystals.	Uraninite p. 386 UO_2 Isometric
		4.7	$5\frac{1}{2}$–6	May be slightly magnetic. Often associated with magnetite. Massive granular; platy crystals; as sand.	**ILMENITE** p. 383 $FeTiO_3$ Hexagonal
		3.7 to 4.7	5–6	Compact massive, stalactitic, botryoidal. Associated with other manganese minerals and told by greater hardness.	ROMANECHITE p. 395 $BaMn^{2+}Mn^{4+}O_{16}(OH)_4$ Orthorhombic
Dark brown to black	Black	5.3 to 7.3	6	Luster black and shiny on fresh surface. May have slight bluish tarnish. Granular or in stout prismatic crystals.	Columbite-Tantalite p. 392 $(Fe,Mn)(Nb,Ta)_2O_6$ Orthorhombic

TABLE 14.1 (continued)

LUSTER: METALLIC OR SUBMETALLIC
III. Hardness: $>5\frac{1}{2}$
(Cannot be scratched by a knife)

Streak	Color	G	H	Remarks	Name, Composition, Crystal System
Dark brown	Iron-brown to brownish-black	7.0 to 7.5	$5-5\frac{1}{2}$	Cleavage perfect {010}. With greater amounts of Mn streak and color are darker.	WOLFRAMITE p. 431 (Fe,Mn)WO$_4$ Monoclinic
		4.6	$5\frac{1}{2}$	Luster pitchy. Frequently accompanied by green oxidation products. Usually in granular masses in peridotites.	**CHROMITE** p. 391 FeCr$_2$O$_4$ Isometric
		5.15	6	Slightly magnetic. Granular or in octahedral crystals. Common only at Franklin, N.J., associated with zincite and willemite.	FRANKLINITE p. 391 (Fe,Zn,Mn)(Fe,Mn)$_2$O$_4$ Isometric
Red-brown, brown-red	Dark brown to steel-gray to black	4.8 to 5.3	$5\frac{1}{2}-6\frac{1}{2}$	Radiating, reniform, massive, micaceous. Rarely in steel-black rhombohedral crystals. Some varieties softer.	**HEMATITE** p. 380 Fe$_2$O$_3$ Hexagonal
Pale brown	Brown to black	4.18 to 4.25	$6-6\frac{1}{2}$	In prismatic crystals, vertically striated; often slender acicular. Crystals frequently twinned. Found in black sands.	**RUTILE** p. 383 TiO$_2$ Tetragonal
Yellow-brown to yellow-ocher	Dark brown to black	4.37	$5-5\frac{1}{2}$	Cleavage {010}. Radiating, colloform, stalactitic.	**GOETHITE** p. 395 αFeO(OH) Orthorhombic

TABLE 14.1   (continued)

LUSTER: NONMETALLIC

I. Streak definitely colored

Streak	Color	G	H	Remarks	Name, Composition, Crystal System
Dark red	Dark red to vermillion	8.10	$2\frac{1}{2}$	Cleavage $\{10\bar{1}0\}$. Usually granular or earthy. Commonly impure and dark red or brown. When pure, translucent or transparent and bright red.	**CINNABAR** p. 362 HgS Hexagonal
	Red-brown. Ruby-red when transparent	6.0	$3\frac{1}{2}-4$	Massive or in cubes or octahedrons. May be in very slender crystals. Associated with malachite, azurite, native copper.	CUPRITE p. 378 Cu_2O Isometric
Red-brown	Dark brown to steel-gray to black	4.8 to 5.3	$5\frac{1}{2}-6\frac{1}{2}$	Radiating, reniform massive, micaceous. Rarely in steel-black rhombohedral crystals. Some varieties softer.	**HEMATITE** p. 380 Fe_2O_3 Hexagonal
	Deep red to black	5.8	$2\frac{1}{2}$	Cleavage $\{10\bar{1}1\}$. Fusible in candle flame. Shows dark ruby-red color in thin splinters. Associated with other silver minerals.	Pyrargyrite p. 370 Ag_3SbS_3 Hexagonal
Bright red	Ruby-red	5.55	$2-2\frac{1}{2}$	Cleavage $\{10\bar{1}1\}$. Fusible in candle flame. Light "ruby silver." Associated with pyrargyrite.	Proustite p. 370 Ag_3AsS_3 Hexagonal
Pink	Red to pink	2.95	$1\frac{1}{2}-2\frac{1}{2}$	Cleavage perfect $\{010\}$. Usually reniform or as pulverulent or earthy crusts. Found as coatings on cobalt minerals.	Erythrite (cobalt bloom) p. 437 $Co_3(AsO_4)_2 \cdot 8H_2O$ Monoclinic
Yellow-brown to yellow-ocher	Dark brown to black	4.4	$5-5\frac{1}{2}$	Cleavage $\{010\}$. In radiating fibers, mammillary and stalactitic forms. Rarely in crystals.	**GOETHITE** p. 395 $\alpha FeO(OH)$ Orthorhombic
Brown	Dark brown	7.0 to 7.5	$5-5\frac{1}{2}$	Cleavage perfect $\{010\}$. With greater amounts of manganese the streak and color are darker.	WOLFRAMITE p. 431 $(Fe,Mn)WO_4$ Monoclinic
	Light to dark brown	3.83 to 3.88	$3\frac{1}{2}-4$	In cleavable masses or in small curved rhombohedral crystals. Becomes magnetic after heating in candle flame.	**SIDERITE** p. 414 $FeCO_3$ Hexagonal

TABLE 14.1 (continued)

LUSTER: NONMETALLIC
I. Streak definitely colored

Streak	Color	G	H	Remarks	Name, Composition, Crystal System
Light brown	Light to dark brown	3.9 to 4.1	$3\frac{1}{2}$–4	Cleavage perfect {011}. (6 directions).Usually cleavable, granular; may be in tetrahedral crystals. The streak is always of a lighter color than the specimen.	**SPHALERITE** p. 357 ZnS Isometric
	Brown to black	6.8 7.1	6–7	Occurs in twinned crystals. Fibrous, reniform and irregular masses; in rolled grains.	**CASSITERITE** p. 385 SnO_2 Tetragonal
	Reddish-brown to black	4.18 to 4.25	6–$6\frac{1}{2}$	Crystals vertically striated; often acicular. Twinning common.	**RUTILE** p. 383 TiO_2 Tetragonal
Orange-yellow	Deep red to orange-yellow	5.68	4–$4\frac{1}{2}$	Cleavage {0001}. Found only at Franklin, N.J., associated with franklinite and willemite.	ZINCITE p. 378 ZnO Hexagonal
	Bright red	5.9 to 6.1	$2\frac{1}{2}$–3	Luster adamantine. In long slender crystals, often interlacing groups. Decrepitates in candle flame.	Crocoite p. 428 $PbCrO_4$ Monoclinic
	Deep red	3.48	$1\frac{1}{2}$–2	Frequently earthy. Associated with orpiment. Fusible in candle flame.	REALGAR p. 362 AsS Monoclinic
Pale yellow	Lemon-yellow	3.49	$1\frac{1}{2}$–2	Cleavage {010}. Luster resinous. Associated with realgar. Fusible in candle flame.	ORPIMENT p. 363 As_2S_3 Monoclinic
	Pale yellow	2.05 to 2.09	$1\frac{1}{2}$–$2\frac{1}{2}$	Burns with blue flame giving odor of SO_2. Crystallized, granular, earthy.	**SULFUR** p. 346 S Orthorhombic
Light green	Dark emerald-green	3.75 to 3.77	3–$3\frac{1}{2}$	One perfect cleavage {010}. In granular cleavable masses or small prismatic crystals.	Atacamite p. 402 $Cu_2Cl(OH)_3$ Orthorhombic
Light green	Dark emerald-green	3.9±	$3\frac{1}{2}$–4	One good cleavage {010}. In small prismatic crystals or granular masses.	Antlerite p. 430 $Cu_3(SO_4)(OH)_4$ Orthorhombic
	Bright green	3.9 to 4.03	$3\frac{1}{2}$–4	Radiating fibrous, mammillary. Associated with azurite and may alter to it. Effervesces in cold acid.	**MALACHITE** p. 421 $Cu_2CO_3(OH)_2$ Monoclinic
Light blue	Intense azure-blue	3.77	$3\frac{1}{2}$–4	In small crystals, often in groups. Radiating fibrous, usually as alteration from malachite. Effervesces in cold acid.	**AZURITE** p. 421 $Cu_3(CO_3)_2(OH)_2$ Monoclinic
Very light blue	Light green to turquoise-blue	2.0 to 2.4	2–4	Massive compact. Associated with oxidized copper minerals.	CHRYSOCOLLA p. 542 ~$Cu_4H_4Si_4O_{10}(OH)_8$ Amorphous

TABLE 14.1   (continued)

LUSTER: NONMETALLIC
II. Streak colorless
A. Hardness: $\leq 2\frac{1}{2}$
(Can be scratched by fingernail)

Cleavage, Fracture	Color	G	H	Remarks	Name, Composition, Crystal System
Perfect cleavage in one direction The micas or related micaceous minerals possess such a perfect cleavage that they can be split into exceedingly thin sheets. They may occur as aggregates of minute scales, when the micaceous structure may not be readily apparent.	Pale brown, green, yellow, white	2.76 to 2.88	$2-2\frac{1}{2}$	In foliated masses and scales. Crystals tabular with hexagonal or diamond-shaped outline. Cleavage flakes elastic.	**MUSCOVITE** p. 534 $KAl_2(AlSi_3O_{10})(OH)_2$ Monoclinic
	Dark brown, green to black; may be yellow	2.95 to 3	$2\frac{1}{2}-3$	Usually in irregular foliated masses. Crystals have hexagonal outline, but rare. Cleavage flakes elastic.	**BIOTITE** p. 537 $K(Mg,Fe)_3(AlSi_3O_{10})(OH)_2$ Monoclinic
	Yellowish-brown, green white	2.86	$2\frac{1}{2}-3$	Often in six-sided tabular crystals; in irregular foliated masses. May show copperlike reflection from cleavage.	PHLOGOPITE p. 537 $KMg_3(AlSi_3O_{10})(OH)_2$ Monoclinic
	Green of various shades	2.6 to 2.9	$2-2\frac{1}{2}$	Usually in irregular foliated masses. May be in compact masses of minute scales. Thin sheets flexible but not elastic.	**CHLORITE** p. 538 $Mg_3Fe_3(Si,Al)_4O_{10}(OH)_2\cdot(Mg,Fe)_3(OH)_6$ Monoclinic
	White, apple-green, gray. When impure as in soapstone, dark gray, dark green to almost black	2.7 to 2.8 2.8 to 2.9	1 1–2	Greasy feel. Frequently distinctly foliated or micaceous. Cannot be positively identified by physical tests.	**TALC** p. 531 $Mg_3Si_4O_{10}(OH)_2$ Monoclinic Pyrophyllite p. 534 $Al_2Si_4O_{10}(OH)_2$ Monoclinic
	White, gray, green	2.39	$2\frac{1}{2}$	Pearly luster on cleavage face, elsewhere vitreous. Sectile. Commonly foliated massive, may be in broad tabular crystals. Thin sheets flexible but not elastic.	BRUCITE p. 393 $Mg(OH)_2$ Hexagonal
{001} perfect seldom seen. Fracture earthy.	White; may be darker	2.6 to 2.63	$2-2\frac{1}{2}$	Compact, earthy. Breathed upon, gives argillaceous odor. Will adhere to the dry tongue.	**KAOLINITE** p. 530 $Al_2Si_2O_5(OH)_4$ Triclinic
Cubic {100}	Colorless or white	1.99	2	Soluble in water, bitter taste. Resembles halite but softer. In granular cleavable masses or cubic crytals.	Sylvite p. 400 KCl Isometric

TABLE 14.1 (continued)

LUSTER: NONMETALLIC
II. Streak colorless
A. Hardness: $\leq 2\frac{1}{2}$
(Can be scratched by fingernail)

Cleavage, Fracture	Color	G	H	Remarks	Name, Composition, Crystal System
{010} perfect, {100}, {011} good	Colorless, white, gray. May be colored by impurities	2.32	2	Occurs in crystals, broad cleavage flakes. May be compact massive without cleavage, or fibrous with silky luster.	**GYPSUM** p. 429 $CaSO_4 \cdot 2H_2O$ Monoclinic
Rhombohedral {1011} poor		2.29	1–2	Occurs in saline crusts. Readily soluble in water; cooling and salty taste. Fusible in candle flame.	NITRATITE p. 422 $NaNO_3$ Hexagonal
Prismatic {110}, seldom seen. Fracture conchoidal	Colorless or white	2.09 to 2.14	2	Usually in crusts, silky tufts and delicate acicular crystals. Readily soluble in water; cooling and salty taste. Fusible in the candle flame.	Niter, p. 422 KNO_3 Orthorhombic
Fracture uneven	Pearl-gray or colorless. Turns to pale brown on exposure to light	5.5±	2–3	Perfectly sectile. Translucent in thin plates. In irregular masses, rarely in crystals. Distinguished from other silver halides only by chemical tests.	Chlorargyrite p. 400 AgCl Isometric
	Pale yellow	2.05 to 2.09	$1\frac{1}{2}$–$2\frac{1}{2}$	Burns with blue flame, giving odor of SO_2. Crystallized, granular, earthy.	**SULFUR** p. 346 S Orthorhombic
	Yellow, brown, gray, white	2.0 to 2.55	1–3	In rounded grains, often earthy and claylike. Usually harder than $2\frac{1}{2}$.	**BAUXITE** p. 397 A mixture of Al hydroxides
Cleavage seldom seen	White	1.95	1	Usually in rounded masses of fine fibers and acicular crystals.	ULEXITE p. 425 $NaCaB_5O_6(OH) \cdot 5H_2O$ Triclinic

TABLE 14.1   (continued)

LUSTER: NONMETALLIC
II. Streak colorless
B. Hardness: $>2\frac{1}{2}-\leq3$
(Cannot be scratched by fingernail; can be scratched by cent)
1. Cleavage prominent

Cleavage, Fracture		Color	G	H	Remarks	Name, Composition, Crystal System
Perfect cleavage in one direction. See also the minerals of the mica group, p. 594 which may be harder than the fingernail.	{001}	Lilac, grayish white	2.8 to 3.0	$2\frac{1}{2}-4$	Crystals six-sided prismatic. Usually in small irregular sheets and scales. A pegmatite mineral.	LEPIDOLITE p. 538 $K(Li,Al)_{2-3}(AlSi_3O_{10})\text{-}$ $(O,OH,F)_2$ Monoclinic
	{001}	Pink, gray, white	3.0 to 3.1	$3\frac{1}{2}-5$	Usually in irregular foliated masses; folia brittle. Associated with emery.	MARGARITE p. 538 $CaAl_2(Al_2Si_2O_{10})(OH)_2$ Monoclinic
	{010}	Colorless or white	4.3	$3\frac{1}{2}$	Usually massive, with radiating habit. Effervesces in cold acid.	**WITHERITE** p. 418 $BaCO_3$ Orthorhombic
Cleavage in two directions {001} {100}		Colorless or white to gray	1.95	3	Occurs in cleavable crystalline aggregates.	KERNITE p. 422 $Na_2B_4O_6(OH)_2\cdot3H_2O$ Monoclinic
Cleavage in three directions at right angles	Cubic {100}	Colorless, white, red, blue	2.1 to 2.3	$2\frac{1}{2}$	Common salt. Soluble in water, taste salty, fusible in candle flame. In granular masses or in cubic crystals.	**HALITE** p. 398 $NaCl$ Isometric
	Cubic {100}	Colorless or white	1.99	2	Resembles halite, but distinguished from it by more bitter taste and lesser hardness.	Sylvite p. 400 KCl Isometric
	{001} {010} {100}	Colorless, white, blue, gray, red	2.89 to 2.98	$3-3\frac{1}{2}$	Commonly in massive aggregates, not showing cleavage; then distinguished only by chemical tests.	**ANHYDRITE** p. 428 $CaSO_4$ Orthorhombic
Cleavage in three directions, not at right angles. Rhombohedral $\{10\bar{1}1\}$		Colorless, white, and variously tinted	2.71	3	Effervesces in cold acid. Crystals show many forms. Occurs as limestone and marble. Clear varieties show strong double refraction.	**CALCITE** p. 411 $CaCO_3$ Hexagonal
Cleavage in three directions, not at right angles. Rhombohedral $\{10\bar{1}1\}$.		Colorless, white, pink	2.85	$3\frac{1}{2}-4$	Usually harder than 3. Often in curved rhombohedral crystals with pearly luster; as dolomitic limestone and marble. Powdered mineral will effervesce in cold acid.	**DOLOMITE** p. 419 $CaMg(CO_3)_2$ Hexagonal

TABLE 14.1   (continued)

LUSTER: NONMETALLIC
II. Streak colorless
B. Hardness: $>2\frac{1}{2}-\leq3$
(Cannot be scratched by fingernail; can be scratched by cent)
1. Cleavage prominent

Cleavage, Fracture	Color	G	H	Remarks	Name, Composition, Crystal System
Cleavage in three directions. Basal {001} at right angles to prismatic {210}	Colorless, white, blue, yellow red	4.5	$3-3\frac{1}{2}$	Frequently in aggregates of platy crystals. Pearly luster on basal cleavage. Characterized by high specific gravity and thus distinguished from celestite.	**BARITE** p. 425 $BaSO_4$ Orthorhombic
	Colorless, white, blue, red	3.95 to 3.97	$3-3\frac{1}{2}$	Similar to barite but lower specific gravity.	**CELESTITE** p. 427 $SrSO_4$ Orthorhombic
	Colorless or white; gray-brown when impure	6.2 to 6.4	3	Adamantine luster. Usually massive but may be in small tabular crystals. Alteration of galena. When massive may need test for SO_4 to distinguish from cerrusite ($PbCO_3$).	**ANGLESITE** p. 427 $PbSO_4$ Orthorhombic

2. Cleavage not prominent
a. Small splinter is fusible in the candle flame

Color	G	H	Remarks	Name, Composition, Crystal System
Colorless or white	1.7±	$2-2\frac{1}{2}$	Soluble in water. One good cleavage, seldom seen. In crusts and prismatic crystals. In candle, flame swells and then fuses. Sweetish alkaline taste.	**BORAX** p. 423 $Na_2B_4O_5(OH)_4 \cdot 8H_2O$ Monoclinic
	2.95 to 3.0	$2\frac{1}{2}$	Massive with peculiar translucent appearance. Fine powder becomes nearly invisible when placed in water. Ivigtut, Greenland, only important locality. Pseudocubic parting.	CRYOLITE p. 401 Na_3AlF_6 Monoclinic
	6.55	$3-3\frac{1}{2}$	Luster adamantine. In granular masses and platy crystals, usually associated with galena. Effervesces in cold nitric acid.	**CERUSSITE** p. 418 $PbCO_3$ Orthorhombic

TABLE 14.1   (continued)

LUSTER: NONMETALLIC
II. Streak colorless
B. Hardness: $>2\frac{1}{2}$, ≤ 3
(Cannot be scratched by fingernail; can be scratched by cent)
2. Cleavage not prominent
b. Infusible in candle flame

Color	G	H	Remarks	Name, Composition, Crystal System
Colorless or white	4.3	$3\frac{1}{2}$	Often in radiating masses; granular; rarely in pseudohexagonal crystals. Effervesces in cold acid	**WITHERITE** p. 418 $BaCO_3$ Orthorhombic
	2.6 to 2.63	$2-2\frac{1}{2}$	Usually compact, earthy. When breathed upon, gives an argillaceous odor.	**KAOLINITE** p. 530 $Al_2Si_2O_5(OH)_4$ Triclinic
Colorless, white, blue, gray, red	2.89 to 2.98	$3-3\frac{1}{2}$	Commonly in massive fine aggregates, not showing cleavage, and can be distinguished only by chemical tests.	**ANHYDRITE** p. 428 $CaSO_4$ Orthorhombic
Yellow, brown, gray, white	2.0 to 2.55	$1-3$	Usually pisolitic; in rounded grains and earthy masses. Often impure.	**BAUXITE** p. 397 A mixture of aluminum hydroxides
Ruby-red, brown, yellow	6.7 to 7.1	3	Luster resinous. In slender prismatic and cavernous crystals; in barrel-shaped forms.	Vanadinite p. 437 $Pb_5(VO_4)_3Cl$ Hexagonal
Yellow, green, white, brown	2.33	$3\frac{1}{2}-4$	Characteristically in radiating hemispherical globular aggregates. Cleavage seldom seen.	Wavellite p. 438 $Al_3(PO_4)_2(OH)_3 \cdot 5H_2O$ Orthorhombic
Olive- to blackish-green, yellow-green, white	2.3 to 2.66	$2-5$	Massive. Fibrous in the asbestos variety, chrysotile. Frequently mottled green in the massive variety.	**SERPENTINE** p. 528 $Mg_3Si_2O_5(OH)_4$ Monoclinic

TABLE 14.1   (continued)

LUSTER: NONMETALLIC
II. Streak colorless
C. Hardness: $>3-\leq 5\frac{1}{2}$
(Cannot be scratched by cent; can be scratched by knife)
1. Cleavage prominent

Cleavage		Color	G	H	Remarks	Name, Composition, Crystal System
One cleavage direction	{100}	Blue, usually darker at center	3.56 to 3.66	5–7	In bladed aggregates with cleavage parallel to length. Can be scratched by knife parallel to the length of the crystal but not in a direction at right angles to this.	**KYANITE** p. 500 Al_2SiO_5 Triclinic
	{010}	White, yellow, brown, red	2.1 to 2.2	$3\frac{1}{2}-4$	Characteristically in sheaflike crystal aggregates. May be in flat tabular crystals. Luster pearly on cleavage face.	**STILBITE** p. 562 $NaCa_2Al_5Si_{13}O_{36}\cdot 14H_2O$ Monoclinic
	{001}	Colorless, white, pale green, yellow, rose	2.3 to 2.4	$4\frac{1}{2}-5$	In prismatic crystals vertically striated. Crystals often resemble cubes truncated by the octahedron. Luster pearly on base, elsewhere vitreous.	APOPHYLLITE p. 541 $KCa_4(Si_4O_{10})_2F\cdot 8H_2O$ Tetragonal
	{010}	White, yellow, red	2.18 to 2.20	$3\frac{1}{2}-4$	Luster pearly on cleavage, elsewhere vitreous. Crystals often tabular parallel to cleavage plane. Found in cavities in igneous rocks.	HEULANDITE p. 561 $CaAl_2Si_7O_{18}\cdot 6H_2O$ Monoclinic
	{010}	Colorless, white	2.42	$4-4\frac{1}{2}$	In crystals and in cleavable aggregates. Decrepitates violently in the candle flame.	COLEMANITE p. 425 $CaB_3O_4(OH)_3\cdot H_2O$ Monoclinic
	{010}	Colorless, white	4.3	$3\frac{1}{2}$	Often in radiating crystal aggregates; granular. Rarely in pseudohexagonal crystals.	WITHERITE p. 418 $BaCO_3$ Orthorhombic
	{010} {110} poor	Colorless, white	2.95	$3\frac{1}{2}-4$	Effervesces in cold acid. Frequently in radiating groups of acicular crystals; in pseudohexagonal twins.	**ARAGONITE** p. 416 $CaCO_3$ Orthorhombic
Two cleavage directions	{001} {100}	Bluish gray, salmon to clove-brown	3.42 to 3.56	$4\frac{1}{2}-5$	Commonly cleavable, massive. Found in pegmatites with other lithium minerals.	Triphylite-lithiophilite p. 433 $Li(Fe,Mn)PO_4$ Orthorhombic
	{001} {100}	Colorless, white, gray	2.8 to 2.9	$5-5\frac{1}{2}$	Usually cleavable, massive to fibrous. Also compact. Associated with crystalline limestone.	WOLLASTONITE p. 520 $CaSiO_3$ Triclinic
	{001} {100}	Colorless, white, gray	2.7 to 2.8	5	In radiating aggregates of sharp acicular crystals. Associated with zeolites in cavities in igneous rocks.	Pectolite p. 523 $Ca_2NaH(SiO_3)_3$ Triclinic

TABLE 14.1 (continued)

LUSTER: NONMETALLIC
II. Streak colorless
C. Hardness: $>3-\leq5\frac{1}{2}$
(Cannot be scratched by cent; can be scratched by knife)
1. Cleavage prominent

Cleavage		Color	G	H	Remarks	Name, Composition, Crystal System
Two cleavage directions	{110}	Colorless, white	2.25	$5-5\frac{1}{2}$	Slender prismatic crystals, prism faces vertically striated. Often in radiating groups. Found lining cavities in igneous rocks.	**NATROLITE** p. 559 $Na_2Al_2Si_3O_{10}\cdot2H_2O$ Orthorhombic
	{110}	Colorless, white	3.7	$3\frac{1}{2}-4$	In prismatic crystals and pseudohexagonal twins. Also fibrous and massive. Effervesces in cold acid.	**STRONTIANITE** p. 418 $SrCO_3$ Orthorhombic
	{110}	White, pale green, blue	3.4 to 3.5	$4\frac{1}{2}-5$	Often in radiating crystal groups. Also stalactitic, mammillary. Prismatic cleavage seldom seen.	**HEMIMORPHITE** p. 506 $Zn_4(Si_2O_7)(OH)_2\cdot H_2O$ Orthorhombic
	{110}	Brown, gray, green, yellow	3.4 to 3.55	$5-5\frac{1}{2}$	Luster adamantine to resinous. In thin wedge-shaped crystals with sharp edges. Prismatic cleavage seldom seen.	TITANITE p. 504 $CaTiO(SiO_4)$ Monoclinic
	Prismatic at angles of 56° and 124° — White, green, black		3.0 to 3.3	5–6	Crystals usually slender, fibrous; may be asbestiform. Tremolite (white, gray, violet), actinolite (green) common in metamorphic rocks. Hornblende and arfvedsonite (dark green to black) common in igneous and metamorphic rocks.	**AMPHIBOLE GROUP** p. 523 Essentially hydrous Ca-Mg-Fe silicates Monoclinic
	Prismatic at angles of 56° and 124° — Gray, clove-brown, green		2.85 to 3.2	$5\frac{1}{2}-6$	An amphibole. Distinct crystals rare. Commonly in aggregates and fibrous masses.	Anthophyllite p. 523 or grunerite p. 523 $(Mg,Fe)_7Si_8O_{22}(OH)_2$ Orthorhombic or monoclinic
	Prismatic at nearly 90° angles — White, green black, brown		3.1 to 3.5	5–6	In stout prisms with rectangular cross section. Often in granular crystalline masses. Diopside (colorless, white, green), aegirine (brown, green), augite (dark green to black), members of the orthopyroxene series (light brown)	**PYROXENE GROUP** p. 514 Essentially Ca-Mg-Fe silicates Monoclinic and orthorhombic
Prismatic at nearly 90° angles		Rose-red, pink, brown	3.58 to 3.70	$5\frac{1}{2}-6$	Color diagnostic. Usually massive, cleavable to compact, in imbedded grains; in large rough crystals, with rounded edges.	RHODONITE p. 521 $MnSiO_3$ Triclinic

TABLE 14.1 (continued)

LUSTER: NONMETALLIC
II. Streak colorless
C. Hardness: $>3-\leq5\frac{1}{2}$
(Cannot be scratched by cent; can be scratched by knife)
1. Cleavage prominent

Cleavage	Color	G	H	Remarks	Name, Composition, Crystal System
Three cleavage directions — Three directions not at right angles. Rhombohedral $\{10\bar{1}1\}$	Colorless, white, and variously tinted	2.72	3	Effervesces in cold acid. Crystals show many forms. Occurs as limestone and marble. Clear varieties show strong double refraction.	**CALCITE** p. 411 $CaCO_3$ Hexagonal
	Colorless, white, pink	2.85	$3\frac{1}{2}-4$	Often in curved rhombohedral crystals with pearly luster. In coarse masses as dolomitic limestone and marble. Powdered mineral will effervesce in cold acid.	**DOLOMITE** p. 419 $CaMg(CO_3)_2$ Hexagonal
	White, yellow gray, brown	3.0 to 3.2	$3\frac{1}{2}-5$	Commonly in dense compact masses; also in fine to coarse cleavable masses. Effervesces in hot hydrochloric acid.	**MAGNESITE** p. 413 $MgCO_3$ Hexagonal
	Light to dark brown	3.83 to 3.88	$3\frac{1}{2}-4$	In cleavable masses or in small curved rhombohedral crystals. Becomes magnetic after heating.	**SIDERITE** p. 414 $FeCO_3$ Hexagonal
	Pink, rose-red, brown	3.45 to 3.6	$3\frac{1}{2}-4\frac{1}{2}$	In cleavable masses or in small rhombohedral crystals. Characterized by its color.	RHODOCHROSITE p. 415 $MnCO_3$ Hexagonal
	Brown, green, blue, pink, white	4.35 to 4.40	5	In rounded botryoidal aggregates and honeycombed masses. Effervesces in cold hydrochloric acid. Cleavage rarely seen.	SMITHSONITE p. 416 $ZnCO_3$ Hexagonal
$\{10\bar{1}1\}$	White, yellow, flesh-red	2.05 to 2.15	4–5	In small rhombohedral crystals with nearly cubic angles. Found lining cavities in igneous rocks.	CHABAZITE p. 560 $Ca_2Al_2Si_4O_{12}\cdot6H_2O$ Hexagonal
$\{001\}$ $\{010\}$ $\{100\}$	Colorless, white, blue, gray, red	2.89 to 2.98	$3-3\frac{1}{2}$	Commonly in massive fine aggregates, not showing cleavage, and can be distinguished only by chemical tests.	**ANHYDRITE** p. 428 $CaSO_4$ Orthorhombic
$\{001\}$ at right angles to $\{110\}$	Colorless, white, blue, yellow, red	4.5	$3-3\frac{1}{2}$	Frequently in aggregates of platy crystals. Pearly luster on basal cleavage. Characterized by high specific gravity and thus distinguished from celestite.	**BARITE** p. 425 $BaSO_4$ Orthorhombic
	Colorless, white, blue, red	3.95 to 3.97	$3-3\frac{1}{2}$	Similar to barite but lower specific gravity.	**CELESTITE** p. 427 $SrSO_4$ Orthorhombic

TABLE 14.1   (continued)

LUSTER: NONMETALLIC
II. Streak colorless
C. Hardness: $>3-\leq5\frac{1}{2}$
(Cannot be scratched by cent; can be scratched by knife)
1. Cleavage prominent

Cleavage		Color	G	H	Remarks	Name, Composition, Crystal System
Four cleavage directions	{111} Octa-hedral	Colorless, violet, green, yellow, pink	3.18	4	In cubic crystals often in penetration twins. Characterized by cleavage.	**FLUORITE** p. 401 CaF_2 Isometric
	{100} {110}	White, pink gray, green, brown	2.65 to 2.74	5–6	In prismatic crystals, granular or massive. Commonly altered. Prismatic cleavage obscure.	SCAPOLITE p. 557 Essentially Na, Ca aluminum silicate Tetragonal
Six cleavage directions Dodecahedral {011}		Yellow, brown, white	3.9 to 4.1	$3\frac{1}{2}$–4	Luster resinous. Small tetrahedral crystals rare. Usually in cleavable masses. If massive, difficult to determine.	**SPHALERITE** p. 357 ZnS Isometric
		Blue, white, gray, green	2.15 to 2.3	$5\frac{1}{2}$–6	Massive or in embedded grains; rarely in crystals. A feldspathoid associated with nepheline, never with quartz.	SODALITE p. 556 $Na_6(AlSiO_4)_6Cl_2$ Isometric

TABLE 14.1 (continued)

LUSTER: NONMETALLIC
II. Streak colorless
C. Hardness: $>3-\leq5\frac{1}{2}$
(Cannot be scratched by cent; can be scratched by knife)
2. Cleavage not prominent

Color	G	H	Remarks	Name, Composition, Crystal System
Colorless, white	2.25	$5-5\frac{1}{2}$	In slender prismatic crystals, prism faces vertically striated. Often in radiating groups. Found lining cavities in igneous rocks. Poor prismatic cleavage.	NATROLITE p. 559 $Na_2Al_2Si_3O_{10}\cdot2H_2O$ Orthorhombic
	2.95	$3\frac{1}{2}-4$	Effervesces in cold acid. Falls to powder in the candle flame. Frequently in radiating groups of acicular crystals; in pseudohexagonal twins. Cleavage indistinct.	ARAGONITE p. 416 $CaCO_3$ Orthorhombic
	2.27	$5-5\frac{1}{2}$	Usually in trapezohedrons with vitreous luster. Found lining cavities in igneous rocks.	ANALCIME p. 558 $NaAlSi_2O_6\cdot H_2O$ Isometric
	3.7	$3\frac{1}{2}-4$	Occurs in prismatic crystals and pseudohexagonal twins. Also fibrous and massive. Effervesces in cold hydrochloric acid.	STRONTIANITE p. 418 $SrCO_3$ Orthorhombic
	3.0 to 3.2	$3\frac{1}{2}-5$	Commonly in dense compact masses showing no cleavage. Effervesces in hot hydrochloric acid.	MAGNESITE p. 413 $MgCO_3$ Hexagonal
	4.3	$3\frac{1}{2}$	Often in radiating masses; granular; rarely in pseudohexagonal crystals. Effervesces in cold hydrochloric acid.	WITHERITE p. 418 $BaCO_3$ Orthorhombic
	2.7 to 2.8	5	Commonly fibrous in radiating aggregates of sharp acicular crystals. Associated with zeolites in cavities in igneous rocks.	Pectolite p. 523 $Ca_2NaH(SiO_3)_3$ Triclinic
Colorless, pale green, yellow	2.8 to 3.0	$5-5\frac{1}{2}$	Usually in crystals with many brilliant faces. Occurs with zeolites lining cavities in igneous rocks.	DATOLITE p. 504 $CaB(SiO_4)(OH)$ Monoclinic
White, pale green, blue	3.4 to 3.5	$4\frac{1}{2}-5$	Often in radiating crystal groups. Also stalactitic, mammillary. Prismatic cleavage seldom seen.	HEMIMORPHITE p. 506 $Zn_4(Si_2O_7)(OH)_2\cdot H_2O$ Orthorhombic
White, pink, gray, green, brown	2.65 to 2.74	5-6	In prismatic crystals, granular or massive. Commonly altered. Prismatic cleavage obscure.	SCAPOLITE p. 557 Essentially Na, Ca-aluminum silicate Tetragonal
White, grayish, red	2.6 to 2.8	4	May be in rhombohedral crystals. Usually massive granular. Definitely determined only by chemical tests. Cleavage {0001}, poor.	ALUNITE p. 431 $KAl_3(SO_4)_2(OH)_6$ Hexagonal

TABLE 14.1   (continued)

LUSTER: NONMETALLIC
II. Streak colorless
C. Hardness: $>3-\leq5\frac{1}{2}$
(Cannot be scratched by cent; can be scratched by knife)
2. Cleavage not prominent

Color	G	H	Remarks	Name, Composition, Crystal System
Colorless, white, yellow, red, brown	1.9 to 2.2	5–6	Conchoidal fracture. Precious opal shows internal play of colors. Specific gravity and hardness less than fine-grained quartz.	**OPAL** p. 548 $SiO_2 \cdot nH_2O$ Essentially amorphous
Brown, green, blue, pink, white	4.35 to 4.40	5	Usually in rounded botryoidal aggregates and in honeycombed masses. Effervesces in cold hydrochloric acid. Cleavage rarely seen.	**SMITHSONITE** p. 416 $ZnCO_3$ Hexagonal
Brown, gray, green, yellow	3.4 to 3.55	5–5$\frac{1}{2}$	Adamantine to resinous luster. In thin wedge-shaped crystals with sharp edges. Prismatic cleavage seldom seen.	TITANITE p. 504 $CaTiO(SiO_4)$ Monoclinic
Colorless, white, yellow, red, brown	2.72	3	May be fibrous or fine granular, banded in Mexican onyx variety. Effervesces in cold hydrochloric acid.	**CALCITE** p. 411 $CaCO_3$ Hexagonal
Yellowish- to reddish-brown	5.0 to 5.3	5–5$\frac{1}{2}$	In small crystals or as rolled grains. Found in pegmatites.	Monazite p. 434 $(Ce,La,Y,Th)PO_4$ Monoclinic
Light to dark brown	3.38 to 3.88	3$\frac{1}{2}$–4	Usually cleavable but may be in compact concretions in clay or shale—clay ironstone variety. Becomes magnetic on heating.	**SIDERITE** p. 414 $FeCO_3$ Hexagonal
White, yellow, green, brown	5.9 to 6.1	4$\frac{1}{2}$–5	Luster vitreous to adamantine. Massive and in octahedral-like crystals. Frequently associated with quartz. Will fluoresce.	SCHEELITE p. 432 $CaWO_4$ Tetragonal
Yellow, orange, red, gray, green	6.8±	3	Luster adamantine. Usually in square tabular crystals. Also granular massive. Characterized by color and high specific gravity.	WULFENITE p. 432 $PbMoO_4$ Tetragonal
White, yellow brown, gray	2.6 to 2.9	3–5	Occurs massive as rock phosphate. Difficult to identify without chemical tests.	**APATITE** p. 434 $Ca_5(PO_4)_3(F,Cl,OH)$ Appears amorphous
Yellow, brown, gray, white	2.0 to 2.55	1–3	Usually pisolitic; in rounded grains and earthy masses. Often impure.	**BAUXITE** p. 397 A mixture of aluminum hydroxides
Green, blue, violet, brown, colorless	3.15 to 3.20	5	Usually in hexagonal prisms with pyramid. Also massive. Poor basal cleavage.	**APATITE** p. 434 $Ca_5(PO_4)_3(F,Cl,OH)$ Hexagonal
Green, brown, yellow, gray	6.5 to 7.1	3$\frac{1}{2}$–4	In small hexagonal crystals, often curved and barrel-shaped. Crystals may be cavernous. Often globular and botryoidal.	Pyromorphite p. 435 $Pb_5(PO_4)_3Cl$ Hexagonal

TABLE 14.1 (continued)

LUSTER: NONMETALLIC
II. Streak colorless
C. Hardness: $>3-\leq5\frac{1}{2}$
(Cannot be scratched by cent; can be scratched by knife)
2. Cleavage not prominent

Color	G	H	Remarks	Name, Composition, Crystal System
Yellow, green, white, brown	2.33	$3\frac{1}{2}-4$	Characteristically in radiating, hemispherical globular aggregates. Cleavage seldom seen.	Wavellite p. 438 $Al_3(PO_4)_2(OH)_3 \cdot 5H_2O$ Orthorhombic
Olive- to blackish-green, yellow-green, white	2.3 to 2.6	2–5	Massive. Fibrous in the asbestos variety, chrysotile. Frequently mottled green in the massive variety.	**SERPENTINE** p. 528 $Mg_3Si_2O_5(OH)_4$ Monoclinic
Yellow-green, white, blue, gray, brown	3.9 to 4.2	$5\frac{1}{2}$	Massive and in disseminated grains. Rarely in hexagonal prisms. Associated with red zincite and black franklinite at Franklin, N.J. Will fluoresce.	WILLEMITE p. 492 Zn_2SiO_4 Hexagonal
White, gray, blue, green	2.15 to 2.3	$5\frac{1}{2}-6$	Massive or in embedded grains; rarely in crystals. A feldspathoid associated with nepheline, never with quartz. Dodecahedral cleavage poor.	SODALITE p. 556 $Na_6(AlSiO_4)_6Cl_2$ Isometric
Deep azure-blue, greenish-blue	2.4 to 2.45	$5-5\frac{1}{2}$	Usually massive. Associated with feldspathoids and pyrite. Poor dodecahedral cleavage.	Lazurite p. 556 $(Na,Ca)_8(AlSiO_4)_6(SO_4,S,Cl)_2$ Isometric

TABLE 14.1 (continued)

LUSTER: NONMETALLIC
II. Streak colorless
D. Hardness: $>5\frac{1}{2} \leq 7$
(Cannot be scratched by knife; can be scratched by quartz)
1. Cleavage prominent

	Cleavage	Color	G	H	Remarks	Name, Composition, Crystal System
One cleavage direction	{010}	White, gray, pale lavender, yellow-green	3.35 to 3.45	$6\frac{1}{2}$–7	In thin tabular crystals. Luster pearly on cleavage face. Associated with emery, margarite, chlorite.	Diaspore p. 395 αAlO(OH) Orthorhombic
	{010} perfect	Hair-brown, grayish-green	3.23	6–7	In long, prismatic crystals. May be in parallel groups—columnar or fibrous. Found in schists.	SILLIMANITE p. 500 Al_2SiO_5 Orthorhombic
	{001}	Yellowish to blackish-green	3.35 to 3.45	6–7	In prismatic crystals striated parallel to length. Found in metamorphic rocks and crystalline limestones	**EPIDOTE** p. 507 $Ca_2(Al,Fe)Al_2O(SiO_4)$- $(Si_2O_7)(OH)$ Monoclinic
	{100}	Blue; may be gray, or green	3.56 to 3.66	5–7	In bladed aggregates with cleavage parallel to length. H = 5 parallel to length of crystal.	**KYANITE** p. 500 Al_2SiO_5 Triclinic
Two cleavage directions	{100} good {110} poor	White, pale green, or blue	3.0 to 3.1	6	Usually cleavable, resembling fledspar. Found in pegmatites associated with other lithium minerals.	AMBLYGONITE p. 437 $LiAlFPO_4$ Triclinic
	{100} good {001}	Colorless, white, gray	2.8 to 2.9	5–$5\frac{1}{2}$	Usually cleavable, massive to fibrous. Also compact. Associated with crystalline limestone.	WOLLASTONITE p. 520 $CaSiO_3$ Triclinic
	{001} {100}	Grayish-white, green, pink	3.25 to 3.37	6–$6\frac{1}{2}$	In prismatic crystals, deeply striated. Also massive, columnar, compact. Luster pearly on cleavage, elsewhere vitreous.	Clinozoisite p. 507 $Ca_2Al_3O(SiO_4)(Si_2O_7)(OH)$ Monoclinic
	{110}	Colorless, white	2.25	5–$5\frac{1}{2}$	In slender prismatic crystals, prism faces vertically striated. Often in radiating groups lining cavities in volcanic rocks.	**NATROLITE** p. 559 $Na_2Al_2Si_3O_{10} \cdot 2H_2O$ Orthorhombic
	{110}	Brown, gray, green, yellow	3.4 to 3.55	5–$5\frac{1}{2}$	Luster adamantine to resinous. In thin wedge-shaped crystals with sharp edges. Prismatic cleavage seldom seen.	TITANITE p. 504 $CaTiO(SiO_4)$ Monoclinic

TABLE 14.1   (continued)

LUSTER: NONMETALLIC

II. Streak colorless

D. Hardness: $>5\frac{1}{2} \leq 7$

(Cannot be scratched by knife; can be scratched by quartz)

1. Cleavage prominent

	Cleavage	Color	G	H	Remarks	Name, Composition, Crystal System
Two cleavage directions at 90° or nearly 90°	{001} {010}	Colorless, white, gray, cream, red, green	2.54 to 2.56	6	In cleavable masses or in irregular grains as rock constituents. May be in crystals in pegmatites. Distinguished with certainty only with the microscope. Green amazonite is microcline.	**ORTHOCLASE** p. 551 Monoclinic **MICROCLINE** p. 548 Triclinic $KAlSi_3O_8$
	{001} {010}	Colorless, white, gray, bluish. Often shows play of colors	2.62 (ab) to 2.76 (an)	6	In cleavable masses or in irregular grains as a rock constituent. On the better cleavage can be seen a series of fine parallel striations due to albite twinning; these distinguish it from orthoclase.	**PLAGIOCLASE** p. 552 Various proportions of albite, $NaAlSi_3O_8$, and anorthite, $CaAl_2Si_2O_8$ Triclinic
	{110}	White, gray, pink, green	3.15 to 3.20	$6\frac{1}{2}$–7	In flattened prismatic crystals, vertically striated. Also massive, cleavable. Found in pegmatites. Frequently shows good {100} parting.	SPODUMENE p. 519 $LiAlSi_2O_6$ Monoclinic
	{110}	White, green, black	3.1 to 3.5	5–6	In stout prisms with rectangular cross section. Often in granular crystalline masses. Diopside (colorless, white, green), aegirine (brown, green), augite (dark green to black). Characterized by cleavage.	**PYROXENE GROUP** p. 514 Essentially Ca-Mg-Fe silicates Monoclinic
	{110}	Gray-brown, green, bronze-brown, black	3.2 to 3.5	$5\frac{1}{2}$	Crystals are usually prismatic but rare. Commonly massive, fibrous, lamellar. Fe may replace Mg and mineral is darker.	ENSTATITE p. 514 $MgSiO_3$ Orthorhombic
	{110}	Rose-red, pink, brown	3.58 to 3.70	$5\frac{1}{2}$–6	Color diagnostic. Usually massive; cleavable to compact, in embedded grains; in large rough crystals with rounded edges.	RHODONITE p. 521 $MnSiO_3$ Triclinic

TABLE 14.1 (continued)

LUSTER: NONMETALLIC
II. Streak colorless
D. Hardness: $>5\frac{1}{2} \leq 7$
(Cannot be scratched by knife; can be scratched by quartz)
1. Cleavage prominent

Cleavage		Color	G	H	Remarks	Name, Composition, Crystal System
Two cleavage directions at 54° and 126° angles	{110}	White, green black	3.0 to 3.3	5–6	Crystals usually slender fibrous; may be asbestiform. Tremolite (white, gray, violet) and actinolite (green) are common in metamorphic rocks. Hornblende and arfvedsonite (dark green to black) are common in igneous rocks. Characterized by cleavage angle.	**AMPHIBOLE GROUP** p. 523 Essentially hydrous Ca-Mg-Fe silicates Monoclinic
	{110}	Gray, clove-brown, green	2.85 to 3.2	$5\frac{1}{2}$–6	An amphibole. Distinct crystals rare. Commonly in aggregates and fibrous, massive.	Anthophyllite p. 523 and grunerite p. 523 $(Mg,Fe)_7(Si_8O_{22})(OH)_2$ Orthorhombic and monoclinic
Six cleavage directions	{110}	Blue, gray, white, green	2.15 to 2.3	$5\frac{1}{2}$–6	Massive or embedded grains; rarely in crystals. A feldspathoid associated with nepheline, never with quartz.	SODALITE p. 556 $Na_8(AlSiO_4)_6Cl_2$ Isometric

TABLE 14.1   (continued)

LUSTER: NONMETALLIC
II. Streak colorless
D. Hardness: $>5\frac{1}{2} \leq 7$
(Cannot be scratched by knife; can be scratched by quartz)
2. Cleavage not prominent

Color	G	H	Remarks	Name, Composition, Crystal System
Colorless	2.26	7	Occurs as small crystals in cavities in volcanic rocks. Difficult to determine without optical aid.	Tridymite p. 547 SiO_2 Pseudohexagonal
Colorless or white	2.27	$5-5\frac{1}{2}$	Usually in trapezohedrons with vitreous luster. Found lining cavities in igneous rocks.	ANALCIME p. 558 $NaAlSi_2O_6 \cdot H_2O$ Isometric
	2.32	7	Occurs in spherical aggregations in volcanic rocks. Difficult to determine without optical aid.	Cristobalite p. 547 SiO_2 Pseudoisometric
Colorless, yellow, red, brown, green, gray, blue	1.9 to 2.2	5–6	Conchoidal fracture. Precious opal shows internal play of colors. Gravity and hardness less than fine-grained quartz.	**OPAL** p. 548 $SiO_2 \cdot nH_2O$ Essentially amorphous
Gray, white, colorless	2.45 to 2.50	$5\frac{1}{2}-6$	In trapezohedral crystals embedded in dark igneous rock. Does not line cavities as analcime does.	LEUCITE p. 554 $KAlSi_2O_6$ Pseudoisometric
Colorless, pale, green, yellow	2.8 to 3.0	$5-5\frac{1}{2}$	Usually in crystals with many brilliant faces. Occurs with zeolites lining cavities in igneous rocks.	DATOLITE p. 504 $CaB(SiO_4)(OH)$ Monoclinic
Colorless, white, smoky. Variously colored	2.65	7	Crystals usually show horizontally striated prism with rhombohedral terminations.	**QUARTZ** p. 543 SiO_2 Hexagonal
Colorless, gray, greenish, reddish	2.55 to 2.65	$5\frac{1}{2}-6$	Greasy luster. A rock constituent, usually massive; rarely in hexagonal prisms. Poor prismatic cleavage. A feldspathoid.	NEPHELINE p. 555 $(Na,K)AlSiO_4$ Hexagonal
White, gray, light to dark green, brown	2.65 to 2.74	5–6	In prismatic crystals, granular or massive. Commonly altered and prismatic cleavage obscure.	SCAPOLITE p. 557 Essentially Na-Ca aluminum silicate Tetragonal
Light yellow, brown, orange	3.1 to 3.2	$6-6\frac{1}{2}$	Occurs in disseminated crystals and grains. Commonly in crystalline limestones.	Chondrodite p. 503 $Mg_5(SiO_4)_2(F,OH)_2$ Monoclinic
Light brown, yellow, red, green	2.6	7	Chalcedony, waxy luster, commonly colloform; as agate lining cavities. Dull luster, massive: jasper red, flint black, chert gray.	**MICROSRYSTALLINE QUARTZ** p. 545 SiO_2
Blue, bluish-green, green	2.6 to 2.8	6	Usually appears amorphous in reniform and stalactitic masses.	TURQUOISE p. 438 $CuAl_6(PO_4)_4(OH)_8 \cdot 5H_2O$ Triclinic

TABLE 14.1 (continued)

LUSTER: NONMETALLIC
II. Streak colorless
D. Hardness: $>5\frac{1}{2} \leq 7$
(Cannot be scratched by knife; can be scratched by quartz)
2. Cleavage not prominent

Color	G	H	Remarks	Name, Composition, Crystal System
Apple-green, gray, white	2.8 to 2.95	$6-6\frac{1}{2}$	Reniform and stalactitic with crystalline surface. In subparallel groups of tabular crystals.	PREHNITE p. 542 $Ca_2Al(AlSi_3O_{10})(OH)_2$ Orthorhombic
Yellow-green, white, blue, gray, brown	3.9 to 4.2	$5\frac{1}{2}$	Massive and in disseminated grains. Rarely in hexagonal prisms. Associated with red zincite and black franklinite at Franklin, N.J. Will fluoresce.	WILLEMITE p. 492 Zn_2SiO_4 Hexagonal
Olive to grayish-green, brown	3.27 to 4.37	$6\frac{1}{2}-7$	Usually in disseminated grains in basic igneous rocks. May be massive, granular.	OLIVINE p. 492 $(Mg,Fe)_2SiO_4$ Orthorhombic
Black, green, brown, blue, red, pink, white	3.0 to 3.25	$7-7\frac{1}{2}$	In slender prismatic crystals with triangular cross section. Crystals may be in radiating groups. Found usually in pegmatites. Black most common; other colors associated with lithium minerals.	**TOURMALINE** p. 513 Complex boron silicate of Na-Ca-Al-Mg-Fe-Mn Hexagonal
Green, brown, yellow, blue, red	3.35 to 4.45	$6\frac{1}{2}$	In square prismatic crystals, vertically striated. Often columnar and granular massive. Found in crystalline limestones.	VESUVIANITE p. 509 Complex hydrous Ca-Mg-Fe-Al silicate Tetragonal
Clove-brown, gray, green, yellow	3.27 to 3.35	$6\frac{1}{2}-7$	In wedge-shaped crystals with sharp edges. Also lamellar.	AXINITE p. 510 $(Ca,Fe,Mn)_3Al_2B-$ $(Si_4O_{15})(OH)$ Triclinic
Red-brown to brownish-black	3.65 to 3.75	$7-7\frac{1}{2}$	In prismatic crystals; commonly in cruciform penetration twins. Frequently altered on the surface and then soft. Found in schists.	**STAUROLITE** p. 502 $Fe_2Al_9O_6(SiO_4)_4(O,OH)_2$ Pseudo-orthorhombic
Reddish-brown, flesh-red, olive-green	3.16 to 3.20	$7\frac{1}{2}$	Prismatic crystals with nearly square cross section. Cross section may show black cross (chiastolite). May be altered to mica and then soft. Found in schists.	**ANDALUSITE** p. 499 Al_2SiO_5 Orthorhombic
Brown, gray, green, yellow	3.4 to 3.55	$5-5\frac{1}{2}$	Luster adamantine to resinous. In thin wedge-shaped crystals with sharp edges. Prismatic cleavage seldom seen.	**TITANITE** p. 504 $CaTiO(SiO_4)$ Monoclinic
Yellowish- to reddish-brown	5.0 to 5.3	$5-5\frac{1}{2}$	In isolated crystals, granular. Commonly found in pegmatites.	Monazite p. 434 $(Ce,La,Y,Th)PO_4$ Monoclinic
Brown to black	6.8 to 7.1	$6-7$	Rarely in prismatic crystals, twinned. Fibrous, giving reniform surface. Rolled grains. Usually gives light brown streak.	**CASSITERITE** p. 385 SnO_2 Tetragonal

TABLE 14.1 (continued)

LUSTER: NONMETALLIC
II. Streak colorless
D. Hardness: $>5\frac{1}{2} \leq 7$
(Cannot be scratched by knife; can be scratched by quartz)
2. Cleavage not prominent

Color	G	H	Remarks	Name, Composition, Crystal System
Reddish-brown to black	4.18 to 4.25	$6-6\frac{1}{2}$	In prismatic crystals, vertically striated; often slender, acicular. Crystals frequently twinned. A constituent of black sands.	**RUTILE** p. 383 TiO_2 Tetragonal
Brown to pitch-black	3.5 to 4.2	$5\frac{1}{2}-6$	Crystals often tabular. Massive and in embedded grains. An accessory mineral in igneous rocks.	Allanite p. 509 Ce, La, Th epidote Monoclinic
Blue, rarely colorless	2.60 to 2.66	$7-7\frac{1}{2}$	In embedded grains and massive, resembling quartz. Commonly altered and foliated; then softer than a knife.	Cordierite p. 512 $(Mg,Fe)_2Al_4Si_5O_{18} \cdot nH_2O$ Orthorhombic (Pseudohexagonal)
Deep azure-blue, greenish-blue	2.4 to 2.45	$5-5\frac{1}{2}$	Usually massive. Associated with feldspathoids and pyrite. Poor dodecahedral cleavage.	LAZURITE p. 556 $(Na,Ca)_8(AlSiO_4)_6(SO_4,S,Cl)_2$ Isometric
Azure-blue	3.0 to 3.1	$5-5\frac{1}{2}$	Usually in pyramidal crystals, which distinguishes it from massive lazurite. A rare mineral.	LAZULITE p. 438 $(Mg,Fe)Al_2(PO_4)_2(OH)_2$ Monoclinic
Blue, green, white, gray	2.15 to 2.3	$5\frac{1}{2}-6$	Massive or in embedded grains; rarely in crystals. A feldspathoid associated with nepheline, never with quartz. Poor dodecahedral cleavage.	SODALITE p. 556 $Na_8(AlSiO_4)_6Cl_2$ Isometric

TABLE 14.1   (continued)

LUSTER: NONMETALLIC
II. Streak colorless
E. Hardness: >7
(Cannot be scratched by quartz)
1. Cleavage prominent

Cleavage		Color	G	H	Remarks	Name, Composition, Crystal System
One cleavage direction	{001}	Colorless, yellow, pink, bluish, greenish	3.4 to 3.6	8	Usually in crystals, also coarse to fine granular. Found in pegmatites.	**TOPAZ** p. 501 $Al_2SiO_4(F,OH)_2$ Orthorhombic
One cleavage direction	{010}	Brown, gray, greenish-gray	3.23	6–7	Commonly in long slender prismatic crystals. May be in parallel groups, columnar or fibrous. Found in schistose rocks.	**SILLIMANITE** p. 500 Al_2SiO_5 Orthorhombic
Two cleavage directions	{110}	White, gray, pink, green	3.15 to 3.20	$6\frac{1}{2}$–7	In flattened prismatic crystals, vertically striated. Also massive, cleavable. Pink variety, kunzite; green, hiddenite. Found in pegmatites. Frequently shows good {100} parting.	SPODUMENE p. 519 $LiAlSi_2O_6$ Monoclinic
Three cleavage directions	{010} {110}	Colorless pale blue, gray	3.09	8	Commonly in tabular or prismatic crystals in schists.	Lawsonite p. 507 $CaAl_2(Si_2O_7)(OH_2) \cdot H_2O$ Orthorhombic
Four cleavage directions	{111}	Colorless, yellow, red, blue, black	3.5	10	Adamantine luster. In octahedral crystals, frequently twinned. Faces may be curved.	Diamond p. 347 C Isometric
No cleavage. Rhombohedral and basal parting		Colorless, gray, blue, red, yellow, brown, green	3.95 to 4.1	9	Luster adamantine to vitreous. Parting fragments may appear nearly cubic. In rude barrel-shaped crystals.	**CORUNDUM** p. 379 Al_2O_3 Hexagonal

TABLE 14.1 (continued)

LUSTER: NONMETALLIC
II. Streak colorless
E. Hardness: >7
(Cannot be scratched by quartz)
2. Cleavage not prominent

Color	G	H	Remarks	Name, Composition, Crystal System
Colorless, white, smoky, variously colored	2.65	7	Crystals usually show horizontally striated prism with rhombohedral terminations.	**QUARTZ** p. 543 SiO_2 Hexagonal
White, colorless	2.97 to 3.0	$7\frac{1}{2}$–8	In small rhombohedral crystals. A rare mineral.	Phenacite p. 491 Be_2SiO_4 Hexagonal
White and almost any color	3.95 to 4.1	9	Luster adamantine to vitreous. Parting fragments may appear nearly cubic. In rude barrel-shaped crystals.	**CORUNDUM** p. 379 Al_2O_3 Hexagonal
Red, black, blue, green, brown	3.6 to 4.0	8	In octahedrons; twinning common. Associated with crystalline limestones.	**SPINEL** p. 388 $MgAl_2O_4$ Isometric
Bluish-green, yellow, pink, colorless	2.65 to 2.8	$7\frac{1}{2}$–8	Commonly in hexagonal prisms terminated by the base; pyramid faces are rare. Crystals large in places. Poor basal cleavage.	**Beryl** p. 511 $Be_3Al_2(Si_6O_{18})$ Hexagonal
Yellowish to emerald-green	3.65 to 3.8	$8\frac{1}{2}$	In tabular crystals; frequently in pseudohexagonal twins. Found in pegmatites.	CHRYSOBERYL p. 392 $BeAl_2O_4$ Orthorhombic
Green, brown, blue, red, pink, black	3.0 to 3.25	7–$7\frac{1}{2}$	In slender prismatic crystals with triangular cross section. Found usually in pegmatites. Black most common, other colors associated with lithium minerals.	TOURMALINE p. 513 Complex boron silicate of Na-Ca-Al-Mg-Fe-Mn Hexagonal
Green, gray, white	3.3 to 3.5	$6\frac{1}{2}$–7	Massive, closely compact. Poor prismatic cleavage at nearly 90° angles. A pyroxene.	Jadeite p. 518 $NaAlSi_2O_6$ Monoclinic
Olive to grayish-green, brown	3.27 to 4.37	$6\frac{1}{2}$–7	Usually in disseminated grains in basic igneous rocks. May be massive granular.	**OLIVINE** p. 492 $(Mg,Fe)_2SiO_4$ Orthorhombic
Green, brown, yellow, blue, red	3.35 to 3.45	$6\frac{1}{2}$	In square prismatic crystals, vertically striated. Often columnar and granular, massive. Found in crystalline limestones.	VESUVIANITE p. 509 Complex hydrous Ca-Mg-Fe-Al silicate Tetragonal
Dark green	4.55	$7\frac{1}{2}$–8	Usually in octahedrons, characteristically striated. A zinc spinel.	Gahnite p. 389 $ZnAl_2O_4$ Isometric
Reddish-brown to black	6.8 to 7.1	6–7	Rarely in prismatic crystals, twinned. Fibrous, giving reniform surface. Rolled grains. Usually gives light brown streak.	**CASSITERITE** p. 385 SnO_2 Tetragonal

TABLE 14.1 (continued)

LUSTER: NONMETALLIC
II. Streak colorless
E. Hardness: >7
(Cannot be scratched by quartz)
2. Cleavage not prominent

Color	G	H	Remarks	Name, Composition, Crystal System
Reddish-brown, flesh-red, olive-green	3.16 to 3.20	$7\frac{1}{2}$	Prismatic crystals with nearly square cross section. Cross section may show black cross (chiastolite). May be altered to mica and then soft. Found in schists.	**ANDALUSITE** p. 499 Al_2SiO_5 Orthorhombic
Clove-brown, green, yellow, gray	3.27 to 3.35	$6\frac{1}{2}-7$	In wedge-shaped crystals with sharp edges. Also lamellar.	AXINITE p. 510 $Ca_2(Fe,Mn)Al_2(BO_3)$- $(Si_4O_{12})(OH)$ Triclinic
Red-brown to brownish-black	3.65 to 3.75	$7-7\frac{1}{2}$	In prismatic crystals; commonly in cruciform penetration twins. Frequently altered on the surface and then soft. Found in schists.	**STAUROLITE** p. 502 $Fe_2Al_9O_6(SiO_4)_4(O,OH)_2$ Pseudo-orthorhombic
Brown, red, gray, green, colorless	4.68	$7\frac{1}{2}$	Usually in small prisms truncated by the pyramid. An accessory mineral in igneous rocks. Found as rolled grains in sand.	**ZIRCON** p. 498 $ZrSiO_4$ Tetragonal
Usually brown to red. Also yellow, green, pink	3.5 to 4.3	$6\frac{1}{2}-7\frac{1}{2}$	Usually in dodecahedrons or trapezohedrons or in combinations of the two. An accessory mineral in igneous rocks and pegmatites. Commonly in metamorphic rocks. As sand.	**GARNET** p. 495 $A_3B_2(SiO_4)_3$ Isometric

TABLE 14.2   Minerals Arranged According to Increasing Specific Gravity

G	Name	G	Name	G	Name
1.6	Carnallite	2.62–2.76	**Plagioclase**	3.35–3.45	**Epidote**
1.7	**Borax**	2.6–2.9	**Collophane**	3.35–3.45	**Vesuvianite**
1.95	**Kernite**	2.74	**Bytownite**	3.4–3.5	**Hemimorphite**
1.96	**Ulexite**	2.7–2.8	Pectolite	3.45	Arfvedsonite
1.99	Sylvite	2.7–2.8	**Talc**	3.40–3.55	Aegirine
		2.76	**Anorthite**	3.4–3.55	**Titanite**
2.0–2.19				3.48	**Realgar**
		2.8–2.99		3.42–3.56	Triphylite
2.0–2.55	**Bauxite**			3.49	**Orpiment**
2.0–2.4	**Chrysocolla**	2.6–2.9	**Collophane**	3.4–3.6	**Topaz**
2.05–2.09	**Sulfur**	2.8–2.9	Pyrophyllite	3.5	**Diamond**
2.05–2.15	**Chabazite**	2.8–2.9	**Wollastonite**	3.45–3.60	**Rhodochrosite**
1.9–2.2	**Opal**	2.85	**Dolomite**	3.5–4.3	Garnet
2.09–2.14	Niter	2.86	Phlogopite		
2.1–2.2	**Stilbite**	2.76–2.88	**Muscovite**	**3.6–3.79**	
2.16	**Halite**	2.8–2.95	**Prehnite**		
2.18–2.20	**Heulandite**	2.8–3.0	**Datolite**	3.27–4.37	Olivine
		2.8–3.0	**Lepidolite**	3.5–4.2	Allanite
2.2–2.39		2.89–2.98	**Anhydrite**	3.5–4.3	**Garnet**
		2.9–3.0	Boracite	3.6–4.0	**Spinel**
2.0–2.4	**Chrysocolla**	2.95	**Aragonite**	3.56–3.66	**Kyanite**
2.2–2.65	**Serpentine**	2.95	Erythrite	3.58–3.70	**Rhodonite**
2.23	**Graphite**	2.8–3.2	**Biotite**	3.65–3.75	**Staurolite**
2.25	**Natrolite**	2.95–3.0	**Cryolite**	3.7	**Strontianite**
2.26	Tridymite	2.97–3.00	Phenacite	3.65–3.8	**Chrysoberyl**
2.27	**Analcime**			3.75–3.77	Atacamite
2.29	Nitratite	**3.0–3.19**		3.77	**Azurite**
2.30	Cristobalite				
2.30	**Sodalite**	2.97–3.02	Danburite	**3.8–3.99**	
2.32	**Gypsum**	2.85–3.2	Anthophyllite		
2.33	Wavellite	3.0–3.1	**Amblygonite**	3.7–4.7	**Romanechite**
2.3–2.4	**Apophyllite**	3.0–3.1	**Lazulite**	3.6–4.0	**Spinel**
2.39	**Brucite**	3.0–3.2	**Magnesite**	3.6–4.0	Limonite
		3.0–3.1	Margarite	3.83–3.88	**Siderite**
2.4–2.59		3.0–3.25	**Tourmaline**	3.5–4.2	Allanite
		3.0–3.3	**Tremolite**	3.5–4.3	**Garnet**
2.0–2.55	**Bauxite**	3.09	Lawsonite	3.9	Antlerite
2.2–2.65	**Serpentine**	3.1–3.2	Autunite	3.9–4.03	**Malachite**
2.42	**Colemanite**	3.1–3.2	Chondrodite	3.95–3.97	**Celestite**
2.42	Petalite	3.15–3.20	**Apatite**		
2.4–2.45	**Lazurite**	3.15–3.20	**Spodumene**	**4.0–4.19**	
2.45–2.50	**Leucite**	3.16–3.20	**Andalusite**		
2.2–2.8	**Garnierite**	3.18	**Fluorite**	3.9–4.1	**Sphalerite**
2.54–2.57	**Microcline**			4.02	**Corundum**
2.57	**Orthoclase**	**3.2–3.39**		3.9–4.2	**Willemite**
2.6–2.79		3.1–3.3	Scorodite	**4.2–4.39**	
		3.2	**Hornblende**		
2.55–2.65	**Nepheline**	3.23	**Sillimanite**	4.1–4.3	**Chalcopyrite**
2.6–2.63	**Kaolinite**	3.2–3.3	**Diopside**	3.7–4.7	**Romanechite**
2.62	**Albite**	3.2–3.4	**Augite**	4.18–4.25	**Rutile**
2.60–2.66	Cordierite	3.25–3.37	Clinozoisite	4.3	**Manganite**
2.65	**Oligoclase**	3.26–3.36	Dumortierite	4.3	**Witherite**
2.65	**Quartz**	3.27–3.35	Axinite	4.37	**Goethite**
2.69	**Andesine**	3.27–4.37	**Olivine**	4.35–4.40	**Smithsonite**
2.6–2.8	Alunite	3.2–3.5	**Enstatite**		
2.6–2.8	Turquoise			**4.4–4.59**	
2.71	**Labradorite**	**3.4–3.59**			
2.65–2.74	**Scapolite**			4.43–4.45	**Enargite**
2.65–2.8	**Beryl**	3.27–4.27	Olivine	4.5	**Barite**
2.72	**Calcite**	3.3–3.5	Jadeite	4.55	Gahnite
2.6–3.3	**Chlorite**	3.35–3.45	**Diaspore**	4.52–4.62	**Stibnite**

TABLE 14.2   (continued)

G	Name	G	Name	G	Name
4.6–4.79		**5.4–5.59**		**6.5–6.99**	
3.7–4.7	**Romanechite**	5.5	Millerite	6.5	**Skutterudite**
4.6	**Chromite**	5.5±	**Chlorargyrite**	6.55	**Cerussite**
4.58–4.65	**Pyrrhotite**	5.55	**Proustite**	6.78	Bismuthinite
4.7	**Ilmenite**			6.5–7.1	Pyromorphite
4.75	**Pyrolusite**	**5.6–5.79**		6.8	**Wulfenite**
4.6–4.76	Covellite			6.7–7.1	Vanadinite
4.62–4.73	**Molybdenite**	5.5–5.8	**Chalcocite**	6.8–7.1	**Cassiterite**
4.68	**Zircon**	5.68	**Zincite**		
		5.5–6.0	Jamesonite	**7.0–7.49**	
4.8–4.99		5.3–7.3	Columbite		
				7.0–7.5	**Wolframite**
4.6–5.0	Pentlandite	**5.8–5.99**		7.3	**Acanthite**
4.6–5.1	**Tetrahedrite-**				
	Tennantite	5.8–5.9	Bournonite	**7.5–7.99**	
4.89	**Marcasite**	5.85	**Pyrargyrite**		
				7.4–7.6	**Galena**
5.0–5.19		**6.0–6.49**		7.3–7.9	Iron
				7.78	**Nickeline**
5.02	**Pyrite**	5.9 6.1	Crocoite		
4.8–5.3	**Hematite**	5.9–6.1	**Scheelite**	**>8.0**	
5.06–5.08	**Bornite**	6.0	**Cuprite**		
5.15	**Franklinite**	6.07	**Arsenopyrite**	8.10	**Cinnabar**
5.0–5.3	Monazite	6.0–6.2	Polybasite	8.9	**Copper**
5.18	**Magnetite**	6.2–6.4	**Anglesite**	9.0–9.7	Uraninite
		5.3–7.3	Columbite	10.5	**Silver**
5.2–5.39		6.33	**Cobaltite**	15.0–19.3	**Gold**
				14–19	**Platinum**

TABLE 14.3 Nonopaque Minerals and Some Synthetic Compounds Arranged According to Increasing Refractive Index*

n, ω, β	ε, α	γ	Optic Sign	Name	n, ω, β	ε, α	γ	Optic Sign	Name
1.338	1.338	1.339	B+	Cryolite	1.587–1.610	1.575–1.595	1.590–1.622	B−	Amblygonite
1.433			I	Fluorite	1.588	1.544	1.598	B−	Pyrophyllite
1.45			I	Opal	1.589	1.539	1.589	B−	Talc
1.469	1.447	1.472	B−	Borax	1.59			I	Howlite
1.472	1.454	1.485	B−	Kernite	1.592	1.586	1.614	B+	Colemanite
1.475	1.473	1.479	B+	Tridymite	1.598	1.551	1.598	B−	Phlogopite
1.482	1.480		U−	Chabazite	1.594	1.585		U−	Beryl (gem)
1.482	1.480	1.493	B+	Natrolite	1.602	1.556	1.603	B−	Muscovite
1.483			I	Sodalite	1.604	1.595	1.633	B+	Pectolite
1.492	1.490	1.493	B+	Heulandite	1.61–1.70	1.57–1.63	1.61–1.70	B−	Biotite
1.487	1.484		U−	Cristobalite	1.609	1.602	1.621	B+	Brazilianite
1.487			I	Analcime	1.609–1.631	1.606–1.630	1.616–1.638	B+	Topaz
1.490			I	Sylvite	1.61–1.71	1.60–1.69	1.62–1.72	B±	Anthophyllite
1.498	1.494	1.500	B−	Stilbite	1.612	1.598	1.626	B±	Chondrodite
1.50			I	Lazurite	1.61–1.66	1.59–1.65	1.61–1.66	B−	Glaucophane
1.50			≈I	Chrysocolla	1.616	1.600	1.627	B−	Tremolite
1.504	1.332	1.504	B−	Niter	1.617	1.614	1.636	B+	Hemimorphite
1.504	1.491	1.520	B+	Ulexite	1.62–1.72	1.61–1.71	1.63–1.73	B−	Hornblende
1.508	1.509		U+	Leucite	1.62	1.61	1.65	B+	Turquoise
1.511	1.505	1.518	B+	Petalite	1.624	1.622	1.631	B+	Celestite
1.526	1.522	1.530	B−	Microcline	1.627	1.600	1.649	B+	Prehnite
1.523	1.520	1.530	B+	Gypsum	1.630	1.619	1.640	B−	Anthophyllite
1.524	1.518	1.526	B−	Orthoclase	1.63–1.67	1.62–1.67		U−	Apatite
1.524–1.574	1.522–1.560	1.527–1.578	B±	Cordierite	1.632	1.620	1.634	B−	Wollastonite
1.526	1.521	1.528	B−	Sanidine	1.633	1.630	1.636	B−	Danburite
1.532	1.527	1.538	B+	Albite	1.634	1.612	1.643	B−	Lazulite
1.54–1.59	1.52–1.56		U−	Scapolite	1.640–1.60	1.61–1.66		U−	Tourmaline
1.537	1.550		U+	Chalcedony	1.637	1.622	1.649	B−	Actinolite
1.537	1.534		U−	Nepheline	1.637	1.636	1.648	B+	Barite
1.533	1.536		U+	Apophyllite	1.638	1.632	1.643	B−	Andalusite
1.535	1.525	1.550	B+	Wavellite	1.638	1.633	1.652	B+	Anthophyllite
1.540	1.540	1.550	B+	Chrysotile	1.64–1.68	1.63–1.66	1.65–1.70	B+	Cummingtonite
1.542	1.540		U−	Apophyllite	1.643	1.626	1.643	B−	Glaucophane
1.543	1.539	1.547	B−	Oligoclase	1.645	1.634	1.647	B−	Margarite
1.544			I	Halite	1.646	1.622	1.658	U−	Tourmaline (gem)
1.544	1.531		U−	Marialite	1.648	1.584		B−	Biotite
1.544	1.553		U+	Quartz	1.649	1.644		U−	Apatite
1.547	1.540	1.550	B−	Cordierite	1.651	1.703		U+	Dioptase
1.553	1.550	1.557	B+	Andesine	1.65–1.88	1.63–1.88	1.67–1.87	B±	Olivine
1.55			≈I	Antigorite	1.651	1.635	1.670	B+	Forsterite
1.55–1.59	1.53–1.55	1.55–1.59	B−	Lepidolite	1.652	1.624	1.668	B−	Datolite
1.588	1.552	1.561	B−	Beryllonite	1.654	1.670		U+	Phenacite
1.56–1.64	1.53–1.59	1.56–1.64	B−	Phlogopite	1.658	1.486		U−	Calcite
1.563	1.559	1.568	B+	Labradorite	1.658	1.657	1.677	B+	Sillimanite
1.564	1.559	1.565	B−	Kaolinite	1.659	1.654	1.667	B+	Jadeite
1.57–1.60	1.56–1.60		U−	Beryl	1.662	1.636		U−	Tourmaline
1.570	1.590		U+	Brucite	1.661	1.626	1.699	B±	Erythrite
1.57–1.67	1.57–1.66	1.57–1.67	B±	Chlorite	1.665	1.650	1.679	B+	Cummingtonite
1.572	1.567	1.576	B−	Bytownite	1.665	1.660	1.674	B+	Enstatite
1.577	1.595		U+	Alunite	1.666	1.654	1.670	B−	Hornblende
1.575	1.570	1.614	B+	Anhydrite	1.666	1.655	1.672	B+	Spodumene
1.577	1.553		U−	Autunite	1.667	1.520	1.669	B−	Strontianite
1.579	1.579	1.584	B+	Clinochlore	1.683	1.678	1.688	B−	Bronzite (orthopyroxene)
1.58			I	Variscite					
1.581	1.572	1.586	B−	Anorthite	1.672	1.654	1.690	B+	Olivine (peridot)
1.583	1.577		U−	Emerald	1.69–1.72	1.68–1.74	1.71–1.75	B+	Augite
1.587	1.336		U−	Nitratite					

*A continuous listing of minerals is given according to increasing refractive index: n (isotropic), ω (uniaxial), β (biaxial). Under optic sign: I = isotropic; U = uniaxial, + or −; B = biaxial, + or −. Although only a few minerals have constant refractive indices, a single listing is given for most. A range of refractive indices is given when the variation is appreciably greater than ±0.01.

TABLE 14.3   (continued)

n, ω, β	ϵ, α	γ	Optic Sign	Name	n, ω, β	ϵ, α	γ	Optic Sign	Name
1.674	1.665	1.684	B+	Lawsonite	1.80			I	Gahnite
1.676	1.623	1.677	B−	Biotite	1.80			I	Spessartine
1.676	1.529	1.677	B−	Witherite	1.816	1.579		U−	Rhodochrosite
1.676	1.667	1.699	B+	Diopside	1.819	1.776	1.836	B−	Aegirine
1.679	1.676	1.680	B+	Lithiophilite	1.820			I	Almandine
1.680	1.530	1.685	B−	Aragonite	1.833			I	YAG
1.681	1.500		U−	Dolomite	1.861	1.831	1.880	B−	Atacamite
1.688	1.681	1.692	B−	Axinite	1.883	1.877	1.894	B+	Anglesite
1.690	1.670	1.708	B−	Grunerite	1.850	1.623		U−	Smithsonite
1.691	1.719		U+	Willemite	1.875	1.633		U−	Siderite
1.692	1.690	1.694	B±	Triphylite	1.875	1.655	1.909	B−	Malachite
1.694	1.679	1.698	B−	Hornblende	1.877	1.835	1.886	B−	Fayalite
1.694	1.693	1.702	B+	Zoisite	1.868			I	Uvarovite
1.699	1.668	1.707	B−	Sinhalite	1.887			I	Andradite
1.700	1.696	1.711	B−	Riebeckite	1.907	1.900	2.034	B+	Titanite
1.700	1.509		U−	Magnesite	1.920	1.934		U+	Scheelite
1.701	1.680	1.720	B−	Olivine	1.920–1.960	1.967–2.015		U+	Zircon
1.702	1.703	1.728	B+	Pigeonite	1.93	1.75	1.95	B−	Carnotite
1.703–1.752	1.700–1.746		U−	Vesuvianite	1.975			I	GGG
1.704	1.698	1.723	B+	Augite	1.997	2.093		U+	Cassiterite
1.704	1.694	1.707	B−	Hypersthene (orthopyroxene)	2.013	2.029		U+	Zincite
1.714–1.750			I	Pyrope	2.058	2.048		U−	Pyromorphite
1.72–1.74	1.71–1.74	1.72–1.75	B+	Rhodonite	2.077	1.804	2.079	B−	Cerussite
1.720–1.734	1.713–1.730	1.723–1.740	B−	Chloritoid	2.11			I	Chromite
1.720	1.712	1.728	B−	Kyanite	2.16			I	Zirconia (cubic)
1.722	1.702	1.850	B+	Diaspore	2.39	2.26	2.40	B−	Goethite
1.724			I	Spinel	2.35	2.27		U−	Vanadinite
1.72–1.78	1.71–1.75	1.73–1.80	B−	Epidote	2.37			I	Sphalerite
1.725	1.718	1.748	B+	Hedenbergite	2.37	2.31	2.66	B+	Crocoite
1.734			I	Grossular	2.404	2.283		U−	Wulfenite
1.737	1.733	1.747	B+	Rhodonite	2.409			I	Strontium titanate
1.738	1.726	1.789	B+	Antlerite	2.417			I	Diamond
1.748	1.746	1.756	B+	Chrysoberyl	2.554	2.493		U−	Anatase
1.749	1.743	1.757	B+	Staurolite	2.586	2.583	2.704	B+	Brookite
1.755	1.733	1.765	B−	Epidote	2.684	2.538	2.704	B−	Realgar
1.757	1.804		U+	Benitoite	2.61	2.90		U+	Rutile
1.758	1.730	1.838	B+	Azurite	2.81	2.40	3.02	B−	Orpiment
1.769	1.760		U−	Corundum	2.85			I	Cuprite
1.794	1.792	1.845	B+	Monazite	2.85	3.20		U+	Cinnabar
1.78–1.87	1.76–1.78	1.80–1.84	B−	Aegirine	2.98	2.71		U−	Proustite
					3.08	2.88		U−	Pyrargyrite
					3.22	2.94		U−	Hematite

MINERAL INDEX

In this index the mineral name is followed by the commonly sought information: composition, crystal system (XI Sys.), specific gravity (**G**), hardness (**H**), and index of refraction (n). For uniaxial crystals $n = \omega$, for biaxial crystals $n = \beta$. The refractive index is given as a single entry when the range is usually no greater than ± 0.01. Mineral names in boldface (e.g., **Arsenopyrite**) represent species for which complete descriptions are given in the chapters on systematic mineralogy (Chapters 8 to 12); those in lightface (e.g., Arfvedsonite) refer to minerals that are briefly treated or only mentioned in the text.

Name, Page	Composition	XI Sys.	G	H	n	Remarks
A						
Acadialite, 561	....					See chabazite
Acanthite, 351	Ag_2S	Mon / Iso	7.3	$2-2\frac{1}{2}$	—	Low temp. Ag_2S
Achroite, 513						Colorless tourmaline
Acmite, 519	....					See aegirine
Actinolite, 525	$Ca_2(Mg,Fe)_5Si_8O_{22}(OH)_2$	Mon	3.1–3.3	5–6	1.64	Green amphibole
Adularia, 552	$KAlSi_3O_8$	Mon		....		Colorless, translucent K-feldspar
Aegirine, 519	$NaFe^{3+}Si_2O_6$	Mon	3.5	$6-6\frac{1}{2}$	1.82	Na-pyroxene
Aegirine-augite, 519	$(Na,Ca)(Fe^{3+},Fe^{2+},Mg,Al)Si_2O_6$	Mon	3.4–3.5	6	1.71–1.78	Pyroxene
Aenigmatite, 504	$Na_2Fe_5^{2+}TiO_2(Si_2O_6)_3$	Tric	3.75	$5\frac{1}{2}$	1.80	In prismatic crystals, black
Agalmatolite, 534	....					Compact talc substitute for gem jade
Agate, 545	....					Concentric layers of chalcedony
Alabandite, 357, 385	MnS	Iso	4.0	$3\frac{1}{2}-4$	—	Black
Alabaster, 429	....					See gypsum
Albite, 553	$NaAlSi_3O_8(An_0\text{-}An_{10})$	Tric	2.62	6	1.53	Na endmember of plagioclase
Alexandrite, 392	....					Gem chrysoberyl
Alkali feldspar, 554	....					Na- or K-feldspar
Allanite, 509	$(Ca,Ce)_3(Fe^{2+},Fe^{3+})Al_2O(SiO_4)(Si_2O_7)(OH)$	Mon	3.5–4.2	$5\frac{1}{2}-6$	1.70–1.81	Brown-black, pitchy luster
Almandine, 496	$Fe_3Al_2Si_3O_{12}$	Iso	4.32	7	1.83	A garnet
Altaite, 357	$PbTe$	Iso	8.16	3	—	Tin-white
Alumstone, 431	....					See alunite
Alunite, 431	$KAl_3(SO_4)_2(OH)_6$	Hex	2.6–2.8	4	1.57	Usually massive
Amalgam, 344	Ag-Hg			....		See silver
Amazonite, 550	....					Green microcline
Amblygonite, 437	$LiAlFPO_4$	Tric	3.0–3.1	6	1.60	Cleavable masses
Amethyst, 544	SiO_2			....		Purple quartz
Amosite, 524	....					Asbestiform cummingtonite; "brown asbestos"
Amphibole, 523	....					A mineral group
Analcime, 558	$NaAlSi_2O_6 \cdot H_2O$	Iso	2.27	$5-5\frac{1}{2}$	1.48–1.49	Usually in trapezohedrons
Anatase, 384	TiO_2	Tet	3.9	$5\frac{1}{2}-6$	2.6	Adamantine luster
Anauxite, 530	....	Mon	2.6	2	1.56	Si-rich kaolinite
Andalusite, 499	Al_2SiO_5	Orth	3.16–3.20	$7\frac{1}{2}$	1.64	Square cross sections

Name, Page	Composition	XI Sys.	G	H	n	Remarks
Andesine, 554	$Ab_{70}An_{30}$-$Ab_{50}An_{50}$	Tric	2.69	6	1.55	Plagioclase feldspar
Andradite, 496	$Ca_3Fe_2Si_3O_{12}$	Iso	3.86	7	1.89	A garnet
Anglesite, 427	$PbSO_4$	Orth	6.2–6.4	3	1.88	Cl{001} {210}
Anhydrite, 428	$CaSO_4$	Orth	2.89–2.98	$3–3\frac{1}{2}$	1.58	Cl{010} {100} {001}
Ankerite, 419	$CaFe(CO_3)_2$	Hex	2.95–3	$3\frac{1}{2}$	1.70–1.75	Cl{10$\bar{1}$1}
Annabergite, 437	$Ni_3(AsO_4)_2 \cdot 8H_2O$	Mon	3.0	$2\frac{1}{2}–3$	1.68	Nickel bloom; green
Anorthite, 553	$CaAl_2Si_2O_8(An_{90}$-$An_{100})$	Tric	2.76	6	1.58	Ca endmember of plagioclase
Anorthoclase, 552	$(K,Na)AlSi_3O_8$	Tric	2.58	6	1.53	An alkali feldspar between K-spar and Ab
Anthophyllite, 523	$(Mg,Fe)_7Si_8O_{22}(OH)_2$	Orth	2.85–3.2	$5\frac{1}{2}–6$	1.61–1.71	An amphibole
Antigorite, 528	$Mg_3Si_2O_5(OH)_4$	Mon	2.5–2.6	~4	1.55	Platy serpentine
Antimony, 337	Sb	Hex	6.7	3	—	Cl{0001}
Antlerite, 430	$Cu_3SO_4(OH)_4$	Orth	3.9±	$3\frac{1}{2}–4$	1.74	Green
Apatite, 434	$Ca_5(PO_4)_3(F,Cl,OH)$	Hex	3.15–3.20	5	1.63	Cl{0001} poor
Apophyllite, 541	$KCa_4(Si_4O_{10})_2F \cdot 8H_2O$	Tet	2.3–2.4	$4\frac{1}{2}–5$	1.54	Cl{001}
Aquamarine, 511	....					Greenish-blue gem beryl
Aragonite, 416	$CaCO_3$	Orth	2.95	$3\frac{1}{2}–4$	1.68	Cl{010} {110}
Arfvedsonite, 527	$Na_3Fe_4^{2+}Fe^{3+}Si_8O_{22}(OH)_2$	Mon	3.45	6	1.69	Deep green amphibole
Argentite, 351	Ag_2S	....				Now known as acanthite
Arsenopyrite, 368	$FeAsS$	Mon	6.07	$5\frac{1}{2}–6$	—	Pseudo-orthorhombic
Asbestos, 527, 528	....					See amphibole and serpentine
Astrophyllite, 504	$(K,Na)_3(Fe,Mn)_7(Ti,Zr)_2$-$Si_8(O,OH)_{31}$	Tric	3.35	3	1.71	Micaceous cleavage
Atacamite, 402	$Cu_2Cl(OH)_3$	Orth	3.75–3.77	$3–3\frac{1}{2}$	1.86	Green, Cl{010}
Augite, 517	$(Ca,Na)(Mg,Fe,Al)(Si,Al)_2O_6$	Mon	3.2–3.4	5–6	1.67–1.73	Dark green to black pyroxene
Aurichalcite, 422	$(Zn,Cu)_5(CO_3)_2(OH)_3$	Orth	3.64	2	1.74	Green to blue
Autunite, 439	$Ca(UO_2)_2(PO_4)_2 \cdot 10–12\ H_2O$	Tet	3.1–3.2	$2–2\frac{1}{2}$	1.58	Yellow-green
Aventurine, 544, 554	....					Quartz or oligoclase
Axinite, 510	$(Ca,Fe,Mn)_3Al_2BSi_4O_{15}(OH)$	Tric	3.27–3.35	$6\frac{1}{2}–7$	1.69	Crystal angles acute
Azurite, 421	$Cu_3(CO_3)_2(OH)_2$	Mon	3.77	$3\frac{1}{2}–4$	1.76	Always blue

B

Name, Page	Composition	XI Sys.	G	H	n	Remarks
Balas ruby, 389	....					Red gem spinel
Barite, 425	$BaSO_4$	Orth	4.5	$3–3\frac{1}{2}$	1.64	Cl{001} {210}
Bauxite, 397	mixture of diaspore, gibbsite, boehmite	—	2.0–2.55	1–3	—	An earthy rock
Beidellite, 531	$(Ca,Na)_{0.3}Al_2(OH)_2$-$(Al,Si)_4O_{10} \cdot 4H_2O$	Mon	2–3	1–2	—	Member of montmorillonite group
Benitoite, 504	$BaTiSi_3O_9$	Hex	3.6	$6\frac{1}{2}$	1.76	Blue
Bentonite, 531	....					Montmorillonite alteration of volcanic ash
Beryl, 511, 567, 579	$Be_3Al_2(Si_6O_{18})$	Hex	2.65–2.8	$7\frac{1}{2}–8$	1.57–1.61	Usually green
Beryllonite, 438	$NaBePO_4$	Mon	2.81	$5\frac{1}{2}$	1.56	Rare gem mineral
Biotite, 537	$K(Mg,Fe)_3(AlSi_3O_{10})(OH)_2$	Mon	2.8–3.2	$2\frac{1}{2}–3$	1.61–1.70	Black mica
Bismuthinite, 364	Bi_2S_3	Orth	6.78 ± 0.03	2	—	Cl{010}
Black-band ore, 414	....					See siderite
Black jack, 357	....					See sphalerite
Bloodstone, 545	....					Green and red chalcedony chalcedony
Boehmite, 395, 397	$\gamma AlO(OH)$	Orth	3.01–3.06	$3\frac{1}{2}–4$	1.65	In bauxite
Bog-iron ore, 396	....					See limonite
Boracite, 425	$Mg_3ClB_7O_{13}$	Orth	2.9–3.0	7	1.66	Pseudoisometric
Borax, 423	$Na_2B_4O_5(OH)_4 \cdot 8H_2O$	Mon	1.7±	$2–2\frac{1}{2}$	1.47	Cl{100}
Bornite, 354	Cu_5FeS_4	Tet / Iso	5.06–5.08	3	—	Purple-blue tarnish
Bort, 347	C	Iso	3.5	10	2.4	Variety of diamond

Name, Page	Composition	XI Sys.	G	H	n	Remarks
D						
Danburite, 502	$Ca(B_2Si_2O_8)$	Orth	2.97–3.02	7	1.63	In crystals
Datolite, 504	$CaB(SiO_4)(OH)$	Mon	2.8–3.0	$5–5\frac{1}{2}$	1.65	Usually in crystals
Demantoid, 497	....					Green andradite garnet
Diallage, 517	....					Variety of diopside
Diamond, 347, 567, 580	C	Iso	3.51	10	2.42	Adamantine luster
Diaspore, 395, 397	$\alpha AlO(OH)$	Orth	3.35–3.45	$6\frac{1}{2}–7$	1.72	Cl{010} perfect
Diatomaceous earth, 548	....					See opal
Diatomite, 548	....					See opal
Dichroite, 512	....					See cordierite
Dickite, 530	$Al_2Si_2O_5(OH)_4$	Mon	2.6	$2–2\frac{1}{2}$	1.56	Clay mineral
Digenite, 354	Cu_9S_5	Iso	5.6	$2\frac{1}{2}–3$	—	Similar to chalcocite
Diopside, 517	$CaMgSi_2O_6$	Mon	3.2	5–6	1.67	White to light green pyroxene
Dioptase, 543	$Cu_6(Si_6O_{18})\cdot 6H_2O$	Hex	3.3	5	1.65	Green; minor gem
Djurleite, 354	$Cu_{31}S_6$	Mon	5.75	2.5–3	—	Black; dark gray
Dolomite, 419	$CaMg(CO_3)_2$	Hex	2.85	$3\frac{1}{2}–4$	1.68	Cl{10$\bar{1}$1}
Dravite, 513	....					Brown tourmaline
Dry-bone ore, 416	....					See smithsonite
Dumortierite, 500	$Al_7O_3(BO_3)(SiO_4)_3$	Orth	3.26–3.36	7	1.69	Radiating
E						
Edenite, 526	$NaCa_2Mg_5AlSi_7O_{22}(OH)_2$	Mon	3.0	6	1.63	See hornblende
Elbaite, 513	....					See tourmaline
Electrum, 342	....					See gold
Emerald, 511	....					Deep green gem beryl
Emery, 379	....					Corundum with magnetite
Enargite, 369	Cu_3AsS_4	Orth	4.45	3	—	Cl{110}
Endlichite, 437	....					See vanadinite
Enstatite, 514	$MgSiO_3$	Orth	3.2–3.5	$5\frac{1}{2}$	1.65	Cl{210} ~ 90°
Epidote, 507	$Ca_2(Al,Fe)Al_2O(SiO_4)$-$(Si_2O_7)(OH)$	Mon	3.35–3.45	6–7	1.72–1.78	Cl{001}, {100}, green
Epsomite, 431	$MgSO_4\cdot 7H_2O$	Orth	1.75	$2–2\frac{1}{2}$	1.46	Bitter taste
Epsom salt, 431	....					See epsomite
Erythrite, 437	$Co_3(AsO_4)_2\cdot 8H_2O$	Mon	3.06	$1\frac{1}{2}–2\frac{1}{2}$	1.66	Pink; cobalt bloom
Essonite, 496	....					See grossular; garnet
Euclase, 512	$BeAl(SiO_4)(OH)$	Mon	3.1	$7\frac{1}{2}$	1.66	Cl{010}
Eucryptite, 520	$LiAlSiO_4$	Hex	2.67	—	1.55	Spodumene alteration
F						
Fairy stone, 503	....					See staurolite
Famatinite, 369	Cu_3SbS_4	Tet	4.52	$3\frac{1}{2}$	—	Gray
Fayalite, 492	Fe_2SiO_4	Orth	4.39	$6\frac{1}{2}$	1.86	Olivine, Fe endmember
Feldspar, 548	....					A mineral group
Feldspathoid, 554	....					A mineral group
Ferberite, 431	$FeWO_4$	Mon	7.0–7.5	5	—	See wolframite
Fergusonite, 393	$(Y,Er,Ce,Fe)NbO_4$	Tet	5.8	$5\frac{1}{2}–6$	—	Brown-black
Ferrimolybdite, 367	$Fe_2(MoO_4)_3\cdot 8H_2O$	Orth?	2.99	~1	1.73–1.79	Canary-yellow, soft
Ferroactinolite, 525	$Ca_2Fe_5Si_8O_{22}(OH)_2$	Mon	3.2–3.3	5–6	1.68	Dark green amphibole
Ferropseudobrookite, 383	$FeTi_2O_5$					See ilmenite
Ferrosilite, 514	$FeSiO_3$	Orth	3.9	6	1.79	Orthopyroxene endmember
Fersmannite, 504	$Na_4Ca_4Ti_4(SiO_4)_3(O,OH,F)_3$	Mon	3.44	$5\frac{1}{2}$	1.93	Brown; rare; in alkalic rocks
Fibrolite, 500	....					See sillimanite
Fischesserite, 343	Ag_3AuSe_2					Rare gold ore
Flint, 545	SiO_2	—	2.65	7	1.54	Microcryst. quartz
Flos ferri, 417	....					See aragonite
Fluorapatite, 434	$Ca_5(PO_4)_3F$					See apatite

Name, Page	Composition	Xl Sys.	G	H	n	Remarks
Fluorite, 401	CaF_2	Iso	3.18	4	1.43	Cl octahedral
Forsterite, 492	Mg_2SiO_4	Orth	3.2	$6\frac{1}{2}$	1.63	Olivine; Mg endmember
Fowlerite, 521						Zn-bearing rhodonite
Franklinite, 391	$(Zn,Fe,Mn)(Fe,Mn)_2O_4$	Iso	5.15	6	—	Spinel from Franklin, N.J.
Freibergite, 370						Ag-bearing tetrahedrite
Fuchsite, 534						Chrome-rich muscovite

G

Name, Page	Composition	Xl Sys.	G	H	n	Remarks
Gadolinite, 512	$YFeBe_2(SiO_4)_2O_2$	Mon	4.0–4.5	$6\frac{1}{2}$–7	1.79	Black
Gahnite, 389	$ZnAl_2O_4$	Iso	4.55	$7\frac{1}{2}$–8	1.80	Spinel; dark green octahedrons
Galaxite, 389	$MnAl_2O_4$	Iso	4.03	$7\frac{1}{2}$–8	1.92	Mn spinel
Galena, 354	PbS	Iso	7.4–7.6	$2\frac{1}{2}$	—	Cl cubic
Garnet, 495, 571, 582	$A_3B_2(SiO_4)_3$	Iso	3.5–4.3	$6\frac{1}{2}$–$7\frac{1}{2}$	1.71–1.88	A mineral group; commonly in crystals
Garnierite, 530	$(Ni,Mg)_3Si_2O_5(OH)_4$	Mon	2.2–2.8	2–3	1.59	Green Ni serpentine
Gaylussite, 422	$Na_2Ca(CO_3)_2 \cdot 5H_2O$	Mon	1.99	2–3	1.52	Cl{110} perfect
Gedrite, 523	$\sim Na_{0.5}(Mg,Fe)_2(Mg,Fe)_{3.5^-}$ $(Al,Fe^{3+})_{1.5}Si_6Al_2O_{22}(OH)_2$	Orth				See anthophyllite
Geikielite, 383	$MgTiO_3$	Hex	4.05	$5\frac{1}{2}$–6	—	Cl{10$\bar{1}$1}
Gersdorffite, 368	NiAsS	Iso	5.9	$5\frac{1}{2}$	—	See cobaltite
Geyserite, 548						Opal in hot springs
Gibbsite, 394, 397	$Al(OH)_3$	Mon	2.3–2.4	$2\frac{1}{2}$–$3\frac{1}{2}$	1.57	Basal Cl{001}
Glaucodot, 368	$(Co,Fe)AsS$	Orth	6.04	5	—	Tin-white
Glauconite, 537	$(K,Na,Ca)_{0.5-1}(Fe^{3+},Al,$ $Fe^{2+},Mg)_2(Si,Al)_4O_{10}(OH)_2 \cdot nH_2O$	Mon	2.4±	2	1.62	In green sands
Glaucophane, 526	$Na_2Mg_3Al_2Si_8O_{22}(OH)_2$	Mon	3.1–3.3	6–$6\frac{1}{2}$	1.62–1.67	Blue to black amphibole
Gmelinite, 562	$(Na_2,Ca)(Al_2Si_4O_{12}) \cdot 6H_2O$	Hex	2.1±	$4\frac{1}{2}$	1.49	A zeolite
Goethite, 395	$\alpha FeO(OH)$	Orth	4.37	5–$5\frac{1}{2}$	2.39	Cl{010} perfect
Gold, 342	Au	Iso	15–19.3	$2\frac{1}{2}$–3	—	Yellow, soft
Goshenite, 511						Colorless gem beryl
Graphite, 351	C	Hex	2.23	1–2	—	Black, platy
Greenalite, 530	$(Fe,Mg)_3Si_2O_5(OH)_4$	Mon	3.2	—	1.67	In iron-formations
Greenockite, 357	CdS	Hex	4.9	3–$3\frac{1}{2}$	—	Yellow-orange
Grossular, 496	$Ca_3Al_2Si_3O_{12}$	Iso	3.59	$6\frac{1}{2}$	1.73	A garnet
Grunerite, 523	$Fe_7Si_8O_{22}(OH)_2$	Mon	3.6	6	1.71	Light brown, needlelike amphibole
Gypsum, 429	$CaSO_4 \cdot 2H_2O$	Mon	2.32	2	1.52	Cl{010} {100} {011}

H

Name, Page	Composition	Xl Sys.	G	H	n	Remarks
Halite, 399	NaCl	Iso	2.16	$2\frac{1}{2}$	1.54	Cl cubic; salty
Halloysite, 531	$Al_2Si_2O_5(OH)_4$ and $Al_2Si_2O_5(OH)_4 \cdot 2H_2O$	Mon	2.0–2.2	1–2	1.54	Clay mineral
Harmotome, 562	$Ba(Al_2Si_6O_{16}) \cdot 6H_2O$	Mon	2.45	$4\frac{1}{2}$	1.51	A zeolite
Hastingsite, 526	$NaCa_2Fe_4(Al,Fe)Al_2Si_6-$ $O_{22}(OH)_2$	Mon	3.2	6	1.66	See hornblende
Haüynite, 556	$(Na,Ca)_{4-8}(AlSiO_4)_6(SO_4)_{1-2}$	Iso	2.4–2.5	$5\frac{1}{2}$–6	1.50	A feldspathoid
Hectorite, 531	$(Mg,Li)_3Si_4O_{10}(OH)_2Na_{0.3} \cdot 4H_2O$	Mon	2.5	1–$1\frac{1}{2}$	1.52	Li montmorillonite
Hedenbergite, 517	$CaFeSi_2O_6$	Mon	3.55	5–6	1.73	Clinopyroxene endmember
Heliotrope, 545						Green and red chalcedony
Hematite, 380	Fe_2O_3	Hex	5.26	$5\frac{1}{2}$–$6\frac{1}{2}$	—	Red streak
Hemimorphite, 506	$Zn_4(Si_2O_7)(OH)_2 \cdot H_2O$	Orth	3.4–3.5	$4\frac{1}{2}$–5	1.62	Cl{110}
Hercynite, 388	$FeAl_2O_4$	Iso	4.39	$7\frac{1}{2}$–8	1.80	Fe spinel
Heulandite, 561	$CaAl_2Si_7O_{18} \cdot 6H_2O$	Mon	2.18–2.20	$3\frac{1}{2}$–4	1.48	Cl{010} perfect
Hiddenite, 519						Green gem spodumene
Hollandite, 395	$Ba_2Mn_8O_{16}$	Tet / Mon	4.9	6	—	Silver-gray to black

Name, Page	Composition	Xl Sys.	G	H	n	Remarks
Holmquistite, 459	$Li_2(Mg,Fe)_3(Al,Fe^{3+})_2Si_8O_{22}(OH)_2$	Orth	3.06–3.13	5–6	1.64–1.66	Blue to violet-blue; near Li-pegmatites
Hornblende, 526	$(Ca,Na)_{2-3}(Mg,Fe,Al)_5Si_6(Si,Al)_2O_{22}(OH)_2$	Mon	3.0–3.4	5–6	1.62–1.72	Green to black amphibole
Horn silver, 400						See chlorargyrite
Huebnerite, 431	$MnWO_4$	Mon	7.0	5	—	See wolframite
Humite, 503	$Mg_7(SiO_4)_3(F,OH)_2$	Orth	3.1–3.2	6	1.64	See chondrodite
Hyalite, 548						Colorless, globular opal
Hyalophane, 476	$(K,Ba)(Al,Si)_2Si_2O_8$	Mon	2.8	6	1.54	A feldspar
Hydroboracite, 425	$CaMgB_6O_8(OH)_6\cdot3H_2O$	Mon	2.17	2	1.53	Clear, colorless to white
Hydrogrossular, 496	$Ca_3Al_2(Si_2O_8(SiO_4)_{1-m}(OH)_{4m}$	Iso	3.13–3.59	6–7	1.67–1.73	Hydrous garnet
Hydroxylapatite, 434	$Ca_5(PO_4)_3(OH)$					See apatite
Hydrozincite, 416	$Zn_5(CO_3)_2(OH)_6$	Mon	3.6–3.8	$2-2\frac{1}{2}$	1.74	Secondary mineral
Hypersthene, 514	$(Mg,Fe)SiO_3$	Orth	3.4–3.5	5–6	1.68–1.73	An intermediate member of the orthopyroxene series

I

Name, Page	Composition	Xl Sys.	G	H	n	Remarks
Iceland spar, 413						See calcite
Idocrase, 509						See vesuvianite
Illite, 531					1.57–1.61	Micalike clay mineral
Ilmenite, 383	$FeTiO_3$	Hex	4.7	$5\frac{1}{2}$–6	—	May be slightly magnetic
Ilvaite, 507	$CaFe_2^{2+}Fe^{3+}O(Si_2O_7)(OH)$	Orth	4.0	$5\frac{1}{2}$–6	1.91	Black
Indialite, 512	$(Mg,Fe)_2Al_4Si_5O_{18}\cdot nH_2O$	Hex	~2.6	7	—	High T form of cordierite
Indicolite, 513						Blue tourmaline
Inyoite, 425	$CaB_3O_3(OH)_5\cdot4H_2O$	Mon	1.88	2	1.51	Colorless, transparent
Iodargyrite, 400	AgI	Hex	5.5–5.7	$1-1\frac{1}{2}$	2.18	Sectile
Iodobromite, 400	$Ag(Cl,Br,I)$	Iso	5.71	$1-1\frac{1}{2}$	2.20	Sectile; same as iodian bromargyrite
Iolite, 512						See cordierite
Iridium, 345	Ir	Iso	22.7	6–7	—	A platinum metal
Iridosmine, 345	$Ir-Os$	Hex	19.3–21.1	6–7	—	See platinum
Iron, 345	Fe	Iso	7.3–7.9	$4\frac{1}{2}$	—	Very rare

J

Name, Page	Composition	Xl Sys.	G	H	n	Remarks
Jacobsite, 390	$MnFe_2O_4$	Iso	5.1	$5\frac{1}{2}-6\frac{1}{2}$	2.3	A spinel
Jade, 518, 569, 581						See nephrite and jadeite
Jadeite, 518	$NaAlSi_2O_6$	Mon	3.3–3.5	$6\frac{1}{2}$–7	1.66	Green, compact
Jargon, 498						See zircon
Jarosite, 431	$KFe_3(SO_4)_2(OH)_6$	Hex	3.2±	3	1.82	Yellow-brown
Jasper, 545						Red, microcryst. quartz
Jimthompsonite, 461	$(Mg,Fe)_{10}Si_{12}O_{32}(OH)_4$	Orth				Microscopic alteration of anthophyllite; a "biopyribole"
Johannsenite, 459	$CaMnSi_2O_6$	Mon	3.4–3.5	6	1.71–1.73	Mn analogue of diopside

K

Name, Page	Composition	Xl Sys.	G	H	n	Remarks
Kainite, 400	$KMg(Cl,SO_4)\cdot2\frac{3}{4}H_2O$	Mon	2.1	3	1.51	Cl{001}; salty, bitter
Kalsilite, 555	$KAlSiO_4$	Hex	2.61	6	1.54	Isostructural with nepheline
Kamacite, 345	$Fe-Ni$	Iso	7.3–7.9	4	—	In meteorites
Kaolin, 530						Mixture of clay minerals
Kaolinite, 530	$Al_2Si_2O_5(OH)_4$	Tric	2.6	2	1.55	Earthy
Keatite, 475	SiO_2	Tet	2.50	7	1.52	Synthetic
Kernite, 422	$Na_2B_4O_6(OH)_2\cdot3H_2O$	Mon	1.9	3	1.47	Cl{001} {100}
K-feldspar, 548						See microcline, orthoclase, sanidine

Name, Page	Composition	XI Sys.	G	H	n	Remarks
Kieserite, 398	$MgSO_4 \cdot H_2O$	Mon	2.57	$3\frac{1}{2}$	1.53	Massive, granular; whitish gray
Kirsteinite, 492	$CaFeSiO_4$					An olivine
Kostovite, 343	$CuAuTe_4$					Rare gold ore
Krennerite, 343	$AuTe_2$	Orth	8.62	2–3	—	Basal Cl{001}
Kunzite, 519						Pink gem spodumene
Kutnahorite, 421	$CaMn(CO_3)_2$	Hex	3.12	$3\frac{1}{2}$–4	1.74	Mn-dolomite
Kyanite, 500	Al_2SiO_5	Tric	3.55–3.66	5–7	1.72	Blue, bladed

L

Name, Page	Composition	XI Sys.	G	H	n	Remarks
Labradorite, 554	$Ab_{50}An_{50}\text{-}Ab_{30}An_{70}$	Tric	2.71	6	1.56	A member of the plagioclase feldspar series
Lamprophyllite, 504	$Na_3Sr_2Ti_3(Si_2O_7)_2(O,OH,F)_2$	Orth	3.45	4	1.75	Platy
Langbeinite, 398	$K_2Mg_2(SO_4)_3$	Iso	2.83	$3\frac{1}{2}$–4	1.53	Colorless; in evaporites
Lapis lazuli, 556						See lazurite
Laumontite, 562	$Ca(Al_2Si_4O_{12}) \cdot 4H_2O$	Mon	2.28	4	1.52	A zeolite
Lawsonite, 507	$CaAl_2(Si_2O_7)(OH)_2 \cdot H_2O$	Orth	3.09	8	1.67	In gneisses and schists
Lazulite, 438	$(Mg,Fe)Al_2(PO_4)_2(OH)_2$	Mon	3.0–3.1	5–$5\frac{1}{2}$	1.64	Blue
Lazurite, 556	$(Na,Ca)_8(AlSiO_4)_6(SO_4,S,Cl)_2$	Iso	2.4–2.45	5–$5\frac{1}{2}$	1.50	Blue; associated with pyrite
Lechatelierite, 547	SiO_2	Amor	2.2	6–7	1.46	Fused silica
Lepidochrosite, 396	$\gamma FeO(OH)$	Orth	4.09	5	2.2	Red
Lepidolite, 538	$K(Li,Al)_{2-3}(AlSi_3O_{10})(O,OH,F)_2$	Mon	2.8–2.9	$2\frac{1}{2}$–4	1.55–1.59	Pink, lilac, gray mica
Leucite, 554	$KAlSi_2O_6$	{Tet Iso	2.47	$5\frac{1}{2}$–6	1.51	In trapezohedrons
Leucoxene, 384						Brownish alteration of Ti minerals
Liddicoatite, 513						See tourmaline
Ligure, 566						Possible ancient name for zircon
Limonite, 396	$FeO \cdot OH \cdot nH_2O$	Amor	3.6–4.0	5–$5\frac{1}{2}$	—	Streak yellow-brown
Linnaeite, 369	Co_3S_4	Iso	4.8	$4\frac{1}{2}$–$5\frac{1}{2}$	—	Steel-gray
Litharge, 356	PbO	Tet	9.14	2	2.66	Red
Lithiophilite, 433	$Li(Mn,Fe)PO_4$	Orth	3.5	5	1.67	Cl{001} {010}
Lizardite, 528	$Mg_3Si_2O_5(OH)_4$					Massive; a polymorph of serpentine
Lodestone, 389						Natural magnet
Lonsdaelite, 349	C	Hex	3.3+	10	2.42	Hexagonal diamond
Luzonite, 369	Cu_3AsS_4	Tet	4.4	3–4	—	Low T form of enargite

M

Name, Page	Composition	XI Sys.	G	H	n	Remarks
Maghemite, 383	γFe_2O_3					Hematite dimorph
Magnesiochromite, 392	$MgCr_2O_4$	Iso	4.2	$5\frac{1}{2}$	—	A spinel
Magnesioferrite, 390	$MgFe_2O_4$	Iso	4.5–4.6	$5\frac{1}{2}$–$6\frac{1}{2}$	—	A spinel
Magnesite, 413	$MgCO_3$	Hex	3.0–3.2	$3\frac{1}{2}$–5	1.70	Commonly massive
Magnetite, 389	Fe_3O_4	Iso	5.18	6	—	Strongly magnetic; a spinel
Malachite, 421	$Cu_2(CO_3)(OH)_2$	Mon	3.9–4.03	$3\frac{1}{2}$–4	1.88	Green
Manganite, 394	$MnO(OH)$	Mon	4.3	4	—	Prismatic, pseudo-orthorhombic crystals
Manganotantalite, 392	$(Mn,Fe)Ta_2O_6$	Orth	6.6±	$4\frac{1}{2}$	—	See columbite
Manjiroite, 395	$(Na,K)Mn_8O_{16} \cdot nH_2O$					See romanechite
Marcasite, 366	FeS_2	Orth	4.89	6–$6\frac{1}{2}$	—	Polymorphous with pyrite
Margarite, 538	$CaAl_2(Al_2Si_2)_{10}(OH)_2$	Mon	3.0–3.1	$3\frac{1}{2}$–5	1.65	Brittle mica
Marialite, 558	$Na_4(AlSi_3O_8)_3(Cl_2,CO_3,SO_4)$	Tet	2.60±	$5\frac{1}{2}$–6	1.55	See scapolite
Martite, 380						See hematite
Meionite, 558	$Ca_4(Al_2Si_2O_8)_3(Cl_2,CO_3,SO_4)$	Tet	2.69	$5\frac{1}{2}$–6	1.59	See scapolite
Melanite, 497	$Ca_3Fe_2(SiO_4)_3$	Iso	3.7	7	1.94	Black andradite garnet
Melanterite, 367, 431	$FeSO_4 \cdot 7H_2O$	Mon	1.90	2	1.48	Green-blue

Name, Page	Composition	Xl Sys.	G	H	n	Remarks
Metacinnabar, 362	$Hg_{1-x}S$	Iso	7.65	3	—	Grayish-black
Mica, 534						A mineral group
Microcline, 548	$KAlSi_3O_8$	Tric	2.54–2.57	6	1.53	Low T K-feldspar
Microlite, 393	$Ca_2Ta_2O_6(O,OH,F)$	Iso	5.48–5.56	$5\frac{1}{2}$	1.92–1.99	Streak yellow to brown
Microperthite, 549						Intergrowth of K-spar and albite
Millerite, 360	NiS	Hex	5.5 ± 0.2	$3–3\frac{1}{2}$	—	Capillary crystals
Mimetite, 436, 437	$Pb_5(AsO_4)_3Cl$	Hex	7.0–7.2	$3\frac{1}{2}$	2.1–2.2	Pale yellow, yellow-brown
Minium, 356	Pb_3O_4	?	8.9–9.2	$2\frac{1}{2}$	2.42	Earthy, brownish-red
Minnesotaite, 532	$Fe_3Si_4O_{10}(OH)_2$	Mon	3.01	1 ?	~1.60	In iron-formations
Molybdenite, 367	MoS_2	Hex	4.62–4.73	$1–1\frac{1}{2}$	—	Lead-gray, platy
Monalbite, 481	$NaAlSi_3O_8$	Mon				Monoclinic, high T albite
Monazite, 434	$(Ce,La,Y,Th)PO_4$	Mon	4.6–5.4	$5–5\frac{1}{2}$	1.79	Parting {001}
Montebrasite, 437	$(Li,Na)Al(PO_4)(OH,F)$	Tric	~2.98	$5\frac{1}{2}–6$	1.61	See amblygonite
Montbrayite, 343	$(Au,Sb)_2Te_3$					Rare gold ore
Monticellite, 492	$CaMgSiO_4$	Orth	3.2	5	1.65	An olivine
Montmorillonite, 531	$(Al,Mg)_8(Si_4O_{10})_4(OH)_8 \cdot 12H_2O$	Mon	2.5	$1–1\frac{1}{2}$	1.50–1.64	A clay mineral
Moonstone, 552						See adularia, albite, and orthoclase
Morganite, 511						Rose gem beryl
Moss agate, 545						Agate with mosslike patterns
Mountain cork, 525						Felted tremolite
Mountain leather, 525						Felted tremolite
Mullite, 500	$~Al_6Si_3O_{15}$	Orth	3.23	6–7	1.67	Cl{010}
Muscovite, 534	$KAl_2(AlSi_3O_{10})(OH)_2$	Mon	2.76–2.88	$2–2\frac{1}{2}$	1.60	Cl{001} perfect
Muthmannite, 343	$(Ag,Au)Te$					Rare gold ore

N

Name, Page	Composition	Xl Sys.	G	H	n	Remarks
Nacrite, 530	$Al_2Si_2O_5(OH)_4$	Mon	2.6	$2–2\frac{1}{2}$	1.56	Clay mineral
Nagyagite, 343	$Pb_5Au(Te,Sb)_4S_{5-8}$	Hex	7.4	$1–1\frac{1}{2}$	—	Blackish lead-gray
Natroalunite, 431	$(Na,K)Al_3(SO_4)_2(OH)_6$	Hex	2.6–2.9	$3\frac{1}{2}–4$	1.57	White, grayish, massive
Natrolite, 559	$Na_2Al_2Si_3O_{10} \cdot 2H_2O$	Orth	2.25	$5–5\frac{1}{2}$	1.48	Cl{110} perfect
Nepheline, 555	$(Na,K)AlSiO_4$	Hex	2.60–2.65	$5\frac{1}{2}–6$	1.54	Greasy luster
Nephrite, 526						Tough, compact tremolite
Neptunite, 504	$KNa_2Li(Fe,Mn)_2TiO_2(Si_4O_{11})_2$	Mon	3.23	5–6	1.70	Black
Nickeline, 360	NiAs	Hex	7.78	$5–5\frac{1}{2}$	—	Copper-red
Nickel bloom, 437						See annabergite
Nickel iron, 345	Fe-Ni	Iso	7.8–8.2	5	—	Kamacite, taenite
Nickel skutterudite, 369	$(Ni,Co)As_3$	Iso	6.5 ± 0.4	$5\frac{1}{2}–6$	—	Tin-white
Niter, 422	KNO_3	Orth	2.09–2.14	2	1.50	Cooling taste
Nitratite, 422	$NaNO_3$	Hex	2.29	1–2	1.59	Cooling taste
Nontronite, 531	$Fe_2(Al,Si)_4O_{10}(OH)_2Na_{0.3} \cdot nH_2O$	Mon	2.5	$1–1\frac{1}{2}$	1.60	Clay mineral
Norbergite, 503	$Mg_3(SiO_4)(F,OH)_2$	Orth	3.1–3.2	6	1.57	See chondrodite
Noselite, 556	$Na_8(AlSiO_4)_6SO_4$	Iso	2.3±	6	1.50	A feldspathoid

O

Name, Page	Composition	Xl Sys.	G	H	n	Remarks
Oligoclase, 554	$Ab_{90}An_{10}$-$Ab_{70}An_{30}$	Tric	2.65	6	1.54	A member of the plagioclase feldspar series
Olivine, 492, 493, 571	$(Mg,Fe)_2SiO_4$	Orth	3.27–4.37	$6\frac{1}{2}–7$	1.69	Green rock-forming mineral
Omphacite, 519	$(Ca,Na)(Mg,Fe,Al)Si_2O_6$	Mon	3.2–3.4	5–6	1.67–1.70	Green pyx. in eclogite
Onyx, 545						Layered chalcedony
Onyx marble, 412						See calcite and aragonite
Opal, 548, 569, 581	$SiO_2 \cdot nH_2O$	Amor	2.0–2.25	5–6	1.44	Conchoidal fracture
Orpiment, 363	As_2S_3	Mon	3.49	$1\frac{1}{2}–2$	2.8	Cl{010}; yellow
Orthite, 509						See allanite
Orthoclase, 551	$KAlSi_3O_8$	Mon	2.57	6	1.52	Medium T K-feldspar

Name, Page	Composition	Xl Sys.	G	H	n	Remarks
Orthoferrosilite, 514	$FeSiO_3$	Orth	3.9	6	1.79	Orthopyroxene endmember (= ferrosilite)

P

Name, Page	Composition	Xl Sys.	G	H	n	Remarks
Palladium, 345	Pd	Iso	11.9	$4\frac{1}{2}$–5	—	See platinum
Paragonite, 535	$NaAl_2(AlSi_3O_{10})(OH)_2$	Mon	2.85	2	1.60	Similar to muscovite
Pargasite, 526	$NaCa_2Fe_4(Al,Fe)Al_2Si_6O_{22}(OH)_2$	Mon	3–3.5	$5\frac{1}{2}$	1.62	See hornblende
Patronite, 437	VS_4	Mon				Ore of vanadium
Pectolite, 523	$Ca_2NaH(SiO_3)_3$	Tric	2.8	5	1.60	Crystals acicular
Pennantite, 541					1.66	See chlorite
Pentlandite, 360	$(Fe,Ni)_9S_8$	Iso	4.6–5.0	$3\frac{1}{2}$–4	—	Generally with pyrrhotite
Periclase, 379	MgO	Iso	3.56	$5\frac{1}{2}$	1.73	Cl{001}, cubic
Peridot, 495						Gem olivine
Peristerite, 482						Intergrowths between Ab_{98} and Ab_{85}; "moonstone"
Perovskite, 383	$CaTiO_3$	Orth	4.03	$5\frac{1}{2}$	2.38	Pseudoisometric crystals
Perthite, 549						Coarse K-feldspar–albite intergrowth
Petalite, 557	$Li(AlSi_4O_{10})$	Mon	2.4	6–$6\frac{1}{2}$	1.51	Cl{001} {201}
Petzite, 343	$(Ag,Au)_2Te$	Iso	8.7–9.0	$2\frac{1}{2}$–3	—	Steel-gray to iron-black
Phenacite, 491	Be_2SiO_4	Hex	2.97–3.0	$7\frac{1}{2}$–8	1.65	In pegmatites
Phillipsite, 562	$KCa(Al_3Si_5O_{16})\cdot6H_2O$	Mon	2.2	$4\frac{1}{2}$–5	1.50	A zeolite
Phlogopite, 537	$KMg_3(AlSi_3O_{10})(OH)_2$	Mon	2.86	$2\frac{1}{2}$–3	1.56–1.64	Yellow-brown mica
Phosgenite, 419	$Pb_2CO_3Cl_2$	Tet	6.0–6.3	3	2.12	Adamantine luster
Phosphorite, 435						See apatite
Piemontite, 507	$Ca_2MnAl_2O(SiO_4)(Si_2O_7)(OH)$	Mon	3.4	$6\frac{1}{2}$	1.75–1.81	Reddish-brown
Pigeonite, 515	$\sim Ca_{0.25}(Mg,Fe)_{1.75}Si_2O_6$	Mon	3.30–3.46	6	1.64–1.72	High T pyroxene
Pitchblende, 386						Massive UO_2
Plagioclase, 552	$Ab_{100}An_0$-Ab_0An_{100}	Tric	2.62–2.76	6	1.53–1.59	The Na-Ca-feldspar series
Plancheite, 543	$Cu_8(Si_4O_{11})_2(OH)_2\cdot H_2O$	Orth	3.3	$5\frac{1}{2}$	1.66	Fibrous, mammillary, blue
Platinum, 345	Pt	Iso	14–19	4–$4\frac{1}{2}$	—	Steel-gray with bright luster; malleable
Pleonaste, 388						Ferroan spinel
Pollucite, 555	$CsAlSi_2O_6\cdot H_2O$	Iso	2.9	$6\frac{1}{2}$	1.52	Colorless; in pegmatites
Polyhalite, 400	$K_2Ca_2Mg(SO_4)_2\cdot 2H_2O$	Tric	2.78	$2\frac{1}{2}$–3	1.56	Bitter taste
Potash feldspar, 548	$KAlSi_3O_8$					Microcline, orthoclase, sanidine
Powellite, 432	$CaMoO_4$	Tet	4.23	$3\frac{1}{2}$–4	1.97	Fluoresces yellow
Prase, 545						Dull green, microcrystalline quartz
Prehnite, 542	$Ca_2Al(AlSi_3O_{10})(OH)_2$	Orth	2.8–2.95	6–$6\frac{1}{2}$	1.63	Tabular crystals, green
Proustite, 370	Ag_3AsS_3	Hex	5.57	2–$2\frac{1}{2}$	3.09	Light ruby silver
Pseudobrookite, 383	Fe_2TiO_5					See ilmenite
Pseudoleucite, 555						See leucite
Pseudophite, 581						Massive form of chlorite; substitute for jade
Pseudowollastonite, 521	$CaSiO_3$	Tric				High T form of wollastonite
Psilomelane, 395						See romanechite
Pyrargyrite, 370	Ag_3SbS_2	Hex	5.85	2–$2\frac{1}{2}$	3.08	Dark ruby-silver
Pyrite, 364	FeS_2	Iso	5.02	6–$6\frac{1}{2}$	—	Crystals striated
Pyrochlore, 393	$(Ca,Na)_2(Nb,Ta)_2O_6(O,OH,F)$	Iso	4.3±	5	—	Usually metamict
Pyrolusite, 384	MnO_2	Tet	4.75	1–2	—	Sooty black
Pyromorphite, 435	$Pb_5(PO_4)_3Cl$	Hex	7.04	$3\frac{1}{2}$–4	2.06	Adamantine luster
Pyrope, 496	$Mg_3Al_2Si_3O_{12}$	Iso	3.58	7	1.71	A garnet
Pyrophanite, 383	$MnTiO_3$	Hex	4.54	5–6	2.48	Cl{0221}
Pyrophyllite, 534	$Al_2Si_4O_{10}(OH)_2$	Mon	2.8	1–2	1.59	Smooth feel; micaceous habit
Pyroxene, 514						A mineral group
Pyroxenoid, 520						A mineral group

Name, Page	Composition	XI Sys.	G	H	n	Remarks
Serpentine, 528	$Mg_3Si_2O_5(OH)_4$	Mon / Orth	2.3	3–5	1.55	Green to yellow; waxlike to silky (when fibrous)
Shattuckite, 543	$Cu_5(SiO_3)_4(OH)_2$	Orth	3.8	—	1.78	Blue
Siderite, 414	$FeCO_3$	Hex	3.96	$3\frac{1}{2}$–4	1.88	Cl{$10\bar{1}1$}
Sillimanite, 500	Al_2SiO_5	Orth	3.23	6–7	1.66	Cl{010} perfect
Silver, 343	Ag	Iso	10.5	$2\frac{1}{2}$–3	—	White, malleable
Sinhalite, 425	$Mg(Al,Fe)BO_4$	Orth	3.48	$6\frac{1}{2}$	1.70	Brown gem mineral
Skutterudite, 369	$(Co,Ni)As_3$	Iso	6.5 ± 0.4	$5\frac{1}{2}$–6	—	Tin-white
Smaltite, 369	$(Co,Ni)As_{3-x}$	Iso	6.5 ± 0.4	$5\frac{1}{2}$–6	—	See skutterudite
Smectite, 531						Clay mineral
Smithsonite, 416	$ZnCO_3$	Hex	4.30–4.45	4–$4\frac{1}{2}$	1.85	Reniform; many colors
Smoky quartz, 544	SiO_2					Brown to black
Soapstone, 531						See talc
Sodalite, 556	$Na_8(AlSiO_4)_6Cl_2$	Iso	2.15–2.30	$5\frac{1}{2}$–6	1.48	Usually blue
Soda niter, 422						See nitratite
Specularite, 380						Platy, metallic hematite
Sperrylite, 345	$PtAs_2$	Iso	10.50	6–7	—	See platinum
Spessartine, 496	$Mn_3Al_2Si_3O_{12}$	Iso	4.19	7	1.80	A garnet
Sphalerite, 357	ZnS	Iso	3.9–4.1	$3\frac{1}{2}$–4	2.37	Cl{011}; 6 directions
Sphene, 504						See titanite
Spinel, 388, 582	$MgAl_2O_4$	Iso	3.5–4.1	8	1.72	In octahedrons
Spodumene, 519	$LiAlSi_2O_6$	Mon	3.15–3.20	$6\frac{1}{2}$–7	1.67	Cl{110} ~ 90°; in pegmatites
Staetite, 531						See talc
Stannite, 359, 386	Cu_2FeSnS_4	Tet	4.3–4.5	4	—	Metallic, gray to black
Starlite, 571						Blue, heat treated zircon; a gem
Staurolite, 502	$Fe_2Al_9O_6(SiO_4)_4(O,OH)_2$	Mon	3.65–3.75	7–$7\frac{1}{2}$	1.75	Pseudo-orthorhombic; cruciform twins
Stibnite, 363	Sb_2S_3	Orth	4.52–4.62	2	—	Cl{010} perfect; prismatic crystals
Stilbite, 562	$NaCa_2Al_5Si_{13}O_{36}\cdot14H_2O$	Mon	2.1–2.2	$3\frac{1}{2}$–4	1.50	A zeolite; sheaflike aggregates
Stilpnomelane, 537	$\sim K_{0.6}(Mg,Fe^{2+},Fe^{3+})_6$- $Si_8Al(O,OH)_{27}\cdot2$–$4H_4O$	Mon	2.59–2.96	3–4	1.58–1.74	Brownish; in micalike plates
Stishovite, 476	SiO_2	Tet	4.35	7	1.80	In meteorite craters; high P form of quartz
Stolzite, 410	$PbWO_4$	Tet	7.9–8.3	$2\frac{1}{2}$–3	2.27	Cl{001} {011}
Stream tin, 386	SnO_2					See cassiterite
Strontianite, 418	$SrCO_3$	Orth	3.7	$3\frac{1}{2}$–4	1.67	Efferv. in HCl
Sudoite, 541	$Mg_2(Al,Fe^{3+})_3Si_3AlO_{10}(OH)_8$	Mon				Member of the chlorite group
Sulfur, 346	S	Orth	2.05–2.09	$1\frac{1}{2}$–$2\frac{1}{2}$	2.04	Burns with blue flame
Sunstone, 554						See oligoclase
Sylvanite, 343	$(Au,Ag)Te_2$	Mon	8.0–8.2	$1\frac{1}{2}$–2	—	Cl{010} perfect
Sylvite, 400	KCl	Iso	1.99	2	1.49	Cl cubic; bitter

T

Name, Page	Composition	XI Sys.	G	H	n	Remarks
Taenite, 345	Fe-Ni	Iso	7.8–8.2	5	—	In meteorites
Talc, 531	$Mg_3Si_4O_{10}(OH)_2$	Mon	2.7–2.8	1	1.59	Greasy feel; sectile
Tantalite, 392	$(Fe,Mn)Ta_2O_6$	Orth	6.5±	6	—	See columbite
Tanzanite, 509						Blue gem zoisite
Tennantite, 370	$Cu_{12}As_4S_{13}$	Iso	4.6–5.1	3–$4\frac{1}{2}$	—	In tetrahedrons
Tenorite, 522	CuO	Tric	6.5	3–4	—	Black
Tephroite, 492	Mn_2SiO_4	Orth	4.1	6	1.70–1.80	An olivine
Tetrahedrite, 370	$Cu_{12}Sb_4S_{13}$	Iso	4.6–5.1	3–$4\frac{1}{2}$	—	In tetrahedrons
Thomsonite, 562	$NaCa_2(Al_5Si_5O_{20})\cdot6H_2O$	Orth	2.3	5	1.52	A zeolite
Thorianite, 386, 387	ThO_2	Iso	9.7	$6\frac{1}{2}$	—	Dark gray, brownish-black
Thorite, 434, 499	$ThSiO_4$	Tet	5.3	5	1.8	Brown to black; radioactive

Name, Page	Composition	XI Sys.	G	H	n	Remarks
Tiger's eye, 527, 544						Brown chatoyant gem material; see crocidolite and quartz
Tincalconite, 423	$Na_2B_4O_5(OH)_4 \cdot 3H_2O$	Hex	1.88	1	1.46	Alteration of borax
Titanite, 504	$CaTiO(SiO_4)$	Mon	3.40–3.55	$5-5\frac{1}{2}$	1.91	Wedge-shaped crystals; formerly known as sphene
Todorokite, 395	$(Mn,Ca,Mg)Mn_3O_7 \cdot H_2O$					See romanechite
Topaz, 501, 570	$Al_2SiO_4(F,OH)_2$	Orth	3.4–3.6	8	1.61–1.63	Cl{001} perfect
Torbernite, 439	$Cu(UO_2)_2(PO_4)_2 \cdot 8-12H_2O$	Tet	3.22	$2-2\frac{1}{2}$	1.59	Green
Tourmaline, 513, 570	$(Na,Ca)(Li,Mg,Al)_3$-$(Al,Fe,Mn)_6(BO_3)_3$-$(Si_6O_{18})(OH)_4$	Hex	3.0–3.25	$7-7\frac{1}{2}$	1.64–1.68	Trigonal cross sections and conchoidal fracture
Travertine, 412						See calcite
Tremolite, 525	$Ca_2Mg_5Si_8O_{22}(OH)_2$	Mon	3.0–3.2	5–6	1.61	Cl{110}; white to light green
Tridymite, 547	SiO_2	Mon Orth	2.26	7	1.47	In volcanic rocks
Triphylite, 433	$Li(Fe,Mn)PO_4$	Orth	3.42–3.56	$4\frac{1}{2}-5$	1.69	Cl{001} {010}
Troilite, 360	FeS	Hex	4.7	4	—	In meteorites
Trona, 422	$Na_3H(CO_3)_2 \cdot 2H_2O$	Mon	2.13	3	1.49	Alkaline taste
Troostite, 492						Manganiferous willemite
Tsavorite, 571						Emerald green gem garnet
Tschermakite, 526	$Ca_2Mg_3(Al,Fe)_2Al_2Si_6O_{22}(OH)_2$					See hornblende
Tufa, 412						See calcite
Turquoise, 438, 571, 582	$CuAl_6(PO_4)_4(OH)_8 \cdot 5H_2O$	Tric	2.6–2.8	6	1.62	Blue-green; gem material
Tyuyamunite, 440	$Ca(UO_2)_2(VO_4)_2 \cdot 5-8\frac{1}{2}H_2O$	Orth	3.7–4.3	2	1.86	Yellow; radioactive; secondary alteration

U

Name, Page	Composition	XI Sys.	G	H	n	Remarks
Ulexite, 425	$NaCaB_5O_6(OH)_6 \cdot 5H_2O$	Tric	1.96	$1-2\frac{1}{2}$	1.50	"Cotton balls"
Ulvöspinel, 390	Fe_2TiO_4	Iso	4.78	$7\frac{1}{2}-8$	—	Commonly as exsolution in magnetite
Uraninite, 386	UO_2	Iso	7.5–9.7	$5\frac{1}{2}$	—	Pitchy luster; radioactive
Uvarovite, 496	$Ca_3Cr_2Si_3O_{12}$	Iso	3.90	$7\frac{1}{2}$	1.87	Rare green garnet
Uytenbogaardite, 343	Ag_3AuS_3					Rare gold ore

V

Name, Page	Composition	XI Sys.	G	H	n	Remarks
Vanadinite, 437	$Pb_5(VO_4)_3Cl$	Hex	6.9	3	2.25–2.42	Adamantine luster; red to yellow
Variscite, 439	$Al(PO_4) \cdot 2H_2O$	Orth	2.57	$3\frac{1}{2}-4\frac{1}{2}$	1.58	Green, massive
Verde antique, 529						Marble and green serpentine
Verdelite, 513						Green tourmaline
Verdite, 581						Green rock material containing fuchsite; substitute for gem jade
Vermiculite, 537	$(Mg,Ca)_{0.3}(Mg,Fe,Al)_{3.0}$-$(Al,Si)_4O_{10}(OH)_4 \cdot 8H_2O$	Mon	2.4	$1\frac{1}{2}$	1.55–1.58	Altered biotite
Vesuvianite, 509	$Ca_{10}(Mg,Fe)_2Al_4(SiO_4)_5$-$(Si_2O_7)_2(OH)_4$	Tet	3.35–3.45	$6\frac{1}{2}$	1.70–1.75	Prismatic crystals; formerly known as idocrase
Vivianite, 437	$Fe_3(PO_4)_2 \cdot 8H_2O$	Mon	2.58–2.68	$1\frac{1}{2}-2$	1.60	Cl{010} perfect

W

Name, Page	Composition	XI Sys.	G	H	n	Remarks
Wad						Manganese ore; mixture of manganese minerals
Wadsleyite, 494	Mg_2SiO_4					High-pressure polymorph of olivine

Name, Page	Composition	Xl Sys.	G	H	n	Remarks
Wavellite, 438	$Al_3(PO_4)_2(OH)_3 \cdot 5H_2O$	Orth	2.36	$3\frac{1}{2}$–4	1.54	Radiating aggregates
Wernerite, 558					1.55–1.60	See scapolite
White iron pyrites, 366						See marcasite
Willemite, 492	Zn_2SiO_4	Hex	3.9–4.2	$5\frac{1}{2}$	1.69	From Franklin, N.J.
Witherite, 418	$BaCO_3$	Orth	4.3	$3\frac{1}{2}$	1.68	Efferv. in HCl
Wolframite, 431	$(Fe,Mn)WO_4$	Mon	7.0–7.5	4–$4\frac{1}{2}$	—	Cl{010} perfect
Wollastonite, 520	$CaSiO_3$	Tric	2.8–2.9	5–$5\frac{1}{2}$	1.63	Cl{100}, {001} perfect
Wood opal, 548						Fossil wood with opal
Wood tin, 385						See cassiterite
Wulfenite, 432	$PbMoO_4$	Tet	6.8±	3	2.40	Orange-red
Wurtzite, 357	ZnS	Hex	3.98	4	2.35	Polymorph of sphalerite

X

Xanthophyllite, 538	$Ca(Mg,Al)_{3-2}(Al_2Si_2O_{10})(OH)_2$	Mon	3–3.1	$3\frac{1}{2}$	1.65	Brittle mica

Z

Zeolite, 559						A mineral group
Zinc blende, 357						See sphalerite
Zincite, 378	ZnO	Hex	5.68	4	2.01	From Franklin, N.J.
Zircon, 498, 571	$ZrSiO_4$	Tet	4.68	$7\frac{1}{2}$	1.92–1.96	Commonly in small crystals
Zoisite, 509	$Ca_2Al_3O(SiO_4)(Si_2O_7)(OH)$	Orth	3.35	6	1.69	Gray, green-brown, metamorphic

SUBJECT INDEX

LOCATIONS OF SOME IMPORTANT TABLES AND ILLUSTRATIONS

SOME UNITS, SYMBOLS, AND CONVERSION FACTORS

Length

meter (m)	3.28083 feet = 39.370 inches = 100 cm = 10^{10} Å
centimeter (cm)	1 cm = 10^{-2} m, or 0.01 m
millimeter (mm)	1 mm = 10^{-3} m = 0.0394 inches
micrometer (μm)	1 μm = 10^{-6} m = 10^{-3} mm = 10^4 Å
nanometer (nm)	1 nm = 10^{-9} m = 10^{-7} cm = 10 Å
angstrom (Å)	1 Å = 10^{-8} cm = 10^{-4} μm = 10^{-1} nm, or 0.1 nm
(inch)	1 inch = 2.54 cm
(foot)	1 foot = 30.48 cm

Volume

liter (l)	1 liter = 1000 cm^3 = 1.0567 quarts (U.S.)
cubic centimeters (cm^3)	
cubic angstroms ($Å^3$)	

Chemical Concentration
weight percent (wt %)
molecular percent (mole %)
volume percent (vol %)
parts per million (ppm)
parts per billion (ppb)

Temperature

degrees Celsius (°C)	5/9 (°F − 32); F = Fahrenheit
kelvins (K)	K = °C + 273.15; C = Celsius; absolute zero = −273.15°C

Pressure

bar	1 bar = 0.9869 atm = 10^5 Pa
kilobar (kbar)	986.9 atm = 10^3 bars = 10^8 Pa
pascal (Pa)	1 pascal = 10^{-5} bars
atmosphere (atm)	1 atm = 760 mm Hg

Density = mass per unit volume;
e.g. grams per cubic centimeter (g/cm^3)

Miller Indices and Diffraction Notation
face symbol: ($h\,k\,l$)
form symbol: {$h\,k\,l$}
edge or zone symbol: [$h\,k\,l$]
diffraction symbol: $h\,k\,l$
interplanar spacing (d); e.g. $d_{h\,k\,l}$

Unit Cell Measurements
edge lengths (a, b, c)
angles (α, β, γ)

Optical Parameters
refractive indices (n; ε, ω; α, β, γ)
optic axial angle (2V)
extinction angle (X, Y or Z ∧ a or c)

Angle
radian (ra)　　1 radian = 57.296 degrees

Some Metric Units and Their Prefixes

Prefix	Multiple or Submultiple	
mega	1,000,000	10^6
kilo	1,000	10^3
hecto	100	10^2
deka	10	10^1
	1	10^0
deci	0.1	10^{-1}
centi	0.01	10^{-2}
milli	0.001	10^{-3}
micro	0.000 001	10^{-6}
nano	0.000 000 001	10^{-9}

THE EIGHT MOST COMMON ELEMENTS IN THE EARTH'S CRUST (See Figure 3.2)

Weight Percentage		Volume Percentage	
O	46.60	O	~94
Si	27.72	K	
Al	8.13	Na	
Fe	5.00	Ca	
Ca	3.63	Si	~6
Na	2.83	Al	in total
K	2.59	Fe	
Mg	2.09	Mg	
Total	98.59		

MOLECULAR WEIGHTS OF COMPONENTS COMMON IN SILICATES (See also Table 3.3)

SiO_2	60.08	MnO	70.94
TiO_2	79.90	MgO	40.30
Al_2O_3	101.96	Na_2O	61.98
FeO	71.85	K_2O	94.20
Fe_2O_3	159.69	H_2O	18.02
CaO	56.08		

SOME OF THE MOST COMMON IONS, THEIR COORDINATION, AND RADII

(The number in brackets is **C.N.** = coordination number.) See Table 3.11 for a complete listing.

Ion	Coordination Number with Oxygen	Ionic Radius Å
O^{2-}		1.36 [3]
K^+	8-12	1.51 [8]–1.64 [12]
Na^+	8-6 } cubic to	1.18 [8]–1.02 [6]
Ca^{2+}	8-6 } octahedral	1.12 [8]–1.00 [6]
Mn^{2+}	6	0.83 [6]
Fe^{2+}	6	0.78 [6]
Mg^{2+}	6 } octahedral	0.72 [6]
Fe^{3+}	6	0.65 [6]
Ti^{4+}	6	0.61 [6]
Al^{3+}	6	0.54 [6]
Al^{3+}	4	0.39 [4]
Si^{4+}	4 } tetrahedral	0.26 [4]
P^{5+}	4	0.17 [4]
S^{6+}	4	0.12 [4]

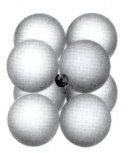

Cubic coordination

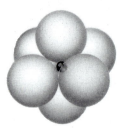

Octahedral coordination

Tetrahedral coordination

STEREOGRAPHIC NET WITH 10CM RADIUS

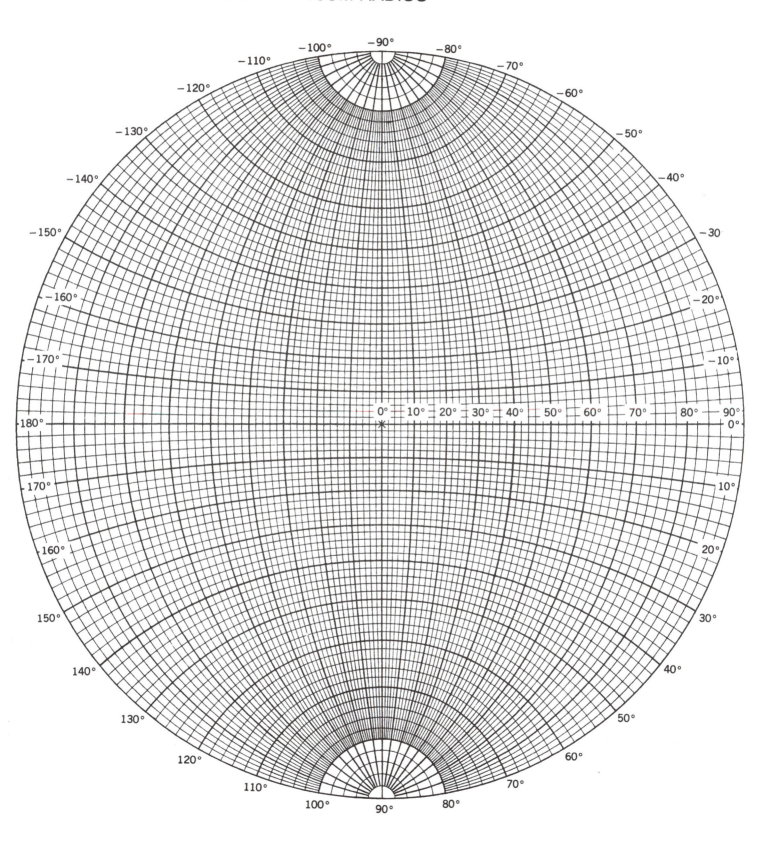

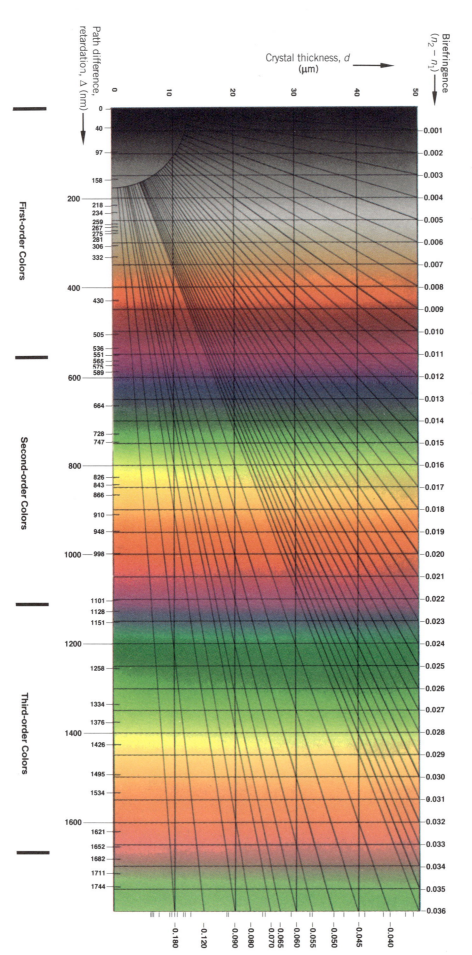

Interference Color Chart

Graphical representation of the relationship between retardation, birefringence, and crystal thickness: $\Delta = d(n_2 - n_1)$

Crystal thickness, d (μm)

Birefringence ($n_2 - n_1$)

Path difference, retardation, Δ (nm)

First-order Colors

Second-order Colors

Third-order Colors